Statistical Abstract of the World

TABLE OF CONTENTS

Statistical Abstract of the World

Marlita A. Reddy, Editor

Gale Research Inc.
An International Thomson Publishing Company

I(T)P

NEW YORK • LONDON • BONN • BOSTON • DETROIT • MADRID
MELBOURNE • MEXICO CITY • PARIS • SINGAPORE • TOKYO
TORONTO • WASHINGTON • ALBANY NY • BELMONT CA • CINCINNATI OH

Marlita A. Reddy, *Editor*

Editorial Code and Data, Inc. Staff
Gary Alampi, *Programmer/Analyst*
Arsen J. Darnay, *Senior Editor*
Kenneth J. Muth, *Cartography*
Nancy Ratliff and Sherae R. Fowler, *Data Entry Associates*

Gale Research, Inc. Staff
Peg Bessette and Kelle S. Sisung, *Developmental Editors*
Jolen Gedridge and Camille Killens, *Associate Developmental Editors*
Lawrence W. Baker, *Senior Developmental Editor*

Mary Beth Trimper, *Production Director*
Mary Kelley, *Production Assistant*
Cynthia Baldwin, *Art Director*
Mary Krzewinsky, *Cover Design*

∞™ This book is printed on acid-free paper that meets the minimum requirements of American National Standard for Information Sciences—Permanence Paper for Printed Library Materials, ANSI Z39.48-1984.

♲ This book is printed on recycled paper that meets Environmental Protection Agency standards.

Copyright © 1994
Gale Research Inc.
835 Penobscot Building
Detroit, MI 48226-4094

ISBN 0-8103-9199-6
ISSN 1077-1360
Printed in the United States of America

Published simultaneously in the United Kingdom
by Gale Research International Limited
(An affiliated company of Gale Research Inc.)

10 9 8 7 6 5 4 3 2

I(T)P™ Gale Research Inc., an International Thomson Publishing Company.
ITP logo is a trademark under license.

INTRODUCTION

Statistical Abstract of the World (*SAW*) is a comprehensive presentation of data on 185 countries of the world in a uniform format. *SAW* was conceived as a general reference resource for all types of users, including students, educators, business and government analysts and officials, journalists and other communicators, and the public at large. *SAW* covers the full range of topics on the countries of the world, including physical characteristics; demographic subjects: ethnicities, religion, and language; education; science and technology; government and defense; labor; energy, production, and manufacturing; and finance, economics, and trade.

This introduction presents brief discussions on the (1) general design of *SAW*, (2) data characteristics, (3) content, (4) coverage of countries, (5) sources, and (6) arrangement of the book, which also covers the contents of appended materials.

General Design

Most books that cover the countries of the world tend to do so in selective fashion—featuring heavy coverage of large, well-known countries and leaving out or covering in brief summary only smaller and poorer countries that are relatively hidden from view by turmoil or political alignment.

The editors' intent in the preparation of *SAW* was to counter this tendency and to present available statistical information on the countries of the world in a standardized, uniform format—even at the cost of highlighting the absence of information.

Arrangement of Data

To meet this design objective, countries were arranged in alphabetical order and information for each country is presented in a uniform manner. Data are shown in 5 or 6 pages, beginning with a regional and country map and then followed by 42 panels. The panels are grouped by topic. All panels are numbered and titled (even if empty) and may hold tabular material or blocks of text. If no information is available for a panel, it is shaded gray.

Variant Pages

Detailed information on *energy* and *manufacturing* is lacking for a large number of countries. Rather than presenting large blocks of empty space, the fourth page of a presentation is redesigned to hold an *Energy Resource Summary* if detailed energy consumption and production data are unavailable. Similarly, for those countries with limited manufacturing data, the fifth and sixth pages of the presentation are combined. In these cases, the manufacturing page is replaced by a single panel, *Industrial Summary*.

It thus develops that each country is covered in either 5 or 6 pages. Panel numbers, however, remain the same; the general location of the panels follows the same pattern whether variant pages are used or not.

Variant Contents

Where at all possible, information for each panel is drawn from the same source. However, since statistics for many panels were unavailable—while alternate sources provided comparable data—alternative data were used rather than leaving the panel blank. This has resulted in variant contents in some of the panels.

Sources, Notes, and Footnotes

Sources, notes, and footnote texts are grouped in a separate *Annotated Source Appendix* accessible by panel number and also by the table of contents provided at the beginning of the source appendix. This ensures a more attractive, less cluttered look.

Data Characteristics

Variability of Statistical Information

The preparation of a volume like *SAW* forces the recognition of the diverse meaning hidden by words such as "country," "state," or "sovereignty." The range of development is enormous. *SAW* thus covers very large industrial societies side-by-side with very small island or city states that—in population, resources, or by any other measure—are often smaller than an urban county in the United States.

In this first edition, the editors had to rely on data published by a number of U.S. or international agencies which collect data from individual countries and present these in various "predigested" formats. These sources of data clearly show that the statistics available for a country often parallel the country's size and development. International bodies that collect statistics have different objectives, and their aims dictate the information they collect. Yet even such bodies, with all of their well-developed machinery of intelligence-gathering, fail to capture accu-

rately (or at all) the numerical measures of some political entities.

For these reasons, statistical data for the world are much less uniform— and thus also less comparable, country-to-country—than U.S. Census statistics are state-to-state. Even demographic data are diverse, often based on very rough estimates; they are also collected with widely different dates. Comparability of economic data for manufacturing, the production sectors, international trade, military expenditures, etc. is further complicated by the great diversity of the world's economies. Other difficulties arise in translating local values to a global standard or in comparing "cost of living" between two or more countries with very different standards of living.

Data Gaps

For these reasons, data gaps are common. In *SAW*, the absence of data is highlighted rather than obscured, partly to preserve uniformity of presentation, partly in the hope of stimulating data collection by international agencies or data reporting by countries.

The absence of data, signalled by a gray box, is most frequently encountered in very small countries (typically the island states); in the successor states to the former U.S.S.R., the former Czechoslovakia, the former Yugoslavia; and in countries that, for one reason or other (political alignment, lack of resources) are not providing data to international agencies.

Dated Information

Statistics are, by their very nature, retrospective. The "most current" statistics, even in an industrially advanced nation, are often several years old. This phenomenon is equally true of international statistics, with the added problem that collecting and processing such data takes even more time and effort.

Events typically out-pace statistics. Thus information regarding Rwanda, Serbia and Herzegovina,

Bosnia and Montenegro, Somalia, and other trouble spots around the world are of historical use only.

Similarly, statistics on Germany are sometimes available only for the former Federal Republic of Germany or only for the former German Democratic Republic (the former East Germany).

Military data are still strongly reflective of the Cold War era. Economic data relating to the successor states of the U.S.S.R. do not adequately reflect the realities of Russia and other states. The new shape of things in these countries will emerge gradually and will be reflected in future editions of *SAW*.

Content

The contents of *SAW* reflect the availability of statistical information. The main topics covered under 8 major headings are shown in the inset.

In general, coverage is quite comprehensive. One area where information is particularly difficult to find is the Arts. Information on Libraries, Newspapers, and Cinema were available (spottily). These topics are grouped under the heading Education.

Coverage

SAW covers all members of the United Nations and a selection of non-UN members that, by their size and importance, properly belong in a book of this type. The non-UN countries are **Hong-Kong**, **Switzerland**, and **Taiwan**.

A new African nation, **Eritrea**, is excluded from the listing because data on this state are altogether unavailable. It is expected that Eritrea will be included in future editions as data become available.

Geography

Human Factors
- Demographics
- Health
- Health Personnel
- Health Expenditures
- Health Care Ratios
- Infants and Malnutrition
- Ethnic Division
- Religion
- Major Languages

Education
- Public Education Expenditures
- Educational Attainment
- Literacy Rate
- Libraries
- Daily Newspapers
- Cinema

Science and Technology
- Scientific/Technical Forces
- R&D Expenditures
- U.S. Patents Issued

Government and Law
- Organization of Government
- Political Parties
- Government Budget

- Military Affairs
- Crime
- Human Rights

Labor Force
- Total Labor Force
- Labor Force by Occupation
- Unemployment Rate

Production Sectors
- Energy Production
- Energy Consumption
- Telecommunications
- Transportation
- Top Agricultural Products
- Top Mining Products
- Tourism

Manufacturing Sector

Finance, Economics, Trade
- Economic Indicators
- Balance of Payments
- Exchange Rates
- Top Import Origins
- Top Export Destinations
- Foreign Aid
- Import/Export Commodities

Sources

Data for *SAW* were provided by a number of international and U.S. federal agencies. Because the goal of this book is to provide a uniform format for as many of the 185 countries as possible, the editors chose sources that supplied meaningful and consistent information for a large number of countries.

The major entities that collect statistical data on countries worldwide are the United Nations, the International Bank for Reconstruction and Development (World Bank), the International Monetary Fund (IMF), the Central Intelligence Agency (CIA), and the U.S. State Department.

While these agencies provide a broad range of statistical data, various branches of the United Nations, such as regional and specialized agencies, and executive branches of the United States government also compile statistics relating to subject matter within their jurisdiction. International agencies include organizations such as the Organization for Economic Community and Development (OECD), International Civil Aviation Organization (ICAO), Organization of American States (OAS), Asian Development Bank, and specialized agencies of the United Nations including the International Labor Organization (ILO), the World Health Organization (WHO), the United Nations Educational, Scientific, and Cultural Organization (UNESCO), and the Economic Commission for Latin America and the Caribbean (ECLAC).

Federal agencies of the U.S. government that publish international statistics include the U.S. Department of Commerce (Bureau of the Census, International Trade Administration, Office of Patents and Trademarks), the U.S. Department of Labor (Bureau of Labor Statistics), the U.S. Department of Agriculture, the U.S. Department of Defense, the U.S. Department of Energy, the U.S. Department of Education, the State Department, the U.S. Department of the Interior (Bureau of Mines), and the U.S. National Science Foundation. Agencies not listed here also publish some international statistics from time to time in their areas of expertise.

Specialized and regional international agencies offer excellent resources for information in their own right. The OECD, for instance, has a wide range of reports on a host of subjects. These publications are timely and informative. ECLAC also reports on a variety of important issues. However, regional reports tend to reflect the issues that are of direct concern to a particular geographic or economic area. Also, methods of statistical collection may vary among agencies, depending on the purpose of the report.

With the exception of tourism information, no data from associations or private organizations were used.

The editors compiled information from agencies that reported on large numbers of countries generally using the same method of data collection and reporting for all countries cited. This approach should provide the reader with a more consistent presentation of statistics than would be the case if the editors attempted to reconcile statistics from variant sources.

The editors compared many publications from such sources as noted above and chose those reports that could best provide tabular data for a large range of countries. By far the most comprehensive source in terms of geographical range is the *CIA World Factbook*, which provides textual (non-tabular) information for all countries included in *SAW*. This information, with a great deal of editorial and computer processing, was very helpful in the production of this book.

All sources used in the preparation of *SAW* are cited in the *Annotated Source Appendix*, which begins on page 1035. The citations are arranged in display order by panel number. In the event that more than one source is used for a single panel, both sources are cited with explanatory notes specifying when each source has been used.

Arrangement of the Book

Access to *SAW* is provided by a **Table of Contents**. It lists each country, in alphabetical order, by the page on which the country's presentation starts.

The **Introduction** follows the Table of Contents.

Two pages follow showing the **Regions of the World**. These are regional maps on which the location of every country in *SAW* is indicated.

A brief section showing **Abbreviations and Acronyms** follows. Additional explanations of many abbreviations are also provided in the Annotated Source Appendix (see below).

The body of the book begins immediately after the Abbreviations and Acronyms. Countries are arranged in alphabetical order. Each country is shown in 5 or 6 standard pages. Information is presented in the order depicted in the inset on the previous page.

Annotated Source Appendix

The **Annotated Source Appendix** follows the last country. This section documents the source or sources used, presents technical or informational notes, and presents the texts of footnotes sequentially.

The Source Appendix is largely drawn, with minor editorial changes, from the primary sources used in the compilation of this book. One or more sources are cited. If two sources were used, a listing of countries covered by the second source is provided. Notes, if any, are shown next. Finally, footnotes are presented, beginning with general notes and followed by numbered footnote texts.

Explanations of technical phrases and expansions of some acronyms are provided in the notes.

Keyword Index

The last item in *SAW* is the **Keyword Index**. The index holds more than 3,000 subject references or the names of languages, religions, ethnicities, and political parties. The names of political parties are sometimes difficult to identify; for this reason, each such entry is marked with an informational tag, *pol*. In all cases, page references are provided. In most cases, the name of the country is also provided with the page reference.

Acknowledgments

Many people helped in the compilation of *SAW* by providing data, suggestions, and advice. For expeditious handling of permissions issues, special thanks to Mai Stewart of the International Monetary Fund; to the self-effacing "Operator 2" and to Carole Rosen of the World Bank. Thanks to John B. Forbes of the USDA's National Agriculture Library and Shelley G. Altenstadter of INTERPOL for providing statistical reports used in this edition; to Gene Alloway and Jim Ottariani of the University of Michigan Engineering Library for assistance with the U.S. patents data; to Pat McGibbon of the Association of Manufacturing Technology, Dr. Lawrence M. Rausch of the U.S. National Science Foundation, and Dr. Ebraham Shekarchi of the U.S. Bureau of Mines for expert advice in locating statistical materials; and to Gary Alampi and Annemarie Muth for creative effort and ingenuity lavished on the original layout of the sample pages. An especially big "Thank you!" goes to Ms. Peg Bessette of Gale Research Inc. for her patient advice and assistance throughout the preparation of this book.

Comments and Suggestions

Comments on *SAW* or suggestions for improvement of its usefulness, format, and coverage are always welcome. Although every effort has been made to maintain accuracy, errors may have occurred; the editors will be grateful if these are called to their attention. Please contact:

Editors, *Statistical Abstract of the World*
Gale Research Inc.
835 Penobscot Building
Detroit, MI 48226-4094
Phone: (313) 961-2242 or (800) 347-GALE
Fax: (313) 961- 6815

Regions of the World

North America

Europe

South America

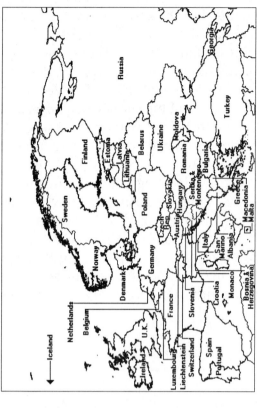

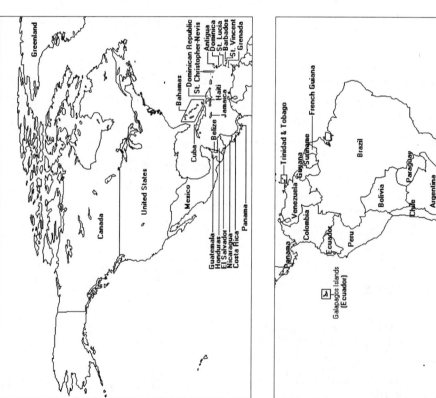

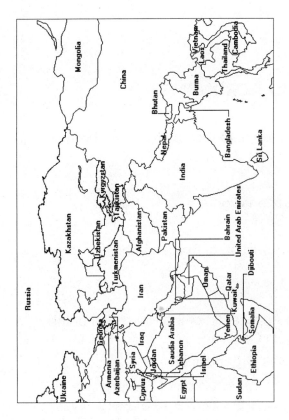

Africa

Southwest Asia

Southeast Asia

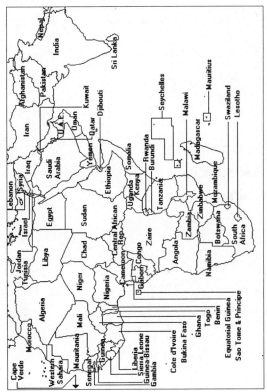

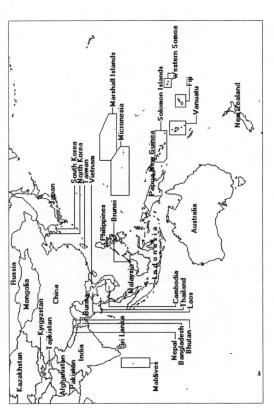

ABBREVIATIONS AND ACRONYMS

Additional explanations of some of the terms defined here may be found in the *Annotated Source Appendix*, page 1035.

ACHR	American Convention on Human Rights
Admin.	Administrative
AFL	Convention Concerning the Abolition of Forced Labor
AG	Silver
AM	Amplitude modulation
APROBC	Convention Concerning the Application of the Principles of the Right to Organize and Bargain Collectively
ARABSAT	Arab Satellite Communications Organization
ASST	Supplementary Convention on the Abolition of Slavery, the Slave Trade, and Institutions and Practices Similar to Slavery
AU	Gold
avg.	Average
bd. ft.	Board foot
bll.	Billion
BTU	British thermal unit
C	Canadian
CARICOM	Caribbean Community and Common Market
CEMA	Council for Mutual Economic Assistance
C.F.A.	Communaute Financiere Afraicaine
CGE	Central government expenditure
CIA	Central Intelligence Agency
c.i.f.	Cost, insurance, and freight
Circ.	Circulation
CIS	Commonwealth of Independent States
CO	Cobalt
const.	Constant
CPI	Consumer price index
CPR	International Covenant on Civil and Political Rights

CU	Copper
curr.	Current
Dec	December
DM	Deutsche Marks
dom.	Domestic
DPT3	Diphtheria, pertussis, tetanus, and measles (vaccine)
DWT	Deadweight ton
EAFDAW	Convention on the Elimination of All Forms of Discrimination Against Women
EAFRD	International Convention on the Elimination of All Forms of Racial Discrimination
EC	European Community
ECOWAS	Economic Community of West African States
ed.	Education
EFTA	European Free Trade Association
equiv.	Equivalent
ESCR	Internation Covenant on Economic, Social, and Cultural Rights
est.	Estimate
EUTELSAT	European Telecommunications Satellite Organization
Ex-Im	Export-Import Bank of the United States
Expend.	Expenditure
FAPRO	Convention Concerning Freedom of Association and Protection of the Right to Organize
FL	Convention Concerning Forced Labor
f.o.b.	Free on board
FM	Frequency modulation
FRG	Federal Republic of Germany
FY	Fiscal year
gal.	Gallon
GDP	Gross domestic product
GDR	[former] German Democratic Republic (East Germany)
GFCF	Gross fixed capital formation
GNP	Gross national product
Govt.	Government
GRT	Gross register ton
GSP	Generalized system of preferences
hl	Hectoliter

ICJ	International Court of Justice
ILO	International Labor Organization
IMF	International Monetary Fund
incl.	Including
INMARSAT	International Maritime Satellite Organization
INTELSAT	International Telecommunications Satellite Organization
Jan	January
kg	Kilogram
km	Kilometer
km^2	Square kilometer
kW	Kilowatt
kWh	Kilowatt hour
lb.	Pound
LDCs	Less-developed countries
m	Meter
MAAE	Convention Concerning the Minimum Age for Admission to Employment
MARACS	Maritime Communications Satellite
MARISAT	Maritime Satellite System
max.	Maximum
mfg.	Manufacturing
Mg	Magnesia
mkt	Market
mi.	Mile
mil.	Million
Mn	Manganese
Mo	Molybdenum
Mt.	Metric tons
N	Nitrogen
NA	Not available
NEGL	Negligible
NG	Natural gas
Ni	Nickel
nm	Nautical mile
NMT	Nordic Mobile Telecommunications
nom.	Nominal
NZ	New Zealand

ODA	Official development assistance
OECD	Organization for Economic Cooperation and Development
OECS	Organization of Eastern Caribbean States
OOF	Other official flows
OPEC	Organization of Petroleum Exporting Countries
Pb	Lead
pct	Percent
PCPTW	Geneva Convention Relative to the Protection of Civilian Persons in Time of War
PDRY	People's Democratic Republic of Yemen (South Yemen)
PHRFF	European Convention for the Protection of Human Rights and Fundamental Freedoms
Pop.	Population
PPCG	Convention on the Prevention and Punishment of the Crime of Genocide
prod.	Production
PRW	Convention on the Political Rights of Women
PVIAC	Protocol Additional to the Geneva Conventions and Relating to the protection of Victims of International Armed Conflicts
PVNAC	Protocol Additional to the Geneva Conventions and Relating to the protection of Victims of Non-International Armed Conflicts
RC	Convention on the Rights of the Child
Reg.	Registered
RMB	Renminbi
SAAR	Seasonally adjusted annual rate
Sb	Antimony
SHF	Super High Frequency
Sn	Tin
sq.	Square
SR	Protocol Relating to the Status of Refugees
SSTS	Convention to Suppress Slavery Trade and Slavery
STPEP	Convention for the Suppression of the Traffic in Persons and of the exploitation and Prostitution of Others
Svc. Pts.	Service points
svgs.	Savings
TCIDTP	Convention Against Torture and Other Cruel, Inhuman or Degrading Treatment or Punishment
TPW	Geneva Convention Relative to the Protection of Prisoners of War

tril.	Trillion
TT	Trinidad and Tobago
TV	Television
UHF	Ultrahigh frequency
VHF	Very High Frequency
Vols.	Volumes
UAE	United Arab Emirates
UK	United Kingdom
UN	United Nations
US	United States
USSR	[former] Union of Soviet Socialist Republics (Soviet Union)
W	Tungsten
WHO	World Health Organization
WPI	Wholesale price index
YAR	Yemen Arab Republic (North Yemen)
Zn	Zinc

Statistical Abstract of the World

Afghanistan

Geography [1]

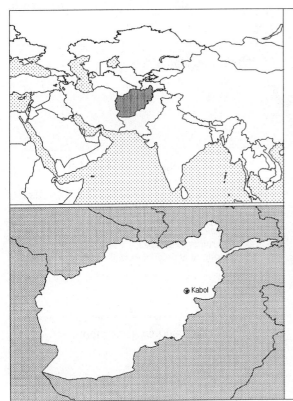

Total area:
647,500 km2
Land area:
647,500 km2
Comparative area:
Slightly smaller than Texas
Land boundaries:
Total 5,529 km, China 76 km, Iran 936 km, Pakistan 2,430 km, Tajikistan 1,206 km, Turkmenistan 744 km, Uzbekistan 137 km
Coastline:
0 km (landlocked)
Climate:
Arid to semiarid; cold winters and hot summers
Terrain:
Mostly rugged mountains; plains in north and southwest
Natural resources:
Natural gas, petroleum, coal, copper, talc, barites, sulphur, lead, zinc, iron ore, salt, precious and semiprecious stones
Land use:
Arable land:
12%
Permanent crops:
0%
Meadows and pastures:
46%
Forest and woodland:
3%
Other:
39%

Demographics [2]

	1960	1970	1980	1990	1991[1]	1994	2000	2010	2020
Population	9,829	12,431	14,985	15,332	16,450	16,903	25,725	32,889	41,518
Population density (persons per sq. mi.)	39	50	60	62	66	NA	100	129	160
Births	NA	NA	NA	NA	719	735	NA	NA	NA
Deaths	NA	NA	NA	NA	329	320	NA	NA	NA
Life expectancy - males	NA	NA	NA	NA	44	46	NA	NA	NA
Life expectancy - females	NA	NA	NA	NA	43	44	NA	NA	NA
Birth rate (per 1,000)	NA	NA	NA	NA	44	43	NA	NA	NA
Death rate (per 1,000)	NA	NA	NA	NA	20	19	NA	NA	NA
Women of reproductive age (15-44 yrs.)	NA	NA	NA	3,452	NA	3,821	5,878	NA	NA
of which are currently married	NA	NA	NA	2,778	NA	3,079	4,747	NA	NA
Fertility rate	NA	NA	NA	6.5	6.32	6.3	5.9	5.1	4.3

Population values are in thousands, life expectancy in years, and other items as indicated.

Health

Health Personnel [3]

Doctors per 1,000 pop., 1988-92	0.11
Nurse-to-doctor ratio, 1988-92	0.8
Hospital beds per 1,000 pop., 1985-90	0.3
Percentage of children immunized (age 1 yr. or less)	
Third dose of DPT, 1990-91	NA
Measles, 1990-91	NA

Health Expenditures [4]

Total health expenditure, 1990 (official exchange rate)	
Millions of dollars	NA
Millions of dollars per capita	NA
Health expenditures as a percentage of GDP	
Total	NA
Public sector	NA
Private sector	NA
Development assistance for health	
Total aid flows (millions of dollars)[1]	53
Aid flows per capita (millions of dollars)	2.6
Aid flows as a percentage of total health expenditure	NA

For sources, notes, and explanations, see Annotated Source Appendix, page 1035.

1

Human Factors

Health Care Ratios [5]	Infants and Malnutrition [6]

Ethnic Division [7]

Minor ethnic groups (Chahar Aimaks, Turkmen, Baloch, and others).	
Pashtun	38%
Tajik	25%
Uzbek	6%
Hazara	19%

Religion [8]

Sunni Muslim	84%
Shi'a Muslim	15%
Other	1%

Major Languages [9]

Much bilingualism.	
Pashtu	35%
Afghan Persian (Dari)	50%
Turkic languages (primarily Uzbek Turkmen)	11%
30 minor languages (primarily Balochi Pashai)	4%

Education

Public Education Expenditures [10]

Million Afghani	1980	1985	1987	1988	1989[6]	1990
Total education expenditure	3,205	NA	3,810	NA	3,540	5,667
as percent of GNP	NA	NA	NA	NA	NA	NA
as percent of total govt. expend.	12.7	NA	4.0	NA	NA	NA
Current education expenditures	2,886	NA	3,150	NA	3,474	5,282
as percent of GNP	NA	NA	NA	NA	NA	NA
as percent of current govt. expend.	14.4	NA	3.7	NA	NA	NA
Capital expenditures	319	NA	660	NA	66	385

Educational Attainment [11]

Literacy Rate [12]

In thousands and percent	1985[a]	1991[a]	2000[a]
Illiterate population +15 years	6,414	6,781	8,969
Illiteracy rate - total pop. (%)	75.9	70.6	59.1
Illiteracy rate - males (%)	62.0	55.9	44.0
Illiteracy rate - females (%)	90.7	86.1	74.8

Libraries [13]

Daily Newspapers [14]

	1975	1980	1985	1990
Number of papers	15	13	13	14
Circ. (000)	80[e]	90[e]	110[e]	180[e]

Cinema [15]

Science and Technology

Scientific/Technical Forces [16]	R&D Expenditures [17]	U.S. Patents Issued [18]

Government and Law

Organization of Government [19]

Long-form name:
Islamic State of Afghanistan
Type:
transitional government
Independence:
19 August 1919 (from UK)
Constitution:
the old Communist-era constitution has been suspended; a new Islamic constitution has yet to be ratified
Legal system:
a new legal system has not been adopted but the transitional government has declared it will follow Islamic law (Shari'a)
National holiday:
Victory of the Muslim Nation, 28 April; Remembrance Day for Martyrs and Disabled, 4 May; Independence Day, 19 August
Executive branch:
president, prime minister
Legislative branch:
a unicameral parliament consisting of 205 members was chosen by the shura in January 1993
Judicial branch:
an interim Chief Justice of the Supreme Court has been appointed, but a new court system has not yet been organized

Crime [23]

Elections [20]

The former ruling Watan Party has been disbanded.

Government Budget [21]

Revenues	NA
Expenditures	NA
Capital expenditures	NA

Military Expenditures and Arms Transfers [22]

	1985	1986	1987	1988	1989
Military expenditures					
Current dollars (mil.)	NA	NA	NA	NA	NA
1989 constant dollars (mil.)	NA	NA	NA	NA	NA
Armed forces (000)	55	60[e]	55[e]	55	55
Gross national product (GNP)					
Current dollars (mil.)	2,918[e]	2,774[e]	NA	NA	NA
1989 constant dollars (mil.)	3,322[e]	3,079[e]	NA	NA	NA
Central government expenditures (CGE)					
1989 constant dollars (mil.)	NA	NA	NA	NA	NA
People (mil.)	13.9	14.0	14.2	14.5	14.8
Military expenditure as % of GNP	NA	NA	NA	NA	NA
Military expenditure as % of CGE	NA	NA	NA	NA	NA
Military expenditure per capita	NA	NA	NA	NA	NA
Armed forces per 1,000 people	4.0	4.3	3.9	3.8	3.7
GNP per capita	239	220	NA	NA	NA
Arms imports[6]					
Current dollars (mil.)	650	1,300	1,400	2,600	3,800
1989 constant dollars (mil.)	740	1,443	1,506	2,707	3,800
Arms exports[6]					
Current dollars (mil.)	0	0	0	0	0
1989 constant dollars (mil.)	0	0	0	0	0
Total imports[7]					
Current dollars (mil.)	1,194	1,404	996	900	822
1989 constant dollars	1,359	1,558	1,070	937	822
Total exports[7]					
Current dollars (mil.)	567	552	512	395	236
1989 constant dollars	645	613	551	411	236
Arms as percent of total imports[8]	54.4	92.6	140.6	288.9	462.3
Arms as percent of total exports[8]	0	0	0	0	0

Human Rights [24]

	SSTS	FL	FAPRO	PPCG	APROBC	TPW	PCPTW	STPEP	PHRFF	PRW	ASST	AFL
Observes	P			P		P	P	P		P	P	P
		EAFRD	CPR	ESCR	SR	ACHR	MAAE	PVIAC	PVNAC	EAFDAW	TCIDTP	RC
Observes		P	P	P						S	P	S

P = Party; S = Signatory; see Appendix for meaning of abbreviations.

Labor Force

Total Labor Force [25]

4.98 million

Labor Force by Occupation [26]

Agriculture and animal husbandry	67.8%
Industry	10.2
Construction	6.3
Commerce	5.0
Services and other	10.7

Date of data: 1980 est.

Unemployment Rate [27]

Production Sectors

Energy Resource Summary [28]

Energy Resources: Natural gas, petroleum, coal. **Electricity**: 480,000 kW capacity; 1,000 million kWh produced, 60 kWh per capita (1992).
Pipelines: Petroleum products—Uzbekistan to Bagram and Turkmenistan to Shindand; natural gas 180 km.

Telecommunications [30]

- Limited telephone, telegraph, and radiobroadcast services
- Television introduced in 1980
- 31,200 telephones
- Broadcast stations - 5 AM, no FM, 1 TV
- 1 satellite earth station

Top Agricultural Products [32]

Largely subsistence farming and nomadic animal husbandry; cash products—wheat, fruits, nuts, karakul pelts, wool, mutton.

Transportation [31]

Railroads. 9.6 km (single track) 1.524-meter gauge from Kushka (Turkmenistan) to Towraghondi and 15.0 km from Termez (Uzbekistan) to Kheyrabad transshipment point on south bank of Amu Darya

Highways. 21,000 km total (1984); 2,800 km hard surface, 1,650 km bituminous-treated gravel and improved earth, 16,550 km unimproved earth and tracks

Airports

Total:	41
Usable:	36
With permanent-surface runways:	9
With runways over 3,659 m:	0
With runways 2,440-3,659 m:	11
With runways 1,220-2,439 m:	16

Top Mining Products [33]

Metric tons unless otherwise specified	M.t.[2]
Barite	2,000
Cement, hydraulic	112,000
Coal, bituminous	170,000
Copper, mine output, Cu content	5,000
Gas, natural (gross; million cubic meters)	2,500
Gypsum	3,000
Natural gas liquids (000 42-gal. barrels)	35
Nitrogen, N content of ammonia	40,000
Salt, rock	12,000

Tourism [34]

Finance, Economics, and Trade

GDP and Manufacturing Summary [35]

	1980	1985	1990	1991	1992
Gross Domestic Product					
Millions of 1980 dollars	3,852	4,298	3,221	3,382	3,483[e]
Growth rate in percent	-3.72	0.20	-6.15	5.00	3.00[e]
Manufacturing Value Added					
Millions of 1980 dollars	272	258	254	256	267[e]
Growth rate in percent	-6.06	0.63	-2.50	1.15	3.98[e]
Manufacturing share in percent of current prices	NA	NA	NA	NA	NA

Economic Indicators [36]

National product: GDP—exchange rate conversion—$3 billion (1989 est.). **National product real growth rate**: NA%. **National product per capita**: $200 (1989 est.). **Inflation rate (consumer prices)**: over 90% (1991 est.). **External debt**: $2.3 billion (March 1991 est.).

Balance of Payments Summary [37]

Values in millions of dollars.

	1985	1986	1987	1988	1989
Exports of goods (f.o.b.)	628.2	497.0	538.7	453.8	252.3
Imports of goods (f.o.b.)	-921.6	-1,138.8	-904.5	-731.8	-623.5
Trade balance	-293.4	-641.8	-365.8	-278.0	-371.2
Services - debits	-162.7	-215.9	-167.6	-131.5	-111.3
Services - credits	69.2	53.1	54.8	92.9	28.3
Private transfers (net)	-	-	-	-	-1.2
Government transfers (net)	143.7	267.4	311.7	342.8	312.1
Long term capital (net)	77.6	217.6	113.6	22.4	-186.4
Short term capital (net)	23.2	84.5	-147.5	-26.5	126.8
Errors and omissions	168.4	216.4	211.6	-47.9	182.7
Overall balance	26.0	-18.7	10.8	-25.8	-20.2

Exchange Rates [38]

Currency: **afghanis.**
Symbol: **Af.**

These rates reflect the free market exchange rates rather than the official exchange rates. Data are currency units per $1.

March 1993	1,019
November 1991	900
1991	850
1989-90	700
1988-89	220

Imports and Exports

Top Import Origins [39]

$874 million (c.i.f., FY91 est.).

Origins	%
Former USSR	NA
Pakistan	NA

Top Export Destinations [40]

$236 million (f.o.b., FY91 est.).

Destinations	%
Former USSR	NA
Pakistan	NA

Foreign Aid [41]

	U.S. $	
US commitments, including Ex-Im (FY70-89)	380	million
Western (non-US) countries, ODA and OOF bilateral commitments (1970-89)	510	million
OPEC bilateral aid (1979-89)	57	million
Communist countries (1970-89)	4.1	billion
Net official Western disbursements (1985-89)	270	million

Import and Export Commodities [42]

Import Commodities

Food
Petroleum products

Export Commodities

Natural gas 55%
Fruits & nuts 24%
Handwoven carpets
Wool
Cotton
Hides
Pelts

Albania

Geography [1]

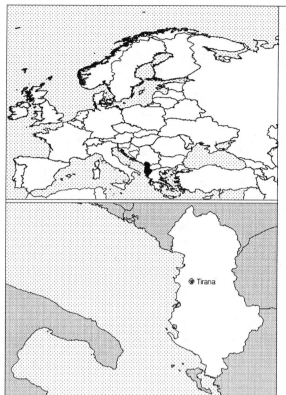

Total area:
 28,750 km2
Land area:
 27,400 km2
Comparative area:
 Slightly larger than Maryland
Land boundaries:
 Total 720 km, Greece 282 km, Macedonia 151 km, Serbia and Montenegro 287 km (114 km with Serbia, 173 km with Montenegro)
Coastline:
 362 km
Climate:
 Mild temperate; cool, cloudy, wet winters; hot, clear, dry summers; interior is cooler and wetter
Terrain:
 Mostly mountains and hills; small plains along coast
Natural resources:
 Petroleum, natural gas, coal, chromium, copper, timber, nickel
Land use:
 Arable land:
 21%
 Permanent crops:
 4%
 Meadows and pastures:
 15%
 Forest and woodland:
 38%
 Other:
 22%

Demographics [2]

	1960	1970	1980	1990	1991[1]	1994	2000	2010	2020
Population	1,607	2,136	2,673	3,249	3,335	3,374	3,610	4,016	4,424
Population density (persons per sq. mi.)	152	202	253	309	315	NA	361	403	442
Births	NA	NA	NA	NA	79	76	NA	NA	NA
Deaths	NA	NA	NA	NA	18	18	NA	NA	NA
Life expectancy - males	NA	NA	NA	NA	NA	70	NA	NA	NA
Life expectancy - females	NA	NA	NA	NA	NA	77	NA	NA	NA
Birth rate (per 1,000)	NA	NA	NA	NA	24	22	NA	NA	NA
Death rate (per 1,000)	NA	NA	NA	NA	5	5	NA	NA	NA
Women of reproductive age (15-44 yrs.)	NA	NA	NA	814	NA	844	926	NA	NA
of which are currently married	NA	NA	NA	599	NA	627	690	NA	NA
Fertility rate	NA	NA	NA	3.1	2.87	2.8	2.4	2.1	1.9

Population values are in thousands, life expectancy in years, and other items as indicated.

Health

Health Personnel [3]

Doctors per 1,000 pop., 1988-92	1.39
Nurse-to-doctor ratio, 1988-92	2.5
Hospital beds per 1,000 pop., 1985-90	4.1
Percentage of children immunized (age 1 yr. or less)	
Third dose of DPT, 1990-91	94.0
Measles, 1990-91	87.0

Health Expenditures [4]

Total health expenditure, 1990 (official exchange rate)	
Millions of dollars	84
Millions of dollars per capita	26
Health expenditures as a percentage of GDP	
Total	4.0
Public sector	3.4
Private sector	0.6
Development assistance for health	
Total aid flows (millions of dollars)[1]	NA
Aid flows per capita (millions of dollars)	NA
Aid flows as a percentage of total health expenditure	NA

For sources, notes, and explanations, see Annotated Source Appendix, page 1035.

Human Factors	
Health Care Ratios [5]	Infants and Malnutrition [6]

Ethnic Division [7]		Religion [8]		Major Languages [9]

Ethnic Division [7]

Other includes Vlachs, Gypsies, Serbs, and Bulgarians.

Albanian	90%
Greeks	8%
Other	2%

Religion [8]

Muslim	70%
Greek Orthodox	20%
Roman Catholic	10%

Major Languages [9]

Albanian (Tosk is the official dialect), Greek.

Education

Public Education Expenditures [10]

Million Lek	1980	1985	1987	1988	1989	1990
Total education expenditure	767	NA	946	952	937	984
as percent of GNP	NA	NA	NA	NA	NA	NA
as percent of total govt. expend.	10.3	NA	11.2	11.1	NA	9.1
Current education expenditure	NA	NA	NA	NA	NA	NA
as percent of GNP	NA	NA	NA	NA	NA	NA
as percent of current govt. expend.	NA	NA	NA	NA	NA	NA
Capital expenditure	NA	NA	NA	NA	NA	NA

Educational Attainment [11]

Literacy Rate [12]

Libraries [13]

	Admin. Units	Svc. Pts.	Vols. (000)	Shelving (meters)	Vols. Added	Reg. Users
National (1986)	1	NA	883	17,049	19,701	8,710
Non-specialized (1986)	40	NA	2,344	NA	NA	116,755
Public (1988)	45	NA	4,072	NA	NA	228,786[26]
Higher ed.	NA	NA	NA	NA	NA	NA
School	NA	NA	NA	NA	NA	NA

Daily Newspapers [14]

	1975	1980	1985	1990
Number of papers	2	2	2	2
Circ. (000)	115	145	135	135

Cinema [15]

Data for 1989.

Cinema seats per 1,000	8.9
Annual attendance per person	2.2
Gross box office receipts (mil. Lek)	5.5

Science and Technology

Scientific/Technical Forces [16]	R&D Expenditures [17]	U.S. Patents Issued [18]

For sources, notes, and explanations, see Annotated Source Appendix, page 1035.

7

Government and Law

Organization of Government [19]

Long-form name:
Republic of Albania
Type:
nascent democracy
Independence:
28 November 1912 (from Ottoman Empire)
Constitution:
an interim basic law was approved by the People's Assembly on 29 April 1991; a new constitution was to be drafted for adoption in 1992, but is still in process
Legal system:
has not accepted compulsory ICJ jurisdiction
National holiday:
Liberation Day, 29 November (1944)
Executive branch:
president, prime minister of the Council of Ministers, two deputy prime ministers of the Council of Ministers
Legislative branch:
unicameral People's Assembly (Kuvendi Popullor)
Judicial branch:
Supreme Court

Crime [23]

Political Parties [20]

People's Assembly	% of votes
Democratic Party (DP)	62.3
Albanian Socialist Party (ASP)	25.6
Social Democratic Party (SDP)	4.3
Albanian Republican Party (RP)	3.2
Unity for Human Rights (UHP)	2.9
Other	1.7

Government Budget [21]

For 1991 est.

Revenues	1.10
Expenditures	1.40
Capital expenditures	0.07

Military Expenditures and Arms Transfers [22]

	1985	1986	1987	1988	1989
Military expenditures					
Current dollars (mil.)	143	NA	151[e]	157[e]	157[e]
1989 constant dollars (mil.)	163	NA	162[e]	163[e]	157[e]
Armed forces (000)	42	42	42	42	43
Gross national product (GNP)					
Current dollars (mil.)	2,700	2,800	2,940[e]	3,087[e]	3,800[e]
1989 constant dollars (mil.)	3,074	3,107	3,162[e]	3,214[e]	3,800[e]
Central government expenditures (CGE)					
1989 constant dollars (mil.)	1,491	NA	1,463[e]	1,434[e]	NA
People (mil.)	3.0	3.0	3.1	3.1	3.2
Military expenditure as % of GNP	5.3	NA	5.1	5.1	4.1
Military expenditure as % of CGE	10.9	NA	11.1	11.4	NA
Military expenditure per capita	55	NA	53	52	49
Armed forces per 1,000 people	14.1	13.9	13.7	13.3	13.5
GNP per capita	1,037	1,028	1,026	1,021	1,184
Arms imports[6]					
Current dollars (mil.)	0	0	0	0	0
1989 constant dollars (mil.)	0	0	0	0	0
Arms exports[6]					
Current dollars (mil.)	0	0	0	0	0
1989 constant dollars (mil.)	0	0	0	0	0
Total imports[7]					
Current dollars (mil.)	335	363	255	NA	NA
1989 constant dollars	381	403	274	NA	NA
Total exports[7]					
Current dollars (mil.)	345	428	378	NA	NA
1989 constant dollars	393	475	407	NA	NA
Arms as percent of total imports[8]	0	0	0	NA	NA
Arms as percent of total exports[8]	0	0	0	NA	NA

Human Rights [24]

	SSTS	FL	FAPRO	PPCG	APROBC	TPW	PCPTW	STPEP	PHRFF	PRW	ASST	AFL
Observes	P	P	P	P	P	P	P	P		P	P	
	EAFRD	CPR	ESCR	SR	ACHR	MAAE	PVIAC	PVNAC	EAFDAW	TCIDTP		RC
Observes			P	P	P							P

P = Party; S = Signatory; see Appendix for meaning of abbreviations.

Labor Force

Total Labor Force [25]

1.5 million (1987)

Labor Force by Occupation [26]

Agriculture	60%
Industry and commerce	40

Date of data: 1986

Unemployment Rate [27]

40% (1992 est.)

For sources, notes, and explanations, see Annotated Source Appendix, page 1035.

Production Sectors

Commercial Energy Production and Consumption

Production [28]

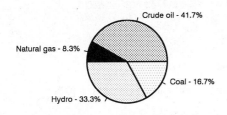

Crude oil - 41.7%
Natural gas - 8.3%
Coal - 16.7%
Hydro - 33.3%

Consumption [29]

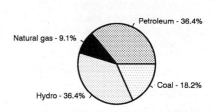

Petroleum - 36.4%
Natural gas - 9.1%
Coal - 18.2%
Hydro - 36.4%

Data are shown in quadrillion (10^{15}) BTUs and percent for 1991

Crude oil[1]	0.05
Natural gas liquids	0.00
Dry natural gas	0.01
Coal[2]	0.02
Net hydroelectric power[3]	0.04
Net nuclear power[3]	0.00
Total	0.12

Petroleum	0.04
Dry natural gas	0.01
Coal[2]	0.02
Net hydroelectric power[3]	0.04
Net nuclear power[3]	0.00
Total	0.11

Telecommunications [30]

- Inadequate service
- 15,000 telephones
- Broadcast stations - 13 AM, 1 TV
- 514,000 radios, 255,000 TVs (1987 est.)

Transportation [31]

Railroads. 543 km total; 509 km 1.435-meter standard gauge, single track and 34 km narrow gauge, single track (1990); line connecting Titograd (Serbia and Montenegro) and Shkoder (Albania) completed August 1986

Highways. 16,700 km total; 6,700 km highways, 10,000 km forest and agricultural cart roads (1990)

Merchant Marine. 11 cargo ships (1,000 GRT or over) totaling 52,967 GRT/76,887 DWT

Airports

Total:	12
Usable:	10
With permanent-surface runways:	3
With runways over 3,659 m:	0
With runways 2,440-3,659 m:	6
With runways 1,220-2,439 m:	4

Top Agricultural Products [32]

	1987	1988
Cereals	1,0110	1,024
Sugarbeets	360	360
Vegetables (including melons)	236	236
Fruit	210	216
Potatoes	135	137

Values shown are 1,000 metric tons.

Top Mining Products [33]

Estimated metric tons unless otherwise specified   M.t.

Chromite, concentrate (000 tons)	75,000
Nickel, mine output, Ni content	7,500
Copper ore, Cu in concentrate (000)	6,100
Coal, lignite (000 tons)	1,100
Dolomite	350,000
Cement, hydraulic (000 tons)	700

Tourism [34]

For sources, notes, and explanations, see Annotated Source Appendix, page 1035.

9

Finance, Economics, and Trade

GDP and Manufacturing Summary [35]

	1980	1985	1990	1991	1992
Gross Domestic Product					
Millions of 1980 dollars	2,373	2,711	3,160	2,718	2,283[e]
Growth rate in percent	6.29	1.48	2.92	-14.00	-16.00[e]
Manufacturing Value Added					
Millions of 1980 dollars	912	1,111	1,350	776	621[e]
Growth rate in percent	6.08	1.57	3.50	-42.50	-20.00[e]
Manufacturing share in percent of current prices	36.1	34.8	33.8	NA	NA

Economic Indicators [36]

National product: GDP—purchasing power equivalent—$2.5 billion (1992 est.). **National product real growth rate**: - 10% (1992 est.). **National product per capita**: $760 (1992 est.). **Inflation rate (consumer prices)**: 210% (1992 est.). **External debt**: $500 million (1992 est.).

Balance of Payments Summary [37]

Exchange Rates [38]

Currency: **leke.**
Symbol: **L.**

Data are currency units per $1.

January 1993	97
January 1992	50
September 1991	25

Imports and Exports

Top Import Origins [39]

$120 million (f.o.b., 1992 est.).

Origins	%
Italy	NA
Macedonia	NA
Germany	NA
Czechoslovakia	NA
Romania	NA
Poland	NA
Hungary	NA
Bulgaria	NA
Greece	NA

Top Export Destinations [40]

$45 million (f.o.b., 1992 est.).

Destinations	%
Italy	NA
Macedonia	NA
Germany	NA
Greece	NA
Czechoslovakia	NA
Poland	NA
Romania	NA
Bulgaria	NA
Hungary	NA

Foreign Aid [41]

	U.S. $	
Recipient - humanitarian aid	190	million
Loans/guarantees/credits	94	million

Import and Export Commodities [42]

Import Commodities

Machinery
Consumer goods
Grains

Export Commodities

Asphalt
Metals and metallic ores
Electricity
Crude oil
Vegetables
Fruits
Tobacco

10

For sources, notes, and explanations, see Annotated Source Appendix, page 1035.

Algeria

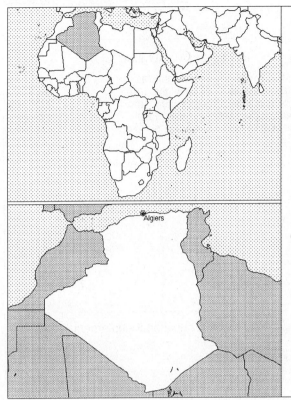

Geography [1]

Total area:
2,381,740 km2
Land area:
2,381,740 km2
Comparative area:
Slightly less than 3.5 times the size of Texas
Land boundaries:
Total 6,343 km, Libya 982 km, Mali 1,376 km, Mauritania 463 km, Morocco 1,559 km, Niger 956 km, Tunisia 965 km, Western Sahara 42 km
Coastline:
998 km
Climate:
Arid to semiarid; mild, wet winters with hot, dry summers along coast; drier with cold winters and hot summers on high plateau; sirocco is a hot, dust/sand-laden wind especially common in summer
Terrain:
Mostly high plateau and desert; some mountains; narrow, discontinuous coastal plain
Natural resources:
Petroleum, natural gas, iron ore, phosphates, uranium, lead, zinc
Land use:
Arable land: 3%
Permanent crops: 0%
Meadows and pastures: 13%
Forest and woodland: 2%
Other: 82%

Demographics [2]

	1960	1970	1980	1990	1991[1]	1994	2000	2010	2020
Population	10,909	13,932	18,862	25,352	26,022	27,895	31,743	38,186	44,096
Population density (persons per sq. mi.)	12	15	21	28	28	NA	35	43	50
Births	NA	NA	NA	NA	826	829	NA	NA	NA
Deaths	NA	NA	NA	NA	177	174	NA	NA	NA
Life expectancy - males	NA	NA	NA	NA	66	67	NA	NA	NA
Life expectancy - females	NA	NA	NA	NA	68	69	NA	NA	NA
Birth rate (per 1,000)	NA	NA	NA	NA	32	30	NA	NA	NA
Death rate (per 1,000)	NA	NA	NA	NA	7	6	NA	NA	NA
Women of reproductive age (15-44 yrs.)	NA	NA	NA	5,740	NA	6,651	8,255	NA	NA
of which are currently married	NA	NA	NA	3,310	NA	3,873	4,829	NA	NA
Fertility rate	NA	NA	NA	4.4	4.24	3.8	3.2	2.5	2.2

Population values are in thousands, life expectancy in years, and other items as indicated.

Health

Health Personnel [3]

Doctors per 1,000 pop., 1988-92	0.26
Nurse-to-doctor ratio, 1988-92	4.7
Hospital beds per 1,000 pop., 1985-90	2.6
Percentage of children immunized (age 1 yr. or less)	
Third dose of DPT, 1990-91	89.0
Measles, 1990-91	83.0

Health Expenditures [4]

Total health expenditure, 1990 (official exchange rate)	
Millions of dollars	4159
Millions of dollars per capita	166
Health expenditures as a percentage of GDP	
Total	7.0
Public sector	5.4
Private sector	1.6
Development assistance for health	
Total aid flows (millions of dollars)[1]	2
Aid flows per capita (millions of dollars)	0.1
Aid flows as a percentage of total health expenditure	0.1

For sources, notes, and explanations, see Annotated Source Appendix, page 1035.

11

Human Factors

Health Care Ratios [5]

Population per physician, 1970	8100
Population per physician, 1990	2330
Population per nursing person, 1970	NA
Population per nursing person, 1990	330
Percent of births attended by health staff, 1985	NA

Infants and Malnutrition [6]

Percent of babies with low birth weight, 1985	9
Infant mortality rate per 1,000 live births, 1970	139
Infant mortality rate per 1,000 live births, 1991	64
Years of life lost per 1,000 population, 1990	27
Prevalence of malnutrition (under age 5), 1990	NA

Ethnic Division [7]

Arab-Berber	99%
European	<1%

Religion [8]

Sunni Muslim (state religion)	99%
Christian and Jewish	1%

Major Languages [9]

Arabic (official), French, Berber dialects.

Education

Public Education Expenditures [10]

Million Dinar	1985	1986	1987	1988	1989	1990
Total education expenditure	24,248	26,487	30,064	30,515	32,826	NA
as percent of GNP	8.5	9.1	9.8	9.1	8.1	NA
as percent of total govt. expend.	20.7	20.7	27.8	27.0	27.0	NA
Current education expenditure	16,814	19,629	22,210	23,415	25,696	NA
as percent of GNP	5.9	6.7	7.2	7.0	6.3	NA
as percent of current govt. expend.	26.2	29.3	35.3	35.7	35.7	NA
Capital expenditure	7,434	6,858	7,854	7,100	7,130	NA

Educational Attainment [11]

Literacy Rate [12]

In thousands and percent	1985[a]	1991[a]	2000[a]
Illiterate population +15 years	6,062	6,004	5,578
Illiteracy rate - total pop. (%)	51.4	42.6	28.1
Illiteracy rate - males (%)	37.3	30.2	19.3
Illiteracy rate - females (%)	64.9	54.5	36.7

Libraries [13]

	Admin. Units	Svc. Pts.	Vols. (000)	Shelving (meters)	Vols. Added	Reg. Users[1]
National (1987)	1	1	1,020	NA	7,307	13,130
Non-specialized	NA	NA	NA	NA	NA	NA
Public	NA	NA	NA	NA	NA	NA
Higher ed.	NA	NA	NA	NA	NA	NA
School	NA	NA	NA	NA	NA	NA

Daily Newspapers [14]

	1975	1980	1985	1990
Number of papers	4	4	5	10
Circ. (000)	285	448	570	1,274

Cinema [15]

Data for 1988.

Cinema seats per 1,000	NA
Annual attendance per person	0.9
Gross box office receipts (mil. Dinar)	181

Science and Technology

Scientific/Technical Forces [16]

R&D Expenditures [17]

U.S. Patents Issued [18]

Government and Law

Organization of Government [19]

Long-form name:
Democratic and Popular Republic of Algeria
Type:
republic
Independence:
5 July 1962 (from France)
Constitution:
19 November 1976, effective 22 November 1976; revised February 1989
Legal system:
socialist, based on French and Islamic law; judicial review of legislative acts in ad hoc Constitutional Council composed of various public officials, including several Supreme Court justices; has not accepted compulsory ICJ jurisdiction
National holiday:
Anniversary of the Revolution, 1 November (1954)
Executive branch:
President of the High State Committee, prime minister, Council of Ministers (cabinet)
Legislative branch:
unicameral National People's Assembly (Al-Majlis Ech-Chaabi Al-Watani)
Judicial branch:
Supreme Court (Cour Supreme)

Crime [23]

Elections [20]

National People's Assembly. The government established a multiparty system in September 1989 and, as of 31 December 1990, over 30 legal parties existed. First round held on 26 December 1991 (second round canceled by the military after President Bendjedid resigned 11 January 1992); results - percent of vote by party NA; seats - (281 total); the fundamentalist FIS won 188 of the 231 seats contested in the first round.

Government Budget [21]

For 1992 est.

Revenues	14.4
Expenditures	14.6
Capital expenditures	3.5

Military Expenditures and Arms Transfers [22]

	1985	1986	1987	1988	1989
Military expenditures					
Current dollars (mil.)[e,2]	1,040	1,271	1,264	1,595	2,313
1989 constant dollars (mil.)[e,2]	1,183	1,411	1,359	1,660	2,313
Armed forces (000)	170	180	170	126	126
Gross national product (GNP)					
Current dollars (mil.)	40,020	41,530	42,590	42,390	45,290
1989 constant dollars (mil.)	45,550	46,090	45,800	44,130	45,290
Central government expenditures (CGE)					
1989 constant dollars (mil.)	18,840	18,850	17,790	17,480	NA
People (mil.)	22.2	22.8	23.5	24.1	24.7
Military expenditure as % of GNP	2.6	3.1	3.0	3.8	5.1
Military expenditure as % of CGE	6.3	7.5	7.6	9.5	NA
Military expenditure per capita	53	62	58	69	94
Armed forces per 1,000 people	7.7	7.9	7.2	5.2	5.1
GNP per capita	2,054	2,018	1,950	1,831	1,831
Arms imports[6]					
Current dollars (mil.)	480	625	700	850	575
1989 constant dollars (mil.)	546	694	753	885	575
Arms exports[6]					
Current dollars (mil.)	0	0	0	0	0
1989 constant dollars (mil.)	0	0	0	0	0
Total imports[7]					
Current dollars (mil.)	9,841	9,228	7,042	7,342	8,818
1989 constant dollars	11,200	10,240	7,574	7,643	8,818
Total exports[7]					
Current dollars (mil.)	12,840	7,430	8,605	7,680	9,100
1989 constant dollars	14,620	8,245	9,255	7,995	9,100
Arms as percent of total imports[8]	4.9	6.8	9.9	11.6	6.5
Arms as percent of total exports[8]	0	0	0	0	0

Human Rights [24]

	SSTS	FL	FAPRO	PPCG	APROBC	TPW	PCPTW	STPEP	PHRFF	PRW	ASST	AFL
Observes	P	P	P	P	P	P	P	P			P	P
	EAFRD	CPR	ESCR	SR	ACHR	MAAE	PVIAC	PVNAC	EAFDAW	TCIDTP	RC
Observes		P	P	P	P		P	P	P		P	S

P = Party; S = Signatory; see Appendix for meaning of abbreviations.

Labor Force

Total Labor Force [25]

6.2 million (1992 est.)

Labor Force by Occupation [26]

Government	29.5%
Agriculture	22
Construction and public works	16.2
Industry	13.6
Commerce and services	13.5
Transportation and communication	5.2

Date of data: 1989

Unemployment Rate [27]

35% (1992 est.)

For sources, notes, and explanations, see Annotated Source Appendix, page 1035.

13

Production Sectors

Commercial Energy Production and Consumption

Production [28]

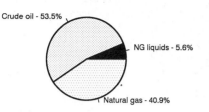

Crude oil - 53.5%
NG liquids - 5.6%
Natural gas - 40.9%

Consumption [29]

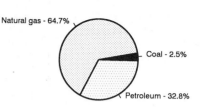

Natural gas - 64.7%
Coal - 2.5%
Petroleum - 32.8%

Data are shown in quadrillion (10^{15}) BTUs and percent for 1991

Crude oil[1]	2.49
Natural gas liquids	0.26
Dry natural gas	1.90
Coal[2]	(s)
Net hydroelectric power[3]	(s)
Net nuclear power[3]	0.00
Total	4.65

Petroleum	0.39
Dry natural gas	0.77
Coal[2]	0.03
Net hydroelectric power[3]	(s)
Net nuclear power[3]	0.00
Total	1.19

Telecommunications [30]

- Excellent domestic and international service in the north, sparse in the south
- 822,000 telephones
- Broadcast stations - 26 AM, no FM, 18 TV
- 1,600,000 TV sets
- 5,200,000 radios
- 5 submarine cables
- Microwave radio relay to Italy, France, Spain, Morocco, and Tunisia
- Coaxial cable to Morocco and Tunisia
- Satellite earth stations - 1 Atlantic Ocean INTELSAT, 1 Indian Ocean INTELSAT, 1 Intersputnik, I ARABSAT, and 12 domestic
- 20 additional satellite earth stations are planned

Transportation [31]

Railroads. 4,060 km total; 2,616 km standard gauge (1.435 m), 1,188 km 1.055-meter gauge, 256 km 1.000-meter gauge; 300 km electrified; 215 km double track

Highways. 90,031 km total; 58,868 km concrete or bituminous, 31,163 km gravel, crushed stone, unimproved earth (1990)

Merchant Marine. 75 ships (1,000 GRT or over) totaling 903,179 GRT/1,064,211 DWT; includes 5 short-sea passenger, 27 cargo, 12 roll-on/roll-off cargo, 5 oil tanker, 9 liquefied gas, 7 chemical tanker, 9 bulk, 1 specialized tanker

Airports

Total:	141
Usable:	124
With permanent-surface runways:	53
With runways over 3,659 m:	2
With runways 2,440-3,659 m:	32
With runways 1,220-2,439 m:	65

Top Agricultural Products [32]

	1989	1990
Vegetables	1,463	1,188
Barley	789	833
Potatoes	1,007	809
Wheat	1,152	750
Citrus	200	280
Dates	210	206

Values shown are 1,000 metric tons.

Top Mining Products [33]

Metric tons unless otherwise specified	M.t.
Phosphate rock (000 tons)	1,090[1]
Mercury	431[1]
Gas, natural (mil. cubic meters)	140,000
Iron ore (000 tons)	2,344
Cement, hydraulic (000 tons)	6,319
Salt (000 tons)	211[1]

Tourism [34]

	1987	1988	1989	1990	1991
Tourists[1]	778	967	1,207	1,137	1,193
Tourism receipts	100	85	64	64	
Tourism expenditures	239	294	212	149	
Fare receipts	74	63	82	86	
Fare expenditures	81	94	149	124	

Tourists are in thousands, money in million U.S. dollars.

Manufacturing Sector

GDP and Manufacturing Summary [35]

	1980	1985	1989	1990	% change 1980-1990	% change 1989-1990
GDP (million 1980 $)	42,342	53,959	56,895	56,964	34.5	0.1
GDP per capita (1980 $)	2,259	2,477	2,342	2,282	1.0	-2.6
Manufacturing as % of GDP (current prices)	10.9	11.6	11.5[e]	9.1	-16.5	-20.9
Gross output (million $)	9,122	13,978[e]	13,765[e]	13,238[e]	45.1	-3.8
Value added (million $)	3,644	6,157	5,997[e]	5,739[e]	57.5	-4.3
Value added (million 1980 $)	3,286	5,029	6,090[e]	5,326	62.1	-12.5
Industrial production index	100	154	161[e]	157	57.0	-2.5
Employment (thousands)	312	413[e]	467[e]	470[e]	50.6	0.6

Note: GDP stands for Gross Domestic Product. 'e' stands for estimated value.

Profitability and Productivity

	1980	1985	1989	1990	% change 1980-1990	% change 1989-1990
Intermediate input (%)	60	56[e]	56[e]	57[e]	-5.0	1.8
Wages, salaries, and supplements (%)	22	25[e]	28[e]	26[e]	18.2	-7.1
Gross operating surplus (%)	18	19[e]	16[e]	17[e]	-5.6	6.3
Gross output per worker ($)	29,246	33,067[e]	29,445[e]	2,763[e]	-90.6	-90.6
Value added per worker ($)	11,682	14,740[e]	12,828[e]	12,019[e]	2.9	-6.3
Average wage (incl. benefits) ($)	6,523	8,303[e]	8,199[e]	7,377[e]	13.1	-10.0

Profitability is in percent of gross output. Productivity is in U.S. $. 'e' stands for estimated value.

Profitability - 1990

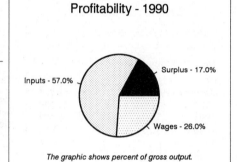

Inputs - 57.0%
Surplus - 17.0%
Wages - 26.0%

The graphic shows percent of gross output.

Value Added in Manufacturing

	1980 $ mil.	1980 %	1985 $ mil.	1985 %	1989 $ mil.	1989 %	1990 $ mil.	1990 %	% change 1980-1990	% change 1989-1990
311 Food products	655	18.0	852	13.8	847[e]	14.1	815[e]	14.2	24.4	-3.8
313 Beverages	135	3.7	176	2.9	185[e]	3.1	172[e]	3.0	27.4	-7.0
314 Tobacco products	176	4.8	229	3.7	211[e]	3.5	216[e]	3.8	22.7	2.4
321 Textiles	291	8.0	450	7.3	399[e]	6.7	421[e]	7.3	44.7	5.5
322 Wearing apparel	234	6.4	362	5.9	319[e]	5.3	370[e]	6.4	58.1	16.0
323 Leather and fur products	52	1.4	80	1.3	84[e]	1.4	73[e]	1.3	40.4	-13.1
324 Footwear	90	2.5	140	2.3	146[e]	2.4	127[e]	2.2	41.1	-13.0
331 Wood and wood products	120	3.3	205	3.3	216[e]	3.6	189[e]	3.3	57.5	-12.5
332 Furniture and fixtures	57	1.6	97	1.6	101[e]	1.7	89[e]	1.6	56.1	-11.9
341 Paper and paper products	143	3.9	242	3.9	266[e]	4.4	223[e]	3.9	55.9	-16.2
342 Printing and publishing	16	0.4	27	0.4	30[e]	0.5	25[e]	0.4	56.3	-16.7
351 Industrial chemicals	14	0.4	25	0.4	26[e]	0.4	22[e]	0.4	57.1	-15.4
352 Other chemical products	93	2.6	167	2.7	172[e]	2.9	170[e]	3.0	82.8	-1.2
353 Petroleum refineries	83	2.3	150	2.4	149[e]	2.5	162[e]	2.8	95.2	8.7
354 Miscellaneous petroleum and coal products	4	0.1	7	0.1	8[e]	0.1	7[e]	0.1	75.0	-12.5
355 Rubber products	17	0.5	30	0.5	30[e]	0.5	25[e]	0.4	47.1	-16.7
356 Plastic products	34	0.9	61	1.0	66[e]	1.1	58[e]	1.0	70.6	-12.1
361 Pottery, china and earthenware	10	0.3	14	0.2	15[e]	0.3	14[e]	0.2	40.0	-6.7
362 Glass and glass products	36	1.0	51	0.8	55[e]	0.9	52[e]	0.9	44.4	-5.5
369 Other non-metal mineral products	355	9.7	497	8.1	543[e]	9.1	510[e]	8.9	43.7	-6.1
371 Iron and steel	323	8.9	727	11.8	795[e]	13.3	574[e]	10.0	77.7	-27.8
372 Non-ferrous metals	19	0.5	42	0.7	47[e]	0.8	37[e]	0.6	94.7	-21.3
381 Metal products	265	7.3	598	9.7	496[e]	8.3	576[e]	10.0	117.4	16.1
382 Non-electrical machinery	46	1.3	105	1.7	87[e]	1.5	89[e]	1.6	93.5	2.3
383 Electrical machinery	123	3.4	278	4.5	231[e]	3.9	265[e]	4.6	115.4	14.7
384 Transport equipment	181	5.0	407	6.6	338[e]	5.6	317[e]	5.5	75.1	-6.2
385 Professional and scientific equipment	30	0.8	67	1.1	56[e]	0.9	69[e]	1.2	130.0	23.2
390 Other manufacturing industries	42	1.2	72	1.2	79[e]	1.3	72[e]	1.3	71.4	-8.9

Note: The industry codes shown are International Standard Industry codes (ISIC). Percentages are percent of total Value Added. 'e' stands for estimated value

For sources, notes, and explanations, see Annotated Source Appendix, page 1035.

15

Finance, Economics, and Trade

Economic Indicators [36]

Billions of Algerian Dinars unless otherwise stated.

	1989[e]	1990	1991
Real GDP (1980 Prices)	208.1	210.4	214.6
Real GDP Growth Rate (%. Chg.)	3.4	1.1	2.0
Money Supply (M1)	250.0	275.2	295.0
Commercial Interest Rates (Short-Term)	8.5	13.0	17.0
Savings Rate (%. of GDP)	31.0	40.0	35.0
CPI (%. Chg.)	17.0	30.0	50.0
WPI	12.0	25.0	40.0
External Debt	25,325	26,100	26,700

Balance of Payments Summary [37]

Values in millions of dollars.

	1986	1987	1988	1989	1990
Exports of goods (f.o.b.)	8,065.0	9,029.0	7,620.0	9,534.0	12,964.0
Imports of goods (f.o.b.)	-7,879.0	-6,616.0	-6,675.0	-8,372.0	-8,777.0
Trade balance	186.0	2,413.0	945.0	1,162.0	1,487.0
Services - debits	-3,900.0	-3,464.0	-3,917.0	-3,390.0	-3,671.0
Services - credits	721.0	675.0	542.0	607.0	571.0
Private transfers (net)	765.0	522.0	385.0	535.0	332.0
Government transfers (net)	-1.0	-5.0	5.0	6.0	1.0
Long term capital (net)	364.0	21.0	767.0	715.0	-926.0
Short term capital (net)	226.0	289.0	-23.0	40.0	-74.0
Errors and omissions	142.0	-802.0	337.0	-448.0	-336.0
Overall balance	-1,497.0	-351.0	-959.0	-773.0	84.0

Exchange Rates [38]

Currency: **Algerian dinars.**
Symbol: **DA.**

Data are currency units per $1.

January 1993	22.787
1992	21.836
1991	18.473
1990	8.958
1989	7.609
1988	5.915

Imports and Exports

Top Import Origins [39]

$8.2 billion (f.o.b., 1992 est.).

Origins	%
France	NA
Italy	NA
Germany	NA
US	NA
Spain	NA

Top Export Destinations [40]

$11.6 billion (f.o.b., 1992 est.).

Destinations	%
Italy	NA
France	NA
US	NA
Germany	NA
Spain	NA

Foreign Aid [41]

	U.S. $	
US commitments, including Ex-Im (FY70-85)	1.4	billion
Western (non-US) countries, ODA and OOF bilateral commitments (1970-89)	925	million
OPEC bilateral aid (1979-89)	1.8	billion
Communist countries (1970-89)	2.7	billion
Net official disbursements (1985-89)	375	million

Import and Export Commodities [42]

Import Commodities

Capital goods 39.7%
Food and beverages 21.7%
Consumer goods 11.8%

Export Commodities

Petroleum & natural gas 97%

                                        For sources, notes, and explanations, see Annotated Source Appendix, page 1035.

Angola

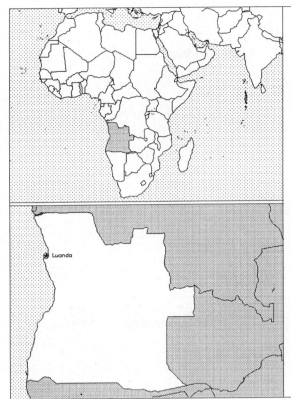

Geography [1]

Total area:
 1,246,700 km2
Land area:
 1,246,700 km2
Comparative area:
 Slightly less than twice the size of Texas
Land boundaries:
 Total 5,198 km, Congo 201 km, Namibia 1,376 km, Zaire 2,511 km, Zambia 1,110 km
Coastline:
 1,600 km
Climate:
 Semiarid in south and along coast to Luanda; north has cool, dry season (May to October) and hot, rainy season (November to April)
Terrain:
 Narrow coastal plain rises abruptly to vast interior plateau
Natural resources:
 Petroleum, diamonds, iron ore, phosphates, copper, feldspar, gold, bauxite, uranium
Land use:
 Arable land: 2%
 Permanent crops: 0%
 Meadows and pastures: 23%
 Forest and woodland: 43%
 Other: 32%

Demographics [2]

	1960	1970	1980	1990	1991[1]	1994	2000	2010	2020
Population	4,797	5,606	6,794	8,430	558,377	9,804	11,513	14,982	19,272
Population density (persons per sq. mi.)	10	12	14	18	18	NA	24	31	40
Births	NA	NA	NA	NA	403	445	NA	NA	NA
Deaths	NA	NA	NA	NA	172	182	NA	NA	NA
Life expectancy - males	NA	NA	NA	NA	61	44	NA	NA	NA
Life expectancy - females	NA	NA	NA	NA	64	48	NA	NA	NA
Birth rate (per 1,000)	NA	NA	NA	NA	47	45	NA	NA	NA
Death rate (per 1,000)	NA	NA	NA	NA	20	19	NA	NA	NA
Women of reproductive age (15-44 yrs.)	NA	NA	NA	1,895	NA	2,186	2,578	NA	NA
of which are currently married	NA	NA	NA	1,396	NA	1,614	1,903	NA	NA
Fertility rate	NA	NA	NA	6.7	6.66	6.5	6.1	5.2	4.2

Population values are in thousands, life expectancy in years, and other items as indicated.

Health

Health Personnel [3]

Doctors per 1,000 pop., 1988-92	0.07
Nurse-to-doctor ratio, 1988-92	16.4
Hospital beds per 1,000 pop., 1985-90	1.2
Percentage of children immunized (age 1 yr. or less)	
Third dose of DPT, 1990-91	26.0
Measles, 1990-91	39.0

Health Expenditures [4]

Total health expenditure, 1990 (official exchange rate)	
Millions of dollars	NA
Millions of dollars per capita	NA
Health expenditures as a percentage of GDP	
Total	NA
Public sector	NA
Private sector	NA
Development assistance for health	
Total aid flows (millions of dollars)[1]	28
Aid flows per capita (millions of dollars)	2.8
Aid flows as a percentage of total health expenditure	NA

For sources, notes, and explanations, see Annotated Source Appendix, page 1035.

17

Human Factors	
Health Care Ratios [5]	Infants and Malnutrition [6]

Ethnic Division [7]		Religion [8]		Major Languages [9]
Ovimbundu	37%	Indigenous beliefs	47%	Portuguese (official), Bantu dialects.
Kimbundu	25%	Roman Catholic	38%	
Bakongo	13%	Protestant	15%	
Mestico	2%			
European	1%			
Other	22%			

Education

Public Education Expenditures [10]

Million Kwansa	1985	1986	1987	1988	1989	1990[1]
Total education expenditure	9,643	12,883	12,854	NA	NA	12,076
as percent of GNP	NA	NA	NA	NA	NA	NA
as percent of total govt. expend.	10.8	12.6	13.8	NA	NA	10.7
Current education expenditure	9,419	12,455	11,567	NA	NA	10,856
as percent of GNP	NA	NA	NA	NA	NA	NA
as percent of current govt. expend.	14.0	15.6	15.5	NA	NA	NA
Capital expenditure	224	428	1,287	NA	NA	1,220

Educational Attainment [11]

Literacy Rate [12]

In thousands and percent	1985[a]	1991[a]	2000[a]
Illiterate population + 15 years	3,117	3,221	3,395
Illiteracy rate - total pop. (%)	64.3	58.3	46.6
Illiteracy rate - males (%)	50.4	44.4	33.6
Illiteracy rate - females (%)	77.4	71.5	59.1

Libraries [13]

Daily Newspapers [14]

	1975	1980	1985	1990
Number of papers	4	4	4	4
Circ. (000)	85[e]	143	103	115[e]

Cinema [15]

Science and Technology

Scientific/Technical Forces [16]	R&D Expenditures [17]	U.S. Patents Issued [18]

Government and Law

Organization of Government [19]

Long-form name:
Republic of Angola
Type:
transitional government nominally a multiparty democracy with a strong presidential system
Independence:
11 November 1975 (from Portugal)
Constitution:
11 November 1975; revised 7 January 1978, 11 August 1980, and 6 March 1991
Legal system:
based on Portuguese civil law system and customary law; recently modified to accommodate political pluralism and increased use of free markets
National holiday:
Independence Day, 11 November (1975)
Executive branch:
president, prime minister, Council of Ministers (cabinet)
Legislative branch:
unicameral National Assembly (Assembleia Nacional)
Judicial branch:
Supreme Court (Tribunal da Relacrao)

Crime [23]

Elections [20]

Popular Movement for the Liberation of Angola (MPLA) is the ruling party and has been in power since 1975; National Union for the Total Independence of Angola (UNITA) remains a legal party despite its returned to armed resistance to the government; five minor parties have small numbers of seats in the National Assembly. First nationwide, multiparty elections were held in late September 1992 with disputed results; further elections are being discussed.

Government Budget [21]

For 1991 est.

Revenues	2.100
Expenditures	3.600
Capital expenditures	0.963

Military Expenditures and Arms Transfers [22]

	1985	1986	1987	1988	1989
Military expenditures					
Current dollars (mil.)	NA	1,127[e]	NA	NA	NA
1989 constant dollars (mil.)	NA	1,250[e]	NA	NA	NA
Armed forces (000)	66[e]	70[e]	74[e]	107	107
Gross national product (GNP)					
Current dollars (mil.)	4,018	4,710	5,581	5,861	6,031
1989 constant dollars (mil.)	4,575	5,227	6,003	6,101	6,031
Central government expenditures (CGE)					
1989 constant dollars (mil.)	NA	NA	NA	3,322[e]	NA
People (mil.)	7.6	7.7	7.9	8.1	8.3
Military expenditure as % of GNP	NA	23.9	NA	NA	NA
Military expenditure as % of CGE	NA	NA	NA	NA	NA
Military expenditure per capita	NA	161	NA	NA	NA
Armed forces per 1,000 people	8.7	9.0	9.4	13.2	12.9
GNP per capita	604	674	759	754	727
Arms imports[6]					
Current dollars (mil.)	800	1,300	1,700	1,400	750
1989 constant dollars (mil.)	911	1,443	1,828	1,457	750
Arms exports[6]					
Current dollars (mil.)	0	0	0	0	0
1989 constant dollars (mil.)	0	0	0	0	0
Total imports[7]					
Current dollars (mil.)	1,700	1,400	NA	NA	2,500
1989 constant dollars	1,935	1,554	NA	NA	2,500
Total exports[7]					
Current dollars (mil.)	2,238	1,303	NA	NA	2,900
1989 constant dollars	2,548	1,446	NA	NA	2,900
Arms as percent of total imports[8]	47.1	92.9	NA	NA	30.0
Arms as percent of total exports[8]	0	0	NA	NA	0

Human Rights [24]

	SSTS	FL	FAPRO	PPCG	APROBC	TPW	PCPTW	STPEP	PHRFF	PRW	ASST	AFL
Observes		P			P	P	P			P		P
	EAFRD	CPR	ESCR	SR	ACHR	MAAE	PVIAC	PVNAC	EAFDAW	TCIDTP		RC
Observes		P	P	P			P			P		P

P = Party; S = Signatory; see Appendix for meaning of abbreviations.

Labor Force

Total Labor Force [25]

2.783 million economically active

Labor Force by Occupation [26]

Agriculture	85%
Industry	15

Date of data: 1985 est.

Unemployment Rate [27]

For sources, notes, and explanations, see Annotated Source Appendix, page 1035.

19

Production Sectors

Commercial Energy Production and Consumption

Production [28]

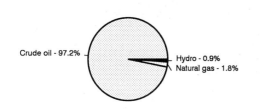

Crude oil - 97.2%
Hydro - 0.9%
Natural gas - 1.8%

Consumption [29]

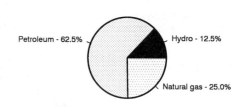

Petroleum - 62.5%
Hydro - 12.5%
Natural gas - 25.0%

Data are shown in quadrillion (10^{15}) BTUs and percent for 1991

Crude oil[1]	1.06
Natural gas liquids	0.00
Dry natural gas	0.02
Coal[2]	0.00
Net hydroelectric power[3]	0.01
Net nuclear power[3]	0.00
Total	1.09

Petroleum	0.05
Dry natural gas	0.02
Coal[2]	0.00
Net hydroelectric power[3]	0.01
Net nuclear power[3]	0.00
Total	0.09

Telecommunications [30]

- Limited system of wire, microwave radio relay, and troposcatter routes
- High frequency radio used extensively for military links
- 40,300 telephones
- Broadcast stations - 17 AM, 13 FM, 6 TV
- 2 Atlantic Ocean INTELSAT earth stations

Transportation [31]

Railroads. 3,189 km total; 2,879 km 1.067-meter gauge, 310 km 0.600-meter gauge; limited trackage in use because of landmines still in place from the civil war; majority of the Benguela Railroad also closed because of civil war

Highways. 73,828 km total; 8,577 km bituminous-surface treatment, 29,350 km crushed stone, gravel, or improved earth, remainder unimproved earth

Merchant Marine. 12 ships (1,000 GRT or over) totaling 66,348 GRT/102,825 DWT; includes 11 cargo, 1 oil tanker

Airports

Total:	302
Usable:	173
With permanent-surface runways:	32
With runways over 3,659 m:	2
With runways 2,440-3,659 m:	17
With runways 1,220-2,439 m:	57

Top Agricultural Products [32]

Cash crops—coffee, sisal, corn, cotton, sugar cane, manioc, tobacco; food crops—cassava, corn, vegetables, plantains, bananas; livestock production Agriculture accounts for 20%, fishing 4%, forestry 2% of total agricultural output; disruptions caused by civil war and marketing deficiencies require food imports.

Top Mining Products [33]

Estimated metric tons unless otherwise specified   M.t.

Petroleum, crude (000 barrels)	184,000
Salt	70,000
Gypsum	57,000
Cement, hydraulic (000 tons)	1,000
Diamonds, gem & industrial (000 carats)	961
Asphalt and bitumen (natural metric tons)	13,000

Tourism [34]

For sources, notes, and explanations, see Annotated Source Appendix, page 1035.

Finance, Economics, and Trade

Industrial Summary [35]

Industrial Production: Growth rate NA%; accounts for about 60% of GDP, including petroleum output. **Industries**: Petroleum; mining diamonds, iron ore, phosphates, feldspar, bauxite, uranium, and gold; fish processing; food processing; brewing; tobacco; sugar; textiles; cement; basic metal products.

Economic Indicators [36]

Millions of U.S. Dollars unless otherwise stated.

	1988	1989	1990[1]
Nominal GDP	6,926	7,724	9,073
Real GDP growth rate (%)	14.3	1.8	2.8
Nominal GDP per capita ($)	739.8	796.3	907.3
Money supply (M2)	317.1	370.8	289.6
Savings Rate	NA	NA	NA
Investment rate	NA	NA	NA
CPI (official prices)	21.7	NA	6.1
External debt			
(incl. interest arrears)	5,928.0	6,533.0	7,281.0

Balance of Payments Summary [37]

Values in millions of dollars.

	1987	1988	1989	1990	1991
Exports of goods (f.o.b.)	2,322.0	2,492.0	3,014.0	3,883.9	3,449.3
Imports of goods (f.o.b.)	-1,303.0	-1,372.0	-1,338.0	-1,578.2	-1,347.2
Trade balance	1,019.0	1,120.0	1,676.0	2,305.7	2,102.1
Services - debits	-717.0	-1,749.0	-1,954.0	-2,583.2	-2,896.2
Services - credits	93.0	128.0	150.0	119.1	186.3
Private transfers (net)	-8.0	-6.0	-68.0	-139.7	-30.0
Government transfers (net)	60.0	38.0	64.0	62.6	58.2
Long term capital (net)	255.0	-192.0	1,699.0	60.9	-532.9
Short term capital (net)	149.0	967.0	-895.0	189.2	1,133.9
Errors and omissions	-834.0	-257.0	-678.0	-19.1	26.9
Overall balance	17.0	49.0	-6.0	-4.5	48.3

Exchange Rates [38]

Currency: **kwanza.**
Symbol: **Kz.**

Data are currency units per $1.

Official rate	4,000
April 30, 1993 black	
market rate	17,000

Imports and Exports

Top Import Origins [39]

$1.5 billion (f.o.b., 1991 est.).

Origins	%
Portugal	NA
Brazil	NA
US	NA
France	NA
Spain	NA

Top Export Destinations [40]

$3.7 billion (f.o.b., 1991 est.).

Destinations	%
US	NA
France	NA
Germany	NA
Netherlands	NA
Brazil	NA

Foreign Aid [41]

	U.S. $	
US commitments, including Ex-Im (FY70-89)	265	million
Western (non-US) countries, ODA and OOF		
bilateral commitments (1970-89)	1,105	million
Communist countries (1970-89)	1.3	billion
Net official disbursements (1985-89)	750	million

Import and Export Commodities [42]

Import Commodities

Machinery & electrical equipment
Food
Vehicles and spare parts
Textiles and clothing
Medicines
Military deliveries

Export Commodities

Oil
Liquefied petroleum gas
Diamonds
Coffee
Sisal
Fish and fish products
Timber
Cotton

For sources, notes, and explanations, see Annotated Source Appendix, page 1035.

21

Antigua and Barbuda

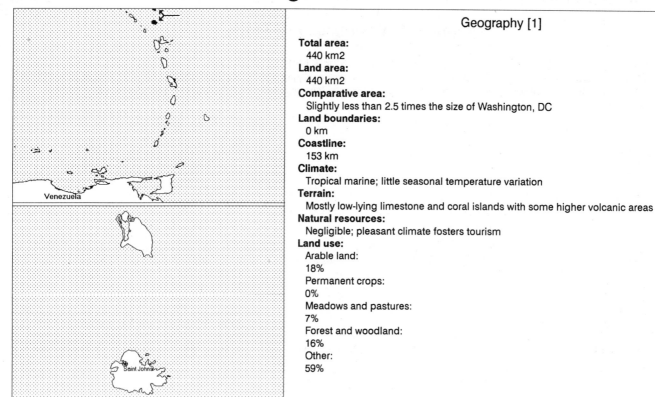

Geography [1]

Total area:
440 km2
Land area:
440 km2
Comparative area:
Slightly less than 2.5 times the size of Washington, DC
Land boundaries:
0 km
Coastline:
153 km
Climate:
Tropical marine; little seasonal temperature variation
Terrain:
Mostly low-lying limestone and coral islands with some higher volcanic areas
Natural resources:
Negligible; pleasant climate fosters tourism
Land use:
Arable land:
18%
Permanent crops:
0%
Meadows and pastures:
7%
Forest and woodland:
16%
Other:
59%

Demographics [2]

	1960	1970	1980	1990	1991[1]	1994	2000	2010	2020
Population	55	66	69	64	64	65	68	74	80
Population density (persons per sq. mi.)	321	386	403	375	376	NA	403	444	478
Births	NA	NA	NA	NA	1	1	NA	NA	NA
Deaths	NA	NA	NA	NA	(Z)	(Z)	NA	NA	NA
Life expectancy - males	NA	NA	NA	NA	70	71	NA	NA	NA
Life expectancy - females	NA	NA	NA	NA	74	75	NA	NA	NA
Birth rate (per 1,000)	NA	NA	NA	NA	18	17	NA	NA	NA
Death rate (per 1,000)	NA	NA	NA	NA	6	5	NA	NA	NA
Women of reproductive age (15-44 yrs.)	NA	NA	NA	NA	NA	NA	NA	NA	NA
of which are currently married	NA	NA	NA	NA	NA	NA	NA	NA	NA
Fertility rate	NA	NA	NA	1.7	1.70	1.7	1.7	1.7	1.7

Population values are in thousands, life expectancy in years, and other items as indicated.

Health

Health Personnel [3]

Health Expenditures [4]

For sources, notes, and explanations, see Annotated Source Appendix, page 1035.

Human Factors	
Health Care Ratios [5]	Infants and Malnutrition [6]

Ethnic Division [7]		Religion [8]	Major Languages [9]
Black African	NA	Anglican (predominant), other Protestant sects, some Roman Catholic.	English (official), local dialects.
British	NA		
Portuguese	NA		
Lebanese	NA		
Syrian	NA		

Education

Public Education Expenditures [10]

Mil. $ E. Caribbean	1980	1985	1987	1988	1989	1990
Total education expenditure	8,621	NA	NA	NA	NA	NA
as percent of GNP	2.9	NA	NA	NA	NA	NA
as percent of total govt. expend.	NA	NA	NA	NA	NA	NA
Current education expenditures	8,529	NA	19,082	30,777	NA	NA
as percent of GNP	2.8	NA	2.6	4.0	NA	NA
as percent of current govt. expend.	NA	NA	NA	NA	NA	NA
Capital expenditures	92	NA	NA	NA	NA	NA

Educational Attainment [11]

Literacy Rate [12]

Libraries [13]

Daily Newspapers [14]

	1975	1980	1985	1990
Number of papers	1	1	1	1
Circ. (000)	4	6	6	6

Cinema [15]

Science and Technology

Scientific/Technical Forces [16]	R&D Expenditures [17]	U.S. Patents Issued [18]

Government and Law

Organization of Government [19]

Long-form name:
none
Type:
parliamentary democracy
Independence:
1 November 1981 (from UK)
Constitution:
1 November 1981
Legal system:
based on English common law
National holiday:
Independence Day, 1 November (1981)
Executive branch:
British monarch, governor general, prime minister, Cabinet
Legislative branch:
bicameral Parliament consists of an upper house or Senate and a lower house or House of Representatives
Judicial branch:
Eastern Caribbean Supreme Court

Political Parties [20]

House of Representatives	% of seats
Antigua Labor Party (ALP)	88.2
United Progressive Party (UPP)	5.9
Independent	5.9

Government Budget [21]

For 1992.

Revenues	105
Expenditures	161
Capital expenditures	56

Defense Summary [22]

Branches: Royal Antigua and Barbuda Defense Force, Royal Antigua and Barbuda Police Force (including the Coast Guard)

Manpower Availability: No information available.

Defense Expenditures: Exchange rate conversion - $1.4 million, 1% of GDP (FY90/91)

Crime [23]

Human Rights [24]

	SSTS	FL	FAPRO	PPCG	APROBC	TPW	PCPTW	STPEP	PHRFF	PRW	ASST	AFL
Observes	P	P	P	P	P	P	P			P	P	P
	EAFRD	CPR	ESCR	SR	ACHR	MAAE	PVIAC	PVNAC	EAFDAW	TCIDTP	RC
Observes		P					P	P	P	P		S

P = Party; S = Signatory; see Appendix for meaning of abbreviations.

Labor Force

Total Labor Force [25]

30,000

Labor Force by Occupation [26]

Commerce and services	82%
Agriculture	11
Industry	7

Date of data: 1983

Unemployment Rate [27]

5% (1988 est.)

For sources, notes, and explanations, see Annotated Source Appendix, page 1035.

Production Sectors

Energy Resource Summary [28]

Energy Resources: None. **Electricity**: 52,100 kW capacity; 95 million kWh produced, 1,482 kWh per capita (1992).

Telecommunications [30]

- Good automatic telephone system
- 6,700 telephones
- Tropospheric scatter links with Saba and Guadeloupe
- Broadcast stations - 4 AM, 2 FM, 2 TV, 2 shortwave
- 1 coaxial submarine cable
- 1 Atlantic Ocean INTELSAT earth station

Transportation [31]

Railroads. 64 km 0.760-meter narrow gauge and 13 km 0.610-meter gauge used almost exclusively for handling sugarcane

Highways. 240 km

Merchant Marine. 149 ships (1,000 GRT or over) totaling 529,202 GRT/778,506 DWT; includes 96 cargo, 3 refrigerated cargo, 21 container, 5 roll-on/roll-off cargo, 1 multifunction large-load carrier, 2 oil tanker, 19 chemical tanker, 2 bulk; note - a flag of convenience registry

Airports

Total:	3
Usable:	3
With permanent-surface runways:	2
With runways 3,659 m:	0
With runways 2,440-3,659 m:	1
With runways 1,220-2,439 m:	0

Top Agricultural Products [32]

Agriculture accounts for 4% of GDP; expanding output of cotton, fruits, vegetables, and livestock; other crops—bananas, coconuts, cucumbers, mangoes, sugarcane; not self-sufficient in food.

Top Mining Products [33]

Detailed information is not available. A summary of mineral resources available follows. **Mineral Resources**: Negligible.

Tourism [34]

	1987	1988	1989	1990	1991
Visitors	347	417	420	458	477
Tourists[2]	173	187	189	197	197
Cruise passengers[3]	174	229	231	260	281
Tourism receipts	191	242	267	298	314
Tourism expenditures	15	16	16	17	18
Fare expenditures		6	7	8	8

Tourists are in thousands, money in million U.S. dollars.

Finance, Economics, and Trade

Industrial Summary [35]

Industrial Production: Growth rate 3% (1989 est.); accounts for 5% of GDP. **Industries**: Tourism, construction, light manufacturing (clothing, alcohol, household appliances).

Economic Indicators [36]

National product: GDP—exchange rate conversion— $424 million (1991 est.). **National product real growth rate**: 1.4% (1991 est.). **National product per capita**: $6,600 (1991 est.). **Inflation rate (consumer prices)**: 6.5% (1991 est.). **External debt**: $250 million (1990 est.).

Balance of Payments Summary [37]

Values in millions of dollars.

	1987	1988	1989	1990	1991
Exports of goods (f.o.b.)	28.5	30.1	31.6	33.2	32.0
Imports of goods (f.o.b.)	-258.3	-274.8	-317.0	-325.9	-317.5
Trade balance	-229.7	-244.7	-285.4	-292.7	-285.5
Services - debits	-66.6	-123.5	-136.6	-149.1	-116.8
Services - credits	194.3	266.9	303.0	349.4	349.4
Private transfers (net)	17.0	18.9	22.8	19.8	15.8
Government transfers (net)	1.7	1.8	0.7	0.8	2.4
Long term capital (net)	76.6	66.2	51.4	71.2	14.0
Short term capital (net)	2.4	14.9	47.1	26.3	18.8
Errors and omissions	8.8	-0.1	0.1	-2.2	6.9
Overall balance	4.5	0.4	3.1	-0.2	5.0

Exchange Rates [38]

Currency: **East Caribbean dollars.**
Symbol: **EC$.**

Data are currency units per $1.

Fixed rate since 1976	2.70

Imports and Exports

Top Import Origins [39]

$317.5 million (c.i.f., 1991).

Origins	%
US	27
UK	16
Canada	4
OECS	3
Other	50

Top Export Destinations [40]

$32 million (f.o.b., 1991).

Destinations	%
OECS	26.0
Barbados	15.0
Guyana	4.0
Trinidad and Tobago	2.0
US	0.3

Foreign Aid [41]

	U.S. $	
US commitments	10	million
Western (non-US) countries, ODA and OOF bilateral commitments (1970-89)	50	million

Import and Export Commodities [42]

Import Commodities	Export Commodities
Food and live animals	Petroleum products 48%
Machinery & transport equipment	Manufactures 23%
Manufactures	Food and live animals 4%
Chemicals	Machinery, transp. equip. 17%
Oil	

For sources, notes, and explanations, see Annotated Source Appendix, page 1035.

Argentina

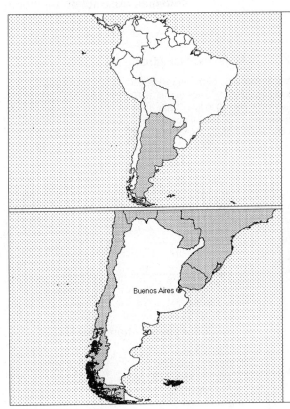

Geography [1]

Total area:
2,766,890 km2
Land area:
2,736,690 km2
Comparative area:
Slightly less than three-tenths the size of the US
Land boundaries:
Total 9,665 km, Bolivia 832 km, Brazil 1,224 km, Chile 5,150 km, Paraguay 1,880 km, Uruguay 579 km
Coastline:
4,989 km
Climate:
Mostly temperate; arid in southeast; subantarctic in southwest
Terrain:
Rich plains of the Pampas in northern half, flat to rolling plateau of Patagonia in south, rugged Andes along western border
Natural resources:
Fertile plains of the pampas, lead, zinc, tin, copper, iron ore, manganese, petroleum, uranium
Land use:
Arable land: 9%
Permanent crops: 4%
Meadows and pastures: 52%
Forest and woodland: 22%
Other: 13%

Demographics [2]

	1960	1970	1980	1990	1991[1]	1994	2000	2010	2020
Population	20,616	23,962	28,237	32,386	32,664	33,913	36,202	39,947	43,190
Population density (persons per sq. mi.)	20	23	27	31	31	NA	34	38	41
Births	NA	NA	NA	NA	648	665	NA	NA	NA
Deaths	NA	NA	NA	NA	284	293	NA	NA	NA
Life expectancy - males	NA	NA	NA	NA	68	68	NA	NA	NA
Life expectancy - females	NA	NA	NA	NA	74	75	NA	NA	NA
Birth rate (per 1,000)	NA	NA	NA	NA	20	20	NA	NA	NA
Death rate (per 1,000)	NA	NA	NA	NA	9	9	NA	NA	NA
Women of reproductive age (15-44 yrs.)	NA	NA	NA	7,705	NA	8,256	9,001	NA	NA
of which are currently married	NA	NA	NA	4,793	NA	5,079	5,552	NA	NA
Fertility rate	NA	NA	NA	2.8	2.74	2.7	2.5	2.3	2.1

Population values are in thousands, life expectancy in years, and other items as indicated.

Health

Health Personnel [3]

Doctors per 1,000 pop., 1988-92	2.99
Nurse-to-doctor ratio, 1988-92	0.2
Hospital beds per 1,000 pop., 1985-90	4.8
Percentage of children immunized (age 1 yr. or less)	
Third dose of DPT, 1990-91	84.0
Measles, 1990-91	99.0

Health Expenditures [4]

Total health expenditure, 1990 (official exchange rate)	
Millions of dollars	4441
Millions of dollars per capita	138
Health expenditures as a percentage of GDP	
Total	4.2
Public sector	2.5
Private sector	1.7
Development assistance for health	
Total aid flows (millions of dollars)[1]	11
Aid flows per capita (millions of dollars)	0.3
Aid flows as a percentage of total health expenditure	0.2

For sources, notes, and explanations, see Annotated Source Appendix, page 1035.

Human Factors

Health Care Ratios [5]

Population per physician, 1970	530
Population per physician, 1990	NA
Population per nursing person, 1970	960
Population per nursing person, 1990	NA
Percent of births attended by health staff, 1985	NA

Infants and Malnutrition [6]

Percent of babies with low birth weight, 1985	6
Infant mortality rate per 1,000 live births, 1970	52
Infant mortality rate per 1,000 live births, 1991	25
Years of life lost per 1,000 population, 1990	12
Prevalence of malnutrition (under age 5), 1990	NA

Ethnic Division [7]

White	85%
Mestizo, Indian, other	15%

Religion [8]

Less than 20% of Catholics are practicing.

Nominally Roman Catholic	90%
Protestant	2%
Jewish	2%
Other	6%

Major Languages [9]

Spanish (official)	NA
English	NA
Italian	NA
German	NA
French	NA

Education

Public Education Expenditures [10]

Billion Austral	1980	1985[1]	1986[1]	1987[1]	1989[1]	1990[1]
Total education expenditure	0.102	0.740	1.29	3.10	348.17	7,356.6
as percent of GNP	3.6	2.0	1.8	1.9	1.5	1.5
as percent of total govt. expend.	15.1	8.6	7.5	8.9	8.0	10.9
Current education expenditure	0.001	0.67	1.13	2.71	328.7	7,063
as percent of GNP	3.1	1.8	1.6	1.7	1.5	1.5
as percent of current govt. expend.	18.8	8.9	7.5	9.5	8.8	12.4
Capital expenditure	0.0002	0.073	0.159	0.39	19.5	293.6

Educational Attainment [11]

Age group	25+
Total population	14,913,575
Highest level attained (%)	
No schooling	7.1
First level	
Incompleted	33.4
Completed	33.0
Entered second level	
S-1	20.4
S-2	NA
Post secondary	6.1

Literacy Rate [12]

In thousands and percent	1985[a]	1991[a]	2000[a]
Illiterate population +15 years	1,097	1,065	976
Illiteracy rate - total pop. (%)	5.2	4.7	3.7
Illiteracy rate - males (%)	4.9	4.5	3.6
Illiteracy rate - females (%)	5.6	4.9	3.8

Libraries [13]

	Admin. Units	Svc. Pts.	Vols. (000)	Shelving (meters)	Vols. Added	Reg. Users
National (1989)	3	11	1,950	NA	115,419	-
Non-specialized	NA	NA	NA	NA	NA	NA
Public	NA	NA	NA	NA	NA	NA
Higher ed.	NA	NA	NA	NA	NA	NA
School	NA	NA	NA	NA	NA	NA

Daily Newspapers [14]

	1975	1980	1985	1990
Number of papers	164	220	218	159
Circ. (000)	3,000[e]	4,000[e]	3,940[e]	4,000[e]

Cinema [15]

Data for 1991.

Cinema seats per 1,000	2.1
Annual attendance per person	0.6[1]
Gross box office receipts (mil. Austral)	NA

Science and Technology

Scientific/Technical Forces [16]

Potential scientists/engineers	695,000
Number female	278,000
Potential technicians	245,000
Number female	200,000
Total	940,000

R&D Expenditures [17]

	Austral (000) 1988
Total expenditure	3,466,700
Capital expenditure	741,900
Current expenditure	2,724,800
Percent current	78.6

U.S. Patents Issued [18]

Values show patents issued to citizens of the country by the U.S. Patents Office.

	1990	1991	1992
Number of patents	19	19	23

For sources, notes, and explanations, see Annotated Source Appendix, page 1035.

Government and Law

Organization of Government [19]

Long-form name:
Argentine Republic
Type:
republic
Independence:
9 July 1816 (from Spain)
Constitution:
1 May 1853
Legal system:
mixture of US and West European legal systems; has not accepted compulsory ICJ jurisdiction
National holiday:
Revolution Day, 25 May (1810)
Executive branch:
president, vice president, Cabinet
Legislative branch:
bicameral National Congress (Congreso Nacional) consists of an upper chamber or Senate (Senado) and a lower chamber or Chamber of Deputies (Camara de Diputados)
Judicial branch:
Supreme Court (Corte Suprema)

Political Parties [20]

Chamber of Deputies	% of seats
Justicialist Party (JP)	48.0
Radical Civic Union (UCR)	33.5
Union Democratic Center (UCD)	3.9
Other	14.6

Government Expenditures [21]

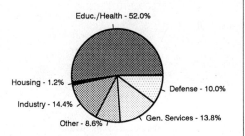

Educ./Health - 52.0%
Housing - 1.2%
Industry - 14.4%
Other - 8.6%
Defense - 10.0%
Gen. Services - 13.8%

% distribution for 1989. Expend. in 1992: 35.8 ($ bil.)

Crime [23]

Crime volume (for 1990)	
Cases known to police	58,429
Attempts (percent)	0.20
Percent cases solved	9.05
Crimes per 100,000 persons	177.05
Persons responsible for offenses	
Total number offenders	6,132
Percent female	11.46
Percent juvenile[2]	5.83
Percent foreigners	NA

Military Expenditures and Arms Transfers [22]

	1985	1986	1987	1988	1989
Military expenditures					
Current dollars (mil.)	1,847[e]	2,075[e]	2,017	1,989	1,858
1989 constant dollars (mil.)	2,103[e]	2,302[e]	2,169	2,070	1,858
Armed forces (000)	129	104	118	95	95
Gross national product (GNP)					
Current dollars (mil.)	48,480	53,360	56,240	56,460	54,080
1989 constant dollars (mil.)	55,190	59,210	60,490	58,780	54,080
Central government expenditures (CGE)					
1989 constant dollars (mil.)	17,000	14,470	14,540	9,364	NA
People (mil.)	30.4	30.8	31.1	31.5	31.9
Military expenditure as % of GNP	3.8	3.9	3.6	3.5	3.4
Military expenditure as % of CGE	12.4	15.9	14.9	22.1	NA
Military expenditure per capita	69	75	70	66	58
Armed forces per 1,000 people	4.3	3.4	3.8	3.0	3.0
GNP per capita	1,818	1,926	1,942	1,864	1,694
Arms imports[6]					
Current dollars (mil.)	200	50	140	300	40
1989 constant dollars (mil.)	228	55	151	312	40
Arms exports[6]					
Current dollars (mil.)	80	40	70	70	60
1989 constant dollars (mil.)	91	44	75	73	60
Total imports[7]					
Current dollars (mil.)	3,814	4,724	5,818	5,322	4,201
1989 constant dollars	4,342	5,242	6,258	5,540	4,201
Total exports[7]					
Current dollars (mil.)	8,396	6,852	6,360	9,135	9,579
1989 constant dollars	9,558	7,604	6,840	9,509	9,579
Arms as percent of total imports[8]	5.2	1.1	2.4	5.6	1.0
Arms as percent of total exports[8]	1.0	0.6	1.1	0.8	0.6

Human Rights [24]

	SSTS	FL	FAPRO	PPCG	APROBC	TPW	PCPTW	STPEP	PHRFF	PRW	ASST	AFL
Observes		P	P	P	P	P	P	P		P	P	P
	EAFRD	CPR	ESCR	SR	ACHR	MAAE	PVIAC	PVNAC	EAFDAW	TCIDTP	RC	
Observes		P	P	P	P		P	P	P	P	P	

P = Party; S = Signatory; see Appendix for meaning of abbreviations.

Labor Force

Total Labor Force [25]

10.9 million

Labor Force by Occupation [26]

Agriculture	12%
Industry	31
Services	57

Date of data: 1985 est.

Unemployment Rate [27]

6.9% (1992)

Production Sectors

Commercial Energy Production and Consumption

Production [28]

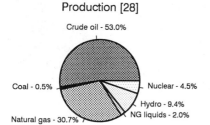

Crude oil - 53.0%

Coal - 0.5%   Nuclear - 4.5%

                              Hydro - 9.4%
Natural gas - 30.7%   NG liquids - 2.0%

Consumption [29]

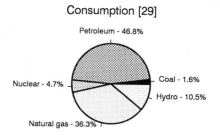

Petroleum - 46.8%

Nuclear - 4.7%   Coal - 1.6%

                              Hydro - 10.5%
Natural gas - 36.3%

Data are shown in quadrillion (10^{15}) BTUs and percent for 1991

Crude oil[1]	1.07
Natural gas liquids	0.04
Dry natural gas	0.62
Coal[2]	0.01
Net hydroelectric power[3]	0.19
Net nuclear power[3]	0.09
Total	2.02

Petroleum	0.89
Dry natural gas	0.69
Coal[2]	0.03
Net hydroelectric power[3]	0.20
Net nuclear power[3]	0.09
Total	1.90

Telecommunications [30]

- Extensive modern system
- 2,650,000 telephones (12,000 public telephones)
- Microwave widely used
- Broadcast stations - 171 AM, no FM, 231 TV, 13 shortwave
- 2 Atlantic Ocean INTELSAT earth stations
- Domestic satellite network has 40 earth stations

Transportation [31]

Railroads. 34,172 km total (includes 209 km electrified); includes a mixture of 1.435-meter standard gauge, 1.676-meter broad gauge, 1.000-meter narrow gauge, and 0.750-meter narrow gauge

Highways. 208,350 km total; 47,550 km paved, 39,500 km gravel, 101,000 km improved earth, 20,300 km unimproved earth

Merchant Marine. 60 ships (1,000 GRT or over) totaling 1,695,420 GRT/1,073,904 DWT; includes 30 cargo, 5 refrigerated cargo, 4 container, 1 railcar carrier, 14 oil tanker, 1 chemical tanker, 4 bulk, 1 roll-on/roll-off

Airports

Total:	1,700
Usable:	1,451
With permanet-surface runways:	137
With runways over 3,659 m:	1
With runways 2,440-3,659 m:	31
With runways 1,220-2,439 m:	326

Top Agricultural Products [32]

	89-90[9]	90-91[9]
Sugar cane	10,600	12,250
Soybean	10,075	11,000
Wheat	10,150	10,500
Corn	5,200	7,600
Sunflowerseed	3,820	3,900
Sorghum	2,000	2,250

Values shown are 1,000 metric tons.

Top Mining Products [33]

Estimated metric tons unless otherwise specified   M.t.

Boron materials, crude	250,000
Cadmium: smelter	49
Coal, bituminous (000 tons)	292[1]
Copper, mine output, Cu content	408[1]
Gypsum, crude	450,000
Lead, mine output, Pb content	23,697[1]
Petroleum, crude	178,379[1]
Silver, mine output, Ag content (kilograms)	56,359[1]
Uranium, mine output, U3O8 content	60,000
Zinc, mine output, Zn content	39,253[1]

Tourism [34]

	1987	1988	1989	1990	1991
Tourists[4]	472	2,119	2,492	2,728	2,870
Tourism receipts	615	634	790	1,975	2,336
Tourism expenditures	890	975	1,014	1,171	1,739
Fare receipts	237	269	323	389[5]	375
Fare expenditures	287	230	219	270[5]	418

Tourists are in thousands, money in million U.S. dollars.

Manufacturing Sector

GDP and Manufacturing Summary [35]

	1980	1985	1989	1990	% change 1980-1990	% change 1989-1990
GDP (million 1980 $)	60,917	54,708	53,638	55,101	-9.5	2.7
GDP per capita (1980 $)	2,157	1,804	1,680	1,704	-21.0	1.4
Manufacturing as % of GDP (current prices)	25.0	27.3	19.2[e]	18.3[e]	-26.8	-4.7
Gross output (million $)	55,936[e]	48,084[e]	43,284[e]	79,001[e]	41.2	82.5
Value added (million $)	24,511	28,891	24,712[e]	31,156	27.1	26.1
Value added (million 1980 $)	15,224	12,506	12,136	11,586	-23.9	-4.5
Industrial production index	100	86	82	90	-10.0	9.8
Employment (thousands)	1,346[e]	1,127[e]	1,068[e]	942[e]	-30.0	-11.8

Note: GDP stands for Gross Domestic Product. 'e' stands for estimated value.

Profitability and Productivity

	1980	1985	1989	1990	% change 1980-1990	% change 1989-1990
Intermediate input (%)	56[e]	40[e]	43[e]	61[e]	8.9	41.9
Wages, salaries, and supplements (%)	10[e]	11[e]	9[e]	8[e]	-20.0	-11.1
Gross operating surplus (%)	33[e]	49[e]	48[e]	31[e]	-6.1	-35.4
Gross output per worker ($)	41,552[e]	42,656[e]	40,519[e]	83,878[e]	101.9	107.0
Value added per worker ($)	18,208[e]	25,630[e]	23,134[e]	33,080[e]	81.7	43.0
Average wage (incl. benefits) ($)	4,301[e]	4,596[e]	3,806[e]	6,767[e]	57.3	77.8

Profitability is in percent of gross output. Productivity is in U.S. $. 'e' stands for estimated value.

Profitability - 1990

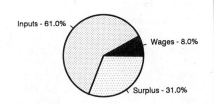

Inputs - 61.0%
Wages - 8.0%
Surplus - 31.0%

The graphic shows percent of gross output.

Value Added in Manufacturing

	1980 $ mil.	1980 %	1985 $ mil.	1985 %	1989 $ mil.	1989 %	1990 $ mil.	1990 %	% change 1980-1990	% change 1989-1990
311 Food products	3,544	14.5	4,912	17.0	4,256[e]	17.2	4,695	15.1	32.5	10.3
313 Beverages	703	2.9	942	3.3	1,033[e]	4.2	932	3.0	32.6	-9.8
314 Tobacco products	498	2.0	719	2.5	435[e]	1.8	480	1.5	-3.6	10.3
321 Textiles	1,703	6.9	1,832	6.3	1,960[e]	7.9	2,209	7.1	29.7	12.7
322 Wearing apparel	919	3.7	558	1.9	403[e]	1.6	492	1.6	-46.5	22.1
323 Leather and fur products	284	1.2	350	1.2	283[e]	1.1	336	1.1	18.3	18.7
324 Footwear	245	1.0	240	0.8	146[e]	0.6	190	0.6	-22.4	30.1
331 Wood and wood products	363	1.5	283	1.0	237[e]	1.0	255	0.8	-29.8	7.6
332 Furniture and fixtures	225	0.9	185	0.6	232[e]	0.9	246	0.8	9.3	6.0
341 Paper and paper products	554	2.3	763	2.6	636[e]	2.6	882	2.8	59.2	38.7
342 Printing and publishing	679	2.8	800	2.8	561[e]	2.3	695	2.2	2.4	23.9
351 Industrial chemicals	914	3.7	1,367	4.7	1,271[e]	5.1	1,844	5.9	101.8	45.1
352 Other chemical products	1,206	4.9	1,916	6.6	1,409[e]	5.7	1,791	5.7	48.5	27.1
353 Petroleum refineries	3,647	14.9	5,120	17.7	3,295[e]	13.3	6,069	19.5	66.4	84.2
354 Miscellaneous petroleum and coal products	86	0.4	121	0.4	153[e]	0.6	122	0.4	41.9	-20.3
355 Rubber products	331	1.4	327	1.1	272[e]	1.1	368	1.2	11.2	35.3
356 Plastic products	424	1.7	485	1.7	371[e]	1.5	436	1.4	2.8	17.5
361 Pottery, china and earthenware	189	0.8	130	0.4	149[e]	0.6	156	0.5	-17.5	4.7
362 Glass and glass products	199	0.8	153	0.5	210[e]	0.8	249	0.8	25.1	18.6
369 Other non-metal mineral products	659	2.7	587	2.0	740[e]	3.0	932	3.0	41.4	25.9
371 Iron and steel	900	3.7	1,239	4.3	1,523[e]	6.2	1,651	5.3	83.4	8.4
372 Non-ferrous metals	235	1.0	257	0.9	229[e]	0.9	305	1.0	29.8	33.2
381 Metal products	1,272	5.2	1,499	5.2	1,593[e]	6.4	1,611	5.2	26.7	1.1
382 Non-electrical machinery	1,358	5.5	930	3.2	640[e]	2.6	835	2.7	-38.5	30.5
383 Electrical machinery	902	3.7	936	3.2	914[e]	3.7	1,025	3.3	13.6	12.1
384 Transport equipment	2,289	9.3	2,054	7.1	1,586[e]	6.4	2,140	6.9	-6.5	34.9
385 Professional and scientific equipment	86	0.4	95	0.3	91[e]	0.4	112	0.4	30.2	23.1
390 Other manufacturing industries	96	0.4	92	0.3	84[e]	0.3	97	0.3	1.0	15.5

Note: The industry codes shown are International Standard Industry codes (ISIC). Percentages are percent of total Value Added. 'e' stands for estimated value

Finance, Economics, and Trade

Economic Indicators [36]

	1989	1990	1991[e]
GDP (1991 US$ billions)	134	133	140
Real GDP growth rate (%)	-4.6	-0.7	5.0
Real per capita GDP (1991 US$)	4,133	4,048	4,203
Money supply (M1) (% yr-end to yr-end)	4,960	886	115
Commercial interest rates (% of GDP)	40.0	6.7	1.2
Savings rate (% of GDP)	7.6	10.2	11.0
Investment rate (% of GDP)	9.4	8.6	13.8
CPI (% yr-end to yr-end)	4,923	1,344	75
WPI (% yr-end to yr-end)	5,386	798	63
External public debt[3]	58.4	56.2	57.5

Balance of Payments Summary [37]

Values in millions of dollars.

	1987	1988	1989	1990	1991
Exports of goods (f.o.b.)	6,360.0	9,134.0	9,573.0	12,354.0	11,972.0
Imports of goods (f.o.b.)	-5,343.0	-4,892.0	-3,864.0	-3,726.0	-7,400.0
Trade balance	1,017.0	4,242.0	5,709.0	8,628.0	4,572.0
Services - debits	-7,290.0	-8,040.0	-9,491.0	-9,540.0	-10,175.0
Services - credits	2,046.0	2,226.0	2,469.0	2,744.0	2,742.0
Private transfers (net)	-8.0	-	8.0	71.0	29.0
Government transfers (net)	-	-	-	-	-
Long term capital (net)	2,459.0	1,228.0	4,747.0	1,246.0	3,688.0
Short term capital (net)	-155.0	2,429.0	-4,466.0	-451.0	2,093.0
Errors and omissions	-112.0	-165.0	-249.0	715.0	-341.0
Overall balance	-2,043.0	1,920.0	-1,273.0	3,413.0	2,608.0

Exchange Rates [38]

Currency: **Pesos.**
Symbol: **P.**

Data are currency units per $1.

January 1993	0.99000
1992	0.99064
1991	0.95355
1990	0.48759
1989	0.04233
1988	0.00088

Imports and Exports

Top Import Origins [39]

$14.0 billion (c.i.f., 1992 est.).

Origins	%
US	22
Brazil	NA
Germany	NA
Bolivia	NA
Japan	NA
Italy	NA
Netherlands	NA

Top Export Destinations [40]

$12.3 billion (f.o.b., 1992 est.).

Destinations	%
US	12
Brazil	NA
Italy	NA
Japan	NA
Netherlands	NA

Foreign Aid [41]

	U.S. $	
US commitments, including Ex-Im (FY70-89)	1.0	billion
Western (non-US) countries, ODA and OOF bilateral commitments (1970-89)	4.4	billion
Communist countries (1970-89)	718	million

Import and Export Commodities [42]

Import Commodities

Machinery & equipment
Chemicals
Metals
Fuels and lubricants
Agricultural products

Export Commodities

Meat
Wheat
Corn
Oilseed
Hides
Wool

32

For sources, notes, and explanations, see Annotated Source Appendix, page 1035.

Armenia

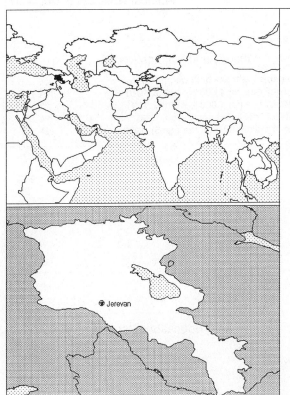

Geography [1]

Total area:
29,800 km2
Land area:
28,400 km2
Comparative area:
Slightly larger than Maryland
Land boundaries:
Total 1,254 km, Azerbaijan (east) 566 km, Azerbaijan (south) 221 km, Georgia 164 km, Iran 35 km, Turkey 268 km
Coastline:
0 km (landlocked)
Climate:
Continental, hot, and subject to drought
Terrain:
High Armenian Plateau with mountains; little forest land; fast flowing rivers; good soil in Aras River valley
Natural resources:
Small deposits of gold, copper, molybdenum, zinc, alumina
Land use:
Arable land:
29%
Permanent crops:
0%
Meadows and pastures:
15%
Forest and woodland:
0%
Other:
56%

Demographics [2]

	1960	1970	1980	1990	1991[1]	1994	2000	2010	2020
Population	1,869	2,520	3,115	3,363	NA	3,522	3,685	3,854	3,959
Population density (persons per sq. mi.)	NA	NA	NA	NA	NA	NA	NA	NA	NA
Births	NA	NA	NA	NA	NA	85	NA	NA	NA
Deaths	NA	NA	NA	NA	NA	24	NA	NA	NA
Life expectancy - males	NA	NA	NA	NA	NA	69	NA	NA	NA
Life expectancy - females	NA	NA	NA	NA	NA	76	NA	NA	NA
Birth rate (per 1,000)	NA	NA	NA	NA	NA	24	NA	NA	NA
Death rate (per 1,000)	NA	NA	NA	NA	NA	7	NA	NA	NA
Women of reproductive age (15-44 yrs.)	NA	NA	NA	844	NA	876	926	NA	NA
of which are currently married	NA	NA	NA	574	NA	600	626	NA	NA
Fertility rate	NA	NA	NA	2.9	NA	3.2	2.6	2.2	2.0

Population values are in thousands, life expectancy in years, and other items as indicated.

Health

Health Personnel [3]

Doctors per 1,000 pop., 1988-92	4.28
Nurse-to-doctor ratio, 1988-92	2.5
Hospital beds per 1,000 pop., 1985-90	9.0
Percentage of children immunized (age 1 yr. or less)	
Third dose of DPT, 1990-91	88.0
Measles, 1990-91	92.0

Health Expenditures [4]

Total health expenditure, 1990 (official exchange rate)	
Millions of dollars	506
Millions of dollars per capita	152
Health expenditures as a percentage of GDP	
Total	4.2
Public sector	2.5
Private sector	1.7
Development assistance for health	
Total aid flows (millions of dollars)[1]	NA
Aid flows per capita (millions of dollars)	NA
Aid flows as a percentage of total health expenditure	NA

For sources, notes, and explanations, see Annotated Source Appendix, page 1035.

33

Human Factors

Health Care Ratios [5]

Population per physician, 1970	NA
Population per physician, 1990	250
Population per nursing person, 1970	NA
Population per nursing person, 1990	NA
Percent of births attended by health staff, 1985	NA

Infants and Malnutrition [6]

Percent of babies with low birth weight, 1985	NA
Infant mortality rate per 1,000 live births, 1970	NA
Infant mortality rate per 1,000 live births, 1991	22
Years of life lost per 1,000 population, 1990	14
Prevalence of malnutrition (under age 5), 1990	NA

Ethnic Division [7]

Armenian	93%
Azeri	3%
Russian	2%
Other	2%

Religion [8]

Armenian Orthodox	94%

Major Languages [9]

Armenian	96%
Russian	2%
Other	2%

Education

Public Education Expenditures [10]

Educational Attainment [11]

Literacy Rate [12]

Libraries [13]

Daily Newspapers [14]

Cinema [15]

Science and Technology

Scientific/Technical Forces [16]

R&D Expenditures [17]

U.S. Patents Issued [18]

Government and Law

Organization of Government [19]

Long-form name:
Republic of Armenia
Type:
republic
Independence:
23 September 1991 (from Soviet Union)
Constitution:
adopted NA April 1978; post-Soviet
constitution not yet adopted
Legal system:
based on civil law system
National holiday:
NA
Executive branch:
president, council of ministers, prime
minister
Legislative branch:
unicameral Supreme Soviet
Judicial branch:
Supreme Court

Political Parties [20]

Supreme Soviet	% of seats
Non-aligned	62.1
Armenian National Movement	21.7
Armenian Democratic	5.8
Dashnatktsutyan	5.0
National Democratic Union	3.8
Christian Democratic Union	0.4
Constitutional Rights Union	0.4
National Self-Determination	0.4
Republican Party	0.4

Government Budget [21]

For 1992.

Revenues	NA
Expenditures	NA
Capital expenditures	NA

Defense Summary [22]

Branches: Army, Air Force, National Guard, Security Forces (internal and border troops)

Manpower Availability: Males age 15-49 848,223; fit for military service 681,058; reach military age (18) annually 28,101 (1993 est.)

Defense Expenditures: 250 million rubles; note - conversion of the military budget into US dollars using the current exchange rate could produce misleading results

Crime [23]

Human Rights [24]

Labor Force

Total Labor Force [25]

1.63 million

Labor Force by Occupation [26]

Industry and construction	42%
Agriculture and forestry	18
Other	40

Date of data: 1990

Unemployment Rate [27]

2% of officially registered unemployed but large numbers of underemployed

For sources, notes, and explanations, see Annotated Source Appendix, page 1035.

35

Production Sectors

Energy Resource Summary [28]

Energy Resources: None. **Electricity**: 2,875,000 kW capacity; 9,000 million kWh produced, 2,585 kWh per capita (1992). **Pipelines**: Natural gas 900 km (1991).

Telecommunications [30]

- Progress on installation of fiber optic cable and construction of facilities for mobile cellular phone service remains in the negotiation phase for joint venture agreement
- Armenia has about 260,000 telephones, of which about 110,000 are in Yerevan
- Average telephone density is 8 per 100 persons
- International connections to other former republics of the USSR are by landline or microwave and to other countries by satellite and by leased connection through the Moscow international gateway switch
- Broadcast stations - 100% of population receives Armenian and Russian TV programs
- Satellite earth station - INTELSAT

Transportation [31]

Railroads. 840 km; does not include industrial lines (1990)

Highways. 11,300 km total; 10,500 km hard surfaced, 800 km earth (1990)

Airports

Total:	12
Useable:	10
With permanent-surface runways:	6
With runways over 3,659 m:	1
With runways 2,440-3,659 m:	4
With runways 1,220-2,439 m:	3

Top Agricultural Products [32]

	1990	1991
Vegetables	390	457
Grains	254	340
Potatoes	213	316
Grapes	144	200
Fruits and berries	156	120

Values shown are 1,000 metric tons.

Top Mining Products [33]

Detailed information is not available. A summary of mineral resources available follows. **Mineral Resources**: Small deposits of gold, copper, molybdenum, zinc, alumina.

Tourism [34]

Finance, Economics, and Trade

Industrial Summary [35]

Industrial Production: Growth rate - 50% (1992 est.). **Industries**: Diverse, including (in percent of output of former USSR) metalcutting machine tools (5.5%), forging-pressing machines (1.9%), electric motors (9%), tires (1.5%), knitted wear (4.4%), hosiery (3.0%), shoes (2.2%), silk fabric (0.8%), washing machines (2.0%), chemicals, trucks, watches, instruments, and microelectronics (1990).

Economic Indicators [36]

National product: GDP not available. **National product real growth rate**: - 34% (1992). **National product per capita**: not available. **Inflation rate (consumer prices)**: 20% per month (first quarter 1993). **External debt**: $650 million (December 1991 est.).

Balance of Payments Summary [37]

Exchange Rates [38]

Currency: **rubles.**
Symbol: **R.**

Subject to wide fluctuations. Data are currency units per $1.

December 24, 1994	415

Imports and Exports

Top Import Origins [39]

$300 million from outside the successor statees of the former USSR (c.i.f., 1992).

Origins	%
NA	NA

Top Export Destinations [40]

$30 million to outside the successor states of the former USSR (f.o.b., 1992).

Destinations	%
NA	NA

Foreign Aid [41]

	U.S. $
Wheat from US, Turkey	NA

Import and Export Commodities [42]

Import Commodities	Export Commodities
Machinery	Machinery & transport equip.
Energy	Light industrial products
Consumer goods	Processed food items

For sources, notes, and explanations, see Annotated Source Appendix, page 1035.

37

Australia

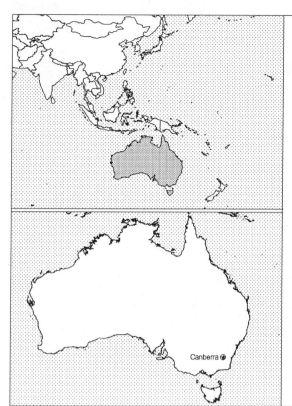

Geography [1]

Total area:
  7,686,850 km2
Land area:
  7,617,930 km2
Comparative area:
  Slightly smaller than the US
Land boundaries:
  0 km
Coastline:
  25,760 km
Climate:
  Generally arid to semiarid; temperate in south and east; tropical in north
Terrain:
  Mostly low plateau with deserts; fertile plain in southeast
Natural resources:
  Bauxite, coal, iron ore, copper, tin, silver, uranium, nickel, tungsten, mineral sands, lead, zinc, diamonds, natural gas, petroleum
Land use:
  Arable land:
  6%
  Permanent crops:
  0%
  Meadows and pastures:
  58%
  Forest and woodland:
  14%
  Other:
  22%

Demographics [2]

	1960	1970	1980	1990	1991[1]	1994	2000	2010	2020
Population	10,361	12,660	14,616	17,071	17,288	18,077	19,386	21,151	22,724
Population density (persons per sq. mi.)	4	4	5	6	6	NA	7	7	8
Births	NA	NA	NA	NA	254	258	NA	NA	NA
Deaths	NA	NA	NA	NA	127	133	NA	NA	NA
Life expectancy - males	NA	NA	NA	NA	74	74	NA	NA	NA
Life expectancy - females	NA	NA	NA	NA	80	81	NA	NA	NA
Birth rate (per 1,000)	NA	NA	NA	NA	15	14	NA	NA	NA
Death rate (per 1,000)	NA	NA	NA	NA	7	7	NA	NA	NA
Women of reproductive age (15-44 yrs.)	NA	NA	NA	4,489	NA	4,742	4,928	NA	NA
of which are currently married	NA	NA	NA	2,568	NA	2,785	2,951	NA	NA
Fertility rate	NA	NA	NA	1.8	1.84	1.8	1.8	1.8	1.8

Population values are in thousands, life expectancy in years, and other items as indicated.

Health

Health Personnel [3]

Doctors per 1,000 pop., 1988-92	2.29
Nurse-to-doctor ratio, 1988-92	3.8
Hospital beds per 1,000 pop., 1985-90	5.6
Percentage of children immunized (age 1 yr. or less)	
Third dose of DPT, 1990-91	90.0
Measles, 1990-91	68.0

Health Expenditures [4]

Total health expenditure, 1990 (official exchange rate)	
Millions of dollars	22736
Millions of dollars per capita	1331
Health expenditures as a percentage of GDP	
Total	7.7
Public sector	5.4
Private sector	2.3
Development assistance for health	
Total aid flows (millions of dollars)[1]	NA
Aid flows per capita (millions of dollars)	NA
Aid flows as a percentage of total health expenditure	NA

For sources, notes, and explanations, see Annotated Source Appendix, page 1035.

Human Factors

Health Care Ratios [5]

Population per physician, 1970	830
Population per physician, 1990	NA
Population per nursing person, 1970	NA
Population per nursing person, 1990	NA
Percent of births attended by health staff, 1985	99

Infants and Malnutrition [6]

Percent of babies with low birth weight, 1985	6
Infant mortality rate per 1,000 live births, 1970	18
Infant mortality rate per 1,000 live births, 1991	8
Years of life lost per 1,000 population, 1990	9
Prevalence of malnutrition (under age 5), 1990	NA

Ethnic Division [7]

Caucasian	95%
Asian	4%
Aboriginal, other	1%

Religion [8]

Anglican	26.1%
Roman Catholic	26.0%
Other Christian	24.3%

Major Languages [9]

English, Native languages.

Education

Public Education Expenditures [10]

Million Australian Dollars	1980	1985	1987	1988	1989	1990
Total education expenditure	7,592	12,925	14,726	16,113	18,001	19,364
as percent of GNP	5.5	5.6	5.1	4.9	5.1	5.4
as percent of total govt. expend.	14.8	12.8	12.5	12.8	12.7	14.8
Current education expenditure	6,899	11,848	13,586	14,954	16,628	17,889
as percent of GNP	5.0	5.1	4.7	4.6	4.7	5.0
as percent of current govt. expend.	16.8	14.2	13.4	13.7	14.1	14.8
Capital expenditure	693	1,077	1,140	1,159	1,373	1,475

Educational Attainment [11]

Literacy Rate [12]

Libraries [13]

	Admin. Units	Svc. Pts.	Vols. (000)	Shelving (meters)	Added	Users
National (1991)	1	NA	4,625[42a]	NA	NA	NA
Non-specialized	NA	NA	NA	NA	NA	NA
Public (1987)	497	1,804	27,000	NA	118,800	NA
Higher ed. (1987)[42b]	341	574	24,215	NA	1 mil.	NA
School	NA	NA	NA	NA	NA	NA

Daily Newspapers [14]

	1975	1980	1985	1990
Number of papers	70	62	62	62
Circ. (000)	5,336[e]	4,700[e]	4,300[e]	4,200[e]

Cinema [15]

Data for 1989.

Cinema seats per 1,000	NA
Annual attendance per person	2.4
Gross box office receipts (mil. Dollars)	258

Science and Technology

Scientific/Technical Forces [16]

Potential scientists/engineers	521,027
Number female	184,827
Potential technicians	317,112
Number female	158,908
Total	838,139

R&D Expenditures [17]

	Dollar (000) 1988
Total expenditure	4,187,100
Capital expenditure	590,200
Current expenditure	3,596,900
Percent current	85.9

U.S. Patents Issued [18]

Values show patents issued to citizens of the country by the U.S. Patents Office.

	1990	1991	1992
Number of patents	517	571	498

Government and Law

Organization of Government [19]

Long-form name:
Commonwealth of Australia

Type:
federal parliamentary state

Independence:
1 January 1901 (federation of UK colonies)

Constitution:
9 July 1900, effective 1 January 1901

Legal system:
based on English common law; accepts compulsory ICJ jurisdiction, with reservations

National holiday:
Australia Day, 26 January

Executive branch:
British monarch, governor general, prime minister, deputy prime minister, Cabinet

Legislative branch:
bicameral Federal Parliament consists of an upper house or Senate and a lower house or House of Representatives

Judicial branch:
High Court

Political Parties [20]

House of Representatives	% of seats
Labor	54.4
Liberal-National	44.2
Independent	1.4

Government Expenditures [21]

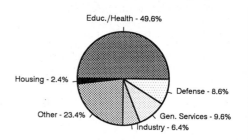

Educ./Health - 49.6%
Housing - 2.4%
Defense - 8.6%
Other - 23.4%
Gen. Services - 9.6%
Industry - 6.4%

% distribution for 1991. Expend. in FY93: 78.0 ($ bil.)

Military Expenditures and Arms Transfers [22]

	1985	1986	1987	1988	1989
Military expenditures					
Current dollars (mil.)	5,856	5,946	6,543	6,093	6,153
1989 constant dollars (mil.)	6,666	6,599	7,038	6,342	6,153
Armed forces (000)	70	71	70	71	70
Gross national product (GNP)					
Current dollars (mil.)	207,500	217,300	232,700	247,600	270,800
1989 constant dollars (mil.)	236,200	241,100	250,300	257,800	270,800
Central government expenditures (CGE)					
1989 constant dollars (mil.)	70,320	72,050	72,040	69,160	66,620[e]
People (mil.)	15.8	16.0	16.2	16.5	16.7
Military expenditure as % of GNP	2.8	2.7	2.8	2.5	2.3
Military expenditure as % of CGE	9.5	9.2	9.8	9.2	9.2
Military expenditure per capita	422	412	433	385	368
Armed forces per 1,000 people	4.4	4.4	4.3	4.3	4.2
GNP per capita	14,960	15,050	15,410	15,650	16,220
Arms imports[6]					
Current dollars (mil.)	900	1,000	725	1,300	675
1989 constant dollars (mil.)	1,025	1,110	780	1,353	675
Arms exports[6]					
Current dollars (mil.)	140	50	80	80	80
1989 constant dollars (mil.)	159	55	86	83	80
Total imports[7]					
Current dollars (mil.)	25,890	26,100	29,320	36,090	44,930
1989 constant dollars	29,470	28,970	31,530	37,570	44,930
Total exports[7]					
Current dollars (mil.)	22,850	22,670	26,490	33,070	37,760
1989 constant dollars	26,010	25,160	28,490	34,430	37,760
Arms as percent of total imports[8]	3.5	3.8	2.5	3.6	1.5
Arms as percent of total exports[8]	0.6	0.2	0.3	0.2	0.2

Crime [23]

Crime volume	
Cases known to police	1,049,708
Attempts (percent)	NA
Percent cases solved	NA
Crimes per 100,000 persons	26,088.412
Persons responsible for offenses	
Total number offenders	142,672
Percent female	NA
Percent juvenile	NA
Percent foreigners	NA

Human Rights [24]

	SSTS	FL	FAPRO	PPCG	APROBC	TPW	PCPTW	STPEP	PHRFF	PRW	ASST	AFL
Observes	P	P	P	P	P	P	P			P	P	P
	EAFRD	CPR	ESCR	SR	ACHR	MAAE	PVIAC	PVNAC	EAFDAW	TCIDTP	RC
Observes	P	P	P	P			P	P	P	P	P

P = Party; S = Signatory; see Appendix for meaning of abbreviations.

Labor Force

Total Labor Force [25]

8.63 million (September 1991)

Labor Force by Occupation [26]

Finance and services	33.8%
Public and community services	22.3
Wholesale and retail trade	20.1
Manufacturing and industry	16.2
Agriculture	6.1

Date of data: 1987

Unemployment Rate [27]

11.3% (December 1992)

40

For sources, notes, and explanations, see Annotated Source Appendix, page 1035.

Production Sectors

Commercial Energy Production and Consumption

Production [28]

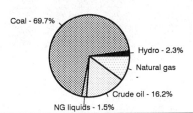

Coal - 69.7%
Hydro - 2.3%
Natural gas
Crude oil - 16.2%
NG liquids - 1.5%

Consumption [29]

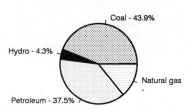

Coal - 43.9%
Hydro - 4.3%
Natural gas
Petroleum - 37.5%

Data are shown in quadrillion (10^{15}) BTUs and percent for 1991

Crude oil[1]	1.11
Natural gas liquids	0.10
Dry natural gas	0.70
Coal[2]	4.77
Net hydroelectric power[3]	0.16
Net nuclear power[3]	0.00
Total	6.83

Petroleum	1.39
Dry natural gas	0.53
Coal[2]	1.63
Net hydroelectric power[3]	0.16
Net nuclear power[3]	0.00
Total	3.71

Telecommunications [30]

- Good international and domestic service
- 8.7 million telephones
- Broadcast stations - 258 AM, 67 FM, 134 TV
- Submarine cables to New Zealand, Papua New Guinea, and Indonesia
- Domestic satellite service
- Satellite stations - 4 Indian Ocean INTELSAT, 6 Pacific Ocean INTELSAT earth stations

Transportation [31]

Railroads. 40,478 km total; 7,970 km 1.600-meter gauge, 16,201 km 1.435-meter standard gauge, 16,307 km 1.067-meter gauge; 183 km dual gauge; 1,130 km electrified; government owned (except for a few hundred kilometers of privately owned track) (1985)

Highways. 837,872 km total; 243,750 km paved, 228,396 km gravel, crushed stone, or stabilized soil surface, 365,726 km unimproved earth

Merchant Marine. 82 ships (1,000 GRT or over) totaling 2,347,271 GRT/3,534,926 DWT; includes 2 short-sea passenger, 8 cargo, 7 container, 8 roll-on/roll-off, 1 vehicle carrier, 17 oil tanker, 3 chemical tanker, 4 liquefied gas, 30 bulk, 2 combination bulk

Top Agricultural Products [32]

	89-90[1]	90-91[1]
Sugar cane	27,622	27,567
Wheat	14,121	15,068
Barley	4,096	4,055
Oats	1,638	1,500
Potatoes	1,166	1,075
Sorghum	920	883

Values shown are 1,000 metric tons.

Airports

Total:	481
Usable:	439
With permanent-surface runways:	243
With runways over 3,659 m:	1
With runways 2,440-3,659 m:	20
With runways 1,220-2,439 m:	268

Top Mining Products [33]

Metric tons unless otherwise specified	M.t.
Alumina	11,713[1]
Bauxite, gross weight (000 tons)	40,503[1]
Diamond (000 carats)	35,956[1]
Ilmenite (000 tons)	1,363[1]
Lead, mine output, Pb content (000 tons)	571[1]
Monazite concentrate, gross weight	5,000[e]
Opal (value, $ thousands)	85,000
Rutile	201,000
Sapphire (value, $ thousands)	10,000
Zirconium concentrate, gross weight (000 tons)	292

Tourism [34]

	1987	1988	1989	1990	1991
Visitors[4]	1,785	2,249	2,080	2,215	2,370
Tourism receipts	2,105	3,315	3,157	3,660	4,183
Tourism expenditures	2,427	2,933	3,859	4,148	3,940
Fare receipts	920	1,276	1,131	1,326	1,518
Fare expenditures	993	1,310	1,663	1,802	1,794

Tourists are in thousands, money in million U.S. dollars.

Manufacturing Sector

GDP and Manufacturing Summary [35]

	1980	1985	1989	1990	% change 1980-1990	% change 1989-1990
GDP (million 1980 $)	159,241	185,531	212,701	211,350	32.7	-0.6
GDP per capita (1980 $)	10,836	11,773	12,772	12,516	15.5	-2.0
Manufacturing as % of GDP (current prices)	19.0	17.4	16.9	15.4	-18.9	-8.9
Gross output (million $)	75,474	69,328	120,330	127,086	68.4	5.6
Value added (million $)	29,173	26,900	44,957	47,351[e]	62.3	5.3
Value added (million 1980 $)	30,815	31,682	37,077	34,798	12.9	-6.1
Industrial production index	100	104	111	111	11.0	0.0
Employment (thousands)	1,139	1,012	1,060	1,045[e]	-8.3	-1.4

Note: GDP stands for Gross Domestic Product. 'e' stands for estimated value.

Profitability and Productivity

	1980	1985	1989	1990	% change 1980-1990	% change 1989-1990
Intermediate input (%)	61	61	63	63[e]	3.3	0.0
Wages, salaries, and supplements (%)	20	19	17	17[e]	-15.0	0.0
Gross operating surplus (%)	18	20	21	21[e]	16.7	0.0
Gross output per worker ($)	65,402	67,896	113,488	119,912[e]	83.3	5.7
Value added per worker ($)	25,280	26,346	42,401	44,848[e]	77.4	5.8
Average wage (incl. benefits) ($)	13,356	12,999	19,134	20,370[e]	52.5	6.5

Profitability is in percent of gross output. Productivity is in U.S. $. 'e' stands for estimated value.

Profitability - 1990

Wages - 16.8%
Inputs - 62.4%
Surplus - 20.8%

The graphic shows percent of gross output.

Value Added in Manufacturing

	1980 $ mil.	1980 %	1985 $ mil.	1985 %	1989 $ mil.	1989 %	1990 $ mil.	1990 %	% change 1980-1990	% change 1989-1990
311 Food products	3,993	13.7	3,764	14.0	6,306	14.0	7,048[e]	14.9	76.5	11.8
313 Beverages	785	2.7	847	3.1	1,391	3.1	1,675[e]	3.5	113.4	20.4
314 Tobacco products	248	0.9	179	0.7	279	0.6	289[e]	0.6	16.5	3.6
321 Textiles	1,050	3.6	955	3.6	1,624	3.6	1,714	3.6	63.2	5.5
322 Wearing apparel	821	2.8	722	2.7	1,204	2.7	1,006[e]	2.1	22.5	-16.4
323 Leather and fur products	93	0.3	76	0.3	132	0.3	137[e]	0.3	47.3	3.8
324 Footwear	223	0.8	205	0.8	300	0.7	247[e]	0.5	10.8	-17.7
331 Wood and wood products	1,052	3.6	1,028	3.8	1,652	3.7	1,776[e]	3.8	68.8	7.5
332 Furniture and fixtures	505	1.7	507	1.9	937	2.1	1,023[e]	2.2	102.6	9.2
341 Paper and paper products	744	2.6	703	2.6	1,362	3.0	1,458[e]	3.1	96.0	7.0
342 Printing and publishing	1,818	6.2	2,131	7.9	3,749	8.3	4,000[e]	8.4	120.0	6.7
351 Industrial chemicals	969	3.3	982	3.7	1,590	3.5	1,799[e]	3.8	85.7	13.1
352 Other chemical products	1,186	4.1	1,191	4.4	1,995	4.4	2,155[e]	4.6	81.7	8.0
353 Petroleum refineries	323	1.1	285	1.1	277	0.6	325[e]	0.7	0.6	17.3
354 Miscellaneous petroleum and coal products	30	0.1	25	0.1	27	0.1	29[e]	0.1	-3.3	7.4
355 Rubber products	341	1.2	264	1.0	437	1.0	515[e]	1.1	51.0	17.8
356 Plastic products	831	2.8	808	3.0	1,535	3.4	1,599[e]	3.4	92.4	4.2
361 Pottery, china and earthenware	46	0.2	41	0.2	73	0.2	69[e]	0.1	50.0	-5.5
362 Glass and glass products	246	0.8	254	0.9	428	1.0	403[e]	0.9	63.8	-5.8
369 Other non-metal mineral products	1,183	4.1	1,085	4.0	1,755	3.9	1,635[e]	3.5	38.2	-6.8
371 Iron and steel	1,920	6.6	1,391	5.2	2,002	4.5	2,194[e]	4.6	14.3	9.6
372 Non-ferrous metals	1,473	5.0	1,409	5.2	2,287	5.1	2,318[e]	4.9	57.4	1.4
381 Metal products	2,467	8.5	2,041	7.6	3,740	8.3	4,144	8.8	68.0	10.8
382 Non-electrical machinery	2,091	7.2	1,575	5.9	2,635	5.9	2,749[e]	5.8	31.5	4.3
383 Electrical machinery	1,351	4.6	1,329	4.9	2,385	5.3	2,477[e]	5.2	83.3	3.9
384 Transport equipment	2,830	9.7	2,579	9.6	3,967	8.8	3,636	7.7	28.5	-8.3
385 Professional and scientific equipment	290	1.0	279	1.0	494	1.1	517[e]	1.1	78.3	4.7
390 Other manufacturing industries	263	0.9	246	0.9	395	0.9	416[e]	0.9	58.2	5.3

Note: The industry codes shown are International Standard Industry codes (ISIC). Percentages are percent of total Value Added. 'e' stands for estimated value

Finance, Economics, and Trade

Economic Indicators [36]

Millions of Australian Dollars, unless otherwise noted[4].

	1989	1990	1991[e]
Real GDP[5]	255,762	259,928	257,500
GDP growth rate (percent)	4.4	1.6	-1.3
Money supply (MI)	40,820.0	42,373.0	44,744.0
Saving rate	7.9	7.6	4.6
CPI (June 1984 = 100)	194.6	208.7	215.3
WPI	NA	NA	NA
Gross external public debt	67,224	73,043	67,500

Balance of Payments Summary [37]

Values in millions of dollars.

	1987	1988	1989	1990	1991
Exports of goods (f.o.b.)	27,014	33,196	36,883	39,332	42,025
Imports of goods (f.o.b.)	-26,749	-33,892	-40,329	-38,966	-38,495
Trade balance	265	-696	-3,446	366	3,530
Services - debits	-17,317	-22,270	-28,146	-30,341	-28,843
Services - credits	8,717	11,395	12,319	13,409	13,595
Private transfers (net)	1,196	1,706	2,124	1,939	2,062
Government transfers (net)	-178	-232	-182	-98	-198
Long term capital (net)	8,736	18,538	18,748	9,597	11,502
Short term capital (net)	-515	-248	-3,314	1,100	448
Errors and omissions	-533	-2,943	2,525	5,755	-2,414
Overall balance	371	5,250	628	1,727	-318

Exchange Rates [38]

Currency: **Australian dollars.**
Symbol: **$A.**

Data are currency units per $1.

January 1993	1.4837
1992	1.3600
1991	1.2836
1990	1.2799
1989	1.2618
1988	1.2752

Imports and Exports

Top Import Origins [39]

$37.8 billion (f.o.b., FY91). Data are for 1990.

Origins	%
US	24
Japan	19
UK	6
FRG	7
NZ	4

Top Export Destinations [40]

$41.7 billion (f.o.b., FY91).

Destinations	%
Japan	26
US	11
NZ	6
South Korea	4
Singapore	4
UK	NA
Taiwan	NA
Hong Kong	NA

Foreign Aid [41]

	U.S. $	
Donor - ODA and OOF commitments (1970-89)	10.4	billion

Import and Export Commodities [42]

Import Commodities	**Export Commodities**
Machinery & transport equipment	Coal
Computers and office machines	Gold
Crude oil & petroleum products	Meat
	Wool
	Alumina
	Wheat
	Machinery
	Transport equip.

Austria

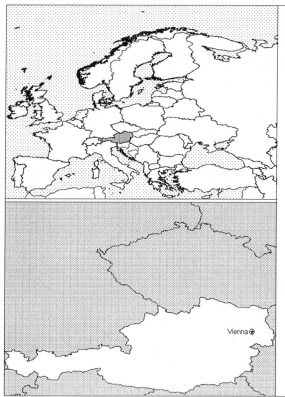

Geography [1]

Total area:
83,850 km2
Land area:
82,730 km2
Comparative area:
Slightly smaller than Maine
Land boundaries:
Total 2,496 km, Czech Republic 362 km, Germany 784 km, Hungary 366 km, Italy 430 km, Liechtenstein 37 km, Slovakia 91 km, Slovenia 262 km, Switzerland 164 km
Coastline:
0 km (landlocked)
Climate:
Temperate; continental, cloudy; cold winters with frequent rain in lowlands and snow in mountains; cool summers with occasional showers
Terrain:
In the west and south mostly mountains (Alps); along the eastern and Northern margins mostly flat or gently sloping
Natural resources:
iron ore, petroleum, timber, magnesite, aluminum, lead, coal, lignite, copper, hydropower
Land use:
Arable land: 17%
Permanent crops: 1%
Meadows and pastures: 24%
Forest and woodland: 39%
Other: 19%

Demographics [2]

	1960	1970	1980	1990	1991[1]	1994	2000	2010	2020
Population	7,047	7,467	7,549	7,718	7,666	7,955	8,108	8,259	8,329
Population density (persons per sq. mi.)	221	234	236	239	240	NA	243	240	234
Births	NA	NA	NA	NA	88	91	NA	NA	NA
Deaths	NA	NA	NA	NA	82	82	NA	NA	NA
Life expectancy - males	NA	NA	NA	NA	74	73	NA	NA	NA
Life expectancy - females	NA	NA	NA	NA	81	80	NA	NA	NA
Birth rate (per 1,000)	NA	NA	NA	NA	12	11	NA	NA	NA
Death rate (per 1,000)	NA	NA	NA	NA	11	10	NA	NA	NA
Women of reproductive age (15-44 yrs.)	NA	NA	NA	1,966	NA	1,981	1,992	NA	NA
of which are currently married	NA	NA	NA	1,220	NA	1,267	1,292	NA	NA
Fertility rate	NA	NA	NA	1.5	1.46	1.5	1.5	1.5	1.6

Population values are in thousands, life expectancy in years, and other items as indicated.

Health

Health Personnel [3]

Doctors per 1,000 pop., 1988-92	4.34
Nurse-to-doctor ratio, 1988-92	2.4
Hospital beds per 1,000 pop., 1985-90	10.8
Percentage of children immunized (age 1 yr. or less)	
Third dose of DPT, 1990-91	90.0
Measles, 1990-91	60.0

Health Expenditures [4]

Total health expenditure, 1990 (official exchange rate)	
Millions of dollars	13193
Millions of dollars per capita	1711
Health expenditures as a percentage of GDP	
Total	8.3
Public sector	5.5
Private sector	2.8
Development assistance for health	
Total aid flows (millions of dollars)[1]	NA
Aid flows per capita (millions of dollars)	NA
Aid flows as a percentage of total health expenditure	NA

For sources, notes, and explanations, see Annotated Source Appendix, page 1035.

Human Factors

Health Care Ratios [5]

Population per physician, 1970	540
Population per physician, 1990	230
Population per nursing person, 1970	300
Population per nursing person, 1990	70
Percent of births attended by health staff, 1985	NA

Infants and Malnutrition [6]

Percent of babies with low birth weight, 1985	6
Infant mortality rate per 1,000 live births, 1970	26
Infant mortality rate per 1,000 live births, 1991	8
Years of life lost per 1,000 population, 1990	11
Prevalence of malnutrition (under age 5), 1990	NA

Ethnic Division [7]

German	99.4%
Croatian	0.3%
Slovene	0.2%
Other	0.1%

Religion [8]

Roman Catholic	85%
Protestant	6%
Other	9%

Major Languages [9]

German.

Education

Public Education Expenditures [10]

Million Schilling	1980	1985	1987	1988	1989	1990
Total education expenditure	55,016	78,639	NA	88,355	91,624	97,301
as percent of GNP	5.6	5.9	NA	5.7	5.6	5.5
as percent of total govt. expend.	8.0	7.9	NA	7.6	7.6	7.6
Current education expendenditures	46,955	70,847	NA	80,422	84,911	89,858
as percent of GNP	4.8	5.3	NA	5.2	5.1	5.1
as percent of current govt. expend.	8.4	8.6	NA	8.7	8.9	8.8
Capital expenditures	8,061	7,792	NA	7,933	6,713	7,443

Educational Attainment [11]

Age group	25+
Total population	4,558,681
Highest level attained (%)	
No schooling	-
First level	
Incompleted	49.3
Completed	NA
Entered second level	
S-1	NA
S-2	47.5
Post secondary	3.3

Literacy Rate [12]

Libraries [13]

	Admin. Units	Svc. Pts.	Vols. (000)	Shelving (meters)	Vols. Added	Reg. Users
National (1989)	1	1	2,686	NA	83,520	326,694
Non-specialized (1989)	8	NA	1,544	NA	37,489	68,556
Public (1990)	2,374	2,374	8,195	NA	NA	869,292
Higher ed. (1990)[27]	21	NA	15,446	NA	371,591	3.2mil.
School	NA	NA	NA	NA	NA	NA

Daily Newspapers [14]

	1975	1980	1985	1990
Number of papers	30	30	33	25
Circ. (000)	2,405	2,651	2,729	2,706

Cinema [15]

Data for 1991.

Cinema seats per 1,000	10.4
Annual attendance per person	1.4
Gross box office receipts (mil. Schilling)	654

Science and Technology

Scientific/Technical Forces [16]

Potential scientists/engineers	134,336
Number female	34,516
Potential technicians	28,467
Number female	19,790
Total	162,803

R&D Expenditures [17]

	Schilling (000) 1985
Total expenditure	17,182,272
Capital expenditure	3,284,207
Current expenditure	13,898,065
Percent current	80.9

U.S. Patents Issued [18]

Values show patents issued to citizens of the country by the U.S. Patents Office.

	1990	1991	1992
Number of patents	423	388	402

For sources, notes, and explanations, see Annotated Source Appendix, page 1035.

45

Government and Law

Organization of Government [19]

Long-form name:
Republic of Austria

Type:
federal republic

Independence:
12 November 1918 (from Austro-Hungarian Empire)

Constitution:
1920; revised 1929 (reinstated 1945)

Legal system:
civil law system with Roman law origin; judicial review of legislative acts by a Constitutional Court; separate administrative and civil/penal supreme courts; has not accepted compulsory ICJ jurisdiction

National holiday:
National Day, 26 October (1955)

Executive branch:
president, chancellor, vice chancellor, Council of Ministers (cabinet)

Legislative branch:
bicameral Federal Assembly

Judicial branch:
Supreme Judicial Court (Oberster Gerichtshof) for civil and criminal cases, Administrative Court (Verwaltungsgerichtshof) for bureaucratic cases, Constitutional Court (Verfassungsgerichtshof) for constitutional cases

Political Parties [20]

National Council	% of votes
Social Democratic Party (SPO)	43.0
Austrian People's Party (OVP)	32.1
Freedom Party of Austria (FPO)	16.6
Green Alternative List (GAL)	4.5
Communist Party (KPO)	0.7
Other	0.3

Government Expenditures [21]

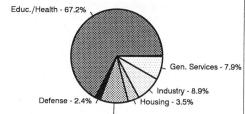

Educ./Health - 67.2%
Gen. Services - 7.9%
Industry - 8.9%
Housing - 3.5%
Defense - 2.4%
Other - 10.1%

% distribution for 1991. Expend. in 1992 est.: 53.0 ($ bil.)

Military Expenditures and Arms Transfers [22]

	1985	1986	1987	1988	1989
Military expenditures					
Current dollars (mil.)	1,322[e]	1,402[e]	1,323	1,310	1,402
1989 constant dollars (mil.)	1,505[e]	1,556[e]	1,423	1,364	1,402
Armed forces (000)	40	39	70	55	48
Gross national product (GNP)					
Current dollars (mil.)	99,040	102,400	107,700	115,900	125,200
1989 constant dollars (mil.)	112,700	113,700	115,800	120,600	125,200
Central government expenditures (CGE)					
1989 constant dollars (mil.)	45,390	47,170	47,340	49,320	49,340
People (mil.)	7.6	7.6	7.6	7.6	7.6
Military expenditure as % of GNP	1.3	1.4	1.2	1.1	1.1
Military expenditure as % of CGE	3.3	3.3	3.0	2.8	2.8
Military expenditure per capita	199	206	188	180	184
Armed forces per 1,000 people	5.3	5.2	9.2	7.2	6.3
GNP per capita	14,920	15,020	15,290	15,880	16,430
Arms imports[6]					
Current dollars (mil.)	20	10	20	90	100
1989 constant dollars (mil.)	23	11	22	94	100
Arms exports[6]					
Current dollars (mil.)	160	270	60	70	40
1989 constant dollars (mil.)	182	300	65	73	40
Total imports[7]					
Current dollars (mil.)	20,990	26,860	32,720	36,220	38,980
1989 constant dollars	23,890	29,810	35,200	37,710	38,980
Total exports[7]					
Current dollars (mil.)	17,110	22,420	27,080	31,000	32,450
1989 constant dollars	19,480	24,880	29,130	32,270	32,450
Arms as percent of total imports[8]	0.1	0	0.1	0.2	0.3
Arms as percent of total exports[8]	0.9	1.2	0.2	0.2	0.1

Crime [23]

Crime volume (for 1990)	
Cases known to police	457,623
Attempts (percent)	4.8
Percent cases solved	44.2
Crimes per 100,000 persons	6,002.7
Persons responsible for offenses	
Total number offenders	176,649
Percent female	19.1
Percent juvenile[3]	10.8
Percent foreigners	18.4

Human Rights [24]

	SSTS	FL	FAPRO	PPCG	APROBC	TPW	PCPTW	STPEP	PHRFF	PRW	ASST	AFL
Observes	P	P	P	P	P	P	P		P	P	P	P
	EAFRD	CPR	ESCR	SR	ACHR	MAAE	PVIAC	PVNAC	EAFDAW	TCIDTP	RC
Observes		P	P	P	P			P	P	P	P	P

P = Party; S = Signatory; see Appendix for meaning of abbreviations.

Labor Force

Total Labor Force [25]

3.47 million (1989)

Labor Force by Occupation [26]

Services	56.4%
Industry and crafts	35.4
Agriculture and forestry	8.1

Unemployment Rate [27]

6.4% (1992 est.)

For sources, notes, and explanations, see Annotated Source Appendix, page 1035.

Production Sectors

Commercial Energy Production and Consumption

Production [28]

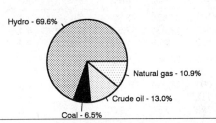

Hydro - 69.6%
Natural gas - 10.9%
Crude oil - 13.0%
Coal - 6.5%

Consumption [29]

Petroleum - 40.8%
Nuclear - 1.7%
Coal - 12.5%
Natural gas - 20.8%
Hydro - 24.2%

Data are shown in quadrillion (10^{15}) BTUs and percent for 1991

Crude oil[1]	0.06
Natural gas liquids	(s)
Dry natural gas	0.05
Coal[2]	0.03
Net hydroelectric power[3]	0.32
Net nuclear power[3]	0.00
Total	0.45

Petroleum	0.49
Dry natural gas	0.25
Coal[2]	0.15
Net hydroelectric power[3]	0.29
Net nuclear power[3]	0.02
Total	1.21

Telecommunications [30]

- Highly developed and efficient
- 4,014,000 telephones
- Broadcast stations - 6 AM, 21 (545 repeaters) FM, 47 (870 repeaters) TV
- Satellite ground stations for Atlantic Ocean INTELSAT, Indian Ocean INTELSAT, and EUTELSAT systems

Transportation [31]

Railroads. 5,749 km total; 5,652 km government owned and 97 km privately owned (0.760-, 1.435- and 1.000-meter gauge); 5,394 km 1.435-meter standard gauge of which 3,154 km is electrified and 1,520 km is double tracked; 339 km 0.760-meter narrow gauge of which 84 km is electrified

Highways. 95,412 km total; 34,612 km are the primary network (including 1,012 km of autobahn, 10,400 km of federal, and 23,200 km of provincial roads); of this number, 21,812 km are paved and 12,800 km are unpaved; in addition, there are 60,800 km of communal roads (mostly gravel, crushed stone, earth)

Merchant Marine. 29 ships (1,000 GRT or over) totaling 154,159 GRT/256,765 DWT; includes 23 cargo, 1 refrigerated cargo, 1 oil tanker, 1 chemical tanker, 3 bulk

Airports

Total:	55
Usable:	55
With permanent-surface runways:	20
With runways over 3,659 m:	0
With runways 2,440-3,659 m:	6
With runways 1,220-2,439 m:	4

Top Agricultural Products [32]

	1990[2]	1991[2]
Sugarbeets	2,494	2,406
Corn	1,544	1,500
Barley	1,512	1,424
Wheat	1,403	1,375
Apples	268	243
Rapeseed	85	111

Values shown are 1,000 metric tons.

Top Mining Products [33]

Metric tons, unless otherwise specified	M.t.
Iron ore and concentrate (000 tons)	2,130
Lead, mine output (Pb content of concentrate)	1,100
Tungsten, mine output, W content of concentrate	1,300
Zinc, mine output, Zn content of concentrate	14,800
Graphite, crude	654,594
Magnesite, crude	961
Coal, brown and lignite (000 tons)	2,081

Tourism [34]

	1987	1988	1989	1990	1991
Tourists[6]	15,761	16,571	18,202	19,011	19,092
Tourism receipts	8,863	10,090[7]	10,717[7]	13,410[7]	13,956[7]
Tourism expenditures	5,592	6,307[8]	6,266[8]	7,723[8]	7,449[8]
Fare receipts	451				
Fare expenditures	140				

Tourists are in thousands, money in million U.S. dollars.

For sources, notes, and explanations, see Annotated Source Appendix, page 1035.

47

Manufacturing Sector

GDP and Manufacturing Summary [35]

	1980	1985	1989	1990	% change 1980-1990	% change 1989-1990
GDP (million 1980 $)	76,882	82,071	91,586	95,394	24.1	4.2
GDP per capita (1980 $)	10,183	10,859	12,084	12,578	23.5	4.1
Manufacturing as % of GDP (current prices)	29.1	28.3	26.6	27.2	-6.5	2.3
Gross output (million $)	54,666	45,959	80,980	105,001	92.1	29.7
Value added (million $)	17,987	15,108	28,903	36,828	104.7	27.4
Value added (million 1980 $)	21,384	23,001	26,524	27,704	29.6	4.4
Industrial production index	100	112	126	137	37.0	8.7
Employment (thousands)	824	783	758	770	-6.6	1.6

Note: GDP stands for Gross Domestic Product. 'e' stands for estimated value.

Profitability and Productivity

	1980	1985	1989	1990	% change 1980-1990	% change 1989-1990
Intermediate input (%)	67	67	64	65	-3.0	1.6
Wages, salaries, and supplements (%)	24	23	18^e	23^e	-4.2	27.8
Gross operating surplus (%)	9	10	17^e	12^e	33.3	-29.4
Gross output per worker ($)	66,339	58,531	106,842	135,118	103.7	26.5
Value added per worker ($)	25,579	22,905	381,133	56,942	122.6	-85.1
Average wage (incl. benefits) ($)	15,686	13,342	19,579^e	30,941^e	97.3	58.0

Profitability is in percent of gross output. Productivity is in U.S. $. 'e' stands for estimated value.

Profitability - 1990

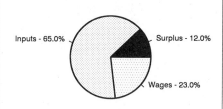

Inputs - 65.0% Surplus - 12.0% Wages - 23.0%

The graphic shows percent of gross output.

Value Added in Manufacturing

	1980 $ mil.	1980 %	1985 $ mil.	1985 %	1989 $ mil.	1989 %	1990 $ mil.	1990 %	% change 1980-1990	% change 1989-1990
311 Food products	1,752	9.7	1,472	9.7	2,549^e	8.8	3,086^e	8.4	76.1	21.1
313 Beverages	474	2.6	380	2.5	665^e	2.3	797^e	2.2	68.1	19.8
314 Tobacco products	807	4.5	725	4.8	1,234	4.3	1,491	4.0	84.8	20.8
321 Textiles	904	5.0	657	4.3	1,132	3.9	1,456	4.0	61.1	28.6
322 Wearing apparel	512	2.8	353	2.3	552	1.9	619	1.7	20.9	12.1
323 Leather and fur products	63	0.4	45	0.3	61^e	0.2	100	0.3	58.7	63.9
324 Footwear	223	1.2	157	1.0	221^e	0.8	231	0.6	3.6	4.5
331 Wood and wood products	192	1.1	298	2.0	545^e	1.9	698^e	1.9	263.5	28.1
332 Furniture and fixtures	965	5.4	733	4.9	1,645^e	5.7	2,030^e	5.5	110.4	23.4
341 Paper and paper products	645	3.6	509	3.4	990^e	3.4	1,282^e	3.5	98.8	29.5
342 Printing and publishing	726	4.0	626	4.1	1,257^e	4.3	1,609^e	4.4	121.6	28.0
351 Industrial chemicals	663	3.7	584	3.9	1,123^e	3.9	1,307^e	3.5	97.1	16.4
352 Other chemical products	534	3.0	398	2.6	786^e	2.7	919^e	2.5	72.1	16.9
353 Petroleum refineries	80	0.4	72	0.5	472	1.6	607	1.6	658.7	28.6
354 Miscellaneous petroleum and coal products	35	0.2	27	0.2	55^e	0.2	64^e	0.2	82.9	16.4
355 Rubber products	258	1.4	200	1.3	514^e	1.8	578^e	1.6	124.0	12.5
356 Plastic products	281	1.6	215	1.4	491^e	1.7	559^e	1.5	98.9	13.8
361 Pottery, china and earthenware	63	0.4	42	0.3	92	0.3	114^e	0.3	81.0	23.9
362 Glass and glass products	244	1.4	237	1.6	391	1.4	616	1.7	152.5	57.5
369 Other non-metal mineral products	894	5.0	724	4.8	1,343	4.6	1,646	4.5	84.1	22.6
371 Iron and steel	1,225	6.8	1,055	7.0	1,843	6.4	2,194	6.0	79.1	19.0
372 Non-ferrous metals	280	1.6	241	1.6	396	1.4	518	1.4	85.0	30.8
381 Metal products	1,542	8.6	1,170	7.7	2,534	8.8	3,035	8.2	96.8	19.8
382 Non-electrical machinery	1,765	9.8	1,502	9.9	2,690	9.3	4,052	11.0	129.6	50.6
383 Electrical machinery	1,615	9.0	1,472	9.7	2,985	10.3	4,343	11.8	168.9	45.5
384 Transport equipment	943	5.2	941	6.2	1,812	6.3	2,224	6.0	135.8	22.7
385 Professional and scientific equipment	161	0.9	144	1.0	270^e	0.9	327^e	0.9	103.1	21.1
390 Other manufacturing industries	143	0.8	130	0.9	252^e	0.9	327^e	0.9	128.7	29.8

Note: The industry codes shown are International Standard Industry codes (ISIC). Percentages are percent of total Value Added. 'e' stands for estimated value

For sources, notes, and explanations, see Annotated Source Appendix, page 1035.

Finance, Economics, and Trade

Economic Indicators [36]

Billions of Austrian Schillings (AS) unless otherwise stated.

	1989	1990	1991[e]
Real GDP (1983 prices)	1,384.2	1,452.5	1,495.6
Real GDP growth rate (%)	3.7	4.9	3.0
Money supply (M1; year-end)	249.2	262.7	NA
Savings rate			
(% of disposable income)	12.5	13.2	12.5
Investment rate (as % of GDP)	24.0	24.3	24.8
CPI (%)	2.5	3.3	3.3
Wholesale prices index (%)	1.7	2.9	1.5
External federal			
government debt (year-end)	125.8	135.8	147.4

Balance of Payments Summary [37]

Values in millions of dollars.

	1987	1988	1989	1990	1991
Exports of goods (f.o.b.)	26,558	30,056	31,832	40,252	40,136
Imports of goods (f.o.b.)	-31,702	-36,358	-38,437	-48,234	-52,186
Trade balance	-5,144	-6,302	-6,605	-7,982	-12,050
Services - debits	-15,263	-17,469	-18,897	-24,088	-24,217
Services - credits	20,059	23,349	25,690	33,028	35,957
Private transfers (net)	-10	38	-57	108	166
Government transfers (net)	-71	-74	-72	-109	-108
Long term capital (net)	1,879	599	399	-2,313	-2,551
Short term capital (net)	-1,403	552	432	2,094	2,568
Errors and omissions	356	-297	106	-775	1,107
Overall balance	403	396	996	-37	872

Exchange Rates [38]

Currency: **Austrian schillings.**
Symbol: **S.**

Data are currency units per $1.

January 1993	11.363
1992	10.989
1991	11.676
1990	11.370
1989	13.231
1988	12.348

Imports and Exports

Top Import Origins [39]

$50.7 billion (1992 est.). Data are for 1991.

Origins	%
EC	67.8
EFTA	6.9
Eastern Europe/ former USSR	6.0
Japan	4.8
US	3.9

Top Export Destinations [40]

$43.5 billion (1992 est.). Data are for 1991.

Destinations	%
EC	65.8
EFTA	9.1
Eastern Europe/ former USSR	9.0
Japan	1.7
US	2.8

Foreign Aid [41]

	U.S. $	
Donor - ODA and OOF commitments (1970-89)	2.4	billion

Import and Export Commodities [42]

Import Commodities

Petroleum
Foodstuffs
Machinery & equipment
Vehicles
Chemicals
Textiles and clothing
Pharmaceuticals

Export Commodities

Machinery & equipment
Iron and steel
Lumber
Textiles
Paper products
Chemicals

For sources, notes, and explanations, see Annotated Source Appendix, page 1035.

49

Azerbaijan

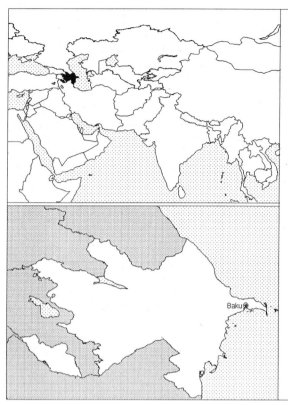

Geography [1]

Total area:
86,600 km2
Land area:
86,100 km2
Comparative area:
Slightly larger than Maine
Land boundaries:
Total 2,013 km, Armenia (west) 566 km, Armenia (southwest) 221 km, Georgia 322 km, Iran (south) 432 km, Iran (southwest) 179 km, Russia 284 km, Turkey 9 km
Coastline:
0 km (landlocked)
Climate:
Dry, semiarid steppe; subject to drought
Terrain:
Large, flat Kura-Aras Lowland (much of it below sea level) with Great Caucasus Mountains to the north, Karabakh Upland in west; Baku lies on Aspheson Peninsula that juts into Caspian Sea
Natural resources:
Petroleum, natural gas, iron ore, nonferrous metals, alumina
Land use:
Arable land: 18%
Permanent crops: 0%
Meadows and pastures: 25%
Forest and woodland: 0%
Other: 57%

Demographics [2]

	1960	1970	1980	1990	1991[1]	1994	2000	2010	2020
Population	3,882	5,169	6,173	7,216	NA	7,684	8,243	8,995	9,689
Population density (persons per sq. mi.)	NA	NA	NA	NA	NA	NA	NA	NA	NA
Births	NA	NA	NA	NA	NA	177	NA	NA	NA
Deaths	NA	NA	NA	NA	NA	51	NA	NA	NA
Life expectancy - males	NA	NA	NA	NA	NA	67	NA	NA	NA
Life expectancy - females	NA	NA	NA	NA	NA	75	NA	NA	NA
Birth rate (per 1,000)	NA	NA	NA	NA	NA	23	NA	NA	NA
Death rate (per 1,000)	NA	NA	NA	NA	NA	7	NA	NA	NA
Women of reproductive age (15-44 yrs.)	NA	NA	NA	1,828	NA	1,967	2,199	NA	NA
of which are currently married	NA	NA	NA	1,126	NA	1,250	1,403	NA	NA
Fertility rate	NA	NA	NA	2.9	NA	2.7	2.4	2.1	2.0

Population values are in thousands, life expectancy in years, and other items as indicated.

Health

Health Personnel [3]

Doctors per 1,000 pop., 1988-92	3.93
Nurse-to-doctor ratio, 1988-92	2.4
Hospital beds per 1,000 pop., 1985-90	10.2
Percentage of children immunized (age 1 yr. or less)	
Third dose of DPT, 1990-91	89.0
Measles, 1990-91	91.0

Health Expenditures [4]

Total health expenditure, 1990 (official exchange rate)	
Millions of dollars	785
Millions of dollars per capita	98
Health expenditures as a percentage of GDP	
Total	4.3
Public sector	2.6
Private sector	1.7
Development assistance for health	
Total aid flows (millions of dollars)[1]	NA
Aid flows per capita (millions of dollars)	NA
Aid flows as a percentage of total health expenditure	NA

For sources, notes, and explanations, see Annotated Source Appendix, page 1035.

Human Factors

Health Care Ratios [5]

Population per physician, 1970	NA
Population per physician, 1990	250
Population per nursing person, 1970	NA
Population per nursing person, 1990	NA
Percent of births attended by health staff, 1985	NA

Infants and Malnutrition [6]

Percent of babies with low birth weight, 1985	NA
Infant mortality rate per 1,000 live births, 1970	NA
Infant mortality rate per 1,000 live births, 1991	33
Years of life lost per 1,000 population, 1990	16
Prevalence of malnutrition (under age 5), 1990	NA

Ethnic Division [7]

Armenian share may be less than 5.6% because many Armenians have fled the ethnic violence since 1989 census.

Azeri	82.7%
Russian	5.6%
Armenian	5.6%
Daghestanis	3.2%
Other	2.9%

Religion [8]

Moslem	87.0%
Russian Orthodox	5.6%
Armenian Orthodox	5.6%
Other	1.8%

Major Languages [9]

Azeri	82%
Russian	7%
Armenian	5%
Other	6%

Education

Public Education Expenditures [10]

Educational Attainment [11]

Literacy Rate [12]

Libraries [13]

Daily Newspapers [14]

Cinema [15]

Science and Technology

Scientific/Technical Forces [16]

R&D Expenditures [17]

U.S. Patents Issued [18]

Government and Law

Organization of Government [19]

Long-form name:
Republic of Azerbaijan
Type:
republic
Independence:
30 August 1991 (from Soviet Union)
Constitution:
adopted NA April 1978; writing a new
constitution mid-1993
Legal system:
based on civil law system
National holiday:
NA
Executive branch:
president, council of ministers
Legislative branch:
National Parliament (National Assembly or
Milli Mejlis)
Judicial branch:
Supreme Court

Political Parties [20]

National Council	% of seats
Popular Front	50.0
Opposition elements	50.0

Government Budget [21]

For 1992.

Revenues	NA
Expenditures	NA
Capital expenditures	NA

Defense Summary [22]

Branches: Army, Air Force, Navy, National Guard, Security Forces (internal and border troops)

Manpower Availability: Males age 15-49 1,842,917; fit for military service 1,497,640; reach military age (18) annually 66,928 (1993 est.)

Defense Expenditures: 2,848 million rubles; note - conversion of the military budget into US dollars using the current exchange rate could produce misleading results

Crime [23]

Human Rights [24]

	SSTS	FL	FAPRO	PPCG	APROBC	TPW	PCPTW	STPEP	PHRFF	PRW	ASST	AFL
Observes		P	P		P							
	EAFRD	CPR	ESCR	SR	ACHR	MAAE	PVIAC	PVNAC	EAFDAW	TCIDTP	RC
Observes		P	P			P					P

P = Party; S = Signatory; see Appendix for meaning of abbreviations.

Labor Force

Total Labor Force [25]

2.789 million

Labor Force by Occupation [26]

Agriculture and forestry	32%
Industry and construction	26
Other	42

Date of data: 1990

Unemployment Rate [27]

0.2% includes officially registered unemployed; also large numbers of underemployed workers

Production Sectors

Energy Resource Summary [28]

Energy Resources: Petroleum, natural gas. **Electricity**: 6,025,000 kW capacity; 22,300 million kWh produced, 2,990 kWh per capita (1992).
Pipelines: Crude oil 1,130 km, petroleum products 630 km, natural gas 1,240 km.

Telecommunications [30]

- Domestic telephone service is of poor quality and inadequate
- 644,000 domestic telephone lines (density—9 lines per 100 persons (1991)), 202,000 persons waiting for telephone installations (January 1991)
- Connections to former USSR republics by cable and microwave and elsewhere via Moscow's international gateway switch
- INTELSAT earth station installed in late 1992 in Baku to 200 countries through Turkey
- Domestic and Russian TV programs received locally; Turkish and Iranian TV received from an INTELSAT satellite through a receive-only earth station

Transportation [31]

Railroads. 2,090 km; does not include industrial lines (1990)

Highways. 36,700 km total (1990); 31,800 km hard surfaced; 4,900 km earth

Airports

Total:	65
Useable:	33
With permanent-surface runways:	26
With runways over 3,659 m:	0
With runways 2,440-3,659 m:	8
With runways 1,220-2,439 m:	23

Top Agricultural Products [32]

	1990	1991
Grains	1,400	1,300
Vegetables	856	839
Cotton	543	500
Potatoes	185	180
Fruits and berries	319	-

Values shown are 1,000 metric tons.

Top Mining Products [33]

Detailed information is not available. A summary of mineral resources available follows. **Mineral Resources**: Petroleum, natural gas, iron ore, nonferrous metals, alumina.

Tourism [34]

For sources, notes, and explanations, see Annotated Source Appendix, page 1035.

53

Finance, Economics, and Trade

Industrial Summary [35]

Industrial Production: Growth rate - 27% (1992). **Industries**: Petroleum and natural gas, petroleum products, oilfield equipment; steel, iron ore, cement; chemicals and petrochemicals; textiles.

Economic Indicators [36]

National product: GDP not available. **National product real growth rate**: - 25% (1992). **National product per capita**: not available. **Inflation rate (consumer prices)**: 20% per month (1992 est.). **External debt**: $1.3 billion (1991 est.).

Balance of Payments Summary [37]

Exchange Rates [38]

Imports and Exports

Top Import Origins [39]

$300 million from outside the successor states of the former USSR (c.i.f., 1992 est.).

Origins	%
European countries	NA

Top Export Destinations [40]

$821 million to outside the successor states of the former USSR (f.o.b., 1992 est.).

Destinations	%
Mostly CIS and European countries	NA

Foreign Aid [41]

	U.S. $
Wheat from Turkey	NA

Import and Export Commodities [42]

Import Commodities	Export Commodities
Machinery & parts	Oil and gas
Consumer durables	Chemicals
Foodstuffs	Oilfield equipment
Textiles	Textiles
	Cotton

54

For sources, notes, and explanations, see Annotated Source Appendix, page 1035.

The Bahamas

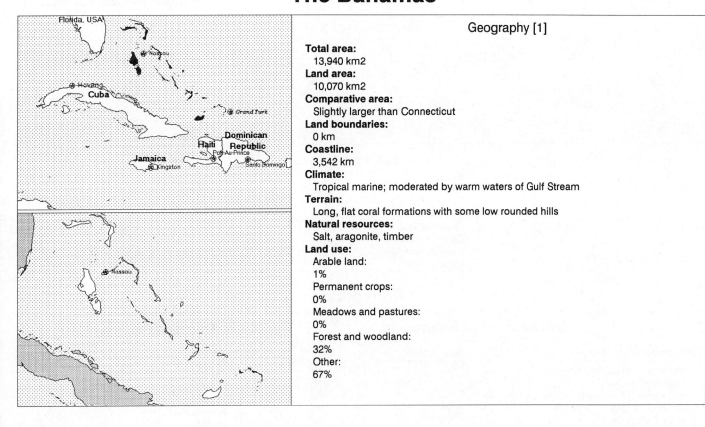

Geography [1]

Total area:
13,940 km2
Land area:
10,070 km2
Comparative area:
Slightly larger than Connecticut
Land boundaries:
0 km
Coastline:
3,542 km
Climate:
Tropical marine; moderated by warm waters of Gulf Stream
Terrain:
Long, flat coral formations with some low rounded hills
Natural resources:
Salt, aragonite, timber
Land use:
Arable land:
1%
Permanent crops:
0%
Meadows and pastures:
0%
Forest and woodland:
32%
Other:
67%

Demographics [2]

	1960	1970	1980	1990	1991[1]	1994	2000	2010	2020
Population	112	170	210	255	252	273	298	332	356
Population density (persons per sq. mi.)	29	44	54	64	65	NA	73	80	86
Births	NA	NA	NA	NA	5	5	NA	NA	NA
Deaths	NA	NA	NA	NA	1	1	NA	NA	NA
Life expectancy - males	NA	NA	NA	NA	69	68	NA	NA	NA
Life expectancy - females	NA	NA	NA	NA	76	75	NA	NA	NA
Birth rate (per 1,000)	NA	NA	NA	NA	19	19	NA	NA	NA
Death rate (per 1,000)	NA	NA	NA	NA	5	5	NA	NA	NA
Women of reproductive age (15-44 yrs.)	NA	NA	NA	74	NA	82	91	NA	NA
of which are currently married	NA	NA	NA	40	NA	46	52	NA	NA
Fertility rate	NA	NA	NA	1.9	2.18	1.9	1.8	1.8	1.8

Population values are in thousands, life expectancy in years, and other items as indicated.

Health

Health Personnel [3]

Health Expenditures [4]

For sources, notes, and explanations, see Annotated Source Appendix, page 1035.

55

Human Factors

Health Care Ratios [5]	Infants and Malnutrition [6]

Ethnic Division [7]

Black	85%
White	15%

Religion [8]

Baptist	32%
Anglican	20%
Roman Catholic	19%
Methodist	6%
Church of God	6%
Other Protestant	12%
None or unknown	3%
Other	2%

Major Languages [9]

English, Creole spoken among Haitian immigrants.

Education

Public Education Expenditures [10]

Million Bahamian Dollars	1980	1985	1987	1988	1989	1990
Total education expenditure	NA	NA	NA	NA	NA	NA
as percent of GNP	NA	NA	NA	NA	NA	NA
as percent of total govt. expend.	NA	NA	NA	NA	NA	NA
Current education expenditure	53	NA	NA	99	105	107
as percent of GNP	4.4	NA	NA	3.7	3.8	3.6
as percent of current govt. expend.	22.1	NA	NA	19.3	18.4	18.9
Capital expenditure	NA	NA	NA	NA	NA	NA

Educational Attainment [11]

Age group	25+
Total population	104,472
Highest level attained (%)	
No schooling	3.5
First level	
Incompleted	25.4
Completed	NA
Entered second level	
S-1	57.7
S-2	NA
Post secondary	13.5

Literacy Rate [12]

Libraries [13]

	Admin. Units	Svc. Pts.	Vols. (000)	Shelving (meters)	Vols. Added	Reg. Users
National	NA	NA	NA	NA	NA	NA
Non-specialized	NA	NA	NA	NA	NA	NA
Public	NA	NA	NA	NA	Na	NA
Higher ed. (1987)	2	2	68	NA	300	2,500
School	NA	NA	NA	NA	NA	NA

Daily Newspapers [14]

	1975	1980	1985	1990
Number of papers	2	3	3	3
Circ. (000)	31	33	39	35

Cinema [15]

Science and Technology

Scientific/Technical Forces [16]

R&D Expenditures [17]

U.S. Patents Issued [18]

Values show patents issued to citizens of the country by the U.S. Patents Office.

	1990	1991	1992
Number of patents	2	8	2

56

For sources, notes, and explanations, see Annotated Source Appendix, page 1035.

Government and Law

Organization of Government [19]

Long-form name:
The Commonwealth of The Bahamas
Type:
commonwealth
Independence:
10 July 1973 (from UK)
Constitution:
10 July 1973
Legal system:
based on English common law
National holiday:
National Day, 10 July (1973)
Executive branch:
British monarch, governor general, prime minister, deputy prime minister, Cabinet
Legislative branch:
bicameral Parliament consists of an appointed upper house or Senate and a directly elected lower house or House of Assembly
Judicial branch:
Supreme Court

Political Parties [20]

House of Assembly	% of seats
Free National Movement (FNM)	65.3
Progressive Liberal Party (PLP)	34.7

Government Budget [21]

For 1992 est.

Revenues	627.5
Expenditures	727.5
Capital expenditures	100.0

Defense Summary [22]

Branches: Royal Bahamas Defense Force (Coast Guard only), Royal Bahamas Police Force

Manpower Availability: Males age 15-49 68,020 (1993 est.)

Defense Expenditures: Exchange rate conversion-$65 million, 2.7% of GDP (1990)

Crime [23]

Crime volume (for 1990)	
Cases known to police	17,409
Attempts (percent)	2.19
Percent cases solved	29.39
Crimes per 100,000 persons	6,835.50
Persons responsible for offenses	
Total number offenders	5,003
Percent female	10.13
Percent juvenile[4]	10.11
Percent foreigners	NA

Human Rights [24]

	SSTS	FL	FAPRO	PPCG	APROBC	TPW	PCPTW	STPEP	PHRFF	PRW	ASST	AFL
Observes	P	P		P	P	P	P			P	P	P
	EAFRD	CPR	ESCR	SR	ACHR	MAAE	PVIAC	PVNAC	EAFDAW	TCIDTP	RC
Observes		1		1			P	P			P

P = Party; S = Signatory; see Appendix for meaning of abbreviations.

Labor Force

Total Labor Force [25]

127,400

Labor Force by Occupation [26]

Government	30%
Hotels and restaurants	25
Business services	10
Agriculture	5

Date of data: 1989

Unemployment Rate [27]

16% (1991 est.)

For sources, notes, and explanations, see Annotated Source Appendix, page 1035.

57

Production Sectors

Energy Resource Summary [28]

Energy Resources: None. **Electricity**: 424,000 kW capacity; 929 million kWh produced, 3,599 kWh per capita (1992).

Telecommunications [30]

- Highly developed
- 99,000 telephones in totally automatic system
- Tropospheric scatter and submarine cable links to Florida
- Broadcast stations-3 AM, 2 FM, 1 TV
- 3 coaxial submarine cables
- 1 Atlantic Ocean INTELSAT earth station

Top Agricultural Products [32]

Agriculture accounts for 5% of GDP; dominated by small-scale producers; principal products—citrus fruit, vegetables, poultry; large net importer of food.

Top Mining Products [33]

Metric tons unless otherwise specified	M.t.
Salt	1,096
Sand: aragonite	1,211

Transportation [31]

Highways. 2,400 km total; 1,350 km paved, 1,050 km gravel

Merchant Marine. 853 ships (1,000 GRT or over) totaling 20,136,078 GRT/33,119,750 DWT; includes 53 passenger, 18 short-sea passenger, 159 cargo, 40 roll-on/roll-off cargo, 48 container, 6 vehicle carrier, 181 oil tanker, 14 liquefied gas, 22 combination ore/oil, 43 chemical tanker, 1 specialized tanker, 159 bulk, 7 combination bulk, 102 refrigerated cargo; note-a flag of convenience registry

Airports

Total:	60
Usable:	55
With permanent-surface runways:	31
With runways over 3,659 m:	0
With runways 2,440-3, 659 m:	3
With runways 1,220-2,439 m:	26

Tourism [34]

	1987	1988	1989	1990	1991
Visitors	3,078	3,158	3,398	3,629	3,622
Tourists	1,480	1,475	1,575	1,562	1,427
Cruise passengers	1,434	1,505	1,645	1,854	2,020
Tourism receipts	1,146	1,150	1,310	1,333	1,222
Tourism expenditures	153	172	184	196	205
Fare receipts	9	9	10	12	9
Fare expenditures	31	34	19	21	20

Tourists are in thousands, money in million U.S. dollars.

58

For sources, notes, and explanations, see Annotated Source Appendix, page 1035.

Finance, Economics, and Trade

GDP and Manufacturing Summary [35]

	1980	1985	1990	1991	1992
Gross Domestic Product					
Millions of 1980 dollars	1,320	1,709	2,036	1,991	1,970[e]
Growth rate in percent	-3.77	13.51	3.00	-2.20	-1.09[e]
Manufacturing Value Added					
Millions of 1980 dollars	NA	NA	NA	NA	NA
Growth rate in percent	NA	NA	NA	NA	NA
Manufacturing share in percent of current prices	7.5	NA	NA	NA	NA

Economic Indicators [36]

Millions of U.S. Dollars.

	1989	1990	1991
Real GDP	2,497	2,522	NA
GDP growth rate	2.0	1.0	NA
GDP per capita (U.S. Dollars)	9,987	9,902	NA
Money supply (M1)	300.7	328.4	324.6
Commercial interest rate (%)	9.0	9.0	9.0
Savings rate	NA	NA	NA
Investment rate	NA	NA	NA
CPI	112.6	120.4	129.0
WPI	NA	NA	NA
External public debt	151.5	188.2	201.5

Balance of Payments Summary [37]

Values in millions of dollars.

	1987	1988	1989	1990	1991
Exports of goods (f.o.b.)	273.1	273.6	259.2	307.6	319.8
Imports of goods (f.o.b.)	-1,154.7	-1,058.9	-1,203.5	-1,228.8	-1,154.2
Trade balance	-881.6	-785.3	-944.3	-921.2	-834.4
Services - debits	-670.2	-683.3	-721.2	-784.0	-792.1
Services - credits	1,389.4	1,354.8	1,506.7	1,514.4	1,425.8
Private transfers (net)	-17.8	-28.9	-17.9	-10.7	-6.9
Government transfers (net)	14.2	14.4	18.9	21.3	27.4
Long term capital (net)	-27.0	24.0	46.5	25.1	165.0
Short term capital (net)	8.2	40.7	50.0	43.7	11.9
Errors and omissions	125.2	64.4	48.1	123.1	14.2
Overall balance	-59.6	0.8	-13.2	11.7	10.9

Exchange Rates [38]

Currency: **Bahamian dollar.**
Symbol: **B$.**

Data are currency units per $1.

Fixed rate	1.00

Imports and Exports

Top Import Origins [39]

$1.14 billion (c.i.f., 1991 est.).

Origins	%
US	35
Nigeria	21
Japan	13
Angola	11

Top Export Destinations [40]

$306 million (f.o.b., 1991 est.).

Destinations	%
US	41
Norway	30
Denmark	4

Foreign Aid [41]

	U.S. $	
US commitments, including Ex-Im (FY85-89)	1.0	million
Western (non-US) countries, ODA and OOF bilateral commitments (1970-89)	345	million

Import and Export Commodities [42]

Import Commodities	Export Commodities
Foodstuffs	Pharmaceuticals
Manufactured goods	Cement
Mineral fuels	Rum
Crude oil	Crawfish

For sources, notes, and explanations, see Annotated Source Appendix, page 1035.

59

Bahrain

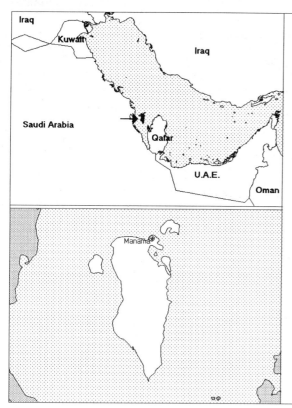

Geography [1]

Total area:
620 km2
Land area:
620 km2
Comparative area:
Slightly less than 3.5 times the size of Washington, DC
Land boundaries:
0 km
Coastline:
161 km
Climate:
Arid; mild, pleasant winters; very hot, humid summers
Terrain:
Mostly low desert plain rising gently to low central escarpment
Natural resources:
Oil, associated and nonassociated natural gas, fish
Land use:
Arable land:
2%
Permanent crops:
2%
Meadows and pastures:
6%
Forest and woodland:
0%
Other:
90%

Demographics [2]

	1960	1970	1980	1990	1991[1]	1994	2000	2010	2020
Population	157	220	348	518	537	586	687	849	1,008
Population density (persons per sq. mi.)	655	919	1,454	2,177	2,247	NA	2,902	3,609	4,313
Births	NA	NA	NA	NA	15	16	NA	NA	NA
Deaths	NA	NA	NA	NA	2	2	NA	NA	NA
Life expectancy - males	NA	NA	NA	NA	71	71	NA	NA	NA
Life expectancy - females	NA	NA	NA	NA	76	76	NA	NA	NA
Birth rate (per 1,000)	NA	NA	NA	NA	27	27	NA	NA	NA
Death rate (per 1,000)	NA	NA	NA	NA	3	4	NA	NA	NA
Women of reproductive age (15-44 yrs.)	NA	NA	NA	111	NA	126	149	NA	NA
of which are currently married	NA	NA	NA	72	NA	83	99	NA	NA
Fertility rate	NA	NA	NA	4.1	4.04	4.0	3.7	3.3	3.0

Population values are in thousands, life expectancy in years, and other items as indicated.

Health

Health Personnel [3]

Health Expenditures [4]

For sources, notes, and explanations, see Annotated Source Appendix, page 1035.

Human Factors

Health Care Ratios [5]	Infants and Malnutrition [6]

Ethnic Division [7]

Bahraini	63%
Asian	13%
Other Arab	10%
Iranian	8%
Other	6%

Religion [8]

Shi'a Muslim	70%
Sunni Muslim	30%

Major Languages [9]

Arabic	NA
English	NA
Farsi	NA
Urdu	NA

Education

Public Education Expenditures [10]

Million Dinar	1980	1985	1987[6]	1988[6]	1989[6]	1990[6]
Total education expenditure	33	52	56	60	62	65
as percent of GNP	2.9	4.0	5.2	5.4	5.0	4.8
as percent of total govt. expend.	10.3	NA	NA	NA	NA	NA
Current education expenditure	28	49	53	56	58	61
as percent of GNP	2.5	3.8	4.9	5.1	4.6	4.5
as percent of current govt. expend.	14.7	14.1	12.0	13.5	NA	NA
Capital expenditure	4	3	2	4	4	4

Educational Attainment [11]

Literacy Rate [12]

In thousands and percent	1985[a]	1991[a]	2000[a]
Illiterate population +15 years	78	79	74
Illiteracy rate - total pop. (%)	27.1	22.6	15.4
Illiteracy rate - males (%)	21.5	17.9	12.3
Illiteracy rate - females (%)	36.7	30.7	20.7

Libraries [13]

	Admin. Units	Svc. Pts.	Vols. (000)	Shelving (meters)	Vols. Added	Reg. Users
National	NA	NA	NA	NA	NA	NA
Non-specialized	NA	NA	NA	NA	NA	NA
Public (1989)	1	10	218	NA	NA	NA
Higher ed. (1987)	2	9	100	3,500	9,615	4,200
School (1990 est.)	184	184	166	5,520	100,000	80,000

Daily Newspapers [14]

	1975	1980	1985	1990
Number of papers	-	3	2	2
Circ. (000)	-	14[e]	19	29

Cinema [15]

Science and Technology

Scientific/Technical Forces [16]

Potential scientists/engineers	10,747[1]
Number female	3,184[1]
Potential technicians	11,615[1]
Number female	4,215[1]
Total	22,362

R&D Expenditures [17]

U.S. Patents Issued [18]

Government and Law

Organization of Government [19]

Long-form name:
State of Bahrain
Type:
traditional monarchy
Independence:
15 August 1971 (from UK)
Constitution:
26 May 1973, effective 6 December 1973
Legal system:
based on Islamic law and English common law
National holiday:
Independence Day, 16 December
Executive branch:
amir, crown prince and heir apparent, prime minister, Cabinet
Legislative branch:
unicameral National Assembly was dissolved 26 August 1975 and legislative powers were assumed by the Cabinet; appointed Advisory Council established 16 December 1992
Judicial branch:
High Civil Appeals Court

Elections [20]

Political parties prohibited; several small, clandestine leftist and Islamic fundamentalist groups are active. No elections.

Government Expenditures [21]

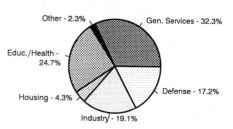

Other - 2.3%
Gen. Services - 32.3%
Educ./Health - 24.7%
Defense - 17.2%
Housing - 4.3%
Industry - 19.1%

% distribution for 1992. Expend. in 1989: 1.32 ($ bil.)

Military Expenditures and Arms Transfers [22]

	1985	1986	1987	1988	1989
Military expenditures					
Current dollars (mil.)	151	161	160	187	196
1989 constant dollars (mil.)	171	178	173	195	196
Armed forces (000)	3	4[e]	4[e]	5	5
Gross national product (GNP)					
Current dollars (mil.)	3,414	2,887	2,884	2,913	2,999[e]
1989 constant dollars (mil.)[e]	3,886	3,203	3,102	3,032	2,999
Central government expenditures (CGE)					
1989 constant dollars (mil.)	1,598	1,511	1,534	967	1,491
People (mil.)	0.4	0.4	0.5	0.5	0.5
Military expenditure as % of GNP	4.4	5.6	5.6	6.4	6.5
Military expenditure as % of CGE	10.7	11.8	11.2	20.2	13.1
Military expenditure per capita	396	396	369	402	389
Armed forces per 1,000 people	6.9	7.8	8.6	10.3	10.1
GNP per capita	8,977	7,122	6,642	6,256	5,966
Arms imports[6]					
Current dollars (mil.)	10	50	370	30	50
1989 constant dollars (mil.)	11	55	398	31	50
Arms exports[6]					
Current dollars (mil.)	0	0	0	0	0
1989 constant dollars (mil.)	0	0	0	0	0
Total imports[7]					
Current dollars (mil.)	3,107	2,408	2,717	2,669	2,803
1989 constant dollars	3,537	2,672	2,922	2,778	2,803
Total exports[7]					
Current dollars (mil.)	2,787	2,342	2,384	2,354	2,685
1989 constant dollars	3,173	2,599	2,564	2,450	2,685
Arms as percent of total imports[8]	0.3	2.1	13.6	1.1	1.8
Arms as percent of total exports[8]	0	0	0	0	0

Crime [23]

Crime volume (for 1989)
Cases known to police	9,175
Attempts (percent)	NA
Percent cases solved	NA
Crimes per 100,000 persons	1,878.02
Persons responsible for offenses	
Total number offenders	NA
Percent female	NA
Percent juvenile[5]	NA
Percent foreigners	NA

Human Rights [24]

	SSTS	FL	FAPRO	PPCG	APROBC	TPW	PCPTW	STPEP	PHRFF	PRW	ASST	AFL
Observes	P	P		P		P	P				P	
	EAFRD	CPR	ESCR	SR	ACHR	MAAE	PVIAC	PVNAC	EAFDAW	TCIDTP	RC	
Observes		P						P	P			P

P = Party; S = Signatory; see Appendix for meaning of abbreviations.

Labor Force

Total Labor Force [25]

140,000

Labor Force by Occupation [26]

Industry and commerce	85%
Agriculture	5
Services	5
Government	3

Date of data: 1982

Unemployment Rate [27]

8%-10% (1989)

Production Sectors

Commercial Energy Production and Consumption

Production [28]

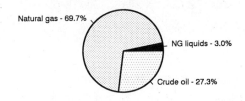

Natural gas - 69.7%

NG liquids - 3.0%

Crude oil - 27.3%

Consumption [29]

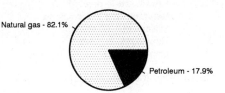

Natural gas - 82.1%

Petroleum - 17.9%

Data are shown in quadrillion (10^{15}) BTUs and percent for 1991

Crude oil[1]	0.09
Natural gas liquids	0.01
Dry natural gas	0.23
Coal[2]	0.00
Net hydroelectric power[3]	0.00
Net nuclear power[3]	0.00
Total	0.32

Petroleum	0.05
Dry natural gas	0.23
Coal[2]	0.00
Net hydroelectric power[3]	0.00
Net nuclear power[3]	0.00
Total	0.28

Telecommunications [30]

- Modern system
- Good domestic services
- 98,000 telephones (1 for every 6 persons)
- Excellent international connections
- Tropospheric scatter to Qatar, UAE
- Microwave radio relay to Saudi Arabia
- Submarine cable to Qatar, UAE, and Saudi Arabia
- Satellite earth stations - 1 Atlantic Ocean INTELSAT, 1 Indian Ocean INTELSAT, 1 ARABSAT
- Broadcast stations - 2 AM, 3 FM, 2 TV

Transportation [31]

Highways. 200 km bituminous surfaced, including 25 km bridge-causeway to Saudi Arabia opened in November 1986; NA km natural surface tracks

Merchant Marine. 9 ships (1,000 GRT or over) totaling 186,331 GRT/249,490 DWT; includes 5 cargo, 2 container, 1 liquefied gas, 1 bulk

Airports

Total:	3
Usable:	3
With permanent-surface runways:	2
With runways over 3,659 m:	2
With runways 2,440-3,659 m:	0
With runways 1,220-2,439 m:	1

Top Agricultural Products [32]

Including fishing, accounts for less than 2% of GDP; not self-sufficient in food production; heavily subsidized sector produces fruit, vegetables, poultry, dairy products, shrimp, fish; fish catch 9,000 metric tons in 1987.

Top Mining Products [33]

Metric tons unless otherwise specified	M.t.
Aluminum, primary metal	215,000
Cement	150,000
Gas, natural (mil. cubic meters)	8,087[1]
Petroleum, crude (000 barrels)	15,434[1]
Butane (000 barrels)	1,102[1]
Propane (000 barrels)	1,474

Tourism [34]

	1987	1988	1989	1990	1991
Visitors	2,488	2,049	2,100	2,051	2,416
Tourists	178	1,171	1,342	1,376	1,674
Tourism receipts	101	114	116	135	162
Tourism expenditures	80	77	77	94	98

Tourists are in thousands, money in million U.S. dollars.

Finance, Economics, and Trade

GDP and Manufacturing Summary [35]

	1980	1985	1990	1991	1992
Gross Domestic Product					
Millions of 1980 dollars	3,072	2,902	3,334	3,423	3,469[e]
Growth rate in percent	0.24	-3.94	1.24	2.68	1.34[e]
Manufacturing Value Added					
Millions of 1980 dollars	498	457	577	591	600[e]
Growth rate in percent	-3.38	-16.13	5.23	2.35	1.64[e]
Manufacturing share in percent of current prices	14.8	8.5	15.9	NA	NA

Economic Indicators [36]

In Millions of Bahraini Dinars (BD) unless otherwise indicated.

	1989	1990	1991
GDP at Current Prices	1,347	1,497	1,542
GDP Growth (nominal)	6.7	11.1	3.0
Per Capita GNP (U.S.$)	6,780	NA	NA
Money Supply			
(M1, end of period)	234.7	295.7	NA
Savings Rate (one month %)	7.0	7.0	NA
Investment Rate (%)	8.3	8.2	NA
CPI (%)	1.5	1.3	NA
External Public Debt	58	NA	NA

Balance of Payments Summary [37]

Values in millions of dollars.

	1987	1988	1989	1990	1991
Exports of goods (f.o.b.)	2,429.5	2,411.4	2,831.1	3,760.6	3,468.9
Imports of goods (f.o.b.)	-2,419.1	-2,309.8	-2,793.4	-3,340.4	-3,662.2
Trade balance	10.4	101.6	37.7	420.2	-193.3
Services - debits	-1,203.5	-1,223.4	-1,294.4	-1,557.7	-1,624.2
Services - credits	1,146.3	1,162.5	1,250.5	1,196.0	1,280.9
Private transfers (net)	-243.6	-193.1	-198.9	-272.3	-303.5
Government transfers (net)	113.3	366.5	102.1	458.8	101.9
Long term capital (net)	-63.3	205.1	93.6	-96.5	-58.2
Short term capital (net)	9.6	-419.4	-359.3	552.9	-283.0
Errors and omissions	-111.8	92.8	180.2	-132.1	974.1
Overall balance	-342.6	92.6	-188.5	569.3	-105.3

Exchange Rates [38]

Currency: **Bahraini dinars.**
Symbol: **BD.**

Data are currency units per $1.

Fixed rate	0.3760

Imports and Exports

Top Import Origins [39]

$3.7 billion (f.o.b., 1991).

Origins	%
Saudi Arabia	41
US	14
UK	7
Japan	5

Top Export Destinations [40]

$3.5 billion (f.o.b., 1991).

Destinations	%
Japan	13
UAE	12
India	10
Pakistan	8

Foreign Aid [41]

	U.S. $	
US commitments, including Ex-Im (FY70-79)	24	million
Western (non-US) countries, ODA and OOF bilateral commitments (1970-89)	45	million
OPEC bilateral aid (1979-89)	9.8	billion

Import and Export Commodities [42]

Import Commodities

Nonoil 59%
Crude oil 41%

Export Commodities

Petroleum & petr. products 80%
Aluminum 7%

Bangladesh

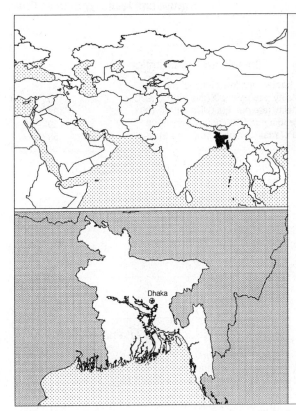

Geography [1]

Total area:
144,000 km2
Land area:
133,910 km2
Comparative area:
Slightly smaller than Wisconsin
Land boundaries:
Total 4,246 km, Burma 193 km, India 4,053 km
Coastline:
580 km
Climate:
Tropical; cool, dry winter (October to March); hot, humid summer (March to June); cool, rainy monsoon (June to October)
Terrain:
Mostly flat alluvial plain; hilly in southeast
Natural resources:
Natural gas, arable land, timber
Land use:
Arable land:
67%
Permanent crops:
2%
Meadows and pastures:
4%
Forest and woodland:
16%
Other:
11%

Demographics [2]

	1960	1970	1980	1990	1991[1]	1994	2000	2010	2020
Population	54,622	67,403	88,077	114,023	116,601	125,149	143,548	176,902	210,248
Population density (persons per sq. mi.)	883	1,304	1,704	2,204	2,255	NA	2,770	3,415	4,060
Births	NA	NA	NA	NA	4,219	735	NA	NA	NA
Deaths	NA	NA	NA	NA	1,516	320	NA	NA	NA
Life expectancy - males	NA	NA	NA	NA	54	55	NA	NA	NA
Life expectancy - females	NA	NA	NA	NA	52	55	NA	NA	NA
Birth rate (per 1,000)	NA	NA	NA	NA	36	43	NA	NA	NA
Death rate (per 1,000)	NA	NA	NA	NA	13	19	NA	NA	NA
Women of reproductive age (15-44 yrs.)	NA	NA	NA	25,926	NA	29,574	35,675	NA	NA
of which are currently married	NA	NA	NA	21,658	NA	24,734	29,938	NA	NA
Fertility rate	NA	NA	NA	4.8	4.72	4.5	4.0	3.3	2.8

Population values are in thousands, life expectancy in years, and other items as indicated.

Health

Health Personnel [3]

Doctors per 1,000 pop., 1988-92	0.15
Nurse-to-doctor ratio, 1988-92	0.8
Hospital beds per 1,000 pop., 1985-90	0.3
Percentage of children immunized (age 1 yr. or less)	
Third dose of DPT, 1990-91	87.0
Measles, 1990-91	83.0

Health Expenditures [4]

Total health expenditure, 1990 (official exchange rate)	
Millions of dollars	715
Millions of dollars per capita	7
Health expenditures as a percentage of GDP	
Total	3.2
Public sector	1.4
Private sector	1.8
Development assistance for health	
Total aid flows (millions of dollars)[1]	128
Aid flows per capita (millions of dollars)	1.2
Aid flows as a percentage of total health expenditure	17.9

For sources, notes, and explanations, see Annotated Source Appendix, page 1035.

Human Factors

Health Care Ratios [5]

Population per physician, 1970	8450
Population per physician, 1990	NA
Population per nursing person, 1970	65780
Population per nursing person, 1990	NA
Percent of births attended by health staff, 1985	NA

Infants and Malnutrition [6]

Percent of babies with low birth weight, 1985	31
Infant mortality rate per 1,000 live births, 1970	140
Infant mortality rate per 1,000 live births, 1991	103
Years of life lost per 1,000 population, 1990	69
Prevalence of malnutrition (under age 5), 1990	60

Ethnic Division [7]

Bengali	98%
Biharis	250,000
Tribals	<1 million

Religion [8]

Muslim	83%
Hindu	16%
Buddhist, Christian, other	1%

Major Languages [9]

Bangla (official), English.

Education

Public Education Expenditures [10]

Million Taka	1980[1]	1985[1]	1987[1]	1988[1]	1989[1]	1990[1]
Total education expenditure	3,009	7,782	10,688	12,276	14,125	14,942
as percent of GNP	1.5	1.9	2.0	2.1	2.2	2.0
as percent of total govt. expend.	7.8	9.7	9.9	10.3	10.5	10.3
Current education expenditure	2,010	6,005	8,200	9,485	10,939	11,820
as percent of GNP	1.0	1.5	1.5	1.6	1.7	1.6
as percent of current govt. expend.	13.6	15.3	15.1	13.7	14.3	14.4
Capital expenditure	999	1,777	2,488	2,791	3,186	3,122

Educational Attainment [11]

Age group	25+
Total population	31,593,122
Highest level attained (%)	
No schooling	70.4
First level	
Incompleted	16.7
Completed	NA
Entered second level	
S-1	7.4
S-2	4.2
Post secondary	1.3

Literacy Rate [12]

In thousands and percent	1985[a]	1991[a]	2000[a]
Illiterate population +15 years	37,226	41,961	52,164
Illiteracy rate - total pop. (%)	67.8	64.7	58.3
Illiteracy rate - males (%)	55.5	52.9	45.8
Illiteracy rate - females (%)	81.0	78.0	71.6

Libraries [13]

	Admin. Units	Svc. Pts.	Vols. (000)	Shelving (meters)	Vols. Added	Reg. Users
National (1989)	1	1	15	328	-	-
Non-specialized	NA	NA	NA	NA	NA	NA
Public (1989)	57	61	521	NA	26,600	-
Higher ed. (1987)	946	983	4,076	NA	19,848	680,639
School	NA	NA	NA	NA	NA	NA

Daily Newspapers [14]

	1975	1980	1985	1990
Number of papers	30	44	60	52
Circ. (000)	250[e]	274	591	700[e]

Cinema [15]

Science and Technology

Scientific/Technical Forces [16]

R&D Expenditures [17]

U.S. Patents Issued [18]

Government and Law

Organization of Government [19]

Long-form name:
People's Republic of Bangladesh
Type:
republic
Independence:
16 December 1971 (from Pakistan)
Constitution:
4 November 1972, effective 16 December 1972, suspended following coup of 24 March 1982, restored 10 November 1986, amended NA March 1991
Legal system:
based on English common law
National holiday:
Independence Day, 26 March (1971)
Executive branch:
president, prime minister, Cabinet
Legislative branch:
unicameral National Parliament (Jatiya Sangsad)
Judicial branch:
Supreme Court

Political Parties [20]

National Parliament	% of seats
Nationalist Party (BNP)	50.9
Awami League (AL)	28.2
Jatiyo Party (JP)	10.6
Jamaat-E-Islami (JI)	6.1
Communist Party (BCP)	1.5
National Awami Party	0.3
Workers Party	0.3
Jatiyo Samajtantik Dal (JSD)	0.3
Ganotantri Party	0.3
Other	1.5

Government Budget [21]

For FY92.
Revenues	2.5
Expenditures	3.7
Capital expenditures	NA

Crime [23]

Crime volume (for 1990)	
Cases known to police	18,438
Attempts (percent)	100
Percent cases solved	98.21
Crimes per 100,000 persons	16.76
Persons responsible for offenses	
Total number offenders	32,963
Percent female	NA
Percent juvenile[6]	NA
Percent foreigners	NA

Military Expenditures and Arms Transfers [22]

	1985	1986	1987	1988	1989
Military expenditures					
Current dollars (mil.)	262	283	326	NA	323
1989 constant dollars (mil.)	298	314	351	NA	323
Armed forces (000)	91	95	100[e]	102[e]	103
Gross national product (GNP)					
Current dollars (mil.)	15,460	16,610	17,810	18,780	20,020
1989 constant dollars (mil.)	17,600	18,430	19,150	19,550	20,020
Central government expenditures (CGE)					
1989 constant dollars (mil.)	2,294	NA	NA	NA	2,139
People (mil.)	101.4	103.8	106.3	108.8	111.3
Military expenditure as % of GNP	1.7	1.7	1.8	NA	1.6
Military expenditure as % of CGE	13.0	NA	NA	NA	15.1
Military expenditure per capita	3	3	3	NA	3
Armed forces per 1,000 people	0.9	0.9	0.9	0.9	0.9
GNP per capita	174	178	180	180	180
Arms imports[6]					
Current dollars (mil.)	60	70	50	50	120
1989 constant dollars (mil.)	68	78	54	52	120
Arms exports[6]					
Current dollars (mil.)	0	0	0	0	0
1989 constant dollars (mil.)	0	0	0	0	0
Total imports[7]					
Current dollars (mil.)	2,772	2,486	2,680	3,046	3,650
1989 constant dollars	3,156	2,759	2,882	3,171	3,650
Total exports[7]					
Current dollars (mil.)	999	887	1,067	1,291	1,305
1989 constant dollars	1,137	984	1,148	1,344	1,305
Arms as percent of total imports[8]	2.2	2.8	1.9	1.6	3.3
Arms as percent of total exports[8]	0	0	0	0	0

Human Rights [24]

	SSTS	FL	FAPRO	PPCG	APROBC	TPW	PCPTW	STPEP	PHRFF	PRW	ASST	AFL
Observes	P	P	P		P	P	P	P			P	P
	EAFRD	CPR	ESCR	SR	ACHR	MAAE	PVIAC	PVNAC	EAFDAW	TCIDTP	RC	
Observes		P					P	P	P		P	

P = Party; S = Signatory; see Appendix for meaning of abbreviations.

Labor Force

Total Labor Force [25]

35.1 million

Labor Force by Occupation [26]

Agriculture	74%
Services	15
Industry and commerce	11

Date of data: FY86

Unemployment Rate [27]

For sources, notes, and explanations, see Annotated Source Appendix, page 1035.

67

Production Sectors

Energy Resource Summary [28]

Energy Resources: Natural gas. **Electricity**: 2,400,000 kW capacity; 9,000 million kWh produced, 75 kWh per capita (1992). **Pipelines**: Natural gas 1,220 km.

Telecommunications [30]

- Adequate international radio communications and landline service
- Fair domestic wire and microwave service
- Fair broadcast service
- 241,250 telephones
- Broadcast stations - 9 AM, 6 FM, 11 TV
- 2 Indian Ocean INTELSAT satellite earth stations

Transportation [31]

Railroads. 2,892 km total (1986); 1,914 km 1.000 meter gauge, 978 km 1.676 meter broad gauge

Highways. 7,240 km total (1985); 3,840 km paved, 3,400 km unpaved

Merchant Marine. 42 ships (1,000 GRT or over) totaling 314,228 GRT/461,607 DWT; includes 34 cargo, 2 oil tanker, 3 refrigerated cargo, 3 bulk

Airports

Total:	16
Usable:	12
With permanent-surface runways:	12
With runways over 3,659 m:	0
With runways 2,440-3,659 m:	4
With runways 1,220-2,439 m:	6

Top Agricultural Products [32]

	89-90[1]	90-91[1]
Rice	17,860	17,860
Sugar cane	7,423	8,380
Potatoes	1,066	1,100
Wheat	890	1,004
Jute	842	952
Sweet potatoes	512	550

Values shown are 1,000 metric tons.

Top Mining Products [33]

Detailed information is not available. A summary of mineral resources available follows. **Mineral Resources**: Natural gas.

Tourism [34]

	1987	1988	1989	1990	1991
Tourists[4]	107	121	128	115	113
Tourism receipts	12	13	18	11	9
Tourism expenditures	53	99	123	78	83

Tourists are in thousands, money in million U.S. dollars.

Manufacturing Sector

GDP and Manufacturing Summary [35]

	1980	1985	1989	1990	% change 1980-1990	% change 1989-1990
GDP (million 1980 $)	15,806	19,043	20,032	22,765	44.0	13.6
GDP per capita (1980 $)	179	188	178	197	10.1	10.7
Manufacturing as % of GDP (current prices)	9.4	7.8	7.4	7.1	-24.5	-4.1
Gross output (million $)	2,253	2,498	3,121^e	3,025^e	34.3	-3.1
Value added (million $)	834	863	1,226^e	1,107^e	32.7	-9.7
Value added (million 1980 $)	1,479	1,612	1,851	1,985	34.2	7.2
Industrial production index	100	98	122	113	13.0	-7.4
Employment (thousands)	412	469	466^e	522^e	26.7	12.0

Note: GDP stands for Gross Domestic Product. 'e' stands for estimated value.

Profitability and Productivity

	1980	1985	1989	1990	% change 1980-1990	% change 1989-1990
Intermediate input (%)	63	65	61^e	64^e	1.6	4.9
Wages, salaries, and supplements (%)	12^e	10^e	12^e	18^e	50.0	50.0
Gross operating surplus (%)	25^e	24^e	27^e	19^e	-24.0	-29.6
Gross output per worker ($)	5,466	5,192	6,694^e	5,416^e	-0.9	-19.1
Value added per worker ($)	2,023	1,794	2,628^e	1,972^e	-2.5	-25.0
Average wage (incl. benefits) ($)	369^e	557^e	810^e	1,036^e	180.8	27.9

Profitability is in percent of gross output. Productivity is in U.S. $. 'e' stands for estimated value.

Profitability - 1990

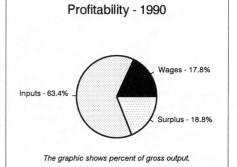

The graphic shows percent of gross output.

Wages - 17.8%
Inputs - 63.4%
Surplus - 18.8%

Value Added in Manufacturing

	1980 $ mil.	1980 %	1985 $ mil.	1985 %	1989 $ mil.	1989 %	1990 $ mil.	1990 %	% change 1980-1990	% change 1989-1990
311 Food products	78	9.4	98	11.4	101	8.2	94^e	8.5	20.5	-6.9
313 Beverages	7	0.8	6	0.7	12	1.0	7^e	0.6	0.0	-41.7
314 Tobacco products	111	13.3	109	12.6	158	12.9	114^e	10.3	2.7	-27.8
321 Textiles	336	40.3	230	26.7	370	30.2	366^e	33.1	8.9	-1.1
322 Wearing apparel	NA	0.0	8	0.9	11^e	0.9	17^e	1.5	NA	54.5
323 Leather and fur products	18	2.2	14	1.6	22^e	1.8	22^e	2.0	22.2	0.0
324 Footwear	4	0.5	10	1.2	13^e	1.1	13^e	1.2	225.0	0.0
331 Wood and wood products	3	0.4	10	1.2	10^e	0.8	10^e	0.9	233.3	0.0
332 Furniture and fixtures	1	0.1	2	0.2	3^e	0.2	2^e	0.2	100.0	-33.3
341 Paper and paper products	23	2.8	19	2.2	23	1.9	29^e	2.6	26.1	26.1
342 Printing and publishing	6	0.7	8	0.9	11^e	0.9	10^e	0.9	66.7	-9.1
351 Industrial chemicals	33	4.0	70	8.1	113^e	9.2	84^e	7.6	154.5	-25.7
352 Other chemical products	97	11.6	85	9.8	154^e	12.6	118^e	10.7	21.6	-23.4
353 Petroleum refineries	2	0.2	75	8.7	70	5.7	66^e	6.0	3,200.0	-5.7
354 Miscellaneous petroleum and coal products	1	0.1	2	0.2	3^e	0.2	6^e	0.5	500.0	100.0
355 Rubber products	4	0.5	1	0.1	6	0.5	3^e	0.3	-25.0	-50.0
356 Plastic products	NA	0.0	2	0.2	3^e	0.2	4^e	0.4	NA	33.3
361 Pottery, china and earthenware	2	0.2	4	0.5	4^e	0.3	5^e	0.5	150.0	25.0
362 Glass and glass products	4	0.5	4	0.5	5	0.4	5^e	0.5	25.0	0.0
369 Other non-metal mineral products	14	1.7	7	0.8	12^e	1.0	12^e	1.1	-14.3	0.0
371 Iron and steel	39	4.7	35	4.1	29	2.4	29^e	2.6	-25.6	0.0
372 Non-ferrous metals	NA	0.0	NA	0.0	NA	0.0	NA	0.0	NA	NA
381 Metal products	9	1.1	13	1.5	16^e	1.3	15^e	1.4	66.7	-6.3
382 Non-electrical machinery	4	0.5	17^e	2.0	19^e	1.5	13^e	1.2	225.0	-31.6
383 Electrical machinery	19	2.3	18	2.1	24	2.0	30^e	2.7	57.9	25.0
384 Transport equipment	11	1.3	10	1.2	28	2.3	23^e	2.1	109.1	-17.9
385 Professional and scientific equipment	NA	0.0	NA	0.0	NA	0.0	NA	0.0	NA	NA
390 Other manufacturing industries	8	1.0	7	0.8	7^e	0.6	8^e	0.7	0.0	14.3

Note: The industry codes shown are International Standard Industry codes (ISIC). Percentages are percent of total Value Added. 'e' stands for estimated value

Finance, Economics, and Trade

Economic Indicators [36]

Millions of U.S. Dollars unless otherwise noted[6].

	FY88/89[7]		90/91[e]
GDP (1991 mil. $)	21,100	22,400	23,100
Real GDP growth (%)	3.01	6.22	3.20
GDP/capita (1991 $)	190.03	197.11	198.65
Money supply (Taka bil.)	191	223	251
Ntl Savings/GDP (%)	4.1	4.3	4.8
Investment/GDP (%)	11.0	11.1	11.4
CPI (% growth)	8.00	9.29	8.95
External public debt	9,625	10,255	11,087

Balance of Payments Summary [37]

Values in millions of dollars.

	1987	1988	1989	1990	1991
Exports of goods (f.o.b.)	1,076.9	1,291.0	1,304.8	1,672.4	1,688.6
Imports of goods (f.o.b.)	-2,445.6	-2,734.5	-3,300.1	-3,259.4	-3,088.6
Trade balance	-1,368.7	-1,443.5	-1,995.3	-1,587.0	-1,400.0
Services - debits	-666.7	-793.9	-923.4	-880.2	-861.2
Services - credits	295.4	332.5	423.1	455.8	503.5
Private transfers (net)	788.3	827.2	806.8	828.3	901.8
Government transfers (net)	714.6	804.8	589.1	785.8	905.6
Long term capital (net)	596.4	631.6	889.8	849.6	550.7
Short term capital (net)	-36.3	-232.6	-56.5	-151.8	-118.5
Errors and omissions	-123.8	6.6	-43.1	-76.3	-48.2
Overall balance	199.2	132.7	-309.5	224.2	433.7

Exchange Rates [38]

Currency: **taka.**
Symbol: **Tk.**

Data are currency units per $1.

January 1993	39.000
1992	38.951
1991	36.596
1990	34.569
1989	32.270
1988	31.733

Imports and Exports

Top Import Origins [39]

$3.4 billion (FY91/92). Data are for FY91.

Origins	%
Japan	10.0
Western Europe	17
US	5.0

Top Export Destinations [40]

$2.0 billion (FY92) Data are for FY91.

Destinations	%
US	28
Western Europe	39

Foreign Aid [41]

	U.S. $	
US commitments, including Ex-Im (FY70-89)	3.4	billion
Western (non-US) countries, ODA and OOF bilateral commitments (1980-89)	11.65	million
OPEC bilateral aid (1979-89)	6.52	million
Communist countries (1970-89)	1.5	billion

Import and Export Commodities [42]

Import Commodities	Export Commodities
Capital goods	Garments
Petroleum	Jute and jute goods
Food	Leather
Textiles	Shrimp

Barbados

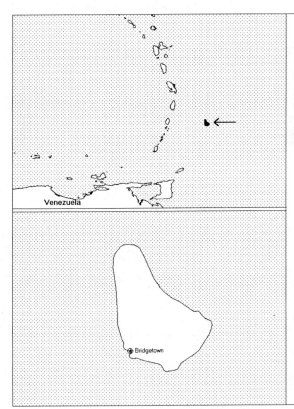

Geography [1]

Total area:
430 km2
Land area:
430 km2
Comparative area:
Slightly less than 2.5 times the size of Washington, DC
Land boundaries:
0 km
Coastline:
97 km
Climate:
Tropical; rainy season (June to October)
Terrain:
Relatively flat; rises gently to central highland region
Natural resources:
Petroleum, fishing, natural gas
Land use:
Arable land:
77%
Permanent crops:
0%
Meadows and pastures:
9%
Forest and woodland:
0%
Other:
14%

Demographics [2]

	1960	1970	1980	1990	1991[1]	1994	2000	2010	2020
Population	232	239	252	254	255	256	260	272	284
Population density (persons per sq. mi.)	1,400	1,438	1,518	1,533	1,534	NA	1,568	1,638	1,710
Births	NA	NA	NA	NA	4	4	NA	NA	NA
Deaths	NA	NA	NA	NA	2	2	NA	NA	NA
Life expectancy - males	NA	NA	NA	NA	70	71	NA	NA	NA
Life expectancy - females	NA	NA	NA	NA	76	77	NA	NA	NA
Birth rate (per 1,000)	NA	NA	NA	NA	16	16	NA	NA	NA
Death rate (per 1,000)	NA	NA	NA	NA	9	8	NA	NA	NA
Women of reproductive age (15-44 yrs.)	NA	NA	NA	70	NA	72	74	NA	NA
of which are currently married	NA	NA	NA	33	NA	35	37	NA	NA
Fertility rate	NA	NA	NA	1.8	1.77	1.8	1.8	1.8	1.8

Population values are in thousands, life expectancy in years, and other items as indicated.

Health

Health Personnel [3]

Health Expenditures [4]

For sources, notes, and explanations, see Annotated Source Appendix, page 1035.

71

Human Factors

Health Care Ratios [5]	Infants and Malnutrition [6]

Ethnic Division [7]

African	80%
Mixed	16%
European	4%

Religion [8]

Protestant	67%
Anglican	40%
Pentecostal	8%
Methodist	7%
Other Protestant	12%
Roman Catholic	4%
Unknown	3%
Other or none	26%

Major Languages [9]

English.

Education

Public Education Expenditures [10]

Million Barbados Dollars	1980	1985	1987	1988	1989	1990
Total education expenditure	109	NA	NA	186	232	269
as percent of GNP	6.5	NA	NA	6.2	6.9	7.9
as percent of total govt. expend.	20.5	NA	NA	NA	NA	NA
Current education expenditure	90	NA	NA	171	196	218
as percent of GNP	5.4	NA	Na	5.7	5.9	6.4
as percent of current govt. expend.	NA	NA	NA	NA	NA	NA
Capital expenditure	19	NA	NA	15	36	51

Educational Attainment [11]

Age group	25+
Total population	116,874
Highest level attained (%)	
No schooling	0.8
First level	
Incompleted	63.5
Completed	NA
Entered second level	
S-1	32.3
S-2	NA
Post secondary	3.3

Literacy Rate [12]

	1970[b]	1980[b]	1990[b]
Illiterate population +15 years	1,093	NA	NA
Illiteracy rate - total pop. (%)[3]	0.7	NA	NA
Illiteracy rate - males (%)	0.7	NA	NA
Illiteracy rate - females (%)	0.7	NA	NA

Libraries [13]

	Admin. Units	Svc. Pts.	Vols. (000)	Shelving (meters)	Vols. Added	Reg. Users
National (1989)	1	10	159	3,928	11,228	59,198
Non-specialized	NA	NA	NA	NA	NA	NA
Public	NA	NA	NA	NA	NA	NA
Higher ed. (1987)[12]	2	4	27	1,613	1,334	1,208
School	NA	NA	NA	NA	NA	NA

Daily Newspapers [14]

	1975	1980	1985	1990
Number of papers	1	2	2	2
Circ. (000)	24	39	40	30

Cinema [15]

Data for 1991.

Cinema seats per 1,000	NA
Annual attendance per person	0.0
Gross box office receipts (mil. Dollars)	NA

Science and Technology

Scientific/Technical Forces [16]	R&D Expenditures [17]	U.S. Patents Issued [18]

U.S. Patents Issued [18]

Values show patents issued to citizens of the country by the U.S. Patents Office.

	1990	1991	1992
Number of patents	0	0	1

Government and Law

Organization of Government [19]

Long-form name:
none
Type:
parliamentary democracy
Independence:
30 November 1966 (from UK)
Constitution:
30 November 1966
Legal system:
English common law; no judicial review of legislative acts
National holiday:
Independence Day, 30 November (1966)
Executive branch:
British monarch, governor general, prime minister, deputy prime minister, Cabinet
Legislative branch:
bicameral Parliament consists of an upper house or Senate and a lower house or House of Assembly
Judicial branch:
Supreme Court of Judicature

Political Parties [20]

House of Assembly	% of seats
Democratic Labor Party (DLP)	64.3
Barbados Labor Party (BLP)	35.7

Government Expenditures [21]

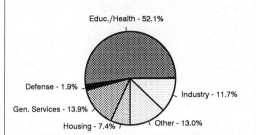

Educ./Health - 52.1%
Defense - 1.9%
Gen. Services - 13.9%
Housing - 7.4%
Other - 13.0%
Industry - 11.7%

% distribution for 1989. Expend. in FY92: 620 ($ mil.)

Military Expenditures and Arms Transfers [22]

	1985	1986	1987	1988	1989
Military expenditures					
Current dollars (mil.)	13	9	8	11	10
1989 constant dollars (mil.)	14	10	9	11	10
Armed forces (000)	1	1	1	NA	0
Gross national product (GNP)					
Current dollars (mil.)	1,244	1,393	1,440	1,579	1,669
1989 constant dollars (mil.)	1,416	1,546	1,549	1,644	1,669
Central government expenditures (CGE)					
1989 constant dollars (mil.)	485	514	580	601	536
People (mil.)	0.3	0.3	0.3	0.3	0.3
Military expenditure as % of GNP	1.0	0.6	0.6	0.7	0.6
Military expenditure as % of CGE	3.0	1.8	1.6	1.9	1.8
Military expenditure per capita	57	37	36	44	38
Armed forces per 1,000 people	3.9	2.0	2.0	NA	1.4
GNP per capita	5,580	6,082	6,091	6,464	6,562
Arms imports[6]					
Current dollars (mil.)	0	5	20	0	0
1989 constant dollars (mil.)	0	6	22	0	0
Arms exports[6]					
Current dollars (mil.)	0	0	0	0	0
1989 constant dollars (mil.)	0	0	0	0	0
Total imports[7]					
Current dollars (mil.)	607	587	515	582	673
1989 constant dollars	691	651	554	606	673
Total exports[7]					
Current dollars (mil.)	352	275	154	173	186
1989 constant dollars	401	305	166	180	186
Arms as percent of total imports[8]	0	0.9	3.9	0	0
Arms as percent of total exports[8]	0	0	0	0	0

Crime [23]

Crime volume (for 1990)	
Cases known to police	11,615
Attempts (percent)	NA
Percent cases solved	29.42
Crimes per 100,000 persons	4,519.46
Persons responsible for offenses	
Total number offenders	3,768
Percent female	7.94
Percent juvenile[7]	2.12
Percent foreigners	NA

Human Rights [24]

	SSTS	FL	FAPRO	PPCG	APROBC	TPW	PCPTW	STPEP	PHRFF	PRW	ASST	AFL
Observes	P	P	P	P	P	P	P			P	P	P
		EAFRD	CPR	ESCR	SR	ACHR	MAAE	PVIAC	PVNAC	EAFDAW	TCIDTP	RC
Observes		P	P	P		P		P	P	P		P

P = Party; S = Signatory; see Appendix for meaning of abbreviations.

Labor Force

Total Labor Force [25]

120,900 (1991)

Labor Force by Occupation [26]

Services and government	37%
Commerce	22
Manufacturing and construction	22
Transp., commun., finance	9
Agriculture	8
Utilities	2

Date of data: 1985 est.

Unemployment Rate [27]

23% (1992)

Production Sectors

Energy Resource Summary [28]

Energy Resources: Petroleum, natural gas. **Electricity**: 152,100 kW capacity; 540 million kWh produced, 2,118 kWh per capita (1992).

Telecommunications [30]

- Islandwide automatic telephone system with 89,000 telephones
- Tropospheric scatter link to Trinidad and Saint Lucia
- Broadcast stations - 3 AM, 2 FM, 2 (1 is pay) TV
- 1 Atlantic Ocean INTELSAT earth station

Transportation [31]

Highways. 1,570 km total; 1,475 km paved, 95 km gravel and earth

Merchant Marine. 3 ships (1,000 GRT or over) totaling 48,710 GRT79,263 DWT; includes 1 cargo, 2 oil tanker

Airports

Total:	1
Usable:	1
With permanent-surface runways:	1
With runways over 3,659 m:	0
With runways 2,440-3,659 m:	1
With runways 1,220-2,439 m:	0

Top Agricultural Products [32]

Agriculture accounts for 8% of GDP; major cash crop is sugarcane; other crops—vegetables, cotton; not self-sufficient in food.

Top Mining Products [33]

Estimated metric tons unless otherwise specified   M.t.

Cement, hydraulic	200
Gas, liquefied petroleum (42-gal. barrels)	18,500
Gas, natural, gross (mil. cubic meters)	35
Petroleum	
Crude (000 42-gal. barrels)	470
Refinery products (000 42-gal. barrels)	2,200

Tourism [34]

	1987	1988	1989	1990	1991
Visitors	651	742	798	795	766
Tourists	422	451	461	432	394
Cruise passengers	229	291	337	363	372
Tourism receipts	379	460	528	494	453
Tourism expenditures	36	37	45	47	44
Fare receipts					2
Fare expenditures	26	26	29	25	26

Tourists are in thousands, money in million U.S. dollars.

Manufacturing Sector

GDP and Manufacturing Summary [35]

	1980	1985	1989	1990	% change 1980-1990	% change 1989-1990
GDP (million 1980 $)	861	837	NA	1,008	17.1	NA
GDP per capita (1980 $)	3,442	3,268	NA	3,937	14.4	NA
Manufacturing as % of GDP (current prices)	11.9	10.6	NA	7.8	-34.5	NA
Gross output (million $)	241	383	NA	398	65.1	NA
Value added (million $)	53	90	NA	90	69.8	NA
Value added (million 1980 $)	91	79	NA	83	-8.8	NA
Industrial production index	100	89	NA	95	-5.0	NA
Employment (thousands)	8	9	NA	6	-25.0	NA

Note: GDP stands for Gross Domestic Product. 'e' stands for estimated value.

Profitability and Productivity

	1980	1985	1989	1990	% change 1980-1990	% change 1989-1990
Intermediate input (%)	78	77	NA	77	-1.3	NA
Wages, salaries, and supplements (%)	14	18	NA	15	7.1	NA
Gross operating surplus (%)	8	5	NA	7	-12.5	NA
Gross output per worker ($)	31,301	41,558	NA	62,043	98.2	NA
Value added per worker ($)	6,854	9,725	NA	13,998	104.2	NA
Average wage (incl. benefits) ($)	4,337	7,726	NA	9,449	117.9	NA

Profitability is in percent of gross output. Productivity is in U.S. $. 'e' stands for estimated value.

Profitability - 1990

Inputs - 77.8%
Surplus - 7.1%
Wages - 15.2%

The graphic shows percent of gross output.

Value Added in Manufacturing

	1980 $ mil.	1980 %	1985 $ mil.	1985 %	1989 $ mil.	1989 %	1990 $ mil.	1990 %	% change 1980-1990	% change 1989-1990
311 Food products	12	22.6	25	27.8	NA	NA	30	33.3	150.0	NA
313 Beverages	6	11.3	12	13.3	NA	NA	10	11.1	66.7	NA
314 Tobacco products	1	1.9	2	2.2	NA	NA	1	1.1	0.0	NA
321 Textiles	NA	0.0	NA	0.0	NA	NA	1	1.1	NA	NA
322 Wearing apparel	6	11.3	7	7.8	NA	NA	5	5.6	-16.7	NA
323 Leather and fur products	NA	0.0	NA	0.0	NA	NA	NA	0.0	NA	NA
324 Footwear	NA	0.0	NA	0.0	NA	NA	NA	0.0	NA	NA
331 Wood and wood products	NA	0.0	NA	0.0	NA	NA	NA	0.0	NA	NA
332 Furniture and fixtures	1	1.9	2	2.2	NA	NA	2	2.2	100.0	NA
341 Paper and paper products	NA	0.0	1	1.1	NA	NA	1	1.1	NA	NA
342 Printing and publishing	4	7.5	8	8.9	NA	NA	8	8.9	100.0	NA
351 Industrial chemicals	NA	0.0	NA	0.0	NA	NA	NA	0.0	NA	NA
352 Other chemical products	1	1.9	3	3.3	NA	NA	4	4.4	300.0	NA
353 Petroleum refineries	2	3.8	3	3.3	NA	NA	5	5.6	150.0	NA
354 Miscellaneous petroleum and coal products	NA	0.0	NA	0.0	NA	NA	NA	0.0	NA	NA
355 Rubber products	1[e]	1.9	1[e]	1.1	NA	NA	2[e]	2.2	100.0	NA
356 Plastic products	1[e]	1.9	1[e]	1.1	NA	NA	2[e]	2.2	100.0	NA
361 Pottery, china and earthenware	NA	0.0	NA	0.0	NA	NA	NA	0.0	NA	NA
362 Glass and glass products	NA	0.0	NA	0.0	NA	NA	1	1.1	NA	NA
369 Other non-metal mineral products	3	5.7	-3	-3.3	NA	NA	5	5.6	66.7	NA
371 Iron and steel	NA	0.0	NA	0.0	NA	NA	NA	0.0	NA	NA
372 Non-ferrous metals	NA	0.0	NA	0.0	NA	NA	NA	0.0	NA	NA
381 Metal products	3	5.7	5	5.6	NA	NA	8	8.9	166.7	NA
382 Non-electrical machinery	5	9.4	11	12.2	NA	NA	2	2.2	-60.0	NA
383 Electrical machinery	3	5.7	8	8.9	NA	NA	1	1.1	-66.7	NA
384 Transport equipment	1	1.9	2	2.2	NA	NA	1	1.1	0.0	NA
385 Professional and scientific equipment	NA	0.0	NA	0.0	NA	NA	NA	0.0	NA	NA
390 Other manufacturing industries	3	5.7	1	1.1	NA	NA	NA	0.0	NA	NA

Note: The industry codes shown are International Standard Industry codes (ISIC). Percentages are percent of total Value Added. 'e' stands for estimated value

For sources, notes, and explanations, see Annotated Source Appendix, page 1035.

75

Finance, Economics, and Trade

Economic Indicators [36]

Millions of U.S. Dollars.

	1988	1989	1990
Real GDP (1974 prices)	438.75	454.55	440.45
Real GDP growth rate (%)	3.5	3.6	-3.1
Money supply (M1)	288.35	257.3	300.8
Commercial interest rate	9.0	11.0	10.25
Savings rate	NA	NA	NA
Investment rate (%)	20.3	22.6	21.9
CPI (1980 = 100)	164.4	174.6	179.9
WPI	NA	NA	NA
External public debt	576.75	551.8	539.85

Balance of Payments Summary [37]

Values in millions of dollars.

	1987	1988	1989	1990	1991
Exports of goods (f.o.b.)	131.4	144.8	146.9	151.0	143.7
Imports of goods (f.o.b.)	-458.4	-517.9	-599.3	-623.0	-617.7
Trade balance	-327.0	-373.1	-452.4	-472.0	-474.0
Services - debits	-289.9	-300.9	-331.0	-272.5	-277.8
Services - credits	556.9	658.4	774.8	685.6	688.9
Private transfers (net)	19.0	29.7	31.9	34.8	32.0
Government transfers (net)	-12.5	-12.0	-26.1	7.8	1.0
Long term capital (net)	70.2	31.4	1.1	31.3	-3.7
Short term capital (net)	30.8	10.4	-1.2	13.7	2.8
Errors and omissions	-41.4	-6.1	-39.5	-65.6	-9.2
Overall balance	6.1	37.8	-42.4	-36.9	-40.0

Exchange Rates [38]

Currency: **Barbadian dollars.**
Symbol: **Bds$.**

Data are currency units per $1.

Fixed rate	2.0113

Imports and Exports

Top Import Origins [39]

$697 million (c.i.f., 1991).

Origins	%
US	34
CARICOM	16
UK	11
Canada	6

Top Export Destinations [40]

$205.8 million (f.o.b., 1991).

Destinations	%
CARICOM	31
US	16
UK	13

Foreign Aid [41]

	U.S. $	
US commitments, including Ex-Im (FY70-89)	15	million
Western (non-US) countries, ODA and OOF bilateral commitments (1970-89)	171	million

Import and Export Commodities [42]

Import Commodities	**Export Commodities**
Foodstuffs	Sugar and molasses
Consumer durables	Chemicals
Raw materials	Electrical components
Machinery	Clothing
Crude oil	Rum
Construction materials	Machinery & transport equip.
Chemicals	

Belarus

Geography [1]

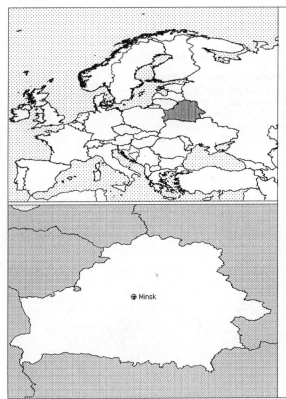

Total area:
207,600 km2
Land area:
207,600 km2
Comparative area:
Slightly smaller than Kansas
Land boundaries:
Total 3,098 km, Latvia 141 km, Lithuania 502 km, Poland 605 km, Russia 959 Km, Ukraine 891 km
Coastline:
0 km (landlocked)
Climate:
Mild and moist; transitional between continental and maritime
Terrain:
Generally flat and contains much marshland
Natural resources:
Forest land, peat deposits
Land use:
Arable land:
29%
Permanent crops:
0%
Meadows and pastures:
15%
Forest and woodland:
0%
Other:
56%

Demographics [2]

	1960	1970	1980	1990	1991[1]	1994	2000	2010	2020
Population	8,168	9,027	9,644	10,248	NA	10,405	10,576	10,864	11,047
Population density (persons per sq. mi.)	NA	NA	NA	NA	NA	NA	NA	NA	NA
Births	NA	NA	NA	NA	NA	137	NA	NA	NA
Deaths	NA	NA	NA	NA	NA	116	NA	NA	NA
Life expectancy - males	NA	NA	NA	NA	NA	66	NA	NA	NA
Life expectancy - females	NA	NA	NA	NA	NA	76	NA	NA	NA
Birth rate (per 1,000)	NA	NA	NA	NA	NA	13	NA	NA	NA
Death rate (per 1,000)	NA	NA	NA	NA	NA	11	NA	NA	NA
Women of reproductive age (15-44 yrs.)	NA	NA	NA	2,466	NA	2,550	2,667	NA	NA
of which are currently married	NA	NA	NA	1,692	NA	1,757	1,815	NA	NA
Fertility rate	NA	NA	NA	2.0	NA	1.9	1.8	1.8	1.7

Population values are in thousands, life expectancy in years, and other items as indicated.

Health

Health Personnel [3]

Doctors per 1,000 pop., 1988-92	4.05
Nurse-to-doctor ratio, 1988-92	NA
Hospital beds per 1,000 pop., 1985-90	13.2
Percentage of children immunized (age 1 yr. or less)	
Third dose of DPT, 1990-91	90.0
Measles, 1990-91	97.0

Health Expenditures [4]

Total health expenditure, 1990 (official exchange rate)	
Millions of dollars	1613
Millions of dollars per capita	157
Health expenditures as a percentage of GDP	
Total	3.2
Public sector	2.2
Private sector	1.0
Development assistance for health	
Total aid flows (millions of dollars)[1]	NA
Aid flows per capita (millions of dollars)	NA
Aid flows as a percentage of total health expenditure	NA

For sources, notes, and explanations, see Annotated Source Appendix, page 1035.

77

Human Factors

Health Care Ratios [5]

Population per physician, 1970	NA
Population per physician, 1990	250
Population per nursing person, 1970	NA
Population per nursing person, 1990	NA
Percent of births attended by health staff, 1985	NA

Infants and Malnutrition [6]

Percent of babies with low birth weight, 1985	NA
Infant mortality rate per 1,000 live births, 1970	NA
Infant mortality rate per 1,000 live births, 1991	15
Years of life lost per 1,000 population, 1990	14
Prevalence of malnutrition (under age 5), 1990	NA

Ethnic Division [7]

Belarusian	77.9%
Russian	13.2%
Polish	4.1%
Ukrainian	2.9%
Other	1.9%

Religion [8]

Eastern Orthodox and other.

Major Languages [9]

Byelorussian, Russian, other.

Education

Public Education Expenditures [10]

Million Roubles	1980	1985	1987	1988	1989	1990
Total education expenditure	1,253	1,499	NA	1,858	1,967	2,093
as percent of GNP	NA	NA	NA	NA	NA	NA
as percent of total govt. expend.	NA	NA	NA	NA	NA	NA
Current education expenditure	1,051	1,264	NA	1,488	1,551	1,758
as percent of GNP	NA	NA	NA	NA	NA	NA
as percent of current govt. expend.	NA	NA	NA	NA	NA	NA
Capital expenditure	202	235	NA	371	416	335

Educational Attainment [11]

Literacy Rate [12]

	1970[b]	1979[b]	1989[b]
Illiterate population + 15 years	NA	NA	NA
Illiteracy rate - total pop. (%)	0.2	0.1	2.1
Illiteracy rate - males (%)	0.2	0.1	0.6
Illiteracy rate - females (%)	0.2	0.1	3.4

Libraries [13]

	Admin. Units	Svc. Pts.	Vols. (000)	Shelving (meters)	Vols. Added	Reg. Users
National	NA	NA	NA	NA	NA	NA
Non-specialized	NA	NA	NA	NA	NA	NA
Public	NA	NA	NA	NA	NA	NA
Higher ed. (1990)	33	NA	19,570	NA	1 mil +	203,600
School (1990)	5,198	NA	76,281	NA	NA	NA

Daily Newspapers [14]

	1975	1980	1985	1990
Number of papers	25	27	28	25
Circ. (000)	2,214	2,343	2,446	2,937

Cinema [15]

Science and Technology

Scientific/Technical Forces [16]

Potential scientists/engineers[2]	580,000
Number female	325,000
Potential technicians[2]	730,000
Number female	479,000
Total	1,310,000

R&D Expenditures [17]

U.S. Patents Issued [18]

Government and Law

Organization of Government [19]

Long-form name:
  Republic of Belarus
Type:
  republic
Independence:
  25 August 1991 (from Soviet Union)
Constitution:
  adopted NA April 1978
Legal system:
  based on civil law system
National holiday:
  24 August (1991)
Executive branch:
  chairman of the Supreme Soviet,
  chairman of the Council of Ministers; note
  - Belarus has approved a directly elected
  presidency but so far no elections have
  been scheduled
Legislative branch:
  unicameral Supreme Soviet
Judicial branch:
  Supreme Court

Elections [20]

Supreme Soviet. last held 4 April 1990
(next to be held NA); results -
Communists 87%; seats - (360 total)
number of seats by party NA; note - 50
seats are for public bodies; the
Communist Party obtained an
overwhelming majority.

Government Expenditures [21]

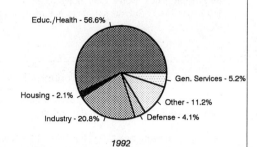

1992

Defense Summary [22]

Branches: Army, Air Forces, Air Defense Forces, Security Forces (internal and border troops)

Manpower Availability: Males age 15-49 2,491,039; fit for military service 1,964,577; reach
military age (18) annually 71,875 (1993 est.)

Defense Expenditures: 56.5 billion rubles; note - conversion of the military budget into US
dollars using the current exchange rate could produce misleading results

Crime [23]

Human Rights [24]

	SSTS	FL	FAPRO	PPCG	APROBC	TPW	PCPTW	STPEP	PHRFF	PRW	ASST	AFL
Observes	P	P	P	P	P	P	P	P		P	P	
	EAFRD	CPR	ESCR	SR	ACHR	MAAE	PVIAC	PVNAC	EAFDAW	TCIDTP	RC
Observes		P	P	P		P	P	P	P	P	P

P = Party; S = Signatory; see Appendix for meaning of abbreviations.

Labor Force

Total Labor Force [25]

5.418 million

Labor Force by Occupation [26]

Industry and construction	42%
Agriculture and forestry	20
Other	38

Date of data: 1990

Unemployment Rate [27]

0.5% of officially registered unemployed;
large numbers of underemployed workers

Production Sectors

Energy Resource Summary [28]

Energy Resources: None. **Electricity**: 8,025,000 kW capacity; 37,600 million kWh produced, 3,626 kWh per capita (1992). **Pipelines**: Crude oil 1,470 km, refined products 1,100 km, natural gas 1,980 km (1992).

Telecommunications [30]

- Construction of NMT-450 analog cellular network proceeding in Minsk, in addition to installation of 300 km of fiber optic cable in the city network
- Telephone network has 1.7 million lines, 15% of which are switched automatically
- Minsk has 450,000 lines
- Telephone density: 17 per 100 persons
- International connections to other former Soviet republics are by landline or microwave and to other countries by leased connection through the Moscow international gateway switch
- Belarus has no ground stations for international telecommunications via satellite

Top Agricultural Products [32]

	1990	1991
Potatoes	8.6	8.9
Grains	7.0	6.3
Vegetables	0.7	0.7
Fruits and berries	0.4	0.3

Values shown are 1,000 metric tons.

Top Mining Products [33]

Detailed information is not available. A summary of mineral resources available follows. **Mineral Resources**: Peat deposits.

Transportation [31]

Railroads. 5,570 km; does not include industrial lines (1990)

Highways. 98,200 km total; 66,100 km hard surfaced, 32,100 km earth (1990)

Merchant Marine. claims 5% of former Soviet fleet

Airports

Total:	124
Useable:	55
With permanent-surface runways:	31
With runways over 3,659 m:	1
With runways 2,440-3,659 m:	28
With runways 1,220-2,439 m:	20

Tourism [34]

Finance, Economics, and Trade

Industrial Summary [35]

Industrial Production: Growth rate - 9.6%; accounts for about 50% of GDP (1992). **Industries**: Employ about 27% of labor force and produce a wide variety of products essential to the other states; products include (in percent share of total output of former Soviet Union): tractors (12%); metal-cutting machine tools (11%); off-highway dump trucks up to 110-metric-ton load capacity (100%); wheel-type earthmovers for construction and mining (100%); eight- wheel-drive, high-flotation trucks with cargo capacity of 25 metric tons for use in tundra and roadless areas (100%); equipment for animal husbandry and livestock feeding (25%); motorcycles (21.3%); television sets (11%); chemical fibers (28%); fertilizer (18%); linen fabric (11%); wool fabric (7%); radios; refrigerators; and other consumer goods.

Economic Indicators [36]

National product: GDP not available. **National product real growth rate**: - 13% (1992 est.). **National product per capita**: not available. **Inflation rate (consumer prices)**: 30% per month (first quarter 1993). **External debt**: $2.6 billion (end of 1991).

Balance of Payments Summary [37]

Exchange Rates [38]

Imports and Exports

Top Import Origins [39]

$751 million from outside the successor states of the former USSR (c.i.f., 1992).

Origins	%
NA	NA

Top Export Destinations [40]

$1.1 billion to outside of the successor states of the former USSR (f.o.b., 1992).

Destinations	%
NA	NA

Foreign Aid [41]

Import and Export Commodities [42]

Import Commodities	**Export Commodities**
Machinery	Machinery & transport equip.
Chemicals	Chemicals
Textiles	Foodstuffs

For sources, notes, and explanations, see Annotated Source Appendix, page 1035.

81

Belgium

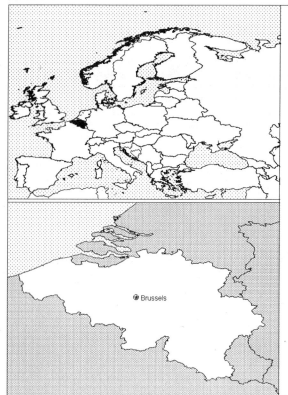

Geography [1]

Total area:
30,510 km2
Land area:
30,230 km2
Comparative area:
Slightly larger than Maryland
Land boundaries:
Total 1,385 km, France 620 km, Germany 167 km, Luxembourg 148 km, Netherlands 450 km
Coastline:
64 km
Climate:
Temperate; mild winters, cool summers; rainy, humid, cloudy
Terrain:
Flat coastal plains in northwest, central rolling hills, rugged mountains of Ardennes Forest in southeast
Natural resources:
Coal, natural gas
Land use:
Arable land:
24%
Permanent crops:
1%
Meadows and pastures:
20%
Forest and woodland:
21%
Other:
34%

Demographics [2]

	1960	1970	1980	1990	1991[1]	1994	2000	2010	2020
Population	9,119	9,638	9,847	9,962	9,922	10,063	10,144	10,135	10,015
Population density (persons per sq. mi.)	781	826	844	849	850	NA	856	847	830
Births	NA	NA	NA	NA	118	118	NA	NA	NA
Deaths	NA	NA	NA	NA	105	103	NA	NA	NA
Life expectancy - males	NA	NA	NA	NA	74	74	NA	NA	NA
Life expectancy - females	NA	NA	NA	NA	81	80	NA	NA	NA
Birth rate (per 1,000)	NA	NA	NA	NA	12	12	NA	NA	NA
Death rate (per 1,000)	NA	NA	NA	NA	11	10	NA	NA	NA
Women of reproductive age (15-44 yrs.)	NA	NA	NA	2,438	NA	2,467	2,418	NA	NA
of which are currently married	NA	NA	NA	1,712	NA	1,767	1,742	NA	NA
Fertility rate	NA	NA	NA	1.6	1.59	1.6	1.6	1.6	1.6

Population values are in thousands, life expectancy in years, and other items as indicated.

Health

Health Personnel [3]

Doctors per 1,000 pop., 1988-92	3.21
Nurse-to-doctor ratio, 1988-92	0.1
Hospital beds per 1,000 pop., 1985-90	8.3
Percentage of children immunized (age 1 yr. or less)	
Third dose of DPT, 1990-91	94.0
Measles, 1990-91	75.0

Health Expenditures [4]

Total health expenditure, 1990 (official exchange rate)	
Millions of dollars	14428
Millions of dollars per capita	1449
Health expenditures as a percentage of GDP	
Total	7.5
Public sector	6.2
Private sector	1.3
Development assistance for health	
Total aid flows (millions of dollars)[1]	NA
Aid flows per capita (millions of dollars)	NA
Aid flows as a percentage of total health expenditure	NA

For sources, notes, and explanations, see Annotated Source Appendix, page 1035.

Human Factors

Health Care Ratios [5]

Population per physician, 1970	650
Population per physician, 1990	310
Population per nursing person, 1970	NA
Population per nursing person, 1990	NA
Percent of births attended by health staff, 1985	100

Infants and Malnutrition [6]

Percent of babies with low birth weight, 1985	5
Infant mortality rate per 1,000 live births, 1970	21
Infant mortality rate per 1,000 live births, 1991	8
Years of life lost per 1,000 population, 1990	11
Prevalence of malnutrition (under age 5), 1990	NA

Ethnic Division [7]

Fleming	55%
Walloon	33%
Mixed or other	12%

Religion [8]

Roman Catholic	75%
Protestant or other	25%

Major Languages [9]

Flemish (Dutch)	56%
French	32%
German	1%
Legally bilingual (divided along ethnic lines)	11%

Education

Public Education Expenditures [10]

Million Francs	1980[1]	1985[1]	1987[1]	1988[1]	1989[1]	1990[1]
Total education expenditure	208,469	287,388	NA	274,028	310,629	325,282
as percent of GNP	6.1	6.2	NA	5.0	5.2	5.1
as percent of total govt. expend.	16.3	15.2	NA	NA	NA	NA
Current education expenditure	206,227	272,843	NA	271,902	299,473	321,427
as percent of GNP	6.0	5.9	NA	5.0	5.0	5.1
as percent of current govt. expend.	18.6	16.0	NA	NA	NA	NA
Capital expenditure	2,243	14,545	NA	2,126	11,156	3,855

Educational Attainment [11]

Literacy Rate [12]

Libraries [13]

	Admin. Units	Svc. Pts.	Vols. (000)	Shelving (meters)	Vols. Added	Reg. Users
National (1989 est.)[28]	1	1	4,000	100,000	41,229	12,411
Non-specialized	NA	NA	NA	NA	NA	NA
Public	NA	NA	NA	NA	NA	NA
Higher ed.	NA	NA	NA	NA	NA	NA
School	NA	NA	NA	NA	NA	NA

Daily Newspapers [14]

	1975	1980	1985	1990
Number of papers	30	26	24	33
Circ. (000)	2,340	2,289	2,171	3,000[e]

Cinema [15]

Science and Technology

Scientific/Technical Forces [16]

R&D Expenditures [17]

	Franc[26] (000) 1988
Total expenditure	91,265,100
Capital expenditure	NA
Current expenditure	NA
Percent current	NA

U.S. Patents Issued [18]

Values show patents issued to citizens of the country by the U.S. Patents Office.

	1990	1991	1992
Number of patents	352	366	370

Government and Law

Organization of Government [19]

Long-form name:
Kingdom of Belgium
Type:
constitutional monarchy
Independence:
4 October 1830 (from the Netherlands)
Constitution:
7 February 1831, last revised 8-9 August 1980; the government is in the process of revising the Constitution with the aim of federalizing the Belgian state
Legal system:
civil law system influenced by English constitutional theory; judicial review of legislative acts; accepts compulsory ICJ jurisdiction, with reservations
National holiday:
National Day, 21 July (ascension of King Leopold to the throne in 1831)
Executive branch:
monarch, prime minister, three deputy prime ministers, Cabinet
Legislative branch:
Parliament with a Senate and a Chamber of Representatives
Judicial branch:
Supreme Court of Justice (Flemish - Hof van Cassatie, French - Cour de Cassation)

Crime [23]

Crime volume (for 1990)	
Cases known to police	332,041
Attempts (percent)	NA
Percent cases solved	18
Crimes per 100,000 persons	3.337.8
Persons responsible for offenses	
Total number offenders	NA
Percent female	NA
Percent juvenile[8]	NA
Percent foreigners	NA

Political Parties [20]

Chamber of Representatives	% of votes
Flemish Social Christian (CVP)	16.7
Walloon Socialist (PS)	13.6
Flemish Socialist (SP)	12.0
Flemish Liberals and Dem.'s (VLD)	11.9
Walloon Liberal (PRL)	8.2
Walloon Social Christian (PSC)	7.8
Vlaams Blok (VB)	6.6
Volksunie (VU)	5.9
ECOLO	5.1
Others	12.2

Government Budget [21]

For 1989.

Revenues	97.8
Expenditures	109.3
Capital expenditures	NA

Military Expenditures and Arms Transfers [22]

	1985	1986	1987	1988	1989
Military expenditures					
Current dollars (mil.)	3,592	3,752	3,879	3,811	3,881
1989 constant dollars (mil.)	4,089	4,163	4,172	3,968	3,881
Armed forces (000)	107	107	109	110	110
Gross national product (GNP)					
Current dollars (mil.)	119,400	124,900	131,700	141,800	154,600
1989 constant dollars (mil.)	135,900	138,600	141,700	147,600	154,600
Central government expenditures (CGE)					
1989 constant dollars (mil.)	77,760	75,870	74,540	75,860	53,690[e]
People (mil.)	9.9	9.9	9.9	9.9	9.9
Military expenditure as % of GNP	3.0	3.0	2.9	2.7	2.5
Military expenditure as % of CGE	5.3	5.5	5.6	5.2	7.2
Military expenditure per capita	415	422	423	401	392
Armed forces per 1,000 people	10.9	10.8	11.0	11.1	11.1
GNP per capita	13,780	14,060	14,350	14,930	15,620
Arms imports[6]					
Current dollars (mil.)	300	180	160	600	220
1989 constant dollars (mil.)	342	200	172	625	220
Arms exports[6]					
Current dollars (mil.)	380	220	90	40	20
1989 constant dollars (mil.)	433	244	97	42	20
Total imports[7]					
Current dollars (mil.)	56,180	68,600	83,230	92,440	98,460
1989 constant dollars	63,960	76,130	89,520	96,230	98,460
Total exports[7]					
Current dollars (mil.)	53,740	68,820	83,100	92,130	100,000
1989 constant dollars	61,180	76,370	89,380	95,910	100,000
Arms as percent of total imports[8]	0.5	0.3	0.2	0.6	0.2
Arms as percent of total exports[8]	0.7	0.3	0.1	0	0

Human Rights [24]

	SSTS	FL	FAPRO	PPCG	APROBC	TPW	PCPTW	STPEP	PHRFF	PRW	ASST	AFL
Observes	P	P	P	P	P	P	P	P	P	P	P	P
	EAFRD	CPR	ESCR	SR	ACHR	MAAE	PVIAC	PVNAC	EAFDAW	TCIDTP	RC
Observes	P	P	P	P		P	P	P	P	S	P

P = Party; S = Signatory; see Appendix for meaning of abbreviations.

Labor Force

Total Labor Force [25]

4.126 million

Labor Force by Occupation [26]

Services	63.6%
Industry	28
Construction	6.1
Agriculture	2.3

Date of data: 1988

Unemployment Rate [27]

9.8% (end 1992)

For sources, notes, and explanations, see Annotated Source Appendix, page 1035.

Production Sectors

Commercial Energy Production and Consumption

Production [28]	Consumption [29]

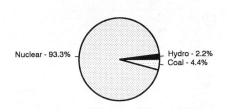

Nuclear - 93.3%   Hydro - 2.2%   Coal - 4.4%

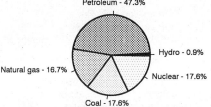

Petroleum - 47.3%   Hydro - 0.9%   Natural gas - 16.7%   Nuclear - 17.6%   Coal - 17.6%

Data are shown in quadrillion (10^{15}) BTUs and percent for 1991

Crude oil[1]	0.00		Petroleum	1.05
Natural gas liquids	0.00		Dry natural gas	0.37
Dry natural gas	0.00		Coal[2]	0.39
Coal[2]	0.02		Net hydroelectric power[3]	0.02
Net hydroelectric power[3]	0.01		Net nuclear power[3]	0.39
Net nuclear power[3]	0.42		Total	2.21
Total	0.45			

Telecommunications [30]

- Highly developed, technologically advanced, and completely automated domestic and international telephone and telegraph facilities
- Extensive cable network
- Limited microwave radio relay network
- 4,720,000 telephones
- Broadcast stations - 3 AM, 39 FM, 32 TV
- 5 submarine cables
- 2 satellite earth stations - Atlantic Ocean INTELSAT and EUTELSAT systems
- Nationwide mobile phone system

Transportation [31]

Railroads. Belgian National Railways (SNCB) operates 3,568 km 1.435-meter standard gauge, government owned; 2,563 km double track; 2,207 km electrified

Highways. 103,396 km total; 1,317 km limited access, divided autoroute; 11,717 km national highway; 1,362 km provincial road; about 38,000 km paved and 51,000 km unpaved rural roads

Merchant Marine. 23 ships (1,000 GRT or over) totaling 96,949 GRT/133,658 DWT; includes 10 cargo, 5 oil tanker, 2 liquefied gas, 5 chemical tanker, 1 bulk

Airports

Total:	42
Usable:	42
With permanent-surface runways:	24
With runways over 3,659 m:	0
With runways 2,440-3,659 m:	14
With runways 1,220-2,439 m:	3

Top Agricultural Products [32]

Agriculture accounts for 2.3% of GDP; emphasis on livestock production—beef, veal, pork, milk; major crops are sugar beets, fresh vegetables, fruits, grain, tobacco; net importer of farm products.

Top Mining Products [33]

Metric tons unless otherwise specified	M.t.
Coal, bituminous (000 tons)	634
Quicklime (000 tons)	2,021[1]
Limestone (000 tons)	33,255[1]
Sand and gravel (000 tons)	20,899[1]
Marble	
In blocks (cubic meters)	358[1]
Crushed and other	1,340[1]
Silica sand	NA

Tourism [34]

Manufacturing Sector

GDP and Manufacturing Summary [35]

	1980	1985	1989	1990	% change 1980-1990	% change 1989-1990
GDP (million 1980 $)	118,016	122,611	141,464	143,454	21.6	1.4
GDP per capita (1980 $)	11,979	12,436	14,366	14,573	21.7	1.4
Manufacturing as % of GDP (current prices)	25.5	24.6	23.3	23.7	-7.1	1.7
Gross output (million $)	84,723[e]	59,419	116,888	146,712	73.2	25.5
Value added (million $)	28,130	18,232	33,454	42,392	50.7	26.7
Value added (million 1980 $)	25,772	29,229	37,645	34,819	35.1	-7.5
Industrial production index	100	107	123	125	25.0	1.6
Employment (thousands)	872	755	718[e]	735	-15.7	2.4

Note: GDP stands for Gross Domestic Product. 'e' stands for estimated value.

Profitability and Productivity

	1980	1985	1989	1990	% change 1980-1990	% change 1989-1990
Intermediate input (%)	67[e]	69	71	71	6.0	0.0
Wages, salaries, and supplements (%)	17[e]	13[e]	12[e]	11[e]	-35.3	-8.3
Gross operating surplus (%)	17[e]	17[e]	16[e]	17[e]	0.0	6.3
Gross output per worker ($)	91,198[e]	78,700	162,831[e]	199,636	118.9	22.6
Value added per worker ($)	30,345	24,149	46,603[e]	57,684	90.1	23.8
Average wage (incl. benefits) ($)	16,066[e]	10,618[e]	19,824[e]	22,774[e]	41.8	14.9

Profitability is in percent of gross output. Productivity is in U.S. $. 'e' stands for estimated value.

Profitability - 1990

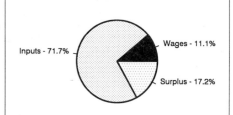

Inputs - 71.7%　　Wages - 11.1%　　Surplus - 17.2%

The graphic shows percent of gross output.

Value Added in Manufacturing

	1980 $ mil.	1980 %	1985 $ mil.	1985 %	1989 $ mil.	1989 %	1990 $ mil.	1990 %	% change 1980-1990	% change 1989-1990
311 Food products	3,991	14.2	2,863	15.7	5,588	16.7	6,015	14.2	50.7	7.6
313 Beverages	549	2.0	359	2.0	627	1.9	717	1.7	30.6	14.4
314 Tobacco products	199	0.7	123	0.7	188	0.6	257	0.6	29.1	36.7
321 Textiles	1,445	5.1	937	5.1	1,612	4.8	1,786	4.2	23.6	10.8
322 Wearing apparel	671	2.4	392	2.2	694	2.1	960	2.3	43.1	38.3
323 Leather and fur products	136	0.5	93	0.5	81	0.2	183	0.4	34.6	125.9
324 Footwear	67	0.2	35	0.2	41	0.1	61	0.1	-9.0	48.8
331 Wood and wood products	226	0.8	131	0.7	266[e]	0.8	458	1.1	102.7	72.2
332 Furniture and fixtures	1,123	4.0	614	3.4	1,257[e]	3.8	1,514	3.6	34.8	20.4
341 Paper and paper products	612	2.2	441	2.4	823[e]	2.5	1,095	2.6	78.9	33.0
342 Printing and publishing	926	3.3	602	3.3	1,111	3.3	1,677	4.0	81.1	50.9
351 Industrial chemicals	2,401	8.5	2,253	12.4	3,603[e]	10.8	4,483	10.6	86.7	24.4
352 Other chemical products	665	2.4	467	2.6	960[e]	2.9	1,186	2.8	78.3	23.5
353 Petroleum refineries	517	1.8	212	1.2	467	1.4	305	0.7	-41.0	-34.7
354 Miscellaneous petroleum and coal products	72	0.3	36	0.2	39	0.1	69	0.2	-4.2	76.9
355 Rubber products	193	0.7	130	0.7	308[e]	0.9	314	0.7	62.7	1.9
356 Plastic products	819	2.9	633	3.5	1,431[e]	4.3	2,197	5.2	168.3	53.5
361 Pottery, china and earthenware	107	0.4	61	0.3	147	0.4	151	0.4	41.1	2.7
362 Glass and glass products	516	1.8	289	1.6	464	1.4	770	1.8	49.2	65.9
369 Other non-metal mineral products	654	2.3	307	1.7	722	2.2	881	2.1	34.7	22.0
371 Iron and steel	2,294	8.2	985	5.4	1,548	4.6	2,510	5.9	9.4	62.1
372 Non-ferrous metals	487	1.7	417	2.3	738	2.2	1,140	2.7	134.1	54.5
381 Metal products	2,071	7.4	1,228	6.7	2,374	7.1	2,835	6.7	36.9	19.4
382 Non-electrical machinery	2,490	8.9	1,556	8.5	2,615	7.8	3,673	8.7	47.5	40.5
383 Electrical machinery	2,303	8.2	1,451	8.0	2,433	7.3	2,913	6.9	26.5	19.7
384 Transport equipment	1,892	6.7	1,217	6.7	2,443	7.3	3,196	7.5	68.9	30.8
385 Professional and scientific equipment	170	0.6	106	0.6	148	0.4	271	0.6	59.4	83.1
390 Other manufacturing industries	537	1.9	294	1.6	725[e]	2.2	772	1.8	43.8	6.5

Note: The industry codes shown are International Standard Industry codes (ISIC). Percentages are percent of total Value Added. 'e' stands for estimated value

　　　　　　　　　　　　　　For sources, notes, and explanations, see Annotated Source Appendix, page 1035.

Finance, Economics, and Trade

Economic Indicators [36]

Billions of Belgian Francs (BF) unless otherwise noted.

	1989[8]	1990[8]	1991[P,8]
Real GNP	5,495	5,688	5,818
Real GDP growth ratio (%.)	3.8	3.7	1.9
Money Supply (M1)	1,308	1,282	1,360
Savings rate (%.)	14.4	14.1	14.7
Investment rate (%.)	16.9	12.6	3.2
CPI (%.)	3.1	3.5	3.3
WPI (%.)	5.7	0.6	2.4
External Public Debt	1,129	1,111	1,124

Balance of Payments Summary [37]

Values in millions of dollars.

	1987	1988	1989	1990	1991
Exports of goods (f.o.b.)	76,088	85,496	89,988	107,654	106,019
Imports of goods (f.o.b.)	-76,268	-84,273	-89,020	-107,064	-106,085
Trade balance	-180	1,223	968	590	-66
Services - debits	-43,864	-51,986	-67,902	-90,646	-101,703
Services - credits	48,264	56,112	71,850	96,971	108,251
Private transfers (net)	-114	41	47	-597	-280
Government transfers (net)	-1,313	-1,796	-1,765	-1,369	-1,470
Direct investments	2,355	5,212	7,057	8,056	9,377
Errors and omissions	-11	59	-86	-2,844	-992
Overall balance	2,272	-105	-2,092	451	505

Exchange Rates [38]

Currency: **Belgian francs.**
Symbol: **BF.**

Data are currency units per $1.

January 1993	33.256
1992	32.150
1991	34.148
1990	33.418
1989	39.404
1988	36.768

Imports and Exports

Top Import Origins [39]

$121 billion (c.i.f., 1991) Belgium-Luxembourg Economic Union. Data are for 1991.

Origins	%
EC	73
US	4.8
Oil-exporting LDCs	4
Former Communist countries	1.8

Top Export Destinations [40]

$118 billion (f.o.b., 1991). Data are for 1991.

Destinations	%
EC	75.5
US	3.7
Former Communist countries	1.4

Foreign Aid [41]

	U.S. $	
Donor - ODA and OOF commitments (1970-89)	5.8	billion

Import and Export Commodities [42]

Import Commodities	Export Commodities
Fuels	Iron and steel
Grains	Transportation equipment
Chemicals	Tractors
Foodstuffs	Diamonds
	Petroleum products

Belize

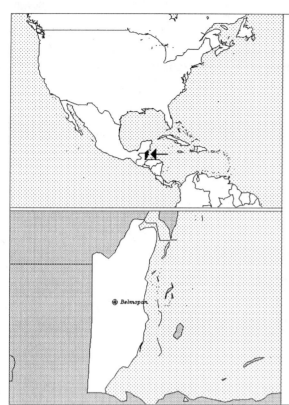

Geography [1]

Total area:
22,960 km2
Land area:
22,800 km2
Comparative area:
Slightly larger than Massachusetts
Land boundaries:
Total 516 km, Guatemala 266 km, Mexico 250 km
Coastline:
386 km
Climate:
Tropical; very hot and humid; rainy season (May to February)
Terrain:
Flat, swampy coastal plain; low mountains in south
Natural resources:
Arable land potential, timber, fish
Land use:
Arable land:
2%
Permanent crops:
0%
Meadows and pastures:
2%
Forest and woodland:
44%
Other:
52%

Demographics [2]

	1960	1970	1980	1990	1991[1]	1994	2000	2010	2020
Population	92	122	144	190	228	209	242	299	356
Population density (persons per sq. mi.)	10	14	17	25	26	NA	34	43	51
Births	NA	NA	NA	NA	9	7	NA	NA	NA
Deaths	NA	NA	NA	NA	1	1	NA	NA	NA
Life expectancy - males	NA	NA	NA	NA	67	66	NA	NA	NA
Life expectancy - females	NA	NA	NA	NA	72	70	NA	NA	NA
Birth rate (per 1,000)	NA	NA	NA	NA	38	35	NA	NA	NA
Death rate (per 1,000)	NA	NA	NA	NA	5	6	NA	NA	NA
Women of reproductive age (15-44 yrs.)	NA	NA	NA	42	NA	48	59	NA	NA
of which are currently married	NA	NA	NA	18	NA	21	26	NA	NA
Fertility rate	NA	NA	NA	5.0	4.67	4.4	3.6	2.7	2.3

Population values are in thousands, life expectancy in years, and other items as indicated.

Health

Health Personnel [3]

Health Expenditures [4]

For sources, notes, and explanations, see Annotated Source Appendix, page 1035.

Human Factors

Health Care Ratios [5]	Infants and Malnutrition [6]

Ethnic Division [7]

Mestizo	44%
Creole	30%
Maya	11%
Garifuna	7%
Other	8%

Religion [8]

Roman Catholic	62%
Protestant	30%
Anglican	12%
Methodist	6%
Mennonite	4%
Seventh-Day Adventist	3%
Pentecostal	2%
Jehovah's Witnesses	1%

Major Languages [9]

English (official)	NA
Spanish	NA
Maya	NA
Garifuna (Carib)	NA

Education

Public Education Expenditures [10]

Million Belize Dollars	1985	1986	1987	1988	1989	1990
Total education expenditure	NA	NA	NA	NA	NA	NA
as percent of GNP	NA	NA	NA	NA	NA	NA
as percent of total govt. expend.	NA	NA	NA	NA	NA	NA
Current education expenditure	NA	18	NA	NA	NA	NA
as percent of GNP	NA	4.4	NA	NA	NA	NA
as percent of current govt. expend.	NA	NA	NA	NA	NA	NA
Capital expenditure	NA	NA	NA	NA	NA	NA

Educational Attainment [11]

Age group	25+
Total population	67,604
Highest level attained (%)	
No schooling	12.8
First level	
Incompleted	63.3
Completed	NA
Entered second level	
S-1	14.7
S-2	NA
Post secondary	9.2

Literacy Rate [12]

	1970[b]	1980[b]	1990[b]
Illiterate population +15 years	5,353	NA	NA
Illiteracy rate - total pop. (%)[3]	8.8	NA	NA
Illiteracy rate - males (%)	8.8	NA	NA
Illiteracy rate - females (%)	8.8	NA	NA

Libraries [13]

	Admin. Units	Svc. Pts.	Vols. (000)	Shelving (meters)	Vols. Added	Reg. Users
National	NA	NA	NA	NA	NA	NA
Non-specialized	NA	NA	NA	NA	NA	NA
Public (1989)[13]	1	34	125	NA	15,000	20,000
Higher ed. (1990)	1	1	6	NA	-	312
School	NA	NA	NA	NA	NA	NA

Daily Newspapers [14]

	1975	1980	1985	1990
Number of papers	1	1	1	-
Circ. (000)	4	3	3	-

Cinema [15]

Science and Technology

Scientific/Technical Forces [16]	R&D Expenditures [17]	U.S. Patents Issued [18]

U.S. Patents Issued [18]

Values show patents issued to citizens of the country by the U.S. Patents Office.

	1990	1991	1992
Number of patents	0	1	0

Government and Law

Organization of Government [19]

Long-form name:
none
Type:
parliamentary democracy
Independence:
21 September 1981 (from UK)
Constitution:
21 September 1981
Legal system:
English law
National holiday:
Independence Day, 21 September
Executive branch:
British monarch, governor general, prime minister, deputy prime minister, Cabinet
Legislative branch:
bicameral National Assembly consists of an upper house or Senate and a lower house or House of Representatives
Judicial branch:
Supreme Court

Political Parties [20]

National Assembly	% of seats
People's United Party (PUP)	57.1
United Democratic Party (UDP)	42.9

Government Expenditures [21]

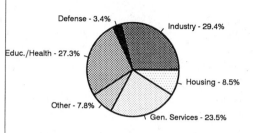

Defense - 3.4%
Industry - 29.4%
Educ./Health - 27.3%
Housing - 8.5%
Other - 7.8%
Gen. Services - 23.5%

% distribution for 1992. Expend. in FY91 est.: 123.1 ($ mil.)

Defense Summary [22]

Branches: British Forces Belize, Belize Defense Force (including Army, Navy, Air Force, and Volunteer Guard), Belize National Police

Manpower Availability: Males age 15-49 47,135; fit for military service 28,070; reach military age (18) annually 2,066 (1993 est.)

Defense Expenditures: Exchange rate conversion - $5.4 million, 2% of GDP (1992)

Crime [23]

Human Rights [24]

	SSTS	FL	FAPRO	PPCG	APROBC	TPW	PCPTW	STPEP	PHRFF	PRW	ASST	AFL
Observes	1	P	P		P	P	P			P	1	P
	EAFRD	CPR	ESCR	SR	ACHR	MAAE	PVIAC	PVNAC	EAFDAW	TCIDTP	RC
Observes		1	1	P			P	P	P	P	P

P = Party; S = Signatory; see Appendix for meaning of abbreviations.

Labor Force

Total Labor Force [25]

51,500

Labor Force by Occupation [26]

Agriculture	30%
Services	16
Government	15.4
Commerce	11.2
Manufacturing	10.3

Unemployment Rate [27]

12% (1991 est.)

Production Sectors

Energy Resource Summary [28]

Energy Resources: None. **Electricity**: 34,532 kW capacity; 90 million kWh produced, 393 kWh per capita (1992).

Telecommunications [30]

- 8,650 telephones
- Above-average system based on microwave radio relay
- Broadcast stations - 6 AM, 5 FM, 1 TV, 1 shortwave
- 1 Atlantic Ocean INTELSAT earth station

Transportation [31]

Highways. 2,710 km total; 500 km paved, 1,600 km gravel, 300 km improved earth, and 310 km unimproved earth

Merchant Marine. 4 ships (1,000 GRT or over) totaling 9,768 GRT/ 12,721 DWT; includes 3 cargo, 1 roll-on/roll-off

Airports

Total:	42
Usable:	32
With permanent-surface runways:	3
With runways over 3,659 m:	0
With runways 2,440-3,659 m:	1
With runways 1,229-2,439 mr:	2

Top Agricultural Products [32]

Agriculture accounts for 22% of GDP (including fish and forestry); commercial crops include sugarcane, bananas, coca, citrus fruits; expanding output of lumber and cultured shrimp; net importer of basic foods.

Top Mining Products [33]

Estimated metric tons unless otherwise specified   M.t.

Clays	2,000,000
Dolomite	100,000
Gold (kilograms)	5
Limestone	300,000
Marl	1,000
Sand and gravel	200,000

Tourism [34]

	1987	1988	1989	1990	1991
Tourists	108	151	182	230	223
Tourism receipts	47	48	79	91	95
Tourism expenditures	5	7	8	7	8
Fare expenditures	5	6	8	8	8

Tourists are in thousands, money in million U.S. dollars.

Finance, Economics, and Trade

GDP and Manufacturing Summary [35]

	1980	1985	1990	1991	1992
Gross Domestic Product					
Millions of 1980 dollars	171	183	267	278	277[e]
Growth rate in percent	4.39	2.25	7.60	4.00	-0.27[e]
Manufacturing Value Added					
Millions of 1980 dollars	22	21	27	28	28[e]
Growth rate in percent	14.91	0.89	4.97	4.06	1.38[e]
Manufacturing share in percent of current prices	17.9	12.3	11.3	10.0[e]	NA

Economic Indicators [36]

National product: GDP—exchange rate conversion—$373 million (1990 est.). **National product real growth rate**: 10% (1990). **National product per capita**: $1,635 (1990 est.). **Inflation rate (consumer prices)**: 5.5% (1991). **External debt**: $143.7 million (1991).

Balance of Payments Summary [37]

Values in millions of dollars.

	1987	1988	1989	1990	1991
Exports of goods (f.o.b.)	102.8	119.4	124.4	129.2	119.8
Imports of goods (f.o.b.)	-126.9	-161.3	-188.5	-188.4	-223.6
Trade balance	-24.1	-41.9	-64.1	-59.2	-103.8
Services - debits	-59.3	-69.0	-81.6	-80.8	-88.7
Services - credits	61.8	82.4	95.5	125.9	116.9
Private transfers (net)	15.2	15.2	20.7	16.3	15.2
Government transfers (net)	15.7	10.6	10.4	13.0	11.3
Long term capital (net)	7.4	23.6	29.9	29.3	22.5
Short term capital (net)	-4.7	3.7	-4.4	-4.2	-7.7
Errors and omissions	-0.2	-2.9	9.1	-25.0	13.7
Overall balance	11.8	21.7	15.5	15.3	-20.6

Exchange Rates [38]

Currency: **Belizean dollars.**
Symbol: **Bz$.**

Data are currency units per $1.

Fixed rate	2.00

Imports and Exports

Top Import Origins [39]

$194 million (c.i.f., 1991 est.).

Origins	%
US	60
UK	NA
EC	NA
Mexico 1991	NA

Top Export Destinations [40]

$95.6 million (f.o.b., 1991).

Destinations	%
US	49
UK	NA
EC	NA
Mexico 1991	NA

Foreign Aid [41]

	U.S. $	
US commitments, including Ex-Im (FY70-89)	104	million
Western (non-US) countries, ODA and OOF bilateral commitments (1970-89)	215	million

Import and Export Commodities [42]

Import Commodities	**Export Commodities**
Machinery & transport equipment	Sugar
Food	Citrus
Manufactured goods	Clothing
Fuels	Bananas
Chemicals	Fish products
Pharmaceuticals	Molasses

Benin

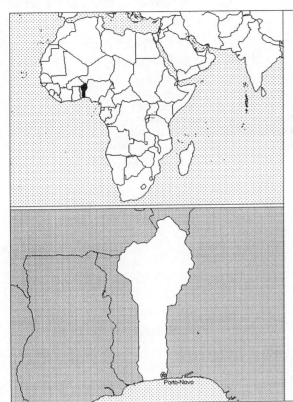

Geography [1]

Total area:
112,620 km2
Land area:
110,620 km2
Comparative area:
Slightly smaller than Pennsylvania
Land boundaries:
Total 1,989 km, Burkina 306 km, Niger 266 km, Nigeria 773 km, Togo 644 km
Coastline:
121 km
Climate:
Tropical; hot, humid in south; semiarid in north
Terrain:
Mostly flat to undulating plain; some hills and low mountains
Natural resources:
Small offshore oil deposits, limestone, marble, timber
Land use:
Arable land:
12%
Permanent crops:
4%
Meadows and pastures:
4%
Forest and woodland:
35%
Other:
45%

Demographics [2]

	1960	1970	1980	1990	1991[1]	1994	2000	2010	2020
Population	2,055	2,620	3,444	4,676	4,832	5,342	6,517	8,955	11,920
Population density (persons per sq. mi.)	48	61	81	109	113	NA	152	209	179
Births	NA	NA	NA	NA	237	255	NA	NA	NA
Deaths	NA	NA	NA	NA	76	77	NA	NA	NA
Life expectancy - males	NA	NA	NA	NA	49	50	NA	NA	NA
Life expectancy - females	NA	NA	NA	NA	52	54	NA	NA	NA
Birth rate (per 1,000)	NA	NA	NA	NA	49	48	NA	NA	NA
Death rate (per 1,000)	NA	NA	NA	NA	16	14	NA	NA	NA
Women of reproductive age (15-44 yrs.)	NA	NA	NA	1,064	NA	1,209	1,475	NA	NA
of which are currently married	NA	NA	NA	857	NA	975	1,191	NA	NA
Fertility rate	NA	NA	NA	7.1	7.00	6.8	6.3	5.4	4.3

Population values are in thousands, life expectancy in years, and other items as indicated.

Health

Health Personnel [3]

Doctors per 1,000 pop., 1988-92	0.07
Nurse-to-doctor ratio, 1988-92	5.8
Hospital beds per 1,000 pop., 1985-90	NA
Percentage of children immunized (age 1 yr. or less)	
Third dose of DPT, 1990-91	67.0
Measles, 1990-91	70.0

Health Expenditures [4]

Total health expenditure, 1990 (official exchange rate)	
Millions of dollars	79
Millions of dollars per capita	17
Health expenditures as a percentage of GDP	
Total	4.3
Public sector	2.8
Private sector	1.6
Development assistance for health	
Total aid flows (millions of dollars)[1]	33
Aid flows per capita (millions of dollars)	7.0
Aid flows as a percentage of total health expenditure	41.8

For sources, notes, and explanations, see Annotated Source Appendix, page 1035.

93

Human Factors

Health Care Ratios [5]

Population per physician, 1970	28570
Population per physician, 1990	NA
Population per nursing person, 1970	2600
Population per nursing person, 1990	NA
Percent of births attended by health staff, 1985	34

Infants and Malnutrition [6]

Percent of babies with low birth weight, 1985	10
Infant mortality rate per 1,000 live births, 1970	155
Infant mortality rate per 1,000 live births, 1991	111
Years of life lost per 1,000 population, 1990	89
Prevalence of malnutrition (under age 5), 1990	35

Ethnic Division [7]

African includes 42 ethnic groups, most important being Fon, Adja, Yoruba, Bariba.

African	99%
Europeans	5,500

Religion [8]

Indigenous beliefs	70%
Muslim	15%
Christian	15%

Major Languages [9]

French (official), Fon and Yoruba (most common vernaculars in south), tribal languages (at least six major ones in north).

Education

Public Education Expenditures [10]

Million Francs C.F.A.	1980	1985	1987	1988	1989	1990
Total education expenditure	NA	NA	NA	NA	NA	NA
as percent of GNP	NA	NA	NA	NA	NA	NA
as percent of total govt. expend.	NA	NA	NA	NA	NA	NA
Current education expenditure	12,426	NA	NA	NA	NA	NA
as percent of GNP	4.2	NA	NA	NA	NA	NA
as percent of current govt. expend.	36.8	NA	NA	NA	NA	NA
Capital expenditure	NA	NA	NA	NA	NA	NA

Educational Attainment [11]

Literacy Rate [12]

In thousands and percent	1985[a]	1991[a]	2000[a]
Illiterate population +15 years	1,754	1,904	2,251
Illiteracy rate - total pop. (%)	81.3	76.6	65.8
Illiteracy rate - males (%)	74.0	68.3	56.3
Illiteracy rate - females (%)	88.3	84.4	74.8

Libraries [13]

	Admin. Units	Svc. Pts.	Vols. (000)	Shelving (meters)	Vols. Added	Reg. Users
National (1989)	1	1	6	177	1,102	158
Non-specialized (1989)	4	4	40	1,100	NA	NA
Public (1989)	12	12	28	777	2,008	689
Higher ed. (1987)	10	10	95	2,611	NA	NA
School (1987)	5	5	16	797	NA	NA

Daily Newspapers [14]

	1975	1980	1985	1990
Number of papers	1	1	1	1
Circ. (000)	1	1	1	12

Cinema [15]

Science and Technology

Scientific/Technical Forces [16]

Potential scientists/engineers	1,341
Number female	263
Potential technicians	NA
Number female	NA
Total	NA

R&D Expenditures [17]

	Francs[1] (000) 1989
Total expenditure	NA
Capital expenditure	NA
Current expenditure	3,347,695
Percent current	NA

U.S. Patents Issued [18]

Government and Law

Organization of Government [19]

Long-form name:
Republic of Benin
Type:
republic under multiparty democratic rule
dropped Marxism-Leninism December
1989; democratic reforms adopted
February 1990; transition to multiparty
system completed 4 April 1991
Independence:
1 August 1960 (from France)
Constitution:
2 December 1990
Legal system:
based on French civil law and customary
law; has not accepted compulsory ICJ
jurisdiction
National holiday:
National Day, 1 August (1990)
Executive branch:
president, cabinet
Legislative branch:
unicameral National Assembly
(Assemblee Nationale)
Judicial branch:
Supreme Court (Cour Supreme)

Crime [23]

Political Parties [20]

National Assembly	% of seats
UDFP-MDPS-ULD	18.8
PNDD/PRD	14.1
PSD/UNSP	12.5
Our Common Cause (NCC)	10.9
National Rally for Democracy (RND)	10.9
MNDD/MSUP/UDRN	9.4
Union for Democracy (UDS)	7.8
Assembly Liberal Democrats (RDL)	6.3
ASD/BSD	4.7
Others	4.7

Government Budget [21]

For 1990 est.
Revenues	194
Expenditures	390
Capital expenditures	104

Military Expenditures and Arms Transfers [22]

	1985	1986	1987	1988	1989
Military expenditures					
Current dollars (mil.)	NA	30	NA	34	33
1989 constant dollars (mil.)	NA	33	NA	36	33
Armed forces (000)	6	4	4	5	5
Gross national product (GNP)					
Current dollars (mil.)	1,451	1,476	1,464	1,551	1,637
1989 constant dollars (mil.)	1,652	1,638	1,575	1,615	1,637
Central government expenditures (CGE)					
1989 constant dollars (mil.)	NA	NA	303[e]	202	168
People (mil.)	4.0	4.1	4.2	4.4	4.5
Military expenditure as % of GNP	NA	2.0	NA	2.2	2.0
Military expenditure as % of CGE	NA	NA	NA	17.7	19.4
Military expenditure per capita	NA	8	NA	8	7
Armed forces per 1,000 people	1.5	1.0	0.9	1.1	1.0
GNP per capita	415	399	371	369	362
Arms imports[6]					
Current dollars (mil.)	10	10	0	10	0
1989 constant dollars (mil.)	11	11	0	10	0
Arms exports[6]					
Current dollars (mil.)	0	0	0	0	0
1989 constant dollars (mil.)	0	0	0	0	0
Total imports[7]					
Current dollars (mil.)	298	314	352	388	309
1989 constant dollars	339	348	379	404	309
Total exports[7]					
Current dollars (mil.)	123	100	104	114	98
1989 constant dollars	140	111	112	119	98
Arms as percent of total imports[8]	3.4	3.2	0	2.6	0
Arms as percent of total exports[8]	0	0	0	0	0

Human Rights [24]

	SSTS	FL	FAPRO	PPCG	APROBC	TPW	PCPTW	STPEP	PHRFF	PRW	ASST	AFL
Observes	2	P	P		P	P	P					P
		EAFRD	CPR	ESCR	SR	ACHR	MAAE	PVIAC	PVNAC	EAFDAW	TCIDTP	RC
Observes		S	P	P	P			P	P	P	P	P

P = Party; S = Signatory; see Appendix for meaning of abbreviations.

Labor Force

Total Labor Force [25]

1.9 million (1987)

Labor Force by Occupation [26]

Agriculture	60%
Transp., commerce, pub. svcs.	38
Industry less than	2

Unemployment Rate [27]

For sources, notes, and explanations, see Annotated Source Appendix, page 1035.

95

Production Sectors

Energy Resource Summary [28]

Energy Resources: Small offshore oil deposits. **Electricity**: 30,000 kW capacity; 25 million kWh produced, 5 kWh per capita (1991).

Telecommunications [30]

- Fair system of open wire, submarine cable, and radio relay microwave
- Broadcast stations - 2 AM, 2 FM, 2 TV
- 1 Atlantic Ocean INTELSAT earth station

Top Agricultural Products [32]

Agriculture accounts for 35% of GDP; small farms produce 90% of agricultural output; production is dominated by food crops—corn, sorghum, cassava, beans, rice; cash crops include cotton, palm oil, peanuts; poultry and livestock output has not kept up with consumption.

Top Mining Products [33]

Estimated metric tons unless otherwise specified	M.t.
Cement, hydraulic	275,000
Steel, crude	8,000
Petroleum, crude (000 barrels)	1,353
Salt, marine	100

Transportation [31]

Railroads. 578 km, all 1.000-meter gauge, single track

Highways. 5,050 km total; 920 km paved, 2,600 laterite, 1,530 km improved earth

Airports

Total:	7
Usable:	5
With permanent-surface runways:	1
With runways over 3,659 m:	0
With runways 2,439-3,659 m:	1
With runways 1,220-2,439 m:	2

Tourism [34]

	1987	1988	1989	1990	1991
Visitors	156	157	189	247	401
Tourists	81	75	75	110	117
Excursionists	75	82	114	137	284
Tourism receipts	39	40	20	28	29
Tourism expenditures	12	13	10	12	10
Fare expenditures	12	13	11	11	11

Tourists are in thousands, money in million U.S. dollars.

Finance, Economics, and Trade

GDP and Manufacturing Summary [35]

	1980	1985	1990	1991	1992
Gross Domestic Product					
Millions of 1980 dollars	1,163	1,225	1,268	1,306	1,343[e]
Growth rate in percent	10.16	-2.47	3.30	3.03	2.80[e]
Manufacturing Value Added					
Millions of 1980 dollars	78	129	91	94	98[e]
Growth rate in percent	-3.47	1.34	5.83	3.26	4.99[e]
Manufacturing share in percent of current prices	12.9	8.2	9.2	9.1	NA

Economic Indicators [36]

National product: GDP—exchange rate conversion—$2 billion (1991). **National product real growth rate:** 3% (1991). **National product per capita:** $410 (1991). **Inflation rate (consumer prices):** 3.4% (1990). **External debt:** $1 billion (December 1990 est.).

Balance of Payments Summary [37]

Values in millions of dollars.

	1987	1988	1989	1990	1991
Exports of goods (f.o.b.)	363.4	379.1	178.4	264.8	291.4
Imports of goods (f.o.b.)	-483.8	-511.0	-316.6	-427.9	-482.4
Trade balance	-120.4	-131.9	-138.2	-163.1	-191.0
Services - debits	-228.9	-238.7	-149.8	-174.1	-175.5
Services - credits	111.8	118.9	87.8	114.6	122.6
Private transfers (net)	50.9	51.7	65.8	69.8	70.2
Government transfers (net)	112.8	117.2	121.3	86.7	84.4
Long term capital (net)	-1.0	75.2	321.9	105.4	190.7
Short term capital (net)	118.8	19.5	-202.5	43.0	-32.6
Errors and omissions	-41.6	-7.8	-112.9	-22.7	35.3
Overall balance	2.4	4.1	-6.6	59.6	104.1

Exchange Rates [38]

Currency: **Communaute Financiere Africaine francs.**
Symbol: **CFAF.**

Data are currency units per $1.

January 1993	274.06
1992	264.69
1991	282.11
1990	272.26
1989	319.01
1988	297.85

Imports and Exports

Top Import Origins [39]

$428 million (f.o.b., 1990 est.).

Origins	%
France	34
Netherlands	10
Japan	7
Italy	6
US	4

Top Export Destinations [40]

$263.3 million (f.o.b., 1990 est.).

Destinations	%
FRG	36
France	16
Spain	14
Italy	8
UK	4

Foreign Aid [41]

	U.S. $	
US commitments, including Ex-Im (FY70-89)	46	million
Western (non-US) countries, ODA and OOF bilateral commitments (1970-89)	1,300	million
OPEC bilateral aid (1979-89)	19	million
Communist countries (1970-89)	101	million

Import and Export Commodities [42]

Import Commodities	Export Commodities
Foodstuffs	Crude oil
Beverages	Cotton
Tobacco	Palm products
Petroleum products	Cocoa
Parts	
Capital goods	
Light consumer goods	

For sources, notes, and explanations, see Annotated Source Appendix, page 1035.

97

Bhutan

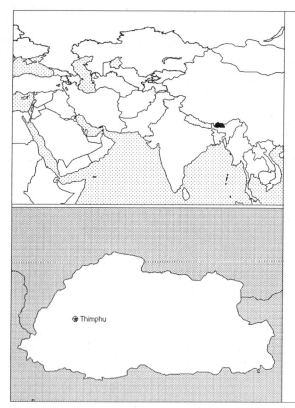

Geography [1]

Total area:
47,000 km2
Land area:
47,000 km2
Comparative area:
Slightly more than half the size of Indiana
Land boundaries:
Total 1,075 km, China 470 km, India 605 km
Coastline:
0 km (landlocked)
Climate:
Varies; tropical in southern plains; cool winters and hot summers in central valleys; severe winters and cool summers in Himalayas
Terrain:
Mostly mountainous with some fertile valleys and savanna
Natural resources:
Timber, hydropower, gypsum, calcium carbide, tourism potential
Land use:
Arable land:
2%
Permanent crops:
0%
Meadows and pastures:
5%
Forest and woodland:
70%
Other:
23%

Demographics [2]

	1960	1970	1980	1990	1991[1]	1994	2000	2010	2020
Population	867	1,045	1,281	1,585	1,598	1,739	1,996	2,474	3,035
Population density (persons per sq. mi.)	48	58	71	86	88	NA	105	125	147
Births	NA	NA	NA	NA	59	68	NA	NA	NA
Deaths	NA	NA	NA	NA	26	28	NA	NA	NA
Life expectancy - males	NA	NA	NA	NA	50	51	NA	NA	NA
Life expectancy - females	NA	NA	NA	NA	48	50	NA	NA	NA
Birth rate (per 1,000)	NA	NA	NA	NA	37	39	NA	NA	NA
Death rate (per 1,000)	NA	NA	NA	NA	17	16	NA	NA	NA
Women of reproductive age (15-44 yrs.)	NA	NA	NA	NA	NA	NA	NA	NA	NA
of which are currently married	NA	NA	NA	NA	NA	NA	NA	NA	NA
Fertility rate	NA	NA	NA	5.5	4.93	5.4	5.1	4.5	3.8

Population values are in thousands, life expectancy in years, and other items as indicated.

Health

Health Personnel [3]

Health Expenditures [4]

For sources, notes, and explanations, see Annotated Source Appendix, page 1035.

Human Factors

Health Care Ratios [5]

Population per physician, 1970	NA
Population per physician, 1990	13110
Population per nursing person, 1970	NA
Population per nursing person, 1990	NA
Percent of births attended by health staff, 1985	3

Infants and Malnutrition [6]

Percent of babies with low birth weight, 1985	NA
Infant mortality rate per 1,000 live births, 1970	182
Infant mortality rate per 1,000 live births, 1991	132
Years of life lost per 1,000 population, 1990	NA
Prevalence of malnutrition (under age 5), 1990	NA

Ethnic Division [7]

Bhote	50%
Ethnic Nepalese	35%
Indigenous/migrant tribes	15%

Religion [8]

Lamaistic Buddhism	75%
Indian- and Nepalese-influenced Hinduism	25%

Major Languages [9]

Dzongkha (official), Bhotes speak various Tibetan dialects; Nepalese speak various Nepalese dialects.

Education

Public Education Expenditures [10]

Million Ngultrum	1985	1986	1987	1988	1989	1990
Total education expenditure	NA	NA	122	NA	NA	NA
as percent of GNP	NA	NA	3.7	NA	NA	NA
as percent of total govt. expend.	NA	NA	NA	NA	NA	NA
Current education expenditure	NA	NA	NA	NA	NA	NA
as percent of GNP	NA	NA	NA	NA	NA	NA
as percent of current govt. expend.	NA	NA	NA	NA	NA	NA
Capital expenditure	NA	NA	NA	NA	NA	NA

Educational Attainment [11]

Literacy Rate [12]

In thousands and percent	1985[a]	1991[a]	2000[a]
Illiterate population +15 years	553	564	565
Illiteracy rate - total pop. (%)	67.8	61.6	49.4
Illiteracy rate - males (%)	55.2	48.7	37.0
Illiteracy rate - females (%)	81.1	75.4	62.8

Libraries [13]

Daily Newspapers [14]

Cinema [15]

Science and Technology

Scientific/Technical Forces [16]

R&D Expenditures [17]

U.S. Patents Issued [18]

Government and Law

Organization of Government [19]

Long-form name:
Kingdom of Bhutan
Type:
monarchy; special treaty relationship with India
Independence:
8 August 1949 (from India)
Constitution:
no written constitution or bill of rights
Legal system:
based on Indian law and English common law; has not accepted compulsory ICJ jurisdiction
National holiday:
National Day, 17 December (1907) (Ugyen Wangchuck became first hereditary king)
Executive branch:
monarch, chairman of the Royal Advisory Council, Royal Advisory Council (Lodoi Tsokde), chairman of the Council of Ministers, Council of Ministers (Lhengye Shungtsog)
Legislative branch:
unicameral National Assembly (Tshogdu)
Judicial branch:
High Court

Elections [20]

No legal parties; no national elections.

Government Expenditures [21]

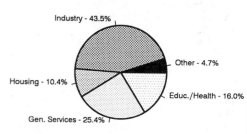

Industry - 43.5%
Other - 4.7%
Housing - 10.4%
Educ./Health - 16.0%
Gen. Services - 25.4%

% distribution for 1991. Expend. in FY91 est.: 121 ($ mil.)

Defense Summary [22]

Branches: Royal Bhutan Army, Palace Guard, Militia

Manpower Availability: Males age 15-49 415,315; fit for military service 222,027; reach military age (18) annually 17,344 (1993 est.)

Defense Expenditures: No information available.

Crime [23]

Human Rights [24]

	SSTS	FL	FAPRO	PPCG	APROBC	TPW	PCPTW	STPEP	PHRFF	PRW	ASST	AFL
Observes												

	EAFRD	CPR	ESCR	SR	ACHR	MAAE	PVIAC	PVNAC	EAFDAW	TCIDTP	RC
Observes	S								P		P

P = Party; S = Signatory; see Appendix for meaning of abbreviations.

Labor Force

Total Labor Force [25]

Labor Force by Occupation [26]

Agriculture	93%
Services	5
Industry and commerce	2

Unemployment Rate [27]

For sources, notes, and explanations, see Annotated Source Appendix, page 1035.

Production Sectors

Energy Resource Summary [28]

Energy Resources: Hydropower. **Electricity**: 336,000 kW capacity; 1,542.2 million kWh produced, 2,203 kWh per capita (25.8% is exported to India, leaving only 1,633 kWh per capita) (1990-91).

Telecommunications [30]

- Domestic telephone service is very poor with very few telephones in use
- International telephone and telegraph service is by land line through India
- A satellite earth station was planned (1990)
- Broadcast stations - 1 AM, 1 FM, no TV (1990)

Top Agricultural Products [32]

Agriculture accounts for 45% of GDP; based on subsistence farming and animal husbandry; self-sufficient in food except for foodgrains; other production—rice, corn, root crops, citrus fruit, dairy products, eggs.

Top Mining Products [33]

Estimated metric tons unless otherwise specified   M.t.[3]

Limestone	100,000
Cement	75,000
Dolomite	50,000
Gypsum	10,000

Transportation [31]

Highways. 2,165 km total; 1,703 km surfaced

Airports

Total:	2
Usable:	2
With permanent-surface runways:	1
With runways over 3,659 m:	0
With runways 2,440-3,659 m:	0
With runways 1,220-2,439 m:	2

Tourism [34]

	1987	1988	1989	1990	1991
Tourists[9]	3	2	2	2	2
Tourism receipts	1	1	2[9]	2	

Tourists are in thousands, money in million U.S. dollars.

Finance, Economics, and Trade

GDP and Manufacturing Summary [35]

	1980	1985	1990	1991	1992
Gross Domestic Product					
Millions of 1980 dollars	142	196	283	291	306
Growth rate in percent	17.63	3.69	4.56	2.96	5.00
Manufacturing Value Added					
Millions of 1980 dollars	5	10	20	21	22[e]
Growth rate in percent	-11.49	12.20	21.44	4.00	6.00[e]
Manufacturing share in percent of current prices	3.2	5.3	9.7	9.7	NA

Economic Indicators [36]

National product: GDP—exchange rate conversion—$320 million (1991 est.). **National product real growth rate**: 3.1% (1991 est.). **National product per capita**: $200 (1991 est.). **Inflation rate (consumer prices)**: 10% (FY91 est.). **External debt**: $120 million (June 91).

Balance of Payments Summary [37]

Exchange Rates [38]

Currency: **ngultrum.**
Symbol: **Nu.**

The Bhutanese ngultrum is at par with the Indian rupee. Data are currency units per $1.

January 1993	26.156
1992	25.918
1991	22.742
1990	17.504
1989	16.226
1988	13.917

Imports and Exports

Top Import Origins [39]

$106.4 million (c.i.f., FY91 est.).

Origins	%
India	83

Top Export Destinations [40]

$74 million (f.o.b., FY91 est.).

Destinations	%
India	90

Foreign Aid [41]

	U.S. $	
Western (non-US) countries, ODA and OOF bilateral commitments (1970-89)	115	million
OPEC bilateral aid (1979-89)	11	million

Import and Export Commodities [42]

Import Commodities	**Export Commodities**
Fuel and lubricants	Cardamon
Grain	Gypsum
Machinery & parts	Timber
Vehicles	Handicrafts
Fabrics	Cement
	Fruit
	Electricity (to India)

Bolivia

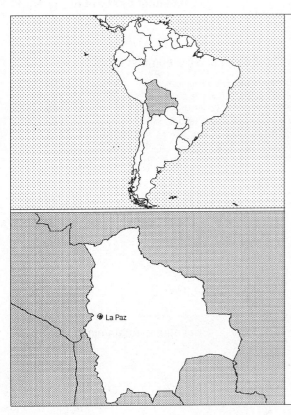

Geography [1]

Total area:
1,098,580 km2
Land area:
1,084,390 km2
Comparative area:
Slightly less than three times the size of Montana
Land boundaries:
Total 6,743 km, Argentina 832 km, Brazil 3,400 km, Chile 861 km, Paraguay 750 km, Peru 900 km
Coastline:
0 km (landlocked)
Climate:
Varies with altitude; humid and tropical to cold and semiarid
Terrain:
Rugged Andes Mountains with a highland plateau (Altiplano), hills, lowland plains of the Amazon basin
Natural resources:
Tin, natural gas, petroleum, zinc, tungsten, antimony, silver, iron ore, lead, gold, timber
Land use:
Arable land: 3%
Permanent crops: 0%
Meadows and pastures: 25%
Forest and woodland: 52%
Other: 20%

Demographics [2]

	1960	1970	1980	1990	1991[1]	1994	2000	2010	2020
Population	3,402	4,272	5,470	7,029	7,157	7,719	8,801	10,671	12,547
Population density (persons per sq. mi.)	8	10	13	17	17	NA	21	25	30
Births	NA	NA	NA	NA	243	249	NA	NA	NA
Deaths	NA	NA	NA	NA	66	65	NA	NA	NA
Life expectancy - males	NA	NA	NA	NA	59	61	NA	NA	NA
Life expectancy - females	NA	NA	NA	NA	64	66	NA	NA	NA
Birth rate (per 1,000)	NA	NA	NA	NA	34	32	NA	NA	NA
Death rate (per 1,000)	NA	NA	NA	NA	9	8	NA	NA	NA
Women of reproductive age (15-44 yrs.)	NA	NA	NA	1,695	NA	1,896	2,219	NA	NA
of which are currently married	NA	NA	NA	993	NA	1,118	1,322	NA	NA
Fertility rate	NA	NA	NA	4.6	4.62	4.2	3.6	2.9	2.5

Population values are in thousands, life expectancy in years, and other items as indicated.

Health

Health Personnel [3]

Doctors per 1,000 pop., 1988-92	0.48
Nurse-to-doctor ratio, 1988-92	0.7
Hospital beds per 1,000 pop., 1985-90	1.3
Percentage of children immunized (age 1 yr. or less)	
Third dose of DPT, 1990-91	58.0
Measles, 1990-91	73.0

Health Expenditures [4]

Total health expenditure, 1990 (official exchange rate)	
Millions of dollars	181
Millions of dollars per capita	25
Health expenditures as a percentage of GDP	
Total	4.0
Public sector	2.4
Private sector	1.6
Development assistance for health	
Total aid flows (millions of dollars)[1]	37
Aid flows per capita (millions of dollars)	5.1
Aid flows as a percentage of total health expenditure	20.3

For sources, notes, and explanations, see Annotated Source Appendix, page 1035.

103

Human Factors

Health Care Ratios [5]

Population per physician, 1970	2020
Population per physician, 1990	NA
Population per nursing person, 1970	3070
Population per nursing person, 1990	NA
Percent of births attended by health staff, 1985	36

Infants and Malnutrition [6]

Percent of babies with low birth weight, 1985	15
Infant mortality rate per 1,000 live births, 1970	153
Infant mortality rate per 1,000 live births, 1991	83
Years of life lost per 1,000 population, 1990	59
Prevalence of malnutrition (under age 5), 1990	18

Ethnic Division [7]

Quechua	30%
Aymara	25%
Mixed	25-30%
European	5-15%

Religion [8]

Roman Catholic 95%, Protestant (Evangelical Methodist).

Major Languages [9]

Spanish (official), Quechua (official), Aymara (official).

Education

Public Education Expenditures [10]

Million Bolivianos	1980	1985	1987	1988	1989[11]	1990[1,11]
Total education expenditure	0.0051	NA	NA	298	269	363
as percent of GNP	4.4	NA	NA	3.0	2.3	2.7
as percent of total govt. expend.	25.3	NA	NA	20.1	NA	NA
Current education expenditure	0.0049	NA	NA	NA	268	NA
as percent of GNP	4.2	NA	NA	NA	2.3	NA
as percent of current govt. expend.	27.0	NA	NA	NA	NA	NA
Capital expenditure	0.0002	NA	NA	NA	1	NA

Educational Attainment [11]

Literacy Rate [12]

In thousands and percent	1985[a]	1991[a]	2000[a]
Illiterate population +15 years	985	923	780
Illiteracy rate - total pop. (%)	27.5	22.5	14.2
Illiteracy rate - males (%)	19.1	15.3	9.4
Illiteracy rate - females (%)	35.5	29.3	18.8

Libraries [13]

Daily Newspapers [14]

	1975	1980	1985	1990
Number of papers	14	14	14	17
Circ. (000)	199	226	290[e]	400[e]

Cinema [15]

Data for 1989.

Cinema seats per 1,000	NA
Annual attendance per person	0.7
Gross box office receipts (mil. Boliviano)	9,905

Science and Technology

Scientific/Technical Forces [16]

Potential scientists/engineers[3]	64,300
Number female	13,650
Potential technicians	3,000
Number female	1,000
Total	67,300

R&D Expenditures [17]

U.S. Patents Issued [18]

Government and Law

Organization of Government [19]

Long-form name:
Republic of Bolivia
Type:
republic
Independence:
6 August 1825 (from Spain)
Constitution:
2 February 1967
Legal system:
based on Spanish law and Code Napoleon; has not accepted compulsory ICJ jurisdiction
National holiday:
Independence Day, 6 August (1825)
Executive branch:
president, vice president, Cabinet
Legislative branch:
bicameral National Congress (Congreso Nacional) consists of an upper chamber or Chamber of Senators (Camara de Senadores) and a lower chamber or Chamber of Deputies (Camara de Diputados)
Judicial branch:
Supreme Court (Corte Suprema)

Political Parties [20]

Chamber of Deputies	% of seats
Nationalist Revolutionary Movement (MNR)	30.8
Nationalist Democratic Action (ADN)	26.9
Movement of the Revolutionary Left (MIR)	25.4
United Left (IU)	7.7
Conscience of the Fatherland (CONDEPA)	6.9
Christian Democratic Party (PDC)	2.3

Government Expenditures [21]

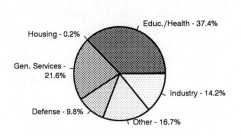

Educ./Health - 37.4%
Housing - 0.2%
Gen. Services - 21.6%
Defense - 9.8%
Industry - 14.2%
Other - 16.7%

% distribution for 1992. Expend. in 1993 est.: 1.570 ($ bil.)

Crime [23]

Military Expenditures and Arms Transfers [22]

	1985	1986	1987	1988	1989
Military expenditures					
Current dollars (mil.)	NA	124[e]	141[e]	157[e]	182[e]
1989 constant dollars (mil.)	NA	138[e]	152[e]	163[e]	182[e]
Armed forces (000)	28	30[e]	30[e]	28	30
Gross national product (GNP)					
Current dollars (mil.)	3,442	3,456	3,658	3,929	4,226
1989 constant dollars (mil.)	3,918	3,835	3,934	4,090	4,226
Central government expenditures (CGE)					
1989 constant dollars (mil.)	NA	611	573	653	717
People (mil.)	6.2	6.3	6.5	6.7	6.8
Military expenditure as % of GNP	NA	3.6	3.9	4.0	4.3
Military expenditure as % of CGE	NA	22.6	26.6	25.0	25.3
Military expenditure per capita	NA	22	23	25	27
Armed forces per 1,000 people	4.5	4.7	4.6	4.2	4.4
GNP per capita	635	606	606	614	619
Arms imports[6]					
Current dollars (mil.)	5	10	0	10	10
1989 constant dollars (mil.)	6	11	0	10	10
Arms exports[6]					
Current dollars (mil.)	0	0	0	0	0
1989 constant dollars (mil.)	0	0	0	0	0
Total imports[7]					
Current dollars (mil.)	552	716	776	604	786
1989 constant dollars	628	795	835	629	786
Total exports[7]					
Current dollars (mil.)	623	564	566	601	634
1989 constant dollars	709	626	609	626	634
Arms as percent of total imports[8]	0.9	1.4	0	1.7	1.3
Arms as percent of total exports[8]	0	0	0	0	0

Human Rights [24]

	SSTS	FL	FAPRO	PPCG	APROBC	TPW	PCPTW	STPEP	PHRFF	PRW	ASST	AFL
Observes	P		P	S	P	P	P	P		P	P	P
	EAFRD	CPR	ESCR	SR	ACHR	MAAE	PVIAC	PVNAC	EAFDAW	TCIDTP	RC	
Observes	P	P	P	S	P		P	P	P	S	P	

P = Party; S = Signatory; see Appendix for meaning of abbreviations.

Labor Force

Total Labor Force [25]

1.7 million

Labor Force by Occupation [26]

Agriculture	50%
Services and utilities	26
Manufacturing	10
Mining	4
Other	10

Date of data: 1983

Unemployment Rate [27]

5% (1992)

For sources, notes, and explanations, see Annotated Source Appendix, page 1035.

Production Sectors

Energy Resource Summary [28]

Energy Resources: Natural gas, petroleum. **Electricity**: 865,000 kW capacity; 1,834 million kWh produced, 250 kWh per capita (1992).
Pipelines: Crude oil 1,800 km; petroleum products 580 km; natural gas 1,495 km.

Telecommunications [30]

- Microwave radio relay system being expanded
- Improved international services
- 144,300 telephones
- Broadcast stations - 129 AM, no FM, 43 TV, 68 shortwave
- 1 Atlantic Ocean INTELSAT earth station

Transportation [31]

Railroads. 3,684 km total, all narrow gauge; 3,652 km 1.000-meter gauge and 32 km 0.760-meter gauge, all government owned, single track

Highways. 38,836 km total; 1,300 km paved, 6,700 km gravel, 30,836 km improved and unimproved earth

Merchant Marine. 2 cargo ships (1,000 GRT or over) totaling 14,051 GRT/22,155 DWT

Airports

Total:	1,225
Usable:	1,043
With permanent-surface runways:	9
With runways over 3,659 m:	2
With runways 2,440-3,659 m:	7
With runways 1,220-2,439 m:	161

Top Agricultural Products [32]

	1990	1991
Sugar cane	2,883.7	3,705.3
Potatoes	582.3	682.4
Bananas and plantains	435.1	460.9
Yucca	323.6	364.6
Corn	284.6	328.6
Rice	217.2	228.6

Values shown are 1,000 metric tons.

Top Mining Products [33]

Preliminary metric tons unless otherwise specified M.t.

Antimony, mine output, Sb content	7,287
Gold, mine output, Au content (kilograms)	3,500
Lead, mine output, Pb content	20,810
Natural gas, gross (mil. cubic meters)	5,432
Silver, mine output, Ag content (kilograms)	375,702[10]
Tin, mine output, Sn content	16,830
Tungsten, mine output, W content	1,065
Zinc, mine output, Zn content	129,778

Tourism [34]

	1987	1988	1989	1990	1991
Tourists[10]	147	167	194	217	221
Tourism receipts	40	67	75	85	90
Tourism expenditures	56	53	59	60	63
Fare receipts	18	20	21	21	23
Fare expenditures	21	22	22	22	24

Tourists are in thousands, money in million U.S. dollars.

Manufacturing Sector

GDP and Manufacturing Summary [35]

	1980	1985	1989	1990	% change 1980-1990	% change 1989-1990
GDP (million 1980 $)	5,018	4,556	4,709	4,952	-1.3	5.2
GDP per capita (1980 $)	901	715	662	677	-24.9	2.3
Manufacturing as % of GDP (current prices)	15.2	11.8	12.9	13.3[e]	-12.5	3.1
Gross output (million $)	2,465	1,956[e]	1,840[e]	1,953[e]	-20.8	6.1
Value added (million $)	834	818	883[e]	886[e]	6.2	0.3
Value added (million 1980 $)	734	551	510	664	-9.5	30.2
Industrial production index	100	69	65	79	-21.0	21.5
Employment (thousands)	102	137[e]	161[e]	160[e]	56.9	-0.6

Note: GDP stands for Gross Domestic Product. 'e' stands for estimated value.

Profitability and Productivity

	1980	1985	1989	1990	% change 1980-1990	% change 1989-1990
Intermediate input (%)	66	58[e]	52[e]	55[e]	-16.7	5.8
Wages, salaries, and supplements (%)	10	11[e]	12[e]	12[e]	20.0	0.0
Gross operating surplus (%)	24	31[e]	36[e]	33[e]	37.5	-8.3
Gross output per worker ($)	24,222	14,325[e]	11,434[e]	12,178[e]	-49.7	6.5
Value added per worker ($)	8,200	28,350	5,487[e]	31,454[e]	283.6	473.2
Average wage (incl. benefits) ($)	2,438	1,555[e]	1,408[e]	1,506[e]	-38.2	7.0

Profitability is in percent of gross output. Productivity is in U.S. $. 'e' stands for estimated value.

Profitability - 1990

Inputs - 55.0%
Wages - 12.0%
Surplus - 33.0%

The graphic shows percent of gross output.

Value Added in Manufacturing

	1980 $ mil.	1980 %	1985 $ mil.	1985 %	1989 $ mil.	1989 %	1990 $ mil.	1990 %	% change 1980-1990	% change 1989-1990
311 Food products	243	29.1	261	31.9	270[e]	30.6	273[e]	30.8	12.3	1.1
313 Beverages	62	7.4	37[e]	4.5	58[e]	6.6	57[e]	6.4	-8.1	-1.7
314 Tobacco products	21	2.5	4	0.5	5[e]	0.6	4[e]	0.5	-81.0	-20.0
321 Textiles	37	4.4	35	4.3	13[e]	1.5	12[e]	1.4	-67.6	-7.7
322 Wearing apparel	47	5.6	30	3.7	26[e]	2.9	28[e]	3.2	-40.4	7.7
323 Leather and fur products	5	0.6	4	0.5	4[e]	0.5	4[e]	0.5	-20.0	0.0
324 Footwear	24	2.9	20	2.4	18[e]	2.0	19[e]	2.1	-20.8	5.6
331 Wood and wood products	24	2.9	21	2.6	22[e]	2.5	22[e]	2.5	-8.3	0.0
332 Furniture and fixtures	21	2.5	18	2.2	18[e]	2.0	19[e]	2.1	-9.5	5.6
341 Paper and paper products	1	0.1	1	0.1	2[e]	0.2	2[e]	0.2	100.0	0.0
342 Printing and publishing	15	1.8	15	1.8	14[e]	1.6	17[e]	1.9	13.3	21.4
351 Industrial chemicals	3	0.4	7	0.9	7[e]	0.8	7[e]	0.8	133.3	0.0
352 Other chemical products	31	3.7	45	5.5	44[e]	5.0	46[e]	5.2	48.4	4.5
353 Petroleum refineries	159	19.1	152	18.6	233[e]	26.4	222[e]	25.1	39.6	-4.7
354 Miscellaneous petroleum and coal products	NA	0.0	NA	0.0	NA	0.0	NA	0.0	NA	NA
355 Rubber products	2	0.2	3	0.4	3[e]	0.3	4[e]	0.5	100.0	33.3
356 Plastic products	12	1.4	9	1.1	7[e]	0.8	8[e]	0.9	-33.3	14.3
361 Pottery, china and earthenware	4	0.5	3	0.4	3[e]	0.3	3[e]	0.3	-25.0	0.0
362 Glass and glass products	11	1.3	9	1.1	9[e]	1.0	10[e]	1.1	-9.1	11.1
369 Other non-metal mineral products	25	3.0	35	4.3	18[e]	2.0	20[e]	2.3	-20.0	11.1
371 Iron and steel	12	1.4	9	1.1	9[e]	1.0	9[e]	1.0	-25.0	0.0
372 Non-ferrous metals	37	4.4	68	8.3	69[e]	7.8	71[e]	8.0	91.9	2.9
381 Metal products	14	1.7	11	1.3	11[e]	1.2	10[e]	1.1	-28.6	-9.1
382 Non-electrical machinery	6	0.7	6	0.7	6[e]	0.7	5[e]	0.6	-16.7	-16.7
383 Electrical machinery	3	0.4	3	0.4	3[e]	0.3	3[e]	0.3	0.0	0.0
384 Transport equipment	5	0.6	5	0.6	4[e]	0.5	3[e]	0.3	-40.0	-25.0
385 Professional and scientific equipment	1	0.1	1	0.1	1[e]	0.1	1[e]	0.1	0.0	0.0
390 Other manufacturing industries	9	1.1	5	0.6	5[e]	0.6	5[e]	0.6	-44.4	0.0

Note: The industry codes shown are International Standard Industry codes (ISIC). Percentages are percent of total Value Added. 'e' stands for estimated value

For sources, notes, and explanations, see Annotated Source Appendix, page 1035.

107

Finance, Economics, and Trade

Economic Indicators [36]

Millions of U.S. Dollars Except Where indicated.

	1989	1990	1991[9]
Real GDP percent change	2.7	2.7	4.0
Real GDP per capita			
Nominal GDP	5,480.8	5,581.8	5,978.8
Money supply (M1)			
(mil. Bolivianos)	706.0	988.0	1,231.9
Inflation (12 months)	16.5	18.0	15.0
Total foreign debt[10]	3,491.6	3,768.5	3,366.1[11]

Balance of Payments Summary [37]

Values in millions of dollars.

	1987	1988	1989	1990	1991
Exports of goods (f.o.b.)	518.7	542.5	723.5	830.8	760.3
Imports of goods (f.o.b.)	-646.3	-590.9	-729.5	-775.6	-804.2
Trade balance	-127.6	-48.4	-6.0	55.2	-43.9
Services - debits	-564.1	-537.8	-581.2	-578.0	-582.8
Services - credits	147.7	146.5	167.2	164.7	181.6
Private transfers (net)	18.0	12.7	20.6	21.6	23.0
Government transfers (net)	103.3	171.6	135.7	145.0	160.0
Long term capital (net)	23.2	266.2	228.0	319.0	320.4
Short term capital (net)	222.3	-128.9	-193.1	-62.0	-36.3
Errors and omissions	174.6	46.6	-32.1	-11.4	53.3
Overall balance	-2.6	-71.5	-260.9	54.1	75.3

Exchange Rates [38]

Currency: **bolivianos.**
Symbol: **$B.**

Data are currency units per $1.

August 1992	3.9437
1992	3.8500
1991	3.5806
1990	3.1727
1989	2.6917
1988	2.3502
1987	2.0549

Imports and Exports

Top Import Origins [39]

$1.185 billion (c.i.f., 1992).

Origins	%
US	22

Top Export Destinations [40]

$609 million (f.o.b., 1992).

Destinations	%
US	15
Argentina	NA

Foreign Aid [41]

	U.S. $	
US commitments, including Ex-Im (FY70-89)	990	million
Western (non-US) countries, ODA and OOF bilateral commitments (1970-89)	2,025	million
Communist countries (1970-89)	340	million

Import and Export Commodities [42]

Import Commodities

Food
Petroleum
Consumer goods
Capital goods

Export Commodities

Metals 46%
Hydrocarbons 21%
Other 33%
(coffee,
soybeans
sugar
cotton
timber)

Bosnia and Herzegovina

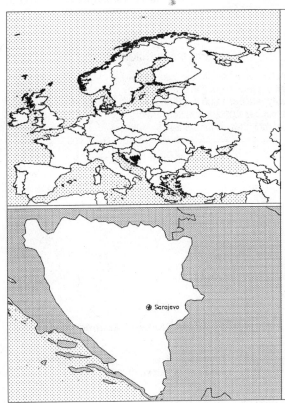

Geography [1]

Total area:
51,233 km2
Land area:
51,233 km2
Comparative area:
Slightly larger than Tennessee
Land boundaries:
Total 1,369 km, Croatia (northwest) 751 km, Croatia (south) 91 km, Serbia and Montenegro 527 km (312 km with Serbia; 215 km with Montenegro)
Coastline:
20 km
Climate:
Hot summers and cold winters; areas of high elevation have short, cool summers and long, severe winters; mild, rainy winters along coast
Terrain:
Mountains and valleys
Natural resources:
Coal, iron, bauxite, manganese, timber, wood products, copper, chromium, lead, zinc
Land use:
Arable land: 20%
Permanent crops: 2%
Meadows and pastures: 25%
Forest and woodland: 36%
Other: 17%

Demographics [2]

	1960	1970	1980	1990	1991[1]	1994	2000	2010	2020
Population	3,240	3,703	4,092	4,517	NA	4,651	4,828	5,039	5,117
Population density (persons per sq. mi.)	NA	NA	NA	NA	NA	NA	NA	NA	NA
Births	NA	NA	NA	NA	NA	62	NA	NA	NA
Deaths	NA	NA	NA	NA	NA	30	NA	NA	NA
Life expectancy - males	NA	NA	NA	NA	NA	72	NA	NA	NA
Life expectancy - females	NA	NA	NA	NA	NA	78	NA	NA	NA
Birth rate (per 1,000)	NA	NA	NA	NA	NA	13	NA	NA	NA
Death rate (per 1,000)	NA	NA	NA	NA	NA	6	NA	NA	NA
Women of reproductive age (15-44 yrs.)	NA	NA	NA	1,195	NA	1,241	1,296	NA	NA
of which are currently married	NA	NA	NA	818	NA	865	918	NA	NA
Fertility rate	NA	NA	NA	1.6	NA	1.6	1.6	1.6	1.6

Population values are in thousands, life expectancy in years, and other items as indicated.

Health

Health Personnel [3]

Doctors per 1,000 pop., 1988-92	2.63 [2]
Nurse-to-doctor ratio, 1988-92	1.9 [2]
Hospital beds per 1,000 pop., 1985-90	6.0 [2]
Percentage of children immunized (age 1 yr. or less)	
Third dose of DPT, 1990-91	79.0 [2]
Measles, 1990-91	75.0 [2]

Health Expenditures [4]

Total health expenditure, 1990 (official exchange rate)	
Millions of dollars	4512
Millions of dollars per capita	205
Health expenditures as a percentage of GDP	
Total	3.0
Public sector	4.0
Private sector	1.0
Development assistance for health	
Total aid flows (millions of dollars)[1]	NA
Aid flows per capita (millions of dollars)	NA
Aid flows as a percentage of total health expenditure	NA

For sources, notes, and explanations, see Annotated Source Appendix, page 1035.

109

Human Factors

Health Care Ratios [5]

Population per physician, 1970	1000 [2]
Population per physician, 1990	530 [2]
Population per nursing person, 1970	420 [2]
Population per nursing person, 1990	110 [2]
Percent of births attended by health staff, 1985	NA

Infants and Malnutrition [6]

Percent of babies with low birth weight, 1985	7
Infant mortality rate per 1,000 live births, 1970	56
Infant mortality rate per 1,000 live births, 1991	21
Years of life lost per 1,000 population, 1990	16
Prevalence of malnutrition (under age 5), 1990	NA

Ethnic Division [7]

Muslim	44%
Serb	31%
Croat	17%
Other	8%

Religion [8]

Muslim	40%
Orthodox	31%
Catholic	15%
Protestant	4%
Other	10%

Major Languages [9]

Serbo-Croatian	99%

Education

Public Education Expenditures [10]

Million Dinar	1980[22]	1985[22]	1987[22]	1988[22]	1989[22]	1990[22]
Total education expenditure	8	42	202	544	9,555	60,318
as percent of GNP	4.7	3.4	4.2	3.6	4.3	6.1
as percent of total govt. expend.	32.5	NA	NA	NA	NA	NA
Current education expenditure	7	39	185	506	9,002	55,911
as percent of GNP	4.0	3.1	3.9	3.3	4.1	5.7
as percent of current govt. expend.	NA	NA	NA	NA	NA	NA
Capital expenditure	1	3	17	38	553	4,407

Educational Attainment [11]

Age group	25+ [1]
Total population	13,083,762 [1]
Highest level attained (%)	
No schooling	15.8 [1]
First level	
Incompleted	53.9 [1]
Completed	NA [1]
Entered second level	
S-1	23.4 [1]
S-2	NA [1]
Post secondary	6.8 [1]

Literacy Rate [12]

In thousands and percent	1985[a,1]	1991[a,1]	2000[a,1]
Illiterate population + 15 years	1,614	1,342	942
Illiteracy rate - total pop. (%)	9.2	7.3	4.7
Illiteracy rate - males (%)	3.5	2.6	1.3
Illiteracy rate - females (%)	14.6	11.9	7.9

Libraries [13]

	Admin. Units[a]	Svc. Pts.[a]	Vols. (000)[a]	Shelving (meters)[a]	Vols. Added[a]	Reg. Users[a]
National (1989)	8	8	12,316	303,555	305,462	163,169
Non-specialized (1989)	20	20	3,488	60,443	72,503	49,262
Public (1989)	808	1,937	30,238	552,866	1.3mil	19.5mil
Higher ed. (1989)	409	421	14,462	319,329	469,142	529,549
School (1989)	7,784	7,784	38,430	NA	1,680	NA

Daily Newspapers [14]

	1975[b]	1980[b]	1985[b]	1990[b]
Number of papers	26	27	27	34
Circ. (000)	1,896	2,649	2,451	2,281

Cinema [15]

Data for 1989.

Cinema seats per 1,000	16.9[a]
Annual attendance per person	1.9[a]
Gross box office receipts (mil. Dinar)	643,490[a]

Science and Technology

Scientific/Technical Forces [16]

Potential scientists/engineers	563,312
Number female	NA
Potential technicians	432,380
Number female	NA
Total	995,692

R&D Expenditures [17]

	Dinar (000) 1989
Total expenditure[a,4,12]	2,152,032
Capital expenditure	815,082
Current expenditure	1,336,950
Percent current	62.1

U.S. Patents Issued [18]

Values show patents issued to citizens of the country by the U.S. Patents Office.

	1990	1991	1992
Number of patents	24	25	18

For sources, notes, and explanations, see Annotated Source Appendix, page 1035.

Government and Law

Organization of Government [19]

Long-form name:
Republic of Bosnia and Herzegovina
Type:
emerging democracy
Independence:
NA April 1992 (from Yugoslavia)
Constitution:
NA
Legal system:
based on civil law system
National holiday:
NA
Executive branch:
collective presidency, prime minister,
deputy prime ministers, cabinet
Legislative branch:
bicameral National Assembly consists of
an upper house or Chamber of
Municipalities (Vijece Opeina) and a lower
house or Chamber of Citizens (Vijece
Gradanstvo)
Judicial branch:
Supreme Court, Constitutional Court

Political Parties [20]

Chamber of Citizens	% of seats
Party of Democratic Action (SDA)	33.1
Serbian Democ. Party (SDS BiH)	26.2
Croatian Democ. Union (HDZ BiH)	16.2
Party of Democratic Changes	11.5
Alliance of Reform Forces (SRSJ BiH)	9.2
Muslim-Bosnian Organization (MBO)	1.5
Democratic Party of Socialists (DSS)	0.8
Democratic League of Greens (DSZ)	0.8
Liberal Party (LS)	0.8

Government Budget [21]

For 1993 est.

Revenues	NA
Expenditures	NA
Capital expenditures	NA

Defense Summary [22]

Branches: Army

Manpower Availability: Males age 15-49 1,283,576; fit for military service 1,045,512; reach military age (19) annually 37,827 (1993 est.)

Defense Expenditures: Information is not available.

Crime [23]

Human Rights [24]

Labor Force

Total Labor Force [25]

1,026,254

Labor Force by Occupation [26]

Agriculture	2%
Industry, mining	45

Date of data: 1991 est.

Unemployment Rate [27]

28% (February 1992 est.)

Production Sectors

Commercial Energy Production and Consumption

Production [28]

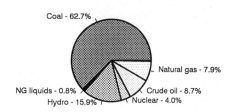

Coal - 62.7%
Natural gas - 7.9%
NG liquids - 0.8%
Crude oil - 8.7%
Hydro - 15.9%
Nuclear - 4.0%

Consumption [29]

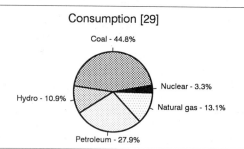

Coal - 44.8%
Nuclear - 3.3%
Hydro - 10.9%
Natural gas - 13.1%
Petroleum - 27.9%

Data are shown in quadrillion (10^{15}) BTUs and percent for 1991

Crude oil[1]	0.11[a]		Petroleum	0.51[a]
Natural gas liquids	0.01[a]		Dry natural gas	0.24[a]
Dry natural gas	0.10[a]		Coal[2]	0.82[a]
Coal[2]	0.79[a]		Net hydroelectric power[3]	0.20[a]
Net hydroelectric power[3]	0.20[a]		Net nuclear power[3]	0.06[a]
Net nuclear power[3]	0.05[a]		Total	1.83[a]
Total	1.26[a]			

Telecommunications [30]

- Telephone and telegraph network is in need of modernization and expansion, many urban areas being below average compared with services in other former Yugoslav republics
- 727,000 telephones
- Broadcast stations - 9 AM, 2 FM, 6 TV
- 840,000 radios
- 1,012,094 TVs
- NA submarine coaxial cables
- Satellite ground stations - none

Transportation [31]

Railroads. NA km

Highways. 21,168 km total (1991); 11,436 km paved, 8,146 km gravel, 1,586 km earth; note - highways now disrupted

Airports

Total:	27
Useable:	22
With permanent-surface runways:	8
With runways over 3659:	0
With runways 2440-3659 m:	4
With runways 1220-2439 m:	5

Top Agricultural Products [32]

Accounted for 9.0% of GDP in 1989; regularly produces less than 50% of food needs; the foothills of northern Bosnia support orchards, vineyards, livestock, and some wheat and corn; long winters and heavy precipitation leach soil fertility reducing agricultural output in the mountains; farms are mostly privately held, small, and not very productive.

Top Mining Products [33]

Metric tons unless otherwise specified	M.t.
Magnesite, crude	175,000
Bauxite (000 tons)	2,850
Copper, Cu content of concentrate	140,000
Gypsum, crude	400,000
Pumice, volcanic tuff	380,000
Cement, hydraulic (000 tons)	7,100

Tourism [34]

	1987	1988	1989	1990	1991
Visitors[74]	26,151	29,635	34,118	39,573	
Tourists[6,74]	8,907	9,018	8,644	7,880	1,459
Tourism receipts[74]	1,668	2,024	2,230	2,774	468
Tourism expenditures[74]	90	109	131	149	103
Fare receipts[74]	340	430	475	620	105

Tourists are in thousands, money in million U.S. dollars.

Manufacturing Sector

GDP and Manufacturing Summary [35]

	1980	1985	1989	1990	% change 1980-1990	% change 1989-1990
GDP (million 1980 $)	69,958	71,058	72,234	66,371	-5.1	-8.1
GDP per capita (1980 $)	3,136	3,073	3,050	2,786	-11.2	-8.7
Manufacturing as % of GDP (current prices)	30.6	37.2	39.5	42.0	37.3	6.3
Gross output (million $)	72,629	57,020	65,078	62,136[e]	-14.4	-4.5
Value added (million $)	21,750	17,171	30,245	27,660[e]	27.2	-8.5
Value added (million 1980 $)	19,526	22,283	24,021	21,703	11.1	-9.6
Industrial production index	100	116	120	108	8.0	-10.0
Employment (thousands)	2,106	2,467	2,658	2,537[e]	20.5	-4.6

Note: GDP stands for Gross Domestic Product. 'e' stands for estimated value.

Profitability and Productivity

	1980	1985	1989	1990	% change 1980-1990	% change 1989-1990
Intermediate input (%)	70	70	54	55[e]	-21.4	1.9
Wages, salaries, and supplements (%)	14	12	12[e]	18[e]	28.6	50.0
Gross operating surplus (%)	15	18	34[e]	26[e]	73.3	-23.5
Gross output per worker ($)	34,487	23,113	24,484	24,248[e]	-29.7	-1.0
Value added per worker ($)	10,328	6,960	11,379	10,796[e]	4.5	-5.1
Average wage (incl. benefits) ($)	4,991	2,703	2,986[e]	4,488[e]	-10.1	50.3

Profitability is in percent of gross output. Productivity is in U.S. $. 'e' stands for estimated value.

Profitability - 1990

Inputs - 55.6%
Wages - 18.2%
Surplus - 26.3%

The graphic shows percent of gross output.

Value Added in Manufacturing

	1980 $ mil.	1980 %	1985 $ mil.	1985 %	1989 $ mil.	1989 %	1990 $ mil.	1990 %	% change 1980-1990	% change 1989-1990
311 Food products	1,897	8.7	1,458	8.5	3,916	12.9	3,484[e]	12.6	83.7	-11.0
313 Beverages	459	2.1	353	2.1	663	2.2	589[e]	2.1	28.3	-11.2
314 Tobacco products	184	0.8	221	1.3	344	1.1	308[e]	1.1	67.4	-10.5
321 Textiles	1,759	8.1	1,428	8.3	2,881	9.5	2,663[e]	9.6	51.4	-7.6
322 Wearing apparel	903	4.2	718	4.2	1,593	5.3	1,427[e]	5.2	58.0	-10.4
323 Leather and fur products	226	1.0	231	1.3	383	1.3	340[e]	1.2	50.4	-11.2
324 Footwear	482	2.2	503	2.9	1,022	3.4	899[e]	3.3	86.5	-12.0
331 Wood and wood products	977	4.5	530	3.1	794	2.6	706[e]	2.6	-27.7	-11.1
332 Furniture and fixtures	730	3.4	438	2.6	1,030	3.4	1,065[e]	3.9	45.9	3.4
341 Paper and paper products	529	2.4	394	2.3	759	2.5	674[e]	2.4	27.4	-11.2
342 Printing and publishing	876	4.0	462	2.7	761	2.5	678[e]	2.5	-22.6	-10.9
351 Industrial chemicals	694	3.2	631	3.7	1,107	3.7	992[e]	3.6	42.9	-10.4
352 Other chemical products	681	3.1	525	3.1	1,419	4.7	1,315[e]	4.8	93.1	-7.3
353 Petroleum refineries	454	2.1	415	2.4	260	0.9	233[e]	0.8	-48.7	-10.4
354 Miscellaneous petroleum and coal products	101	0.5	101	0.6	104	0.3	91[e]	0.3	-9.9	-12.5
355 Rubber products	276	1.3	269	1.6	479	1.6	456[e]	1.6	65.2	-4.8
356 Plastic products	413	1.9	258	1.5	397	1.3	350[e]	1.3	-15.3	-11.8
361 Pottery, china and earthenware	128	0.6	72	0.4	162	0.5	144[e]	0.5	12.5	-11.1
362 Glass and glass products	163	0.7	113	0.7	224	0.7	204[e]	0.7	25.2	-8.9
369 Other non-metal mineral products	906	4.2	513	3.0	683	2.3	604[e]	2.2	-33.3	-11.6
371 Iron and steel	1,221	5.6	1,000	5.8	1,343	4.4	1,171[e]	4.2	-4.1	-12.8
372 Non-ferrous metals	480	2.2	509	3.0	944	3.1	927[e]	3.4	93.1	-1.8
381 Metal products	2,105	9.7	1,577	9.2	1,293	4.3	1,130[e]	4.1	-46.3	-12.6
382 Non-electrical machinery	1,828	8.4	1,463	8.5	2,372	7.8	2,378[e]	8.6	30.1	0.3
383 Electrical machinery	1,600	7.4	1,544	9.0	2,640	8.7	2,334[e]	8.4	45.9	-11.6
384 Transport equipment	1,441	6.6	1,263	7.4	2,389	7.9	2,241[e]	8.1	55.5	-6.2
385 Professional and scientific equipment	101	0.5	93	0.5	154	0.5	146[e]	0.5	44.6	-5.2
390 Other manufacturing industries	134	0.6	88	0.5	128	0.4	114[e]	0.4	-14.9	-10.9

Note: The industry codes shown are International Standard Industry codes (ISIC). Percentages are percent of total Value Added. 'e' stands for estimated value

For sources, notes, and explanations, see Annotated Source Appendix, page 1035.

113

Finance, Economics, and Trade

Economic Indicators [36]

Yugoslav Dinars or U.S. Dollars as indicated[12].

	1989	1990	1991[e]
GSP[13]			
(billions current dinars)	221.9	910.9	1,490
Real GSP growth rate (%)	0.6	-7.6	-20
Money supply (M1)			
(yr-end mil.)	51,216	127,241	NA
Commercial interest			
rates (avg.)	4,354	45	80
Investment rate (%)	19.4	18.9	NA
CPI	1,356	688	310
WPI	1,406	534	320
External public debt	18,569	17,791	15,760

Balance of Payments Summary [37]

Exchange Rates [38]

Imports and Exports

Top Import Origins [39]

$1,891 million (1990).

Origins	%
Principally the other former Yugoslav republics	NA

Top Export Destinations [40]

$2,054 million (1990).

Destinations	%
Principally the other former Yugoslav republics	NA

Foreign Aid [41]

Import and Export Commodities [42]

Import Commodities

Fuels and lubricants 32%
Machinery, transp. equip. 23.3%
Other manufactures 21.3%
Chemicals 10%
Raw materials 6.7%
Food and live animals 5.5%
Beverages and tobacco 1.9%

Export Commodities

Manufactured goods 31%
Machinery, transp. equip. 20.8%
Raw materials 18%
Misc. articles 17.3%
Chemicals 9.4%
Fuel and lubricants 1.4%
Food and live animals 1.2%

Botswana

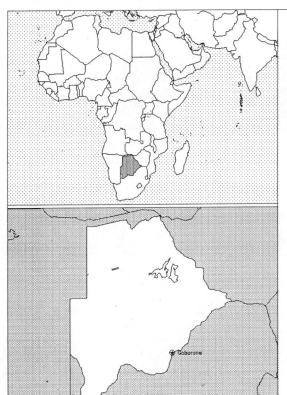

Geography [1]

Total area:
600,370 km2
Land area:
585,370 km2
Comparative area:
Slightly smaller than Texas
Land boundaries:
Total 4,013 km, Namibia 1,360 km, South Africa 1,840 km, Zimbabwe 813 km
Coastline:
0 km (landlocked)
Climate:
Semiarid; warm winters and hot summers
Terrain:
Predominately flat to gently rolling tableland; Kalahari Desert in southwest
Natural resources:
Diamonds, copper, nickel, salt, soda ash, potash, coal, iron ore, silver
Land use:
Arable land:
2%
Permanent crops:
0%
Meadows and pastures:
75%
Forest and woodland:
2%
Other:
21%

Demographics [2]

	1960	1970	1980	1990	1991[1]	1994	2000	2010	2020
Population	497	584	903	1,224	1,258	1,359	1,554	1,871	2,187
Population density (persons per sq. mi.)	2	3	4	5	6	NA	7	8	10
Births	NA	NA	NA	NA	45	44	NA	NA	NA
Deaths	NA	NA	NA	NA	11	10	NA	NA	NA
Life expectancy - males	NA	NA	NA	NA	59	60	NA	NA	NA
Life expectancy - females	NA	NA	NA	NA	65	66	NA	NA	NA
Birth rate (per 1,000)	NA	NA	NA	NA	36	32	NA	NA	NA
Death rate (per 1,000)	NA	NA	NA	NA	9	8	NA	NA	NA
Women of reproductive age (15-44 yrs.)	NA	NA	NA	294	NA	339	418	NA	NA
of which are currently married	NA	NA	NA	122	NA	142	177	NA	NA
Fertility rate	NA	NA	NA	4.8	4.64	4.1	3.1	2.3	2.1

Population values are in thousands, life expectancy in years, and other items as indicated.

Health

Health Personnel [3]

Health Expenditures [4]

For sources, notes, and explanations, see Annotated Source Appendix, page 1035.

115

Human Factors

Health Care Ratios [5]

Population per physician, 1970	15220
Population per physician, 1990	5150
Population per nursing person, 1970	1900
Population per nursing person, 1990	NA
Percent of births attended by health staff, 1985	52

Infants and Malnutrition [6]

Percent of babies with low birth weight, 1985	8
Infant mortality rate per 1,000 live births, 1970	101
Infant mortality rate per 1,000 live births, 1991	36
Years of life lost per 1,000 population, 1990	NA
Prevalence of malnutrition (under age 5), 1990	15

Ethnic Division [7]

Batswana	95%
Kalanga, Basarwa, Kgalagadi	4%
White	1%

Religion [8]

Indigenous beliefs	50%
Christian	50%

Major Languages [9]

English (official), Setswana.

Education

Public Education Expenditures [10]

Million Pula	1985	1986	1987	1988[1]	1989[1]	1990
Total education expenditure	111	NA	204	267	359	NA
as percent of GNP	6.9	NA	8.2	8.3	7.5	NA
as percent of total govt. expend.	15.4	NA	15.9	15.4	16.3	NA
Current education expenditure	88	NA	148	183	249	NA
as percent of GNP	5.5	NA	6.0	5.7	5.2	NA
as percent of current govt. expend.	18.6	NA	20.4	19.6	20.9	NA
Capital expenditure	23	NA	57	84	110	NA

Educational Attainment [11]

Age group	25+
Total population	310,303
Highest level attained (%)	
No schooling	54.7
First level	
Incompleted	31.1
Completed	9.4
Entered second level	
S-1	3.1
S-2	1.3
Post secondary	0.5

Literacy Rate [12]

In thousands and percent	1985[a]	1991[a]	2000[a]
Illiterate population +15 years	168	175	189
Illiteracy rate - total pop. (%)	30.0	26.4	19.9
Illiteracy rate - males (%)	18.5	16.3	12.4
Illiteracy rate - females (%)	39.6	34.9	26.5

Libraries [13]

	Admin. Units	Svc. Pts.	Vols. (000)	Shelving (meters)	Vols. Added	Reg. Users
National (1988)	NA	NA	NA	NA	NA	NA
Non-specialized	NA	NA	NA	NA	NA	NA
Public	NA	NA	NA	NA	NA	NA
Higher ed.[2]	8	8	9	NA	2,834	680
School	NA	NA	NA	NA	NA	NA

Daily Newspapers [14]

	1975	1980	1985	1990
Number of papers	1	1	1	1
Circ. (000)	14	19	18	18

Cinema [15]

Science and Technology

Scientific/Technical Forces [16]

R&D Expenditures [17]

U.S. Patents Issued [18]

For sources, notes, and explanations, see Annotated Source Appendix, page 1035.

Government and Law

Organization of Government [19]

Long-form name:
Republic of Botswana
Type:
parliamentary republic
Independence:
30 September 1966 (from UK)
Constitution:
March 1965, effective 30 September 1966
Legal system:
based on Roman-Dutch law and local customary law; judicial review limited to matters of interpretation; has not accepted compulsory ICJ jurisdiction
National holiday:
Independence Day, 30 September (1966)
Executive branch:
president, vice president, Cabinet
Legislative branch:
bicameral National Assembly consists of an upper house or House of Chiefs and a lower house or National Assembly
Judicial branch:
High Court, Court of Appeal

Political Parties [20]

National Assembly	% of votes
Botswana Democratic Party (BDP)	92.1
Botswana National Front (BNF)	7.9

Government Expenditures [21]

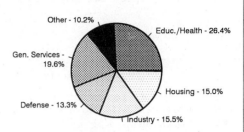

Other - 10.2%
Educ./Health - 26.4%
Gen. Services - 19.6%
Housing - 15.0%
Defense - 13.3%
Industry - 15.5%

% distribution for 1991. Expend. in FY94: 1.990 ($ bil.)

Military Expenditures and Arms Transfers [22]

	1985	1986	1987	1988	1989
Military expenditures					
Current dollars (mil.)	38	54	102	115	62[e]
1989 constant dollars (mil.)	43	60	110	119	62[e]
Armed forces (000)	3	4	4	4	6
Gross national product (GNP)					
Current dollars (mil.)	1,272	1,475	1,626	1,844	2,188
1989 constant dollars (mil.)	1,448	1,637	1,749	1,920	2,188
Central government expenditures (CGE)					
1989 constant dollars (mil.)	743	812	1,013	1,138	1,035[e]
People (mil.)	1.1	1.1	1.1	1.2	1.2
Military expenditure as % of GNP	3.0	3.7	6.3	6.2	2.8
Military expenditure as % of CGE	5.8	7.4	10.9	10.5	6.0
Military expenditure per capita	41	55	98	103	52
Armed forces per 1,000 people	2.8	3.7	3.6	3.5	5.0
GNP per capita	1,369	1,502	1,557	1,660	1,838
Arms imports[6]					
Current dollars (mil.)	5	0	0	30	10
1989 constant dollars (mil.)	6	0	0	31	10
Arms exports[6]					
Current dollars (mil.)	0	0	0	0	0
1989 constant dollars (mil.)	0	0	0	0	0
Total imports[7]					
Current dollars (mil.)	583	713	936	1,031	NA
1989 constant dollars	664	791	1,007	1,418	NA
Total exports[7]					
Current dollars (mil.)	744	865	1,587	1,418	1,445
1989 constant dollars	847	960	1,707	1,476	1,445
Arms as percent of total imports[8]	0.9	0	0	2.9	NA
Arms as percent of total exports[8]	0	0	0	0	0

Crime [23]

Crime volume (for 1989)	
Cases known to police	87,009
Attempts (percent)	NA
Percent cases solved	85.8
Crimes per 100,000 persons	6,693
Persons responsible for offenses	
Total number offenders	52,916
Percent female	9.9
Percent juvenile[9]	1.1
Percent foreigners	NA

Human Rights [24]

	SSTS	FL	FAPRO	PPCG	APROBC	TPW	PCPTW	STPEP	PHRFF	PRW	ASST	AFL
Observes	1					P	P				1	
	EAFRD	CPR	ESCR	SR	ACHR	MAAE	PVIAC	PVNAC	EAFDAW	TCIDTP	RC
Observes	P			P			P	P			1

P = Party; S = Signatory; see Appendix for meaning of abbreviations.

Labor Force

Total Labor Force [25]

400,000

Labor Force by Occupation [26]

198,500 formal sector employees
14,600 in S. Africa mines
Date of data: 1990

Unemployment Rate [27]

25% (1989)

Production Sectors

Energy Resource Summary [28]

Energy Resources: Coal. **Electricity**: 220,000 kW capacity; 1,123 million kWh produced, 846 kWh per capita (1991).

Telecommunications [30]

- The small system is a combination of open-wire lines, microwave radio relay links, and a few radio-communications stations
- 26,000 telephones
- Broadcast stations - 7 AM, 13 FM, no TV
- 1 Indian Ocean INTELSAT earth station

Transportation [31]

Railroads. 712 km 1.067-meter gauge

Highways. 11,514 km total; 1,600 km paved; 1,700 km crushed stone or gravel, 5,177 km improved earth, 3,037 km unimproved earth

Airports

Total:	100
Usable:	87
With permanent-surface runways:	8
With runways over 3,659 m:	0
With runways 2,440-3,659 m:	1
With runways 1,220-2,439 m:	29

Top Agricultural Products [32]

Agriculture accounts for only 5% of GDP; subsistence farming predominates; cattle raising supports 50% of the population; must import up to of 80% of food needs.

Top Mining Products [33]

Estimated metric tons unless otherwise specified   M.t.

Nickel, mine output, Ni content of ore milled	23,500
Copper, mine output, Cu content of ore milled	24,800
Diamonds, gem, near gem and industrial (000 carats)	16,506
Sand (cubic meters)	340,825
Stone, crushed (cubic meters)	783,487
Coal, bituminous	783,873

Tourism [34]

	1987	1988	1989	1990	1991
Visitors[11]	426	379	682	829	899
Tourists	299	268	448	543	412[9]
Excursionists	127	111	234	286	487
Tourism receipts	49	38	54	65	
Tourism expenditures	26	32	34	39	
Fare receipts	6	7	5	6	
Fare expenditures	9	7	37	50	60

Tourists are in thousands, money in million U.S. dollars.

Manufacturing Sector

GDP and Manufacturing Summary [35]

	1980	1985	1989	1990	% change 1980-1990	% change 1989-1990
GDP (million 1980 $)	913	1,559	NA	2,411	164.1	NA
GDP per capita (1980 $)	1,012	1,438	NA	1,846	82.4	NA
Manufacturing as % of GDP (current prices)	4.0	5.1	NA	4.1	2.5	NA
Gross output (million $)	149	169	NA	550[e]	269.1	NA
Value added (million $)	41	46	NA	148[e]	261.0	NA
Value added (million 1980 $)	38	46	NA	77	102.6	NA
Industrial production index	100	171	NA	256	156.0	NA
Employment (thousands)	5	10	NA	24	380.0	NA

Note: GDP stands for Gross Domestic Product. 'e' stands for estimated value.

Profitability and Productivity

	1980	1985	1989	1990	% change 1980-1990	% change 1989-1990
Intermediate input (%)	73	73	NA	73[e]	0.0	NA
Wages, salaries, and supplements (%)	14[e]	11[e]	NA	9[e]	-35.7	NA
Gross operating surplus (%)	14[e]	16[e]	NA	17[e]	21.4	NA
Gross output per worker ($)	27,107	16,581	NA	22,642[e]	-16.5	NA
Value added per worker ($)	7,446	4,518	NA	6,092[e]	-18.2	NA
Average wage (incl. benefits) ($)	3,666[e]	1,880[e]	NA	2,133[e]	-41.8	NA

Profitability is in percent of gross output. Productivity is in U.S. $. 'e' stands for estimated value.

Profitability - 1990

Inputs - 73.7%
Wages - 9.1%
Surplus - 17.2%

The graphic shows percent of gross output.

Value Added in Manufacturing

	1980 $ mil.	1980 %	1985 $ mil.	1985 %	1989 $ mil.	1989 %	1990 $ mil.	1990 %	% change 1980-1990	% change 1989-1990
311 Food products	13	31.7	14	30.4	NA	NA	46[e]	31.1	253.8	NA
313 Beverages	4	9.8	10	21.7	NA	NA	36[e]	24.3	800.0	NA
314 Tobacco products	NA	0.0	NA	0.0	NA	NA	NA	0.0	NA	NA
321 Textiles	4[e]	9.8	2[e]	4.3	NA	NA	5[e]	3.4	25.0	NA
322 Wearing apparel	2[e]	4.9	1[e]	2.2	NA	NA	4[e]	2.7	100.0	NA
323 Leather and fur products	NA	0.0	NA	0.0	NA	NA	NA	0.0	NA	NA
324 Footwear	1[e]	2.4	NA	0.0	NA	NA	2[e]	1.4	100.0	NA
331 Wood and wood products	NA	0.0	NA	0.0	NA	NA	2[e]	1.4	NA	NA
332 Furniture and fixtures	NA	0.0	NA	0.0	NA	NA	2[e]	1.4	NA	NA
341 Paper and paper products	NA	0.0	1[e]	2.2	NA	NA	2[e]	1.4	NA	NA
342 Printing and publishing	NA	0.0	1[e]	2.2	NA	NA	2[e]	1.4	NA	NA
351 Industrial chemicals	NA	0.0	1[e]	2.2	NA	NA	4[e]	2.7	NA	NA
352 Other chemical products	NA	0.0	1[e]	2.2	NA	NA	4[e]	2.7	NA	NA
353 Petroleum refineries	NA	0.0	NA	0.0	NA	NA	NA	0.0	NA	NA
354 Miscellaneous petroleum and coal products	NA	0.0	NA	0.0	NA	NA	NA	0.0	NA	NA
355 Rubber products	NA	0.0	1[e]	2.2	NA	NA	3[e]	2.0	NA	NA
356 Plastic products	NA	0.0	NA	0.0	NA	NA	2[e]	1.4	NA	NA
361 Pottery, china and earthenware	NA	0.0	NA	0.0	NA	NA	NA	0.0	NA	NA
362 Glass and glass products	NA	0.0	NA	0.0	NA	NA	NA	0.0	NA	NA
369 Other non-metal mineral products	NA	0.0	NA	0.0	NA	NA	NA	0.0	NA	NA
371 Iron and steel	NA	0.0	NA	0.0	NA	NA	NA	0.0	NA	NA
372 Non-ferrous metals	NA	0.0	NA	0.0	NA	NA	NA	0.0	NA	NA
381 Metal products	1[e]	2.4	3[e]	6.5	NA	NA	8[e]	5.4	700.0	NA
382 Non-electrical machinery	1[e]	2.4	1[e]	2.2	NA	NA	3[e]	2.0	200.0	NA
383 Electrical machinery	NA	0.0	1[e]	2.2	NA	NA	2[e]	1.4	NA	NA
384 Transport equipment	1[e]	2.4	NA	0.0	NA	NA	1[e]	0.7	0.0	NA
385 Professional and scientific equipment	NA	0.0	NA	0.0	NA	NA	NA	0.0	NA	NA
390 Other manufacturing industries	12	29.3	8	17.4	NA	NA	21[e]	14.2	75.0	NA

Note: The industry codes shown are International Standard Industry codes (ISIC). Percentages are percent of total Value Added. 'e' stands for estimated value

Finance, Economics, and Trade

Economic Indicators [36]

National product: GDP—purchasing power equivalent—$3.6 billion (FY92 est.). **National product real growth rate**: 5.8% (FY92 est.). **National product per capita**: $2,450 (FY92 est.). **Inflation rate (consumer prices)**: 16.5% (December 1992). **External debt**: $344 million (December 1991).

Balance of Payments Summary [37]

Values in millions of dollars.

	1986	1987	1988	1989	1990
Exports of goods (f.o.b.)	852.5	1,586.6	1,468.9	1,819.7	1,753.2
Imports of goods (f.o.b.)	-608.4	-803.9	-986.9	-1,185.1	-1,606.2
Trade balance	244.1	782.7	482.0	634.6	147.0
Services - debits	-400.9	-589.6	-791.9	-711.5	-782.1
Services - credits	216.6	296.9	330.4	355.3	497.3
Private transfers (net)	-2.6	6.8	-17.5	-30.6	-40.8
Government transfers (net)	53.8	166.7	184.6	250.5	316.2
Long term capital (net)	99.6	-84.3	39.9	162.4	252.7
Short term capital (net)	6.0	-5.5	-65.2	-49.5	-61.3
Errors and omissions	90.5	-12.2	220.0	-34.8	-21.7
Overall balance	307.1	561.5	382.3	576.4	307.3

Exchange Rates [38]

Currency: **pula.**
Symbol: **P.**

Data are currency units per $1.

February 1993	2.3100
1992	2.1327
1991	2.0173
1990	1.8601
1989	2.0125
1988	1.8159

Imports and Exports

Top Import Origins [39]

$1.7 billion (c.i.f., 1991).

Origins	%
Switzerland	NA
SACU (Southern African Customs Union)	NA
UK	NA
US	NA

Top Export Destinations [40]

$1.6 billion (f.o.b. 1991).

Destinations	%
Switzerland	NA
UK	NA
SACU (Southern African Customs Union)	NA

Foreign Aid [41]

	U.S. $	
US aid	13	million
US commitments, including Ex-Im (FY70-89)	257	million
Western (non-US) countries, ODA and OOF bilateral commitments (1970-89)	1,875	million
OPEC bilateral aid (1979-89)	43	million
Communist countries (1970-89)	29	million
Norway ((1992) largest donor)	16	million
Sweden	15.5	million
Germany	3.6	million
EC/Lome-IV, in grants	3-6	million
EC/Lome-IV, in long-term projects	28.7	million

Import and Export Commodities [42]

Import Commodities

Foodstuffs
Vehicles and transport equipment
Textiles
Petroleum products

Export Commodities

Diamonds 78%
Copper and nickel 8%
Meat 4%

Brazil

Geography [1]

Total area:
   8,511,965 km2
Land area:
   8,456,510 km2
Comparative area:
   Slightly smaller than the US
Land boundaries:
   Total 14,691 km, Argentina 1,224 km, Bolivia 3,400 km, Colombia 1,643 km, French Guiana 673 km, Guyana 1,119 km, Paraguay 1,290 km, Peru 1,560 km, Suriname 597 km, Uruguay 985 km, Venezuela 2,200 km
Coastline:
   7,491 km
Climate:
   Mostly tropical, but temperate in south
Terrain:
   Mostly flat to rolling lowlands in north; some plains, hills, mountains, and narrow coastal belt
Natural resources:
   Iron ore, manganese, bauxite, nickel, uranium, phosphates, tin, hydropower, gold, platinum, petroleum, timber
Land use:
   Arable land: 7%
   Permanent crops: 1%
   Meadows and pastures: 19%
   Forest and woodland: 67%
   Other: 6%

Demographics [2]

	1960	1970	1980	1990	1991[1]	1994	2000	2010	2020
Population	71,695	95,684	122,830	150,062	155,356	158,739	169,543	183,742	197,466
Population density (persons per sq. mi.)	22	29	38	47	48	NA	55	64	71
Births	NA	NA	NA	NA	4,002	3,410	NA	NA	NA
Deaths	NA	NA	NA	NA	1,153	1,370	NA	NA	NA
Life expectancy - males	NA	NA	NA	NA	62	57	NA	NA	NA
Life expectancy - females	NA	NA	NA	NA	68	67	NA	NA	NA
Birth rate (per 1,000)	NA	NA	NA	NA	26	21	NA	NA	NA
Death rate (per 1,000)	NA	NA	NA	NA	7	9	NA	NA	NA
Women of reproductive age (15-44 yrs.)	NA	NA	NA	39,466	NA	43,219	47,806	NA	NA
of which are currently married	NA	NA	NA	23,660	NA	26,252	29,531	NA	NA
Fertility rate	NA	NA	NA	2.6	3.07	2.4	2.1	1.9	1.8

Population values are in thousands, life expectancy in years, and other items as indicated.

Health

Health Personnel [3]

Doctors per 1,000 pop., 1988-92	1.46
Nurse-to-doctor ratio, 1988-92	0.1
Hospital beds per 1,000 pop., 1985-90	3.5
Percentage of children immunized (age 1 yr. or less)	
Third dose of DPT, 1990-91	75.0
Measles, 1990-91	83.0

Health Expenditures [4]

Total health expenditure, 1990 (official exchange rate)	
Millions of dollars	19871
Millions of dollars per capita	132
Health expenditures as a percentage of GDP	
Total	4.2
Public sector	2.8
Private sector	1.4
Development assistance for health	
Total aid flows (millions of dollars)[1]	84
Aid flows per capita (millions of dollars)	0.6
Aid flows as a percentage of total health expenditure	0.4

For sources, notes, and explanations, see Annotated Source Appendix, page 1035.

121

Human Factors

Health Care Ratios [5]

Population per physician, 1970	2030
Population per physician, 1990	NA
Population per nursing person, 1970	4140
Population per nursing person, 1990	NA
Percent of births attended by health staff, 1985	73

Infants and Malnutrition [6]

Percent of babies with low birth weight, 1985	8
Infant mortality rate per 1,000 live births, 1970	95
Infant mortality rate per 1,000 live births, 1991	58
Years of life lost per 1,000 population, 1990	26
Prevalence of malnutrition (under age 5), 1990	13

Ethnic Division [7]

Black	6%
White	55%
Mixed	38%
Other	1%

Religion [8]

Roman Catholic (nominal) 90%.

Major Languages [9]

Portuguese (official)	NA
Spanish	NA
English	NA
French	NA

Education

Public Education Expenditures [10]

Million Cruzeiros	1980	1985	1986	1988	1989	1990
Total education expenditure	0.431	50	164	3,550	56,101	NA
as percent of GNP	3.6	3.8	4.7	4.3	4.6	NA
as percent of total govt. expend.	NA	NA	NA	NA	NA	NA
Current education expenditure	NA	NA	NA	NA	NA	NA
as percent of GNP	NA	NA	NA	NA	NA	NA
as percent of current govt. expend.	NA	NA	NA	NA	NA	NA
Capital expenditure	NA	NA	NA	NA	NA	NA

Educational Attainment [11]

Age group	$10+^2$
Total population	$110,157,487^2$
Highest level attained (%)	
No schooling	18.7^2
First level	
Incompleted	57.0^2
Completed	6.9^2
Entered second level	
S-1	11.9^2
S-2	5.5^2
Post secondary	NA^2

Literacy Rate [12]

In thousands and percent	1985[a]	1991[a]	2000[a]
Illiterate population + 15 years	18,533	18,407	17,395
Illiteracy rate - total pop. (%)	21.5	18.9	14.2
Illiteracy rate - males (%)	19.7	17.5	13.4
Illiteracy rate - females (%)	23.3	20.2	15.0

Libraries [13]

	Admin. Units	Svc. Pts.	Vols. (000)	Shelving (meters)	Vols. Added	Reg. Users
National (1989)	3	5	5,024	38,548	129,707	-
Non-specialized	NA	NA	NA	NA	NA	NA
Public	NA	NA	NA	NA	NA	NA
Higher ed.	NA	NA	NA	NA	NA	NA
School	NA	NA	NA	NA	NA	NA

Daily Newspapers [14]

	1975	1980	1985	1990
Number of papers	289	343	322	356
Circ. (000)	4,653	5,482	6,534	8,100[e]

Cinema [15]

Science and Technology

Scientific/Technical Forces [16]

Potential scientists/engineers	1,362,206
Number female	668,911
Potential technicians	NA
Number female	NA
Total	NA

R&D Expenditures [17]

	Cruzado (000) 1985
Total expenditure[12,15]	5,390,540
Capital expenditure	NA
Current expenditure	NA
Percent current	NA

U.S. Patents Issued [18]

Values show patents issued to citizens of the country by the U.S. Patents Office.

	1990	1991	1992
Number of patents	45	65	43

Government and Law

Organization of Government [19]

Long-form name:
Federative Republic of Brazil
Type:
federal republic
Independence:
7 September 1822 (from Portugal)
Constitution:
5 October 1988
Legal system:
based on Roman codes; has not accepted compulsory ICJ jurisdiction
National holiday:
Independence Day, 7 September (1822)
Executive branch:
president, vice president, Cabinet
Legislative branch:
bicameral National Congress (Congresso Nacional) consists of an upper chamber or Federal Senate (Senado Federal) and a lower chamber or Chamber of Deputies (Camara dos Deputados)
Judicial branch:
Supreme Federal Tribunal

Political Parties [20]

Chamber of Deputies	% of seats
Democratic Movement (PMDB)	21.5
Liberal Front Party (PFL)	17.3
Democratic Labor Party (PDT)	9.1
PDS	8.5
National Reconstruction (PRN)	8.0
Brazilian Labor Party (PTB)	7.0
Workers' Party (PT)	7.0
Other	21.7

Government Expenditures [21]

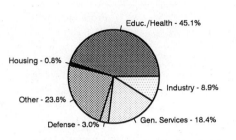

Educ./Health - 45.1%
Housing - 0.8%
Industry - 8.9%
Other - 23.8%
Gen. Services - 18.4%
Defense - 3.0%

% distribution for 1991. Expend. in 1990: 170.6 ($ bil.)

Crime [23]

Military Expenditures and Arms Transfers [22]

	1985	1986	1987	1988	1989
Military expenditures					
Current dollars (mil.)	2,793	3,597	4,185	5,731	NA
1989 constant dollars (mil.)	3,180	3,992	4,502	5,966	NA
Armed forces (000)	496	527	541	319	319
Gross national product (GNP)					
Current dollars (mil.)	348,600	388,400	419,000	430,500	462,300
1989 constant dollars (mil.)	396,900	431,100	450,700	448,200	462,300
Central government expenditures (CGE)					
1989 constant dollars (mil.)	154,600	160,200	212,000	256,000	NA
People (mil.)	138.1	141.0	143.9	146.8	149.6
Military expenditure as % of GNP	0.8	0.9	1.0	1.3	NA
Military expenditure as % of CGE	2.1	2.5	2.1	2.3	NA
Military expenditure per capita	23	28	31	41	NA
Armed forces per 1,000 people	3.6	3.7	3.8	2.2	2.1
GNP per capita	2,874	3,058	3,133	3,054	3,090
Arms imports[6]					
Current dollars (mil.)	50	110	160	360	160
1989 constant dollars (mil.)	57	122	172	375	160
Arms exports[6]					
Current dollars (mil.)	370	270	650	420	50
1989 constant dollars (mil.)	421	300	699	437	50
Total imports[7]					
Current dollars (mil.)	14,330	15,560	16,580	16,050	20,020
1989 constant dollars	16,320	17,260	17,830	16,710	20,020
Total exports[7]					
Current dollars (mil.)	25,640	22,350	26,220	33,780	34,390
1989 constant dollars	29,190	24,800	28,210	35,170	34,390
Arms as percent of total imports[8]	0.3	0.7	1.0	2.2	0.8
Arms as percent of total exports[8]	1.4	1.2	2.5	1.2	0.1

Human Rights [24]

	SSTS	FL	FAPRO	PPCG	APROBC	TPW	PCPTW	STPEP	PHRFF	PRW	ASST	AFL
Observes	2	P		P	P	P	P	P		P	P	P
	EAFRD	CPR	ESCR	SR	ACHR	MAAE	PVIAC	PVNAC	EAFDAW	TCIDTP	RC	
Observes	P	P	P	P			P	P	P	P	P	

P = Party; S = Signatory; see Appendix for meaning of abbreviations.

Labor Force

Total Labor Force [25]

57 million (1989 est.)

Labor Force by Occupation [26]

Services	42%
Agriculture	31
Industry	27

Unemployment Rate [27]

5.9% (1992)

Production Sectors

Commercial Energy Production and Consumption

Production [28]

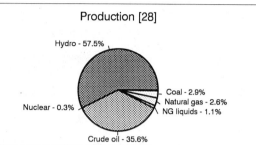

Hydro - 57.5%
Nuclear - 0.3%
Crude oil - 35.6%
Coal - 2.9%
Natural gas - 2.6%
NG liquids - 1.1%

Consumption [29]

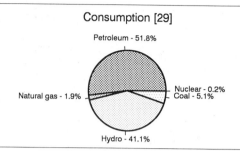

Petroleum - 51.8%
Natural gas - 1.9%
Nuclear - 0.2%
Coal - 5.1%
Hydro - 41.1%

Data are shown in quadrillion (10^{15}) BTUs and percent for 1991

Crude oil[1]	1.35
Natural gas liquids	0.04
Dry natural gas	0.10
Coal[2]	0.11
Net hydroelectric power[3]	2.18
Net nuclear power[3]	0.01
Total	3.79

Petroleum	2.76
Dry natural gas	0.10
Coal[2]	0.27
Net hydroelectric power[3]	2.19
Net nuclear power[3]	0.01
Total	5.34

Telecommunications [30]

- Good system
- Extensive microwave radio relay facilities
- 9.86 million telephones
- Broadcast stations - 1,223 AM, no FM, 112 TV, 151 shortwave
- 3 coaxial submarine cables, 3 Atlantic Ocean INTELSAT earth stations and 64 domestic satellite earth stations

Transportation [31]

Railroads. 28,828 km total; 24,864 km 1.000-meter gauge, 3,877 km 1.600-meter gauge, 74 km mixed 1.600-1.000-meter gauge, 13 km 0.760-meter gauge; 2,360 km electrified

Highways. 1,448,000 km total; 48,000 km paved, 1,400,000 km gravel or earth

Merchant Marine. 232 ships (1,000 GRT or over) totaling 5,335,234 GRT/8,986,734 DWT; includes 5 passenger-cargo, 42 cargo, 1 refrigerated cargo, 10 container, 11 roll-on/roll-off, 58 oil tanker, 15 chemical tanker, 12 combination ore/oil, 65 bulk, 2 combination bulk, 11 vehicle carrier; in addition, 1 naval tanker is sometimes used commercially

Airports

Total:	3,613
Usable:	3,031
With permanent-surface runways:	431
With runways over 3,659 m:	2
With runways 2,440-3,659 m:	22
With runways 1,220-2,439 m:	584

Top Agricultural Products [32]

	1990	1991
Sugar cane	220.0	235.0
Corn	21.8	23.5
Soybeans	20.3	15.5
Oranges	12.0	12.0
Rice	7.2	9.3
Wheat	3.2	3.2

Values shown are 1,000 metric tons.

Top Mining Products [33]

Estimated metric tons unless otherwise specified   M.t.

Bauxite, dry basis, gross weight	10,310,000[1]
Columbium-tantalum ores and concentrates, gross weight	28,779[1]
Diamonds, gems and industrial (000 carats)	1,500
Other gemstones (ex. ruby and sapphire)	8,440
Gold (kilograms)	89,109[1,11]
Iron ore and concentrate, gross wt. (000 tons)[12]	150,000
Kaolin, marketable product[12]	13,100
Manganese ore and concentrate, gross wt.[12]	2,500,000
Tin, mine output, Sn content	29,300

Tourism [34]

	1987	1988	1989	1990	1991
Tourists	1,929	1,743	1,403	1,091	1,352
Cruise passengers	18				
Tourism receipts[12]	1,502	1,643	1,225	1,444	1,559
Tourism expenditures	1,249	1,084	751	1,559	1,224
Fare receipts	141	55	21	32	
Fare expenditures	123	234	358	346	

Tourists are in thousands, money in million U.S. dollars.

Manufacturing Sector

GDP and Manufacturing Summary [35]

	1980	1985	1989	1990	% change 1980-1990	% change 1989-1990
GDP (million 1980 $)	233,962	254,528	296,273	280,406	19.9	-5.4
GDP per capita (1980 $)	1,929	1,878	2,010	1,863	-3.4	-7.3
Manufacturing as % of GDP (current prices)	31.1	30.0	27.4e	23.3	-25.1	-15.0
Gross output (million $)	176,175	174,241	330,962e	291,993e	65.7	-11.8
Value added (million $)	71,690	77,082	148,882e	73,294	2.2	-50.8
Value added (million 1980 $)	70,679	68,069	73,225	69,529	-1.6	-5.0
Industrial production index	100	98	112	103	3.0	-8.0
Employment (thousands)	4,449	5,501	4,165e	5,213	17.2	25.2

Note: GDP stands for Gross Domestic Product. 'e' stands for estimated value.

Profitability and Productivity

	1980	1985	1989	1990	% change 1980-1990	% change 1989-1990
Intermediate input (%)	59	56	55e	75e	27.1	36.4
Wages, salaries, and supplements (%)	7	9e	9e	8e	14.3	-11.1
Gross operating surplus (%)	34	36e	36e	17e	-50.0	-52.8
Gross output per worker ($)	39,599	31,674	79,462e	56,015e	41.5	-29.5
Value added per worker ($)	16,114	14,012	35,745e	14,061	-12.7	-60.7
Average wage (incl. benefits) ($)	2,773	2,753e	7,008e	4,334e	56.3	-38.2

Profitability is in percent of gross output. Productivity is in U.S. $. 'e' stands for estimated value.

Profitability - 1990

Inputs - 75.0%
Wages - 8.0%
Surplus - 17.0%

The graphic shows percent of gross output.

Value Added in Manufacturing

	1980 $ mil.	1980 %	1985 $ mil.	1985 %	1989 $ mil.	1989 %	1990 $ mil.	1990 %	% change 1980-1990	% change 1989-1990
311 Food products	7,996	11.2	9,259	12.0	14,703e	9.9	8,687	11.9	8.6	-40.9
313 Beverages	1,375	1.9	957	1.2	1,418e	1.0	1,388	1.9	0.9	-2.1
314 Tobacco products	495	0.7	587	0.8	962e	0.6	726	1.0	46.7	-24.5
321 Textiles	4,860	6.8	4,586	5.9	10,016e	6.7	3,862	5.3	-20.5	-61.4
322 Wearing apparel	2,307	3.2	2,639	3.4	4,198e	2.8	2,425e	3.3	5.1	-42.2
323 Leather and fur products	309	0.4	464	0.6	1,023e	0.7	371	0.5	20.1	-63.7
324 Footwear	985	1.4	1,353	1.8	3,572e	2.4	1,243e	1.7	26.2	-65.2
331 Wood and wood products	1,903	2.7	1,220	1.6	1,619e	1.1	951	1.3	-50.0	-41.3
332 Furniture and fixtures	1,087	1.5	949	1.2	1,206e	0.8	843	1.2	-22.4	-30.1
341 Paper and paper products	2,238	3.1	2,260	2.9	4,904e	3.3	2,556	3.5	14.2	-47.9
342 Printing and publishing	1,901	2.7	1,496	1.9	3,172e	2.1	2,305	3.1	21.3	-27.3
351 Industrial chemicals	3,428	4.8	5,933e	7.7	9,823e	6.6	3,930e	5.4	14.6	-60.0
352 Other chemical products	3,544	4.9	6,465e	8.4	9,773e	6.6	4,560e	6.2	28.7	-53.3
353 Petroleum refineries	3,075	4.3	1,956e	2.5	9,355e	6.3	1,343e	1.8	-56.3	-85.6
354 Miscellaneous petroleum and coal products	1,216	1.7	990e	1.3	1,295e	0.9	714e	1.0	-41.3	-44.9
355 Rubber products	941	1.3	1,420	1.8	3,154e	2.1	1,059	1.4	12.5	-66.4
356 Plastic products	1,994	2.8	1,742	2.3	3,911e	2.6	1,847	2.5	-7.4	-52.8
361 Pottery, china and earthenware	190	0.3	844	1.1	323e	0.2	761e	1.0	300.5	135.6
362 Glass and glass products	558	0.8	525	0.7	1,022e	0.7	447e	0.6	-19.9	-56.3
369 Other non-metal mineral products	3,447	4.8	1,941	2.5	4,956e	3.3	1,930e	2.6	-44.0	-61.1
371 Iron and steel	4,128	5.8	4,927	6.4	10,707e	7.2	5,811e	7.9	40.8	-45.7
372 Non-ferrous metals	1,115	1.6	1,564	2.0	3,476e	2.3	2,172e	3.0	94.8	-37.5
381 Metal products	3,599	5.0	3,063	4.0	5,893e	4.0	3,714e	5.1	3.2	-37.0
382 Non-electrical machinery	7,171	10.0	7,092	9.2	13,879e	9.3	6,282e	8.6	-12.4	-54.7
383 Electrical machinery	4,536	6.3	5,831	7.6	11,893e	8.0	6,341	8.7	39.8	-46.7
384 Transport equipment	3,625	5.1	4,954	6.4	9,578e	6.4	5,652	7.7	55.9	-41.0
385 Professional and scientific equipment	453	0.6	532e	0.7	1,422e	1.0	637e	0.9	40.6	-55.2
390 Other manufacturing industries	1,216	1.7	1,532e	2.0	1,628e	1.1	1,738e	2.4	42.9	6.8

Note: The industry codes shown are International Standard Industry codes (ISIC). Percentages are percent of total Value Added. 'e' stands for estimated value

For sources, notes, and explanations, see Annotated Source Appendix, page 1035.

125

Finance, Economics, and Trade

Economic Indicators [36]

	1989	1990	1991
GDP (bil. $)	370.0	355.0	358.0[e]
Real GDP Growth (percent)	3.6	-4.0	0.8[14]
GDP Per Capita (dollars)	2,593.0	2,489.0	NA
M-1 (bil. cruzeiros yr-end)	103.1	1,651.0	6,139[15]
Comm. Interest Rate			
(Avg. Monthly Rate)	62.2	28.5	24.0[16]
Gross Savings Rate (% of GDP)	22.5	21.0	NA
Gross Domestic Investment			
(% of GDP)	22.4	20.5	NA
Total Foreign Debt			
(yr-end bil. $)	115.1	124.0	118.4[17]

Balance of Payments Summary [37]

Values in millions of dollars.

	1986	1987	1988	1989	1990
Exports of goods (f.o.b.)	22,348.0	26,210.0	33,773.0	34,375.0	31,408.0
Imports of goods (f.o.b.)	-14,044.0	-15,052.0	-14,605.0	-18,263.0	-20,661.0
Trade balance	8,304.0	11,158.0	19,168.0	16,112.0	10,747.0
Services - debits	-16,478.0	-15,198.0	-18,153.0	-19,773.0	-20,288.0
Services - credits	2,783.0	2,520.0	3,050.0	4,442.0	4,919.0
Private transfers (net)	89.0	113.0	107.0	226.0	813.0
Government transfers (net)	-2.0	-43.0	-13.0	18.0	21.0
Long term capital (net)	763.0	-995.0	451.0	-3,025.0	-4,359.0
Short term capital (net)	847.0	4,887.0	-2,145.0	3,421.0	9,563.0
Errors and omissions	66.0	-802.0	-827.0	-819.0	-296.0
Overall balance	-3,628.0	1,640.0	1,638.0	602.0	1,120.0

Exchange Rates [38]

Currency: **cruzeiros.**
Symbol: **Cr$.**

Data are currency units per $1.

January 1993	13,827.06
1992	4,506.45
1991	406.61
1990	68.30
1989	2.83
1988	0.26

Imports and Exports

Top Import Origins [39]

$20.0 billion (1992). Data are for 1991.

Origins	%
Middle East	12.4
US	23.5
EC	21.8
Latin America	18.8
Japan	6

Top Export Destinations [40]

$35.0 billion (1992) Data are for 1991.

Destinations	%
EC	32.3
US	20.3
Latin America	11.6
Japan	9

Foreign Aid [41]

	U.S. $	
US commitments, including Ex-Im (FY70-89)	2.5	billion
Western (non-US) countries, ODA and OOF bilateral commitments (1970-89)	10.2	million
OPEC bilateral aid (1979-89)	284	million
Former Communist countries (1970-89)	1.3	billion

Import and Export Commodities [42]

Import Commodities	Export Commodities
Crude oil	Iron ore
Capital goods	Soybean bran
Chemical products	Orange juice
Foodstuffs	Footwear
Coal	Coffee
	Motor vehicle parts

For sources, notes, and explanations, see Annotated Source Appendix, page 1035.

Brunei

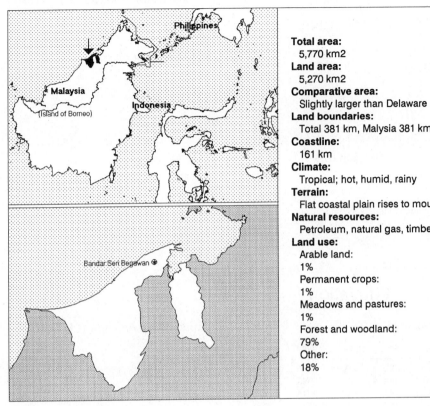

Geography [1]

Total area:
5,770 km2
Land area:
5,270 km2
Comparative area:
Slightly larger than Delaware
Land boundaries:
Total 381 km, Malysia 381 km
Coastline:
161 km
Climate:
Tropical; hot, humid, rainy
Terrain:
Flat coastal plain rises to mountains in east; hilly lowland in west
Natural resources:
Petroleum, natural gas, timber
Land use:
Arable land:
1%
Permanent crops:
1%
Meadows and pastures:
1%
Forest and woodland:
79%
Other:
18%

Demographics [2]

	1960	1970	1980	1990	1991[1]	1994	2000	2010	2020
Population	83	128	185	254	398	285	331	410	491
Population density (persons per sq. mi.)	41	63	91	183	195	NA	276	324	365
Births	NA	NA	NA	NA	9	7	NA	NA	NA
Deaths	NA	NA	NA	NA	2	1	NA	NA	NA
Life expectancy - males	NA	NA	NA	NA	74	69	NA	NA	NA
Life expectancy - females	NA	NA	NA	NA	77	73	NA	NA	NA
Birth rate (per 1,000)	NA	NA	NA	NA	22	26	NA	NA	NA
Death rate (per 1,000)	NA	NA	NA	NA	4	5	NA	NA	NA
Women of reproductive age (15-44 yrs.)	NA	NA	NA	63	NA	72	84	NA	NA
of which are currently married	NA	NA	NA	40	NA	46	54	NA	NA
Fertility rate	NA	NA	NA	3.5	2.88	3.4	3.3	3.1	3.0

Population values are in thousands, life expectancy in years, and other items as indicated.

Health

Health Personnel [3]

Health Expenditures [4]

For sources, notes, and explanations, see Annotated Source Appendix, page 1035.

127

Human Factors

Health Care Ratios [5]	Infants and Malnutrition [6]

Ethnic Division [7]

Malay	64%
Chinese	20%
Other	16%

Religion [8]

Muslim (official)	63%
Buddhism	14%
Christian	8%
Indigenous beliefs and other	15%

Major Languages [9]

Malay (official), English, Chinese.

Education

Public Education Expenditures [10]

Million Brunei Dollars	1980	1985	1986	1988	1989	1990
Total education expenditure	129	NA	243	296	275	253
as percent of GNP	NA	NA	NA	NA	NA	NA
as percent of total govt. expend.	11.8	NA	NA	NA	NA	NA
Current education expenditure	115	NA	217	272	244	229
as percent of GNP	NA	NA	NA	NA	NA	NA
as percent of current govt. expend.	12.5	NA	NA	NA	NA	NA
Capital expenditure	15	NA	25	24	31	17

Educational Attainment [11]

Age group	25+
Total population	75,283
Highest level attained (%)	
No schooling	32.1
First level	
Incompleted	32.1
Completed	28.3
Entered second level	
S-1	NA
S-2	30.1
Post secondary	9.4

Literacy Rate [12]

Libraries [13]

	Admin. Units	Svc. Pts.	Vols. (000)	Shelving (meters)	Vols. Added	Reg. Users
National	NA	NA	NA	NA	NA	NA
Non-specialized	NA	NA	NA	NA	NA	NA
Public	NA	NA	NA	NA	NA	NA
Higher ed. (1990)	2	3	150	NA	10,854	1,504
School (1991)	23	NA	287	2,228	18,941	36,890

Daily Newspapers [14]

	1975	1980	1985	1990
Number of papers	-	-	-	1
Circ. (000)	-	-	-	10

Cinema [15]

Science and Technology

Scientific/Technical Forces [16]

Potential scientists/engineers[4]	2,214
Number female[4]	390
Potential technicians[5]	4,301
Number female	1,770
Total[5]	6,515

R&D Expenditures [17]

	Dollar[19] (000) 1984
Total expenditure	10,880
Capital expenditure	2,660
Current expenditure	8,220
Percent current	75.6

U.S. Patents Issued [18]

For sources, notes, and explanations, see Annotated Source Appendix, page 1035.

Government and Law

Organization of Government [19]

Long-form name:
Negara Brunei Darussalam
Type:
constitutional sultanate
Independence:
1 January 1984 (from UK)
Constitution:
29 September 1959 (some provisions
suspended under a State of Emergency
since December 1962, others since
independence on 1 January 1984)
Legal system:
based on Islamic law
National holiday:
23 February (1984)
Executive branch:
sultan, prime minister, Council of Cabinet
Ministers
Legislative branch:
unicameral Legislative Council (Majlis
Masyuarat Megeri)
Judicial branch:
Supreme Court

Elections [20]

Legislative Council. The Brunei
National Democratic Party, the first legal
political party, is now banned. Elections
were last held in March 1962; in 1970
the Council was changed to an
appointive body by decree of the sultan
and no elections are planned.

Government Budget [21]

For 1989 est.

Revenues	1.300
Expenditures	1.500
Capital expenditures	0.255

Defense Summary [22]

Branches: Ground Force, Navy, Air Force, Royal Brunei Police

Manpower Availability: Males age 15-49 77,407; fit for military service 45,112; reach military age (18) annually 2,676 (1993 est.)

Defense Expenditures: Exchange rate conversion - $300 million, 9% of GDP (1990)

Crime [23]

Crime volume (for 1989)

Cases known to police	833
Attempts (percent)	NA
Percent cases solved	NA
Crimes per 100,000 persons	334.54
Persons responsible for offenses	
Total number offenders	NA
Percent female	NA
Percent juvenile[10]	NA
Percent foreigners	NA

Human Rights [24]

	SSTS	FL	FAPRO	PPCG	APROBC	TPW	PCPTW	STPEP	PHRFF	PRW	ASST	AFL
Observes	1					P	P			P	1	P
	EAFRD	CPR	ESCR	SR	ACHR	MAAE	PVIAC	PVNAC	EAFDAW	TCIDTP	RC
Observes			P				P	P			P

P = Party; S = Signatory; see Appendix for meaning of abbreviations.

Labor Force

Total Labor Force [25]

89,000 (includes members of the Army)

Labor Force by Occupation [26]

Government	47.5%
Oil, natural gas, services, construction	41.9
Agriculture, forestry, and fishing	3.8

Date of data: 1986

Unemployment Rate [27]

3.7% (1989)

Production Sectors

Commercial Energy Production and Consumption

Production [28]

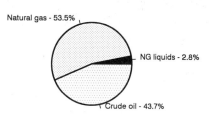

Natural gas - 53.5%

NG liquids - 2.8%

Crude oil - 43.7%

Consumption [29]

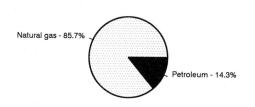

Natural gas - 85.7%

Petroleum - 14.3%

Data are shown in quadrillion (10^{15}) BTUs and percent for 1991

Crude oil[1]	0.31
Natural gas liquids	0.02
Dry natural gas	0.38
Coal[2]	0.00
Net hydroelectric power[3]	0.00
Net nuclear power[3]	0.00
Total	0.71

Petroleum	0.01
Dry natural gas	0.06
Coal[2]	0.00
Net hydroelectric power[3]	0.00
Net nuclear power[3]	0.00
Total	0.07

Telecommunications [30]

- Service throughout country is adequate for present needs
- International service good to adjacent Malaysia
- Radiobroadcast coverage good
- 33,000 telephones (1987)
- Broadcast stations - 4 AM/FM, 1 TV
- 74,000 radio receivers (1987)
- Satellite earth stations - 1 Indian Ocean INTELSAT and 1 Pacific Ocean INTELSAT

Transportation [31]

Railroads. 13 km 0.610-meter narrow-gauge private line

Highways. 1,090 km total; 370 km paved (bituminous treated) and another 52 km under construction, 720 km gravel or unimproved

Merchant Marine. 7 liquefied gas carriers (1,000 GRT or over) totaling 348,476 GRT/340,635 DWT

Airports

Total:	2
Usable:	2
With permanent-surface runways:	1
With runway over 3,659 m:	1
With runway 2,440-3,659 m:	0
With runway 1,220-2,439 m:	1

Top Agricultural Products [32]

Imports about 80% of its food needs; principal crops and livestock include rice, cassava, bananas, buffaloes, and pigs.

Top Mining Products [33]

Estimated metric tons unless otherwise specified   M.t.

Natural gas, gross (million cubic meters)	9,200
Natural gas liquids condensate (000 42-gal. barrels)	4,000
Petroleum, crude (000 42-gal. barrels)	53,290
Petroleum, refined products (000 42-gal. barrels)	1,435

Tourism [34]

	1987	1988	1989	1990	1991
Visitors	523	457	393	377	
Tourism receipts	25	30	32		

Tourists are in thousands, money in million U.S. dollars.

Finance, Economics, and Trade

GDP and Manufacturing Summary [35]

	1980	1985	1990	1991	1992
Gross Domestic Product					
Millions of 1980 dollars	4,848	4,115	3,778	3,846	3,923
Growth rate in percent	-7.00	0.73	-1.15	3,846	3,923
Manufacturing Value Added					
Millions of 1980 dollars	573	339	367	376	385[e]
Growth rate in percent	-8.35	-5.42	0.09	2.33	2.44[e]
Manufacturing share in percent of current prices	11.7	10.0	NA	NA	NA

Economic Indicators [36]

National product: GDP—exchange rate conversion—$3.5 billion (1990 est.). **National product real growth rate**: 1% (1990 est.). **National product per capita**: $8,800 (1990 est.). **Inflation rate (consumer prices)**: 1.3% (1989). **External debt**: $0.

Balance of Payments Summary [37]

Exchange Rates [38]

Currency: **Bruneian dollars.**
Symbol: **B$.**

The Bruneian dollar is at par with the Singapore dollar. Data are currency units per $1.

January 1993	1.6531
1992	1.6290
1991	1.7276
1990	1.8125
1989	1.9503
1988	2.0124

Imports and Exports

Top Import Origins [39]

$1.7 billion (c.i.f., 1990 est.). Data are for 1990.

Origins	%
Singapore	35
UK	26
Switzerland	9
US	9
Japan	5

Top Export Destinations [40]

$2.2 billion (f.o.b., 1990 est.). Data are for 1990.

Destinations	%
Japan	53
UK	12
South Korea	9
Thailand	7
Singapore	5

Foreign Aid [41]

	U.S. $	
US commitments, including Ex-Im (FY70-87)	20.6	million
Western (non-US) countries, ODA and OOF bilateral commitments (1970-89)	153	million

Import and Export Commodities [42]

Import Commodities	**Export Commodities**
Machinery & transport equipment	Crude oil
Manufactured goods	Liquefied natural gas
Food	Petroleum products
Chemicals	

Bulgaria

Geography [1]

Total area:
110,910 km2
Land area:
110,550 km2
Comparative area:
Slightly larger than Tennessee
Land boundaries:
Total 1,808 km, Greece 494 km, Macedonia 148 km, Romania 608 km, Serbia and Montenegro 318 km (all with Serbia), Turkey 240 km
Coastline:
354 km
Climate:
Temperate; cold, damp winters; hot, dry summers
Terrain:
Mostly mountains with lowlands in north and south
Natural resources:
Bauxite, copper, lead, zinc, coal, timber, arable land
Land use:
Arable land:
34%
Permanent crops:
3%
Meadows and pastures:
18%
Forest and woodland:
35%
Other:
10%

Demographics [2]

	1960	1970	1980	1990	1991[1]	1994	2000	2010	2020
Population	7,867	8,490	8,844	8,966	8,911	8,800	8,742	8,757	8,642
Population density (persons per sq. mi.)	184	199	207	209	209	NA	211	213	213
Births	NA	NA	NA	NA	116	103	NA	NA	NA
Deaths	NA	NA	NA	NA	104	100	NA	NA	NA
Life expectancy - males	NA	NA	NA	NA	69	70	NA	NA	NA
Life expectancy - females	NA	NA	NA	NA	76	77	NA	NA	NA
Birth rate (per 1,000)	NA	NA	NA	NA	13	12	NA	NA	NA
Death rate (per 1,000)	NA	NA	NA	NA	12	11	NA	NA	NA
Women of reproductive age (15-44 yrs.)	NA	NA	NA	2,144	NA	2,130	2,059	NA	NA
of which are currently married	NA	NA	NA	1,644	NA	1,624	1,588	NA	NA
Fertility rate	NA	NA	NA	1.7	1.91	1.7	1.7	1.7	1.7

Population values are in thousands, life expectancy in years, and other items as indicated.

Health

Health Personnel [3]

Doctors per 1,000 pop., 1988-92	3.19
Nurse-to-doctor ratio, 1988-92	2.1
Hospital beds per 1,000 pop., 1985-90	9.8
Percentage of children immunized (age 1 yr. or less)	
Third dose of DPT, 1990-91	99.0
Measles, 1990-91	97.0

Health Expenditures [4]

Total health expenditure, 1990 (official exchange rate)	
Millions of dollars	1154
Millions of dollars per capita	131
Health expenditures as a percentage of GDP	
Total	5.4
Public sector	4.4
Private sector	1.0
Development assistance for health	
Total aid flows (millions of dollars)[1]	NA
Aid flows per capita (millions of dollars)	NA
Aid flows as a percentage of total health expenditure	NA

Human Factors

Health Care Ratios [5]

Population per physician, 1970	540
Population per physician, 1990	320
Population per nursing person, 1970	240
Population per nursing person, 1990	NA
Percent of births attended by health staff, 1985	100

Infants and Malnutrition [6]

Percent of babies with low birth weight, 1985	NA
Infant mortality rate per 1,000 live births, 1970	27
Infant mortality rate per 1,000 live births, 1991	17
Years of life lost per 1,000 population, 1990	15
Prevalence of malnutrition (under age 5), 1990	NA

Ethnic Division [7]

Bulgarian	85.3%
Turk	8.5%
Gypsy	2.6%
Macedonian	2.5%
Armenian	0.3%
Russian	0.2%
Other	0.6%

Religion [8]

Bulgarian Orthodox	85.0%
Muslim	13.0%
Jewish	0.8%
Roman Catholic	0.5%
Uniate Catholic	0.2%
Other	0.5%

Major Languages [9]

Bulgarian; secondary languages closely correspond to ethnic breakdown.

Education

Public Education Expenditures [10]

Million Leva	1980	1985	1987	1988	1989	1990
Total education expenditure	1,145	1,784	NA	2,044	2,112	2,357
as percent of GNP	4.5	5.5	NA	5.4	5.5	5.4
as percent of total govt. expend.	NA	NA	NA	NA	NA	NA
Current education expenditure	1,098	1,598	NA	1,869	1,946	2,183
as percent of GNP	4.3	4.9	NA	5.0	5.1	5.0
as percent of current govt. expend.	NA	NA	NA	NA	NA	NA
Capital expenditure	47	186	NA	175	166	174

Educational Attainment [11]

Literacy Rate [12]

Libraries [13]

	Admin. Units	Svc. Pts.	Vols. (000)	Shelving (meters)	Vols. Added	Reg. Users
National (1989)	1	NA	2,161	NA	36,639	26,063
Non-specialized (1989)	27	NA	10,426	NA	360,851	298,200
Public (1989)	5,356	NA	46,148	NA	1.8mil	1.6mil
Higher ed. (1990)	56	NA	7,043	NA	220,555	144,688
School (1990)	3,208	NA	16,625	NA	520,286	761,532

Daily Newspapers [14]

	1975	1980	1985	1990
Number of papers	14	14	17	24
Circ. (000)	2,109	2,244	2,626	4,065

Cinema [15]

Data for 1991.

Cinema seats per 1,000	30.2
Annual attendance per person	2.9
Gross box office receipts (mil. Leva)	57

Science and Technology

Scientific/Technical Forces [16]

Potential scientists/engineers	323,575
Number female	161,918
Potential technicians[7]	690,501
Number female[7]	391,789
Total[7]	1,014,076

R&D Expenditures [17]

	Leva (000) 1989
Total expenditure	1,042,400
Capital expenditure	109,500
Current expenditure	932,900
Percent current	89.5

U.S. Patents Issued [18]

Values show patents issued to citizens of the country by the U.S. Patents Office.

	1990	1991	1992
Number of patents	27	10	5

For sources, notes, and explanations, see Annotated Source Appendix, page 1035.

133

Government and Law

Organization of Government [19]

Long-form name:
Republic of Bulgaria
Type:
emerging democracy
Independence:
22 September 1908 (from Ottoman Empire)
Constitution:
adopted 12 July 1991
Legal system:
based on civil law system, with Soviet law influence; has accepted compulsory ICJ jurisdiction
National holiday:
3 March (1878)
Executive branch:
president, chairman of the Council of Ministers (prime minister), three deputy chairmen of the Council of Ministers, Council of Ministers
Legislative branch:
unicameral National Assembly (Narodno Sobranie)
Judicial branch:
Supreme Court, Constitutional Court

Political Parties [20]

National Assembly	% of votes
Union of Democratic Forces (UDF)	34.0
Bulgarian Socialist Party (BSP)	33.0
MRF	7.5

Government Expenditures [21]

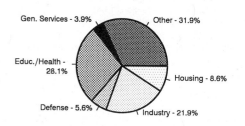

Gen. Services - 3.9%
Other - 31.9%
Educ./Health - 28.1%
Housing - 8.6%
Defense - 5.6%
Industry - 21.9%

% distribution for 1990. Expend. in 1991 est.: 5 ($ bil.)

Crime [23]

Crime volume (for 1990)	
Cases known to police	66,429
Attempts (percent)	NA
Number cases solved	30,457
Crimes per 100,000 persons	736.99
Persons responsible for offenses	
Total number offenders	17,400
Percent female	11.5
Percent juvenile	68.6
Percent foreigners	1.1

Military Expenditures and Arms Transfers [22]

	1985	1986	1987	1988	1989
Military expenditures					
Current dollars (mil.)[e]	5,808	6,215	6,451	5,868	5,885
1989 constant dollars (mil.)[e]	6,612	6,897	6,938	6,109	5,885
Armed forces (000)	189	190	191	160	150
Gross national product (GNP)					
Current dollars (mil.)	41,220	43,410	44,870	47,970	49,590
1989 constant dollars (mil.)	46,920	48,180	48,260	49,940	49,590
Central government expenditures (CGE)					
1989 constant dollars (mil.)	20,320	23,850	23,540	19,690	19,800
People (mil.)	8.9	9.0	9.0	9.0	9.0
Military expenditure as % of GNP[3]	14.1	14.3	14.4	12.2	11.9
Military expenditure as % of CGE[3]	32.5	28.9	29.5	31.0	29.7
Military expenditure per capita	739	770	773	680	656
Armed forces per 1,000 people	21.2	21.2	21.3	17.8	16.7
GNP per capita	5,246	5,378	5,379	5,560	5,530
Arms imports[6]					
Current dollars (mil.)	950	1,300	700	400	290
1989 constant dollars (mil.)	1,081	1,442	753	416	290
Arms exports[6]					
Current dollars (mil.)	550	470	575	430	160
1989 constant dollars (mil.)	626	522	618	448	160
Total imports[7]					
Current dollars (mil.)	13,090	15,580	17,060	20,980	14,970
1989 constant dollars	14,900	17,290	18,350	21,840	14,970
Total exports[7]					
Current dollars (mil.)	12,780	14,520	16,660	20,350	16,040
1989 constant dollars	14,550	16,110	17,920	21,180	16,040
Arms as percent of total imports[8]	7.3	8.3	4.1	1.9	1.9
Arms as percent of total exports[8]	4.3	3.2	3.5	2.1	1.0

Human Rights [24]

	SSTS	FL	FAPRO	PPCG	APROBC	TPW	PCPTW	STPEP	PHRFF	PRW	ASST	AFL
Observes	P	P	P	P	P	P	P	P	P	P	P	

	EAFRD	CPR	ESCR	SR	ACHR	MAAE	PVIAC	PVNAC	EAFDAW	TCIDTP	RC
Observes		P	P	P			P	P	P	P	P

P = Party; S = Signatory; see Appendix for meaning of abbreviations.

Labor Force

Total Labor Force [25]

4.3 million

Labor Force by Occupation [26]

Industry	33%
Agriculture	20
Other	47

Date of data: 1987

Unemployment Rate [27]

15% (1992)

For sources, notes, and explanations, see Annotated Source Appendix, page 1035.

Production Sectors

Commercial Energy Production and Consumption

Production [28]

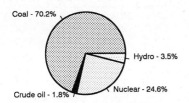

Coal - 70.2%
Hydro - 3.5%
Crude oil - 1.8%
Nuclear - 24.6%

Consumption [29]

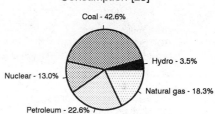

Coal - 42.6%
Hydro - 3.5%
Nuclear - 13.0%
Natural gas - 18.3%
Petroleum - 22.6%

Data are shown in quadrillion (10^{15}) BTUs and percent for 1991

Crude oil[1]	0.01
Natural gas liquids	0.00
Dry natural gas	0.00
Coal[2]	0.40
Net hydroelectric power[3]	0.02
Net nuclear power[3]	0.14
Total	0.57

Petroleum	0.26
Dry natural gas	0.21
Coal[2]	0.49
Net hydroelectric power[3]	0.04
Net nuclear power[3]	0.15
Total	1.14

Telecommunications [30]

- Extensive but antiquated transmission system of coaxial cable and mirowave radio relay
- 2.6 million phones; direct dialing to 36 countries; phone density: 29 phones per 100 persons (1992)
- 67% of Sofia households have phones (November 1988) and telephone service is available in most villages
- Broadcast stations - 20 AM, 15 FM, and 29 TV, with 1 Soviet TV repeater in Sofia
- 2.1 million TV sets (1990); 92% of country receives No. 1 television program (May 1990)
- 1 satellite ground station using Intersputnik; INTELSAT is used through a Greek earth station

Top Agricultural Products [32]

	90-91	91-92[3]
Wheat	5,290	4,549
Corn	1,220	2,718
Barley	1,390	1,495
Vegetables	1,662	1,334
Wine grapes	543	644
Potatoes	430	503

Values shown are 1,000 metric tons.

Top Mining Products [33]

Metric tons unless otherwise specified	M.t.
Copper (000 tons)	6,000
Lead, mine output, Pb content	50,000
Zinc, mine output, Zn content	50,000
Manganese ore, Mn content	14,800
Gypsum (000 tons)	450
Quicklime (000 tons)	1,300

Transportation [31]

Railroads. 4,300 km total, all government owned (1987); 4,055 km 1.435-meter standard gauge, 245 km narrow gauge; 917 km double track; 2,640 km electrified

Highways. 36,908 km total; 33,535 km hard surface (including 242 km superhighways); 3,373 km earth roads (1987)

Merchant Marine. 112 ships (1,000 GRT and over) totaling 1,262,320 GRT/1,887,729 DWT; includes 2 short-sea passenger, 30 cargo, 2 container, 1 passenger-cargo training, 6 roll-on/roll-off, 15 oil tanker, 4 chemical carrier, 2 railcar carrier, 50 bulk; Bulgaria owns 1 ship (1,000 GRT or over) totaling 8,717 DWT operating under Liberian registry

Airports

Total:	380
Usable:	380
With permanent-surface runways:	120
With runways over 3659 m:	0
With runways 2,440-3,659 m:	20
With runways 1,220-2,439 m:	20

Tourism [34]

	1987	1988	1989	1990	1991
Visitors	7,594	8,295	8,221	10,330	6,818
Tourism receipts	494	484	495	320	
Tourism expenditures	69	74	113	189	

Tourists are in thousands, money in million U.S. dollars.

For sources, notes, and explanations, see Annotated Source Appendix, page 1035.

135

Manufacturing Sector

GDP and Manufacturing Summary [35]

	1980	1985	1989	1990	% change 1980-1990	% change 1989-1990
GDP (million 1980 $)	19,993	23,992	25,776	23,907	19.6	-7.3
GDP per capita (1980 $)	2,256	2,677	2,863	2,652	17.6	-7.4
Manufacturing as % of GDP (current prices)	38.5	46.5	58.1[e]	40.1	4.2	-31.0
Gross output (million $)	24,310[e]	39,270[e]	20,969[e]	18,225[e]	-25.0	-13.1
Value added (million $)	11,771	14,745	17,410	14,535[e]	23.5	-16.5
Value added (million 1980 $)	8,447	12,050	NA	12,059	42.8	NA
Industrial production index	100	125	148	123	23.0	-16.9
Employment (thousands)	1,260	1,316	1,315[e]	1,374	9.0	4.5

Note: GDP stands for Gross Domestic Product. 'e' stands for estimated value.

Profitability and Productivity

	1980	1985	1989	1990	% change 1980-1990	% change 1989-1990
Intermediate input (%)	NA	NA	NA	NA	NA	NA
Wages, salaries, and supplements (%)	6[e]	5[e]	NA	8[e]	33.3	NA
Gross operating surplus (%)	NA	NA	NA	NA	NA	NA
Gross output per worker ($)	30,586[e]	48,256	25,749[e]	21,446	-29.9	-16.7
Value added per worker ($)	9,675	11,738	13,733[e]	11,012[e]	13.8	-19.8
Average wage (incl. benefits) ($)	1,737[e]	2,677[e]	1,558[e]	1,666[e]	-4.1	6.9

Profitability is in percent of gross output. Productivity is in U.S. $. 'e' stands for estimated value.

Profitability - 1990

Value Added in Manufacturing

	1980 $ mil.	1980 %	1985 $ mil.	1985 %	1989 $ mil.	1989 %	1990 $ mil.	1990 %	% change 1980-1990	% change 1989-1990
311 Food products	1,870	15.9	1,945	13.2	2,094	12.0	1,842[e]	12.7	-1.5	-12.0
313 Beverages	308	2.6	357	2.4	351	2.0	339	2.3	10.1	-3.4
314 Tobacco products	426	3.6	472	3.2	438	2.5	349	2.4	-18.1	-20.3
321 Textiles	904	7.7	1,003	6.8	1,193	6.9	1,175	8.1	30.0	-1.5
322 Wearing apparel	517	4.4	626	4.2	843	4.8	915	6.3	77.0	8.5
323 Leather and fur products	84	0.7	110	0.7	134	0.8	112	0.8	33.3	-16.4
324 Footwear	156	1.3	218	1.5	329	1.9	284	2.0	82.1	-13.7
331 Wood and wood products	248	2.1	258	1.7	270	1.6	240	1.7	-3.2	-11.1
332 Furniture and fixtures	233	2.0	347	2.4	405	2.3	352	2.4	51.1	-13.1
341 Paper and paper products	119	1.0	141	1.0	140	0.8	112	0.8	-5.9	-20.0
342 Printing and publishing	83	0.7	91	0.6	108	0.6	108	0.7	30.1	0.0
351 Industrial chemicals	404	3.4	573	3.9	549	3.2	500	3.4	23.8	-8.9
352 Other chemical products	291	2.5	486	3.3	623	3.6	483	3.3	66.0	-22.5
353 Petroleum refineries	NA	0.0	NA	0.0	NA	0.0	NA	0.0	NA	NA
354 Miscellaneous petroleum and coal products	126	1.1	134	0.9	181[e]	1.0	156	1.1	23.8	-13.8
355 Rubber products	227	1.9	323	2.2	405	2.3	332	2.3	46.3	-18.0
356 Plastic products	110	0.9	150[e]	1.0	162[e]	0.9	175[e]	1.2	59.1	8.0
361 Pottery, china and earthenware	45	0.4	40	0.3	52	0.3	58	0.4	28.9	11.5
362 Glass and glass products	121	1.0	140	0.9	137	0.8	133	0.9	9.9	-2.9
369 Other non-metal mineral products	469	4.0	507	3.4	479	2.8	390	2.7	-16.8	-18.6
371 Iron and steel	447	3.8	513	3.5	531	3.0	304	2.1	-32.0	-42.7
372 Non-ferrous metals	189	1.6	180	1.2	192[e]	1.1	102	0.7	-46.0	-46.9
381 Metal products	484	4.1	600	4.1	551	3.2	489	3.4	1.0	-11.3
382 Non-electrical machinery	1,491[e]	12.7	2,427[e]	16.5	3,475[e]	20.0	2,511[e]	17.3	68.4	-27.7
383 Electrical machinery	743	6.3	1,241	8.4	1,479	8.5	1,152	7.9	55.0	-22.1
384 Transport equipment	567	4.8	726	4.9	828	4.8	687	4.7	21.2	-17.0
385 Professional and scientific equipment	172[e]	1.5	283[e]	1.9	448[e]	2.6	299[e]	2.1	73.8	-33.3
390 Other manufacturing industries	937	8.0	853	5.8	1,012	5.8	937	6.4	0.0	-7.4

Note: The industry codes shown are International Standard Industry codes (ISIC). Percentages are percent of total Value Added. 'e' stands for estimated value

Finance, Economics, and Trade

Economic Indicators [36]

Bil. Dollars or Leva as noted	1989	1990	1991[e]
Real GDP (bil. 1990 Lev)	47.6	42.0	32.8
Real GDP Growth Rate (%)	-0.3	-11.8	-22
Money Supply (M1, bil. Lev)	21.9	28.7	50.8
Commercial interest rate (%)[19]	4.7	5.6	60.0
GDS Rate	0.29	0.24	0.08
GDI Rate	0.32	0.29	0.27
CPI (Dec. 1990 equals 100)	60	100	520
WPI	NA	NA	NA
External Public Debt (US$ bil.)	9.2	10.0	12.0

Balance of Payments Summary [37]

Values in millions of dollars.

	1986	1987	1988	1989	1990
Exports of goods (f.o.b.)	8,862	10,297	9,283	8,268	6,113
Imports of goods (f.o.b.)	-10,045	-11,308	-9,889	-8,960	-7,427
Trade balance	-1,183	-1,011	-606	-692	-1,314
Services - debits	-931	-1,090	-1,167	-1,504	-1,478
Services - credits	1,096	1,273	1,268	1,350	957
Private transfers (net)	67	108	103	77	125
Government transfers (net)	-	-	-	-	-
Long term capital (net)	597	-69	1,554	319	-3,535
Short term capital (net)	-185	549	-9	-359	4,297
Errors and omissions	-346	-257	-486	375	70
Overall balance	-885	-497	657	-434	-878

Exchange Rates [38]

Currency: **leva.**
Symbol: **Lv.**

Floating exchange rate since February 1991. Data are currency units per $1.

January 1993	24.56
January 1992	17.18
March 1991	16.13
November 1990	0.74
1989	0.84
1988	0.82
1987	0.90

Imports and Exports

Top Import Origins [39]

$2.8 billion (f.o.b., 1991).

Origins	%
Former CEMA countries	51.0
Former USSR	43.2
Poland	3.7
Developed countries	32.8
Germany	7.0
Austria	4.7
Less developed	16.2
Iran	2.8
Lybia	2.5

Top Export Destinations [40]

$3.5 billion (f.o.b., 1991). Data are for 1991.

Destinations	%
USSR	48.6
Germany	4.8
Greece	2.2
Libya	2.1
Poland	2.1
Czechoslovakia	0.9
Iran	0.7

Foreign Aid [41]

	U.S. $	
Donor - in bilateral aid to non-Communist less developed countries (1956-89)	1.6	billion

Import and Export Commodities [42]

Import Commodities

Fuels
Minerals
And raw materials 58.7%
Machinery & equipment 15.8%
Manufactured consumer goods 4.4%
Agricultural products 15.2%
Other 5.9%

Export Commodities

Machinery & equipment 30.6%
Agricultural products 24%
Consumer goods 22.2%
Fuels
Minerals
Raw materials
And metals 10.5%
Other 12.7%

For sources, notes, and explanations, see Annotated Source Appendix, page 1035.

137

Burkina

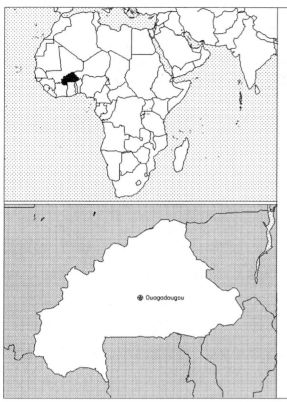

Geography [1]

Total area:
274,200 km2
Land area:
273,800 km2
Comparative area:
Slightly larger than Colorado
Land boundaries:
Total 3,192 km, Benin 306 km, Ghana 548 km, Cote d'Ivoire 584 km, Mali 1,000 km, Niger 628 km, Togo 126 km
Coastline:
0 km (landlocked)
Climate:
Tropical; warm, dry winters; hot, wet summers
Terrain:
Mostly flat to dissected, undulating plains; hills in west and southeast
Natural resources:
Manganese, limestone, marble; small deposits of gold, antimony, copper, nickel, bauxite, lead, phosphates, zinc, silver
Land use:
Arable land:
10%
Permanent crops:
0%
Meadows and pastures:
37%
Forest and woodland:
26%
Other:
27%

Demographics [2]

	1960	1970	1980	1990	1991[1]	1994	2000	2010	2020
Population	4,866	5,626	6,939	9,042	9,360	10,135	11,871	14,478	18,123
Population density (persons per sq. mi.)	46	53	66	86	89	NA	118	162	218
Births	NA	NA	NA	NA	465	491	NA	NA	NA
Deaths	NA	NA	NA	NA	152	184	NA	NA	NA
Life expectancy - males	NA	NA	NA	NA	52	46	NA	NA	NA
Life expectancy - females	NA	NA	NA	NA	53	48	NA	NA	NA
Birth rate (per 1,000)	NA	NA	NA	NA	50	48	NA	NA	NA
Death rate (per 1,000)	NA	NA	NA	NA	16	18	NA	NA	NA
Women of reproductive age (15-44 yrs.)	NA	NA	NA	2,035	NA	2,272	2,647	NA	NA
of which are currently married	NA	NA	NA	1,692	NA	1,888	2,192	NA	NA
Fertility rate	NA	NA	NA	7.2	7.12	6.9	6.5	5.4	4.3

Population values are in thousands, life expectancy in years, and other items as indicated.

Health

Health Personnel [3]

Doctors per 1,000 pop., 1988-92	0.03
Nurse-to-doctor ratio, 1988-92	8.2
Hospital beds per 1,000 pop., 1985-90	0.3
Percentage of children immunized (age 1 yr. or less)	
Third dose of DPT, 1990-91	37.0
Measles, 1990-91	42.0

Health Expenditures [4]

Total health expenditure, 1990 (official exchange rate)	
Millions of dollars	219
Millions of dollars per capita	24
Health expenditures as a percentage of GDP	
Total	8.5
Public sector	7.0
Private sector	1.5
Development assistance for health	
Total aid flows (millions of dollars)[1]	42
Aid flows per capita (millions of dollars)	4.7
Aid flows as a percentage of total health expenditure	19.4

For sources, notes, and explanations, see Annotated Source Appendix, page 1035.

Human Factors

Health Care Ratios [5]

Population per physician, 1970	97120
Population per physician, 1990	57320
Population per nursing person, 1970	NA
Population per nursing person, 1990	1680
Percent of births attended by health staff, 1985	NA

Infants and Malnutrition [6]

Percent of babies with low birth weight, 1985	18
Infant mortality rate per 1,000 live births, 1970	178
Infant mortality rate per 1,000 live births, 1991	133
Years of life lost per 1,000 population, 1990	114
Prevalence of malnutrition (under age 5), 1990	46

Ethnic Division [7]

Mossi	2.5 million
Gurunsi	NA
Senufo	NA
Lobi	NA
Bobo	NA
Mande	NA
Fulani	NA

Religion [8]

Indigenous beliefs	65%
Muslim	25%
Christian (mainly Roman Catholic)	10%

Major Languages [9]

French (official), tribal languages belong to Sudanic family, spoken by 90% of the population.

Education

Public Education Expenditures [10]

Million Francs C.F.A.	1980	1985	1987	1988[3]	1989	1990
Total education expenditure	7,994	12,901	14,683	11,196	18,780	NA
as percent of GNP	2.2	2.0	2.0	1.4	2.3	NA
as percent of total govt. expend.	19.8	21.0	14.9	NA	17.5	NA
Current education expenditure	7,436	12,292	14,305	11,096	18,727	NA
as percent of GNP	2.1	1.9	1.9	1.4	2.3	NA
as percent of current govt. expend.	21.1	22.4	18.8	NA	21.9	NA
Capital expenditure	558	609	378	100	53	NA

Educational Attainment [11]

Literacy Rate [12]

In thousands and percent	1985[a]	1991[a]	2000[a]
Illiterate population + 15 years	3,791	4,137	4,813
Illiteracy rate - total pop. (%)	85.5	81.8	72.3
Illiteracy rate - males (%)	77.0	72.1	60.9
Illiteracy rate - females (%)	93.8	91.1	83.3

Libraries [13]

Daily Newspapers [14]

	1975	1980	1985	1990
Number of papers	1	1	2	1
Circ. (000)	2	2	4	3

Cinema [15]

Science and Technology

Scientific/Technical Forces [16]

R&D Expenditures [17]

U.S. Patents Issued [18]

For sources, notes, and explanations, see Annotated Source Appendix, page 1035.

139

Government and Law

Organization of Government [19]

Long-form name:
Burkina Faso
Type:
parliamentary
Independence:
5 August 1960 (from France)
Constitution:
June 1991
Legal system:
based on French civil law system and
customary law
National holiday:
Anniversary of the Revolution, 4 August
(1983)
Executive branch:
president, Council of Ministers
Legislative branch:
Assembly of People's Deputies
Judicial branch:
Appeals Court

Political Parties [20]

Assembly of People's Deputies	% of seats
Organization People's Dem. (ODP-MT)	72.9
National Convention (CNPP-PSD)	11.2
Democratic Assembly (RDA)	5.6
Alliance for Democracy (ADF)	3.7
Other	6.5

Government Budget [21]

For 1991.

Revenues	495
Expenditures	786
Capital expenditures	NA

Crime [23]

Military Expenditures and Arms Transfers [22]

	1985	1986	1987	1988	1989
Military expenditures					
Current dollars (mil.)	35	53	45	52	NA
1989 constant dollars (mil.)	40	59	49	54	NA
Armed forces (000)	9	9[e]	9	8	8
Gross national product (GNP)					
Current dollars (mil.)	1,836	2,048	2,156	2,437	2,552
1989 constant dollars (mil.)	2,090	2,273	2,319	2,537	2,552
Central government expenditures (CGE)					
1989 constant dollars (mil.)	212	267	281	307	NA
People (mil.)	7.9	8.1	8.3	8.6	8.8
Military expenditure as % of GNP	1.9	2.6	2.1	2.1	NA
Military expenditure as % of CGE	18.7	22.1	17.3	17.5	NA
Military expenditure per capita	5	7	6	6	NA
Armed forces per 1,000 people	1.1	1.1	1.1	0.9	0.9
GNP per capita	266	281	279	297	290
Arms imports[6]					
Current dollars (mil.)	20	30	0	10	10
1989 constant dollars (mil.)	23	33	0	10	10
Arms exports[6]					
Current dollars (mil.)	0	0	0	0	0
1989 constant dollars (mil.)	0	0	0	0	0
Total imports[7]					
Current dollars (mil.)	332	405	434	489	NA
1989 constant dollars	378	449	467	509	NA
Total exports[7]					
Current dollars (mil.)	71	83	155	142	NA
1989 constant dollars	81	92	167	148	NA
Arms as percent of total imports[8]	6.0	7.4	0	2.0	NA
Arms as percent of total exports[8]	0	0	0	0	NA

Human Rights [24]

	SSTS	FL	FAPRO	PPCG	APROBC	TPW	PCPTW	STPEP	PHRFF	PRW	ASST	AFL
Observes	P	P	P	P	P	P	P					
	EAFRD	CPR	ESCR	SR	ACHR	MAAE	PVIAC	PVNAC	EAFDAW	TCIDTP	RC
Observes	P		P				P	P	P		P

P = Party; S = Signatory; see Appendix for meaning of abbreviations.

Labor Force

Total Labor Force [25]

3.3 million residents; 30,000 are wage
earners. 20% of male labor force migrates
annually to neighboring countries for
seasonal employment (1984); 44% of
population was of working age (1985).

Labor Force by Occupation [26]

Agriculture	82%
Industry	13
Commerce, services, and government	5

Unemployment Rate [27]

Not available.

Production Sectors

Energy Resource Summary [28]

Energy Resources: None. **Electricity**: 120,000 kW capacity; 320 million kWh produced, 40 kWh per capita (1991).

Telecommunications [30]

- Microwave radio relay, wire, and radio communication stations in use
- Broadcast stations - 2 AM, 1 FM, 2 TV
- 1 Atlantic Ocean INTELSAT earth station

Top Agricultural Products [32]

Agriculture accounts for about 30% of GDP; cash crops—peanuts, shea nuts, sesame, cotton; food crops—sorghum, millet, corn, rice; livestock; not self-sufficient in food grains.

Top Mining Products [33]

Metric tons unless otherwise specified	M.t.
Gold (kilograms)	5,300[30]
Phosphate rock (000 tons)	3
Pumice	10,000
Salt	6,500
Marble (000)	100

Transportation [31]

Railroads. 620 km total; 520 km Ouagadougou to Cote d'Ivoire border and 100 km Ouagadougou to Kaya; all 1.00-meter gauge and single track.

Highways. 16,500 km total; 1,300 km paved, 7,400 km improved, 7,800 km unimproved (1985).

Airports

Total:	48
Usable:	38
With permanent-surface runways:	2
With runways over 3,659 m:	0
With runways 2,440-3,659 m:	2
With runways 1,220-2,439 m:	8

Tourism [34]

	1987	1988	1989	1990	1991
Tourists[10]	51	82	79	74	80
Tourism receipts	4	5	6	8	8
Tourism expenditures	28	30	30	35	34
Fare expenditures	27	28	26	30	29

Tourists are in thousands, money in million U.S. dollars.

For sources, notes, and explanations, see Annotated Source Appendix, page 1035.

141

Manufacturing Sector

GDP and Manufacturing Summary [35]

	1980	1985	1989	1990	% change 1980-1990	% change 1989-1990
GDP (million 1980 $)	1,287	1,435	1,767	1,607	24.9	-9.1
GDP per capita (1980 $)	185	182	202	179	-3.2	-11.4
Manufacturing as % of GDP (current prices)	11.4	11.2	12.5[e]	11.5	0.9	-8.0
Gross output (million $)	391	318[e]	524[e]	596[e]	52.4	13.7
Value added (million $)	144	121[e]	218[e]	206[e]	43.1	-5.5
Value added (million 1980 $)	141	144	183[e]	163	15.6	-10.9
Industrial production index	100	110	140[e]	129	29.0	-7.9
Employment (thousands)	8	9[e]	10[e]	9[e]	12.5	-10.0

Note: GDP stands for Gross Domestic Product. 'e' stands for estimated value.

Profitability and Productivity

	1980	1985	1989	1990	% change 1980-1990	% change 1989-1990
Intermediate input (%)	63	62[e]	58[e]	65[e]	3.2	12.1
Wages, salaries, and supplements (%)	8	7[e]	7[e]	8[e]	0.0	14.3
Gross operating surplus (%)	28	31[e]	35[e]	27[e]	-3.6	-22.9
Gross output per worker ($)	47,326	36,452[e]	54,487[e]	64,078[e]	35.4	17.6
Value added per worker ($)	17,465	13,890[e]	22,679[e]	22,195[e]	27.1	-2.1
Average wage (incl. benefits) ($)	4,021	2,712[e]	3,669[e]	5,088[e]	26.5	38.7

Profitability is in percent of gross output. Productivity is in U.S. $. 'e' stands for estimated value.

Profitability - 1990

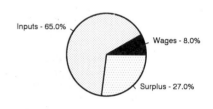

Inputs - 65.0%
Wages - 8.0%
Surplus - 27.0%

The graphic shows percent of gross output.

Value Added in Manufacturing

	1980 $ mil.	1980 %	1985 $ mil.	1985 %	1989 $ mil.	1989 %	1990 $ mil.	1990 %	% change 1980-1990	% change 1989-1990
311 Food products	55	38.2	55[e]	45.5	108[e]	49.5	98[e]	47.6	78.2	-9.3
313 Beverages	29	20.1	21[e]	17.4	33[e]	15.1	36[e]	17.5	24.1	9.1
314 Tobacco products	1	0.7	1[e]	0.8	2[e]	0.9	2[e]	1.0	100.0	0.0
321 Textiles	20	13.9	18[e]	14.9	34[e]	15.6	28[e]	13.6	40.0	-17.6
322 Wearing apparel	2	1.4	2[e]	1.7	4[e]	1.8	3[e]	1.5	50.0	-25.0
323 Leather and fur products	2	1.4	1[e]	0.8	2[e]	0.9	3[e]	1.5	50.0	50.0
324 Footwear	3	2.1	3[e]	2.5	4[e]	1.8	5[e]	2.4	66.7	25.0
331 Wood and wood products	NA	0.0	NA	0.0	NA	0.0	NA	0.0	NA	NA
332 Furniture and fixtures	2	1.4	1[e]	0.8	3[e]	1.4	3[e]	1.5	50.0	0.0
341 Paper and paper products	NA	0.0	NA	0.0	NA	0.0	NA	0.0	NA	NA
342 Printing and publishing	1	0.7	1[e]	0.8	2[e]	0.9	1[e]	0.5	0.0	-50.0
351 Industrial chemicals	1	0.7	1[e]	0.8	1[e]	0.5	1[e]	0.5	0.0	0.0
352 Other chemical products	NA	0.0	NA	0.0	NA	0.0	NA	0.0	NA	NA
353 Petroleum refineries	NA	0.0	NA	0.0	NA	0.0	NA	0.0	NA	NA
354 Miscellaneous petroleum and coal products	NA	0.0	NA	0.0	NA	0.0	NA	0.0	NA	NA
355 Rubber products	4	2.8	2[e]	1.7	3[e]	1.4	3[e]	1.5	-25.0	0.0
356 Plastic products	2	1.4	1[e]	0.8	2[e]	0.9	1[e]	0.5	-50.0	-50.0
361 Pottery, china and earthenware	NA	0.0	NA	0.0	NA	0.0	NA	0.0	NA	NA
362 Glass and glass products	NA	0.0	NA	0.0	NA	0.0	NA	0.0	NA	NA
369 Other non-metal mineral products	NA	0.0	NA	0.0	NA	0.0	NA	0.0	NA	NA
371 Iron and steel	1[e]	0.7	1[e]	0.8	2[e]	0.9	1[e]	0.5	0.0	-50.0
372 Non-ferrous metals	1[e]	0.7	NA	0.0	1[e]	0.5	1[e]	0.5	0.0	0.0
381 Metal products	1	0.7	NA	0.0	1[e]	0.5	1[e]	0.5	0.0	0.0
382 Non-electrical machinery	1	0.7	NA	0.0	NA	0.0	NA	0.0	NA	NA
383 Electrical machinery	1	0.7	NA	0.0	1[e]	0.5	1[e]	0.5	0.0	0.0
384 Transport equipment	3	2.1	1[e]	0.8	2[e]	0.9	3[e]	1.5	0.0	50.0
385 Professional and scientific equipment	NA	0.0	NA	0.0	NA	0.0	NA	0.0	NA	NA
390 Other manufacturing industries	12	8.3	9[e]	7.4	12[e]	5.5	15[e]	7.3	25.0	25.0

Note: The industry codes shown are International Standard Industry codes (ISIC). Percentages are percent of total Value Added. 'e' stands for estimated value

142

For sources, notes, and explanations, see Annotated Source Appendix, page 1035.

Finance, Economics, and Trade

Economic Indicators [36]

National product: GDP—exchange rate conversion—$3.3 billion (1991). **National product real growth rate**: 1.3% (1990 est.). **National product per capita**: $350 (1991). **Inflation rate (consumer prices)**: - 1% (1990). **External debt**: $865 million (December 1991 est.).

Balance of Payments Summary [37]

Values in millions of dollars.

	1987	1988	1989	1990	1991
Exports of goods (f.o.b.)	229.9	249.1	215.7	272.2	283.2
Imports of goods (f.o.b.)	-475.2	-486.8	-501.6	-593.2	-601.5
Trade balance	-245.3	-237.7	-285.9	-321.0	-318.3
Services - debits	-213.3	-235.4	-237.3	-272.9	-277.6
Services - credits	48.9	49.0	52.7	65.0	63.5
Private transfers (net)	123.1	113.4	97.5	113.9	106.4
Government transfers (net)	235.6	261.4	460.9	312.9	335.6
Long term capital (net)	85.1	70.5	-206.3	75.8	113.4
Short term capital (net)	10.9	20.2	52.4	33.5	-180.2
Errors and omissions	-5.3	-2.4	5.1	-5.2	192.8
Overall balance	39.7	39.0	-60.9	2.0	35.6

Exchange Rates [38]

Currency: **CFA francs.**
Symbol: **CFAF.**

Data are currency units per $1.

January 1993	274.06
1992	264.69
1991	282.11
1990	272.26
1989	319.01
1988	297.85

Imports and Exports

Top Import Origins [39]

$593 million (f.o.b., 1990). Data are for 1987.

Origins	%
EC	51
Africa	25
US	6

Top Export Destinations [40]

$304.8 million (f.o.b., 1990) Data are for 1987.

Destinations	%
EC	45
Taiwan	15
Cote d'Ivoire	15

Foreign Aid [41]

	U.S. $	
US commitments, including Ex-Im (FY70-89)	294	million
Western (non-US) countries, ODA and OOF bilateral commitments (1970-89)	2.9	billion
Communist countries (1970-89)	113	million

Import and Export Commodities [42]

Import Commodities	Export Commodities
Machinery	Cotton
Food products	Gold
Petroleum	Animal products

For sources, notes, and explanations, see Annotated Source Appendix, page 1035.

143

Burma

Geography [1]

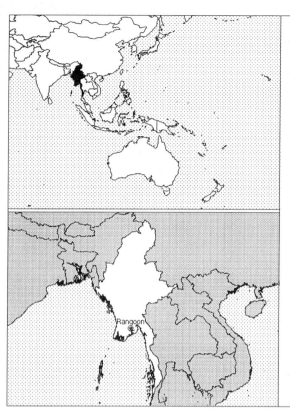

Total area:
678,500 km2
Land area:
657,740 km2
Comparative area:
Slightly smaller than Texas
Land boundaries:
Total 5,876 km, Bangladesh 193 km, China 2,185 km, India 1,463 km, Laos 235 km, Thailand 1,800 km
Coastline:
1,930 km
Climate:
Tropical monsoon; cloudy, rainy, hot, humid summers (southwest monsoon, June to September); less cloudy, scant rainfall, mild temperatures, lower humidity during winter (northeast monsoon, December to April)
Terrain:
Central lowlands ringed by steep, rugged highlands
Natural resources:
Petroleum, timber, tin, antimony, zinc, copper, tungsten, lead, coal, some marble, limestone, precious stones, natural gas
Land use:
Arable land: 15%
Permanent crops: 1%
Meadows and pastures: 1%
Forest and woodland: 49%
Other: 34%

Demographics [2]

	1960	1970	1980	1990	1991[1]	1994	2000	2010	2020
Population	22,836	27,386	33,578	41,044	42,112	44,277	49,300	57,720	65,914
Population density (persons per sq. mi.)	90	108	132	163	166	NA	196	231	267
Births	NA	NA	NA	NA	1,365	1,260	NA	NA	NA
Deaths	NA	NA	NA	NA	528	436	NA	NA	NA
Life expectancy - males	NA	NA	NA	NA	74	58	NA	NA	NA
Life expectancy - females	NA	NA	NA	NA	77	62	NA	NA	NA
Birth rate (per 1,000)	NA	NA	NA	NA	32	28	NA	NA	NA
Death rate (per 1,000)	NA	NA	NA	NA	13	10	NA	NA	NA
Women of reproductive age (15-44 yrs.)	NA	NA	NA	10,244	NA	11,200	12,875	NA	NA
of which are currently married	NA	NA	NA	6,337	NA	7,011	8,103	NA	NA
Fertility rate	NA	NA	NA	3.9	4.08	3.6	3.3	2.8	2.4

Population values are in thousands, life expectancy in years, and other items as indicated.

Health

Health Personnel [3]

Health Expenditures [4]

Human Factors

Health Care Ratios [5]	Infants and Malnutrition [6]

Ethnic Division [7]

Burman	68%
Shan	9%
Karen	7%
Rakhine	4%
Chinese	3%
Mon	2%
Indian	2%
Other	5%

Religion [8]

Buddhist	89%
Christian	4%
Baptist	3%
Roman Catholic	1%
Muslim	4%
Animist beliefs	1%
Other	2%

Major Languages [9]

Burmese; minority ethnic groups have their own languages.

Education

Public Education Expenditures [10]	Educational Attainment [11]
Literacy Rate [12]	Libraries [13]
Daily Newspapers [14]	Cinema [15]

Science and Technology

Scientific/Technical Forces [16]	R&D Expenditures [17]	U.S. Patents Issued [18]

For sources, notes, and explanations, see Annotated Source Appendix, page 1035.

Government and Law

Organization of Government [19]

Long-form name:
Union of Burma
Type:
military regime
Independence:
4 January 1948 (from UK)
Constitution:
3 January 1974 (suspended since 18 September 1988); National Convention started on 9 January 1993 to draft chapter headings for a new constitution
Legal system:
has not accepted compulsory ICJ jurisdiction
National holiday:
Independence Day, 4 January (1948)
Executive branch:
chairman of the State Law and Order Restoration Council, State Law and Order Restoration Council
Legislative branch:
unicameral People's Assembly (Pyithu Hluttaw) was dissolved after the coup of 18 September 1988
Judicial branch:
none; Council of People's Justices was abolished after the coup of 18 September 1988

Crime [23]

Political Parties [20]

People's Assembly	% of votes
Nat. League for Democr. (NLD)	80.0
Other	20.0

Government Budget [21]

For 1992.

Revenues	8.1
Expenditures	11.6
Capital expenditures	NA

Military Expenditures and Arms Transfers [22]

	1985	1986	1987	1988	1989
Military expenditures					
Current dollars (mil.)	483	476[e]	482[e]	436[e]	611[e]
1989 constant dollars (mil.)	550	528[e]	518[e]	454[e]	611[e]
Armed forces (000)	210	210[e]	210[e]	186	200
Gross national product (GNP)					
Current dollars (mil.)	15,750	15,980	15,880	14,590	16,330
1989 constant dollars (mil.)	17,930	17,730	17,080	15,190	16,330
Central government expenditures (CGE)					
1989 constant dollars (mil.)	2,800	2,845	2,326	1,847	NA
People (mil.)	37.2	38.0	38.8	39.6	40.4
Military expenditure as % of GNP	3.1	3.0	3.0	3.0	3.7
Military expenditure as % of CGE	19.6	18.6	22.3	24.6	NA
Military expenditure per capita	15	14	13	11	15
Armed forces per 1,000 people	5.6	5.5	5.4	4.7	4.9
GNP per capita	482	466	440	383	404
Arms imports[6]					
Current dollars (mil.)	50	30	20	20	20
1989 constant dollars (mil.)	57	33	22	21	20
Arms exports[6]					
Current dollars (mil.)	0	0	0	0	0
1989 constant dollars (mil.)	0	0	0	0	0
Total imports[7]					
Current dollars (mil.)	283	304	268	244	210
1989 constant dollars	322	337	288	254	210
Total exports[7]					
Current dollars (mil.)	303	288	219	147	215
1989 constant dollars	345	320	236	153	215
Arms as percent of total imports[8]	17.7	9.9	7.5	8.2	9.5
Arms as percent of total exports[8]	0	0	0	0	0

Human Rights [24]

	SSTS	FL	FAPRO	PPCG	APROBC	TPW	PCPTW	STPEP	PHRFF	PRW	ASST	AFL
Observes	P	P	P	P				S		S		
	EAFRD	CPR	ESCR	SR	ACHR	MAAE	PVIAC	PVNAC	EAFDAW	TCIDTP		RC
Observes												P

P = Party; S = Signatory; see Appendix for meaning of abbreviations.

Labor Force

Total Labor Force [25]

16.007 million (1992)

Labor Force by Occupation [26]

Agriculture	65.2%
Industry	14.3
Trade	10.1
Government	6.3
Other	4.1

Date of data: FY89 est.

Unemployment Rate [27]

9.6% (FY89 est.) in urban areas

For sources, notes, and explanations, see Annotated Source Appendix, page 1035.

Production Sectors

Energy Resource Summary [28]

Energy Resources: Petroleum, coal, natural gas. **Electricity**: 1,100,000 kW capacity; 2,800 million kWh produced, 65 kWh per capita (1992).
Pipelines: Crude oil 1,343 km; natural gas 330 km.

Telecommunications [30]

- Meets minimum requirements for local and intercity service for business and government
- International service is good
- 53,000 telephones (1986)
- Radiobroadcast coverage is limited to the most populous areas
- Broadcast stations - 2 AM, 1 FM, 1 TV (1985)
- 1 Indian Ocean INTELSAT earth station

Transportation [31]

Railroads. 3,991 km total, all government owned; 3,878 km 1.000-meter gauge, 113 km narrow-gauge industrial lines; 362 km double track

Highways. 27,000 km total; 3,200 km bituminous, 17,700 km improved earth or gravel, 6,100 km unimproved earth

Merchant Marine. 62 ships (1,000 GRT or over) totaling 940,264 GRT/1,315,156 DWT; includes 3 passenger-cargo, 18 cargo, 5 refrigerated cargo, 4 vehicle carrier, 2 container, 2 oil tanker, 3 chemical, 1 combination ore/oil, 23 bulk, 1 combination bulk

Airports

Total:	83
Usable:	78
With permanent-surface runways:	26
With runways over 3,659 m:	0
With runways 2,440-3,659 m:	3
With runways 1,220-2,439 m:	38

Top Agricultural Products [32]

	1989	1990
Rice (rough)	13,500	13,700
Sugar cane	2,008	1,998
Pulses and beans	347	430
Corn	194	186
Sorghum	116	138
Wheat	124	138

Values shown are 1,000 metric tons.

Top Mining Products [33]

Detailed information is not available. A summary of mineral resources available follows. **Mineral Resources**: Petroleum, tin, antimony, zinc, copper, tungsten, lead, coal, some marble, limestone, precious stones, natural gas.

Tourism [34]

Finance, Economics, and Trade

Industrial Summary [35]

Industrial Production: Growth rate 2.6% (FY90 est.); accounts for 10% of GDP. **Industries**: Agricultural processing; textiles and footwear; wood and wood products; petroleum refining; mining of copper, tin, tungsten, iron; construction materials; pharmaceuticals; fertilizer.

Economic Indicators [36]

National product: GDP—exchange rate conversion—$28 billion (1992). **National product real growth rate**: 1.3% (1992). **National product per capita**: $660 (1992). **Inflation rate (consumer prices)**: 50% (1992). **External debt**: $4 billion (1992).

Balance of Payments Summary [37]

Exchange Rates [38]

Currency: **kyats.**
Symbol: **K.**

Unofficial rate - 105. Data are currency units per $1.

January 1992	6.0963
1991	6.2837
1990	6.3386
1989	6.7049
1988	6.4600
1987	6.6535

Imports and Exports

Top Import Origins [39]

$907.0 million (FY92).

Origins	%
Japan	NA
China	NA
Singapore	NA

Top Export Destinations [40]

$535.1 million (FY92).

Destinations	%
China	NA
India	NA
Thailand	NA
Singapore	NA

Foreign Aid [41]

	U.S. $	
US commitments, including Ex-Im (FY70-89)	158	million
Western (non-US) countries, ODA and OOF bilateral commitments (1970-89)	3.9	billion
Communist countries (1970-89)	424	million

Import and Export Commodities [42]

Import Commodities	**Export Commodities**
Machinery	Teak
Transport equipment	Rice
Chemicals	Oilseed
Food products	Metals
	Rubber
	Gems

Burundi

Geography [1]

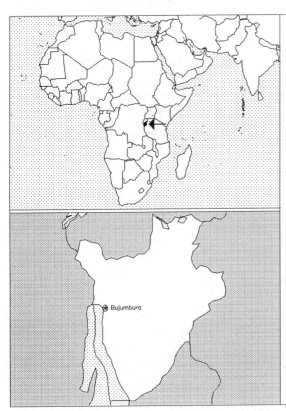

Total area:
27,830 km2
Land area:
25,650 km2
Comparative area:
Slightly larger than Maryland
Land boundaries:
Total 974 km, Rwanda 290 km, Tanzania 451 km, Zaire 233 km
Coastline:
0 km (landlocked)
Climate:
Temperate; warm; occasional frost in uplands
Terrain:
Mostly rolling to hilly highland; some plains
Natural resources:
Nickel, uranium, rare earth oxide, peat, cobalt, copper, platinum (not yet Exploited), vanadium
Land use:
Arable land:
43%
Permanent crops:
8%
Meadows and pastures:
35%
Forest and woodland:
2%
Other:
12%

Demographics [2]

	1960	1970	1980	1990	1991[1]	1994	2000	2010	2020
Population	2,812	3,513	4,138	5,558	5,831	6,125	6,939	8,382	10,734
Population density (persons per sq. mi.)	284	355	418	570	589	NA	781	1,053	1,386
Births	NA	NA	NA	NA	272	270	NA	NA	NA
Deaths	NA	NA	NA	NA	85	131	NA	NA	NA
Life expectancy - males	NA	NA	NA	NA	50	38	NA	NA	NA
Life expectancy - females	NA	NA	NA	NA	54	42	NA	NA	NA
Birth rate (per 1,000)	NA	NA	NA	NA	47	44	NA	NA	NA
Death rate (per 1,000)	NA	NA	NA	NA	15	21	NA	NA	NA
Women of reproductive age (15-44 yrs.)	NA	NA	NA	1,268	NA	1,372	1,582	NA	NA
of which are currently married	NA	NA	NA	818	NA	881	972	NA	NA
Fertility rate	NA	NA	NA	7.0	6.89	6.7	6.3	5.3	4.4

Population values are in thousands, life expectancy in years, and other items as indicated.

Health

Health Personnel [3]

Doctors per 1,000 pop., 1988-92	0.06
Nurse-to-doctor ratio, 1988-92	4.3
Hospital beds per 1,000 pop., 1985-90	1.3
Percentage of children immunized (age 1 yr. or less)	
Third dose of DPT, 1990-91	83.0
Measles, 1990-91	75.0

Health Expenditures [4]

Total health expenditure, 1990 (official exchange rate)	
Millions of dollars	36
Millions of dollars per capita	7
Health expenditures as a percentage of GDP	
Total	3.3
Public sector	1.7
Private sector	1.6
Development assistance for health	
Total aid flows (millions of dollars)[1]	15
Aid flows per capita (millions of dollars)	2.8
Aid flows as a percentage of total health expenditure	42.7

For sources, notes, and explanations, see Annotated Source Appendix, page 1035.

149

Human Factors

Health Care Ratios [5]

Population per physician, 1970	58570
Population per physician, 1990	NA
Population per nursing person, 1970	6870
Population per nursing person, 1990	NA
Percent of births attended by health staff, 1985	12

Infants and Malnutrition [6]

Percent of babies with low birth weight, 1985	14
Infant mortality rate per 1,000 live births, 1970	138
Infant mortality rate per 1,000 live births, 1991	107
Years of life lost per 1,000 population, 1990	81
Prevalence of malnutrition (under age 5), 1990	38

Ethnic Division [7]

No detailed information available.

Religion [8]

Christian	67%
Roman Catholic	62%
Protestant	5%
Indigenous beliefs	32%
Muslim	1%

Major Languages [9]

Kirundi (official), French (official), Swahili (along Lake Tanganyika and in the Bujumbura area).

Education

Public Education Expenditures [10]

Million Francs	1985[1]	1986[1]	1987[1]	1988[1]	1989[1]	1990[1]
Total education expenditure	3,467	NA	4,275	4,876	NA	6,570
as percent of GNP	2.5	NA	3.1	3.3	NA	3.5
as percent of total govt. expend.	15.5	NA	18.1	16.7	NA	16.7
Current education expenditure	3,212	NA	4,092	4,797	NA	6,370
as percent of GNP	2.4	NA	3.0	3.2	NA	3.4
as percent of current govt. expend.	17.1	NA	20.0	19.0	NA	19.4
Capital expenditure	254	NA	183	79	NA	200

Educational Attainment [11]

Literacy Rate [12]

In thousands and percent	1985[a]	1991[a]	2000[a]
Illiterate population +15 years	1,508	1,482	1,386
Illiteracy rate - total pop. (%)	57.9	50.0	34.6
Illiteracy rate - males (%)	46.6	39.1	25.4
Illiteracy rate - females (%)	68.2	60.2	43.3

Libraries [13]

	Admin. Units	Svc. Pts.	Vols. (000)	Shelving (meters)	Vols. Added	Reg. Users
National	NA	NA	NA	NA	NA	NA
Non-specialized (1989)	174	NA	174	NA	NA	500
Public (1989)	60	60	NA	NA	NA	20,000
Higher ed. (1987)[3]	1	1	0.2	NA	224	NA
School	NA	NA	NA	NA	NA	NA

Daily Newspapers [14]

	1975	1980	1985	1990
Number of papers	1	1	1	1
Circ. (000)	2	1	2	20

Cinema [15]

Science and Technology

Scientific/Technical Forces [16]

R&D Expenditures [17]

	Francs[2] (000) 1989
Total expenditure	536,187
Capital expenditure	NA
Current expenditure	NA
Percent current	NA

U.S. Patents Issued [18]

For sources, notes, and explanations, see Annotated Source Appendix, page 1035.

Government and Law

Organization of Government [19]

Long-form name:
Republic of Burundi
Type:
republic
Independence:
1 July 1962 (from UN trusteeship under Belgian administration)
Constitution:
13 March 1992 draft provides for establishment of plural political system
Legal system:
based on German and Belgian civil codes and customary law; has not accepted compulsory ICJ jurisdiction
National holiday:
Independence Day, 1 July (1962)
Executive branch:
president; chairman of the Central Committee of the National Party of Unity and Progress (UPRONA), prime minister
Legislative branch:
unicameral National Assembly (Assemblee Nationale) was dissolved following the coup of 3 September 1987; now ruled by the Military Committee for National Salvation
Judicial branch:
Supreme Court (Cour Supreme)

Elections [20]

National Assembly. The National Unity Charter outlining the principles for constitutional government was adopted by a national referendum on 5 February 1991; new elections to the National Assembly are to take place 29 June 1993; presidential elections are to take place 1 June 1993.

Government Budget [21]

For 1991 est.

Revenues	318
Expenditures	326
Capital expenditures	150

Military Expenditures and Arms Transfers [22]

	1985	1986	1987	1988	1989
Military expenditures					
Current dollars (mil.)	24	29	30	NA	28
1989 constant dollars (mil.)	27	32	33	NA	28
Armed forces (000)	9[e]	10[e]	10[e]	11	11
Gross national product (GNP)					
Current dollars (mil.)	802	841	921	1,001	1,058
1989 constant dollars (mil.)	914	933	991	1,043	1,058
Central government expenditures (CGE)					
1989 constant dollars (mil.)	129	136	156	166	195
People (mil.)	4.8	5.0	5.1	5.3	5.5
Military expenditure as % of GNP	2.9	3.4	3.3	NA	2.6
Military expenditure as % of CGE	20.8	23.3	21.0	NA	14.2
Military expenditure per capita	6	6	6	NA	5
Armed forces per 1,000 people	1.9	2.0	1.9	2.1	2.0
GNP per capita	190	188	193	197	194
Arms imports[6]					
Current dollars (mil.)	5	10	20	10	10
1989 constant dollars (mil.)	6	11	22	10	10
Arms exports[6]					
Current dollars (mil.)	0	0	0	0	0
1989 constant dollars (mil.)	0	0	0	0	0
Total imports[7]					
Current dollars (mil.)	189	202	212	204	187
1989 constant dollars	215	224	228	212	187
Total exports[7]					
Current dollars (mil.)	110	154	91	133	78
1989 constant dollars	125	171	98	138	78
Arms as percent of total imports[8]	2.6	5.0	9.4	4.9	5.3
Arms as percent of total exports[8]	0	0	0	0	0

Crime [23]

Crime volume (for 1990)	
Cases known to police	4,916
Attempts (percent)	1.3
Percent cases solved	NA
Crimes per 100,000 persons	87.29
Persons responsible for offenses	
Total number offenders	NA
Percent female	NA
Percent juvenile[11]	NA
Percent foreigners	NA

Human Rights [24]

	SSTS	FL	FAPRO	PPCG	APROBC	TPW	PCPTW	STPEP	PHRFF	PRW	ASST	AFL
Observes		P				P	P					P
	EAFRD	CPR	ESCR	SR	ACHR	MAAE	PVIAC	PVNAC	EAFDAW	TCIDTP		RC
Observes	P	P	P	P					P			P

P = Party; S = Signatory; see Appendix for meaning of abbreviations.

Labor Force

Total Labor Force [25]

1.9 million (1983 est.)

Labor Force by Occupation [26]

Agriculture	93.0%
Government	4.0
Industry and commerce	1.5
Services	1.5

Unemployment Rate [27]

For sources, notes, and explanations, see Annotated Source Appendix, page 1035.

151

Production Sectors

Energy Resource Summary [28]

Energy Resources: Uranium. **Electricity**: 55,000 kW capacity; 105 million kWh produced, 20 kWh per capita (1991).

Telecommunications [30]

- Sparse system of wire, radiocommunications, and low-capacity microwave radio relay links
- 8,000 telephones
- Broadcast stations - 2 AM, 2 FM, 1 TV
- 1 Indian Ocean INTELSAT earth station

Transportation [31]

Highways. 5,900 km total; 400 km paved, 2,500 km gravel or laterite, 3,000 km improved or unimproved earth

Airports

Total:	5
Usable:	4
With permanent-surface runways:	1
With runways over 3,659 m:	0
With runways 2,440-3,659 m:	1
With runways 1,220-2,439 m:	4

Top Agricultural Products [32]

Agriculture accounts for 60% of GDP; 90% of population dependent on subsistence farming; marginally self-sufficient in food production; cash crops—coffee, cotton, tea; food crops—corn, sorghum, sweet potatoes, bananas, manioc; livestock - meat, milk, hides and skins.

Top Mining Products [33]

Estimated metric tons unless otherwise specified   M.t.

Clay, kaolin	6,682
Lime	86
Peat	10,026
Gold (kilograms)	25[31]
Tin, mine output, Sn content	74

Tourism [34]

	1987	1988	1989	1990	1991
Visitors	132	191	152	212	246
Tourists[1]	80	99	82	109	125
Excursionists	52	92	70	103	121
Tourism receipts	1	2	3	4	4
Tourism expenditures	17	15	14	16	17
Fare receipts	1	1	1	2	2

Tourists are in thousands, money in million U.S. dollars.

152

For sources, notes, and explanations, see Annotated Source Appendix, page 1035.

Manufacturing Sector

GDP and Manufacturing Summary [35]

	1980	1985	1989	1990	% change 1980-1990	% change 1989-1990
GDP (million 1980 $)	951	1,208	1,374	1,421	49.4	3.4
GDP per capita (1980 $)	230	255	259	260	13.0	0.4
Manufacturing as % of GDP (current prices)	9.0	13.6	10.0[e]	14.0	55.6	40.0
Gross output (million $)	95	137[e]	152[e]	175[e]	84.2	15.1
Value added (million $)	56	84[e]	96[e]	109[e]	94.6	13.5
Value added (million 1980 $)	77	80	101[e]	94	22.1	-6.9
Industrial production index	100	144[e]	152[e]	178[e]	78.0	17.1
Employment (thousands)	3	4[e]	5[e]	5[e]	66.7	0.0

Note: GDP stands for Gross Domestic Product. 'e' stands for estimated value.

Profitability and Productivity

	1980	1985	1989	1990	% change 1980-1990	% change 1989-1990
Intermediate input (%)	41	39[e]	37[e]	38[e]	-7.3	2.7
Wages, salaries, and supplements (%)	9[e]	10[e]	6[e]	9[e]	0.0	50.0
Gross operating surplus (%)	51[e]	51[e]	57[e]	54[e]	5.9	-5.3
Gross output per worker ($)	27,640	27,581[e]	29,956[e]	30,394[e]	10.0	1.5
Value added per worker ($)	16,370	17,238[e]	18,869[e]	19,680[e]	20.2	4.3
Average wage (incl. benefits) ($)	2,357[e]	3,378[e]	1,822[e]	3,008[e]	27.6	65.1

Profitability is in percent of gross output. Productivity is in U.S. $. 'e' stands for estimated value.

Profitability - 1990

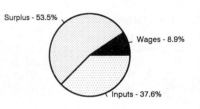

Surplus - 53.5%
Wages - 8.9%
Inputs - 37.6%

The graphic shows percent of gross output.

Value Added in Manufacturing

	1980 $ mil.	1980 %	1985 $ mil.	1985 %	1989 $ mil.	1989 %	1990 $ mil.	1990 %	% change 1980-1990	% change 1989-1990
311 Food products	27[e]	48.2	40[e]	47.6	41[e]	42.7	56[e]	51.4	107.4	36.6
313 Beverages	13[e]	23.2	19[e]	22.6	26[e]	27.1	25[e]	22.9	92.3	-3.8
314 Tobacco products	4[e]	7.1	6[e]	7.1	13[e]	13.5	7[e]	6.4	75.0	-46.2
321 Textiles	2	3.6	3[e]	3.6	2[e]	2.1	3[e]	2.8	50.0	50.0
322 Wearing apparel	3	5.4	4[e]	4.8	3[e]	3.1	4[e]	3.7	33.3	33.3
323 Leather and fur products	1	1.8	1[e]	1.2	NA	0.0	NA	0.0	NA	NA
324 Footwear	NA	0.0	NA	0.0	NA	0.0	NA	0.0	NA	NA
331 Wood and wood products	NA	0.0	NA	0.0	NA	0.0	NA	0.0	NA	NA
332 Furniture and fixtures	NA	0.0	NA	0.0	NA	0.0	NA	0.0	NA	NA
341 Paper and paper products	NA	0.0	NA	0.0	NA	0.0	NA	0.0	NA	NA
342 Printing and publishing	1	1.8	1[e]	1.2	1[e]	1.0	1[e]	0.9	0.0	0.0
351 Industrial chemicals	1	1.8	3[e]	3.6	3[e]	3.1	3[e]	2.8	200.0	0.0
352 Other chemical products	NA	0.0	1[e]	1.2	1[e]	1.0	1[e]	0.9	NA	0.0
353 Petroleum refineries	NA	0.0	NA	0.0	NA	0.0	NA	0.0	NA	NA
354 Miscellaneous petroleum and coal products	NA	0.0	NA	0.0	NA	0.0	NA	0.0	NA	NA
355 Rubber products	NA	0.0	NA	0.0	NA	0.0	NA	0.0	NA	NA
356 Plastic products	NA	0.0	NA	0.0	NA	0.0	NA	0.0	NA	NA
361 Pottery, china and earthenware	NA	0.0	NA	0.0	NA	0.0	NA	0.0	NA	NA
362 Glass and glass products	NA	0.0	NA	0.0	NA	0.0	NA	0.0	NA	NA
369 Other non-metal mineral products	1	1.8	2[e]	2.4	1[e]	1.0	2[e]	1.8	100.0	100.0
371 Iron and steel	NA	0.0	NA	0.0	NA	0.0	NA	0.0	NA	NA
372 Non-ferrous metals	NA	0.0	NA	0.0	NA	0.0	NA	0.0	NA	NA
381 Metal products	2	3.6	4[e]	4.8	4[e]	4.2	5[e]	4.6	150.0	25.0
382 Non-electrical machinery	NA	0.0	NA	0.0	NA	0.0	NA	0.0	NA	NA
383 Electrical machinery	NA	0.0	NA	0.0	NA	0.0	NA	0.0	NA	NA
384 Transport equipment	NA	0.0	NA	0.0	NA	0.0	NA	0.0	NA	NA
385 Professional and scientific equipment	NA	0.0	NA	0.0	NA	0.0	NA	0.0	NA	NA
390 Other manufacturing industries	NA	0.0	NA	0.0	NA	0.0	NA	0.0	NA	NA

Note: The industry codes shown are International Standard Industry codes (ISIC). Percentages are percent of total Value Added. 'e' stands for estimated value

Finance, Economics, and Trade

Economic Indicators [36]

National product: GDP—exchange rate conversion—$1.23 billion (1991 est.). **National product real growth rate**: 5% (1991 est.). **National product per capita**: $205 (1991 est.). **Inflation rate (consumer prices)**: 9% (1991 est.). **External debt**: $1 billion (1990 est.).

Balance of Payments Summary [37]

Values in millions of dollars.

	1987	1988	1989	1990	1991
Exports of goods (f.o.b.)	98.3	124.4	93.2	72.9	91.5
Imports of goods (f.o.b.)	-159.2	-166.1	-151.4	-189.0	-195.9
Trade balance	-60.9	-41.7	-58.2	-116.1	-104.4
Services - debits	-163.4	-140.7	-119.1	-148.5	-157.7
Services - credits	14.8	14.8	24.2	24.8	35.1
Private transfers (net)	7.2	9.9	8.6	10.0	13.2
Government transfers (net)	105.7	87.2	132.4	163.6	182.5
Long term capital (net)	115.9	79.3	64.0	62.3	56.8
Short term capital (net)	15.2	4.9	0.4	15.7	13.7
Errors and omissions	-37.3	-6.6	-14.3	-15.1	-101.7
Overall balance	-2.8	7.1	38.0	-3.3	-62.5

Exchange Rates [38]

Currency: **Burundi francs.**
Symbol: **FBu.**

Data are currency units per $1.

January 1993	235.75
1992	208.30
1991	181.51
1990	171.26
1989	158.67
1988	140.40

Imports and Exports

Top Import Origins [39]

$246 million (c.i.f., 1991).

Origins	%
EC	57
Asia	23
US	3

Top Export Destinations [40]

$91.7 million (f.o.b., 1991).

Destinations	%
EC	83
US	5
Asia	2

Foreign Aid [41]

	U.S. $	
US commitments, including Ex-Im (FY70-89)	71	million
Western (non-US) countries, ODA and OOF bilateral commitments (1970-89)	10.2	billion
OPEC bilateral aid (1979-89)	32	million
Communist countries (1970-89)	175	million

Import and Export Commodities [42]

Import Commodities	**Export Commodities**
Capital goods 31%	Coffee 81%
Petroleum products 15%	Tea
Foodstuffs	Hides
Consumer goods	And skins

Cambodia

Geography [1]

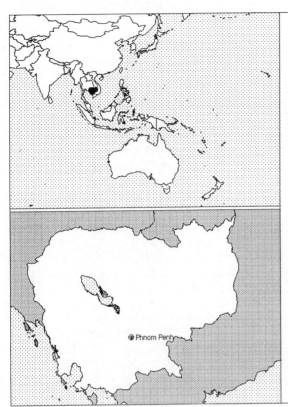

Total area:
181,040 km2
Land area:
176,520 km2
Comparative area:
Slightly smaller than Oklahoma
Land boundaries:
Total 2,572 km, Laos 541 km, Thailand 803 km, Vietnam 1,228 km
Coastline:
443 km
Climate:
Tropical; rainy, monsoon season (May to October); dry season (December to March); little seasonal temperature variation
Terrain:
Mostly low, flat plains; mountains in southwest and north
Natural resources:
Timber, gemstones, some iron ore, manganese, phosphates, hydropower potential
Land use:
Arable land:
16%
Permanent crops:
1%
Meadows and pastures:
3%
Forest and woodland:
76%
Other:
4%

Demographics [2]

	1960	1970	1980	1990	1991[1]	1994	2000	2010	2020
Population	5,364	6,996	6,499	8,731	7,146	10,265	12,098	15,679	20,208
Population density (persons per sq. mi.)	79	104	84	103	105	NA	125	147	175
Births	NA	NA	NA	NA	269	463	NA	NA	NA
Deaths	NA	NA	NA	NA	114	168	NA	NA	NA
Life expectancy - males	NA	NA	NA	NA	48	48	NA	NA	NA
Life expectancy - females	NA	NA	NA	NA	51	51	NA	NA	NA
Birth rate (per 1,000)	NA	NA	NA	NA	38	45	NA	NA	NA
Death rate (per 1,000)	NA	NA	NA	NA	16	16	NA	NA	NA
Women of reproductive age (15-44 yrs.)	NA	NA	NA	NA	NA	NA	NA	NA	NA
of which are currently married	NA	NA	NA	NA	NA	NA	NA	NA	NA
Fertility rate	NA	NA	NA	5.8	4.47	5.8	5.8	5.2	4.6

Population values are in thousands, life expectancy in years, and other items as indicated.

Health

Health Personnel [3]

Doctors per 1,000 pop., 1988-92	0.04
Nurse-to-doctor ratio, 1988-92	8.0
Hospital beds per 1,000 pop., 1985-90	2.2
Percentage of children immunized (age 1 yr. or less)	
Third dose of DPT, 1990-91	38.0
Measles, 1990-91	38.0

Health Expenditures [4]

Total health expenditure, 1990 (official exchange rate)	
Millions of dollars	NA
Millions of dollars per capita	NA
Health expenditures as a percentage of GDP	
Total	NA
Public sector	NA
Private sector	NA
Development assistance for health	
Total aid flows (millions of dollars)[1]	NA
Aid flows per capita (millions of dollars)	NA
Aid flows as a percentage of total health expenditure	NA

For sources, notes, and explanations, see Annotated Source Appendix, page 1035.

155

Human Factors

Health Care Ratios [5]	Infants and Malnutrition [6]

Ethnic Division [7]

Khmer	90%
Vietnamese	5%
Chinese	1%
Other	4%

Religion [8]

Theravada Buddhism	95%
Other	5%

Major Languages [9]

Khmer (official), French.

Education

Public Education Expenditures [10]	Educational Attainment [11]

Literacy Rate [12]

In thousands and percent	1985[a]	1991[a]	2000[a]
Illiterate population + 15 years	3,498	3,479	3,213
Illiteracy rate - total pop. (%)	71.2	64.8	52.0
Illiteracy rate - males (%)	58.7	51.8	38.9
Illiteracy rate - females (%)	83.4	77.6	64.9

Libraries [13]

Daily Newspapers [14]

Cinema [15]

Science and Technology

Scientific/Technical Forces [16]	R&D Expenditures [17]	U.S. Patents Issued [18]

For sources, notes, and explanations, see Annotated Source Appendix, page 1035.

Government and Law

Organization of Government [19]

Long-form name:
none
Type:
transitional government currently administered by the Supreme National Council (SNC)
Independence:
9 November 1949 (from France)
Constitution:
a new constitution will be drafted after the national election in 1993
Legal system:
NA
Executive branch:
a 12 member Supreme National Council (SNC), chaired by Prince NORODOM SIHANOUK
Legislative branch:
transitional legislative body
Judicial branch:
Supreme People's Court pending a national election in 1993

Elections [20]

UN-supervised election for a 120-member constituent assembly based on proportional representation within each province is scheduled for 23-27 May 1993; the assembly will draft and approve a constitution and then transform itself into a legislature that will create a new Cambodian Government.

Government Budget [21]

For 1992 est.
Revenues	120
Expenditures	NA
Capital expenditures	NA

Crime [23]

Military Expenditures and Arms Transfers [22]

	1985	1986	1987	1988	1989
Military expenditures					
Current dollars (mil.)	NA	NA	NA	NA	NA
1989 constant dollars (mil.)	NA	NA	NA	NA	NA
Armed forces (000)	35	40[e]	50[e]	60	99
Gross national product (GNP)					
Current dollars (mil.)	NA	NA	NA	NA	NA
1989 constant dollars (mil.)	NA	NA	NA	NA	NA
Central government expenditures (CGE)					
1989 constant dollars (mil.)	NA	NA	NA	NA	NA
People (mil.)	6.2	6.4	6.5	6.7	6.8
Military expenditure as % of GNP	NA	NA	NA	NA	NA
Military expenditure as % of CGE	NA	NA	NA	NA	NA
Military expenditure per capita	NA	NA	NA	NA	NA
Armed forces per 1,000 people	5.6	6.3	7.7	9.0	14.5
GNP per capita	NA	NA	NA	NA	NA
Arms imports[6]					
Current dollars (mil.)	280	150	460	240	490
1989 constant dollars (mil.)	319	166	495	250	490
Arms exports[6]					
Current dollars (mil.)	0	0	0	0	0
1989 constant dollars (mil.)	0	0	0	0	0
Total imports[7]					
Current dollars (mil.)	28	17	NA	NA	NA
1989 constant dollars	31	19	NA	NA	NA
Total exports[7]					
Current dollars (mil.)	3	3	NA	NA	NA
1989 constant dollars	3	3	NA	NA	NA
Arms as percent of total imports[8]	1,014.5	882.4	NA	NA	NA
Arms as percent of total exports[8]	0	0	NA	NA	NA

Human Rights [24]

	SSTS	FL	FAPRO	PPCG	APROBC	TPW	PCPTW	STPEP	PHRFF	PRW	ASST	AFL
Observes	P			P		P	P				P	
	EAFRD	CPR	ESCR	SR	ACHR	MAAE	PVIAC	PVNAC	EAFDAW	TCIDTP		RC
Observes	P	P	P						S			P

P = Party; S = Signatory; see Appendix for meaning of abbreviations.

Labor Force

Total Labor Force [25]

2,500,000 to 3,000,000

Labor Force by Occupation [26]

Agriculture   80Agriculture
Date of data: 1988 est.

Unemployment Rate [27]

Production Sectors

Energy Resource Summary [28]

Energy Resources: Hydropower potential. **Electricity**: 35,000 kW capacity; 70 million kWh produced, 9 kWh per capita (1990).

Telecommunications [30]

- Service barely adequate for government requirements and virtually nonexistent for general public
- International service limited to Vietnam and other adjacent countries
- Broadcast stations - 1 AM, no FM, 1 TV

Transportation [31]

Railroads. 612 km 1.000-meter gauge, government owned

Highways. 13,351 km total; 2,622 km bituminous; 7,105 km crushed stone, gravel, or improved earth; 3,624 km unimproved earth; some roads in disrepair

Airports

Total:	15
Usable:	9
With permanent-surface runways:	5
With runways over 3,659 m:	0
With runways 2,440-3,659 m:	2
With runways 1,220-2,439 m:	4

Top Agricultural Products [32]

Mainly subsistence farming except for rubber plantations; main crops—rice, rubber, corn; food shortages—rice, meat, vegetables, dairy products, sugar, flour.

Top Mining Products [33]

Metric tons unless otherwise specified	M.t.
Brick	NA
Clays	NA
Gravel	NA
Stone	NA
Cement	NA

Tourism [34]

Finance, Economics, and Trade

Industrial Summary [35]

Industrial Production: Growth rate NA%. **Industries**: Rice milling, fishing, wood and wood products, rubber, cement, gem mining.

Economic Indicators [36]

National product: GDP—exchange rate conversion—$2 billion (1991 est.). **National product real growth rate**: NA%. **National product per capita**: $280 (1991 est.). **Inflation rate (consumer prices)**: 250-300% (1992 est.). **External debt**: $717 million (1990).

Balance of Payments Summary [37]

Exchange Rates [38]

Currency: **riels.**
Symbol: **CR.**

Data are currency units per $1.

September 1992	2,800
December 1991	500
1990	560
1988	159
1987	100

Imports and Exports

Top Import Origins [39]

$170 million (c.i.f., 1990 est.).

Origins	%
Vietnam	NA
USSR	NA
Eastern Europe	NA
Japan	NA
India	NA

Top Export Destinations [40]

$59 million (f.o.b., 1990 est.).

Destinations	%
Vietnam	NA
USSR	NA
Eastern Europe	NA
Japan	NA
India	NA

Foreign Aid [41]

	U.S. $	
US commitments, including Ex-Im (FY70-89)	725	million
Western (non-US countries) (1970-89)	300	million
Communist countries (1970-89)	1.8	billion

Import and Export Commodities [42]

Import Commodities	Export Commodities
International food aid	Natural rubber
Fuels	Rice
Consumer goods	Pepper
Machinery	Wood

Cameroon

Geography [1]

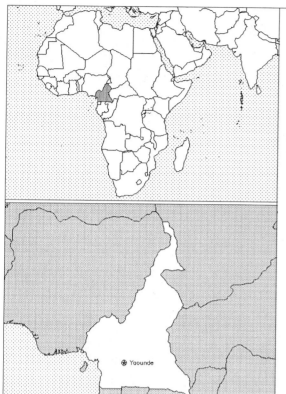

Total area:
475,440 km2
Land area:
469,440 km2
Comparative area:
Slightly larger than California
Land boundaries:
Total 4,591 km, Central African Republic 797 km, Chad 1,094 km, Congo 523 km, Equatorial Guinea 189 km, Gabon 298 km, Nigeria 1,690 km
Coastline:
402 km
Climate:
Varies with terrain from tropical along coast to semiarid and hot in north
Terrain:
Diverse, with coastal plain in southwest, dissected plateau in center, mountains in west, plains in north
Natural resources:
Petroleum, bauxite, iron ore, timber, hydropower potential
Land use:
Arable land:
13%
Permanent crops:
2%
Meadows and pastures:
18%
Forest and woodland:
54%
Other:
13%

Demographics [2]

	1960	1970	1980	1990	1991[1]	1994	2000	2010	2020
Population	5,609	6,727	8,756	11,697	11,390	13,132	15,677	21,165	28,329
Population density (persons per sq. mi.)	31	37	47	61	63	NA	80	103	130
Births	NA	NA	NA	NA	471	532	NA	NA	NA
Deaths	NA	NA	NA	NA	169	150	NA	NA	NA
Life expectancy - males	NA	NA	NA	NA	49	55	NA	NA	NA
Life expectancy - females	NA	NA	NA	NA	53	59	NA	NA	NA
Birth rate (per 1,000)	NA	NA	NA	NA	41	41	NA	NA	NA
Death rate (per 1,000)	NA	NA	NA	NA	15	11	NA	NA	NA
Women of reproductive age (15-44 yrs.)	NA	NA	NA	2,590	NA	2,952	3,590	NA	NA
of which are currently married	NA	NA	NA	1,772	NA	1,999	2,428	NA	NA
Fertility rate	NA	NA	NA	6.0	5.63	5.8	5.6	5.2	4.7

Population values are in thousands, life expectancy in years, and other items as indicated.

Health

Health Personnel [3]

Doctors per 1,000 pop., 1988-92	0.08
Nurse-to-doctor ratio, 1988-92	6.4
Hospital beds per 1,000 pop., 1985-90	2.7
Percentage of children immunized (age 1 yr. or less)	
Third dose of DPT, 1990-91	56.0
Measles, 1990-91	56.0

Health Expenditures [4]

Total health expenditure, 1990 (official exchange rate)	
Millions of dollars	286
Millions of dollars per capita	24
Health expenditures as a percentage of GDP	
Total	2.6
Public sector	1.0
Private sector	1.6
Development assistance for health	
Total aid flows (millions of dollars)[1]	38
Aid flows per capita (millions of dollars)	3.3
Aid flows as a percentage of total health expenditure	13.4

For sources, notes, and explanations, see Annotated Source Appendix, page 1035.

Human Factors

Health Care Ratios [5]

Population per physician, 1970	28920
Population per physician, 1990	12190
Population per nursing person, 1970	2560
Population per nursing person, 1990	1690
Percent of births attended by health staff, 1985	NA

Infants and Malnutrition [6]

Percent of babies with low birth weight, 1985	13
Infant mortality rate per 1,000 live births, 1970	126
Infant mortality rate per 1,000 live births, 1991	64
Years of life lost per 1,000 population, 1990	67
Prevalence of malnutrition (under age 5), 1990	NA

Ethnic Division [7]

Cameroon Highlanders	31%
Equatorial Bantu	19%
Kirdi	11%
Fulani	10%
Northwestern Bantu	8%
Eastern Nigritic	7%
Other African	13%
Non-African	<1%

Religion [8]

Indigenous beliefs	51%
Christian	33%
Muslim	16%

Major Languages [9]

24 major African language groups, English (official), French (official).

Education

Public Education Expenditures [10]

Million Francs C.F.A.	1980	1985[1]	1987[1]	1988[1]	1989[1]	1990[1]
Total education expenditure	45,099	109,344	NA	96,541	115,479	107,968
as percent of GNP	3.2	3.0	NA	2.7	3.4	3.4
as percent of total govt. expend.	20.3	14.8	NA	16.1	19.2	19.6
Current education expenditure	36,653	90,045	NA	86,759	103,319	97,948
as percent of GNP	2.6	2.5	NA	2.4	3.0	3.1
as percent of current govt. expend.	NA	20.9	NA	23.1	24.3	26.9
Capital expenditure	8,446	19,299	NA	9,782	12,160	10,020

Educational Attainment [11]

Literacy Rate [12]

In thousands and percent	1985[a]	1991[a]	2000[a]
Illiterate population +15 years	2,911	2,912	2,858
Illiteracy rate - total pop. (%)	52.0	45.9	34.0
Illiteracy rate - males (%)	38.9	33.7	24.3
Illiteracy rate - females (%)	64.4	57.4	43.4

Libraries [13]

	Admin. Units	Svc. Pts.	Vols. (000)	Shelving (meters)	Added	Reg. Users
National (1989)	1	7	40	1,000	25,000	5,000
Non-specialized (1989)	1	14	50	NA	3,500	3,000
Public	NA	NA	NA	NA	NA	NA
Higher ed. (1988)	3	3	13	NA	2,939	617
School	NA	NA	NA	NA	NA	NA

Daily Newspapers [14]

	1975	1980	1985	1990
Number of papers	2	2	1	2
Circ. (000)	25	65	35	80[e]

Cinema [15]

Data for 1991.

Cinema seats per 1,000	3.4
Annual attendance per person	NA
Gross box office receipts (mil. Franc C.F.A.)	NA

Science and Technology

Scientific/Technical Forces [16]

R&D Expenditures [17]

U.S. Patents Issued [18]

For sources, notes, and explanations, see Annotated Source Appendix, page 1035.

Government and Law

Organization of Government [19]

Long-form name:
Republic of Cameroon
Type:
unitary republic; multiparty presidential regime (opposition parties legalized 1990)
Independence:
1 January 1960 (from UN trusteeship under French administration)
Constitution:
20 May 1972
Legal system:
based on French civil law system, with common law influence; has not accepted compulsory ICJ jurisdiction
National holiday:
National Day, 20 May (1972)
Executive branch:
president, Cabinet
Legislative branch:
unicameral National Assembly (Assemblee Nationale)
Judicial branch:
Supreme Court

Political Parties [20]

National Assembly	% of seats
People's Democratic Movement	48.9
National Union	37.8
Democratic Union	10.0
MDR	3.3

Government Budget [21]

For FY90 est.

Revenues	1.700
Expenditures	2.400
Capital expenditures	0.422

Crime [23]

Military Expenditures and Arms Transfers [22]

	1985	1986	1987	1988	1989
Military expenditures					
Current dollars (mil.)	209	244[e]	224	160	148
1989 constant dollars (mil.)	238	270[e]	241	166	148
Armed forces (000)	15	15	15[e]	21	21
Gross national product (GNP)					
Current dollars (mil.)	10,520	11,770	11,720	11,260	11,100
1989 constant dollars (mil.)	11,980	13,060	12,600	11,720	11,100
Central government expenditures (CGE)					
1989 constant dollars (mil.)	2,849	2,993	2,876	NA	2,232
People (mil.)	9.7	10.0	10.3	10.6	10.8
Military expenditure as % of GNP	2.0	2.1	1.9	1.4	1.3
Military expenditure as % of CGE	8.3	9.0	8.4	NA	6.6
Military expenditure per capita	24	27	23	16	14
Armed forces per 1,000 people	1.5	1.5	1.5	1.9	1.9
GNP per capita	1,230	1,303	1,224	1,109	1,025
Arms imports[6]					
Current dollars (mil.)	20	10	10	5	5
1989 constant dollars (mil.)	23	11	11	5	5
Arms exports[6]					
Current dollars (mil.)	0	0	0	0	0
1989 constant dollars (mil.)	0	0	0	0	0
Total imports[7]					
Current dollars (mil.)	1,151	1,704	1,723	1,272	NA
1989 constant dollars	1,310	1,891	1,853	1,324	NA
Total exports[7]					
Current dollars (mil.)	722	782	806	928	1,000
1989 constant dollars	822	868	867	966	1,000
Arms as percent of total imports[8]	1.7	0.6	0.6	0.4	NA
Arms as percent of total exports[8]	0	0	0	0	0

Human Rights [24]

	SSTS	FL	FAPRO	PPCG	APROBC	TPW	PCPTW	STPEP	PHRFF	PRW	ASST	AFL
Observes	P	P	P		P	P	P	P			P	P
	EAFRD	CPR	ESCR	SR	ACHR	MAAE	PVIAC	PVNAC	EAFDAW	TCIDTP	RC	
Observes		P	P	P	P				P	s	P	S

P = Party; S = Signatory; see Appendix for meaning of abbreviations.

Labor Force

Total Labor Force [25]

Labor Force by Occupation [26]

Agriculture	74.4%
Industry and transport	11.4
Other services	14.2

Date of data: 1983

Unemployment Rate [27]

25% (1990 est.)

For sources, notes, and explanations, see Annotated Source Appendix, page 1035.

Production Sectors

Energy Resource Summary [28]

Energy Resources: Petroleum, hydropower potential. **Electricity**: 755,000 kW capacity; 2,190 million kWh produced, 190 kWh per capita (1991).

Telecommunications [30]

- Good system of open wire, cable, troposcatter, and microwave radio relay
- 26,000 telephones, 2 telephones per 1,000 persons, available only to business and government
- Broadcast stations - 11 AM, 11 FM, 1 TV
- 2 Atlantic Ocean INTELSAT earth stations

Transportation [31]

Railroads. 1,003 km total; 858 km 1.000-meter gauge, 145 km 0.600-meter gauge

Highways. about 65,000 km total; includes 2,682 km paved, 32,318 km gravel and improved earth, and 30,000 km of unimproved earth

Merchant Marine. 2 cargo ships (1,000 GRT or over) totaling 24,122 GRT/33,509 DWT

Airports

Total:	59
Usable:	51
With permanent-surface runways:	11
With runways over 3,659 m:	0
With runways 2,440-3,659 m:	6
With runways 1,220-2,439 m:	51

Top Agricultural Products [32]

The agriculture and forestry sectors provide employment for the majority of the population, contributing nearly 25% to GDP and providing a high degree of self-sufficiency in staple foods; commercial and food crops include coffee, cocoa, timber, cotton, rubber, bananas, oilseed, grains, livestock, root starches.

Top Mining Products [33]

Metric tons unless otherwise specified	M.t.
Aluminum metal, primary	82,516
Gold, mine output, Au content (kilograms)	8
Pozzolana	130,000
Limestone	57,000
Tin ore, Sn content of concentrate (kilograms)	3,050
Marble	200

Tourism [34]

	1987	1988	1989	1990	1991
Tourists[10]	118	100	87	100	71
Tourism receipts	25	21	17	21	15
Tourism expenditures	244	283			
Fare receipts	54	55	55	54	54
Fare expenditures	69	59	64	65	64

Tourists are in thousands, money in million U.S. dollars.

Manufacturing Sector

GDP and Manufacturing Summary [35]

	1980	1985	1989	1990	% change 1980-1990	% change 1989-1990
GDP (million 1980 $)	6,674	9,911	NA	8,597	28.8	NA
GDP per capita (1980 $)	771	986	NA	727	-5.7	NA
Manufacturing as % of GDP (current prices)	9.2	11.4	NA	11.1	20.7	NA
Gross output (million $)	1,708[e]	1,650[e]	NA	2,607	52.6	NA
Value added (million $)	707[e]	705[e]	NA	826	16.8	NA
Value added (million 1980 $)	587	1,433	NA	1,399	138.3	NA
Industrial production index	100	153	NA	156	56.0	NA
Employment (thousands)	51[e]	68[e]	NA	50	-2.0	NA

Note: GDP stands for Gross Domestic Product. 'e' stands for estimated value.

Profitability and Productivity

	1980	1985	1989	1990	% change 1980-1990	% change 1989-1990
Intermediate input (%)	59[e]	57[e]	NA	68	15.3	NA
Wages, salaries, and supplements (%)	14[e]	13[e]	NA	14[e]	0.0	NA
Gross operating surplus (%)	27[e]	30[e]	NA	18[e]	-33.3	NA
Gross output per worker ($)	33,434[e]	24,107[e]	NA	51,631	54.4	NA
Value added per worker ($)	13,838[e]	10,410[e]	NA	16,357	18.2	NA
Average wage (incl. benefits) ($)	4,794[e]	3,075[e]	NA	7,281[e]	51.9	NA

Profitability is in percent of gross output. Productivity is in U.S. $. 'e' stands for estimated value.

Profitability - 1990

Wages - 14.0%
Inputs - 68.0%
Surplus - 18.0%

The graphic shows percent of gross output.

Value Added in Manufacturing

	1980 $ mil.	1980 %	1985 $ mil.	1985 %	1989 $ mil.	1989 %	1990 $ mil.	1990 %	% change 1980-1990	% change 1989-1990
311 Food products	187[e]	26.4	128[e]	18.2	NA	NA	185	22.4	-1.1	NA
313 Beverages	183[e]	25.9	179[e]	25.4	NA	NA	294	35.6	60.7	NA
314 Tobacco products	24[e]	3.4	20[e]	2.8	NA	NA	23	2.8	-4.2	NA
321 Textiles	36[e]	5.1	44[e]	6.2	NA	NA	-87	-10.5	-341.7	NA
322 Wearing apparel	10[e]	1.4	14[e]	2.0	NA	NA	-27	-3.3	-370.0	NA
323 Leather and fur products	7[e]	1.0	3[e]	0.4	NA	NA	3	0.4	-57.1	NA
324 Footwear	10[e]	1.4	4[e]	0.6	NA	NA	5	0.6	-50.0	NA
331 Wood and wood products	30[e]	4.2	57[e]	8.1	NA	NA	61	7.4	103.3	NA
332 Furniture and fixtures	13[e]	1.8	24[e]	3.4	NA	NA	26	3.1	100.0	NA
341 Paper and paper products	17[e]	2.4	7[e]	1.0	NA	NA	11	1.3	-35.3	NA
342 Printing and publishing	20[e]	2.8	8[e]	1.1	NA	NA	6	0.7	-70.0	NA
351 Industrial chemicals	10[e]	1.4	18[e]	2.6	NA	NA	17	2.1	70.0	NA
352 Other chemical products	12[e]	1.7	19[e]	2.7	NA	NA	21	2.5	75.0	NA
353 Petroleum refineries	15[e]	2.1	54[e]	7.7	NA	NA	114[e]	13.8	660.0	NA
354 Miscellaneous petroleum and coal products	NA	0.0	NA	0.0	NA	NA	NA	0.0	NA	NA
355 Rubber products	2[e]	0.3	2[e]	0.3	NA	NA	1[e]	0.1	-50.0	NA
356 Plastic products	16[e]	2.3	20[e]	2.8	NA	NA	25[e]	3.0	56.3	NA
361 Pottery, china and earthenware	6[e]	0.8	4[e]	0.6	NA	NA	8	1.0	33.3	NA
362 Glass and glass products	4[e]	0.6	3[e]	0.4	NA	NA	6	0.7	50.0	NA
369 Other non-metal mineral products	12[e]	1.7	9[e]	1.3	NA	NA	16	1.9	33.3	NA
371 Iron and steel	24[e]	3.4	30[e]	4.3	NA	NA	30	3.6	25.0	NA
372 Non-ferrous metals	19[e]	2.7	25[e]	3.5	NA	NA	23	2.8	21.1	NA
381 Metal products	13[e]	1.8	9[e]	1.3	NA	NA	17	2.1	30.8	NA
382 Non-electrical machinery	18[e]	2.5	13[e]	1.8	NA	NA	32	3.9	77.8	NA
383 Electrical machinery	4[e]	0.6	3[e]	0.4	NA	NA	9	1.1	125.0	NA
384 Transport equipment	3[e]	0.4	5[e]	0.7	NA	NA	3	0.4	0.0	NA
385 Professional and scientific equipment	NA	0.0	NA	0.0	NA	NA	NA	0.0	NA	NA
390 Other manufacturing industries	11[e]	1.6	4[e]	0.6	NA	NA	6	0.7	-45.5	NA

Note: The industry codes shown are International Standard Industry codes (ISIC). Percentages are percent of total Value Added. 'e' stands for estimated value

For sources, notes, and explanations, see Annotated Source Appendix, page 1035.

Finance, Economics, and Trade

Economic Indicators [36]

National product: GDP—exchange rate conversion—$11.5 billion (1990 est.). **National product real growth rate**: 3% (1990 est.). **National product per capita**: $1,040 (1990 est.). **Inflation rate (consumer prices)**: 3% (1990 est.). **External debt**: $6 billion (1991).

Balance of Payments Summary [37]

Values in millions of dollars.

	1980	1985	1986	1987	1988
Exports of goods (f.o.b.)	1,657.5	1,626.3	2,077.0	1,688.7	1,841.2
Imports of goods (f.o.b.)	-1,620.3	-1,135.9	-1,634.5	-1,434.8	-1,220.8
Trade balance	37.2	490.4	442.5	253.9	620.4
Services - debits	-931.4	-1,570.1	-1,419.3	-1,471.4	-1,416.4
Services - credits	439.3	536.8	518.5	423.8	473.0
Private transfers (net)	-73.4	-99.9	-122.8	-125.7	-134.9
Government transfers (net)	82.9	81.3	30.1	26.6	29.0
Long term capital (net)	532.0	468.3	90.4	430.0	304.8
Short term capital (net)	19.4	21.6	156.6	375.9	-27.8
Errors and omissions	-6.4	108.9	-21.7	89.3	166.2
Overall balance	99.6	37.3	-325.8	2.4	14.3

Exchange Rates [38]

Currency: **Communaute Financiere Africaine francs.**
Symbol: **CFAF.**

Data are currency units per $1.

January 1993	274.06
1992	264.69
1991	282.11
1990	272.26
1989	319.01
1988	297.85

Imports and Exports

Top Import Origins [39]

$1.2 billion (c.i.f., 1991).

Origins	%
EC	60
France	41
Germany	9
African countries	NA
Japan	NA
US	4

Top Export Destinations [40]

$1.8 billion (f.o.b., 1991).

Destinations	%
EC (particularly France)	50
US	NA
African countries	NA

Foreign Aid [41]

	U.S. $	
US commitments, including Ex-Im (FY70-90)	479	million
Western (non-US) countries, ODA and OOF bilateral commitments (1970-90)	4.75	billion
OPEC bilateral aid (1979-89)	29	million
Communist countries (1970-89)	125	million

Import and Export Commodities [42]

Import Commodities	Export Commodities
Machines and electrical equipment	Petroleum products 51%
Food	Coffee
Consumer goods	Beans
Transport equipment	Cocoa
	Aluminum products
	Timber

For sources, notes, and explanations, see Annotated Source Appendix, page 1035.

165

Canada

Geography [1]

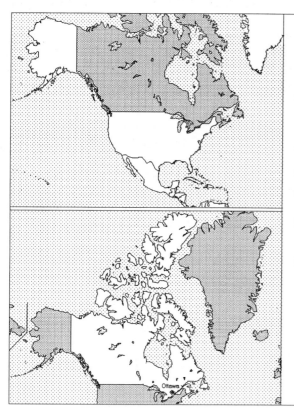

Total area:
9,976,140 km2
Land area:
9,220,970 km2
Comparative area:
Slightly larger than US
Land boundaries:
Total 8,893 km, US 8,893 km (includes 2,477 km with Alaska)
Coastline:
243,791 km
Climate:
Varies from temperate in south to subarctic and arctic in north
Terrain:
Mostly plains with mountains in west and lowlands in southeast
Natural resources:
Nickel, zinc, copper, gold, lead, molybdenum, potash, silver, fish, timber, wildlife, coal, petroleum, natural gas
Land use:
Arable land:
5%
Permanent crops:
0%
Meadows and pastures:
3%
Forest and woodland:
35%
Other:
57%

Demographics [2]

	1960	1970	1980	1990	1991[1]	1994	2000	2010	2020
Population	17,909	21,324	24,070	26,620	26,835	28,114	29,867	32,265	34,347
Population density (persons per sq. mi.)	5	6	7	7	8	NA	8	9	9
Births	NA	NA	NA	NA	370	396	NA	NA	NA
Deaths	NA	NA	NA	NA	200	208	NA	NA	NA
Life expectancy - males	NA	NA	NA	NA	74	75	NA	NA	NA
Life expectancy - females	NA	NA	NA	NA	81	82	NA	NA	NA
Birth rate (per 1,000)	NA	NA	NA	NA	14	14	NA	NA	NA
Death rate (per 1,000)	NA	NA	NA	NA	7	7	NA	NA	NA
Women of reproductive age (15-44 yrs.)	NA	NA	NA	7,154	NA	7,453	7,630	NA	NA
of which are currently married	NA	NA	NA	4,389	NA	4,654	4,779	NA	NA
Fertility rate	NA	NA	NA	1.8	1.68	1.8	1.8	1.8	1.8

Population values are in thousands, life expectancy in years, and other items as indicated.

Health

Health Personnel [3]

Doctors per 1,000 pop., 1988-92	2.22
Nurse-to-doctor ratio, 1988-92	4.7
Hospital beds per 1,000 pop., 1985-90	16.1
Percentage of children immunized (age 1 yr. or less)	
Third dose of DPT, 1990-91	85.0
Measles, 1990-91	85.0

Health Expenditures [4]

Total health expenditure, 1990 (official exchange rate)	
Millions of dollars	51594
Millions of dollars per capita	1945
Health expenditures as a percentage of GDP	
Total	9.1
Public sector	6.8
Private sector	2.4
Development assistance for health	
Total aid flows (millions of dollars)[1]	NA
Aid flows per capita (millions of dollars)	NA
Aid flows as a percentage of total health expenditure	NA

For sources, notes, and explanations, see Annotated Source Appendix, page 1035.

Human Factors

Health Care Ratios [5]

Population per physician, 1970	680
Population per physician, 1990	450
Population per nursing person, 1970	140
Population per nursing person, 1990	NA
Percent of births attended by health staff, 1985	99

Infants and Malnutrition [6]

Percent of babies with low birth weight, 1985	6
Infant mortality rate per 1,000 live births, 1970	19
Infant mortality rate per 1,000 live births, 1991	7
Years of life lost per 1,000 population, 1990	9
Prevalence of malnutrition (under age 5), 1990	NA

Ethnic Division [7]

British Isles origin	40.0%
French origin	27.0%
Other European	20.0%
Indigenous Indian/Eskimo	1.5%

Religion [8]

Roman Catholic	46%
United Church	16%
Anglican	10%
Other	28%

Major Languages [9]

English (official), French (official).

Education

Public Education Expenditures [10]

Million Canadian Dollars	1980[7]	1985[7]	1987[7]	1988[7]	1989[7]	1990[7,8]
Total education expenditure	22,100	32,429	38,809	42,024	44,187	47,708
as percent of GNP	7.4	7.1	7.3	7.2	7.1	7.4
as percent of total govt. expend.	17.3	12.7	15.4	15.9	15.3	15.6
Current education expenditure	20,451	30,176	36,226	39,073	41,091	44,319
as percent of GNP	6.8	6.6	6.8	6.7	6.6	6.9
as percent of current govt. expend.	NA	NA	NA	NA	NA	NA
Capital expenditure	1,649	2,253	2,583	2,951	3,096	3,389

Educational Attainment [11]

Age group	25+
Total population	13,971,280
Highest level attained (%)	
No schooling	2.0
First level	
Incompleted	14.2
Completed	9.5
Entered second level	
S-1	36.8
S-2	NA
Post secondary	37.4

Literacy Rate [12]

	1980[b]	1986[b]	1990[b]
Illiterate population +15 years	NA	659,745	NA
Illiteracy rate - total pop. (%)	NA	3.4	
Illiteracy rate - males (%)	NA	NA	NA
Illiteracy rate - females (%)	NA	NA	NA

Libraries [13]

	Admin. Units	Svc. Pts.	Vols. (000)	Shelving (meters)	Vols. Added	Reg. Users
National (1989)	1	NA	7,200	NA	84,740	NA
Non-specialized	NA	NA	NA	NA	NA	NA
Public (1988)[14]	993	6,157[14]	59,581	NA	4.4mil	NA
Higher ed.	NA	NA	NA	NA	NA	NA
School	NA	NA	NA	NA	NA	NA

Daily Newspapers [14]

	1975	1980	1985	1990[2]
Number of papers	121	123	117	107
Circ. (000)	4,900[e]	5,425	5,566	5,993

Cinema [15]

Data for 1990.

Cinema seats per 1,000	27.1
Annual attendance per person	3.0
Gross box office receipts (mil. Dollar)	439[1]

Science and Technology

Scientific/Technical Forces [16]

Potential scientists/engineers	1,623,045
Number female	638,035
Potential technicians	3,063,745
Number female	1,433,685
Total	4,686,790

R&D Expenditures [17]

	Dollar[7] (000) 1989
Total expenditure	8,658,000[e]
Capital expenditure	NA
Current expenditure	NA
Percent current	NA

U.S. Patents Issued [18]

Values show patents issued to citizens of the country by the U.S. Patents Office.

	1990	1991	1992
Number of patents	2,090	2,302	2,218

Government and Law

Organization of Government [19]

Long-form name:
none
Type:
confederation with parliamentary democracy
Independence:
1 July 1867 (from UK)
Constitution:
amended British North America Act 1867 patriated to Canada 17 April 1982; charter of rights and unwritten customs
Legal system:
based on English common law, except in Quebec, where civil law system based on French law prevails; accepts compulsory ICJ jurisdiction, with reservations
National holiday:
Canada Day, 1 July (1867)
Executive branch:
British monarch, governor general, prime minister, deputy prime minister, Cabinet
Legislative branch:
bicameral Parliament (Parlement) consists of an upper house or Senate (Senat) and a lower house or House of Commons (Chambre des Communes)
Judicial branch:
Supreme Court

Political Parties [20]

House of Commons	% of votes
Progressive Conservative Party	43.0
Liberal Party	32.0
New Democratic Party	20.0
Other	5.0

Government Expenditures [21]

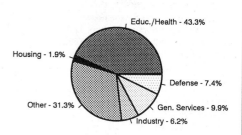

Educ./Health - 43.3%
Housing - 1.9%
Defense - 7.4%
Other - 31.3%
Gen. Services - 9.9%
Industry - 6.2%

% distribution for 1989. Expend. in FY90 est.: 138.3 ($ bil.)

Crime [23]

Crime volume (for 1990)	
Cases known to police	3,320,155
Attempts (percent)	NA
Percent cases solved	44.65
Crimes per 100,000 persons	11,913
Persons responsible for offenses	
Total number offenders	898,824
Percent female	12.58
Percent juvenile[12]	17.43
Percent foreigners	NA

Military Expenditures and Arms Transfers [22]

	1985	1986	1987	1988	1989
Military expenditures					
Current dollars (mil.)	8,943	9,509	10,040	10,500	10,840
1989 constant dollars (mil.)	10,180	10,550	10,800	10,930	10,840
Armed forces (000)	83	85	86	88	88
Gross national product (GNP)					
Current dollars (mil.)	401,300	423,100	457,500	496,500	531,000
1989 constant dollars (mil.)	456,900	469,500	492,000	516,800	531,000
Central government expenditures (CGE)					
1989 constant dollars (mil.)	118,000	114,800	115,800	121,400	123,000
People (mil.)	25.2	25.4	25.6	25.9	26.2
Military expenditure as % of GNP	2.2	2.2	2.2	2.1	2.0
Military expenditure as % of CGE	8.6	9.2	9.3	9.0	8.8
Military expenditure per capita	404	416	421	422	413
Armed forces per 1,000 people	3.3	3.4	3.4	3.4	3.4
GNP per capita	18,140	18,500	19,190	19,930	20,240
Arms imports[6]					
Current dollars (mil.)	100	130	170	210	190
1989 constant dollars (mil.)	114	144	183	219	190
Arms exports[6]					
Current dollars (mil.)	550	525	600	650	410
1989 constant dollars (mil.)	626	583	645	677	410
Total imports[7]					
Current dollars (mil.)	80,640	85,070	92,590	112,200	121,200
1989 constant dollars	91,800	94,400	99,590	116,800	121,200
Total exports[7]					
Current dollars (mil.)	90,950	90,190	97,850	117,800	121,400
1989 constant dollars	103,500	100,100	105,200	122,600	121,400
Arms as percent of total imports[8]	0.1	0.2	0.2	0.2	0.2
Arms as percent of total exports[8]	0.6	0.6	0.6	0.6	0.3

Human Rights [24]

	SSTS	FL	FAPRO	PPCG	APROBC	TPW	PCPTW	STPEP	PHRFF	PRW	ASST	AFL
Observes	P		P	P		P	P			P	P	P
	EAFRD	CPR	ESCR	SR	ACHR	MAAE	PVIAC	PVNAC	EAFDAW	TCIDTP	RC	
Observes		P	P	P	P			S	S	P	P	P

P = Party; S = Signatory; see Appendix for meaning of abbreviations.

Labor Force

Total Labor Force [25]

13.38 million

Labor Force by Occupation [26]

Services	75%
Manufacturing	14
Agriculture	4
Construction	3
Other	4

Date of data: 1988

Unemployment Rate [27]

11.5% (December 1992)

For sources, notes, and explanations, see Annotated Source Appendix, page 1035.

Production Sectors

Commercial Energy Production and Consumption

Production [28]

NG liquids - 4.7%
Natural gas - 28.9%
Crude oil - 24.0%
Coal - 12.2%
Nuclear - 7.0%
Hydro - 23.2%

Consumption [29]

Petroleum - 30.2%
Coal - 9.9%
Nuclear - 8.9%
Hydro - 27.9%
Natural gas - 23.0%

Data are shown in quadrillion (10^{15}) BTUs and percent for 1991

Crude oil[1]	3.28
Natural gas liquids	0.64
Dry natural gas	3.95
Coal[2]	1.67
Net hydroelectric power[3]	3.16
Net nuclear power[3]	0.95
Total	13.66

Petroleum	3.23
Dry natural gas	2.46
Coal[2]	1.06
Net hydroelectric power[3]	2.98
Net nuclear power[3]	0.95
Total	10.67

Telecommunications [30]

- Excellent service provided by modern media
- 18.0 million telephones
- Broadcast stations - 900 AM, 29 FM, 53 (1,400 repeaters) TV
- 5 coaxial submarine cables
- Over 300 earth stations operating in INTELSAT (including 4 Atlantic Ocean and 1 Pacific Ocean) and domestic systems

Transportation [31]

Railroads. 146,444 km total; two major transcontinental freight railway systems - Canadian National (government owned) and Canadian Pacific Railway; passenger service - VIA (government operated); 158 km is electrified

Highways. 884,272 km total; 712,936 km surfaced (250,023 km paved), 171,336 km earth

Merchant Marine. 63 ships (1,000 GRT or over) totaling 454,582 GRT/646,329 DWT; includes 1 passenger, 3 short-sea passenger, 2 passenger-cargo, 8 cargo, 2 railcar carrier, 1 refrigerated cargo, 7 roll-on/roll-off, 1 container, 24 oil tanker, 4 chemical tanker, 1 specialized tanker, 9 bulk; note - does not include ships used exclusively in the Great Lakes

Airports

Total:	1,420
Useable:	1,142
With permanent-surface runways:	457
With runways over 3,659 m:	4
With runways 2,440-3,659 m:	30
With runways 1,220-2,439 m:	330

Top Agricultural Products [32]

	89-90	90-91[3]
Wheat	32.7	32.8
Barley	13.9	12.5
Corn	7.1	7.3
Canola	3.3	4.3
Soybeans	1.3	1.4

Values shown are 1,000 metric tons.

Top Mining Products [33]

Preliminary metric tons unless otherwise specified M.t.

Petroleum crude (000 42-gal. barrels)[13]	563,985
Natural gas, gross (mil. cubic meters)	144,987
Gold (kilograms)	178,712
Natural gas liquids (000 42-gal. barrels)	157,973
Copper, mine output, recoverable Cu content[14]	797,603
Coal (000 tons)	71,130
Nickel, mine output, Ni content[15]	196,868
Zinc, mine output, Zn content	1,148,149
Iron ore, gross weight (000 tons)[16]	35,961
Potash, K2O eqivalent (000 tons)	7,012

Tourism [34]

	1987	1988	1989	1990	1991
Visitors	39,595	39,253	37,982	37,990	36,818
Tourists	14,975	15,485	15,111	15,258	14,989
Excursionists	24,620	23,768	22,871	22,732	21,829
Tourism receipts[13]	3,961	4,603	5,014	5,231	5,537
Tourism expenditures[13]	5,304	6,316	7,370	9,974	10,526
Fare receipts	765	969	1,060	1,143	1,209
Fare expenditures	1,348	1,470	1,635	2,265	2,391

Tourists are in thousands, money in million U.S. dollars.

For sources, notes, and explanations, see Annotated Source Appendix, page 1035.

169

Manufacturing Sector

GDP and Manufacturing Summary [35]

	1980	1985	1989	1990	% change 1980-1990	% change 1989-1990
GDP (million 1980 $)	263,242	303,726	NA	349,214	32.7	NA
GDP per capita (1980 $)	10,949	11,968	NA	13,158	20.2	NA
Manufacturing as % of GDP (current prices)	19.5	19.0	NA	18.3[e]	-6.2	NA
Gross output (million $)	167,211	211,017	NA	305,886	82.9	NA
Value added (million $)	59,803	74,209	NA	115,821	93.7	NA
Value added (million 1980 $)	47,086	53,990	NA	56,928	20.9	NA
Industrial production index	100	111	NA	119	19.0	NA
Employment (thousands)	1,853	1,765	NA	1,828	-1.3	NA

Note: GDP stands for Gross Domestic Product. 'e' stands for estimated value.

Profitability and Productivity

	1980	1985	1989	1990	% change 1980-1990	% change 1989-1990
Intermediate input (%)	64	65	NA	62	-3.1	NA
Wages, salaries, and supplements (%)	17	16	NA	17[e]	0.0	NA
Gross operating surplus (%)	19	19	NA	21[e]	10.5	NA
Gross output per worker ($)	89,995	119,212	NA	167,206	85.8	NA
Value added per worker ($)	32,187	41,950	NA	63,338	96.8	NA
Average wage (incl. benefits) ($)	15,296	19,168	NA	27,410[e]	79.2	NA

Profitability is in percent of gross output. Productivity is in U.S. $. 'e' stands for estimated value.

Profitability - 1990

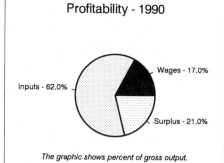

Wages - 17.0%
Inputs - 62.0%
Surplus - 21.0%

The graphic shows percent of gross output.

Value Added in Manufacturing

	1980 $ mil.	1980 %	1985 $ mil.	1985 %	1989 $ mil.	1989 %	1990 $ mil.	1990 %	% change 1980-1990	% change 1989-1990
311 Food products	6,142	10.3	8,001	10.8	NA	NA	12,352	10.7	101.1	NA
313 Beverages	1,660	2.8	2,189	2.9	NA	NA	3,211	2.8	93.4	NA
314 Tobacco products	479	0.8	608	0.8	NA	NA	928	0.8	93.7	NA
321 Textiles	2,130	3.6	2,152	2.9	NA	NA	2,978	2.6	39.8	NA
322 Wearing apparel	1,694	2.8	1,933	2.6	NA	NA	2,679	2.3	58.1	NA
323 Leather and fur products	154	0.3	154	0.2	NA	NA	172	0.1	11.7	NA
324 Footwear	299	0.5	344	0.5	NA	NA	343	0.3	14.7	NA
331 Wood and wood products	2,968	5.0	3,236	4.4	NA	NA	5,014[e]	4.3	68.9	NA
332 Furniture and fixtures	1,044	1.7	1,332	1.8	NA	NA	1,888[e]	1.6	80.8	NA
341 Paper and paper products	5,714	9.6	5,410	7.3	NA	NA	10,124	8.7	77.2	NA
342 Printing and publishing	3,054	5.1	4,517	6.1	NA	NA	7,088	6.1	132.1	NA
351 Industrial chemicals	2,164	3.6	2,570	3.5	NA	NA	5,865	5.1	171.0	NA
352 Other chemical products	2,421	4.0	3,755	5.1	NA	NA	5,929	5.1	144.9	NA
353 Petroleum refineries	1,531	2.6	1,867	2.5	NA	NA	1,471	1.3	-3.9	NA
354 Miscellaneous petroleum and coal products	111	0.2	132	0.2	NA	NA	307	0.3	176.6	NA
355 Rubber products	873	1.5	1,069	1.4	NA	NA	1,388[e]	1.2	59.0	NA
356 Plastic products	873	1.5	1,654	2.2	NA	NA	2,717[e]	2.3	211.2	NA
361 Pottery, china and earthenware	43	0.1	29	0.0	NA	NA	76[e]	0.1	76.7	NA
362 Glass and glass products	385	0.6	578	0.8	NA	NA	706	0.6	83.4	NA
369 Other non-metal mineral products	1,497	2.5	1,713	2.3	NA	NA	3,048	2.6	103.6	NA
371 Iron and steel	2,652	4.4	2,906	3.9	NA	NA	4,006	3.5	51.1	NA
372 Non-ferrous metals	2,190	3.7	2,284	3.1	NA	NA	3,902	3.4	78.2	NA
381 Metal products	4,414	7.4	4,363	5.9	NA	NA	6,551	5.7	48.4	NA
382 Non-electrical machinery	3,952	6.6	4,912	6.6	NA	NA	8,060	7.0	103.9	NA
383 Electrical machinery	3,849	6.4	4,531	6.1	NA	NA	7,648	6.6	98.7	NA
384 Transport equipment	5,911	9.9	10,088	13.6	NA	NA	14,818	12.8	150.7	NA
385 Professional and scientific equipment	667	1.1	659	0.9	NA	NA	891[e]	0.8	33.6	NA
390 Other manufacturing industries	932	1.6	1,223	1.6	NA	NA	1,664	1.4	78.5	NA

Note: The industry codes shown are International Standard Industry codes (ISIC). Percentages are percent of total Value Added. 'e' stands for estimated value

Finance, Economics, and Trade

Economic Indicators [36]

Millions of Canadian Dollars unless otherwise stated.

	1989	1990	1991
Real GDP (SAAR, bil. $C)[20]	565.0	567.5	562.1[21]
Real GDP Growth Rate (%)	2.5	0.5	-1.0[21]
Money Supply (M1)	39,478	38,997	NA
Personal Savings Rate (%)	10.4	10.2	10.9[21]
Annual CPI (1986 = 100)	114.0	119.5	126.2[22]
Gross External Debt: (bil. $C)	297.0	329.7	NA

Balance of Payments Summary [37]

Values in millions of dollars.

	1987	1988	1989	1990	1991
Exports of goods (f.o.b.)	98,052	115,432	123,185	129,075	127,459
Imports of goods (f.o.b.)	-89,092	-107,274	-116,893	-119,894	-121,530
Trade balance	8,960	8,158	6,292	9,181	5,929
Services - debits	-35,368	-44,812	-49,660	-55,736	-56,228
Services - credits	17,709	23,844	23,796	24,656	25,087
Private transfers (net)	394	663	767	769	612
Government transfers (net)	-449	-416	-500	-835	-929
Long term capital (net)	8,163	8,422	18,359	13,562	17,859
Short term capital (net)	52,540	5,750	30,110	11,190	-20,710
Errors and omissions	-2,403	-560	-65	229	-5,110
Overall balance	3,342	7,559	293	626	-2,486

Exchange Rates [38]

Currency: **Canadian dollars.**
Symbol: **Can$.**

Data are currency units per $1.

January 1993	1.2776
1992	1.2087
1991	1.1457
1990	1.1668
1989	1.1840
1988	1.2307

Imports and Exports

Top Import Origins [39]

$118 billion (c.i.f., 1991).

Origins	%
US	NA
Japan	NA
UK	NA
Germany	NA
France	NA
Mexico	NA
Taiwan	NA
South Korea	NA

Top Export Destinations [40]

$124.0 billion (f.o.b., 1991).

Destinations	%
US	NA
Japan	NA
UK	NA
Germany	NA
South Korea	NA
Netherlands	NA
China	NA

Foreign Aid [41]

	U.S. $	
Donor - ODA and OOF commitments (1970-89)	7.2	billion

Import and Export Commodities [42]

Import Commodities	Export Commodities
Crude oil	Newsprint
Chemicals	Wood pulp
Motor vehicles and parts	Timber
Durable consumer goods	Crude petroleum
Electronic computers	Machinery
Telecom equipment and parts	Natural gas
	Aluminum
	Motor vehicles and parts
	Telecom equipment

For sources, notes, and explanations, see Annotated Source Appendix, page 1035.

171

Cape Verde

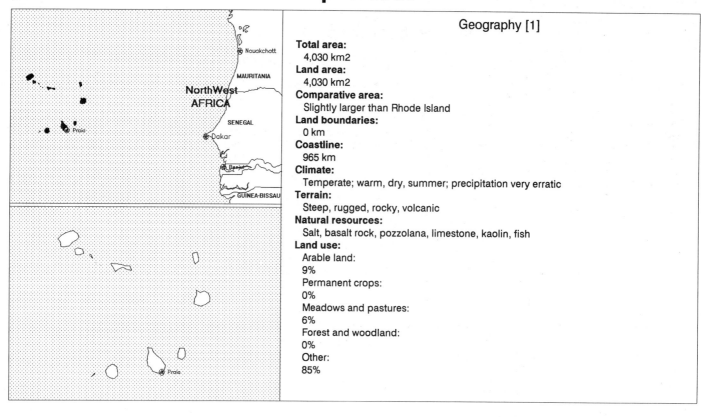

Geography [1]

Total area:
4,030 km2
Land area:
4,030 km2
Comparative area:
Slightly larger than Rhode Island
Land boundaries:
0 km
Coastline:
965 km
Climate:
Temperate; warm, dry, summer; precipitation very erratic
Terrain:
Steep, rugged, rocky, volcanic
Natural resources:
Salt, basalt rock, pozzolana, limestone, kaolin, fish
Land use:
Arable land:
9%
Permanent crops:
0%
Meadows and pastures:
6%
Forest and woodland:
0%
Other:
85%

Demographics [2]

	1960	1970	1980	1990	1991[1]	1994	2000	2010	2020
Population	197	269	296	375	387	423	503	646	812
Population density (persons per sq. mi.)	126	173	190	241	248	NA	324	417	528
Births	NA	NA	NA	NA	19	20	NA	NA	NA
Deaths	NA	NA	NA	NA	4	4	NA	NA	NA
Life expectancy - males	NA	NA	NA	NA	60	61	NA	NA	NA
Life expectancy - females	NA	NA	NA	NA	63	65	NA	NA	NA
Birth rate (per 1,000)	NA	NA	NA	NA	48	46	NA	NA	NA
Death rate (per 1,000)	NA	NA	NA	NA	10	9	NA	NA	NA
Women of reproductive age (15-44 yrs.)	NA	NA	NA	86	NA	97	118	NA	NA
of which are currently married	NA	NA	NA	23	NA	26	32	NA	NA
Fertility rate	NA	NA	NA	6.7	6.60	6.3	5.7	4.5	3.5

Population values are in thousands, life expectancy in years, and other items as indicated.

Health

Health Personnel [3]

Health Expenditures [4]

For sources, notes, and explanations, see Annotated Source Appendix, page 1035.

Human Factors

Health Care Ratios [5]	Infants and Malnutrition [6]

Ethnic Division [7]

Creole (mulatto)	71%
African	28%
European	1%

Religion [8]

Roman Catholicism fused with indigenous beliefs.

Major Languages [9]

Portuguese, Crioulo, a blend of Portuguese and West African words.

Education

Public Education Expenditures [10]

Million Escudos	1985	1986	1987	1988	1989	1990
Total education expenditure	341	NA	493	NA	NA	NA
as percent of GNP	3.6	NA	3.8	NA	NA	NA
as percent of total govt. expend.	NA	NA	14.8	NA	NA	NA
Current education expenditure	325	NA	472	NA	NA	NA
as percent of GNP	3.5	NA	3.7	NA	NA	NA
as percent of current govt. expend.	15.2	NA	15.3	NA	NA	NA
Capital expenditure	16	NA	21	NA	NA	NA

Educational Attainment [11]

Literacy Rate [12]

	1970[b]	1980[b,4]	1989[b,4]
Illiterate population +15 years	134,633	NA	64,510
Illiteracy rate - total pop. (%)	63.1	52.6	33.5
Illiteracy rate - males (%)	NA	38.6	NA
Illiteracy rate - females (%)	NA	61.4	NA

Libraries [13]

Daily Newspapers [14]

Cinema [15]

Science and Technology

Scientific/Technical Forces [16]	R&D Expenditures [17]	U.S. Patents Issued [18]

For sources, notes, and explanations, see Annotated Source Appendix, page 1035.

173

Government and Law

Organization of Government [19]

Long-form name:
Republic of Cape Verde
Type:
republic
Independence:
5 July 1975 (from Portugal)
Constitution:
7 September 1980; amended 12 February 1981, December 1988, and 28 September 1990 (legalized opposition parties)
Legal system:
NA
National holiday:
Independence Day, 5 July (1975)
Executive branch:
president, prime minister, deputy minister, secretaries of state, Council of Ministers (cabinet)
Legislative branch:
unicameral People's National Assembly (Assembleia Nacional Popular)
Judicial branch:
Supreme Tribunal of Justice (Supremo Tribunal de Justia)

Political Parties [20]

People's National Assembly	% of seats
Movement for Democracy (MPD)	70.9
African Party for Independence (PAICV)	29.1

Government Budget [21]

For 1991 est.

Revenues	104
Expenditures	133
Capital expenditures	72

Crime [23]

Military Expenditures and Arms Transfers [22]

	1985	1986	1987	1988	1989
Military expenditures					
Current dollars (mil.)	NA	NA	NA	NA	NA
1989 constant dollars (mil.)	NA	NA	NA	NA	NA
Armed forces (000)	6	4	4	3	3
Gross national product (GNP)					
Current dollars (mil.)	203	216	239	257	282
1989 constant dollars (mil.)	231	240	257	268	282
Central government expenditures (CGE)					
1989 constant dollars (mil.)	NA	NA	NA	NA	278[e]
People (mil.)	0.3	0.3	0.3	0.4	0.4
Military expenditure as % of GNP	NA	NA	NA	NA	NA
Military expenditure as % of CGE	NA	NA	NA	NA	NA
Military expenditure per capita	NA	NA	NA	NA	NA
Armed forces per 1,000 people	18.3	11.9	11.6	8.5	8.2
GNP per capita	706	714	747	757	775
Arms imports[6]					
Current dollars (mil.)	10	5	5	5	5
1989 constant dollars (mil.)	11	6	5	5	5
Arms exports[6]					
Current dollars (mil.)	10	0	0	0	0
1989 constant dollars (mil.)	11	0	0	0	0
Total imports[7]					
Current dollars (mil.)	81	80	101	106	112
1989 constant dollars	92	89	109	110	112
Total exports[7]					
Current dollars (mil.)	5	4	8	3	7
1989 constant dollars	6	4	9	3	7
Arms as percent of total imports[8]	12.3	6.3	5.0	4.7	4.5
Arms as percent of total exports[8]	200.0	0	0	0	0

Human Rights [24]

	SSTS	FL	FAPRO	PPCG	APROBC	TPW	PCPTW	STPEP	PHRFF	PRW	ASST	AFL
Observes		P			P	P	P					P
	EAFRD	CPR	ESCR	SR	ACHR	MAAE	PVIAC	PVNAC	EAFDAW	TCIDTP	RC	
Observes		P		P					P	P	P	

P = Party; S = Signatory; see Appendix for meaning of abbreviations.

Labor Force

Total Labor Force [25]

102,000 (1985 est.); 51% of population of working age (1985).

Labor Force by Occupation [26]

Agriculture (mostly subsistence)	57%
Services	29
Industry	14

Date of data: 1981

Unemployment Rate [27]

25% (1988)

For sources, notes, and explanations, see Annotated Source Appendix, page 1035.

Production Sectors

Energy Resource Summary [28]

Energy Resources: None. **Electricity**: 15,000 kW capacity; 15 million kWh produced, 40 kWh per capita (1991).

Telecommunications [30]

- Interisland microwave radio relay system, high-frequency radio to Senegal and Guinea-Bissau
- Over 1,700 telephones
- Broadcast stations - 1 AM, 6 FM, 1 TV
- 2 coaxial submarine cables
- 1 Atlantic Ocean INTELSAT earth station

Top Agricultural Products [32]

Agriculture accounts for 20% of GDP (including fishing); largely subsistence farming; bananas are the only export crop; other crops—corn, beans, sweet potatoes, coffee; growth potential of agricultural sector limited by poor soils and scanty rainfall; annual food imports required; fish catch provides for both domestic consumption and small exports.

Top Mining Products [33]

Detailed information is not available. A summary of mineral resources available follows. **Mineral Resources**: Salt, basalt rock, pozzolana, limestone, kaolin.

Transportation [31]

Ports. Mindelo, Praia.

Merchant marine. 7 cargo ships (1,000 GRT or over) totaling 11,717 GRT/19,000 DWT

Airports

Total:	6
Usable:	6
With permanent-surface runways:	6
With runways over 3,659 m:	0
With runways 2,440-3,659 m:	1
With runways 1,220-2,439 m:	2

Tourism [34]

Finance, Economics, and Trade

GDP and Manufacturing Summary [35]

	1980	1985	1990	1991	1992
Gross Domestic Product					
Millions of 1980 dollars	142	196	254	263	277[e]
Growth rate in percent	3.32	8.63	4.00	3.56	5.05[e]
Manufacturing Value Added					
Millions of 1980 dollars	7	11	14	15	16[e]
Growth rate in percent	7.14	31.07	4.98	6.75	6.50[e]
Manufacturing share in percent of current prices	4.8	5.8	6.0[e]	5.9[e]	NA

Economic Indicators [36]

National product: GDP—exchange rate conversion—$310 million (1990 est.). **National product real growth rate**: 4% (1990 est.). **National product per capita**: $800 (1990 est.). **Inflation rate (consumer prices)**: 8.7% (1991 est.). **External debt**: $156 million (1991).

Balance of Payments Summary [37]

Values in millions of dollars.

	1987	1988	1989	1990	1991
Exports of goods (f.o.b.)	7.8	3.3	6.7	5.6	6.3
Imports of goods (f.o.b.)	-92.8	-101.8	-106.9	-119.5	-123.5
Trade balance	-85.0	-98.5	-100.1	-113.8	-117.2
Services - debits	-25.5	-24.2	-30.3	-32.4	-21.2
Services - credits	43.3	45.0	57.3	62.7	21.1
Private transfers (net)	34.0	39.5	43.2	52.1	45.5
Government transfers (net)	46.9	38.7	27.6	25.5	31.4
Long term capital (net)	5.4	-1.8	1.6	3.8	3.1
Short term capital (net)	-2.0	-1.7	-1.2	-	-
Errors and omissions	15.7	5.7	-19.7	-3.1	28.8
Overall balance	32.8	2.7	-21.6	-5.3	-8.5

Exchange Rates [38]

Currency: **Cape Verdean escudos.**
Symbol: **CVEsc.**

Data are currency units per $1.

January 1993	75.47
1992	73.10
1991	71.41
November 1990	64.10
December 1989	74.86
1988	72.01

Imports and Exports

Top Import Origins [39]

$120 million (c.i.f., 1990 est.).

Origins	%
Sweden	33
Spain	11
Germany	5
Portugal	3
France	3
Netherlands	NA
US 1990 est.	NA

Top Export Destinations [40]

$5.7 million (f.o.b., 1990 est.).

Destinations	%
Portugal	40
Algeria	31
Angola	NA
Netherlands	NA

Foreign Aid [41]

	U.S. $	
US commitments, including Ex-Im (FY75-90)	93	million
Western (non-US) countries, ODA and OOF bilateral commitments (1970-90)	586	million
OPEC bilateral aid (1979-89)	12	million
Communist countries (1970-89)	36	million

Import and Export Commodities [42]

Import Commodities	**Export Commodities**
Foodstuffs	Fish
Consumer goods	Bananas
Industrial products	Hides and skins
Transport equipment	

For sources, notes, and explanations, see Annotated Source Appendix, page 1035.

Central African Republic

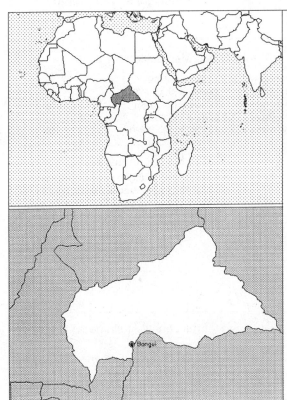

Geography [1]

Total area:
622,980 km2
Land area:
622,980 km2
Comparative area:
Slightly smaller than Texas
Land boundaries:
Total 5,203 km, Cameroon 797 km, Chad 1,197 km, Congo 467 km, Sudan 1,165 km, Zaire 1,577 km
Coastline:
0 km (landlocked)
Climate:
Tropical; hot, dry winters; mild to hot, wet summers
Terrain:
Vast, flat to rolling, monotonous plateau; scattered hills in northeast and southwest
Natural resources:
Diamonds, uranium, timber, gold, oil
Land use:
Arable land:
3%
Permanent crops:
0%
Meadows and pastures:
5%
Forest and woodland:
64%
Other:
28%

Demographics [2]

	1960	1970	1980	1990	1991[1]	1994	2000	2010	2020
Population	1,467	1,827	2,269	2,866	2,952	3,142	3,511	3,898	4,561
Population density (persons per sq. mi.)	6	8	9	12	12	NA	15	20	25
Births	NA	NA	NA	NA	129	133	NA	NA	NA
Deaths	NA	NA	NA	NA	53	65	NA	NA	NA
Life expectancy - males	NA	NA	NA	NA	45	41	NA	NA	NA
Life expectancy - females	NA	NA	NA	NA	49	44	NA	NA	NA
Birth rate (per 1,000)	NA	NA	NA	NA	44	42	NA	NA	NA
Death rate (per 1,000)	NA	NA	NA	NA	18	21	NA	NA	NA
Women of reproductive age (15-44 yrs.)	NA	NA	NA	NA	NA	NA	NA	NA	NA
of which are currently married	NA	NA	NA	NA	NA	NA	NA	NA	NA
Fertility rate	NA	NA	NA	5.6	5.58	5.4	5.1	4.4	3.7

Population values are in thousands, life expectancy in years, and other items as indicated.

Health

Health Personnel [3]

Doctors per 1,000 pop., 1988-92	0.04
Nurse-to-doctor ratio, 1988-92	4.5
Hospital beds per 1,000 pop., 1985-90	0.9
Percentage of children immunized (age 1 yr. or less)	
Third dose of DPT, 1990-91	82.0
Measles, 1990-91	82.0

Health Expenditures [4]

Total health expenditure, 1990 (official exchange rate)	
Millions of dollars	55
Millions of dollars per capita	18
Health expenditures as a percentage of GDP	
Total	4.2
Public sector	2.6
Private sector	1.6
Development assistance for health	
Total aid flows (millions of dollars)[1]	20
Aid flows per capita (millions of dollars)	6.5
Aid flows as a percentage of total health expenditure	35.8

For sources, notes, and explanations, see Annotated Source Appendix, page 1035.

177

Human Factors

Health Care Ratios [5]

Population per physician, 1970	44740
Population per physician, 1990	25930
Population per nursing person, 1970	2460
Population per nursing person, 1990	NA
Percent of births attended by health staff, 1985	NA

Infants and Malnutrition [6]

Percent of babies with low birth weight, 1985	15
Infant mortality rate per 1,000 live births, 1970	139
Infant mortality rate per 1,000 live births, 1991	106
Years of life lost per 1,000 population, 1990	74
Prevalence of malnutrition (under age 5), 1990	NA

Ethnic Division [7]

Of Europeans, 3,600 are French.

Baya	34%
Banda	27%
Sara	10%
Mandjia	21%
Mboum	4%
M'Baka	4%
Europeans	6,500

Religion [8]

Indigenous beliefs	24%
Protestant	25%
Roman Catholic	25%
Muslim	15%
Other	11%

Major Languages [9]

French (official), Sangho (lingua franca and national language), Arabic, Hunsa, Swahili.

Education

Public Education Expenditures [10]

Million Francs C.F.A.	1985	1986	1987	1988[1]	1989[1]	1990
Total education expenditure	NA	9,553	9,197	8,475	8,305	9,862
as percent of GNP	NA	2.8	3.0	2.6	2.5	2.8
as percent of total govt. expend.	NA	NA	16.8	NA	NA	NA
Current education expenditure	NA	9,313	9,002	8,227	8,097	9,622
as percent of GNP	NA	2.8	2.9	2.6	2.4	2.8
as percent of current govt. expend.	NA	25.6	24.7	21.7	12.8	NA
Capital expenditure	NA	240	195	248	208	240

Educational Attainment [11]

Literacy Rate [12]

In thousands and percent	1985[a]	1991[a]	2000[a]
Illiterate population +15 years	1,014	1,028	1,062
Illiteracy rate - total pop. (%)	68.5	62.3	49.9
Illiteracy rate - males (%)	55.0	48.2	36.5
Illiteracy rate - females (%)	80.7	75.1	62.4

Libraries [13]

Daily Newspapers [14]

	1975	1980	1985	1990
Number of papers	-	-	-	1
Circ. (000)	-	-	-	2e

Cinema [15]

Science and Technology

Scientific/Technical Forces [16]

R&D Expenditures [17]

	(000) 1984
Total expenditure	680,791
Capital expenditure	NA
Current expenditure	NA
Percent current	NA

U.S. Patents Issued [18]

      For sources, notes, and explanations, see Annotated Source Appendix, page 1035.

Government and Law

Organization of Government [19]

Long-form name:
Central African Republic
Type:
republic; one-party presidential regime since 1986
Independence:
13 August 1960 (from France)
Constitution:
21 November 1986
Legal system:
based on French law
National holiday:
National Day, 1 December (1958) (proclamation of the republic)
Executive branch:
president, prime minister, Council of Ministers (cabinet)
Legislative branch:
unicameral National Assembly (Assemblee Nationale) advised by the Economic and Regional Council (Conseil Economique et Regional); when they sit together this is known as the Congress (Congres)
Judicial branch:
Supreme Court (Cour Supreme)

Elections [20]

National Assembly. Elections last held 25 October 1992; widespread irregularities at some polls led to dismissal of results by Supreme Court; elections are rescheduled for 17 October 1993.

Government Budget [21]

For 1991 est.	
Revenues	175
Expenditures	312
Capital expenditures	122

Military Expenditures and Arms Transfers [22]

	1985	1986	1987	1988	1989
Military expenditures					
Current dollars (mil.)	NA	NA	17[e]	NA	18
1989 constant dollars (mil.)	NA	NA	18[e]	NA	18
Armed forces (000)	5	5	5	5	5
Gross national product (GNP)					
Current dollars (mil.)	928	957	957	1,005	1,088
1989 constant dollars (mil.)	1,057	1,062	1,030	1,046	1,088
Central government expenditures (CGE)					
1989 constant dollars (mil.)	296	278	283	268	274[e]
People (mil.)	2.6	2.6	2.7	2.7	2.8
Military expenditure as % of GNP	NA	NA	1.8	NA	1.7
Military expenditure as % of CGE	NA	NA	6.4	NA	6.6
Military expenditure per capita	NA	NA	7	NA	6
Armed forces per 1,000 people	1.9	1.9	1.9	1.9	1.8
GNP per capita	411	405	385	382	388
Arms imports[6]					
Current dollars (mil.)	0	0	0	0	0
1989 constant dollars (mil.)	0	0	0	0	0
Arms exports[6]					
Current dollars (mil.)	0	0	0	0	0
1989 constant dollars (mil.)	0	0	0	0	0
Total imports[7]					
Current dollars (mil.)	113	252	204	201	150
1989 constant dollars	129	280	219	209	150
Total exports[7]					
Current dollars (mil.)	92	131	130	130	134
1989 constant dollars	105	145	140	135	134
Arms as percent of total imports[8]	0	0	0	0	0
Arms as percent of total exports[8]	0	0	0	0	0

Crime [23]

Crime volume (for 1989)	
Cases known to police	4,076
Attempts (percent)	157.9
Percent cases solved	NA
Crimes per 100,000 persons	135.8
Persons responsible for offenses	
Total number offenders	4,685
Percent female	94.6
Percent juvenile[13]	99.4
Percent foreigners	NA

Human Rights [24]

	SSTS	FL	FAPRO	PPCG	APROBC	TPW	PCPTW	STPEP	PHRFF	PRW	ASST	AFL
Observes	2	P	P		P	P	P	P		P	P	P
	EAFRD	CPR	ESCR	SR	ACHR	MAAE	PVIAC	PVNAC	EAFDAW	TCIDTP	RC	
Observes	P	P	P				P	P	P		P	

P = Party; S = Signatory; see Appendix for meaning of abbreviations.

Labor Force

Total Labor Force [25]

775,413 (1986 est.)

Labor Force by Occupation [26]

Agriculture	85%
Commerce and services	9
Industry	3
Government	3

Unemployment Rate [27]

30% (1988 est.) in Bangui

For sources, notes, and explanations, see Annotated Source Appendix, page 1035.

179

Production Sectors

Energy Resource Summary [28]

Energy Resources: Uranium, oil. **Electricity**: 40,000 kW capacity; 95 million kWh produced, 30 kWh per capita (1991).

Telecommunications [30]

- Fair system
- Network relies primarily on radio relay links, with low-capacity, low-powered radiocommunication also used
- Broadcast stations - 1 AM, 1 FM, 1 TV
- 1 Atlantic Ocean INTELSAT earth station

Transportation [31]

Highways. 22,000 km total; 458 km bituminous, 10,542 km improved earth, 11,000 unimproved earth

Airports

Total:	66
Usable:	51
With permanent-surface runways:	3
With runways over 3,659 m:	0
With runways 2,440-3,659 m:	2
With runways 1,220-2,439 m:	20

Top Agricultural Products [32]

Agriculture accounts for 40% of GDP; self-sufficient in food production except for grain; commercial crops—cotton, coffee, tobacco, timber; food crops - manioc, yams, millet, corn, bananas.

Top Mining Products [33]

Metric tons unless otherwise specified	M.t.
Diamonds, gem and industrial (carats)	378,643
Gold (kilograms)	176

Tourism [34]

Manufacturing Sector

GDP and Manufacturing Summary [35]

	1980	1985	1989	1990	% change 1980-1990	% change 1989-1990
GDP (million 1980 $)	797	883	1,033	958	20.2	-7.3
GDP per capita (1980 $)	343	334	350	315	-8.2	-10.0
Manufacturing as % of GDP (current prices)	8.8	7.1	7.4[e]	9.6[e]	9.1	29.7
Gross output (million $)	98[e]	108	104[e]	151[e]	54.1	45.2
Value added (million $)	35[e]	33	24[e]	48[e]	37.1	100.0
Value added (million 1980 $)	71	83	94[e]	97	36.6	3.2
Industrial production index	100	112	118[e]	131	31.0	11.0
Employment (thousands)	6[e]	8	5[e]	4[e]	-33.3	-20.0

Note: GDP stands for Gross Domestic Product. 'e' stands for estimated value.

Profitability and Productivity

	1980	1985	1989	1990	% change 1980-1990	% change 1989-1990
Intermediate input (%)	64[e]	70	77[e]	68[e]	6.3	-11.7
Wages, salaries, and supplements (%)	16[e]	18[e]	18[e]	17[e]	6.3	-5.6
Gross operating surplus (%)	19[e]	12[e]	5[e]	15[e]	-21.1	200.0
Gross output per worker ($)	16,369[e]	13,101	20,988[e]	32,976[e]	101.5	57.1
Value added per worker ($)	5,932[e]	4,157	4,799[e]	10,774[e]	81.6	124.5
Average wage (incl. benefits) ($)	2,691[e]	2,415[e]	3,822[e]	6,022[e]	123.8	57.6

Profitability is in percent of gross output. Productivity is in U.S. $. 'e' stands for estimated value.

Profitability - 1990

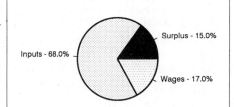

Surplus - 15.0%
Inputs - 68.0%
Wages - 17.0%

The graphic shows percent of gross output.

Value Added in Manufacturing

	1980 $ mil.	1980 %	1985 $ mil.	1985 %	1989 $ mil.	1989 %	1990 $ mil.	1990 %	% change 1980-1990	% change 1989-1990
311 Food products	5	14.3	8	24.2	8[e]	33.3	13[e]	27.1	160.0	62.5
313 Beverages	3	8.6	4	12.1	4[e]	16.7	6[e]	12.5	100.0	50.0
314 Tobacco products	4	11.4	6	18.2	7[e]	29.2	10[e]	20.8	150.0	42.9
321 Textiles	5[e]	14.3	NA	0.0	-7[e]	-29.2	1[e]	2.1	-80.0	-114.3
322 Wearing apparel	1[e]	2.9	NA	0.0	-1[e]	-4.2	NA	0.0	NA	NA
323 Leather and fur products	NA	0.0	NA	0.0	NA	0.0	NA	0.0	NA	NA
324 Footwear	NA	0.0	NA	0.0	NA	0.0	NA	0.0	NA	NA
331 Wood and wood products	11[e]	31.4	8	24.2	6	25.0	9[e]	18.8	-18.2	50.0
332 Furniture and fixtures	NA	0.0	1	3.0	NA	0.0	1[e]	2.1	NA	NA
341 Paper and paper products	NA	0.0	NA	0.0	NA	0.0	NA	0.0	NA	NA
342 Printing and publishing	1	2.9	2	6.1	1[e]	4.2	2[e]	4.2	100.0	100.0
351 Industrial chemicals	1	2.9	1	3.0	1[e]	4.2	1[e]	2.1	0.0	0.0
352 Other chemical products	2	5.7	1	3.0	2[e]	8.3	2[e]	4.2	0.0	0.0
353 Petroleum refineries	NA	0.0	NA	0.0	NA	0.0	NA	0.0	NA	NA
354 Miscellaneous petroleum and coal products	NA	0.0	NA	0.0	NA	0.0	NA	0.0	NA	NA
355 Rubber products	NA	0.0	NA	0.0	NA	0.0	NA	0.0	NA	NA
356 Plastic products	NA	0.0	NA	0.0	NA	0.0	NA	0.0	NA	NA
361 Pottery, china and earthenware	NA	0.0	NA	0.0	NA	0.0	NA	0.0	NA	NA
362 Glass and glass products	NA	0.0	NA	0.0	NA	0.0	NA	0.0	NA	NA
369 Other non-metal mineral products	NA	0.0	NA	0.0	NA	0.0	NA	0.0	NA	NA
371 Iron and steel	NA	0.0	NA	0.0	NA	0.0	NA	0.0	NA	NA
372 Non-ferrous metals	NA	0.0	NA	0.0	NA	0.0	NA	0.0	NA	NA
381 Metal products	1	2.9	NA	0.0	NA	0.0	NA	0.0	NA	NA
382 Non-electrical machinery	NA	0.0	NA	0.0	NA	0.0	NA	0.0	NA	NA
383 Electrical machinery	NA	0.0	NA	0.0	NA	0.0	NA	0.0	NA	NA
384 Transport equipment	2	5.7	1	3.0	1[e]	4.2	2[e]	4.2	0.0	100.0
385 Professional and scientific equipment	NA	0.0	NA	0.0	NA	0.0	NA	0.0	NA	NA
390 Other manufacturing industries	NA	0.0	1	3.0	2[e]	8.3	1[e]	2.1	NA	-50.0

Note: The industry codes shown are International Standard Industry codes (ISIC). Percentages are percent of total Value Added. 'e' stands for estimated value

Finance, Economics, and Trade

Economic Indicators [36]

National product: GDP—exchange rate conversion—$1.3 billion (1990 est.). **National product real growth rate**: -3% (1990 est.). **National product per capita**: $440 (1990 est.). **Inflation rate (consumer prices)**: - 3% (1990 est.). **External debt**: $859 million (1991).

Balance of Payments Summary [37]

Values in millions of dollars.

	1986	1987	1988	1989	1990
Exports of goods (f.o.b.)	129.5	128.9	133.7	148.1	150.5
Imports of goods (f.o.b.)	-201.0	-197.7	-178.2	-186.0	-241.6
Trade balance	-71.5	-68.8	-44.5	-37.9	-91.1
Services - debits	-156.8	-176.4	NA	-165.9	-190.9
Services - credits	58.8	70.6	NA	66.2	69.8
Private transfers (net)	-19.5	-23.7	NA	-24.9	-32.9
Government transfers (net)	102.4	124.8	-	129.0	155.9
Long term capital (net)	94.0	85.1	-	47.9	90.5
Short term capital (net)	-2.1	11.9	-3.8	-0.3	-9.3
Errors and omissions	7.2	-1.7	-	1.3	1.1
Overall balance	12.5	21.8	-48.3	15.4	-6.9

Exchange Rates [38]

Currency: **Communaute Financiere Africaine francs.**
Symbol: **CFAF.**

Data are currency units per $1.

January 1993	274.06
1992	264.69
1991	282.11
1990	272.26
1989	319.01
1988	297.85

Imports and Exports

Top Import Origins [39]

$205 million (1991 est.).

Origins	%
France	NA
Other EC countries	NA
Japan	NA
Algeria	NA

Top Export Destinations [40]

$138 million (1991 est.).

Destinations	%
France	NA
Belgium	NA
Italy	NA
Japan	NA
US	NA

Foreign Aid [41]

	U.S. $	
US commitments, including Ex-Im (FY70-90)	52	million
Western (non-US) countries, ODA and OOF bilateral commitments (1970-90)	1.6	billion
OPEC bilateral aid (1979-89)	6	million
Communist countries (1970-89)	38	million

Import and Export Commodities [42]

Import Commodities

Food
Textiles
Petroleum products
Machinery
Electrical equipment
Motor vehicles
Chemicals
Pharmaceuticals
Consumer goods
Industrial products

Export Commodities

Diamonds
Cotton
Coffee
Timber
Tobacco

Chad

Geography [1]

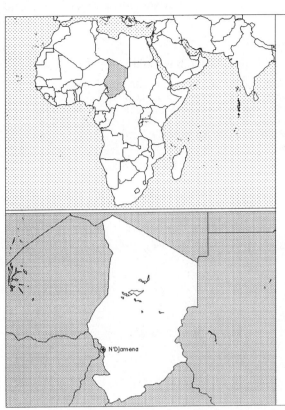

Total area:
  1.284 million km2
Land area:
  1,259,200 km2
Comparative area:
  Slightly more than three times the size of California
Land boundaries:
  Total 5,968 km, Cameroon 1,094 km, Central African Republic 1,197 km, Libya 1,055 km, Niger 1,175 km, Nigeria 87 km, Sudan 1,360 km
Coastline:
  0 km (landlocked)
Climate:
  Tropical in south, desert in north
Terrain:
  Broad, arid plains in center, desert in north, mountains in northwest, lowlands in south
Natural resources:
  Petroleum (unexploited but exploration under way), uranium, natron, kaolin, fish (Lake Chad)
Land use:
  Arable land: 2%
  Permanent crops: 0%
  Meadows and pastures: 36%
  Forest and woodland: 11%
  Other: 51%

Demographics [2]

	1960	1970	1980	1990	1991[1]	1994	2000	2010	2020
Population	3,106	3,557	4,024	5,024	5,122	5,467	6,221	7,680	9,396
Population density (persons per sq. mi.)	6	7	8	10	11	NA	13	16	19
Births	NA	NA	NA	NA	217	230	NA	NA	NA
Deaths	NA	NA	NA	NA	111	113	NA	NA	NA
Life expectancy - males	NA	NA	NA	NA	39	40	NA	NA	NA
Life expectancy - females	NA	NA	NA	NA	41	42	NA	NA	NA
Birth rate (per 1,000)	NA	NA	NA	NA	42	42	NA	NA	NA
Death rate (per 1,000)	NA	NA	NA	NA	22	21	NA	NA	NA
Women of reproductive age (15-44 yrs.)	NA	NA	NA	1,232	NA	1,326	1,498	NA	NA
of which are currently married	NA	NA	NA	1,012	NA	1,089	1,230	NA	NA
Fertility rate	NA	NA	NA	5.3	5.33	5.3	5.1	4.5	3.8

Population values are in thousands, life expectancy in years, and other items as indicated.

Health

Health Personnel [3]

Doctors per 1,000 pop., 1988-92	0.03
Nurse-to-doctor ratio, 1988-92	0.9
Hospital beds per 1,000 pop., 1985-90	NA
Percentage of children immunized (age 1 yr. or less)	
Third dose of DPT, 1990-91	18.0
Measles, 1990-91	28.0

Health Expenditures [4]

Total health expenditure, 1990 (official exchange rate)	
Millions of dollars	76
Millions of dollars per capita	13
Health expenditures as a percentage of GDP	
Total	6.3
Public sector	4.7
Private sector	1.6
Development assistance for health	
Total aid flows (millions of dollars)[1]	33
Aid flows per capita (millions of dollars)	5.8
Aid flows as a percentage of total health expenditure	43.0

For sources, notes, and explanations, see Annotated Source Appendix, page 1035.

183

Human Factors

Health Care Ratios [5]

Population per physician, 1970	61900
Population per physician, 1990	30030
Population per nursing person, 1970	8010
Population per nursing person, 1990	NA
Percent of births attended by health staff, 1985	NA

Infants and Malnutrition [6]

Percent of babies with low birth weight, 1985	11
Infant mortality rate per 1,000 live births, 1970	171
Infant mortality rate per 1,000 live births, 1991	124
Years of life lost per 1,000 population, 1990	106
Prevalence of malnutrition (under age 5), 1990	35

Ethnic Division [7]

No detailed information available.

Religion [8]

Muslim	44%
Christian	33%
Indigenous beliefs and animism	23%

Major Languages [9]

More than 100 different languages and dialects are spoken.

French (official)	NA
Arabic (official)	NA
Sara (in south)	NA
Sango (in south)	NA

Education

Public Education Expenditures [10]

Educational Attainment [11]

Literacy Rate [12]

In thousands and percent	1985[a]	1991[a]	2000[a]
Illiterate population + 15 years	2,230	2,280	2,354
Illiteracy rate - total pop. (%)	77.0	70.2	56.6
Illiteracy rate - males (%)	66.0	57.8	43.3
Illiteracy rate - females (%)	87.5	82.1	69.4

Libraries [13]

	Admin. Units	Svc. Pts.	Vols. (000)	Shelving (meters)	Vols. Added	Reg. Users
National	NA	NA	NA	NA	NA	NA
Non-specialized	NA	NA	NA	NA	NA	NA
Public	NA	NA	NA	NA	NA	NA
Higher ed. (1987)[4]	1	1	10	128	324	350
School	NA	NA	NA	NA	NA	NA

Daily Newspapers [14]

	1975	1980	1985	1990
Number of papers	4	1	1	1
Circ. (000)	NA	1	1	2

Cinema [15]

Science and Technology

Scientific/Technical Forces [16]

R&D Expenditures [17]

U.S. Patents Issued [18]

Government and Law

Organization of Government [19]

Long-form name:
Republic of Chad
Type:
republic
Independence:
11 August 1960 (from France)
Constitution:
22 December 1989, suspended 3
December 1990; Provisional National
Charter 1 March 1991; national
conference drafting new constitution to
submit to referendum January 1993
Legal system:
based on French civil law system and
Chadian customary law; has not accepted
compulsory ICJ jurisdiction
National holiday:
11 August
Executive branch:
president, Council of State (cabinet)
Legislative branch:
unicameral National Consultative Council
(Conseil National Consultatif) was
disbanded 3 December 1990 and
replaced by the Provisional Council of the
Republic, with 30 members appointed by
President DEBY on 8 March 1991
Judicial branch:
Court of Appeal

Elections [20]

National Consultative Council. Last
held 8 July 1990; disbanded 3
December 1990 President Deby has
promised political pluralism, a new
constitution, and free elections by
September 1993; numerous dissident
groups; 26 opposition political parties.

Government Budget [21]

For 1991 est.

Revenues	115
Expenditures	412
Capital expenditures	218

Crime [23]

Military Expenditures and Arms Transfers [22]

	1985	1986	1987	1988	1989
Military expenditures					
Current dollars (mil.)	14	29	34	NA	NA
1989 constant dollars (mil.)	16	32	36	NA	NA
Armed forces (000)	16	22	30	33	33
Gross national product (GNP)					
Current dollars (mil.)	805	788	785	954	1,002
1989 constant dollars (mil.)	916	875	845	993	1,002
Central government expenditures (CGE)					
1989 constant dollars (mil.)	49[e]	86	90	83	NA
People (mil.)	4.4	4.5	4.6	4.8	4.9
Military expenditure as % of GNP	1.7	3.6	4.3	NA	NA
Military expenditure as % of CGE	32.3	37.1	40.1	NA	NA
Military expenditure per capita	4	7	8	NA	NA
Armed forces per 1,000 people	3.6	4.9	6.5	6.9	6.7
GNP per capita	207	194	182	208	205
Arms imports[6]					
Current dollars (mil.)	20	40	100	50	10
1989 constant dollars (mil.)	23	44	108	52	10
Arms exports[6]					
Current dollars (mil.)	0	0	0	0	0
1989 constant dollars (mil.)	0	0	0	0	0
Total imports[7]					
Current dollars (mil.)	240	288	366	419	NA
1989 constant dollars	273	320	394	436	NA
Total exports[7]					
Current dollars (mil.)	88	99	111	141	NA
1989 constant dollars	100	110	119	147	NA
Arms as percent of total imports[8]	8.3	13.9	27.3	11.9	NA
Arms as percent of total exports[8]	0	0	0	0	NA

Human Rights [24]

	SSTS	FL	FAPRO	PPCG	APROBC	TPW	PCPTW	STPEP	PHRFF	PRW	ASST	AFL
Observes		P	P		P	P	P					P
	EAFRD	CPR	ESCR	SR	ACHR	MAAE	PVIAC	PVNAC	EAFDAW	TCIDTP	RC	
Observes		P			P							P

P = Party; S = Signatory; see Appendix for meaning of abbreviations.

Labor Force

Total Labor Force [25]

Labor Force by Occupation [26]

Agriculture   85Agriculture

Unemployment Rate [27]

Production Sectors

Energy Resource Summary [28]

Energy Resources: Petroleum (unexploited but exploration under way), uranium. **Electricity**: 40,000 kW capacity; 70 million kWh produced, 15 kWh per capita (1991).

Telecommunications [30]

- Fair system of radiocommunication stations for intercity links
- Broadcast stations - 6 AM, 1 FM, limited TV service
- Many facilities are inoperative
- 1 Atlantic Ocean INTELSAT earth station

Transportation [31]

Highways. 31,322 km total; 32 km bituminous; 7,300 km gravel and laterite; remainder unimproved earth

Airports

Total:	69
Usable:	55
With permanent-surface runways:	5
With runways over 3,659 m:	0
With runways 2,440-3,659 m:	4
With runways 1,220-2,439 m:	24

Top Agricultural Products [32]

Agriculture accounts for about 45% of GDP; largely subsistence farming; cotton most important cash crop; food crops include sorghum, millet, peanuts, rice, potatoes, manioc; livestock—cattle, sheep, goats, camels; self-sufficient in food in years of adequate rainfall.

Top Mining Products [33]

Detailed information is not available. A summary of mineral resources available follows. **Mineral Resources**: Petroleum (unexploited but exploration under way), uranium, natron, kaolin.

Tourism [34]

	1987	1988	1989	1990	1991
Tourists	27	21	12	9	21
Tourism receipts	6	7	9	12	10
Tourism expenditures	47	32	33	36	32
Fare receipts	1	1	1	4	3
Fare expenditures	6	5	9	12	11

Tourists are in thousands, money in million U.S. dollars.

Finance, Economics, and Trade

GDP and Manufacturing Summary [35]

	1980	1985	1990	1991	1992
Gross Domestic Product					
Millions of 1980 dollars	1,005	804	952	1,017	1,048[e]
Growth rate in percent	-7.40	6.86	-2.70	6.81	3.04[e]
Manufacturing Value Added					
Millions of 1980 dollars	92	69	83	68	70[e]
Growth rate in percent	-12.00	5.39	-1.90	-18.00	2.61[e]
Manufacturing share in percent of current prices	10.7[e]	11.1	15.4	11.1	NA

Economic Indicators [36]

National product: GDP—exchange rate conversion—$1.1 billion (1991 est.). **National product real growth rate**: 8.4% (1991 est.). **National product per capita**: $215 (1991 est.). **Inflation rate (consumer prices)**: 2%- 3% (1991 est.). **External debt**: $492 million (December 1990 est.).

Balance of Payments Summary [37]

Values in millions of dollars.

	1987	1988	1989	1990	1991
Exports of goods (f.o.b.)	109.4	145.9	155.4	230.3	193.9
Imports of goods (f.o.b.)	-225.9	-228.4	-240.3	-259.5	-294.1
Trade balance	-116.5	-82.5	-84.9	-29.2	-100.2
Services - debits	-211.0	-233.4	-220.7	-252.0	-280.4
Services - credits	73.3	80.9	43.6	44.0	47.5
Private transfers (net)	-9.8	-17.1	-20.2	-12.9	-13.8
Government transfers (net)	238.5	277.7	230.9	266.1	266.6
Long term capital (net)	55.8	68.9	81.0	128.4	121.3
Short term capital (net)	-28.9	-23.2	-19.0	-119.3	3.7
Errors and omissions	16.5	-83.7	23.7	-33.3	-52.9
Overall balance	17.9	-12.4	34.4	-8.2	-8.2

Exchange Rates [38]

Currency: **Communaute Financi- ere Africaine francs.**
Symbol: **CFAF.**

Data are currency units per $1.

January 1993	274.06
1992	264.69
1991	282.11
1990	272.26
1989	319.01
1988	297.85

Imports and Exports

Top Import Origins [39]

$294.1 million (f.o.b., 1991).

Origins	%
US	NA
France	NA
Nigeria	NA
Cameroon	NA

Top Export Destinations [40]

$193.9 million (f.o.b., 1991).

Destinations	%
France	NA
Nigeria	NA
Cameroon	NA

Foreign Aid [41]

	U.S. $	
US commitments, including Ex-Im (FY70-89)	198	million
Western (non-US) countries, ODA and OOF bilateral commitments (1970-89)	1.5	billion
OPEC bilateral aid (1979-89)	28	million
Communist countries (1970-89)	80	million

Import and Export Commodities [42]

Import Commodities	**Export Commodities**
Machinery, transp. equip. 39%	Cotton 48%
Industrial goods 20%	Cattle 35%
Petroleum products 13%	Textiles 5%
Foodstuffs 9%	Fish

For sources, notes, and explanations, see Annotated Source Appendix, page 1035.

187

Chile

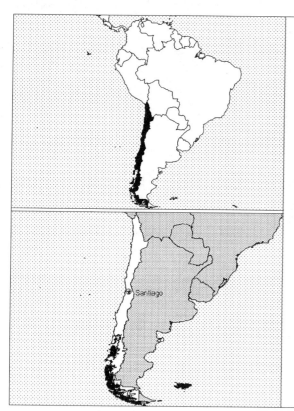

Geography [1]

Total area:
756,950 km2
Land area:
748,800 km2
Comparative area:
Slightly smaller than twice the size of Montana
Land boundaries:
Total 6,171 km, Argentina 5,150 km, Bolivia 861 km, Peru 160 km
Coastline:
6,435 km
Climate:
Temperate; desert in north; cool and damp in south
Terrain:
Low coastal mountains; fertile central valley; rugged Andes in east
Natural resources:
Copper, timber, iron ore, nitrates, precious metals, molybdenum
Land use:
Arable land:
7%
Permanent crops:
0%
Meadows and pastures:
16%
Forest and woodland:
21%
Other:
56%

Demographics [2]

	1960	1970	1980	1990	1991[1]	1994	2000	2010	2020
Population	7,585	9,369	11,094	13,108	13,287	13,951	15,207	17,266	19,225
Population density (persons per sq. mi.)	26	32	38	45	46	NA	52	58	64
Births	NA	NA	NA	NA	280	287	NA	NA	NA
Deaths	NA	NA	NA	NA	77	77	NA	NA	NA
Life expectancy - males	NA	NA	NA	NA	70	72	NA	NA	NA
Life expectancy - females	NA	NA	NA	NA	77	78	NA	NA	NA
Birth rate (per 1,000)	NA	NA	NA	NA	21	21	NA	NA	NA
Death rate (per 1,000)	NA	NA	NA	NA	6	5	NA	NA	NA
Women of reproductive age (15-44 yrs.)	NA	NA	NA	3,486	NA	3,685	3,978	NA	NA
of which are currently married	NA	NA	NA	2,027	NA	2,166	2,345	NA	NA
Fertility rate	NA	NA	NA	2.6	2.49	2.5	2.4	2.3	2.2

Population values are in thousands, life expectancy in years, and other items as indicated.

Health

Health Personnel [3]

Doctors per 1,000 pop., 1988-92	0.46
Nurse-to-doctor ratio, 1988-92	0.8
Hospital beds per 1,000 pop., 1985-90	3.3
Percentage of children immunized (age 1 yr. or less)	
Third dose of DPT, 1990-91	91.0
Measles, 1990-91	93.0

Health Expenditures [4]

Total health expenditure, 1990 (official exchange rate)	
Millions of dollars	1315
Millions of dollars per capita	100
Health expenditures as a percentage of GDP	
Total	4.7
Public sector	3.4
Private sector	1.4
Development assistance for health	
Total aid flows (millions of dollars)[1]	10
Aid flows per capita (millions of dollars)	0.7
Aid flows as a percentage of total health expenditure	0.7

For sources, notes, and explanations, see Annotated Source Appendix, page 1035.

Human Factors

Health Care Ratios [5]

Population per physician, 1970	2160
Population per physician, 1990	2150
Population per nursing person, 1970	460
Population per nursing person, 1990	340
Percent of births attended by health staff, 1985	97

Infants and Malnutrition [6]

Percent of babies with low birth weight, 1985	7
Infant mortality rate per 1,000 live births, 1970	78
Infant mortality rate per 1,000 live births, 1991	17
Years of life lost per 1,000 population, 1990	13
Prevalence of malnutrition (under age 5), 1990	2

Ethnic Division [7]

European/Euro-Indian	95%
Indian	3%
Other	2%

Religion [8]

Roman Catholic 89%, Protestant 11%, Jewish.

Major Languages [9]

Spanish.

Education

Public Education Expenditures [10]

Million Pesos	1980	1985	1987	1988	1989	1990[12]
Total education expenditure	47,961	101,493	137,863	181,378	NA	232,516
as percent of GNP	4.6	4.4	3.6	3.7	NA	2.9
as percent of total govt. expend.	11.9	15.3	NA	NA	NA	10.4
Current education expenditure	45,504	NA	NA	164,054	225,620	297,975
as percent of GNP	4.4	NA	NA	3.3	2.8	2.9
as percent of current govt. expend.	13.4	NA	NA	NA	NA	NA
Capital expenditure	2,457	NA	NA	17,324	6,896	9,745

Educational Attainment [11]

Age group	25+
Total population	5,204,698
Highest level attained (%)	
No schooling	9.4
First level	
Incompleted	56.6
Completed	NA
Entered second level	
S-1	26.9
S-2	NA
Post secondary	7.2

Literacy Rate [12]

In thousands and percent	1985[a]	1991[a]	2000[a]
Illiterate population +15 years	648	603	518
Illiteracy rate - total pop. (%)	7.8	6.6	4.8
Illiteracy rate - males (%)	7.4	6.5	4.9
Illiteracy rate - females (%)	8.1	6.8	4.8

Libraries [13]

	Admin. Units	Svc. Pts.	Vols. (000)	Shelving (meters)	Vols. Added	Reg. Users
National (1989)	1	2	3,514	NA	427	NA
Non-specialized	NA	NA	NA	NA	NA	NA
Public (1989)	293	293	1,054	48,263	42,104	27,450
Higher ed. (1989)	178	NA	5,669	NA	NA	NA
School (1989)	821	NA	3,820	NA	NA	NA

Daily Newspapers [14]

	1975	1980	1985	1990
Number of papers	47	34	38e|	45e|
Circ. (000)	NA	NA	NA	6000e|

Cinema [15]

Data for 1991.

Cinema seats per 1,000	6.5[1]
Annual attendance per person	0.7[1]
Gross box office receipts (mil. Peso)	NA

Science and Technology

Scientific/Technical Forces [16]

R&D Expenditures [17]

	Peso[4] (000) 1988
Total expenditure	23,161,300
Capital expenditure	NA
Current expenditure	NA
Percent current	NA

U.S. Patents Issued [18]

Values show patents issued to citizens of the country by the U.S. Patents Office.

	1990	1991	1992
Number of patents	2	8	5

For sources, notes, and explanations, see Annotated Source Appendix, page 1035.

Government and Law

Organization of Government [19]

Long-form name:
Republic of Chile
Type:
republic
Independence:
18 September 1810 (from Spain)
Constitution:
11 September 1980, effective 11 March 1981; amended 30 July 1989
Legal system:
based on Code of 1857 derived from Spanish law and subsequent codes influenced by French and Austrian law; judicial review of legislative acts in the Supreme Court; has not accepted compulsory ICJ jurisdiction
National holiday:
Independence Day, 18 September (1810)
Executive branch:
president, Cabinet
Legislative branch:
bicameral National Congress (Congreso Nacional) consisting of an upper house or Senate (Senado) and a lower house or Chamber of Deputies (Camara de Diputados)
Judicial branch:
Supreme Court (Corte Suprema)

Political Parties [20]

Chamber of Deputies	% of seats
Concertation of Parties for Democracy	59.2
PDC	31.7
PPD	14.2
PR	4.2
Other	9.2
National Renovation (RN)	24.2
Independent Democratic Union (UDI)	9.2
Right-wing independents	7.5

Government Expenditures [21]

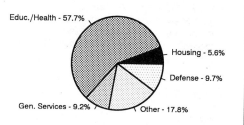

Educ./Health - 57.7%
Housing - 5.6%
Defense - 9.7%
Gen. Services - 9.2%
Other - 17.8%

% distribution for 1992. Expend. in 1993: 10.9 ($ bil.)

Military Expenditures and Arms Transfers [22]

	1985	1986	1987	1988	1989
Military expenditures					
Current dollars (mil.)[4]	619	596	820[e]	803[e]	790[e]
1989 constant dollars (mil.)[4]	704	662	882[e]	836[e]	790[e]
Armed forces (000)	124	127	127	96	95
Gross national product (GNP)					
Current dollars (mil.)	15,240	16,530	18,400	20,220	23,300
1989 constant dollars (mil.)	17,350	18,340	19,790	21,050	23,300
Central government expenditures (CGE)					
1989 constant dollars (mil.)	4,417	4,322	4,439	6,600	NA
People (mil.)	12.1	12.3	12.5	12.7	12.9
Military expenditure as % of GNP	4.1	3.6	4.5	4.0	3.4
Military expenditure as % of CGE	15.9	15.3	19.9	12.7	NA
Military expenditure per capita	58	54	71	66	61
Armed forces per 1,000 people	10.3	10.3	10.1	7.6	7.4
GNP per capita	1,438	1,496	1,588	1,661	1,809
Arms imports[6]					
Current dollars (mil.)	80	80	120	90	120
1989 constant dollars (mil.)	91	89	129	94	120
Arms exports[6]					
Current dollars (mil.)	80	20	170	280	160
1989 constant dollars (mil.)	91	22	183	291	160
Total imports[7]					
Current dollars (mil.)	2,743	2,914	3,793	4,731	6,496
1989 constant dollars	3,123	3,234	4,080	4,925	6,496
Total exports[7]					
Current dollars (mil.)	3,823	4,222	5,091	7,046	8,191
1989 constant dollars	4,352	4,685	5,476	7,335	8,191
Arms as percent of total imports[8]	2.9	2.7	3.2	1.9	1.8
Arms as percent of total exports[8]	2.1	0.5	3.3	4.0	2.0

Crime [23]

Crime volume (for 1990)

Cases known to police	183,210
Attempts (percent)	NA
Percent cases solved	44.92
Crimes per 100,000 persons	1,347.18
Persons responsible for offenses	
Total number offenders	30,963
Percent female	14.46
Percent juvenile[14]	8.17
Percent foreigners	0.62

Human Rights [24]

	SSTS	FL	FAPRO	PPCG	APROBC	TPW	PCPTW	STPEP	PHRFF	PRW	ASST	AFL
Observes		P		P		P	P			P		
	EAFRD	CPR	ESCR	SR	ACHR	MAAE	PVIAC	PVNAC	EAFDAW	TCIDTP	RC	
Observes	P	P	P	P	P		P	P	P	P	P	

P = Party; S = Signatory; see Appendix for meaning of abbreviations.

Labor Force

Total Labor Force [25]

4.728 million

Labor Force by Occupation [26]

Services, government	38.3%
Industry and commerce	33.8
Agriculture, forestry, and fishing	19.2
Mining	2.3
Construction	6.4

Date of data: 1990

Unemployment Rate [27]

4.9% (1992)

For sources, notes, and explanations, see Annotated Source Appendix, page 1035.

Production Sectors

Commercial Energy Production and Consumption

Production [28]

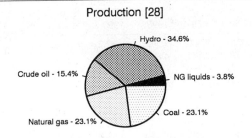

Hydro - 34.6%
NG liquids - 3.8%
Coal - 23.1%
Natural gas - 23.1%
Crude oil - 15.4%

Consumption [29]

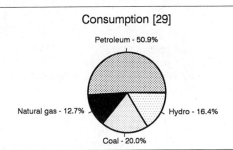

Petroleum - 50.9%
Hydro - 16.4%
Coal - 20.0%
Natural gas - 12.7%

Data are shown in quadrillion (10^{15}) BTUs and percent for 1991

Crude oil[1]	0.04
Natural gas liquids	0.01
Dry natural gas	0.06
Coal[2]	0.06
Net hydroelectric power[3]	0.09
Net nuclear power[3]	0.00
Total	0.26

Petroleum	0.28
Dry natural gas	0.07
Coal[2]	0.11
Net hydroelectric power[3]	0.09
Net nuclear power[3]	0.00
Total	0.55

Telecommunications [30]

- Modern telephone system based on extensive microwave radio relay facilities
- 768,000 telephones
- Broadcast stations - 159 AM, no FM, 131 TV, 11 shortwave
- Satellite ground stations - 2 Atlantic Ocean INTELSAT and 3 domestic

Transportation [31]

Railroads. 7,766 km total; 3,974 km 1.676-meter gauge, 150 km 1.435-meter standard gauge, 3,642 km 1.000-meter gauge; 1,865 km 1.676-meter gauge and 80 km 1.000-meter gauge electrified

Highways. 79,025 km total; 9,913 km paved, 33,140 km gravel, 35,972 km improved and unimproved earth (1984)

Merchant Marine. 31 ships (1,000 GRT or over) totaling 445,330 GRT/756,018 DWT; includes 8 cargo, 1 refrigerated cargo, 3 roll-on/roll-off cargo, 2 oil tanker, 3 chemical tanker, 3 liquefied gas tanker, 3 combination ore/oil, 8 bulk; note - in addition, 1 naval tanker and 1 military transport are sometimes used commercially

Airports

Total:	396
Usable:	351
With permanent-surface runways:	48
With runways over 3,659 m:	0
With runways 2,440-3,659 m:	13
With runways 1,220-2,439 m:	57

Top Agricultural Products [32]

	1990	1991
Grains	2,974	2,856
Fruits	2,070	2,270
Sugar beets	2,594	2,150
Potatoes	829	844
Beans	87	117
Oils	80	90

Values shown are 1,000 metric tons.

Top Mining Products [33]

Preliminary metric tons unless otherwise specified M.t.

Copper, mine output, Cu content	1,814,300
Potassium nitrate	250,000[e]
Sodium nitrate	600,000[e]
Iodine, elemental	5,700
Lithium carbonate	8,575
Molybdenum, mine output, Mo content	14,434
Rhenium, mine output, Re content (kilograms)	6,500[e]
Gold, mine output, Au content (kilograms)	28,879
Silver (kilograms)	676,339
Selenium (kilograms)	50,600

Tourism [34]

	1987	1988	1989	1990	1991
Tourists	575	624	797	943	1,349
Tourism receipts	182	205	407	540	700
Tourism expenditures	353	423	397	426	409
Fare receipts	85	86	80	135	172
Fare expenditures	156	191	178	174	196

Tourists are in thousands, money in million U.S. dollars.

Manufacturing Sector

GDP and Manufacturing Summary [35]

	1980	1985	1989	1990	% change 1980-1990	% change 1989-1990
GDP (million 1980 $)	21,489	21,075	27,360	28,402	32.2	3.8
GDP per capita (1980 $)	1,928	1,739	2,111	2,155	11.8	2.1
Manufacturing as % of GDP (current prices)	19.9	22.7	20.8e	20.9e	5.0	0.5
Gross output (million $)	10,790	10,477	15,605e	21,215	96.6	36.0
Value added (million $)	4,991	4,713	6,345e	8,757	75.5	38.0
Value added (million 1980 $)	4,830	4,482	5,872	6,107	26.4	4.0
Industrial production index	100	100	130	128	28.0	-1.5
Employment (thousands)	206	185	222e	298	44.7	34.2

Note: GDP stands for Gross Domestic Product. 'e' stands for estimated value.

Profitability and Productivity

	1980	1985	1989	1990	% change 1980-1990	% change 1989-1990
Intermediate input (%)	54	55	59e	59	9.3	0.0
Wages, salaries, and supplements (%)	9e	6	7e	7	-22.2	0.0
Gross operating surplus (%)	38e	39	34e	34	-10.5	0.0
Gross output per worker ($)	51,994	56,369	70,365e	70,919	36.4	0.8
Value added per worker ($)	24,050	25,359	28,610e	29,274	21.7	2.3
Average wage (incl. benefits) ($)	4,444e	3,498	4,816e	4,861	9.4	0.9

Profitability is in percent of gross output. Productivity is in U.S. $. 'e' stands for estimated value.

Profitability - 1990

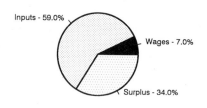

Inputs - 59.0%
Wages - 7.0%
Surplus - 34.0%

The graphic shows percent of gross output.

Value Added in Manufacturing

	1980 $ mil.	1980 %	1985 $ mil.	1985 %	1989 $ mil.	1989 %	1990 $ mil.	1990 %	% change 1980-1990	% change 1989-1990
311 Food products	827	16.6	805	17.1	1,182e	18.6	1,543	17.6	86.6	30.5
313 Beverages	289	5.8	177	3.8	261e	4.1	374	4.3	29.4	43.3
314 Tobacco products	214	4.3	205	4.3	99e	1.6	303	3.5	41.6	206.1
321 Textiles	234	4.7	162	3.4	252e	4.0	333	3.8	42.3	32.1
322 Wearing apparel	111	2.2	83	1.8	122e	1.9	163	1.9	46.8	33.6
323 Leather and fur products	22	0.4	18	0.4	20e	0.3	37	0.4	68.2	85.0
324 Footwear	77	1.5	51	1.1	77e	1.2	121	1.4	57.1	57.1
331 Wood and wood products	153	3.1	143	3.0	159e	2.5	270	3.1	76.5	69.8
332 Furniture and fixtures	37	0.7	14	0.3	34e	0.5	53	0.6	43.2	55.9
341 Paper and paper products	281	5.6	278	5.9	479e	7.5	561	6.4	99.6	17.1
342 Printing and publishing	182	3.6	104	2.2	154e	2.4	224	2.6	23.1	45.5
351 Industrial chemicals	55	1.1	94	2.0	88e	1.4	247	2.8	349.1	180.7
352 Other chemical products	324	6.5	289	6.1	411e	6.5	617	7.0	90.4	50.1
353 Petroleum refineries	184	3.7	277	5.9	234e	3.7	480	5.5	160.9	105.1
354 Miscellaneous petroleum and coal products	27	0.5	47	1.0	78e	1.2	69	0.8	155.6	-11.5
355 Rubber products	60	1.2	49	1.0	65e	1.0	72	0.8	20.0	10.8
356 Plastic products	50	1.0	63	1.3	99e	1.6	178	2.0	256.0	79.8
361 Pottery, china and earthenware	14	0.3	9	0.2	14e	0.2	9	0.1	-35.7	-35.7
362 Glass and glass products	38	0.8	27	0.6	46e	0.7	51	0.6	34.2	10.9
369 Other non-metal mineral products	146	2.9	115	2.4	182e	2.9	218	2.5	49.3	19.8
371 Iron and steel	188	3.8	226	4.8	300e	4.7	284	3.2	51.1	-5.3
372 Non-ferrous metals	965	19.3	1,175	24.9	1,540e	24.3	1,716	19.6	77.8	11.4
381 Metal products	181	3.6	130	2.8	208e	3.3	366	4.2	102.2	76.0
382 Non-electrical machinery	96	1.9	50	1.1	72e	1.1	168	1.9	75.0	133.3
383 Electrical machinery	90	1.8	61	1.3	96e	1.5	125	1.4	38.9	30.2
384 Transport equipment	127	2.5	50	1.1	64e	1.0	153	1.7	20.5	139.1
385 Professional and scientific equipment	5	0.1	4	0.1	5e	0.1	9	0.1	80.0	80.0
390 Other manufacturing industries	13	0.3	7	0.1	7e	0.1	14	0.2	7.7	100.0

Note: The industry codes shown are International Standard Industry codes (ISIC). Percentages are percent of total Value Added. 'e' stands for estimated value

192

For sources, notes, and explanations, see Annotated Source Appendix, page 1035.

Finance, Economics, and Trade

Economic Indicators [36]

Billions of 1977 Chilean Pesos, unless otherwise noted[23].

	1989	1990	1991
Real GDP	470	480	504[P]
Real GDP Growth Rate	10	2.1	5[P]
Money Supply (M1) (bil. pesos)	412	484	688[P]
Gross Domestic Saving	16.20	17.20	17.50
Investment Rate (% of GDP)	17.2	17.5	16.2[P]
CPI	21.4	27.30	20.0[P]
WPI	22.8	25.7	18.0[P]
External Public Debt	16.3	17.5	17.4[24]

Balance of Payments Summary [37]

Values in millions of dollars.

	1987	1988	1989	1990	1991
Exports of goods (f.o.b.)	5,224.0	7,052.0	8,080.0	8,310.0	8,929.0
Imports of goods (f.o.b.)	-3,994.0	-4,833.0	-6,502.0	-7,037.0	-7,354.0
Trade balance	1,230.0	2,219.0	1,578.0	1,273.0	1,575.0
Services - debits	-3,432.0	-3,962.0	-4,337.0	-4,426.0	-4,506.0
Services - credits	1,268.0	1,399.0	1,777.0	2,355.0	2,733.0
Private transfers (net)	65.0	63.0	58.0	54.0	40.0
Government transfers (net)	61.0	114.0	157.0	146.0	300.0
Long term capital (net)	849.0	1,310.0	666.0	2,108.0	1,083.0
Short term capital (net)	95.0	-301.0	612.0	1,184.0	-149.0
Errors and omissions	-78.0	-109.0	-71.0	-326.0	161.0
Overall balance	58.0	733.0	440.0	2,368.0	1,237.0

Exchange Rates [38]

Currency: **Chilean pesos.**
Symbol: **Ch$.**

Data are currency units per $1.

January 1993	384.04
1992	362.59
1991	349.37
1990	305.06
1989	267.16
1988	245.05

Imports and Exports

Top Import Origins [39]

$9.2 billion (f.o.b., 1992). Data are for 1991.

Origins	%
US	21
EC	18
Brazil	9
Japan	8

Top Export Destinations [40]

$10 billion (f.o.b., 1992) Data are for 1991.

Destinations	%
EC	32
US	18
Japan	18
Brazil	5

Foreign Aid [41]

	U.S. $	
US commitments, including Ex-Im (FY70-89)	521	million
Western (non-US) countries, ODA and OOF bilateral commitments (1970-89)	1.6	billion
Communist countries (1970-89)	386	million

Import and Export Commodities [42]

Import Commodities

Capital goods 25.2%
Spare parts 24.8%
Raw materials 15.4%
Petroleum 10%
Foodstuffs 5.7%

Export Commodities

Copper 41%
Other metals and minerals 8.7%
Wood products 7.1%
Fish and fishmeal 9.8%
Fruits 8.4%

China

Geography [1]

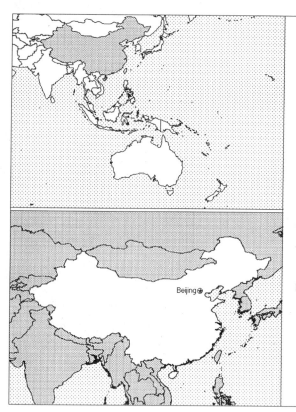

Total area:
9,596,960 km2
Land area:
9,326,410 km2
Comparative area:
Slightly larger than the US
Land boundaries:
Total 22,143.34 km, Afghanistan 76 km, Bhutan 470 km, Burma 2,185 km, Hong Kong 30 km, India 3,380 km, Kazakhstan 1,533 km, North Korea 1,416 km, Kyrgyzstan 858 km, Laos 423 km, Macau 0.34 km, Mongolia 4,673 km, Nepal 1,236 km, Pakistan 523 km, Russia (northeast) 3,605 km, Russia (northwest) 40 km, Tajikistan 414 km, Vietnam 1,281 km
Coastline:
14,500 km
Climate:
Extremely diverse; tropical in south to subarctic in north
Terrain:
Mostly mountains, high plateaus, deserts in west; plains, deltas, and hills in east
Land use:
Arable land: 10%
Permanent crops: 0%
Meadows and pastures: 31%
Forest and woodland: 14%
Other: 45%

Demographics [2]

	1960	1970	1980	1990	1991[1]	1994	2000	2010	2020
Population	650,661	820,403	984,736	1,136,626	1,151,487	1,190,431	1,260,154	1,348,429	1,424,725
Population density (persons per sq. mi.)	181	228	273	315	320	NA	362	394	428
Births	NA	NA	NA	NA	25,655	21,547	NA	NA	NA
Deaths	NA	NA	NA	NA	7,703	8,750	NA	NA	NA
Life expectancy - males	NA	NA	NA	NA	68	67	NA	NA	NA
Life expectancy - females	NA	NA	NA	NA	72	69	NA	NA	NA
Birth rate (per 1,000)	NA	NA	NA	NA	22	18	NA	NA	NA
Death rate (per 1,000)	NA	NA	NA	NA	7	7	NA	NA	NA
Women of reproductive age (15-44 yrs.)	NA	NA	NA	308,049	NA	325,810	345,163	NA	NA
of which are currently married	NA	NA	NA	191,654	NA	219,057	250,908	NA	NA
Fertility rate	NA	NA	NA	2.2	2.30	1.8	1.8	1.8	1.8

Population values are in thousands, life expectancy in years, and other items as indicated.

Health

Health Personnel [3]

Doctors per 1,000 pop., 1988-92	1.37
Nurse-to-doctor ratio, 1988-92	0.5
Hospital beds per 1,000 pop., 1985-90	2.6
Percentage of children immunized (age 1 yr. or less)	
Third dose of DPT, 1990-91	95.0
Measles, 1990-91	96.0

Health Expenditures [4]

Total health expenditure, 1990 (official exchange rate)	
Millions of dollars	12969
Millions of dollars per capita	11
Health expenditures as a percentage of GDP	
Total	3.5
Public sector	2.1
Private sector	1.4
Development assistance for health	
Total aid flows (millions of dollars)[1]	77
Aid flows per capita (millions of dollars)	0.1
Aid flows as a percentage of total health expenditure	0.6

For sources, notes, and explanations, see Annotated Source Appendix, page 1035.

Human Factors

Health Care Ratios [5]

Population per physician, 1970	NA
Population per physician, 1990	NA
Population per nursing person, 1970	2500
Population per nursing person, 1990	NA
Percent of births attended by health staff, 1985	NA

Infants and Malnutrition [6]

Percent of babies with low birth weight, 1985	6
Infant mortality rate per 1,000 live births, 1970	69
Infant mortality rate per 1,000 live births, 1991	38
Years of life lost per 1,000 population, 1990	NA
Prevalence of malnutrition (under age 5), 1990	NA

Ethnic Division [7]

Other includes Zhuang, Uygur, Hui, Yi, Tibetan, Miao, Manchu, Mongol, Buyi, Korean.

Han Chinese	91.9%
Other	8.1%

Religion [8]

Daoism (Taoism), Buddhism, Muslim 2-3%, Christian 1% (est.).

Major Languages [9]

Standard Chinese (Putonghua) or Mandarin (based on the Beijing dialect), Yue (Cantonese), Wu (Shanghainese), Minbei (Fuzhou), Minnan (Hokkien-Taiwanese), Xiang, Gan, Hakka dialects, minority languages (see Ethnic divisions entry).

Education

Public Education Expenditures [10]

Million Yuan	1980	1985	1987	1988	1989	1990
Total education expenditure	11,319	22,489	NA	32,322	37,299	NA
as percent of GNP	2.5	2.6	NA	2.3	2.3	NA
as percent of total govt. expend.	9.3	12.2	NA	12.1	12.4	NA
Current education expenditure	10,263	19,770	NA	29,673	34,254	NA
as percent of GNP	2.3	2.3	NA	2.1	2.2	NA
as percent of current govt. expend.	NA	NA	NA	NA	NA	NA
Capital expenditure	1,056	2,719	NA	2,649	3,045	NA

Educational Attainment [11]

Age group	25+[3]
Total population	466,915,380[3]
Highest level attained (%)	
No schooling	44.5[3]
First level	
Incompleted	NA[3]
Completed	32.7[3]
Entered second level	
S-1	16.1[3]
S-2	5.6[3]
Post secondary	1.0[3]

Literacy Rate [12]

In thousands and percent	1985[a]	1991[a]	2000[a]
Illiterate population +15 years	236,741	223,727	188,263
Illiteracy rate - total pop. (%)	31.8	26.7	19.7
Illiteracy rate - males (%)	19.6	15.9	10.9
Illiteracy rate - females (%)	44.7	38.2	29.0

Libraries [13]

	Admin. Units	Svc. Pts.	Vols. (000)	Shelving (meters)	Vols. Added	Reg. Users
National (1989)	1	NA	13,768	290,000	476,796	166,861
Non-specialized	NA	NA	NA	NA	NA	NA
Public (1989)	2,512	2,512	283,680	7 mil.	4.9mil	5.9mil
Higher ed. (1990)	1,064	NA	356,415	NA	419,850	3.2mil
School	NA	NA	NA	NA	NA	NA

Daily Newspapers [14]

	1975	1980	1985	1990
Number of papers	NA	50	70e	44
Circ. (000)	NA	34,375	39000e	NA

Cinema [15]

Data for 1989.

Cinema seats per 1,000	NA
Annual attendance per person	14.1
Gross box office receipts (mil. Yuan)	1,950[e]

Science and Technology

Scientific/Technical Forces [16]

Potential scientists/engineers	NA
Number female	NA
Potential technicians	NA
Number female	NA
Total	9,661,000

R&D Expenditures [17]

U.S. Patents Issued [18]

Values show patents issued to citizens of the country by the U.S. Patents Office.

	1990	1991	1992
Number of patents	48	54	41

Government and Law

Organization of Government [19]

Long-form name:
People's Republic of China
Type:
Communist state
Independence:
221 BC (unification under the Qin or Ch'in Dynasty 221 BC; Qing or Ch'ing Dynasty replaced by the Republic on 12 February 1912; People's Republic established 1 October 1949)
Constitution:
most recent promulgated 4 December 1982
Legal system:
a complex amalgam of custom and statute, largely criminal law; rudimentary civil code in effect since 1 January 1987; new legal codes since 1 January 1980
National holiday:
National Day, 1 October (1949)
Executive branch:
president, vice president, premier, four vice premiers, State Council
Legislative branch:
unicameral National People's Congress (Quanguo Renmin Daibiao Dahui)
Judicial branch:
Supreme People's Court

Elections [20]

National People's Congress. Last held March 1993 (next to be held March 1998); results - CCP is the only party but there are also independents; seats - (2,977 total) (elected at county or xian level).

Government Budget [21]

For 1992.
Deficit   16.3

Crime [23]

Crime volume (for 1990)	
Cases known to police	2,216,997
Attempts (percent)	NA
Percent cases solved	57.1
Crimes per 100,000 persons	200.9
Persons responsible for offenses	
Total number offenders	1,176,882
Percent female	2.6
Percent juvenile[15]	69.7
Percent foreigners	0.02

Military Expenditures and Arms Transfers [22]

	1985	1986	1987	1988	1989
Military expenditures					
Current dollars (mil.)[e]	19,850	19,800	20,480	21,830	22,330
1989 constant dollars (mil.)[e]	22,600	21,970	22,030	22,720	22,330
Armed forces (000)	4,100	4,030	3,530	3,783	3,903
Gross national product (GNP)					
Current dollars (mil.)	386,300	427,700	486,400	559,100	603,500
1989 constant dollars (mil.)	439,800	474,600	523,100	582,000	603,500
Central government expenditures (CGE)					
1989 constant dollars (mil.)	94,900	113,800	113,100	113,900	117,100
People (mil.)	1,042.9	1,055.6	1,071.3	1,087.2	1,102.4
Military expenditure as % of GNP	5.1	4.6	4.2	3.9	3.7
Military expenditure as % of CGE	23.8	19.3	19.5	20.0	19.1
Military expenditure per capita	22	21	21	21	20
Armed forces per 1,000 people	3.9	3.8	3.3	3.5	3.5
GNP per capita	422	450	488	535	547
Arms imports[6]					
Current dollars (mil.)	650	550	625	300	110
1989 constant dollars (mil.)	740	610	672	312	110
Arms exports[6]					
Current dollars (mil.)	675	1,200	1,800	2,600	2,000
1989 constant dollars (mil.)	768	1,332	1,936	2,707	2,000
Total imports[7]					
Current dollars (mil.)	42,530	43,170	43,390	55,280	58,280
1989 constant dollars	48,410	47,910	46,670	57,540	58,280
Total exports[7]					
Current dollars (mil.)	27,330	31,150	39,540	47,540	51,630
1989 constant dollars	31,110	34,570	42,530	49,490	61,530
Arms as percent of total imports[8]	1.5	1.3	1.4	0.5	0.2
Arms as percent of total exports[8]	2.5	3.9	4.6	5.5	3.9

Human Rights [24]

	SSTS	FL	FAPRO	PPCG	APROBC	TPW	PCPTW	STPEP	PHRFF	PRW	ASST	AFL
Observes				P		P	P					
	EAFRD	CPR	ESCR	SR	ACHR	MAAE	PVIAC	PVNAC	EAFDAW	TCIDTP	RC	
Observes	P			P			P	P	P	P	P	

P = Party; S = Signatory; see Appendix for meaning of abbreviations.

Labor Force

Total Labor Force [25]

567.4 million

Labor Force by Occupation [26]

Agriculture and forestry	60%
Industry and commerce	25
Construction and mining	5
Social services	5
Other	5

Date of data: 1990 est.

Unemployment Rate [27]

2.3% in urban areas (1992)

Production Sectors

Commercial Energy Production and Consumption

Production [28]

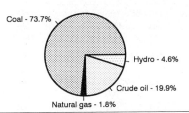

Coal - 73.7%
Hydro - 4.6%
Crude oil - 19.9%
Natural gas - 1.8%

Consumption [29]

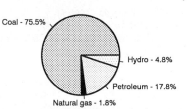

Coal - 75.5%
Hydro - 4.8%
Petroleum - 17.8%
Natural gas - 1.8%

Data are shown in quadrillion (10^{15}) BTUs and percent for 1991

Crude oil[1]	6.02
Natural gas liquids	0.00
Dry natural gas	0.54
Coal[2]	22.24
Net hydroelectric power[3]	1.39
Net nuclear power[3]	0.00
Total	30.19

Petroleum	5.20
Dry natural gas	0.54
Coal[2]	22.05
Net hydroelectric power[3]	1.40
Net nuclear power[3]	(s)
Total	29.19

Telecommunications [30]

- Domestic and international services are increasingly available for private use
- Unevenly distributed internal system serves principal cities, industrial centers, and most townships
- 11,000,000 telephones (December 1989)
- Broadcast stations - 274 AM, unknown FM, 202 (2,050 repeaters) TV
- More than 215 million radio receivers
- 75 million TVs
- Satellite earth stations - 4 Pacific Ocean INTELSAT, 1 Indian Ocean INTELSAT, 1 INMARSAT, and 55 domestic

Transportation [31]

Railroads. total about 64,000 km; 54,000 km of common carrier lines, of which 53,400 km are 1.435-meter gauge (standard) and 600 km are 1.000-meter gauge (narrow); 11,200 km of standard gauge common carrier route are double tracked and 6,900 km are electrified (1990); an additional 10,000 km of varying gauges (0.762 to 1.067-meter) are dedicated industrial lines

Highways. about 1,029,000 km (1990) total; 170,000 km (est.) paved roads, 648,000 km (est.) gravel/improved earth roads, 211,000 km (est.) unimproved earth roads and tracks

Merchant Marine. 1,478 ships (1,000 GRT or over) totaling 14,029,320 GRT/21,120,522 DWT; includes 25 passenger, 42 short-sea passenger, 18 passenger-cargo, 6 cargo/training, 811 cargo, 11 refrigerated cargo, 81 container, 18 roll-on/roll-off cargo, 1 multifunction/barge carrier, 177 oil tanker, 11 chemical tanker, 263 bulk, 3 liquefied gas, 1 vehicle carrier, 9 combination bulk, 1 barge carrier; note - China beneficially owns an additional 227 ships (1,000 GRT or over) totaling approximately 6,187,117 DWT that operate under Panamanian, British, Hong Kong, Maltese, Liberian, Vanuatu, Cypriot, Saint Vincent, Bahamian, and Romanian registry

Airports

Total:	330
Usable:	330
With permanent-surface runways:	260
With runways over 3,500 m:	fewer than 10
With runways 2,440-3,659 m:	90
With runways 1,220-2,439 m:	200

Top Agricultural Products [32]

	1990	1991[4]
Rice	189.3	187.0
Wheat	98.2	96.0
Corn	96.8	95.0
Sugar cane	57.6	61.0
Sugar beets	14.5	15.0
Soybeans	11.0	10.1

Values shown are 1,000 metric tons.

Top Mining Products [33]

Estimated metric tons unless otherwise specified	M.t.
Antimony, mine output, Sb content	45,000
Barite (000 tons)	1,800
Flurospar	1,600,000
Magnesite (000 tons)	2,600
Rare earths	NA
Tungsten, mine output, W content	25,000

Tourism [34]

	1987	1988	1989	1990	1991
Visitors[14]	26,902	31,695	24,501	27,462	33,350
Tourists[15]	1,728	1,842	1,461	1,747	2,710
Tourism receipts	1,862	2,247	1,861	2,218	2,845
Tourism expenditures	387	633	429	448	417
Fare receipts	152	450	372	480	494

Tourists are in thousands, money in million U.S. dollars.

For sources, notes, and explanations, see Annotated Source Appendix, page 1035.

197

Manufacturing Sector

GDP and Manufacturing Summary [35]

	1980	1985	1989	1990	% change 1980-1990	% change 1989-1990
GDP (million 1980 $)	286,716	459,012	NA	662,193	131.0	NA
GDP per capita (1980 $)	293	441	NA	592	102.0	NA
Manufacturing as % of GDP (current prices)	40.1	36.1	NA	35.7	-11.0	NA
Gross output (million $)	232,460[e]	246,331	NA	349,604	50.4	NA
Value added (million $)	88,577[e]	78,380	NA	90,259	1.9	NA
Value added (million 1980 $)	107,146[e]	171,761	NA	275,032	156.7	NA
Industrial production index	NA	NA	NA	NA	NA	NA
Employment (thousands)	24,390	29,743	NA	33,950	39.2	NA

Note: GDP stands for Gross Domestic Product. 'e' stands for estimated value.

Profitability and Productivity

	1980	1985	1989	1990	% change 1980-1990	% change 1989-1990
Intermediate input (%)	62[e]	68	NA	74	19.4	NA
Wages, salaries, and supplements (%)	6[e]	5	NA	5[e]	-16.7	NA
Gross operating surplus (%)	32[e]	27	NA	21[e]	-34.4	NA
Gross output per worker ($)	9,531[e]	8,282	NA	10,298	8.0	NA
Value added per worker ($)	3,632[e]	2,635	NA	2,659	-26.8	NA
Average wage (incl. benefits) ($)	548	384	NA	500[e]	-8.8	NA

Profitability is in percent of gross output. Productivity is in U.S. $. 'e' stands for estimated value.

Profitability - 1990

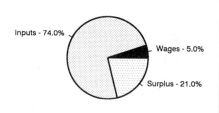

Inputs - 74.0%
Wages - 5.0%
Surplus - 21.0%

The graphic shows percent of gross output.

Value Added in Manufacturing

	1980 $ mil.	1980 %	1985 $ mil.	1985 %	1989 $ mil.	1989 %	1990 $ mil.	1990 %	% change 1980-1990	% change 1989-1990
311 Food products	3,764[e]	4.2	3,433	4.4	NA	NA	4,489	5.0	19.3	NA
313 Beverages	1,587[e]	1.8	1,696	2.2	NA	NA	2,414	2.7	52.1	NA
314 Tobacco products	3,545	4.0	3,999	5.1	NA	NA	6,220	6.9	75.5	NA
321 Textiles	13,409	15.1	8,587	11.0	NA	NA	10,299	11.4	-23.2	NA
322 Wearing apparel	1866a	0.0	1716a	0.0	NA	NA	2109a	0.0	NA	NA
323 Leather and fur products	911	1.0	747	1.0	NA	NA	944	1.0	3.6	NA
324 Footwear	a	0.0	a	0.0	NA	NA	a	0.0	NA	NA
331 Wood and wood products	751[e]	0.8	591	0.8	NA	NA	502	0.6	-33.2	NA
332 Furniture and fixtures	653[e]	0.7	514	0.7	NA	NA	455	0.5	-30.3	NA
341 Paper and paper products	1,929[e]	2.2	1,532	2.0	NA	NA	1,949	2.2	1.0	NA
342 Printing and publishing	1,042	1.2	960	1.2	NA	NA	1,036	1.1	-0.6	NA
351 Industrial chemicals	7,125	8.0	5,584	7.1	NA	NA	8,459	9.4	18.7	NA
352 Other chemical products	2,429	2.7	2,292	2.9	NA	NA	3,372	3.7	38.8	NA
353 Petroleum refineries	4,223	4.8	3,676	4.7	NA	NA	2,714	3.0	-35.7	NA
354 Miscellaneous petroleum and coal products	154[e]	0.2	183	0.2	NA	NA	208	0.2	35.1	NA
355 Rubber products	2,175	2.5	1,593	2.0	NA	NA	1,603	1.8	-26.3	NA
356 Plastic products	1,256	1.4	1,317	1.7	NA	NA	1,736	1.9	38.2	NA
361 Pottery, china and earthenware	439	0.5	431	0.5	NA	NA	504	0.6	14.8	NA
362 Glass and glass products	838	0.9	822	1.0	NA	NA	705	0.8	-15.9	NA
369 Other non-metal mineral products	4,425	5.0	4,340	5.5	NA	NA	4,524	5.0	2.2	NA
371 Iron and steel	6,538[e]	7.4	5,810	7.4	NA	NA	6,571	7.3	0.5	NA
372 Non-ferrous metals	1,868[e]	2.1	1,730	2.2	NA	NA	2,050	2.3	9.7	NA
381 Metal products	4,861	5.5	2,582	3.3	NA	NA	2,946	3.3	-39.4	NA
382 Non-electrical machinery	13,418	15.1	10,941	14.0	NA	NA	10,116	11.2	-24.6	NA
383 Electrical machinery	3,216	3.6	6,458	8.2	NA	NA	7,445	8.2	131.5	NA
384 Transport equipment	3,013	3.4	4,134	5.3	NA	NA	3,918	4.3	30.0	NA
385 Professional and scientific equipment	810	0.9	1,021	1.3	NA	NA	843	0.9	4.1	NA
390 Other manufacturing industries	1,838[e]	2.1	1,691	2.2	NA	NA	2,125	2.4	15.6	NA

Note: The industry codes shown are International Standard Industry codes (ISIC). Percentages are percent of total Value Added. 'e' stands for estimated value

For sources, notes, and explanations, see Annotated Source Appendix, page 1035.

Finance, Economics, and Trade

Economic Indicators [36]

	1989	1990	1991[P]
GNP (bil. RMB)[25]	1,568	1,740	2,001
Real GNP Growth (pct)	3.9	4.0	7.0
GNP Per Capita (RMB)	1,410	1,522	1,760
Money Supply (M-1)[5]	744	840	920
Commercial Interest Rates	NA	NA	NA
Savings Rate[27]	34.8	35.7	40.0
General Retail Price Index (CPI) (% chg.)[28]	6.4	5.4	5.5
Wholesale Price Index	NA	NA	NA
External Debt, Year-End	41.3	52.5	55.0

Balance of Payments Summary [37]

Values in millions of dollars.

	1987	1988	1989	1990	1991
Exports of goods (f.o.b.)	34,734.0	41,054.0	43,220.0	51,519.0	58,919.0
Imports of goods (f.o.b.)	-36,395.0	-46,369.0	-48,840.0	-42,354.0	-50,176.0
Trade balance	-1,661.0	-5,315.0	-5,620.0	9,165.0	8,743.0
Services - debits	-3,676.0	-5,233.0	-5,575.0	-6,314.0	-7,000.0
Services - credits	5,413.0	6,327.0	6,497.0	8,873.0	11,191.0
Private transfers (net)	249.0	416.0	238.0	222.0	444.0
Government transfers (net)	-25.0	3.0	143.0	52.0	387.0
Long term capital (net)	5,869.0	7,139.0	5,240.0	5,963.0	-123.0
Short term capital (net)	212.0	77.0	-1,518.0	2,242.0	362.0
Errors and omissions	-1,598.0	-1,040.0	117.0	-8,161.0	533.0
Overall balance	4,783.0	2,374.0	-478.0	12,042.0	14,537.0

Exchange Rates [38]

Currency: **yuan.**
Symbol: **Y.**

Data are currency units per $1.

January 1993	5.7640
1992	5.5146
1991	5.3234
1990	4.7832
1989	3.7651
1988	3.7221

Imports and Exports

Top Import Origins [39]

$80.6 billion (c.i.f., 1992).

Origins	%
Hong Kong and Macau	NA
Japan	NA
US	NA
Taiwan	NA
Germany	NA
Russia 1992	NA

Top Export Destinations [40]

$85.0 billion (f.o.b., 1992).

Destinations	%
Hong Kong and Macau	NA
Japan	NA
US	NA
Germany	NA
South Korea	NA
Russia 1992	NA

Foreign Aid [41]

	U.S. $	
Donor - to less developed countries (1970-89)	7.0	billion
US commitments, including Ex-Im (FY70-87)	220.7	million
Western (non-US) countries, ODA and OOF bilateral commitments (1970-87)	13.5	billion

Import and Export Commodities [42]

Import Commodities	Export Commodities
Specialized industrial machinery	Textiles
Chemicals	Garments
Manufactured goods	Telecom & recording equip.
Steel	Petroleum
Textile yarn	Minerals
Fertilizer	

Colombia

Geography [1]

Total area:
 1,138,910 km2
Land area:
 1,038,700 km2
Comparative area:
 Slightly less than three times the size of Montana
Land boundaries:
 Total 7,408 km, Brazil 1,643 km, Ecuador 590 km, Panama 225 km, Peru 2,900 km, Venezuela 2,050 km
Coastline:
 3,208 km (Caribbean Sea 1,760 km, North Pacific Ocean 1,448 km)
Climate:
 Tropical along coast and eastern plains; cooler in highlands
Terrain:
 Flat coastal lowlands, central highlands, high Andes mountains, eastern lowland plains
Natural resources:
 Petroleum, natural gas, coal, iron ore, nickel, gold, copper, emeralds
Land use:
 Arable land:
 4%
 Permanent crops:
 2%
 Meadows and pastures:
 29%
 Forest and woodland:
 49%
 Other:
 16%

Santa Fe de Bogata

Demographics [2]

	1960	1970	1980	1990	1991[1]	1994	2000	2010	2020
Population	15,953	21,430	26,580	32,983	33,778	35,578	39,172	44,504	49,266
Population density (persons per sq. mi.)	40	53	66	82	84	NA	99	114	128
Births	NA	NA	NA	NA	881	805	NA	NA	NA
Deaths	NA	NA	NA	NA	169	169	NA	NA	NA
Life expectancy - males	NA	NA	NA	NA	68	69	NA	NA	NA
Life expectancy - females	NA	NA	NA	NA	74	75	NA	NA	NA
Birth rate (per 1,000)	NA	NA	NA	NA	26	23	NA	NA	NA
Death rate (per 1,000)	NA	NA	NA	NA	5	5	NA	NA	NA
Women of reproductive age (15-44 yrs.)	NA	NA	NA	9,022	NA	9,830	10,991	NA	NA
of which are currently married	NA	NA	NA	4,937	NA	5,530	6,310	NA	NA
Fertility rate	NA	NA	NA	2.8	2.83	2.5	2.2	1.9	1.9

Population values are in thousands, life expectancy in years, and other items as indicated.

Health

Health Personnel [3]

Doctors per 1,000 pop., 1988-92	0.87
Nurse-to-doctor ratio, 1988-92	0.6
Hospital beds per 1,000 pop., 1985-90	1.5
Percentage of children immunized (age 1 yr. or less)	
Third dose of DPT, 1990-91	84.0
Measles, 1990-91	75.0

Health Expenditures [4]

Total health expenditure, 1990 (official exchange rate)	
Millions of dollars	1604
Millions of dollars per capita	50
Health expenditures as a percentage of GDP	
Total	4.0
Public sector	1.8
Private sector	2.2
Development assistance for health	
Total aid flows (millions of dollars)[1]	26
Aid flows per capita (millions of dollars)	0.8
Aid flows as a percentage of total health expenditure	1.6

Human Factors

Health Care Ratios [5]

Population per physician, 1970	2260
Population per physician, 1990	NA
Population per nursing person, 1970	NA
Population per nursing person, 1990	NA
Percent of births attended by health staff, 1985	51

Infants and Malnutrition [6]

Percent of babies with low birth weight, 1985	15
Infant mortality rate per 1,000 live births, 1970	77
Infant mortality rate per 1,000 live births, 1991	23
Years of life lost per 1,000 population, 1990	11
Prevalence of malnutrition (under age 5), 1990	12

Ethnic Division [7]

Mestizo	58%
White	20%
Mulatto	14%
Black	4%
Mixed black-Indian	3%
Indian	1%

Religion [8]

Roman Catholic	95%

Major Languages [9]

Spanish.

Education

Public Education Expenditures [10]

Million Pesos	1980[1]	1985[1]	1987[1]	1988[1]	1989[1]	1990[1]
Total education expenditure	29,240	136,570	219,485	NA	409,215	526,686
as percent of GNP	1.9	2.9	2.6	NA	2.9	2.7
as percent of total govt. expend.	14.3	NA	22.4	NA	21.4	12.4
Current education expenditure	27,286	127,908	206,992	NA	389,489	NA
as percent of GNP	1.7	2.7	2.5	NA	2.7	NA
as percent of current govt. expend.	19.9	NA	31.2	NA	38.5	NA
Capital expenditure	1,954	8,662	12,493	NA	19,726	NA

Educational Attainment [11]

Literacy Rate [12]

In thousands and percent	1985[a]	1991[a]	2000[a]
Illiterate population +15 years	2,761	2,702	2,532
Illiteracy rate - total pop. (%)	15.3	13.3	9.9
Illiteracy rate - males (%)	14.2	12.5	9.6
Illiteracy rate - females (%)	16.3	14.1	10.2

Libraries [13]

Daily Newspapers [14]

	1975	1980	1985	1990
Number of papers	40	36	46[e]	45
Circ. (000)	1,450[e]	1,400[e]	1,800[e]	2,000[e]

Cinema [15]

Data for 1988.

Cinema seats per 1,000	7.5
Annual attendance per person	1.3
Gross box office receipts (mil. Peso)	12,078

Science and Technology

Scientific/Technical Forces [16]

R&D Expenditures [17]

	Peso[1] (000) 1982
Total expenditure	27,542,730
Capital expenditure	NA
Current expenditure	NA
Percent current	NA

U.S. Patents Issued [18]

Values show patents issued to citizens of the country by the U.S. Patents Office.

	1990	1991	1992
Number of patents	6	4	6

Government and Law

Organization of Government [19]

Long-form name:
Republic of Colombia
Type:
republic; executive branch dominates
government structure
Independence:
20 July 1810 (from Spain)
Constitution:
5 July 1991
Legal system:
based on Spanish law; judicial review of
executive and legislative acts; accepts
compulsory ICJ jurisdiction, with
reservations
National holiday:
Independence Day, 20 July (1810)
Executive branch:
president, presidential designate, Cabinet
Legislative branch:
bicameral Congress (Congreso) consists
of a nationally elected upper chamber or
Senate (Senado) and a nationally elected
lower chamber or House of
Representatives (Camara de
Representantes)
Judicial branch:
Supreme Court of Justice (Corte Suprema
de Justical), Constitutional Court, Council
of State

Crime [23]

Political Parties [20]

House of Representatives	% of seats
Liberal	54.0
Conservative	19.3
Democratic Alliance M-19 (AD/M-19)	8.1
National Salvation Movement (MSN)	6.2
Patriotic Union (UP)	1.9
Other	10.6

Government Budget [21]

For 1991 est.

Revenues	5.000
Current expenditures	5.100
Capital expenditures	0.964

Military Expenditures and Arms Transfers [22]

	1985	1986	1987	1988	1989
Military expenditures					
Current dollars (mil.)	424[e]	425[e]	495[e]	614[e]	758[e]
1989 constant dollars (mil.)	482[e]	472[e]	533[e]	640[e]	758[e]
Armed forces (000)	66	76[e]	86[e]	76	91
Gross national product (GNP)					
Current dollars (mil.)	27,250	29,230	31,770	34,050	36,890
1989 constant dollars (mil.)	31,020	32,440	34,170	35,440	36,890
Central government expenditures (CGE)					
1989 constant dollars (mil.)	4,798	4,642	5,012	4,454	4,627
People (mil.)	29.7	30.3	31.0	31.7	32.4
Military expenditure as % of GNP	1.6	1.5	1.6	1.8	2.1
Military expenditure as % of CGE	10.1	10.2	10.6	14.4	16.4
Military expenditure per capita	16	16	17	20	23
Armed forces per 1,000 people	2.2	2.5	2.8	2.4	2.8
GNP per capita	1,046	1,070	1,102	1,118	1,139
Arms imports[6]					
Current dollars (mil.)	20	50	10	60	150
1989 constant dollars (mil.)	23	55	11	62	150
Arms exports[6]					
Current dollars (mil.)	0	0	0	0	0
1989 constant dollars (mil.)	0	0	0	0	0
Total imports[7]					
Current dollars (mil.)	4,141	3,862	4,322	5,002	5,004
1989 constant dollars	4,741	4,286	4,649	5,207	5,004
Total exports[7]					
Current dollars (mil.)	3,552	5,102	4,642	5,037	5,717
1989 constant dollars	4,044	5,662	4,993	5,243	5,717
Arms as percent of total imports[8]	0.5	1.3	0.2	1.2	3.0
Arms as percent of total exports[8]	0	0	0	0	0

Human Rights [24]

	SSTS	FL	FAPRO	PPCG	APROBC	TPW	PCPTW	STPEP	PHRFF	PRW	ASST	AFL
Observes		P	P	P	P	P	P			P		P
	EAFRD	CPR	ESCR	SR	ACHR	MAAE	PVIAC	PVNAC	EAFDAW	TCIDTP	RC	
Observes		P	P	P	P	P			P	P	P	

P = Party; S = Signatory; see Appendix for meaning of abbreviations.

Labor Force

Total Labor Force [25]

12 million (1990)

Labor Force by Occupation [26]

Services	46%
Agriculture	30
Industry	24

Date of data: 1990

Unemployment Rate [27]

10% (1992)

For sources, notes, and explanations, see Annotated Source Appendix, page 1035.

Production Sectors

Commercial Energy Production and Consumption

Production [28]

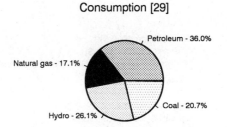

Crude oil - 47.9%
NG liquids - 0.5%
Natural gas - 9.3%
Hydro - 14.9%
Coal - 27.3%

Consumption [29]

Petroleum - 36.0%
Natural gas - 17.1%
Coal - 20.7%
Hydro - 26.1%

Data are shown in quadrillion (10^{15}) BTUs and percent for 1991

Crude oil[1]	0.93
Natural gas liquids	0.01
Dry natural gas	0.18
Coal[2]	0.53
Net hydroelectric power[3]	0.29
Net nuclear power[3]	0.00
Total	1.94

Petroleum	0.40
Dry natural gas	0.19
Coal[2]	0.23
Net hydroelectric power[3]	0.29
Net nuclear power[3]	(s)
Total	1.10

Telecommunications [30]

- Nationwide radio relay system
- 1,890,000 telephones
- Broadcast stations - 413 AM, no FM, 33 TV, 28 shortwave
- 2 Atlantic Ocean INTELSAT earth stations and 11 domestic satellite earth stations

Transportation [31]

Railroads. 3,386 km; 3,236 km 0.914-meter gauge, single track (2,611 km in use), 150 km 1.435-meter gauge

Highways. 75,450 km total; 9,350 km paved, 66,100 km earth and gravel surfaces

Merchant Marine. 27 ships (1,000 GRT or over) totaling 227,719 GRT/356,665 DWT; includes 9 cargo, 3 oil tanker, 8 bulk, 7 container

Airports

Total:	1,233
Usable:	1,059
With permanent-surface:	69
With runways over 3,659 m:	1
With runways 2,440-2,459 m:	9
With runways 1,220-2,439 m:	200

Top Agricultural Products [32]

	1990	1991[4]
Plantains	2,684	2,748
Potatoes	2,400	2,372
Yucca	1,854	1,966
Sugar (raw)[5]	1,611	1,602
Bananas	1,391	1,413
Corn	1,200	1,274

Values shown are 1,000 metric tons.

Top Mining Products [33]

Preliminary metric tons unless otherwise specified M.t.

Coal (000 tons)	20,200
Clay, common and kaolin (000 tons)	1,984
Platinum-group metals (kilograms)	1,603
Asbestos, mine output	160,332
Cement, hydraulic (000 tons)	6,277
Ferronickel, Ni content	8,425
Gold (kilograms)	34,844
Salt (000 tons)	701
Petroleum, cude (42-gal. barrels)	155,329
Natural gas, gross (mil. cubic meters)	5,202

Tourism [34]

	1987	1988	1989	1990	1991
Tourists	541	829	733	813	857
Tourism receipts	213	280	252	270	410
Tourism expenditures	454	469	383	411	593
Fare receipts	151	179	160	168	176
Fare expenditures	89	104	130	137	144

Tourists are in thousands, money in million U.S. dollars.

Manufacturing Sector

GDP and Manufacturing Summary [35]

	1980	1985	1989	1990	% change 1980-1990	% change 1989-1990
GDP (million 1980 $)	33,400	37,325	44,145	46,589	39.5	5.5
GDP per capita (1980 $)	1,241	1,249	1,365	1,412	13.8	3.4
Manufacturing as % of GDP (current prices)	23.3	21.4	20.3	20.2	-13.3	-0.5
Gross output (million $)	16,453	16,814	20,177^e	21,249^e	29.1	5.3
Value added (million $)	7,131	6,711	7,871^e	8,250^e	15.7	4.8
Value added (million 1980 $)	7,772	8,230	9,691	10,354	33.2	6.8
Industrial production index	100	108	127	130	30.0	2.4
Employment (thousands)	508	440	478^e	486^e	-4.3	1.7

Note: GDP stands for Gross Domestic Product. 'e' stands for estimated value.

Profitability and Productivity

	1980	1985	1989	1990	% change 1980-1990	% change 1989-1990
Intermediate input (%)	57	60	61^e	61^e	7.0	0.0
Wages, salaries, and supplements (%)	8	7	6^e	6^e	-25.0	0.0
Gross operating surplus (%)	35	33	33^e	33^e	-5.7	0.0
Gross output per worker ($)	31,860	37,616	42,230^e	42,936^e	34.8	1.7
Value added per worker ($)	13,809	15,013	16,474^e	16,671^e	20.7	1.2
Average wage (incl. benefits) ($)	2,583	2,708	2,364^e	2,467^e	-4.5	4.4

Profitability is in percent of gross output. Productivity is in U.S. $. 'e' stands for estimated value.

Profitability - 1990

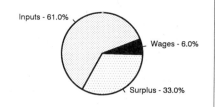

Inputs - 61.0%
Wages - 6.0%
Surplus - 33.0%

The graphic shows percent of gross output.

Value Added in Manufacturing

	1980 $ mil.	1980 %	1985 $ mil.	1985 %	1989 $ mil.	1989 %	1990 $ mil.	1990 %	% change 1980-1990	% change 1989-1990
311 Food products	951	13.3	1,166	17.4	1,326^e	16.8	1,324^e	16.0	39.2	-0.2
313 Beverages	1,021	14.3	1,032	15.4	969^e	12.3	1,002^e	12.1	-1.9	3.4
314 Tobacco products	160	2.2	224	3.3	153^e	1.9	175^e	2.1	9.4	14.4
321 Textiles	803	11.3	619	9.2	772^e	9.8	838^e	10.2	4.4	8.5
322 Wearing apparel	241	3.4	206	3.1	228^e	2.9	247^e	3.0	2.5	8.3
323 Leather and fur products	59	0.8	47	0.7	53^e	0.7	61^e	0.7	3.4	15.1
324 Footwear	50	0.7	54	0.8	95^e	1.2	98^e	1.2	96.0	3.2
331 Wood and wood products	50	0.7	46	0.7	57^e	0.7	62^e	0.8	24.0	8.8
332 Furniture and fixtures	34	0.5	29	0.4	39^e	0.5	40^e	0.5	17.6	2.6
341 Paper and paper products	227	3.2	274	4.1	293^e	3.7	313^e	3.8	37.9	6.8
342 Printing and publishing	185	2.6	180	2.7	240^e	3.0	250^e	3.0	35.1	4.2
351 Industrial chemicals	303	4.2	405	6.0	495^e	6.3	552^e	6.7	82.2	11.5
352 Other chemical products	419	5.9	457	6.8	555^e	7.1	592^e	7.2	41.3	6.7
353 Petroleum refineries	773	10.8	90	1.3	104^e	1.3	107^e	1.3	-86.2	2.9
354 Miscellaneous petroleum and coal products	17	0.2	28	0.4	33^e	0.4	46^e	0.6	170.6	39.4
355 Rubber products	117	1.6	138	2.1	135^e	1.7	131^e	1.6	12.0	-3.0
356 Plastic products	141	2.0	169	2.5	226^e	2.9	238^e	2.9	68.8	5.3
361 Pottery, china and earthenware	44	0.6	46	0.7	60^e	0.8	61^e	0.7	38.6	1.7
362 Glass and glass products	76	1.1	92	1.4	108^e	1.4	106^e	1.3	39.5	-1.9
369 Other non-metal mineral products	232	3.3	264	3.9	327^e	4.2	378^e	4.6	62.9	15.6
371 Iron and steel	217	3.0	205	3.1	387^e	4.9	377^e	4.6	73.7	-2.6
372 Non-ferrous metals	34	0.5	36	0.5	52^e	0.7	44^e	0.5	29.4	-15.4
381 Metal products	260	3.6	242	3.6	244^e	3.1	278^e	3.4	6.9	13.9
382 Non-electrical machinery	120	1.7	114	1.7	143^e	1.8	148^e	1.8	23.3	3.5
383 Electrical machinery	244	3.4	211	3.1	281^e	3.6	309^e	3.7	26.6	10.0
384 Transport equipment	256	3.6	221	3.3	358^e	4.5	306^e	3.7	19.5	-14.5
385 Professional and scientific equipment	26	0.4	38	0.6	55^e	0.7	77^e	0.9	196.2	40.0
390 Other manufacturing industries	72	1.0	78	1.2	83^e	1.1	92^e	1.1	27.8	10.8

Note: The industry codes shown are International Standard Industry codes (ISIC). Percentages are percent of total Value Added. 'e' stands for estimated value

Finance, Economics, and Trade

Economic Indicators [36]

Billions of 1975 pesos; 1975 peso rate: 32.96 = US $1[29].

	1989	1990	1991[P]
Real GDP	703.8	733.4	744.4
Real GDP Growth Rate	3.2	4.2	1.5
Real Per Capita GDP (pesos)	21,718	21,894	22,152
Money Supply Growth (M1)	29.1	25.8	30.0
Commercial Interest Rates	36-42	34-38	35-39
Savings Rate	33.9	36.7	37.8
Investment Rate	19.6	19.0	NA
CPI Increase	26.1	32.4	27.0
WPI Increase	25.6	25.0	24.0
External Public Debt	13,296	13,680	13,700

Balance of Payments Summary [37]

Values in millions of dollars.

	1987	1988	1989	1990	1991
Exports of goods (f.o.b.)	5,661.0	5,343.0	6,031.0	7,079.0	7,507.0
Imports of goods (f.o.b.)	-3,793.0	-4,516.0	-4,557.0	-5,108.0	-4,548.0
Trade balance	1,868.0	827.0	1,474.0	1,971.0	2,959.0
Services - debits	-3,901.0	-3,672.0	-4,151.0	-4,402.0	-4,292.0
Services - credits	1,368.0	1,665.0	1,578.0	1,947.0	1,984.0
Private transfers (net)	1,009.0	975.0	912.0	1,041.0	1,712.0
Government transfers (net)	-8.0	-11.0	-14.0	-15.0	-14.0
Long term capital (net)	191.0	834.0	653.0	196.0	143.0
Short term capital (net)	-203.0	106.0	-246.0	-170.0	-928.0
Errors and omissions	67.0	-530.0	157.0	70.0	269.0
Overall balance	391.0	194.0	363.0	638.0	1,833.0

Exchange Rates [38]

Currency: **Colombian pesos.**
Symbol: **Col$.**

Data are currency units per $1.

January 1993	820.08
1992	759.28
1991	633.05
1990	502.26
1989	382.57
1988	299.17

Imports and Exports

Top Import Origins [39]

$5.5 billion (c.i.f., 1992 est.). Data are for 1991.

Origins	%
US	36
EC	16
Brazil	4
Venezuela	3
Japan	3

Top Export Destinations [40]

$7.4 billion (f.o.b., 1992 est.). Data are for 1991.

Destinations	%
US	44
EC	21
Japan	5
Netherlands	4
Sweden	3

Foreign Aid [41]

	U.S. $	
US commitments, including Ex-Im (FY70-89)	1.6	billion
Western (non-US) countries, ODA and OOF bilateral commitments (1970-89)	3.3	billion
Communist countries (1970-89)	399	million

Import and Export Commodities [42]

Import Commodities	Export Commodities
Industrial equipment	Petroleum
Transport equipment	Coffee
Consumer goods	Coal
Chemicals	Bananas
Paper products	Fresh cut flowers

For sources, notes, and explanations, see Annotated Source Appendix, page 1035.

205

Comoros

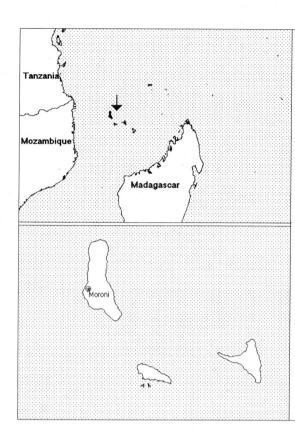

Geography [1]

Total area:
  2,170 km2
Land area:
  2,170 km2
Comparative area:
  Slightly more than 12 times the size of Washington, DC
Land boundaries:
  0 km
Coastline:
  340 km
Climate:
  Tropical marine; rainy season (November to May)
Terrain:
  Volcanic islands, interiors vary from steep mountains to low hills
Natural resources:
  Negligible
Land use:
  Arable land:
  35%
  Permanent crops:
  8%
  Meadows and pastures:
  7%
  Forest and woodland:
  16%
  Other:
  34%

Demographics [2]

	1960	1970	1980	1990	1991[1]	1994	2000	2010	2020
Population	183	236	3,334	460	477	530	656	919	1,249
Population density (persons per sq. mi.)	218	282	398	549	569	NA	782	1,098	1,497
Births	NA	NA	NA	NA	23	25	NA	NA	NA
Deaths	NA	NA	NA	NA	6	6	NA	NA	NA
Life expectancy - males	NA	NA	NA	NA	54	56	NA	NA	NA
Life expectancy - females	NA	NA	NA	NA	59	60	NA	NA	NA
Birth rate (per 1,000)	NA	NA	NA	NA	47	46	NA	NA	NA
Death rate (per 1,000)	NA	NA	NA	NA	12	11	NA	NA	NA
Women of reproductive age (15-44 yrs.)	NA	NA	NA	101	NA	116	144	NA	NA
of which are currently married	NA	NA	NA	67	NA	77	96	NA	NA
Fertility rate	NA	NA	NA	7.1	6.99	6.8	6.3	5.4	4.3

Population values are in thousands, life expectancy in years, and other items as indicated.

Health

Health Personnel [3]

Health Expenditures [4]

For sources, notes, and explanations, see Annotated Source Appendix, page 1035.

Human Factors

Health Care Ratios [5]	Infants and Malnutrition [6]

Ethnic Division [7]

Antalote	NA
Cafre	NA
Makoa	NA
Oimatsaha	NA
Sakalava	NA

Religion [8]

Sunni Muslim	86%
Roman Catholic	14%

Major Languages [9]

Arabic (official), French (official), Comoran (a blend of Swahili and Arabic).

Education

Public Education Expenditures [10]

Million Francs C.F.A.	1985	1986	1987	1988	1989	1990
Total education expenditure	NA	3,653	NA	NA	NA	NA
as percent of GNP	NA	6.5	NA	NA	NA	NA
as percent of total govt. expend.	NA	13.2	NA	NA	NA	NA
Current education expenditure	NA	2,427	NA	NA	NA	NA
as percent of GNP	NA	4.3	NA	NA	NA	NA
as percent of current govt. expend.	NA	23.4	NA	NA	NA	NA
Capital expenditure	NA	1,226	NA	NA	NA	NA

Educational Attainment [11]

Literacy Rate [12]

	1980[b]	1984[b]	1990[b]
Illiterate population +15 years	88,780	NA	NA
Illiteracy rate - total pop. (%)	52.1	NA	NA
Illiteracy rate - males (%)	44.0	NA	NA
Illiteracy rate - females (%)	60.0	NA	NA

Libraries [13]

Daily Newspapers [14]

Cinema [15]

Science and Technology

Scientific/Technical Forces [16]	R&D Expenditures [17]	U.S. Patents Issued [18]

Government and Law

Organization of Government [19]

Long-form name:
Federal Islamic Republic of the Comoros
Type:
independent republic
Independence:
6 July 1975 (from France)
Constitution:
7 June 1992
Legal system:
French and Muslim law in a new
consolidated code
National holiday:
Independence Day, 6 July (1975)
Executive branch:
president, Council of Ministers (cabinet),
prime minister
Legislative branch:
unicameral Federal Assembly (Assemblee
Federale)
Judicial branch:
Supreme Court (Cour Supreme)

Political Parties [20]

Federal Assembly	% of seats
Union for Democracy (UNDC)	16.7
Islands' Fraternity (CHUMA)	7.1
ADP	4.8
MDP/NGDC	11.9
FDC	4.8
MAECHA BORA	4.8
Comoran Popular Front (FPC)	4.8
Other	45.3

Government Budget [21]

For 1991 est.

Revenues	96
Expenditures	88
Capital expenditures	33

Defense Summary [22]

Branches: Comoran Defense Force (FDC)

Manpower Availability: Males age 15-49 108,867; fit for military service 65,106 (1993 est.)

Defense Expenditures: Information is not available.

Crime [23]

Human Rights [24]

	SSTS	FL	FAPRO	PPCG	APROBC	TPW	PCPTW	STPEP	PHRFF	PRW	ASST	AFL
Observes		P	P		P	P	P					P
	EAFRD	CPR	ESCR	SR	ACHR	MAAE	PVIAC	PVNAC	EAFDAW	TCIDTP	RC
Observes							P	P			S

P = Party; S = Signatory; see Appendix for meaning of abbreviations.

Labor Force

Total Labor Force [25]

140,000 (1982)

Labor Force by Occupation [26]

Agriculture	80%
Government	3

Unemployment Rate [27]

over 16% (1988 est.)

Production Sectors

Energy Resource Summary [28]

Energy Resources: None. **Electricity**: 16,000 kW capacity; 25 million kWh produced, 50 kWh per capita (1991).

Telecommunications [30]

- Sparse system of radio relay and high-frequency radio communication stations for interisland and external communications to Madagascar and Reunion
- Over 1,800 telephones
- Broadcast stations - 2 AM, 1 FM, no TV

Transportation [31]

Highways. 750 km total; about 210 km bituminous, remainder crushed stone or gravel

Airports

Total:	4
Usable:	4
With permanent-surface runways:	4
With runways over 3,659 m:	0
With runways 2,440-3,659 m:	1
With runways 1,220-2,439 m:	3

Top Agricultural Products [32]

Agriculture accounts for 40% of GDP; most of population works in subsistence agriculture and fishing; plantations produce cash crops for export—vanilla, cloves, perfume essences, copra; principal food crops—coconuts, bananas, cassava; world's leading producer of essence of ylang-ylang (for perfumes) and second-largest producer of vanilla; large net food importer.

Top Mining Products [33]

Detailed information is not available. A summary of mineral resources available follows. **Mineral Resources**: Negligible.

Tourism [34]

	1987	1988	1989	1990	1991
Tourists[16]	8	8	13	8	17
Tourism receipts	3	3	3	2	9
Tourism expenditures	5	5	5	6	7
Fare receipts		1		2	2
Fare expenditures	5	6	6	7	5

Tourists are in thousands, money in million U.S. dollars.

For sources, notes, and explanations, see Annotated Source Appendix, page 1035.

209

Finance, Economics, and Trade

Industrial Summary [35]

Industrial Production: Growth rate - 6.5% (1989 est.); accounts for 10% of GDP. **Industries**: Perfume distillation, textiles, furniture, jewelry, construction materials, soft drinks.

Economic Indicators [36]

National product: GDP—exchange rate conversion—$260 million (1991 est.). **National product real growth rate**: 2.7% (1991 est.). **National product per capita**: $540 (1991 est.). **Inflation rate (consumer prices)**: 4% (1991 est.). **External debt**: $196 million (1991 est.).

Balance of Payments Summary [37]

Values in millions of dollars.

	1987	1988	1989	1990	1991
Exports of goods (f.o.b.)	11.6	21.5	18.1	17.9	24.4
Imports of goods (f.o.b.)	-44.2	-44.3	-35.7	-45.2	-53.6
Trade balance	-32.5	-22.8	-17.6	-27.3	-29.2
Services - debits	-44.7	-46.2	-42.6	-46.9	-48.1
Services - credits	16.2	18.6	21.8	20.2	27.5
Private transfers (net)	0.9	3.1	2.7	5.5	3.7
Government transfers (net)	38.7	40.8	41.1	39.2	37.2
Long term capital (net)	18.1	5.8	6.9	0.2	13.8
Short term capital (net)	11.4	-1.6	0.7	13.5	-9.6
Errors and omissions	0.7	-1.4	-7.6	-9.2	1.7
Overall balance	8.8	-3.7	5.5	-4.8	-3.0

Exchange Rates [38]

Currency: **Comoran francs.**
Symbol: **CF.**

Linked to the French franc at 50 to 1 French franc. Data are currency units per $1.

January 1993	274.06
1992	264.69
1991	282.11
1990	272.26
1989	319.01
1988	297.85

Imports and Exports

Top Import Origins [39]

$41 million (f.o.b., 1990 est.).

Origins	%
Europe	62
Africa	5
Pakistan	NA
China 1988	NA

Top Export Destinations [40]

$16 million (f.o.b., 1990 est.). Data are for 1988.

Destinations	%
US	53
France	41
Africa	4
FRG	2

Foreign Aid [41]

	U.S. $	
US commitments, including Ex-Im (FY80-89)	10	million
Western (non-US) countries, ODA and OOF bilateral commitments (1970-89)	435	million
OPEC bilateral aid (1979-89)	22	million
Communist countries (1970-89)	18	million

Import and Export Commodities [42]

Import Commodities	Export Commodities
Rice and other foodstuffs	Vanilla
Cement	Cloves
Petroleum products	Perfume oil
Consumer goods	Copra
	Ylang-ylang

Congo

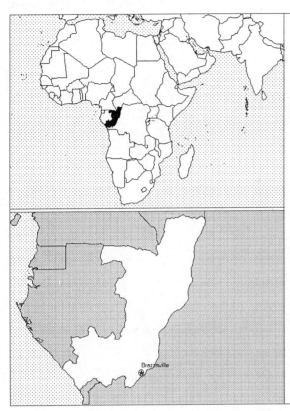

Geography [1]

Total area:
 342,000 km2
Land area:
 341,500 km2
Comparative area:
 Slightly smaller than Montana
Land boundaries:
 Total 5,504 km, Angola 201 km, Cameroon 523 km, Central African Republic 467
 Km, Gabon 1,903 km, Zaire 2,410 km
Coastline:
 169 km
Climate:
 Tropical; rainy season (March to June); dry season (June to October);
 Constantly high temperatures and humidity; particularly enervating climate
 Astride the Equator
Terrain:
 Coastal plain, southern basin, central plateau, northern basin
Natural resources:
 Petroleum, timber, potash, lead, zinc, uranium, copper, phosphates, natural
 Gas
Land use:
 Arable land: 2%
 Permanent crops: 0%
 Meadows and pastures: 29%
 Forest and woodland: 62%
 Other: 7%

Demographics [2]

	1960	1970	1980	1990	1991[1]	1994	2000	2010	2020
Population	931	1,183	1,620	2,215	2,309	2,447	2,784	3,219	3,775
Population density (persons per sq. mi.)	7	9	12	17	18	NA	23	30	38
Births	NA	NA	NA	NA	99	99	NA	NA	NA
Deaths	NA	NA	NA	NA	31	40	NA	NA	NA
Life expectancy - males	NA	NA	NA	NA	52	46	NA	NA	NA
Life expectancy - females	NA	NA	NA	NA	56	49	NA	NA	NA
Birth rate (per 1,000)	NA	NA	NA	NA	43	40	NA	NA	NA
Death rate (per 1,000)	NA	NA	NA	NA	13	16	NA	NA	NA
Women of reproductive age (15-44 yrs.)	NA	NA	NA	NA	NA	NA	NA	NA	NA
of which are currently married	NA	NA	NA	NA	NA	NA	NA	NA	NA
Fertility rate	NA	NA	NA	5.6	5.73	5.3	4.8	4.0	3.2

Population values are in thousands, life expectancy in years, and other items as indicated.

Health

Health Personnel [3]

Health Expenditures [4]

For sources, notes, and explanations, see Annotated Source Appendix, page 1035.

211

Human Factors

Health Care Ratios [5]

Population per physician, 1970	9510
Population per physician, 1990	NA
Population per nursing person, 1970	780
Population per nursing person, 1990	NA
Percent of births attended by health staff, 1985	NA

Infants and Malnutrition [6]

Percent of babies with low birth weight, 1985	12
Infant mortality rate per 1,000 live births, 1970	126
Infant mortality rate per 1,000 live births, 1991	115
Years of life lost per 1,000 population, 1990	NA
Prevalence of malnutrition (under age 5), 1990	24

Ethnic Division [7]

No detailed information available.

Religion [8]

Christian	50%
Animist	48%
Muslim	2%

Major Languages [9]

French (official)	NA
African languages	NA
Lingala	NA
Kikongo	NA
Other	NA

Education

Public Education Expenditures [10]

Million Francs C.F.A.	1980	1985	1987	1988	1989	1990
Total education expenditure	22,942	NA	NA	NA	38,989	37,899
as percent of GNP	7.0	NA	NA	NA	5.9	5.6
as percent of total govt. expend.	23.6	NA	NA	NA	16.5	14.4
Current education expenditure	21,517	NA	NA	NA	38,033	36,906
as percent of GNP	6.6	NA	NA	NA	5.7	5.5
as percent of current govt. expend.	24.1	NA	NA	NA	20.2	18.0
Capital expenditure	1,424	NA	NA	NA	956	993

Educational Attainment [11]

Age group	25+
Total population	646,626
Highest level attained (%)	
No schooling	58.7
First level	
Incompleted	12.9
Completed	8.4
Entered second level	
S-1	13.2
S-2	3.7
Post secondary	3.0

Literacy Rate [12]

In thousands and percent	1985[a]	1991[a]	2000[a]
Illiterate population +15 years	473	485	502
Illiteracy rate - total pop. (%)	48.3	43.4	34.1
Illiteracy rate - males (%)	34.0	30.0	23.0
Illiteracy rate - females (%)	61.8	56.1	44.8

Libraries [13]

	Admin. Units	Svc. Pts.	Vols. (000)	Shelving (meters)	Vols. Added	Reg. Users
National (1989)	1	1	9	125	2,000	995
Non-specialized	NA	NA	NA	NA	NA	NA
Public (1989)	1	4	15	360	1,200	22,365
Higher ed. (1987)	1	9	78	2,446	650	12,000
School (1990)	1	1	10	96	155	492

Daily Newspapers [14]

	1975	1980	1985	1990
Number of papers	3	1	1	5
Circ. (000)	1	3	8	17[e]

Cinema [15]

Science and Technology

Scientific/Technical Forces [16]

R&D Expenditures [17]

	(000) 1984
Total expenditure	25,530
Capital expenditure	14,263
Current expenditure	11,267
Percent current	44.1

U.S. Patents Issued [18]

For sources, notes, and explanations, see Annotated Source Appendix, page 1035.

Government and Law

Organization of Government [19]

Long-form name:
Republic of the Congo
Type:
republic
Independence:
15 August 1960 (from France)
Constitution:
8 July 1979, currently being modified
Legal system:
based on French civil law system and
customary law
National holiday:
Congolese National Day, 15 August (1960)
Executive branch:
president, prime minister, Council of
Ministers (cabinet)
Legislative branch:
unicameral National Assembly
(Assemblee Nationale) was dissolved on
NA November 1992
Judicial branch:
Supreme Court (Cour Supreme)

Elections [20]

National Assembly. National Assembly
dissolved in November 1992; next
election to be held May 1993.

Government Budget [21]

For 1990.
Revenues	765
Expenditures	952
Capital expenditures	65

Military Expenditures and Arms Transfers [22]

	1985	1986	1987	1988	1989
Military expenditures					
Current dollars (mil.)	73[e]	115[e]	95[e]	NA	NA
1989 constant dollars (mil.)	84[e]	128[e]	102[e]	NA	NA
Armed forces (000)	15	13	14	15	15
Gross national product (GNP)					
Current dollars (mil.)	1,819	1,780	1,844	1,900	2,008
1989 constant dollars (mil.)	2,070	1,975	1,983	1,978	2,008
Central government expenditures (CGE)					
1989 constant dollars (mil.)	909	1,136	817	861	752
People (mil.)	1.9	2.0	2.1	2.1	2.2
Military expenditure as % of GNP	4.0	6.5	5.1	NA	NA
Military expenditure as % of CGE	9.2	11.2	12.5	NA	NA
Military expenditure per capita	43	64	50	NA	NA
Armed forces per 1,000 people	7.7	6.5	6.8	7.0	6.8
GNP per capita	1,069	991	966	936	922
Arms imports[6]					
Current dollars (mil.)	40	30	10	20	0
1989 constant dollars (mil.)	46	33	11	21	0
Arms exports[6]					
Current dollars (mil.)	0	0	0	0	0
1989 constant dollars (mil.)	0	0	0	0	0
Total imports[7]					
Current dollars (mil.)	598	528	529	564	524
1989 constant dollars	681	586	556	587	524
Total exports[7]					
Current dollars (mil.)	1,145	673	517	752	912
1989 constant dollars	1,303	747	556	783	912
Arms as percent of total imports[8]	6.7	5.7	1.9	3.5	0
Arms as percent of total exports[8]	0	0	0	0	0

Crime [23]

Crime volume	
Cases known to police	649
Attempts (percent)	21
Percent cases solved	4.28
Crimes per 100,000 persons	32.45
Persons responsible for offenses	
Total number offenders	404
Percent female	40
Percent juvenile[17]	10
Percent foreigners	88

Human Rights [24]

	SSTS	FL	FAPRO	PPCG	APROBC	TPW	PCPTW	STPEP	PHRFF	PRW	ASST	AFL
Observes	P	P	P			P	P	P		P	P	
	EAFRD	CPR	ESCR	SR	ACHR	MAAE	PVIAC	PVNAC	EAFDAW	TCIDTP	RC	
Observes	P	P	P	P			P	P	P	P	P	

P = Party; S = Signatory; see Appendix for meaning of abbreviations.

Labor Force

Total Labor Force [25]

79,100 wage earners

Labor Force by Occupation [26]

Agriculture	75%
Commerce, industry, and government	25

Unemployment Rate [27]

Production Sectors

Energy Resource Summary [28]

Energy Resources: Petroleum, uranium, natural gas. **Electricity**: 140,000 kW capacity; 315 million kWh produced, 135 kWh per capita (1991).
Pipelines: Crude oil 25 km.

Telecommunications [30]

- Services adequate for government use
- Primary network is composed of radio relay routes and coaxial cables
- Key centers are Brazzaville, Pointe-Noire, and Loubomo
- 18,100 telephones
- Broadcast stations - 4 AM, 1 FM, 4 TV
- 1 Atlantic Ocean satellite earth station

Top Agricultural Products [32]

Agriculture accounts for 13% of GDP (including fishing and forestry); cassava accounts for 90% of food output; other crops—rice, corn, peanuts, vegetables; cash crops include coffee and cocoa; forest products important export earner; imports over 90% of food needs.

Top Mining Products [33]

Estimated metric tons unless otherwise specified   M.t.

Cement, hydraulic	102,571
Gas, natural (mil. cubic meters)	368
Gold, mine output, Au content (kilograms)	12[1]
Lime (kilograms)	300[1]
Petroleum, crude (000 barrels)	56,575

Transportation [31]

Railroads. 797 km, 1.067-meter gauge, single track (includes 285 km that are privately owned)

Highways. 11,960 km total; 560 km paved; 850 km gravel and laterite; 5,350 km improved earth; 5,200 km unimproved earth

Airports

Total:	44
Usable:	41
With permanent-surface runways:	5
With runways over 3,659 m:	0
With runways 2,440-3,659 m:	1
With runways 1,220-2,439 m:	16

Tourism [34]

	1987	1988	1989	1990	1991
Tourists	39[17]	39[17]	33	33	31
Tourism receipts	7	6	7	8	7
Tourism expenditures	79	126	86	114	93
Fare expenditures	25	25	27	33	32

Tourists are in thousands, money in million U.S. dollars.

Manufacturing Sector

GDP and Manufacturing Summary [35]

	1980	1985	1989	1990	% change 1980-1990	% change 1989-1990
GDP (million 1980 $)	1,706	2,860	2,455	2,784	63.2	13.4
GDP per capita (1980 $)	1,022	1,474	1,116	1,227	20.1	9.9
Manufacturing as % of GDP (current prices)	7.7	5.7	7.6[e]	7.9	2.6	3.9
Gross output (million $)	193[e]	154	281[e]	276[e]	43.0	-1.8
Value added (million $)	69[e]	57	80[e]	104[e]	50.7	30.0
Value added (million 1980 $)	128	256	231[e]	272	112.5	17.7
Industrial production index	100	183	121[e]	168	68.0	38.8
Employment (thousands)	5[e]	9	9[e]	8[e]	60.0	-11.1

Note: GDP stands for Gross Domestic Product. 'e' stands for estimated value.

Profitability and Productivity

	1980	1985	1989	1990	% change 1980-1990	% change 1989-1990
Intermediate input (%)	64[e]	63	72[e]	62[e]	-3.1	-13.9
Wages, salaries, and supplements (%)	15[e]	17[e]	17[e]	15[e]	0.0	-11.8
Gross operating surplus (%)	20[e]	20[e]	11[e]	22[e]	10.0	100.0
Gross output per worker ($)	16,482[e]	17,590	30,826[e]	34,628[e]	110.1	12.3
Value added per worker ($)	5,895[e]	6,525	8,736[e]	13,059[e]	121.5	49.5
Average wage (incl. benefits) ($)	5,463[e]	3,032[e]	5,234[e]	5,320[e]	-2.6	1.6

Profitability is in percent of gross output. Productivity is in U.S. $. 'e' stands for estimated value.

Profitability - 1990

Inputs - 62.6%
Wages - 15.2%
Surplus - 22.2%

The graphic shows percent of gross output.

Value Added in Manufacturing

	1980 $ mil.	1980 %	1985 $ mil.	1985 %	1989 $ mil.	1989 %	1990 $ mil.	1990 %	% change 1980-1990	% change 1989-1990
311 Food products	11[e]	15.9	10	17.5	11[e]	13.8	23[e]	22.1	109.1	109.1
313 Beverages	12[e]	17.4	11	19.3	13[e]	16.3	24[e]	23.1	100.0	84.6
314 Tobacco products	3[e]	4.3	3	5.3	5[e]	6.3	9[e]	8.7	200.0	80.0
321 Textiles	4[e]	5.8	2[e]	3.5	6[e]	7.5	2[e]	1.9	-50.0	-66.7
322 Wearing apparel	1[e]	1.4	1[e]	1.8	2[e]	2.5	1[e]	1.0	0.0	-50.0
323 Leather and fur products	NA	0.0	NA	0.0	NA	0.0	NA	0.0	NA	NA
324 Footwear	3[e]	4.3	2	3.5	5[e]	6.3	2[e]	1.9	-33.3	-60.0
331 Wood and wood products	7[e]	10.1	5[e]	8.8	7[e]	8.8	7[e]	6.7	0.0	0.0
332 Furniture and fixtures	4[e]	5.8	3[e]	5.3	4[e]	5.0	4[e]	3.8	0.0	0.0
341 Paper and paper products	1[e]	1.4	NA	0.0	1[e]	1.3	1[e]	1.0	0.0	0.0
342 Printing and publishing	1[e]	1.4	NA	0.0	1[e]	1.3	1[e]	1.0	0.0	0.0
351 Industrial chemicals	6[e]	8.7	4[e]	7.0	1[e]	1.3	8[e]	7.7	33.3	700.0
352 Other chemical products	3[e]	4.3	2[e]	3.5	3[e]	3.8	4[e]	3.8	33.3	33.3
353 Petroleum refineries	1[e]	1.4	1[e]	1.8	1[e]	1.3	2[e]	1.9	100.0	100.0
354 Miscellaneous petroleum and coal products	NA	0.0	NA	0.0	NA	0.0	NA	0.0	NA	NA
355 Rubber products	1[e]	1.4	1[e]	1.8	1[e]	1.3	1[e]	1.0	0.0	0.0
356 Plastic products	NA	0.0	NA	0.0	NA	0.0	1[e]	1.0	NA	NA
361 Pottery, china and earthenware	1[e]	1.4	2[e]	3.5	NA	0.0	1[e]	1.0	0.0	NA
362 Glass and glass products	NA	0.0	NA	0.0	NA	0.0	NA	0.0	NA	NA
369 Other non-metal mineral products	NA	0.0	NA	0.0	2[e]	2.5	NA	0.0	NA	NA
371 Iron and steel	NA	0.0	NA	0.0	NA	0.0	NA	0.0	NA	NA
372 Non-ferrous metals	NA	0.0	NA	0.0	NA	0.0	NA	0.0	NA	NA
381 Metal products	4[e]	5.8	4[e]	7.0	7[e]	8.8	5[e]	4.8	25.0	-28.6
382 Non-electrical machinery	1[e]	1.4	1[e]	1.8	3[e]	3.8	2[e]	1.9	100.0	-33.3
383 Electrical machinery	2[e]	2.9	2[e]	3.5	3[e]	3.8	2[e]	1.9	0.0	-33.3
384 Transport equipment	3[e]	4.3	2	3.5	4[e]	5.0	3[e]	2.9	0.0	-25.0
385 Professional and scientific equipment	NA	0.0	NA	0.0	NA	0.0	NA	0.0	NA	NA
390 Other manufacturing industries	NA	0.0	NA	0.0	NA	0.0	NA	0.0	NA	NA

Note: The industry codes shown are International Standard Industry codes (ISIC). Percentages are percent of total Value Added. 'e' stands for estimated value

For sources, notes, and explanations, see Annotated Source Appendix, page 1035.

215

Finance, Economics, and Trade

Economic Indicators [36]

National product: GDP—exchange rate conversion—$2.5 billion (1991 est.). **National product real growth rate**: 0.6% (1991 est.). **National product per capita**: $1,070 (1991 est.). **Inflation rate (consumer prices)**: - 0.6% (1991 est.). **External debt**: $4.1 billion (1991).

Balance of Payments Summary [37]

Values in millions of dollars.

	1987	1988	1989	1990	1991
Exports of goods (f.o.b.)	876.7	843.2	1,160.5	1,388.7	1,135.7
Imports of goods (f.o.b.)	-419.9	-522.7	-532.0	-512.7	-458.3
Trade balance	456.8	320.5	628.5	876.0	677.4
Services - debits	-818.3	-873.5	-857.3	-1,243.3	-940.4
Services - credits	127.7	99.7	97.5	113.9	92.2
Private transfers (net)	-56.8	-56.6	-46.7	-62.8	-60.3
Government transfers (net)	67.9	64.4	93.1	65.7	62.4
Long term capital (net)	173.2	8.8	-279.9	-128.2	-175.8
Short term capital (net)	62.0	410.1	361.1	537.7	312.6
Errors and omissions	27.6	40.6	8.5	-44.7	3.3
Overall balance	40.1	14.0	4.8	114.3	-28.6

Exchange Rates [38]

Currency: **Communaute Financiere Africaine francs.**
Symbol: **CFAF.**

Data are currency units per $1.

January 1993	274.06
1992	264.69
1991	282.11
1990	272.26
1989	319.01
1988	297.85

Imports and Exports

Top Import Origins [39]

$704 million (c.i.f., 1990).

Origins	%
France	NA
Italy	NA
Other EC countries	NA
US	NA
Germany	NA
Spain	NA
Japan	NA
Brazil	NA

Top Export Destinations [40]

$1.1 billion (f.o.b., 1990).

Destinations	%
US	NA
France	NA
Other EC countries	NA

Foreign Aid [41]

	U.S. $	
US commitments, including Ex-Im (FY70-90)	63	million
Western (non-US) countries, ODA and OOF bilateral commitments (1970-90)	2.5	billion
OPEC bilateral aid (1979-89)	15	million
Communist countries (1970-89)	338	million

Import and Export Commodities [42]

Import Commodities

Foodstuffs
Consumer goods
Intermediate manufactures
Capital equipment

Export Commodities

Crude oil 72%
Lumber
Plywood
Coffee
Cocoa
Sugar
Diamonds

Costa Rica

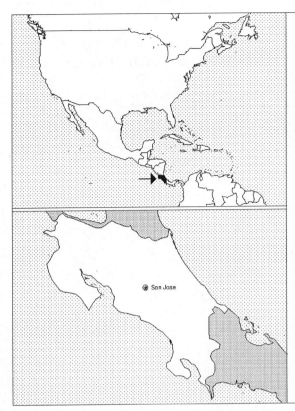

Geography [1]

Total area:
51,100 km2
Land area:
50,660 km2
Comparative area:
Slightly smaller than West Virginia
Land boundaries:
Total 639 km, Nicaragua 309 km, Panama 330 km
Coastline:
1,290 km
Climate:
Tropical; dry season (December to April); rainy season (May to November)
Terrain:
Coastal plains separated by rugged mountains
Natural resources:
Hydropower potential
Land use:
Arable land:
6%
Permanent crops:
7%
Meadows and pastures:
45%
Forest and woodland:
34%
Other:
8%

Demographics [2]

	1960	1970	1980	1990	1991[1]	1994	2000	2010	2020
Population	1,248	1,736	2,307	3,031	3,111	3,342	3,797	4,537	5,257
Population density (persons per sq. mi.)	64	89	118	155	159	NA	194	232	271
Births	NA	NA	NA	NA	85	85	NA	NA	NA
Deaths	NA	NA	NA	NA	12	12	NA	NA	NA
Life expectancy - males	NA	NA	NA	NA	75	76	NA	NA	NA
Life expectancy - females	NA	NA	NA	NA	79	80	NA	NA	NA
Birth rate (per 1,000)	NA	NA	NA	NA	27	25	NA	NA	NA
Death rate (per 1,000)	NA	NA	NA	NA	4	4	NA	NA	NA
Women of reproductive age (15-44 yrs.)	NA	NA	NA	773	NA	855	984	NA	NA
of which are currently married	NA	NA	NA	466	NA	524	603	NA	NA
Fertility rate	NA	NA	NA	3.3	3.23	3.1	2.8	2.5	2.3

Population values are in thousands, life expectancy in years, and other items as indicated.

Health

Health Personnel [3]

Health Expenditures [4]

For sources, notes, and explanations, see Annotated Source Appendix, page 1035.

217

Human Factors

Health Care Ratios [5]

Population per physician, 1970	1620
Population per physician, 1990	1030
Population per nursing person, 1970	460
Population per nursing person, 1990	NA
Percent of births attended by health staff, 1985	93

Infants and Malnutrition [6]

Percent of babies with low birth weight, 1985	9
Infant mortality rate per 1,000 live births, 1970	62
Infant mortality rate per 1,000 live births, 1991	14
Years of life lost per 1,000 population, 1990	NA
Prevalence of malnutrition (under age 5), 1990	NA

Ethnic Division [7]

White/mestizo	96%
Black	2%
Indian	1%
Chinese	1%

Religion [8]

Roman Catholic	95%

Major Languages [9]

Spanish (official), English; spoken around Puerto Limon.

Education

Public Education Expenditures [10]

Million Colon	1980	1985	1987	1988	1989	1990
Total education expenditure	3,069	8,181	NA	14,216	17,611	22,907
as percent of GNP	7.8	4.5	NA	4.4	4.5	4.6
as percent of total govt. expend.	22.2	22.7	NA	20.8	20.9	20.8
Current education expenditure	2,802	7,787	NA	13,607	17,126	22,188
as percent of GNP	7.1	4.2	NA	4.2	4.4	4.4
as percent of current govt. expend.	26.7	26.2	NA	26.6	27.0	26.3
Capital expenditure	267	394	NA	609	485	719

Educational Attainment [11]

Literacy Rate [12]

In thousands and percent	1985[a]	1991[a]	2000[a]
Illiterate population +15 years	136	139	133
Illiteracy rate - total pop. (%)	8.2	7.2	5.3
Illiteracy rate - males (%)	8.4	7.4	5.7
Illiteracy rate - females (%)	8.0	6.9	5.0

Libraries [13]

	Admin. Units	Svc. Pts.	Vols. (000)	Shelving (meters)	Vols. Added	Reg. Users
National (1986)	1	1	7	92	11,982	190,419
Non-specialized	NA	NA	NA	NA	NA	NA
Public (1986)	81	87	321	NA	16,588	293,615
Higher ed. (1990)	4	4	633	NA	25,145	NA
School	NA	NA	NA	NA	NA	NA

Daily Newspapers [14]

	1975	1980	1985	1990[2]
Number of papers	6	4	6	4
Circ. (000)	174	251	280[e]	314

Cinema [15]

Science and Technology

Scientific/Technical Forces [16]

Potential scientists/engineers	48,010
Number female	NA
Potential technicians	NA
Number female	NA
Total	NA

R&D Expenditures [17]

	Colon (000) 1986
Total expenditure	612,000
Capital expenditure	NA
Current expenditure	NA
Percent current	NA

U.S. Patents Issued [18]

Values show patents issued to citizens of the country by the U.S. Patents Office.

	1990	1991	1992
Number of patents	2	4	1

Government and Law

Organization of Government [19]

Long-form name:
Republic of Costa Rica
Type:
democratic republic
Independence:
15 September 1821 (from Spain)
Constitution:
9 November 1949
Legal system:
based on Spanish civil law system;
judicial review of legislative acts in the
Supreme Court; has not accepted
compulsory ICJ jurisdiction
National holiday:
Independence Day, 15 September (1821)
Executive branch:
president, two vice presidents, Cabinet
Legislative branch:
unicameral Legislative Assembly
(Asamblea Legislativa)
Judicial branch:
Supreme Court (Corte Suprema)

Political Parties [20]

Legislative Assembly	% of seats
Social Christian Unity (PUSC)	50.9
National Liberation Party (PLN)	43.9
PVP/PPC	1.8
Regional parties	3.5

Government Expenditures [21]

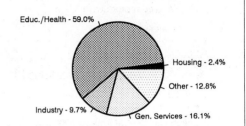

% distribution for 1990. Expend. in 1991 est.: 1.340 ($ bil.)

Military Expenditures and Arms Transfers [22]

	1985	1986	1987	1988	1989
Military expenditures					
Current dollars (mil.)	24	24	25[e]	21	22
1989 constant dollars (mil.)	27	27	27[e]	22	22
Armed forces (000)	8	8	8	8	8
Gross national product (GNP)					
Current dollars (mil.)	3,597	3,888	4,176	4,433	4,899
1989 constant dollars (mil.)	4,095	4,314	4,491	4,615	4,899
Central government expenditures (CGE)					
1989 constant dollars (mil.)	969	1,223	1,325	1,249	1,394
People (mil.)	2.6	2.7	2.8	2.9	3.0
Military expenditure as % of GNP	0.7	0.6	0.6	0.5	0.5
Military expenditure as % of CGE	2.8	2.2	2.0	1.8	1.6
Military expenditure per capita	10	10	10	8	8
Armed forces per 1,000 people	3.0	2.9	2.9	2.8	2.7
GNP per capita	1,551	1,587	1,606	1,605	1,658
Arms imports[6]					
Current dollars (mil.)	20	10	0	0	0
1989 constant dollars (mil.)	23	11	0	0	0
Arms exports[6]					
Current dollars (mil.)	0	0	0	0	0
1989 constant dollars (mil.)	0	0	0	0	0
Total imports[7]					
Current dollars (mil.)	1,098	1,148	1,383	1,410	1,743
1989 constant dollars	1,250	1,274	1,487	1,468	1,743
Total exports[7]					
Current dollars (mil.)	976	1,121	1,158	1,246	1,404
1989 constant dollars	1,111	1,244	1,245	1,297	1,404
Arms as percent of total imports[8]	1.8	0.9	0	0	0
Arms as percent of total exports[8]	0	0	0	0	0

Crime [23]

Crime volume	
Cases known to police	25,692
Attempts (percent)	1.72
Percent cases solved	NA
Crimes per 100,000 persons	868.21
Persons responsible for offenses	
Total number offenders	NA
Percent female	NA
Percent juvenile	NA
Percent foreigners	NA

Human Rights [24]

	SSTS	FL	FAPRO	PPCG	APROBC	TPW	PCPTW	STPEP	PHRFF	PRW	ASST	AFL
Observes		P	P	P	P	P	P			P		P
	EAFRD	CPR	ESCR	SR	ACHR	MAAE	PVIAC	PVNAC	EAFDAW	TCIDTP	RC
Observes	P	P	P	P	P	P	P	P	P	S	P

P = Party; S = Signatory; see Appendix for meaning of abbreviations.

Labor Force

Total Labor Force [25]

868,300

Labor Force by Occupation [26]

Industry and commerce	35.1%
Government and services	33
Agriculture	27
Other	4.9

Date of data: 1985 est.

Unemployment Rate [27]

4% (1992)

For sources, notes, and explanations, see Annotated Source Appendix, page 1035.

219

Production Sectors

Energy Resource Summary [28]

Energy Resources: Hydropower potential. **Electricity**: 927,000 kW capacity; 3,612 million kWh produced, 1,130 kWh per capita (1992).
Pipelines: Petroleum products 176 km.

Telecommunications [30]

- Very good domestic telephone service
- 292,000 telephones
- Connection into Central American Microwave System
- Broadcast stations - 71 AM, no FM, 18 TV, 13 shortwave
- 1 Atlantic Ocean INTELSAT earth station

Transportation [31]

Railroads. 950 km total, all 1.067-meter gauge; 260 km electrified

Highways. 15,400 km total; 7,030 km paved, 7,010 km gravel, 1,360 km unimproved earth

Merchant Marine. 1 cargo ship (1,000 GRT or over) totaling 2,878 GRT/4,506 DWT

Airports

Total:	162
Usable:	144
With permanent-surface runways:	28
With runways over 3,659 m:	0
With runways 2,440-3,659 m:	2
With runways 1,220-2,439 m:	8

Top Agricultural Products [32]

	1990[6]	1991[6]
Bananas	1,362	1,472
Sugar	245	265
Rice	225	194
Coffee	147	154
Beef	85	91
Corn	80	68

Values shown are 1,000 metric tons.

Top Mining Products [33]

Preliminary metric tons unless otherwise specified M.t.

Limestone and other calcerous materials (000 tons)	1,300
Cement	700,000[1]
Gold (kilograms)	550
Diatomite	12,000[e]
Iron and steel: semimanufactures	65,000
Petroleum refinery products (000 42-gal. barrels)	2,507

Tourism [34]

	1987	1988	1989	1990	1991
Tourists	278	329	376	435	505
Cruise passengers		56	49	55	68
Tourism receipts	136	165	207	275	331
Tourism expenditures	71	72	114	148	149
Fare receipts	25	37	41	45	45
Fare expenditures	22	10	10	11	11

Tourists are in thousands, money in million U.S. dollars.

220

For sources, notes, and explanations, see Annotated Source Appendix, page 1035.

Manufacturing Sector

GDP and Manufacturing Summary [35]

	1980	1985	1989	1990	% change 1980-1990	% change 1989-1990
GDP (million 1980 $)	4,832	4,900	5,884	6,136	27.0	4.3
GDP per capita (1980 $)	2,114	1,854	2,001	2,031	-3.9	1.5
Manufacturing as % of GDP (current prices)	18.4	21.9	20.6	19.9	8.2	-3.4
Gross output (million $)	2,743	2,468	3,145	3,183	16.0	1.2
Value added (million $)	788	762	944	926	17.5	-1.9
Value added (million 1980 $)	899	908	1,090	1,114	23.9	2.2
Industrial production index	100	89	106	110	10.0	3.8
Employment (thousands)	64[e]	104	114[e]	134	109.4	17.5

Note: GDP stands for Gross Domestic Product. 'e' stands for estimated value.

Profitability and Productivity

	1980	1985	1989	1990	% change 1980-1990	% change 1989-1990
Intermediate input (%)	71	69	70	71	0.0	1.4
Wages, salaries, and supplements (%)	12	11	9[e]	11	-8.3	22.2
Gross operating surplus (%)	17	20	21[e]	18	5.9	-14.3
Gross output per worker ($)	35,932[e]	23,757	25,703[e]	23,827	-33.7	-7.3
Value added per worker ($)	10,532[e]	7,338	7,772[e]	6,928	-34.2	-10.9
Average wage (incl. benefits) ($)	4,915[e]	2,569	2,420[e]	2,711	-44.8	12.0

Profitability is in percent of gross output. Productivity is in U.S. $. 'e' stands for estimated value.

Profitability - 1990

Inputs - 71.0%
Wages - 11.0%
Surplus - 18.0%

The graphic shows percent of gross output.

Value Added in Manufacturing

	1980 $ mil.	1980 %	1985 $ mil.	1985 %	1989 $ mil.	1989 %	1990 $ mil.	1990 %	% change 1980-1990	% change 1989-1990
311 Food products	241	30.6	247	32.4	297	31.5	286	30.9	18.7	-3.7
313 Beverages	96	12.2	94	12.3	116	12.3	122	13.2	27.1	5.2
314 Tobacco products	24	3.0	28	3.7	16	1.7	30	3.2	25.0	87.5
321 Textiles	33	4.2	23	3.0	28	3.0	31	3.3	-6.1	10.7
322 Wearing apparel	31	3.9	34	4.5	29	3.1	31	3.3	0.0	6.9
323 Leather and fur products	7	0.9	5	0.7	6	0.6	5	0.5	-28.6	-16.7
324 Footwear	10	1.3	9	1.2	10	1.1	7	0.8	-30.0	-30.0
331 Wood and wood products	30	3.8	25	3.3	21	2.2	21	2.3	-30.0	0.0
332 Furniture and fixtures	26	3.3	14	1.8	24	2.5	21	2.3	-19.2	-12.5
341 Paper and paper products	20	2.5	22	2.9	39	4.1	43	4.6	115.0	10.3
342 Printing and publishing	18	2.3	21	2.8	36	3.8	32	3.5	77.8	-11.1
351 Industrial chemicals	19	2.4	26	3.4	34	3.6	32	3.5	68.4	-5.9
352 Other chemical products	40	5.1	42	5.5	52	5.5	48	5.2	20.0	-7.7
353 Petroleum refineries	40	5.1	45	5.9	55	5.8	33	3.6	-17.5	-40.0
354 Miscellaneous petroleum and coal products	NA	0.0	NA	0.0	NA	0.0	NA	0.0	NA	NA
355 Rubber products	14	1.8	15	2.0	19	2.0	16	1.7	14.3	-15.8
356 Plastic products	19	2.4	26	3.4	36	3.8	34	3.7	78.9	-5.6
361 Pottery, china and earthenware	1	0.1	2	0.3	2	0.2	3	0.3	200.0	50.0
362 Glass and glass products	3	0.4	7	0.9	8	0.8	10	1.1	233.3	25.0
369 Other non-metal mineral products	25	3.2	19	2.5	30	3.2	34	3.7	36.0	13.3
371 Iron and steel	4	0.5	NA	0.0	NA	0.0	NA	0.0	NA	NA
372 Non-ferrous metals	1[e]	0.1	NA	0.0	NA	0.0	1	0.1	0.0	NA
381 Metal products	18	2.3	12	1.6	20	2.1	19	2.1	5.6	-5.0
382 Non-electrical machinery	8	1.0	10	1.3	15	1.6	13	1.4	62.5	-13.3
383 Electrical machinery	25	3.2	21	2.8	33	3.5	32	3.5	28.0	-3.0
384 Transport equipment	31	3.9	10	1.3	14	1.5	16	1.7	-48.4	14.3
385 Professional and scientific equipment	NA	0.0	1[e]	0.1	2[e]	0.2	2[e]	0.2	NA	0.0
390 Other manufacturing industries	2	0.3	3	0.4	3	0.3	4	0.4	100.0	33.3

Note: The industry codes shown are International Standard Industry codes (ISIC). Percentages are percent of total Value Added. 'e' stands for estimated value

Finance, Economics, and Trade

Economic Indicators [36]

Millions 1966 Colones unless otherwise indicated.

	1989	1990	1991[30]
Real GDP	11,827.3	12,228.8	12,408.6
Real GDP Growth (%)	5.7	3.4	1.5
Real GDP per capita ($US/1990 rate)	1,825	1,892	NA
Money Supply (M1) (mil. colones)	63,057.0	67,804.0	77,787.0
GDI (percent of GDP)	23.1	22.1	20.8
CPI (percent change Dec-Dec)	10.0	27.3	25.0
WPI (percent change Dec-Dec)	10.7	25.9	23.0
Foreign public debt	3,800.9	3,269.2	NA

Balance of Payments Summary [37]

Values in millions of dollars.

	1987	1988	1989	1990	1991
Exports of goods (f.o.b.)	1,106.7	1,180.7	1,333.4	1,354.2	1,490.5
Imports of goods (f.o.b.)	-1,245.2	-1,278.6	-1,572.0	-1,796.7	-1,697.8
Trade balance	-138.5	-97.9	-238.6	-442.5	-207.3
Services - debits	-729.5	-814.1	-985.5	-912.7	-793.0
Services - credits	385.6	478.1	617.8	739.3	784.1
Private transfers (net)	38.7	40.0	39.2	55.4	51.4
Government transfers (net)	187.3	215.4	152.2	136.5	83.1
Long term capital (net)	-363.1	-98.0	59.7	548.0	438.4
Short term capital (net)	531.1	286.1	296.1	-370.5	-128.1
Errors and omissions	131.2	224.6	208.9	56.4	112.7
Overall balance	42.8	234.2	149.8	-190.1	341.3

Exchange Rates [38]

Currency: **Costa Rican colones.**
Symbol: **C.**

Data are currency units per $1.

January 1993	137.72
1992	134.51
1991	122.43
1990	91.58
1989	81.50
1988	75.81

Imports and Exports

Top Import Origins [39]

$1.8 billion (c.i.f., 1992 est.).

Origins	%
US	45
Japan	NA
Guatemala	NA
Germany	NA

Top Export Destinations [40]

$1.7 billion (f.o.b., 1992 est.).

Destinations	%
US	75
Germany	NA
Guatemala	NA
Netherlands	NA
UK	NA
Japan	NA

Foreign Aid [41]

	U.S. $	
US commitments, including Ex-Im (FY70-89)	1.4	billion
Western (non-US) countries, ODA and OOF bilateral commitments (1970-89)	935	million
Communist countries (1971-89)	27	million

Import and Export Commodities [42]

Import Commodities

Raw materials
Consumer goods
Capital equipment
Petroleum

Export Commodities

Coffee
Bananas
Textiles
Sugar

Cote d'Ivoire

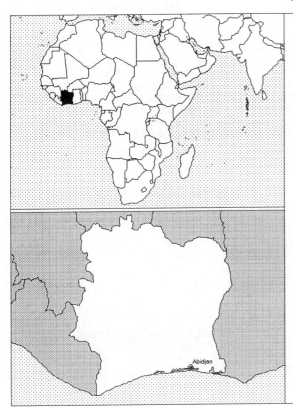

Geography [1]

Total area:
322,460 km2
Land area:
318,000 km2
Comparative area:
Slightly larger than New Mexico
Land boundaries:
Total 3,110 km, Burkina 584 km, Ghana 668 km, Guinea 610 km, Liberia 716 km, Mali 532 km
Coastline:
515 km
Climate:
Tropical along coast, semiarid in far north; three seasons - warm and dry (November to March), hot and dry (March to May), hot and wet (June to October)
Terrain:
Mostly flat to undulating plains; mountains in northwest
Natural resources:
Petroleum, diamonds, manganese, iron ore, cobalt, bauxite, copper
Land use:
Arable land: 9%
Permanent crops: 4%
Meadows and pastures: 9%
Forest and woodland: 26%
Other: 52%

Demographics [2]

	1960	1970	1980	1990	1991[1]	1994	2000	2010	2020
Population	3,565	5,427	8,418	12,399	12,978	14,296	17,371	22,924	29,705
Population density (persons per sq. mi.)	29	44	69	102	106	NA	148	206	274
Births	NA	NA	NA	NA	620	665	NA	NA	NA
Deaths	NA	NA	NA	NA	161	215	NA	NA	NA
Life expectancy - males	NA	NA	NA	NA	52	47	NA	NA	NA
Life expectancy - females	NA	NA	NA	NA	56	51	NA	NA	NA
Birth rate (per 1,000)	NA	NA	NA	NA	48	47	NA	NA	NA
Death rate (per 1,000)	NA	NA	NA	NA	12	15	NA	NA	NA
Women of reproductive age (15-44 yrs.)	NA	NA	NA	2,691	NA	3,076	3,710	NA	NA
of which are currently married	NA	NA	NA	2,053	NA	2,348	2,831	NA	NA
Fertility rate	NA	NA	NA	6.9	6.84	6.7	6.2	5.1	3.9

Population values are in thousands, life expectancy in years, and other items as indicated.

Health

Health Personnel [3]

Doctors per 1,000 pop., 1988-92	0.06
Nurse-to-doctor ratio, 1988-92	4.8
Hospital beds per 1,000 pop., 1985-90	0.8
Percentage of children immunized (age 1 yr. or less)	
Third dose of DPT, 1990-91	48.0
Measles, 1990-91	42.0

Health Expenditures [4]

Total health expenditure, 1990 (official exchange rate)	
Millions of dollars	332
Millions of dollars per capita	28
Health expenditures as a percentage of GDP	
Total	3.3
Public sector	1.7
Private sector	1.6
Development assistance for health	
Total aid flows (millions of dollars)[1]	11
Aid flows per capita (millions of dollars)	0.9
Aid flows as a percentage of total health expenditure	3.4

For sources, notes, and explanations, see Annotated Source Appendix, page 1035.

223

Human Factors

Health Care Ratios [5]

Population per physician, 1970	15520
Population per physician, 1990	NA
Population per nursing person, 1970	1930
Population per nursing person, 1990	NA
Percent of births attended by health staff, 1985	20

Infants and Malnutrition [6]

Percent of babies with low birth weight, 1985	14
Infant mortality rate per 1,000 live births, 1970	135
Infant mortality rate per 1,000 live births, 1991	95
Years of life lost per 1,000 population, 1990	50
Prevalence of malnutrition (under age 5), 1990	12

Ethnic Division [7]

Baoule	23%
Bete	18%
Senoufou	15%
Malinke	11%
Agni	NA
Foreign Africans	NA

Religion [8]

Indigenous	63%
Muslim	25%
Christian	12%

Major Languages [9]

French (official), 60 native dialects—Dioula is the most widely spoken.

Education

Public Education Expenditures [10]

Million Francs C.F.A.	1980	1985	1987	1988	1989	1990
Total education expenditure	147,478	NA	NA	NA	NA	NA
as percent of GNP	7.0	NA	NA	NA	NA	NA
as percent of total govt. expend.	22.6	NA	NA	NA	NA	NA
Current education expenditure	123,196	179,447	NA	NA	NA	NA
as percent of GNP	5.9	6.3	NA	NA	NA	NA
as percent of current govt. expend.	36.4	NA	NA	NA	NA	NA
Capital expenditure	24,282	NA	NA	NA	NA	NA

Educational Attainment [11]

Literacy Rate [12]

In thousands and percent	1985[a]	1991[a]	2000[a]
Illiterate population +15 years	2,687	2,941	3,397
Illiteracy rate - total pop. (%)	51.3	46.2	36.5
Illiteracy rate - males (%)	37.5	33.1	25.1
Illiteracy rate - females (%)	65.7	59.8	48.2

Libraries [13]

Daily Newspapers [14]

	1975	1980	1985	1990
Number of papers	3	2	1	1
Circ. (000)	55[e]	81	90	90

Cinema [15]

Science and Technology

Scientific/Technical Forces [16]

R&D Expenditures [17]

U.S. Patents Issued [18]

Government and Law

Organization of Government [19]

Long-form name:
Republic of Cote d'Ivoire
Type:
republic multiparty presidential regime
established 1960
Independence:
7 August 1960 (from France)
Constitution:
3 November 1960
Legal system:
based on French civil law system and
customary law; judicial review in the
Constitutional Chamber of the Supreme
Court; has not accepted compulsory ICJ
jurisdiction
National holiday:
National Day, 7 December
Executive branch:
president, Council of Ministers (cabinet)
Legislative branch:
unicameral National Assembly
(Assemblee Nationale)
Judicial branch:
Supreme Court (Cour Supreme)

Political Parties [20]

National Assembly	% of seats
Democratic Party (PDCI)	93.1
Ivorian Popular Front (FPI)	5.1
Ivorian Worker's Party (PIT)	0.6
Independents	1.1

Government Budget [21]

For 1990 est.

Revenues	2.300
Expenditures	3.600
Capital expenditures	0.274

Crime [23]

Crime volume	
Cases known to police	15,502
Attempts (percent)	NA
Percent cases solved	50.58
Crimes per 100,000 persons	124.92
Persons responsible for offenses	
Total number offenders	11,120
Percent female	NA
Percent juvenile	NA
Percent foreigners	NA

Military Expenditures and Arms Transfers [22]

	1985	1986	1987	1988	1989
Military expenditures					
Current dollars (mil.)	95	110[e]	152	173	130
1989 constant dollars (mil.)	105	122[e]	163	180	130
Armed forces (000)	8[e]	8	8	8	8
Gross national product (GNP)					
Current dollars (mil.)	7,365	8,037	8,255	8,714	8,460
1989 constant dollars (mil.)	8,385	8,919	8,879	9,072	8,460
Central government expenditures (CGE)					
1989 constant dollars (mil.)	NA	2,459[e]	NA	1,959	NA
People (mil.)	10.2	10.7	11.1	11.5	12.0
Military expenditure as % of GNP	1.2	1.4	1.8	2.0	1.5
Military expenditure as % of CGE	NA	4.9	NA	9.2	NA
Military expenditure per capita	10	11	15	16	11
Armed forces per 1,000 people	0.8	0.8	0.7	0.7	0.7
GNP per capita	819	837	801	787	705
Arms imports[6]					
Current dollars (mil.)	20	0	5	0	0
1989 constant dollars (mil.)	23	0	5	0	0
Arms exports[6]					
Current dollars (mil.)	0	0	0	0	0
1989 constant dollars (mil.)	0	0	0	0	0
Total imports[7]					
Current dollars (mil.)	1,749	2,055	2,370	2,081	1,641
1989 constant dollars	1,991	2,280	2,549	2,167	1,641
Total exports[7]					
Current dollars (mil.)	3,198	3,354	3,110	2,770	1,684
1989 constant dollars	3,641	3,722	3,345	2,884	1,684
Arms as percent of total imports[8]	1.1	0	0.2	0	0
Arms as percent of total exports[8]	0	0	0	0	0

Human Rights [24]

	SSTS	FL	FAPRO	PPCG	APROBC	TPW	PCPTW	STPEP	PHRFF	PRW	ASST	AFL
Observes	P	P	P		P	P	P				P	P
		EAFRD	CPR	ESCR	SR	ACHR	MAAE	PVIAC	PVNAC	EAFDAW	TCIDTP	RC
Observes		P	P	P	P			P	P	S		P

P = Party; S = Signatory; see Appendix for meaning of abbreviations.

Labor Force

Total Labor Force [25]

5.718 million

Labor Force by Occupation [26]

Agriculture, forestry, livestock	85%
Other	15

Unemployment Rate [27]

14% (1985)

Production Sectors

Energy Resource Summary [28]

Energy Resources: Petroleum. **Electricity**: 1,210,000 kW capacity; 1,970 million kWh produced, 150 kWh per capita (1991).

Telecommunications [30]

- Well-developed by African standards but operating well below capacity
- Consists of open-wire lines and radio relay microwave links
- 87,700 telephones
- Broadcast stations - 3 AM, 17 FM, 13 TV, 1 Atlantic Ocean and 1 Indian Ocean INTELSAT earth statio
- 2 coaxial submarine cables

Top Agricultural Products [32]

	1990[7]	1991[7]
Yams	2,528	2,540
Cassava	1,393	1,410
Plantains	1,086	1,100
Cocoa	710	785
Rice (paddy)	687	725
Corn	484	490

Values shown are 1,000 metric tons.

Top Mining Products [33]

Detailed information is not available. A summary of mineral resources available follows. **Mineral Resources**: Petroleum, diamonds, manganese, iron ore, cobalt, bauxite, copper.

Transportation [31]

Railroads. 660 km (Burkina border to Abidjan, 1.00-meter gauge, single track, except 25 km Abidjan-Anyama section is double track)

Highways. 46,600 km total; 3,600 km paved; 32,000 km gravel, crushed stone, laterite, and improved earth; 11,000 km unimproved

Merchant Marine. 7 ships (1,000 GRT or over) totaling 71,945 GRT/ 90,684 DWT; includes 1 oil tanker, 1 chemical tanker, 3 container, 2 roll-on/roll-off

Airports

Total:	42
Usable:	37
With permanent-surface runways:	7
With runways over 3,659 m:	0
With runways 2,440-3,659 m:	3
With runways 1,220-2,439 m:	15

Tourism [34]

	1987	1988	1989	1990	1991
Tourists[16]	175	181	192	196	200
Tourism receipts	59	60	44	48	46
Tourism expenditures	233	235	216	246	223
Fare receipts	2	2	2	2	2
Fare expenditures	110	107	151	202	184

Tourists are in thousands, money in million U.S. dollars.

Manufacturing Sector

GDP and Manufacturing Summary [35]

	1980	1985	1989	1990	% change 1980-1990	% change 1989-1990
GDP (million 1980 $)	10,176	10,660	9,962	10,184	0.1	2.2
GDP per capita (1980 $)	1,242	1,073	862	849	-31.6	-1.5
Manufacturing as % of GDP (current prices)	11.7	13.8	13.9[e]	14.1[e]	20.5	1.4
Gross output (million $)	4,006	2,869[e]	6,031[e]	5,423[e]	35.4	-10.1
Value added (million $)	1,273	719[e]	1,334[e]	1,409[e]	10.7	5.6
Value added (million 1980 $)	1,141	1,226	970	1,178	3.2	21.4
Industrial production index	100	100	111	97	-3.0	-12.6
Employment (thousands)	67	55[e]	50[e]	51[e]	-23.9	2.0

Note: GDP stands for Gross Domestic Product. 'e' stands for estimated value.

Profitability and Productivity

	1980	1985	1989	1990	% change 1980-1990	% change 1989-1990
Intermediate input (%)	68	75[e]	78[e]	74[e]	8.8	-5.1
Wages, salaries, and supplements (%)	10[e]	9[e]	9[e]	7[e]	-30.0	-22.2
Gross operating surplus (%)	22[e]	16[e]	14[e]	19[e]	-13.6	35.7
Gross output per worker ($)	59,631	51,722[e]	124,054[e]	104,503[e]	75.2	-15.8
Value added per worker ($)	18,950	12,964[e]	27,435[e]	27,184[e]	43.5	-0.9
Average wage (incl. benefits) ($)	5,744[e]	4,926[e]	10,586[e]	7,859[e]	36.8	-25.8

Profitability is in percent of gross output. Productivity is in U.S. $. 'e' stands for estimated value.

Profitability - 1990

Inputs - 74.0%
Wages - 7.0%
Surplus - 19.0%

The graphic shows percent of gross output.

Value Added in Manufacturing

	1980 $ mil.	1980 %	1985 $ mil.	1985 %	1989 $ mil.	1989 %	1990 $ mil.	1990 %	% change 1980-1990	% change 1989-1990
311 Food products	303[e]	23.8	171[e]	23.8	278[e]	20.8	339[e]	24.1	11.9	21.9
313 Beverages	75	5.9	35[e]	4.9	85[e]	6.4	68[e]	4.8	-9.3	-20.0
314 Tobacco products	66[e]	5.2	32[e]	4.5	40[e]	3.0	59[e]	4.2	-10.6	47.5
321 Textiles	169[e]	13.3	97[e]	13.5	138[e]	10.3	162[e]	11.5	-4.1	17.4
322 Wearing apparel	8[e]	0.6	5[e]	0.7	7[e]	0.5	8[e]	0.6	0.0	14.3
323 Leather and fur products	5[e]	0.4	6[e]	0.8	10[e]	0.7	12[e]	0.9	140.0	20.0
324 Footwear	5[e]	0.4	6[e]	0.8	23[e]	1.7	7[e]	0.5	40.0	-69.6
331 Wood and wood products	67[e]	5.3	24[e]	3.3	59[e]	4.4	32[e]	2.3	-52.2	-45.8
332 Furniture and fixtures	21[e]	1.6	8[e]	1.1	18[e]	1.3	11[e]	0.8	-47.6	-38.9
341 Paper and paper products	14[e]	1.1	7[e]	1.0	...	0.0	10[e]	0.7	-28.6	NA
342 Printing and publishing	22[e]	1.7	9[e]	1.3	...	0.0	13[e]	0.9	-40.9	NA
351 Industrial chemicals	22[e]	1.7	10[e]	1.4	18[e]	1.3	18[e]	1.3	-18.2	0.0
352 Other chemical products	53[e]	4.2	29[e]	4.0	82[e]	6.1	80[e]	5.7	50.9	-2.4
353 Petroleum refineries	181[e]	14.2	119[e]	16.6	201[e]	15.1	233[e]	16.5	28.7	15.9
354 Miscellaneous petroleum and coal products	NA	0.0	NA	0.0	NA	0.0	NA	0.0	NA	NA
355 Rubber products	4	0.3	2[e]	0.3	2[e]	0.1	4[e]	0.3	0.0	100.0
356 Plastic products	1[e]	0.1	NA	0.0	NA	0.0	NA	0.0	NA	NA
361 Pottery, china and earthenware	2[e]	0.2	1[e]	0.1	4[e]	0.3	2[e]	0.1	0.0	-50.0
362 Glass and glass products	NA	0.0	NA	0.0	NA	0.0	NA	0.0	NA	NA
369 Other non-metal mineral products	27[e]	2.1	12[e]	1.7	24[e]	1.8	26[e]	1.8	-3.7	8.3
371 Iron and steel	5[e]	0.4	1[e]	0.1	2[e]	0.1	3[e]	0.2	-40.0	50.0
372 Non-ferrous metals	3[e]	0.2	1[e]	0.1	2[e]	0.1	2[e]	0.1	-33.3	0.0
381 Metal products	70	5.5	33[e]	4.6	85[e]	6.4	56[e]	4.0	-20.0	-34.1
382 Non-electrical machinery	3	0.2	1[e]	0.1	3[e]	0.2	2[e]	0.1	-33.3	-33.3
383 Electrical machinery	20	1.6	9[e]	1.3	22[e]	1.6	16[e]	1.1	-20.0	-27.3
384 Transport equipment	106	8.3	88[e]	12.2	209[e]	15.7	223[e]	15.8	110.4	6.7
385 Professional and scientific equipment	NA	0.0	NA	0.0	NA	0.0	NA	0.0	NA	NA
390 Other manufacturing industries	20	1.6	14[e]	1.9	21[e]	1.6	23[e]	1.6	15.0	9.5

Note: The industry codes shown are International Standard Industry codes (ISIC). Percentages are percent of total Value Added. 'e' stands for estimated value

Finance, Economics, and Trade

Economic Indicators [36]

National product: GDP—exchange rate conversion—$10 billion (1991). **National product real growth rate**: - 0.6% (1991). **National product per capita**: $800 (1991). **Inflation rate (consumer prices)**: 1% (1991 est.). **External debt**: $15 billion (1990 est.).

Balance of Payments Summary [37]

Values in millions of dollars.

	1987	1988	1989	1990	1991
Exports of goods (f.o.b.)	2,949.7	2,691.3	2,807.8	3,120.1	2,803.9
Imports of goods (f.o.b.)	-1,863.3	-1,769.4	-1,720.3	-1,701.7	-1,641.6
Trade balance	1,086.4	921.9	1,087.5	1,418.4	1,162.3
Services - debits	-2,309.9	-2,335.1	-2,377.7	-2,753.2	-2,811.0
Services - credits	613.2	627.5	513.8	563.8	525.7
Private transfers (net)	-501.1	-514.4	-469.6	-539.9	-491.3
Government transfers (net)	143.1	61.1	91.2	106.5	163.1
Long term capital (net)	344.7	307.5	200.9	228.1	647.6
Short term capital (net)	765.0	1,013.8	892.0	899.9	627.1
Errors and omissions	12.6	2.3	174.9	-28.4	241.6
Overall balance	154.0	84.6	113.5	-104.8	65.1

Exchange Rates [38]

Currency: **Communaute Financiere Africaine francs.**
Symbol: **CFAF.**

Data are currency units per $1.

January 1993	274.06
1992	264.69
1991	282.11
1990	272.26
1989	319.01
1988	297.85

Imports and Exports

Top Import Origins [39]

$1.6 billion (f.o.b., 1990). Data are for 1989.

Origins	%
France	29
Other EC	29
Nigeria	16
US	4
Japan	3

Top Export Destinations [40]

$2.8 billion (f.o.b., 1990).

Destinations	%
France	NA
FRG	NA
Netherlands	NA
US	NA
Belgium	NA
Spain (1985)	NA

Foreign Aid [41]

	U.S. $	
US commitments, including Ex-Im (FY70-89)	356	million
Western (non-US) countries, ODA and OOF bilateral commitments (1970-88)	5.2	billion

Import and Export Commodities [42]

Import Commodities	**Export Commodities**
Food	Cocoa 30%
Capital goods	Coffee 20%
Consumer goods	Tropical woods 11%
Fuel	Petroleum
	Cotton
	Bananas
	Pineapples
	Palm oil
	Cotton

For sources, notes, and explanations, see Annotated Source Appendix, page 1035.

Croatia

Geography [1]

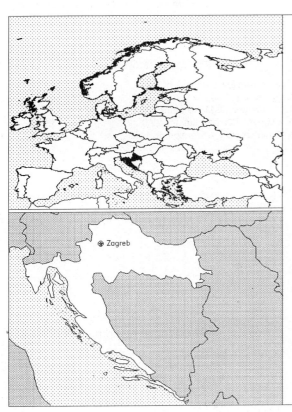

Total area:
56,538 km2
Land area:
56,410 km2
Comparative area:
Slightly smaller than West Virginia
Land boundaries:
Total 1,843 km, Bosnia and Herzegovina (east) 751 km, Bosnia and Herzegovina (southeast) 91 km, Hungary 292 km, Serbia and Montenegro 254 km (239 km with Serbia; 15 km with Montenego), Slovenia 455 km
Coastline:
5,790 km (mainland 1,778 km, islands 4,012 km)
Climate:
Mediterranean and continental; continental climate predominant with hot summers and cold winters; mild winters, dry summers along coast
Terrain:
Geographically diverse; flat plains along Hungarian border, low mountains and highlands near Adriatic coast, coastline, and islands
Natural resources:
Oil, some coal, bauxite, low-grade iron ore, calcium, natural asphalt, silica, mica, clays, salt
Land use:
Arable land: 32%
Permanent crops: 20%
Meadows and pastures: 18%
Forest and woodland: 15%
Other: 15%

Demographics [2]

	1960	1970	1980	1990	1991[1]	1994	2000	2010	2020
Population	4,140	4,411	4,593	4,686	NA	4,717	4,717	4,729	4,647
Population density (persons per sq. mi.)	NA	NA	NA	NA	NA	NA	NA	NA	NA
Births	NA	NA	NA	NA	NA	53	NA	NA	NA
Deaths	NA	NA	NA	NA	NA	50	NA	NA	NA
Life expectancy - males	NA	NA	NA	NA	NA	70	NA	NA	NA
Life expectancy - females	NA	NA	NA	NA	NA	77	NA	NA	NA
Birth rate (per 1,000)	NA	NA	NA	NA	NA	12	NA	NA	NA
Death rate (per 1,000)	NA	NA	NA	NA	NA	11	NA	NA	NA
Women of reproductive age (15-44 yrs.)	NA	NA	NA	1,136	NA	1,145	1,153	NA	NA
of which are currently married	NA	NA	NA	802	NA	810	815	NA	NA
Fertility rate	NA	NA	NA	1.7	NA	1.7	1.6	1.6	1.6

Population values are in thousands, life expectancy in years, and other items as indicated.

Health

Health Personnel [3]

Doctors per 1,000 pop., 1988-92	2.63 [2]
Nurse-to-doctor ratio, 1988-92	1.9 [2]
Hospital beds per 1,000 pop., 1985-90	6.0 [2]
Percentage of children immunized (age 1 yr. or less)	
Third dose of DPT, 1990-91	79.0 [2]
Measles, 1990-91	75.0 [2]

Health Expenditures [4]

For sources, notes, and explanations, see Annotated Source Appendix, page 1035.

229

Human Factors

Health Care Ratios [5]

Population per physician, 1970	1000 [2]
Population per physician, 1990	530 [2]
Population per nursing person, 1970	420 [2]
Population per nursing person, 1990	110 [2]
Percent of births attended by health staff, 1985	NA

Infants and Malnutrition [6]

Percent of babies with low birth weight, 1985	7
Infant mortality rate per 1,000 live births, 1970	56
Infant mortality rate per 1,000 live births, 1991	21
Years of life lost per 1,000 population, 1990	16
Prevalence of malnutrition (under age 5), 1990	NA

Ethnic Division [7]

Croat	78.0%
Serb	12.0%
Muslim	0.9%
Hungarian	0.5%
Slovenian	0.5%
Others	8.1%

Religion [8]

Catholic	76.5%
Orthodox	11.1%
Slavic Muslim	1.2%
Protestant	1.4%
Others and unknown	9.8%

Major Languages [9]

Serbo-Croatian	96%
Other	4%

Education

Public Education Expenditures [10]

Million Dinar	1980[22]	1985[22]	1987[22]	1988[22]	1989[22]	1990[22]
Total education expenditure	8	42	202	544	9,555	60,318
as percent of GNP	4.7	3.4	4.2	3.6	4.3	6.1
as percent of total govt. expend.	32.5	NA	NA	NA	NA	NA
Current education expenditure	7	39	185	506	9,002	55,911
as percent of GNP	4.0	3.1	3.9	3.3	4.1	5.7
as percent of current govt. expend.	NA	NA	NA	NA	NA	NA
Capital expenditure	1	3	17	38	553	4,407

Educational Attainment [11]

Age group	25+ [1]
Total population	13,083,762 [1]
Highest level attained (%)	
No schooling	15.8 [1]
First level	
Incompleted	53.9 [1]
Completed	NA [1]
Entered second level	
S-1	23.4 [1]
S-2	NA [1]
Post secondary	6.8 [1]

Literacy Rate [12]

In thousands and percent	1985[a,1]	1991[a,1]	2000[a,1]
Illiterate population +15 years	1,614	1,342	942
Illiteracy rate - total pop. (%)	9.2	7.3	4.7
Illiteracy rate - males (%)	3.5	2.6	1.3
Illiteracy rate - females (%)	14.6	11.9	7.9

Libraries [13]

	Admin. Units[a]	Svc. Pts.[a]	Vols. (000)[a]	Shelving (meters)[a]	Vols. Added[a]	Reg. Users[a]
National (1989)	8	8	12,316	303,555	305,462	163,169
Non-specialized (1989)	20	20	3,488	60,443	72,503	49,262
Public (1989)	808	1,937	30,238	552,866	1.3mil	19.5mil
Higher ed. (1989)	409	421	14,462	319,329	469,142	529,549
School (1989)	7,784	7,784	38,430	NA	1,680	NA

Daily Newspapers [14]

	1975[b]	1980[b]	1985[b]	1990[b]
Number of papers	26	27	27	34
Circ. (000)	1,896	2,649	2,451	2,281

Cinema [15]

Data for 1989.

Cinema seats per 1,000	16.9[a]
Annual attendance per person	1.9[a]
Gross box office receipts (mil. Dinar)	643,490[a]

Science and Technology

Scientific/Technical Forces [16]

Potential scientists/engineers	563,312
Number female	NA
Potential technicians	432,380
Number female	NA
Total	995,692

R&D Expenditures [17]

	Dinar (000) 1989
Total expenditure[a,4,12]	2,152,032
Capital expenditure	815,082
Current expenditure	1,336,950
Percent current	62.1

U.S. Patents Issued [18]

For sources, notes, and explanations, see Annotated Source Appendix, page 1035.

Government and Law

Organization of Government [19]

Long-form name:
Republic of Croatia
Type:
parliamentary democracy
Independence:
NA June 1991 (from Yugoslavia)
Constitution:
adopted on 2 December 1991
Legal system:
based on civil law system
National holiday:
Statehood Day, 30 May (1990)
Executive branch:
president, prime minister, deputy prime ministers, cabinet
Legislative branch:
bicameral Parliament consists of an upper house or House of Parishes (Zupanije Dom) and a lower house or Chamber of Deputies (Predstavnicke Dom)
Judicial branch:
Supreme Court, Constitutional Court

Political Parties [20]

Chamber of Deputies	% of seats
Croatian Democratic Union (HDZ)	63.0
Other	37.0

Government Budget [21]

For 1990 est.

Revenues	NA
Expenditures	NA
Capital expenditures	NA

Defense Summary [22]

Branches: Ground Forces, Naval Forces, Air and Air Defense Forces

Manpower Availability: Males age 15-49 1,177,029; fit for military service 943,259; reach military age (19) annually 32,873 (1993 est.)

Defense Expenditures: 337-393 billion Croatian dinars; note - conversion of defense expenditures into US dollars using the current exchange rate could produce misleading results

Crime [23]

Human Rights [24]

	SSTS	FL	FAPRO	PPCG	APROBC	TPW	PCPTW	STPEP	PHRFF	PRW	ASST	AFL
Observes	P			P		P	P			P	P	
	EAFRD	CPR	ESCR	SR	ACHR	MAAE	PVIAC	PVNAC	EAFDAW	TCIDTP	RC
Observes	P	P	P	P			P	P	P	P	P

P = Party; S = Signatory; see Appendix for meaning of abbreviations.

Labor Force

Total Labor Force [25]

1,509,489

Labor Force by Occupation [26]

Industry and mining	37%
Agriculture	16
Other	47

Unemployment Rate [27]

20% (December 1991 est.)

Production Sectors

Energy Resource Summary [28]

Energy Resources: Oil, some coal. **Electricity**: 3,570,000 kW capacity; 11,500 million kWh produced, 2,400 kWh per capita (1992). **Pipelines**: Crude oil 670 km, petroleum products 20 km, natural gas 310 km (1992); note: now disrupted because of territorial dispute.

Telecommunications [30]

- 350,000 telephones
- Broadcast stations - 14 AM, 8 FM, 12 (2 repeaters) TV
- 1,100,000 radios
- 1,027,000 TVs
- NA submarine coaxial cables
- Satellite ground stations - none

Top Agricultural Products [32]

Croatia normally produces a food surplus; most agricultural land in private hands and concentrated in Croat-majority districts in Slavonia and Istria; much of Slavonia's land has been put out of production by fighting; wheat, corn, sugar beets, sunflowers, alfalfa, and clover are main crops in Slavonia; central Croatian highlands are less fertile but support cereal production, orchards, vineyards, livestock breeding, and dairy farming.

Top Mining Products [33]

Metric tons unless otherwise specified	M.t.
Magnesite, crude	175,000
Bauxite (000 tons)	2,850
Copper, Cu content of concentrate	140,000
Gypsum, crude	400,000
Pumice, volcanic tuff	380,000
Cement, hydraulic (000 tons)	7,100

Transportation [31]

Railroads. 2,592 km of standard guage (1.435 m) of which 864 km are electrified (1992); note - disrupted by territorial dispute

Highways. 32,071 km total; 23,305 km paved, 8,439 km gravel, 327 km earth (1990); note - key highways note disrupted because of territorial dispute

Merchant Marine. 18 ships (1,000 GRT or over) totaling 77,074 GRT/93,052 DWT; includes 4 cargo, 1 roll-on/roll-off, 10 passenger ferries, 2 bulk, 1 oil tanker; note - also controlled by Croatian shipowners are 198 ships (1,000 GRT or over) under flags of convenience - primarily Malta and St. Vincent - totaling 2,602,678 GRT/4,070,852 DWT; includes 89 cargo, 9 roll-on/ roll-off, 6 refrigerated cargo, 14 container, 3 multifunction large load carriers, 51 bulk, 5 passenger, 11 oil tanker, 4 chemical tanker, 6 service vessel

Airports

Total:	75
Usable:	72
With permanent-surface runways:	15
With runways over 3,659 m:	0
With runways 2,440-3,659 m:	10
With runways 1,220-2,439 m:	5

Tourism [34]

	1987	1988	1989	1990	1991
Visitors[74]	26,151	29,635	34,118	39,573	
Tourists[6,74]	8,907	9,018	8,644	7,880	1,459
Tourism receipts[74]	1,668	2,024	2,230	2,774	468
Tourism expenditures[74]	90	109	131	149	103
Fare receipts[74]	340	430	475	620	105

Tourists are in thousands, money in million U.S. dollars.

Manufacturing Sector

GDP and Manufacturing Summary [35]

	1980	1985	1989	1990	% change 1980-1990	% change 1989-1990
GDP (million 1980 $)	69,958	71,058	72,234	66,371	-5.1	-8.1
GDP per capita (1980 $)	3,136	3,073	3,050	2,786	-11.2	-8.7
Manufacturing as % of GDP (current prices)	30.6	37.2	39.5	42.0	37.3	6.3
Gross output (million $)	72,629	57,020	65,078	62,136^e	-14.4	-4.5
Value added (million $)	21,750	17,171	30,245	27,660^e	27.2	-8.5
Value added (million 1980 $)	19,526	22,283	24,021	21,703	11.1	-9.6
Industrial production index	100	116	120	108	8.0	-10.0
Employment (thousands)	2,106	2,467	2,658	2,537^e	20.5	-4.6

Note: GDP stands for Gross Domestic Product. 'e' stands for estimated value.

Profitability and Productivity

	1980	1985	1989	1990	% change 1980-1990	% change 1989-1990
Intermediate input (%)	70	70	54	55^e	-21.4	1.9
Wages, salaries, and supplements (%)	14	12	12^e	18^e	28.6	50.0
Gross operating surplus (%)	15	18	34^e	26^e	73.3	-23.5
Gross output per worker ($)	34,487	23,113	24,484	24,248^e	-29.7	-1.0
Value added per worker ($)	10,328	6,960	11,379	10,796^e	4.5	-5.1
Average wage (incl. benefits) ($)	4,991	2,703	2,986^e	4,488^e	-10.1	50.3

Profitability is in percent of gross output. Productivity is in U.S. $. 'e' stands for estimated value.

Profitability - 1990

Wages - 18.2%
Inputs - 55.6%
Surplus - 26.3%

The graphic shows percent of gross output.

Value Added in Manufacturing

	1980 $ mil.	1980 %	1985 $ mil.	1985 %	1989 $ mil.	1989 %	1990 $ mil.	1990 %	% change 1980-1990	% change 1989-1990
311 Food products	1,897	8.7	1,458	8.5	3,916	12.9	3,484^e	12.6	83.7	-11.0
313 Beverages	459	2.1	353	2.1	663	2.2	589^e	2.1	28.3	-11.2
314 Tobacco products	184	0.8	221	1.3	344	1.1	308^e	1.1	67.4	-10.5
321 Textiles	1,759	8.1	1,428	8.3	2,881	9.5	2,663^e	9.6	51.4	-7.6
322 Wearing apparel	903	4.2	718	4.2	1,593	5.3	1,427^e	5.2	58.0	-10.4
323 Leather and fur products	226	1.0	231	1.3	383	1.3	340^e	1.2	50.4	-11.2
324 Footwear	482	2.2	503	2.9	1,022	3.4	899^e	3.3	86.5	-12.0
331 Wood and wood products	977	4.5	530	3.1	794	2.6	706^e	2.6	-27.7	-11.1
332 Furniture and fixtures	730	3.4	438	2.6	1,030	3.4	1,065^e	3.9	45.9	3.4
341 Paper and paper products	529	2.4	394	2.3	759	2.5	674^e	2.4	27.4	-11.2
342 Printing and publishing	876	4.0	462	2.7	761	2.5	678^e	2.5	-22.6	-10.9
351 Industrial chemicals	694	3.2	631	3.7	1,107	3.7	992^e	3.6	42.9	-10.4
352 Other chemical products	681	3.1	525	3.1	1,419	4.7	1,315^e	4.8	93.1	-7.3
353 Petroleum refineries	454	2.1	415	2.4	260	0.9	233^e	0.8	-48.7	-10.4
354 Miscellaneous petroleum and coal products	101	0.5	101	0.6	104	0.3	91^e	0.3	-9.9	-12.5
355 Rubber products	276	1.3	269	1.6	479	1.6	456^e	1.6	65.2	-4.8
356 Plastic products	413	1.9	258	1.5	397	1.3	350^e	1.3	-15.3	-11.8
361 Pottery, china and earthenware	128	0.6	72	0.4	162	0.5	144^e	0.5	12.5	-11.1
362 Glass and glass products	163	0.7	113	0.7	224	0.7	204^e	0.7	25.2	-8.9
369 Other non-metal mineral products	906	4.2	513	3.0	683	2.3	604^e	2.2	-33.3	-11.6
371 Iron and steel	1,221	5.6	1,000	5.8	1,343	4.4	1,171^e	4.2	-4.1	-12.8
372 Non-ferrous metals	480	2.2	509	3.0	944	3.1	927^e	3.4	93.1	-1.8
381 Metal products	2,105	9.7	1,577	9.2	1,293	4.3	1,130^e	4.1	-46.3	-12.6
382 Non-electrical machinery	1,828	8.4	1,463	8.5	2,372	7.8	2,378^e	8.6	30.1	0.3
383 Electrical machinery	1,600	7.4	1,544	9.0	2,640	8.7	2,334^e	8.4	45.9	-11.6
384 Transport equipment	1,441	6.6	1,263	7.4	2,389	7.9	2,241^e	8.1	55.5	-6.2
385 Professional and scientific equipment	101	0.5	93	0.5	154	0.5	146^e	0.5	44.6	-5.2
390 Other manufacturing industries	134	0.6	88	0.5	128	0.4	114^e	0.4	-14.9	-10.9

Note: The industry codes shown are International Standard Industry codes (ISIC). Percentages are percent of total Value Added. 'e' stands for estimated value

Finance, Economics, and Trade

Economic Indicators [36]

Yugoslav Dinars or U.S. Dollars as indicated[12].

	1989	1990	1991[e]
GSP[13]			
(billions current dinars)	221.9	910.9	1,490
Real GSP growth rate (%)	0.6	-7.6	-20
Money supply (M1)			
(yr-end mil.)	51,216	127,241	NA
Commercial interest			
rates (avg.)	4,354	45	80
Investment rate (%)	19.4	18.9	NA
CPI	1,356	688	310
WPI	1,406	534	320
External public debt	18,569	17,791	15,760

Balance of Payments Summary [37]

Exchange Rates [38]

Currency: **Croatian dinar.**
Symbol: **CD.**

Data are currency units per $1.

April 1992	60.00

Imports and Exports

Top Import Origins [39]

$4.4 billion (1990).

Origins	%
Other former Yugoslav republics	NA

Top Export Destinations [40]

$2.9 billion (1990).

Destinations	%
Principally the other former Yugoslav republics	NA

Foreign Aid [41]

Import and Export Commodities [42]

Import Commodities

Machinery, transp. equip. 21%
Fuels and lubricants 19%
Food and live animals 16%
Chemicals 14%
Manufactured goods 13%
Manufactured articles 9%
Raw materials 6.5%
Beverages and tobacco 1%

Export Commodities

Machinery, transp. equip. 30%
Other manufacturers 37%
Chemicals 11%
Food and live animals 9%
Raw materials 6.5%
Fuels and lubricants 5%

234

For sources, notes, and explanations, see Annotated Source Appendix, page 1035.

Cuba

Geography [1]

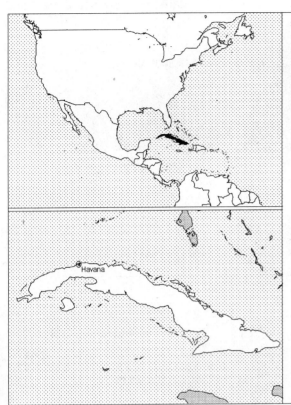

Total area:
110,860 km2
Land area:
110,860 km2
Comparative area:
Slightly smaller than Pennsylvania
Land boundaries:
Total 29 km, US Naval Base at Guantanamo 29 km
Coastline:
3,735 km
Climate:
Tropical; moderated by trade winds; dry season (November to April); rainy season (May to October)
Terrain:
Mostly flat to rolling plains with rugged hills and mountains in the southeast
Natural resources:
Cobalt, nickel, iron ore, copper, manganese, salt, timber, silica, petroleum
Land use:
Arable land:
23%
Permanent crops:
6%
Meadows and pastures:
23%
Forest and woodland:
17%
Other:
31%

Demographics [2]

	1960	1970	1980	1990	1991[1]	1994	2000	2010	2020
Population	7,027	8,543	9,653	10,622	10,732	11,064	11,617	12,274	12,755
Population density (persons per sq. mi.)	164	200	226	248	251	NA	271	287	299
Births	NA	NA	NA	NA	192	184	NA	NA	NA
Deaths	NA	NA	NA	NA	72	72	NA	NA	NA
Life expectancy - males	NA	NA	NA	NA	73	75	NA	NA	NA
Life expectancy - females	NA	NA	NA	NA	78	79	NA	NA	NA
Birth rate (per 1,000)	NA	NA	NA	NA	18	17	NA	NA	NA
Death rate (per 1,000)	NA	NA	NA	NA	7	7	NA	NA	NA
Women of reproductive age (15-44 yrs.)	NA	NA	NA	3,003	NA	3,065	3,075	NA	NA
of which are currently married	NA	NA	NA	1,958	NA	2,076	2,142	NA	NA
Fertility rate	NA	NA	NA	1.9	1.86	1.8	1.8	1.8	1.8

Population values are in thousands, life expectancy in years, and other items as indicated.

Health

Health Personnel [3]

Doctors per 1,000 pop., 1988-92	3.75
Nurse-to-doctor ratio, 1988-92	1.7
Hospital beds per 1,000 pop., 1985-90	5.0
Percentage of children immunized (age 1 yr. or less)	
Third dose of DPT, 1990-91	99.0
Measles, 1990-91	99.0

Health Expenditures [4]

Total health expenditure, 1990 (official exchange rate)	
Millions of dollars	NA
Millions of dollars per capita	NA
Health expenditures as a percentage of GDP	
Total	NA
Public sector	NA
Private sector	NA
Development assistance for health	
Total aid flows (millions of dollars)[1]	3
Aid flows per capita (millions of dollars)	0.3
Aid flows as a percentage of total health expenditure	NA

For sources, notes, and explanations, see Annotated Source Appendix, page 1035.

235

Human Factors

Health Care Ratios [5]	Infants and Malnutrition [6]

Ethnic Division [7]

Mulatto	51%
White	37%
Black	11%
Chinese	1%

Religion [8]

Nominally Roman Catholic 85% prior to Castro assuming power.

Major Languages [9]

Spanish.

Education

Public Education Expenditures [10]

Million Pesos	1980[9]	1985[9]	1987[9]	1988[9]	1989[9]	1990[9]
Total education expenditure	1,267	1,690	1,732	1,767	1,778	1,748
as percent of GNP	7.2	6.3	6.8	6.7	6.7	NA
as percent of total govt. expend.	NA	NA	18.4	14.1	12.8	12.3
Current education expenditure	1,135	1,588	1,600	1,652	1,659	1,627
as percent of GNP	6.4	5.9	6.3	6.3	6.2	NA
as percent of current govt. expend.	NA	NA	23.5	16.7	15.4	14.4
Capital expenditure	133	103	132	115	119	121

Educational Attainment [11]

Age group	25-49
Total population	3,013,315
Highest level attained (%)	
No schooling	3.7
First level	
Incompleted	22.6
Completed	27.6
Entered second level	
S-1	40.2
S-2	NA
Post secondary	5.9

Literacy Rate [12]

In thousands and percent	1985[a]	1991[a]	2000[a]
Illiterate population + 15 years	562	484	334
Illiteracy rate - total pop. (%)	7.6	6.0	3.8
Illiteracy rate - males (%)	6.3	5.0	3.3
Illiteracy rate - females (%)	8.9	7.0	4.4

Libraries [13]

	Admin. Units	Svc. Pts.	Vols. (000)	Shelving (meters)	Vols. Added	Reg. Users
National (1989)	1	1	2,431	NA	84,620	13,384
Non-specialized	NA	NA	NA	NA	NA	NA
Public (1989)	332	4,671	4,334	NA	367,371	334,007
Higher ed. (1990)[15]	85	85	2,415	NA	4,104	287,102
School (1987)	3,860	3,860	15,415	NA	1.2mil	1.6mil

Daily Newspapers [14]

	1975	1980	1985	1990
Number of papers	15	17	17	19
Circ. (000)	600[e]	1,050	1,207	1,824

Cinema [15]

Data for 1991.

Cinema seats per 1,000	20.0
Annual attendance per person	2.8
Gross box office receipts (mil. Peso)	12

Science and Technology

Scientific/Technical Forces [16]

Potential scientists/engineers	139,469
Number female	55,924
Potential technicians	NA
Number female	NA
Total	NA

R&D Expenditures [17]

	Peso[8] (000) 1989
Total expenditure	222,000
Capital expenditure	63,309
Current expenditure	158,691
Percent current	71.5

U.S. Patents Issued [18]

For sources, notes, and explanations, see Annotated Source Appendix, page 1035.

Government and Law

Organization of Government [19]

Long-form name:
Republic of Cuba
Type:
Communist state
Independence:
20 May 1902 (from Spain 10 December 1898; admin. by the US 1898-1902)
Constitution:
24 February 1976
Legal system:
based on Spanish and American law, with large elements of Communist legal theory; does not accept compulsory ICJ jurisdiction
National holiday:
Rebellion Day, 26 July (1953)
Executive branch:
president of the Council of State (CS), first vice president of CS, CS, president of the Council of Ministers (CM), first vice president of CM, Executive Committee of CM, CM
Legislative branch:
unicameral National Assembly of the People's Power (Asamblea Nacional del Poder Popular)
Judicial branch:
People's Supreme Court (Tribunal Supremo Popular)

Crime [23]

Elections [20]

National Assembly of People's Power.
Last held December 1986 (next to be held February 1993); results - PCC (Cuban Communist Party) is the only party; seats - (510 total; after the February election, the National Assembly will have 590 seats) indirectly elected from slates approved by special candidacy commissions.

Government Budget [21]

For 1990 est.
Revenues	12.46
Expenditures	14.45
Capital expenditures	NA

Military Expenditures and Arms Transfers [22]

	1985	1986	1987	1988	1989
Military expenditures					
Current dollars (mil.)[4]	1,335	1,307	1,306	1,350	1,377
1989 constant dollars (mil.)[4]	1,520	1,450	1,405	1,405	1,377
Armed forces (000)	297	297[e]	297[e]	297	297
Gross national product (GNP)					
Current dollars (mil.)[e]	29,520	31,420	33,700	34,720	35,460
1989 constant dollars (mil.)[e]	33,610	34,870	36,250	36,140	35,460
Central government expenditures (CGE)					
1989 constant dollars (mil.)	NA	NA	NA	NA	13,530
People (mil.)	10.1	10.2	10.3	10.4	10.5
Military expenditure as % of GNP	4.5	4.2	3.9	3.9	3.9
Military expenditure as % of CGE	NA	NA	NA	NA	10.2
Military expenditure per capita	151	142	137	135	131
Armed forces per 1,000 people	29.4	29.2	28.9	28.6	28.3
GNP per capita	3,332	3,422	3,523	3,477	3,375
Arms imports[6]					
Current dollars (mil.)	2,400	1,600	1,800	1,700	1,200
1989 constant dollars (mil.)	2,732	1,776	1,936	1,770	1,200
Arms exports[6]					
Current dollars (mil.)	5	0	0	0	5
1989 constant dollars (mil.)	6	0	0	0	5
Total imports[7]					
Current dollars (mil.)	8,677	9,158	7,612	7,580	8,124
1989 constant dollars	9,878	10,160	8,187	7,891	8,124
Total exports[7]					
Current dollars (mil.)	6,503	6,438	5,401	5,500	5,392
1989 constant dollars	7,403	7,144	5,809	5,725	5,392
Arms as percent of total imports[8]	27.7	17.5	23.6	22.4	14.8
Arms as percent of total exports[8]	0.1	0	0	0	0.1

Human Rights [24]

	SSTS	FL	FAPRO	PPCG	APROBC	TPW	PCPTW	STPEP	PHRFF	PRW	ASST	AFL
Observes	P	P	P	P	P	P	P	P		P	P	P
		EAFRD	CPR	ESCR	SR	ACHR	MAAE	PVIAC	PVNAC	EAFDAW	TCIDTP	RC
Observes		P					P			P	S	P

P = Party; S = Signatory; see Appendix for meaning of abbreviations.

Labor Force

Total Labor Force [25]

4,620,800 economically active population (1988) 3,578,800 in state sector

Labor Force by Occupation [26]

Services and government	30%
Industry	22
Agriculture	20
Commerce	11
Construction	10
Transportation and communications	7

Date of data: June 1990

Unemployment Rate [27]

For sources, notes, and explanations, see Annotated Source Appendix, page 1035.

237

Production Sectors

Commercial Energy Production and Consumption

Production [28]

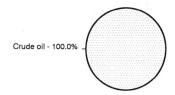

Crude oil - 100.0%

Consumption [29]

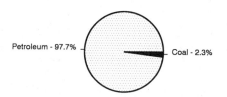

Petroleum - 97.7%   Coal - 2.3%

Data are shown in quadrillion (10^{15}) BTUs and percent for 1991

Crude oil[1]	0.04
Natural gas liquids	(s)
Dry natural gas	0.00
Coal[2]	0.00
Net hydroelectric power[3]	(s)
Net nuclear power[3]	0.00
Total	0.04

Petroleum	0.43
Dry natural gas	0.00
Coal[2]	0.01
Net hydroelectric power[3]	(s)
Net nuclear power[3]	0.00
Total	0.44

Telecommunications [30]

- Broadcast stations - 150 AM, 5 FM, 58 TV
- 1,530,000 TVs
- 2,140,000 radios
- 229,000 telephones
- 1 Atlantic Ocean INTELSAT earth station

Transportation [31]

Railroads. 12,947 km total; Cuban National Railways operates 5,053 km of 1.435-meter gauge track; 151.7 km electrified; 7,742 km of sugar plantation lines of 0.914-m and 1.435-m gauge

Highways. 26,477 km total; 14,477 km paved, 12,000 km gravel and earth surfaced (1989 est.)

Merchant Marine. 73 ships (1,000 GRT or over) totaling 511,522 GRT/720,270 DWT; includes 42 cargo, 10 refrigerated cargo, 1 cargo/training, 11 oil tanker, 1 chemical tanker, 4 liquefied gas, 4 bulk; note - Cuba beneficially owns an additional 38 ships (1,000 GRT and over) totaling 529,090 DWT under the registry of Panama, Cyprus, and Malta

Airports

Total:	186
Usable:	166
With permanent-surface runways:	73
With runways over 3,659 m:	3
With runways 2,440-3,659 m:	12
With runways 1,220-2,439 m:	19

Top Agricultural Products [32]

Agriculture accounts for 11% of GNP (including fishing and forestry); key commercial crops—sugarcane, tobacco, and citrus fruits; other products - coffee, rice, potatoes, meat, beans; world's largest sugar exporter; not self-sufficient in food (excluding sugar); sector hurt by growing shortages of fuels and parts.

Top Mining Products [33]

Estimated metric tons unless otherwise specified   M.t.

Gypsum (000 tons)	130
Lime (000 tons)	180
Nickel, metallurgical products, Ni content[23]	36,400
Copper, mine output, Cu content	3,000
Steel, crude (000 tons)	270
Cement, hydraulic (000 tons)	4,000
Nitrogen content of anhydrous ammonia (000)	140
Sulfur, byproduct of petroleum (000 tons)	5[e]

Tourism [34]

	1987	1988	1989	1990	1991
Visitors	293	309	326	340	424
Tourists	282	298	314	327	408
Excursionists[18]	11	11	12	13	16
Cruise passengers	3	5	5		
Tourism receipts[19]	145	189	204	246	300

Tourists are in thousands, money in million U.S. dollars.

For sources, notes, and explanations, see Annotated Source Appendix, page 1035.

Manufacturing Sector

GDP and Manufacturing Summary [35]

	1980	1985	1989	1990	% change 1980-1990	% change 1989-1990
GDP (million 1980 $)	16,653	24,937	23,969	23,051	38.4	-3.8
GDP per capita (1980 $)	1,721	2,474	2,282	2,172	26.2	-4.8
Manufacturing as % of GDP (current prices)	35.8	34.1	33.6e	33.8e	-5.6	0.6
Gross output (million $)	9,725	12,032	13,485	15,644e	60.9	16.0
Value added (million $)	4,882	5,120	5,255	5,990e	22.7	14.0
Value added (million 1980 $)	5,735	10,905	10,207e	9,523	66.1	-6.7
Industrial production index	100	131	125e	113	13.0	-9.6
Employment (thousands)	501	654	700	709e	41.5	1.3

Note: GDP stands for Gross Domestic Product. 'e' stands for estimated value.

Profitability and Productivity

	1980	1985	1989	1990	% change 1980-1990	% change 1989-1990
Intermediate input (%)	50	57	61	62e	24.0	1.6
Wages, salaries, and supplements (%)	13e	14e	14e	14e	7.7	0.0
Gross operating surplus (%)	37e	29e	25e	24e	-35.1	-4.0
Gross output per worker ($)	18,444	17,225	19,270	20,658e	12.0	7.2
Value added per worker ($)	6,135	4,757	7,510	8,011e	30.6	6.7
Average wage (incl. benefits) ($)	2,606e	2,514e	2,770e	3,199e	22.8	15.5

Profitability is in percent of gross output. Productivity is in U.S. $. 'e' stands for estimated value.

Profitability - 1990

Inputs - 62.0%
Wages - 14.0%
Surplus - 24.0%

The graphic shows percent of gross output.

Value Added in Manufacturing

	1980 $ mil.	1980 %	1985 $ mil.	1985 %	1989 $ mil.	1989 %	1990 $ mil.	1990 %	% change 1980-1990	% change 1989-1990
311 Food products	655	13.4	957	18.7	905	17.2	1,033e	17.2	57.7	14.1
313 Beverages	246	5.0	273	5.3	314	6.0	358e	6.0	45.5	14.0
314 Tobacco products	1,805	37.0	2,004	39.1	2,303	43.8	2,629e	43.9	45.7	14.2
321 Textiles	50	1.0	40	0.8	95	1.8	109e	1.8	118.0	14.7
322 Wearing apparel	146	3.0	98	1.9	79	1.5	88e	1.5	-39.7	11.4
323 Leather and fur products	53	1.1	32	0.6	27	0.5	29e	0.5	-45.3	7.4
324 Footwear	79	1.6	48	0.9	41	0.8	45e	0.8	-43.0	9.8
331 Wood and wood products	58	1.2	53	1.0	44	0.8	52e	0.9	-10.3	18.2
332 Furniture and fixtures	48	1.0	43	0.8	37	0.7	42e	0.7	-12.5	13.5
341 Paper and paper products	46	0.9	44	0.9	10	0.2	13e	0.2	-71.7	30.0
342 Printing and publishing	96	2.0	59	1.2	72	1.4	81e	1.4	-15.6	12.5
351 Industrial chemicals	79e	1.6	54e	1.1	55e	1.0	68e	1.1	-13.9	23.6
352 Other chemical products	331e	6.8	225e	4.4	237e	4.5	286e	4.8	-13.6	20.7
353 Petroleum refineries	NA	0.0	NA	0.0	NA	0.0	NA	0.0	NA	NA
354 Miscellaneous petroleum and coal products	NA	0.0	NA	0.0	NA	0.0	NA	0.0	NA	NA
355 Rubber products	96e	2.0	65e	1.3	72e	1.4	83e	1.4	-13.5	15.3
356 Plastic products	83e	1.7	57e	1.1	69e	1.3	72e	1.2	-13.3	4.3
361 Pottery, china and earthenware	8	0.2	6	0.1	7	0.1	8e	0.1	0.0	14.3
362 Glass and glass products	17	0.3	13	0.3	16	0.3	19e	0.3	11.8	18.8
369 Other non-metal mineral products	188	3.9	104	2.0	104	2.0	112e	1.9	-40.4	7.7
371 Iron and steel	27	0.6	44	0.9	36	0.7	37e	0.6	37.0	2.8
372 Non-ferrous metals	41	0.8	48	0.9	57	1.1	64e	1.1	56.1	12.3
381 Metal products	108	2.2	92	1.8	68	1.3	78e	1.3	-27.8	14.7
382 Non-electrical machinery	153e	3.1	219e	4.3	149e	2.8	176e	2.9	15.0	18.1
383 Electrical machinery	60	1.2	58	1.1	52	1.0	57e	1.0	-5.0	9.6
384 Transport equipment	197e	4.0	281e	5.5	205e	3.9	225e	3.8	14.2	9.8
385 Professional and scientific equipment	111e	2.3	16e	0.3	13e	0.2	13e	0.2	-88.3	0.0
390 Other manufacturing industries	201	4.1	188	3.7	187	3.6	213e	3.6	6.0	13.9

Note: The industry codes shown are International Standard Industry codes (ISIC). Percentages are percent of total Value Added. 'e' stands for estimated value

Finance, Economics, and Trade

Economic Indicators [36]

National product: GNP—exchange rate conversion—$14.9 billion (1992 est.). **National product real growth rate**: - 15% (1992 est.). **National product per capita**: $1,370 (1992 est.). **Inflation rate (consumer prices)**: NA%. **External debt**: $6.8 billion (convertible currency, July 1989).

Balance of Payments Summary [37]

Exchange Rates [38]

Currency: **Cuban pesos.**
Symbol: **Cu$.**

Data are currency units per $1.

linked to the US dollar	1.0000

Imports and Exports

Top Import Origins [39]

$2.2 billion (c.i.f., 1992 est.).

Russia	10
China	9
Spain	9
Mexico	5
Italy	5
Canada	4
France	4

Top Export Destinations [40]

$2.1 billion (f.o.b., 1992 est.).

Destinations	%
Russia	30
Canada	10
China	9
Japan	6
Spain	4

Foreign Aid [41]

	U.S. $	
Western (non-US) countries, ODA and OOF bilateral commitments (1970-89)	710	million
Communist countries (1970-89)	18.5	billion

Import and Export Commodities [42]

Import Commodities

Petroleum
Food
Machinery
Chemicals

Export Commodities

Sugar
Nickel
Shellfish
Tobacco
Medical products
Citrus
Coffee

240

For sources, notes, and explanations, see Annotated Source Appendix, page 1035.

Cyprus

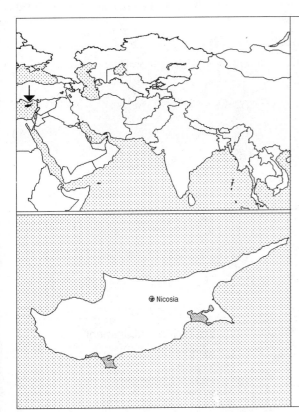

Geography [1]

Total area:
9,250 km2
Land area:
9,240 km2
Comparative area:
About 0.7 times the size of Connecticut
Land boundaries:
0 km
Coastline:
648 km
Climate:
Temperate, Mediterranean with hot, dry summers and cool, wet winters
Terrain:
Central plain with mountains to north and south
Natural resources:
Copper, pyrites, asbestos, gypsum, timber, salt, marble, clay earth pigment
Land use:
Arable land:
40%
Permanent crops:
7%
Meadows and pastures:
10%
Forest and woodland:
18%
Other:
25%

⊕ Nicosia

Demographics [2]

	1960	1970	1980	1990	1991[1]	1994	2000	2010	2020
Population	573	615	627	702	709	730	768	829	883
Population density (persons per sq. mi.)	161	172	176	197	199	NA	215	233	249
Births	NA	NA	NA	NA	13	12	NA	NA	NA
Deaths	NA	NA	NA	NA	6	6	NA	NA	NA
Life expectancy - males	NA	NA	NA	NA	73	74	NA	NA	NA
Life expectancy - females	NA	NA	NA	NA	78	79	NA	NA	NA
Birth rate (per 1,000)	NA	NA	NA	NA	18	17	NA	NA	NA
Death rate (per 1,000)	NA	NA	NA	NA	8	8	NA	NA	NA
Women of reproductive age (15-44 yrs.)	NA	NA	NA	177	NA	181	188	NA	NA
of which are currently married	NA	NA	NA	122	NA	126	128	NA	NA
Fertility rate	NA	NA	NA	2.4	2.38	2.3	2.2	2.1	2.0

Population values are in thousands, life expectancy in years, and other items as indicated.

Health

Health Personnel [3]

Health Expenditures [4]

For sources, notes, and explanations, see Annotated Source Appendix, page 1035.

241

Human Factors

Health Care Ratios [5]	Infants and Malnutrition [6]

Ethnic Division [7]

Greek	78%
Turkish	18%
Other	4%

Religion [8]

Greek Orthodox	78%
Muslim	18%
Maronite, Armenian, Apostolic, and other	4%

Major Languages [9]

Greek, Turkish, English.

Education

Public Education Expenditures [10]

Million Cypriot Pounds	1980[13]	1985[13]	1987[13]	1988[13]	1989[13]	1990[13]
Total education expenditure	27	55	NA	70	79	89
as percent of GNP	3.5	3.7	NA	3.5	3.5	3.5
as percent of total govt. expend.	12.9	12.2	NA	11.5	11.7	11.3
Current education expenditure	25	53	NA	68	76	84
as percent of GNP	3.3	3.6	NA	3.4	3.3	3.3
as percent of current govt. expend.	16.2	13.4	NA	12.7	12.8	12.3
Capital expenditure	2	3	NA	3	4	5

Educational Attainment [11]

Age group	20+
Total population	NA
Highest level attained (%)	
No schooling	6.0
First level	
Incompleted	46.0
Completed	NA
Entered second level	
S-1	34.0
S-2	NA
Post secondary	14.0

Literacy Rate [12]

Libraries [13]

Daily Newspapers [14]

	1975	1980	1985	1990
Number of papers	12	12	10	11
Circ. (000)	80[e]	80[e]	83	78

Cinema [15]

Science and Technology

Scientific/Technical Forces [16]

Potential scientists/engineers	23,222[e]
Number female	7,352[e]
Potential technicians	19,644[e]
Number female	10,350[e]
Total	42,866[e]

R&D Expenditures [17]

	Pound[5] (000) 1984
Total expenditure	1,173
Capital expenditure	NA
Current expenditure	NA
Percent current	NA

U.S. Patents Issued [18]

Values show patents issued to citizens of the country by the U.S. Patents Office.

	1990	1991	1992
Number of patents	3	0	0

For sources, notes, and explanations, see Annotated Source Appendix, page 1035.

Government and Law

Organization of Government [19]

Long-form name:
Republic of Cyprus
Type:
republic
Independence:
16 August 1960 (from UK)
Constitution:
16 August 1960; Turkish area May 1985
Legal system:
based on common law, with civil law modifications
National holiday:
Independence Day, 1 October (15 November is celebrated as Independence Day in the Turkish area)
Executive branch:
president, Council of Ministers (cabinet)
Legislative branch:
unicameral House of Representatives (Vouli Antiprosopon)
Judicial branch:
Supreme Court; note - there is also a Supreme Court in the Turkish area

Elections [20]

House of Representatives. Last held 19 May 1991; seats - (56 total) DISY 20, AKEL (Communist) 18, DIKO 11, EDEK 7; Turkish Area **Assembly of the Republic**: Last held 6 May 1990 (next to be held May 1995); seats as of July 1992 UBP 34, SPD 1, HDP 1, YDP 2, DP 10, independents 2.

Government Expenditures [21]

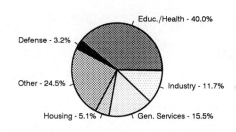

% distribution for 1991. Expend. in 1993: 2.200 ($ bil.)

Military Expenditures and Arms Transfers [22]

	1985	1986	1987	1988	1989
Military expenditures					
Current dollars (mil.)	39	28	34	41	41
1989 constant dollars (mil.)	44	31	37	43	41
Armed forces (000)	13	13	13	13	14
Gross national product (GNP)					
Current dollars (mil.)	3,087	3,283	3,623	4,063	4,468
1989 constant dollars (mil.)	3,515	6,343	3,896	4,229	4,468
Central government expenditures (CGE)					
1989 constant dollars (mil.)	1,078	1,104	1,175	1,269	1,345
People (mil.)	0.7	0.7	0.7	0.7	0.7
Military expenditure as % of GNP	1.2	0.9	0.9	1.0	0.9
Military expenditure as % of CGE	4.1	2.8	3.1	3.4	3.0
Military expenditure per capita	66	46	54	63	59
Armed forces per 1,000 people	19.5	19.3	19.1	18.9	19.9
GNP per capita	5,283	5,412	5,727	6,151	6,431
Arms imports[6]					
Current dollars (mil.)	10	20	90	10	40
1989 constant dollars (mil.)	11	22	97	10	40
Arms exports[6]					
Current dollars (mil.)	0	0	0	0	0
1989 constant dollars (mil.)	0	0	0	0	0
Total imports[7]					
Current dollars (mil.)	1,247	1,272	1,484	1,859	2,288
1989 constant dollars	1,420	1,412	1,596	1,935	2,288
Total exports[7]					
Current dollars (mil.)	476	451	568	705	790
1989 constant dollars	542	500	611	734	790
Arms as percent of total imports[8]	0.8	1.6	6.1	0.5	1.7
Arms as percent of total exports[8]	0	0	0	0	0

Crime [23]

Crime volume (for 1990)	
Cases known to police	3,832
Attempts (percent)	NA
Percent cases solved	90.8
Crimes per 100,000 persons	670.9
Persons responsible for offenses	
Total number offenders	1,063
Percent female	8.0
Percent juvenile[16]	16.1
Percent foreigners	16.0

Human Rights [24]

	SSTS	FL	FAPRO	PPCG	APROBC	TPW	PCPTW	STPEP	PHRFF	PRW	ASST	AFL
Observes	P	P	P	P	P	P	P	P	P	P	P	P
	EAFRD	CPR	ESCR	SR	ACHR	MAAE	PVIAC	PVNAC	EAFDAW	TCIDTP	RC	
Observes	P	P	P	P			P		P	P	P	

P = Party; S = Signatory; see Appendix for meaning of abbreviations.

Labor Force

Total Labor Force [25]

Labor force total is not available.

Labor Force by Occupation [26]

Services	57%
Industry	29
Agriculture	14

Date of data: 1991

Unemployment Rate [27]

Unemployment figures are not available.

For sources, notes, and explanations, see Annotated Source Appendix, page 1035.

243

Production Sectors

Energy Resource Summary [28]

Energy Resources: None. **Electricity**: 620,000 kW capacity; 1,770 million kWh produced, 2,530 kWh per capita (1991).

Telecommunications [30]

- Excellent in both the area controlled by the Cypriot Government (Greek area), and in the Turkish-Cypriot administered area
- 210,000 telephones
- Largely open-wire and microwave radio relay
- Broadcast stations - 11 AM, 8 FM, 1 (34 repeaters) TV in Greek sector and 2 AM, 6 FM and 1 TV in Turkish sector
- International service by tropospheric scatter, 3 submarine cables, and satellite earth stations - 1 Atlantic Ocean INTELSAT, 1 Indian Ocean INTELSAT and EUTELSAT earth stations

Top Agricultural Products [32]

Contributes 6% to GDP and employs 14% of labor force in the south; major crops—potatoes, vegetables, barley, grapes, olives, citrus fruits; vegetables and fruit provide 25% of export revenues.

Top Mining Products [33]

Estimated metric tons unless otherwise specified   M.t.

Cement, hydraulic (000 tons)	1,134
Limestone, crushed, Havara (000 tons)	2,700
Clay, bentonite, crude	58,500
Gypsum, crude	37,000
Umber	5,800
Lime, hydrated	6,566

Transportation [31]

Highways. 10,780 km total; 5,170 km paved; 5,610 km gravel, crushed stone, and earth

Merchant Marine. 1,299 ships (1,000 GRT or over) totaling 21,045,037 GRT/37,119,933 DWT; includes 10 short-sea passenger, 1 passenger-cargo, 463 cargo, 77 refrigerated cargo, 24 roll-on/roll-off, 70 container, 4 multifunction large load carrier, 110 oil tanker, 3 specialized tanker, 3 liquefied gas, 26 chemical tanker, 32 combination ore/oil, 422 bulk, 3 vehicle carrier, 48 combination bulk, 1 railcar carrier, 2 passenger; note - a flag of convenience registry; Cuba owns 27 of these ships, Russia owns 36, Latvia also has 7 ships, Croatia owns 2, and Romania 5

Airports

Total:	13
Usable:	13
With permanent-surface runways:	10
With runways over 3,659 m:	0
With runways 2,440-3,659 m:	7
With runways 1,220-2,439 m:	1

Tourism [34]

	1987	1988	1989	1990	1991
Visitors	1,157	1,312	1,540	1,676	1,473
Tourists	949	1,112	1,378	1,561	1,385
Excursionists[20]	208	200	162	114	88
Tourism receipts	666	782	990	1,258	1,026
Tourism expenditures	67	81	78	111	113
Fare receipts	141	144	133	163	200
Fare expenditures	53	62	64	64	82

Tourists are in thousands, money in million U.S. dollars.

For sources, notes, and explanations, see Annotated Source Appendix, page 1035.

Manufacturing Sector

GDP and Manufacturing Summary [35]

	1980	1985	1989	1990	% change 1980-1990	% change 1989-1990
GDP (million 1980 $)	2,154	2,831	3,660	3,935	82.7	7.5
GDP per capita (1980 $)	3,419	4,251	5,274	5,614	64.2	6.4
Manufacturing as % of GDP (current prices)	18.2	16.4	14.8e	15.0	-17.6	1.4
Gross output (million $)	1,134	1,122	1,855	2,196	93.7	18.4
Value added (million $)	406	378	670	792	95.1	18.2
Value added (million 1980 $)	378	473	601e	586	55.0	-2.5
Industrial production index	100	118	140	147	47.0	5.0
Employment (thousands)	34	39	43	43	26.5	0.0

Note: GDP stands for Gross Domestic Product. 'e' stands for estimated value.

Profitability and Productivity

	1980	1985	1989	1990	% change 1980-1990	% change 1989-1990
Intermediate input (%)	64	66	64	64	0.0	0.0
Wages, salaries, and supplements (%)	15e	18	17	19	26.7	11.8
Gross operating surplus (%)	21e	16	19	17	-19.0	-10.5
Gross output per worker ($)	29,417	25,804	43,605	46,057	56.6	5.6
Value added per worker ($)	10,525	8,697	15,761	16,606	57.8	5.4
Average wage (incl. benefits) ($)	5,063e	5,143	7,283	9,738	92.3	33.7

Profitability is in percent of gross output. Productivity is in U.S. $. 'e' stands for estimated value.

Profitability - 1990

Surplus - 17.0%
Inputs - 64.0%
Wages - 19.0%

The graphic shows percent of gross output.

Value Added in Manufacturing

	1980 $ mil.	1980 %	1985 $ mil.	1985 %	1989 $ mil.	1989 %	1990 $ mil.	1990 %	% change 1980-1990	% change 1989-1990
311 Food products	42	10.3	49	13.0	85	12.7	101	12.8	140.5	18.8
313 Beverages	37	9.1	29	7.7	62	9.3	73	9.2	97.3	17.7
314 Tobacco products	36	8.9	26	6.9	37	5.5	41	5.2	13.9	10.8
321 Textiles	16	3.9	14	3.7	28	4.2	32	4.0	100.0	14.3
322 Wearing apparel	53	13.1	54	14.3	97	14.5	118	14.9	122.6	21.6
323 Leather and fur products	5	1.2	6	1.6	9	1.3	11	1.4	120.0	22.2
324 Footwear	21	5.2	19	5.0	26	3.9	30	3.8	42.9	15.4
331 Wood and wood products	19	4.7	23	6.1	35	5.2	39	4.9	105.3	11.4
332 Furniture and fixtures	17	4.2	22	5.8	30	4.5	36	4.5	111.8	20.0
341 Paper and paper products	11	2.7	8	2.1	12	1.8	17	2.1	54.5	41.7
342 Printing and publishing	15	3.7	18	4.8	30	4.5	37	4.7	146.7	23.3
351 Industrial chemicals	3	0.7	2	0.5	5	0.7	3	0.4	0.0	-40.0
352 Other chemical products	12	3.0	12	3.2	24	3.6	28	3.5	133.3	16.7
353 Petroleum refineries	6	1.5	5	1.3	6	0.9	7	0.9	16.7	16.7
354 Miscellaneous petroleum and coal products	NA	0.0	NA	0.0	NA	0.0	NA	0.0	NA	NA
355 Rubber products	3	0.7	2	0.5	3	0.4	3	0.4	0.0	0.0
356 Plastic products	11	2.7	11	2.9	21	3.1	25	3.2	127.3	19.0
361 Pottery, china and earthenware	NA	0.0	1	0.3	2	0.3	2	0.3	NA	0.0
362 Glass and glass products	NA	0.0	NA	0.0	1	0.1	1	0.1	NA	0.0
369 Other non-metal mineral products	44	10.8	24	6.3	59	8.8	69	8.7	56.8	16.9
371 Iron and steel	NA	0.0	NA	0.0	NA	0.0	NA	0.0	NA	NA
372 Non-ferrous metals	NA	0.0	NA	0.0	NA	0.0	NA	0.0	NA	NA
381 Metal products	23	5.7	26	6.9	45	6.7	55	6.9	139.1	22.2
382 Non-electrical machinery	11	2.7	12	3.2	20	3.0	24	3.0	118.2	20.0
383 Electrical machinery	5	1.2	6	1.6	11	1.6	12	1.5	140.0	9.1
384 Transport equipment	8	2.0	4	1.1	8	1.2	9	1.1	12.5	12.5
385 Professional and scientific equipment	NA	0.0	NA	0.0	NA	0.0	NA	0.0	NA	NA
390 Other manufacturing industries	7	1.7	7	1.9	16	2.4	19	2.4	171.4	18.8

Note: The industry codes shown are International Standard Industry codes (ISIC). Percentages are percent of total Value Added. 'e' stands for estimated value

Finance, Economics, and Trade

Economic Indicators [36]

National product: Greek area: GDP—purchasing power equivalent—$6.3 billion (1992). Turkish area: GDP—purchasing power equivalent—$600 million (1990).
National product real growth rate: Greek area: 6.5% (1992). Turkish area: 5.9% (1990). **National product per capita**: Greek area: $11,000 (1992). Turkish area: $4,000 (1990). **Inflation rate (consumer prices)**: Greek area: 5.1% (1991). Turkish area: 69.4% (1990) **External debt**: $1.9 billion (1991).

Balance of Payments Summary [37]

Values in millions of dollars.

	1987	1988	1989	1990	1991
Exports of goods (f.o.b.)	566.4	645.5	717.3	846.7	875.0
Imports of goods (f.o.b.)	-1,326.8	-1,666.8	-2,072.1	-2,308.9	-2,363.1
Trade balance	-760.4	-1,021.3	-1,354.8	-1,462.2	-1,488.1
Services - debits	-567.5	-650.6	-696.6	-859.4	-892.0
Services - credits	1,375.8	1,628.3	1,878.6	2,345.1	2,159.1
Private transfers (net)	26.4	24.6	22.9	21.9	20.6
Government transfers (net)	17.5	27.4	17.0	20.8	21.0
Long term capital (net)	45.9	51.2	376.7	111.3	206.6
Short term capital (net)	34.9	97.7	33.6	353.1	54.5
Errors and omissions	-109.0	-86.6	-48.9	-233.3	-171.6
Overall balance	63.6	70.7	228.5	297.3	-89.9

Exchange Rates [38]

Currency: **Cypriot pound.**
Symbol: **#C.**

Currencies in use are the Cypriot pound (Greek area) and the Turkish lira, Turkish area. No exchange rate data are available.

Imports and Exports

Top Import Origins [39]

$2.4 billion (f.o.b., 1991).

Origins	%
UK	13.0
Japan	12.0
Italy	10.0
Germany	9.1

Top Export Destinations [40]

$875 million (f.o.b., 1991).

Destinations	%
UK	23
Greece	10
Lebanon	10
Germany	5

Foreign Aid [41]

	U.S. $	
US commitments, including Ex-Im (FY70-89)	292	million
Western (non-US) countries, ODA and OOF bilateral commitments (1970-89)	250	million
OPEC bilateral aid (1979-89)	62	million
Communist countries (1970-89)	24	million

Import and Export Commodities [42]

Import Commodities

Consumer goods
Petroleum and lubricants
Food and feed grains
Machinery

Export Commodities

Citrus
Potatoes
Grapes
Wine
Cement
Clothing and shoes

Czech Republic

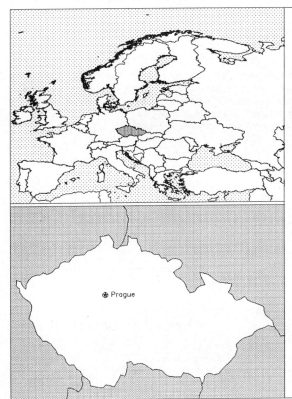

Geography [1]

Total area:
48,845 km2
Land area:
48,800 km2
Comparative area:
About twice the size of New Hampshire
Land boundaries:
Total 1,355 km, Austria 91 km, Czech Republic 215 km, Hungary 515 km, Poland 444 km, Ukraine 90 km
Coastline:
0 km (landlocked)
Climate:
Temperate; cool summers; cold, cloudy, humid winters
Terrain:
Rugged mountains in the central and northern part and lowlands in the south
Natural resources:
Brown coal and lignite; small amounts of iron ore, copper and manganese ore; salt; gas
Land use:
Arable land:
NA%
Permanent crops:
NA%
Meadows and pastures:
NA%
Forest and woodland:
NA%
Other:
NA%

Demographics [2]

	1960	1970	1980	1990	1991[1]	1994	2000	2010	2020
Population	9,660	9,795	10,289	10,363	5,133	10,408	10,607	10,892	10,991
Population density (persons per sq. mi.)	282	296	315	324	325	NA	337	347	351
Births	NA	NA	NA	NA	213	138	NA	NA	NA
Deaths	NA	NA	NA	NA	169	116	NA	NA	NA
Life expectancy - males	NA	NA	NA	NA	69	69	NA	NA	NA
Life expectancy - females	NA	NA	NA	NA	77	77	NA	NA	NA
Birth rate (per 1,000)	NA	NA	NA	NA	14	13	NA	NA	NA
Death rate (per 1,000)	NA	NA	NA	NA	11	11	NA	NA	NA
Women of reproductive age (15-44 yrs.)	NA	NA	NA	2,594	NA	2,669	2,591	NA	NA
of which are currently married	NA	NA	NA	1,740	NA	1,765	1,771	NA	NA
Fertility rate	NA	NA	NA	1.9	1.95	1.8	1.8	1.8	1.8

Population values are in thousands, life expectancy in years, and other items as indicated.

Health

Health Personnel [3]

Doctors per 1,000 pop., 1988-92	3.23 [3]
Nurse-to-doctor ratio, 1988-92	2.4 [3]
Hospital beds per 1,000 pop., 1985-90	7.9 [3]
Percentage of children immunized (age 1 yr. or less)	
Third dose of DPT, 1990-91	99.0 [3]
Measles, 1990-91	98.0 [3]

Health Expenditures [4]

Total health expenditure, 1990 (official exchange rate)	
Millions of dollars	2711
Millions of dollars per capita	173
Health expenditures as a percentage of GDP	
Total	5.9
Public sector	5.0
Private sector	0.9
Development assistance for health	
Total aid flows (millions of dollars)[1]	NA
Aid flows per capita (millions of dollars)	NA
Aid flows as a percentage of total health expenditure	NA

For sources, notes, and explanations, see Annotated Source Appendix, page 1035.

247

Human Factors

Health Care Ratios [5]

Population per physician, 1970	470 [3]
Population per physician, 1990	310 [3]
Population per nursing person, 1970	170 [3]
Population per nursing person, 1990	NA [3]
Percent of births attended by health staff, 1985	100 [3]

Infants and Malnutrition [6]

Percent of babies with low birth weight, 1985	6
Infant mortality rate per 1,000 live births, 1970	22
Infant mortality rate per 1,000 live births, 1991	11
Years of life lost per 1,000 population, 1990	16
Prevalence of malnutrition (under age 5), 1990	NA

Ethnic Division [7]

Slovak 85.6%, Hungarian 10.8%, Gypsy 1.5% (the 1992 census figures underreport the Gypsy/Romany community, which could reach 500,000 or more), Czech 1.1%, Ruthenian 15,000, Ukrainian 13,000, Moravian 6,000, German 5,000, Polish 3,000.

Religion [8]

Roman Catholic	60.3%
Atheist	9.7%
Protestant	8.4%
Orthodox	4.1%
Other	17.5%

Major Languages [9]

Slovak (official), Hungarian.

Education

Public Education Expenditures [10]

Million Koruna	1980[1,15]	1985[1,15]	1987[1,15]	1988[1,15]	1989[1,15]	1990[1,15]
Total education expenditure	23,181	28,201	30,549	32,254	NA	37,323
as percent of GNP	4.0	4.2	4.3	4.4	NA	4.6
as percent of total govt. expend.	NA	7.9	8.0	8.0	NA	8.2
Current education expenditure	21,802	26,959	29,247	31,005	NA	35,482
as percent of GNP	3.8	4.0	4.1	4.2	NA	4.4
as percent of current govt. expend.	NA	8.6	8.7	8.7	NA	8.7
Capital expenditure	1,379	1,242	1,302	1,249	NA	1,841

Educational Attainment [11]

Age group	25+[4]
Total population	9,274,694[4]
Highest level attained (%)	
No schooling	0.4[4]
First level	
Incompleted	47.6[4]
Completed	NA[4]
Entered second level	
S-1	45.9[4]
S-2	NA[4]
Post secondary	6.0[4]

Literacy Rate [12]

Libraries [13]

	Admin. Units[b]	Svc. Pts.[b]	Vols. (000)[b]	Shelving (meters)[b]	Vols. Added[b]	Reg. Users[b]
National (1989)	19	NA	26,962	NA	1.2mil[29]	455,748
Non-specialized	NA	NA	NA	NA	NA	NA
Public (1989)	8,398	11,454	58,627	NA	3.1mil[29]	2.8mil
Higher ed. (1990)	1,590	1,590	12,720	NA	366,500	269,591
School	NA	NA	NA	NA	NA	NA

Daily Newspapers [14]

	1975	1980	1985	1990
Number of papers	29	30	30	48
Circ. (000)	4,436	4,798	5,124	7,943

Cinema [15]

Data for 1990.

Cinema seats per 1,000	51.2[b]
Annual attendance per person	3.2[b]
Gross box office receipts (mil. Koruna)	386[b]

Science and Technology

Scientific/Technical Forces [16]

Potential scientists/engineers	542,706
Number female	191,256
Potential technicians	NA
Number female	NA
Total	NA

R&D Expenditures [17]

	Koruny (000) 1989
Total expenditure[b,27]	24,721,000
Capital expenditure	2,621,000
Current expenditure	22,100,000
Percent current	89.4

U.S. Patents Issued [18]

Values show patents issued to citizens of the country by the U.S. Patents Office.

	1990	1991	1992
Number of patents	39	27	18

For sources, notes, and explanations, see Annotated Source Appendix, page 1035.

Government and Law

Organization of Government [19]

Long-form name:
Slovak Republic
Type:
parliamentary democracy
Independence:
1 January 1993 (from Czechoslovakia)
Constitution:
ratified 3 September 1992; fully effective 1
January 1993
Legal system:
civil law system based on Austro-
Hungarian codes; has not accepted
compulsory ICJ jurisdiction; legal code
modified to comply with the obligations of
Conference on Security and Cooperation
in Europe (CSCE) and to expunge
Marxist-Leninist legal theory
National holiday:
Slovak National Uprising, August 29
(1944)
Executive branch:
president, prime minister, Cabinet
Legislative branch:
unicameral National Council (Narodni
Rada)
Judicial branch:
Supreme Court

Political Parties [20]

Chamber of Deputies	% of seats
Civic Democratic Party	38.0
Left Bloc	17.5
Czechoslovak Social Democracy	8.0
Liberal Social Union	8.0
Christian Democratic Union	7.5
Assembly for the Republic	7.0
Civic Democratic Alliance	7.0
Self-Governing Moravia & Silesia	7.0

Government Expenditures [21]

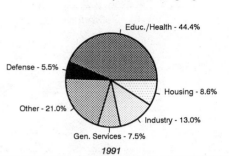

Educ./Health - 44.4%
Defense - 5.5%
Housing - 8.6%
Other - 21.0%
Industry - 13.0%
Gen. Services - 7.5%

1991

Military Expenditures and Arms Transfers [22]

	1985	1986	1987	1988	1989
Military expenditures					
Current dollars (mil.)	7,998	8,460	8,840	9,131	8,361
1989 constant dollars (mil.)	9,105	9,388	9,508	9,505	8,361
Armed forces (000)	210	214	215	211	175
Gross national product (GNP)					
Current dollars (mil.)	97,790	102,400	106,600	117,100	123,200
1989 constant dollars (mil.)	111,300	113,600	114,600	121,900	123,200
Central government expenditures (CGE)					
1989 constant dollars (mil.)	34,390	34,520	35,490	38,260	38,540
People (mil.)	15.5	15.5	15.6	15.6	15.6
Military expenditure as % of GNP	8.2	8.3	8.3	7.8	6.8
Military expenditure as % of CGE	26.5	27.2	26.8	24.8	21.7
Military expenditure per capita	587	604	611	609	534
Armed forces per 1,000 people	13.5	13.8	13.8	13.5	11.2
GNP per capita	7,182	7,313	7,362	7,812	7,876
Arms imports[6]					
Current dollars (mil.)	800	1,100	925	210	460
1989 constant dollars (mil.)	911	1,221	995	219	460
Arms exports[6]					
Current dollars (mil.)	1,600	1,400	1,300	925	875
1989 constant dollars (mil.)	1,821	1,554	1,398	963	875
Total imports[7]					
Current dollars (mil.)	28,200	34,810	37,260	40,160	13,010
1989 constant dollars	32,100	38,630	40,080	41,800	13,010
Total exports[7]					
Current dollars (mil.)	29,370	34,770	36,660	38,450	13,180
1989 constant dollars	33,430	38,580	39,430	40,030	13,180
Arms as percent of total imports[8]	2.8	3.2	2.5	0.5	3.5
Arms as percent of total exports[8]	5.4	4.0	3.5	2.4	6.6

Crime [23]

Crime volume	
Cases known to police	286,724
Attempts (percent)	11,381
Percent cases solved	124,633
Crimes per 100,000 persons	1,911.49
Persons responsible for offenses	
Total number offenders	107,032
Percent female	9,699
Percent juvenile[55]	10,775
Percent foreigners	2,133

Human Rights [24]

	SSTS	FL	FAPRO	PPCG	APROBC	TPW	PCPTW	STPEP	PHRFF	PRW	ASST	AFL
Observes	P	P	P	P	P	P	P	P	P	P	P	
		EAFRD	CPR	ESCR	SR	ACHR	MAAE	PVIAC	PVNAC	EAFDAW	TCIDTP	RC
Observes		P	P	P	P			P	P	P	P	P

P = Party; S = Signatory; see Appendix for meaning of abbreviations.

Labor Force

Total Labor Force [25]

5.389 million

Labor Force by Occupation [26]

Industry	37.9%
Agriculture	8.1
Construction	8.8
Communications and other	45.2

Date of data: 1990

Unemployment Rate [27]

3.1% (1992 est.)

For sources, notes, and explanations, see Annotated Source Appendix, page 1035.

249

Production Sectors

Commercial Energy Production and Consumption

Production [28]	Consumption [29]
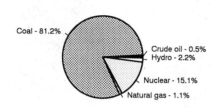	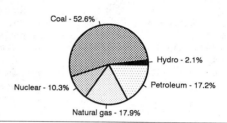
Coal - 81.2% Crude oil - 0.5% Hydro - 2.2% Nuclear - 15.1% Natural gas - 1.1%	Coal - 52.6% Hydro - 2.1% Petroleum - 17.2% Nuclear - 10.3% Natural gas - 17.9%

Data are shown in quadrillion (10^{15}) BTUs and percent for 1991

Crude oil[1]	0.01[b]	Petroleum	0.50[b]	
Natural gas liquids	(s)[b]	Dry natural gas	0.52[b]	
Dry natural gas	0.02[b]	Coal[2]	1.53[b]	
Coal[2]	1.51[b]	Net hydroelectric power[3]	0.06[b]	
Net hydroelectric power[3]	0.04[b]	Net nuclear power[3]	0.30[b]	
Net nuclear power[3]	0.28[b]	Total	2.92[b]	
Total	1.85[b]			

Telecommunications [30]

Transportation [31]

Railroads. 3,669 km total (1990)

Highways. 17,650 km total (1990)

Merchant Marine. the former Czechoslovakia had 22 ships (1,000 GRT or over) totaling 290,185 GRT/437,291 DWT; includes 13 cargo, 9 bulk; may be shared with the Czech Republic

Airports

Total:	34
Usable:	34
With permanent-surface runways:	9
With runways over 3,659 m:	0
With runways 2,440-3,659 m:	1
With runways 1,220-2,439 m:	5

Top Agricultural Products [32]

	1990[8]	1991[8]
Wheat	6,707	6,216
Sugar beets	5,608	6,050
Barley	4,071	3,798
Potatoes	2,534	2,964
Corn	468	867
Rye	736	446

Values shown are 1,000 metric tons.

Top Mining Products [33]

Estimated metric tons unless otherwise specified   M.t.

Gypsum, crude	624,000
Magnesite, crude	328,000
Clay, kaolin	705,000
Zinc, mine output, Zn content of concentrate	533,000
Iron ore, Fe content (000 tons)	460
Limestone (000 tons)	7,442

Tourism [34]

	1987	1988	1989	1990	1991
Visitors[21]	21,756	24,593	29,683	46,607	64,801
Tourists[21]	6,126	14,028	8,036		
Excursionists[21]	15,630	10,565	21,647		
Tourism receipts	493	608	581	470	825
Tourism expenditures	409	399	431	636[22]	393

Tourists are in thousands, money in million U.S. dollars.

Manufacturing Sector

GDP and Manufacturing Summary [35]

	1980	1985	1989	1990	% change 1980-1990	% change 1989-1990
GDP (million 1980 $)	40,327	43,826	62,060	46,773	16.0	-24.6
GDP per capita (1980 $)	2,634	2,827	3,970	2,986	13.4	-24.8
Manufacturing as % of GDP (current prices)	45.6	42.3	47.9[e]	40.1	-12.1	-16.3
Gross output (million $)	41,415	45,108	53,480	44,915	8.5	-16.0
Value added (million $)	17,194	13,083	15,456	12,471	-27.5	-19.3
Value added (million 1980 $)	22,261	24,404	35,035	26,006	16.8	-25.8
Industrial production index	100	121	133	130	30.0	-2.3
Employment (thousands)	2,518	2,588	2,572	2,448	-2.8	-4.8

Note: GDP stands for Gross Domestic Product. 'e' stands for estimated value.

Profitability and Productivity

	1980	1985	1989	1990	% change 1980-1990	% change 1989-1990
Intermediate input (%)	58	71	71	72	24.1	1.4
Wages, salaries, and supplements (%)	15	13	12	13	-13.3	8.3
Gross operating surplus (%)	27	16	17	15	-44.4	-11.8
Gross output per worker ($)	16,448	17,430	20,793	18,348	11.6	-11.8
Value added per worker ($)	6,828	5,055	6,010	5,094	-25.4	-15.2
Average wage (incl. benefits) ($)	2,438	2,264	2,522	2,396	-1.7	-5.0

Profitability is in percent of gross output. Productivity is in U.S. $. 'e' stands for estimated value.

Profitability - 1990

Inputs - 72.0%
Wages - 13.0%
Surplus - 15.0%

The graphic shows percent of gross output.

Value Added in Manufacturing

	1980 $ mil.	1980 %	1985 $ mil.	1985 %	1989 $ mil.	1989 %	1990 $ mil.	1990 %	% change 1980-1990	% change 1989-1990
311 Food products	1,257	7.3	911	7.0	1,126	7.3	916	7.3	-27.1	-18.7
313 Beverages	285	1.7	209	1.6	272	1.8	258	2.1	-9.5	-5.1
314 Tobacco products	33	0.2	23	0.2	27	0.2	24	0.2	-27.3	-11.1
321 Textiles	1,100	6.4	848	6.5	1,046	6.8	790	6.3	-28.2	-24.5
322 Wearing apparel	271	1.6	236	1.8	281	1.8	223	1.8	-17.7	-20.6
323 Leather and fur products	94	0.5	69	0.5	94	0.6	66	0.5	-29.8	-29.8
324 Footwear	299	1.7	244	1.9	306	2.0	256	2.1	-14.4	-16.3
331 Wood and wood products	387	2.3	259	2.0	318	2.1	289	2.3	-25.3	-9.1
332 Furniture and fixtures	210	1.2	162	1.2	183	1.2	154	1.2	-26.7	-15.8
341 Paper and paper products	391	2.3	287	2.2	391	2.5	255	2.0	-34.8	-34.8
342 Printing and publishing	136	0.8	103	0.8	141	0.9	127	1.0	-6.6	-9.9
351 Industrial chemicals	1,262	7.3	862	6.6	875	5.7	698	5.6	-44.7	-20.2
352 Other chemical products	178	1.0	130	1.0	162	1.0	177	1.4	-0.6	9.3
353 Petroleum refineries	497	2.9	390	3.0	429	2.8	316	2.5	-36.4	-26.3
354 Miscellaneous petroleum and coal products	120	0.7	74	0.6	92	0.6	209	1.7	74.2	127.2
355 Rubber products	214	1.2	158	1.2	203	1.3	131	1.1	-38.8	-35.5
356 Plastic products	50	0.3	34	0.3	32	0.2	49	0.4	-2.0	53.1
361 Pottery, china and earthenware	45	0.3	39	0.3	40	0.3	46	0.4	2.2	15.0
362 Glass and glass products	422	2.5	263	2.0	361	2.3	298	2.4	-29.4	-17.5
369 Other non-metal mineral products	773	4.5	488	3.7	574	3.7	411	3.3	-46.8	-28.4
371 Iron and steel	1,753	10.2	1,312	10.0	1,910	12.4	1,271	10.2	-27.5	-33.5
372 Non-ferrous metals	327	1.9	214	1.6	264	1.7	236	1.9	-27.8	-10.6
381 Metal products	792	4.6	590	4.5	710	4.6	602	4.8	-24.0	-15.2
382 Non-electrical machinery	3,452	20.1	2,827	21.6	2,866	18.5	2,597	20.8	-24.8	-9.4
383 Electrical machinery	853	5.0	828	6.3	1,025	6.6	894	7.2	4.8	-12.8
384 Transport equipment	1,677	9.8	1,315	10.1	1,441	9.3	903	7.2	-46.2	-37.3
385 Professional and scientific equipment	94	0.5	67	0.5	94	0.6	84	0.7	-10.6	-10.6
390 Other manufacturing industries	223	1.3	140	1.1	194	1.3	192	1.5	-13.9	-1.0

Note: The industry codes shown are International Standard Industry codes (ISIC). Percentages are percent of total Value Added. 'e' stands for estimated value

Finance, Economics, and Trade

Economic Indicators [36]

In Crowns or U.S. Dollars as indicated[31].

	1989	1990	1991
Real GDP (Bil. Crowns)	736.6	730.5	NA
Real GDP Growth Rate	1.4	-0.3	NA
Money Supply (MI) (Bil. Crowns, est.)	311.1	291.2	346.5
Commercial Interest Rate (Jan) (%)	5.20	6.16	15.76
Savings Rate (%)	3.5	-0.2	7.3
Investment Rate (%)	2.6	5.7	NA
CPI (1989 equals 100)	100.0	110.0	160.7[32]
WPI (1985 equals 100)	99.4	103.8	165.7[33]
External Public Debt (Bil. $US)	7.92	8.1	9.27[34]

Balance of Payments Summary [37]

Exchange Rates [38]

Currency: **koruny.**
Symbol: **Kcs.**

Data are currency units per $1.

December 1992	28.59
1992	28.26
1991	29.53
1990	17.95
1989	15.05
1988	14.36
1987	13.69

Imports and Exports

Top Import Origins [39]

$8.9 billion (f.o.b., 1992).

Origins	%
Slovakia	NA
CIS republics	NA
Germany Austria	NA
Poland	NA
Switzerland	NA
Hungary	NA
UK	NA
Italy	NA

Top Export Destinations [40]

$8.2 billion (f.o.b., 1992).

Destinations	%
Slovakia	NA
Germany	NA
Poland	NA
Austria	NA
Hungary	NA
Italy	NA
France	NA
US	NA
UK	NA
CIS republics	NA

Foreign Aid [41]

	U.S. $	
The former Czechoslovakia was a donor of bilateral aid to non-Communist less developed countries (1954-89)	4.2	billion

Import and Export Commodities [42]

Import Commodities	**Export Commodities**
Machinery & transport equipment	Manufactured goods
Fuels and lubricants	Machinery & transport equip.
Manfactured goods	Chemicals
Raw materials	Fuels
Chemicals	Minerals
Agricultural products	And metals

Denmark

Geography [1]

Total area:
43,070 km2
Land area:
42,370 km2
Comparative area:
Slightly more than twice the size of Massachusetts
Land boundaries:
Total 68 km, Germany 68 km
Coastline:
3,379 km
Climate:
Temperate; humid and overcast; mild, windy winters and cool summers
Terrain:
Low and flat to gently rolling plains
Natural resources:
Petroleum, natural gas, fish, salt, limestone
Land use:
Arable land:
61%
Permanent crops:
0%
Meadows and pastures:
6%
Forest and woodland:
12%
Other:
21%

Demographics [2]

	1960	1970	1980	1990	1991[1]	1994	2000	2010	2020
Population	4,581	4,929	5,123	5,141	5,133	5,188	5,255	5,311	5,307
Population density (persons per sq. mi.)	280	301	313	314	314	NA	315	311	304
Births	NA	NA	NA	NA	60	65	NA	NA	NA
Deaths	NA	NA	NA	NA	59	59	NA	NA	NA
Life expectancy - males	NA	NA	NA	NA	73	73	NA	NA	NA
Life expectancy - females	NA	NA	NA	NA	79	79	NA	NA	NA
Birth rate (per 1,000)	NA	NA	NA	NA	12	12	NA	NA	NA
Death rate (per 1,000)	NA	NA	NA	NA	11	11	NA	NA	NA
Women of reproductive age (15-44 yrs.)	NA	NA	NA	1,310	NA	1,303	1,232	NA	NA
of which are currently married	NA	NA	NA	671	NA	683	669	NA	NA
Fertility rate	NA	NA	NA	1.7	1.57	1.7	1.7	1.7	1.7

Population values are in thousands, life expectancy in years, and other items as indicated.

Health

Health Personnel [3]

Doctors per 1,000 pop., 1988-92	2.56
Nurse-to-doctor ratio, 1988-92	5.6
Hospital beds per 1,000 pop., 1985-90	5.7
Percentage of children immunized (age 1 yr. or less)	
Third dose of DPT, 1990-91	95.0
Measles, 1990-91	86.0

Health Expenditures [4]

Total health expenditure, 1990 (official exchange rate)	
Millions of dollars	8160
Millions of dollars per capita	1588
Health expenditures as a percentage of GDP	
Total	6.3
Public sector	5.3
Private sector	1.0
Development assistance for health	
Total aid flows (millions of dollars)[1]	NA
Aid flows per capita (millions of dollars)	NA
Aid flows as a percentage of total health expenditure	NA

For sources, notes, and explanations, see Annotated Source Appendix, page 1035.

253

Human Factors

Health Care Ratios [5]

Population per physician, 1970	690
Population per physician, 1990	390
Population per nursing person, 1970	NA
Population per nursing person, 1990	NA
Percent of births attended by health staff, 1985	NA

Infants and Malnutrition [6]

Percent of babies with low birth weight, 1985	6
Infant mortality rate per 1,000 live births, 1970	14
Infant mortality rate per 1,000 live births, 1991	8
Years of life lost per 1,000 population, 1990	12
Prevalence of malnutrition (under age 5), 1990	NA

Ethnic Division [7]

Scandinavian	NA
Eskimo	NA
Faroese	NA
German	NA

Religion [8]

Evangelical Lutheran	91%
Other Protestant and Roman Catholic	2%
Other	7%

Major Languages [9]

Danish	NA
Faroese	NA
Greenlandic (an Eskimo dialect)	NA
German (small minority)	NA

Education

Public Education Expenditures [10]

Million Krone	1980	1985	1986	1987	1988	1989
Total education expenditure	25,020	42,672	45,108	50,042	54,305	55,448
as percent of GNP	6.9	7.2	7.1	7.4	7.7	7.5
as percent of total govt. expend.	9.5	NA	12.7	13.7	13.1	13.0
Current education expenditure	22,188	NA	43,054	47,597	51,299	52,270
as percent of GNP	6.1	NA	6.7	7.1	7.3	7.1
as percent of current govt. expend.	9.0	NA	12.9	NA	NA	12.9
Capital expenditure	2,832	NA	2,054	2,445	3,006	3,178

Educational Attainment [11]

Literacy Rate [12]

Libraries [13]

	Admin. Units	Svc. Pts.	Vols. (000)	Shelving (meters)	Vols. Added	Reg. Users
National (1989)	1	4	3,348	89,238	72,426	NA
Non-specialized (1989)	5	14	4,903	123,071	124,199	NA
Public (1990)	250	NA	34,285	NA	NA	NA
Higher ed. (1990)	19	54	11,247	268,996	288,109	NA
School (1990)	275	1,773	32,235	NA	1.2mil	NA

Daily Newspapers [14]

	1975	1980	1985	1990
Number of papers	49	48	47	47
Circ. (000)	1,723	1,874	1,855	1,810

Cinema [15]

Data for 1991.

Cinema seats per 1,000	10.7
Annual attendance per person	1.8
Gross box office receipts (mil. Krone)	271

Science and Technology

Scientific/Technical Forces [16]

Potential scientists/engineers	116,882
Number female	33,798
Potential technicians	320,245
Number female	204,355
Total	437,127

R&D Expenditures [17]

	Kroner (000) 1989
Total expenditure	11,892,000
Capital expenditure	1,375,000
Current expenditure	10,517,000
Percent current	88.4

U.S. Patents Issued [18]

Values show patents issued to citizens of the country by the U.S. Patents Office.

	1990	1991	1992
Number of patents	204	285	275

Government and Law

Organization of Government [19]

Long-form name:
Kingdom of Denmark
Type:
constitutional monarchy
Independence:
1849 (became a constitutional monarchy)
Constitution:
5 June 1953
Legal system:
civil law system; judicial review of legislative acts; accepts compulsory ICJ jurisdiction, with reservations
National holiday:
Birthday of the Queen, 16 April (1940)
Executive branch:
monarch, heir apparent, prime minister, Cabinet
Legislative branch:
unicameral parliament (Folketing)
Judicial branch:
Supreme Court

Political Parties [20]

Parliament	% of votes
Social Democratic Party	37.4
Conservative Party	16.0
Liberal	15.8
Socialist People's Party	8.3
Progress Party	6.4
Center Democratic Party	5.1
Radical Liberal Party	3.5
Christian People's Party	2.3
Other	5.2

Government Budget [21]

For 1992.
Revenues	48.8
Expenditures	55.3
Capital expenditures	NA

Crime [23]

Crime volume	
Cases known to police	527,422
Attempts (percent)	NA
Percent cases solved	21.38
Crimes per 100,000 persons	10,270.30
Persons responsible for offenses	
Total number offenders	NA
Percent female	NA
Percent juvenile	NA
Percent foreigners	NA

Military Expenditures and Arms Transfers [22]

	1985	1986	1987	1988	1989
Military expenditures					
Current dollars (mil.)	1,909	1,870	2,019	2,135	2,184
1989 constant dollars (mil.)	2,173	2,076	2,171	2,222	2,184
Armed forces (000)	29	28	28	30	31
Gross national product (GNP)					
Current dollars (mil.)	84,250	89,640	92,050	95,050	100,400
1989 constant dollars (mil.)	95,910	99,470	99,010	98,940	100,400
Central government expenditures (CGE)					
1989 constant dollars (mil.)	41,570	39,780	39,700	41,380	42,090
People (mil.)	5.1	5.1	5.1	5.1	5.1
Military expenditure as % of GNP	2.3	2.1	2.2	2.2	2.2
Military expenditure as % of CGE	5.2	5.2	5.5	5.4	5.2
Military expenditure per capita	425	405	423	433	426
Armed forces per 1,000 people	5.7	5.5	5.5	5.8	6.0
GNP per capita	18,750	19,430	19,310	19,290	19,580
Arms imports[6]					
Current dollars (mil.)	100	50	110	200	110
1989 constant dollars (mil.)	114	55	118	208	110
Arms exports[6]					
Current dollars (mil.)	20	20	60	40	20
1989 constant dollars (mil.)	23	22	65	42	20
Total imports[7]					
Current dollars (mil.)	18,240	22,880	25,500	25,920	26,690
1989 constant dollars	20,770	25,390	27,430	26,980	26,690
Total exports[7]					
Current dollars (mil.)	17,090	21,290	25,670	27,650	28,110
1989 constant dollars	19,460	23,620	27,610	28,790	28,110
Arms as percent of total imports[8]	0.5	0.2	0.4	0.8	0.4
Arms as percent of total exports[8]	0.1	0.1	0.2	0.1	0.1

Human Rights [24]

	SSTS	FL	FAPRO	PPCG	APROBC	TPW	PCPTW	STPEP	PHRFF	PRW	ASST	AFL
Observes	P	P	P	P	P	P	P	S	P	P	P	P
	EAFRD	CPR	ESCR	SR	ACHR	MAAE	PVIAC	PVNAC	EAFDAW	TCIDTP	RC	
Observes		P	P	P	P			P	P	P	P	P

P = Party; S = Signatory; see Appendix for meaning of abbreviations.

Labor Force

Total Labor Force [25]

2,553,900

Labor Force by Occupation [26]

Private services	37.1%
Government services	30.4
Manufacturing and mining	20.0
Construction	6.3
Agriculture, forestry, and fishing	5.6
Electricity/gas/water	0.6
Date of data: 1991	

Unemployment Rate [27]

11.4% (1992)

For sources, notes, and explanations, see Annotated Source Appendix, page 1035.

255

Production Sectors

Commercial Energy Production and Consumption

Production [28]

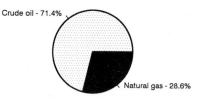

Crude oil - 71.4%

Natural gas - 28.6%

Consumption [29]

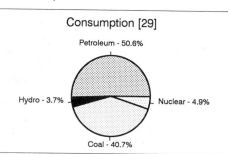

Petroleum - 50.6%

Hydro - 3.7%

Nuclear - 4.9%

Coal - 40.7%

Data are shown in quadrillion (10^{15}) BTUs and percent for 1991

Crude oil[1]	0.30	Petroleum	0.41
Natural gas liquids	0.00	Dry natural gas	0.00
Dry natural gas	0.12	Coal[2]	0.33
Coal[2]	0.00	Net hydroelectric power[3]	0.03
Net hydroelectric power[3]	(s)	Net nuclear power[3]	0.04
Net nuclear power[3]	0.00	Total	0.81
Total	0.42		

Telecommunications [30]

- Excellent telephone, telegraph, and broadcast services
- 4,509,000 telephones
- Buried and submarine cables and microwave radio relay support trunk network
- Broadcast stations - 3 AM, 2 FM, 50 TV
- 19 submarine coaxial cables
- 7 earth stations operating in INTELSAT, EUTELSAT, and INMARSAT

Transportation [31]

Railroads. 2,770 km; Danish State Railways (DSB) operate 2,120 km (1,999 km rail line and 121 km rail ferry services); 188 km electrified, 730 km double tracked; 650 km of standard-gauge lines are privately owned and operated

Highways. 66,482 km total; 64,551 km concrete, bitumen, or stone block; 1,931 km gravel, crushed stone, improved earth

Merchant Marine. 328 ships (1,000 GRT or over) totaling 5,043,277 GRT/7,230,634 DWT; includes 13 short-sea passenger, 102 cargo, 19 refrigerated cargo, 47 container, 37 roll-on/roll-off, 1 railcar carrier, 33 oil tanker, 18 chemical tanker, 36 liquefied gas, 4 livestock carrier, 17 bulk, 1 combination bulk; note - Denmark has created its own internal register, called the Danish International Ship register (DIS); DIS ships do not have to meet Danish manning regulations, and they amount to a flag of convenience within the Danish register; by the end of 1990, 258 of the Danish-flag ships belonged to the DIS

Airports

Total:	118
Usable:	109
With permanent-surface runways:	28
With runways over 3,659 m:	0
With runways 2,440-3,659 m:	9
With runways 1,220-2,439 m:	7

Top Agricultural Products [32]

	1990	1991
Pasture and grass/feed	20,411	18,926
Root crops for feed	6,827	5,790
Barley	4,984	4,978
Wheat	3,953	3,629
Sugar beets	3,533	3,039
Potatoes	1,483	1,557

Values shown are 1,000 metric tons.

Top Mining Products [33]

Estimated metric tons unless otherwise specified	M.t.
Cement, hydraulic (000 tons)	2,016[1]
Chalk (000 tons)	2,000
Moler	84,000
Clay, kaolin	17,000
Sand and gravel (000 cubic meters)	31,000
Limestone, agricultural (000 tons)	1,450

Tourism [34]

Manufacturing Sector

GDP and Manufacturing Summary [35]

	1980	1985	1989	1990	% change 1980-1990	% change 1989-1990
GDP (million 1980 $)	66,321	75,577	78,631	81,529	22.9	3.7
GDP per capita (1980 $)	12,943	14,752	15,301	15,853	22.5	3.6
Manufacturing as % of GDP (current prices)	19.7	19.6	15.8	18.6	-5.6	17.7
Gross output (million $)	31,526	27,652	46,393	56,712	79.9	22.2
Value added (million $)	12,774	11,184	20,189	24,593	92.5	21.8
Value added (million 1980 $)	11,411	12,938	13,060	12,746	11.7	-2.4
Industrial production index	100	121	130	131	31.0	0.8
Employment (thousands)	381	405	390	391	2.6	0.3

Note: GDP stands for Gross Domestic Product. 'e' stands for estimated value.

Profitability and Productivity

	1980	1985	1989	1990	% change 1980-1990	% change 1989-1990
Intermediate input (%)	59	60	56	57	-3.4	1.8
Wages, salaries, and supplements (%)	24	22	22	23[e]	-4.2	4.5
Gross operating surplus (%)	17	18	21	21[e]	23.5	0.0
Gross output per worker ($)	82,313	68,092	118,806	144,739	75.8	21.8
Value added per worker ($)	33,351	27,541	51,701	62,773	88.2	21.4
Average wage (incl. benefits) ($)	19,697	15,021	26,400	33,096[e]	68.0	25.4

Profitability is in percent of gross output. Productivity is in U.S. $. 'e' stands for estimated value.

Profitability - 1990

Surplus - 20.8%
Inputs - 56.4%
Wages - 22.8%

The graphic shows percent of gross output.

Value Added in Manufacturing

	1980 $ mil.	1980 %	1985 $ mil.	1985 %	1989 $ mil.	1989 %	1990 $ mil.	1990 %	% change 1980-1990	% change 1989-1990
311 Food products	2,344	18.3	2,022	18.1	3,609	17.9	4,276	17.4	82.4	18.5
313 Beverages	490	3.8	386	3.5	625	3.1	766	3.1	56.3	22.6
314 Tobacco products	109	0.9	96	0.9	225	1.1	205	0.8	88.1	-8.9
321 Textiles	423	3.3	375	3.4	622	3.1	706	2.9	66.9	13.5
322 Wearing apparel	231	1.8	199	1.8	247	1.2	293	1.2	26.8	18.6
323 Leather and fur products	30	0.2	20	0.2	25	0.1	15	0.1	-50.0	-40.0
324 Footwear	62	0.5	43	0.4	54	0.3	74	0.3	19.4	37.0
331 Wood and wood products	285	2.2	219	2.0	447	2.2	549	2.2	92.6	22.8
332 Furniture and fixtures	330	2.6	371	3.3	618	3.1	781	3.2	136.7	26.4
341 Paper and paper products	315	2.5	275	2.5	557	2.8	659	2.7	109.2	18.3
342 Printing and publishing	941	7.4	752	6.7	1,454	7.2	1,773	7.2	88.4	21.9
351 Industrial chemicals	551	4.3	498	4.5	924	4.6	1,142	4.6	107.3	23.6
352 Other chemical products	586	4.6	618	5.5	1,259	6.2	1,586	6.4	170.6	26.0
353 Petroleum refineries	65	0.5	55	0.5	170	0.8	127	0.5	95.4	-25.3
354 Miscellaneous petroleum and coal products	99	0.8	63	0.6	184	0.9	223	0.9	125.3	21.2
355 Rubber products	79	0.6	59	0.5	109	0.5	129	0.5	63.3	18.3
356 Plastic products	267	2.1	297	2.7	546	2.7	714	2.9	167.4	30.8
361 Pottery, china and earthenware	87	0.7	41	0.4	56	0.3	74	0.3	-14.9	32.1
362 Glass and glass products	98	0.8	60	0.5	96	0.5	118	0.5	20.4	22.9
369 Other non-metal mineral products	627	4.9	478	4.3	892	4.4	1,040	4.2	65.9	16.6
371 Iron and steel	175	1.4	124	1.1	242	1.2	295	1.2	68.6	21.9
372 Non-ferrous metals	71	0.6	45	0.4	65	0.3	76	0.3	7.0	16.9
381 Metal products	912	7.1	882	7.9	1,636	8.1	2,030	8.3	122.6	24.1
382 Non-electrical machinery	1,718	13.4	1,475	13.2	2,572	12.7	3,242	13.2	88.7	26.0
383 Electrical machinery	712	5.6	631	5.6	1,082	5.4	1,337	5.4	87.8	23.6
384 Transport equipment	663	5.2	589	5.3	861	4.3	1,162	4.7	75.3	35.0
385 Professional and scientific equipment	284	2.2	304	2.7	568	2.8	643	2.6	126.4	13.2
390 Other manufacturing industries	219	1.7	211	1.9	444	2.2	559	2.3	155.3	25.9

Note: The industry codes shown are International Standard Industry codes (ISIC). Percentages are percent of total Value Added. 'e' stands for estimated value

Finance, Economics, and Trade

Economic Indicators [36]

Millions of Danish Kroner (DKK) unless otherwise stated.

	1989	1990	1991[P]
Real GDP (1989 prices)	776,015	792,045	805,500
Real GDP Growth (%)	1.2	2.1	1.7
Real GDP, Per Capita (DKK)	151,223	154,097	156,286
Money Supply M1 (mid-year)[35]	260,141	276,632	254,418
Commercial Interest Rate (%)	13.4	14.2	13.2
Personal Savings Rate[36]	0.6	2.6	2.0
Investment Rate (%/GDP)	18.2	17.7	17.3
CPI (1980:100)	172.9	177.4	181.8
WPI (1980:100)	151.1	153.0	155.0
Govt. External Debt[37]	116,031	119,101	91,100

Balance of Payments Summary [37]

Values in millions of dollars.

	1987	1988	1989	1990	1991
Exports of goods (f.o.b.)	25,695	27,537	28,728	36,072	36,877
Imports of goods (f.o.b.)	-24,900	-25,654	-26,304	-31,197	-31,996
Trade balance	795	1,883	2,424	4,875	4,881
Services - debits	-14,051	-16,303	-17,688	-22,544	-24,679
Services - credits	10,475	13,299	14,290	19,075	22,497
Private transfers (net)	-56	-88	80	-46	-150
Government transfers (net)	-164	-131	-223	-10	-346
Long term capital (net)	8,153	2,469	-3,566	6,256	-1,718
Short term capital (net)	-793	805	1,207	-1,575	-1,364
Errors and omissions	85	-619	-347	-2,658	-2,424
Overall balance	4,444	1,315	-3,823	3,373	-3,303

Exchange Rates [38]

Currency: **Danish kroner.**
Symbol: **DKr.**

Data are currency units per $1.

January 1993	6.236
1992	6.036
1991	6.396
1990	6.189
1989	7.310
1988	6.732

Imports and Exports

Top Import Origins [39]

$30.3 billion (c.i.f., 1992). Data are for 1992.

Origins	%
EC	53.4
UK	8.2
France	5.6
Sweden	10.8
Norway	5.4
US	5.7
Japan	4.1

Top Export Destinations [40]

$37.3 billion (f.o.b., 1992) Data are for 1992.

Destinations	%
EC	54.3
UK	10.1
France	5.7
Sweden	10.5
Norway	5.8
US	4.9
Japan	3.6

Foreign Aid [41]

	U.S. $	
Donor - ODA and OOF commitments (1970-89)	5.9	billion

Import and Export Commodities [42]

Import Commodities

Petroleum
Machinery & equipment
Chemicals
Grain and foodstuffs
Textiles
Paper

Export Commodities

Meat and meat products
Dairy products
Transport equip. (shipbuilding)
Fish
Chemicals
Industrial machinery

Djibouti

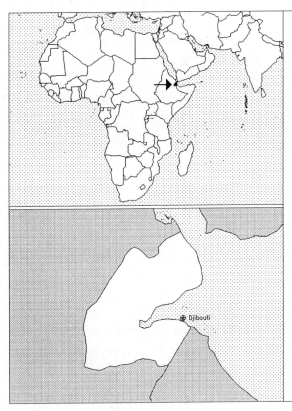

Geography [1]

Total area:
22,000 km2
Land area:
21,980 km2
Comparative area:
Slightly larger than Massachusetts
Land boundaries:
Total 508 km, Erithea 113 km, Ethiopia 337 km, Somalia 58 km
Coastline:
314 km
Climate:
Desert; torrid, dry
Terrain:
Coastal plain and plateau separated by central mountains
Natural resources:
Geothermal areas
Land use:
Arable land:
0%
Permanent crops:
0%
Meadows and pastures:
9%
Forest and woodland:
0%
Other:
91%

Demographics [2]

	1960	1970	1980	1990	1991[1]	1994	2000	2010	2020
Population	78	158	279	370	346	413	454	588	751
Population density (persons per sq. mi.)	9	19	33	40	41	NA	52	67	86
Births	NA	NA	NA	NA	15	18	NA	NA	NA
Deaths	NA	NA	NA	NA	6	7	NA	NA	NA
Life expectancy - males	NA	NA	NA	NA	46	47	NA	NA	NA
Life expectancy - females	NA	NA	NA	NA	50	51	NA	NA	NA
Birth rate (per 1,000)	NA	NA	NA	NA	43	43	NA	NA	NA
Death rate (per 1,000)	NA	NA	NA	NA	16	16	NA	NA	NA
Women of reproductive age (15-44 yrs.)	NA	NA	NA	NA	NA	NA	NA	NA	NA
of which are currently married	NA	NA	NA	NA	NA	NA	NA	NA	NA
Fertility rate	NA	NA	NA	6.4	6.38	6.2	5.8	5.0	4.1

Population values are in thousands, life expectancy in years, and other items as indicated.

Health

Health Personnel [3]

Health Expenditures [4]

For sources, notes, and explanations, see Annotated Source Appendix, page 1035.

259

Human Factors

Health Care Ratios [5]	Infants and Malnutrition [6]

Ethnic Division [7]

Other includes French, Arab, Ethiopian, and Italian.

Somali	60%
Afar	35%
Other	5%

Religion [8]

Muslim	94%
Christian	6%

Major Languages [9]

French (official)	NA
Arabic (official)	NA
Somali	NA
Afar	NA

Education

Public Education Expenditures [10]

Million Francs	1985[1]	1986	1987	1988	1989	1990
Total education expenditure	1,690	NA	NA	2,532	2,596	2,614
as percent of GNP	2.7	NA	NA	3.5	3.4	3.3
as percent of total govt. expend.	7.5	NA	NA	10.5	10.7	10.5
Current education expenditure	1,690	NA	NA	2,532	2,596	2,614
as percent of GNP	2.7	NA	NA	3.5	3.4	3.3
as percent of current govt. expend.	NA	NA	NA	11.0	10.9	10.5
Capital expenditure	-	NA	NA	-	-	-

Educational Attainment [11]

Literacy Rate [12]

Libraries [13]

Daily Newspapers [14]

Cinema [15]

Science and Technology

Scientific/Technical Forces [16]	R&D Expenditures [17]	U.S. Patents Issued [18]

Government and Law

Organization of Government [19]

Long-form name:
Republic of Djibouti
Type:
republic
Independence:
27 June 1977 (from France)
Constitution:
multiparty constitution approved in referendum September 1992
Legal system:
based on French civil law system, traditional practices, and Islamic law
National holiday:
Independence Day, 27 June (1977)
Executive branch:
president, prime minister, Council of Ministers
Legislative branch:
unicameral Chamber of Deputies (Chambre des Deputes)
Judicial branch:
Supreme Court (Cour Supreme)

Political Parties [20]

National Assembly	% of seats
People's Progress Assembly	100.0

Government Budget [21]

For 1991 est.

Revenues	170
Expenditures	203
Capital expenditures	70

Defense Summary [22]

Branches: Djibouti National Army (including Navy and Air Force), National Security Force (Force Nationale de Securite), National Police Force

Manpower Availability: Males age 15-49 97,943; fit for military service 57,187 (1993 est.)

Defense Expenditures: Exchange rate conversion - $26 million (1989)

Crime [23]

Crime volume	
Cases known to police	2,436
Attempts (percent)	41
Percent cases solved	72
Crimes per 100,000 persons	487.2
Persons responsible for offenses	
Total number offenders	2,285
Percent female	11
Percent juvenile[19]	13
Percent foreigners	60

Human Rights [24]

	SSTS	FL	FAPRO	PPCG	APROBC	TPW	PCPTW	STPEP	PHRFF	PRW	ASST	AFL
Observes		P	P		P	P	P	P			P	P
	EAFRD	CPR	ESCR	SR	ACHR	MAAE	PVIAC	PVNAC	EAFDAW	TCIDTP	RC	
Observes				P			P	P			P	

P = Party; S = Signatory; see Appendix for meaning of abbreviations.

Labor Force

Total Labor Force [25]

Labor Force by Occupation [26]

Unemployment Rate [27]

over 30% (1989)

For sources, notes, and explanations, see Annotated Source Appendix, page 1035.

261

Production Sectors

Energy Resource Summary [28]

Energy Resources: Geothermal areas. **Electricity**: 115,000 kW capacity; 200 million kWh produced, 580 kWh per capita (1991).

Telecommunications [30]

- Telephone facilities in the city of Djibouti are adequate as are the microwave radio relay connections to outlying areas of the country
- International connections via submarine cable to Saudi Arabia and by satellite to other countries
- One ground station each for Indian Ocean INTELSAT and ARABSAT
- Broadcast stations - 2 AM, 2 FM, 1 TV

Transportation [31]

Railroads. the Ethiopian-Djibouti railroad extends for 97 km through Djibouti

Highways. 2,900 km total; 280 km paved; 2,620 km improved or unimproved earth (1982)

Merchant Marine. 1 cargo ship (1,000 GRT or over) totaling 1,369 GRT/3,030 DWT

Airports

Total:	13
Usable:	11
With permanent-surface runways:	2
With runways over 3,659 m:	0
With runways 2,440-3,659 m:	2
With runways 1,220-2,439 m:	5

Top Agricultural Products [32]

Agriculture accounts for only 3% of GDP; scanty rainfall limits crop production to mostly fruit and vegetables; half of population pastoral nomads herding goats, sheep, and camels; imports bulk of food needs.

Top Mining Products [33]

Detailed information is not available. A summary of mineral resources available follows. **Mineral Resources**: None.

Tourism [34]

For sources, notes, and explanations, see Annotated Source Appendix, page 1035.

Finance, Economics, and Trade

GDP and Manufacturing Summary [35]

	1980	1985	1990	1991	1992
Gross Domestic Product					
Millions of 1980 dollars	339	357	397	404	415[e]
Growth rate in percent	4.72	0.85	2.00	1.60	2.91[e]
Manufacturing Value Added					
Millions of 1980 dollars	34	36	43	45	47[e]
Growth rate in percent	2.98	0.49	5.10	4.22	3.98[e]
Manufacturing share in percent of current prices	9.7	9.4[e]	NA	NA	NA

Economic Indicators [36]

National product: GDP—exchange rate conversion— $358 million (1990 est.). **National product real growth rate**: 1.2% (1990 est.). **National product per capita**: $1,030 (1990 est.). **Inflation rate (consumer prices)**: 7.7% (1991 est.). **External debt**: $355 million (December 1990).

Balance of Payments Summary [37]

Exchange Rates [38]

Currency: **Djiboutian francs.**
Symbol: **DF.**

Data are currency units per $1.

Fixed rate since 1973   177.721

Imports and Exports

Top Import Origins [39]

$360 million (f.o.b., 1991 est.).

Origins	%
Western Europe	54
Middle East	20
Asia	19

Top Export Destinations [40]

$186 million (f.o.b., 1991 est.).

Destinations	%
Africa	50
Middle East	40
Western Europe	9

Foreign Aid [41]

	U.S. $	
US commitments, including Ex-Im (FY78-89)	39	million
Western (non-US) countries, ODA and OOF bilateral commitments (1970-89)	1.1	billion
OPEC bilateral aid (1979-89)	149	million
Communist countries (1970-89)	35	million

Import and Export Commodities [42]

Import Commodities	**Export Commodities**
Foods	Hides and skins
Beverages	Coffee (in transit)
Transport equipment	
Chemicals	
Petroleum products	

Dominica

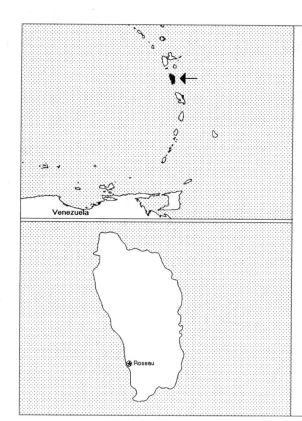

Geography [1]

Total area:
750 km2
Land area:
750 km2
Comparative area:
Slightly more than four times the size of Washington, DC
Land boundaries:
0 km
Coastline:
148 km
Climate:
Tropical; moderated by northeast trade winds; heavy rainfall
Terrain:
Rugged mountains of volcanic origin
Natural resources:
Timber
Land use:
Arable land:
9%
Permanent crops:
13%
Meadows and pastures:
3%
Forest and woodland:
41%
Other:
34%

Demographics [2]

	1960	1970	1980	1990	1991[1]	1994	2000	2010	2020
Population	60	71	75	83	86	88	95	107	118
Population density (persons per sq. mi.)	208	244	259	293	298	NA	348	406	463
Births	NA	NA	NA	NA	2	2	NA	NA	NA
Deaths	NA	NA	NA	NA	(Z)	(Z)	NA	NA	NA
Life expectancy - males	NA	NA	NA	NA	73	74	NA	NA	NA
Life expectancy - females	NA	NA	NA	NA	79	80	NA	NA	NA
Birth rate (per 1,000)	NA	NA	NA	NA	26	20	NA	NA	NA
Death rate (per 1,000)	NA	NA	NA	NA	5	5	NA	NA	NA
Women of reproductive age (15-44 yrs.)	NA	NA	NA	23	NA	25	28	NA	NA
of which are currently married	NA	NA	NA	11	NA	13	16	NA	NA
Fertility rate	NA	NA	NA	2.1	2.59	2.0	1.9	1.8	1.8

Population values are in thousands, life expectancy in years, and other items as indicated.

Health

Health Personnel [3]

Health Expenditures [4]

For sources, notes, and explanations, see Annotated Source Appendix, page 1035.

Human Factors

Health Care Ratios [5]	Infants and Malnutrition [6]

Ethnic Division [7]

Black	NA
Carib Indians	NA

Religion [8]

Roman Catholic	77%
Protestant	15%
Methodist	5%
Pentecostal	3%
Seventh-Day Adventist	3%
Baptist	2%
Other Protestant	2%
None, unkown, other	8%

Major Languages [9]

English (official), French patois.

Education

Public Education Expenditures [10]

Educational Attainment [11]

Age group	25+
Total population	27,508
Highest level attained (%)	
No schooling	6.6
First level	
Incompleted	80.5
Completed	NA
Entered second level	
S-1	11.1
S-2	NA
Post secondary	1.7

Literacy Rate [12]

	1970[b]	1980[b]	1990[b]
Illiterate population +15 years	2,083	NA	NA
Illiteracy rate - total pop. (%)[3]	5.9	NA	NA
Illiteracy rate - males (%)	6.0	NA	NA
Illiteracy rate - females (%)	5.8	NA	NA

Libraries [13]

Daily Newspapers [14]

Cinema [15]

Science and Technology

Scientific/Technical Forces [16]

R&D Expenditures [17]

U.S. Patents Issued [18]

Values show patents issued to citizens of the country by the U.S. Patents Office.

	1990	1991	1992
Number of patents	1	2	0

For sources, notes, and explanations, see Annotated Source Appendix, page 1035.

Government and Law

Organization of Government [19]

Long-form name:
Commonwealth of Dominica
Type:
parliamentary democracy
Independence:
3 November 1978 (from UK)
Constitution:
3 November 1978
Legal system:
based on English common law
National holiday:
Independence Day, 3 November (1978)
Executive branch:
president, prime minister, Cabinet
Legislative branch:
unicameral House of Assembly
Judicial branch:
Eastern Caribbean Supreme Court

Political Parties [20]

House of Assembly	% of seats
Dominica Freedom Party (DFP)	36.7
United Workers Party (UWP)	20.0
Dominica Labor Party (DLP)	13.3

Government Budget [21]

For FY91 est.

Revenues	70
Expenditures	84
Capital expenditures	26

Defense Summary [22]

Branches: Commonwealth of Dominica Police Force

Manpower Availability: No information available.

Defense Expenditures: No information available.

Crime [23]

Human Rights [24]

	SSTS	FL	FAPRO	PPCG	APROBC	TPW	PCPTW	STPEP	PHRFF	PRW	ASST	AFL
Observes	1	P	P		P	P	P			1	1	P
	EAFRD	CPR	ESCR	SR	ACHR	MAAE	PVIAC	PVNAC	EAFDAW	TCIDTP	RC
Observes						P			P	P	P

P = Party; S = Signatory; see Appendix for meaning of abbreviations.

Labor Force

Total Labor Force [25]

25,000

Labor Force by Occupation [26]

Agriculture	40%
Industry and commerce	32
Services	28

Date of data: 1984

Unemployment Rate [27]

15% (1991)

For sources, notes, and explanations, see Annotated Source Appendix, page 1035.

Production Sectors

Energy Resource Summary [28]

Energy Resources: None. **Electricity**: 7,000 kW capacity; 16 million kWh produced, 185 kWh per capita (1992).

Telecommunications [30]

- 4,600 telephones in fully automatic network
- VHF and UHF link to Saint Lucia
- New SHF links to Martinique and Guadeloupe
- Broadcast stations - 3 AM, 2 FM, 1 cable TV

Transportation [31]

Highways. 750 km total; 370 km paved, 380 km gravel and earth

Airports

Total:	2
Usable:	2
With permanent-surface runways:	2
With runways over 3,659 m:	0
With runways 2,440-3,659 m:	0
With runways 1,220-2,439 m:	1

Top Agricultural Products [32]

Agriculture accounts for 26% of GDP; principal crops—bananas, citrus, mangoes, root crops, coconuts; bananas provide the bulk of export earnings; forestry and fisheries potential not exploited.

Top Mining Products [33]

Metric tons unless otherwise specified	M.t.
Clay	NA
Limestone	NA
Pumice	NA
Volcanic ash	NA
Sand and gravel	NA

Tourism [34]

	1987	1988	1989	1990	1991
Visitors	40	45	46	59	120
Tourists	27	34	37	45	46
Excursionists	1	2	2	7	9
Cruise passengers	12	9	7	7	65
Tourism receipts	13	14	19	25	28
Tourism expenditures	2	2	4	4	
Fare expenditures			4	3	

Tourists are in thousands, money in million U.S. dollars.

For sources, notes, and explanations, see Annotated Source Appendix, page 1035.

267

Finance, Economics, and Trade

Industrial Summary [35]

Industrial Production: Growth rate 4.5% in manufacturing (1988 est.); accounts for 18% of GDP. **Industries**: Soap, coconut oil, tourism, copra, furniture, cement blocks, shoes.

Economic Indicators [36]

National product: GDP—purchasing power equivalent—$174 million (1991 est.). **National product real growth rate**: 2.1% (1991 est.). **National product per capita**: $2,100 (1991 est.). **Inflation rate (consumer prices)**: 4.5% (1991). **External debt**: $87 million (1991).

Balance of Payments Summary [37]

Values in millions of dollars.

	1986	1987	1988	1989	1990
Exports of goods (f.o.b.)	44.6	49.3	57.0	46.3	59.9
Imports of goods (f.o.b.)	-49.2	-58.7	-77.0	-94.2	-103.9
Trade balance	-4.4	-9.4	-20.0	-47.9	-44.1
Services - debits	-19.7	-22.0	-24.3	-33.9	-40.9
Services - credits	15.3	18.1	23.8	26.1	37.4
Private transfers (net)	6.7	7.6	8.4	11.1	13.0
Government transfers (net)	9.5	10.6	9.5	7.7	8.8
Long term capital (net)	8.9	17.4	15.3	26.5	22.1
Short term capital (net)	-1.5	-12.0	-9.8	9.8	11.2
Errors and omissions	-8.6	-1.7	-3.9	0.7	-2.8
Overall balance	6.2	8.6	-1.0	0.1	4.7

Exchange Rates [38]

Currency: **East Caribbean dollars.**
Symbol: **EC$.**

Data are currency units per $1.

Fixed rate since 1976	2.70

Imports and Exports

Top Import Origins [39]

$110.0 million (c.i.f., 1991).

Origins	%
US	27
CARICOM	NA
UK	NA
Canada	NA

Top Export Destinations [40]

$66.0 million (c.i.f., 1991).

Destinations	%
UK	50
CARICOM countries	NA
US	NA
Italy	NA

Foreign Aid [41]

	U.S. $	
Western (non-US) countries, ODA and OOF bilateral commitments (1970-89)	120	million

Import and Export Commodities [42]

Import Commodities	Export Commodities
Manufactured goods	Bananas
Machinery & equipment	Soap
Food	Bay oil
Chemicals	Vegetables
	Grapefruit
	Oranges

For sources, notes, and explanations, see Annotated Source Appendix, page 1035.

Dominican Republic

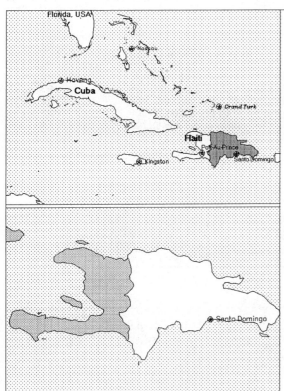

Geography [1]

Total area:
48,730 km2
Land area:
48,380 km2
Comparative area:
Slightly more than twice the size of New Hampshire
Land boundaries:
Total 275 km, Haiti 275 km
Coastline:
1,288 km
Climate:
Tropical maritime; little seasonal temperature variation
Terrain:
Rugged highlands and mountains with fertile valleys interspersed
Natural resources:
Nickel, bauxite, gold, silver
Land use:
Arable land:
23%
Permanent crops:
7%
Meadows and pastures:
43%
Forest and woodland:
13%
Other:
14%

Demographics [2]

	1960	1970	1980	1990	1991[1]	1994	2000	2010	2020
Population	3,159	4,373	5,847	7,249	7,385	7,826	8,644	9,931	11,153
Population density (persons per sq. mi.)	169	234	313	388	395	NA	464	540	612
Births	NA	NA	NA	NA	201	195	NA	NA	NA
Deaths	NA	NA	NA	NA	50	49	NA	NA	NA
Life expectancy - males	NA	NA	NA	NA	65	66	NA	NA	NA
Life expectancy - females	NA	NA	NA	NA	69	71	NA	NA	NA
Birth rate (per 1,000)	NA	NA	NA	NA	27	25	NA	NA	NA
Death rate (per 1,000)	NA	NA	NA	NA	7	6	NA	NA	NA
Women of reproductive age (15-44 yrs.)	NA	NA	NA	1,841	NA	2,037	2,331	NA	NA
of which are currently married	NA	NA	NA	1,186	NA	1,342	1,570	NA	NA
Fertility rate	NA	NA	NA	3.2	3.12	2.8	2.4	2.1	2.0

Population values are in thousands, life expectancy in years, and other items as indicated.

Health

Health Personnel [3]

Doctors per 1,000 pop., 1988-92	1.08
Nurse-to-doctor ratio, 1988-92	0.7
Hospital beds per 1,000 pop., 1985-90	2.0
Percentage of children immunized (age 1 yr. or less)	
Third dose of DPT, 1990-91	47.0
Measles, 1990-91	69.0

Health Expenditures [4]

Total health expenditure, 1990 (official exchange rate)	
Millions of dollars	263
Millions of dollars per capita	37
Health expenditures as a percentage of GDP	
Total	3.7
Public sector	2.1
Private sector	1.6
Development assistance for health	
Total aid flows (millions of dollars)[1]	11
Aid flows per capita (millions of dollars)	1.5
Aid flows as a percentage of total health expenditure	4.1

For sources, notes, and explanations, see Annotated Source Appendix, page 1035.

269

Human Factors

Health Care Ratios [5]

Population per physician, 1970	NA
Population per physician, 1990	NA
Population per nursing person, 1970	1400
Population per nursing person, 1990	NA
Percent of births attended by health staff, 1985	57

Infants and Malnutrition [6]

Percent of babies with low birth weight, 1985	16
Infant mortality rate per 1,000 live births, 1970	90
Infant mortality rate per 1,000 live births, 1991	54
Years of life lost per 1,000 population, 1990	24
Prevalence of malnutrition (under age 5), 1990	13

Ethnic Division [7]

Mixed	73%
White	16%
Black	11%

Religion [8]

Roman Catholic	95%

Major Languages [9]

Spanish.

Education

Public Education Expenditures [10]

Million Pesos	1980	1985	1986	1987	1988	1989
Total education expenditure	139	234	228	NA	NA	NA
as percent of GNP	2.2	1.8	1.5	NA	NA	NA
as percent of total govt. expend.	16.0	14.0	10.0	NA	NA	NA
Current education expenditure[10]	104	204	219	NA	NA	NA
as percent of GNP	1.6	1.5	1.5	NA	NA	NA
as percent of current govt. expend.	NA	NA	NA	NA	NA	NA
Capital expenditure[10]	10	2	5	NA	NA	NA

Educational Attainment [11]

Literacy Rate [12]

In thousands and percent	1985[a]	1991[a]	2000[a]
Illiterate population + 15 years	759	744	690
Illiteracy rate - total pop. (%)	19.6	16.7	12.1
Illiteracy rate - males (%)	17.8	15.2	11.3
Illiteracy rate - females (%)	21.5	18.2	12.9

Libraries [13]

Daily Newspapers [14]

	1975	1980	1985	1990
Number of papers	10	7	7	12
Circ. (000)	200[e]	220	216	230[e]

Cinema [15]

Science and Technology

Scientific/Technical Forces [16]

R&D Expenditures [17]

U.S. Patents Issued [18]

For sources, notes, and explanations, see Annotated Source Appendix, page 1035.

Government and Law

Organization of Government [19]

Long-form name:
Dominican Republic
Type:
republic
Independence:
27 February 1844 (from Haiti)
Constitution:
28 November 1966
Legal system:
based on French civil codes
National holiday:
Independence Day, 27 February (1844)
Executive branch:
president, vice president, Cabinet
Legislative branch:
bicameral National Congress (Congreso Nacional) consists of an upper chamber or Senate (Senado) and lower chamber or Chamber of Deputies (Camara de Diputados)
Judicial branch:
Supreme Court (Corte Suprema)

Political Parties [20]

Chamber of Deputies	% of seats
Dominican Liberation Party	36.7
Social Christian Reformist Party	34.2
Dominican Revolutionary Party	27.5
Independent Revolutionary Party	1.7

Government Expenditures [21]

Industry - 36.9%
Defense - 4.9%
Gen. Services - 20.7%
Educ./Health - 20.7%
Housing - 16.8%

% distribution for 1990. Expend. in 1993 est.: 1.8 ($ bil.)

Crime [23]

Military Expenditures and Arms Transfers [22]

	1985	1986	1987	1988	1989
Military expenditures					
Current dollars (mil.)	57	67	77	60	52
1989 constant dollars (mil.)	65	75	83	63	52
Armed forces (000)	22	21	21	20	21
Gross national product (GNP)					
Current dollars (mil.)	4,781	5,077	5,545	5,735	6,422
1989 constant dollars (mil.)	5,442	5,634	5,963	5,970	6,422
Central government expenditures (CGE)					
1989 constant dollars (mil.)	817	840	1,144[e]	1,219	1,043[e]
People (mil.)	6.5	6.7	6.8	7.0	7.1
Military expenditure as % of GNP	1.2	1.3	1.4	1.1	0.8
Military expenditure as % of CGE	8.0	8.9	7.3	5.1	5.0
Military expenditure per capita	10	11	12	9	7
Armed forces per 1,000 people	3.4	3.1	3.1	2.9	3.0
GNP per capita	833	844	875	858	905
Arms imports[6]					
Current dollars (mil.)	5	5	5	5	5
1989 constant dollars (mil.)	6	6	5	5	5
Arms exports[6]					
Current dollars (mil.)	0	0	0	0	0
1989 constant dollars (mil.)	0	0	0	0	0
Total imports[7]					
Current dollars (mil.)	1,487	1,433	1,830	1,849	2,280
1989 constant dollars	1,693	1,590	1,968	1,925	2,280
Total exports[7]					
Current dollars (mil.)	735	718	711	890	928
1989 constant dollars	837	797	765	926	928
Arms as percent of total imports[8]	0.3	0.3	0.3	0.3	0.2
Arms as percent of total exports[8]	0	0	0	0	0

Human Rights [24]

	SSTS	FL	FAPRO	PPCG	APROBC	TPW	PCPTW	STPEP	PHRFF	PRW	ASST	AFL
Observes		P	P	S	P	P	P			P	P	P
	EAFRD	CPR	ESCR	SR	ACHR	MAAE	PVIAC	PVNAC	EAFDAW	TCIDTP	RC	
Observes	P	P	P	P	P				P	S	P	

P = Party; S = Signatory; see Appendix for meaning of abbreviations.

Labor Force

Total Labor Force [25]

2,300,000 to 2,600,000

Labor Force by Occupation [26]

Agriculture	49%
Services	33
Industry	18

Date of data: 1986

Unemployment Rate [27]

30% (1992 est.)

For sources, notes, and explanations, see Annotated Source Appendix, page 1035.

271

Production Sectors

Energy Resource Summary [28]

Energy Resources: None. **Electricity**: 2,283,000 kW capacity; 5,000 million kWh produced, 660 kWh per capita (1992). **Pipelines**: Crude oil 96 km; petroleum products 8 km.

Telecommunications [30]

- Relatively efficient domestic system based on islandwide microwave relay network
- 190,000 telephones
- Broadcast stations - 120 AM, no FM, 18 TV, 6 shortwave
- 1 coaxial submarine cable
- 1 Atlantic Ocean INTELSAT earth station

Top Agricultural Products [32]

	1990	1991
Beans	732	656
Sugar	627	650
Cassava (Yucca)	756	608
Plantains (mil. units)	277	270
Tobacco	98	93
Rice	73	70

Values shown are 1,000 metric tons.

Top Mining Products [33]

Preliminary metric tons unless otherwise specified M.t.

Gold (kilograms)	3,160
Nickel, mine output, Ni content	29,062
Silver (kilograms)	21,954
Bauxite, dry equiv., gross weight (000 tons)	7
Cement, hydraulic (000 tons)	1,231
Gypsum (000 tons)	118
Limestone	448,654
Petroleum refinery products (000 42-gal. barrels)	318
Salt[24]	11,400[e]
Steel (kilograms)	21,954

Transportation [31]

Railroads. 1,655 km total in numerous segments; 4 different gauges from 0.558 m to 1.435 m

Highways. 12,000 km total; 5,800 km paved, 5,600 km gravel and improved earth, 600 km unimproved

Merchant Marine. 1 cargo ship (1,000 GRT or over) totaling 1,587 GRT/1,165 DWT

Airports

Total:	36
Usable:	30
With permanent-surface runways:	12
With runways over 3,659 m:	0
With runways 2,440-3,659 m:	4
With runways 1,220-2,439 m:	8

Tourism [34]

	1987	1988	1989	1990	1991
Visitors	1,069	1,216	1,500	1,633	1,421
Tourists[23]	902	1,116	1,400	1,533	1,321
Cruise passengers[24]	167	100	100	100	100
Tourism receipts	571	768	818	900	877
Tourism expenditures	95	127	136	144	154
Fare expenditures	15	15	15	15	20

Tourists are in thousands, money in million U.S. dollars.

For sources, notes, and explanations, see Annotated Source Appendix, page 1035.

Manufacturing Sector

GDP and Manufacturing Summary [35]

	1980	1985	1989	1990	% change 1980-1990	% change 1989-1990
GDP (million 1980 $)	6,631	7,159	8,358	7,913	19.3	-5.3
GDP per capita (1980 $)	1,164	1,116	1,191	1,102	-5.3	-7.5
Manufacturing as % of GDP (current prices)	15.3	13.6	15.6[e]	13.5	-11.8	-13.5
Gross output (million $)	2,376	1,822[e]	2,887[e]	3,034[e]	27.7	5.1
Value added (million $)	1,013	783[e]	1,252[e]	1,298[e]	28.1	3.7
Value added (million 1980 $)	1,015	986	1,176	1,067	5.1	-9.3
Industrial production index	100	99	117	102	2.0	-12.8
Employment (thousands)	146	131	133[e]	139[e]	-4.8	4.5

Note: GDP stands for Gross Domestic Product. 'e' stands for estimated value.

Profitability and Productivity

	1980	1985	1989	1990	% change 1980-1990	% change 1989-1990
Intermediate input (%)	57	57[e]	57[e]	57[e]	0.0	0.0
Wages, salaries, and supplements (%)	11	7[e]	6[e]	6[e]	-45.5	0.0
Gross operating surplus (%)	31	36[e]	38[e]	37[e]	19.4	-2.6
Gross output per worker ($)	16,284	13,877[e]	21,786[e]	21,898[e]	34.5	0.5
Value added per worker ($)	6,940	5,966[e]	9,451[e]	9,373[e]	35.1	-0.8
Average wage (incl. benefits) ($)	1,867	998	1,230[e]	1,348[e]	-27.8	9.6

Profitability is in percent of gross output. Productivity is in U.S. $. 'e' stands for estimated value.

Profitability - 1990

Inputs - 57.0%
Wages - 6.0%
Surplus - 37.0%

The graphic shows percent of gross output.

Value Added in Manufacturing

	1980 $ mil.	1980 %	1985 $ mil.	1985 %	1989 $ mil.	1989 %	1990 $ mil.	1990 %	% change 1980-1990	% change 1989-1990
311 Food products	510	50.3	293[e]	37.4	400[e]	31.9	414[e]	31.9	-18.8	3.5
313 Beverages	103	10.2	110[e]	14.0	167[e]	13.3	179[e]	13.8	73.8	7.2
314 Tobacco products	50	4.9	42[e]	5.4	81[e]	6.5	67[e]	5.2	34.0	-17.3
321 Textiles	29	2.9	26[e]	3.3	53[e]	4.2	45[e]	3.5	55.2	-15.1
322 Wearing apparel	13	1.3	9[e]	1.1	11[e]	0.9	16[e]	1.2	23.1	45.5
323 Leather and fur products	11	1.1	8[e]	1.0	15[e]	1.2	14[e]	1.1	27.3	-6.7
324 Footwear	13	1.3	13[e]	1.7	26[e]	2.1	25[e]	1.9	92.3	-3.8
331 Wood and wood products	2	0.2	3[e]	0.4	3[e]	0.2	2[e]	0.2	0.0	-33.3
332 Furniture and fixtures	11	1.1	11[e]	1.4	23[e]	1.8	19[e]	1.5	72.7	-17.4
341 Paper and paper products	19	1.9	21[e]	2.7	38[e]	3.0	37[e]	2.9	94.7	-2.6
342 Printing and publishing	14	1.4	13[e]	1.7	24[e]	1.9	22[e]	1.7	57.1	-8.3
351 Industrial chemicals	18	1.8	16[e]	2.0	20[e]	1.6	21[e]	1.6	16.7	5.0
352 Other chemical products	41	4.0	27[e]	3.4	49[e]	3.9	44[e]	3.4	7.3	-10.2
353 Petroleum refineries	66	6.5	81[e]	10.3	138[e]	11.0	209[e]	16.1	216.7	51.4
354 Miscellaneous petroleum and coal products	1	0.1	NA	0.0	NA	0.0	1[e]	0.1	0.0	NA
355 Rubber products	6	0.6	6[e]	0.8	11[e]	0.9	10[e]	0.8	66.7	-9.1
356 Plastic products	21	2.1	12[e]	1.5	28[e]	2.2	21[e]	1.6	0.0	-25.0
361 Pottery, china and earthenware	1	0.1	1[e]	0.1	1[e]	0.1	1[e]	0.1	0.0	0.0
362 Glass and glass products	3	0.3	5[e]	0.6	8[e]	0.6	8[e]	0.6	166.7	0.0
369 Other non-metal mineral products	32	3.2	29[e]	3.7	47[e]	3.8	46[e]	3.5	43.8	-2.1
371 Iron and steel	10	1.0	15[e]	1.9	25[e]	2.0	24[e]	1.8	140.0	-4.0
372 Non-ferrous metals	1	0.1	1[e]	0.1	2[e]	0.2	3[e]	0.2	200.0	50.0
381 Metal products	21	2.1	28[e]	3.6	55[e]	4.4	48[e]	3.7	128.6	-12.7
382 Non-electrical machinery	5	0.5	3[e]	0.4	8[e]	0.6	6[e]	0.5	20.0	-25.0
383 Electrical machinery	7	0.7	6[e]	0.8	13[e]	1.0	11[e]	0.8	57.1	-15.4
384 Transport equipment	NA	0.0	NA	0.0	NA	0.0	1[e]	0.1	NA	NA
385 Professional and scientific equipment	1	0.1	1[e]	0.1	1[e]	0.1	2[e]	0.2	100.0	100.0
390 Other manufacturing industries	2	0.2	1[e]	0.1	3[e]	0.2	3[e]	0.2	50.0	0.0

Note: The industry codes shown are International Standard Industry codes (ISIC). Percentages are percent of total Value Added. 'e' stands for estimated value

For sources, notes, and explanations, see Annotated Source Appendix, page 1035.

273

Finance, Economics, and Trade

Economic Indicators [36]

	1989	1990	1991[P]
Real GDP (1970 pesos)	3,655.7	3,468.4	3,329.7
Real GDP growth rate	4.1	-5.1	-2.0
Money Supply (M1)			
(mil. pesos)	5,911.7	8,304.8	8,362.9
Com'l Interest Rates (Prime)	30.0	43.0	34.0
Gross Nat'l Savings Rate	15.6	13.3	13.0
Investment Rate	NA	NA	NA
CPI (percent change)	41.2	100.6	9.0
WPI	NA	NA	NA
External Public Debt	4,068.5	4,481.8	4,673.1

Balance of Payments Summary [37]

Values in millions of dollars.

	1987	1988	1989	1990	1991
Exports of goods (f.o.b.)	711.3	889.7	924.4	734.5	658.3
Imports of goods (f.o.b.)	-1,591.3	-1,608.0	-1,963.8	-1,792.8	-1,728.8
Trade balance	-880.2	-718.3	-1,039.4	-1,058.3	-1,070.5
Services - debits	-678.1	-676.1	-706.4	-603.4	-712.5
Services - credits	863.6	1,021.9	1,162.8	1,282.4	1,338.1
Private transfers (net)	273.1	288.8	300.5	314.8	329.5
Government transfers (net)	57.5	64.8	83.9	55.8	57.0
Long term capital (net)	59.5	239.4	243.1	106.2	133.1
Short term capital (net)	-20.7	-57.4	9.9	116.5	14.0
Errors and omissions	248.9	-46.1	-192.6	-128.2	288.1
Overall balance	-76.4	117.0	-138.2	85.8	376.8

Exchange Rates [38]

Currency: **Dominican pesos.**
Symbol: **RD$.**

Data are currency units per $1.

1992	12.700
1991	12.692
1990	8.525
1989	6.340
1988	6.113

Imports and Exports

Top Import Origins [39]

$2 billion (c.i.f., 1992 est.).

Origins	%
US	50

Top Export Destinations [40]

$600 million (f.o.b., 1992) Data are for 1990.

Destinations	%
US	60
EC	19
Puerto Rico	8

Foreign Aid [41]

	U.S. $	
US commitments, including Ex-Im (FY85-89)	575	million
Western (non-US) countries, ODA and OOF		
bilateral commitments (1970-89)	655	million

Import and Export Commodities [42]

Import Commodities	**Export Commodities**
Foodstuffs	Ferronickel
Petroleum	Sugar
Cotton and fabrics	Gold
Chemicals and pharmaceuticals	Coffee
	Cocoa

Ecuador

Geography [1]

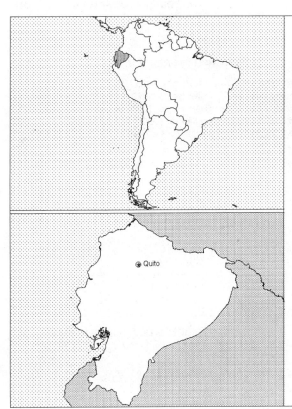

Total area:
283,560 km2
Land area:
276,840 km2
Comparative area:
Slightly smaller than Nevada
Land boundaries:
Total 2,010 km, Colombia 590 km, Peru 1,420 km
Coastline:
2,237 km
Climate:
Tropical along coast becoming cooler inland
Terrain:
Coastal plain (Costa), inter-Andean central highlands (Sierra), and flat to rolling eastern jungle (Oriente)
Natural resources:
Petroleum, fish, timber
Land use:
Arable land:
6%
Permanent crops:
3%
Meadows and pastures:
17%
Forest and woodland:
51%
Other:
23%

Demographics [2]

	1960	1970	1980	1990	1991[1]	1994	2000	2010	2020
Population	4,413	6,051	8,123	9,806	10,752	10,677	11,945	13,990	15,894
Population density (persons per sq. mi.)	41	57	76	98	101	NA	122	145	169
Births	NA	NA	NA	NA	320	276	NA	NA	NA
Deaths	NA	NA	NA	NA	74	61	NA	NA	NA
Life expectancy - males	NA	NA	NA	NA	64	67	NA	NA	NA
Life expectancy - females	NA	NA	NA	NA	68	73	NA	NA	NA
Birth rate (per 1,000)	NA	NA	NA	NA	30	26	NA	NA	NA
Death rate (per 1,000)	NA	NA	NA	NA	7	6	NA	NA	NA
Women of reproductive age (15-44 yrs.)	NA	NA	NA	2,462	NA	2,769	3,238	NA	NA
of which are currently married	NA	NA	NA	1,510	NA	1,718	2,044	NA	NA
Fertility rate	NA	NA	NA	3.5	3.70	3.1	2.6	2.2	2.1

Population values are in thousands, life expectancy in years, and other items as indicated.

Health

Health Personnel [3]

Doctors per 1,000 pop., 1988-92	1.04
Nurse-to-doctor ratio, 1988-92	0.3
Hospital beds per 1,000 pop., 1985-90	1.7
Percentage of children immunized (age 1 yr. or less)	
Third dose of DPT, 1990-91	89.0
Measles, 1990-91	54.0

Health Expenditures [4]

Total health expenditure, 1990 (official exchange rate)	
Millions of dollars	441
Millions of dollars per capita	43
Health expenditures as a percentage of GDP	
Total	4.1
Public sector	2.6
Private sector	1.6
Development assistance for health	
Total aid flows (millions of dollars)[1]	31
Aid flows per capita (millions of dollars)	3.0
Aid flows as a percentage of total health expenditure	7.0

For sources, notes, and explanations, see Annotated Source Appendix, page 1035.

275

Human Factors

Health Care Ratios [5]

Population per physician, 1970	2910
Population per physician, 1990	980
Population per nursing person, 1970	2680
Population per nursing person, 1990	620
Percent of births attended by health staff, 1985	27

Infants and Malnutrition [6]

Percent of babies with low birth weight, 1985	10
Infant mortality rate per 1,000 live births, 1970	100
Infant mortality rate per 1,000 live births, 1991	47
Years of life lost per 1,000 population, 1990	21
Prevalence of malnutrition (under age 5), 1990	38

Ethnic Division [7]

Mestizo	55%
Indian	25%
Spanish	10%
Black	10%

Religion [8]

Roman Catholic	95%

Major Languages [9]

Spanish (official), Indian languages (especially Quechua).

Education

Public Education Expenditures [10]

Million Sucre	1980	1985	1987	1988	1989	1990
Total education expenditure	15,580	38,009	57,914	84,328	134,554	NA
as percent of GNP	5.6	3.7	3.5	3.0	2.8	NA
as percent of total govt. expend.	33.3	20.6	21.3	21.3	19.1	NA
Current education expenditure	14,649	35,611	53,016	77,569	127,394	NA
as percent of GNP	5.3	3.5	3.2	2.8	2.7	NA
as percent of current govt. expend.	36.0	25.8	26.9	26.5	24.7	NA
Capital expenditure	931	2,397	4,898	6,759	7,160	NA

Educational Attainment [11]

Age group	25+
Total population	3,180,447
Highest level attained (%)	
No schooling	2.2
First level	
Incompleted	54.3
Completed	NA
Entered second level	
S-1	28.0
S-2	NA
Post secondary	15.5

Literacy Rate [12]

In thousands and percent	1985[a]	1991[a]	2000[a]
Illiterate population +15 years	928	909	843
Illiteracy rate - total pop. (%)	17.0	14.2	9.8
Illiteracy rate - males (%)	14.5	12.2	8.6
Illiteracy rate - females (%)	19.5	16.2	11.0

Libraries [13]

	Admin. Units	Svc. Pts.	Vols. (000)	Shelving (meters)	Vols. Added	Reg. Users
National	NA	NA	NA	NA	NA	NA
Non-specialized	NA	NA	NA	NA	NA	NA
Public (1988)	210	210	142	NA	55,200	NA
Higher ed. (1987)	128	128	531	NA	110,029	224,517
School	NA	NA	NA	NA	NA	NA

Daily Newspapers [14]

	1975	1980	1985	1990
Number of papers	29	18	26	25
Circ. (000)	332	558	800[e]	920[e]

Cinema [15]

Data for 1991.

Cinema seats per 1,000	7.0
Annual attendance per person	0.6
Gross box office receipts (mil. Sucre)	4,949

Science and Technology

Scientific/Technical Forces [16]

R&D Expenditures [17]

U.S. Patents Issued [18]

Values show patents issued to citizens of the country by the U.S. Patents Office.

	1990	1991	1992
Number of patents	2	3	1

                                             For sources, notes, and explanations, see Annotated Source Appendix, page 1035.

Government and Law

Organization of Government [19]

Long-form name:
Republic of Ecuador
Type:
republic
Independence:
24 May 1822 (from Spain)
Constitution:
10 August 1979
Legal system:
based on civil law system; has not accepted compulsory ICJ jurisdiction
National holiday:
Independence Day, 10 August (1809) (independence of Quito)
Executive branch:
president, vice president, Cabinet
Legislative branch:
unicameral National Congress (Congreso Nacional)
Judicial branch:
Supreme Court (Corte Suprema)

Political Parties [20]

National Congress	% of seats
Social Christian Party	26.0
Roldista Party	19.5
Republican Unity Party	15.6
Democratic Left	9.1
PC	7.8
Popular Democracy	6.5
Ecuadorian Socialist Party	3.9
Popular Democratic Movement	3.9
Ecuadorian Radical Liberal Party	2.6
Other	5.2

Government Expenditures [21]

Housing - 1.0%
Other - 33.0%
Educ./Health - 31.1%
Gen. Services - 11.3%
Industry - 10.6%
Defense - 13.0%

% distribution for 1990. Expend. in 1992: 1.9 ($ bil.)

Crime [23]

Military Expenditures and Arms Transfers [22]

	1985	1986	1987	1988	1989
Military expenditures					
Current dollars (mil.)	208	228	229	220	163
1989 constant dollars (mil.)	237	253	246	229	163
Armed forces (000)	43	44	44	46	46
Gross national product (GNP)					
Current dollars (mil.)	7,532	7,951	7,698	9,264	9,668
1989 constant dollars (mil.)	8,575	8,823	8,280	9,644	9,668
Central government expenditures (CGE)					
1989 constant dollars (mil.)	1,397	1,511	1,375	1,368	1,402
People (mil.)	9.3	9.5	9.8	10.0	10.3
Military expenditure as % of GNP	2.8	2.9	3.0	2.4	1.7
Military expenditure as % of CGE	16.9	16.7	17.9	16.7	11.6
Military expenditure per capita	25	26	25	23	16
Armed forces per 1,000 people	4.6	4.6	4.5	4.6	4.5
GNP per capita	922	925	847	962	942
Arms imports[6]					
Current dollars (mil.)	70	100	90	50	20
1989 constant dollars (mil.)	80	111	97	52	20
Arms exports[6]					
Current dollars (mil.)	0	0	0	0	0
1989 constant dollars (mil.)	0	0	0	0	0
Total imports[7]					
Current dollars (mil.)	1,767	1,810	2,252	1,714	1,855
1989 constant dollars	2,012	2,009	2,422	1,784	1,855
Total exports[7]					
Current dollars (mil.)	2,905	2,172	1,928	2,192	2,354
1989 constant dollars	3,307	2,410	2,074	2,282	2,354
Arms as percent of total imports[8]	4.0	5.5	4.0	2.9	1.1
Arms as percent of total exports[8]	0	0	0	0	0

Human Rights [24]

	SSTS	FL	FAPRO	PPCG	APROBC	TPW	PCPTW	STPEP	PHRFF	PRW	ASST	AFL
Observes	P	P	P	P	P	P	P	P		P	P	P
		EAFRD	CPR	ESCR	SR	ACHR	MAAE	PVIAC	PVNAC	EAFDAW	TCIDTP	RC
Observes		P	P	P	P	P		P	P	P	P	P

P = Party; S = Signatory; see Appendix for meaning of abbreviations.

Labor Force

Total Labor Force [25]

2.8 million

Labor Force by Occupation [26]

Agriculture	35%
Manufacturing	21
Commerce	16
Services and other activities	28

Date of data: 1982

Unemployment Rate [27]

8% (1992)

For sources, notes, and explanations, see Annotated Source Appendix, page 1035.

277

Production Sectors

Commercial Energy Production and Consumption

Production [28]

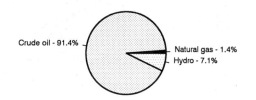

Crude oil - 91.4%
Natural gas - 1.4%
Hydro - 7.1%

Consumption [29]

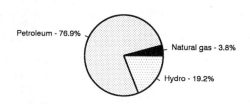

Petroleum - 76.9%
Natural gas - 3.8%
Hydro - 19.2%

Data are shown in quadrillion (10^{15}) BTUs and percent for 1991

Crude oil[1]	0.64
Natural gas liquids	(s)
Dry natural gas	0.01
Coal[2]	0.00
Net hydroelectric power[3]	0.05
Net nuclear power[3]	0.00
Total	0.70

Petroleum	0.20
Dry natural gas	0.01
Coal[2]	0.00
Net hydroelectric power[3]	0.05
Net nuclear power[3]	0.00
Total	0.26

Telecommunications [30]

- Domestic facilities generally adequate
- 318,000 telephones
- Broadcast stations - 272 AM, no FM, 33 TV, 39 shortwave
- 1 Atlantic Ocean INTELSAT earth station

Transportation [31]

Railroads. 965 km total; all 1.067-meter-gauge single track

Highways. 28,000 km total; 3,600 km paved, 17,400 km gravel and improved earth, 7,000 km unimproved earth

Merchant Marine. 45 ships (1,000 GRT or over) totaling 333,380 GRT/483,862 DWT; includes 2 passenger, 4 cargo, 17 refrigerated cargo, 4 container, 1 roll-on/roll-off, 15 oil tanker, 1 liquefied gas, 1 bulk

Airports

Total:	174
Usable:	173
With permanent-surface runways:	52
With runway over 3,659 m:	1
With runways 2,440-3,659 m:	6
With runways 1,220-2,439 m:	21

Top Agricultural Products [32]

	1990	1991[4]
Bananas	3,054	3,500
Sugar cane	3,256	3,400
Rice (paddy)	840	840
Corn (soft and hard)	465	615
Potatoes	368	367
Soybeans	166	168

Values shown are 1,000 metric tons.

Top Mining Products [33]

Preliminary metric tons unless otherwise specified M.t.

Gold, mine output, Au content (kilograms)	200
Lead concentrate, Pb content	200[e]
Silver, mine output, Ag content (kilograms)	60
Zinc, mine output, Zn content	100[e]
Clay, bentonite	200[e]
Kaolin	26,000
Limestone for cement manufacture	3,885
Marble	22,200
Sulfur	14,000[e]

Tourism [34]

	1987	1988	1989	1990	1991
Visitors	274	347	335	362	365
Tourism receipts	167	173	187	188	189
Tourism expenditures	170	167	169	175	177
Fare receipts	37	57	59	62	63
Fare expenditures	51	56	60	62	64

Tourists are in thousands, money in million U.S. dollars.

For sources, notes, and explanations, see Annotated Source Appendix, page 1035.

Manufacturing Sector

GDP and Manufacturing Summary [35]

	1980	1985	1989	1990	% change 1980-1990	% change 1989-1990
GDP (million 1980 $)	11,733	13,040	14,184	14,383	22.6	1.4
GDP per capita (1980 $)	1,444	1,399	1,374	1,357	-6.0	-1.2
Manufacturing as % of GDP (current prices)	17.8	19.3	21.8	24.2	36.0	11.0
Gross output (million $)	3,571	4,379	3,894[e]	3,934	10.2	1.0
Value added (million $)	1,289	1,322	1,022[e]	860	-33.3	-15.9
Value added (million 1980 $)	2,072	2,219	2,156	2,278	9.9	5.7
Industrial production index	100	110	117	127	27.0	8.5
Employment (thousands)	112	97	113[e]	112	0.0	-0.9

Note: GDP stands for Gross Domestic Product. 'e' stands for estimated value.

Profitability and Productivity

	1980	1985	1989	1990	% change 1980-1990	% change 1989-1990
Intermediate input (%)	64	70	74[e]	78	21.9	5.4
Wages, salaries, and supplements (%)	16[e]	13	9[e]	9[e]	-43.8	0.0
Gross operating surplus (%)	20[e]	18	17[e]	13[e]	-35.0	-23.5
Gross output per worker ($)	31,623	45,072	34,409[e]	35,083	10.9	2.0
Value added per worker ($)	11,414	13,606	9,030[e]	7,666	-32.8	-15.1
Average wage (incl. benefits) ($)	5,092[e]	5,677	3,118[e]	3,106[e]	-39.0	-0.4

Profitability is in percent of gross output. Productivity is in U.S. $. 'e' stands for estimated value.

Profitability - 1990

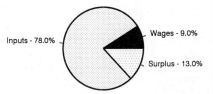

Inputs - 78.0%
Wages - 9.0%
Surplus - 13.0%

The graphic shows percent of gross output.

Value Added in Manufacturing

	1980 $ mil.	1980 %	1985 $ mil.	1985 %	1989 $ mil.	1989 %	1990 $ mil.	1990 %	% change 1980-1990	% change 1989-1990
311 Food products	294	22.8	328	24.8	302[e]	29.5	228	26.5	-22.4	-24.5
313 Beverages	96	7.4	65	4.9	38[e]	3.7	33	3.8	-65.6	-13.2
314 Tobacco products	46	3.6	17	1.3	12[e]	1.2	1	0.1	-97.8	-91.7
321 Textiles	134	10.4	146	11.0	100[e]	9.8	95	11.0	-29.1	-5.0
322 Wearing apparel	20	1.6	15	1.1	10[e]	1.0	10	1.2	-50.0	0.0
323 Leather and fur products	7	0.5	6	0.5	4[e]	0.4	4	0.5	-42.9	0.0
324 Footwear	6	0.5	7	0.5	6[e]	0.6	6	0.7	0.0	0.0
331 Wood and wood products	35	2.7	18	1.4	19[e]	1.9	16	1.9	-54.3	-15.8
332 Furniture and fixtures	28	2.2	23	1.7	12[e]	1.2	9	1.0	-67.9	-25.0
341 Paper and paper products	42	3.3	41	3.1	33[e]	3.2	34	4.0	-19.0	3.0
342 Printing and publishing	40	3.1	35	2.6	34[e]	3.3	27	3.1	-32.5	-20.6
351 Industrial chemicals	25	1.9	32	2.4	33[e]	3.2	17	2.0	-32.0	-48.5
352 Other chemical products	90	7.0	76	5.7	55[e]	5.4	75	8.7	-16.7	36.4
353 Petroleum refineries	29	2.2	38	2.9	32[e]	3.1	37	4.3	27.6	15.6
354 Miscellaneous petroleum and coal products	4	0.3	14	1.1	12[e]	1.2	4	0.5	0.0	-66.7
355 Rubber products	25	1.9	29	2.2	12[e]	1.2	17	2.0	-32.0	41.7
356 Plastic products	34	2.6	57	4.3	45[e]	4.4	42	4.9	23.5	-6.7
361 Pottery, china and earthenware	7	0.5	15	1.1	10[e]	1.0	7	0.8	0.0	-30.0
362 Glass and glass products	9	0.7	15	1.1	10[e]	1.0	8	0.9	-11.1	-20.0
369 Other non-metal mineral products	100	7.8	101	7.6	78[e]	7.6	60	7.0	-40.0	-23.1
371 Iron and steel	25	1.9	56	4.2	20[e]	2.0	19	2.2	-24.0	-5.0
372 Non-ferrous metals	5	0.4	10	0.8	7[e]	0.7	2	0.2	-60.0	-71.4
381 Metal products	93	7.2	78	5.9	54[e]	5.3	44	5.1	-52.7	-18.5
382 Non-electrical machinery	4	0.3	7	0.5	4[e]	0.4	3	0.3	-25.0	-25.0
383 Electrical machinery	59	4.6	58	4.4	51[e]	5.0	32	3.7	-45.8	-37.3
384 Transport equipment	23	1.8	23	1.7	17[e]	1.7	22	2.6	-4.3	29.4
385 Professional and scientific equipment	2	0.2	9	0.7	8[e]	0.8	3	0.3	50.0	-62.5
390 Other manufacturing industries	7	0.5	5	0.4	6[e]	0.6	3	0.3	-57.1	-50.0

Note: The industry codes shown are International Standard Industry codes (ISIC). Percentages are percent of total Value Added. 'e' stands for estimated value

Finance, Economics, and Trade

Economic Indicators [36]

	1989	1990	1991
GDP (bil. $U.S.)[38]	10.1	10.9	11.5
GDP Growth Rate (percent)	0.6	2.3	2.5
GDP/Capita ($U.S.)[38]	963	1,007	1,038
Money Supply (M1, %. growth)	32.2	44.6	50.0
Commercial Interest Rates[e]	50.0	53.0	53.0
Savings Rate (% GDP)	17.1	19.0	19.0
Inflation (CPI, year-end)	54.2	49.5	49.0
External Public Debt (bil.)	11.2	11.7	12.4

Balance of Payments Summary [37]

Values in millions of dollars.

	1987	1988	1989	1990	1991
Exports of goods (f.o.b.)	2,021.0	2,202.0	2,354.0	2,714.0	2,851.0
Imports of goods (f.o.b.)	-2,054.0	-1,583.0	-1,693.0	-1,711.0	-2,207.0
Trade balance	-33.0	619.0	661.0	1,003.0	644.0
Services - debits	-1,672.0	-1,709.0	-1,808.0	-1,839.0	-1,808.0
Services - credits	449.0	457.0	536.0	563.0	587.0
Private transfers (net)	-	-	-	-	-
Government transfers (net)	132.0	97.0	97.0	107.0	110.0
Long term capital (net)	358.0	157.0	356.0	-422.0	-506.0
Short term capital (net)	924.0	599.0	294.0	716.0	932.0
Errors and omissions	-221.6	-144.2	66.6	124.1	182.7
Overall balance	-63.6	75.8	202.6	252.1	141.7

Exchange Rates [38]

Currency: **sucres.**
Symbol: **S/.**

Data are currency units per $1.

August 1992	1,453.80
1991	1,046.25
December 1990	869.54
1990	767.75
1989	526.35
1988	301.61

Imports and Exports

Top Import Origins [39]

$2.4 billion (f.o.b., 1992).

Origins	%
US	32.7
Latin America	NA
Caribbean	NA
EC countries	NA
Japan	NA

Top Export Destinations [40]

$3.0 billion (f.o.b., 1992).

Destinations	%
US	53.4
Latin America	NA
Caribbean	NA
EC countries	NA

Foreign Aid [41]

	U.S. $	
US commitments, including Ex-Im (FY70-89)	498	million
Western (non-US) countries, ODA and OOF bilateral commitments (1970-89)	2.15	billion
Communist countries (1970-89)	64	million

Import and Export Commodities [42]

Import Commodities	Export Commodities
Transport equipment	Petroleum 42%
Vehicles	Bananas
Machinery	Shrimp
Chemicals	Cocoa
	Coffee

Egypt

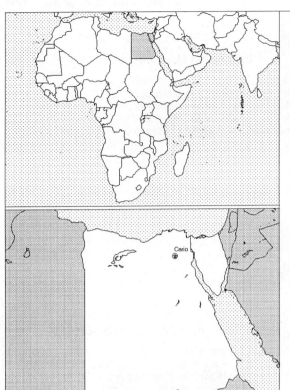

Geography [1]

Total area:
1,001,450 km2
Land area:
995,450 km2
Comparative area:
Slightly more than three times the size of New Mexico
Land boundaries:
Total 2,689 km, Gaza Strip 11 km, Israel 255 km, Libya 1,150 km, Sudan 1,273 km
Coastline:
2,450 km
Climate:
Desert; hot, dry summers with moderate winters
Terrain:
Vast desert plateau interrupted by Nile valley and delta
Natural resources:
Petroleum, natural gas, iron ore, phosphates, manganese, limestone, gypsum, talc, asbestos, lead, zinc
Land use:
Arable land:
3%
Permanent crops:
2%
Meadows and pastures:
0%
Forest and woodland:
0%
Other:
95%

Demographics [2]

	1960	1970	1980	1990	1991[1]	1994	2000	2010	2020
Population	26,847	33,574	41,663	53,993	54,452	59,325	67,542	82,478	97,434
Population density (persons per sq. mi.)	70	87	108	138	142	NA	173	213	254
Births	NA	NA	NA	NA	1,780	1,879	NA	NA	NA
Deaths	NA	NA	NA	NA	517	538	NA	NA	NA
Life expectancy - males	NA	NA	NA	NA	60	59	NA	NA	NA
Life expectancy - females	NA	NA	NA	NA	61	63	NA	NA	NA
Birth rate (per 1,000)	NA	NA	NA	NA	33	32	NA	NA	NA
Death rate (per 1,000)	NA	NA	NA	NA	10	9	NA	NA	NA
Women of reproductive age (15-44 yrs.)	NA	NA	NA	12,467	NA	14,184	16,742	NA	NA
of which are currently married	NA	NA	NA	8,507	NA	9,544	11,309	NA	NA
Fertility rate	NA	NA	NA	4.6	4.52	4.3	3.8	3.1	2.7

Population values are in thousands, life expectancy in years, and other items as indicated.

Health

Health Personnel [3]

Doctors per 1,000 pop., 1988-92	0.77
Nurse-to-doctor ratio, 1988-92	1.2
Hospital beds per 1,000 pop., 1985-90	1.9
Percentage of children immunized (age 1 yr. or less)	
Third dose of DPT, 1990-91	86.0
Measles, 1990-91	89.0

Health Expenditures [4]

Total health expenditure, 1990 (official exchange rate)	
Millions of dollars	921
Millions of dollars per capita	18
Health expenditures as a percentage of GDP	
Total	2.6
Public sector	1.0
Private sector	1.6
Development assistance for health	
Total aid flows (millions of dollars)[1]	111
Aid flows per capita (millions of dollars)	2.1
Aid flows as a percentage of total health expenditure	12.1

For sources, notes, and explanations, see Annotated Source Appendix, page 1035.

281

Human Factors

Health Care Ratios [5]

Population per physician, 1970	1900
Population per physician, 1990	1320
Population per nursing person, 1970	2320
Population per nursing person, 1990	490
Percent of births attended by health staff, 1985	24

Infants and Malnutrition [6]

Percent of babies with low birth weight, 1985	7
Infant mortality rate per 1,000 live births, 1970	158
Infant mortality rate per 1,000 live births, 1991	59
Years of life lost per 1,000 population, 1990	33
Prevalence of malnutrition (under age 5), 1990	13

Ethnic Division [7]

Eastern Hamitic stock	90%
Other	10%
(Greek,	
Italian,	
Syro-Lebanese)	

Religion [8]

Muslim (mostly Sunni)	94%
Coptic Christian and other	6%

Major Languages [9]

Arabic (official), English and French widely understood by educated classes.

Education

Public Education Expenditures [10]

Million Egyptian Pounds	1985[3]	1986[3]	1987[3]	1988[3]	1989[3]	1990[1]
Total education expenditure	1,878	NA	2,330	2,834	4,020	2,664
as percent of GNP	6.3	NA	5.7	5.7	6.7	3.8
as percent of total govt. expend.	NA	NA	NA	NA	NA	Na
Current education expenditure	1,775	NA	1,960	2,412	3,599	2,406
as percent of GNP	5.9	NA	4.8	4.8	6.0	3.4
as percent of current govt. expend.	10.8	NA	8.5	NA	NA	NA
Capital expenditure	103	NA	371	422	421	258

Educational Attainment [11]

Age group	25+
Total population	19,441,903
Highest level attained (%)	
No schooling	64.1
First level	
Incompleted	16.5
Completed	NA
Entered second level[5]	
S-1	14.8
S-2	NA
Post secondary	4.6

Literacy Rate [12]

In thousands and percent	1985[a]	1991[a]	2000[a]
Illiterate population +15 years	15,686	16,492	18,535
Illiteracy rate - total pop. (%)	55.4	51.6	43.3
Illiteracy rate - males (%)	40.4	37.1	30.4
Illiteracy rate - females (%)	70.5	66.2	56.4

Libraries [13]

	Admin. Units	Svc. Pts.	Vols. (000)	Shelving (meters)	Vols. Added	Reg. Users
National (1990)	1	NA	2,117	NA	NA	1.8mil
Non-specialized	NA	NA	NA	NA	NA	NA
Public (1988)	836	NA	8,523	NA	NA	1,644
Higher ed. (1989)	272	272	35,790	NA	NA	65,900
School	NA	NA	NA	NA	NA	NA

Daily Newspapers [14]

	1975	1980	1985	1990
Number of papers	12	12	12	14
Circ. (000)	1,095	1,701	2,383	3,000[e]

Cinema [15]

Data for 1989.

Cinema seats per 1,000	2.5
Annual attendance per person	0.5
Gross box office receipts (mil. Pound)	22

Science and Technology

Scientific/Technical Forces [16]

Potential scientists/engineers	1,573,118
Number female	NA
Potential technicians	8,863,831
Number female	NA
Total	10,436,949

R&D Expenditures [17]

	Pound[4] (000) 1982
Total expenditure	40,378
Capital expenditure	6,136
Current expenditure	34,242
Percent current	84.8

U.S. Patents Issued [18]

Values show patents issued to citizens of the country by the U.S. Patents Office.

	1990	1991	1992
Number of patents	1	3	1

                           For sources, notes, and explanations, see Annotated Source Appendix, page 1035.

Government and Law

Organization of Government [19]

Long-form name:
Arab Republic of Egypt
Type:
republic
Independence:
28 February 1922 (from UK)
Constitution:
11 September 1971
Legal system:
based on English common law, Islamic law, and Napoleonic codes; judicial review by Supreme Court and Council of State (oversees validity of administrative decisions); accepts compulsory ICJ jurisdiction, with reservations
National holiday:
Anniversary of the Revolution, 23 July (1952)
Executive branch:
president, prime minister, Cabinet
Legislative branch:
unicameral People's Assembly (Majlis al-Cha'b); note - there is an Advisory Council (Majlis al-Shura) that functions in a consultative role
Judicial branch:
Supreme Constitutional Court

Political Parties [20]

People's Assembly	% of seats
National Democratic Party (NDP)	78.4
Socialist Labor Party (NPUG)	1.4
Independents	18.7

Government Expenditures [21]

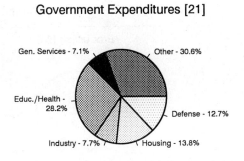

Gen. Services - 7.1%
Other - 30.6%
Educ./Health - 28.2%
Defense - 12.7%
Industry - 7.7%
Housing - 13.8%

% distribution for 1989. Expend. in FY92 est.: 15.2 ($ bil.)

Crime [23]

Crime volume	
Cases known to police	1,785,838
Attempts (percent)	0.02
Percent cases solved	95
Crimes per 100,000 persons	3,314.41
Persons responsible for offenses	
Total number offenders	1,509,090
Percent female	1
Percent juvenile[20]	1.41
Percent foreigners	0.02

Military Expenditures and Arms Transfers [22]

	1985	1986	1987	1988	1989
Military expenditures					
Current dollars (mil.)	6,785	6,460	6,664	5,559	3,499
1989 constant dollars (mil.)	7,724	7,168	7,167	5,786	3,499
Armed forces (000)	466	400	450	452	450
Gross national product (GNP)					
Current dollars (mil.)	53,040	55,250	59,480	64,920	69,780
1989 constant dollars (mil.)	60,380	61,310	63,970	67,580	69,780
Central government expenditures (CGE)					
1989 constant dollars (mil.)	34,900	36,730	32,100	33,760	32,590
People (mil.)	47.3	48.4	49.6	50.8	52.0
Military expenditure as % of GNP	12.8	11.7	11.2	8.6	5.0
Military expenditure as % of CGE	22.1	19.5	22.3	17.1	10.7
Military expenditure per capita	163	148	145	114	67
Armed forces per 1,000 people	9.9	8.3	9.1	8.9	8.7
GNP per capita	1,277	1,267	1,290	1,331	1,342
Arms imports[6]					
Current dollars (mil.)	1,500	1,200	1,700	775	600
1989 constant dollars (mil.)	1,708	1,332	1,828	807	600
Arms exports[6]					
Current dollars (mil.)	90	110	100	210	370
1989 constant dollars (mil.)	102	122	108	219	370
Total imports[7]					
Current dollars (mil.)	9,961	8,680	7,596	8,657	7,434
1989 constant dollars	11,340	9,632	8,170	9,012	7,434
Total exports[7]					
Current dollars (mil.)	3,714	2,214	2,037	2,120	2,565
1989 constant dollars	4,228	2,457	2,191	2,207	2,565
Arms as percent of total imports[8]	15.1	13.8	22.4	9.0	8.1
Arms as percent of total exports[8]	2.4	5.0	4.9	9.9	14.4

Human Rights [24]

	SSTS	FL	FAPRO	PPCG	APROBC	TPW	PCPTW	STPEP	PHRFF	PRW	ASST	AFL
Observes	P	P	P	P	P	P	P	P		P	P	P
	EAFRD	CPR	ESCR	SR	ACHR	MAAE	PVIAC	PVNAC	EAFDAW	TCIDTP	RC
Observes		P	P	P	P			S	S	P	P	P

P = Party; S = Signatory; see Appendix for meaning of abbreviations.

Labor Force

Total Labor Force [25]

15 million (1989 est.)

Labor Force by Occupation [26]

Government, pub. enterprises, military	36%
Agriculture	34
Privately enterprises	20

Date of data: 1984

Unemployment Rate [27]

20% (1992 est.)

Production Sectors

Commercial Energy Production and Consumption

Production [28]

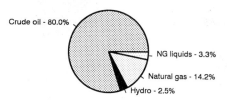

Crude oil - 80.0%
NG liquids - 3.3%
Natural gas - 14.2%
Hydro - 2.5%

Consumption [29]

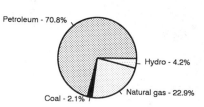

Petroleum - 70.8%
Hydro - 4.2%
Natural gas - 22.9%
Coal - 2.1%

Data are shown in quadrillion (10^{15}) BTUs and percent for 1991

Crude oil[1]	1.92
Natural gas liquids	0.08
Dry natural gas	0.34
Coal[2]	0.00
Net hydroelectric power[3]	0.06
Net nuclear power[3]	0.00
Total	2.40

Petroleum	1.02
Dry natural gas	0.33
Coal[2]	0.03
Net hydroelectric power[3]	0.06
Net nuclear power[3]	0.00
Total	1.44

Telecommunications [30]

- Large system but inadequate for present needs
- About 600,000 telephones (est.)—11 phones per 1,000 persons
- Principal centers at Alexandria, Cairo, Al Mansurah, Ismailia Suez, and Tanta are connected by coaxial cable and microwave radio relay
- International traffic is carried by satellite—one earth station each for INTELSAT (Atlantic and Indian Ocean), ARABSAT, and INMARSAT; by 5 coaxial submarine cables; microwave troposcatter (to Sudan), and microwave radio relay (to Libya, Israel, and Jordan)
- Broadcast stations - 39 AM, 6 FM, and 41 TV

Transportation [31]

Railroads. 5,110 km total; 4,763 km 1,435-meter standard gauge, 347 km 0.750-meter gauge; 951 km double track; 25 km electrified

Highways. 51,925 km total; 17,900 km paved, 2,500 km gravel, 13,500 km improved earth, 18,025 km unimproved earth

Merchant Marine. 168 ships (1,000 GRT or over) totaling 1,097,707 GRT/1,592,885 DWT; includes 25 passenger, 6 short-sea passenger, 2 passenger-cargo, 88 cargo, 3 refrigerated cargo, 14 roll-on/roll-off, 13 oil tanker, 16 bulk, 1 container

Airports

Total:	92
Usable:	82
With permanent-surface runways:	66
With runways over 3,659 m:	2
With runways 2,440-3,659 m:	44
With runways 1,220-2,439 m:	24

Top Agricultural Products [32]

	1990[9]	1991[9]
Corn	4.6	4.7
Wheat	4.3	4.5
Tomatoes	4.2	4.3
Rice (paddy)	2.8	3.2
Potatoes	1.6	1.7
Oranges	1.6	1.6

Values shown are 1,000 metric tons.

Top Mining Products [33]

Metric tons unless otherwise specified	M.t.
Phosphate rock (000 tons)	700
Clay, kaolin	49,000
Gypsum, crude	480,000
Cement, hydraulic (000 tons)	15,000
Sand, including glass sand (000 cubic meters)	23,000
Salt (000 tons)	900

Tourism [34]

	1987	1988	1989	1990	1991
Visitors	1,795	1,969	2,503	2,600	2,214
Tourists	1,671	1,833	2,351	2,411	2,112
Excursionists	124	136	152	189	102
Tourism receipts	1,586	1,784	1,646	1,994	2,029
Tourism expenditures	78	43	87	129	225
Fare receipts	309	375	370	430	300
Fare expenditures	70	72	52	46	57

Tourists are in thousands, money in million U.S. dollars.

For sources, notes, and explanations, see Annotated Source Appendix, page 1035.

Manufacturing Sector

GDP and Manufacturing Summary [35]

	1980	1985	1989	1990	% change 1980-1990	% change 1989-1990
GDP (million 1980 $)	17,433	24,646	47,546	28,538	63.7	-40.0
GDP per capita (1980 $)	426	530	928	544	27.7	-41.4
Manufacturing as % of GDP (current prices)	12.9	15.0	14.5[e]	19.5	51.2	34.5
Gross output (million $)	6,986	10,141	26,751[e]	15,976[e]	128.7	-40.3
Value added (million $)	1,769	2,938	7,121[e]	5,036[e]	184.7	-29.3
Value added (million 1980 $)	2,173	3,021	5,381[e]	3,508	61.4	-34.8
Industrial production index	100	162	159	156	56.0	-1.9
Employment (thousands)	868	917	968[e]	1,019[e]	17.4	5.3

Note: GDP stands for Gross Domestic Product. 'e' stands for estimated value.

Profitability and Productivity

	1980	1985	1989	1990	% change 1980-1990	% change 1989-1990
Intermediate input (%)	75	71	73[e]	68[e]	-9.3	-6.8
Wages, salaries, and supplements (%)	17[e]	18[e]	15[e]	13[e]	-23.5	-13.3
Gross operating surplus (%)	8[e]	11[e]	11[e]	18[e]	125.0	63.6
Gross output per worker ($)	7,984	10,977	27,648[e]	15,532[e]	94.5	-43.8
Value added per worker ($)	2,023	3,180	7,360[e]	4,897[e]	142.1	-33.5
Average wage (incl. benefits) ($)	1,360[e]	2,008[e]	4,257[e]	2,061[e]	51.5	-51.6

Profitability is in percent of gross output. Productivity is in U.S. $. 'e' stands for estimated value.

Profitability - 1990

Wages - 13.1%
Inputs - 68.7%
Surplus - 18.2%

The graphic shows percent of gross output.

Value Added in Manufacturing

	1980 $ mil.	1980 %	1985 $ mil.	1985 %	1989 $ mil.	1989 %	1990 $ mil.	1990 %	% change 1980-1990	% change 1989-1990
311 Food products	308	17.4	421	14.3	1,354[e]	19.0	1,102[e]	21.9	257.8	-18.6
313 Beverages	14	0.8	71	2.4	270[e]	3.8	97[e]	1.9	592.9	-64.1
314 Tobacco products	21	1.2	131	4.5	300[e]	4.2	181[e]	3.6	761.9	-39.7
321 Textiles	506	28.6	509	17.3	1,170[e]	16.4	763[e]	15.2	50.8	-34.8
322 Wearing apparel	6	0.3	15	0.5	36[e]	0.5	28[e]	0.6	366.7	-22.2
323 Leather and fur products	3	0.2	7	0.2	30[e]	0.4	12[e]	0.2	300.0	-60.0
324 Footwear	22	1.2	9	0.3	27[e]	0.4	22[e]	0.4	0.0	-18.5
331 Wood and wood products	9	0.5	24	0.8	33[e]	0.5	29[e]	0.6	222.2	-12.1
332 Furniture and fixtures	7	0.4	19	0.6	29[e]	0.4	64[e]	1.3	814.3	120.7
341 Paper and paper products	42	2.4	76	2.6	135[e]	1.9	59[e]	1.2	40.5	-56.3
342 Printing and publishing	39	2.2	101	3.4	199[e]	2.8	75[e]	1.5	92.3	-62.3
351 Industrial chemicals	69	3.9	145	4.9	292[e]	4.1	162[e]	3.2	134.8	-44.5
352 Other chemical products	87	4.9	205	7.0	506[e]	7.1	311[e]	6.2	257.5	-38.5
353 Petroleum refineries	40	2.3	59	2.0	141[e]	2.0	106[e]	2.1	165.0	-24.8
354 Miscellaneous petroleum and coal products	61	3.4	78	2.7	174[e]	2.4	54[e]	1.1	-11.5	-69.0
355 Rubber products	12	0.7	28	1.0	52[e]	0.7	21[e]	0.4	75.0	-59.6
356 Plastic products	33	1.9	-21	-0.7	143[e]	2.0	103[e]	2.0	212.1	-28.0
361 Pottery, china and earthenware	6	0.3	12	0.4	75[e]	1.1	43[e]	0.9	616.7	-42.7
362 Glass and glass products	17	1.0	22	0.7	74[e]	1.0	37[e]	0.7	117.6	-50.0
369 Other non-metal mineral products	78	4.4	167	5.7	525[e]	7.4	383[e]	7.6	391.0	-27.0
371 Iron and steel	88	5.0	98	3.3	123[e]	1.7	264[e]	5.2	200.0	114.6
372 Non-ferrous metals	64	3.6	279	9.5	421[e]	5.9	414[e]	8.2	546.9	-1.7
381 Metal products	42	2.4	95	3.2	248[e]	3.5	174[e]	3.5	314.3	-29.8
382 Non-electrical machinery	54	3.1	83	2.8	129[e]	1.8	185[e]	3.7	242.6	43.4
383 Electrical machinery	69	3.9	181	6.2	308[e]	4.3	117[e]	2.3	69.6	-62.0
384 Transport equipment	65	3.7	106	3.6	284[e]	4.0	187[e]	3.7	187.7	-34.2
385 Professional and scientific equipment	4	0.2	13	0.4	27[e]	0.4	39[e]	0.8	875.0	44.4
390 Other manufacturing industries	1	0.1	6	0.2	13[e]	0.2	5[e]	0.1	400.0	-61.5

Note: The industry codes shown are International Standard Industry codes (ISIC). Percentages are percent of total Value Added. 'e' stands for estimated value

Finance, Economics, and Trade

Economic Indicators [36]

In Millions of Egyptian Pounds (LE) unless otherwise noted[39].

	1989	1990	1991[40]
GDP[174]	45,648.0	48,228.0	50,177.0
GDP Growth Rate/(%)	5.5	5.6	4.0
Per Capita GDP			
(LE curr. prices)	1,588.0	NA	1,511.0
Money Supply (M1)	21,562.0	25,572.0	NA
Bank Lending Interest Rates	18.3	19.0	21.0
Savings Rate	11.7	12.0	15.5
Fixed Investment	9,282.0	9,745.0	9,040.0
CPI (%)	16.7	21.2	14.7
WPI (%)	25.2	21.8	NA

Balance of Payments Summary [37]

Values in millions of dollars.

	1987	1988	1989	1990	1991
Exports of goods (f.o.b.)	3,115.0	2,770.0	2,907.0	3,604.0	3,856.0
Imports of goods (f.o.b.)	-8,095.0	-9,378.0	-8,841.0	-10,303.0	-9,831.0
Trade balance	-4,980.0	-6,608.0	-5,934.0	-6,699.0	-5,975.0
Services - debits	-3,725.0	-3,858.0	-4,672.0	-5,667.0	-5,507.0
Services - credits	4,130.0	4,982.0	5,123.0	7,147.0	7,951.0
Private transfers (net)	3,604.0	3,770.0	3,293.0	4,284.0	4,054.0
Government transfers (net)	991.0	674.0	880.0	13,871.0	2,854.0
Long term capital (net)	1,102.0	1,486.0	1,870.0	-9,344.0	462.0
Short term capital (net)	-1,434.0	-178.0	-1,387.0	-1,667.0	-1,793.0
Errors and omissions	892.0	-362.0	414.0	631.0	730.0
Overall balance	580.0	-94.0	-413.0	2,556.0	2,776.0

Exchange Rates [38]

Currency: **Egyptian pounds.**
Symbol: **#E.**

Data are currency units per $1.

November 1992	3.3450
1990	2.7072
1989	2.5171
1988	2.2233
1987	1.5183

Imports and Exports

Top Import Origins [39]

$10.0 billion (c.i.f., FY92 est.).

Origins	%
EC	NA
US	NA
Japan	NA
Eastern Europe	NA

Top Export Destinations [40]

$3.6 billion (f.o.b., FY92 est.).

Destinations	%
EC	NA
Eastern Europe	NA
US	NA
Japan	NA

Foreign Aid [41]

	U.S. $	
US commitments, including Ex-Im (FY70-89)	15.7	billion
Western (non-US) countries, ODA and OOF bilateral commitments (1970-88)	10.1	billion
OPEC bilateral aid (1979-89)	2.9	billion
Communist countries (1970-89)	2.4	billion

Import and Export Commodities [42]

Import Commodities

Machinery & equipment
Foods
Fertilizers
Wood products
Durable consumer goods
Capital goods

Export Commodities

Crude oil, petroleum products
Cotton yarn
Raw cotton
Textiles
Metal products
Chemicals

El Salvador

Geography [1]

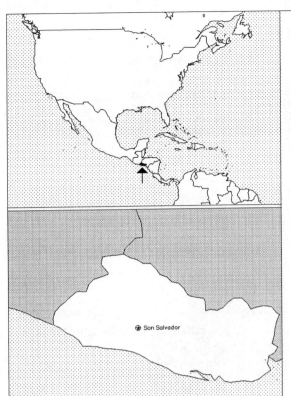

Total area:
21,040 km2
Land area:
20,720 km2
Comparative area:
Slightly smaller than Massachusetts
Land boundaries:
Total 545 km, Guatemala 203 km, Honduras 342 km
Coastline:
307 km
Climate:
Tropical; rainy season (May to October); dry season (November to April)
Terrain:
Mostly mountains with narrow coastal belt and central plateau
Natural resources:
Hydropower, geothermal power, petroleum
Land use:
Arable land:
27%
Permanent crops:
8%
Meadows and pastures:
29%
Forest and woodland:
6%
Other:
30%

Demographics [2]

	1960	1970	1980	1990	1991[1]	1994	2000	2010	2020
Population	2,574	3,583	4,655	5,303	5,419	5,753	6,459	7,603	8,763
Population density (persons per sq. mi.)	322	448	582	664	677	NA	809	953	1,101
Births	NA	NA	NA	NA	182	189	NA	NA	NA
Deaths	NA	NA	NA	NA	37	37	NA	NA	NA
Life expectancy - males	NA	NA	NA	NA	63	64	NA	NA	NA
Life expectancy - females	NA	NA	NA	NA	68	70	NA	NA	NA
Birth rate (per 1,000)	NA	NA	NA	NA	34	33	NA	NA	NA
Death rate (per 1,000)	NA	NA	NA	NA	7	6	NA	NA	NA
Women of reproductive age (15-44 yrs.)	NA	NA	NA	1,236	NA	1,415	1,625	NA	NA
of which are currently married	NA	NA	NA	735	NA	841	1,015	NA	NA
Fertility rate	NA	NA	NA	4.1	4.05	3.8	3.3	2.7	2.4

Population values are in thousands, life expectancy in years, and other items as indicated.

Health

Health Personnel [3]

Doctors per 1,000 pop., 1988-92	0.64
Nurse-to-doctor ratio, 1988-92	1.5
Hospital beds per 1,000 pop., 1985-90	1.5
Percentage of children immunized (age 1 yr. or less)	
Third dose of DPT, 1990-91	60.0
Measles, 1990-91	53.0

Health Expenditures [4]

Total health expenditure, 1990 (official exchange rate)	
Millions of dollars	317
Millions of dollars per capita	61
Health expenditures as a percentage of GDP	
Total	5.9
Public sector	2.6
Private sector	3.3
Development assistance for health	
Total aid flows (millions of dollars)[1]	44
Aid flows per capita (millions of dollars)	8.5
Aid flows as a percentage of total health expenditure	13.9

For sources, notes, and explanations, see Annotated Source Appendix, page 1035.

287

Human Factors

Health Care Ratios [5]

Population per physician, 1970	4100
Population per physician, 1990	NA
Population per nursing person, 1970	890
Population per nursing person, 1990	NA
Percent of births attended by health staff, 1985	35

Infants and Malnutrition [6]

Percent of babies with low birth weight, 1985	15
Infant mortality rate per 1,000 live births, 1970	103
Infant mortality rate per 1,000 live births, 1991	42
Years of life lost per 1,000 population, 1990	28
Prevalence of malnutrition (under age 5), 1990	NA

Ethnic Division [7]

Mestizo	94%
Indian	5%
White	1%

Religion [8]

Roman Catholic	75%

Major Languages [9]

Spanish, Nahua (among some Indians).

Education

Public Education Expenditures [10]

Million Colon	1980	1985	1987	1988	1989	1990
Total education expenditure	340	NA	NA	NA	625	722
as percent of GNP	3.9	NA	NA	NA	2.0	1.8
as percent of total govt. expend.	17.1	NA	NA	NA	26.9	28.1
Current education expenditure	320	NA	NA	NA	615	715
as percent of GNP	3.7	NA	NA	NA	2.0	1.8
as percent of current govt. expend.	22.9	NA	NA	NA	NA	NA
Capital expenditure	20	NA	NA	NA	10	7

Educational Attainment [11]

Age group	10+
Total population	3,132,400
Highest level attained (%)	
No schooling	30.2[6]
First level	
Incompleted	60.7
Completed	NA
Entered second level	
S-1	6.9
S-2	NA
Post secondary	2.3

Literacy Rate [12]

In thousands and percent	1985[a]	1991[a]	2000[a]
Illiterate population +15 years	803	787	780
Illiteracy rate - total pop. (%)	31.2	27.0	19.8
Illiteracy rate - males (%)	27.4	23.8	17.4
Illiteracy rate - females (%)	34.7	30.0	21.9

Libraries [13]

	Admin. Units	Svc. Pts.	Vols. (000)	Shelving (meters)	Vols. Added	Reg. Users
National (1989)	1	3	10	334	12,309	NA
Non-specialized	NA	NA	NA	NA	NA	NA
Public (1989)	44	83	55	NA	8,145	21,490
Higher ed. (1987)	110	695	220	NA	7,000	23,000
School (1987)	360	360	NA	NA	1,320	3,900

Daily Newspapers [14]

	1975	1980	1985	1990
Number of papers	10	7	4	5
Circ. (000)	211	291	243	457

Cinema [15]

Science and Technology

Scientific/Technical Forces [16]

R&D Expenditures [17]

	Colon[9] (000) 1989
Total expenditure	290,881
Capital expenditure	94,917
Current expenditure	195,964
Percent current	67.4

U.S. Patents Issued [18]

For sources, notes, and explanations, see Annotated Source Appendix, page 1035.

Government and Law

Organization of Government [19]

Long-form name:
Republic of El Salvador
Type:
republic
Independence:
15 September 1821 (from Spain)
Constitution:
20 December 1983
Legal system:
based on civil and Roman law, with traces of common law; judicial review of legislative acts in the Supreme Court; accepts compulsory ICJ jurisdiction, with reservations
National holiday:
Independence Day, 15 September (1821)
Executive branch:
president, vice president, Council of Ministers (cabinet)
Legislative branch:
unicameral Legislative Assembly (Asamblea Legislativa)
Judicial branch:
Supreme Court (Corte Suprema)

Crime [23]

Political Parties [20]

Legislative Assembly	% of votes
National Republican Alliance (ARENA)	44.3
Christian Democratic Party (PDC)	28.0
Democratic Convergence (CD)	12.2
National Conciliation Party (PCN)	9.0
Authentic Christian Movement (MAC)	3.2
Democratic Nationalist Union (UDN)	2.7

Government Expenditures [21]

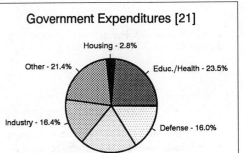

Housing - 2.8%
Other - 21.4%
Educ./Health - 23.5%
Industry - 16.4%
Defense - 16.0%
Gen. Services - 19.9%

% distribution for 1992. Expend. in 1992 est.: 890 ($ mil.)

Military Expenditures and Arms Transfers [22]

	1985	1986	1987	1988	1989
Military expenditures					
Current dollars (mil.)	276[e]	268	258[e]	237[e]	252[e]
1989 constant dollars (mil.)	314[e]	298	277[e]	247[e]	252[e]
Armed forces (000)	48	48	49	45	45
Gross national product (GNP)					
Current dollars (mil.)	5,161	5,324	5,647	6,017	6,335
1989 constant dollars (mil.)	5,876	5,908	6,073	6,263	6,335
Central government expenditures (CGE)					
1989 constant dollars (mil.)	1,082	845	783	709	675
People (mil.)	4.8	4.9	5.0	5.1	5.2
Military expenditure as % of GNP	5.3	5.0	4.6	3.9	4.0
Military expenditure as % of CGE	29.1	35.2	35.4	34.9	37.3
Military expenditure per capita	66	61	56	48	48
Armed forces per 1,000 people	10.0	9.8	9.8	8.8	8.6
GNP per capita	1,225	1,208	1,216	1,228	1,217
Arms imports[6]					
Current dollars (mil.)	90	80	50	60	70
1989 constant dollars (mil.)	102	89	54	62	70
Arms exports[6]					
Current dollars (mil.)	0	0	0	0	0
1989 constant dollars (mil.)	0	0	0	0	0
Total imports[7]					
Current dollars (mil.)	961	923	994	1,049	1,100
1989 constant dollars	1,094	1,024	1,069	1,092	1,100
Total exports[7]					
Current dollars (mil.)	679	757	591	566	497
1989 constant dollars	773	840	636	589	497
Arms as percent of total imports[8]	9.4	8.7	5.0	5.7	6.4
Arms as percent of total exports[8]	0	0	0	0	0

Human Rights [24]

	SSTS	FL	FAPRO	PPCG	APROBC	TPW	PCPTW	STPEP	PHRFF	PRW	ASST	AFL
Observes				P		P	P			S	S	P
	EAFRD	CPR	ESCR	SR	ACHR	MAAE	PVIAC	PVNAC	EAFDAW	TCIDTP	RC	
Observes		P	P	P	P		P	P	P		P	

P = Party; S = Signatory; see Appendix for meaning of abbreviations.

Labor Force

Total Labor Force [25]

1.7 million (1982 est.)

Labor Force by Occupation [26]

Agriculture	40%
Commerce	16
Manufacturing	15
Government	13
Financial services	9
Transportation	6
Other	1

Unemployment Rate [27]

7.5% (1991)

For sources, notes, and explanations, see Annotated Source Appendix, page 1035.

Production Sectors

Energy Resource Summary [28]

Energy Resources: Hydropower, geothermal power, petroleum. **Electricity**: 713,800 kW capacity; 2,190 million kWh produced, 390 kWh per capita (1992).

Telecommunications [30]

- Nationwide trunk microwave radio relay system
- Connection into Central American Microwave System
- 116,000 telephones (21 telephones per 1,000 persons)
- Broadcast stations - 77 AM, no FM, 5 TV, 2 shortwave
- 1 Atlantic Ocean INTELSAT earth station

Transportation [31]

Railroads. 602 km 0.914-meter gauge, single track; 542 km in use

Highways. 10,000 km total; 1,500 km paved, 4,100 km gravel, 4,400 km improved and unimproved earth

Airports

Total:	105
Usable:	74
With permanent-surface runways:	5
With runways over 3,659 m:	0
With runways 2,440-3,659 m:	1
With runways 1,220-2,439 m:	5

Top Agricultural Products [32]

		1[10]
Coffee (000 60-kg. bags)	2,787	2,402
Tobacco (metric tons)	680	727
Corn	582	596
Sugar cane	211	270
Sorghum	148	159
Rice (milled)	42	40

Values shown are 1,000 metric tons.

Top Mining Products [33]

Preliminary metric tons unless otherwise specified M.t.

Gypsum	5,000[e]
Limestone	1,900,000
Salt, marine	15,000
Cement	679,723
Petroleum refinery products (000 42-gal. barrels)	5,662
Steel, crude	11,000[e]

Tourism [34]

	1987	1988	1989	1990	1991
Tourists	125	134	131	194	199
Tourism receipts	43	61	63	145	157
Tourism expenditures	76	75	60	55	60
Fare receipts	49	58	59	49	51
Fare expenditures	4	4	8	10	11

Tourists are in thousands, money in million U.S. dollars.

Manufacturing Sector

GDP and Manufacturing Summary [35]

	1980	1985	1989	1990	% change 1980-1990	% change 1989-1990
GDP (million 1980 $)	3,567	3,247	3,375	3,562	-0.1	5.5
GDP per capita (1980 $)	788	681	657	681	-13.6	3.7
Manufacturing as % of GDP (current prices)	15.0	16.4	18.1	18.6	24.0	2.8
Gross output (million $)	1,130	860	1,587[e]	1,274[e]	12.7	-19.7
Value added (million $)	448	393	735[e]	603[e]	34.6	-18.0
Value added (million 1980 $)	536	471	525	541	0.9	3.0
Industrial production index	100	83	93	95	-5.0	2.2
Employment (thousands)	39	25	25[e]	26[e]	-33.3	4.0

Note: GDP stands for Gross Domestic Product. 'e' stands for estimated value.

Profitability and Productivity

	1980	1985	1989	1990	% change 1980-1990	% change 1989-1990
Intermediate input (%)	60	54	54[e]	53[e]	-11.7	-1.9
Wages, salaries, and supplements (%)	15[e]	12[e]	9[e]	12[e]	-20.0	33.3
Gross operating surplus (%)	24[e]	34[e]	38[e]	35[e]	45.8	-7.9
Gross output per worker ($)	28,857	34,129	63,062[e]	47,118[e]	63.3	-25.3
Value added per worker ($)	11,426	15,595	29,222[e]	22,472[e]	96.7	-23.1
Average wage (incl. benefits) ($)	4,383[e]	3,990[e]	5,427[e]	6,024[e]	37.4	11.0

Profitability is in percent of gross output. Productivity is in U.S. $. 'e' stands for estimated value.

Profitability - 1990

Inputs - 53.0%
Wages - 12.0%
Surplus - 35.0%

The graphic shows percent of gross output.

Value Added in Manufacturing

	1980 $ mil.	1980 %	1985 $ mil.	1985 %	1989 $ mil.	1989 %	1990 $ mil.	1990 %	% change 1980-1990	% change 1989-1990
311 Food products	78	17.4	55	14.0	81[e]	11.0	63[e]	10.4	-19.2	-22.2
313 Beverages	63	14.1	59	15.0	108[e]	14.7	105[e]	17.4	66.7	-2.8
314 Tobacco products	26	5.8	29	7.4	47[e]	6.4	42[e]	7.0	61.5	-10.6
321 Textiles	62	13.8	40	10.2	59[e]	8.0	55[e]	9.1	-11.3	-6.8
322 Wearing apparel	16	3.6	10	2.5	19[e]	2.6	15[e]	2.5	-6.3	-21.1
323 Leather and fur products	5	1.1	5	1.3	8[e]	1.1	7[e]	1.2	40.0	-12.5
324 Footwear	13	2.9	1	0.3	1[e]	0.1	3[e]	0.5	-76.9	200.0
331 Wood and wood products	1	0.2	NA	0.0	NA	0.0	NA	0.0	NA	NA
332 Furniture and fixtures	3	0.7	4	1.0	7[e]	1.0	6[e]	1.0	100.0	-14.3
341 Paper and paper products	40	8.9	24	6.1	51[e]	6.9	39[e]	6.5	-2.5	-23.5
342 Printing and publishing	8	1.8	8	2.0	12[e]	1.6	16[e]	2.7	100.0	33.3
351 Industrial chemicals	4	0.9	7	1.8	11[e]	1.5	11[e]	1.8	175.0	0.0
352 Other chemical products	46	10.3	57	14.5	129[e]	17.6	87[e]	14.4	89.1	-32.6
353 Petroleum refineries	14	3.1	20	5.1	47[e]	6.4	30[e]	5.0	114.3	-36.2
354 Miscellaneous petroleum and coal products	2	0.4	NA	0.0	2[e]	0.3	2[e]	0.3	0.0	0.0
355 Rubber products	4	0.9	3	0.8	4[e]	0.5	4[e]	0.7	0.0	0.0
356 Plastic products	13	2.9	15	3.8	36[e]	4.9	25[e]	4.1	92.3	-30.6
361 Pottery, china and earthenware	NA	0.0	NA	0.0	NA	0.0	NA	0.0	NA	NA
362 Glass and glass products	NA	0.0	NA	0.0	NA	0.0	NA	0.0	NA	NA
369 Other non-metal mineral products	11	2.5	13	3.3	23[e]	3.1	22[e]	3.6	100.0	-4.3
371 Iron and steel	9	2.0	7	1.8	14[e]	1.9	11[e]	1.8	22.2	-21.4
372 Non-ferrous metals	1	0.2	1	0.3	1[e]	0.1	1[e]	0.2	0.0	0.0
381 Metal products	10	2.2	12	3.1	26[e]	3.5	20[e]	3.3	100.0	-23.1
382 Non-electrical machinery	6	1.3	7	1.8	15[e]	2.0	11[e]	1.8	83.3	-26.7
383 Electrical machinery	9	2.0	12	3.1	27[e]	3.7	21[e]	3.5	133.3	-22.2
384 Transport equipment	1	0.2	NA	0.0	1[e]	0.1	NA	0.0	NA	NA
385 Professional and scientific equipment	NA	0.0	1	0.3	1[e]	0.1	1[e]	0.2	NA	0.0
390 Other manufacturing industries	4	0.9	2	0.5	5[e]	0.7	4[e]	0.7	0.0	-20.0

Note: The industry codes shown are International Standard Industry codes (ISIC). Percentages are percent of total Value Added. 'e' stands for estimated value

Finance, Economics, and Trade

Economic Indicators [36]

In Colones or U.S. Dollars as indicated.

	1989	1990	1991[e]
GDP (mil. current colones)	32,230.0	41,057.2	48,093.0
Real GDP growth (% change)	1.1	3.4	2.7
GDP per capita (const 1988 $)	913.3	926.0	932.5
Money (M1)			
(mil. current cols)	3,137.0	3,854.9	3,876.1
Savings rate (% GDP)	14.5	9.9	10.5
Investment rate (% GDP)	12.1	11.8	13.8
CPI (% change)	23.5	19.3	12.0
WPI (% change)	18.5	18.0	16.0
External public sector			
debt (bil. $)[41]	1.9	2.0	2.0

Balance of Payments Summary [37]

Values in millions of dollars.

	1987	1988	1989	1990	1991
Exports of goods (f.o.b.)	598.6	610.6	497.8	580.2	588.0
Imports of goods (f.o.b.)	-938.7	-966.5	-1,089.5	-1,180.0	-1,294.1
Trade balance	-349.1	-355.9	-591.7	-599.8	-706.1
Services - debits	-415.0	-471.0	-463.8	-428.8	-474.9
Services - credits	361.1	352.2	336.7	323.2	341.9
Private transfers (net)	180.5	202.1	207.8	324.0	469.9
Government transfers (net)	358.4	298.5	337.4	244.6	201.4
Long term capital (net)	-36.7	29.2	193.8	26.9	62.0
Short term capital (net)	-22.9	32.7	-43.3	-6.9	-90.0
Errors and omissions	7.0	-107.2	126.3	270.3	125.8
Overall balance	83.3	-19.4	103.2	153.5	-70.0

Exchange Rates [38]

Currency: **Salvadoran colones.**
Symbol: **C.**

Data are currency units per $1.

January 1993	8.76
1992	9.17
1991	8.03
1986-1989	
fixed rate	5.00

Imports and Exports

Top Import Origins [39]

$1.47 billion (c.i.f., 1992 est.).

Origins	%
US	43
Guatemala	NA
Mexico	NA
Venezuela	NA
Germany	NA

Top Export Destinations [40]

$693 million (f.o.b., 1992 est.).

Destinations	%
US	33
Guatemala	NA
Germany	NA
Costa Rica	NA

Foreign Aid [41]

	U.S. $	
US commitments, including Ex-Im (FY70-90)	2.95	billion
US commitments, including Ex-Im for 1992-96	250	million
Western (non-US) countries, ODA and OOF bilateral commitments (1970-89)	525	million

Import and Export Commodities [42]

Import Commodities	Export Commodities
Raw materials	Coffee 45%
Consumer goods	Sugar
Capital goods	Shrimp
	Cotton

Equatorial Guinea

Geography [1]

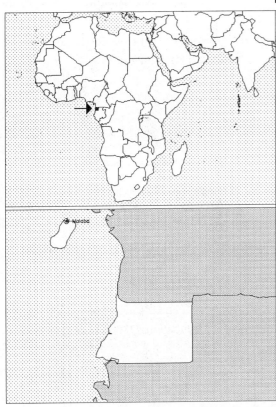

Total area:
28,050 km2
Land area:
28,050 km2
Comparative area:
Slightly larger than Maryland
Land boundaries:
Total 539 km, Cameroon 189 km, Gabon 350 km
Coastline:
296 km
Climate:
Tropical; always hot, humid
Terrain:
Coastal plains rise to interior hills; islands are volcanic
Natural resources:
Timber, petroleum, small unexploited deposits of gold, manganese, uranium
Land use:
Arable land:
8%
Permanent crops:
4%
Meadows and pastures:
4%
Forest and woodland:
51%
Other:
33%

Demographics [2]

	1960	1970	1980	1990	1991[1]	1994	2000	2010	2020
Population	244	270	256	369	379	410	478	615	783
Population density (persons per sq. mi.)	23	25	24	34	35	NA	44	57	73
Births	NA	NA	NA	NA	16	17	NA	NA	NA
Deaths	NA	NA	NA	NA	6	6	NA	NA	NA
Life expectancy - males	NA	NA	NA	NA	49	50	NA	NA	NA
Life expectancy - females	NA	NA	NA	NA	53	54	NA	NA	NA
Birth rate (per 1,000)	NA	NA	NA	NA	42	41	NA	NA	NA
Death rate (per 1,000)	NA	NA	NA	NA	16	15	NA	NA	NA
Women of reproductive age (15-44 yrs.)	NA	NA	NA	NA	NA	NA	NA	NA	NA
of which are currently married	NA	NA	NA	NA	NA	NA	NA	NA	NA
Fertility rate	NA	NA	NA	5.5	5.44	5.3	4.9	4.4	3.8

Population values are in thousands, life expectancy in years, and other items as indicated.

Health

Health Personnel [3]

Health Expenditures [4]

For sources, notes, and explanations, see Annotated Source Appendix, page 1035.

293

Human Factors

Health Care Ratios [5]	Infants and Malnutrition [6]

Ethnic Division [7]

Bioko (primarily Bubi, some Fernandinos), Rio Muni (primarily Fang), Europeans less than 1,000, mostly Spanish.

Religion [8]

Nominally Christian and predominantly Roman Catholic, pagan practices.

Major Languages [9]

Spanish (official)	NA
Pidgin English	NA
Fang	NA
Bubi	NA
Ibo	NA

Education

Public Education Expenditures [10]

Million Francs C.F.A.	1985	1986	1987	1988	1989	1990
Total education expenditure	NA	NA	NA	620	NA	NA
as percent of GNP	NA	NA	NA	1.7	NA	NA
as percent of total govt. expend.	NA	NA	NA	3.9	NA	NA
Current education expenditure	NA	NA	NA	527	NA	NA
as percent of GNP	NA	NA	NA	1.4	NA	NA
as percent of current govt. expend.	NA	NA	NA	3.5	NA	NA
Capital expenditure	NA	NA	NA	93	NA	NA

Educational Attainment [11]

Literacy Rate [12]

In thousands and percent	1985[a]	1991[a]	2000[a]
Illiterate population + 15 years	127	127	128
Illiteracy rate - total pop. (%)	55.1	49.8	39.5
Illiteracy rate - males (%)	40.6	35.9	27.4
Illiteracy rate - females (%)	68.8	63.0	51.1

Libraries [13]

Daily Newspapers [14]

	1975	1980	1985	1990
Number of papers	2	2	2	2
Circ. (000)	1[e]	2[e]	2[e]	2[e]

Cinema [15]

Science and Technology

Scientific/Technical Forces [16]	R&D Expenditures [17]	U.S. Patents Issued [18]

Government and Law

Organization of Government [19]

Long-form name:
Republic of Equatorial Guinea
Type:
republic in transition to multiparty democracy
Independence:
12 October 1968 (from Spain)
Constitution:
new constitution 17 November 1991
Legal system:
partly based on Spanish civil law and tribal custom
National holiday:
Independence Day, 12 October (1968)
Executive branch:
president, prime minister, deputy prime minister, Council of Ministers (cabinet)
Legislative branch:
unicameral House of Representatives of the People (Camara de Representantes del Pueblo)
Judicial branch:
Supreme Tribunal

Political Parties [20]

Chamber of People's Rep's	% of eats
Democratic Party for Equatorial Guinea (PDGE)	100.0

Government Budget [21]

For 1991 est.

Revenues	26
Expenditures	30
Capital expenditures	3

Military Expenditures and Arms Transfers [22]

	1985	1986	1987	1988	1989
Military expenditures					
Current dollars (mil.)	NA	NA	NA	NA	NA
1989 constant dollars (mil.)	NA	NA	NA	NA	NA
Armed forces (000)	3	2	2	1	1
Gross national product (GNP)					
Current dollars (mil.)	96	99	107	119	125
1989 constant dollars (mil.)	109	110	115	124	125
Central government expenditures (CGE)					
1989 constant dollars (mil.)	NA	NA	NA	27	27
People (mil.)	0.3	0.3	0.3	0.4	0.4
Military expenditure as % of GNP	NA	NA	NA	NA	NA
Military expenditure as % of CGE	NA	NA	NA	NA	NA
Military expenditure per capita	NA	NA	NA	NA	NA
Armed forces per 1,000 people	9.2	6.0	5.9	4.0	3.9
GNP per capita	336	329	337	354	347
Arms imports[6]					
Current dollars (mil.)	0	0	0	0	10
1989 constant dollars (mil.)	0	0	0	0	10
Arms exports[6]					
Current dollars (mil.)	0	0	0	0	0
1989 constant dollars (mil.)	0	0	0	0	0
Total imports[7]					
Current dollars (mil.)	25	41	50	50	NA
1989 constant dollars	28	45	54	52	NA
Total exports[7]					
Current dollars (mil.)	23	39	39	49	41
1989 constant dollars	26	43	42	51	41
Arms as percent of total imports[8]	0	0	0	0	NA
Arms as percent of total exports[8]	0	0	0	0	0

Crime [23]

Human Rights [24]

	SSTS	FL	FAPRO	PPCG	APROBC	TPW	PCPTW	STPEP	PHRFF	PRW	ASST	AFL
Observes						P	P					
	EAFRD	CPR	ESCR	SR	ACHR	MAAE	PVIAC	PVNAC	EAFDAW	TCIDTP	RC	
Observes		P	P	P		P	P	P	P		P	

P = Party; S = Signatory; see Appendix for meaning of abbreviations.

Labor Force

Total Labor Force [25]

172,000 (1986 est.)

Labor Force by Occupation [26]

Agriculture	66%
Services	23
Industry	11
Date of data: 1980	

Unemployment Rate [27]

For sources, notes, and explanations, see Annotated Source Appendix, page 1035.

295

Production Sectors

Energy Resource Summary [28]

Energy Resources: Petroleum, uranium. **Electricity**: 23,000 kW capacity; 60 million kWh produced, 160 kWh per capita (1991).

Telecommunications [30]

- Poor system with adequate government services
- International communications from Bata and Malabo to African and European countries
- 2,000 telephones
- Broadcast stations - 2 AM, no FM, 1 TV
- 1 Indian Ocean INTELSAT earth station

Transportation [31]

Highways. Rio Muni - 2,460 km; Bioko - 300 km

Merchant Marine. 2 ships (1,000 GRT or over) totaling 6,413 GRT/ 6,699 DWT; includes 1 cargo and 1 passenger-cargo

Airports

Total:	3
Usable:	3
With permanent-surface runways:	2
With runways over 3,659 m:	0
With runways 2,440-3,659 m:	1
With runways 1,220-2,439 m:	1

Top Agricultural Products [32]

Cash crops—timber and coffee from Rio Muni, cocoa from Bioko; food crops - rice, yams, cassava, bananas, oil palm nuts, manioc, livestock.

Top Mining Products [33]

Detailed information is not available. A summary of mineral resources available follows. **Mineral Resources**: Petroleum, small unexploited deposits of gold, manganese, uranium.

Tourism [34]

Finance, Economics, and Trade

GDP and Manufacturing Summary [35]

	1980	1985	1990	1991	1992
Gross Domestic Product					
Millions of 1980 dollars	43	49	54	53	51[e]
Growth rate in percent	-9.14	7.31	3.14	-0.93	-4.30[e]
Manufacturing Value Added					
Millions of 1980 dollars	0	0	0	0	0[e]
Growth rate in percent	-9.33	4.27	3.90	-4.66	-4.86[e]
Manufacturing share in percent of current prices	3.0	1.9	1.3[e]	NA	NA

Economic Indicators [36]

National product: GDP—exchange rate conversion—$144 million (1991 est.). **National product real growth rate**: - 1% (1991 est.). **National product per capita**: $380 (1991 est.). **Inflation rate (consumer prices)**: 1.4% (1990). **External debt**: $213 million (1990).

Balance of Payments Summary [37]

Values in millions of dollars.

	1987	1988	1989	1990	1991
Exports of goods (f.o.b.)	38.5	44.7	32.7	37.8	35.8
Imports of goods (f.o.b.)	-47.9	-56.5	-43.6	-53.2	-59.6
Trade balance	-9.4	-11.9	-10.9	-15.3	-23.8
Services - debits	-47.2	-56.6	-39.8	-46.0	-52.4
Services - credits	6.1	5.9	5.8	4.5	6.2
Private transfers (net)	-2.5	-3.8	-13.0	-17.1	-16.6
Government transfers (net)	28.8	49.9	39.3	59.0	64.4
Long term capital (net)	13.1	13.8	16.6	21.5	49.2
Short term capital (net)	7.8	8.1	7.0	-1.1	5.2
Errors and omissions	0.8	-1.7	-4.5	-2.4	-30.7
Overall balance	-2.5	3.7	0.6	3.1	1.5

Exchange Rates [38]

Currency: **Communaute Financiere Africaine francs.**
Symbol: **CFAF.**

Data are currency units per $1.

January 1993	274.06
1992	264.69
1991	282.11
1990	272.26
1989	319.01
1988	297.85

Imports and Exports

Top Import Origins [39]

$63.0 million (c.i.f., 1990). Data are for 1988.

Origins	%
France	25.9
Spain	21.0
Italy	16.0
US	12.8
Netherlands	8.0
FRG	3.1
Gabon	2.9
Nigeria	1.8

Top Export Destinations [40]

$37 million (f.o.b., 1990 est.). Data are for 1988.

Destinations	%
Spain	38.2
Italy	12.2
Netherlands	11.4
FRG	6.9
Nigeria	12.4

Foreign Aid [41]

	U.S. $	
US commitments, including Ex-Im (FY81-89)	14	million
Western (non-US) countries, ODA and OOF bilateral commitments (1970-89)	130	million
Communist countries (1970-89)	55	million

Import and Export Commodities [42]

Import Commodities	**Export Commodities**
Petroleum	Coffee
Food	Timber
Beverages	Cocoa beans
Clothing	
Machinery	

For sources, notes, and explanations, see Annotated Source Appendix, page 1035.

297

Estonia

Geography [1]

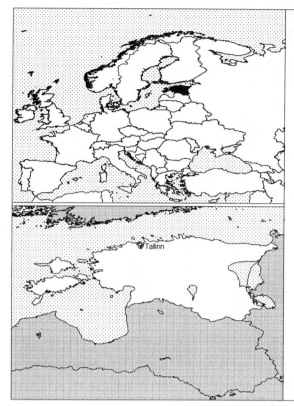

Total area:
45,100 km2
Land area:
43,200 km2
Comparative area:
Slightly larger than New Hampshire and Vermont combined
Land boundaries:
Total 557 km, Latvia 267 km, Russia 290 km
Coastline:
1,393 km
Climate:
Maritime, wet, moderate winters
Terrain:
Marshy, lowlands
Natural resources:
Shale oil, peat, phosphorite, amber
Land use:
Arable land:
22%
Permanent crops:
0%
Meadows and pastures:
11%
Forest and woodland:
31%
Other:
36%

Demographics [2]

	1960	1970	1980	1990	1991[1]	1994	2000	2010	2020
Population	1,211	1,363	1,482	1,583	NA	1,617	1,670	1,776	1,880
Population density (persons per sq. mi.)	NA	NA	NA	NA	NA	NA	NA	NA	NA
Births	NA	NA	NA	NA	NA	23	NA	NA	NA
Deaths	NA	NA	NA	NA	NA	19	NA	NA	NA
Life expectancy - males	NA	NA	NA	NA	NA	65	NA	NA	NA
Life expectancy - females	NA	NA	NA	NA	NA	75	NA	NA	NA
Birth rate (per 1,000)	NA	NA	NA	NA	NA	14	NA	NA	NA
Death rate (per 1,000)	NA	NA	NA	NA	NA	12	NA	NA	NA
Women of reproductive age (15-44 yrs.)	NA	NA	NA	384	NA	397	417	NA	NA
of which are currently married	NA	NA	NA	242	NA	249	260	NA	NA
Fertility rate	NA	NA	NA	2.1	NA	2.0	1.9	1.8	1.8

Population values are in thousands, life expectancy in years, and other items as indicated.

Health

Health Personnel [3]

Health Expenditures [4]

Human Factors

Health Care Ratios [5]

Population per physician, 1970	NA
Population per physician, 1990	210
Population per nursing person, 1970	NA
Population per nursing person, 1990	NA
Percent of births attended by health staff, 1985	NA

Infants and Malnutrition [6]

Percent of babies with low birth weight, 1985	NA
Infant mortality rate per 1,000 live births, 1970	20
Infant mortality rate per 1,000 live births, 1991	14
Years of life lost per 1,000 population, 1990	NA
Prevalence of malnutrition (under age 5), 1990	NA

Ethnic Division [7]

Estonian	61.50%
Russian	30.30%
Ukrainian	3.17%
Belarusian	1.80%
Finn	1.10%
Other	2.13%

Religion [8]

Lutheran.

Major Languages [9]

Estonian (official)	NA
Latvian	NA
Lithuanian	NA
Russian	NA
Other	NA

Education

Public Education Expenditures [10]

Educational Attainment [11]

Age group	25+
Total population	1,001,198
Highest level attained (%)	
No schooling	2.2
First level	
Incompleted	NA
Completed	39.0
Entered second level	
S-1	NA
S-2	45.1
Post secondary	13.7

Literacy Rate [12]

	1980[b]	1989[b]	1990[b]
Illiterate population +15 years	NA	3,329	NA
Illiteracy rate - total pop. (%)	NA	0.3	NA
Illiteracy rate - males (%)	NA	0.1	NA
Illiteracy rate - females (%)	NA	0.4	NA

Libraries [13]

Daily Newspapers [14]

Cinema [15]

Science and Technology

Scientific/Technical Forces [16]

R&D Expenditures [17]

U.S. Patents Issued [18]

For sources, notes, and explanations, see Annotated Source Appendix, page 1035.

299

Government and Law

Organization of Government [19]

Long-form name:
Republic of Estonia

Type:
republic

Independence:
6 September 1991 (from Soviet Union)

Constitution:
adopted 28 June 1992

Legal system:
based on civil law system; no judicial review of legislative acts

National holiday:
Independence Day, 24 February (1918)

Executive branch:
president, prime minister, cabinet

Legislative branch:
unicameral Parliament (Riigikogu)

Judicial branch:
Supreme Court

Political Parties [20]

Parliament	% of votes
Fatherland	21.0
Safe Home	14.0
Popular Front	13.0
Moderates	10.0
National Independence Party	8.0
Royalists	7.0
Estonian Citizen	7.0
Estonian Entrepreneurs	2.0
Other	18.0

Government Expenditures [21]

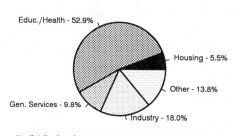

Educ./Health - 52.9%
Housing - 5.5%
Other - 13.8%
Gen. Services - 9.8%
Industry - 18.0%

% distribution for 1991. Expend. in 1992: 142 ($ mil.)

Defense Summary [22]

Branches: Ground Forces, Maritime Border Guard, National Guard (Kaitseliit), Security Forces (internal and border troops)

Manpower Availability: Males age 15-49 387,733; fit for military service 306,056; reach military age (18) annually 11,570 (1993 est.)

Defense Expenditures: 124.4 million kroons; note - conversion of the military budget into US dollars using the current exchange rate could produce misleading results

Crime [23]

Human Rights [24]

	SSTS	FL	FAPRO	PPCG	APROBC	TPW	PCPTW	STPEP	PHRFF	PRW	ASST	AFL
Observes	2			P								
	EAFRD	CPR	ESCR	SR	ACHR	MAAE	PVIAC	PVNAC	EAFDAW	TCIDTP	RC	
Observes	P	P	P						P	P	P	

P = Party; S = Signatory; see Appendix for meaning of abbreviations.

Labor Force

Total Labor Force [25]

796,000

Labor Force by Occupation [26]

Industry and construction	42%
Agriculture and forestry	20
Other	38

Date of data: 1990

Unemployment Rate [27]

3% (March 1993); but large number of underemployed workers

For sources, notes, and explanations, see Annotated Source Appendix, page 1035.

Production Sectors

Energy Resource Summary [28]

Energy Resources: Shale oil. **Electricity**: 3,700,000 kW capacity; 22,900 million kWh produced, 14,245 kWh per capita (1992). **Pipelines**: Natural gas 420 km (1992).

Telecommunications [30]

- 300,000 telephone subscribers (1990)—international and direct dial to Finland, Germany, Austria, UK and France
- 21 telephone lines per 100 persons (1991)
- Broadcast stations - 3 TV
- International traffic is carried to former USSR republics by landline or microwave and elsewhere by leased connection to the Moscow international gateway switch via 19 incoming/20 outgoing international channels, by the Finnish cellular net, and by an old copper submarine cable to Finland
- There is also a new international telephone exchange in Tallinn handling 60 channels via Helsinki
- 2 analog mobile cellular networks

Transportation [31]

Railroads. 1,030 km (includes NA km electrified); does not include industrial lines (1990)

Highways. 30,300 km total (1990); 29,200 km hard surfaced; 1,100 km earth

Merchant Marine. 68 ships (1,000 GRT or over) totaling 394,501 GRT/526,502 DWT; includes 52 cargo, 6 roll-on/roll-off, 2 short-sea passenger, 6 bulk, 2 container

Airports

Total:	29
Useable:	18
With permanent-surface runways:	11
With runways over 3,659 m:	0
With runways 2,440-3,659 m:	10
With runways 1,220-2,439 m:	8

Top Agricultural Products [32]

Employs 20% of work force; very efficient; net exports of meat, fish, dairy products, and potatoes; imports of feedgrains for livestock; fruits and vegetables.

Top Mining Products [33]

Detailed information is not available. A summary of mineral resources available follows. **Mineral Resources**: Shale oil, peat, phosphorite, amber.

Tourism [34]

Finance, Economics, and Trade

Industrial Summary [35]

Industrial Production: Growth rate - 40% (1992). **Industries**: Accounts for 30% of labor force; oil shale, shipbuilding, phosphates, electric motors, excavators, cement, furniture, clothing, textiles, paper, shoes, apparel.

Economic Indicators [36]

Millions of Rubles unless otherwise noted.

	1989	1990	1991
GNP (current prices)	4,478	4,543	NA
Money Supply	NA	NA	NA
Investment rate (% of GDP)	NA	NA	NA
Savings rate	NA	NA	NA
CPI	NA	NA	NA
WPI	NA	NA	NA
External public debt	NA	NA	NA

Balance of Payments Summary [37]

Exchange Rates [38]

Currency: **kroons.**
Symbol: **EEK.**

Data are currency units per $1.

January 1993	12

Imports and Exports

Top Import Origins [39]

Amount not available.

Origins	%
Finland	15
Russia	18

Top Export Destinations [40]

Amounts not available. Data are for 1992.

Destinations	%
Russia	NA
Former Soviet republics	50
West	50

Foreign Aid [41]

	U.S. $	
US commitments, including Ex-Im (1992)	10	million

Import and Export Commodities [42]

Import Commodities

Machinery 45%
Oil 13%
Chemicals 12%

Export Commodities

Textile 11%
Wood products and timber 9%
Dairy products 9%

Ethiopia

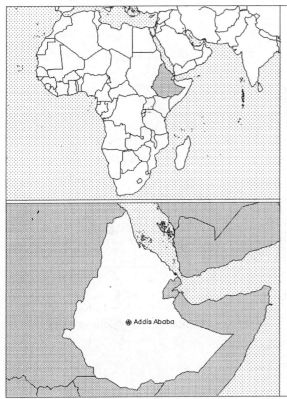

Geography [1]

Total area:
1,127,127 km2
Land area:
1,119,683 km2
Comparative area:
Slightly less than twice the size of Texas
Land boundaries:
Total 5,311 km, Djibouti 337 km, Erithea 912 km, Kenya 830 km, Somalia 1,626 km, Sudan 1,606 km
Coastline:
None - landlocked
Climate:
Tropical monsoon with wide topographic-induced variation; some areas prone to extended droughts
Terrain:
High plateau with central mountain range divided by Great Rift Valley
Natural resources:
Small reserves of gold, platinum, copper, potash
Land use:
Arable land:
12%
Permanent crops:
1%
Meadows and pastures:
41%
Forest and woodland:
24%
Other:
22%

Demographics [2]

	1960	1970	1980	1990	1991[1]	1994	2000	2010	2020
Population	25,864	31,826	38,967	51,507	53,191	58,710	70,340	94,496	124,294
Population density (persons per sq. mi.)	59	73	91	121	125	NA	163	220	291
Births	NA	NA	NA	NA	2,399	2,643	NA	NA	NA
Deaths	NA	NA	NA	NA	779	815	NA	NA	NA
Life expectancy - males	NA	NA	NA	NA	50	51	NA	NA	NA
Life expectancy - females	NA	NA	NA	NA	53	54	NA	NA	NA
Birth rate (per 1,000)	NA	NA	NA	NA	45	45	NA	NA	NA
Death rate (per 1,000)	NA	NA	NA	NA	15	14	NA	NA	NA
Women of reproductive age (15-44 yrs.)	NA	NA	NA	11,478	NA	13,092	15,754	NA	NA
of which are currently married	NA	NA	NA	9,130	NA	10,412	12,504	NA	NA
Fertility rate	NA	NA	NA	7.1	6.97	6.8	6.4	5.4	4.4

Population values are in thousands, life expectancy in years, and other items as indicated.

Health

Health Personnel [3]

Doctors per 1,000 pop., 1988-92	0.03
Nurse-to-doctor ratio, 1988-92	2.4
Hospital beds per 1,000 pop., 1985-90	0.3
Percentage of children immunized (age 1 yr. or less)	
Third dose of DPT, 1990-91	44.0
Measles, 1990-91	37.0

Health Expenditures [4]

Total health expenditure, 1990 (official exchange rate)	
Millions of dollars	229
Millions of dollars per capita	4
Health expenditures as a percentage of GDP	
Total	3.8
Public sector	2.3
Private sector	1.5
Development assistance for health	
Total aid flows (millions of dollars)[1]	43
Aid flows per capita (millions of dollars)	0.8
Aid flows as a percentage of total health expenditure	18.8

For sources, notes, and explanations, see Annotated Source Appendix, page 1035.

303

Human Factors

Health Care Ratios [5]

Population per physician, 1970	86120
Population per physician, 1990	32650
Population per nursing person, 1970	NA
Population per nursing person, 1990	NA
Percent of births attended by health staff, 1985	58

Infants and Malnutrition [6]

Percent of babies with low birth weight, 1985	NA
Infant mortality rate per 1,000 live births, 1970	158
Infant mortality rate per 1,000 live births, 1991	130
Years of life lost per 1,000 population, 1990	107
Prevalence of malnutrition (under age 5), 1990	NA

Ethnic Division [7]

Oromo	40%
Amhara Tigrean	32%
Sidamo	9%
Shankella	6%
Somali	6%
Afar	4%
Gurage	2%
Other	1%

Religion [8]

Muslim	45-50%
Ethiopian Orthodox	35-40%
Animist	12%
Other	5%

Major Languages [9]

English is the major foreign language taught in schools.

Amharic (official)	NA
Tigrinya	NA
Orominga	NA
Guaraginga	NA
Somali	NA
Arabic	NA

Education

Public Education Expenditures [10]

Million Birr	1980	1985	1987	1988	1989	1990
Total education expenditure	279	420	476	509	577	600
as percent of GNP	3.3	4.3	4.2	4.3	4.7	4.9
as percent of total govt. expend.	10.4	9.5	9.3	8.4	8.8	9.4
Current education expenditure	222	354	406	437	471	494
as percent of GNP	2.6	3.6	3.6	3.7	3.8	4.0
as percent of current govt. expend.	12.7	14.3	13.1	11.6	11.3	11.1
Capital expenditure	57	66	70	72	107	105

Educational Attainment [11]

Literacy Rate [12]

	1970[b]	1980[b,4]	1989[b,4]
Illiterate population +15 years	134,633	NA	64,510
Illiteracy rate - total pop. (%)	63.1	52.6	33.5
Illiteracy rate - males (%)	NA	38.6	NA
Illiteracy rate - females (%)	NA	61.4	NA

Libraries [13]

	Admin. Units	Svc. Pts.	Vols. (000)	Shelving (meters)	Vols. Added	Reg. Users
National (1986)	1	10	100	NA	22,608	-
Non-specialized	NA	NA	NA	NA	NA	NA
Public (1986)	4	17	124	NA	12,197	11,680
Higher ed. (1986)	5	13	530	10,100	25,000	9,400
School	NA	NA	NA	NA	NA	NA

Daily Newspapers [14]

	1975	1980	1985	1990[1]
Number of papers	3	3	3	3
Circ. (000)	44	40	41	42

Cinema [15]

Science and Technology

Scientific/Technical Forces [16]

Potential scientists/engineers	47,113
Number female	12,476
Potential technicians	NA
Number female	NA
Total	NA

R&D Expenditures [17]

U.S. Patents Issued [18]

         For sources, notes, and explanations, see Annotated Source Appendix, page 1035.

Government and Law

Organization of Government [19]

Long-form name:
none
Type:
transitional government
Independence:
oldest independent country in Africa and one of the oldest in the world - at least 2,000 years
Constitution:
to be redrafted by 1993
Legal system:
NA
National holiday:
National Day, 28 May (1991) (defeat of Mengistu regime)
Executive branch:
president, prime minister, Council of Ministers
Legislative branch:
unicameral Constituent Assembly
Judicial branch:
Supreme Court

Crime [23]

Elections [20]

Constituent Assembly. Now planned for January 1994 (to ratify constitution to be drafted by end of 1993).

Government Expenditures [21]

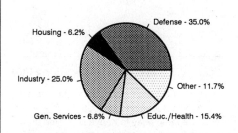

Defense - 35.0%
Housing - 6.2%
Industry - 25.0%
Other - 11.7%
Gen. Services - 6.8%
Educ./Health - 15.4%

% distribution for 1989. Expend. in FY91: 2.300 ($ bil.)

Military Expenditures and Arms Transfers [22]

	1985	1986	1987	1988	1989
Military expenditures					
Current dollars (mil.)	382[e]	388	432	NA	763
1989 constant dollars (mil.)	435[e]	430	464	NA	763
Armed forces (000)	240	300	300	250	250
Gross national product (GNP)					
Current dollars (mil.)	4,253	4,661	5,245	5,596	5,959
1989 constant dollars (mil.)	4,842	5,172	5,641	5,825	5,959
Central government expenditures (CGE)					
1989 constant dollars (mil.)	1,857	1,914	1,991	2,242	2,507
People (mil.)	43.5	44.7	46.3	48.0	49.8
Military expenditure as % of GNP	9.0	8.3	8.2	NA	12.8
Military expenditure as % of CGE	23.4	22.5	23.3	NA	30.4
Military expenditure per capita	10	10	10	NA	15
Armed forces per 1,000 people	5.5	6.7	6.5	5.2	5.0
GNP per capita	111	116	122	121	120
Arms imports[6]					
Current dollars (mil.)	775	330	1,000	700	925
1989 constant dollars (mil.)	882	366	1,076	729	925
Arms exports[6]					
Current dollars (mil.)	0	0	0	0	0
1989 constant dollars (mil.)	0	0	0	0	0
Total imports[7]					
Current dollars (mil.)	993	1,102	1,066	1,129	951
1989 constant dollars	1,130	1,223	1,147	1,175	951
Total exports[7]					
Current dollars (mil.)	333	455	355	429	440
1989 constant dollars	379	505	382	447	440
Arms as percent of total imports[8]	78.0	29.9	93.8	62.0	97.3
Arms as percent of total exports[8]	0	0	0	0	0

Human Rights [24]

	SSTS	FL	FAPRO	PPCG	APROBC	TPW	PCPTW	STPEP	PHRFF	PRW	ASST	AFL
Observes	P		P	P	P	P	P	P		P	P	
		EAFRD	CPR	ESCR	SR	ACHR	MAAE	PVIAC	PVNAC	EAFDAW	TCIDTP	RC
Observes		P			P					P		P

P = Party; S = Signatory; see Appendix for meaning of abbreviations.

Labor Force

Total Labor Force [25]

18 million

Labor Force by Occupation [26]

Agriculture and animal husbandry	80%
Government and services	12
Industry and construction	8

Date of data: 1985

Unemployment Rate [27]

For sources, notes, and explanations, see Annotated Source Appendix, page 1035.

305

Production Sectors

Energy Resource Summary [28]

Energy Resources: None. **Electricity**: 330,000 kW capacity; 650 million kWh produced, 10 kWh per capita (1991).

Telecommunications [30]

- Open-wire and radio relay system adequate for government use
- Open-wire to Sudan and Djibouti
- Microwave radio relay to Kenya and Djibouti
- Broadcast stations - 4 AM, no FM, 1 TV
- 100,000 TV sets
- 9,000,000 radios
- Satellite earth stations - 1 Atlantic Ocean INTELSAT and 2 Pacific Ocean INTELSAT

Top Agricultural Products [32]

Agriculture accounts for 47% of GDP and is the most important sector of the economy even though frequent droughts and poor cultivation practices keep farm output low; famines not uncommon; export crops of coffee and oilseeds grown partly on state farms; estimated 50% of agricultural production at subsistence level; principal crops and livestock—cereals, pulses, coffee, oilseeds, sugarcane, potatoes and other vegetables, hides and skins, cattle, sheep, goats.

Top Mining Products [33]

Estimated metric tons unless otherwise specified   M.t.

Construction stone, crushed (000 tons)	1,300
Cement, hydraulic	290,000
Lime	3,400
Salt, marine	85,000
Pumice	20,000[36]
Sand (000 tons)	1,000

Transportation [31]

Railroads. 781 km total; 781 km 1.000-meter gauge; 307 km 0.950-meter gauge linking Addis Ababa (Ethiopia) to Djibouti; control of railroad is shared between Djibouti and Ethiopia

Highways. 39,150 km total; 2,776 km paved, 7,504 km gravel, 2,054 km improved earth, 26,816 km unimproved earth (1993 est.)

Merchant Marine. none; landlocked

Airports

Total:	121
Usable:	82
With permanent-surface runways:	9
With runways over 3,659 m:	1
With runways 2,440-3,659 m:	13
With runways 1,220-2,439 m:	83 (1993 est.)

Tourism [34]

	1987	1988	1989	1990	1991
Tourists[26]	73	76	77	79	82
Tourism receipts	15	19	21	26	20
Tourism expenditures	6	6	10	11	
Fare receipts	79	91	117	138	

Tourists are in thousands, money in million U.S. dollars.

Manufacturing Sector

GDP and Manufacturing Summary [35]

	1980	1985	1989	1990	% change 1980-1990	% change 1989-1990
GDP (million 1980 $)	4,106	4,064	4,847	4,826	17.5	-0.4
GDP per capita (1980 $)	106	94	101	98	-7.5	-3.0
Manufacturing as % of GDP (current prices)	10.8	11.5	11.1[e]	11.3	4.6	1.8
Gross output (million $)	1,016	1,375	1,848[e]	1,787[e]	75.9	-3.3
Value added (million $)	459	577[e]	868[e]	861[e]	87.6	-0.8
Value added (million 1980 $)	400	479	591[e]	526	31.5	-11.0
Industrial production index	100	121	152[e]	135	35.0	-11.2
Employment (thousands)	77	88[e]	102[e]	105[e]	36.4	2.9

Note: GDP stands for Gross Domestic Product. 'e' stands for estimated value.

Profitability and Productivity

	1980	1985	1989	1990	% change 1980-1990	% change 1989-1990
Intermediate input (%)	55	58[e]	53[e]	52[e]	-5.5	-1.9
Wages, salaries, and supplements (%)	8	9	9[e]	9[e]	12.5	0.0
Gross operating surplus (%)	37	33[e]	38[e]	39[e]	5.4	2.6
Gross output per worker ($)	13,215	15,558	18,083[e]	16,984[e]	28.5	-6.1
Value added per worker ($)	6,009	6,528[e]	8,490[e]	8,191[e]	36.3	-3.5
Average wage (incl. benefits) ($)	1,079	1,332[e]	1,631[e]	1,549[e]	43.6	-5.0

Profitability is in percent of gross output. Productivity is in U.S. $. 'e' stands for estimated value.

Profitability - 1990

Inputs - 52.0%
Wages - 9.0%
Surplus - 39.0%

The graphic shows percent of gross output.

Value Added in Manufacturing

	1980 $ mil.	1980 %	1985 $ mil.	1985 %	1989 $ mil.	1989 %	1990 $ mil.	1990 %	% change 1980-1990	% change 1989-1990
311 Food products	110	24.0	114[e]	19.8	152[e]	17.5	143[e]	16.6	30.0	-5.9
313 Beverages	83	18.1	141	24.4	198[e]	22.8	196[e]	22.8	136.1	-1.0
314 Tobacco products	30	6.5	35	6.1	69[e]	7.9	63[e]	7.3	110.0	-8.7
321 Textiles	106	23.1	69	12.0	114[e]	13.1	115[e]	13.4	8.5	0.9
322 Wearing apparel	3	0.7	11	1.9	12[e]	1.4	11[e]	1.3	266.7	-8.3
323 Leather and fur products	14	3.1	13	2.3	32[e]	3.7	36[e]	4.2	157.1	12.5
324 Footwear	10	2.2	10	1.7	13[e]	1.5	12[e]	1.4	20.0	-7.7
331 Wood and wood products	8	1.7	6	1.0	9[e]	1.0	9[e]	1.0	12.5	0.0
332 Furniture and fixtures	2	0.4	4	0.7	6[e]	0.7	5[e]	0.6	150.0	-16.7
341 Paper and paper products	9	2.0	9	1.6	6[e]	0.7	7[e]	0.8	-22.2	16.7
342 Printing and publishing	11	2.4	17	2.9	21[e]	2.4	18[e]	2.1	63.6	-14.3
351 Industrial chemicals	1	0.2	1	0.2	2[e]	0.2	2[e]	0.2	100.0	0.0
352 Other chemical products	13	2.8	21	3.6	25[e]	2.9	27[e]	3.1	107.7	8.0
353 Petroleum refineries	20	4.4	54	9.4	112[e]	12.9	129[e]	15.0	545.0	15.2
354 Miscellaneous petroleum and coal products	NA	0.0	NA	0.0	NA	0.0	NA	0.0	NA	NA
355 Rubber products	8	1.7	13	2.3	14[e]	1.6	13[e]	1.5	62.5	-7.1
356 Plastic products	3	0.7	11	1.9	13[e]	1.5	12[e]	1.4	300.0	-7.7
361 Pottery, china and earthenware	NA	0.0	NA	0.0	NA	0.0	NA	0.0	NA	NA
362 Glass and glass products	2	0.4	4	0.7	5[e]	0.6	5[e]	0.6	150.0	0.0
369 Other non-metal mineral products	8	1.7	19	3.3	22[e]	2.5	21[e]	2.4	162.5	-4.5
371 Iron and steel	9	2.0	8	1.4	13[e]	1.5	10[e]	1.2	11.1	-23.1
372 Non-ferrous metals	NA	0.0	NA	0.0	NA	0.0	NA	0.0	NA	NA
381 Metal products	7	1.5	12	2.1	14[e]	1.6	13[e]	1.5	85.7	-7.1
382 Non-electrical machinery	NA	0.0	NA	0.0	NA	0.0	NA	0.0	NA	NA
383 Electrical machinery	NA	0.0	1	0.2	1[e]	0.1	1[e]	0.1	NA	0.0
384 Transport equipment	NA	0.0	7	1.2	14[e]	1.6	13[e]	1.5	NA	-7.1
385 Professional and scientific equipment	NA	0.0	NA	0.0	NA	0.0	NA	0.0	NA	NA
390 Other manufacturing industries	NA	0.0	NA	0.0	NA	0.0	NA	0.0	NA	NA

Note: The industry codes shown are International Standard Industry codes (ISIC). Percentages are percent of total Value Added. 'e' stands for estimated value

For sources, notes, and explanations, see Annotated Source Appendix, page 1035.

307

Finance, Economics, and Trade

Economic Indicators [36]

National product: GDP—exchange rate conversion—$6.6 billion (FY92 est.). **National product real growth rate**: 6% (FY92 est.). **National product per capita**: $130 (FY92 est.). **Inflation rate (consumer prices)**: 7.8% (1989). **External debt**: $3.48 billion (1991).

Balance of Payments Summary [37]

Values in millions of dollars.

	1986	1987	1988	1989	1990
Exports of goods (f.o.b.)	477.1	355.2	400.0	451.3	301.7
Imports of goods (f.o.b.)	-932.6	-932.7	-956.0	-817.9	-912.1
Trade balance	-455.5	-577.5	-556.0	-366.6	-610.4
Services - debits	-332.9	-366.4	-402.9	-408.4	-436.6
Services - credits	277.9	318.7	288.9	301.8	313.8
Private transfers (net)	69.4	129.6	180.5	145.7	229.1
Government transfers (net)	113.7	278.0	261.6	190.5	220.0
Long term capital (net)	240.6	292.8	292.9	170.8	132.0
Short term capital (net)	-1.0	0.4	6.8	42.6	274.8
Errors and omissions	201.6	-182.8	-94.0	-39.4	-144.3
Overall balance	113.8	-107.2	-22.2	37.0	-21.6

Exchange Rates [38]

Currency: **birr.**
Symbol: **Br.**

Data are currency units per $1.

Fixed rate	5.0000

Imports and Exports

Top Import Origins [39]

$1.0 billion (c.i.f., FY90).

Origins	%
EC	NA
Eastern Europe	NA
Japan	NA
US	NA

Top Export Destinations [40]

$276 million (f.o.b., FY90).

Destinations	%
EC	NA
Djibouti	NA
Japan	NA
Saudi Arabia	NA
US	NA

Foreign Aid [41]

	U.S. $	
US commitments, including Ex-Im (FY70-89)	504	million
Western (non-US) countries, ODA and OOF bilateral commitments (1970-89)	3.4	billion
OPEC bilateral aid (1979-89)	8	million
Communist countries (1970-89)	2.0	billion

Import and Export Commodities [42]

Import Commodities	Export Commodities
Capital goods	Coffee
Consumer goods	Leather products
Fuel	Gold
	Petroleum products

Fiji

Geography [1]

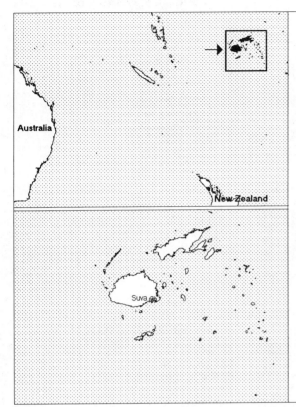

Total area:
18,270 km2
Land area:
18,270 km2
Comparative area:
Slightly smaller than New Jersey
Land boundaries:
0 km
Coastline:
1,129 km
Climate:
Tropical marine; only slight seasonal temperature variation
Terrain:
Mostly mountains of volcanic origin
Natural resources:
Timber, fish, gold, copper, offshore oil potential
Land use:
Arable land:
8%
Permanent crops:
5%
Meadows and pastures:
3%
Forest and woodland:
65%
Other:
19%

Demographics [2]

	1960	1970	1980	1990	1991[1]	1994	2000	2010	2020
Population	393	521	635	738	744	764	823	933	1,037
Population density (persons per sq. mi.)	56	74	90	105	105	NA	117	132	147
Births	NA	NA	NA	NA	19	18	NA	NA	NA
Deaths	NA	NA	NA	NA	5	5	NA	NA	NA
Life expectancy - males	NA	NA	NA	NA	62	63	NA	NA	NA
Life expectancy - females	NA	NA	NA	NA	67	68	NA	NA	NA
Birth rate (per 1,000)	NA	NA	NA	NA	26	24	NA	NA	NA
Death rate (per 1,000)	NA	NA	NA	NA	7	7	NA	NA	NA
Women of reproductive age (15-44 yrs.)	NA	NA	NA	188	NA	198	221	NA	NA
of which are currently married	NA	NA	NA	124	NA	130	144	NA	NA
Fertility rate	NA	NA	NA	3.1	3.09	2.9	2.7	2.4	2.2

Population values are in thousands, life expectancy in years, and other items as indicated.

Health

Health Personnel [3]

Health Expenditures [4]

For sources, notes, and explanations, see Annotated Source Appendix, page 1035.

309

Human Factors

Health Care Ratios [5]	Infants and Malnutrition [6]

Ethnic Division [7]	Religion [8]		Major Languages [9]
Other includes European, other Pacific Islanders, overseas Chinese.	Christian	52%	English (official), Fijian, Hindustani.
	Methodist	37%	
Fijian 49%	Roman Catholic	9%	
Indian 46%	Hindu	38%	
Other 5%	Muslim	8%	
	Other	2%	

Education

Public Education Expenditures [10]

Million Fiji Dollars	1980	1985	1986	1988	1989	1990
Total education expenditure	50	NA	85	NA	86	NA
as percent of GNP	5.1	NA	6.0	NA	5.0	NA
as percent of total govt. expend.	NA	NA	NA	NA	15.4	NA
Current education expenditure	48	NA	83	NA	82	NA
as percent of GNP	4.9	NA	5.9	NA	4.7	NA
as percent of current govt. expend.	21.5	NA	NA	NA	19.4	NA
Capital expenditure	2	NA	2	NA	4	NA

Educational Attainment [11]

Age group	25+
Total population	287,175
Highest level attained (%)	
No schooling	10.9
First level	
Incompleted	35.9
Completed	23.9
Entered second level	
S-1	24.9
S-2	NA
Post secondary	4.5

Literacy Rate [12]

	1976[b]	1980[b]	1989[b]
Illiterate population + 15 years	65,957	NA	NA
Illiteracy rate - total pop. (%)	21.0	NA	NA
Illiteracy rate - males (%)	16.0	NA	NA
Illiteracy rate - females (%)	26.0	NA	NA

Libraries [13]

	Admin. Units	Svc. Pts.	Vols. (000)	Shelving (meters)	Vols. Added	Reg. Users
National	NA	NA	NA	NA	NA	NA
Non-specialized	NA	NA	NA	NA	NA	NA
Public (1986)	1	4	71	NA	3,374	24,677
Higher ed. (1989)	7	7	65	1,905	2,250	800
School (1987)	1	2	18	1,005	4,500	17,000

Daily Newspapers [14]

	1975	1980	1985	1990
Number of papers	1	3	3	1
Circ. (000)	20	64	68	27

Cinema [15]

Science and Technology

Scientific/Technical Forces [16]	R&D Expenditures [17]	U.S. Patents Issued [18]

R&D Expenditures [17]

	Dollar[35] (000) 1986
Total expenditure	3,800
Capital expenditure	800
Current expenditure	3,000
Percent current	78.9

For sources, notes, and explanations, see Annotated Source Appendix, page 1035.

Government and Law

Organization of Government [19]

Long-form name:
Republic of Fiji
Type:
republic
Independence:
10 October 1970 (from UK)
Constitution:
10 October 1970 (suspended 1 October 1987); a new Constitution was proposed on 23 September 1988 and promulgated on 25 July 1990; the 1990 Constitution is currently still under review (February 1993)
Legal system:
based on British system
National holiday:
Independence Day, 10 October (1970)
Executive branch:
president, prime minister, Cabinet, Great Councils of Chiefs (highest ranking members of the traditional chiefly system)
Legislative branch:
bicameral Parliament (Senate, House), was dissolved following the coup of 14 May 1987; the Constitution of 23 September 1988 provides for a bicameral Parliament
Judicial branch:
Supreme Court

Elections [20]

House of Representatives. Last held 23-29 May 1992 (next to be held NA 1997); results - percent of vote by party NA; seats - (70 total, with ethnic Fijians allocated 37 seats, ethnic Indians 27 seats, and independents and other 6 seats) number of seats by party NA.

Government Expenditures [21]

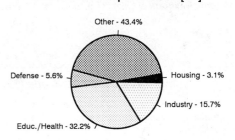

% distribution for 1993. Expend. in 1993 est.: 546 ($ mil.)

Military Expenditures and Arms Transfers [22]

	1985	1986	1987	1988	1989
Military expenditures					
Current dollars (mil.)	12	12	22	24	26
1989 constant dollars (mil.)	13	13	24	25	26
Armed forces (000)	4	3	3	4	4
Gross national product (GNP)					
Current dollars (mil.)	926	1,037	985	995	1,164
1989 constant dollars (mil.)	1,054	1,151	1,059	1,036	1,164
Central government expenditures (CGE)					
1989 constant dollars (mil.)	312	335	317	286	NA
People (mil.)	0.7	0.7	0.7	0.7	0.8
Military expenditure as % of GNP	1.3	1.2	2.3	2.4	2.2
Military expenditure as % of CGE	4.3	4.0	7.6	8.8	NA
Military expenditure per capita	19	19	33	34	35
Armed forces per 1,000 people	5.8	3.8	3.7	4.7	4.6
GNP per capita	1,517	1,624	1,462	1,399	1,540
Arms imports[6]					
Current dollars (mil.)	0	0	0	0	0
1989 constant dollars (mil.)	0	0	0	0	0
Arms exports[6]					
Current dollars (mil.)	0	0	0	0	0
1989 constant dollars (mil.)	0	0	0	0	0
Total imports[7]					
Current dollars (mil.)	442	437	380	454	615
1989 constant dollars	503	485	409	473	615
Total exports[7]					
Current dollars (mil.)	237	274	299	312	370
1989 constant dollars	270	304	322	325	370
Arms as percent of total imports[8]	0	0	0	0	0
Arms as percent of total exports[8]	0	0	0	0	0

Crime [23]

Crime volume	
Cases known to police	14,169
Attempts (percent)	NA
Percent cases solved	51
Crimes per 100,000 persons	1,914.7
Persons responsible for offenses	
Total number offenders	6,203
Percent female	101
Percent juvenile[24]	206
Percent foreigners	19

Human Rights [24]

	SSTS	FL	FAPRO	PPCG	APROBC	TPW	PCPTW	STPEP	PHRFF	PRW	ASST	AFL
Observes	P	P		P	P	P	P			P	P	P
	EAFRD	CPR	ESCR	SR	ACHR	MAAE	PVIAC	PVNAC	EAFDAW	TCIDTP	RC
Observes	1			P	P	P	P				P	P

P = Party; S = Signatory; see Appendix for meaning of abbreviations.

Labor Force

Total Labor Force [25]

235,000

Labor Force by Occupation [26]

Subsistence agriculture	67%
Wage earners	18
Salary earners	15

Date of data: 1987

Unemployment Rate [27]

5.9% (1991 est.)

Production Sectors

Energy Resource Summary [28]

Energy Resources: Offshore oil potential. **Electricity**: 215,000 kW capacity; 420 million kWh produced, 560 kWh per capita (1992).

Telecommunications [30]

- Modern local, interisland, and international (wire/radio integrated) public and special-purpose telephone, telegraph, and teleprinter facilities
- Regional radio center
- Important COMPAC cable link between US-Canada and New Zealand-Australia
- 53,228 telephones (71 telephones per 1,000 persons)
- Broadcast stations - 7 AM, 1 FM, no TV
- 1 Pacific Ocean INTELSAT earth station

Top Agricultural Products [32]

Agriculture accounts for 23% of GDP; principal cash crop is sugarcane; coconuts, cassava, rice, sweet potatoes, bananas; small livestock sector includes cattle, pigs, horses, and goats; fish catch nearly 33,000 tons (1989).

Top Mining Products [33]

Preliminary metric tons unless otherwise specified M.t.[4]

Cement, hydraulic	90,000[e]
Gold, mine output, Au content	2,827
Limestone	NA
Silver, mine output, Ag content	494
Stone, sand and gravel (cubic meters)	964,801

Transportation [31]

Railroads. 644 km 0.610-meter narrow gauge, belonging to the government-owned Fiji Sugar Corporation

Highways. 3,300 km total; 1,590 km paved; 1,290 km gravel, crushed stone, or stabilized soil surface; 420 unimproved earth (1984)

Merchant Marine. 7 ships (1,000 GRT or over) totaling 40,072 GRT/47,187 DWT; includes 2 roll-on/roll-off, 2 container, 1 oil tanker, 1 chemical tanker, 1 cargo

Airports

Total:	25
Usable:	22
With permanent-surface runways:	2
With runways over 3,659 m:	0
With runways 2,440-3,659 m:	1
With runways 1,220-2,439 m:	2

Tourism [34]

	1987	1988	1989	1990	1991
Visitors	222	228	281	307	287
Tourists	190	208	251	279	259
Cruise passengers	33	20	31	28	27
Tourism receipts	121	131	199	227	211
Tourism expenditures	53	35	41	41	45
Fare receipts	22	32	54	72	84
Fare expenditures	11	12	19	21	15

Tourists are in thousands, money in million U.S. dollars.

Manufacturing Sector

GDP and Manufacturing Summary [35]

	1980	1985	1989	1990	% change 1980-1990	% change 1989-1990
GDP (million 1980 $)	1,204	1,266	1,424	1,506	25.1	5.8
GDP per capita (1980 $)	1,900	1,809	1,894	1,966	3.5	3.8
Manufacturing as % of GDP (current prices)	11.5	9.1	11.0[e]	10.1[e]	-12.2	-8.2
Gross output (million $)	489	395	461[e]	642	31.3	39.3
Value added (million $)	121	90	98[e]	142	17.4	44.9
Value added (million 1980 $)	127	124	155	154	21.3	-0.6
Industrial production index	100	89	101	98	-2.0	-3.0
Employment (thousands)	13	13	15[e]	21	61.5	40.0

Note: GDP stands for Gross Domestic Product. 'e' stands for estimated value.

Profitability and Productivity

	1980	1985	1989	1990	% change 1980-1990	% change 1989-1990
Intermediate input (%)	75	77	79[e]	78	4.0	-1.3
Wages, salaries, and supplements (%)	11	13	13[e]	11	0.0	-15.4
Gross operating surplus (%)	14	9	9[e]	12	-14.3	33.3
Gross output per worker ($)	37,145	28,851	31,654[e]	30,433	-18.1	-3.9
Value added per worker ($)	9,230	6,605	6,767[e]	6,731	-27.1	-0.5
Average wage (incl. benefits) ($)	4,114	3,990	4,061[e]	3,253	-20.9	-19.9

Profitability is in percent of gross output. Productivity is in U.S. $. 'e' stands for estimated value.

Profitability - 1990

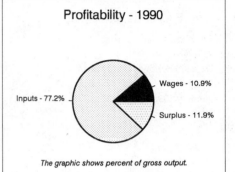

Wages - 10.9%
Inputs - 77.2%
Surplus - 11.9%

The graphic shows percent of gross output.

Value Added in Manufacturing

	1980 $ mil.	1980 %	1985 $ mil.	1985 %	1989 $ mil.	1989 %	1990 $ mil.	1990 %	% change 1980-1990	% change 1989-1990
311 Food products	71	58.7	37	41.1	36[e]	36.7	60	42.3	-15.5	66.7
313 Beverages	6	5.0	7	7.8	9[e]	9.2	11	7.7	83.3	22.2
314 Tobacco products	2	1.7	2	2.2	2[e]	2.0	3	2.1	50.0	50.0
321 Textiles	NA	0.0	NA	0.0	NA	0.0	NA	0.0	NA	NA
322 Wearing apparel	2	1.7	4	4.4	5[e]	5.1	16	11.3	700.0	220.0
323 Leather and fur products	NA	0.0	NA	0.0	NA	0.0	NA	0.0	NA	NA
324 Footwear	NA	0.0	NA	0.0	NA	0.0	1	0.7	NA	NA
331 Wood and wood products	7	5.8	6	6.7	6[e]	6.1	11	7.7	57.1	83.3
332 Furniture and fixtures	3	2.5	3	3.3	4[e]	4.1	3	2.1	0.0	-25.0
341 Paper and paper products	2	1.7	2[e]	2.2	3[e]	3.1	5	3.5	150.0	66.7
342 Printing and publishing	4	3.3	5	5.6	6[e]	6.1	6	4.2	50.0	0.0
351 Industrial chemicals	NA	0.0	NA	0.0	NA	0.0	NA	0.0	NA	NA
352 Other chemical products	4	3.3	5	5.6	6[e]	6.1	7	4.9	75.0	16.7
353 Petroleum refineries	NA	0.0	NA	0.0	NA	0.0	NA	0.0	NA	NA
354 Miscellaneous petroleum and coal products	NA	0.0	NA	0.0	NA	0.0	NA	0.0	NA	NA
355 Rubber products	1	0.8	1	1.1	1[e]	1.0	1	0.7	0.0	0.0
356 Plastic products	2	1.7	2	2.2	2[e]	2.0	3	2.1	50.0	50.0
361 Pottery, china and earthenware	NA	0.0	NA	0.0	NA	0.0	NA	0.0	NA	NA
362 Glass and glass products	NA	0.0	NA	0.0	NA	0.0	NA	0.0	NA	NA
369 Other non-metal mineral products	6	5.0	7	7.8	8[e]	8.2	5	3.5	-16.7	-37.5
371 Iron and steel	NA	0.0	NA	0.0	NA	0.0	NA	0.0	NA	NA
372 Non-ferrous metals	NA	0.0	NA	0.0	NA	0.0	NA	0.0	NA	NA
381 Metal products	6	5.0	4[e]	4.4	NA	0.0	5	3.5	-16.7	NA
382 Non-electrical machinery	1	0.8	1	1.1	1[e]	1.0	1	0.7	0.0	0.0
383 Electrical machinery	NA	0.0	1	1.1	1[e]	1.0	NA	0.0	NA	NA
384 Transport equipment	4	3.3	3	3.3	4[e]	4.1	1	0.7	-75.0	-75.0
385 Professional and scientific equipment	NA	0.0	NA	0.0	NA	0.0	NA	0.0	NA	NA
390 Other manufacturing industries	NA	0.0	1	1.1	1[e]	1.0	1	0.7	NA	0.0

Note: The industry codes shown are International Standard Industry codes (ISIC). Percentages are percent of total Value Added. 'e' stands for estimated value

For sources, notes, and explanations, see Annotated Source Appendix, page 1035.

313

Finance, Economics, and Trade

Economic Indicators [36]

National product: GDP—exchange rate conversion—$1.4 billion (1992 est.). **National product real growth rate**: 3% (1992 est.). **National product per capita**: $1,900 (1992 est.). **Inflation rate (consumer prices)**: 5% (1992 est.). **External debt**: $428 million (December 1990 est.).

Balance of Payments Summary [37]

Values in millions of dollars.

	1987	1988	1989	1990	1991
Exports of goods (f.o.b.)	303.6	345.6	399.4	470.1	434.8
Imports of goods (f.o.b.)	-329.0	-395.3	-495.2	-642.2	-553.2
Trade balance	-25.4	-49.7	-95.8	-172.1	-118.4
Services - debits	-225.5	-222.6	-274.5	-336.1	-332.9
Services - credits	243.4	272.9	384.1	446.9	469.8
Private transfers (net)	-20.8	-3.6	-13.1	-23.3	-24.9
Government transfers (net)	10.3	33.7	28.4	24.9	25.3
Long term capital (net)	-17.8	55.2	6.0	62.5	1.7
Short term capital (net)	-37.6	17.7	0.4	26.7	-3.6
Errors and omissions	3.1	16.5	-44.7	12.6	-11.8
Overall balance	-70.3	120.1	-9.2	42.1	5.2

Exchange Rates [38]

Currency: **Fijian dollars.**
Symbol: **F$.**

Data are currency units per $1.

January 1993	1.5809
1992	1.5029
1991	1.4756
1990	1.4809
1989	1.4833
1988	1.4303

Imports and Exports

Top Import Origins [39]

$553 million (c.i.f., 1991).

Origins	%
Australia	30
NZ	17
Japan	13
EC	6
US	6

Top Export Destinations [40]

$435 million (f.o.b., 1991).

Destinations	%
EC	31
Australia	21
Japan	8
US	6

Foreign Aid [41]

	U.S. $	
Western (non-US) countries, ODA and OOF bilateral commitments (1980-89)	815	million

Import and Export Commodities [42]

Import Commodities	**Export Commodities**
Machinery & transport equipment 32%	Sugar 40%
Food 15%	Gold
Petroleum products	Clothing
Consumer goods	Copra
Chemicals	Processed fish
	Lumber

Finland

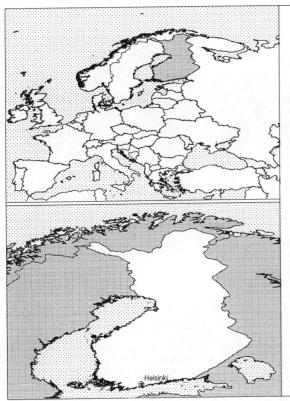

Geography [1]

Total area:
337,030 km2
Land area:
305,470 km2
Comparative area:
Slightly smaller than Montana
Land boundaries:
Total 2,628 km, Norway 729 km, Sweden 586 km, Russia 1,313 km
Coastline:
1,126 km (excludes islands and coastal indentations)
Climate:
Cold temperate; potentially subarctic, but comparatively mild because of moderating influence of the North Atlantic Current, Baltic Sea, and more than 60,000 lakes
Terrain:
Mostly low, flat to rolling plains interspersed with lakes and low hills
Natural resources:
Timber, copper, zinc, iron ore, silver
Land use:
Arable land:
8%
Permanent crops:
0%
Meadows and pastures:
0%
Forest and woodland:
76%
Other:
16%

Demographics [2]

	1960	1970	1980	1990	1991[1]	1994	2000	2010	2020
Population	4,430	4,606	4,780	4,986	4,991	5,069	5,153	5,246	5,283
Population density (persons per sq. mi.)	38	39	41	42	42	NA	43	43	43
Births	NA	NA	NA	NA	62	63	NA	NA	NA
Deaths	NA	NA	NA	NA	49	50	NA	NA	NA
Life expectancy - males	NA	NA	NA	NA	71	72	NA	NA	NA
Life expectancy - females	NA	NA	NA	NA	80	80	NA	NA	NA
Birth rate (per 1,000)	NA	NA	NA	NA	12	12	NA	NA	NA
Death rate (per 1,000)	NA	NA	NA	NA	10	10	NA	NA	NA
Women of reproductive age (15-44 yrs.)	NA	NA	NA	1,258	NA	1,271	1,216	NA	NA
of which are currently married	NA	NA	NA	675	NA	689	649	NA	NA
Fertility rate	NA	NA	NA	1.8	1.71	1.8	1.8	1.8	1.8

Population values are in thousands, life expectancy in years, and other items as indicated.

Health

Health Personnel [3]

Doctors per 1,000 pop., 1988-92	2.47
Nurse-to-doctor ratio, 1988-92	4.3
Hospital beds per 1,000 pop., 1985-90	10.8
Percentage of children immunized (age 1 yr. or less)	
Third dose of DPT, 1990-91	95.0
Measles, 1990-91	97.0

Health Expenditures [4]

Total health expenditure, 1990 (official exchange rate)	
Millions of dollars	10200
Millions of dollars per capita	2046
Health expenditures as a percentage of GDP	
Total	7.4
Public sector	6.2
Private sector	1.2
Development assistance for health	
Total aid flows (millions of dollars)[1]	NA
Aid flows per capita (millions of dollars)	NA
Aid flows as a percentage of total health expenditure	NA

For sources, notes, and explanations, see Annotated Source Appendix, page 1035.

315

Human Factors

Health Care Ratios [5]

Population per physician, 1970	960
Population per physician, 1990	410
Population per nursing person, 1970	130
Population per nursing person, 1990	NA
Percent of births attended by health staff, 1985	NA

Infants and Malnutrition [6]

Percent of babies with low birth weight, 1985	4
Infant mortality rate per 1,000 live births, 1970	13
Infant mortality rate per 1,000 live births, 1991	6
Years of life lost per 1,000 population, 1990	11
Prevalence of malnutrition (under age 5), 1990	NA

Ethnic Division [7]

Finn	NA
Swede	NA
Lapp	NA
Gypsy	NA
Tatar	NA

Religion [8]

Evangelical Lutheran	89%
Greek Orthodox	1%
None	9%
Other	1%

Major Languages [9]

Small Lapp- and Russian-speaking minorities.

Finnish (official)	93.5%
Swedish (official)	6.3%

Education

Public Education Expenditures [10]

Million Markka	1980[7]	1985[7]	1987[7]	1988[7]	1989[7]	1990[7]
Total education expenditure	10,435	18,967	NA	25,040	28,060	31,362
as percent of GNP	5.5	5.8	NA	5.8	5.8	6.2
as percent of total govt. expend.	NA	12.9	NA	13.3	NA	NA
Current education expenditure	9,565	17,604	NA	23,677	26,326	29,300
as percent of GNP	5.1	5.4	NA	5.5	5.4	5.8
as percent of current govt. expend.	12.5	13.9	NA	14.9	15.1	14.9
Capital expenditure	870	1,363	NA	1,363	1,734	2,062

Educational Attainment [11]

Age group	25+
Total population	3,387,384
Highest level attained (%)	
No schooling	-
First level	
Incompleted	49.4
Completed	NA
Entered second level	
S-1	35.3
S-2	NA
Post secondary	15.4

Literacy Rate [12]

Libraries [13]

	Admin. Units	Svc. Pts.	Vols. (000)	Shelving (meters)	Vols. Added	Reg. Users
National (1990)	1	NA	2,800	NA	NA	NA
Non-specialized	NA	NA	NA	NA	NA	NA
Public (1989)	460	1,429	34,900	NA	2.3mil	2.4mil
Higher ed. (1990)	31	NA	14,399	410,047	505,442	NA
School (1990)	5,349	5,349	7,428	NA	194,000	697,329

Daily Newspapers [14]

	1975	1980	1985	1990
Number of papers	60	58	65	66
Circ. (000)	2,100	2,414	2,661	2,780

Cinema [15]

Data for 1991.

Cinema seats per 1,000	12.6
Annual attendance per person	1.2
Gross box office receipts (mil. Markka)	184

Science and Technology

Scientific/Technical Forces [16]

Potential scientists/engineers	274,644
Number female	112,191
Potential technicians	244,407
Number female	126,673
Total	519,051

R&D Expenditures [17]

	Markkaa[28] (000) 1989
Total expenditure	8,887,800[e]
Capital expenditure	1,093,400
Current expenditure	6,114,400
Percent current	84.8

U.S. Patents Issued [18]

Values show patents issued to citizens of the country by the U.S. Patents Office.

	1990	1991	1992
Number of patents	315	343	380

Government and Law

Organization of Government [19]

Long-form name:
Republic of Finland
Type:
republic
Independence:
6 December 1917 (from Soviet Union)
Constitution:
17 July 1919
Legal system:
civil law system based on Swedish law; Supreme Court may request legislation interpreting or modifying laws; accepts compulsory ICJ jurisdiction, with reservations
National holiday:
Independence Day, 6 December (1917)
Executive branch:
president, prime minister, deputy prime minister, Council of State (Valtioneuvosto)
Legislative branch:
unicameral Parliament (Eduskunta)
Judicial branch:
Supreme Court (Korkein Oikeus)

Political Parties [20]

Parliament	% of votes
Center Party	24.8
Social Democratic Party	22.1
National Coalition Party	19.3
Leftist Alliance (Communist)	10.1
Green League	6.8
Swedish People's Party	5.5
Rural	4.8
Finnish Christian League	3.1
Liberal People's Party	0.8

Government Expenditures [21]

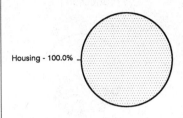

Housing - 100.0%

Crime [23]

Crime volume
Cases known to police	848,978
Attempts (percent)	NA
Percent cases solved	73.3
Crimes per 100,000 persons	16,984.7

Persons responsible for offenses
Total number offenders	697,918
Percent female	12.3
Percent juvenile[23]	20
Percent foreigners	NA

Military Expenditures and Arms Transfers [22]

	1985	1986	1987	1988	1989
Military expenditures					
Current dollars (mil.)	1,435	1,608	1,574	1,687	1,788
1989 constant dollars (mil.)	1,633	1,784	1,693	1,756	1,788
Armed forces (000)	40[e]	37	37	40	39
Gross national product (GNP)					
Current dollars (mil.)	84,440	88,960	94,810	103,100	112,800
1989 constant dollars (mil.)	96,130	98,720	102,000	107,400	112,800
Central government expenditures (CGE)					
1989 constant dollars (mil.)	30,220	31,630	32,910	33,710	33,850
People (mil.)	4.9	4.9	4.9	4.9	5.0
Military expenditure as % of GNP	1.7	1.8	1.7	1.6	1.6
Military expenditure as % of CGE	5.4	5.6	5.1	5.2	5.3
Military expenditure per capita	333	363	343	355	360
Armed forces per 1,000 people	8.2	7.4	7.4	8.1	7.9
GNP per capita	19,610	20,070	20,670	21,710	22,740
Arms imports[6]					
Current dollars (mil.)	140	220	200	90	20
1989 constant dollars (mil.)	159	244	215	94	20
Arms exports[6]					
Current dollars (mil.)	0	0	0	10	0
1989 constant dollars (mil.)	0	0	0	10	0
Total imports[7]					
Current dollars (mil.)	13,230	15,340	19,630	21,130	24,440
1989 constant dollars	15,060	17,020	21,120	22,000	24,440
Total exports[7]					
Current dollars (mil.)	13,620	16,360	20,040	21,750	23,300
1989 constant dollars	15,500	18,150	21,550	22,640	23,300
Arms as percent of total imports[8]	1.1	1.4	1.0	0.4	0.1
Arms as percent of total exports[8]	0	0	0	0	0

Human Rights [24]

	SSTS	FL	FAPRO	PPCG	APROBC	TPW	PCPTW	STPEP	PHRFF	PRW	ASST	AFL
Observes	P	P	P	P	P	P	PP	P	P	P	P	P
		EAFRD	CPR	ESCR	SR	ACHR	MAAE	PVIAC	PVNAC	EAFDAW	TCIDTP	RC
Observes		P	P	P	P		P	P	P	P	P	P

P = Party; S = Signatory; see Appendix for meaning of abbreviations.

Labor Force

Total Labor Force [25]

2.533 million

Labor Force by Occupation [26]

Public services
Industry
Commerce
Finance, insurance, and business services
Agriculture and forestry
Transport and communications
Construction

Unemployment Rate [27]

13.1% (1992)

Production Sectors

Commercial Energy Production and Consumption

Production [28]

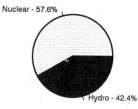

Nuclear - 57.6%

Hydro - 42.4%

Consumption [29]

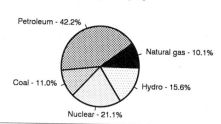

Petroleum - 42.2%

Natural gas - 10.1%

Coal - 11.0%

Hydro - 15.6%

Nuclear - 21.1%

Data are shown in quadrillion (10^{15}) BTUs and percent for 1991

Crude oil[1]	0.00
Natural gas liquids	0.00
Dry natural gas	0.00
Coal[2]	0.00
Net hydroelectric power[3]	0.14
Net nuclear power[3]	0.19
Total	0.33

Petroleum	0.46
Dry natural gas	0.11
Coal[2]	0.12
Net hydroelectric power[3]	0.17
Net nuclear power[3]	0.23
Total	1.09

Telecommunications [30]

- Good service from cable and microwave radio relay network
- 3,140,000 telephones
- Broadcast stations - 6 AM, 105 FM, 235 TV
- 1 submarine cable
- INTELSAT satellite transmission service via Swedish earth station and a receive-only INTELSAT earth station near Helsinki

Transportation [31]

Railroads. 5,924 km total; Finnish State Railways (VR) operate a total of 5,863 km 1.524-meter gauge, of which 480 km are multiple track and 1,445 km are electrified

Highways. about 103,000 km total, including 35,000 km paved (bituminous, concrete, bituminous-treated surface) and 38,000 km unpaved (stabilized gravel, gravel, earth); additional 30,000 km of private (state-subsidized) roads

Merchant Marine. 87 ships (1,000 GRT or over) totaling 935,260 GRT/973,995 DWT; includes 3 passenger, 11 short-sea passenger, 17 cargo, 1 refrigerated cargo, 26 roll-on/roll-off, 14 oil tanker, 6 chemical tanker, 2 liquefied gas, 7 bulk

Airports

Total:	160
Usable:	157
With permanent-surface runways:	66
With runways over 3,659 m:	0
With runways 2,440-3,659 m:	25
With runways 1,220-2,439 m:	22

Top Agricultural Products [32]

	1990	1991
Barley	1,720	1,749
Oats	1,662	1,109
Sugar beets	1,126	1,035
Potatoes	881	672
Wheat	627	419
Rapeseed	125	120

Values shown are 1,000 metric tons.

Top Mining Products [33]

Metric tons unless otherwise specified	M.t.
Nickel, mine output, Ni content	9,900
Zinc, mine output, Zn content	55,500
Copper, mine output, Cu content	11,700
Feldspar	53,000
Lime (000 tons)	225

Tourism [34]

For sources, notes, and explanations, see Annotated Source Appendix, page 1035.

Manufacturing Sector

GDP and Manufacturing Summary [35]

	1980	1985	1989	1990	% change 1980-1990	% change 1989-1990
GDP (million 1980 $)	51,637	59,582	70,546	70,565	36.7	0.0
GDP per capita (1980 $)	10,803	12,152	14,212	14,172	31.2	-0.3
Manufacturing as % of GDP (current prices)	27.4	24.5	20.3	21.0	-23.4	3.4
Gross output (million $)	40,839	36,968	65,600	74,422	82.2	13.4
Value added (million $)	14,343	13,598	24,435	16,433	14.6	-32.7
Value added (million 1980 $)	12,998	15,184	17,203	17,233	32.6	0.2
Industrial production index	100	115	132	131	31.0	-0.8
Employment (thousands)	531	496	443	432	-18.6	-2.5

Note: GDP stands for Gross Domestic Product. 'e' stands for estimated value.

Profitability and Productivity

	1980	1985	1989	1990	% change 1980-1990	% change 1989-1990
Intermediate input (%)	65	63	63	64	-1.5	1.6
Wages, salaries, and supplements (%)	19	20	16	21	10.5	31.3
Gross operating surplus (%)	16	17	21	14	-12.5	-33.3
Gross output per worker ($)	76,435	74,040	148,110	171,270	124.1	15.6
Value added per worker ($)	26,845	27,234	55,170	60,849	126.7	10.3
Average wage (incl. benefits) ($)	14,694	14,601	23,559	36,654	149.4	55.6

Profitability is in percent of gross output. Productivity is in U.S. $. 'e' stands for estimated value.

Profitability - 1990

Inputs - 64.6%
Surplus - 14.1%
Wages - 21.2%

The graphic shows percent of gross output.

Value Added in Manufacturing

	1980 $ mil.	1980 %	1985 $ mil.	1985 %	1989 $ mil.	1989 %	1990 $ mil.	1990 %	% change 1980-1990	% change 1989-1990
311 Food products	1,402	9.8	1,418	10.4	2,321	9.5	2,582	15.7	84.2	11.2
313 Beverages	225	1.6	227	1.7	568	2.3	667	4.1	196.4	17.4
314 Tobacco products	46	0.3	58	0.4	119	0.5	177	1.1	284.8	48.7
321 Textiles	469	3.3	310	2.3	412	1.7	386	2.3	-17.7	-6.3
322 Wearing apparel	499	3.5	434	3.2	422	1.7	428	2.6	-14.2	1.4
323 Leather and fur products	54	0.4	37	0.3	45	0.2	47	0.3	-13.0	4.4
324 Footwear	134	0.9	106	0.8	98	0.4	88	0.5	-34.3	-10.2
331 Wood and wood products	1,196	8.3	652	4.8	1,297	5.3	1,053	6.4	-12.0	-18.8
332 Furniture and fixtures	257	1.8	215	1.6	475	1.9	515	3.1	100.4	8.4
341 Paper and paper products	2,088	14.6	1,846	13.6	3,890	15.9	3,618	22.0	73.3	-7.0
342 Printing and publishing	1,080	7.5	1,223	9.0	1,853	7.6	2,120	12.9	96.3	14.4
351 Industrial chemicals	555	3.9	561	4.1	1,244	5.1	1,371	8.3	147.0	10.2
352 Other chemical products	349	2.4	371	2.7	623	2.5	707	4.3	102.6	13.5
353 Petroleum refineries	445	3.1	384	2.8	536	2.2	675	4.1	51.7	25.9
354 Miscellaneous petroleum and coal products	46	0.3	47	0.3	93	0.4	121	0.7	163.0	30.1
355 Rubber products	105	0.7	84	0.6	124	0.5	129	0.8	22.9	4.0
356 Plastic products	164	1.1	168	1.2	322	1.3	407	2.5	148.2	26.4
361 Pottery, china and earthenware	46	0.3	40	0.3	72	0.3	73	0.4	58.7	1.4
362 Glass and glass products	105	0.7	77	0.6	144	0.6	167	1.0	59.0	16.0
369 Other non-metal mineral products	434	3.0	432	3.2	940	3.8	1,050	6.4	141.9	11.7
371 Iron and steel	544	3.8	463	3.4	942	3.9	849	5.2	56.1	-9.9
372 Non-ferrous metals	142	1.0	103	0.8	445	1.8	359	2.2	152.8	-19.3
381 Metal products	756	5.3	766	5.6	1,618	6.6	1,754	10.7	132.0	8.4
382 Non-electrical machinery	1,469	10.2	1,618	11.9	2,622	10.7	3,339	20.3	127.3	27.3
383 Electrical machinery	694	4.8	763	5.6	1,556	6.4	1,831	11.1	163.8	17.7
384 Transport equipment	823	5.7	915	6.7	1,206	4.9	1,406	8.6	70.8	16.6
385 Professional and scientific equipment	110	0.8	166	1.2	293	1.2	344	2.1	212.7	17.4
390 Other manufacturing industries	107	0.7	111	0.8	157	0.6	169	1.0	57.9	7.6

Note: The industry codes shown are International Standard Industry codes (ISIC). Percentages are percent of total Value Added. 'e' stands for estimated value

For sources, notes, and explanations, see Annotated Source Appendix, page 1035.

319

Finance, Economics, and Trade

Economic Indicators [36]

Billions of Finnmarks (FIM) unless otherwise noted.

	1989	1990	1991[e]
Real GDP (1985 = 100)	395.1	396.7	377.0
Real GDP growth rate (%)	5.2	0.4	-5.0
Money supply (M1)[42]	124.3	141.5	129.0
Commercial interest rates[43]	12.53	13.99	13.5
Household savings rate (% DI)	2.0	3.3	6.0
Investment rate (% of GDP)	27.7	26.3	23.5
CPI (1985 = 100)	120.0	127.3	132.3
WPI (1985 = 100)	107.3	110.8	111.3
External public debt (central and local govts)	23.1	25.0	33.0

Balance of Payments Summary [37]

Values in millions of dollars.

	1987	1988	1989	1990	1991
Exports of goods (f.o.b.)	19,079	21,826	22,882	26,089	22,557
Imports of goods (f.o.b.)	-17,700	-20,686	-23,101	-25,322	-20,299
Trade balance	1,379	1,140	-219	767	2,258
Services - debits	-7,675	-9,686	-11,425	-15,157	-15,130
Services - credits	5,057	6,363	6,611	8,784	7,915
Private transfers (net)	-162	-87	-252	-342	-310
Government transfers (net)	-328	-425	-510	-735	-736
Long term capital (net)	-418	725	1,655	7,771	9,707
Short term capital (net)	7,691	1,481	1,845	4,254	-6,210
Errors and omissions	-1,523	745	1,238	-1,407	617
Overall balance	4,021	256	-1,057	3,935	-1,889

Exchange Rates [38]

Currency: **markkaa.**
Symbol: **FMk.**

Data are currency units per $1.

January 1993	5.4193
1992	4.4794
1991	4.0440
1990	3.8235
1989	4.2912
1988	4.1828

Imports and Exports

Top Import Origins [39]

$21.2 billion (c.i.f., 1992). Data are for 1992.

Origins	%
EC	47.2
UK	8.7
EFTA	19.0
US	6.1
Japan	5.5
Russia	7.1

Top Export Destinations [40]

$24.0 billion (f.o.b., 1992).

Destinations	%
EC	53.2
UK	10.7
EFTA	19.5
US	5.9
Japan	1.3
Russia	2.8

Foreign Aid [41]

	U.S. $	
Donor - ODA and OOF commitments (1970-89)	2.7	billion

Import and Export Commodities [42]

Import Commodities	Export Commodities
Foodstuffs	Timber
Petroleum and petroleum products	Paper and pulp
Chemicals	Ships
Transport equipment	Machinery
Iron and steel	Clothing and footwear
Machinery	
Textile yarn and fabrics	
Fodder grains	

France

Geography [1]

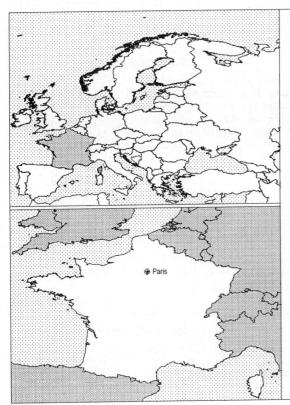

Total area:
547,030 km2
Land area:
545,630 km2
Comparative area:
Slightly more than twice the size of Colorado
Land boundaries:
Total 2,892.4 km, Andorra 60 km, Belgium 620 km, Germany 451 km, Italy 488 km, Luxembourg 73 km, Monaco 4.4 km, Spain 623 km, Switzerland 573 km
Coastline:
3,427 km (mainland 2,783 km, Corsica 644 km)
Climate:
Generally cool winters and mild summers, but mild winters and hot summers along the Mediterranean
Terrain:
Mostly flat plains or gently rolling hills in north and west; remainder is mountainous, especially Pyrenees in south, Alps in east
Natural resources:
Coal, iron ore, bauxite, fish, timber, zinc, potash
Land use:
Arable land: 32%
Permanent crops: 2%
Meadows and pastures: 23%
Forest and woodland: 27%
Other: 16%

Demographics [2]

	1960	1970	1980	1990	1991[1]	1994	2000	2010	2020
Population	45,670	50,787	53,870	56,720	56,596	57,840	59,354	61,001	61,793
Population density (persons per sq. mi.)	217	241	256	268	269	NA	278	183	286
Births	NA	NA	NA	NA	766	759	NA	NA	NA
Deaths	NA	NA	NA	NA	531	538	NA	NA	NA
Life expectancy - males	NA	NA	NA	NA	74	74	NA	NA	NA
Life expectancy - females	NA	NA	NA	NA	82	82	NA	NA	NA
Birth rate (per 1,000)	NA	NA	NA	NA	14	13	NA	NA	NA
Death rate (per 1,000)	NA	NA	NA	NA	9	9	NA	NA	NA
Women of reproductive age (15-44 yrs.)	NA	NA	NA	14,155	NA	14,458	14,124	NA	NA
of which are currently married	NA	NA	NA	8,750	NA	9,194	9,124	NA	NA
Fertility rate	NA	NA	NA	1.8	1.82	1.8	1.8	1.8	1.8

Population values are in thousands, life expectancy in years, and other items as indicated.

Health

Health Personnel [3]

Doctors per 1,000 pop., 1988-92	2.89
Nurse-to-doctor ratio, 1988-92	1.6
Hospital beds per 1,000 pop., 1985-90	9.3
Percentage of children immunized (age 1 yr. or less)	
Third dose of DPT, 1990-91	95.0
Measles, 1990-91	69.0

Health Expenditures [4]

Total health expenditure, 1990 (official exchange rate)	
Millions of dollars	105467
Millions of dollars per capita	1869
Health expenditures as a percentage of GDP	
Total	8.9
Public sector	6.6
Private sector	2.3
Development assistance for health	
Total aid flows (millions of dollars)[1]	NA
Aid flows per capita (millions of dollars)	NA
Aid flows as a percentage of total health expenditure	NA

For sources, notes, and explanations, see Annotated Source Appendix, page 1035.

321

Human Factors

Health Care Ratios [5]

Population per physician, 1970	750
Population per physician, 1990	350
Population per nursing person, 1970	270
Population per nursing person, 1990	NA
Percent of births attended by health staff, 1985	NA

Infants and Malnutrition [6]

Percent of babies with low birth weight, 1985	5
Infant mortality rate per 1,000 live births, 1970	18
Infant mortality rate per 1,000 live births, 1991	7
Years of life lost per 1,000 population, 1990	10
Prevalence of malnutrition (under age 5), 1990	NA

Ethnic Division [7]

Celtic and Latin with Teutonic, Slavic, North African, Indochinese, Basque minorities.

Religion [8]

Roman Catholic	90%
Protestant	2%
Jewish	1%
Muslim (North African workers)	1%
Unaffiliated	6%

Major Languages [9]

Rapidly declining regional dialects and languages (Provencal, Breton, Alsatian, Corsican, Catalan, Basque, Flemish).

French	100%

Education

Public Education Expenditures [10]

Million Francs	1980[16]	1985[16]	1987[16]	1988[16]	1989[16]	1990[16]
Total education expenditure	142,099	269,191	NA	310,485	329,051	351,867
as percent of GNP	5.0	5.8	NA	5.4	5.4	5.4
as percent of total govt. expend.	NA	NA	NA	NA	NA	NA
Current education expenditure	131,441	254,433	NA	288,647	305,806	327,427
as percent of GNP	4.7	5.4	NA	5.1	5.0	5.1
as percent of current govt. expend.	NA	NA	NA	NA	NA	NA
Capital expenditure	10,658	14,758	NA	21,838	23,245	24,440

Educational Attainment [11]

Literacy Rate [12]

Libraries [13]

	Admin. Units[30]	Svc. Pts.[30]	Vols. (000)[30]	Shelving (meters)[30]	Vols. Added[30]	Reg. Users[30]
National	NA	NA	NA	NA	NA	NA
Non-specialized (1991)[31]	1	1	400	15,500	19,500	3.7mil
Public (1987)	1,462	2,740	78,474	NA	4.9mil	16.5mil
Higher ed. (1990)	67	220	21,400	1 mil	522,000	794,000
School	NA	NA	NA	NA	NA	NA

Daily Newspapers [14]

	1975	1980	1985	1990
Number of papers	95	90	92	79
Circ. (000)	11,000[e]	10,332	10,670	11,792

Cinema [15]

Data for 1991.

Cinema seats per 1,000	17.3
Annual attendance per person	2.1
Gross box office receipts (mil. Franc)	3,880

Science and Technology

Scientific/Technical Forces [16]

R&D Expenditures [17]

	Franc (000) 1988
Total expenditure	130,631,000
Capital expenditure	13,902,000
Current expenditure	116,729,000
Percent current	89.4

U.S. Patents Issued [18]

Values show patents issued to citizens of the country by the U.S. Patents Office.

	1990	1991	1992
Number of patents	3,093	3,249	3,282

For sources, notes, and explanations, see Annotated Source Appendix, page 1035.

Government and Law

Organization of Government [19]

Long-form name:
French Republic
Type:
republic
Independence:
486 (unified by Clovis)
Constitution:
28 September 1958, amended concerning
election of president in 1962, ammended
to comply with provisions of EC
Maastricht Treaty in 1992
Legal system:
civil law system with indigenous
concepts; review of administrative but not
legislative acts
National holiday:
National Day, Taking of theBastille, 14
July (1789)
Executive branch:
president, prime minister, Council of
Ministers (cabinet)
Legislative branch:
bicameral Parliament (Parlement)
consists of an upper house or Senate
(Senat) and a lower house or National
Assembly (Assemblee Nationale)
Judicial branch:
Constitutional Court (Cour
Constitutionnelle)

Political Parties [20]

National Assembly	% of seats
Rally for the Republic (RPR)	42.8
Union for French Democracy (UDF)	36.9
Socialist Party (PS)	11.6
Communist Party (PCF)	4.2
Independents	4.5

Government Budget [21]

For 1993 budget.
Revenues	220.5
Expenditures	249.1
Capital expenditures	47.0

Crime [23]

Crime volume	
Cases known to police	3,492,712
Attempts (percent)	NA
Percent cases solved	37.50
Crimes per 100,000 persons	6,169.29
Persons responsible for offenses	
Total number offenders	754,161
Percent female	17.27
Percent juvenile[25]	13.03
Percent foreigners	16.97

Military Expenditures and Arms Transfers [22]

	1985	1986	1987	1988	1989
Military expenditures					
Current dollars (mil.)	29,600	30,500	32,530	33,490	35,260
1989 constant dollars (mil.)	33,690	33,840	34,980	34,860	35,260
Armed forces (000)	563	558	559	558	554
Gross national product (GNP)					
Current dollars (mil.)	740,900	779,200	823,600	884,200	954,100
1989 constant dollars (mil.)	843,500	864,700	885,800	920,500	954,100
Central government expenditures (CGE)					
1989 constant dollars (mil.)	382,900	389,600	386,800	401,300	410,500
People (mil.)	55.2	55.4	55.6	55.9	56.1
Military expenditure as % of GNP	4.0	3.9	3.9	3.8	3.7
Military expenditure as % of CGE	8.8	8.7	9.0	8.7	8.6
Military expenditure per capita	611	611	629	624	628
Armed forces per 1,000 people	10.2	10.1	10.0	10.0	9.9
GNP per capita	15,290	15,610	15,920	16,470	17,000
Arms imports[6]					
Current dollars (mil.)	150	210	290	140	210
1989 constant dollars (mil.)	171	233	312	146	210
Arms exports[6]					
Current dollars (mil.)	5,400	4,900	3,000	2,300	2,700
1989 constant dollars (mil.)	6,147	5,438	3,227	2,394	2,700
Total imports[7]					
Current dollars (mil.)	108,300	129,400	158,500	178,900	193,000
1989 constant dollars	123,200	143,600	170,400	186,200	193,000
Total exports[7]					
Current dollars (mil.)	101,700	124,900	148,400	167,800	179,400
1989 constant dollars	115,700	138,700	159,600	174,700	179,400
Arms as percent of total imports[8]	0.1	0.2	0.2	0.1	0.1
Arms as percent of total exports[8]	5.3	3.9	2.0	1.4	1.5

Human Rights [24]

	SSTS	FL	FAPRO	PPCG	APROBC	TPW	PCPTW	STPEP	PHRFF	PRW	ASST	AFL
Observes	P	P	P	P	P	P	P	P	P	P	P	P
		EAFRD	CPR	ESCR	SR	ACHR	MAAE	PVIAC	PVNAC	EAFDAW	TCIDTP	RC
Observes		P	P	P	P		P		P	P	P	P

P = Party; S = Signatory; see Appendix for meaning of abbreviations.

Labor Force

Total Labor Force [25]

24.17 million

Labor Force by Occupation [26]

Services	61.5%
Industry	31.3
Agriculture	7.2
Date of data: 1987	

Unemployment Rate [27]

10.5% (end 1992)

For sources, notes, and explanations, see Annotated Source Appendix, page 1035.

Production Sectors

Commercial Energy Production and Consumption

Production [28]

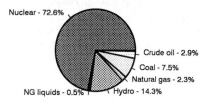

Nuclear - 72.6%
Crude oil - 2.9%
Coal - 7.5%
Natural gas - 2.3%
NG liquids - 0.5%
Hydro - 14.3%

Consumption [29]

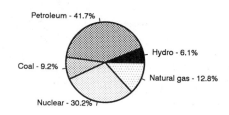

Petroleum - 41.7%
Hydro - 6.1%
Coal - 9.2%
Natural gas - 12.8%
Nuclear - 30.2%

Data are shown in quadrillion (10^{15}) BTUs and percent for 1991

Crude oil[1]	0.13
Natural gas liquids	0.02
Dry natural gas	0.10
Coal[2]	0.33
Net hydroelectric power[3]	0.63
Net nuclear power[3]	3.20
Total	4.40

Petroleum	3.90
Dry natural gas	1.20
Coal[2]	0.86
Net hydroelectric power[3]	0.57
Net nuclear power[3]	2.83
Total	9.35

Telecommunications [30]

- Highly developed. Extensive cable and microwave radio relay networks
- Large-scale introduction of optical-fiber systems
- Satellite systems for domestic traffic
- 39,200,000 telephones
- Broadcast stations - 41 AM, 800 (mostly repeaters) FM, 846 (mostly repeaters) TV
- 24 submarine coaxial cables
- 2 INTELSAT earth stations (with total of 5 antennas - 2 for the Indian Ocean INTELSAT and 3 for the Atlantic Ocean INTELSAT)
- HF radio communications with more than 20 countries
- INMARSAT service; EUTELSAT TV service

Top Agricultural Products [32]

	1990	1991
Grain	55,200	60,200
Wine grapes (000 hl.)	65,529	43,500
Sugar beets	26,369	26,103
Pulses	4,580	5,104
Sunflowerseed	2,370	2,654
Rapeseed	1,930	2,269

Values shown are 1,000 metric tons.

Top Mining Products [33]

Estimated metric tons unless otherwise specified   M.t.

Bauxite (000 tons)	400
Barite	95,000
Talc, crude	310,000
Diatomite (000 tons)	250
Iron ore and concentrate (000 tons)	7,472[1]
Coal, anthracite, lignite, and bituminous (000 tons)	12,328[1]

Transportation [31]

Railroads. French National Railways (SNCF) operates 34,322 km 1.435-meter standard gauge; 12,434 km electrified, 15,132 km double or multiple track; 99 km of various gauges (1.000-meter), privately owned and operated

Highways. 1,551,400 km total; 33,400 km national highway; 347,000 km departmental highway; 421,000 km community roads; 750,000 km rural roads; 5,401 km of controlled-access divided autoroutes; about 803,000 km paved

Merchant Marine. 130 ships (1,000 GRT or over) totaling 3,224,945 GRT/5,067,252 DWT; includes 7 short-sea passenger, 10 cargo, 20 container, 1 multifunction large-load carrier, 27 roll-on/roll-off, 36 oil tanker, 11 chemical tanker, 6 liquefied gas, 2 specialized tanker, 10 bulk; note - France also maintains a captive register for French-owned ships in the Kerguelen Islands (French Southern and Antarctic Lands) and French Polynesia

Airports

Total:	471
Usable:	461
With permanent-surface runways:	256
With runways over 3,659 m:	3
With runways 2,440-3,659 m:	37
With runways 1,220-2,439 m:	136

Tourism [34]

	1987	1988	1989	1990	1991
Visitors	74,722	76,489	88,714	93,992	98,155
Tourists	41,734	42,721	49,549	52,497	54,822
Excursionists	32,988	33,768	39,165	41,495	43,333
Tourism receipts	11,870	13,786	16,245	20,185	21,300
Tourism expenditures	8,493	9,715	10,031	12,424	12,338

Tourists are in thousands, money in million U.S. dollars.

324

For sources, notes, and explanations, see Annotated Source Appendix, page 1035.

Manufacturing Sector

GDP and Manufacturing Summary [35]

	1980	1985	1989	1990	% change 1980-1990	% change 1989-1990
GDP (million 1980 $)	664,529	716,607	807,893	835,089	25.7	3.4
GDP per capita (1980 $)	12,333	12,989	14,439	14,873	20.6	3.0
Manufacturing as % of GDP (current prices)	25.5	23.1	21.3	21.8	-14.5	2.3
Gross output (million $)	453,636	326,412	550,124	676,345	49.1	22.9
Value added (million $)	161,552	115,430	205,249	256,663	58.9	25.0
Value added (million 1980 $)	160,795	158,170	176,468	174,014	8.2	-1.4
Industrial production index	100	94	106	108	8.0	1.9
Employment (thousands)	5,103	4,470	4,185	4,243	-16.9	1.4

Note: GDP stands for Gross Domestic Product. 'e' stands for estimated value.

Profitability and Productivity

	1980	1985	1989	1990	% change 1980-1990	% change 1989-1990
Intermediate input (%)	64	65	63	62	-3.1	-1.6
Wages, salaries, and supplements (%)	24	23	23	23	-4.2	0.0
Gross operating surplus (%)	11	12	14	15	36.4	7.1
Gross output per worker ($)	84,523	69,176	131,464	151,193	78.9	15.0
Value added per worker ($)	30,101	24,463	49,049	57,376	90.6	17.0
Average wage (incl. benefits) ($)	21,643	17,129	30,708	36,111	66.8	17.6

Profitability is in percent of gross output. Productivity is in U.S. $. 'e' stands for estimated value.

Profitability - 1990

Inputs - 62.0%
Surplus - 15.0%
Wages - 23.0%

The graphic shows percent of gross output.

Value Added in Manufacturing

	1980 $ mil.	1980 %	1985 $ mil.	1985 %	1989 $ mil.	1989 %	1990 $ mil.	1990 %	% change 1980-1990	% change 1989-1990
311 Food products	15,952	9.9	12,825	11.1	21,240	10.3	25,508	9.9	59.9	20.1
313 Beverages	3,486	2.2	2,268	2.0	3,588	1.7	5,856	2.3	68.0	63.2
314 Tobacco products	1,497	0.9	948	0.8	1,518	0.7	1,862	0.7	24.4	22.7
321 Textiles	6,130	3.8	4,239	3.7	6,839	3.3	7,149	2.8	16.6	4.5
322 Wearing apparel	4,742	2.9	3,104	2.7	4,850	2.4	5,581	2.2	17.7	15.1
323 Leather and fur products	757	0.5	527	0.5	828	0.4	1,066	0.4	40.8	28.7
324 Footwear	1,411	0.9	929	0.8	1,190	0.6	1,480	0.6	4.9	24.4
331 Wood and wood products	2,888	1.8	1,704	1.5	2,700	1.3	3,812	1.5	32.0	41.2
332 Furniture and fixtures	2,846	1.8	1,632	1.4	2,958	1.4	3,841	1.5	35.0	29.9
341 Paper and paper products	3,592	2.2	2,817	2.4	5,053	2.5	6,513	2.5	81.3	28.9
342 Printing and publishing	6,660	4.1	5,069	4.4	10,309	5.0	12,991	5.1	95.1	26.0
351 Industrial chemicals	6,462	4.0	4,669	4.0	9,403	4.6	10,047	3.9	55.5	6.8
352 Other chemical products	6,302	3.9	4,996	4.3	9,365	4.6	12,860	5.0	104.1	37.3
353 Petroleum refineries	9,973	6.2	8,127	7.0	12,952	6.3	15,153	5.9	51.9	17.0
354 Miscellaneous petroleum and coal products	118	0.1	78	0.1	144	0.1	180	0.1	52.5	25.0
355 Rubber products	2,483	1.5	1,544	1.3	2,767	1.3	3,285	1.3	32.3	18.7
356 Plastic products	3,083	1.9	2,415	2.1	4,934	2.4	6,530	2.5	111.8	32.3
361 Pottery, china and earthenware	639	0.4	367	0.3	657	0.3	821	0.3	28.5	25.0
362 Glass and glass products	2,170	1.3	1,365	1.2	2,572	1.3	3,124	1.2	44.0	21.5
369 Other non-metal mineral products	5,653	3.5	3,153	2.7	6,425	3.1	7,852	3.1	38.9	22.2
371 Iron and steel	6,741	4.2	3,788	3.3	7,026	3.4	9,209	3.6	36.6	31.1
372 Non-ferrous metals	2,479	1.5	2,340	2.0	4,949	2.4	4,891	1.9	97.3	-1.2
381 Metal products	12,119	7.5	7,792	6.8	14,213	6.9	19,445	7.6	60.5	36.8
382 Non-electrical machinery	16,245	10.1	11,998	10.4	19,822	9.7	25,094	9.8	54.5	26.6
383 Electrical machinery	14,411	8.9	11,491	10.0	19,148	9.3	25,777	10.0	78.9	34.6
384 Transport equipment	17,733	11.0	11,316	9.8	24,029	11.7	28,442	11.1	60.4	18.4
385 Professional and scientific equipment	2,206	1.4	1,752	1.5	3,431	1.7	4,125	1.6	87.0	20.2
390 Other manufacturing industries	2,772	1.7	2,178	1.9	2,337	1.1	4,172	1.6	50.5	78.5

Note: The industry codes shown are International Standard Industry codes (ISIC). Percentages are percent of total Value Added. 'e' stands for estimated value

Finance, Economics, and Trade

Economic Indicators [36]

Billions of French Francs (FF) unless otherwise stated.

	1989	1990	1991[44]
Real GDP (FF 1980)	3,437.1	3,534.0	3,611.7
Real GDP Growth (%)	3.9	2.9	2.0
Money Supply (M1)	1,538.7	1,597.8	1,566.8[45]
Savings Rate (households)(%)	12.7	11.9	12.4[46]
Investment Growth corporate (%)	11.5	6.9	2.0
CPI (%, yr end)	3.6	3.4	2.8
WPI[47]	NA	NA	NA
External Public Debt (FF billions)	-29.4	NA	NA

Balance of Payments Summary [37]

Values in millions of dollars.

	1987	1988	1989	1990	1991
Exports of goods (f.o.b.)	141,658	160,188	170,761	206,672	207,084
Imports of goods (f.o.b.)	-150,325	-168,726	-181,412	-220,339	-217,223
Trade balance	-8,667	-8,538	-10,651	-13,667	-10,139
Services - debits	-69,713	-79,114	-89,158	-121,357	-140,051
Services - credits	79,344	89,530	102,721	134,726	152,021
Private transfers (net)	-2,297	-2,437	-2,668	-3,957	-3,025
Government transfers (net)	-3,114	-4,237	-5,863	-9,517	-4,955
Long term capital (net)	2,211	-1,017	6,845	6,559	7,098
Short term capital (net)	-6,945	4,805	-1,889	12,535	-14,062
Errors and omissions	850	940	-1,688	6,501	7,905
Overall balance	-8,331	-68	-2,351	11,823	-5,208

Exchange Rates [38]

Currency: **French francs.**
Symbol: **F.**

Data are currency units per $1.

January 1993	5.4812
1992	5.2938
1991	5.6421
1990	5.4453
1989	6.3801
1988	5.9569

Imports and Exports

Top Import Origins [39]

$230.3 billion (c.i.f., 1991). Data are for 1991 est.

Origins	%
Germany	17.8
Italy	10.9
US	9.5
Netherlands	8.9
Spain	8.8
Belgium-Luxembourg	8.5
UK	7.5
Japan	4.1
Former USSR	1.3

Top Export Destinations [40]

$212.7 billion (f.o.b., 1991) Data are for 1991 est.

Destinations	%
Germany	18.6
Italy	11.0
Spain	11.0
Belgium-Luxembourg	9.1
UK	8.8
Netherlands	7.9
US	6.4
Japan	2.0
Former USSR	0.7

Foreign Aid [41]

	U.S. $	
Donor - ODA and OOF commitments (1970-89)	75.1	billion

Import and Export Commodities [42]

Import Commodities
Crude oil
Machinery & equipment
Agricultural products
Chemicals
Iron and steel products

Export Commodities
Machinery & transport equip.
Chemicals
Foodstuffs
Agricultural products
Iron and steel products
Textiles and clothing

Gabon

Geography [1]

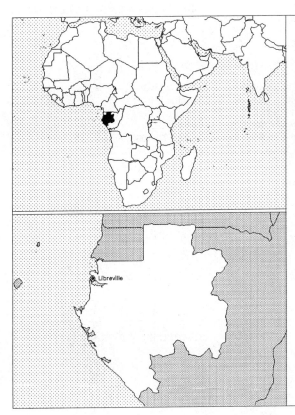

Total area:
267,670 km2
Land area:
257,670 km2
Comparative area:
Slightly smaller than Colorado
Land boundaries:
Total 2,551 km, Cameroon 298 km, Congo 1,903 km, Equatorial Guinea 350 km
Coastline:
885 km
Climate:
Tropical; always hot, humid
Terrain:
Narrow coastal plain; hilly interior; savanna in east and south
Natural resources:
Petroleum, manganese, uranium, gold, timber, iron ore
Land use:
Arable land:
1%
Permanent crops:
1%
Meadows and pastures:
18%
Forest and woodland:
78%
Other:
2%

Demographics [2]

	1960	1970	1980	1990	1991[1]	1994	2000	2010	2020
Population	446	514	808	1,078	1,080	1,139	1,244	1,445	1,675
Population density (persons per sq. mi.)	4	5	8	11	11	NA	12	14	16
Births	NA	NA	NA	NA	30	32	NA	NA	NA
Deaths	NA	NA	NA	NA	16	16	NA	NA	NA
Life expectancy - males	NA	NA	NA	NA	51	52	NA	NA	NA
Life expectancy - females	NA	NA	NA	NA	56	58	NA	NA	NA
Birth rate (per 1,000)	NA	NA	NA	NA	28	28	NA	NA	NA
Death rate (per 1,000)	NA	NA	NA	NA	14	14	NA	NA	NA
Women of reproductive age (15-44 yrs.)	NA	NA	NA	259	NA	268	288	NA	NA
of which are currently married	NA	NA	NA	214	NA	221	238	NA	NA
Fertility rate	NA	NA	NA	4.2	3.97	4.0	3.7	3.4	3.0

Population values are in thousands, life expectancy in years, and other items as indicated.

Health

Health Personnel [3]

Health Expenditures [4]

For sources, notes, and explanations, see Annotated Source Appendix, page 1035.

327

Human Factors

Health Care Ratios [5]

Population per physician, 1970	5250
Population per physician, 1990	NA
Population per nursing person, 1970	570
Population per nursing person, 1990	NA
Percent of births attended by health staff, 1985	92

Infants and Malnutrition [6]

Percent of babies with low birth weight, 1985	16
Infant mortality rate per 1,000 live births, 1970	138
Infant mortality rate per 1,000 live births, 1991	95
Years of life lost per 1,000 population, 1990	NA
Prevalence of malnutrition (under age 5), 1990	25

Ethnic Division [7]

Bantu tribes including four major tribal groupings (Fang, Eshira, Bapounou, Bateke), Africans and Europeans 100,000, including 27,000 French.

Religion [8]

Christian 55-75%, Muslim less than 1%, animist.

Major Languages [9]

French (official)	NA
Fang	NA
Myene	NA
Bateke	NA
Bapounou/Eschira	NA
Bandjabi	NA

Education

Public Education Expenditures [10]

Million Francs C.F.A.	1980	1985	1986	1987	1988	1989
Total education expenditure	22,204	69,500	78,305	53,372	NA	NA
as percent of GNP	2.7	4.5	7.0	5.7	NA	NA
as percent of total govt. expend.	NA	9.4	NA	NA	NA	NA
Current education expenditure	16,055	47,500	55,766	47,774	NA	NA
as percent of GNP	2.0	3.1	5.0	5.1	NA	NA
as percent of current govt. expend.	NA	21.7	NA	NA	NA	NA
Capital expenditure	6,149	22,000	22,539	5,598	NA	NA

Educational Attainment [11]

Literacy Rate [12]

In thousands and percent	1985[a]	1991[a]	2000[a]
Illiterate population +15 years	284	311	297
Illiteracy rate - total pop. (%)	43.9	39.3	30.7
Illiteracy rate - males (%)	30.1	26.5	20.3
Illiteracy rate - females (%)	56.9	51.5	40.7

Libraries [13]

	Admin. Units	Svc. Pts.	Vols. (000)	Shelving (meters)	Vols. Added	Reg. Users
National (1989)	1	1	NA	922	185	5,308
Non-specialized	NA	NA	NA	NA	NA	NA
Public	NA	NA	NA	NA	NA	NA
Higher ed.	NA	NA	NA	NA	NA	NA
School	NA	NA	NA	NA	NA	NA

Daily Newspapers [14]

	1975	1980	1985	1990
Number of papers	1	1	1	1
Circ. (000)	3	15	20	20

Cinema [15]

Science and Technology

Scientific/Technical Forces [16]

R&D Expenditures [17]

	Franc[5] (000) 1986
Total expenditure	380,000
Capital expenditure	130,000
Current expenditure	250,000
Percent current	65.8

U.S. Patents Issued [18]

Government and Law

Organization of Government [19]

Long-form name:
Gabonese Republic
Type:
republic; multiparty presidential regime
(opposition parties legalized 1990)
Independence:
17 August 1960 (from France)
Constitution:
21 February 1961, revised 15 April 1975
Legal system:
based on French civil law system and
customary law; judicial review of
legislative acts in Constitutional Chamber
of the Supreme Court; compulsory ICJ
jurisdiction not accepted
National holiday:
Renovation Day, 12 March (1968)
(Gabonese Democratic Party established)
Executive branch:
president, prime minister, Cabinet
Legislative branch:
unicameral National Assembly
(Assemblee Nationale)
Judicial branch:
Supreme Court (Cour Supreme)

Political Parties [20]

National Assembly	% of seats
Gabonese Democratic Party	51.7
NRM - Lumberjacks	15.8
Gabonese Party for Progress	15.0
National Recovery Movement	5.8
Association for Socialism in Gabon	5.0
Gabonese Socialist Union	3.3
Circle for Renewal and Progress	0.8
Independents	2.5

Government Budget [21]

For 1990 est.
Revenues	1.400
Expenditures	1.400
Capital expenditures	0.247

Military Expenditures and Arms Transfers [22]

	1985	1986	1987	1988	1989
Military expenditures					
Current dollars (mil.)	81	149[e]	128	163[e]	140[e]
1989 constant dollars (mil.)	92	166[e]	137	170[e]	140E
Armed forces (000)	7	9	9	8	10
Gross national product (GNP)					
Current dollars (mil.)	2,897	3,311	2,786	2,882	3,119
1989 constant dollars (mil.)	3,297	3,674	2,997	3,000	3,119
Central government expenditures (CGE)					
1989 constant dollars (mil.)	1,389	2,167[e]	NA	1,108	1,127
People (mil.)	1.0	1.0	1.0	1.1	1.1
Military expenditure as % of GNP	2.8	4.5	4.6	5.7	4.5
Military expenditure as % of CGE	6.6	7.6	NA	15.3	12.4
Military expenditure per capita	91	161	132	161	132
Armed forces per 1,000 people	6.9	8.7	8.6	7.4	9.0
GNP per capita	3,262	3,561	2,872	2,853	2,944
Arms imports[6]					
Current dollars (mil.)	100	0	0	20	20
1989 constant dollars (mil.)	114	0	0	21	20
Arms exports[6]					
Current dollars (mil.)	0	0	0	0	0
1989 constant dollars (mil.)	0	0	0	0	0
Total imports[7]					
Current dollars (mil.)	855	866	732	930	634
1989 constant dollars	973	961	787	968	634
Total exports[7]					
Current dollars (mil.)	1,951	1,271	1,288	NA	1,140
1989 constant dollars	2,221	1,410	1,385	NA	1,140
Arms as percent of total imports[8]	11.7	0	0	2.2	3.0
Arms as percent of total exports[8]	0	0	0	NA	0

Crime [23]

Crime volume
Cases known to police	3,231
Attempts (percent)	176
Percent cases solved	1,759
Crimes per 100,000 persons	323.1
Persons responsible for offenses	
Total number offenders	2,180
Percent female	143
Percent juvenile[26]	49
Percent foreigners	1,011

Human Rights [24]

	SSTS	FL	FAPRO	PPCG	APROBC	TPW	PCPTW	STPEP	PHRFF	PRW	ASST	AFL
Observes		P	P	P	P	P	P			P		P
	EAFRD	CPR	ESCR	SR	ACHR	MAAE	PVIAC	PVNAC	EAFDAW	TCIDTP	RC
Observes	P	P	P	P	S		P	P	S	S

P = Party; S = Signatory; see Appendix for meaning of abbreviations.

Labor Force

Total Labor Force [25]

120,000 salaried

Labor Force by Occupation [26]

Agriculture	65.0%
Industry and commerce	30.0
Services	2.5
Government	2.5

Unemployment Rate [27]

For sources, notes, and explanations, see Annotated Source Appendix, page 1035.

Production Sectors

Commercial Energy Production and Consumption

Production [28]	Consumption [29]
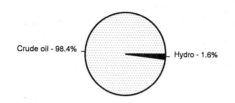	
Crude oil - 98.4% Hydro - 1.6%	Petroleum - 75.0% Hydro - 25.0%

Data are shown in quadrillion (10^{15}) BTUs and percent for 1991

Crude oil[1]	0.63
Natural gas liquids	0.00
Dry natural gas	(s)
Coal[2]	0.00
Net hydroelectric power[3]	0.01
Net nuclear power[3]	0.00
Total	0.64

Petroleum	0.03
Dry natural gas	(s)
Coal[2]	0.00
Net hydroelectric power[3]	0.01
Net nuclear power[3]	0.00
Total	0.04

Telecommunications [30]

- Adequate system of cable, radio relay, tropospheric scatter links and radiocommunication stations
- 15,000 telephones
- Broadcast stations - 6 AM, 6 FM, 3 (5 repeaters) TV
- Satellite earth stations - 3 Atlantic Ocean INTELSAT and 12 domestic satellite

Transportation [31]

Railroads. 649 km 1.437-meter standard-gauge single track (Transgabonese Railroad)

Highways. 7,500 km total; 560 km paved, 960 km laterite, 5,980 km earth

Merchant Marine. 2 cargo ships (1,000 GRT or over) totaling 18,563 GRT/25,330 DWT

Airports

Total:	68
Usable:	56
With permanent-surface runways:	10
With runways over 3,659 m:	0
With runways 2,440-3,659 m:	2
With runways 1,220-2,439 m:	22

Top Agricultural Products [32]

Agriculture accounts for 10% of GDP (including fishing and forestry); cash crops—cocoa, coffee, palm oil; livestock not developed; importer of food; small fishing operations provide a catch of about 20,000 metric tons; okoume (a tropical softwood) is the most important timber product.

Top Mining Products [33]

Metric tons unless otherwise specified	M.t.
Cement, hydraulic	117,000[38]
Manganese, ore & pellets	1,620,388
Uranium oxide, content of concentrate	700
Clinker	125,677
Gold, mine output, Au content (kilograms)	50[39]
Petroleum, crude (000 barrels)	109,500[1]

Tourism [34]

	1987	1988	1989	1990	1991
Tourists	21	20	113[27]	108[27]	128
Tourism receipts	5	7	4	4	8
Tourism expenditures	132	134	124	143	152
Fare receipts	37	44	43	30	32
Fare expenditures	33	36	36	42	44

Tourists are in thousands, money in million U.S. dollars.

Manufacturing Sector

GDP and Manufacturing Summary [35]

	1980	1985	1989	1990	% change 1980-1990	% change 1989-1990
GDP (million 1980 $)	4,281	4,459	3,989	3,076	-28.1	-22.9
GDP per capita (1980 $)	5,305	4,522	3,521	2,622	-50.6	-25.5
Manufacturing as % of GDP (current prices)	5.1	5.6	8.6[e]	5.1	0.0	-40.7
Gross output (million $)	690	615[e]	978[e]	843[e]	22.2	-13.8
Value added (million $)	224	182[e]	275[e]	268[e]	19.6	-2.5
Value added (million 1980 $)	239	245	NA	152	-36.4	NA
Industrial production index	100	102	94	90	-10.0	-4.3
Employment (thousands)	18[e]	18[e]	17[e]	15[e]	-16.7	-11.8

Note: GDP stands for Gross Domestic Product. 'e' stands for estimated value.

Profitability and Productivity

	1980	1985	1989	1990	% change 1980-1990	% change 1989-1990
Intermediate input (%)	68[e]	70[e]	72[e]	68[e]	0.0	-5.6
Wages, salaries, and supplements (%)	16[e]	17[e]	17[e]	19[e]	18.8	11.8
Gross operating surplus (%)	16[e]	13[e]	12[e]	13[e]	-18.8	8.3
Gross output per worker ($)	38,481[e]	34,305[e]	57,509[e]	53,905[e]	40.1	-6.3
Value added per worker ($)	12,470[e]	10,360[e]	16,189[e]	17,264[e]	38.4	6.6
Average wage (incl. benefits) ($)	6,283[e]	5,783[e]	9,520[e]	10,333[e]	64.5	8.5

Profitability is in percent of gross output. Productivity is in U.S. $. 'e' stands for estimated value.

Profitability - 1990

Inputs - 68.0%
Surplus - 13.0%
Wages - 19.0%

The graphic shows percent of gross output.

Value Added in Manufacturing

	1980 $ mil.	1980 %	1985 $ mil.	1985 %	1989 $ mil.	1989 %	1990 $ mil.	1990 %	% change 1980-1990	% change 1989-1990
311 Food products	18[e]	8.0	17[e]	9.3	31[e]	11.3	26[e]	9.7	44.4	-16.1
313 Beverages	19	8.5	13[e]	7.1	14[e]	5.1	20[e]	7.5	5.3	42.9
314 Tobacco products	17	7.6	12[e]	6.6	12[e]	4.4	17[e]	6.3	0.0	41.7
321 Textiles	3	1.3	2[e]	1.1	1[e]	0.4	3[e]	1.1	0.0	200.0
322 Wearing apparel	5	2.2	3[e]	1.6	3[e]	1.1	5[e]	1.9	0.0	66.7
323 Leather and fur products	1	0.4	NA	0.0	NA	0.0	1[e]	0.4	0.0	NA
324 Footwear	1	0.4	NA	0.0	NA	0.0	1[e]	0.4	0.0	NA
331 Wood and wood products	64	28.6	36[e]	19.8	27[e]	9.8	53[e]	19.8	-17.2	96.3
332 Furniture and fixtures	9	4.0	5[e]	2.7	4[e]	1.5	7[e]	2.6	-22.2	75.0
341 Paper and paper products	2	0.9	1[e]	0.5	3[e]	1.1	2[e]	0.7	0.0	-33.3
342 Printing and publishing	3	1.3	3[e]	1.6	3[e]	1.1	4[e]	1.5	33.3	33.3
351 Industrial chemicals	6	2.7	6[e]	3.3	6[e]	2.2	7[e]	2.6	16.7	16.7
352 Other chemical products	3	1.3	2[e]	1.1	6[e]	2.2	3[e]	1.1	0.0	-50.0
353 Petroleum refineries	18	8.0	15[e]	8.2	40[e]	14.5	31[e]	11.6	72.2	-22.5
354 Miscellaneous petroleum and coal products	NA	0.0	NA	0.0	NA	0.0	NA	0.0	NA	NA
355 Rubber products	NA	0.0	NA	0.0	NA	0.0	NA	0.0	NA	NA
356 Plastic products	NA	0.0	NA	0.0	NA	0.0	NA	0.0	NA	NA
361 Pottery, china and earthenware	NA	0.0	NA	0.0	NA	0.0	NA	0.0	NA	NA
362 Glass and glass products	1	0.4	2[e]	1.1	2[e]	0.7	3[e]	1.1	200.0	50.0
369 Other non-metal mineral products	8	3.6	14[e]	7.7	30[e]	10.9	17[e]	6.3	112.5	-43.3
371 Iron and steel	3	1.3	3[e]	1.6	5[e]	1.8	4[e]	1.5	33.3	-20.0
372 Non-ferrous metals	3	1.3	3[e]	1.6	5[e]	1.8	4[e]	1.5	33.3	-20.0
381 Metal products	13	5.8	15[e]	8.2	25[e]	9.1	20[e]	7.5	53.8	-20.0
382 Non-electrical machinery	2	0.9	2[e]	1.1	4[e]	1.5	3[e]	1.1	50.0	-25.0
383 Electrical machinery	8	3.6	9[e]	4.9	21[e]	7.6	12[e]	4.5	50.0	-42.9
384 Transport equipment	11	4.9	12[e]	6.6	20[e]	7.3	17[e]	6.3	54.5	-15.0
385 Professional and scientific equipment	1	0.4	1[e]	0.5	2[e]	0.7	1[e]	0.4	0.0	-50.0
390 Other manufacturing industries	5	2.2	5[e]	2.7	9[e]	3.3	7[e]	2.6	40.0	-22.2

Note: The industry codes shown are International Standard Industry codes (ISIC). Percentages are percent of total Value Added. 'e' stands for estimated value

For sources, notes, and explanations, see Annotated Source Appendix, page 1035.

331

Finance, Economics, and Trade

Economic Indicators [36]

Millions of CFA Francs unless otherwise noted.

	1989	1990	1991[48]
GDP, current prices (bil. CFA)	1,125.8	1,284.8	1,305.7
GDP, % change, nominal	16.5	14.1	1.6
M1 (bil. CFA)	258.3	267.5	272.1
Commercial Lending Rate (%)	7.5	18.5	NA
Savings Rate (%)	18.5	23.9	24.4
Investment Rate (%)	23.6	18.9	19.0
CPI[49]	NA	NA	NA
WPI[49]	NA	NA	NA
External Public Debt	3,263.0	3,499.0	NA

Balance of Payments Summary [37]

Values in millions of dollars.

	1987	1988	1989	1990	1991
Exports of goods (f.o.b.)	1,286.4	1,195.6	1,626.0	2,481.6	2,272.9
Imports of goods (f.o.b.)	-731.8	-791.2	-751.7	-772.0	-827.0
Trade balance	554.6	404.4	874.3	1,709.6	1,445.9
Services - debits	-1,011.1	-1,103.7	-1,248.6	-1,622.6	-1,737.6
Services - credits	130.8	228.1	308.0	261.7	246.0
Private transfers (net)	-147.8	-155.5	-135.2	-158.5	-139.3
Government transfers (net)	24.4	11.2	9.3	24.2	24.8
Long term capital (net)	555.9	652.7	258.4	67.7	536.3
Short term capital (net)	-149.9	130.3	-145.2	-208.8	-294.6
Errors and omissions	-51.0	-101.9	35.0	-45.4	-21.3
Overall balance	-94.1	65.6	-44.0	27.9	60.2

Exchange Rates [38]

Currency: **Communaute Financiere Africaine francs.**
Symbol: **CFAF.**

Data are currency units per $1.

January 1993	274.06
1992	264.69
1991	282.11
1990	272.26
1989	319.01
1988	297.85

Imports and Exports

Top Import Origins [39]

$702 million (c.i.f., 1991 est.).

Origins	%
France	64
African countries	7
US	5
Japan	3

Top Export Destinations [40]

$2.2 billion (f.o.b., 1991).

Destinations	%
France	48
US	15
Germany	2
Japan	2

Foreign Aid [41]

	U.S. $	
US commitments, including Ex-Im (FY70-90)	68	million
Western (non-US) countries, ODA and OOF bilateral commitments (1970-90)	2,342	million
Communist countries (1970-89)	27	million

Import and Export Commodities [42]

Import Commodities	**Export Commodities**
Foodstuffs	Crude oil 80%
Chemical products	Manganese 7%
Petroleum products	Wood 7%
Construction materials	Uranium 2%
Manufactures	
Machinery	

The Gambia

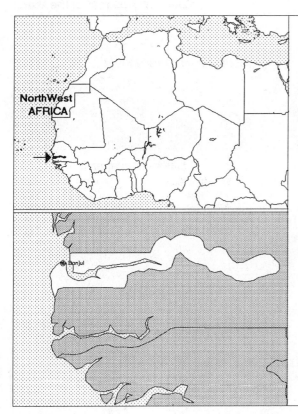

Geography [1]

Total area:
11,300 km2
Land area:
10,000 km2
Comparative area:
Slightly more than twice the size of Delaware
Land boundaries:
Total 740 km, Senegal 740 km
Coastline:
80 km
Climate:
Tropical; hot, rainy season (June to November); cooler, dry season (November to May)
Terrain:
Flood plain of the Gambia River flanked by some low hills
Natural resources:
Fish
Land use:
Arable land:
16%
Permanent crops:
0%
Meadows and pastures:
9%
Forest and woodland:
20%
Other:
55%

Demographics [2]

	1960	1970	1980	1990	1991[1]	1994	2000	2010	2020
Population	391	502	644	848	875	959	1,154	1,561	2,073
Population density (persons per sq. mi.)	101	130	167	220	227	NA	298	402	533
Births	NA	NA	NA	NA	42	45	NA	NA	NA
Deaths	NA	NA	NA	NA	15	15	NA	NA	NA
Life expectancy - males	NA	NA	NA	NA	47	48	NA	NA	NA
Life expectancy - females	NA	NA	NA	NA	51	52	NA	NA	NA
Birth rate (per 1,000)	NA	NA	NA	NA	48	46	NA	NA	NA
Death rate (per 1,000)	NA	NA	NA	NA	17	16	NA	NA	NA
Women of reproductive age (15-44 yrs.)	NA	NA	NA	NA	NA	NA	NA	NA	NA
of which are currently married	NA	NA	NA	NA	NA	NA	NA	NA	NA
Fertility rate	NA	NA	NA	6.5	6.47	6.3	5.9	5.1	4.4

Population values are in thousands, life expectancy in years, and other items as indicated.

Health

Health Personnel [3]

Health Expenditures [4]

For sources, notes, and explanations, see Annotated Source Appendix, page 1035.

333

Human Factors	
Health Care Ratios [5]	Infants and Malnutrition [6]

Ethnic Division [7]

African	99%
Mandinka	42%
Fula	18%
Wolof	16%
Jola	10%
Serahuli	9%
other	4%
Non-Gambian	1%

Religion [8]

Muslim	90%
Christian	9%
Indigenous beliefs	1%

Major Languages [9]

English (official)	NA
Mandinka	NA
Wolof	NA
Fula	NA
Other indigenous vernaculars	NA

Education

Public Education Expenditures [10]

Million Dalasi	1980	1985	1987	1988	1989	1990
Total education expenditure	13	31	NA	53	NA	95
as percent of GNP	3.3	3.2	NA	3.0	NA	3.8
as percent of total govt. expend.	NA	NA	NA	8.8	NA	11.0
Current education expenditure	11	25	NA	37	61	73
as percent of GNP	2.9	2.6	NA	2.1	2.8	2.9
as percent of current govt. expend.	NA	16.4	NA	9.3	NA	11.6
Capital expenditure	2	6	NA	16	NA	22

Educational Attainment [11]

Literacy Rate [12]

In thousands and percent	1985[a]	1991[a]	2000[a]
Illiterate population +15 years	336	350	368
Illiteracy rate - total pop. (%)	79.7	72.8	58.7
Illiteracy rate - males (%)	69.6	61.0	45.5
Illiteracy rate - females (%)	89.5	84.0	71.3

Libraries [13]

	Admin. Units	Svc. Pts.	Vols. (000)	Shelving (meters)	Vols. Added	Reg. Users
National (1986)	1	5	3	220	48	1,000
Non-specialized	NA	NA	NA	NA	NA	NA
Public (1986)	4	8	89	NA	11,800	1,000
Higher ed.	NA	NA	NA	NA	NA	NA
School	NA	NA	NA	NA	NA	NA

Daily Newspapers [14]

	1975	1980	1985	1990
Number of papers	-	-	6	2
Circ. (000)	-	-	4	2[e]

Cinema [15]

Science and Technology

Scientific/Technical Forces [16]	R&D Expenditures [17]	U.S. Patents Issued [18]

Government and Law

Organization of Government [19]

Long-form name:
Republic of The Gambia

Type:
republic under multiparty democratic rule

Independence:
18 February 1965 (from UK; The Gambia and Senegal signed an agreement on 12 December 1981 that called for the creation of a loose confederation to be known as Senegambia, but the agreement was dissolved on 30 September 1989)

Constitution:
24 April 1970

Legal system:
based on a composite of English common law, Koranic law, and customary law; accepts compulsory ICJ jurisdiction, with reservations

National holiday:
Independence Day, 18 February (1965)

Executive branch:
president, vice president, Cabinet

Legislative branch:
unicameral House of Representatives

Judicial branch:
Supreme Court

Political Parties [20]

House of Representatives	% of votes
People's Progressive Party (PPP)	56.6
National Convention Party (NCP)	27.6
Gambian People's Party (GPP)	14.7
People's Democratic Organization (PDOIS)	1.0

Government Expenditures [21]

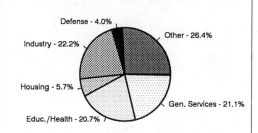

Defense - 4.0%
Other - 26.4%
Industry - 22.2%
Housing - 5.7%
Gen. Services - 21.1%
Educ./Health - 20.7%

% distribution for 1990. Expend. in FY91 est.: 80 ($ mil.)

Crime [23]

Military Expenditures and Arms Transfers [22]

	1985	1986	1987	1988	1989
Military expenditures					
Current dollars (mil.)	NA	NA	1[e]	NA	1[e]
1989 constant dollars (mil.)	NA	NA	1[e]	NA	1[e]
Armed forces (000)	1	1	1	1	2
Gross national product (GNP)					
Current dollars (mil.)	143	148	170	190	210
1989 constant dollars (mil.)	162	164	183	197	210
Central government expenditures (CGE)					
1989 constant dollars (mil.)	NA	NA	NA	NA	74[e]
People (mil.)	0.7	0.8	0.8	0.8	0.8
Military expenditure as % of GNP	NA	NA	0.7	NA	0.7
Military expenditure as % of CGE	NA	NA	NA	NA	2.0
Military expenditure per capita	NA	NA	2	NA	2
Armed forces per 1,000 people	1.4	1.3	1.3	1.3	1.8
GNP per capita	222	218	236	247	255
Arms imports[6]					
Current dollars (mil.)	10	10	10	10	0
1989 constant dollars (mil.)	11	11	11	10	0
Arms exports[6]					
Current dollars (mil.)	0	0	0	0	0
1989 constant dollars (mil.)	0	0	0	0	0
Total imports[7]					
Current dollars (mil.)	93	104	127	137	161
1989 constant dollars	106	115	137	143	161
Total exports[7]					
Current dollars (mil.)	43	35	40	NA	NA
1989 constant dollars	49	39	43	NA	NA
Arms as percent of total imports[8]	10.8	9.6	7.9	7.3	0
Arms as percent of total exports[8]	0	0	0	NA	NA

Human Rights [24]

	SSTS	FL	FAPRO	PPCG	APROBC	TPW	PCPTW	STPEP	PHRFF	PRW	ASST	AFL
Observes	1			P		P	P				1	
	EAFRD	CPR	ESCR	SR	ACHR	MAAE	PVIAC	PVNAC	EAFDAW	TCIDTP	RC	
Observes	P	P	P	P			P	P	S	S	P	

P = Party; S = Signatory; see Appendix for meaning of abbreviations.

Labor Force

Total Labor Force [25]

400,000 (1986 est.)

Labor Force by Occupation [26]

Agriculture	75.0%
Industry, commerce, and services	18.9
Government	6.1

Unemployment Rate [27]

Production Sectors

Energy Resource Summary [28]

Energy Resources: None. **Electricity**: 30,000 kW capacity; 65 million kWh produced, 75 kWh per capita (1991).

Telecommunications [30]

- Adequate network of radio relay and wire
- 3,500 telephones
- Broadcast stations - 3 AM, 2 FM
- 1 Atlantic Ocean INTELSAT earth station

Transportation [31]

Highways. 3,083 km total; 431 km paved, 501 km gravel/laterite, and 2,151 km unimproved earth

Airports

Total:	1
Usable:	1
With permanent-surface runways:	1
With runways over 3,659 m:	0
With runways 2,440-3,659 m:	1
With runways 1,220-2,439 m:	0

Top Agricultural Products [32]

Agriculture accounts for 30% of GDP and employs about 75% of the population; imports one-third of food requirements; major export crop is peanuts; other principal crops - millet, sorghum, rice, corn, cassava, palm kernels; livestock - cattle, sheep, goats; forestry and fishing resources not fully exploited.

Top Mining Products [33]

Detailed information is not available. A summary of mineral resources available follows. **Mineral Resources**: None.

Tourism [34]

	1987	1988	1989	1990	1991
Tourists	97	102	86	101	114
Excursionists[18]	1	2	1	1	1
Tourism receipts	14	23	19	26	26
Tourism expenditures	3	5	5	8	
Fare expenditures	2	2	1	2	

Tourists are in thousands, money in million U.S. dollars.

For sources, notes, and explanations, see Annotated Source Appendix, page 1035.

Manufacturing Sector

GDP and Manufacturing Summary [35]

	1980	1985	1989	1990	% change 1980-1990	% change 1989-1990
GDP (million 1980 $)	239	320	NA	376	57.3	NA
GDP per capita (1980 $)	374	430	NA	436	16.6	NA
Manufacturing as % of GDP (current prices)	3.6	7.7	NA	5.8	61.1	NA
Gross output (million $)	30	40[e]	NA	52[e]	73.3	NA
Value added (million $)	11	9[e]	NA	13[e]	18.2	NA
Value added (million 1980 $)	16	23	NA	28	75.0	NA
Industrial production index	100	107	NA	125	25.0	NA
Employment (thousands)	2	3[e]	NA	2[e]	0.0	NA

Note: GDP stands for Gross Domestic Product. 'e' stands for estimated value.

Profitability and Productivity

	1980	1985	1989	1990	% change 1980-1990	% change 1989-1990
Intermediate input (%)	62	78[e]	NA	76[e]	22.6	NA
Wages, salaries, and supplements (%)	10	7[e]	NA	8[e]	-20.0	NA
Gross operating surplus (%)	28	14[e]	NA	17[e]	-39.3	NA
Gross output per worker ($)	16,115	13,431[e]	NA	15,916[e]	-1.2	NA
Value added per worker ($)	6,094	3,052[e]	NA	4,230[e]	-30.6	NA
Average wage (incl. benefits) ($)	1,566	1,111[e]	NA	1,628[e]	4.0	NA

Profitability is in percent of gross output. Productivity is in U.S. $. 'e' stands for estimated value.

Profitability - 1990

Inputs - 75.2%
Wages - 7.9%
Surplus - 16.8%

The graphic shows percent of gross output.

Value Added in Manufacturing

	1980 $ mil.	1980 %	1985 $ mil.	1985 %	1989 $ mil.	1989 %	1990 $ mil.	1990 %	% change 1980-1990	% change 1989-1990
311 Food products	3	27.3	4[e]	44.4	NA	NA	5[e]	38.5	66.7	NA
313 Beverages	1	9.1	1[e]	11.1	NA	NA	2[e]	15.4	100.0	NA
314 Tobacco products	NA	0.0	NA	0.0	NA	NA	NA	0.0	NA	NA
321 Textiles	NA	0.0	NA	0.0	NA	NA	NA	0.0	NA	NA
322 Wearing apparel	NA	0.0	NA	0.0	NA	NA	NA	0.0	NA	NA
323 Leather and fur products	NA	0.0	NA	0.0	NA	NA	NA	0.0	NA	NA
324 Footwear	NA	0.0	NA	0.0	NA	NA	NA	0.0	NA	NA
331 Wood and wood products	NA	0.0	NA	0.0	NA	NA	NA	0.0	NA	NA
332 Furniture and fixtures	1	9.1	NA	0.0	NA	NA	1[e]	7.7	0.0	NA
341 Paper and paper products	NA	0.0	NA	0.0	NA	NA	NA	0.0	NA	NA
342 Printing and publishing	NA	0.0	NA	0.0	NA	NA	NA	0.0	NA	NA
351 Industrial chemicals	NA	0.0	NA	0.0	NA	NA	NA	0.0	NA	NA
352 Other chemical products	NA	0.0	NA	0.0	NA	NA	NA	0.0	NA	NA
353 Petroleum refineries	NA	0.0	NA	0.0	NA	NA	NA	0.0	NA	NA
354 Miscellaneous petroleum and coal products	NA	0.0	NA	0.0	NA	NA	NA	0.0	NA	NA
355 Rubber products	NA	0.0	NA	0.0	NA	NA	NA	0.0	NA	NA
356 Plastic products	NA	0.0	NA	0.0	NA	NA	NA	0.0	NA	NA
361 Pottery, china and earthenware	NA	0.0	NA	0.0	NA	NA	NA	0.0	NA	NA
362 Glass and glass products	NA	0.0	NA	0.0	NA	NA	NA	0.0	NA	NA
369 Other non-metal mineral products	NA	0.0	NA	0.0	NA	NA	NA	0.0	NA	NA
371 Iron and steel	NA	0.0	NA	0.0	NA	NA	NA	0.0	NA	NA
372 Non-ferrous metals	NA	0.0	NA	0.0	NA	NA	NA	0.0	NA	NA
381 Metal products	NA	0.0	NA	0.0	NA	NA	NA	0.0	NA	NA
382 Non-electrical machinery	NA	0.0	NA	0.0	NA	NA	NA	0.0	NA	NA
383 Electrical machinery	NA	0.0	NA	0.0	NA	NA	NA	0.0	NA	NA
384 Transport equipment	NA	0.0	NA	0.0	NA	NA	NA	0.0	NA	NA
385 Professional and scientific equipment	NA	0.0	NA	0.0	NA	NA	NA	0.0	NA	NA
390 Other manufacturing industries	6	54.5	2[e]	22.2	NA	NA	3[e]	23.1	-50.0	NA

Note: The industry codes shown are International Standard Industry codes (ISIC). Percentages are percent of total Value Added. 'e' stands for estimated value

For sources, notes, and explanations, see Annotated Source Appendix, page 1035.

337

Finance, Economics, and Trade

Economic Indicators [36]

National product: GDP—exchange rate conversion—$292 million (1991 est.). **National product real growth rate**: 3% (1991). **National product per capita**: $325 (1991 est.). **Inflation rate (consumer prices)**: 12% (1992 est.). **External debt**: $336 million (December 1990 est.).

Balance of Payments Summary [37]

Values in millions of dollars.

	1986	1987	1988	1989	1990
Exports of goods (f.o.b.)	64.9	74.5	83.1	100.2	110.6
Imports of goods (f.o.b.)	-84.6	-95.0	-105.9	-125.4	-140.5
Trade balance	-19.7	-20.4	-22.9	-25.1	-29.9
Services - debits	-22.4	-61.9	-61.4	-66.4	-64.9
Services - credits	28.7	50.2	63.7	67.7	71.4
Private transfers (net)	10.7	13.5	12.8	6.7	14.1
Government transfers (net)	6.9	36.2	41.9	33.9	45.0
Long term capital (net)	-9.8	33.8	11.5	12.9	6.7
Short term capital (net)	35.2	-39.4	-6.6	-16.2	-19.6
Errors and omissions	-26.7	5.5	-11.3	-20.8	-24.0
Overall balance	2.9	17.4	27.7	-7.3	-1.1

Exchange Rates [38]

Currency: **dalasi.**
Symbol: **D.**

Data are currency units per $1.

October 1992	8.673
1991	8.803
1990	7.883
1989	7.585
1988	6.709
1987	7.0744

Imports and Exports

Top Import Origins [39]

$174 million (f.o.b., FY91 est.). Data are for 1989.

Origins	%
Europe	57
Asia	25
USSR and Eastern Europe	9
US	6
Other	3

Top Export Destinations [40]

$133 million (f.o.b., FY91 est.). Data are for 1989.

Destinations	%
Japan	60
Europe	29
Africa	5
US	1
Other	5

Foreign Aid [41]

	U.S. $	
US commitments, including Ex-Im (FY70-89)	93	million
Western (non-US) countries, ODA and OOF bilateral commitments (1970-89)	535	million
Communist countries (1970-89)	39	million

Import and Export Commodities [42]

Import Commodities

Foodstuffs
Manufactures
Raw materials
Fuel
Machinery & transport equipment

Export Commodities

Peanuts and peanut products
Fish
Cotton lint
Palm kernels

Georgia

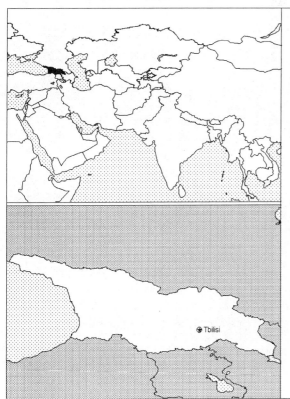

Geography [1]

Total area:
69,700 km2
Land area:
69,700 km2
Comparative area:
Slightly larger than South Carolina
Land boundaries:
Total 1,461 km, Armenia 164 km, Azerbaijan 322 km, Russia 723 km, Turkey 252 km
Coastline:
310 km
Climate:
Warm and pleasant; Mediterranean-like on Black Sea coast
Terrain:
Largely mountainous with Great Caucasus Mountains in the north and Lesser Caucasus Mountains in the south; Kolkhida Lowland opens to the Black Sea in the west; Kura River Basin in the east; good soils in river valley flood plains, foothills of Kolkhida lowland
Natural resources:
Forest lands, hydropower, manganese deposits, iron ores, copper, minor coal and oil deposits; coastal climate and soils allow for important tea and citrus growth
Land use:
Arable land: NA%
Permanent crops: NA%
Meadows and pastures: NA%
Forest and woodland: NA%
Other: NA%

Demographics [2]

	1960	1970	1980	1990	1991[1]	1994	2000	2010	2020
Population	4,147	4,694	5,048	5,484	NA	5,681	5,925	6,253	6,506
Population density (persons per sq. mi.)	NA	NA	NA	NA	NA	NA	NA	NA	NA
Births	NA	NA	NA	NA	NA	92	NA	NA	NA
Deaths	NA	NA	NA	NA	NA	49	NA	NA	NA
Life expectancy - males	NA	NA	NA	NA	NA	69	NA	NA	NA
Life expectancy - females	NA	NA	NA	NA	NA	77	NA	NA	NA
Birth rate (per 1,000)	NA	NA	NA	NA	NA	16	NA	NA	NA
Death rate (per 1,000)	NA	NA	NA	NA	NA	9	NA	NA	NA
Women of reproductive age (15-44 yrs.)	NA	NA	NA	1,353	NA	1,404	1,480	NA	NA
of which are currently married	NA	NA	NA	882	NA	924	975	NA	NA
Fertility rate	NA	NA	NA	2.3	NA	2.2	2.1	1.9	1.9

Population values are in thousands, life expectancy in years, and other items as indicated.

Health

Health Personnel [3]

Health Expenditures [4]

For sources, notes, and explanations, see Annotated Source Appendix, page 1035.

339

Human Factors

Health Care Ratios [5]	Infants and Malnutrition [6]

Ethnic Division [7]

Georgian	70.1%
Armenian	8.1%
Russian	6.3%
Azeri	5.7%
Ossetian	3.0%
Abkhaz	1.8%
Other	5.0%

Religion [8]

Georgian Orthodox	65%
Russian Orthodox	10%
Muslim	11%
Armenian Orthodox	8%
Unknown	6%

Major Languages [9]

Georgian (official)	71%
Russian	9%
Armenian	7%
Azerbaijani	6%
Other	7%

Education

Public Education Expenditures [10]	Educational Attainment [11]
Literacy Rate [12]	Libraries [13]
Daily Newspapers [14]	Cinema [15]

Science and Technology

Scientific/Technical Forces [16]	R&D Expenditures [17]	U.S. Patents Issued [18]

　　　　　　　　　　For sources, notes, and explanations, see Annotated Source Appendix, page 1035.

Government and Law

Organization of Government [19]

Long-form name:
Republic of Georgia
Type:
republic
Independence:
9 April 1991 (from Soviet Union)
Constitution:
adopted NA 1921; currently amending
constitution for Parliamentary and popular
review by late 1995
Legal system:
based on civil law system
National holiday:
Independence Day, 9 April 1991
Executive branch:
chairman of Parliament, Council of
Ministers, prime minister
Legislative branch:
unicameral Parliament
Judicial branch:
Supreme Court

Elections [20]

Georgian Parliament (Supreme Soviet). last held 11 October 1992 (next to be held NA); results—percent of vote by party NA; seats - (225 total) number of seats by party NA; note—representatives of 26 parties elected; Peace Bloc, October 11, Unity, National Democratic Party, and the Greens Party won the largest representation.

Government Budget [21]

For FY88.

Revenues	NA
Expenditures	NA
Capital expenditures	NA

Defense Summary [22]

Branches: Army, National Guard, Interior Ministry Troops

Manpower Availability: Males age 15-49 1,338,606; fit for military service 1,066,309; reach military age (18) annually 43,415 (1993 est.)

Defense Expenditures: No information available.

Note: Georgian forces are poorly organized and not fully under the government's control

Crime [23]

Human Rights [24]

Labor Force

Total Labor Force [25]

2.763 million

Labor Force by Occupation [26]

Industry and construction	31%
Agriculture and forestry	25
Other	44

Date of data: 1990

Unemployment Rate [27]

3% but large numbers of underemployed workers

For sources, notes, and explanations, see Annotated Source Appendix, page 1035.

341

Production Sectors

Energy Resource Summary [28]

Energy Resources: Hydropower, minor coal and oil deposits. **Electricity**: 4,875,000 kW capacity; 15,800 million kWh produced, about 2,835 kWh per capita (1992). **Pipelines**: Crude oil 370 km, refined products 300 km, natural gas 440 km (1992).

Telecommunications [30]

- Poor telephone service
- As of 1991, 672,000 republic telephone lines providing 12 lines per 100 persons
- 339,000 unsatisfied applications for telephones (31 January 1992)
- International links via landline to CIS members and Turkey
- Low capacity satellite earth station and leased international connections via the Moscow international gateway switch
- International electronic mail and telex service established

Transportation [31]

Railroads. 1,570 km, does not include industrial lines (1990)

Highways. 33,900 km total; 29,500 km hard surfaced, 4,400 km earth (1990)

Merchant Marine. 47 ships (1,000 GRT or over) totaling 658,192 GRT/1,014,056 DWT; includes 16 bulk cargo, 30 oil tanker, and 1 specialized liquid carrier

Airports

Total:	37
Useable:	26
With permanent-surface runways:	19
With runways over 3,659 m:	0
With runways 2,440-3,659 m:	10
With runways 1,220-2,439 m:	9

Top Agricultural Products [32]

Accounted for 97% of former USSR citrus fruits and 93% of former USSR tea; berries and grapes; sugar; vegetables, grains, potatoes; cattle, pigs, sheep, goats, poultry; tobacco.

Top Mining Products [33]

Detailed information is not available. A summary of mineral resources available follows. **Mineral Resources**: manganese deposits, iron ores, copper, minor coal and oil deposits.

Tourism [34]

Finance, Economics, and Trade

Industrial Summary [35]

Industrial Production: Growth rate - 50% (1992). **Industries**: Heavy industrial products include raw steel, rolled steel, cement, lumber; machine tools, foundry equipment, electric mining locomotives, tower cranes, electric welding equipment, machinery for food preparation, meat packing, dairy, and fishing industries; air-conditioning electric motors up to 100 kW in size, electric motors for cranes, magnetic starters for motors; devices for control of industrial processes; trucks, tractors, and other farm machinery; light industrial products, including cloth, hosiery, and shoes.

Economic Indicators [36]

National product: GDP not available. **National product real growth rate**: - 35% (1992 est.). **National product per capita**: not available. **Inflation rate (consumer prices)**: 50% per month (January 1993 est.). **External debt**: $650 million (1991 est.).

Balance of Payments Summary [37]

Exchange Rates [38]

Currency: **rubles.**
Symbol: **R.**

Subject to wide fluctuations. Data are currency units per $1.

December 24, 1992	415

Imports and Exports

Top Import Origins [39]

Amount not available.

Origins	%
Russia	NA
Ukraine (1992)	NA

Top Export Destinations [40]

Amount not available.

Destinations	%
Russia	NA
Turkey	NA
Armenia	NA
Azerbaijan (1992)	NA

Foreign Aid [41]

Import and Export Commodities [42]

Import Commodities	Export Commodities
Machinery & parts	Citrus fruits
Fuel	Tea
Transport equipment	Other agricultural products
Textiles	Diverse types of machinery
	Ferrous and nonferrous metals
	Textiles

Germany

Geography [1]

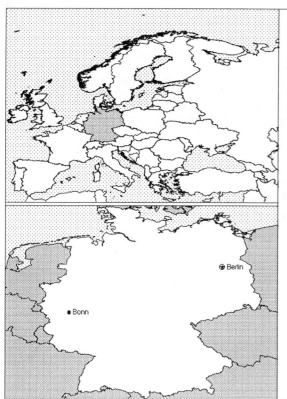

Total area:
356,910 km2
Land area:
349,520 km2
Comparative area:
Slightly smaller than Montana
Land boundaries:
Total 3,621 km, Austria 784 km, Belgium 167 km, Czech Republic 646 km, Denmark 68 km, France 451 km, Luxembourg 138 km, Netherlands 577 km, Poland 456 km, Switzerland 334 km
Coastline:
2,389 km
Climate:
Temperate and marine; cool, cloudy, wet winters and summers; occasional warm, tropical foehn wind; high relative humidity
Terrain:
Lowlands in north, uplands in center, Bavarian Alps in south
Natural resources:
Iron ore, coal, potash, timber, lignite, uranium, copper, natural gas, salt, nickel
Land use:
Arable land: 34%
Permanent crops: 1%
Meadows and pastures: 16%
Forest and woodland: 30%
Other: 19%

Demographics [2]

	1960	1970	1980	1990	1991[1]	1994	2000	2010	2020
Population	72,481	77,783	78,298	79,357	79,548	81,088	82,239	82,837	82,385
Population density (persons per sq. mi.)	536	575	579	585	588	NA	603	608	605
Births	NA	NA	NA	NA	895	895	NA	NA	NA
Deaths	NA	NA	NA	NA	888	883	NA	NA	NA
Life expectancy - males	NA	NA	NA	NA	73	73	NA	NA	NA
Life expectancy - females	NA	NA	NA	NA	79	80	NA	NA	NA
Birth rate (per 1,000)	NA	NA	NA	NA	11	11	NA	NA	NA
Death rate (per 1,000)	NA	NA	NA	NA	11	11	NA	NA	NA
Women of reproductive age (15-44 yrs.)	NA	NA	NA	19,399	NA	19,255	19,288	NA	NA
of which are currently married	NA	NA	NA	12,390	NA	12,743	12,982	NA	NA
Fertility rate	NA	NA	NA	1.5	1.45	1.5	1.5	1.5	1.6

Population values are in thousands, life expectancy in years, and other items as indicated.

Health

Health Personnel [3]

Doctors per 1,000 pop., 1988-92	2.73
Nurse-to-doctor ratio, 1988-92	1.7
Hospital beds per 1,000 pop., 1985-90	8.7
Percentage of children immunized (age 1 yr. or less)	
Third dose of DPT, 1990-91	80.0
Measles, 1990-91	90.0

Health Expenditures [4]

Total health expenditure, 1990 (official exchange rate)	
Millions of dollars	120072
Millions of dollars per capita	1511
Health expenditures as a percentage of GDP	
Total	8.0
Public sector	5.8
Private sector	2.2
Development assistance for health	
Total aid flows (millions of dollars)[1]	NA
Aid flows per capita (millions of dollars)	NA
Aid flows as a percentage of total health expenditure	NA

For sources, notes, and explanations, see Annotated Source Appendix, page 1035.

Human Factors

Health Care Ratios [5]

Population per physician, 1970	580 [1]
Population per physician, 1990	370 [1]
Population per nursing person, 1970	NA
Population per nursing person, 1990	NA
Percent of births attended by health staff, 1985	NA

Infants and Malnutrition [6]

Percent of babies with low birth weight, 1985	5
Infant mortality rate per 1,000 live births, 1970	23
Infant mortality rate per 1,000 live births, 1991	7
Years of life lost per 1,000 population, 1990	12
Prevalence of malnutrition (under age 5), 1990	NA

Ethnic Division [7]

German	95.1%
Turkish	2.3%
Italians	0.7%
Greeks	0.4%
Poles	0.4%
Other	1.1%

Religion [8]

Protestant	45%
Roman Catholic	37%
Unaffiliated or other	18%

Major Languages [9]

German.

Education

Public Education Expenditures [10]

Million Deutschemarks	1980[21]	1985[21]	1987[21]	1988[21]	1989[21]	1990[21]
Total education expenditure	70,099	83,691	88,445	89,747	92,631	98,412
as percent of GNP	4.7	4.6	4.4	4.3	4.1	4.1
as percent of total govt. expend.	9.5	9.2	9.0	8.8	8.8	8.6
Current education expenditure	60,558	75,566	79,906	81,123	83,481	88,499
as percent of GNP	4.1	4.1	4.0	3.8	3.7	3.6
as percent of current govt. expend.	NA	9.4	9.2	8.9	8.9	8.7
Capital expenditure	9,541	8,125	8,539	8,624	9,150	9,913

Educational Attainment [11]

Age group	25+
Total population	10,714,841
Highest level attained (%)	
No schooling	-
First level	
Incompleted	30.1
Completed	NA
Entered second level	
S-1	52.6
S-2	NA
Post secondary	17.3

Literacy Rate [12]

Libraries [13]

	Admin. Units	Svc. Pts.	Vols. (000)	Shelving (meters)	Vols. Added	Reg. Users
National	NA	NA	NA	NA	NA	NA
Non-specialized	NA	NA	NA	NA	NA	NA
Public (1990)	18,284	18,284	148,683	NA	NA	9.4mil
Higher ed. (1990)[32]	237	NA	118,695	NA	3.6mil	1.7mil
School	NA	NA	NA	NA	NA	NA

Daily Newspapers [14]

	1975[c]	1980[c]	1985[c]	1990[c,2]
Number of papers	350	329	319	315
Circ. (000)	20,200	20,611	21,108	10,677

Cinema [15]

Data for 1991.

Cinema seats per 1,000	9.6[2]
Annual attendance per person	1.5[2]
Gross box office receipts (mil. Deutsche Mark)	981[2]

Science and Technology

Scientific/Technical Forces [16]

Potential scientists/engineers	3,040,000
Number female	883,000
Potential technicians	2,150,000
Number female	414,000
Total	5,190,000

R&D Expenditures [17]

	DM[29] (000) 1987
Total expenditure	57,240,000
Capital expenditure	7,018,000
Current expenditure	49,578,000
Percent current	87.6

U.S. Patents Issued [18]

Values show patents issued to citizens of the country by the U.S. Patents Office.

	1990	1991	1992
Number of patents	7,858	7,984	7,608

For sources, notes, and explanations, see Annotated Source Appendix, page 1035.

345

Government and Law

Organization of Government [19]

Long-form name:
Federal Republic of Germany
Type:
federal republic
Independence:
18 January 1871 (German Empire unification); Federal Republic of Germany (West) proclaimed 23 May 1949; German Democratic Republic (East) proclaimed 7 October 1949; unification took place 3 October 1990
Constitution:
23 May 1949, provisional constitution known as Basic Law
Legal system:
civil law system with indigenous concepts; judicial review of legislative acts; has not accepted compulsory ICJ jurisdiction
National holiday:
German Unity Day, 3 October (1990)
Executive branch:
president, chancellor, Cabinet
Legislative branch:
bicameral parliament with a Federal Council (Bundesrat) and a lower chamber (Bundestag)
Judicial branch:
Federal Constitutional Court

Crime [23]

Crime volume (for 1990)	
Cases known to police	4,455333
Attempts (percent)	7.9
Percent cases solved	47
Crimes per 100,000 persons	7,108.2
Persons responsible for offenses	
Total number offenders	1,437,923
Percent female	23.5
Percent juvenile	9.8[1]
Percent foreigners	26.7

Political Parties [20]

Federal Diet	% of votes
Christian Democratic Union (CDU)	36.7
Social Democratic Party (SPD)	33.5
Free Democratic Party (FDP)	11.0
Christian Social Union (CSU)	7.1
Green Party (West Germany)	3.9
Party of Democratic Socialism (PDS)	2.4
Republikaner	2.1
Alliance 90/Green Party (East)	1.2
Other	2.1

Government Expenditures [21]

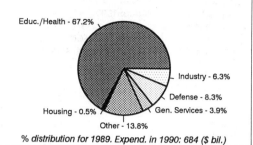

Educ./Health - 67.2%
Industry - 6.3%
Defense - 8.3%
Gen. Services - 3.9%
Housing - 0.5%
Other - 13.8%

% distribution for 1989. Expend. in 1990: 684 ($ bil.)

Military Expenditures and Arms Transfers [22]

	1985	1986	1987	1988	1989
Military expenditures					
Current dollars (mil.)	29,980	30,590	31,570	32,300	33,600
1989 constant dollars (mil.)	34,130	33,950	33,960	33,630	33,600
Armed forces (000)	495	495	495	495	503
Gross national product (GNP)					
Current dollars (bil.)	942,900	989.7	1,038	1,112	1,207
1989 constant dollars (bil.)	1,073,000	1,098	1,117	1,158	1,207
Central government expenditures (CGE)					
1989 constant dollars (mil.)	332,900	332,700	339,200	351,400	353,200
People (mil.)	61.0	61.1	61.2	61.5	61.8
Military expenditure as % of GNP	3.2	3.1	3.0	2.9	2.8
Military expenditure as % of CGE	10.3	10.2	10.0	9.6	9.5
Military expenditure per capita	559	556	555	547	544
Armed forces per 1,000 people	8.1	8.1	8.1	8.1	8.1
GNP per capita	17,590	17,980	18,240	18,830	19,520
Arms imports[6]					
Current dollars (mil.)	490	440	625	1,100	875
1989 constant dollars (mil.)	558	488	672	1,145	875
Arms exports[6]					
Current dollars (mil.)	1,400	1,100	1,500	1,200	1,200
1989 constant dollars (mil.)	1,594	1,221	1,613	1,249	1,200
Total imports[7]					
Current dollars (mil.)	158,500	190,900	228,400	250,500	269,700
1989 constant dollars	180,400	211,800	245,700	260,700	269,700
Total exports[7]					
Current dollars (mil.)	183,900	243,300	294,400	323,300	341,200
1989 constant dollars	209,400	270,000	316,600	336,600	341,200
Arms as percent of total imports[8]	0.3	0.2	0.3	0.4	0.3
Arms as percent of total exports[8]	0.8	0.5	0.5	0.4	0.4

Human Rights [24]

	SSTS	FL	FAPRO	PPCG	APROBC	TPW	PCPTW	STPEP	PHRFF	PRW	ASST	AFL
Observes		P	P	P	P	P	P		P	P	P	P

	EAFRD	CPR	ESCR	SR	ACHR	MAAE	PVIAC	PVNAC	EAFDAW	TCIDTP	RC
Observes	P	P	P	P		P	P	P	P	S	P

P = Party; S = Signatory; see Appendix for meaning of abbreviations.

Labor Force

Total Labor Force [25]

36.75 million

Labor Force by Occupation [26]

Industry	41%
Agriculture	6
Other	53

Date of data: 1987

Unemployment Rate [27]

Production Sectors

Commercial Energy Production and Consumption

Production [28]

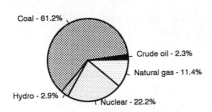

Coal - 61.2%
Crude oil - 2.3%
Natural gas - 11.4%
Hydro - 2.9%
Nuclear - 22.2%

Consumption [29]

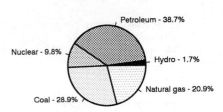

Petroleum - 38.7%
Nuclear - 9.8%
Hydro - 1.7%
Coal - 28.9%
Natural gas - 20.9%

Data are shown in quadrillion (10^{15}) BTUs and percent for 1991

Crude oil[1]	0.15
Natural gas liquids	0.00
Dry natural gas	0.75
Coal[2]	4.02
Net hydroelectric power[3]	0.19
Net nuclear power[3]	1.46
Total	6.58

Petroleum	5.82
Dry natural gas	3.14
Coal[2]	4.35
Net hydroelectric power[3]	0.26
Net nuclear power[3]	1.47
Total	15.06

Telecommunications [30]

- *Western*: highly developed, modern telecommunication service to all parts of the country, fully adequate in all respects
- 40,300,000 telephones
- Well developed cable and microwave radio relay networks, all completely automatic
- Broadcast stations—80 AM, 470 FM, 225 (6,000 repeaters) TV
- 6 submarine coaxial cables; satellite earth stations—12 Atlantic Ocean INTELSAT antennas, 2 Indian Ocean INTELSAT antennas, EUTELSAT, and domestic systems; 2 HF radiocommunication centers; tropospheric links
- *Eastern*: badly needs modernization; 3,970,000 telephones; broadcast stations—23 AM, 17 FM, 21 TV (15 Soviet TV repeaters); 6,181,860 TVs; 6,700,000 radios; 1 satellite earth station operating in INTELSAT and Intersputnik systems

Transportation [31]

Merchant Marine. 565 ships (1,000 GRT or over) totaling 4,928,759 GRT/6,292,193 DWT; includes 5 short-sea passenger, 3 passenger, 303 cargo, 10 refrigerated cargo, 134 container, 28 roll-on/roll-off cargo, 5 railcar carrier, 7 barge carrier, 9 oil tanker, 21 chemical tanker, 17 liquefied gas tanker, 5 combination ore/oil, 6 combination bulk, 12 bulk; note - the German register includes ships of the former East and West Germany; during 1991 the fleet underwent major restructuring as surplus ships were sold off

Airports

Total:	499
Usable:	492
With permanent-surface runways:	271
With runways over 3,659 m:	5
With runways 2,440-3,659 m:	59
With runways 1,220-2,439 m:	67

Top Agricultural Products [32]

	1990	1991[11]
Grains	37,579	39,280
Sugar beets	30,366	26,000
Potatoes	14,039	9,856

Values shown are 1,000 metric tons.

Top Mining Products [33]

Estimated metric tons unless otherwise specified   M.t.

Coal, anthracite, lignite & bituminous (000 tons)	279,403
Steel, crude (000 tons)	42,169
Cement, hydraulic (000 tons)	34,396
Potash, crude (000 tons)	41,322
Salt (000 tons)	16,025
Limestone, industrial (000 tons)	58,106

Tourism [34]

	1987	1988	1989	1990	1991
Tourists[28]	14,045	14,501	16,115	17,045[29]	15,648[29]
Tourism receipts	7,678	8,449	8,658	10,683	10,947
Tourism expenditures	23,341	25,036	23,727	29,836	31,650
Fare receipts	2,964	3,650	3,940	5,172	5,076
Fare expenditures	3,481	4,098	4,016	5,325	5,532

Tourists are in thousands, money in million U.S. dollars.

For sources, notes, and explanations, see Annotated Source Appendix, page 1035.

347

Manufacturing Sector

GDP and Manufacturing Summary [35]

	1980	1985	1989	1990	% change 1980-1990	% change 1989-1990
GDP (million 1980 $)	813,498	858,550	961,194	998,198	22.7	3.8
GDP per capita (1980 $)	13,213	14,069	15,687	16,273	23.2	3.7
Manufacturing as % of GDP (current prices)	33.6	32.6	31.1	31.9	-5.1	2.6
Gross output (million $)	632,161	490,047	879,040	1,095,480	73.3	24.6
Value added (million $)	265,588	224,215	433,240	543,666	104.7	25.5
Value added (million 1980 $)	265,589	273,829	305,461	305,136	14.9	-0.1
Industrial production index	100	104	118	124	24.0	5.1
Employment (thousands)	7,229	6,614	6,910	7,119	-1.5	3.0

Note: GDP stands for Gross Domestic Product. 'e' stands for estimated value.

Profitability and Productivity

	1980	1985	1989	1990	% change 1980-1990	% change 1989-1990
Intermediate input (%)	58	54	51	50	-13.8	-2.0
Wages, salaries, and supplements (%)	26	24	20	25e	-3.8	25.0
Gross operating surplus (%)	16	22	29	25e	56.3	-13.8
Gross output per worker ($)	87,448	74,092	127,213	153,881	76.0	21.0
Value added per worker ($)	36,739	33,926	62,698	76,368	107.9	21.8
Average wage (incl. benefits) ($)	22,606	17,567	25,605	38,440e	70.0	50.1

Profitability is in percent of gross output. Productivity is in U.S. $. 'e' stands for estimated value.

Profitability - 1990

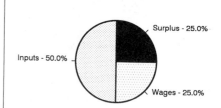

Surplus - 25.0%
Inputs - 50.0%
Wages - 25.0%

The graphic shows percent of gross output.

Value Added in Manufacturing

	1980 $ mil.	1980 %	1985 $ mil.	1985 %	1989 $ mil.	1989 %	1990 $ mil.	1990 %	% change 1980-1990	% change 1989-1990
311 Food products	18,570	7.0	10,829	4.8	21,142	4.9	28,236	5.2	52.1	33.6
313 Beverages	6,452	2.4	5,048	2.3	9,178	2.1	12,715	2.3	97.1	38.5
314 Tobacco products	6,909	2.6	5,720	2.6	9,629	2.2	12,885	2.4	86.5	33.8
321 Textiles	6,964	2.6	5,510	2.5	9,706	2.2	11,996	2.2	72.3	23.6
322 Wearing apparel	4,934	1.9	2,802	1.2	4,487	1.0	5,689	1.0	15.3	26.8
323 Leather and fur products	935	0.4	499	0.2	740	0.2	956	0.2	2.2	29.2
324 Footwear	1,205	0.5	727	0.3	1,046	0.2	1,158	0.2	-3.9	10.7
331 Wood and wood products	4,485	1.7	2,429	1.1	4,246e	1.0	5,378e	1.0	19.9	26.7
332 Furniture and fixtures	5,548	2.1	3,084	1.4	6,749e	1.6	8,615e	1.6	55.3	27.6
341 Paper and paper products	5,099	1.9	5,221	2.3	10,141	2.3	12,781	2.4	150.7	26.0
342 Printing and publishing	6,150	2.3	4,141	1.8	7,846	1.8	10,309	1.9	67.6	31.4
351 Industrial chemicals	13,944	5.3	16,569	7.4	32,552e	7.5	39,324e	7.2	182.0	20.8
352 Other chemical products	8,003	3.0	11,596	5.2	23,060e	5.3	28,030e	5.2	250.2	21.6
353 Petroleum refineries	14,637	5.5	10,126	4.5	14,303e	3.3	19,676	3.6	34.4	37.6
354 Miscellaneous petroleum and coal products	990	0.4	985	0.4	1,348e	0.3	2,250	0.4	127.3	66.9
355 Rubber products	3,201	1.2	2,880	1.3	5,975e	1.4	7,170e	1.3	124.0	20.0
356 Plastic products	6,095	2.3	5,639	2.5	12,530e	2.9	15,612e	2.9	156.1	24.6
361 Pottery, china and earthenware	1,304	0.5	669	0.3	1,219	0.3	1,537e	0.3	17.9	26.1
362 Glass and glass products	2,492	0.9	1,916	0.9	4,131	1.0	4,887	0.9	96.1	18.3
369 Other non-metal mineral products	7,937	3.0	4,874	2.2	9,970	2.3	12,016	2.2	51.4	20.5
371 Iron and steel	18,872	7.1	9,538	4.3	16,377	3.8	19,685	3.6	4.3	20.2
372 Non-ferrous metals	2,508	0.9	3,414	1.5	6,720	1.6	8,284	1.5	230.3	23.3
381 Metal products	14,455	5.4	14,161	6.3	29,594	6.8	38,187	7.0	164.2	29.0
382 Non-electrical machinery	34,263	12.9	33,811	15.1	65,194	15.0	82,039	15.1	139.4	25.8
383 Electrical machinery	30,501	11.5	28,329	12.6	61,328	14.2	74,395	13.7	143.9	21.3
384 Transport equipment	31,232	11.8	29,076	13.0	55,334	12.8	68,629	12.6	119.7	24.0
385 Professional and scientific equipment	6,205	2.3	3,448	1.5	6,337	1.5	8,167e	1.5	31.6	28.9
390 Other manufacturing industries	1,700	0.6	1,175	0.5	2,357e	0.5	3,063e	0.6	80.2	30.0

Note: The industry codes shown are International Standard Industry codes (ISIC). Percentages are percent of total Value Added. 'e' stands for estimated value

For sources, notes, and explanations, see Annotated Source Appendix, page 1035.

Finance, Economics, and Trade

Economic Indicators [36]

Billions of Deutschemarks (DM) unless Otherwise noted.

	1990[50]	1991[e,50]	1991[e]
Real GNP (1985 prices)	2,138.7	2,206.6	2,390.0
Real GNP Growth Rate	4.5	3.2	1.2
Real GNP Per Capita Income (in DM)	33,823	34,450	29,880
Money Supply (M1)	584.2[51]	-	581.1[52]
Commercial Interest Rate[53]	10.28	-	11.31[54]
Savings Rate[6]	14.7	14.5	NA
Investment Rate (% nom GNP)	21.0	21.8	22.9
CPI, (1985 = 100)	107.0	110.7	NA
WPI, (1985 = 100)	95.2	96.7[55]	NA
External Public Debt	224.4	-	239.6[56]

Balance of Payments Summary [37]

Exchange Rates [38]

Currency: **deutsche marks.**
Symbol: **DM.**

Data are currency units per $1.

January 1993	1.6158
1992	1.5617
1991	1.6595
1990	1.6157
1989	1.8800
1988	1.7562

Imports and Exports

Top Import Origins [39]

$354.5 billion (f.o.b., 1991). Data are for 1992.

Origins	%
EC	52.0
France	12.0
Netherlands	9.6
Italy	9.2
UK	6.8
Belgium-Luxembourg	7.0
Other Western Europe	15.2
US	6.6
Eastern Europe	5.5
OPEC	2.4

Top Export Destinations [40]

$378.0 billion (f.o.b., 1991) Data are for 1992.

Destinations	%
EC	54.3
Netherlands	8.3
Italy	9.3
UK	7.7
Belgium-Luxembourg	7.4
Other Western Europe	17.0
US	6.4
Eastern Europe	5.6
OPEC	3.4

Foreign Aid [41]

	U.S. $	
Western: donor - ODA and OOF commitments (1970-89)	75.5	billion
Eastern: donor - extended bilaterally to non-Communist less developed countries (1956-89)	4.0	billion

Import and Export Commodities [42]

Import Commodities

Manufactures 68.5%
Agricultural products 12.0%
Fuels 9.7%
Raw materials 7.1%

Export Commodities

Manufactures 86.6% including:
Machines
Machine tools
Chemicals
Motor vehicles
Iron & steel products
Agricultural products 4.9%
Raw materials 2.3%
Fuels 1.3%

For sources, notes, and explanations, see Annotated Source Appendix, page 1035.

349

Ghana

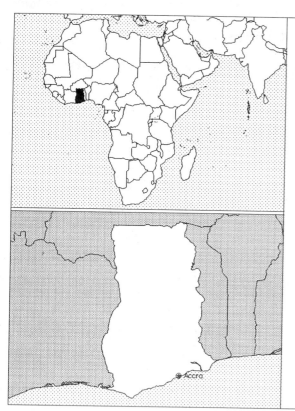

Geography [1]

Total area:
238,540 km2
Land area:
230,020 km2
Comparative area:
Slightly smaller than Oregon
Land boundaries:
Total 2,093 km, Burkina 548 km, Cote d'Ivoire 668 km, Togo 877 km
Coastline:
539 km
Climate:
Tropical; warm and comparatively dry along southeast coast; hot and humid in southwest; hot and dry in north
Terrain:
Mostly low plains with dissected plateau in south-central area
Natural resources:
Gold, timber, industrial diamonds, bauxite, manganese, fish, rubber
Land use:
Arable land:
5%
Permanent crops:
7%
Meadows and pastures:
15%
Forest and woodland:
37%
Other:
36%

Demographics [2]

	1960	1970	1980	1990	1991[1]	1994	2000	2010	2020
Population	6,958	8,789	10,777	15,195	15,617	17,225	20,608	27,305	35,877
Population density (persons per sq. mi.)	78	99	121	170	176	NA	231	306	401
Births	NA	NA	NA	NA	713	760	NA	NA	NA
Deaths	NA	NA	NA	NA	203	211	NA	NA	NA
Life expectancy - males	NA	NA	NA	NA	53	54	NA	NA	NA
Life expectancy - females	NA	NA	NA	NA	56	58	NA	NA	NA
Birth rate (per 1,000)	NA	NA	NA	NA	46	44	NA	NA	NA
Death rate (per 1,000)	NA	NA	NA	NA	13	12	NA	NA	NA
Women of reproductive age (15-44 yrs.)	NA	NA	NA	NA	NA	NA	NA	NA	NA
of which are currently married	NA	NA	NA	NA	NA	NA	NA	NA	NA
Fertility rate	NA	NA	NA	6.4	6.34	6.2	5.8	5.0	4.3

Population values are in thousands, life expectancy in years, and other items as indicated.

Health

Health Personnel [3]

Doctors per 1,000 pop., 1988-92	0.04
Nurse-to-doctor ratio, 1988-92	9.1
Hospital beds per 1,000 pop., 1985-90	1.5
Percentage of children immunized (age 1 yr. or less)	
Third dose of DPT, 1990-91	39.0
Measles, 1990-91	39.0

Health Expenditures [4]

Total health expenditure, 1990 (official exchange rate)	
Millions of dollars	204
Millions of dollars per capita	14
Health expenditures as a percentage of GDP	
Total	3.5
Public sector	1.7
Private sector	1.8
Development assistance for health	
Total aid flows (millions of dollars)[1]	29
Aid flows per capita (millions of dollars)	1.9
Aid flows as a percentage of total health expenditure	14.2

For sources, notes, and explanations, see Annotated Source Appendix, page 1035.

Human Factors

Health Care Ratios [5]

Population per physician, 1970	12910
Population per physician, 1990	22970
Population per nursing person, 1970	690
Population per nursing person, 1990	1670
Percent of births attended by health staff, 1985	73

Infants and Malnutrition [6]

Percent of babies with low birth weight, 1985	17
Infant mortality rate per 1,000 live births, 1970	11
Infant mortality rate per 1,000 live births, 1991	83
Years of life lost per 1,000 population, 1990	55
Prevalence of malnutrition (under age 5), 1990	36

Ethnic Division [7]

Black African	99.8%
Akan	44.0%
Moshi-Dagomba	16.0%
Ewe	13.0%
Ga	8.0%
European/other	0.2%

Religion [8]

Indigenous beliefs	38%
Muslim	30%
Christian	24%
Other	8%

Major Languages [9]

English (official)	NA
African languages	NA
Akan	NA
Moshi-Dagomba	NA
Ewe	NA
Ga	NA

Education

Public Education Expenditures [10]

Million Cedi	1980	1985[1]	1987[1]	1988[1]	1989	1990
Total education expenditure	1,319	8,675	24,795	33,656	47,791	61,900
as percent of GNP	3.1	2.6	3.4	3.3	3.5	3.3
as percent of total govt. expend.	17.1	19.0	24.3	23.4	24.3	24.3
Current education expenditure	NA	NA	23,006	30,914	43,857	53,664
as percent of GNP	NA	NA	3.2	3.0	3.2	2.9
as percent of current govt. expend.	NA	NA	28.1	27.8	29.5	27.1
Capital expenditure	NA	NA	1,789	2,742	3,934	8,236

Educational Attainment [11]

Literacy Rate [12]

In thousands and percent	1985[a]	1991[a]	2000[a]
Illiterate population +15 years	3,316	3,258	2,974
Illiteracy rate - total pop. (%)	47.2	39.7	26.4
Illiteracy rate - males (%)	36.3	30.0	19.5
Illiteracy rate - females (%)	57.8	49.0	33.2

Libraries [13]

	Admin. Units	Svc. Pts.	Vols. (000)	Shelving (meters)	Vols. Added	Reg. Users
National	NA	NA	NA	NA	NA	NA
Non-specialized	NA	NA	NA	NA	NA	NA
Public (1989)	13	47	1,576	NA	32,200	56,211
Higher ed.	NA	NA	NA	NA	NA	NA
School	NA	NA	NA	NA	NA	NA

Daily Newspapers [14]

	1975	1980	1985	1990
Number of papers	4	5	5	2
Circ. (000)	500	500[e]	510[e]	200

Cinema [15]

Science and Technology

Scientific/Technical Forces [16]

R&D Expenditures [17]

U.S. Patents Issued [18]

For sources, notes, and explanations, see Annotated Source Appendix, page 1035.

351

Government and Law

Organization of Government [19]

Long-form name:
Republic of Ghana
Type:
constitutional democracy
Independence:
6 March 1957 (from UK)
Constitution:
new constitution approved 28 April 1992
Legal system:
based on English common law and customary law; has not accepted compulsory ICJ jurisdiction
National holiday:
Independence Day, 6 March (1957)
Executive branch:
president, cabinet
Legislative branch:
unicameral National Assembly
Judicial branch:
Supreme Court

Elections [20]

National Assembly. Last held 29 December 1992 (next to be held NA).

Government Expenditures [21]

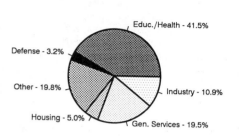

% distribution for 1988. Expend. in 1991 est.: 0.905 ($ bil.)

Military Expenditures and Arms Transfers [22]

	1985	1986	1987	1988	1989
Military expenditures					
Current dollars (mil.)	37	36	39	21	30[e]
1989 constant dollars (mil.)	42	40	42	22	30[e]
Armed forces (000)	15	9	11	16	16
Gross national product (GNP)					
Current dollars (mil.)	3,639	3,913	4,228	4,649	5,134
1989 constant dollars (mil.)	4,143	4,342	4,548	4,839	5,134
Central government expenditures (CGE)					
1989 constant dollars (mil.)	588	638	671	703	NA
People (mil.)	12.9	13.4	13.8	14.2	14.7
Military expenditure as % of GNP	1.0	0.9	0.9	0.4	0.6
Military expenditure as % of CGE	7.2	6.3	6.2	3.1	NA
Military expenditure per capita	3	3	3	2	2
Armed forces per 1,000 people	1.2	0.7	0.8	1.1	1.1
GNP per capita	322	324	329	340	350
Arms imports[6]					
Current dollars (mil.)	0	10	20	20	30
1989 constant dollars (mil.)	0	11	22	21	30
Arms exports[6]					
Current dollars (mil.)	0	0	0	0	0
1989 constant dollars (mil.)	0	0	0	0	0
Total imports[7]					
Current dollars (mil.)	731	783	988	907	NA
1989 constant dollars	832	869	1,063	944	NA
Total exports[7]					
Current dollars (mil.)	617	863	977	1,014	NA
1989 constant dollars	702	958	1,051	1,056	NA
Arms as percent of total imports[8]	0	1.3	2.0	2.2	NA
Arms as percent of total exports[8]	0	0	0	0	NA

Crime [23]

Crime volume	
Cases known to police	120,933
Attempts (percent)	NA
Percent cases solved	93
Crimes per 100,000 persons	863.81
Persons responsible for offenses	
Total number offenders	112,400
Percent female	2.05
Percent juvenile	NA
Percent foreigners	NA

Human Rights [24]

	SSTS	FL	FAPRO	PPCG	APROBC	TPW	PCPTW	STPEP	PHRFF	PRW	ASST	AFL
Observes	P	P	P	P	P	P	P			P	P	P
	EAFRD	CPR	ESCR	SR	ACHR	MAAE	PVIAC	PVNAC	EAFDAW	TCIDTP	RC
Observes		P		P			P		P	P		P

P = Party; S = Signatory; see Appendix for meaning of abbreviations.

Labor Force

Total Labor Force [25]

3.7 million

Labor Force by Occupation [26]

Agriculture and fishing
Industry
Sales and clerical
Services, transportation, and communications
Professional

Unemployment Rate [27]

10% (1991)

For sources, notes, and explanations, see Annotated Source Appendix, page 1035.

Production Sectors

Energy Resource Summary [28]

Energy Resources: None. **Electricity**: 1,180,000 kW capacity; 4,490 million kWh produced, 290 kWh per capita (1991). **Pipelines**: None.

Telecommunications [30]

- Poor to fair system handled primarily by microwave radio relay links
- 42,300 telephones
- Broadcast stations - 4 AM, 1 FM, 4 (8 translators) TV
- 1 Atlantic Ocean INTELSAT earth station

Transportation [31]

Railroads. 953 km, all 1.067-meter gauge; 32 km double track; railroads undergoing major renovation

Highways. 32,250 km total; 6,084 km concrete or bituminous surface, 26,166 km gravel, laterite, and improved earth surfaces

Merchant Marine. 6 ships (1,000 GRT or over) totaling 59,293 GRT/ 78,246 DWT; includes 5 cargo, 1 refrigerated cargo

Airports

Total:	10
Usable:	9
With permanent-surface runways:	5
With runways over 3,659 m:	0
With runways 2,440-3,659 m:	2
With runways 1,220-2,439 m:	6

Top Agricultural Products [32]

	1990	1991
Cassava	2,717	5,701
Yams	877	2,631
Plantains	798	1,178
Corn	553	931
Sorghum and millet	210	353
Cocoa	295	290

Values shown are 1,000 metric tons.

Top Mining Products [33]

Estimated metric tons unless otherwise specified   M.t.

Cement, hydraulic (000 tons)	750[1]
Diamond, gem & industrial (000 carats)	700[40]
Manganese, Mn content of ore and concentrate	125,000[41]
Gold (kilograms)	26,311
Salt	50,000
Bauxite	380,000

Tourism [34]

	1987	1988	1989	1990	1991
Tourists	103	114	125	146	172
Tourism receipts	36	55	72	81	118
Tourism expenditures	12	12	13	13	
Fare receipts	12	12	12	10	
Fare expenditures	25	27	27	27	

Tourists are in thousands, money in million U.S. dollars.

For sources, notes, and explanations, see Annotated Source Appendix, page 1035.

353

Manufacturing Sector

GDP and Manufacturing Summary [35]

	1980	1985	1989	1990	% change 1980-1990	% change 1989-1990
GDP (million 1980 $)	4,788	4,686	5,795	5,893	23.1	1.7
GDP per capita (1980 $)	446	365	398	392	-12.1	-1.5
Manufacturing as % of GDP (current prices)	7.8	11.5	10.4[e]	9.9	26.9	-4.8
Gross output (million $)	505	696	950[e]	1,309[e]	159.2	37.8
Value added (million $)	244	338	498[e]	620[e]	154.1	24.5
Value added (million 1980 $)	374	299	449[e]	409	9.4	-8.9
Industrial production index	100	70	111[e]	104	4.0	-6.3
Employment (thousands)	80	61	68[e]	95[e]	18.8	39.7

Note: GDP stands for Gross Domestic Product. 'e' stands for estimated value.

Profitability and Productivity

	1980	1985	1989	1990	% change 1980-1990	% change 1989-1990
Intermediate input (%)	52	51	48[e]	53[e]	1.9	10.4
Wages, salaries, and supplements (%)	10	6	7[e]	7[e]	-30.0	0.0
Gross operating surplus (%)	39	42	46[e]	40[e]	2.6	-13.0
Gross output per worker ($)	6,293	11,306	14,045[e]	13,685[e]	117.5	-2.6
Value added per worker ($)	3,034	5,495	7,360[e]	6,501[e]	114.3	-11.7
Average wage (incl. benefits) ($)	606	711	961[e]	970[e]	60.1	0.9

Profitability is in percent of gross output. Productivity is in U.S. $. 'e' stands for estimated value.

Profitability - 1990

Inputs - 53.0%
Wages - 7.0%
Surplus - 40.0%

The graphic shows percent of gross output.

Value Added in Manufacturing

	1980 $ mil.	1980 %	1985 $ mil.	1985 %	1989 $ mil.	1989 %	1990 $ mil.	1990 %	% change 1980-1990	% change 1989-1990
311 Food products	20	8.2	35	10.4	45[e]	9.0	76[e]	12.3	280.0	68.9
313 Beverages	38	15.6	51	15.1	72[e]	14.5	87[e]	14.0	128.9	20.8
314 Tobacco products	32	13.1	68	20.1	77[e]	15.5	70[e]	11.3	118.8	-9.1
321 Textiles	22	9.0	18	5.3	27[e]	5.4	50[e]	8.1	127.3	85.2
322 Wearing apparel	3	1.2	1	0.3	1[e]	0.2	1[e]	0.2	-66.7	0.0
323 Leather and fur products	1	0.4	NA	0.0	1[e]	0.2	1[e]	0.2	0.0	0.0
324 Footwear	1	0.4	NA	0.0	1[e]	0.2	1[e]	0.2	0.0	0.0
331 Wood and wood products	16	6.6	41	12.1	78[e]	15.7	56[e]	9.0	250.0	-28.2
332 Furniture and fixtures	2	0.8	2	0.6	3[e]	0.6	7[e]	1.1	250.0	133.3
341 Paper and paper products	1	0.4	2	0.6	3[e]	0.6	4[e]	0.6	300.0	33.3
342 Printing and publishing	5	2.0	4	1.2	6[e]	1.2	10[e]	1.6	100.0	66.7
351 Industrial chemicals	2	0.8	1	0.3	1[e]	0.2	5[e]	0.8	150.0	400.0
352 Other chemical products	9	3.7	26	7.7	33[e]	6.6	27[e]	4.4	200.0	-18.2
353 Petroleum refineries	37	15.2	34	10.1	63[e]	12.7	45[e]	7.3	21.6	-28.6
354 Miscellaneous petroleum and coal products	NA	0.0	NA	0.0	NA	0.0	NA	0.0	NA	NA
355 Rubber products	5	2.0	2	0.6	2[e]	0.4	5[e]	0.8	0.0	150.0
356 Plastic products	1	0.4	2	0.6	3[e]	0.6	3[e]	0.5	200.0	0.0
361 Pottery, china and earthenware	1	0.4	NA	0.0	NA	0.0	NA	0.0	NA	NA
362 Glass and glass products	NA	0.0	1	0.3	2[e]	0.4	1[e]	0.2	NA	-50.0
369 Other non-metal mineral products	6	2.5	16	4.7	16[e]	3.2	23[e]	3.7	283.3	43.8
371 Iron and steel	1	0.4	1	0.3	1[e]	0.2	3[e]	0.5	200.0	200.0
372 Non-ferrous metals	29	11.9	16	4.7	45[e]	9.0	121[e]	19.5	317.2	168.9
381 Metal products	7	2.9	8	2.4	11[e]	2.2	14[e]	2.3	100.0	27.3
382 Non-electrical machinery	NA	0.0	NA	0.0	NA	0.0	2[e]	0.3	NA	NA
383 Electrical machinery	2	0.8	3	0.9	4[e]	0.8	5[e]	0.8	150.0	25.0
384 Transport equipment	3	1.2	2	0.6	2[e]	0.4	4[e]	0.6	33.3	100.0
385 Professional and scientific equipment	1	0.4	1	0.3	1[e]	0.2	1[e]	0.2	0.0	0.0
390 Other manufacturing industries	NA	0.0	NA	0.0	NA	0.0	1[e]	0.2	NA	NA

Note: The industry codes shown are International Standard Industry codes (ISIC). Percentages are percent of total Value Added. 'e' stands for estimated value

For sources, notes, and explanations, see Annotated Source Appendix, page 1035.

Finance, Economics, and Trade

Economic Indicators [36]

Millions of Cedis unless otherwise stated.

	1989	1990	1991[e]
GDP, curr. prices (bil. Cedis)	1,346.9	1,609.1	NA
GDP, curr. prices (% chg.)	27.3	19.5	NA
Per capita GDP/current dollars	331.0	325.0	330.0
Money Supply (M2, yr-end)	204.1	265.2	NA
Commercial Lending Rate			
(%, yr-end)	23-31	23-31	NA
Savings Growth Rate	6.5	4.6	6.5
Investment (percent/GDP)	14.0	14.4	15.6
CPI (1980 - 100)	2,601	3,511	4,213
WPI (1977 - 100)	18,084	NA	NA
External Public Debt (yr-end)	2,862.4	3,043.1	NA

Balance of Payments Summary [37]

Values in millions of dollars.

	1986	1987	1988	1989	1990
Exports of goods (f.o.b.)	773.4	826.8	881.0	807.2	890.6
Imports of goods (f.o.b.)	-712.5	-951.5	-993.4	-1,002.2	-1,198.9
Trade balance	60.9	-124.7	-112.4	-195.0	-308.3
Services - debits	-343.6	-376.3	-399.6	-407.8	-429.0
Services - credits	45.1	79.3	77.7	81.9	93.1
Private transfers (net)	72.1	201.6	172.4	202.1	201.9
Government transfers (net)	122.5	123.2	196.1	220.2	213.8
Long term capital (net)	146.3	232.4	183.9	180.2	334.2
Short term capital (net)	-41.0	-38.6	-48.1	-61.0	-43.2
Errors and omissions	-81.2	-18.7	37.9	40.6	8.8
Overall balance	-18.9	78.2	107.9	61.2	71.3

Exchange Rates [38]

Currency: **ceolis.**
Symbol: **Ce.**

Data are currency units per $1.

July 1992	437

Imports and Exports

Top Import Origins [39]

$1.4 billion (c.i.f., 1992 est.).

Origins	%
UK	23
US	11
Germany	10
Japan	6

Top Export Destinations [40]

$1.1 billion (f.o.b., 1992).

Destinations	%
Germany	29
UK	12
US	12
Japan	5

Foreign Aid [41]

	U.S. $	
US commitments, including Ex-Im (FY70-89)	455	million
Western (non-US) countries, ODA and OOF bilateral commitments (1970-89)	2.6	billion
OPEC bilateral aid (1979-89)	78	million
Communist countries (1970-89)	106	million

Import and Export Commodities [42]

Import Commodities

Petroleum 16%
Consumer goods
Foods
Intermediate goods
Capital equipment

Export Commodities

Cocoa 45%
Gold
Timber
Tuna
Bauxite
And aluminum

Greece

Geography [1]

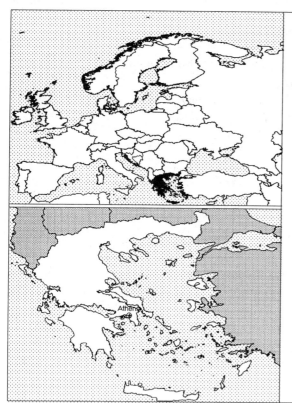

Total area:
131,940 km2
Land area:
130,800 km2
Comparative area:
Slightly smaller than Alabama
Land boundaries:
Total 1,210 km, Albania 282 km, Bulgaria 494 km, Turkey 206 km, Macedonia 228 km
Coastline:
13,676 km
Climate:
Temperate; mild, wet winters; hot, dry summers
Terrain:
Mostly mountains with ranges extending into sea as peninsulas or chains of islands
Natural resources:
Bauxite, lignite, magnesite, petroleum, marble
Land use:
Arable land:
23%
Permanent crops:
8%
Meadows and pastures:
40%
Forest and woodland:
20%
Other:
9%

Demographics [2]

	1960	1970	1980	1990	1991[1]	1994	2000	2010	2020
Population	8,327	8,793	9,643	10,123	10,043	10,565	10,878	10,920	10,689
Population density (persons per sq. mi.)	165	174	191	199	199	NA	201	201	196
Births	NA	NA	NA	NA	110	111	NA	NA	NA
Deaths	NA	NA	NA	NA	95	98	NA	NA	NA
Life expectancy - males	NA	NA	NA	NA	75	75	NA	NA	NA
Life expectancy - females	NA	NA	NA	NA	80	80	NA	NA	NA
Birth rate (per 1,000)	NA	NA	NA	NA	11	11	NA	NA	NA
Death rate (per 1,000)	NA	NA	NA	NA	9	9	NA	NA	NA
Women of reproductive age (15-44 yrs.)	NA	NA	NA	2,397	NA	2,583	2,660	NA	NA
of which are currently married	NA	NA	NA	1,677	NA	1,822	1,901	NA	NA
Fertility rate	NA	NA	NA	1.4	1.54	1.5	1.5	1.5	1.6

Population values are in thousands, life expectancy in years, and other items as indicated.

Health

Health Personnel [3]

Doctors per 1,000 pop., 1988-92	1.73
Nurse-to-doctor ratio, 1988-92	1.6
Hospital beds per 1,000 pop., 1985-90	5.1
Percentage of children immunized (age 1 yr. or less)	
Third dose of DPT, 1990-91	54.0
Measles, 1990-91	76.0

Health Expenditures [4]

Total health expenditure, 1990 (official exchange rate)	
Millions of dollars	3609
Millions of dollars per capita	358
Health expenditures as a percentage of GDP	
Total	5.5
Public sector	4.2
Private sector	1.3
Development assistance for health	
Total aid flows (millions of dollars)[1]	NA
Aid flows per capita (millions of dollars)	NA
Aid flows as a percentage of total health expenditure	NA

For sources, notes, and explanations, see Annotated Source Appendix, page 1035.

Human Factors

Health Care Ratios [5]

Population per physician, 1970	620
Population per physician, 1990	580
Population per nursing person, 1970	990
Population per nursing person, 1990	NA
Percent of births attended by health staff, 1985	NA

Infants and Malnutrition [6]

Percent of babies with low birth weight, 1985	6
Infant mortality rate per 1,000 live births, 1970	30
Infant mortality rate per 1,000 live births, 1991	10
Years of life lost per 1,000 population, 1990	10
Prevalence of malnutrition (under age 5), 1990	NA

Ethnic Division [7]

Greek	98%
Other	2%

Religion [8]

Greek Orthodox	98.0%
Muslim	1.3%
Other	0.7%

Major Languages [9]

Greek (official), English, French.

Education

Public Education Expenditures [10]

Million Drachma	1985	1986	1987	1988	1989	1990
Total education expenditure	133,091	147,976	166,808	206,794	265,351	NA
as percent of GNP	2.9	2.7	2.7	2.7	3.0	NA
as percent of total govt. expend.	7.5	6.8	6.1	5.6	NA	NA
Current education expenditure	126,749	139,730	156,807	194,338	248,040	NA
as percent of GNP	2.7	2.5	2.5	2.6	2.8	NA
as percent of current govt. expend.	8.5	7.6	6.5	5.8	NA	NA
Capital expenditure	6,342	8,246	10,001	12,456	17,311	NA

Educational Attainment [11]

Age group	25+
Total population	5,966,511
Highest level attained (%)	
No schooling	11.5
First level	
Incompleted	16.9
Completed	44.3
Entered second level	
S-1	6.1
S-2	13.6
Post secondary	7.6

Literacy Rate [12]

In thousands and percent	1985[a]	1991[a]	2000[a]
Illiterate population +15 years	672	548	338
Illiteracy rate - total pop. (%)	8.6	6.8	4.0
Illiteracy rate - males (%)	3.2	2.4	1.5
Illiteracy rate - females (%)	13.6	10.9	6.4

Libraries [13]

	Admin. Units	Svc. Pts.	Vols. (000)	Shelving (meters)	Vols. Added	Reg. Users
National (1989)	1	1	2,500	32,600	11,830	1,234
Non-specialized (1989)	758	758	7,492	194,430	216,221	1.2mil
Public (1988)[33]	758	758	9,962	235,341	228,051	1.7mil
Higher ed.	NA	NA	NA	NA	NA	NA
School	NA	NA	NA	NA	NA	NA

Daily Newspapers [14]

	1975	1980	1985	1990[1]
Number of papers	108	128	140	117
Circ. (000)	921	1,160[e]	1,350[e]	1,400[e]

Cinema [15]

Science and Technology

Scientific/Technical Forces [16]

Potential scientists/engineers	329,489
Number female	100,528
Potential technicians	155,966
Number female	73,420
Total	485,455

R&D Expenditures [17]

	Drachma (000) 1986
Total expenditure	18,331,000
Capital expenditure	NA
Current expenditure	NA
Percent current	NA

U.S. Patents Issued [18]

Values show patents issued to citizens of the country by the U.S. Patents Office.

	1990	1991	1992
Number of patents	10	13	9

For sources, notes, and explanations, see Annotated Source Appendix, page 1035.

Government and Law

Organization of Government [19]

Long-form name:
  Hellenic Republic
Type:
  presidential parliamentary government; monarchy rejected by referendum 8 December 1974
Independence:
  1829 (from the Ottoman Empire)
Constitution:
  11 June 1975
Legal system:
  based on codified Roman law; judiciary divided into civil, criminal, and administrative courts
National holiday:
  Independence Day, 25 March (1821) (proclamation of the war of independence)
Executive branch:
  president, prime minister, Cabinet
Legislative branch:
  unicameral Greek Chamber of Deputies (Vouli ton Ellinon)
Judicial branch:
  Supreme Judicial Court, Special Supreme Tribunal

Political Parties [20]

Chamber of Deputies	% of votes
New Democracy (ND)	46.9
Panhellenic Socialist Movement (PASOK)	38.6
Left Alliance	10.3
PASOK/Left Alliance	1.0
Ecologist-Alternative List	0.8
Democratic Renewal (DIANA)	0.7
Muslim independents	0.5

Government Expenditures [21]

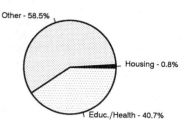

Other - 58.5%
Housing - 0.8%
Educ./Health - 40.7%

% distribution for 1990. Expend. in 1993: 45.1 ($ bil.)

Crime [23]

Crime volume	
Cases known to police	330,803
Attempts (percent)	0.3
Percent cases solved	88.2
Crimes per 100,000 persons	3,306.57
Persons responsible for offenses	
Total number offenders	310,569
Percent female	10.3
Percent juvenile[27]	4.9
Percent foreigners	1.1

Military Expenditures and Arms Transfers [22]

	1985	1986	1987	1988	1989
Military expenditures					
Current dollars (mil.)	3,052	2,791	2,926	3,170	3,097
1989 constant dollars (mil.)	3,474	3,097	3,148	3,300	3,097
Armed forces (000)	201	202	199	199	201
Gross national product (GNP)					
Current dollars (mil.)	43,430	44,630	46,110	49,640	52,930
1989 constant dollars (mil.)	49,440	49,520	49,590	51,680	52,930
Central government expenditures (CGE)					
1989 constant dollars (mil.)	25,210	18,440	19,390	21,240	23,090
People (mil.)	9.9	10.0	10.0	10.0	10.0
Military expenditure as % of GNP	7.0	6.3	6.3	6.4	5.9
Military expenditure as % of CGE	13.8	16.8	16.2	15.5	13.4
Military expenditure per capita	350	311	315	330	309
Armed forces per 1,000 people	20.2	20.3	19.9	19.9	20.1
GNP per capita	4,977	4,970	4,964	5,166	5,286
Arms imports[6]					
Current dollars (mil.)	300	200	310	470	2,000
1989 constant dollars (mil.)	342	222	333	489	2,000
Arms exports[6]					
Current dollars (mil.)	30	40	40	30	0
1989 constant dollars (mil.)	34	44	43	31	0
Total imports[7]					
Current dollars (mil.)	10,130	11,350	13,170	12,320	16,150
1989 constant dollars	11,540	12,600	14,160	12,830	16,150
Total exports[7]					
Current dollars (mil.)	4,539	5,648	6,533	5,429	7,545
1989 constant dollars	5,167	6,268	7,027	5,652	7,545
Arms as percent of total imports[8]	3.0	1.8	2.4	3.8	12.4
Arms as percent of total exports[8]	0.7	0.7	0.6	0.6	0

Human Rights [24]

	SSTS	FL	FAPRO	PPCG	APROBC	TPW	PCPTW	STPEP	PHRFF	PRW	ASST	AFL
Observes	P	P	P	P	P	P	P		P	P	P	P
	EAFRD	CPR	ESCR	SR	ACHR	MAAE	PVIAC	PVNAC	EAFDAW	TCIDTP	RC	
Observes		P		P	P	S	P	P		P	P	S

P = Party; S = Signatory; see Appendix for meaning of abbreviations.

Labor Force

Total Labor Force [25]

3,966,900

Labor Force by Occupation [26]

Services	45%
Agriculture	27
Industry	28
Date of data: 1990	

Unemployment Rate [27]

9.1% (1992)

For sources, notes, and explanations, see Annotated Source Appendix, page 1035.

Production Sectors

Commercial Energy Production and Consumption

Production [28]

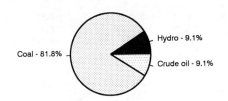

Coal - 81.8%
Hydro - 9.1%
Crude oil - 9.1%

Consumption [29]

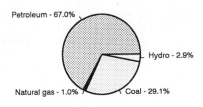

Petroleum - 67.0%
Hydro - 2.9%
Natural gas - 1.0%
Coal - 29.1%

Data are shown in quadrillion (10^{15}) BTUs and percent for 1991

Crude oil[1]	0.03
Natural gas liquids	(s)
Dry natural gas	(s)
Coal[2]	0.27
Net hydroelectric power[3]	0.03
Net nuclear power[3]	0.00
Total	0.34

Petroleum	0.69
Dry natural gas	0.01
Coal[2]	0.30
Net hydroelectric power[3]	0.03
Net nuclear power[3]	0.00
Total	1.02

Telecommunications [30]

- Adequate, modern networks reach all areas
- 4,080,000 telephones
- Microwave radio relay carries most traffic
- Extensive open-wire network
- Submarine cables to off-shore islands
- Broadcast stations - 29 AM, 17 (20 repeaters) FM, 361 TV
- Tropospheric links, 8 submarine cables
- 1 satellite earth station operating in INTELSAT (1 Atlantic Ocean and 1 Indian Ocean antenna), and EUTELSAT systems

Transportation [31]

Railroads. 2,479 km total; 1,565 km 1.435-meter standard gauge, of which 36 km electrified and 100 km double track; 892 km 1.000-meter gauge; 22 km 0.750-meter narrow gauge; all government owned

Highways. 38,938 km total; 16,090 km paved, 13,676 km crushed stone and gravel, 5,632 km improved earth, 3,540 km unimproved earth

Merchant Marine. 998 ships (1,000 GRT or over) totaling 25,483,768 GRT/47,047,285 DWT; includes 14 passenger, 66 short-sea passenger, 2 passenger-cargo, 128 cargo, 26 container, 15 roll-on/roll-off cargo, 14 refrigerated cargo, 1 vehicle carrier, 214 oil tanker, 19 chemical tanker, 7 liquefied gas, 42 combination ore/oil, 3 specialized tanker, 424 bulk, 22 combination bulk, 1 livestock carrier; note - ethnic Greeks also own large numbers of ships under the registry of Liberia, Panama, Cyprus, Malta, and The Bahamas

Airports

Total:	78
Usable:	77
With permanent-surface runways:	63
With runways over 3,659 m:	0
With runways 2,440-3,659 m:	20
With runways 1,220-2,439 m:	24

Top Agricultural Products [32]

	1990[12]	1991[4,12]
Wheat	1,680	2,800
Corn	1,400	1,850
Tomatoes	1,800	1,800
Alfalfa	1,489	1,480
Peaches and nectarines	760	800
Barley	480	750

Values shown are 1,000 metric tons.

Top Mining Products [33]

Estimated metric tons unless otherwise specified   M.t.

Clay, bentonite, crude	600,286
Emery	7,855
Coal, lignite (000 tons)	50,538
Magnesite, crude	590,188
Perlite	369,495
Bauxite (000 tons)	2,134

Tourism [34]

	1987	1988	1989	1990	1991
Visitors	8,004	8,351	8,541	9,310	8,271
Tourists[30]	7,564	7,923	8,082	8,873	8,036
Cruise passengers	440	428	459	437	235
Tourism receipts	2,268	2,396	1,976	2,587	2,566
Tourism expenditures	508	735	816	1,090	1,011
Fare receipts	28	29	16	30	45
Fare expenditures	127	113	114	147	158

Tourists are in thousands, money in million U.S. dollars.

For sources, notes, and explanations, see Annotated Source Appendix, page 1035.

359

Manufacturing Sector

GDP and Manufacturing Summary [35]

	1980	1985	1989	1990	% change 1980-1990	% change 1989-1990
GDP (million 1980 $)	40,147	42,902	46,298	46,588	16.0	0.6
GDP per capita (1980 $)	4,163	4,318	4,616	4,633	11.3	0.4
Manufacturing as % of GDP (current prices)	19.5	18.2	15.3	16.2	-16.9	5.9
Gross output (million $)	25,525	20,633	31,109	37,186	45.7	19.5
Value added (million $)	7,591	5,759	9,261	11,645	53.4	25.7
Value added (million 1980 $)	6,968	7,000	7,353	7,096	1.8	-3.5
Industrial production index	100	98	102	99	-1.0	-2.9
Employment (thousands)	474	441	425[e]	439	-7.4	3.3

Note: GDP stands for Gross Domestic Product. 'e' stands for estimated value.

Profitability and Productivity

	1980	1985	1989	1990	% change 1980-1990	% change 1989-1990
Intermediate input (%)	70	72	70	69	-1.4	-1.4
Wages, salaries, and supplements (%)	13	14	12[e]	15[e]	15.4	25.0
Gross operating surplus (%)	16	14	18[e]	16[e]	0.0	-11.1
Gross output per worker ($)	53,854	46,760	73,235[e]	84,686	57.3	15.6
Value added per worker ($)	20,069	16,348	21,801[e]	33,631	67.6	54.3
Average wage (incl. benefits) ($)	7,266	6,610	8,595[e]	12,941[e]	78.1	50.6

Profitability is in percent of gross output. Productivity is in U.S. $. 'e' stands for estimated value.

Profitability - 1990

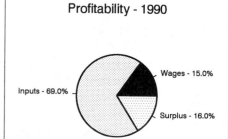

Wages - 15.0%
Inputs - 69.0%
Surplus - 16.0%

The graphic shows percent of gross output.

Value Added in Manufacturing

	1980 $ mil.	1980 %	1985 $ mil.	1985 %	1989 $ mil.	1989 %	1990 $ mil.	1990 %	% change 1980-1990	% change 1989-1990
311 Food products	1,039	13.7	897	15.6	1,514	16.3	1,919	16.5	84.7	26.8
313 Beverages	264	3.5	246	4.3	432	4.7	536	4.6	103.0	24.1
314 Tobacco products	138	1.8	114	2.0	141	1.5	280	2.4	102.9	98.6
321 Textiles	1,063	14.0	820	14.2	1,365	14.7	1,194	10.3	12.3	-12.5
322 Wearing apparel	494	6.5	409	7.1	573	6.2	962	8.3	94.7	67.9
323 Leather and fur products	105	1.4	88	1.5	139	1.5	155	1.3	47.6	11.5
324 Footwear	111	1.5	88	1.5	102	1.1	144	1.2	29.7	41.2
331 Wood and wood products	241	3.2	114	2.0	246[e]	2.7	306	2.6	27.0	24.4
332 Furniture and fixtures	148	1.9	93	1.6	130[e]	1.4	253	2.2	70.9	94.6
341 Paper and paper products	126	1.7	101	1.8	222	2.4	290	2.5	130.2	30.6
342 Printing and publishing	216	2.8	138	2.4	223	2.4	403	3.5	86.6	80.7
351 Industrial chemicals	185	2.4	197	3.4	337[e]	3.6	297	2.6	60.5	-11.9
352 Other chemical products	339	4.5	241	4.2	440[e]	4.8	679	5.8	100.3	54.3
353 Petroleum refineries	153	2.0	140	2.4	220[e]	2.4	218	1.9	42.5	-0.9
354 Miscellaneous petroleum and coal products	37	0.5	22	0.4	42[e]	0.5	32	0.3	-13.5	-23.8
355 Rubber products	77	1.0	58	1.0	128	1.4	111	1.0	44.2	-13.3
356 Plastic products	214	2.8	128	2.2	298	3.2	317	2.7	48.1	6.4
361 Pottery, china and earthenware	67	0.9	48	0.8	65	0.7	79	0.7	17.9	21.5
362 Glass and glass products	53	0.7	24	0.4	41	0.4	53	0.5	0.0	29.3
369 Other non-metal mineral products	483	6.4	321	5.6	540	5.8	748	6.4	54.9	38.5
371 Iron and steel	203	2.7	155	2.7	281	3.0	284	2.4	39.9	1.1
372 Non-ferrous metals	245	3.2	184	3.2	209	2.3	347	3.0	41.6	66.0
381 Metal products	512	6.7	387	6.7	523	5.6	626	5.4	22.3	19.7
382 Non-electrical machinery	181	2.4	116	2.0	193	2.1	258	2.2	42.5	33.7
383 Electrical machinery	334	4.4	248	4.3	318	3.4	499	4.3	49.4	56.9
384 Transport equipment	483	6.4	285	4.9	415	4.5	520	4.5	7.7	25.3
385 Professional and scientific equipment	10	0.1	7	0.1	9	0.1	22	0.2	120.0	144.4
390 Other manufacturing industries	73	1.0	91	1.6	113[e]	1.2	109	0.9	49.3	-3.5

Note: The industry codes shown are International Standard Industry codes (ISIC). Percentages are percent of total Value Added. 'e' stands for estimated value

360

For sources, notes, and explanations, see Annotated Source Appendix, page 1035.

Finance, Economics, and Trade

Economic Indicators [36]

Billions of Drachmas (Dr) unless otherwise noted[57].

	1989	1990[P]	1991[P]
Real GDP (1970 mkt prices)	550.1	549.3	554.8
Real GDP growth rate (%)	3.5	-0.1	1.0
Money supply (M1)			
(end period)	1,517.8	1,880.8	2,000
Commercial interest rate (%)	22-25.5	26-32	26-26.5
Savings interest rate (%)	14.5-15	16-18	18.0
Investm. interest rate (%)[58]	19-20	25-27	25-27
CPI (yr-end) (1982 = 100)	339.6	417.0	487.0
WPI (yr-end) (1980 = 100)	433.0	512.6	590.0
External public debt	18,829.5	21,930.0	21,500.0

Balance of Payments Summary [37]

Values in millions of dollars.

	1987	1988	1989	1990	1991
Exports of goods (f.o.b.)	5,612	5,933	5,994	6,365	6,797
Imports of goods (f.o.b.)	-11,112	-12,005	-13,377	-16,543	-16,909
Trade balance	-5,500	-6,072	-7,383	-10,178	-10,112
Services - debits	-3,362	-3,980	-4,352	-5,045	-5,349
Services - credits	4,604	5,445	5,191	6,968	7,757
Private transfers (net)	1,370	1,713	1,381	1,817	2,149
Government transfers (net)	1,665	1,936	2,602	2,901	4,034
Long term capital (net)	1,387	1,438	1,941	2,975	3,587
Short term capital (net)	419	627	817	787	-170
Errors and omissions	223	41	-538	-185	-236
Overall balance	806	1,148	-341	40	1,660

Exchange Rates [38]

Currency: **drachma.**
Symbol: **Dr.**

Data are currency units per $1.

January 1993	215.82
1992	190.62
1991	182.27
1990	158.51
1989	162.42
1988	141.86

Imports and Exports

Top Import Origins [39]

$21.5 billion (c.i.f., 1991).

Origins	%
Germany	20
Italy	14
France	8
UK	5
US	4

Top Export Destinations [40]

$6.8 billion (f.o.b., 1991).

Destinations	%
Germany	24
France	18
Italy	17
UK	7
US	6

Foreign Aid [41]

	U.S. $	
US commitments, including Ex-Im (FY70-81)	525	million
Western (non-US) countries, ODA and OOF bilateral commitments (1970-89)	1,390	million

Import and Export Commodities [42]

Import Commodities

Manufactured goods 71%
Foodstuffs 14%
Fuels 10%

Export Commodities

Manufactured goods 53%
Foodstuffs 31%
Fuels 9%

Grenada

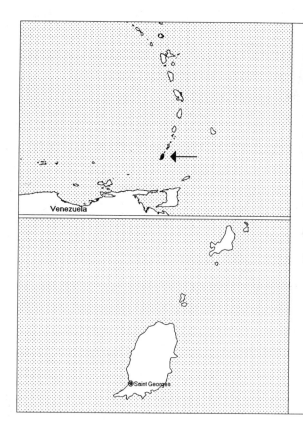

Geography [1]

Total area:
340 km2
Land area:
340 km2
Comparative area:
Slightly less than twice the size of Washington, DC
Land boundaries:
0 km
Coastline:
121 km
Climate:
Tropical; tempered by northeast trade winds
Terrain:
Volcanic in origin with central mountains
Natural resources:
Timber, tropical fruit, deepwater harbors
Land use:
Arable land:
15%
Permanent crops:
26%
Meadows and pastures:
3%
Forest and woodland:
9%
Other:
47%

Demographics [2]

	1960	1970	1980	1990	1991[1]	1994	2000	2010	2020
Population	90	95	90	94	84	94	98	115	141
Population density (persons per sq. mi.)	688	728	688	642	640	NA	637	736	933
Births	NA	NA	NA	NA	3	3	NA	NA	NA
Deaths	NA	NA	NA	NA	1	1	NA	NA	NA
Life expectancy - males	NA	NA	NA	NA	69	68	NA	NA	NA
Life expectancy - females	NA	NA	NA	NA	74	73	NA	NA	NA
Birth rate (per 1,000)	NA	NA	NA	NA	35	30	NA	NA	NA
Death rate (per 1,000)	NA	NA	NA	NA	7	6	NA	NA	NA
Women of reproductive age (15-44 yrs.)	NA	NA	NA	21	NA	21	22	NA	NA
of which are currently married	NA	NA	NA	9	NA	9	10	NA	NA
Fertility rate	NA	NA	NA	4.2	4.72	3.9	3.5	2.9	2.5

Population values are in thousands, life expectancy in years, and other items as indicated.

Health

Health Personnel [3]

Health Expenditures [4]

Human Factors

Health Care Ratios [5]	Infants and Malnutrition [6]

Ethnic Division [7]		Religion [8]	Major Languages [9]
Black African	NA	Roman Catholic, Anglican, other Protestant sects.	English (official), French patois.

Education

Public Education Expenditures [10]

Million E. Carribbean Dollars	1985	1986	1987	1988	1989	1990
Total education expenditure	NA	NA	NA	NA	NA	NA
as percent of GNP	NA	NA	NA	NA	NA	NA
as percent of total govt. expend.	NA	NA	NA	NA	NA	NA
Current education expenditure	14	16	NA	NA	NA	NA
as percent of GNP	4.6	4.7	NA	NA	NA	NA
as percent of current govt. expend.	NA	NA	NA	NA	NA	NA
Capital expenditure	NA	NA	NA	NA	NA	NA

Educational Attainment [11]

Age group	25+
Total population	33,401
Highest level attained (%)	
No schooling	2.2
First level	
Incompleted	87.8
Completed	NA
Entered second level	
S-1	8.5
S-2	NA
Post secondary	1.5

Literacy Rate [12]

	1970[b]	1980[b]	1990[b]
Illiterate population +15 years	1,070	NA	NA
Illiteracy rate - total pop. (%)[3]	2.2	NA	NA
Illiteracy rate - males (%)	2.0	NA	NA
Illiteracy rate - females (%)	2.4	NA	NA

Libraries [13]

Daily Newspapers [14]

	1975	1980	1985	1990
Number of papers	1	1	-	-
Circ. (000)	3[e]	4[e]	-	-

Cinema [15]

Science and Technology

Scientific/Technical Forces [16]	R&D Expenditures [17]	U.S. Patents Issued [18]

For sources, notes, and explanations, see Annotated Source Appendix, page 1035.

363

Government and Law

Organization of Government [19]

Long-form name:
none
Type:
parliamentary democracy
Independence:
7 February 1974 (from UK)
Constitution:
19 December 1973
Legal system:
based on English common law
National holiday:
Independence Day, 7 February (1974)
Executive branch:
British monarch, governor general, prime minister, Ministers of Government (cabinet)
Legislative branch:
bicameral Parliament consists of an upper house or Senate and a lower house or House of Representatives
Judicial branch:
Supreme Court

Political Parties [20]

House of Representatives	% of seats
National Democratic Congress (NDC)	53.3
Grenada United Labor Party (GULP)	20.0
The National Party (TNP)	13.3
New National Party (NNP)	13.3

Government Budget [21]

For 1991 est.

Revenues	78
Expenditures	51
Capital expenditures	22

Defense Summary [22]

Branches: Royal Grenada Police Force, Coast Guard

Manpower Availability: No information available.

Defense Expenditures: No information available.

Crime [23]

Crime volume	
Cases known to police	2,679
Attempts (percent)	50
Percent cases solved	72.68
Crimes per 100,000 persons	2,679
Persons responsible for offenses	
Total number offenders	1,972
Percent female	8.47
Percent juvenile	3.45
Percent foreigners	0.66

Human Rights [24]

	SSTS	FL	FAPRO	PPCG	APROBC	TPW	PCPTW	STPEP	PHRFF	PRW	ASST	AFL
Observes	1	P		1	P	P	P			P	1	P
	EAFRD	CPR	ESCR	SR	ACHR	MAAE	PVIAC	PVNAC	EAFDAW	TCIDTP	RC
Observes		S	P	P		P				P		P

P = Party; S = Signatory; see Appendix for meaning of abbreviations.

Labor Force

Total Labor Force [25]

36,000

Labor Force by Occupation [26]

Services	31%
Agriculture	24
Construction	8
Manufacturing	5
Other	32

Date of data: 1985

Unemployment Rate [27]

25% (1992 est.)

Production Sectors

Energy Resource Summary [28]

Energy Resources: None. **Electricity**: 12,500 kW capacity; 26 million kWh produced, 310 kWh per capita (1992).

Telecommunications [30]

- Automatic, islandwide telephone system with 5,650 telephones
- New SHF radio links to Trinidad and Tobago and Saint Vincent
- VHF and UHF radio links to Trinidad and Carriacou
- Broadcast stations - 1 AM, no FM, 1 TV

Transportation [31]

Highways. 1,000 km total; 600 km paved, 300 km otherwise improved; 100 km unimproved

Airports

Total:	3
Usable:	3
With permanent-surface runways:	2
With runways over 3,659 m:	0
With runways 2,440-3,659 m:	1
With runways 1,220-2,439 m:	1

Top Agricultural Products [32]

Agriculture accounts for 16% of GDP and 80% of exports; bananas, cocoa, nutmeg, and mace account for two-thirds of total crop production; world's second-largest producer and fourth-largest exporter of nutmeg and mace; small-size farms predominate, growing a variety of citrus fruits, avocados, root crops, sugarcane, corn, and vegetables.

Top Mining Products [33]

Detailed information is not available. A summary of mineral resources available follows. **Mineral Resources**: None.

Tourism [34]

	1987	1988	1989	1990	1991
Visitors	187	200	192	265	288
Tourists	57	62	69	76	85
Cruise passengers	127	135	120	183	196
Tourism receipts	30	29	31	38	42
Tourism expenditures	4	5	4	5	
Fare receipts	2	2	2		
Fare expenditures	4	4	5	6	

Tourists are in thousands, money in million U.S. dollars.

Finance, Economics, and Trade

Industrial Summary [35]

Industrial Production: Growth rate 5.8% (1989 est.); accounts for 9% of GDP. **Industries**: Food and beverage, textile, light assembly operations, tourism, construction.

Economic Indicators [36]

National product: GDP—purchasing power equivalent—$250 million (1992 est.). **National product real growth rate**: - 0.4% (1992 est.). **National product per capita**: $3,000 (1992 est.). **Inflation rate (consumer prices)**: 2.6% (1991 est.). **External debt**: $104 million (1990 est.).

Balance of Payments Summary [37]

Values in millions of dollars.

	1986	1987	1988	1989	1990
Exports of goods (f.o.b.)	28.8	31.6	32.8	28.0	26.6
Imports of goods (f.o.b.)	-86.0	-92.8	-94.5	-92.5	-106.3
Trade balance	-57.2	-61.2	-61.7	-64.5	-79.7
Services - debits	-29.5	-31.6	-31.6	-34.4	-50.3
Services - credits	40.5	48.4	52.7	45.1	69.9
Private transfers (net)	10.4	11.8	15.3	17.0	17.0
Government transfers (net)	25.5	8.2	5.2	12.2	15.1
Long term capital (net)	11.0	23.9	23.8	11.5	25.6
Short term capital (net)	3.5	2.8	-7.0	23.0	-11.1
Errors and omissions	-1.8	1.6	-1.9	-10.8	16.1
Overall balance	2.5	3.9	-5.2	-0.8	2.6

Exchange Rates [38]

Currency: **East Caribbean dollars.**
Symbol: **EC$.**

Data are currency units per $1.

Fixed rate since 1976	2.70

Imports and Exports

Top Import Origins [39]

$110 million (f.o.b., 1991 est.).

Origins	%
US	29
UK	NA
Trinidad and Tobago	NA
Japan	NA
Canada 1989	NA

Top Export Destinations [40]

$30 million (f.o.b., 1991 est.).

Destinations	%
US	12
UK	NA
FRG	NA
Netherlands	NA
Trinidad and Tobago (1989)	NA

Foreign Aid [41]

	U.S. $	
US commitments, including Ex-Im (FY84-89)	60	million
Western (non-US) countries, ODA and OOF bilateral commitments (1970-89)	70	million
Communist countries (1970-89)	32	million

Import and Export Commodities [42]

Import Commodities

Food 25%
Manufactured goods 22%
Machinery 20%
Chemicals 10%
Fuel 6%

Export Commodities

Nutmeg 36%
Cocoa beans 9%
Bananas 14%
Mace 8%
Textiles 5%

Guatemala

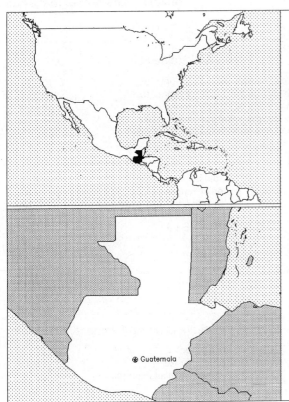

Geography [1]

Total area:
108,890 km2
Land area:
108,430 km2
Comparative area:
Slightly smaller than Tennessee
Land boundaries:
Total 1,687 km, Belize 266 km, El Salvador 203 km, Honduras 256 km, Mexico 962 km
Coastline:
400 km
Climate:
Tropical; hot, humid in lowlands; cooler in highlands
Terrain:
Mostly mountains with narrow coastal plains and rolling limestone plateau (Peten)
Natural resources:
Petroleum, nickel, rare woods, fish, chicle
Land use:
Arable land:
12%
Permanent crops:
4%
Meadows and pastures:
12%
Forest and woodland:
40%
Other:
32%

Demographics [2]

	1960	1970	1980	1990	1991[1]	1994	2000	2010	2020
Population	3,975	5,287	7,232	9,633	9,266	10,721	12,408	15,284	18,131
Population density (persons per sq. mi.)	95	126	164	216	221	270	NA	323	373
Births	NA	NA	NA	NA	327	380	NA	NA	NA
Deaths	NA	NA	NA	NA	76	81	NA	NA	NA
Life expectancy - males	NA	NA	NA	NA	61	62	NA	NA	NA
Life expectancy - females	NA	NA	NA	NA	66	67	NA	NA	NA
Birth rate (per 1,000)	NA	NA	NA	NA	35	35	NA	NA	NA
Death rate (per 1,000)	NA	NA	NA	NA	8	8	NA	NA	NA
Women of reproductive age (15-44 yrs.)	NA	NA	NA	2,176	NA	2,480	2,986	NA	NA
of which are currently married	NA	NA	NA	1,428	NA	1,625	1,971	NA	NA
Fertility rate	NA	NA	NA	5.3	4.78	4.8	4.0	3.0	2.5

Population values are in thousands, life expectancy in years, and other items as indicated.

Health

Health Personnel [3]

Doctors per 1,000 pop., 1988-92	0.44
Nurse-to-doctor ratio, 1988-92	2.5
Hospital beds per 1,000 pop., 1985-90	1.7
Percentage of children immunized (age 1 yr. or less)	
Third dose of DPT, 1990-91	63.0
Measles, 1990-91	48.0

Health Expenditures [4]

Total health expenditure, 1990 (official exchange rate)	
Millions of dollars	283
Millions of dollars per capita	31
Health expenditures as a percentage of GDP	
Total	3.7
Public sector	2.1
Private sector	1.6
Development assistance for health	
Total aid flows (millions of dollars)[1]	32
Aid flows per capita (millions of dollars)	3.4
Aid flows as a percentage of total health expenditure	11.1

For sources, notes, and explanations, see Annotated Source Appendix, page 1035.

367

Human Factors

Health Care Ratios [5]

Population per physician, 1970	3660
Population per physician, 1990	NA
Population per nursing person, 1970	NA
Population per nursing person, 1990	NA
Percent of births attended by health staff, 1985	19

Infants and Malnutrition [6]

Percent of babies with low birth weight, 1985	10
Infant mortality rate per 1,000 live births, 1970	100
Infant mortality rate per 1,000 live births, 1991	60
Years of life lost per 1,000 population, 1990	41
Prevalence of malnutrition (under age 5), 1990	34

Ethnic Division [7]

Ladino	56%
Indian	44%

Religion [8]

Roman Catholic, Protestant, traditional Mayan.

Major Languages [9]

Spanish	60%
Indian language	40%
18 Indian dialects	NA
Quiche	NA
Cakchiquel	NA
Kekchi	NA
Other	NA

Education

Public Education Expenditures [10]

Million Quetzal	1985	1986	1987	1988[1]	1989[1]	1990[1]
Total education expenditure	NA	NA	NA	377	434	468
as percent of GNP	NA	NA	NA	1.9	1.9	1.4
as percent of total govt. expend.	NA	NA	NA	13.1	12.5	11.8
Current education expenditure	NA	NA	NA	NA	NA	NA
as percent of GNP	NA	NA	NA	NA	NA	NA
as percent of current govt. expend.	NA	NA	NA	NA	NA	NA
Capital expenditure	NA	NA	NA	NA	NA	NA

Educational Attainment [11]

Age group	25+
Total population	2,070,399
Highest level attained (%)	
No schooling	54.7
First level	
Incompleted	27.1
Completed	8.6
Entered second level	
S-1	4.8
S-2	2.5
Post secondary	2.2

Literacy Rate [12]

In thousands and percent	1985[a]	1991[a]	2000[a]
Illiterate population +15 years	2,072	2,253	2,685
Illiteracy rate - total pop. (%)	48.1	44.9	38.5
Illiteracy rate - males (%)	40.0	36.9	31.1
Illiteracy rate - females (%)	56.2	52.9	45.9

Libraries [13]

	Admin. Units	Svc. Pts.	Vols. (000)	Shelving (meters)	Vols. Added	Reg. Users
National	NA	NA	NA	NA	NA	NA
Non-specialized	NA	NA	NA	NA	NA	NA
Public	NA	NA	NA	NA	NA	NA
Higher ed. (1987)[16]	1	1	133	NA	2,524	54,496
School	NA	NA	NA	NA	NA	NA

Daily Newspapers [14]

	1975	1980	1985	1990
Number of papers	10	9	9	5
Circ. (000)	249	200[e]	250[e]	190

Cinema [15]

Data for 1988.

Cinema seats per 1,000	6.8
Annual attendance per person	0.9
Gross box office receipts (mil. Quetzal)	13

Science and Technology

Scientific/Technical Forces [16]

Potential scientists/engineers	27,292
Number female	NA
Potential technicians	NA
Number female	NA
Total	NA

R&D Expenditures [17]

	Quetzal[10] (000) 1988
Total expenditure	31,859
Capital expenditure	NA
Current expenditure	NA
Percent current	NA

U.S. Patents Issued [18]

Values show patents issued to citizens of the country by the U.S. Patents Office.

	1990	1991	1992
Number of patents	2	4	0

For sources, notes, and explanations, see Annotated Source Appendix, page 1035.

Government and Law

Organization of Government [19]

Long-form name:
Republic of Guatemala
Type:
republic
Independence:
15 September 1821 (from Spain)
Constitution:
31 May 1985, effective 14 January 1986
Legal system:
civil law system; judicial review of legislative acts; has not accepted compulsory ICJ jurisdiction
National holiday:
Independence Day, 15 September (1821)
Executive branch:
president, vice president, Council of Ministers (cabinet)
Legislative branch:
unicameral Congress of the Republic (Congreso de la Republica)
Judicial branch:
Supreme Court of Justice (Corte Suprema de Justicia)

Political Parties [20]

Congress	% of votes
National Centrist Union (UCN)	25.6
Solidarity Action Movement (MAS)	24.3
Christian Democratic Party (DCG)	17.5
National Advancement Party (PAN)	17.3
National Liberation Movement (MLN)	4.8
PSD/AP-5	3.6
Revolutionary Party (PR)	2.1

Government Expenditures [21]

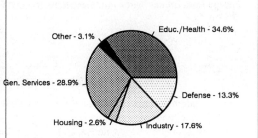

Educ./Health - 34.6%
Other - 3.1%
Gen. Services - 28.9%
Defense - 13.3%
Housing - 2.6%
Industry - 17.6%

% distribution for 1989. Expend. in 1990 est.: 808 ($ mil.)

Military Expenditures and Arms Transfers [22]

	1985	1986	1987	1988	1989
Military expenditures					
Current dollars (mil.)	106	82	127	127	131
1989 constant dollars (mil.)	121	91	136	133	131
Armed forces (000)	43	43	43	36	43
Gross national product (GNP)					
Current dollars (mil.)	6,456	6,575	7,027	7,579	8,208
1989 constant dollars (mil.)	7,349	7,296	7,558	7,890	8,208
Central government expenditures (CGE)					
1989 constant dollars (mil.)	713	736	874	986	1,007
People (mil.)	7.9	8.1	8.4	8.6	8.8
Military expenditure as % of GNP	1.6	1.2	1.8	1.7	1.6
Military expenditure as % of CGE	17.0	12.3	15.6	13.4	13.0
Military expenditure per capita	15	11	16	15	15
Armed forces per 1,000 people	5.4	5.3	5.1	4.2	4.9
GNP per capita	929	897	904	919	932
Arms imports[6]					
Current dollars (mil.)	30	0	5	10	10
1989 constant dollars (mil.)	34	0	5	10	10
Arms exports[6]					
Current dollars (mil.)	0	0	0	0	0
1989 constant dollars (mil.)	0	0	0	0	0
Total imports[7]					
Current dollars (mil.)	1,175	619	1,479	1,557	1,664
1989 constant dollars	1,338	687	1,591	1,621	1,664
Total exports[7]					
Current dollars (mil.)	1,057	1,044	981	1,034	900
1989 constant dollars	1,203	1,159	1,055	1,076	900
Arms as percent of total imports[8]	2.6	0	0.3	0.6	0.6
Arms as percent of total exports[8]	0	0	0	0	0

Crime [23]

Crime volume	
Cases known to police	46,940
Attempts (percent)	28
Percent cases solved	72
Crimes per 100,000 persons	510.36
Persons responsible for offenses	
Total number offenders	NA
Percent female	NA
Percent juvenile	NA
Percent foreigners	NA

Human Rights [24]

	SSTS	FL	FAPRO	PPCG	APROBC	TPW	PCPTW	STPEP	PHRFF	PRW	ASST	AFL
Observes	P	P	P	P	P	P	P			P	P	P
	EAFRD	CPR	ESCR	SR	ACHR	MAAE	PVIAC	PVNAC	EAFDAW	TCIDTP	RC	
Observes	P	P	P	P	P	P	P	P	P	P	P	

P = Party; S = Signatory; see Appendix for meaning of abbreviations.

Labor Force

Total Labor Force [25]

2.5 million

Labor Force by Occupation [26]

Agriculture	60%
Services	13
Manufacturing	12
Commerce	7
Mining and construction	4.4
Transport and utilities	3.8

Date of data: 1985

Unemployment Rate [27]

6.5% (1991 est.), with 30-40% underemployment

Production Sectors

Energy Resource Summary [28]

Energy Resources: Petroleum. **Electricity**: 847,600 kW capacity; 2,500 million kWh produced, 260 kWh per capita (1992). **Pipelines**: Crude oil 275 km.

Telecommunications [30]

- Fairly modern network centered in Guatemala [city]
- 97,670 telephones
- Broadcast stations - 91 AM, no FM, 25 TV, 15 shortwave
- Connection into Central American Microwave System
- 1 Atlantic Ocean INTELSAT earth station

Top Agricultural Products [32]

		1[13]
Coffee (mil. 60-kg. bags)	3,472.0	3,282.0
Corn	1,200.0	1,274.0
Sugar	875.0	1,015.0
Sesame seed	21.7	28.9
Wheat	34.0	23.0
Rubber	15.4	17.8

Values shown are 1,000 metric tons.

Top Mining Products [33]

Preliminary metric tons unless otherwise specified M.t.

Antimony, mine output, Sb content	609
Gold (kilograms)	31
Iron ore, gross weight	5,103
Lead metal including secondary	28
Gypsum	51,519
Marble (block, chips, and fragments)	18,851[e]
Sand and gravel (000 tons)	1,033

Transportation [31]

Railroads. 1,019 km 0.914-meter gauge, single track; 917 km government owned, 102 km privately owned

Highways. 26,429 km total; 2,868 km paved, 11,421 km gravel, and 12,140 unimproved

Merchant Marine. 1 cargo ship (1,000 GRT or over) totaling 4,129 GRT/6,450 DWT

Airports

Total:	474
Usable:	418
With permanent-surface runways:	11
With runways over 3,659 m:	0
With runways 2,440-3,659 m:	3
With runways 1,220-2,439 m:	21

Tourism [34]

	1987	1988	1989	1990	1991
Tourists	353	405	437	509	513
Tourism receipts	103	124	152	185	211
Tourism expenditures	95	109	126	100	67
Fare expenditures	7	8	6	7	8

Tourists are in thousands, money in million U.S. dollars.

Manufacturing Sector

GDP and Manufacturing Summary [35]

	1980	1985	1989	1990	% change 1980-1990	% change 1989-1990
GDP (million 1980 $)	7,879	7,446	8,301	8,599	9.1	3.6
GDP per capita (1980 $)	1,139	935	929	935	-17.9	0.6
Manufacturing as % of GDP (current prices)	11.6	11.1	10.3	9.7[e]	-16.4	-5.8
Gross output (million $)	1,968	2,195	2,250[e]	2,018[e]	2.5	-10.3
Value added (million $)	794	906	919[e]	821[e]	3.4	-10.7
Value added (million 1980 $)	1,312	1,179	1,258	1,291	-1.6	2.6
Industrial production index	100	92	135	101	1.0	-25.2
Employment (thousands)	82	73	90[e]	94[e]	14.6	4.4

Note: GDP stands for Gross Domestic Product. 'e' stands for estimated value.

Profitability and Productivity

	1980	1985	1989	1990	% change 1980-1990	% change 1989-1990
Intermediate input (%)	60	59	59[e]	59[e]	-1.7	0.0
Wages, salaries, and supplements (%)	11[e]	10[e]	8[e]	8[e]	-27.3	0.0
Gross operating surplus (%)	30[e]	31[e]	33[e]	32[e]	6.7	-3.0
Gross output per worker ($)	23,209	28,308	24,871[e]	19,731[e]	-15.0	-20.7
Value added per worker ($)	9,560	11,685	10,161[e]	8,032[e]	-16.0	-21.0
Average wage (incl. benefits) ($)	2,513[e]	3,079[e]	1,973[e]	1,812[e]	-27.9	-8.2

Profitability is in percent of gross output. Productivity is in U.S. $. 'e' stands for estimated value.

Profitability - 1990

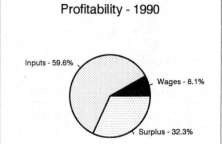

Inputs - 59.6%
Wages - 8.1%
Surplus - 32.3%

The graphic shows percent of gross output.

Value Added in Manufacturing

	1980 $ mil.	1980 %	1985 $ mil.	1985 %	1989 $ mil.	1989 %	1990 $ mil.	1990 %	% change 1980-1990	% change 1989-1990
311 Food products	204	25.7	276	30.5	248[e]	27.0	250[e]	30.5	22.5	0.8
313 Beverages	91	11.5	90	9.9	58[e]	6.3	52[e]	6.3	-42.9	-10.3
314 Tobacco products	14	1.8	15	1.7	24[e]	2.6	24[e]	2.9	71.4	0.0
321 Textiles	45	5.7	71	7.8	64[e]	7.0	51[e]	6.2	13.3	-20.3
322 Wearing apparel	19	2.4	13	1.4	30[e]	3.3	24[e]	2.9	26.3	-20.0
323 Leather and fur products	3	0.4	3	0.3	4[e]	0.4	3[e]	0.4	0.0	-25.0
324 Footwear	15	1.9	12	1.3	8[e]	0.9	7[e]	0.9	-53.3	-12.5
331 Wood and wood products	10	1.3	7	0.8	11[e]	1.2	8[e]	1.0	-20.0	-27.3
332 Furniture and fixtures	4	0.5	3	0.3	6[e]	0.7	5[e]	0.6	25.0	-16.7
341 Paper and paper products	19	2.4	21	2.3	15[e]	1.6	15[e]	1.8	-21.1	0.0
342 Printing and publishing	34	4.3	34	3.8	45[e]	4.9	35[e]	4.3	2.9	-22.2
351 Industrial chemicals	28	3.5	28	3.1	33[e]	3.6	28[e]	3.4	0.0	-15.2
352 Other chemical products	110	13.9	121	13.4	147[e]	16.0	114[e]	13.9	3.6	-22.4
353 Petroleum refineries	14	1.8	8	0.9	8[e]	0.9	9[e]	1.1	-35.7	12.5
354 Miscellaneous petroleum and coal products	2	0.3	NA	0.0	NA	0.0	NA	0.0	NA	NA
355 Rubber products	21	2.6	24	2.6	24[e]	2.6	21[e]	2.6	0.0	-12.5
356 Plastic products	19	2.4	37	4.1	34[e]	3.7	27[e]	3.3	42.1	-20.6
361 Pottery, china and earthenware	2	0.3	8	0.9	9[e]	1.0	8[e]	1.0	300.0	-11.1
362 Glass and glass products	22	2.8	17	1.9	16[e]	1.7	13[e]	1.6	-40.9	-18.8
369 Other non-metal mineral products	34	4.3	41	4.5	36[e]	3.9	38[e]	4.6	11.8	5.6
371 Iron and steel	16	2.0	21	2.3	29[e]	3.2	24[e]	2.9	50.0	-17.2
372 Non-ferrous metals	1	0.1	NA	0.0	NA	0.0	NA	0.0	NA	NA
381 Metal products	23	2.9	23	2.5	26[e]	2.8	24[e]	2.9	4.3	-7.7
382 Non-electrical machinery	6	0.8	4	0.4	7[e]	0.8	6[e]	0.7	0.0	-14.3
383 Electrical machinery	25	3.1	19	2.1	27[e]	2.9	26[e]	3.2	4.0	-3.7
384 Transport equipment	8	1.0	5	0.6	4[e]	0.4	4[e]	0.5	-50.0	0.0
385 Professional and scientific equipment	1	0.1	1	0.1	1[e]	0.1	1[e]	0.1	0.0	0.0
390 Other manufacturing industries	4	0.5	3	0.3	5[e]	0.5	4[e]	0.5	0.0	-20.0

Note: The industry codes shown are International Standard Industry codes (ISIC). Percentages are percent of total Value Added. 'e' stands for estimated value

Finance, Economics, and Trade

Economic Indicators [36]

	1989	1990	1991[e]
GDP (Current $U.S.)	8,364	7,624	8,131
Real GDP Growth Rate	4.0	3.1	3.2
Real Per Capita GDP			
(Curr. $U.S.)	934	828	703
Money Supply (M1)			
(Mil. Quetzals)	2,415	3,224	2,891
Savings rate			
(% GDP, curr. prices)	14.5	10.9	13.8
Investment rate			
(% GDP, curr. prices)	9.6	8.4	9.6
CPI (% change from prev. yr)	17.9	60.6	11.0
WPI (% change from prev. yr)	NA	NA	NA
External Public Debt	2,731	2,602	2,612

Balance of Payments Summary [37]

Values in millions of dollars.

	1987	1988	1989	1990	1991
Exports of goods (f.o.b.)	977.9	1,073.3	1,126.1	1,211.4	1,230.0
Imports of goods (f.o.b.)	-1,333.2	-1,413.2	-1,484.4	-1,428.0	-1,673.0
Trade balance	-355.3	-339.9	-358.3	-216.6	-443.0
Services - debits	-469.9	-525.8	-587.3	-600.3	-523.1
Services - credits	189.4	227.4	328.7	377.0	522.7
Private transfers (net)	101.0	141.7	178.8	205.3	257.7
Government transfers (net)	92.3	82.6	71.0	21.7	2.0
Long term capital (net)	137.0	108.0	126.0	33.0	224.0
Short term capital (net)	324.0	170.0	255.0	114.0	432.0
Errors and omissions	-72.7	-2.4	54.7	36.2	83.3
Overall balance	-54.2	-138.4	68.6	-29.7	555.6

Exchange Rates [38]

Currency: **free market quetzales.**
Symbol: **Q.**

Data are currency units per $1.

December 1993	5.2850
1992	5.1706
1991	5.0289
1989	2.8161
1988	2.6196

Imports and Exports

Top Import Origins [39]

$1.8 billion (c.i.f., 1992).

Origins	%
US	40
Mexico	NA
Venezuela	NA
Japan	NA
Germany	NA

Top Export Destinations [40]

$1.3 billion (f.o.b., 1992).

Destinations	%
US	36
El Salvador	NA
Costa Rica	NA
Germany	NA
Honduras	NA

Foreign Aid [41]

	U.S. $	
US commitments, including Ex-Im (FY70-90)	1.1	billion
Western (non-US) countries, ODA and OOF bilateral commitments (1970-89)	7.92	billion

Import and Export Commodities [42]

Import Commodities	Export Commodities
Fuel and petroleum products	Coffee 26%
Machinery	Sugar 13%
Grain	Bananas 7%
Fertilizers	Beef 3%
Motor vehicles	

Guinea

Geography [1]

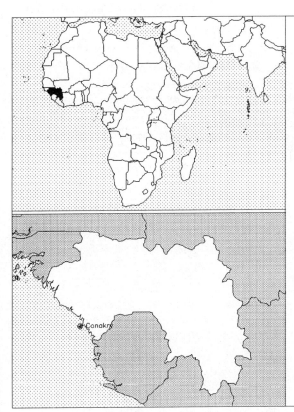

Total area:
245,860 km2
Land area:
245,860 km2
Comparative area:
Slightly smaller than Oregon
Land boundaries:
Total 3,399 km, Guinea-Bissau 386 km, Cote d'Ivoire 610 km, Liberia 563 km, Mali 858 km, Senegal 330 km, Sierra Leone 652 km
Coastline:
320 km
Climate:
Generally hot and humid; monsoonal-type rainy season (June to November) with southwesterly winds; dry season (December to May) with northeasterly Harmattan winds
Terrain:
Generally flat coastal plain, hilly to mountainous interior
Natural resources:
Bauxite, iron ore, diamonds, gold, uranium, hydropower, fish
Land use:
Arable land: 6%
Permanent crops: 0%
Meadows and pastures: 12%
Forest and woodland: 42%
Other: 40%

Demographics [2]

	1960	1970	1980	1990	1991[1]	1994	2000	2010	2020
Population	3,019	3,587	4,320	5,930	7,456	6,392	7,372	9,303	11,664
Population density (persons per sq. mi.)	41	48	58	77	79	NA	97	122	152
Births	NA	NA	NA	NA	347	282	NA	NA	NA
Deaths	NA	NA	NA	NA	160	125	NA	NA	NA
Life expectancy - males	NA	NA	NA	NA	41	42	NA	NA	NA
Life expectancy - females	NA	NA	NA	NA	45	46	NA	NA	NA
Birth rate (per 1,000)	NA	NA	NA	NA	47	44	NA	NA	NA
Death rate (per 1,000)	NA	NA	NA	NA	21	20	NA	NA	NA
Women of reproductive age (15-44 yrs.)	NA	NA	NA	1,386	NA	1,493	1,743	NA	NA
of which are currently married	NA	NA	NA	1,284	NA	1,383	1,615	NA	NA
Fertility rate	NA	NA	NA	6.1	6.02	5.9	5.5	4.7	3.9

Population values are in thousands, life expectancy in years, and other items as indicated.

Health

Health Personnel [3]

Doctors per 1,000 pop., 1988-92	0.02
Nurse-to-doctor ratio, 1988-92	4.3
Hospital beds per 1,000 pop., 1985-90	0.6
Percentage of children immunized (age 1 yr. or less)	
Third dose of DPT, 1990-91	41.0
Measles, 1990-91	39.0

Health Expenditures [4]

Total health expenditure, 1990 (official exchange rate)	
Millions of dollars	106
Millions of dollars per capita	19
Health expenditures as a percentage of GDP	
Total	3.9
Public sector	2.3
Private sector	1.6
Development assistance for health	
Total aid flows (millions of dollars)[1]	20
Aid flows per capita (millions of dollars)	3.5
Aid flows as a percentage of total health expenditure	23.8

For sources, notes, and explanations, see Annotated Source Appendix, page 1035.

373

Human Factors

Health Care Ratios [5]

Population per physician, 1970	50010
Population per physician, 1990	NA
Population per nursing person, 1970	3720
Population per nursing person, 1990	NA
Percent of births attended by health staff, 1985	NA

Infants and Malnutrition [6]

Percent of babies with low birth weight, 1985	18
Infant mortality rate per 1,000 live births, 1970	181
Infant mortality rate per 1,000 live births, 1991	136
Years of life lost per 1,000 population, 1990	125
Prevalence of malnutrition (under age 5), 1990	NA

Ethnic Division [7]

Fulani	35%
Malinke	30%
Soussou	20%
Indigenous tribes	15%

Religion [8]

Muslim	85%
Christian	8%
Indigenous beliefs	7%

Major Languages [9]

French (official); each tribe has its own language.

Education

Public Education Expenditures [10]

Million Syli	1985	1986	1987	1988	1989	1990
Total education expenditure	NA	NA	10,968	19,743	NA	23,483
as percent of GNP	NA	NA	1.4	1.9	NA	1.4
as percent of total govt. expend.	NA	NA	13.0	21.5	NA	NA
Current education expenditure	NA	NA	7,268	15,330	NA	NA
as percent of GNP	NA	NA	NA	NA	NA	NA
as percent of current govt. expend.	NA	NA	10.4	NA	NA	NA
Capital expenditure	NA	NA	3,700	4,413	NA	NA

Educational Attainment [11]

Literacy Rate [12]

In thousands and percent	1985[a]	1991[a]	2000[a]
Illiterate population +15 years	2,879	2,947	3,060
Illiteracy rate - total pop. (%)	83.2	76.0	61.5
Illiteracy rate - males (%)	74.5	65.1	48.5
Illiteracy rate - females (%)	91.6	86.6	73.9

Libraries [13]

	Admin. Units	Svc. Pts.	Vols. (000)	Shelving (meters)	Vols. Added	Reg. Users
National	NA	NA	NA	NA	NA	NA
Non-specialized	NA	NA	NA	NA	NA	NA
Public (1987)	1	NA	9	NA	NA	NA
Higher ed. (1988)	6	7	52	1,235	3,315	10,957
School (1987)	6	6	30	585	4,929	7,756

Daily Newspapers [14]

Cinema [15]

Data for 1991.	
Cinema seats per 1,000	6.9
Annual attendance per person	0.7
Gross box office receipts (mil. Syli)	793

Science and Technology

Scientific/Technical Forces [16]

R&D Expenditures [17]

U.S. Patents Issued [18]

For sources, notes, and explanations, see Annotated Source Appendix, page 1035.

Government and Law

Organization of Government [19]

Long-form name:
Republic of Guinea
Type:
republic
Independence:
2 October 1958 (from France)
Constitution:
23 December 1990 (Loi Fundamentale)
Legal system:
based on French civil law system, customary law, and decree; legal codes currently being revised; has not accepted compulsory ICJ jurisdiction
National holiday:
Anniversary of the Second Republic, 3 April (1984)
Executive branch:
president, Transitional Committee for National Recovery (CTRN) replaced the Military Committee for National Recovery (CMRN); Council of Ministers (cabinet)
Legislative branch:
People's National Assembly (Assemblee Nationale Populaire) dissolved after 3 April 1984 coup; framework established in Dec. 1991 for a new National Assembly with 114 seats
Judicial branch:
Court of Appeal (Cour d'Appel)

Crime [23]

Crime volume	
Cases known to police	1,942
Attempts (percent)	1.97
Percent cases solved	19.78
Crimes per 100,000 persons	32.36
Persons responsible for offenses	
Total number offenders	2,811
Percent female	0.91
Percent juvenile[28]	4.07
Percent foreigners	6.63

Elections [20]

None; political parties were legalized on 1 April 1992.

Government Budget [21]

For 1990 est.	
Revenues	449
Expenditures	708
Capital expenditures	361

Military Expenditures and Arms Transfers [22]

	1985	1986	1987	1988	1989
Military expenditures					
Current dollars (mil.)	NA	NA	NA	29	NA
1989 constant dollars (mil.)	NA	NA	NA	30	NA
Armed forces (000)	28	24	24	15	15
Gross national product (GNP)					
Current dollars (mil.)	2,407	2,157	2,287	2,458	2,550
1989 constant dollars (mil.)	2,740[e]	2,394	2,459	2,558	2,550
Central government expenditures (CGE)					
1989 constant dollars (mil.)	NA	450	532	599	549
People (mil.)	6.4	6.6	6.7	6.9	7.1
Military expenditure as % of GNP	NA	NA	NA	1.2	NA
Military expenditure as % of CGE	NA	NA	NA	5.1	NA
Military expenditure per capita	NA	NA	NA	4	NA
Armed forces per 1,000 people	4.4	3.7	3.6	2.2	2.1
GNP per capita	430	364	365	370	360
Arms imports[6]					
Current dollars (mil.)	70	60	70	20	10
1989 constant dollars (mil.)	80	67	75	21	10
Arms exports[6]					
Current dollars (mil.)	0	0	0	0	0
1989 constant dollars (mil.)	0	0	0	0	0
Total imports[7]					
Current dollars (mil.)	NA	571	560	509	NA
1989 constant dollars	NA	634	602	530	NA
Total exports[7]					
Current dollars (mil.)	NA	538	571	553	NA
1989 constant dollars	NA	597	614	576	NA
Arms as percent of total imports[8]	NA	10.5	12.5	3.9	NA
Arms as percent of total exports[8]	NA	0	0	0	NA

Human Rights [24]

	SSTS	FL	FAPRO	PPCG	APROBC	TPW	PCPTW	STPEP	PHRFF	PRW	ASST	AFL
Observes	P	P	P		P	P	P	P		P	P	P
	EAFRD	CPR	ESCR	SR	ACHR	MAAE	PVIAC	PVNAC	EAFDAW	TCIDTP	RC
Observes		P	P	P			P	P	P	P	P

P = Party; S = Signatory; see Appendix for meaning of abbreviations.

Labor Force

Total Labor Force [25]

2.4 million (1983)

Labor Force by Occupation [26]

Agriculture	82.0%
Industry and commerce	11.0
Services	5.4

Unemployment Rate [27]

For sources, notes, and explanations, see Annotated Source Appendix, page 1035.

375

Production Sectors

Energy Resource Summary [28]

Energy Resources: Uranium, hydropower. **Electricity**: 113,000 kW capacity; 300 million kWh produced, 40 kWh per capita (1989).

Telecommunications [30]

- Poor to fair system of open-wire lines, small radiocommunication stations, and new radio relay system
- 15,000 telephones
- Broadcast stations - 3 AM 1 FM, 1 TV
- 65,000 TV sets
- 200,000 radio receivers
- 1 Atlantic Ocean INTELSAT earth station

Transportation [31]

Railroads. 1,045 km; 806 km 1.000-meter gauge, 239 km 1.435-meter standard gauge

Highways. 30,100 km total; 1,145 km paved, 12,955 km gravel or laterite (of which barely 4,500 km are currently all-weather roads), 16,000 km unimproved earth (1987)

Airports

Total:	15
Usable:	15
With permanent-surface runways:	4
With runways over 3,659 m:	0
With runways 2,440-3,659 m:	3
With runways 1,220-2,439 m:	10

Top Agricultural Products [32]

Agriculture accounts for 40% of GDP (includes fishing and forestry); mostly subsistence farming; principal products—rice, coffee, pineapples, palm kernels, cassava, bananas, sweet potatoes, timber; livestock—cattle, sheep and goats; not self-sufficient in food grains.

Top Mining Products [33]

Estimated metric tons unless otherwise specified　　M.t.

	M.t.
Alumina, hydrated & calcined (000 tons)	1,250
Bauxite, wet basis (000 tons)	15,625[42]
Diamond, gem & industrial (000 tons)	113[42,43]
Gold, kilograms	4,453[44]

Tourism [34]

Finance, Economics, and Trade

GDP and Manufacturing Summary [35]

	1980	1985	1990	1991	1992
Gross Domestic Product					
Millions of 1980 dollars	1,897	1,807	2,190	2,267	2,314[e]
Growth rate in percent	5.60	3.89	4.00	3.50	2.08[e]
Manufacturing Value Added					
Millions of 1980 dollars	60	76	68	70	72[e]
Growth rate in percent	2.70	33.33	2.90	2.93	2.48[e]
Manufacturing share in percent of current prices	2.9[e]	2.0[e]	4.3[e]	NA	NA

Economic Indicators [36]

National product: GDP—exchange rate conversion—$3 billion (1990 est.). **National product real growth rate**: 4.3% (1990 est.). **National product per capita**: $410 (1990 est.). **Inflation rate (consumer prices)**: 19.6% (1990 est.). **External debt**: $2.6 billion (1990 est.).

Balance of Payments Summary [37]

Values in millions of dollars.

	1973	1980	1986	1987
Exports of goods (f.o.b.)	59.2	499.8	506.6	564.7
Imports of goods (f.o.b.)	-158.8	-339.0	-422.7	-380.3
Trade balance	-99.5	160.8	83.9	184.4
Services - debits	NA	-237.9	-293.9	-282.8
Services - credits	-4.6	53.6	60.4	58.2
Private transfers (net)	-1.5	-2.7	-6.0	-13.3
Government transfers (net)	7.9	17.1	31.6	35.3
Long term capital (net)	124.7	45.7	200.9	26.5
Short term capital (net)	5.5	-23.7	-64.4	5.5
Errors and omissions	-	15.1	22.6	-7.6
Overall balance	32.4	28.0	35.1	6.2

Exchange Rates [38]

Currency: **Guinean francs.**
Symbol: **FG.**

Data are currency units per $1.

1990	675
1989	618
1988	515
1987	440
1986	383

Imports and Exports

Top Import Origins [39]

$692 million (c.i.f., 1990 est.).

Origins	%
US	16
France	NA
Brazil	NA

Top Export Destinations [40]

$788 million (f.o.b., 1990 est.).

Destinations	%
US	33
EC	33
USSR and Eastern Europe	20
Canada	NA

Foreign Aid [41]

	U.S. $	
US commitments, including Ex-Im (FY70-89)	227	million
Western (non-US) countries, ODA and OOF bilateral commitments (1970-89)	1,465	million
OPEC bilateral aid (1979-89)	120	million
Communist countries (1970-89)	446	million

Import and Export Commodities [42]

Import Commodities	**Export Commodities**
Petroleum products	Alumina
Metals	Bauxite
Machinery	Diamonds
Transport equipment	Coffee
Foodstuffs	Pineapples
Textiles	Bananas
And other grain	Palm kernels

Guinea-Bissau

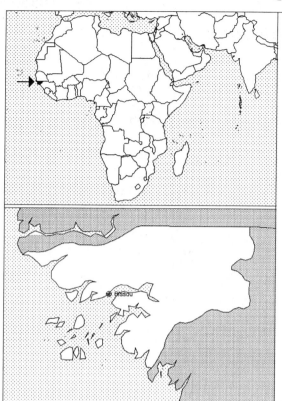

Geography [1]

Total area:
  36,120 km2
Land area:
  28,000 km2
Comparative area:
  Slightly less than three times the size of Connecticut
Land boundaries:
  Total 724 km, Guinea 386 km, Senegal 338 km
Coastline:
  350 km
Climate:
  Tropical; generally hot and humid; monsoonal-type rainy season (June to November) with southwesterly winds; dry season (December to May) with northeasterly harmattan winds
Terrain:
  Mostly low coastal plain rising to savanna in east
Natural resources:
  Unexploited deposits of petroleum, bauxite, phosphates, fish, timber
Land use:
  Arable land:
  11%
  Permanent crops:
  1%
  Meadows and pastures:
  43%
  Forest and woodland:
  38%
  Other:
  7%

Demographics [2]

	1960	1970	1980	1990	1991[1]	1994	2000	2010	2020
Population	617	620	789	998	1,024	1,098	1,263	1,579	1,925
Population density (persons per sq. mi.)	57	57	73	92	95	NA	117	146	178
Births	NA	NA	NA	NA	43	45	NA	NA	NA
Deaths	NA	NA	NA	NA	19	19	NA	NA	NA
Life expectancy - males	NA	NA	NA	NA	45	46	NA	NA	NA
Life expectancy - females	NA	NA	NA	NA	48	49	NA	NA	NA
Birth rate (per 1,000)	NA	NA	NA	NA	42	41	NA	NA	NA
Death rate (per 1,000)	NA	NA	NA	NA	18	17	NA	NA	NA
Women of reproductive age (15-44 yrs.)	NA	NA	NA	NA	NA	NA	NA	NA	NA
of which are currently married	NA	NA	NA	NA	NA	NA	NA	NA	NA
Fertility rate	NA	NA	NA	5.9	5.77	5.5	5.0	4.2	3.4

Population values are in thousands, life expectancy in years, and other items as indicated.

Health

Health Personnel [3]

Health Expenditures [4]

Human Factors

Health Care Ratios [5]

Population per physician, 1970	17500
Population per physician, 1990	NA
Population per nursing person, 1970	2820
Population per nursing person, 1990	NA
Percent of births attended by health staff, 1985	16

Infants and Malnutrition [6]

Percent of babies with low birth weight, 1985	20
Infant mortality rate per 1,000 live births, 1970	185
Infant mortality rate per 1,000 live births, 1991	148
Years of life lost per 1,000 population, 1990	NA
Prevalence of malnutrition (under age 5), 1990	NA

Ethnic Division [7]

African	99%
Balanta	30%
Fula	20%
Manjaca	14%
Mandinga	13%
Papel	7%
European mulatto	<1%

Religion [8]

Indigenous beliefs	65%
Muslim	30%
Christian	5%

Major Languages [9]

Portuguese (official), Criolo, African languages.

Education

Public Education Expenditures [10]

Million Pesos	1980	1985	1987	1988	1989	1990
Total education expenditure	NA	NA	2,533	NA	NA	NA
as percent of GNP	NA	NA	2.8	NA	NA	NA
as percent of total govt. expend.	NA	NA	NA	NA	NA	NA
Current education expenditure	208	NA	2,473	NA	NA	NA
as percent of GNP	4.0	NA	2.8	NA	NA	NA
as percent of current govt. expend.	NA	NA	NA	NA	NA	NA
Capital expenditure	NA	NA	60	NA	NA	NA

Educational Attainment [11]

Literacy Rate [12]

	1979[b]	1980[b]	1990[b]
Illiterate population + 15 years	342,393	NA	367,400
Illiteracy rate - total pop. (%)	80.0	NA	63.5
Illiteracy rate - males (%)	66.7	NA	49.8
Illiteracy rate - females (%)	91.4	NA	76.0

Libraries [13]

	Admin. Units	Svc. Pts.	Vols. (000)	Shelving (meters)	Vols. Added	Reg. Users
National	NA	NA	NA	NA	NA	NA
Non-specialized (1986)	1	1	60	NA	NA	200
Public	NA	NA	NA	NA	NA	NA
Higher ed.	NA	NA	NA	NA	NA	NA
School	NA	NA	NA	NA	NA	NA

Daily Newspapers [14]

	1975	1980	1985	1990
Number of papers	1	1	1	1
Circ. (000)	6	6	6	6

Cinema [15]

Science and Technology

Scientific/Technical Forces [16]

R&D Expenditures [17]

U.S. Patents Issued [18]

Government and Law

Organization of Government [19]

Long-form name:
Republic of Guinea-Bissau
Type:
republic highly centralized multiparty since mid-1991; the African Party for the Independence of Guinea-Bissau and Cape Verde (PAIGC) held a party congress in December 1990 and established a two-year transition program during which the constitution will be revised, allowing for multiple political parties and a presidential election in 1993
Independence:
10 September 1974 (from Portugal)
Constitution:
16 May 1984
Legal system:
NA
National holiday:
Independence Day, 10 September (1974)
Executive branch:
president of the Council of State, vice presidents of the Council of State, Council of State, Council of Ministers (cabinet)
Legislative branch:
unicameral National People's Assembly
Judicial branch:
none; there is a Ministry of Justice in the Council of Ministers

Crime [23]

Elections [20]

National People's Assembly. Last held 15 June 1989 (next to be held 15 June 1994); results - PAIGC is the only party; seats - (150 total) PAIGC 150, appointed by Regional Councils. PAIGC is still the major party (of 10 parties) and controls all aspects of the government.

Government Expenditures [21]

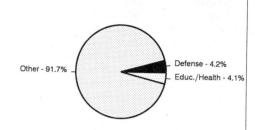

Other - 91.7%
Defense - 4.2%
Educ./Health - 4.1%

% distribution for 1989. Expend. in 1991 est.: 44.8 ($ mil.)

Military Expenditures and Arms Transfers [22]

	1985	1986	1987	1988	1989
Military expenditures					
Current dollars (mil.)	4	4	3	NA	NA
1989 constant dollars (mil.)	5	4	4	NA	NA
Armed forces (000)	11	11	11	10	10
Gross national product (GNP)					
Current dollars (mil.)	135	134	143	154	167
1989 constant dollars (mil.)	154	148	154	160	167
Central government expenditures (CGE)					
1989 constant dollars (mil.)	96	61	90	86	111
People (mil.)	0.9	0.9	0.9	1.0	1.0
Military expenditure as % of GNP	3.0	2.7	2.4	NA	NA
Military expenditure as % of CGE	4.7	6.6	4.1	NA	NA
Military expenditure per capita	5	4	4	NA	NA
Armed forces per 1,000 people	12.4	12.1	11.8	10.5	10.3
GNP per capita	173	163	165	168	171
Arms imports[6]					
Current dollars (mil.)	10	20	20	30	10
1989 constant dollars (mil.)	11	22	22	31	10
Arms exports[6]					
Current dollars (mil.)	0	0	0	0	0
1989 constant dollars (mil.)	0	0	0	0	0
Total imports[7]					
Current dollars (mil.)	NA	NA	49	NA	NA
1989 constant dollars	NA	NA	52	NA	NA
Total exports[7]					
Current dollars (mil.)	12	NA	17	NA	NA
1989 constant dollars	14	NA	18	NA	NA
Arms as percent of total imports[8]	NA	NA	41.0	NA	NA
Arms as percent of total exports[8]	0	NA	0	NA	NA

Human Rights [24]

	SSTS	FL	FAPRO	PPCG	APROBC	TPW	PCPTW	STPEP	PHRFF	PRW	ASST	AFL
Observes		P			P	P	P					P
	EAFRD	CPR	ESCR	SR	ACHR	MAAE	PVIAC	PVNAC	EAFDAW	TCIDTP		RC
Observes			P	P			P	P	P			P

P = Party; S = Signatory; see Appendix for meaning of abbreviations.

Labor Force

Total Labor Force [25]

403,000 (est.)

Labor Force by Occupation [26]

Agriculture	90%
Industry, services, and commerce	5
Government	5

Unemployment Rate [27]

For sources, notes, and explanations, see Annotated Source Appendix, page 1035.

Production Sectors

Energy Resource Summary [28]

Energy Resources: Unexploited deposits of petroleum. **Electricity**: 22,000 kW capacity; 30 million kWh produced, 30 kWh per capita (1991).

Telecommunications [30]

- Poor system of radio relay, open-wire lines, and radiocommunications
- 3,000 telephones
- Broadcast stations - 2 AM, 3 FM, 1 TV

Transportation [31]

Highways. 3,218 km; 2,698 km bituminous, remainder earth

Airports

Total:	33
Usable:	15
With permanent-surface runways:	4
With runways over 3,659 m:	0
With runways 2,440-3,659 m:	1
With runways 1,220-2,439 m:	5

Top Agricultural Products [32]

Agriculture accounts for over 50% of GDP, nearly 100% of exports, and 90% of employment; rice is the staple food; other crops include corn, beans, cassava, cashew nuts, peanuts, palm kernels, and cotton; not self-sufficient in food; fishing and forestry potential not fully exploited.

Top Mining Products [33]

Detailed information is not available. A summary of mineral resources available follows. **Mineral Resources**: Unexploited deposits of petroleum, bauxite, phosphates.

Tourism [34]

For sources, notes, and explanations, see Annotated Source Appendix, page 1035.

381

Finance, Economics, and Trade

GDP and Manufacturing Summary [35]

	1980	1985	1990	1991	1992
Gross Domestic Product					
Millions of 1980 dollars	154	171	218	224	229[e]
Growth rate in percent	-4.19	-2.30	3.04	2.80	2.17[e]
Manufacturing Value Added					
Millions of 1980 dollars	12	11	10	10	10[e]
Growth rate in percent	-5.09	-5.95	1.25	0.14	-0.13[e]
Manufacturing share in percent of current prices	1.6[e]	1.6[e]	NA	NA	NA

Economic Indicators [36]

National product: GDP—exchange rate conversion— $210 million (1991 est.). **National product real growth rate**: 2.3% (1991 est.). **National product per capita**: $210 (1991 est.). **Inflation rate (consumer prices)**: 55% (1991 est.). **External debt**: $462 million (December 1990 est.).

Balance of Payments Summary [37]

Values in millions of dollars.

	1986	1987	1988	1989	1990
Exports of goods (f.o.b.)	9.7	15.4	15.9	14.2	19.3
Imports of goods (f.o.b.)	-51.2	-44.7	-58.9	-68.9	-68.1
Trade balance	-41.5	-29.3	-43.0	-54.7	-48.8
Services - debits	-30.9	-33.2	-37.0	-49.0	-35.5
Services - credits	NA	NA	NA	NA	NA
Private transfers (net)	-1.5	-2.0	1.5	1.2	-2.1
Government transfers (net)	50.9	51.6	49.6	63.9	67.0
Long term capital (net)	17.9	64.2	29.2	29.0	28.4
Short term capital (net)	13.0	-47.0	8.6	11.7	-5.4
Errors and omissions	-3.7	-3.6	3.4	-10.6	1.3
Overall balance	4.2	0.7	12.3	-8.5	4.9

Exchange Rates [38]

Currency: **Guinea-Bissauan pesos.**
Symbol: **PG.**

Data are currency units per $1.

1989	1,987.2
1988	1,363.6
1987	851.65
1986	238.98

Imports and Exports

Top Import Origins [39]

$63.5 million (f.o.b., 1991 est.).

Origins	%
Portugal	NA
Netherlands	NA
Senegal	NA
USSR	NA
Germany	NA

Top Export Destinations [40]

$20.4 million (f.o.b., 1991 est.).

Destinations	%
Portugal	NA
Senegal	NA
France	NA
The Gambia	NA
Netherlands	NA
Spain	NA

Foreign Aid [41]

	U.S. $	
US commitments, including Ex-Im (FY70-89)	49	million
Western (non-US) countries, ODA and OOF bilateral commitments (1970-89)	615	million
OPEC bilateral aid (1979-89)	41	million
Communist countries (1970-89)	68	million

Import and Export Commodities [42]

Import Commodities

Capital equipment
Consumer goods
Semiprocessed goods
Foods
Petroleum

Export Commodities

Cashews
Fish
Peanuts
Palm kernels

382

For sources, notes, and explanations, see Annotated Source Appendix, page 1035.

Guyana

Geography [1]

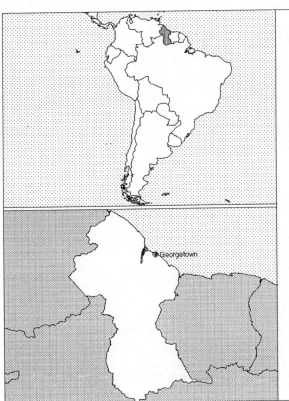

Total area:
214,970 km2
Land area:
196,850 km2
Comparative area:
Slightly smaller than Idaho
Land boundaries:
Total 2,462 km, Brazil 1,119 km, Suriname 600 km, Venezuela 743 km
Coastline:
459 km
Climate:
Tropical; hot, humid, moderated by northeast trade winds; two rainy seasons (May to mid-August, mid-November to mid-January)
Terrain:
Mostly rolling highlands; low coastal plain; savanna in south
Natural resources:
Bauxite, gold, diamonds, hardwood timber, shrimp, fish
Land use:
Arable land:
3%
Permanent crops:
0%
Meadows and pastures:
6%
Forest and woodland:
83%
Other:
8%

Demographics [2]

	1960	1970	1980	1990	1991[1]	1994	2000	2010	2020
Population	571	715	759	748	750	729	710	767	833
Population density (persons per sq. mi.)	8	9	10	10	10	NA	10	11	12
Births	NA	NA	NA	NA	17	15	NA	NA	NA
Deaths	NA	NA	NA	NA	6	5	NA	NA	NA
Life expectancy - males	NA	NA	NA	NA	61	62	NA	NA	NA
Life expectancy - females	NA	NA	NA	NA	68	68	NA	NA	NA
Birth rate (per 1,000)	NA	NA	NA	NA	23	20	NA	NA	NA
Death rate (per 1,000)	NA	NA	NA	NA	7	7	NA	NA	NA
Women of reproductive age (15-44 yrs.)	NA	NA	NA	195	NA	196	199	NA	NA
of which are currently married	NA	NA	NA	99	NA	101	105	NA	NA
Fertility rate	NA	NA	NA	2.5	2.65	2.3	2.1	1.9	1.8

Population values are in thousands, life expectancy in years, and other items as indicated.

Health

Health Personnel [3]

Health Expenditures [4]

For sources, notes, and explanations, see Annotated Source Appendix, page 1035.

383

Human Factors

Health Care Ratios [5]	Infants and Malnutrition [6]

Ethnic Division [7]

East Indian	51%
Black mixed	43%
Amerindian	4%
European Chinese	2%

Religion [8]

Christian	57%
Hindu	33%
Muslim	9%
Other	1%

Major Languages [9]

English, Amerindian dialects.

Education

Public Education Expenditures [10]

Million Guyana Dollars	1985	1986	1987	1988	1989	1990
Total education expenditure	NA	NA	219	322	478	542
as percent of GNP	NA	NA	8.3	9.8	6.0	4.7
as percent of total govt. expend.	NA	NA	NA	NA	8.9	NA
Current education expenditure	NA	NA	160	242	342	435
as percent of GNP	NA	NA	6.1	7.4	4.3	3.8
as percent of current govt. expend.	NA	NA	NA	NA	8.4	5.2
Capital expenditure	NA	NA	58	80	136	107

Educational Attainment [11]

Age group	25+
Total population	270,849
Highest level attained (%)	
No schooling	8.1
First level	
Incompleted	72.9
Completed	NA
Entered second level	
S-1	17.3
S-2	NA
Post secondary	1.8

Literacy Rate [12]

In thousands and percent	1985[a]	1991[a]	2000[a]
Illiterate population +15 years	28	25	18
Illiteracy rate - total pop. (%)	4.6	3.6	2.1
Illiteracy rate - males (%)	3.3	2.5	1.5
Illiteracy rate - females (%)	5.9	4.6	2.7

Libraries [13]

	Admin. Units	Svc. Pts.	Vols. (000)	Shelving (meters)	Vols. Added	Reg. Users
National (1986)	1	37	190	NA	6,693	45,233
Non-specialized	NA	NA	NA	NA	NA	NA
Public	NA	NA	NA	NA	NA	NA
Higher ed. (1987)[18]	1	7	166	6,100	4,237	2,512
School	NA	NA	NA	NA	NA	NA

Daily Newspapers [14]

	1975	1980	1985	1990
Number of papers	2	1	2	2
Circ. (000)	50	58	78	80[e]

Cinema [15]

Science and Technology

Scientific/Technical Forces [16]

Potential scientists/engineers	1,512
Number female	NA
Potential technicians[5]	336
Number female	NA
Total[5]	1,848

R&D Expenditures [17]

	Dollar[4,16] (000) 1982
Total expenditure	2,800
Capital expenditure	NA
Current expenditure	NA
Percent current	NA

U.S. Patents Issued [18]

Values show patents issued to citizens of the country by the U.S. Patents Office.

	1990	1991	1992
Number of patents	0	1	1

For sources, notes, and explanations, see Annotated Source Appendix, page 1035.

Government and Law

Organization of Government [19]

Long-form name:
Co-operative Republic of Guyana
Type:
republic
Independence:
26 May 1966 (from UK)
Constitution:
6 October 1980
Legal system:
based on English common law with
certain admixtures of Roman-Dutch law;
has not accepted compulsory ICJ
jurisdiction
National holiday:
Republic Day, 23 February (1970)
Executive branch:
executive president, first vice president,
prime minister, first deputy prime
minister, Cabinet
Legislative branch:
unicameral National Assembly
Judicial branch:
Supreme Court of Judicature

Political Parties [20]

National Assembly	% of seats
People's Progressive Party (PPP)	53.4
People's National Congress (PNC)	42.3
Working People's Alliance (WPA)	2.0
The United Force (TUF)	1.2

Government Budget [21]

For 1990 est.	
Revenues	121
Expenditures	225
Capital expenditures	50

Crime [23]

Military Expenditures and Arms Transfers [22]

	1985	1986	1987	1988	1989
Military expenditures					
Current dollars (mil.)	39	NA	NA	10	6
1989 constant dollars (mil.)	44	NA	NA	10	6
Armed forces (000)	7	6	5	4	4
Gross national product (GNP)					
Current dollars (mil.)	449	435	202	233	230
1989 constant dollars (mil.)	511	483	217	243	230
Central government expenditures (CGE)					
1989 constant dollars (mil.)	539[e]	572[e]	204[e]	214[e]	168[e]
People (mil.)	0.8	0.8	0.8	0.8	0.8
Military expenditure as % of GNP	8.7	NA	NA	4.3	2.7
Military expenditure as % of CGE	8.3	NA	NA	4.9	3.6
Military expenditure per capita	58	NA	NA	14	8
Armed forces per 1,000 people	9.2	7.4	6.6	4.6	4.6
GNP per capita	672	635	286	321	305
Arms imports[6]					
Current dollars (mil.)	10	0	0	0	0
1989 constant dollars (mil.)	11	0	0	0	0
Arms exports[6]					
Current dollars (mil.)	0	0	0	0	0
1989 constant dollars (mil.)	0	0	0	0	0
Total imports[7]					
Current dollars (mil.)	226	242	254	216	NA
1989 constant dollars	257	269	273	225	NA
Total exports[7]					
Current dollars (mil.)	206	214	242	230	287
1989 constant dollars	235	237	260	239	287
Arms as percent of total imports[8]	4.4	0	0	0	NA
Arms as percent of total exports[8]	0	0	0	0	0

Human Rights [24]

	SSTS	FL	FAPRO	PPCG	APROBC	TPW	PCPTW	STPEP	PHRFF	PRW	ASST	AFL
Observes	1	P	P		P	P	P				1	P
		EAFRD	CPR	ESCR	SR	ACHR	MAAE	PVIAC	PVNAC	EAFDAW	TCIDTP	RC
Observes		P	P	P				P	P	P	P	P

P = Party; S = Signatory; see Appendix for meaning of abbreviations.

Labor Force

Total Labor Force [25]

268,000

Labor Force by Occupation [26]

Industry and commerce	44.5%
Agriculture	33.8
Services	21.7

Unemployment Rate [27]

12%-15% (1991 est.)

Production Sectors

Energy Resource Summary [28]

Energy Resources: None. **Electricity**: 253,500 kW capacity; 276 million kWh produced, 370 kWh per capita (1992).

Telecommunications [30]

- Fair system with radio relay network
- Over 27,000 telephones
- Tropospheric scatter link to Trinidad
- Broadcast stations - 4 AM, 3 FM, no TV, 1 shortwave
- 1 Atlantic Ocean INTELSAT earth station

Top Agricultural Products [32]

Most important sector, accounting for 25% of GDP and about half of exports; sugar and rice are key crops; development potential exists for fishing and forestry; not self-sufficient in food, especially wheat, vegetable oils, and animal products.

Top Mining Products [33]

Metric tons unless otherwise specified	M.t.
Bauxite, dry equiv., gross weight (000 metric tons)	2,204
Diamonds (carats)[26]	18,189
Gold, mine output, Au content (kilograms)	11,000[e]
Stone, crushed	45,000[e]

Transportation [31]

Railroads. 187 km total, all single track 0.914-meter gauge

Highways. 7,665 km total; 550 km paved, 5,000 km gravel, 1,525 km earth, 590 km unimproved

Merchant Marine. 1 cargo ship (1,000 GRT or over) totaling 1,317 GRT/2,558 DWT

Airports

Total:	53
Usable:	48
With permanent-surface runways:	5
With runways over 3,659 m:	0
With runways 2,440-3,659 m:	0
With runways 1,220-2,439 m:	13

Tourism [34]

	1987	1988	1989	1990	1991
Visitors	60	71	67	64	73
Tourism receipts	24	30	30	30	30

Tourists are in thousands, money in million U.S. dollars.

For sources, notes, and explanations, see Annotated Source Appendix, page 1035.

Finance, Economics, and Trade

GDP and Manufacturing Summary [35]

	1980	1985	1990	1991	1992
Gross Domestic Product					
Millions of 1980 dollars	591	494	430	456	491
Growth rate in percent	1.66	1.02	-6.20	6.06	7.77
Manufacturing Value Added					
Millions of 1980 dollars	64	45	29	28	32[e]
Growth rate in percent	0.76	-3.13	-16.67	-1.34	11.78[e]
Manufacturing share in percent of current prices	12.1	13.9	15.9	9.6	9.6[e]

Economic Indicators [36]

National product: GDP—exchange rate conversion—$267.5 million (1992 est.). **National product real growth rate**: 7% (1992 est.). **National product per capita**: $370 (1992 est.). **Inflation rate (consumer prices)**: 15% (1992). **External debt**: $2 billion including arrears (1990).

Balance of Payments Summary [37]

Values in millions of dollars.

	1970	1973	1975	1980	1985
Exports of goods (f.o.b.)	129.0	135.7	351.4	388.9	214.0
Imports of goods (f.o.b.)	-119.9	-159.4	-305.8	-386.4	-209.1
Trade balance	9.1	-23.7	45.6	2.5	4.9
Services - debits	-48.0	-62.0	-84.5	-151.8	-144.3
Services - credits	17.6	21.9	20.4	21.6	48.0
Private transfers (net)	-0.5	-0.8	-4.4	1.0	-2.0
Government transfers (net)	-	0.1	-1.7	-1.8	-3.2
Long term capital (net)	17.1	28.7	97.4	79.8	-36.0
Short term capital (net)	-0.2	-4.1	-3.9	5.3	141.5
Errors and omissions	2.6	13.8	-19.1	0.1	-4.3
Overall balance	-2.3	-26.1	49.8	-43.3	4.6

Exchange Rates [38]

Currency: **Guyanese dollars.**
Symbol: **G$.**

Data are currency units per $1.

January 1993	125.8
1992	125.0
1991	111.8
1990	39.5
1989	27.2
1988	10.0

Imports and Exports

Top Import Origins [39]

$242.4 million (f.o.b., 1990 est.). Data are for 1989.

Origins	%
US	40
Trinidad & Tobago	13
UK	11
Japan	5
Netherland Antilles	3

Top Export Destinations [40]

$268 million (f.o.b., 1992 est.). Data are for 1989.

Destinations	%
UK	28
US	25
FRG	8
Canada	7
Japan	6

Foreign Aid [41]

	U.S. $	
US commitments, including Ex-Im (FY70-89)	116	million
Western (non-US) countries, ODA and OOF bilateral commitments (1970-89)	325	million
Communist countries 1970-89	242	million

Import and Export Commodities [42]

Import Commodities	**Export Commodities**
Manufactures	Sugar
Machinery	Bauxite/alumina
Food	Rice
Petroleum	Gold
	Shrimp
	Molasses
	Timber
	Rum

Haiti

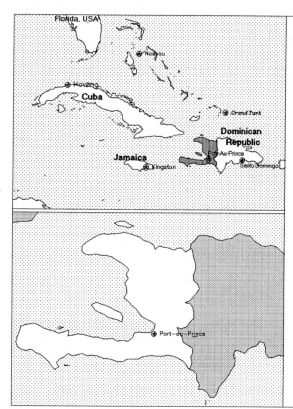

Geography [1]

Total area:
27,750 km2
Land area:
27,560 km2
Comparative area:
Slightly larger than Maryland
Land boundaries:
Total 275 km, Dominican Republic 275 km
Coastline:
1,771 km
Climate:
Tropical; semiarid where mountains in east cut off trade winds
Terrain:
Mostly rough and mountainous
Natural resources:
Bauxite
Land use:
Arable land:
20%
Permanent crops:
13%
Meadows and pastures:
18%
Forest and woodland:
4%
Other:
45%

Demographics [2]

	1960	1970	1980	1990	1991[1]	1994	2000	2010	2020
Population	3,723	4,605	5,473	6,052	6,287	6,491	7,102	8,121	9,499
Population density (persons per sq. mi.)	350	433	514	577	591	NA	719	885	1,069
Births	NA	NA	NA	NA	273	258	NA	NA	NA
Deaths	NA	NA	NA	NA	96	122	NA	NA	NA
Life expectancy - males	NA	NA	NA	NA	52	43	NA	NA	NA
Life expectancy - females	NA	NA	NA	NA	55	47	NA	NA	NA
Birth rate (per 1,000)	NA	NA	NA	NA	43	40	NA	NA	NA
Death rate (per 1,000)	NA	NA	NA	NA	15	19	NA	NA	NA
Women of reproductive age (15-44 yrs.)	NA	NA	NA	1,363	NA	1,426	1,626	NA	NA
of which are currently married	NA	NA	NA	705	NA	729	787	NA	NA
Fertility rate	NA	NA	NA	6.4	6.28	5.9	5.2	3.9	2.9

Population values are in thousands, life expectancy in years, and other items as indicated.

Health

Health Personnel [3]

Doctors per 1,000 pop., 1988-92	0.14
Nurse-to-doctor ratio, 1988-92	0.8
Hospital beds per 1,000 pop., 1985-90	0.8
Percentage of children immunized (age 1 yr. or less)	
Third dose of DPT, 1990-91	41.0
Measles, 1990-91	31.0

Health Expenditures [4]

Total health expenditure, 1990 (official exchange rate)	
Millions of dollars	193
Millions of dollars per capita	30
Health expenditures as a percentage of GDP	
Total	7.0
Public sector	3.2
Private sector	3.8
Development assistance for health	
Total aid flows (millions of dollars)[1]	33
Aid flows per capita (millions of dollars)	5.1
Aid flows as a percentage of total health expenditure	17.0

For sources, notes, and explanations, see Annotated Source Appendix, page 1035.

Human Factors

Health Care Ratios [5]

Population per physician, 1970	12520
Population per physician, 1990	NA
Population per nursing person, 1970	7410
Population per nursing person, 1990	NA
Percent of births attended by health staff, 1985	20

Infants and Malnutrition [6]

Percent of babies with low birth weight, 1985	17
Infant mortality rate per 1,000 live births, 1970	141
Infant mortality rate per 1,000 live births, 1991	94
Years of life lost per 1,000 population, 1990	69
Prevalence of malnutrition (under age 5), 1990	NA

Ethnic Division [7]

Black	95%
Mulatto European	5%

Religion [8]

The overwhelming majority of Catholics also practices Voodoo.

Roman Catholic	80%
Protestant	16%
Baptist	10%
Pentecostal	4%
Adventist	1%
None or other	4%

Major Languages [9]

Creole also spoken.

French (official)	10%

Education

Public Education Expenditures [10]

Million Gourdes	1980	1985	1987	1988	1989	1990
Total education expenditure	107	118	208	199	213	216
as percent of GNP	1.5	1.2	1.9	1.8	1.8	1.8
as percent of total govt. expend.	14.9	16.5	20.6	20.4	19.7	20.0
Current education expenditure	86	117	206	198	213	216
as percent of GNP	1.2	1.2	1.9	1.8	1.8	1.8
as percent of current govt. expend.	17.2	16.7	20.7	20.6	20.0	20.1
Capital expenditure	21	1	2	1	0.0	0.0

Educational Attainment [11]

Age group	25+
Total population	2,103,124
Highest level attained (%)	
No schooling	77.0
First level	
Incompleted	15.2
Completed	NA
Entered second level	
S-1	7.2
S-2	NA
Post secondary	0.7

Literacy Rate [12]

In thousands and percent	1985[a]	1991[a]	2000[a]
Illiterate population +15 years	1,847	1,858	1,812
Illiteracy rate - total pop. (%)	52.1	47.0	37.2
Illiteracy rate - males (%)	45.7	40.9	32.3
Illiteracy rate - females (%)	58.1	52.6	41.7

Libraries [13]

Daily Newspapers [14]

	1975	1980	1985	1990
Number of papers	7	4	5	4
Circ. (000)	93	36[e]	50[e]	45

Cinema [15]

Science and Technology

Scientific/Technical Forces [16]

Potential scientists/engineers[10]	14,189
Number female[10]	4,530
Potential technicians[11]	18,020
Number female[11]	8,639
Total	32,209

R&D Expenditures [17]

U.S. Patents Issued [18]

For sources, notes, and explanations, see Annotated Source Appendix, page 1035.

389

Government and Law

Organization of Government [19]

Long-form name:
Republic of Haiti
Type:
republic
Independence:
1 January 1804 (from France)
Constitution:
27 August 1983, suspended February 1986; draft constitution approved March 1987, suspended June 1988, most articles reinstated March 1989; October 1991, government claims to be observing the Constitution
Legal system:
based on Roman civil law system; accepts compulsory ICJ jurisdiction
National holiday:
Independence Day, 1 January (1804)
Executive branch:
president, Council of Ministers (cabinet)
Legislative branch:
bicameral National Assembly (Assemblee Nationale) consisting of an upper house or Senate and a lower house or Chamber of Deputies
Judicial branch:
Court of Appeal (Cour de Cassation)

Crime [23]

Political Parties [20]

Chamber of Deputies	% of seats
National Front (FNCD)	32.5
ANDP	20.5
Christian Democratic (PDCH)	8.4
Agricultural/Industrial (PAIN)	7.2
Progressive National (RDNP)	7.2
National Development (MDN)	6.0
National Party of Labor (PNT)	3.6
Cooperative Action (MKN)	2.4
Democratic Liberation (MODELH)	2.4
Other	9.6

Government Budget [21]

For 1990 est.
Revenues	300
Expenditures	416
Capital expenditures	145

Military Expenditures and Arms Transfers [22]

	1985	1986	1987	1988	1989
Military expenditures					
Current dollars (mil.)	32	NA	42[e]	35[e]	45[e]
1989 constant dollars (mil.)	37	NA	45[e]	37[e]	45[e]
Armed forces (000)	6	8	8	8	9
Gross national product (GNP)					
Current dollars (mil.)	2,102	2,176	2,229	2,270	2,344
1989 constant dollars (mil.)	2,393	2,415	2,398	2,363	2,344
Central government expenditures (CGE)					
1989 constant dollars (mil.)	489	427	465	364	363
People (mil.)	5.5	5.6	5.7	5.9	6.0
Military expenditure as % of GNP	1.5	NA	1.9	1.6	1.9
Military expenditure as % of CGE	7.5	NA	9.6	10.1	12.4
Military expenditure per capita	7	NA	8	6	8
Armed forces per 1,000 people	1.1	1.3	1.3	1.4	1.5
GNP per capita	436	431	418	403	391
Arms imports[6]					
Current dollars (mil.)	10	0	5	0	0
1989 constant dollars (mil.)	11	0	5	0	0
Arms exports[6]					
Current dollars (mil.)	0	0	0	0	0
1989 constant dollars (mil.)	0	0	0	0	0
Total imports[7]					
Current dollars (mil.)	442	355	374	344	NA
1989 constant dollars	503	394	402	358	NA
Total exports[7]					
Current dollars (mil.)	174	186	216	183	165
1989 constant dollars	198	206	232	191	165
Arms as percent of total imports[8]	2.3	0	1.3	0	NA
Arms as percent of total exports[8]	0	0	0	0	0

Human Rights [24]

	SSTS	FL	FAPRO	PPCG	APROBC	TPW	PCPTW	STPEP	PHRFF	PRW	ASST	AFL
Observes	P	P	P	P	P	P	P	P		P	P	P
	EAFRD	CPR	ESCR	SR	ACHR	MAAE	PVIAC	PVNAC	EAFDAW	TCIDTP	RC
Observes		P	P		P	P				P		S

P = Party; S = Signatory; see Appendix for meaning of abbreviations.

Labor Force

Total Labor Force [25]

2.3 million

Labor Force by Occupation [26]

Agriculture	66%
Services	25
Industry	9

Unemployment Rate [27]

25-50% (1991)

For sources, notes, and explanations, see Annotated Source Appendix, page 1035.

Production Sectors

Energy Resource Summary [28]

Energy Resources: None. **Electricity**: 217,000 kW capacity; 480 million kWh produced, 75 kWh per capita (1992).

Telecommunications [30]

- Domestic facilities barely adequate, international facilities slightly better
- 36,000 telephones
- Broadcast stations - 33 AM, no FM, 4 TV, 2 shortwave
- 1 Atlantic Ocean INTELSAT earth station

Transportation [31]

Railroads. 40 km 0.760-meter narrow gauge, single-track, privately owned industrial line

Highways. 4,000 km total; 950 km paved, 900 km otherwise improved, 2,150 km unimproved

Airports

Total:	13
Usable:	10
With permanent-surface runways:	3
With runways over 3,659 m:	0
With runways 2,440-3,659 m:	1
With runways 1,220-2,439 m:	3

Top Agricultural Products [32]

Agriculture accounts for 28% of GDP and employs around 70% of work force; mostly small-scale subsistence farms; commercial crops—coffee, mangoes, sugarcane, wood; staple crops—rice, corn, sorghum; shortage of wheat flour.

Top Mining Products [33]

Estimated metric tons unless otherwise specified   M.t.

Cement, hydraulic	250,000
Clays, for cement	40,000
Gravel (cubic meters)	3,900,000
Sand (cubic meters)	2,200,000
Limestone, for cement	250,000
Marble (cubic meters)	600

Tourism [34]

	1987	1988	1989	1990	1991
Tourists[9]	122	133	122	120	119
Tourism receipts	60	50	50	46	46
Tourism expenditures	42	34	33	32	33
Fare expenditures	35	40	38	37	39

Tourists are in thousands, money in million U.S. dollars.

For sources, notes, and explanations, see Annotated Source Appendix, page 1035.

391

Finance, Economics, and Trade

GDP and Manufacturing Summary [35]

	1980	1985	1990	1991	1992
Gross Domestic Product					
Millions of 1980 dollars	1,437	1,365	1,389	1,377	1,240
Growth rate in percent	7.34	0.26	-0.70	-0.81	-10.00
Manufacturing Value Added					
Millions of 1980 dollars	274	228	220	215	173[e]
Growth rate in percent	14.69	-2.87	-0.51	-2.37	-19.32[e]
Manufacturing share in percent of current prices	18.3	16.0	NA	NA	NA

Economic Indicators [36]

Gourdes or U.S. Dollars as stated[59].

	FY89	FY90	FY91[P,60]
GDP (nominal) ($ bil.)	2.1	2.4	NA
Real GDP growth rate (%)	-1.5	-3.0	-5.0
Real GDP per capita (1976 $)	156.7	149.3	146.7
Inflation (year-end, %)	9.0	15.5	15.0
Money supply M1 (mil. G.)	2,040	2,083	NA
Interest rates 1 yr CD (%)	15-18	15-22	NA
CPI (1985 = 100)	102.3	116.3	138.8
Investment rate (%. of GDP)	11.6	10.6	10.6
Gross national svgs.(%. of GDP)	5.7	4.5	4.4
External debt	804.8	838.4	861.8

Balance of Payments Summary [37]

Values in millions of dollars.

	1987	1988	1989	1990	1991
Exports of goods (f.o.b.)	210.1	180.4	148.3	160.3	162.9
Imports of goods (f.o.b.)	-311.2	-283.9	-259.3	-247.3	-300.4
Trade balance	-101.1	-103.5	-111.0	-87.0	-137.5
Services - debits	-216.6	-230.4	-219.0	-206.0	-216.3
Services - credits	115.5	100.7	93.1	88.1	92.0
Private transfers (net)	56.2	63.4	59.3	52.8	86.2
Government transfers (net)	114.8	129.5	114.9	113.5	165.1
Long term capital (net)	57.7	23.4	30.0	34.8	56.8
Short term capital (net)	-3.0	31.6	39.9	-17.5	-25.1
Errors and omissions	-17.6	10.7	-5.6	47.8	-42.6
Overall balance	5.9	25.4	1.6	26.5	-21.4

Exchange Rates [38]

Currency: **gourdes.**
Symbol: **G.**

Data are currency units per $1.

December 1991	8.4

Imports and Exports

Top Import Origins [39]

$252 million (f.o.b., 1991 est.). Data are for 1987.

Origins	%
US	64
Netherlands Antilles	5
Japan	5
France	4
Canada	3
Germany	3

Top Export Destinations [40]

$146 million (f.o.b., 1991 est.). Data are for 1987.

Destinations	%
US	84
Italy	4
France	3
Other industrial countries	6
Less developed countries	3

Foreign Aid [41]

	U.S. $	
US commitments, including Ex-Im (1970-89)	700	million
Western (non-US) countries, ODA and OOF bilateral commitments (1970-89)	770	million

Import and Export Commodities [42]

Import Commodities

Machines and manufactures 34%
Food and beverages 22%
Petroleum products 14%
Chemicals 10%
Fats and oils 9%

Export Commodities

Light manufactures 65%
Coffee 19%
Other agriculture 8%
Other 8%

392

For sources, notes, and explanations, see Annotated Source Appendix, page 1035.

Honduras

Geography [1]

Total area:
112,090 km2
Land area:
111,890 km2
Comparative area:
Slightly larger than Tennessee
Land boundaries:
Total 1,520 km, Guatemala 256 km, El Salvador 342 km, Nicaragua 922 km
Coastline:
820 km
Climate:
Subtropical in lowlands, temperate in mountains
Terrain:
Mostly mountains in interior, narrow coastal plains
Natural resources:
Timber, gold, silver, copper, lead, zinc, iron ore, antimony, coal, fish
Land use:
Arable land:
14%
Permanent crops:
2%
Meadows and pastures:
30%
Forest and woodland:
34%
Other:
20%

Demographics [2]

	1960	1970	1980	1990	1991[1]	1994	2000	2010	2020
Population	1,952	2,683	3,625	4,741	4,949	5,315	6,192	7,643	9,042
Population density (persons per sq. mi.)	45	62	84	111	115	NA	145	177	210
Births	NA	NA	NA	NA	189	186	NA	NA	NA
Deaths	NA	NA	NA	NA	36	33	NA	NA	NA
Life expectancy - males	NA	NA	NA	NA	64	65	NA	NA	NA
Life expectancy - females	NA	NA	NA	NA	68	70	NA	NA	NA
Birth rate (per 1,000)	NA	NA	NA	NA	38	35	NA	NA	NA
Death rate (per 1,000)	NA	NA	NA	NA	7	6	NA	NA	NA
Women of reproductive age (15-44 yrs.)	NA	NA	NA	1,077	NA	1,245	1,526	NA	NA
of which are currently married	NA	NA	NA	650	NA	753	931	NA	NA
Fertility rate	NA	NA	NA	5.3	4.98	4.7	3.8	2.8	2.3

Population values are in thousands, life expectancy in years, and other items as indicated.

Health

Health Personnel [3]

Doctors per 1,000 pop., 1988-92	0.32
Nurse-to-doctor ratio, 1988-92	1.0
Hospital beds per 1,000 pop., 1985-90	1.1
Percentage of children immunized (age 1 yr. or less)	
Third dose of DPT, 1990-91	94.0
Measles, 1990-91	86.0

Health Expenditures [4]

Total health expenditure, 1990 (official exchange rate)	
Millions of dollars	134
Millions of dollars per capita	26
Health expenditures as a percentage of GDP	
Total	4.5
Public sector	2.9
Private sector	1.6
Development assistance for health	
Total aid flows (millions of dollars)[1]	20
Aid flows per capita (millions of dollars)	4.0
Aid flows as a percentage of total health expenditure	15.1

For sources, notes, and explanations, see Annotated Source Appendix, page 1035.

393

Human Factors

Health Care Ratios [5]

Population per physician, 1970	3770
Population per physician, 1990	3090
Population per nursing person, 1970	1470
Population per nursing person, 1990	NA
Percent of births attended by health staff, 1985	50

Infants and Malnutrition [6]

Percent of babies with low birth weight, 1985	20
Infant mortality rate per 1,000 live births, 1970	110
Infant mortality rate per 1,000 live births, 1991	49
Years of life lost per 1,000 population, 1990	27
Prevalence of malnutrition (under age 5), 1990	21

Ethnic Division [7]

Mestizo	90%
Indian	7%
Black	2%
White	1%

Religion [8]

Roman Catholic 97%, Protestant minority.

Major Languages [9]

Spanish, Indian dialects.

Education

Public Education Expenditures [10]

Million Lempira	1980	1985	1986	1987[11]	1989	1990
Total education expenditure	155	290	341	376	416	NA
as percent of GNP	3.2	4.2	4.8	4.8	4.3	NA
as percent of total govt. expend.	14.2	13.8	16.4	19.5	15.9	NA
Current education expenditure	141	286	333	370	404	NA
as percent of GNP	2.9	4.1	4.6	4.7	4.2	NA
as percent of current govt. expend.	19.5	NA	NA	NA	NA	NA
Capital expenditure	14	4	8	6	12	NA

Educational Attainment [11]

Age group	25+[8]
Total population	NA[8]
Highest level attained (%)	
No schooling	33.5[8]
First level	
Incompleted	51.3[8]
Completed	NA[8]
Entered second level	
S-1	4.3[8]
S-2	7.6[8]
Post secondary	3.3[8]

Literacy Rate [12]

In thousands and percent	1985[a]	1991[a]	2000[a]
Illiterate population +15 years	752	766	757
Illiteracy rate - total pop. (%)	32.0	26.9	18.8
Illiteracy rate - males (%)	29.0	24.5	17.3
Illiteracy rate - females (%)	35.0	29.4	20.4

Libraries [13]

Daily Newspapers [14]

	1975	1980	1985	1990
Number of papers	8	6	7	5
Circ. (000)	120[e]	212	293	199

Cinema [15]

Science and Technology

Scientific/Technical Forces [16]

R&D Expenditures [17]

U.S. Patents Issued [18]

Values show patents issued to citizens of the country by the U.S. Patents Office.

	1990	1991	1992
Number of patents	1	0	1

For sources, notes, and explanations, see Annotated Source Appendix, page 1035.

Government and Law

Organization of Government [19]

Long-form name:
  Republic of Honduras
Type:
  republic
Independence:
  15 September 1821 (from Spain)
Constitution:
  11 January 1982, effective 20 January 1982
Legal system:
  rooted in Roman and Spanish civil law; some influence of English common law; accepts ICJ jurisdiction, with reservations
National holiday:
  Independence Day, 15 September (1821)
Executive branch:
  president, Council of Ministers (cabinet)
Legislative branch:
  unicameral National Congress (Congreso Nacional)
Judicial branch:
  Supreme Court of Justice (Corte Suprema de Justica)

Political Parties [20]

National Congress	% of votes
PNH	51.0
Liberal Party (PLH)	43.0
Christian Democratic Party (PDCH)	1.9
PINU-SD	1.5
Other	2.6

Government Budget [21]

For 1990 est.

Revenues	1.400
Expenditures	1.900
Capital expenditures	0.511

Crime [23]

Military Expenditures and Arms Transfers [22]

	1985	1986	1987	1988	1989
Military expenditures					
Current dollars (mil.)	128[e,2]	134[e,2]	140[e,2]	124[e,2]	150[e]
1989 constant dollars (mil.)	146[e,2]	149[e,2]	150[e,2]	129[e,2]	150[e]
Armed forces (000)	21	22	22	19	19
Gross national product (GNP)					
Current dollars (mil.)	3,534	3,666	4,029	4,339	4,622
1989 constant dollars (mil.)	4,024	4,068	4,334	4,517	4,622
Central government expenditures (CGE)					
1989 constant dollars (mil.)	1,043	943	900	890	969E
People (mil.)	4.2	4.3	4.4	4.5	4.7
Military expenditure as % of GNP	3.6	3.7	3.5	2.9	3.2
Military expenditure as % of CGE	14.0	15.8	16.7	14.5	15.5
Military expenditure per capita	35	35	34	29	32
Armed forces per 1,000 people	5.0	5.1	4.9	4.2	4.1
GNP per capita	966	951	987	1,003	986
Arms imports[6]					
Current dollars (mil.)	20	70	60	40	30
1989 constant dollars (mil.)	23	78	65	42	30
Arms exports[6]					
Current dollars (mil.)	0	0	0	0	0
1989 constant dollars (mil.)	0	0	0	0	0
Total imports[7]					
Current dollars (mil.)	888	875	897	933	981
1989 constant dollars	1,011	971	965	971	981
Total exports[7]					
Current dollars (mil.)	780	854	769	869	912
1989 constant dollars	888	948	827	905	912
Arms as percent of total imports[8]	2.3	8.0	6.7	4.3	3.1
Arms as percent of total exports[8]	0	0	0	0	0

Human Rights [24]

	SSTS	FL	FAPRO	PPCG	APROBC	TPW	PCPTW	STPEP	PHRFF	PRW	ASST	AFL
Observes		P	P	P	P	P	P	S				P
	EAFRD	CPR	ESCR	SR	ACHR	MAAE	PVIAC	PVNAC	EAFDAW	TCIDTP		RC
Observes		S	P	P	P	P	S	S	P			P

P = Party; S = Signatory; see Appendix for meaning of abbreviations.

Labor Force

Total Labor Force [25]

1.3 million

Labor Force by Occupation [26]

Agriculture	62%
Services	20
Manufacturing	9
Construction	3
Other	6

Date of data: 1985

Unemployment Rate [27]

15% (30-40% underemployed) (1989)

For sources, notes, and explanations, see Annotated Source Appendix, page 1035.

395

Production Sectors

Energy Resource Summary [28]

Energy Resources: Coal. **Electricity**: 575,000 kW capacity; 2,000 million kWh produced, 390 kWh per capita (1992).

Telecommunications [30]

- Inadequate system with only 7 telephones per 1,000 persons
- International services provided by 2 Atlantic Ocean INTELSAT earch stations and the Central American microwave radio relay system
- Broadcast stations - 176 AM, no FM, 7 SW, 28 TV

Top Agricultural Products [32]

		1[14]
Corn	510.0	557.9
Sugar cane	198.7	185.7
Lumber (mil. bd. ft.)	168.6	139.1
Palm oil	71.5	72.7
Bananas (mil. 40-lb. boxes)	59.3	54.9
Coffee (mil. 60-kg. bags)	1.9	1.7

Values shown are 1,000 metric tons.

Top Mining Products [33]

Preliminary metric tons unless otherwise specified M.t.

Lead, mine output, Pb content	8,719
Zinc, mine output, Zn content	38,280
Cadmium, Cd content of lead, zinc concent.	212
Copper, Cu content of lead, zinc concent.	1,000
Gold (kilograms)	179
Silver (kilograms)	39,359
Cement	693,040
Marble (sqaure meters)	95,937
Salt	30,000[e]

Transportation [31]

Railroads. 785 km total; 508 km 1.067-meter gauge, 277 km 0.914-meter gauge

Highways. 8,950 km total; 1,700 km paved, 5,000 km otherwise improved, 2,250 km unimproved earth

Merchant Marine. 252 ships (1,000 GRT or over) totaling 819,100 GRT/1,195,276 DWT; includes 2 passenger-cargo, 162 cargo, 20 refrigerated cargo, 10 container, 6 roll-on/roll-off cargo, 22 oil tanker, 1 chemical tanker, 2 specialized tanker, 22 bulk, 3 passenger, 2 short-sea passenger; note - a flag of convenience registry; Russia owns 10 ships under the Honduran flag

Airports

Total:	165
Usable:	137
With permanent-surface runways:	11
With runways over 3,659 m:	0
With runways 2,440-3,659 m:	4
With runways 1,220-2,439 m:	14

Tourism [34]

	1987	1988	1989	1990	1991
Visitors	194	226	260	290	226
Tourists	158	162	176	202	198
Excursionists	35	64	84	87	28
Tourism receipts[31]	27	28	28	29	31
Tourism expenditures[31]	35	37	38	38	37
Fare receipts	15	15	15	15	15
Fare expenditures	5	5	5	5	5

Tourists are in thousands, money in million U.S. dollars.

For sources, notes, and explanations, see Annotated Source Appendix, page 1035.

Manufacturing Sector

GDP and Manufacturing Summary [35]

	1980	1985	1989	1990	% change 1980-1990	% change 1989-1990
GDP (million 1980 $)	2,544	2,689	3,055	3,101	21.9	1.5
GDP per capita (1980 $)	695	613	613	603	-13.2	-1.6
Manufacturing as % of GDP (current prices)	15.1	15.0	14.0	17.1	13.2	22.1
Gross output (million $)	1,021[e]	1,611	2,468[e]	1,464[e]	43.4	-40.7
Value added (million $)	280[e]	493	715[e]	425[e]	51.8	-40.6
Value added (million 1980 $)	344	363	415	453	31.7	9.2
Industrial production index	100	111	138	134	34.0	-2.9
Employment (thousands)	55[e]	64	69[e]	69[e]	25.5	0.0

Note: GDP stands for Gross Domestic Product. 'e' stands for estimated value.

Profitability and Productivity

	1980	1985	1989	1990	% change 1980-1990	% change 1989-1990
Intermediate input (%)	73[e]	69	71[e]	71[e]	-2.7	0.0
Wages, salaries, and supplements (%)	12[e]	13	11[e]	12[e]	0.0	9.1
Gross operating surplus (%)	16[e]	18	17[e]	17[e]	6.3	0.0
Gross output per worker ($)	18,518[e]	25,167	35,870	20,727[e]	11.9	-42.2
Value added per worker ($)	5,073[e]	7,707	10,386[e]	6,051[e]	19.3	-41.7
Average wage (incl. benefits) ($)	2,147[e]	3,173	4,112[e]	2,443[e]	13.8	-40.6

Profitability is in percent of gross output. Productivity is in U.S. $. 'e' stands for estimated value.

Profitability - 1990

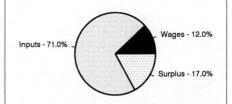

Inputs - 71.0%
Wages - 12.0%
Surplus - 17.0%

The graphic shows percent of gross output.

Value Added in Manufacturing

	1980 $ mil.	1980 %	1985 $ mil.	1985 %	1989 $ mil.	1989 %	1990 $ mil.	1990 %	% change 1980-1990	% change 1989-1990
311 Food products	75[e]	26.8	129	26.2	213[e]	29.8	120[e]	28.2	60.0	-43.7
313 Beverages	57[e]	20.4	78	15.8	111[e]	15.5	70[e]	16.5	22.8	-36.9
314 Tobacco products	19[e]	6.8	42	8.5	47[e]	6.6	27[e]	6.4	42.1	-42.6
321 Textiles	12[e]	4.3	13[e]	2.6	22[e]	3.1	16[e]	3.8	33.3	-27.3
322 Wearing apparel	6[e]	2.1	14	2.8	17[e]	2.4	10[e]	2.4	66.7	-41.2
323 Leather and fur products	2[e]	0.7	2	0.4	4[e]	0.6	2[e]	0.5	0.0	-50.0
324 Footwear	1[e]	0.4	2	0.4	4[e]	0.6	3[e]	0.7	200.0	-25.0
331 Wood and wood products	20[e]	7.1	30	6.1	37[e]	5.2	19[e]	4.5	-5.0	-48.6
332 Furniture and fixtures	5[e]	1.8	8	1.6	10[e]	1.4	6[e]	1.4	20.0	-40.0
341 Paper and paper products	4[e]	1.4	9	1.8	18[e]	2.5	11[e]	2.6	175.0	-38.9
342 Printing and publishing	8[e]	2.9	13	2.6	18[e]	2.5	10[e]	2.4	25.0	-44.4
351 Industrial chemicals	1[e]	0.4	2	0.4	3[e]	0.4	2[e]	0.5	100.0	-33.3
352 Other chemical products	11[e]	3.9	20	4.1	31[e]	4.3	19[e]	4.5	72.7	-38.7
353 Petroleum refineries	9[e]	3.2	38	7.7	41[e]	5.7	23[e]	5.4	155.6	-43.9
354 Miscellaneous petroleum and coal products	NA	0.0	NA	0.0	NA	0.0	NA	0.0	NA	NA
355 Rubber products	5[e]	1.8	8	1.6	12[e]	1.7	7[e]	1.6	40.0	-41.7
356 Plastic products	8[e]	2.9	18	3.7	28[e]	3.9	16[e]	3.8	100.0	-42.9
361 Pottery, china and earthenware	NA	0.0	NA	0.0	NA	0.0	NA	0.0	NA	NA
362 Glass and glass products	NA	0.0	NA	0.0	NA	0.0	NA	0.0	NA	NA
369 Other non-metal mineral products	16[e]	5.7	24	4.9	40[e]	5.6	26[e]	6.1	62.5	-35.0
371 Iron and steel	NA	0.0	1	0.2	4[e]	0.6	3[e]	0.7	NA	-25.0
372 Non-ferrous metals	NA	0.0	1	0.2	1[e]	0.1	1[e]	0.2	NA	0.0
381 Metal products	13[e]	4.6	21	4.3	27[e]	3.8	16[e]	3.8	23.1	-40.7
382 Non-electrical machinery	1[e]	0.4	3	0.6	5[e]	0.7	4[e]	0.9	300.0	-20.0
383 Electrical machinery	3[e]	1.1	8	1.6	9[e]	1.3	6[e]	1.4	100.0	-33.3
384 Transport equipment	NA	0.0	2	0.4	3[e]	0.4	2[e]	0.5	NA	-33.3
385 Professional and scientific equipment	NA	0.0	1	0.2	1[e]	0.1	1[e]	0.2	NA	0.0
390 Other manufacturing industries	1[e]	0.4	5	1.0	8[e]	1.1	5[e]	1.2	400.0	-37.5

Note: The industry codes shown are International Standard Industry codes (ISIC). Percentages are percent of total Value Added. 'e' stands for estimated value

For sources, notes, and explanations, see Annotated Source Appendix, page 1035.

397

Finance, Economics, and Trade

Economic Indicators [36]

In Millions of Lempiras, unless otherwise stated[61].

	1989	1990	1991[e]
Real GDP (1978 prices)	5,030.0	4,979.0	4,964.0
Real GDP Growth (%)	2.3	- 1.0	- 0.3
Money Supply (M1)	1,462.1	1,831.1	2,078.3
Bank's Interest rate	17.0	28.0	28.0
Savings rate/GDP	6.7	2.3	NA
Investment rate/GDP	12.9	16.4	NA
CPI (% change)	11.4	36.4	26.0
WPI (% change)	18.6	29.6	NA
Foreign Public Debt	6,460.0	15,584.4	17,766.0

Balance of Payments Summary [37]

Values in millions of dollars.

	1987	1988	1989	1990	1991
Exports of goods (f.o.b.)	832.8	874.9	883.4	847.8	807.9
Imports of goods (f.o.b.)	-813.0	-870.4	-834.9	-869.7	-863.5
Trade balance	19.8	4.5	48.5	-21.9	-55.6
Services - debits	-444.9	-459.8	-467.2	-469.6	-474.7
Services - credits	143.7	144.4	149.3	149.3	152.6
Private transfers (net)	16.0	17.5	16.0	25.5	9.9
Government transfers (net)	115.3	117.5	56.0	207.6	148.0
Long term capital (net)	79.6	64.5	37.9	107.0	51.0
Short term capital (net)	198.8	164.5	226.1	60.4	43.8
Errors and omissions	-46.9	-28.4	-111.7	-22.1	192.1
Overall balance	81.4	24.7	-45.1	36.2	67.1

Exchange Rates [38]

Currency: **lempiras.**
Symbol: **L.**

The lempira was allowed to float in 1992; current rate about US$1 - 5.65. Data are currency units per $1.

Current rate	5.65
Fixed rate	5.40
November 1990 (black market)	5.70

Imports and Exports

Top Import Origins [39]

$1.3 billion (c.i.f. 1991).

Origins	%
US	45
Japan	9
Netherlands	7
Mexico	7
Venezuela	6

Top Export Destinations [40]

$1.0 billion (f.o.b., 1991).

Destinations	%
US	65
Germany	9
Japan	8
Belgium	7

Foreign Aid [41]

	U.S. $	
US commitments, including Ex-Im (FY70-89)	1.4	billion
Western (non-US) countries, ODA and OOF bilateral commitments (1970-89)	1.1	billion

Import and Export Commodities [42]

Import Commodities

Machinery & transport equipment
Chemical products
Manufactured goods
Fuel and oil
Foodstuffs

Export Commodities

Bananas
Coffee
Shrimp
Lobster
Minerals
Meat
Lumber

Hong Kong

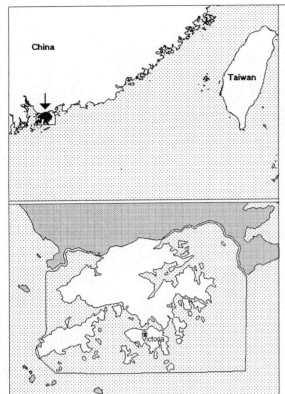

Geography [1]

Total area:
 1,040 km2
Land area:
 990 km2
Comparative area:
 Slightly less than six times the size of Washington, DC
Land boundaries:
 Total 30 km, China 30 km
Coastline:
 733 km
Climate:
 Tropical monsoon; cool and humid in winter, hot and rainy from spring through summer, warm and sunny in fall
Terrain:
 Hilly to mountainous with steep slopes; lowlands in north
Natural resources:
 Outstanding deepwater harbor, feldspar
Land use:
 Arable land:
 7%
 Permanent crops:
 1%
 Meadows and pastures:
 1%
 Forest and woodland:
 12%
 Other:
 79%

Demographics [2]

	1960	1970	1980	1990	1991[1]	1994	2000	2010	2020
Population	3,075	3,959	5,063	5,558	5,856	5,549	5,587	5,734	5,729
Population density (persons per sq. mi.)	8,051	10,364	13,254	15,231	15,329	NA	16,089	16,675	16,856
Births	NA	NA	NA	NA	77	67	NA	NA	NA
Deaths	NA	NA	NA	NA	30	32	NA	NA	NA
Life expectancy - males	NA	NA	NA	NA	77	77	NA	NA	NA
Life expectancy - females	NA	NA	NA	NA	84	84	NA	NA	NA
Birth rate (per 1,000)	NA	NA	NA	NA	13	12	NA	NA	NA
Death rate (per 1,000)	NA	NA	NA	NA	5	6	NA	NA	NA
Women of reproductive age (15-44 yrs.)	NA	NA	NA	1,505	NA	1,539	1,530	NA	NA
of which are currently married	NA	NA	NA	987	NA	1,060	1,085	NA	NA
Fertility rate	NA	NA	NA	1.3	1.42	1.4	1.5	1.5	1.6

Population values are in thousands, life expectancy in years, and other items as indicated.

Health

Health Personnel [3]

Health Expenditures [4]

For sources, notes, and explanations, see Annotated Source Appendix, page 1035.

399

Human Factors

Health Care Ratios [5]	Infants and Malnutrition [6]

Ethnic Division [7]

Chinese	98%
Other	2%

Religion [8]

Eclectic mixture of local religions	90%
Christian	10%

Major Languages [9]

Chinese (Cantonese), English.

Education

Public Education Expenditures [10]	Educational Attainment [11]

Literacy Rate [12]	Libraries [13]

Daily Newspapers [14]

	1975	1980	1985	1990
Number of papers	82	NA	46	38
Circ. (000)	NA	NA	4,100[e]	3,700[e]

Cinema [15]

Science and Technology

Scientific/Technical Forces [16]	R&D Expenditures [17]	U.S. Patents Issued [18]

U.S. Patents Issued [18]

Values show patents issued to citizens of the country by the U.S. Patents Office.

	1990	1991	1992
Number of patents	151	209	159

                                    For sources, notes, and explanations, see Annotated Source Appendix, page 1035.

Government and Law

Organization of Government [19]

Long-form name:
none
Type:
dependent territory of the UK scheduled to revert to China in 1997
Independence:
none (dependent territory of the UK; the UK signed an agreement with China on 19 December 1984 to return Hong Kong to China on 1 July 1997; in the joint declaration, China promises to respect Hong Kong's existing social and economic systems and lifestyle)
Constitution:
unwritten; partly statutes, partly common law and practice; new Basic Law approved in March 1990 in preparation for 1997
Legal system:
based on English common law
National holiday:
Liberation Day, 29 August (1945)
Executive branch:
British monarch, governor, chief secretary of the Executive Council
Legislative branch:
unicameral Legislative Council
Judicial branch:
Supreme Court

Elections [20]

Legislative Council. Indirect elections last held 12 September 1991 and direct elections were held for the first time 15 September 1991 (next to be held in September 1995 when the number of directly-elected seats increases to 20); results - percent of vote by party NA; seats - (60 total; 21 indirectly elected by functional constituencies, 18 directly elected, 18 appointed by governor, 3 ex officio members).

Government Budget [21]

For FY92.

Revenues	17.4
Expenditures	14.7
Capital expenditures	NA

Defense Summary [22]

Branches: Headquarters of British Forces, Royal Navy, Royal Air Force, Royal Hong Kong Auxiliary Air Force, Royal Hong Kong Police Force

Manpower Availability: Males age 15-49 1,635,516; fit for military service 1,256,057; reach military age (18) annually 43,128 (1993 est.)

Defense Expenditures: Exchange rate conversion - $300 million, 0.5% of GDP (1989 est.); this represents one-fourth of the total cost of defending itself, the remainder being paid by the UK

Note: Defense is the responsibility of the UK

Crime [23]

Crime volume	
Cases known to police	88,300
Attempts (percent)	NA
Percent cases solved	45.2
Crimes per 100,000 persons	1,522.3
Persons responsible for offenses	
Total number offenders	44,013
Percent female	12.1
Percent juvenile[29]	15
Percent foreigners	6

Human Rights [24]

Labor Force

Total Labor Force [25]

2.8 million (1990)

Labor Force by Occupation [26]

Manufacturing	28.5%
Wholesale, retail, restaurants, hotels	27.9
Services	17.7
Finance, insurance, and real estate	9.2
Transport and communications	4.5
Construction	2.5
Other	9.7

Date of data: 1989

Unemployment Rate [27]

2% (1992 est.)

Production Sectors

Energy Resource Summary [28]

Energy Resources: None. **Electricity**: 9,566,000 kW capacity; 29,400 million kWh produced, 4,980 kWh per capita (1992).

Telecommunications [30]

- Modern facilities provide excellent domestic and international services
- 3 mil. phones. 1.3 mil. TV sets (1.22 color). 2.5 mil. radio receivers.
- Microwave transmission links, extensive optical fiber transmission network
- Broadcast stations—6 AM, 6 FM, 4 TV
- 1 BBC repeater station and 1 British Forces repeater station
- Satellite earth stations—1 Pacific Ocean INTELSAT and 2 Indian Ocean INTELSAT
- Coaxial cable to Guangzhou, China
- Links to 5 international submarine cables

Top Agricultural Products [32]

Minor role in the economy; rice, vegetables, dairy products; less than 20% self-sufficient; shortages of rice, wheat, water.

Top Mining Products [33]

Detailed information is not available. A summary of mineral resources available follows. **Mineral Resources**: Feldspar.

Transportation [31]

Railroads. 35 km 1.435-meter standard gauge, government owned

Highways. 1,100 km total; 794 km paved, 306 km gravel, crushed stone, or earth

Merchant Marine. 176 ships (1,000 GRT or over), totaling 5,870,007 GRT/10,006,390 DWT; includes 1 passenger, 1 short-sea passenger, 20 cargo, 6 refrigerated cargo, 29 container, 15 oil tanker, 3 chemical tanker, 6 combination ore/oil, 5 liquefied gas, 88 bulk, 2 combination bulk; note - a flag of convenience registry; ships registered in Hong Kong fly the UK flag, and an estimated 500 Hong Kong-owned ships are registered elsewhere

Airports

Total:	2
Useable:	2
With permanent-surface runways:	2
With runways over 3,659 m:	0
With runways 2,440-3,659 m:	1
With runways 1,220-2,439 m:	0

Tourism [34]

	1987	1988	1989	1990	1991
Visitors	4,502	5,589	5,361	5,933	6,032
Cruise passengers[32]	7	8	10	9	
Tourism receipts[33]	3,261	4,273	4,731	5,032	5,078

Tourists are in thousands, money in million U.S. dollars.

402

For sources, notes, and explanations, see Annotated Source Appendix, page 1035.

Manufacturing Sector

GDP and Manufacturing Summary [35]

	1980	1985	1989	1990	% change 1980-1990	% change 1989-1990
GDP (million 1980 $)	27,526	36,134	50,314	52,061	89.1	3.5
GDP per capita (1980 $)	5,463	6,622	8,702	8,873	62.4	2.0
Manufacturing as % of GDP (current prices)	22.0	20.4	20.6[e]	16.1	-26.8	-21.8
Gross output (million $)	22,187	22,835	46,744	41,513	87.1	-11.2
Value added (million $)	7,343	6,582	12,630	12,032	63.9	-4.7
Value added (million 1980 $)	6,134	7,132	9,615	9,996	63.0	4.0
Industrial production index	100	129	185	182	82.0	-1.6
Employment (thousands)	937	908	981[e]	763	-18.6	-22.2

Note: GDP stands for Gross Domestic Product. 'e' stands for estimated value.

Profitability and Productivity

	1980	1985	1989	1990	% change 1980-1990	% change 1989-1990
Intermediate input (%)	67	71	73	71	6.0	-2.7
Wages, salaries, and supplements (%)	18[e]	19	15[e]	17[e]	-5.6	13.3
Gross operating surplus (%)	15[e]	10	12[e]	12[e]	-20.0	0.0
Gross output per worker ($)	23,686	25,140	47,647[e]	54,430	129.8	14.2
Value added per worker ($)	7,840	7,246	12,874[e]	15,775	101.2	22.5
Average wage (incl. benefits) ($)	4,238[e]	4,808	7,260[e]	9,182[e]	116.7	26.5

Profitability is in percent of gross output. Productivity is in U.S. $. 'e' stands for estimated value.

Profitability - 1990

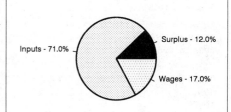

Inputs - 71.0%

Surplus - 12.0%

Wages - 17.0%

The graphic shows percent of gross output.

Value Added in Manufacturing

	1980 $ mil.	1980 %	1985 $ mil.	1985 %	1989 $ mil.	1989 %	1990 $ mil.	1990 %	% change 1980-1990	% change 1989-1990
311 Food products	161	2.2	171	2.6	315	2.5	397	3.3	146.6	26.0
313 Beverages	99	1.3	125	1.9	161	1.3	200	1.7	102.0	24.2
314 Tobacco products	81	1.1	127	1.9	222	1.8	394	3.3	386.4	77.5
321 Textiles	1,027	14.0	964	14.6	1,954	15.5	1,801	15.0	75.4	-7.8
322 Wearing apparel	1,920	26.1	1,594	24.2	2,735	21.7	2,455	20.4	27.9	-10.2
323 Leather and fur products	43	0.6	26	0.4	51[e]	0.4	38	0.3	-11.6	-25.5
324 Footwear	59	0.8	62	0.6	70[e]	0.6	35	0.3	-40.7	-50.0
331 Wood and wood products	45	0.6	32	0.5	41[e]	0.3	38	0.3	-15.6	-7.3
332 Furniture and fixtures	62	0.8	54	0.8	77[e]	0.6	66	0.5	6.5	-14.3
341 Paper and paper products	110	1.5	90	1.4	287	2.3	275	2.3	150.0	-4.2
342 Printing and publishing	290	3.9	350	5.3	713	5.6	877	7.3	202.4	23.0
351 Industrial chemicals	40	0.5	36	0.5	79[e]	0.6	64	0.5	60.0	-19.0
352 Other chemical products	77	1.0	71	1.1	138[e]	1.1	153	1.3	98.7	10.9
353 Petroleum refineries	NA	0.0	NA	0.0	NA	0.0	NA	0.0	NA	NA
354 Miscellaneous petroleum and coal products	NA	0.0	NA	0.0	NA	0.0	13	0.1	NA	NA
355 Rubber products	29	0.4	17	0.3	16	0.1	16	0.1	-44.8	0.0
356 Plastic products	563	7.7	612	9.3	771	6.1	759	6.3	34.8	-1.6
361 Pottery, china and earthenware	5	0.1	3	0.0	9[e]	0.1	6	0.0	20.0	-33.3
362 Glass and glass products	10	0.1	17	0.3	25[e]	0.2	19	0.2	90.0	-24.0
369 Other non-metal mineral products	55	0.7	47[e]	0.7	113[e]	0.9	95	0.8	72.7	-15.9
371 Iron and steel	31	0.4	17	0.3	23[e]	0.2	44	0.4	41.9	91.3
372 Non-ferrous metals	35	0.5	20	0.3	58[e]	0.5	40	0.3	14.3	-31.0
381 Metal products	638	8.7	460	7.0	898	7.1	716	6.0	12.2	-20.3
382 Non-electrical machinery	188	2.6	236	3.6	709	5.6	1,077	9.0	472.9	51.9
383 Electrical machinery	987	13.4	752	11.4	1,835	14.5	1,151	9.6	16.6	-37.3
384 Transport equipment	176	2.4	157	2.4	262	2.1	333	2.8	89.2	27.1
385 Professional and scientific equipment	362	4.9	289	4.4	567	4.5	536	4.5	48.1	-5.5
390 Other manufacturing industries	250	3.4	253	3.8	501	4.0	432	3.6	72.8	-13.8

Note: The industry codes shown are International Standard Industry codes (ISIC). Percentages are percent of total Value Added. 'e' stands for estimated value

For sources, notes, and explanations, see Annotated Source Appendix, page 1035.

403

Finance, Economics, and Trade

Economic Indicators [36]

In Millions of HK Dollars unless Otherwise noted.

	1989	1990	1991[e]
Real GDP (1980 prices)	254,218	261,209	271,135[62]
Real GDP growth rate (%)	2.7	2.8	3.8[62]
Money Supply (M-1)[37]	94,858	107,509	121,485
Commercial Interest Rate (%)[37,63]	10.0	10.0	9.25
Savings Rate (%)[37,64]	5.25	5.5	4.25
Investment Rate (%)[37,65]	6.5	6.75	5.25
CPI[66]	129.4	142.0	159.0
Ext. Public Debt (mil. $U.S.)	-0-	-0-	-0-

Balance of Payments Summary [37]

Exchange Rates [38]

Currency: **Hong Kong dollars.**
Symbol: **HK$.**

Linked to the US dollar at the rate of about 7.8 HK$ per 1 US$ since 1985. Data are currency units per $1.

1992	7.800
1991	7.771
1990	7.790
1989	7.800
1988	7.810
1987	7.760

Imports and Exports

Top Import Origins [39]

$120 billion (c.i.f., 1992 est.). Data are for 1990.

Origins	%
China	37
Japan	16
Taiwan	9
US	8

Top Export Destinations [40]

$118 billion, including reexports of $85.1 billion (f.o.b., 1992 est.). Data are for 1990.

Destinations	%
US	29
China	21
Germany	8
UK	6
Japan	5

Foreign Aid [41]

	U.S. $	
US commitments, including Ex-Im (FY70-87)	152	million
Western (non-US) countries, ODA and OOF bilateral commitments (1970-89)	923	million

Import and Export Commodities [42]

Import Commodities	Export Commodities
Foodstuffs	Clothing
Transport equipment	Textiles
Raw materials	Yarn and fabric
Semimanufactures	Footwear
Petroleum	Electrical appliances
	Watches and clocks
	Toys

Hungary

Geography [1]

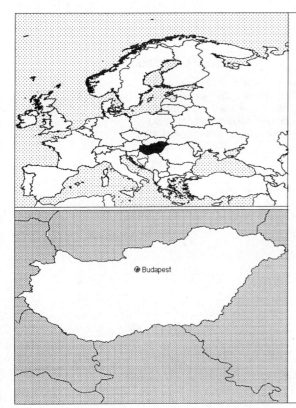

Total area:
93,030 km2
Land area:
92,340 km2
Comparative area:
Slightly smaller than Indiana
Land boundaries:
Total 1,952 km, Austria 366 km, Croatia 292 km, Romania 443 km, Serbia and Montenegro 151 km (all with Serbia), Slovakia 515 km, Slovenia 82 km, Ukraine 103 km
Coastline:
0 km (landlocked)
Climate:
Temperate; cold, cloudy, humid winters; warm summers
Terrain:
Mostly flat to rolling plains
Natural resources:
Bauxite, coal, natural gas, fertile soils
Land use:
Arable land:
50.7%
Permanent crops:
6.1%
Meadows and pastures:
12.6%
Forest and woodland:
18.3%
Other:
12.3%

Demographics [2]

	1960	1970	1980	1990	1991[1]	1994	2000	2010	2020
Population	9,984	10,337	10,711	10,365	10,558	10,319	10,372	10,477	10,449
Population density (persons per sq. mi.)	280	290	300	296	296	NA	297	297	291
Births	NA	NA	NA	NA	124	129	NA	NA	NA
Deaths	NA	NA	NA	NA	133	131	NA	NA	NA
Life expectancy - males	NA	NA	NA	NA	68	67	NA	NA	NA
Life expectancy - females	NA	NA	NA	NA	76	76	NA	NA	NA
Birth rate (per 1,000)	NA	NA	NA	NA	12	12	NA	NA	NA
Death rate (per 1,000)	NA	NA	NA	NA	13	13	NA	NA	NA
Women of reproductive age (15-44 yrs.)	NA	NA	NA	2,542	NA	2,583	2,520	NA	NA
of which are currently married	NA	NA	NA	1,733	NA	1,731	1,743	NA	NA
Fertility rate	NA	NA	NA	1.8	1.77	1.8	1.8	1.8	1.8

Population values are in thousands, life expectancy in years, and other items as indicated.

Health

Health Personnel [3]

Doctors per 1,000 pop., 1988-92	2.98
Nurse-to-doctor ratio, 1988-92	1.1
Hospital beds per 1,000 pop., 1985-90	10.1
Percentage of children immunized (age 1 yr. or less)	
Third dose of DPT, 1990-91	100.0
Measles, 1990-91	100.0

Health Expenditures [4]

Total health expenditure, 1990 (official exchange rate)	
Millions of dollars	1958
Millions of dollars per capita	185
Health expenditures as a percentage of GDP	
Total	6.0
Public sector	5.0
Private sector	0.9
Development assistance for health	
Total aid flows (millions of dollars)[1]	NA
Aid flows per capita (millions of dollars)	NA
Aid flows as a percentage of total health expenditure	NA

For sources, notes, and explanations, see Annotated Source Appendix, page 1035.

405

Human Factors

Health Care Ratios [5]

Population per physician, 1970	510
Population per physician, 1990	340
Population per nursing person, 1970	210
Population per nursing person, 1990	NA
Percent of births attended by health staff, 1985	99

Infants and Malnutrition [6]

Percent of babies with low birth weight, 1985	10
Infant mortality rate per 1,000 live births, 1970	36
Infant mortality rate per 1,000 live births, 1991	16
Years of life lost per 1,000 population, 1990	15
Prevalence of malnutrition (under age 5), 1990	NA

Ethnic Division [7]

Hungarian	89.9%
Gypsy	4.0%
German	2.6%
Serb	2.0%
Slovak	0.8%
Romanian	0.7%

Religion [8]

Roman Catholic	67.5%
Calvinist	20%
Lutheran	5%
Atheist and other	7.5%

Major Languages [9]

Hungarian	98.2%
Other	1.8%

Education

Public Education Expenditures [10]

Million Forintko	1980	1985	1987	1988	1989	1990
Total education expenditure	33,099	54,061	65,488	73,035	98,301	122,120
as percent of GNP	4.7	5.5	5.6	5.2	6.0	6.1
as percent of total govt. expend.	5.2	6.4	6.3	6.4	7.1	7.8
Current education expenditure	27,516	48,125	55,240	63,043	86,887	110,382
as percent of GNP	3.9	4.9	4.7	4.5	5.3	5.5
as percent of current govt. expend.	6.4	7.4	6.9	6.9	7.8	8.6
Capital expenditure	5,583	5,936	10,248	9,992	11,414	11,738

Educational Attainment [11]

Age group	25+
Total population	6,798,765
Highest level attained (%)	
No schooling	1.3
First level	
Incompleted	24.3
Completed	33.6
Entered second level	
S-1	NA
S-2	30.7
Post secondary	10.1

Literacy Rate [12]

	1970[b]	1980[b]	1990[b]
Illiterate population +15 years	163,768	95,542	NA
Illiteracy rate - total pop. (%)	2.0	1.1	NA
Illiteracy rate - males (%)	1.6	0.7	NA
Illiteracy rate - females (%)	2.4	1.5	NA

Libraries [13]

	Admin. Units	Svc. Pts.	Vols. (000)	Shelving (meters)	Vols. Added	Reg. Users
National (1986)	1	4	2,525	NA	30,451	18,061
Non-specialized (1986)	1	1	1,177	NA	23,147	5,147
Public (1987)	4,503	9,049	51,808	NA	NA	2.2mil
Higher ed.	NA	NA	NA	NA	NA	NA
School (1990)	3,956	NA	27,975	NA	NA	994,967

Daily Newspapers [14]

	1975	1980	1985	1990
Number of papers	27	27	28	34
Circ. (000)	2,455	2,648	2,717	2,460

Cinema [15]

Data for 1991.

Cinema seats per 1,000	21.4
Annual attendance per person	2.1
Gross box office receipts (mil. Forint)	1,220

Science and Technology

Scientific/Technical Forces [16]

Potential scientists/engineers[12]	517,650
Number female	232,300
Potential technicians	NA
Number female	NA
Total	517,650

R&D Expenditures [17]

	Forint[27] (000) 1989
Total expenditure	33,441,000
Capital expenditure	4,031,000
Current expenditure	29,410,000
Percent current	87.9

U.S. Patents Issued [18]

Values show patents issued to citizens of the country by the U.S. Patents Office.

	1990	1991	1992
Number of patents	93	86	89

406

For sources, notes, and explanations, see Annotated Source Appendix, page 1035.

Government and Law

Organization of Government [19]

Long-form name:
Republic of Hungary
Type:
republic
Independence:
1001 (unification by King Stephen I)
Constitution:
18 August 1949, effective 20 August 1949, revised 19 April 1972; 18 October 1989 revision ensured legal rights for individuals and constitutional checks on the authority of the prime minister and also established the principle of parliamentary oversight
Legal system:
in process of revision, moving toward rule of law based on Western model
National holiday:
October 23 (1956) (commemorates the Hungarian uprising)
Executive branch:
president, prime minister
Legislative branch:
unicameral National Assembly (Orszaggyules)
Judicial branch:
Constitutional Court

Crime [23]

Political Parties [20]

National Assembly	% of seats
Democratic Forum	42.0
Free Democrats	23.3
Independent Smallholders	11.7
Hungarian Socialist Party (MSP)	8.5
Young Democrats	5.7
Christian Democrats	5.4
Independents	3.4

Government Budget [21]

For 1993 est.
Revenues	13.2
Expenditures	15.4
including capital Expenditures	NA

Military Expenditures and Arms Transfers [22]

	1985	1986	1987	1988	1989
Military expenditures					
Current dollars (mil.)	3,782	3,915	4,055	4,396	4,064
1989 constant dollars (mil.)	4,305	4,345	4,361	4,576	4,064
Armed forces (000)	117	116	116	117	109
Gross national product (GNP)					
Current dollars (mil.)	52,580	55,130	57,850	63,340	64,740
1989 constant dollars (mil.)	59,850	61,180	62,220	65,930	64,740
Central government expenditures (CGE)					
1989 constant dollars (mil.)	28,060	27,490	26,780	27,320	20,260
People (mil.)	10.6	10.6	10.6	10.6	10.6
Military expenditure as % of GNP	7.2	7.1	7.0	6.9	6.3
Military expenditure as % of CGE	15.3	15.8	16.3	16.7	20.1
Military expenditure per capita	404	409	411	432	384
Armed forces per 1,000 people	11.0	10.9	10.9	11.0	10.3
GNP per capita	5,621	5,755	5,863	6,222	6,119
Arms imports[6]					
Current dollars (mil.)	80	180	400	120	30
1989 constant dollars (mil.)	91	200	430	125	30
Arms exports[6]					
Current dollars (mil.)	220	160	240	160	50
1989 constant dollars (mil.)	250	178	258	167	50
Total imports[7]					
Current dollars (mil.)	12,930	16,470	17,360	18,290	18,630
1989 constant dollars	14,720	18,280	18,670	19,040	18,630
Total exports[7]					
Current dollars (mil.)	13,440	16,180	18,050	19,050	20,210
1989 constant dollars	15,300	17,950	19,420	19,840	20,210
Arms as percent of total imports[8]	0.6	1.1	2.3	0.7	0.2
Arms as percent of total exports[8]	1.6	1.0	1.3	0.8	0.2

Human Rights [24]

	SSTS	FL	FAPRO	PPCG	APROBC	TPW	PCPTW	STPEP	PHRFF	PRW	ASST	AFL
Observes	P	P	P	P	P	P	P	P	P	P	P	
	EAFRD	CPR	ESCR	SR	ACHR	MAAE	PVIAC	PVNAC	EAFDAW	TCIDTP	RC
Observes		P	P	P	P			P	P	P	P	P

P = Party; S = Signatory; see Appendix for meaning of abbreviations.

Labor Force

Total Labor Force [25]

5.4 million

Labor Force by Occupation [26]

Services, trade, government, and other	44.8%
Industry	29.7
Agriculture	16.1
Construction	7.0

Date of data: 1991

Unemployment Rate [27]

12.3% (1992)

For sources, notes, and explanations, see Annotated Source Appendix, page 1035.

407

Production Sectors

Commercial Energy Production and Consumption

Production [28]

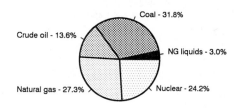

Coal - 31.8%
Crude oil - 13.6%
NG liquids - 3.0%
Natural gas - 27.3%
Nuclear - 24.2%

Consumption [29]

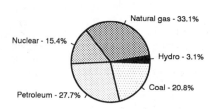

Natural gas - 33.1%
Nuclear - 15.4%
Hydro - 3.1%
Petroleum - 27.7%
Coal - 20.8%

Data are shown in quadrillion (10^{15}) BTUs and percent for 1991

Crude oil[1]	0.09
Natural gas liquids	0.02
Dry natural gas	0.18
Coal[2]	0.21
Net hydroelectric power[3]	(s)
Net nuclear power[3]	0.16
Total	0.66

Petroleum	0.36
Dry natural gas	0.43
Coal[2]	0.27
Net hydroelectric power[3]	0.04
Net nuclear power[3]	0.20
Total	1.30

Telecommunications [30]

- Automatic telephone network based on microwave radio relay system
- 1,128,800 phones (1991)
- Telephone density is at 19.4 per 100 inhabitants
- 49% of all phones are in Budapest
- 608,000 telephones on order (1991)
- 12-15 year wait for a phone
- 14,213 telex lines (1991)
- Broadcast stations - 32 AM, 15 FM, 41 TV (8 Soviet TV repeaters)
- 4.2 million TVs (1990)
- 1 satellite ground station using INTELSAT and Intersputnik

Transportation [31]

Railroads. 7,765 km total; 7,508 km 1.435-meter standard gauge, 222 km narrow gauge (mostly 0.760-meter), 35 km 1.520-meter broad gauge; 1,236 km double track, 2,249 km electrified; all government owned (1990)

Highways. 130,218 km total; 29,919 km national highway system (27,212 km asphalt, 126 km concrete, 50 km stone and road brick, 2,131 km macadam, 400 km unpaved); 58,495 km country roads (66% unpaved), and 41,804 km other roads (70% unpaved) (1988)

Merchant Marine. 12 cargo ships (1,000 GRT or over) and 1 bulk totaling 83,091 GRT/115,950 DWT

Airports

Total:	92
Usable:	92
With permanent-surface runways:	25
With runways over 3,659 m:	1
With runways 2,440-3,659 m:	20
With runways 1,220-2,439 m:	28

Top Agricultural Products [32]

	1989	1990
Wheat	6,540	6,159
Sugar beets	5,301	4,674
Corn	6,996	4,322
Vegetable	1,993	1,938
Barley	1,340	1,359
Potatoes	1,332	1,226

Values shown are 1,000 metric tons.

Top Mining Products [33]

Metric tons unless otherwise specified	M.t.
Bauxite (000 tons)	2,037
Perlite	87,750
Manganese ore, run of mine	54,783
Gypsum	110,000
Alumina	546
Cement, hydraulic (000 tons)	2,529

Tourism [34]

	1987	1988	1989	1990	1991
Visitors[34]	18,953	17,965	24,919	37,632	33,265
Tourists[4,34]	11,826	10,563	14,490	20,510	21,860
Excursionists	3,280	3,622	5,641	10,670	11,405
Tourism receipts	784	758	798	1,000	1,037
Tourism expenditures	250	647	1,008	600	505

Tourists are in thousands, money in million U.S. dollars.

For sources, notes, and explanations, see Annotated Source Appendix, page 1035.

Manufacturing Sector

GDP and Manufacturing Summary [35]

	1980	1985	1989	1990	% change 1980-1990	% change 1989-1990
GDP (million 1980 $)	22,165	24,184	25,114	24,795	11.9	-1.3
GDP per capita (1980 $)	2,069	2,271	2,376	2,351	13.6	-1.1
Manufacturing as % of GDP (current prices)	28.9	28.7	24.5[e]	24.8	-14.2	1.2
Gross output (million $)	24,898	21,690	24,737	25,081	0.7	1.4
Value added (million $)	5,907	5,356	7,109	7,838[e]	32.7	10.3
Value added (million 1980 $)	5,856	7,101	7,031	6,485	10.7	-7.8
Industrial production index	100	111	120	103	3.0	-14.2
Employment (thousands)	1,384	1,278	1,171	1,117	-19.3	-4.6

Note: GDP stands for Gross Domestic Product. 'e' stands for estimated value.

Profitability and Productivity

	1980	1985	1989	1990	% change 1980-1990	% change 1989-1990
Intermediate input (%)	76	75	71	69[e]	-9.2	-2.8
Wages, salaries, and supplements (%)	8	8	10	11	37.5	10.0
Gross operating surplus (%)	16	16	19	20[e]	25.0	5.3
Gross output per worker ($)	17,990	16,972	21,124	22,454	24.8	6.3
Value added per worker ($)	4,268	4,191	6,071	7,017[e]	64.4	15.6
Average wage (incl. benefits) ($)	1,437	1,403	2,159	2,495	73.6	15.6

Profitability is in percent of gross output. Productivity is in U.S. $. 'e' stands for estimated value.

Profitability - 1990

Inputs - 69.0%

Wages - 11.0%

Surplus - 20.0%

The graphic shows percent of gross output.

Value Added in Manufacturing

	1980 $ mil.	1980 %	1985 $ mil.	1985 %	1989 $ mil.	1989 %	1990 $ mil.	1990 %	% change 1980-1990	% change 1989-1990
311 Food products	555	9.4	281	5.2	524	7.4	584[e]	7.5	5.2	11.5
313 Beverages	83	1.4	107	2.0	130	1.8	137[e]	1.7	65.1	5.4
314 Tobacco products	27	0.5	28	0.5	37	0.5	42[e]	0.5	55.6	13.5
321 Textiles	353	6.0	325	6.1	326	4.6	355[e]	4.5	0.6	8.9
322 Wearing apparel	194	3.3	158	2.9	187	2.6	203[e]	2.6	4.6	8.6
323 Leather and fur products	48	0.8	39	0.7	40	0.6	43[e]	0.5	-10.4	7.5
324 Footwear	79	1.3	85	1.6	77	1.1	82[e]	1.0	3.8	6.5
331 Wood and wood products	81	1.4	42	0.8	71	1.0	80[e]	1.0	-1.2	12.7
332 Furniture and fixtures	101	1.7	92	1.7	112	1.6	121[e]	1.5	19.8	8.0
341 Paper and paper products	94	1.6	106	2.0	118	1.7	125[e]	1.6	33.0	5.9
342 Printing and publishing	83	1.4	94	1.8	141	2.0	156[e]	2.0	88.0	10.6
351 Industrial chemicals	417	7.1	320	6.0	517	7.3	563[e]	7.2	35.0	8.9
352 Other chemical products	242	4.1	303	5.7	400	5.6	468[e]	6.0	93.4	17.0
353 Petroleum refineries	153[e]	2.6	193[e]	3.6	272	3.8	314[e]	4.0	105.2	15.4
354 Miscellaneous petroleum and coal products	2[e]	0.0	2[e]	0.0	NA	0.0	4[e]	0.1	100.0	NA
355 Rubber products	55	0.9	71	1.3	100	1.4	124[e]	1.6	125.5	24.0
356 Plastic products	61	1.0	80	1.5	133	1.9	159[e]	2.0	160.7	19.5
361 Pottery, china and earthenware	57	1.0	46	0.9	58	0.8	63[e]	0.8	10.5	8.6
362 Glass and glass products	70	1.2	71	1.3	81	1.1	96[e]	1.2	37.1	18.5
369 Other non-metal mineral products	204	3.5	161	3.0	200	2.8	215[e]	2.7	5.4	7.5
371 Iron and steel	370	6.3	200	3.7	456	6.4	480[e]	6.1	29.7	5.3
372 Non-ferrous metals	215	3.6	54	1.0	305	4.3	415[e]	5.3	93.0	36.1
381 Metal products	214	3.6	215	4.0	281	4.0	303[e]	3.9	41.6	7.8
382 Non-electrical machinery	497	8.4	569	10.6	761	10.7	817[e]	10.4	64.4	7.4
383 Electrical machinery	655	11.1	758	14.2	814	11.5	861[e]	11.0	31.5	5.8
384 Transport equipment	486	8.2	507	9.5	461	6.5	483[e]	6.2	-0.6	4.8
385 Professional and scientific equipment	272	4.6	287	5.4	362	5.1	392[e]	5.0	44.1	8.3
390 Other manufacturing industries	237	4.0	164	3.1	144	2.0	152[e]	1.9	-35.9	5.6

Note: The industry codes shown are International Standard Industry codes (ISIC). Percentages are percent of total Value Added. 'e' stands for estimated value

For sources, notes, and explanations, see Annotated Source Appendix, page 1035.

409

Finance, Economics, and Trade

Economic Indicators [36]

Billions of Forintok (FT) or $U.S. unless otherwise stated.

	1989	1990	1991[e]
GDP (bil. $U.S.)	28.87	27.72	26.33
GDP Growth Rate (%)	(1.0)	(4.3)	(6-8)
GDP per capita ($)	2,645	2,364	2,328
Broad Money (bil. FT)	706	913	NA
Gross Savings (% of GDP)	27.3	27.5	NA
Gross Investment (% of GDP)	25.9	24.5	NA
CPI	17.0	29.0	37.0

Balance of Payments Summary [37]

Values in millions of dollars.

	1987	1988	1989	1990	1991
Exports of goods (f.o.b.)	9,967	9,989	10,493	9,151	9,688
Imports of goods (f.o.b.)	-9,887	-9,406	-9,450	-8,617	-9,330
Trade balance	80	583	1,043	534	358
Services - debits	-2,088	-2,559	-3,283	-4,107	-3,669
Services - credits	1,227	1,287	1,522	3,164	2,847
Private transfers (net)	105	117	130	794	834
Government transfers (net)	-	-	-	-7	34
Long term capital (net)	933	422	1,280	241	2,221
Short term capital (net)	-698	258	-379	-1,693	-747
Errors and omissions	160	50	-141	661	-82
Overall balance	-281	158	172	-413	1,796

Exchange Rates [38]

Currency: **forints.**
Symbol: **HF.**

Data are currency units per $1.

December 1992	83.97
1992	78.99
1991	74.74
1990	63.21
1989	59.07
1988	50.41

Imports and Exports

Top Import Origins [39]

$11.7 billion (f.o.b., 1992 est.). Data are for 1991.

Origins	%
OECD	71.0
EC	45.4
EFTA	20.0
LDCs	3.9
Former CEMA members	23.9
Others	1.2

Top Export Destinations [40]

$10.9 billion (f.o.b., 1992 est.). Data are for 1991.

Destinations	%
OECD	70.7
EC	50.1
EFTA	15.0
Less developed countries	5.1
Former CEMA members	23.2
Others	1.0

Foreign Aid [41]

	U.S. $	
Recipient - assistance from OECD countries (from 1st quarter 1990 to end of 2nd quarter 1991)	9.1	billion

Import and Export Commodities [42]

Import Commodities

Fuels and energy 14.9%
Raw materials
Semi-finished goods
Chemicals 37.6%
Machinery 19.7%
Light industry 21.5%
Food and agricultural 6.3%

Export Commodities

Raw materials
Semi-finished goods
Chemicals 35.5%
Machinery 13.5%
Light industry 23.3%
Food and agricultural 24.8%
Fuels and energy 2.8%

For sources, notes, and explanations, see Annotated Source Appendix, page 1035.

Iceland

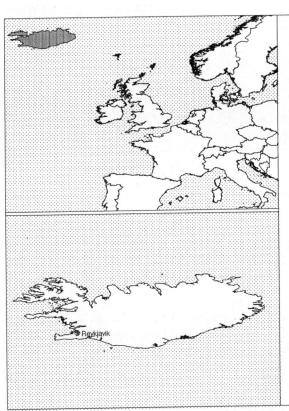

Geography [1]

Total area:
103,000 km2
Land area:
100,250 km2
Comparative area:
Slightly smaller than Kentucky
Land boundaries:
0 km
Coastline:
4,988 km
Climate:
Temperate; moderated by North Atlantic Current; mild, windy winters; damp, cool summers
Terrain:
Mostly plateau interspersed with mountain peaks, icefields; coast deeply indented by bays and fiords
Natural resources:
Fish, hydropower, geothermal power, diatomite
Land use:
Arable land:
1%
Permanent crops:
0%
Meadows and pastures:
20%
Forest and woodland:
1%
Other:
78%

Demographics [2]

	1960	1970	1980	1990	1991[1]	1994	2000	2010	2020
Population	176	204	228	255	260	264	277	293	306
Population density (persons per sq. mi.)	5	5	6	7	7	7	NA	8	8
Births	NA	NA	NA	NA	5	4	NA	NA	NA
Deaths	NA	NA	NA	NA	2	2	NA	NA	NA
Life expectancy - males	NA	NA	NA	NA	75	77	NA	NA	NA
Life expectancy - females	NA	NA	NA	NA	80	81	NA	NA	NA
Birth rate (per 1,000)	NA	NA	NA	NA	17	16	NA	NA	NA
Death rate (per 1,000)	NA	NA	NA	NA	7	7	NA	NA	NA
Women of reproductive age (15-44 yrs.)	NA	NA	NA	65	NA	67	70	NA	NA
of which are currently married	NA	NA	NA	35	NA	37	39	NA	NA
Fertility rate	NA	NA	NA	2.3	2.15	2.1	1.8	1.8	1.8

Population values are in thousands, life expectancy in years, and other items as indicated.

Health

Health Personnel [3]

Health Expenditures [4]

For sources, notes, and explanations, see Annotated Source Appendix, page 1035.

411

Human Factors

Health Care Ratios [5]	Infants and Malnutrition [6]

Ethnic Division [7]
Homogeneous mixture of descendants of Norwegians and Celts.

Religion [8]
Evangelical Lutheran	96%
Other Protestant and Roman Catholic	3%
None	1%

Major Languages [9]
Icelandic.

Education

Public Education Expenditures [10]

Million Krona	1980	1985	1986	1987	1988	1990
Total education expenditure	699	5,684	7,416	9,729	13,357	19,747
as percent of GNP	4.6	5.0	4.9	4.8	5.4	6.0
as percent of total govt. expend.	14.0	13.8	12.8	14.0	14.1	NA
Current education expenditure	NA	NA	NA	NA	NA	14,584
as percent of GNP	NA	NA	NA	NA	NA	4.5
as percent of current govt. expend.	NA	NA	NA	NA	NA	NA
Capital expenditure	NA	NA	NA	NA	NA	5,163

Educational Attainment [11]

Literacy Rate [12]

Libraries [13]

	Admin. Units	Svc. Pts.	Vols. (000)	Shelving (meters)	Vols. Added	Reg. Users
National (1990)	1	NA	415	NA	NA	NA
Non-specialized	NA	NA	NA	NA	NA	NA
Public (1988)	231	NA	1,703	NA	NA	NA
Higher ed. (1990)	2	NA	305	NA	NA	NA
School (1990)	78	78	358	NA	NA	NA

Daily Newspapers [14]

	1975	1980	1985	1990
Number of papers	5	6	6	6
Circ. (000)	93	125	113	NA

Cinema [15]
Data for 1991.
Cinema seats per 1,000	21.8
Annual attendance per person	5.2
Gross box office receipts (mil. Krona)	NA

Science and Technology

Scientific/Technical Forces [16]

R&D Expenditures [17]

U.S. Patents Issued [18]
Values show patents issued to citizens of the country by the U.S. Patents Office.
	1990	1991	1992
Number of patents	3	0	6

For sources, notes, and explanations, see Annotated Source Appendix, page 1035.

Government and Law

Organization of Government [19]

Long-form name:
  Republic of Iceland
Type:
  republic
Independence:
  17 June 1944 (from Denmark)
Constitution:
  16 June 1944, effective 17 June 1944
Legal system:
  civil law system based on Danish law;
  does not accept compulsory ICJ
  jurisdiction
National holiday:
  Anniversary of the Establishment of the
  Republic, 17 June (1944)
Executive branch:
  president, prime minister, Cabinet
Legislative branch:
  unicameral Parliament (Althing)
Judicial branch:
  Supreme Court (Haestirettur)

Political Parties [20]

Althing	% of votes
Independence Party	38.6
Progressive Party	18.9
Social Democratic Party	15.5
People's Alliance	14.4
Womens List	8.3
Liberals	1.2
Other	3.1

Government Expenditures [21]

Educ./Health - 56.2%
Housing - 3.8%
Other - 14.1%
Industry - 17.3%
Gen. Services - 8.6%

% distribution for 1991. Expend. in 1992: 1.900 ($ bil.)

Crime [23]

Military Expenditures and Arms Transfers [22]

	1985	1986	1987	1988	1989
Military expenditures					
Current dollars (mil.)	0	0	0	0	0
1989 constant dollars (mil.)	0	0	0	0	0
Armed forces (000)	0	0	0	0	0
Gross national product (GNP)					
Current dollars (mil.)	3,884	4,312	4,885	4,984	5,000
1989 constant dollars (mil.)	4,422	4,785	5,254	5,188	5,000
Central government expenditures (CGE)					
1989 constant dollars (mil.)	1,402	1,559	1,507	1,536	NA
People (mil.)	0.2	0.2	0.2	0.2	0.3
Military expenditure as % of GNP	0	0	0	0	0
Military expenditure as % of CGE	0	0	0	0	NA
Military expenditure per capita	0	0	0	0	0
Armed forces per 1,000 people	0	0	0	0	0
GNP per capita	18,320	19,690	21,470	20,760	19,700
Arms imports[6]					
Current dollars (mil.)	5	0	0	0	0
1989 constant dollars (mil.)	6	0	0	0	0
Arms exports[6]					
Current dollars (mil.)	0	0	0	0	0
1989 constant dollars (mil.)	0	0	0	0	0
Total imports[7]					
Current dollars (mil.)	905	1,119	1,590	1,597	1,401
1989 constant dollars	1,030	1,242	1,710	1,662	1,401
Total exports[7]					
Current dollars (mil.)	815	1,099	1,375	1,424	1,406
1989 constant dollars	928	1,220	1,479	1,482	1,406
Arms as percent of total imports[8]	0.6	0	0	0	0
Arms as percent of total exports[8]	0	0	0	0	0

Human Rights [24]

	SSTS	FL	FAPRO	PPCG	APROBC	TPW	PCPTW	STPEP	PHRFF	PRW	ASST	AFL
Observes		P	P	P	P	P	P		P	P	P	p
	EAFRD	CPR	ESCR	SR	ACHR	MAAE	PVIAC	PVNAC	EAFDAW	TCIDTP	RC
Observes	P	P	P	P				P	P	P	S	P

P = Party; S = Signatory; see Appendix for meaning of abbreviations.

Labor Force

Total Labor Force [25]

127,900

Labor Force by Occupation [26]

Commerce, transportation, and services
Manufacturing
Fishing and fish processing
Construction
Agriculture
  Date of data: 1990

Unemployment Rate [27]

5% (first quarter 1993)

For sources, notes, and explanations, see Annotated Source Appendix, page 1035.

413

Production Sectors

Energy Resource Summary [28]

Energy Resources: Hydropower, geothermal power. **Electricity**: 1,063,000 kW capacity; 5,165 million kWh produced, 19,940 kWh per capita (1992).

Telecommunications [30]

- Adequate domestic service
- Coaxial and fiber-optical cables and microwave radio relay for trunk network
- 140,000 telephones
- Broadcast stations - 5 AM, 147 (transmitters and repeaters) FM, 202 (transmitters and repeaters) T
- 2 submarine cables
- 1 Atlantic Ocean INTELSAT earth station carries all international traffic
- A second INTELSAT earth station is scheduled to be operational in 1993

Top Agricultural Products [32]

Agriculture accounts for about 25% of GDP; fishing is most important economic activity, contributing nearly 75% to export earnings; principal crops—potatoes, turnips; livestock—cattle, sheep; self-sufficient in crops; fish catch of about 1.4 million metric tons in 1989.

Top Mining Products [33]

Metric tons unless otherwise specified	M.t.
Aluminum metal, primary	88,768
Ferrosilicon	50,299
Diatomite	23,106
Pumice	33,354
Sand, calcareous, shell (000 cubic meters)	106
Stone	
Rhyolite	22,984
Basaltic (000 tons)	116

Transportation [31]

Highways. 11,543 km total; 2,690 km hard surfaced, 8,853 km gravel and earth

Merchant Marine. 10 ships (1,000 GRT or over) totaling 35,832 GRT/53,037 DWT; includes 3 cargo, 3 refrigerated cargo, 2 roll-on/roll-off cargo, 1 oil tanker, 1 chemical tanker

Airports

Total:	90
Usable:	84
With permanent-surface runways:	8
With runways over 3,659 m:	0
With runways 2,440-3,659 m:	1
With runways 1,220-2,439 m:	12

Tourism [34]

	1987	1988	1989	1990	1991
Visitors	272	278	273	284	292
Tourists	129	129	131	142	143
Tourism receipts	86	108	108	122	116
Tourism expenditures	213	200	176	218	295
Fare receipts	118	119	83	69	91

Tourists are in thousands, money in million U.S. dollars.

414

For sources, notes, and explanations, see Annotated Source Appendix, page 1035.

Manufacturing Sector

GDP and Manufacturing Summary [35]

	1980	1985	1989	1990	% change 1980-1990	% change 1989-1990
GDP (million 1980 $)	3,230	3,534	3,994	4,098	26.9	2.6
GDP per capita (1980 $)	14,104	14,605	15,850	16,134	14.4	1.8
Manufacturing as % of GDP (current prices)	20.3	18.6	14.4[e]	18.5[e]	-8.9	28.5
Gross output (million $)	1,969	1,629	2,728[e]	2,956[e]	50.1	8.4
Value added (million $)	765	553	880[e]	939[e]	22.7	6.7
Value added (million 1980 $)	502	506	706[e]	533	6.2	-24.5
Industrial production index	100	101	108[e]	106	6.0	-1.9
Employment (thousands)	28	30	32[e]	26[e]	-7.1	-18.8

Note: GDP stands for Gross Domestic Product. 'e' stands for estimated value.

Profitability and Productivity

	1980	1985	1989	1990	% change 1980-1990	% change 1989-1990
Intermediate input (%)	61	66	68[e]	68[e]	11.5	0.0
Wages, salaries, and supplements (%)	20[e]	19[e]	22[e]	20[e]	0.0	-9.1
Gross operating surplus (%)	19[e]	15[e]	10[e]	12[e]	-36.8	20.0
Gross output per worker ($)	69,708	54,610	85,759[e]	111,225[e]	59.6	29.7
Value added per worker ($)	27,097	18,555	27,667[e]	35,330[e]	30.4	27.7
Average wage (incl. benefits) ($)	13,687[e]	10,407[e]	18,732[e]	22,317[e]	63.1	19.1

Profitability is in percent of gross output. Productivity is in U.S. $. 'e' stands for estimated value.

Profitability - 1990

Inputs - 68.0%
Surplus - 12.0%
Wages - 20.0%

The graphic shows percent of gross output.

Value Added in Manufacturing

	1980 $ mil.	1980 %	1985 $ mil.	1985 %	1989 $ mil.	1989 %	1990 $ mil.	1990 %	% change 1980-1990	% change 1989-1990
311 Food products	330	43.1	221	40.0	323[e]	36.7	402[e]	42.8	21.8	24.5
313 Beverages	11	1.4	10	1.8	24[e]	2.7	14[e]	1.5	27.3	-41.7
314 Tobacco products	NA	0.0	NA	0.0	NA	0.0	NA	0.0	NA	NA
321 Textiles	26	3.4	21	3.8	29[e]	3.3	41[e]	4.4	57.7	41.4
322 Wearing apparel	17	2.2	11	2.0	17[e]	1.9	10[e]	1.1	-41.2	-41.2
323 Leather and fur products	8	1.0	6	1.1	11[e]	1.3	10[e]	1.1	25.0	-9.1
324 Footwear	1	0.1	1	0.2	2[e]	0.2	1[e]	0.1	0.0	-50.0
331 Wood and wood products	NA	0.0	NA	0.0	1[e]	0.1	1[e]	0.1	NA	0.0
332 Furniture and fixtures	53	6.9	32	5.8	50[e]	5.7	38[e]	4.0	-28.3	-24.0
341 Paper and paper products	5	0.7	5	0.9	10[e]	1.1	8[e]	0.9	60.0	-20.0
342 Printing and publishing	36	4.7	37	6.7	69[e]	7.8	84[e]	8.9	133.3	21.7
351 Industrial chemicals	11	1.4	9	1.6	14[e]	1.6	14[e]	1.5	27.3	0.0
352 Other chemical products	11	1.4	9	1.6	15[e]	1.7	13[e]	1.4	18.2	-13.3
353 Petroleum refineries	NA	0.0	NA	0.0	NA	0.0	NA	0.0	NA	NA
354 Miscellaneous petroleum and coal products	NA	0.0	NA	0.0	NA	0.0	NA	0.0	NA	NA
355 Rubber products	5	0.7	6	1.1	14[e]	1.6	11[e]	1.2	120.0	-21.4
356 Plastic products	12	1.6	11	2.0	23[e]	2.6	23[e]	2.4	91.7	0.0
361 Pottery, china and earthenware	1	0.1	NA	0.0	1[e]	0.1	1[e]	0.1	0.0	0.0
362 Glass and glass products	4	0.5	3	0.5	5[e]	0.6	3[e]	0.3	-25.0	-40.0
369 Other non-metal mineral products	22	2.9	19	3.4	37[e]	4.2	33[e]	3.5	50.0	-10.8
371 Iron and steel	6	0.8	11	2.0	9[e]	1.0	25[e]	2.7	316.7	177.8
372 Non-ferrous metals	50	6.5	24	4.3	24[e]	2.7	28[e]	3.0	-44.0	16.7
381 Metal products	24[e]	3.1	16[e]	2.9	29[e]	3.3	29[e]	3.1	20.8	0.0
382 Non-electrical machinery	46[e]	6.0	31[e]	5.6	69[e]	7.8	55[e]	5.9	19.6	-20.3
383 Electrical machinery	15	2.0	16	2.9	30[e]	3.4	27[e]	2.9	80.0	-10.0
384 Transport equipment	65	8.5	47	8.5	64[e]	7.3	59[e]	6.3	-9.2	-7.8
385 Professional and scientific equipment	2	0.3	1	0.2	2[e]	0.2	2[e]	0.2	0.0	0.0
390 Other manufacturing industries	3	0.4	4	0.7	9[e]	1.0	7[e]	0.7	133.3	-22.2

Note: The industry codes shown are International Standard Industry codes (ISIC). Percentages are percent of total Value Added. 'e' stands for estimated value

Finance, Economics, and Trade

Economic Indicators [36]

National product: GDP—purchasing power equivalent—$4.5 billion (1992). **National product real growth rate**: -3.3% (1992). **National product per capita**: $17,400 (1992). **Inflation rate (consumer prices)**: 3.7% (1992 est.). **External debt**: $3.9 billion (1992 est.).

Balance of Payments Summary [37]

Values in millions of dollars.

	1987	1988	1989	1990	1991
Exports of goods (f.o.b.)	1,376	1,425	1,401	1,589	1,551
Imports of goods (f.o.b.)	-1,428	-1,439	-1,267	-1,509	-1,600
Trade balance	-52	-14	134	80	-48
Services - debits	-708	-764	-764	-864	-920
Services - credits	570	558	549	624	652
Private transfers (net)	1	1	1	6	3
Government transfers (net)	-2	-2	-4	-6	-8
Long term capital (net)	178	208	261	278	292
Short term capital (net)	54	19	-136	-68	8
Errors and omissions	-59	-4	14	24	31
Overall balance	-18	1	55	74	10

Exchange Rates [38]

Currency: **Icelandic kronur.**
Symbol: **IKr.**

Data are currency units per $1.

January 1993	63.789
1992	57.546
1991	58.996
1990	58.284
1989	57.042
1988	43.014

Imports and Exports

Top Import Origins [39]

$1.5 billion (c.i.f., 1992).

Origins	%
EC	53
Denmark	10
UK	9
Norway	14
US	9

Top Export Destinations [40]

$1.5 billion (f.o.b., 1992).

Destinations	%
EC	68
Germany	12
US	11
Japan	8

Foreign Aid [41]

	U.S. $	
US commitments, including Ex-Im (FY70-81)	19.1	million

Import and Export Commodities [42]

Import Commodities

Machinery & transport equipment
Petroleum products
Foodstuffs
Textiles

Export Commodities

Fish and fish products
Animal products
Aluminum
Ferrosilicon
Diatomite

India

Geography [1]

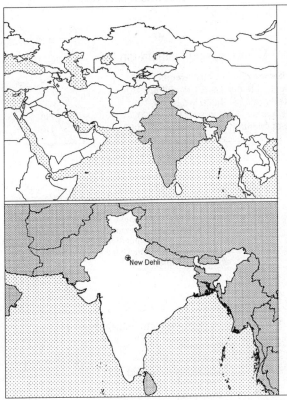

Total area:
  3,287,590 km2
Land area:
  2,973,190 km2
Comparative area:
  Slightly more than one-third the size of the US
Land boundaries:
  Total 14,103 km, Bangladesh 4,053 km, Bhutan 605 km, Burma 1,463 km, China 3,380 km, Nepal 1,690 km, Pakistan 2,912 km
Coastline:
  7,000 km
Climate:
  Varies from tropical monsoon in south to temperate in north
Terrain:
  Upland plain (Deccan Plateau) in south, flat to rolling plain along the Ganges, deserts in west, Himalayas in north
Natural resources:
  Coal (fourth-largest reserves in the world), iron ore, manganese, mica, bauxite, titanium ore, chromite, natural gas, diamonds, petroleum, limestone
Land use:
  Arable land: 55%
  Permanent crops: 1%
  Meadows and pastures: 4%
  Forest and woodland: 23%
  Other: 17%

Demographics [2]

	1960	1970	1980	1990	1991[1]	1994	2000	2010	2020
Population	445,857	555,043	692,394	852,656	869,515	919,903	1,018,105	1,173,621	1,320,746
Population density (persons per sq. mi.)	388	484	603	743	757	NA	887	1,021	1,147
Births	NA	NA	NA	NA	26,390	26,171	NA	NA	NA
Deaths	NA	NA	NA	NA	9,539	9,466	NA	NA	NA
Life expectancy - males	NA	NA	NA	NA	57	58	NA	NA	NA
Life expectancy - females	NA	NA	NA	NA	58	59	NA	NA	NA
Birth rate (per 1,000)	NA	NA	NA	NA	30	28	NA	NA	NA
Death rate (per 1,000)					11	10			
Women of reproductive age (15-44 yrs.)	NA	NA	NA	208,699	NA	227,136	257,723	NA	NA
of which are currently married	NA	NA	NA	167,910	NA	184,190	209,310	NA	NA
Fertility rate	NA	NA	NA	3.8	3.74	3.5	3.0	2.6	2.3

Population values are in thousands, life expectancy in years, and other items as indicated.

Health

Health Personnel [3]

Doctors per 1,000 pop., 1988-92	0.41
Nurse-to-doctor ratio, 1988-92	1.1
Hospital beds per 1,000 pop., 1985-90	0.7
Percentage of children immunized (age 1 yr. or less)	
Third dose of DPT, 1990-91	83.0
Measles, 1990-91	77.0

Health Expenditures [4]

Total health expenditure, 1990 (official exchange rate)	
Millions of dollars	17,740
Millions of dollars per capita	21
Health expenditures as a percentage of GDP	
Total	6.0
Public sector	1.3
Private sector	4.7
Development assistance for health	
Total aid flows (millions of dollars)[1]	286
Aid flows per capita (millions of dollars)	0.3
Aid flows as a percentage of total health expenditure	1.6

For sources, notes, and explanations, see Annotated Source Appendix, page 1035.

417

Human Factors

Health Care Ratios [5]

Population per physician, 1970	4890
Population per physician, 1990	2460
Population per nursing person, 1970	3710
Population per nursing person, 1990	NA
Percent of births attended by health staff, 1985	33

Infants and Malnutrition [6]

Percent of babies with low birth weight, 1985	30
Infant mortality rate per 1,000 live births, 1970	137
Infant mortality rate per 1,000 live births, 1991	90
Years of life lost per 1,000 population, 1990	NA
Prevalence of malnutrition (under age 5), 1990	NA

Ethnic Division [7]

Indo-Aryan	72%
Dravidian	25%
Mongoloid other	3%

Religion [8]

Hindu	82.6%
Muslim	11.4%
Christian	2.4%
Sikh	2.0%
Buddhist	0.7%
Jains	0.5%
Other	0.4%

Major Languages [9]

English enjoys associate status but is the most important language for national, political, and commercial communication, Hindi is the national language and primary tongue of 30% of the people; official languages: Bengali, Telugu, Marathi, Tamil, Urdu, Gujarati, Malayalam, Kannada, Oriya, Punjabi, Assamese, Kashmiri, Sindhi, and Sanskrit; Hindustani, a popular variant of Hindu/Urdu, is spoken widely throughout northern India.

Education

Public Education Expenditures [10]

Million Rupee	1980	1985	1986	1987	1988	1989
Total education expenditure	37,924	87,257	100,410	106,434	122,982	136,197
as percent of GNP	2.8	3.4	3.5	3.2	3.1	3.1
as percent of total govt. expend.	10.0	9.4	NA	8.5	NA	NA
Current education expenditure	37,462	85,198	98,822	104,807	NA	NA
as percent of GNP	2.7	3.3	3.4	3.2	NA	NA
as percent of current govt. expend.	NA	13.0	NA	9.9	NA	NA
Capital expenditure	463	2,059	1,588	1,627	NA	NA

Educational Attainment [11]

Age group	25+
Total population	280,599,720
Highest level attained (%)	
No schooling	72.5
First level	
Incompleted	11.3
Completed	NA
Entered second level	
S-1	13.7
S-2	NA
Post secondary	2.5

Literacy Rate [12]

In thousands and percent	1985[a]	1991[a]	2000[a]
Illiterate population +15 years	266,395	280,732	298,498
Illiteracy rate - total pop. (%)	55.9	51.8	43.7
Illiteracy rate - males (%)	41.8	38.2	31.5
Illiteracy rate - females (%)	70.9	66.3	56.8

Libraries [13]

	Admin. Units	Svc. Pts.	Vols. (000)	Shelving (meters)	Vols. Added	Reg. Users
National (1986)	8	11	1,893	51,488	32,621	33,220
Non-specialized	NA	NA	NA	NA	NA	NA
Public	NA	NA	NA	NA	NA	NA
Higher ed.	NA	NA	NA	NA	NA	NA
School	NA	NA	NA	NA	NA	NA

Daily Newspapers [14]

	1975	1980	1985	1990
Number of papers	835	1,173	1,802	NA
Circ. (000)	9,383	14,531	19,804	NA

Cinema [15]

Data for 1991.

Cinema seats per 1,000	7.8[e,1]
Annual attendance per person	5.0[e,1]
Gross box office receipts (mil. Rupee)	NA

Science and Technology

Scientific/Technical Forces [16]

Potential scientists/engineers	2,471,400
Number female	NA
Potential technicians	639,300
Number female	NA
Total	3,110,700

R&D Expenditures [17]

	Rupee (000) 1988
Total expenditure	34,718,100
Capital expenditure	5,399,930
Current expenditure	29,318,170
Percent current	84.4

U.S. Patents Issued [18]

Values show patents issued to citizens of the country by the U.S. Patents Office.

	1990	1991	1992
Number of patents	23	24	24

For sources, notes, and explanations, see Annotated Source Appendix, page 1035.

Government and Law

Organization of Government [19]

Long-form name:
Republic of India
Type:
federal republic
Independence:
15 August 1947 (from UK)
Constitution:
26 January 1950
Legal system:
based on English common law; limited judicial review of legislative acts; accepts compulsory ICJ jurisdiction, with reservations
National holiday:
Anniversary of the Proclamation of the Republic, 26 January (1950)
Executive branch:
president, vice president, prime minister, Council of Ministers
Legislative branch:
bicameral Parliament (Sansad) consists of an upper house or Council of States (Rajya Sabha) and a lower house or People's Assembly (Lok Sabha)
Judicial branch:
Supreme Court

Political Parties [20]

People's Assembly	% of seats
Congress (I) Party	45.0
Bharatiya Janata Party	21.8
Janata Dal Party	7.2
Janata Dal (Ajit Singh)	3.7
Communist Party of India/Marxist	6.4
Communist Party of India	2.6
Telugu Desam	2.4
All-India Anna Dravida	2.0
Samajwadi Janata Party	0.9
Other (incl. 9 vacant)	7.7

Government Expenditures [21]

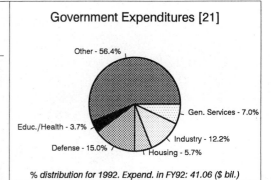

Other - 56.4%
Gen. Services - 7.0%
Industry - 12.2%
Housing - 5.7%
Defense - 15.0%
Educ./Health - 3.7%

% distribution for 1992. Expend. in FY92: 41.06 ($ bil.)

Military Expenditures and Arms Transfers [22]

	1985	1986	1987	1988	1989
Military expenditures					
Current dollars (mil.)	6,662	7,302	8,275	8,377	8,174
1989 constant dollars (mil.)	7,584	8,103	8,900	8,721	8,174
Armed forces (000)	1,515	1,492	1,502	1,362	1,257
Gross national product (GNP)					
Current dollars (mil.)	188,200	201,900	217,700	246,200	267,400
1989 constant dollars (mil.)	214,200	224,100	234,100	256,300	267,400
Central government expenditures (CGE)					
1989 constant dollars (mil.)	48,370	53,720	54,340	59,140	60,120
People (mil.)	767.6	783.8	800.1	816.6	833.2
Military expenditure as % of GNP	3.5	3.6	3.8	3.4	3.1
Military expenditure as % of CGE	15.7	15.1	16.4	14.7	13.6
Military expenditure per capita	10	10	11	11	10
Armed forces per 1,000 people	2.0	1.9	1.9	1.7	1.5
GNP per capita	279	286	293	314	321
Arms imports[6]					
Current dollars (mil.)	2,600	3,200	3,500	3,400	3,500
1989 constant dollars (mil.)	2,960	3,551	3,764	3,539	3,500
Arms exports[6]					
Current dollars (mil.)	10	40	5	0	0
1989 constant dollars (mil.)	11	44	5	0	0
Total imports[7]					
Current dollars (mil.)	16,070	15,410	16,720	19,170	20,430
1989 constant dollars	18,300	17,100	17,990	19,950	20,430
Total exports[7]					
Current dollars (mil.)	9,214	9,499	11,370	13,310	15,820
1989 constant dollars	10,490	10,540	12,230	13,860	15,820
Arms as percent of total imports[8]	16.2	20.8	20.9	17.7	17.1
Arms as percent of total exports[8]	0.1	0.4	0	0	0

Crime [23]

Human Rights [24]

	SSTS	FL	FAPRO	PPCG	APROBC	TPW	PCPTW	STPEP	PHRFF	PRW	ASST	AFL
Observes	P	P		P		P	P	P		P	P	
	EAFRD	CPR	ESCR	SR	ACHR	MAAE	PVIAC	PVNAC	EAFDAW	TCIDTP	RC	
Observes		P	P	P					S	P	P	

P = Party; S = Signatory; see Appendix for meaning of abbreviations.

Labor Force

Total Labor Force [25]

284.4 million

Labor Force by Occupation [26]

Agriculture   67%
Date of data: FY85

Unemployment Rate [27]

For sources, notes, and explanations, see Annotated Source Appendix, page 1035.

419

Production Sectors

Commercial Energy Production and Consumption

Production [28]

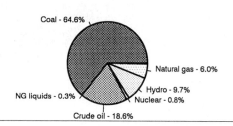

Coal - 64.6%
Natural gas - 6.0%
Hydro - 9.7%
Nuclear - 0.8%
Crude oil - 18.6%
NG liquids - 0.3%

Consumption [29]

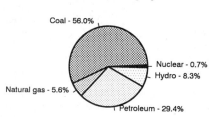

Coal - 56.0%
Nuclear - 0.7%
Hydro - 8.3%
Natural gas - 5.6%
Petroleum - 29.4%

Data are shown in quadrillion (10^{15}) BTUs and percent for 1991

Crude oil[1]	1.33
Natural gas liquids	0.02
Dry natural gas	0.43
Coal[2]	4.62
Net hydroelectric power[3]	0.69
Net nuclear power[3]	0.06
Total	7.15

Petroleum	2.48
Dry natural gas	0.47
Coal[2]	4.72
Net hydroelectric power[3]	0.70
Net nuclear power[3]	0.06
Total	8.43

Telecommunications [30]

- Domestic telephone system is poor providing only one telephone for about 200 persons on average
- Long distance telephoning has been improved by a domestic satellite system which also carries TV
- International service is provided by 3 Indian Ocean INTELSAT earth stations and by submarine cable to Malaysia and the United Arab Emirates
- Broadcast stations - 96 AM, 4 FM, 274 TV (government controlled)

Transportation [31]

Railroads. 61,850 km total (1986); 33,553 km 1.676-meter broad gauge, 24,051 km 1.000-meter gauge, 4,246 km narrow gauge (0.762 meter and 0.610 meter); 12,617 km is double track; 6,500 km is electrified

Highways. 1,970,000 km total (1989); 960,000 km surfaced and 1,010,000 km gravel, crushed stone, or earth

Merchant Marine. 306 ships (1,000 GRT or over) totaling 6,278,672 GRT/10,446,073 DWT; includes 1 short-sea passenger, 6 passenger-cargo, 87 cargo, 1 roll-on/roll-off, 8 container, 63 oil tanker, 10 chemical tanker, 8 combination ore/oil, 114 bulk, 2 combination bulk, 6 liquefied gas

Airports

Total:	336
Usable:	285
With permanent-surface runways:	205
With runways over 3,659 m:	2
With runways 2,440-3,659 m:	58
With runways 1,220-2,439 m:	90

Top Agricultural Products [32]

		[1][9]
Tea	684.1	714.6
Sugar cane	222.6	234.7
Rice	73.6	74.6
Wheat	49.8	54.5
Coarse grains	34.8	33.1
Pulses	12.6	14.0

Values shown are 1,000 metric tons.

Top Mining Products [33]

Preliminary metric tons unless otherwise specified M.t.[5]

Bauxite, gross weight (000 tons)	4,738
Chromite, gross weight	994,674
Iron ore, concentrate (000 tons)	59,915
Manganese ore, concentrate (000 tons)	1,401
Gemstones (diamonds, agate, garnet)	NA
Granite	NA
Graphite	69,922
Mica	5,529
Talc	491,036
Coal, bituminous and lignite (000 tons)	239,849

Tourism [34]

	1987	1988	1989	1990	1991
Visitors	1,498	1,604	1,748	1,721	1,685
Tourists[4]	1,484	1,591	1,736	1,707	1,678
Excursionists	2	1	1	1	1
Cruise passengers	13	12	11	13	6
Tourism receipts	1,430	1,500	1,535	1,437	1,310
Tourism expenditures	352	397	416		

Tourists are in thousands, money in million U.S. dollars.

Manufacturing Sector

GDP and Manufacturing Summary [35]

	1980	1985	1989	1990	% change 1980-1990	% change 1989-1990
GDP (million 1980 $)	172,723	224,026	288,716	302,102	74.9	4.6
GDP per capita (1980 $)	251	291	346	354	41.0	2.3
Manufacturing as % of GDP (current prices)	17.7	17.9	15.8[e]	18.6	5.1	17.7
Gross output (million $)	71,387	88,304	111,915	129,450[e]	81.3	15.7
Value added (million $)	13,086	15,526	19,551	22,598[e]	72.7	15.6
Value added (million 1980 $)	27,526	38,560	53,829	55,238	100.7	2.6
Industrial production index	100	137	173	193	93.0	11.6
Employment (thousands)	6,992	6,578	6,880[e]	7,309[e]	4.5	6.2

Note: GDP stands for Gross Domestic Product. 'e' stands for estimated value.

Profitability and Productivity

	1980	1985	1989	1990	% change 1980-1990	% change 1989-1990
Intermediate input (%)	82	82	83	83[e]	1.2	0.0
Wages, salaries, and supplements (%)	11[e]	10	8[e]	8[e]	-27.3	0.0
Gross operating surplus (%)	8[e]	8	9[e]	9[e]	12.5	0.0
Gross output per worker ($)	10,210	13,423	16,266[e]	17,707[e]	73.4	8.9
Value added per worker ($)	1,872	2,360	2,842[e]	3,098[e]	65.5	9.0
Average wage (incl. benefits) ($)	1,088[e]	1,298	1,340[e]	1,502[e]	38.1	12.1

Profitability is in percent of gross output. Productivity is in U.S. $. 'e' stands for estimated value.

Profitability - 1990

Inputs - 83.0%
Wages - 8.0%
Surplus - 9.0%

The graphic shows percent of gross output.

Value Added in Manufacturing

	1980 $ mil.	1980 %	1985 $ mil.	1985 %	1989 $ mil.	1989 %	1990 $ mil.	1990 %	% change 1980-1990	% change 1989-1990
311 Food products	899	6.9	1,436	9.2	1,699	8.7	2,128[e]	9.4	136.7	25.3
313 Beverages	99	0.8	135	0.9	153	0.8	258[e]	1.1	160.6	68.6
314 Tobacco products	196	1.5	230	1.5	227	1.2	453[e]	2.0	131.1	99.6
321 Textiles	2,642	20.2	2,135	13.8	2,295	11.7	2,443[e]	10.8	-7.5	6.4
322 Wearing apparel	62	0.5	87	0.6	143	0.7	261[e]	1.2	321.0	82.5
323 Leather and fur products	48	0.4	52	0.3	58	0.3	94[e]	0.4	95.8	62.1
324 Footwear	37	0.3	52	0.3	65	0.3	64[e]	0.3	73.0	-1.5
331 Wood and wood products	74	0.6	73	0.5	54	0.3	96[e]	0.4	29.7	77.8
332 Furniture and fixtures	8	0.1	7	0.0	6	0.0	9[e]	0.0	12.5	50.0
341 Paper and paper products	296	2.3	233	1.5	343	1.8	354[e]	1.6	19.6	3.2
342 Printing and publishing	256	2.0	280	1.8	322[e]	1.6	377[e]	1.7	47.3	17.1
351 Industrial chemicals	778	5.9	1,200	7.7	1,595	8.2	1,744[e]	7.7	124.2	9.3
352 Other chemical products	1,062	8.1	1,146	7.4	1,722	8.8	1,677[e]	7.4	57.9	-2.6
353 Petroleum refineries	203	1.6	344	2.2	504	2.6	837[e]	3.7	312.3	66.1
354 Miscellaneous petroleum and coal products	151	1.2	152	1.0	165[e]	0.8	272[e]	1.2	80.1	64.8
355 Rubber products	234	1.8	363	2.3	537	2.7	675[e]	3.0	188.5	25.7
356 Plastic products	93	0.7	166	1.1	157[e]	0.8	221[e]	1.0	137.6	40.8
361 Pottery, china and earthenware	47	0.4	27	0.2	50[e]	0.3	64[e]	0.3	36.2	28.0
362 Glass and glass products	67	0.5	101	0.7	93[e]	0.5	106[e]	0.5	58.2	14.0
369 Other non-metal mineral products	399	3.0	775	5.0	764	3.9	791[e]	3.5	98.2	3.5
371 Iron and steel	1,489	11.4	1,790	11.5	1,825	9.3	2,501[e]	11.1	68.0	37.0
372 Non-ferrous metals	81	0.6	115	0.7	160	0.8	465[e]	2.1	474.1	190.6
381 Metal products	421	3.2	425	2.7	467	2.4	748[e]	3.3	77.7	60.2
382 Non-electrical machinery	1,130	8.6	1,506	9.7	1,634	8.4	1,674[e]	7.4	48.1	2.4
383 Electrical machinery	1,061	8.1	1,201	7.7	2,210	11.3	2,184[e]	9.7	105.8	-1.2
384 Transport equipment	1,088	8.3	1,231	7.9	1,835	9.4	1,821[e]	8.1	67.4	-0.8
385 Professional and scientific equipment	92	0.7	118	0.8	352	1.8	180[e]	0.8	95.7	-48.9
390 Other manufacturing industries	72	0.6	146	0.9	117[e]	0.6	100[e]	0.4	38.9	-14.5

Note: The industry codes shown are International Standard Industry codes (ISIC). Percentages are percent of total Value Added. 'e' stands for estimated value

Finance, Economics, and Trade

Economic Indicators [36]

Billions of Indian Rupees, unless otherwise noted[67].

	1989-90	1990-91[e]	1991-92[p]
Real GNP (1981 prices)	2,211.3	2,321.9	2,391.6
Real GNP growth (%)	5.1	5.0	3.0
Money Supply (M1)	814.2	945.8	1,056.5
Commercial Interest Rate (%)	16.5	17.0	20.0
Gross Savings Rate (%)	21.7	21.8	21.2
Investment Rate (%)	24.1	24.0	23.2
CPI (1982=100)	173	193	216
WPI (1982=100)	165.7	182.7	205.0
External Public Debt (bil. $U.S.)	47.2	53.5	60.0

Balance of Payments Summary [37]

Values in millions of dollars.

	1985	1986	1987	1988	1989
Exports of goods (f.o.b.)	9,465.0	10,248.0	11,884.0	13,510.0	16,144.0
Imports of goods (f.o.b.)	-15,081.0	-15,686.0	-17,661.0	-20,091.0	-22,254.0
Trade balance	-5,616.0	-5,438.0	-5,777.0	-6,581.0	-6,110.0
Services - debits	-5,250.0	-5,526.0	-6,235.0	-7,537.0	-8,372.0
Services - credits	3,913.0	3,746.0	3,813.0	4,218.0	4,586.0
Private transfers (net)	2,456.0	2,223.0	2,636.0	2,295.0	2,567.0
Government transfers (net)	357.0	428.0	391.0	487.0	516.0
Long term capital (net)	3,341.0	4,496.0	4,511.0	5,915.0	6,161.0
Short term capital (net)	-60.0	-504.0	1,223.0	1,328.0	1,188.0
Errors and omissions	500.0	197.0	-409.0	-112.0	-285.0
Overall balance	-359.0	-378.0	153.0	13.0	251.0

Exchange Rates [38]

Currency: **Indian rupees.**
Symbol: **Rs.**

Data are currency units per $1.

January 1993	26.156
1992	25.918
1991	22.742
1990	17.504
1989	16.226
1988	13.917

Imports and Exports

Top Import Origins [39]

$25.5 billion (c.i.f., FY93 est.). Data are for FY91.

Origins	%
US	12.1
West Germany	8.0
Japan	7.5

Top Export Destinations [40]

$19.8 billion (f.o.b., FY93 est.). Data are for FY91.

Destinations	%
USSR	16.1
US	14.7
West Germany	7.8

Foreign Aid [41]

	U.S. $	
US commitments, including Ex-Im (FY70-89)	4.4	billion
Western (non-US) countries, ODA and OOF bilateral commitments (1980-89)	31.7	billion
OPEC bilateral aid (1979-89)	315	million
USSR (1970-89)	11.6	billion
Eastern Europe (1970-89)	105	million

Import and Export Commodities [42]

Import Commodities
Crude & petroleum products
Gems
Fertilizer
Chemicals
Machinery

Export Commodities
Gems and jewelry
Clothing
Engineering goods
Leather manufactures
Cotton yarn
And fabric

For sources, notes, and explanations, see Annotated Source Appendix, page 1035.

Indonesia

Geography [1]

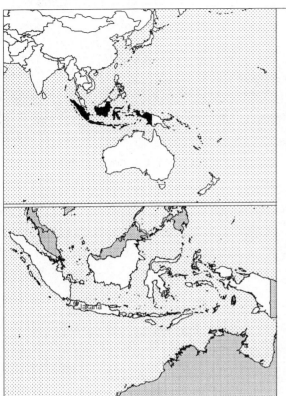

Total area:
 1,919,440 km2
Land area:
 1,826,440 km2
Comparative area:
 Slightly less than three times the size of Texas
Land boundaries:
 Total 2,602 km, Malaysia 1,782 km, Papua New Guinea 820 km
Coastline:
 54,716 km
Climate:
 Tropical; hot, humid; more moderate in highlands
Terrain:
 Mostly coastal lowlands; larger islands have interior mountains
Natural resources:
 Petroleum, tin, natural gas, nickel, timber, bauxite, copper, fertile soils, coal, gold, silver
Land use:
 Arable land:
 8%
 Permanent crops:
 3%
 Meadows and pastures:
 7%
 Forest and woodland:
 67%
 Other:
 15%

Demographics [2]

	1960	1970	1980	1990	1991[1]	1994	2000	2010	2020
Population	100,655	122,671	154,936	187,728	193,560	200,410	219,496	250,033	276,474
Population density (persons per sq. mi.)	143	174	220	270	NA	272	317	364	407
Births	NA	NA	NA	NA	5,017	4,900	NA	NA	NA
Deaths	NA	NA	NA	NA	1,601	1,724	NA	NA	NA
Life expectancy - males	NA	NA	NA	NA	59	59	NA	NA	NA
Life expectancy - females	NA	NA	NA	NA	63	63	NA	NA	NA
Birth rate (per 1,000)	NA	NA	NA	NA	26	24	NA	NA	NA
Death rate (per 1,000)	NA	NA	NA	NA	8	9	NA	NA	NA
Women of reproductive age (15-44 yrs.)	NA	NA	NA	48,926	NA	54,037	61,535	NA	NA
of which are currently married	NA	NA	NA	34,741	NA	38,437	44,478	NA	NA
Fertility rate	NA	NA	NA	3.0	3.03	2.8	2.5	2.3	2.1

Population values are in thousands, life expectancy in years, and other items as indicated.

Health

Health Personnel [3]

Doctors per 1,000 pop., 1988-92	0.14
Nurse-to-doctor ratio, 1988-92	2.8
Hospital beds per 1,000 pop., 1985-90	0.7
Percentage of children immunized (age 1 yr. or less)	
Third dose of DPT, 1990-91	86.0
Measles, 1990-91	80.0

Health Expenditures [4]

Total health expenditure, 1990 (official exchange rate)	
Millions of dollars	2,148
Millions of dollars per capita	12
Health expenditures as a percentage of GDP	
Total	2.0
Public sector	0.7
Private sector	1.3
Development assistance for health	
Total aid flows (millions of dollars)[1]	159
Aid flows per capita (millions of dollars)	0.9
Aid flows as a percentage of total health expenditure	7.4

For sources, notes, and explanations, see Annotated Source Appendix, page 1035.

423

Human Factors

Health Care Ratios [5]

Population per physician, 1970	26820
Population per physician, 1990	7030
Population per nursing person, 1970	4810
Population per nursing person, 1990	NA
Percent of births attended by health staff, 1985	43

Infants and Malnutrition [6]

Percent of babies with low birth weight, 1985	14
Infant mortality rate per 1,000 live births, 1970	118
Infant mortality rate per 1,000 live births, 1991	74
Years of life lost per 1,000 population, 1990	36
Prevalence of malnutrition (under age 5), 1990	14

Ethnic Division [7]

Javanese	45%
Sundanese	14%
Madurese	7.5%
Coastal Malays	7.5%
Other	26%

Religion [8]

Muslim	87%
Protestant	6%
Roman Catholic	3%
Hindu	2%
Buddhist	1%
Other	1%

Major Languages [9]

Bahasa Indonesia (modified form of Malay; official), English, Dutch, local dialects the most widely spoken of which is Javanese.

Education

Public Education Expenditures [10]

Billion Rupiah	1980	1985	1987	1988[1]	1989	1990
Total education expenditure	0.808	NA	NA	1,238.3	NA	NA
as percent of GNP	1.7	NA	NA	0.9	NA	NA
as percent of total govt. expend.	8.9	NA	NA	4.3	NA	NA
Current education expenditure	NA	NA	NA	1,095.5	NA	NA
as percent of GNP	NA	NA	NA	0.8	NA	NA
as percent of current govt. expend.	NA	NA	NA	5.5	NA	NA
Capital expenditure	NA	NA	NA	142.8	NA	NA

Educational Attainment [11]

Age group	25+
Total population	58,441,240
Highest level attained (%)	
No schooling	41.1
First level	
Incompleted	31.6
Completed	16.8
Entered second level	
S-1	4.7
S-2	4.9
Post secondary	0.8

Literacy Rate [12]

In thousands and percent	1985[a]	1991[a]	2000[a]
Illiterate population +15 years	28,810	26,970	22,758
Illiteracy rate - total pop. (%)	28.2	23.0	15.5
Illiteracy rate - males (%)	19.6	15.9	10.5
Illiteracy rate - females (%)	36.5	32.0	20.4

Libraries [13]

	Admin. Units	Svc. Pts.	Vols. (000)	Shelving (meters)	Vols. Added	Reg. Users
National	NA	NA	NA	NA	NA	NA
Non-specialized	NA	NA	NA	NA	NA	NA
Public	NA	NA	NA	NA	NA	NA
Higher ed. (1989)	45	137	1,735	NA	NA	534,798
School	NA	NA	NA	NA	NA	NA

Daily Newspapers [14]

	1975	1980	1985	1990
Number of papers	60[e]	84	97	64
Circ. (000)	2,000[e]	2,281	3,010	5,144

Cinema [15]

Science and Technology

Scientific/Technical Forces [16]

Potential scientists/engineers	639,006
Number female	NA
Potential technicians	NA
Number female	NA
Total	NA

R&D Expenditures [17]

	Rupiah[20] (000) 1988
Total expenditure	259,283,000
Capital expenditure	64,645,000
Current expenditure	194,638,000
Percent current	75.1

U.S. Patents Issued [18]

Values show patents issued to citizens of the country by the U.S. Patents Office.

	1990	1991	1992
Number of patents	3	2	9

Government and Law

Organization of Government [19]

Long-form name:
Republic of Indonesia
Type:
republic
Independence:
17 August 1945 (proclaimed independence; on 27 December 1949, Indonesia became legally independent from the Netherlands)
Constitution:
August 1945, abrogated by Federal Constitution of 1949 and Provisional Constitution of 1950, restored 5 July 1959
Legal system:
based on Roman-Dutch law, substantially modified by indigenous concepts and by new criminal procedures code; has not accepted compulsory ICJ jurisdiction
National holiday:
Independence Day, 17 August (1945)
Executive branch:
president, vice president, Cabinet
Legislative branch:
unicameral House of Representatives (Dewan Perwakilan Rakyat or DPR)
Judicial branch:
Supreme Court (Mahkamah Agung)

Crime [23]

Political Parties [20]

House of Representatives	% of votes
GOLKAR	68.0
Development Unity Party	17.0
Indonesia Democracy Party	15.0

Government Expenditures [21]

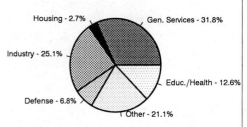

Housing - 2.7%
Gen. Services - 31.8%
Industry - 25.1%
Educ./Health - 12.6%
Defense - 6.8%
Other - 21.1%

% distribution for 1991. Expend. in FY91: 23.4 ($ bil.)

Military Expenditures and Arms Transfers [22]

	1985	1986	1987	1988	1989
Military expenditures					
Current dollars (mil.)	1,446	1,567	1,373	1,402	1,510
1989 constant dollars (mil.)	1,646	1,739	1,477	1,460	1,510
Armed forces (000)	281	278	281	284	285
Gross national product (GNP)					
Current dollars (mil.)	61,070	66,720	72,620	79,810	89,370
1989 constant dollars (mil.)	69,520	74,040	78,110	83,080	89,370
Central government expenditures (CGE)					
1989 constant dollars (mil.)	15,910	18,740	16,950	17,380	18,340
People (mil.)	172.7	176.2	179.7	183.2	186.7
Military expenditure as % of GNP	2.4	2.3	1.9	1.8	1.7
Military expenditure as % of CGE	10.3	9.3	8.7	8.4	8.2
Military expenditure per capita	10	10	8	8	8
Armed forces per 1,000 people	1.6	1.6	1.6	1.6	1.5
GNP per capita	403	420	435	453	479
Arms imports[6]					
Current dollars (mil.)	100	100	220	250	90
1989 constant dollars (mil.)	114	111	237	260	90
Arms exports[6]					
Current dollars (mil.)	5	0	0	5	10
1989 constant dollars (mil.)	6	0	0	5	10
Total imports[7]					
Current dollars (mil.)	10,260	10,720	12,890	13,250	16,440
1989 constant dollars	11,680	11,890	13,860	13,790	16,440
Total exports[7]					
Current dollars (mil.)	18,590	16,070	17,130	19,460	22,160
1989 constant dollars	21,160	17,840	18,430	20,260	22,160
Arms as percent of total imports[8]	1.0	0.9	1.7	1.9	0.5
Arms as percent of total exports[8]	0	0	0	0	0

Human Rights [24]

	SSTS	FL	FAPRO	PPCG	APROBC	TPW	PCPTW	STPEP	PHRFF	PRW	ASST	AFL
Observes		P			P	P	P			P		
	EAFRD	CPR	ESCR	SR	ACHR	MAAE	PVIAC	PVNAC	EAFDAW	TCIDTP	RC	
Observes		S	S	S					P	S	P	

P=Party; S=Signatory; see Appendix for meaning of abbreviations.

Labor Force

Total Labor Force [25]

67 million

Labor Force by Occupation [26]

Agriculture	55%
Manufacturing	10
Construction	4
Transport and communications	3

Date of data: 1985 est.

Unemployment Rate [27]

3% ; underemployment 45% (1991 est.)

For sources, notes, and explanations, see Annotated Source Appendix, page 1035.

425

Production Sectors

Commercial Energy Production and Consumption

Production [28]

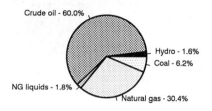

Crude oil - 60.0%
Hydro - 1.6%
Coal - 6.2%
NG liquids - 1.8%
Natural gas - 30.4%

Consumption [29]

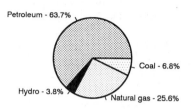

Petroleum - 63.7%
Coal - 6.8%
Hydro - 3.8%
Natural gas - 25.6%

Data are shown in quadrillion (10^{15}) BTUs and percent for 1991

Crude oil[1]	3.38
Natural gas liquids	0.10
Dry natural gas	1.71
Coal[2]	0.35
Net hydroelectric power[3]	0.09
Net nuclear power[3]	0.00
Total	5.64

Petroleum	1.49
Dry natural gas	0.60
Coal[2]	0.16
Net hydroelectric power[3]	0.09
Net nuclear power[3]	0.00
Total	2.35

Telecommunications [30]

- Interisland microwave system and HF police net
- Domestic service fair, international service good
- Radiobroadcast coverage good
- 763,000 telephones (1986)
- Broadcast stations - 618 AM, 38 FM, 9 TV
- Satellite earth stations - 1 Indian Ocean INTELSAT earth station and 1 Pacific Ocean INTELSAT eart station
- And 1 domestic satellite communications system

Transportation [31]

Railroads. 6,964 km total; 6,389 km 1.067-meter gauge, 497 km 0.750-meter gauge; 78 km 0.600-meter gauge; 211 km double track; 101 km electrified; all government owned

Highways. 119,500 km total; 11,812 km state, 34,180 km provincial, and 73,508 km district roads

Merchant Marine. 401 ships (1,000 GRT or over) totaling 1,766,201 GRT/2,642,529 DWT; includes 6 short-sea passenger, 13 passenger-cargo, 238 cargo, 10 container, 4 roll-on/roll-off cargo, 4 vehicle carrier, 78 oil tanker, 6 chemical tanker, 6 liquefied gas, 7 specialized tanker, 1 livestock carrier, 26 bulk, 2 passenger

Airports

Total:	435
Usable:	411
With permanent-surface runways:	119
With runways over 3,659 m:	1
With runways 2,440-3,659 m:	11
With runways 1,220-2,439 m:	67

Top Agricultural Products [32]

	89-90	90-91[4]
Rice (milled basis)[4]	29.40	28.70
Cassava	15.50	17.00
Corn	5.20	5.00
Palm oil	2.25	2.80
Sugar	2.12	2.05
Soybeans	1.32	1.29

Values shown are 1,000 metric tons.

Top Mining Products [33]

Preliminary metric tons unless otherwise specified M.t.

Bauxite, gross weight (000 tons)	1,242
Tin, mine output, Sb content	30,061
Nickel, mine output, Ni content[7]	71,681
Gold, mine output, Au content (kilograms)[8]	16,879
Silver, mine output, Ag content (kilograms)	80,294
Aluminum, primary	173,000
Coal (000 tons)	13,688
Liquid petroleum gas (000 42-gal. barrels)	3,453

Tourism [34]

	1987	1988	1989	1990	1991
Visitors	1,060	1,301	1,626	2,178	2,570
Tourism receipts	924	1,283	1,628	2,153	2,515
Tourism expenditures	511	592	722	836	949

Tourists are in thousands, money in million U.S. dollars.

For sources, notes, and explanations, see Annotated Source Appendix, page 1035.

Manufacturing Sector

GDP and Manufacturing Summary [35]

	1980	1985	1989	1990	% change 1980-1990	% change 1989-1990
GDP (million 1980 $)	78,013	99,479	115,436	134,922	72.9	16.9
GDP per capita (1980 $)	517	595	638	732	41.6	14.7
Manufacturing as % of GDP (current prices)	13.0	16.0	17.9[e]	21.1	62.3	17.9
Gross output (million $)	13,205	23,111	24,834[e]	39,467[e]	198.9	58.9
Value added (million $)	4,368	8,098[e]	9,027[e]	12,268[e]	180.9	35.9
Value added (million 1980 $)	10,133	18,632	20,432	30,905	205.0	51.3
Industrial production index	100	119	155	193	93.0	24.5
Employment (thousands)	963	1,672	1,905[e]	2,378[e]	146.9	24.8

Note: GDP stands for Gross Domestic Product. 'e' stands for estimated value.

Profitability and Productivity

	1980	1985	1989	1990	% change 1980-1990	% change 1989-1990
Intermediate input (%)	69	69	68[e]	71[e]	2.9	4.4
Wages, salaries, and supplements (%)	7	7	7[e]	6[e]	-14.3	-14.3
Gross operating surplus (%)	25	24	25[e]	23[e]	-8.0	-8.0
Gross output per worker ($)	11,219	12,255	12,103[e]	15,322[e]	36.6	26.6
Value added per worker ($)	3,497	3,850	3,912[e]	4,461[e]	27.6	14.0
Average wage (incl. benefits) ($)	743	921	898[e]	941[e]	26.6	4.8

Profitability is in percent of gross output. Productivity is in U.S. $. 'e' stands for estimated value.

Profitability - 1990

Inputs - 71.0%
Wages - 6.0%
Surplus - 23.0%

The graphic shows percent of gross output.

Value Added in Manufacturing

	1980 $ mil.	1980 %	1985 $ mil.	1985 %	1989 $ mil.	1989 %	1990 $ mil.	1990 %	% change 1980-1990	% change 1989-1990
311 Food products	376	8.6	870	10.7	759[e]	8.4	1,287[e]	10.5	242.3	69.6
313 Beverages	51	1.2	77	1.0	92[e]	1.0	81[e]	0.7	58.8	-12.0
314 Tobacco products	649	14.9	741	9.2	981[e]	10.9	1,158[e]	9.4	78.4	18.0
321 Textiles	420	9.6	687	8.5	852[e]	9.4	1,378[e]	11.2	228.1	61.7
322 Wearing apparel	15	0.3	105	1.3	160[e]	1.8	271[e]	2.2	1,706.7	69.4
323 Leather and fur products	5	0.1	14	0.2	29[e]	0.3	15[e]	0.1	200.0	-48.3
324 Footwear	26	0.6	31	0.4	34[e]	0.4	94[e]	0.8	261.5	176.5
331 Wood and wood products	239	5.5	612	7.6	850[e]	9.4	1,230[e]	10.0	414.6	44.7
332 Furniture and fixtures	6	0.1	18	0.2	25[e]	0.3	51[e]	0.4	750.0	104.0
341 Paper and paper products	43	1.0	110	1.4	97[e]	1.1	269[e]	2.2	525.6	177.3
342 Printing and publishing	51	1.2	92	1.1	166[e]	1.8	156[e]	1.3	205.9	-6.0
351 Industrial chemicals	145	3.3	385	4.8	454[e]	5.0	475[e]	3.9	227.6	4.6
352 Other chemical products	241	5.5	430	5.3	387[e]	4.3	488[e]	4.0	102.5	26.1
353 Petroleum refineries	978	22.4	1,611[e]	19.9	1,575[e]	17.4	1,596[e]	13.0	63.2	1.3
354 Miscellaneous petroleum and coal products	NA	0.0	NA	0.0	NA	0.0	NA	0.0	NA	NA
355 Rubber products	164	3.8	328	4.1	252[e]	2.8	516[e]	4.2	214.6	104.8
356 Plastic products	25	0.6	175	2.2	115[e]	1.3	148[e]	1.2	492.0	28.7
361 Pottery, china and earthenware	8	0.2	24	0.3	29[e]	0.3	49[e]	0.4	512.5	69.0
362 Glass and glass products	36	0.8	98	1.2	93[e]	1.0	58[e]	0.5	61.1	-37.6
369 Other non-metal mineral products	200	4.6	262	3.2	253[e]	2.8	232[e]	1.9	16.0	-8.3
371 Iron and steel	107	2.4	469b	0.0	704[e]	7.8	742[e]	6.0	593.5	5.4
372 Non-ferrous metals	NA	0.0	b	0.0	NA	0.0	NA	0.0	NA	NA
381 Metal products	118	2.7	278	3.4	329[e]	3.6	732[e]	6.0	520.3	122.5
382 Non-electrical machinery	53	1.2	76	0.9	65[e]	0.7	108[e]	0.9	103.8	66.2
383 Electrical machinery	180	4.1	246	3.0	224[e]	2.5	292[e]	2.4	62.2	30.4
384 Transport equipment	217	5.0	331	4.1	466[e]	5.2	769[e]	6.3	254.4	65.0
385 Professional and scientific equipment	2	0.0	4	0.0	4[e]	0.0	7[e]	0.1	250.0	75.0
390 Other manufacturing industries	13	0.3	24	0.3	30[e]	0.3	64[e]	0.5	392.3	113.3

Note: The industry codes shown are International Standard Industry codes (ISIC). Percentages are percent of total Value Added. 'e' stands for estimated value

Finance, Economics, and Trade

Economic Indicators [36]

Billions of 1983 Rupiah (RP) unless otherwise noted.

	1989	1990	1991[e]
Real GDP	107,321	114,921	122,391
Real GDP Growth (%)	7.4	7.4	6.0
Money supply (MI) (%)[68]	39.8	18.4	6.1
Interest rates[69]	12.4	14.5	17.7
National savings/GDP (%)	21.4	21.4	22.0
Investment/GDP (%)	23.5	24.6	25.0
CPI[70,71]	337	115	126
WPI[72]	162	178	185

Balance of Payments Summary [37]

Values in millions of dollars.

	1987	1988	1989	1990	1991
Exports of goods (f.o.b.)	17,206.0	19,509.0	22,974.0	26,807.0	29,430.0
Imports of goods (f.o.b.)	-12,532.0	-13,831.0	-16,310.0	-21,455.0	-24,626.0
Trade balance	4,674.0	5,678.0	6,664.0	5,352.0	4,804.0
Services - debits	-8,655.0	-9,190.0	-10,548.0	-11,655.0	-12,570.0
Services - credits	1,626.0	1,861.0	2,437.0	2,897.0	3,424.0
Private transfers (net)	86.0	99.0	167.0	166.0	130.0
Government transfers (net)	171.0	155.0	172.0	252.0	132.0
Long term capital (net)	2,511.0	1,809.0	3,016.0	4,724.0	5,912.0
Short term capital (net)	970.0	408.0	-98.0	-229.0	214.0
Errors and omissions	-753.0	-933.0	-1,315.0	744.0	-517.0
Overall balance	630.0	-113.0	495.0	2,251.0	1,529.0

Exchange Rates [38]

Currency: **Indonesian rupiahs.**
Symbol: **Rp.**

Data are currency units per $1.

January 1993	2,064.7
1992	2,029.9
1991	1,950.3
1990	1,842.8
1989	1,770.1
1988	1,685.7

Imports and Exports

Top Import Origins [39]

$24.6 billion (f.o.b., 1991).

Origins	%
Japan	25
Europe	23
US	13
Singapore	5

Top Export Destinations [40]

$29.4 billion (f.o.b., 1991).

Destinations	%
Japan	37
Europe	13
US	12
Singapore	8

Foreign Aid [41]

	U.S. $	
US commitments, including Ex-Im (FY70-89)	4.4	billion
Western (non-US) countries, ODA and OOF bilateral commitments (1970-89)	25.9	billion
OPEC bilateral aid (1979-89)	213	million
Communist countries (1970-89)	175	million

Import and Export Commodities [42]

Import Commodities

Machinery 39%
Chemical products 19%
Manufactured goods 16%

Export Commodities

Petroleum & LNG 40%
Timber 15%
Textiles 7%
Rubber 5%
Coffee 3%

Iran

Geography [1]

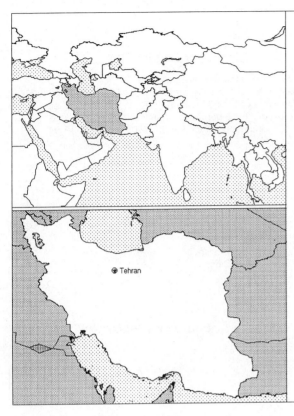

⊕ Tehran

Total area:
1.648 million km2
Land area:
1.636 million km2
Comparative area:
Slightly larger than Alaska
Land boundaries:
Total 5,440 km, Afghanistan 936 km, Armenia 35 km, Azerbaijan (north) 432 Km, Azerbaijan (northwest) 179 km, Iraq 1,458 km, Pakistan 909 km, Turkey 499 km, Turkmenistan 992 km
Coastline:
2,440 km
Climate:
Mostly arid or semiarid, subtropical along Caspian coast
Terrain:
Rugged, mountainous rim; high, central basin with deserts, mountains; small, discontinuous plains along both coasts
Natural resources:
Petroleum, natural gas, coal, chromium, copper, iron ore, lead, manganese, zinc, sulfur
Land use:
Arable land: 8%
Permanent crops: 0%
Meadows and pastures: 27%
Forest and woodland: 11%
Other: 54%

Demographics [2]

	1960	1970	1980	1990	1991[1]	1994	2000	2010	2020
Population	21,577	28,933	38,810	57,003	59,051	65,612	78,347	107,676	143,624
Population density (persons per sq. mi.)	34	46	61	90	93	NA	124	170	227
Births	NA	NA	NA	NA	2,610	2,785	NA	NA	NA
Deaths	NA	NA	NA	NA	505	514	NA	NA	NA
Life expectancy - males	NA	NA	NA	NA	64	65	NA	NA	NA
Life expectancy - females	NA	NA	NA	NA	65	67	NA	NA	NA
Birth rate (per 1,000)	NA	NA	NA	NA	44	42	NA	NA	NA
Death rate (per 1,000)	NA	NA	NA	NA	9	8	NA	NA	NA
Women of reproductive age (15-44 yrs.)	NA	NA	NA	12,011	NA	13,955	17,033	NA	NA
of which are currently married	NA	NA	NA	9,098	NA	10,571	12,858	NA	NA
Fertility rate	NA	NA	NA	6.6	6.55	6.3	5.9	4.9	4.0

Population values are in thousands, life expectancy in years, and other items as indicated.

Health

Health Personnel [3]

Doctors per 1,000 pop., 1988-92	0.32
Nurse-to-doctor ratio, 1988-92	1.1
Hospital beds per 1,000 pop., 1985-90	1.5
Percentage of children immunized (age 1 yr. or less)	
Third dose of DPT, 1990-91	88.0
Measles, 1990-91	84.0

Health Expenditures [4]

Total health expenditure, 1990 (official exchange rate)	
Millions of dollars	3024
Millions of dollars per capita	54
Health expenditures as a percentage of GDP	
Total	2.6
Public sector	1.5
Private sector	1.1
Development assistance for health	
Total aid flows (millions of dollars)[1]	2
Aid flows per capita (millions of dollars)	NA
Aid flows as a percentage of total health expenditure	NA

For sources, notes, and explanations, see Annotated Source Appendix, page 1035.

429

Human Factors

Health Care Ratios [5]

Population per physician, 1970	3270
Population per physician, 1990	3140
Population per nursing person, 1970	1780
Population per nursing person, 1990	1150
Percent of births attended by health staff, 1985	NA

Infants and Malnutrition [6]

Percent of babies with low birth weight, 1985	9
Infant mortality rate per 1,000 live births, 1970	131
Infant mortality rate per 1,000 live births, 1991	68
Years of life lost per 1,000 population, 1990	32
Prevalence of malnutrition (under age 5), 1990	NA

Ethnic Division [7]

Persian	51%
Azerbaijani	24%
Gilaki Mazandarani	8%
Kurd	7%
Arab	3%
Lur	2%
Baloch	2%
Turkmen	2%

Religion [8]

Shi'a Muslim	95%
Sunni Muslim	4%
Zoroastrian, Jewish, Christian, and Baha'i	1%

Major Languages [9]

Persian (Persian dialects)	58%
Turkic (Turkic dialects)	26%
Kurdish	9%
Luri	2%
Baloch	1%
Arabic	1%
Turkish	1%
Other	2%

Education

Public Education Expenditures [10]

Billion Rial	1980	1985	1987	1988	1989	1990
Total education expenditure	498.3	575.5	NA	888.3	1,045.5	1,493.9
as percent of GNP	7.5	3.5	NA	3.8	3.7	4.1
as percent of total govt. expend.	15.7	17.2	NA	19.2	21.9	22.4
Current education expenditure	440.3	510.1	NA	791.9	915.7	1,232.1
as percent of GNP	6.6	3.1	NA	3.4	3.3	3.4
as percent of current govt. expend.	20.1	20.6	NA	20.8	23.9	28.9
Capital expenditure	58	65.4	NA	96.4	129.7	261.8

Educational Attainment [11]

Literacy Rate [12]

In thousands and percent	1985[a]	1991[a]	2000[a]
Illiterate population +15 years	14,155	14,604	14,421
Illiteracy rate - total pop. (%)	52.3	46.0	34.0
Illiteracy rate - males (%)	40.9	35.5	25.6
Illiteracy rate - females (%)	63.7	56.7	42.7

Libraries [13]

	Admin. Units	Svc. Pts.	Vols. (000)	Shelving (meters)	Vols. Added	Reg. Users
National (1989)	1	3	300	NA	7,500[e]	7,200
Non-specialized	NA	NA	NA	NA	NA	NA
Public (1987)	507	507	3,332	NA	NA	7.1mil
Higher ed.	NA	NA	NA	NA	NA	NA
School	NA	NA	NA	NA	NA	NA

Daily Newspapers [14]

	1975	1980	1985	1990
Number of papers	19	45[e]	12[e]	21
Circ. (000)	700[e]	970[e]	1,250[e]	1,500[e]

Cinema [15]

Data for 1991.

Cinema seats per 1,000	2.8
Annual attendance per person	1.1
Gross box office receipts (mil. Rial)	14,632

Science and Technology

Scientific/Technical Forces [16]

Potential scientists/engineers	294,647
Number female	NA
Potential technicians	170,894
Number female	NA
Total	465,541

R&D Expenditures [17]

	Rial[18] (000) 1985
Total expenditure	22,010,713
Capital expenditure	9,464,315
Current expenditure	12,546,398
Percent current	57.0

U.S. Patents Issued [18]

Values show patents issued to citizens of the country by the U.S. Patents Office.

	1990	1991	1992
Number of patents	2	2	1

For sources, notes, and explanations, see Annotated Source Appendix, page 1035.

Government and Law

Organization of Government [19]

Long-form name:
Islamic Republic of Iran
Type:
theocratic republic
Independence:
1 April 1979 (Islamic Republic of Iran proclaimed)
Constitution:
2-3 December 1979; revised 1989 to expand powers of the presidency and eliminate the prime ministership
Legal system:
the Constitution codifies Islamic principles of government
National holiday:
Islamic Republic Day, 1 April (1979)
Executive branch:
supreme leader (velay-t-e faqih), president, Council of Ministers
Legislative branch:
unicameral Islamic Consultative Assembly (Majles-e-Shura-ye-Eslami)
Judicial branch:
Supreme Court

Crime [23]

Elections [20]

Islamic Consultative Assembly. There are at least 18 licensed parties; the three most important are - Tehran Militant Clergy Association, Militant Clerics Association, and Fedaiyin Islam Organization. Last held 8 April 1992 (next to be held April 1996); results - percent of vote by party NA; seats - (270 seats total) number of seats by party NA.

Government Expenditures [21]

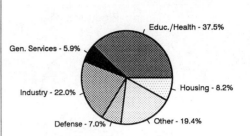

% distribution for 1993. Expend. in FY90 est.: 80 ($ bil.)

Military Expenditures and Arms Transfers [22]

	1985	1986	1987	1988	1989
Military expenditures					
Current dollars (mil.)[e,2]	NA	NA	NA	NA	NA
1989 constant dollars (mil.)[e,2]	NA	NA	NA	NA	NA
Armed forces (000)	345	345[e]	350[e]	654	604
Gross national product (GNP)					
Current dollars (mil.)[e]	71,050	66,780	67,880	71,340	77,540
1989 constant dollars (mil.)[e]	80,880	74,110	73,010	74,260	77,540
Central government expenditures (CGE)					
1989 constant dollars (mil.)[e]	13,790	13,510	12,980	NA	NA
People (mil.)	47.1	48.8	50.4	52.1	53.9
Military expenditure as % of GNP	NA	NA	NA	NA	NA
Military expenditure as % of CGE	NA	NA	NA	NA	NA
Military expenditure per capita	NA	NA	NA	NA	NA
Armed forces per 1,000 people	7.3	7.1	6.9	12.6	11.2
GNP per capita	1,716	1,519	1,447	1,425	1,440
Arms imports[6]					
Current dollars (mil.)	1,900	2,600	2,000	2,300	1,300
1989 constant dollars (mil.)	2,163	2,885	2,151	2,394	1,300
Arms exports[6]					
Current dollars (mil.)	0	0	0	0	0
1989 constant dollars (mil.)	0	0	0	0	0
Total imports[7]					
Current dollars (mil.)	11,630	10,520	9,570	9,454	10,500
1989 constant dollars	13,250	11,680	10,290	9,842	10,500
Total exports[7]					
Current dollars (mil.)	13,330	8,322	9,400	9,500	12,300
1989 constant dollars	15,170	9,235	10,110	9,889	12,300
Arms as percent of total imports[8]	16.3	24.7	20.9	24.3	12.4
Arms as percent of total exports[8]	0	0	0	0	0

Human Rights [24]

	SSTS	FL	FAPRO	PPCG	APROBC	TPW	PCPTW	STPEP	PHRFF	PRW	ASST	AFL
Observes		P		P		P	P	S			P	P
	EAFRD	CPR	ESCR	SR	ACHR	MAAE	PVIAC	PVNAC	EAFDAW	TCIDTP	RC
Observes		P	P	P	P			S	S			S

P = Party; S = Signatory; see Appendix for meaning of abbreviations.

Labor Force

Total Labor Force [25]

15.4 million

Labor Force by Occupation [26]

Agriculture	33%
Manufacturing	21

Unemployment Rate [27]

30% (1991 est.)

For sources, notes, and explanations, see Annotated Source Appendix, page 1035.

431

Production Sectors

Commercial Energy Production and Consumption

Production [28]

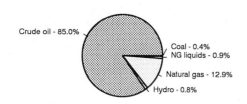

Crude oil - 85.0%
Coal - 0.4%
NG liquids - 0.9%
Natural gas - 12.9%
Hydro - 0.8%

Consumption [29]

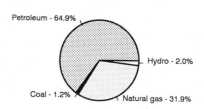

Petroleum - 64.9%
Hydro - 2.0%
Natural gas - 31.9%
Coal - 1.2%

Data are shown in quadrillion (10^{15}) BTUs and percent for 1991

Crude oil[1]	7.17
Natural gas liquids	0.08
Dry natural gas	1.09
Coal[2]	0.03
Net hydroelectric power[3]	0.07
Net nuclear power[3]	0.00
Total	8.44

Petroleum	2.24
Dry natural gas	1.10
Coal[2]	0.04
Net hydroelectric power[3]	0.07
Net nuclear power[3]	0.00
Total	3.44

Telecommunications [30]

- Microwave radio relay extends throughout country
- System centered in Tehran
- 2,143,000 telephones (35 telephones per 1,000 persons)
- Broadcast stations - 77 AM, 3 FM, 28 TV
- Satellite earth stations - 2 Atlantic Ocean INTELSAT and 1 Indian Ocean INTELSAT
- HF radio and microwave radio relay to Turkey, Pakistan, Syria, Kuwait, Tajikistan, and Uzbekistan
- Submarine fiber optic cable to UAE

Transportation [31]

Railroads. 4,852 km total; 4,760 km 1.432-meter gauge, 92 km 1.676-meter gauge; 480 km under construction from Bafq to Bandar-e Abbas, rail construction from Bafq to Sirjan has been completed and is operational; section from Sirjan to Bandar-e Abbas still under construction

Highways. 140,200 km total; 42,694 km paved surfaces; 46,866 km gravel and crushed stone; 49,440 km improved earth; 1,200 km (est.) rural road network

Merchant Marine. 135 ships (1,000 GRT or over) totaling 4,480,726 GRT/8,332,593 DWT; includes 39 cargo, 6 roll-on/roll-off cargo, 32 oil tanker, 4 chemical tanker, 3 refrigerated cargo, 48 bulk, 2 combination bulk, 1 liquefied gas

Airports

Total:	219
Usable:	194
With permanent-surface runways:	83
With runways over 3,659 m:	16
With runways 2,440-3,659 m:	20
With runways 1,220-2,439 m:	70

Top Agricultural Products [32]

Agriculture accounts for about 20% of GDP; principal products—wheat, rice, other grains, sugar beets, fruits, nuts, cotton, dairy products, wool, caviar; not self-sufficient in food.

Top Mining Products [33]

Estimated metric tons unless otherwise specified   M.t.

Petroleum, crude (mil. barrels)	1,217
Cement, hydraulic (000 tons)	15,000
Gypsum (000 tons)	8,050
Salt	900,000
Quartz and silica	832,441
Dimension stone: marble, travertine, granite	
Blocks and slabs (000 tons)	4,940[17]
Crushed (000 tons)	555[17]

Tourism [34]

	1987	1988	1989	1990	1991
Visitors	136	140	143	298	
Tourists	69	67	89	154	
Tourism receipts	27	25	37	62	
Tourism expenditures	246	69			
Fare receipts	52	55			
Fare expenditures	31	103			

Tourists are in thousands, money in million U.S. dollars.

For sources, notes, and explanations, see Annotated Source Appendix, page 1035.

Manufacturing Sector

GDP and Manufacturing Summary [35]

	1980	1985	1989	1990	% change 1980-1990	% change 1989-1990
GDP (million 1980 $)	98,081	135,805	108,798	134,074	36.7	23.2
GDP per capita (1980 $)	2,521	2,852	2,038	2,441	-3.2	19.8
Manufacturing as % of GDP (current prices)	8.5	8.5	7.8[e]	7.6	-10.6	-2.6
Gross output (million $)	15,871	26,458	70,483[e]	81,835[e]	415.6	16.1
Value added (million $)	8,186	13,336[e]	41,280[e]	44,607[e]	444.9	8.1
Value added (million 1980 $)	8,528	11,581	10,733	10,683	25.3	-0.5
Industrial production index	100	140	115[e]	122	22.0	6.1
Employment (thousands)	470	614	737[e]	741[e]	57.7	0.5

Note: GDP stands for Gross Domestic Product. 'e' stands for estimated value.

Profitability and Productivity

	1980	1985	1989	1990	% change 1980-1990	% change 1989-1990
Intermediate input (%)	48	50[e]	41[e]	45[e]	-6.3	9.8
Wages, salaries, and supplements (%)	29	26[e]	23[e]	24[e]	-17.2	4.3
Gross operating surplus (%)	23	24[e]	36[e]	31[e]	34.8	-13.9
Gross output per worker ($)	33,756	43,070[e]	95,619[e]	108,954[e]	222.8	13.9
Value added per worker ($)	17,411	21,709[e]	56,002[e]	61,839[e]	255.2	10.4
Average wage (incl. benefits) ($)	9,668	11,294	21,523[e]	26,415[e]	173.2	22.7

Profitability is in percent of gross output. Productivity is in U.S. $. 'e' stands for estimated value.

Profitability - 1990

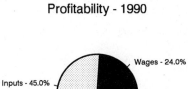

Wages - 24.0%
Inputs - 45.0%
Surplus - 31.0%

The graphic shows percent of gross output.

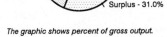

Value Added in Manufacturing

	1980 $ mil.	1980 %	1985 $ mil.	1985 %	1989 $ mil.	1989 %	1990 $ mil.	1990 %	% change 1980-1990	% change 1989-1990
311 Food products	930	11.4	1,259	9.4	3,026[e]	7.3	3,712[e]	8.3	299.1	22.7
313 Beverages	145	1.8	302	2.3	841[e]	2.0	972[e]	2.2	570.3	15.6
314 Tobacco products	190[e]	2.3	103	0.8	3,497[e]	8.5	4,133[e]	9.3	2,075.3	18.2
321 Textiles	1,329	16.2	2,119	15.9	4,872[e]	11.8	5,640[e]	12.6	324.4	15.8
322 Wearing apparel	78	1.0	76	0.6	563[e]	1.4	471[e]	1.1	503.8	-16.3
323 Leather and fur products	36	0.4	67	0.5	349[e]	0.8	379[e]	0.8	952.8	8.6
324 Footwear	100	1.2	165	1.2	434[e]	1.1	540[e]	1.2	440.0	24.4
331 Wood and wood products	68	0.8	120	0.9	501[e]	1.2	587[e]	1.3	763.2	17.2
332 Furniture and fixtures	33	0.4	48	0.4	171[e]	0.4	177[e]	0.4	436.4	3.5
341 Paper and paper products	135	1.6	261	2.0	380[e]	0.9	362[e]	0.8	168.1	-4.7
342 Printing and publishing	80	1.0	97	0.7	438[e]	1.1	517[e]	1.2	546.3	18.0
351 Industrial chemicals	93	1.1	232	1.7	715[e]	1.7	806[e]	1.8	766.7	12.7
352 Other chemical products	278	3.4	606	4.5	1,956[e]	4.7	2,037[e]	4.6	632.7	4.1
353 Petroleum refineries	1,652	20.2	1,977[e]	14.8	9,311[e]	22.6	6,444[e]	14.4	290.1	-30.8
354 Miscellaneous petroleum and coal products	2	0.0	32	0.2	184[e]	0.4	245[e]	0.5	12,150.0	33.2
355 Rubber products	93	1.1	180	1.3	807[e]	2.0	841[e]	1.9	804.3	4.2
356 Plastic products	198	2.4	235	1.8	558[e]	1.4	691[e]	1.5	249.0	23.8
361 Pottery, china and earthenware	45	0.5	76	0.6	134[e]	0.3	191[e]	0.4	324.4	42.5
362 Glass and glass products	115	1.4	167	1.3	331[e]	0.8	485[e]	1.1	321.7	46.5
369 Other non-metal mineral products	819	10.0	1,368	10.3	4,891[e]	11.8	6,030[e]	13.5	636.3	23.3
371 Iron and steel	367	4.5	713	5.3	1,517[e]	3.7	2,056[e]	4.6	460.2	35.5
372 Non-ferrous metals	48	0.6	191	1.4	712[e]	1.7	742[e]	1.7	1,445.8	4.2
381 Metal products	319	3.9	555	4.2	1,159[e]	2.8	1,551[e]	3.5	386.2	33.8
382 Non-electrical machinery	208	2.5	632	4.7	2,130[e]	5.2	2,584[e]	5.8	1,142.3	21.3
383 Electrical machinery	391	4.8	749	5.6	634[e]	1.5	946[e]	2.1	141.9	49.2
384 Transport equipment	399	4.9	927	7.0	996[e]	2.4	1,270[e]	2.8	218.3	27.5
385 Professional and scientific equipment	24	0.3	55	0.4	81[e]	0.2	98[e]	0.2	308.3	21.0
390 Other manufacturing industries	11	0.1	26	0.2	92[e]	0.2	103[e]	0.2	836.4	12.0

Note: The industry codes shown are International Standard Industry codes (ISIC). Percentages are percent of total Value Added. 'e' stands for estimated value

Finance, Economics, and Trade

Economic Indicators [36]

Millions of Iranian Rials (IR) unless otherwise stated.

	1989	1990	1991
Real GDP	NA	NA	NA
Money supply (M1) (bil. rials)	6.14	NA	NA
Commercial interest rates	NA	19[72]	NA
Savings rate	NA	NA	NA
Investment rate	NA	NA	NA
CPI	NA	NA	NA
WPI	NA	NA	NA
External public debt	4,300[73]	NA	NA

Balance of Payments Summary [37]

Values in millions of dollars.

	1980	1985	1986	1987	1988
Exports of goods (f.o.b.)	12,338.0	14,175.0	7,171.0	11,916.0	10,709.0
Imports of goods (f.o.b.)	-10,888.0	-12,006.0	-10,585.0	-12,005.0	-10,608.0
Trade balance	1,450.0	2,169.0	-3,414.0	-89.0	101.0
Services - debits	-5,621.0	-3,408.0	-2,348.0	-2,438.0	-2,436.0
Services - credits	1,735.0	763.0	607.0	437.0	467.0
Private transfers (net)	-	-	-	-	-
Government transfers (net)	-2.0	-	-	-	-
Long term capital (net)	-5,261.0	-160.0	802.0	719.0	-41.0
Short term capital (net)	-2,977.0	704.0	2,325.0	992.0	476.0
Errors and omissions	829.0	487.0	814.0	155.0	421.0
Overall balance	-9,847.0	555.0	-1,214.0	-224.0	-1,012.0

Exchange Rates [38]

Currency: **Iranian rials.**
Symbol: **IR.**

Data are currency units per $1.

Official rate	
(March 1993)	1,538.000
January 1993	67.095
1992	65.552
1991	67.505
1990	68.096
1989	72.015
1988	68.683

Imports and Exports

Top Import Origins [39]

$21.0 billion (c.i.f., FY91 est.).

Origins	%
Germany	NA
Japan	NA
Italy	NA
UK	NA
France	NA

Top Export Destinations [40]

$17.2 billion (f.o.b., FY91 est.).

Destinations	%
Japan	NA
Italy	NA
France	NA
Netherlands	NA
Belgium/Luxembourg	NA
Spain	NA
Germany	NA

Foreign Aid [41]

Note - aid fell sharply following the 1979 revolution.

	U.S. $	
US commitments, including Ex-Im (FY70-80)	1.0	billion
Western (non-US) countries, ODA and OOF bilateral commitments (1970-89)	1.675	billion
Communist countries (1970-89)	976	million

Import and Export Commodities [42]

Import Commodities	**Export Commodities**
Machinery	Petroleum 90%
Military supplies	Carpets
Metal works	Fruits
Foodstuffs	Nuts
Pharmaceuticals	Hides
Technical services	
Refined oil products	

For sources, notes, and explanations, see Annotated Source Appendix, page 1035.

Iraq

Geography [1]

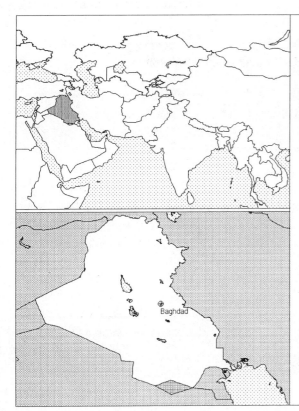

Total area:
437,072 km2
Land area:
432,162 km2
Comparative area:
Slightly more than twice the size of Idaho
Land boundaries:
Total 3,631 km, Iran 1,458 km, Jordan 181 km, Kuwait 242 km, Saudi Arabia 814 km, Syria 605 km, Turkey 331 km
Coastline:
58 km
Climate:
Mostly desert; mild to cool winters with dry, hot, cloudless summers; northernmost regions along Iranian and Turkish borders experience cold winters with occasionally heavy snows
Terrain:
Mostly broad plains; reedy marshes in southeast; mountains along borders with Iran and Turkey
Natural resources:
Petroleum, natural gas, phosphates, sulfur
Land use:
Arable land: 12%
Permanent crops: 1%
Meadows and pastures: 9%
Forest and woodland: 3%
Other: 75%

Demographics [2]

	1960	1970	1980	1990	1991[1]	1994	2000	2010	2020
Population	6,822	9,414	13,233	18,425	19,525	19,890	24,731	34,545	46,260
Population density (persons per sq. mi.)	41	56	79	112	117	NA	162	227	304
Births	NA	NA	NA	NA	893	877	NA	NA	NA
Deaths	NA	NA	NA	NA	140	144	NA	NA	NA
Life expectancy - males	NA	NA	NA	NA	66	65	NA	NA	NA
Life expectancy - females	NA	NA	NA	NA	68	67	NA	NA	NA
Birth rate (per 1,000)	NA	NA	NA	NA	46	44	NA	NA	NA
Death rate (per 1,000)	NA	NA	NA	NA	7	7	NA	NA	NA
Women of reproductive age (15-44 yrs.)	NA	NA	NA	3,915	NA	4,302	5,496	NA	NA
of which are currently married	NA	NA	NA	2,665	NA	2,930	3,786	NA	NA
Fertility rate	NA	NA	NA	7.3	7.16	6.7	5.8	4.8	3.9

Population values are in thousands, life expectancy in years, and other items as indicated.

Health

Health Personnel [3]

Doctors per 1,000 pop., 1988-92	0.58
Nurse-to-doctor ratio, 1988-92	1.2
Hospital beds per 1,000 pop., 1985-90	1.6
Percentage of children immunized (age 1 yr. or less)	
Third dose of DPT, 1990-91	69.0
Measles, 1990-91	73.0

Health Expenditures [4]

Total health expenditure, 1990 (official exchange rate)	
Millions of dollars	NA
Millions of dollars per capita	NA
Health expenditures as a percentage of GDP	
Total	NA
Public sector	NA
Private sector	NA
Development assistance for health	
Total aid flows (millions of dollars)[1]	4
Aid flows per capita (millions of dollars)	0.2
Aid flows as a percentage of total health expenditure	NA

For sources, notes, and explanations, see Annotated Source Appendix, page 1035.

Human Factors

Health Care Ratios [5]	Infants and Malnutrition [6]

Ethnic Division [7]

Arab	75-80%
Kurdish	15-20%
Turkoman, Assyrian, or other	5%

Religion [8]

Muslim	97%
Shi'a	60-65%
Sunni	32-37%
Christian or other	3%

Major Languages [9]

Arabic	NA
Kurdish (official in Kurdish regions)	NA
Assyrian	NA
Armenian	NA

Education

Public Education Expenditures [10]

Million Dinar	1980	1985	1986	1987	1988	1989
Total education expenditure	418	551	555	560	690	NA
as percent of GNP	NA	NA	NA	NA	NA	NA
as percent of total govt. expend.	NA	6.5	6.5	6.4	NA	NA
Current education expenditure	NA	NA	NA	NA	625	NA
as percent of GNP	NA	NA	NA	NA	NA	NA
as percent of current govt. expend.	NA	NA	NA	NA	NA	NA
Capital expenditure	NA	NA	NA	NA	65	NA

Educational Attainment [11]

Literacy Rate [12]

In thousands and percent	1985[a]	1991[a]	2000[a]
Illiterate population +15 years	4,014	4,078	4,012
Illiteracy rate - total pop. (%)	47.6	40.3	27.2
Illiteracy rate - males (%)	36.2	30.2	20.0
Illiteracy rate - females (%)	59.3	50.7	34.9

Libraries [13]

	Admin. Units	Svc. Pts.	Vols. (000)	Shelving (meters)	Vols. Added	Reg. Users
National (1989)	1	3	NA	NA	NA	NA
Non-specialized	NA	NA	NA	NA	NA	NA
Public	NA	NA	NA	NA	NA	NA
Higher ed. (1988)	106	106	2,745	NA	NA	807,942
School	NA	NA	NA	NA	NA	NA

Daily Newspapers [14]

	1975	1980	1985	1990
Number of papers	7	5	6	6
Circ. (000)	230[e]	340[e]	600[e]	650[e]

Cinema [15]

Science and Technology

Scientific/Technical Forces [16]	R&D Expenditures [17]	U.S. Patents Issued [18]

Government and Law

Organization of Government [19]

Long-form name:
Republic of Iraq
Type:
republic
Independence:
3 October 1932 (from League of Nations mandate under British administration)
Constitution:
22 September 1968, effective 16 July 1970 (interim Constitution); new constitution drafted in 1990 but not adopted
Legal system:
based on Islamic law in special religious courts, civil law system elsewhere; has not accepted compulsory ICJ jurisdiction
National holiday:
Anniversary of the Revolution, 17 July (1968)
Executive branch:
president, vice president, chairman of the Revolutionary Command Council, vice chairman of the Revolutionary Command Council, prime minister, first deputy prime minister, Council of Ministers
Legislative branch:
unicameral National Assembly (Majlis al-Watani)
Judicial branch:
Court of Cassation

Crime [23]

Political Parties [20]

National Assembly	% of votes
Sunni Arabs	53.0
Shi'a Arabs	30.0
Kurds	15.0
Christians (est.)	2.0

Government Budget [21]

For FY90 est.

Revenues	NA
Expenditures	NA
Capital expenditures	NA

Military Expenditures and Arms Transfers [22]

	1985	1986	1987	1988	1989
Military expenditures					
Current dollars (mil.) [e,2]	NA	NA	NA	NA	NA
1989 constant dollars (mil.) [e,2]	NA	NA	NA	NA	NA
Armed forces (000)	788	800	900	1,000	1,000
Gross national product (GNP)					
Current dollars (mil.)	41,250[e]	34,930[e]	35,780[e]	39,630[e]	NA
1989 constant dollars (mil.)[e]	46,960	38,760	38,490	41,250	NA
Central government expenditures (CGE)					
1989 constant dollars (mil.)	NA	NA	NA	NA	NA
People (mil.)	15.7	16.2	16.8	17.4	18.1
Military expenditure as % of GNP	NA	NA	NA	NA	NA
Military expenditure as % of CGE[5]	NA	NA	NA	NA	NA
Military expenditure per capita	NA	NA	NA	NA	NA
Armed forces per 1,000 people	50.2	49.2	53.5	57.4	55.3
GNP per capita	2,992	2,386	2,287	2,367	NA
Arms imports[6]					
Current dollars (mil.)	4,600	5,700	5,400	4,900	1,900
1989 constant dollars (mil.)	5,237	6,325	5,808	5,101	1,900
Arms exports[6]					
Current dollars (mil.)	10	10	10	90	60
1989 constant dollars (mil.)	11	11	11	94	60
Total imports[7]					
Current dollars (mil.)	10,560	10,190	7,415	10,270	12,000
1989 constant dollars	12,020	11,310	7,975	10,690	12,000
Total exports[7]					
Current dollars (mil.)	12,220	9,007	9,014	12,500	16,000
1989 constant dollars	13,920	9,995	9,695	13,010	16,000
Arms as percent of total imports[8]	43.6	55.9	72.8	47.7	15.8
Arms as percent of total exports[8]	0.1	0.1	0.1	0.7	0.4

Human Rights [24]

	SSTS	FL	FAPRO	PPCG	APROBC	TPW	PCPTW	STPEP	PHRFF	PRW	ASST	AFL
Observes	P	P		P	P	P	P	P			P	P
	EAFRD	CPR	ESCR	SR	ACHR	MAAE	PVIAC	PVNAC	EAFDAW	TCIDTP	RC
Observes		P	P	P		P			P		P

P = Party; S = Signatory; see Appendix for meaning of abbreviations.

Labor Force

Total Labor Force [25]

4.4 million (1989)

Labor Force by Occupation [26]

Services	48%
Agriculture	30
Industry	22

Unemployment Rate [27]

less than 5% (1989 est.)

For sources, notes, and explanations, see Annotated Source Appendix, page 1035.

437

Production Sectors

Commercial Energy Production and Consumption

Production [28]

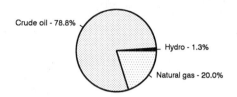

Crude oil - 78.8%

Hydro - 1.3%

Natural gas - 20.0%

Consumption [29]

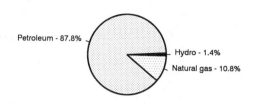

Petroleum - 87.8%

Hydro - 1.4%

Natural gas - 10.8%

Data are shown in quadrillion (10^{15}) BTUs and percent for 1991

Crude oil[1]	0.63
Natural gas liquids	0.00
Dry natural gas	0.16
Coal[2]	0.00
Net hydroelectric power[3]	0.01
Net nuclear power[3]	0.00
Total	0.80

Petroleum	0.65
Dry natural gas	0.08
Coal[2]	(s)
Net hydroelectric power[3]	0.01
Net nuclear power[3]	0.00
Total	0.74

Telecommunications [30]

- Reconstitution of damaged telecommunication facilities began after Desert Storm, most damaged facilities have been rebuilt
- The network consists of coaxial cables and microwave radio relay links
- 632,000 telephones
- Broadcast stations - 16 AM, 1 FM, 13 TV
- Satellite earth stations - 1 Atlantic Ocean INTELSAT, 1 Indian Ocean INTELSAT, 1 Atlantic Ocean GORIZONT in the Intersputnik system and 1 ARABSAT
- Coaxial cable and microwave radio relay to Jordan, Kuwait, Syria, and Turkey, Kuwait line is probably non-operational

Top Agricultural Products [32]

Agriculture accounts for 11% of GNP and 30% of labor force; principal products—wheat, barley, rice, vegetables, dates, other fruit, cotton, wool; livestock—cattle, sheep; not self-sufficient in food output.

Transportation [31]

Railroads. 2,457 km 1.435-meter standard gauge

Highways. 34,700 km total; 17,500 km paved, 5,500 km improved earth, 11,700 km unimproved earth

Merchant Marine. 41 ships (1,000 GRT or over) totaling 930,780 GRT/1,674,878 DWT; includes 1 passenger, 1 passenger-cargo, 15 cargo, 1 refrigerated cargo, 3 roll-on/roll-off cargo, 19 oil tanker, 1 chemical tanker; note - none of the Iraqi flag merchant fleet was trading internationally as of 1 January 1993

Airports

Total:	114
Usable:	99
With permanent-surface runways:	74
With runways over 3,659 m:	9
With runways 2,440-3,659 m:	52
With runways 1,220-2,439 m:	12

Top Mining Products [33]

Estimated metric tons unless otherwise specified	M.t.
Petroleum, crude (000 barrels)	100,000
Cement, hydraulic (000 tons)	5,000
Phosphate rock (000 tons)	400[18]
Salt (000 tons)	300
Gypsum (000 tons)	190[19]

Tourism [34]

	1987	1988	1989	1990	1991
Visitors	739	1,209	1,025	748	268
Tourism receipts	62	61	59		

Tourists are in thousands, money in million U.S. dollars.

For sources, notes, and explanations, see Annotated Source Appendix, page 1035.

Manufacturing Sector

GDP and Manufacturing Summary [35]

	1980	1985	1989	1990	% change 1980-1990	% change 1989-1990
GDP (million 1980 $)	52,749	39,316	16,272	35,145	-33.4	116.0
GDP per capita (1980 $)	3,969	2,473	890	1,857	-53.2	108.7
Manufacturing as % of GDP (current prices)	4.5	9.2	16.6[e]	NA	NA	-100.0
Gross output (million $)	5,393[e]	7,162	11,771[e]	7,056[e]	30.8	-40.1
Value added (million $)	2,068[e]	3,676	6,119[e]	3,807[e]	84.1	-37.8
Value added (million 1980 $)	2,363	2,520	1,584[e]	1,838	-22.2	16.0
Industrial production index	100	107	126[e]	118[e]	18.0	-6.3
Employment (thousands)	177	174	195[e]	169[e]	-4.5	-13.3

Note: GDP stands for Gross Domestic Product. 'e' stands for estimated value.

Profitability and Productivity

	1980	1985	1989	1990	% change 1980-1990	% change 1989-1990
Intermediate input (%)	62[e]	49	48[e]	46[e]	-25.8	-4.2
Wages, salaries, and supplements (%)	12[e]	13	13[e]	13[e]	8.3	0.0
Gross operating surplus (%)	26[e]	39	39[e]	41[e]	57.7	5.1
Gross output per worker ($)	30,443[e]	41,090	60,482[e]	41,407[e]	36.0	-31.5
Value added per worker ($)	11,673[e]	21,088	31,439[e]	22,386[e]	91.8	-28.8
Average wage (incl. benefits) ($)	3,700	5,242	8,126[e]	5,390[e]	45.7	-33.7

Profitability is in percent of gross output. Productivity is in U.S. $. 'e' stands for estimated value.

Profitability - 1990

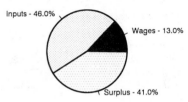

Inputs - 46.0%
Wages - 13.0%
Surplus - 41.0%

The graphic shows percent of gross output.

Value Added in Manufacturing

	1980 $ mil.	1980 %	1985 $ mil.	1985 %	1989 $ mil.	1989 %	1990 $ mil.	1990 %	% change 1980-1990	% change 1989-1990
311 Food products	225[e]	10.9	396	10.8	530[e]	8.7	288[e]	7.6	28.0	-45.7
313 Beverages	74[e]	3.6	125	3.4	162[e]	2.6	112[e]	2.9	51.4	-30.9
314 Tobacco products	105[e]	5.1	140	3.8	226[e]	3.7	124[e]	3.3	18.1	-45.1
321 Textiles	230[e]	11.1	248	6.7	376[e]	6.1	219[e]	5.8	-4.8	-41.8
322 Wearing apparel	30[e]	1.5	53	1.4	74[e]	1.2	45[e]	1.2	50.0	-39.2
323 Leather and fur products	24[e]	1.2	1	0.0	2[e]	0.0	1[e]	0.0	-95.8	-50.0
324 Footwear	19[e]	0.9	81	2.2	88[e]	1.4	49[e]	1.3	157.9	-44.3
331 Wood and wood products	1[e]	0.0	1	0.0	2[e]	0.0	2[e]	0.1	100.0	0.0
332 Furniture and fixtures	10[e]	0.5	13	0.4	30[e]	0.5	14[e]	0.4	40.0	-53.3
341 Paper and paper products	48[e]	2.3	52	1.4	155[e]	2.5	79[e]	2.1	64.6	-49.0
342 Printing and publishing	27[e]	1.3	33	0.9	88[e]	1.4	42[e]	1.1	55.6	-52.3
351 Industrial chemicals	78[e]	3.8	151	4.1	291[e]	4.8	153[e]	4.0	96.2	-47.4
352 Other chemical products	192[e]	9.3	389	10.6	876[e]	14.3	370[e]	9.7	92.7	-57.8
353 Petroleum refineries	392[e]	19.0	868	23.6	1,386[e]	22.7	1,185[e]	31.1	202.3	-14.5
354 Miscellaneous petroleum and coal products	27[e]	1.3	40	1.1	91[e]	1.5	41[e]	1.1	51.9	-54.9
355 Rubber products	6[e]	0.3	10	0.3	21[e]	0.3	11[e]	0.3	83.3	-47.6
356 Plastic products	11[e]	0.5	33	0.9	57[e]	0.9	38[e]	1.0	245.5	-33.3
361 Pottery, china and earthenware	1[e]	0.0	1	0.0	2[e]	0.0	1[e]	0.0	0.0	-50.0
362 Glass and glass products	21[e]	1.0	35	1.0	51[e]	0.8	31[e]	0.8	47.6	-39.2
369 Other non-metal mineral products	190[e]	9.2	565	15.4	828[e]	13.5	587[e]	15.4	208.9	-29.1
371 Iron and steel	5[e]	0.2	20[e]	0.5	25[e]	0.4	19[e]	0.5	280.0	-24.0
372 Non-ferrous metals	NA	0.0	NA	0.0	NA	0.0	NA	0.0	NA	NA
381 Metal products	55[e]	2.7	47	1.3	105[e]	1.7	75[e]	2.0	36.4	-28.6
382 Non-electrical machinery	160[e]	7.7	149	4.1	211[e]	3.4	129[e]	3.4	-19.4	-38.9
383 Electrical machinery	122[e]	5.9	185	5.0	340[e]	5.6	150[e]	3.9	23.0	-55.9
384 Transport equipment	12[e]	0.6	40	1.1	103[e]	1.7	42[e]	1.1	250.0	-59.2
385 Professional and scientific equipment	1[e]	0.0	NA	0.0	NA	0.0	NA	0.0	NA	NA
390 Other manufacturing industries	1[e]	0.0	NA	0.0	NA	0.0	NA	0.0	NA	NA

Note: The industry codes shown are International Standard Industry codes (ISIC). Percentages are percent of total Value Added. 'e' stands for estimated value

For sources, notes, and explanations, see Annotated Source Appendix, page 1035.

439

Finance, Economics, and Trade

Economic Indicators [36]

National product: GNP—exchange rate conversion—$35 billion (1989 est.). **National product real growth rate**: 10% (1989 est.). **National product per capita**: $1,940 (1989 est.). **Inflation rate (consumer prices)**: 200% (1992 est.). **External debt**: $45 billion (1989 est.), excluding debt of about $35 billion owed to Arab.

Balance of Payments Summary [37]

Values in millions of dollars.

	1970	1973	1975
Exports of goods (f.o.b.)	1,098.0	2,204.0	8,301.0
Imports of goods (f.o.b.)	-459.0	-849.0	-4,162.0
Trade balance	639.0	1,355.0	4,139.0
Services - debits	-679.0	-800.0	-1,712.0
Services - credits	143.0	256.0	543.0
Private transfers (net)	1.0	1.0	1.0
Government transfers (net)	1.0	-11.0	-266.0
Direct investments	24.0	297.0	11.0
Errors and omissions	-127.0	-254.0	-726.0
Overall balance	-6.0	663.0	-498.0

Exchange Rates [38]

Currency: **Iraqi dinars.**
Symbol: **ID.**

Data are currency units per $1.

Fixed official rate since 1982	3.20
Black market rate (April 1993)	53.5

Imports and Exports

Top Import Origins [39]

$6.6 billion (c.i.f., 1990).

Origins	%
Germany	NA
US	NA
Turkey	NA
France	NA
UK	NA

Top Export Destinations [40]

$10.4 billion (f.o.b., 1990).

Destinations	%
US	NA
Brazil	NA
Turkey	NA
Japan	NA
Netherlands	NA
Spain	NA

Foreign Aid [41]

	U.S. $	
US commitments, including Ex-Im (FY70-80)	3	million
Western (non-US) countries, ODA and OOF bilateral commitments (1970-89)	647	million
Communist countries (1970-89)	3.9	billion

Import and Export Commodities [42]

Import Commodities
Manufactures
Food

Export Commodities
Crude oil and refined products
Fertilizer
Sulfur

For sources, notes, and explanations, see Annotated Source Appendix, page 1035.

Ireland

Geography [1]

Total area:
70,280 km2
Land area:
68,890 km2
Comparative area:
Slightly larger than West Virginia
Land boundaries:
Total 360 km, UK 360 km
Coastline:
1,448 km
Climate:
Temperate maritime; modified by North Atlantic Current; mild winters, cool summers; consistently humid; overcast about half the time
Terrain:
Mostly level to rolling interior plain surrounded by rugged hills and low mountains; sea cliffs on west coast
Natural resources:
Zinc, lead, natural gas, petroleum, barite, copper, gypsum, limestone, dolomite, peat, silver
Land use:
Arable land: 14%
Permanent crops: 0%
Meadows and pastures: 71%
Forest and woodland: 5%
Other: 10%

Demographics [2]

	1960	1970	1980	1990	1991[1]	1994	2000	2010	2020
Population	2,832	2,950	3,401	3,508	3,489	3,539	3,627	3,846	4,034
Population density (persons per sq. mi.)	106	111	128	132	131	NA	132	139	144
Births	NA	NA	NA	NA	51	50	NA	NA	NA
Deaths	NA	NA	NA	NA	31	30	NA	NA	NA
Life expectancy - males	NA	NA	NA	NA	73	73	NA	NA	NA
Life expectancy - females	NA	NA	NA	NA	79	79	NA	NA	NA
Birth rate (per 1,000)	NA	NA	NA	NA	15	14	NA	NA	NA
Death rate (per 1,000)	NA	NA	NA	NA	9	9	NA	NA	NA
Women of reproductive age (15-44 yrs.)	NA	NA	NA	851	NA	889	940	NA	NA
of which are currently married	NA	NA	NA	471	NA	494	523	NA	NA
Fertility rate	NA	NA	NA	2.1	2.06	2.0	1.8	1.8	1.8

Population values are in thousands, life expectancy in years, and other items as indicated.

Health

Health Personnel [3]

Doctors per 1,000 pop., 1988-92	1.58
Nurse-to-doctor ratio, 1988-92	4.7
Hospital beds per 1,000 pop., 1985-90	3.9
Percentage of children immunized (age 1 yr. or less)	
Third dose of DPT, 1990-91	65.0
Measles, 1990-91	78.0

Health Expenditures [4]

Total health expenditure, 1990 (official exchange rate)	
Millions of dollars	3068
Millions of dollars per capita	876
Health expenditures as a percentage of GDP	
Total	7.1
Public sector	5.8
Private sector	1.4
Development assistance for health	
Total aid flows (millions of dollars)[1]	NA
Aid flows per capita (millions of dollars)	NA
Aid flows as a percentage of total health expenditure	NA

Dublin

For sources, notes, and explanations, see Annotated Source Appendix, page 1035.

441

Human Factors

Health Care Ratios [5]

Population per physician, 1970	980
Population per physician, 1990	630
Population per nursing person, 1970	160
Population per nursing person, 1990	NA
Percent of births attended by health staff, 1985	NA

Infants and Malnutrition [6]

Percent of babies with low birth weight, 1985	4
Infant mortality rate per 1,000 live births, 1970	20
Infant mortality rate per 1,000 live births, 1991	8
Years of life lost per 1,000 population, 1990	11
Prevalence of malnutrition (under age 5), 1990	NA

Ethnic Division [7]

Celtic	NA
English	NA

Religion [8]

Roman Catholic	93%
Anglican	3%
None	1%
Unknown	2%
Other	1%

Major Languages [9]

Irish (Gaelic), spoken mainly in areas located along the western seaboard, English is the language generally used.

Education

Public Education Expenditures [10]

Million Irish Pounds	1980	1985	1987	1988	1989	1990
Total education expenditure	595	1,058	1,263	1,245	1,305	1,372
as percent of GNP	6.6	6.7	6.9	6.5	6.1	6.0
as percent of total govt. expend.	NA	8.9	9.5	8.4	11.2	10.2
Current education expenditure	515	963	1,169	1,184	1,249	1,303
as percent of GNP	5.7	6.1	6.4	6.1	5.9	5.7
as percent of current govt. expend.	NA	10.5	11.2	11.3	12.6	12.1
Capital expenditure	80	95	94	61	56	68

Educational Attainment [11]

Age group	25+
Total population	1,793,855
Highest level attained (%)	
No schooling	52.3
First level	
Incompleted	NA
Completed	NA
Entered second level	
S-1	39.8
S-2	NA
Post secondary	7.9

Literacy Rate [12]

Libraries [13]

	Admin. Units	Svc. Pts.	Vols. (000)	Shelving (meters)	Vols. Added	Reg. Users
National (1986)	1	2	808	14,530	8,000	-
Non-specialized	NA	NA	NA	NA	NA	NA
Public (1989)[34]	31	356	11,259	NA	NA	683,133
Higher ed.	NA	NA	NA	NA	NA	NA
School	NA	NA	NA	NA	NA	NA

Daily Newspapers [14]

	1975	1980	1985	1990
Number of papers	7	7	7	7
Circ. (000)	693	779	685	591

Cinema [15]

Science and Technology

Scientific/Technical Forces [16]

Potential scientists/engineers	139,685
Number female	48,752
Potential technicians[5]	692,105
Number female	237,650
Total[5]	831,790

R&D Expenditures [17]

	Pound (000) 1988
Total expenditure	185,800
Capital expenditure	NA
Current expenditure	NA
Percent current	NA

U.S. Patents Issued [18]

Values show patents issued to citizens of the country by the U.S. Patents Office.

	1990	1991	1992
Number of patents	61	60	61

Government and Law

Organization of Government [19]

Long-form name:
none
Type:
republic
Independence:
6 December 1921 (from UK)
Constitution:
29 December 1937; adopted 1937
Legal system:
based on English common law, substantially modified by indigenous concepts; judicial review of legislative acts in Supreme Court; has not accepted compulsory ICJ jurisdiction
National holiday:
Saint Patrick's Day, 17 March
Executive branch:
president, prime minister, deputy prime minister, Cabinet
Legislative branch:
bicameral Parliament (Oireachtas) consists of an upper house or Senate (Seanad Eireann) and a lower house or House of Representatives (Dail Eireann)
Judicial branch:
Supreme Court

Political Parties [20]

House of Representatives	% of votes
Fianna Fail	39.1
Fine Gael	24.5
Labor Party	19.3
Progressive Democrats	4.7
Democratic Left	2.8
Sinn Fein	1.6
Workers' Party	0.7
Independents	5.9

Government Expenditures [21]

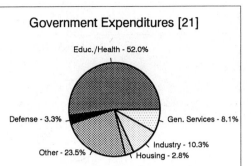

Educ./Health - 52.0%
Defense - 3.3%
Gen. Services - 8.1%
Industry - 10.3%
Housing - 2.8%
Other - 23.5%

% distribution for 1990. Expend. in 1992 est.: 16.6 ($ bil.)

Crime [23]

Crime volume
Cases known to police	87,658
Attempts (percent)	NA
Percent cases solved	33.1
Crimes per 100,000 persons	2,475.76
Persons responsible for offenses	
Total number offenders	NA
Percent female	NA
Percent juvenile	NA
Percent foreigners	NA

Military Expenditures and Arms Transfers [22]

	1985	1986	1987	1988	1989
Military expenditures					
Current dollars (mil.)	419	426	415	431	449
1989 constant dollars (mil.)	477	473	446	448	449
Armed forces (000)	14	14	14	13	13
Gross national product (GNP)					
Current dollars (mil.)	23,110	23,430	25,530	26,700	28,900
1989 constant dollars (mil.)	26,310	26,010	27,460	27,790	28,900
Central government expenditures (CGE)					
1989 constant dollars (mil.)	16,330	15,920	16,240	13,020	12,230
People (mil.)	3.5	3.5	3.5	3.5	3.5
Military expenditure as % of GNP	1.8	1.8	1.6	1.6	1.6
Military expenditure as % of CGE	2.9	3.0	2.7	3.4	3.7
Military expenditure per capita	135	133	126	127	128
Armed forces per 1,000 people	4.0	3.9	4.0	3.7	3.7
GNP per capita	7,431	7,345	7,757	7,874	8,224
Arms imports[6]					
Current dollars (mil.)	10	40	20	20	5
1989 constant dollars (mil.)	11	44	22	21	5
Arms exports[6]					
Current dollars (mil.)	0	0	0	0	0
1989 constant dollars (mil.)	0	0	0	0	0
Total imports[7]					
Current dollars (mil.)	10,020	11,620	13,640	15,570	17,420
1989 constant dollars	11,410	12,900	14,670	16,200	17,420
Total exports[7]					
Current dollars (mil.)	10,360	12,660	16,000	18,720	20,670
1989 constant dollars	11,790	14,050	17,210	19,490	20,670
Arms as percent of total imports[8]	0.1	0.3	0.1	0.1	0
Arms as percent of total exports[8]	0	0	0	0	0

Human Rights [24]

	SSTS	FL	FAPRO	PPCG	APROBC	TPW	PCPTW	STPEP	PHRFF	PRW	ASST	AFL
Observes	P	P	P	P	P	P	P		P	P	P	P
	EAFRD	CPR	ESCR	SR	ACHR	MAAE	PVIAC	PVNAC	EAFDAW	TCIDTP	RC	
Observes	S	P	P	P		P	S	S	P	S	S	

P = Party; S = Signatory; see Appendix for meaning of abbreviations.

Labor Force

Total Labor Force [25]

1.37 million

Labor Force by Occupation [26]

Services	57.0%
Manufacturing and construction	28
Agriculture, forestry, and fishing	13.5
Energy and mining	1.5

Date of data: 1992

Unemployment Rate [27]

22.7% (1992)

Production Sectors

Commercial Energy Production and Consumption

Production [28]

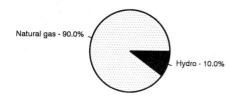

Natural gas - 90.0%

Hydro - 10.0%

Consumption [29]

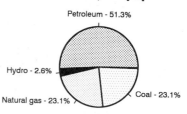

Petroleum - 51.3%

Hydro - 2.6%

Natural gas - 23.1%

Coal - 23.1%

Data are shown in quadrillion (10^{15}) BTUs and percent for 1991

Crude oil[1]	0.00
Natural gas liquids	0.00
Dry natural gas	0.09
Coal[2]	(s)
Net hydroelectric power[3]	0.01
Net nuclear power[3]	0.00
Total	0.10

Petroleum	0.20
Dry natural gas	0.09
Coal[2]	0.09
Net hydroelectric power[3]	0.01
Net nuclear power[3]	0.00
Total	0.38

Telecommunications [30]

- Modern system using cable and digital microwave circuits
- 900,000 telephones
- Broadcast stations - 9 AM, 45 FM, 86 TV
- 2 coaxial submarine cables
- 1 Atlantic Ocean INTELSAT earth station

Transportation [31]

Railroads. Irish National Railways (CIE) operates 1,947 km 1.602-meter gauge, government owned; 485 km double track; 37 km electrified

Highways. 92,294 km total; 87,422 km paved, 4,872 km gravel or crushed stone

Merchant Marine. 57 ships (1,000 GRT or over) totaling 154,647 GRT/186,432 DWT; includes 4 short-sea passenger, 33 cargo, 2 refrigerated cargo, 4 container, 3 oil tanker, 3 specialized tanker, 3 chemical tanker, 5 bulk

Airports

Total:	40
Usable:	39
With permanent-surface runways:	13
With runways over 3,659 m:	0
With runways 2,440-3,659 m:	2
With runways 1,220-2,439 m:	6

Top Agricultural Products [32]

	89-90[1]	90-91[1]
Sugar beets	1,451	1,480
Barley	1,475	1,328
Potatoes	581	633
Wheat	474	601
Turnips	550	471
Oats	103	104

Values shown are 1,000 metric tons.

Top Mining Products [33]

Metric tons unless otherwise specified	M.t.
Zinc, mine output, Zn content	187,500
Alumina (000 tons)	981
Barite (000 tons)	80
Lead, mine output, Pb content	35,000
Peat (000 tons)	249
Lime	110,000

Tourism [34]

	1987	1988	1989	1990	1991
Visitors	10,032	10,355			
Tourists	2,664	3,007	3,484	3,666	3,535
Excursionists	7,368	7,348			
Tourism receipts[35]	839	997	1,070	1,447	1,511
Tourism expenditures	839	961	989	1,159	1,125
Fare receipts	246	282	328	434	442
Fare expenditures	196	236	237	278	292

Tourists are in thousands, money in million U.S. dollars.

Manufacturing Sector

GDP and Manufacturing Summary [35]

	1980	1985	1989	1990	% change 1980-1990	% change 1989-1990
GDP (million 1980 $)	19,261	21,722	24,521	26,899	39.7	9.7
GDP per capita (1980 $)	5,662	6,114	8,654	7,233	27.7	-16.4
Manufacturing as % of GDP (current prices)	26.3	27.1	25.6[e]	31.6	20.2	23.4
Gross output (million $)	15,905	15,394	28,159	33,081	108.0	17.5
Value added (million $)	5,700	5,809[e]	12,416	14,741	158.6	18.7
Value added (million 1980 $)	4,481	5,535	7,911	7,692	71.7	-2.8
Industrial production index	100	122	169	176	76.0	4.1
Employment (thousands)	225	186	188[e]	194[e]	-13.8	3.2

Note: GDP stands for Gross Domestic Product. 'e' stands for estimated value.

Profitability and Productivity

	1980	1985	1989	1990	% change 1980-1990	% change 1989-1990
Intermediate input (%)	64	62[e]	56	55	-14.1	-1.8
Wages, salaries, and supplements (%)	17[e]	14[e]	12[e]	14[e]	-17.6	16.7
Gross operating surplus (%)	19[e]	24[e]	33[e]	31[e]	63.2	-6.1
Gross output per worker ($)	70,068	82,191	149,465[e]	170,065[e]	142.7	13.8
Value added per worker ($)	25,112	31,015[e]	65,903[e]	75,869	202.1	15.1
Average wage (incl. benefits) ($)	11,894[e]	11,580[e]	17,194[e]	23,878[e]	100.8	38.9

Profitability is in percent of gross output. Productivity is in U.S. $. 'e' stands for estimated value.

Profitability - 1990

Inputs - 55.0%
Wages - 14.0%
Surplus - 31.0%

The graphic shows percent of gross output.

Value Added in Manufacturing

	1980 $ mil.	1980 %	1985 $ mil.	1985 %	1989 $ mil.	1989 %	1990 $ mil.	1990 %	% change 1980-1990	% change 1989-1990
311 Food products	1,264	22.2	1,194	20.6	2,420	19.5	3,065	20.8	142.5	26.7
313 Beverages	325	5.7	331	5.7	660	5.3	756	5.1	132.6	14.5
314 Tobacco products	83	1.5	83	1.4	125	1.0	166	1.1	100.0	32.8
321 Textiles	266	4.7	181	3.1	297	2.4	349	2.4	31.2	17.5
322 Wearing apparel	147	2.6	118	2.0	158	1.3	207	1.4	40.8	31.0
323 Leather and fur products	28	0.5	12	0.2	15[e]	0.1	22	0.1	-21.4	46.7
324 Footwear	42	0.7	22	0.4	21[e]	0.2	19	0.1	-54.8	-9.5
331 Wood and wood products	93	1.6	66	1.1	130[e]	1.0	170	1.2	82.8	30.8
332 Furniture and fixtures	59	1.0	40	0.7	58[e]	0.5	86	0.6	45.8	48.3
341 Paper and paper products	105	1.8	75	1.3	154	1.2	187	1.3	78.1	21.4
342 Printing and publishing	265	4.6	219	3.8	447	3.6	561	3.8	111.7	25.5
351 Industrial chemicals	236	4.1	275[e]	4.7	592[e]	4.8	1,050	7.1	344.9	77.4
352 Other chemical products	536	9.4	575[e]	9.9	1,326[e]	10.7	1,391	9.4	159.5	4.9
353 Petroleum refineries	22	0.4	17[e]	0.3	34t	0.0	31	0.2	40.9	NA
354 Miscellaneous petroleum and coal products	NA	0.0	NA	0.0	-t	0.0	NA	0.0	NA	NA
355 Rubber products	52	0.9	56[e]	1.0	96[e]	0.8	118	0.8	126.9	22.9
356 Plastic products	113	2.0	119[e]	2.0	255[e]	2.1	332	2.3	193.8	30.2
361 Pottery, china and earthenware	28	0.5	13	0.2	31[e]	0.2	28	0.2	0.0	-9.7
362 Glass and glass products	109	1.9	113	1.9	181[e]	1.5	144	1.0	32.1	-20.4
369 Other non-metal mineral products	322	5.6	260	4.5	464[e]	3.7	560	3.8	73.9	20.7
371 Iron and steel	31	0.5	37	0.6	84[e]	0.7	92	0.6	196.8	9.5
372 Non-ferrous metals	15	0.3	8	0.1	21[e]	0.2	10	0.1	-33.3	-52.4
381 Metal products	335	5.9	216	3.7	377	3.0	470	3.2	40.3	24.7
382 Non-electrical machinery	449	7.9	854	14.7	1,932	15.6	2,034	13.8	353.0	5.3
383 Electrical machinery	337	5.9	512	8.8	1,697	13.7	1,843	12.5	446.9	8.6
384 Transport equipment	190	3.3	116	2.0	276	2.2	309	2.1	62.6	12.0
385 Professional and scientific equipment	168	2.9	261	4.5	483	3.9	611	4.1	263.7	26.5
390 Other manufacturing industries	79	1.4	39	0.7	82[e]	0.7	132	0.9	67.1	61.0

Note: The industry codes shown are International Standard Industry codes (ISIC). Percentages are percent of total Value Added. 'e' stands for estimated value

For sources, notes, and explanations, see Annotated Source Appendix, page 1035.

445

Finance, Economics, and Trade

Economic Indicators [36]

Millions of Irish Pounds (IP) unless otherwise noted[74].

	1989	1990	1991[e]
Real GDP (1985 prices)	20,577	22,041	22,280
Real GDP growth rate			
(%, 1985 prices)	3.4	6.50	1.0
Money Supply (M1) year-end	2,936	3,175[75]	3,084
Commercial Interest Rates	11.25	10.75	10.50
Savings Interest Rate	8.25	6.00	5.50
CPI[76]	138.9	143.6[77]	106.4
WPI[78]	108.1	105.1	106.4[79]
Gross External			
Government Debt (year-end)	9,168	8,862	8,750

Balance of Payments Summary [37]

Values in millions of dollars.

	1986	1987	1988	1989	1990
Exports of goods (f.o.b.)	12,365	15,569	18,390	20,355	23,357
Imports of goods (f.o.b.)	-11,224	-12,948	-14,569	-16,355	-19,387
Trade balance	1,141	2,621	3,821	4,000	3,970
Services - debits	-5,612	-6,604	-8,328	-9,241	-11,235
Services - credits	2,498	3,048	3,534	4,021	5,568
Private transfers (net)	-61	-162	-121	-94	-66
Government transfers (net)	1,352	1,471	1,659	1,671	2,683
Long term capital (net)	1,050	827	-261	-1,215	-3,001
Short term capital (net)	734	-198	461	-355	1,096
Errors and omissions	-1,198	-118	-173	276	1,734
Overall balance	-96	885	592	-937	749

Exchange Rates [38]

Currency: **Irish pounds.**
Symbol: **#Ir.**

Data are currency units per $1.

January 1993	0.6118
1992	0.5864
1991	0.6190
1990	0.6030
1989	0.7472
1988	0.6553

Imports and Exports

Top Import Origins [39]

$23.3 billion (c.i.f., 1992).

Origins	%
EC	66
Germany	8
Netherlands	4
US	15

Top Export Destinations [40]

$28.3 billion (f.o.b., 1992) Data are for UK 3.

Destinations	%
EC	75
Germany	13
France	10
US	9

Foreign Aid [41]

	U.S. $	
Donor - ODA commitments (1980-89)	90	million

Import and Export Commodities [42]

Import Commodities	Export Commodities
Food	Chemicals
Animal feed	Data processing equipment
Data processing equipment	Industrial machinery
Petroleum and petroleum products	Live animals
Machinery	Animal products
Textiles	
Clothing	

Israel

Geography [1]

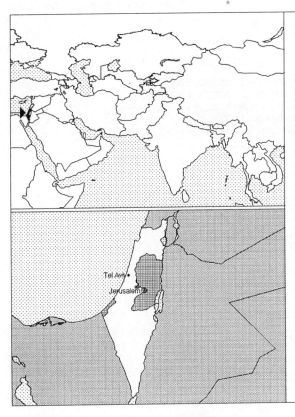

Total area:
  20,770 km2
Land area:
  20,330 km2
Comparative area:
  Slightly larger than New Jersey
Land boundaries:
  Total 1,006 km, Egypt 255 km, Gaza Strip 51 km, Jordan 238 km, Lebanon 79 km, Syria 76 km, West Bank 307 km
Coastline:
  273 km
Climate:
  Temperate; hot and dry in southern and eastern desert areas
Terrain:
  Negev desert in the south; low coastal plain; central mountains; Jordan Rift Valley
Natural resources:
  Copper, phosphates, bromide, potash, clay, sand, sulfur, asphalt, manganese, small amounts of natural gas and crude oil
Land use:
  Arable land: 17%
  Permanent crops: 5%
  Meadows and pastures: 40%
  Forest and woodland: 6%
  Other: 32%

Demographics [2]

	1960	1970	1980	1990	1991[1]	1994	2000	2010	2020
Population	2,141	2,903	3,737	4,303	4,558	5,051	5,507	6,241	6,934
Population density (persons per sq. mi.)	273	370	476	565	581	NA	678	778	873
Births	NA	NA	NA	NA	99	104	NA	NA	NA
Deaths	NA	NA	NA	NA	29	32	NA	NA	NA
Life expectancy - males	NA	NA	NA	NA	75	76	NA	NA	NA
Life expectancy - females	NA	NA	NA	NA	79	80	NA	NA	NA
Birth rate (per 1,000)	NA	NA	NA	NA	22	21	NA	NA	NA
Death rate (per 1,000)	NA	NA	NA	NA	6	6	NA	NA	NA
Women of reproductive age (15-44 yrs.)	NA	NA	NA	1,039	NA	1,257	1,366	NA	NA
of which are currently married	NA	NA	NA	667	NA	816	895	NA	NA
Fertility rate	NA	NA	NA	3.0	2.98	2.8	2.7	2.4	2.2

Population values are in thousands, life expectancy in years, and other items as indicated.

Health

Health Personnel [3]

Doctors per 1,000 pop., 1988-92	2.90
Nurse-to-doctor ratio, 1988-92	2.3
Hospital beds per 1,000 pop., 1985-90	6.3
Percentage of children immunized (age 1 yr. or less)	
Third dose of DPT, 1990-91	88.0
Measles, 1990-91	88.0

Health Expenditures [4]

Total health expenditure, 1990 (official exchange rate)	
Millions of dollars	2301
Millions of dollars per capita	494
Health expenditures as a percentage of GDP	
Total	4.2
Public sector	2.1
Private sector	2.1
Development assistance for health	
Total aid flows (millions of dollars)[1]	3
Aid flows per capita (millions of dollars)	0.6
Aid flows as a percentage of total health expenditure	0.1

For sources, notes, and explanations, see Annotated Source Appendix, page 1035.

447

Human Factors

Health Care Ratios [5]

Population per physician, 1970	410
Population per physician, 1990	NA
Population per nursing person, 1970	NA
Population per nursing person, 1990	NA
Percent of births attended by health staff, 1985	99

Infants and Malnutrition [6]

Percent of babies with low birth weight, 1985	7
Infant mortality rate per 1,000 live births, 1970	25
Infant mortality rate per 1,000 live births, 1991	9
Years of life lost per 1,000 population, 1990	9
Prevalence of malnutrition (under age 5), 1990	NA

Ethnic Division [7]

Jewish	83%
Non-Jewish	17%

Religion [8]

Judaism	82%
Islam (mostly Sunni Muslim)	14%
Christian	2%
Druze and other	2%

Major Languages [9]

Hebrew (official), Arabic used officially for Arab minority, English most commonly used foreign language.

Education

Public Education Expenditures [10]

Million Shekel	1980	1985	1987	1988	1989	1990[7]
Total education expenditure	9	1,876	3,396	4,270	5,168	9,276
as percent of GNP	7.9	6.4	5.7	6.0	6.0	8.9
as percent of total govt. expend.	7.3	8.6	9.2	10.0	10.4	NA
Current education expenditure	8	1,720	3,097	3,926	4,789	NA
as percent of GNP	7.3	5.9	5.2	5.5	5.6	NA
as percent of current govt. expend.	8.9	8.3	9.0	9.9	10.3	NA
Capital expenditure	1	156	299	344	379	NA

Educational Attainment [11]

Age group	25+
Total population	2,003,500
Highest level attained (%)	
No schooling	9.7
First level	
Incompleted	30.6
Completed	NA
Entered second level	
S-1	36.6
S-2	NA
Post secondary	23.1

Literacy Rate [12]

	1972[b]	1983[b,3]	1987[b]
Illiterate population +15 years	238,296	224,080	NA
Illiteracy rate - total pop. (%)	12.1	8.2	3.6
Illiteracy rate - males (%)	7.4	5.0	NA
Illiteracy rate - females (%)	16.7	11.3	NA

Libraries [13]

Daily Newspapers [14]

	1975	1980	1985	1990
Number of papers	23	36	21	30
Circ. (000)	850[e]	1,000[e]	1,100[e]	1,200[e]

Cinema [15]

Science and Technology

Scientific/Technical Forces [16]

Potential scientists/engineers	174,518
Number female	69,500
Potential technicians	174,792
Number female	99,000
Total	349,410

R&D Expenditures [17]

	Shekel[21] (000) 1985
Total expenditure	911,100
Capital expenditure	NA
Current expenditure	NA
Percent current	NA

U.S. Patents Issued [18]

Values show patents issued to citizens of the country by the U.S. Patents Office.

	1990	1991	1992
Number of patents	311	324	350

Government and Law

Organization of Government [19]

Long-form name:
State of Israel
Type:
republic
Independence:
14 May 1948 (from League of Nations mandate under British administration)
Constitution:
no formal constitution
Legal system:
mixture of English common law, British Mandate regulations, and, in personal matters, Jewish, Christian, and Muslim legal systems; does not accept compulsory ICJ jurisdiction
National holiday:
Independence Day, 14 May 1948 (Israel declared independence on 14 May 1948, but the Jewish calendar is lunar and the holiday may occur in April or May)
Executive branch:
president, prime minister, vice prime minister, cabinet
Legislative branch:
unicameral parliament (Knesset)
Judicial branch:
Supreme Court

Political Parties [20]

Knesset	% of seats
Labor Party	36.7
Likud bloc	26.7
Meretz	10.0
Tzomet	6.7
National Religious Party	5.0
Shas	5.0
United Torah Jewry	3.3
Democratic Front	2.5
Moledet	2.5
Arab Democratic Party	1.7

Government Expenditures [21]

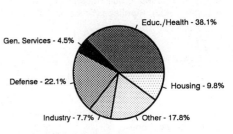

Educ./Health - 38.1%
Gen. Services - 4.5%
Defense - 22.1%
Industry - 7.7%
Other - 17.8%
Housing - 9.8%

% distribution for 1992. Expend. in FY93: 36.8 ($ bil.)

Crime [23]

Crime volume	
Cases known to police	265.498
Attempts (percent)	NA
Percent cases solved	27.6
Crimes per 100,000 persons	5,838.1
Persons responsible for offenses	
Total number offenders	65,107
Percent female	12.7
Percent juvenile[30]	11.3
Percent foreigners	NA

Military Expenditures and Arms Transfers [22]

	1985	1986	1987	1988	1989
Military expenditures					
Current dollars (mil.)	7,027	6,431	5,829	5,585	5,745
1989 constant dollars (mil.)	7,999	7,137	6,269	5,814	5,745
Armed forces (000)	195	180	180	191	191
Gross national product (GNP)					
Current dollars (mil.)	34,370	36,970	40,710	42,940	44,890
1989 constant dollars (mil.)	39,130	41,020	43,780	44,700	44,890
Central government expenditures (CGE)					
1989 constant dollars (mil.)	27,420	25,520	25,880	25,230	22,840
People (mil.)	4.1	4.1	4.2	4.3	4.3
Military expenditure as % of GNP	20.4	17.4	14.3	13.0	12.8
Military expenditure as % of CGE	27.2	28.0	24.2	23.0	25.2
Military expenditure per capita	1,963	1,725	1,491	1,360	1,323
Armed forces per 1,000 people	47.8	43.5	42.8	44.7	44.0
GNP per capita	9,602	9,913	10,410	10,460	10,340
Arms imports[6]					
Current dollars (mil.)	1,000	500	1,800	2,000	725
1989 constant dollars (mil.)	1,138	555	1,936	2,082	725
Arms exports[6]					
Current dollars (mil.)	725	650	725	430	625
1989 constant dollars (mil.)	825	721	780	448	625
Total imports[7]					
Current dollars (mil.)	10,140	10,810	14,360	15,020	14,390
1989 constant dollars	11,540	12,000	15,440	15,640	14,390
Total exports[7]					
Current dollars (mil.)	6,260	7,154	8,454	9,752	10,740
1989 constant dollars	7,126	7,939	9,093	10,150	10,740
Arms as percent of total imports[8]	9.9	4.6	12.5	13.3	5.0
Arms as percent of total exports[8]	11.6	9.1	8.6	4.4	5.8

Human Rights [24]

	SSTS	FL	FAPRO	PPCG	APROBC	TPW	PCPTW	STPEP	PHRFF	PRW	ASST	AFL
Observes	P	P	P	P	P	P	P	P		P	P	P
		EAFRD	CPR	ESCR	SR	ACHR	MAAE	PVIAC	PVNAC	EAFDAW	TCIDTP	RC
Observes		P	P	P	P		P			P	P	P

P = Party; S = Signatory; see Appendix for meaning of abbreviations.

Labor Force

Total Labor Force [25]

1.4 million (1984 est.)

Labor Force by Occupation [26]

Public services	29.3%
Industry, mining, manufacturing	22.8
Commerce	12.8
Finance, business	9.5
Transport, communications, utilities	7.8
Construction and public works	6.5
Services	5.8
Agriculture, forestry, fishing	5.5

Date of data: 1983

Unemployment Rate [27]

11% (1992 est.)

Production Sectors

Energy Resource Summary [28]

Energy Resources: Small amounts of natural gas and crude oil. **Electricity**: 5,835,000 kW capacity; 21,840 million kWh produced, 4,600 kWh per capita (1992). **Pipelines**: Crude oil 708 km; petroleum products 290 km; natural gas 89 km.

Telecommunications [30]

- Most highly developed in the Middle East although not the largest
- Good system of coaxial cable and microwave radio relay
- 1,800,000 telephones
- Broadcast stations - 14 AM, 21 FM, 20 TV
- 3 submarine cables
- Satellite earth stations - 2 Atlantic Ocean INTELSAT and 1 Indian Ocean INTELSAT

Transportation [31]

Railroads. 600 km 1.435-meter gauge, single track; diesel operated

Highways. 4,750 km; majority is bituminous surfaced

Merchant Marine. 35 ships (1,000 GRT or over) totaling 678,584 GRT/785,220 DWT; includes 8 cargo, 24 container, 2 refrigerated cargo, 1 roll-on/roll-off; note - Israel also maintains a significant flag of convenience fleet, which is normally at least as large as the Israeli flag fleet; the Israeli flag of convenience fleet typically includes all of its oil tankers

Airports

Total:	53
Usable:	46
With permanent-surface runways:	28
With runways over 3,659 m:	0
With runways 2,440-3,659 m:	7
With runways 1,220-2,439 m:	12

Top Agricultural Products [32]

	89-90[15]	90-91[15]
Citrus	1,530	1,021
Grapefruit	399	356
Potatoes	214	220
Tomatoes, canning	347	185
Wheat	275	175
Corn on the cob	126	107

Values shown are 1,000 metric tons.

Top Mining Products [33]

Estimated metric tons unless otherwise specified	M.t.
Phosphate rock (000 tons)	1,070
Bromine	135,000
Cement, hydraulic (000 tons)	3,550
Potash (000 tons)	1,320
Magnesia, Mg content	38,600
Stone	
Crushed (000 tons)	17,094[20]
Marble	12,000[20]

Tourism [34]

	1987	1988	1989	1990	1991
Visitors[4]	1,518	1,299	1,425	1,342	9,110
Tourists[4,36]	1,379	1,170	1,177	1,063	943
Cruise passengers	139	129	248	279	167
Tourism receipts	1,342	1,347	1,468	1,382	1,306
Tourism expenditures	1,043	1,161	1,288	1,485	1,783
Fare receipts	330	303	321	357	369
Fare expenditures	188	203	217	195	235

Tourists are in thousands, money in million U.S. dollars.

For sources, notes, and explanations, see Annotated Source Appendix, page 1035.

Manufacturing Sector

GDP and Manufacturing Summary [35]

	1980	1985	1989	1990	% change 1980-1990	% change 1989-1990
GDP (million 1980 $)	23,400	27,066	30,642	32,796	40.2	7.0
GDP per capita (1980 $)	6,034	6,394	6,769	7,123	18.0	5.2
Manufacturing as % of GDP (current prices)	15.5	15.7	13.5[e]	NA	NA	-100.0
Gross output (million $)	14,332	16,351	23,742	29,404	105.2	23.8
Value added (million $)	6,490	6,655	8,869[e]	10,497[e]	61.7	18.4
Value added (million 1980 $)	4,200	5,006	4,321	5,388	28.3	24.7
Industrial production index	100	119	125	133	33.0	6.4
Employment (thousands)	259	292	269	293	13.1	8.9

Note: GDP stands for Gross Domestic Product. 'e' stands for estimated value.

Profitability and Productivity

	1980	1985	1989	1990	% change 1980-1990	% change 1989-1990
Intermediate input (%)	55	59	63[e]	64[e]	16.4	1.6
Wages, salaries, and supplements (%)	30[e]	30[e]	17	21[e]	-30.0	23.5
Gross operating surplus (%)	15[e]	11[e]	21[e]	14[e]	-6.7	-33.3
Gross output per worker ($)	54,619	55,297	88,294	93,942	72.0	6.4
Value added per worker ($)	24,770	22,506	32,982[e]	33,535[e]	35.4	1.7
Average wage (incl. benefits) ($)	16,751[e]	16,850[e]	14,577	21,499[e]	28.3	47.5

Profitability is in percent of gross output. Productivity is in U.S. $. 'e' stands for estimated value.

Profitability - 1990

Inputs - 64.6%
Surplus - 14.1%
Wages - 21.2%

The graphic shows percent of gross output.

Value Added in Manufacturing

	1980 $ mil.	1980 %	1985 $ mil.	1985 %	1989 $ mil.	1989 %	1990 $ mil.	1990 %	% change 1980-1990	% change 1989-1990
311 Food products	706	10.9	748	11.2	1,228[e]	13.8	1,392[e]	13.3	97.2	13.4
313 Beverages	66	1.0	56	0.8	127[e]	1.4	155[e]	1.5	134.8	22.0
314 Tobacco products	24	0.4	10	0.2	16[e]	0.2	17[e]	0.2	-29.2	6.3
321 Textiles	422	6.5	243	3.7	332[e]	3.7	381[e]	3.6	-9.7	14.8
322 Wearing apparel	293	4.5	229	3.4	325[e]	3.7	395[e]	3.8	34.8	21.5
323 Leather and fur products	18	0.3	13	0.2	16[e]	0.2	21[e]	0.2	16.7	31.3
324 Footwear	38	0.6	42	0.6	60[e]	0.7	65[e]	0.6	71.1	8.3
331 Wood and wood products	112	1.7	78	1.2	105[e]	1.2	124[e]	1.2	10.7	18.1
332 Furniture and fixtures	90	1.4	81	1.2	129[e]	1.5	160[e]	1.5	77.8	24.0
341 Paper and paper products	150	2.3	135	2.0	233[e]	2.6	297[e]	2.8	98.0	27.5
342 Printing and publishing	184	2.8	227	3.4	406[e]	4.6	461[e]	4.4	150.5	13.5
351 Industrial chemicals	256	3.9	317	4.8	419[e]	4.7	472[e]	4.5	84.4	12.6
352 Other chemical products	250	3.9	241	3.6	342[e]	3.9	425[e]	4.0	70.0	24.3
353 Petroleum refineries	93	1.4	106	1.6	142[e]	1.6	154[e]	1.5	65.6	8.5
354 Miscellaneous petroleum and coal products	93	1.4	106	1.6	142[e]	1.6	154[e]	1.5	65.6	8.5
355 Rubber products	104	1.6	64	1.0	63[e]	0.7	76[e]	0.7	-26.9	20.6
356 Plastic products	212	3.3	290	4.4	386[e]	4.4	450[e]	4.3	112.3	16.6
361 Pottery, china and earthenware	26	0.4	25	0.4	24[e]	0.3	25[e]	0.2	-3.8	4.2
362 Glass and glass products	30	0.5	23	0.3	19[e]	0.2	24[e]	0.2	-20.0	26.3
369 Other non-metal mineral products	239	3.7	143	2.1	229[e]	2.6	271[e]	2.6	13.4	18.3
371 Iron and steel	148	2.3	118	1.8	82[e]	0.9	97[e]	0.9	-34.5	18.3
372 Non-ferrous metals	61	0.9	36	0.5	55[e]	0.6	70[e]	0.7	14.8	27.3
381 Metal products	1,060	16.3	967	14.5	1,153[e]	13.0	1,382[e]	13.2	30.4	19.9
382 Non-electrical machinery	245	3.8	224	3.4	239[e]	2.7	301[e]	2.9	22.9	25.9
383 Electrical machinery	831	12.8	1,415	21.3	1,846[e]	20.8	2,144[e]	20.4	158.0	16.1
384 Transport equipment	610	9.4	522	7.8	507[e]	5.7	707[e]	6.7	15.9	39.4
385 Professional and scientific equipment	66	1.0	129	1.9	147[e]	1.7	161[e]	1.5	143.9	9.5
390 Other manufacturing industries	63	1.0	67	1.0	95[e]	1.1	114[e]	1.1	81.0	20.0

Note: The industry codes shown are International Standard Industry codes (ISIC). Percentages are percent of total Value Added. 'e' stands for estimated value

For sources, notes, and explanations, see Annotated Source Appendix, page 1035.

451

Finance, Economics, and Trade

Economic Indicators [36]

	1989	1990	1991[e]
GDP, nominal (bil. $U.S.)	44.4	51.2	57.00
GDP Per Capita, nominal[80]	9,460.4	10,622.4	N/A
Money Supply (M1)(bil. NIS)[81]	5.3	7.0	7.8[84]
Commercial Interest Rate[82]	10.6	4.4	N/A
Savings Rate (private)	18.2	17.3	N/A
Investment Rate (gross)	14.5	16.5	N/A
CPI	20.7	17.6	20.00
WPI	127.3	143.4	17.0
External Total Debt[83]	23.79	24.05	N/A

Balance of Payments Summary [37]

Values in millions of dollars.

	1987	1988	1989	1990	1991
Exports of goods (f.o.b.)	9,310	10,355	11,169	12,287	12,180
Imports of goods (f.o.b.)	-12,976	-13,368	-12,876	-15,103	-16,946
Trade balance	-3,666	-3,013	-1,707	-2,816	-4,766
Services - debits	-6,724	-7,340	-7,668	-8,534	-9,016
Services - credits	4,786	5,216	5,708	6,272	6,562
Private transfers (net)	1,388	1,253	1,591	1,846	1,955
Government transfers (net)	3,400	3,370	3,286	3,807	4,436
Long term capital (net)	675	-697	71	-367	-93
Short term capital (net)	538	-193	-1,171	-419	-700
Errors and omissions	265	234	1,287	726	1,583
Overall balance	662	-1,170	1,397	515	-39

Exchange Rates [38]

Currency: **new Israeli shekels.**
Symbol: **NIS.**

Data are currency units per $1.

December 1992	2.8000
1992	2.4591
1991	2.2791
1990	2.0162
1989	1.9164
1988	1.5989
1987	1.5946

Imports and Exports

Top Import Origins [39]

$19.6 billion (c.i.f., 1992 est.).

Origins	%
US	NA
EC	NA
Switzerland	NA
Japan	NA
South Africa	NA
Canada	NA
Hong Kong	NA

Top Export Destinations [40]

$11.8 billion (f.o.b., 1992 est.).

Destinations	%
US	NA
EC	NA
Japan	NA
Hong Kong	NA
Switzerland	NA

Foreign Aid [41]

	U.S. $	
US commitments, including Ex-Im (FY70-90)	18.2	billion
Western (non-US) countries, ODA and OOF bilateral commitments (1970-89)	2.8	billion

Import and Export Commodities [42]

Import Commodities	**Export Commodities**
Military equipment	Polished diamonds
Rough diamonds	Citrus and other fruits
Oil	Textiles and clothing
Chemicals	Processed foods
Machinery	Fertilizer and chemical products
Iron and steel	Military hardware
Cereals	Electronics
Textiles	
Vehicles	
Ships	
Aircraft	

Italy

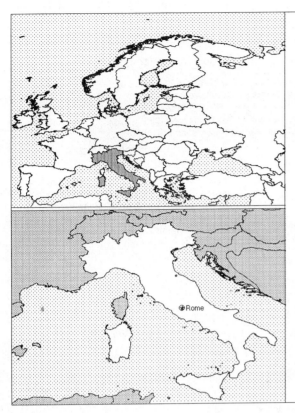

Geography [1]

Total area:
301,230 km2
Land area:
294,020 km2
Comparative area:
Slightly larger than Arizona
Land boundaries:
Total 1,899.2 km, Austria 430 km, France 488 km, Holy See (Vatican City) 3.2 km, San Marino 39 km, Slovenia 199 km, Switzerland 740 km
Coastline:
4,996 km
Climate:
Predominantly Mediterranean; Alpine in far north; hot, dry in south
Terrain:
Mostly rugged and mountainous; some plains, coastal lowlands
Natural resources:
Mercury, potash, marble, sulfur, dwindling natural gas and crude oil reserves, fish, coal
Land use:
Arable land:
32%
Permanent crops:
10%
Meadows and pastures:
17%
Forest and woodland:
22%
Other:
19%

Demographics [2]

	1960	1970	1980	1990	1991[1]	1994	2000	2010	2020
Population	50,198	53,661	56,451	57,661	57,772	58,138	58,865	59,089	57,544
Population density (persons per sq. mi.)	442	473	497	508	509	NA	516	511	494
Births	NA	NA	NA	NA	609	627	NA	NA	NA
Deaths	NA	NA	NA	NA	552	565	NA	NA	NA
Life expectancy - males	NA	NA	NA	NA	75	74	NA	NA	NA
Life expectancy - females	NA	NA	NA	NA	82	81	NA	NA	NA
Birth rate (per 1,000)	NA	NA	NA	NA	11	11	NA	NA	NA
Death rate (per 1,000)	NA	NA	NA	NA	10	10	NA	NA	NA
Women of reproductive age (15-44 yrs.)	NA	NA	NA	14,553	NA	14,640	14,190	NA	NA
of which are currently married	NA	NA	NA	9,284	NA	9,615	9,712	NA	NA
Fertility rate	NA	NA	NA	1.3	1.38	1.4	1.5	1.5	1.6

Population values are in thousands, life expectancy in years, and other items as indicated.

Health

Health Personnel [3]

Doctors per 1,000 pop., 1988-92	4.69
Nurse-to-doctor ratio, 1988-92	0.6
Hospital beds per 1,000 pop., 1985-90	7.5
Percentage of children immunized (age 1 yr. or less)	
Third dose of DPT, 1990-91	95.0
Measles, 1990-91	50.0

Health Expenditures [4]

Total health expenditure, 1990 (official exchange rate)	
Millions of dollars	82214
Millions of dollars per capita	1426
Health expenditures as a percentage of GDP	
Total	7.5
Public sector	5.8
Private sector	1.7
Development assistance for health	
Total aid flows (millions of dollars)[1]	NA
Aid flows per capita (millions of dollars)	NA
Aid flows as a percentage of total health expenditure	NA

For sources, notes, and explanations, see Annotated Source Appendix, page 1035.

453

Human Factors

Health Care Ratios [5]

Population per physician, 1970	550
Population per physician, 1990	120
Population per nursing person, 1970	NA
Population per nursing person, 1990	NA
Percent of births attended by health staff, 1985	NA

Infants and Malnutrition [6]

Percent of babies with low birth weight, 1985	7
Infant mortality rate per 1,000 live births, 1970	30
Infant mortality rate per 1,000 live births, 1991	8
Years of life lost per 1,000 population, 1990	10
Prevalence of malnutrition (under age 5), 1990	NA

Ethnic Division [7]

Italian (includes small clusters of German-, French-, and Slovene-Italians in the north and Albanian-Italians and Greek-Italians in the south), Sicilians, Sardinians.

Religion [8]

Roman Catholic	100%

Major Languages [9]

Italian, German (parts of Trentino-Alto Adige region are predominantly German speaking), French (small French-speaking minority in Valle d'Aosta region), Slovene (Slovene-speaking minority in the Trieste-Gorizia area).

Education

Public Education Expenditures [10]

Billion Lira	1985	1986	1987	1988	1989	1990[6,12]
Total education expenditure	40,533.1	NA	NA	53,349.3	NA	40,846.3
as percent of GNP	5.0	NA	NA	5.0	NA	3.2
as percent of total govt. expend.	8.3	NA	NA	NA	NA	NA
Current education expenditure	36,859.2	NA	NA	49,767.4	NA	40,400.1
as percent of GNP	4.6	NA	NA	4.6	NA	3.1
as percent of current govt. expend.	10.0	NA	NA	NA	NA	NA
Capital expenditure	3,674.0	NA	NA	3,581.9	NA	446.1

Educational Attainment [11]

Age group	25+
Total population	35,596,616
Highest level attained (%)	
No schooling	19.3[6]
First level	
Incompleted	47.4
Completed	NA
Entered second level	
S-1	18.0
S-2	11.2
Post secondary	4.1

Literacy Rate [12]

In thousands and percent	1985[a]	1991[a]	2000[a]
Illiterate population +15 years	1,700	1,378	966
Illiteracy rate - total pop. (%)	3.7	2.9	2.0
Illiteracy rate - males (%)	2.7	2.2	1.6
Illiteracy rate - females (%)	4.5	3.6	2.5

Libraries [13]

	Admin. Units	Svc. Pts.	Vols. (000)	Shelving (meters)	Vols. Added	Reg. Users
National	NA	NA	NA	NA	NA	NA
Non-specialized (1989)	34	34	7,394	321,516	60,264	45,663
Public (1991)[35]	42	NA	18,080	NA	NA	253,457
Higher ed. (1990)[35]	10	NA	5,642	239,009	38,406	826,848
School	NA	NA	NA	NA	NA	NA

Daily Newspapers [14]

	1975	1980	1985	1990[2]
Number of papers	78	82	72	76
Circ. (000)	6,497	4,775	5,511	6,093

Cinema [15]

Data for 1991.

Cinema seats per 1,000	NA
Annual attendance per person	1.5
Gross box office receipts (mil. Lira)	657,890

Science and Technology

Scientific/Technical Forces [16]

Potential scientists/engineers	1,175,418
Number female	420,694
Potential technicians[5]	3,527,943
Number female	1,436,119
Total[5]	4,703,361

R&D Expenditures [17]

	Lire[12] (000) 1988
Total expenditure	13,281,284
Capital expenditure	1,857,879
Current expenditure	11,423,405
Percent current	86.0

U.S. Patents Issued [18]

Values show patents issued to citizens of the country by the U.S. Patents Office.

	1990	1991	1992
Number of patents	1,499	1,388	1,446

454

For sources, notes, and explanations, see Annotated Source Appendix, page 1035.

Government and Law

Organization of Government [19]

Long-form name:
Italian Republic
Type:
republic
Independence:
17 March 1861 (Kingdom of Italy proclaimed)
Constitution:
1 January 1948
Legal system:
based on civil law system, with ecclesiastical law influence; appeals treated as trials de novo; judicial review under certain conditions in Constitutional Court; has not accepted compulsory ICJ jurisdiction
National holiday:
Anniversary of the Republic, 2 June (1946)
Executive branch:
president, prime minister (president of the Council of Ministers)
Legislative branch:
bicameral Parliament (Parlamento) consists of an upper chamber or Senate of the Republic (Senato della Repubblica) and a lower chamber or Chamber of Deputies (Camera dei Deputati)
Judicial branch:
Constitutional Court

Political Parties [20]

Chamber of Deputies	% of votes
Christian Democratic Party (DC)	29.7
Democratic Party of the Left (PDS)	16.1
Socialist Party (PSI)	13.6
Northern League	8.7
Communist Renewal (RC)	5.6
Italian Social Movement (MSI)	5.4
Republican Party (PRI)	4.4
Liberal Party (PLI)	2.8
Social Democratic Party (PSDI)	2.7
Other	11.0

Government Budget [21]

For 1992 est.	
Revenues	447
Expenditures	581
Capital expenditures	46

Crime [23]

Crime volume	
Cases known to police	2,501,640
Attempts (percent)	NA
Percent cases solved	16.96
Crimes per 100,000 persons	4,385.33
Persons responsible for offenses	
Total number offenders	435,751
Percent female	NA
Percent juvenile[31]	3.94
Percent foreigners	NA

Military Expenditures and Arms Transfers [22]

	1985	1986	1987	1988	1989
Military expenditures					
Current dollars (mil.)	15,320	15,780	18,190	19,790	20,720
1989 constant dollars (mil.)	17,440	17,510	19,560	20,600	20,720
Armed forces (000)	531	529	531	533	533
Gross national product (GNP)					
Current dollars (mil.)	665,500	699,200	743,900	800,600	860,000
1989 constant dollars (mil.)	757,600	775,900	800,100	833,400	860,000
Central government expenditures (CGE)					
1989 constant dollars (mil.)	362,900	403,500	418,000	400,500	422,500
People (mil.)	57.1	57.2	57.3	57.4	57.6
Military expenditure as % of GNP	2.3	2.3	2.4	2.5	2.4
Military expenditure as % of CGE	4.8	4.3	4.7	5.1	4.9
Military expenditure per capita	305	306	341	359	360
Armed forces per 1,000 people	9.3	9.2	9.3	9.3	9.3
GNP per capita	13,260	13,550	13,950	14,510	14,940
Arms imports[6]					
Current dollars (mil.)	230	220	200	280	300
1989 constant dollars (mil.)	262	244	215	291	300
Arms exports[6]					
Current dollars (mil.)	1,200	675	525	380	60
1989 constant dollars (mil.)	1,366	749	565	396	60
Total imports[7]					
Current dollars (mil.)	87,690	99,450	125,100	138,600	153,000
1989 constant dollars	99,830	110,400	134,600	144,200	153,000
Total exports[7]					
Current dollars (mil.)	76,720	97,610	116,400	127,900	140,700
1989 constant dollars	87,330	108,300	125,200	133,100	140,700
Arms as percent of total imports[8]	0.3	0.2	0.2	0.2	0.2
Arms as percent of total exports[8]	1.6	0.7	0.5	0.3	0

Human Rights [24]

	SSTS	FL	FAPRO	PPCG	APROBC	TPW	PCPTW	STPEP	PHRFF	PRW	ASST	AFL
Observes	P	P	P	P	P	P	P	P	P	P	P	P
	EAFRD	CPR	ESCR	SR	ACHR	MAAE	PVIAC	PVNAC	EAFDAW	TCIDTP	RC	
Observes		P	P	P		P	P	P	P	P	P	

P = Party; S = Signatory; see Appendix for meaning of abbreviations.

Labor Force

Total Labor Force [25]

23.988 million

Labor Force by Occupation [26]

Services	58%
Industry	32.2
Agriculture	9.8

Date of data: 1988

Unemployment Rate [27]

11% (1992 est.)

For sources, notes, and explanations, see Annotated Source Appendix, page 1035.

Production Sectors

Commercial Energy Production and Consumption

Production [28]

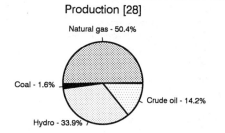

Natural gas - 50.4%
Coal - 1.6%
Crude oil - 14.2%
Hydro - 33.9%

Consumption [29]

Petroleum - 57.0%
Nuclear - 1.8%
Coal - 7.7%
Hydro - 7.4%
Natural gas - 26.1%

Data are shown in quadrillion (10^{15}) BTUs and percent for 1991

Crude oil[1]	0.18
Natural gas liquids	(s)
Dry natural gas	0.64
Coal[2]	0.02
Net hydroelectric power[3]	0.43
Net nuclear power[3]	0.00
Total	1.27

Petroleum	4.07
Dry natural gas	1.86
Coal[2]	0.55
Net hydroelectric power[3]	0.53
Net nuclear power[3]	0.13
Total	7.13

Telecommunications [30]

- Modern, well-developed, fast
- 25,600,000 telephones
- Fully automated telephone, telex, and data services
- High-capacity cable and microwave radio relay trunks
- Broadcast stations - 135 AM, 28 (1,840 repeaters) FM, 83 (1,000 repeaters) TV
- International service by 21 submarine cables, 3 satellite earth stations operating in INTELSAT with 3 Atlantic Ocean antennas and 2 Indian Ocean antennas
- Also participates in INMARSAT and EUTELSAT systems

Transportation [31]

Railroads. 20,011 km total; 16,066 km 1.435-meter government-owned standard gauge (8,999 km electrified); 3,945 km privately owned - 2,100 km 1.435-meter standard gauge (1,155 km electrified) and 1,845 km 0.950-meter narrow gauge (380 km electrified)

Highways. 298,000 km total; autostrada (expressway) 6,000 km, state highways 46,000 km, provincial highways 103,000 km, communal highways 143,000 km; 270,000 km paved, 23,000 km gravel and crushed stone, 5,000 km earth

Merchant Marine. 536 ships (1,000 GRT or over) totaling 6,788,938 GRT/10,128,468 DWT; includes 15 passenger, 36 short-sea passenger, 87 cargo, 4 refrigerated cargo, 21 container, 69 roll-on/roll-off cargo, 8 vehicle carrier, 1 multifunction large-load carrier, 138 oil tanker, 34 chemical tanker, 45 liquefied gas, 10 specialized tanker, 9 combination ore/oil, 57 bulk, 2 combination bulk

Airports

Total:	137
Usable:	133
With permanent-surface runways:	92
With runways over 3,659 m:	2
With runways 2,440-3,659 m:	36
With runways 1,220-2,439 m:	39

Top Agricultural Products [32]

	1989	1990
Sugar beets	16.9	13.8
Wheat	7.4	8.1
Wine grapes	8.0	7.4
Corn	6.4	5.9
Tomatoes	5.7	5.8
Citrus[16]	3.2	2.9

Values shown are 1,000 metric tons.

Top Mining Products [33]

Estimated metric tons unless otherwise specified   M.t.

Feldspar (000 tons)	1,304
Barite	88,486
Bentonite	385
Flurospar	98,518
Marble in blocks, travertine (000 tons)	1,104
Potash, crude salts, (000 tons)	429

Tourism [34]

	1987	1988	1989	1990	1991
Visitors	52,725	55,690	55,131	60,296	51,317
Tourists[4]	25,749	26,155	25,935	26,679	26,840
Excursionists[18]	26,976	29,535	29,196	33,617	24,477
Tourism receipts	12,174	12,403	11,984	19,742	19,668
Tourism expenditures	4,536	5,989	6,772	13,826	13,300
Fare receipts	1,303	1,328	1,416	1,793	1,702
Fare expenditures	797	1,041	1,190	1,427	1,175

Tourists are in thousands, money in million U.S. dollars.

Manufacturing Sector

GDP and Manufacturing Summary [35]

	1980	1985	1989	1990	% change 1980-1990	% change 1989-1990
GDP (million 1980 $)	452,646	485,199	556,069	563,011	24.4	1.2
GDP per capita (1980 $)	8,021	8,491	9,740	9,865	23.0	1.3
Manufacturing as % of GDP (current prices)	28.1	24.4	23.2	22.8	-18.9	-1.7
Gross output (million $)	250,912	212,913	403,868	501,092[e]	99.7	24.1
Value added (million $)	97,032	64,726	132,099	164,069	69.1	24.2
Value added (million 1980 $)	122,239	129,123	157,782	154,259	26.2	-2.2
Industrial production index	100	96	113	113	13.0	0.0
Employment (thousands)	3,333	2,875	2,906[e]	2,858[e]	-14.3	-1.7

Note: GDP stands for Gross Domestic Product. 'e' stands for estimated value.

Profitability and Productivity

	1980	1985	1989	1990	% change 1980-1990	% change 1989-1990
Intermediate input (%)	61	70	67	67[e]	9.8	0.0
Wages, salaries, and supplements (%)	21[e]	18[e]	13[e]	19[e]	-9.5	46.2
Gross operating surplus (%)	18[e]	12[e]	20[e]	14[e]	-22.2	-30.0
Gross output per worker ($)	74,433	73,098	139,000[e]	173,539[e]	133.1	24.8
Value added per worker ($)	28,784	22,227	45,465[e]	56,876[e]	97.6	25.1
Average wage (incl. benefits) ($)	15,647[e]	13,630[e]	17,976[e]	32,527[e]	107.9	80.9

Profitability is in percent of gross output. Productivity is in U.S. $. 'e' stands for estimated value.

Profitability - 1990

Inputs - 67.0%
Surplus - 14.0%
Wages - 19.0%

The graphic shows percent of gross output.

Value Added in Manufacturing

	1980 $ mil.	1980 %	1985 $ mil.	1985 %	1989 $ mil.	1989 %	1990 $ mil.	1990 %	% change 1980-1990	% change 1989-1990
311 Food products	6,362	6.6	3,618	5.6	6,957	5.3	8,793	5.4	38.2	26.4
313 Beverages	1,672	1.7	1,354	2.1	2,782	2.1	3,661	2.2	119.0	31.6
314 Tobacco products	307	0.3	224	0.3	617	0.5	695	0.4	126.4	12.6
321 Textiles	6,716	6.9	5,062	7.8	9,716	7.4	11,734	7.2	74.7	20.8
322 Wearing apparel	3,197	3.3	2,322	3.6	4,465	3.4	5,541	3.4	73.3	24.1
323 Leather and fur products	718	0.7	560	0.9	1,122	0.8	1,394	0.8	94.2	24.2
324 Footwear	1,495	1.5	1,260	1.9	1,895	1.4	2,401	1.5	60.6	26.7
331 Wood and wood products	1,318	1.4	786	1.2	1,464	1.1	1,745[e]	1.1	32.4	19.2
332 Furniture and fixtures	1,936	2.0	1,257	1.9	2,474	1.9	3,215[e]	2.0	66.1	30.0
341 Paper and paper products	2,260	2.3	1,661	2.6	3,172	2.4	3,976	2.4	75.9	25.3
342 Printing and publishing	3,017	3.1	2,271	3.5	5,666	4.3	7,129	4.3	136.3	25.8
351 Industrial chemicals	5,983	6.2	3,994	6.2	7,802	5.9	9,527[e]	5.8	59.2	22.1
352 Other chemical products	4,439	4.6	2,696	4.2	6,095	4.6	7,605	4.6	71.3	24.8
353 Petroleum refineries	1,275	1.3	1,065	1.6	1,580	1.2	2,106	1.3	65.2	33.3
354 Miscellaneous petroleum and coal products	58	0.1	42	0.1	63	0.0	79[e]	0.0	36.2	25.4
355 Rubber products	1,832	1.9	1,107	1.7	2,255[e]	1.7	2,848[e]	1.7	55.5	26.3
356 Plastic products	1,465	1.5	1,729	2.7	4,242[e]	3.2	5,579[e]	3.4	280.8	31.5
361 Pottery, china and earthenware	1,897	2.0	1,139	1.8	2,642	2.0	3,381[e]	2.1	78.2	28.0
362 Glass and glass products	1,116	1.2	666	1.0	1,465	1.1	1,868	1.1	67.4	27.5
369 Other non-metal mineral products	3,667	3.8	2,043	3.2	4,870	3.7	6,327	3.9	72.5	29.9
371 Iron and steel	8,354	8.6	3,846	5.9	7,164	5.4	8,535	5.2	2.2	19.1
372 Non-ferrous metals	1,315	1.4	875	1.4	1,923	1.5	2,286	1.4	73.8	18.9
381 Metal products	5,687	5.9	3,405	5.3	6,915	5.2	8,543	5.2	50.2	23.5
382 Non-electrical machinery	9,326	9.6	8,914	13.8	17,036	12.9	20,985	12.8	125.0	23.2
383 Electrical machinery	8,435	8.7	5,813	9.0	11,934	9.0	14,796	9.0	75.4	24.0
384 Transport equipment	10,280	10.6	6,172	9.5	14,095	10.7	17,597	10.7	71.2	24.8
385 Professional and scientific equipment	2,032	2.1	550	0.8	1,245	0.9	1,229[e]	0.7	-39.5	-1.3
390 Other manufacturing industries	871	0.9	297	0.5	442[e]	0.3	497[e]	0.3	-42.9	12.4

Note: The industry codes shown are International Standard Industry codes (ISIC). Percentages are percent of total Value Added. 'e' stands for estimated value

For sources, notes, and explanations, see Annotated Source Appendix, page 1035.

457

Finance, Economics, and Trade

Economic Indicators [36]

Billions of Italian Lire unless Otherwise stated.

	1989	1990	1991[85]
Real GDP (1985 prices)	922,558	940,574	949,046
Real GDP growth rate	3.0	1.0	0.9
GDP per capita (000s) (1985 prices)	16,033	16,312	16,458
Money supply (M1) (end period)	433,334	467,463	446,522[86]
Public external debt (tril. lire) (year end)[87]	35.1	48.8	NA

Balance of Payments Summary [37]

Values in millions of dollars.

	1987	1988	1989	1990	1991
Exports of goods (f.o.b.)	116,178	127,416	140,118	169,820	168,802
Imports of goods (f.o.b.)	-116,516	-128,778	-142,285	-169,204	-169,700
Trade balance	-338	-1,362	-2,167	616	-898
Services - debits	-39,825	-44,478	-53,621	-89,388	-91,326
Services - credits	39,599	41,195	47,412	77,058	76,823
Private transfers (net)	1,278	1,438	1,326	864	-1,268
Government transfers (net)	-2,272	-2,745	-3,837	-3,573	-4,784
Long term capital (net)	3,078	7,444	22,154	36,354	4,213
Short term capital (net)	6,286	8,880	2,414	5,721	18,189
Errors and omissions	-2,258	-2,008	-2,637	-15,993	-7,355
Overall balance	5,548	8,364	11,044	11,659	-6,406

Exchange Rates [38]

Currency: **Italian lire.**
Symbol: **Lit.**

Data are currency units per $1.

January 1993	1,482.5
1992	1,232.4
1991	1,240.6
1990	1,198.1
1989	1,372.1
1988	1,301.6

Imports and Exports

Top Import Origins [39]

$169.7 million (f.o.b., 1991). Data are for 1992.

Origins	%
EC	58.8
OPEC	6.1
US	5.5

Top Export Destinations [40]

$168.8 million (f.o.b., 1991) Data are for 1992.

Destinations	%
EC	58.3
US	6.8
OPEC	5.1

Foreign Aid [41]

	U.S. $	
Donor - ODA and OOF commitments (1970-89)	25.9	billion

Import and Export Commodities [42]

Import Commodities

Petroleum
Industrial machinery
Chemicals
Metals
Food
Agricultural products

Export Commodities

Textiles
Wearing apparel
Metals
Production machinery
Motor vehicles
Transportation equipment
Chemicals
Other

For sources, notes, and explanations, see Annotated Source Appendix, page 1035.

Jamaica

Geography [1]

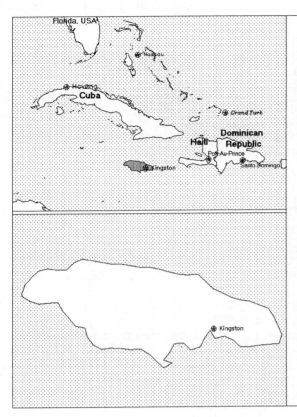

Total area:
10,990 km2
Land area:
10,830 km2
Comparative area:
Slightly smaller than Connecticut
Land boundaries:
0 km
Coastline:
1,022 km
Climate:
Tropical; hot, humid; temperate interior
Terrain:
Mostly mountains with narrow, discontinuous coastal plain
Natural resources:
Bauxite, gypsum, limestone
Land use:
Arable land:
19%
Permanent crops:
6%
Meadows and pastures:
18%
Forest and woodland:
28%
Other:
29%

Demographics [2]

	1960	1970	1980	1990	1991[1]	1994	2000	2010	2020
Population	1,632	1,944	2,229	2,466	2,489	2,555	2,746	3,110	3,446
Population density (persons per sq. mi.)	390	465	533	591	595	NA	661	755	845
Births	NA	NA	NA	NA	59	55	NA	NA	NA
Deaths	NA	NA	NA	NA	15	14	NA	NA	NA
Life expectancy - males	NA	NA	NA	NA	72	72	NA	NA	NA
Life expectancy - females	NA	NA	NA	NA	76	77	NA	NA	NA
Birth rate (per 1,000)	NA	NA	NA	NA	24	22	NA	NA	NA
Death rate (per 1,000)	NA	NA	NA	NA	6	6	NA	NA	NA
Women of reproductive age (15-44 yrs.)	NA	NA	NA	631	NA	671	756	NA	NA
of which are currently married	NA	NA	NA	140	NA	159	194	NA	NA
Fertility rate	NA	NA	NA	2.7	2.63	2.4	2.1	1.9	1.9

Population values are in thousands, life expectancy in years, and other items as indicated.

Health

Health Personnel [3]

Health Expenditures [4]

For sources, notes, and explanations, see Annotated Source Appendix, page 1035.

459

Human Factors

Health Care Ratios [5]

Population per physician, 1970	2630
Population per physician, 1990	NA
Population per nursing person, 1970	530
Population per nursing person, 1990	NA
Percent of births attended by health staff, 1985	89

Infants and Malnutrition [6]

Percent of babies with low birth weight, 1985	8
Infant mortality rate per 1,000 live births, 1970	43
Infant mortality rate per 1,000 live births, 1991	15
Years of life lost per 1,000 population, 1990	NA
Prevalence of malnutrition (under age 5), 1990	8

Ethnic Division [7]

African	76.3%
Afro-European	15.1%
East Indian Afro-East Indian	3.0%
White	3.2%
Chinese Afro-Chinese	1.2%
Other	1.2%

Religion [8]

Protestant	55.9%
Roman Catholic	5.0%
Other, including some spiritual cults	39.1%

Major Languages [9]

English, Creole.

Education

Public Education Expenditures [10]

Million Jamaican Dollars	1980[1]	1985	1987	1988	1989	1990
Total education expenditure	304	550	723	1,104	1,225	NA
as percent of GNP	7.0	5.7	5.2	6.6	6.1	NA
as percent of total govt. expend.	13.1	12.1	11.0	NA	12.9	NA
Current education expenditure	303	515	676	875	1,020	NA
as percent of GNP	7.0	5.4	4.9	5.3	5.1	NA
as percent of current govt. expend.	19.4	15.8	16.1	NA	17.1	NA
Capital expenditure	1	35	47	230	206	NA

Educational Attainment [11]

Age group	25+
Total population	703,714
Highest level attained (%)	
No schooling	3.2
First level	
Incompleted	79.8
Completed	NA
Entered second level	
S-1	15.0
S-2	NA
Post secondary	2.0

Literacy Rate [12]

In thousands and percent	1985[a]	1991[a]	2000[a]
Illiterate population +15 years	30	27	20
Illiteracy rate - total pop. (%)	2.0	1.6	1.0
Illiteracy rate - males (%)	2.2	1.8	1.3
Illiteracy rate - females (%)	1.8	1.4	0.9

Libraries [13]

Daily Newspapers [14]

	1975	1980	1985	1990
Number of papers	3	3	4	3
Circ. (000)	131	109	138[e]	155[e]

Cinema [15]

Science and Technology

Scientific/Technical Forces [16]

R&D Expenditures [17]

	Dollar[11] (000) 1986
Total expenditure	4,016
Capital expenditure	130
Current expenditure	3,886
Percent current	96.8

U.S. Patents Issued [18]

Values show patents issued to citizens of the country by the U.S. Patents Office.

	1990	1991	1992
Number of patents	1	1	1

For sources, notes, and explanations, see Annotated Source Appendix, page 1035.

Government and Law

Organization of Government [19]

Long-form name:
none
Type:
parliamentary democracy
Independence:
6 August 1962 (from UK)
Constitution:
6 August 1962
Legal system:
based on English common law; has not accepted compulsory ICJ jurisdiction
National holiday:
Independence Day (first Monday in August)
Executive branch:
British monarch, governor general, prime minister, Cabinet
Legislative branch:
bicameral Parliament consists of an upper house or Senate and a lower house or House of Representatives
Judicial branch:
Supreme Court

Political Parties [20]

House of Representatives	% of seats
People's National Party (PNP)	86.7
Jamaica Labor Party (JLP)	13.3

Government Budget [21]

For FY91 est.

Revenues	600
Expenditures	736
Capital expenditures	NA

Crime [23]

Crime volume	
Cases known to police	50.101
Attempts (percent)	NA
Percent cases solved	58.51
Crimes per 100,000 persons	1,926.96
Persons responsible for offenses	
Total number offenders	29,314
Percent female	NA
Percent juvenile	NA
Percent foreigners	NA

Military Expenditures and Arms Transfers [22]

	1985	1986	1987	1988	1989
Military expenditures					
Current dollars (mil.)	23	NA	30	34	36
1989 constant dollars (mil.)	26	NA	32	36	36
Armed forces (000)	2	3	3	3	3
Gross national product (GNP)					
Current dollars (mil.)	2,575	2,710	2,940	2,986	3,231
1989 constant dollars (mil.)	2,932	3,008	3,162	3,108	3,231
Central government expenditures (CGE)					
1989 constant dollars (mil.)	1,426	1,246	NA	1,602	NA
People (mil.)	2.4	2.4	2.4	2.4	2.5
Military expenditure as % of GNP	0.9	NA	1.0	1.1	1.1
Military expenditure as % of CGE	1.8	NA	NA	2.2	NA
Military expenditure per capita	11	NA	13	15	15
Armed forces per 1,000 people	0.8	1.3	1.2	1.2	1.2
GNP per capita	1,236	1,256	1,310	1,277	1,318
Arms imports[6]					
Current dollars (mil.)	10	0	5	0	5
1989 constant dollars (mil.)	11	0	5	0	5
Arms exports[6]					
Current dollars (mil.)	0	0	0	0	0
1989 constant dollars (mil.)	0	0	0	0	0
Total imports[7]					
Current dollars (mil.)	1,111	972	1,238	1,440	1,801
1989 constant dollars	1,265	1,079	1,332	1,499	1,801
Total exports[7]					
Current dollars (mil.)	566	589	706	831	967
1989 constant dollars	644	654	759	865	967
Arms as percent of total imports[8]	0.9	0	0.4	0	0.3
Arms as percent of total exports[8]	0	0	0	0	0

Human Rights [24]

	SSTS	FL	FAPRO	PPCG	APROBC	TPW	PCPTW	STPEP	PHRFF	PRW	ASST	AFL
Observes	P	P	P	P	P	P	P			P	P	P
	EAFRD	CPR	ESCR	SR	ACHR	MAAE	PVIAC	PVNAC	EAFDAW	TCIDTP	RC
Observes		P	P	P	P	P		P	P	P		P

P = Party; S = Signatory; see Appendix for meaning of abbreviations.

Labor Force

Total Labor Force [25]

1,062,100

Labor Force by Occupation [26]

Services	41%
Agriculture	22.5
Industry	19
Unemployed	17.5

Date of data: 1989

Unemployment Rate [27]

15.4% (1992)

For sources, notes, and explanations, see Annotated Source Appendix, page 1035.

461

Production Sectors

Energy Resource Summary [28]

Energy Resources: None. **Electricity**: 1,127,000 kW capacity; 2,736 million kWh produced, 1,090 kWh per capita (1992). **Pipelines**: Petroleum products 10 km.

Telecommunications [30]

- Fully automatic domestic telephone network
- 127,000 telephones
- Broadcast stations - 10 AM, 17 FM, 8 TV
- 2 Atlantic Ocean INTELSAT earth stations
- 3 coaxial submarine cables

Transportation [31]

Railroads. 294 km, all 1.435-meter standard gauge, single track

Highways. 18,200 km total; 12,600 km paved, 3,200 km gravel, 2,400 km improved earth

Merchant Marine. 4 ships (1,000 GRT or over) totaling 9,619 GRT/ 16,302 DWT; includes 1 roll-on/roll-off cargo, 1 oil tanker, 2 bulk

Airports

Total:	36
Usable:	23
With permanent-surface runways:	10
With runways over 3,659 m:	0
With runways 2,440-3,659 m:	2
With runways 1,220-2,439 m:	1

Top Agricultural Products [32]

	1990	1991[4]
Sugar cane	237	225
Bananas	63	75
Citrus[17]	76	55
Yams	161	35
Vegetables	108	23
Coffee[17]	7	9

Values shown are 1,000 metric tons.

Top Mining Products [33]

Metric tons unless otherwise specified	M.t.
Bauxite, dry equiv., gross weight (000 tons)	11,550
Alumina (000 tons)	3,015
Cement, hydraulic	395
Lime (000 tons)	95[e]
Salt	14,000[e]
Gypsum	135,844
Marble	12,000
Sand and gravel (000 tons)	1,214
Silica sand	15,622
Steel, crude	25,000[e]

Tourism [34]

	1987	1988	1989	1990	1991
Visitors	1,031	1,020	1,159	1,226	1,335
Tourists	739	649	715	841	845
Cruise passengers	292	368	444	385	490
Tourism receipts	595	525	593	740	764
Tourism expenditures	44	57	54	54	54
Fare receipts	106	108	116	118	89
Fare expenditures	12	13	14	12	12

Tourists are in thousands, money in million U.S. dollars.

Manufacturing Sector

GDP and Manufacturing Summary [35]

	1980	1985	1989	1990	% change 1980-1990	% change 1989-1990
GDP (million 1980 $)	2,667	2,678	2,887	3,187	19.5	10.4
GDP per capita (1980 $)	1,250	1,158	1,189	1,296	3.7	9.0
Manufacturing as % of GDP (current prices)	16.1	19.3	20.2	18.4	14.3	-8.9
Gross output (million $)	1,661	1,464	4,090[e]	2,497[e]	50.3	-38.9
Value added (million $)	436	370	554[e]	734[e]	68.3	32.5
Value added (million 1980 $)	446	469	540	585	31.2	8.3
Industrial production index	100	105	108	122	22.0	13.0
Employment (thousands)	44	46	45[e]	64	45.5	42.2

Note: GDP stands for Gross Domestic Product. 'e' stands for estimated value.

Profitability and Productivity

	1980	1985	1989	1990	% change 1980-1990	% change 1989-1990
Intermediate input (%)	74[e]	75	87[e]	71[e]	-4.1	-18.4
Wages, salaries, and supplements (%)	12	10	7[e]	11[e]	-8.3	57.1
Gross operating surplus (%)	14[e]	16	7[e]	18[e]	28.6	157.1
Gross output per worker ($)	37,512[e]	31,521	90,881[e]	39,259[e]	4.7	-56.8
Value added per worker ($)	9,842[e]	7,959	12,279[e]	11,543[e]	17.3	-6.0
Average wage (incl. benefits) ($)	4,560[e]	3,066	6,169[e]	4,484[e]	-1.7	-27.3

Profitability is in percent of gross output. Productivity is in U.S. $. 'e' stands for estimated value.

Profitability - 1990

Inputs - 71.0%
Wages - 11.0%
Surplus - 18.0%

The graphic shows percent of gross output.

Value Added in Manufacturing

	1980 $ mil.	1980 %	1985 $ mil.	1985 %	1989 $ mil.	1989 %	1990 $ mil.	1990 %	% change 1980-1990	% change 1989-1990
311 Food products	78	17.9	74	20.0	122[e]	22.0	122[e]	16.6	56.4	0.0
313 Beverages	63	14.4	51	13.8	77[e]	13.9	90[e]	12.3	42.9	16.9
314 Tobacco products	61	14.0	48	13.0	71[e]	12.8	84[e]	11.4	37.7	18.3
321 Textiles	3	0.7	2	0.5	3[e]	0.5	5[e]	0.7	66.7	66.7
322 Wearing apparel	15	3.4	12	3.2	15[e]	2.7	28[e]	3.8	86.7	86.7
323 Leather and fur products	2	0.5	3	0.8	4[e]	0.7	6[e]	0.8	200.0	50.0
324 Footwear	8	1.8	5	1.4	5[e]	0.9	11[e]	1.5	37.5	120.0
331 Wood and wood products	3	0.7	2	0.5	3[e]	0.5	4[e]	0.5	33.3	33.3
332 Furniture and fixtures	12	2.8	12	3.2	25[e]	4.5	14[e]	1.9	16.7	-44.0
341 Paper and paper products	7	1.6	6	1.6	21[e]	3.8	10[e]	1.4	42.9	-52.4
342 Printing and publishing	15	3.4	14	3.8	13[e]	2.3	26[e]	3.5	73.3	100.0
351 Industrial chemicals	24[e]	5.5	20	5.4	14[e]	2.5	56[e]	7.6	133.3	300.0
352 Other chemical products	4[e]	0.9	4	1.1	50[e]	9.0	10[e]	1.4	150.0	-80.0
353 Petroleum refineries	55	12.6	28	7.6	46[e]	8.3	78[e]	10.6	41.8	69.6
354 Miscellaneous petroleum and coal products	2[e]	0.5	1	0.3	12[e]	2.2	3[e]	0.4	50.0	-75.0
355 Rubber products	10[e]	2.3	5	1.4	2[e]	0.4	15[e]	2.0	50.0	650.0
356 Plastic products	11[e]	2.5	9	2.4	2[e]	0.4	24[e]	3.3	118.2	1,100.0
361 Pottery, china and earthenware	1	0.2	2	0.5	2[e]	0.4	5[e]	0.7	400.0	150.0
362 Glass and glass products	2	0.5	3	0.8	6[e]	1.1	7[e]	1.0	250.0	16.7
369 Other non-metal mineral products	8	1.8	12	3.2	25[e]	4.5	27[e]	3.7	237.5	8.0
371 Iron and steel	5	1.1	5	1.4	7[e]	1.3	9[e]	1.2	80.0	28.6
372 Non-ferrous metals	NA	0.0	NA	0.0	NA	0.0	NA	0.0	NA	NA
381 Metal products	10	2.3	11	3.0	9[e]	1.6	19[e]	2.6	90.0	111.1
382 Non-electrical machinery	6	1.4	7	1.9	2[e]	0.4	13[e]	1.8	116.7	550.0
383 Electrical machinery	6	1.4	7	1.9	4[e]	0.7	14[e]	1.9	133.3	250.0
384 Transport equipment	23	5.3	23	6.2	12[e]	2.2	44[e]	6.0	91.3	266.7
385 Professional and scientific equipment	NA	0.0	NA	0.0	NA	0.0	NA	0.0	NA	NA
390 Other manufacturing industries	4	0.9	4	1.1	5[e]	0.9	8[e]	1.1	100.0	60.0

Note: The industry codes shown are International Standard Industry codes (ISIC). Percentages are percent of total Value Added. 'e' stands for estimated value

For sources, notes, and explanations, see Annotated Source Appendix, page 1035.

463

Finance, Economics, and Trade

Economic Indicators [36]

Millions of Jamaican Dollars (J$) or U.S. Dollars.

	1989	1990	1991[p]
Real GDP (J$ mil.)			
(1974 base yr)	2,103.9	2,184.2	2,206.2
Real GDP Growth Rate	4.6	3.8	1.0
Real GDP Per Capita			
(J$ 1974 base yr)	884.0	910.1	880
Money Supply (M1) (J$ mil.)	2,739.4	3,516.0	4,651.9[88]
Avg. Commercial Interest Rate	31.0	36.03	31.65[89]
Savings Rate	18.0	18.0	18-21
CPI (Dec.-Dec.)	17.2	29.8	56.3
WPI	NA	NA	NA
External Pub. Debt ($US mil.)	4,038.4	4,152.4	4,147.5

Balance of Payments Summary [37]

Values in millions of dollars.

	1987	1988	1989	1990	1991
Exports of goods (f.o.b.)	708.4	883.0	1,000.4	1,157.5	1,145.2
Imports of goods (f.o.b.)	-1,065.1	-1,240.3	-1,606.4	-1,679.6	-1,551.2
Trade balance	-356.7	-357.3	-606.0	-522.1	-406.0
Services - debits	-876.0	-1,008.9	-1,187.2	-1,248.1	-1,161.2
Services - credits	925.0	890.8	1,008.6	1,170.8	1,096.5
Private transfers (net)	117.2	436.5	299.5	155.4	167.7
Government transfers (net)	54.4	70.0	187.7	116.0	105.0
Long term capital (net)	255.6	12.4	249.5	294.5	230.8
Short term capital (net)	-37.4	131.7	66.4	93.1	-48.3
Errors and omissions	82.8	-44.0	4.6	21.3	-95.3
Overall balance	164.9	131.2	23.1	80.9	-110.8

Exchange Rates [38]

Currency: **Jamaican dollars.**
Symbol: **J$.**

Data are currency units per $1.

September 1992	22.1730
1991	12.1160
1990	7.1840
1989	5.7446
1988	5.4886
1987	5.4867

Imports and Exports

Top Import Origins [39]

$1.6 billion (f.o.b., 1991).

Origins	%
US	51.0
UK	6.0
Venezuela	5.0
Canada	5.0
Japan	4.5

Top Export Destinations [40]

$1.2 billion (f.o.b., 1991).

Destinations	%
US	39
UK	14
Canada	12
Netherlands	8
Norway	7

Foreign Aid [41]

	U.S. $	
US commitments, including Ex-Im (FY70-89)	1.2	billion
Other countries, ODA and OOF bilateral commitments (1970-89)	1.6	billion

Import and Export Commodities [42]

Import Commodities	Export Commodities
Fuel	Alumina
Other raw materials	Bauxite
Construction materials	Sugar
Food	Bananas
Transport equipment	Rum
Other machinery & equipment	

Japan

Geography [1]

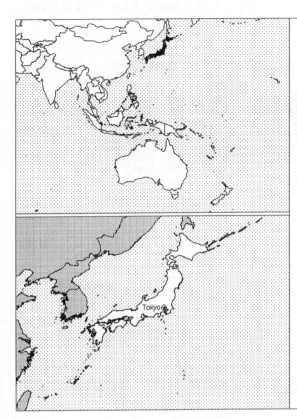

Total area:
377,835 km2
Land area:
374,744 km2
Comparative area:
Slightly smaller than California
Land boundaries:
0 km
Coastline:
29,751 km
Climate:
Varies from tropical in south to cool temperate in north
Terrain:
Mostly rugged and mountainous
Natural resources:
Negligible mineral resources, fish
Land use:
Arable land:
13%
Permanent crops:
1%
Meadows and pastures:
1%
Forest and woodland:
67%
Other:
18%

Demographics [2]

	1960	1970	1980	1990	1991[1]	1994	2000	2010	2020
Population	94,092	104,345	116,807	123,540	124,017	125,107	127,554	129,361	126,062
Population density (persons per sq. mi.)	617	685	766	811	814	NA	841	856	838
Births	NA	NA	NA	NA	1,270	1,312	NA	NA	NA
Deaths	NA	NA	NA	NA	827	915	NA	NA	NA
Life expectancy - males	NA	NA	NA	NA	76	76	NA	NA	NA
Life expectancy - females	NA	NA	NA	NA	82	82	NA	NA	NA
Birth rate (per 1,000)	NA	NA	NA	NA	10	10	NA	NA	NA
Death rate (per 1,000)	NA	NA	NA	NA	7	7	NA	NA	NA
Women of reproductive age (15-44 yrs.)	NA	NA	NA	31,462	NA	31,054	29,368	NA	NA
of which are currently married	NA	NA	NA	18,670	NA	18,306	17,758	NA	NA
Fertility rate	NA	NA	NA	1.5	1.56	1.6	1.6	1.6	1.6

Population values are in thousands, life expectancy in years, and other items as indicated.

Health

Health Personnel [3]

Doctors per 1,000 pop., 1988-92	1.64
Nurse-to-doctor ratio, 1988-92	1.8
Hospital beds per 1,000 pop., 1985-90	15.9
Percentage of children immunized (age 1 yr. or less)	
Third dose of DPT, 1990-91	87.0
Measles, 1990-91	66.0

Health Expenditures [4]

Total health expenditure, 1990 (official exchange rate)	
Millions of dollars	189930
Millions of dollars per capita	1538
Health expenditures as a percentage of GDP	
Total	6.5
Public sector	4.8
Private sector	1.6
Development assistance for health	
Total aid flows (millions of dollars)[1]	NA
Aid flows per capita (millions of dollars)	NA
Aid flows as a percentage of total health expenditure	NA

For sources, notes, and explanations, see Annotated Source Appendix, page 1035.

465

Human Factors

Health Care Ratios [5]

Population per physician, 1970	890
Population per physician, 1990	610
Population per nursing person, 1970	310
Population per nursing person, 1990	NA
Percent of births attended by health staff, 1985	100

Infants and Malnutrition [6]

Percent of babies with low birth weight, 1985	5
Infant mortality rate per 1,000 live births, 1970	13
Infant mortality rate per 1,000 live births, 1991	5
Years of life lost per 1,000 population, 1990	8
Prevalence of malnutrition (under age 5), 1990	NA

Ethnic Division [7]

Japanese	99.4%
Other	0.6%

Religion [8]

Shinto	95.8%
Buddhist	76.3%
Christian	1.4%
Other	12.0%

Major Languages [9]

Japanese.

Education

Public Education Expenditures [10]

Billion Yen	1980	1985	1986	1987	1988	1989
Total education expenditure	13,908	16,143	16,429	16,934	17,461	18,911
as percent of GNP	5.8	5.0	4.9	4.8	4.7	4.7
as percent of total govt. expend.	19.6	17.9	17.7	16.8	16.2	16.5
Current education expenditure[14]	9,417	15,281	15,940	16,448	17,279	NA
as percent of GNP	3.9	4.8	4.7	4.7	4.6	NA
as percent of current govt. expend.	NA	NA	NA	NA	NA	NA
Capital expenditure[14]	3,921	5,144	5,127	5,463	5,533	NA

Educational Attainment [11]

Age group	25+
Total population	80,489,000
Highest level attained (%)	
No schooling	0.0
First level	
Incompleted	NA
Completed	34.3
Entered second level	
S-1	NA
S-2	44.5
Post secondary	21.2

Literacy Rate [12]

Libraries [13]

	Admin. Units	Svc. Pts.	Vols. (000)	Shelving (meters)	Vols. Added	Reg. Users
National (1989)	1	3	4,903[19]	NA	334,000	418,000[19]
Non-specialized	NA	NA	NA	NA	NA	NA
Public (1990)	1,475	1,950	161,694	NA	15 mil	16 mil
Higher ed. (1987)	1,398	1,398	205,237	NA	7.8mil	1.2mil
School (1987)	39,685	39,685	300,827	NA	12 mil	NA

Daily Newspapers [14]

	1975	1980	1985	1990
Number of papers	178	151	124	125
Circ. (000)	60,782	66,258	68,296	75,524

Cinema [15]

Data for 1989.

Cinema seats per 1,000	NA
Annual attendance per person	1.2
Gross box office receipts (mil. Yen)	166,681

Science and Technology

Scientific/Technical Forces [16]

Potential scientists/engineers	8,672,000
Number female	1,228,000
Potential technicians	4,955,000
Number female	3,069,000
Total	13,627,000

R&D Expenditures [17]

	Yen[7, 12] (000) 1988
Total expenditure	10,627,572
Capital expenditure	1,684,476
Current expenditure	8,943,096
Percent current	84.2

U.S. Patents Issued [18]

Values show patents issued to citizens of the country by the U.S. Patents Office.

	1990	1991	1992
Number of patents	20,742	22,403	23,164

For sources, notes, and explanations, see Annotated Source Appendix, page 1035.

Government and Law

Organization of Government [19]

Long-form name:
none
Type:
constitutional monarchy
Independence:
660 BC (traditional founding by Emperor Jimmu)
Constitution:
3 May 1947
Legal system:
modled after European civil law system with English-American influence; judicial review of legislative acts in the Supreme Court; accepts compulsory ICJ jurisdiction, with reservations
National holiday:
Birthday of the Emperor, 23 December (1933)
Executive branch:
Emperor, prime minister, Cabinet
Legislative branch:
bicameral Diet (Kokkai) consists of an upper house or House of Councillors (Sangi-in) and a lower house or House of Representatives (Shugi-in)
Judicial branch:
Supreme Court

Political Parties [20]

House of Representatives	% of seats
Liberal Democratic Party (LDP)	53.5
Social Democratic Party (SDPJ)	26.8
CGP	9.0
Japan Communist Party (JCP)	3.1
Democratic Socialist Party (DSP)	2.5
Others	1.0
Independents	1.2
Vacant	2.9

Government Budget [21]

For FY93.
Revenues	490
Expenditures	579
Public works	68

Crime [23]

Crime volume	
Cases known to police	1,726,188
Attempts (percent)	3.2
Percent cases solved	45.3
Crimes per 100,000 persons	1,396.5
Persons responsible for offenses	
Total number offenders	378.217
Percent female	20.7
Percent juvenile[32]	48.4
Percent foreigners	2.9

Military Expenditures and Arms Transfers [22]

	1985	1986	1987	1988	1989
Military expenditures					
Current dollars (mil.)	20,650	22,170	24,120	26,100	28,410
1989 constant dollars (mil.)	23,510	24,610	25,940	27,170	28,410
Armed forces (000)	241	245	244	245	247
Gross national product (GNP)					
Current dollars (bil.)	2,090	2,197	2,368	2,586	2,820
1989 constant dollars (bil.)	2,379	2,438	2,547	2,692	2,820
Central government expenditures (CGE)					
1989 constant dollars (mil.)	419,400	424,800	438,500	451,900	NA
People (mil.)	120.8	121.5	122.1	122.7	123.2
Military expenditure as % of GNP	1.0	1.0	1.0	1.0	1.0
Military expenditure as % of CGE	5.6	5.8	5.9	6.0	NA
Military expenditure per capita	195	203	212	222	231
Armed forces per 1,000 people	2.0	2.0	2.0	2.0	2.0
GNP per capita	19,700	20,070	20,860	21,950	22,900
Arms imports[6]					
Current dollars (mil.)	1,000	825	1,000	1,200	1,400
1989 constant dollars (mil.)	1,138	916	1,076	1,249	1,400
Arms exports[6]					
Current dollars (mil.)	150	180	120	100	110
1989 constant dollars (mil.)	171	200	129	104	110
Total imports[7]					
Current dollars (mil.)	130,500	127,600	151,000	187,400	209,700
1989 constant dollars	148,500	141,500	162,400	195,100	209,700
Total exports[7]					
Current dollars (mil.)	177,200	210,800	231,300	264,900	273,900
1989 constant dollars	201,700	233,900	248,800	275,700	273,900
Arms as percent of total imports[8]	0.8	0.6	0.7	0.6	0.7
Arms as percent of total exports[8]	0.1	0.1	0.1	0	0

Human Rights [24]

	SSTS	FL	FAPRO	PPCG	APROBC	TPW	PCPTW	STPEP	PHRFF	PRW	ASST	AFL
Observes		P	P		P	P	P	P		P		
	EAFRD	CPR	ESCR	SR	ACHR	MAAE	PVIAC	PVNAC	EAFDAW	TCIDTP	RC	
Observes		P	P	P					P		S	

P = Party; S = Signatory; see Appendix for meaning of abbreviations.

Labor Force

Total Labor Force [25]

63.33 million

Labor Force by Occupation [26]

Trade and services	54%
Manufacturing, mining, and construction	33
Agriculture, forestry, and fishing	7
Government	3

Date of data: 1988

Unemployment Rate [27]

2.2% (1992)

For sources, notes, and explanations, see Annotated Source Appendix, page 1035.

Production Sectors

Commercial Energy Production and Consumption

Production [28]

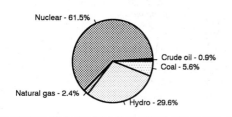

Nuclear - 61.5%
Crude oil - 0.9%
Coal - 5.6%
Natural gas - 2.4%
Hydro - 29.6%

Consumption [29]

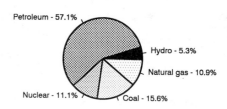

Petroleum - 57.1%
Hydro - 5.3%
Natural gas - 10.9%
Nuclear - 11.1%
Coal - 15.6%

Data are shown in quadrillion (10^{15}) BTUs and percent for 1991

Crude oil[1]	0.03
Natural gas liquids	(s)
Dry natural gas	0.08
Coal[2]	0.19
Net hydroelectric power[3]	1.00
Net nuclear power[3]	2.08
Total	3.39

Petroleum	10.74
Dry natural gas	2.05
Coal[2]	2.93
Net hydroelectric power[3]	1.00
Net nuclear power[3]	2.08
Total	18.81

Telecommunications [30]

- Excellent domestic and international service
- 64,000,000 telephones
- Broadcast stations - 318 AM, 58 FM, 12,350 TV (196 major - 1 kw or greater)
- Satellite earth stations - 4 Pacific Ocean INTELSAT and 1 Indian Ocean INTELSAT
- Submarine cables to US (via Guam), Philippines, China, and Russia

Transportation [31]

Railroads. 27,327 km total; 2,012 km 1.435-meter standard gauge and 25,315 km predominantly 1.067-meter narrow gauge; 5,724 km doubletrack and multitrack sections, 9,038 km 1.067-meter narrow-gauge electrified, 2,012 km 1.435-meter standard-gauge electrified (1987)

Highways. 1,111,974 km total; 754,102 km paved, 357,872 km gravel, crushed stone, or unpaved; 4,400 km national expressways; 46,805 km national highways; 128,539 km prefectural roads; and 930,230 km city, town, and village roads, 6,400 km other

Merchant Marine. 950 ships (1,000 GRT or over) totaling 21,080,149 GRT/32,334,270 DWT; includes 10 passenger, 39 short-sea passenger, 1 passenger cargo, 81 cargo, 43 container, 43 roll-on/roll-off cargo, 87 refrigerated cargo, 97 vehicle carrier, 240 oil tanker, 11 chemical tanker, 39 liquefied gas, 9 combination ore/oil, 2 specialized tanker, 247 bulk, 1 multi-function large load carrier; note - Japan also owns a large flag of convenience fleet, including up to 44% of the total number of ships under the Panamanian flag

Airports

Total:	162
Usable:	159
With permanent-surface runways:	132
With runways over 3,659 m:	2
With runways 2,440-3,659 m:	32
With runways 1,220-2,439 m:	50

Top Agricultural Products [32]

	89-90[18]	90-91[18]
Rice	9,416	9,554
Mikan oranges	2,375	1,988
Wheat	985	952
Raw sugar	983	925
Barley	346	323
Soybeans	272	220

Values shown are 1,000 metric tons.

Top Mining Products [33]

Preliminary metric tons unless otherwise specified M.t.

Cadmium metal, refined	438
Indium metal (kilograms)	51,646
Iodine, elemental	7,502
Electrolytic manganese dioxide	58,526
Pyrophyllite	1,229,287
Tellurium, elemental	57

Tourism [34]

	1987	1988	1989	1990	1991
Visitors[4]	2,155	2,355	2,835	3,226	3,533
Tourists[37]	1,069	1,116	1,499	1,879	2,104
Excursionists	215	209	159	118	85
Tourism receipts	2,097	2,893	3,143	3,578	3,435
Tourism expenditures	10,760	18,682	22,490	24,928	23,983
Fare receipts	878	1,138	1,469	1,264	1,260
Fare expenditures	3,571	4,817	6,538	7,253	7,483

Tourists are in thousands, money in million U.S. dollars.

468

For sources, notes, and explanations, see Annotated Source Appendix, page 1035.

Manufacturing Sector

GDP and Manufacturing Summary [35]

	1980	1985	1989	1990	% change 1980-1990	% change 1989-1990
GDP (billion 1980 $)	1,059.3	1,272.4	1,514.9	1,592.1	50.3	5.1
GDP per capita (1980 $)	9,068	10,530	12,318	12,889	42.1	4.6
Manufacturing as % of GDP (current prices)	28.2	28.4	28.9	27.3	-3.2	-5.5
Gross output (billion $)	970.6	1,114.7	2,177.6	2,245.7	131.4	3.1
Value added (billion $)	339.2	412.5	872.0	891.8	162.9	2.3
Value added (billion 1980 $)	309.7	410.2	553.9	542.5	75.2	-2.1
Industrial production index	100	119	144	151	51.0	4.9
Employment (thousands)	10,253	10,646	10,776	10,980	7.1	1.9

Note: GDP stands for Gross Domestic Product. 'e' stands for estimated value.

Profitability and Productivity

	1980	1985	1989	1990	% change 1980-1990	% change 1989-1990
Intermediate input (%)	65	63	60	60	-7.7	0.0
Wages, salaries, and supplements (%)	12	13	13	13	8.3	0.0
Gross operating surplus (%)	23	24	27	27	17.4	0.0
Gross output per worker ($)	88,443	102,348	202,076	200,997	127.3	-0.5
Value added per worker ($)	30,912	37,876	80,924	79,816	158.2	-1.4
Average wage (incl. benefits) ($)	11,522	13,653	26,840	26,828	132.8	0.0

Profitability is in percent of gross output. Productivity is in U.S. $. 'e' stands for estimated value.

Profitability - 1990

Inputs - 60.0%
Wages - 13.0%
Surplus - 27.0%

The graphic shows percent of gross output.

Value Added in Manufacturing

	1980 $ mil.	1980 %	1985 $ mil.	1985 %	1989 $ mil.	1989 %	1990 $ mil.	1990 %	% change 1980-1990	% change 1989-1990
311 Food products	25,889	7.6	32,032	7.8	66,758	7.7	66,676	7.5	157.5	-0.1
313 Beverages	5,015	1.5	5,307	1.3	10,351	1.2	10,305	1.2	105.5	-0.4
314 Tobacco products	1,888	0.6	700	0.2	2,153	0.2	2,003	0.2	6.1	-7.0
321 Textiles	15,436	4.6	15,259	3.7	27,602	3.2	27,046	3.0	75.2	-2.0
322 Wearing apparel	5,156	1.5	5,622	1.4	11,851	1.4	11,921	1.3	131.2	0.6
323 Leather and fur products	886	0.3	981	0.2	1,841	0.2	1,871	0.2	111.2	1.6
324 Footwear	697	0.2	658	0.2	1,399	0.2	1,478	0.2	112.1	5.6
331 Wood and wood products	8,997	2.7	6,888	1.7	13,852	1.6	14,000	1.6	55.6	1.1
332 Furniture and fixtures	3,788	1.1	3,798	0.9	8,539	1.0	8,730	1.0	130.5	2.2
341 Paper and paper products	9,310	2.7	9,759	2.4	23,014	2.6	22,287	2.5	139.4	-3.2
342 Printing and publishing	17,099	5.0	20,789	5.0	46,289	5.3	47,938	5.4	180.4	3.6
351 Industrial chemicals	13,809	4.1	16,811	4.1	39,519	4.5	38,083	4.3	175.8	-3.6
352 Other chemical products	15,471	4.6	19,758	4.8	47,072	5.4	46,764	5.2	202.3	-0.7
353 Petroleum refineries	6,620	2.0	4,595	1.1	6,516	0.7	4,841	0.5	-26.9	-25.7
354 Miscellaneous petroleum and coal products	1,063	0.3	713	0.2	1,653	0.2	1,540	0.2	44.9	-6.8
355 Rubber products	4,150	1.2	5,077	1.2	10,605	1.2	11,403	1.3	174.8	7.5
356 Plastic products	9,478	2.8	13,570	3.3	30,175	3.5	30,796	3.5	224.9	2.1
361 Pottery, china and earthenware	1,623	0.5	1,627	0.4	3,037	0.3	2,984	0.3	83.9	-1.7
362 Glass and glass products	2,876	0.8	4,029	1.0	8,930	1.0	8,467	0.9	194.4	-5.2
369 Other non-metal mineral products	12,565	3.7	12,321	3.0	26,616	3.1	26,659	3.0	112.2	0.2
371 Iron and steel	26,444	7.8	25,224	6.1	49,993	5.7	48,539	5.4	83.6	-2.9
372 Non-ferrous metals	7,458	2.2	5,236	1.3	11,953	1.4	11,976	1.3	60.6	0.2
381 Metal products	22,409	6.6	26,356	6.4	58,952	6.8	62,905	7.1	180.7	6.7
382 Non-electrical machinery	39,270	11.6	53,580	13.0	116,635	13.4	126,569	14.2	222.3	8.5
383 Electrical machinery	38,868	11.5	63,176	15.3	133,067	15.3	133,877	15.0	244.4	0.6
384 Transport equipment	32,107	9.5	45,158	10.9	88,489	10.1	95,594	10.7	197.7	8.0
385 Professional and scientific equipment	5,685	1.7	6,972	1.7	12,417	1.4	12,798	1.4	125.1	3.1
390 Other manufacturing industries	5,178	1.5	6,510	1.6	12,757	1.5	13,730	1.5	165.2	7.6

Note: The industry codes shown are International Standard Industry codes (ISIC). Percentages are percent of total Value Added. 'e' stands for estimated value

For sources, notes, and explanations, see Annotated Source Appendix, page 1035.

469

Finance, Economics, and Trade

Economic Indicators [36]

	1989	1990	1991
Real GNP (Tril. Y)	383.1	404.7	422.6[90]
Real GNP Growth Rate (%)	4.7	5.6	4.1[90]
Per capita GNP (Dollars)	23,010	23,950	26,753
Money Supply			
(M2+CD Annual Avg) (Tril. Y)	432.7	483.1	498.9[91]
Commercial Interest Rates			
(10-year Govt Bond; Yr-end)	5.73	7.10	5.38
Savings Rate (%)[92]	14.2	14.3	N/A
Investment Rate (%)[93]	31.5	33.0	33.9[90]
CPI (1990 equals 100)	97.0	100.0	102.9[94]
WPI (1985 equals 100)	88.8	90.6	90.8

Balance of Payments Summary [37]

Values in millions of dollars.

	1987	1988	1989	1990	1991
Exports of goods (f.o.b.)	224,620	259,770	269,550	280,350	306,580
Imports of goods (f.o.b.)	-128,200	-164,770	-192,660	-216,770	-203,490
Trade balance	96,420	95,000	76,890	63,580	103,090
Services - debits	-85,380	-123,050	-159,530	-188,150	-206,280
Services - credits	79,660	111,780	143,910	165,960	188,590
Private transfers (net)	-990	-1,120	-990	-1,010	-660
Government transfers (net)	-2,690	-3,000	-3,290	-4,510	-11,830
Long term capital (net)	-133,980	-117,090	-93,760	-53,080	31,390
Short term capital (net)	88,610	50,870	45,830	31,540	-103,240
Errors and omissions	-3,710	3,130	-21,820	-20,920	-7,680
Overall balance	37,940	16,520	-12,760	-6,590	-6,620

Exchange Rates [38]

Currency: **yen.**
Symbol: **Y.**

Data are currency units per $1.

January 1993	125.01
1992	126.65
1991	134.71
1990	144.79
1989	137.96
1988	128.15

Imports and Exports

Top Import Origins [39]

$232.7 billion (c.i.f., 1992).

Origins	%
Southeast Asia	25
US	22
Western Europe	17
Middle East	12
Former Communist countries	NA
China	8

Top Export Destinations [40]

$339.7 billion (f.o.b., 1992).

Destinations	%
Southeast Asia	31
US	29
Western Europe	23
Communist countries	4
Middle East	3

Foreign Aid [41]

	U.S. $	
Donor - ODA and OOF commitments (1970-89)	83.2	billion
ODA outlay in 1990 (est.)	9.1	billion

Import and Export Commodities [42]

Import Commodities

Manufactures 44%
Fossil fuels 33%
Foodstuffs and raw materials 23%

Export Commodities

Manufactures 97% including:
Machinery 40%
Motor vehicles 18%
Consumer electronics 10%

Jordan

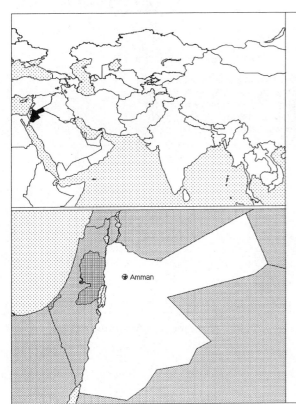

Geography [1]

Total area:
89,213 km2
Land area:
88,884 km2
Comparative area:
Slightly smaller than Indiana
Land boundaries:
Total 1,619 km, Iraq 181 km, Israel 238 km, Saudi Arabia 728 km, Syria 375 km, West Bank 97 km
Coastline:
26 km
Climate:
Mostly arid desert; rainy season in west (November to April)
Terrain:
Mostly desert plateau in east, highland area in west; Great Rift Valley separates East and West Banks of the Jordan River
Natural resources:
Phosphates, potash, shale oil
Land use:
Arable land:
4%
Permanent crops:
0.5%
Meadows and pastures:
1%
Forest and woodland:
0.5%
Other:
94%

Demographics [2]

	1960	1970	1980	1990	1991[1]	1994	2000	2010	2020
Population	849	1,503	2,165	3,305	3,413	3,961	4,814	6,213	7,595
Population density (persons per sq. mi.)	24	43	61	93	97	NA	138	193	254
Births	NA	NA	NA	NA	156	154	NA	NA	NA
Deaths	NA	NA	NA	NA	16	17	NA	NA	NA
Life expectancy - males	NA	NA	NA	NA	70	70	NA	NA	NA
Life expectancy - females	NA	NA	NA	NA	73	74	NA	NA	NA
Birth rate (per 1,000)	NA	NA	NA	NA	46	39	NA	NA	NA
Death rate (per 1,000)	NA	NA	NA	NA	5	4	NA	NA	NA
Women of reproductive age (15-44 yrs.)	NA	NA	NA	723	NA	876	1,091	NA	NA
of which are currently married	NA	NA	NA	480	NA	596	754	NA	NA
Fertility rate	NA	NA	NA	6.3	7.09	5.6	4.6	3.3	2.6

Population values are in thousands, life expectancy in years, and other items as indicated.

Health

Health Personnel [3]

Doctors per 1,000 pop., 1988-92	1.54
Nurse-to-doctor ratio, 1988-92	0.3
Hospital beds per 1,000 pop., 1985-90	1.9
Percentage of children immunized (age 1 yr. or less)	
Third dose of DPT, 1990-91	92.0
Measles, 1990-91	85.0

Health Expenditures [4]

Total health expenditure, 1990 (official exchange rate)	
Millions of dollars	149
Millions of dollars per capita	48
Health expenditures as a percentage of GDP	
Total	3.8
Public sector	1.8
Private sector	2.0
Development assistance for health	
Total aid flows (millions of dollars)[1]	18
Aid flows per capita (millions of dollars)	5.9
Aid flows as a percentage of total health expenditure	12.4

For sources, notes, and explanations, see Annotated Source Appendix, page 1035.

471

Human Factors

Health Care Ratios [5]

Population per physician, 1970	2480
Population per physician, 1990	770
Population per nursing person, 1970	870
Population per nursing person, 1990	500
Percent of births attended by health staff, 1985	75

Infants and Malnutrition [6]

Percent of babies with low birth weight, 1985	7
Infant mortality rate per 1,000 live births, 1970	NA
Infant mortality rate per 1,000 live births, 1991	29
Years of life lost per 1,000 population, 1990	18
Prevalence of malnutrition (under age 5), 1990	NA

Ethnic Division [7]

Arab	98%
Circassian	1%
Armenian	1%

Religion [8]

Sunni Muslim	92%
Christian	8%

Major Languages [9]

Arabic (official), English widely understood among upper and middle classes.

Education

Public Education Expenditures [10]

Million Dinar	1980	1985	1987	1988[11]	1989	1990[6]
Total education expenditure	64	105	NA	95	138	101
as percent of GNP	NA	5.5	NA	4.5	5.9	4.3
as percent of total govt. expend.	11.3	13.0	NA	8.9	13.3	8.5
Current education expenditure	51	92	NA	89	129	92
as percent of GNP	NA	4.8	NA	4.2	5.6	4.0
as percent of current govt. expend.	15.0	18.9	NA	13.5	NA	NA
Capital expenditure	13	14	NA	6	9	9

Educational Attainment [11]

Literacy Rate [12]

In thousands and percent	1985[a]	1991[a]	2000[a]
Illiterate population +15 years	470	442	378
Illiteracy rate - total pop. (%)	25.8	19.9	11.6
Illiteracy rate - males (%)	14.3	10.7	5.8
Illiteracy rate - females (%)	38.0	29.7	17.7

Libraries [13]

	Admin. Units	Svc. Pts.	Vols. (000)	Shelving (meters)	Vols. Added	Reg. Users
National (1986)	1	1	40	NA	1,200	NA
Non-specialized	NA	NA	NA	NA	NA	NA
Public (1986)	5	5	140	NA	14,600	6,865
Higher ed. (1990)	33	44	1,227	NA	36,182	51,369
School	NA	NA	NA	NA	NA	NA

Daily Newspapers [14]

	1975	1980	1985	1990
Number of papers	4	4	4	4
Circ. (000)	58	66	155	225

Cinema [15]

Data for 1989.

Cinema seats per 1,000	24.2[3]
Annual attendance per person	0.3[3]
Gross box office receipts (mil. Dinar)	NA[3]

Science and Technology

Scientific/Technical Forces [16]

Potential scientists/engineers	30,205
Number female	7,280
Potential technicians	NA
Number female	NA
Total	NA

R&D Expenditures [17]

	Dinar[4] (000) 1986
Total expenditure	5,587
Capital expenditure	1,287
Current expenditure	4,300
Percent current	77.0

U.S. Patents Issued [18]

Values show patents issued to citizens of the country by the U.S. Patents Office.

	1990	1991	1992
Number of patents	1	0	1

Government and Law

Organization of Government [19]

Long-form name:
Hashemite Kingdom of Jordan
Type:
constitutional monarchy
Independence:
25 May 1946 (from League of Nations mandate under British administration)
Constitution:
8 January 1952
Legal system:
based on Islamic law and French codes; judicial review of legislative acts in a specially provided High Tribunal; has not accepted compulsory ICJ jurisdiction
National holiday:
Independence Day, 25 May (1946)
Executive branch:
monarch, prime minister, deputy prime minister, Cabinet
Legislative branch:
bicameral National Assembly (Majlis al-'Umma): upper house or House of Notables (Majlis al-A'ayan); lower house or House of Representatives (Majlis al-Nuwaab)
Judicial branch:
Court of Cassation

Crime [23]

Political Parties [20]

House of Representatives	% of seats
Muslim Brotherhood (fund.)	27.5
Independent Islamic bloc (trad.)	7.5
Democratic bloc (mostly leftist)	11.3
Constitutionalist bloc (traditionalist)	21.3
Nationalist bloc (traditionalist)	20.0
Independent	12.5

Government Expenditures [21]

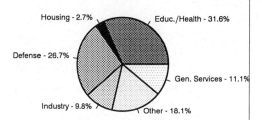

Housing - 2.7%   Educ./Health - 31.6%   Defense - 26.7%   Gen. Services - 11.1%   Industry - 9.8%   Other - 18.1%

% distribution for 1991. Expend. in 1992 est.: 1.900 ($ bil.)

Military Expenditures and Arms Transfers [22]

	1985	1986	1987	1988	1989
Military expenditures					
Current dollars (mil.)	602[e]	633[e]	666[e]	640[e]	548
1989 constant dollars (mil.)	686[e]	702[e]	716[e]	666[e]	548
Armed forces (000)	81	86	100	165	190
Gross national product (GNP)					
Current dollars (mil.)	3,778	3,981	4,224	4,201	4,306
1989 constant dollars (mil.)	4,300	4,418	4,543	4,373	4,306
Central government expenditures (CGE)					
1989 constant dollars (mil.)	1,698	1,785	1,925	1,882	1,678
People (mil.)	2.7	2.8	2.9	3.0	3.1
Military expenditure as % of GNP	15.9	15.9	15.8	15.2	12.7
Military expenditure as % of CGE	40.4	39.3	37.2	35.4	32.7
Military expenditure per capita	258	253	248	221	175
Armed forces per 1,000 people	30.4	31.0	34.6	54.8	60.5
GNP per capita	1,616	1,593	1,573	1,453	1,372
Arms imports[6]					
Current dollars (mil.)	600	450	330	450	190
1989 constant dollars (mil.)	683	499	355	468	190
Arms exports[6]					
Current dollars (mil.)	0	40	70	40	5
1989 constant dollars (mil.)	0	44	75	42	5
Total imports[7]					
Current dollars (mil.)	2,733	2,432	2,710	2,732	2,125
1989 constant dollars	3,111	2,699	2,915	2,844	2,125
Total exports[7]					
Current dollars (mil.)	789	733	930	1,019	1,107
1989 constant dollars	898	813	1,000	1,061	1,107
Arms as percent of total imports[8]	22.0	18.5	12.2	16.5	8.9
Arms as percent of total exports[8]	0	5.5	7.5	3.9	0.5

Human Rights [24]

	SSTS	FL	FAPRO	PPCG	APROBC	TPW	PCPTW	STPEP	PHRFF	PRW	ASST	AFL
Observes	P	P		P	P	P	P	P	P	P	P	P
		EAFRD	CPR	ESCR	SR	ACHR	MAAE	PVIAC	PVNAC	EAFDAW	TCIDTP	RC
Observes		P	P	P				P	P	P	P	P

P = Party; S = Signatory; see Appendix for meaning of abbreviations.

Labor Force

Total Labor Force [25]

572,000 (1988)

Labor Force by Occupation [26]

Agriculture	20%
Manufacturing and mining	20

Date of data: 1987 est.

Unemployment Rate [27]

40% (1991 est.)

Production Sectors

Energy Resource Summary [28]

Energy Resources: Shale oil. **Electricity**: 1,030,000 kW capacity; 3,814 million kWh produced, 1,070 kWh per capita (1992). **Pipelines**: Crude oil 209 km.

Telecommunications [30]

- Adequate telephone system of microwave, cable, and radio links
- 81,500 telephones
- Broadcast stations - 5 AM, 7 FM, 8 TV
- Satellite earth stations - 1 Atlantic Ocean INTELSAT, 1 Indian Ocean INTELSAT, 1 ARABSAT, 1 domestic TV receive-only
- Coaxial cable and microwave to Iraq, Saudi Arabia, and Syria
- Microwave link to Lebanon is inactive
- Participant in MEDARABTEL, a microwave radio relay network linking Syria, Jordan, Egypt, Libya, Tunisia, Algeria, and Morocco

Top Agricultural Products [32]

Agriculture accounts for about 7% of GDP; principal products are wheat, barley, citrus fruit, tomatoes, melons, olives; livestock—sheep, goats, poultry; large net importer of food.

Top Mining Products [33]

Metric tons unless otherwise specified	M.t.
Phosphate, mine output (000 tons)	1,458
Potash	1,364,000[1]
Cement, hydraulic (000 tons)	1,754[1]
Steel, crude	200,300
Marble	180,000[1]
Salt	57,000[1]

Transportation [31]

Railroads. 789 km 1.050-meter gauge, single track

Highways. 7,500 km; 5,500 km asphalt, 2,000 km gravel and crushed stone

Merchant Marine. 2 ships (1,000 GRT or over) totaling 60,378 GRT/ 113,557 DWT; includes 1 cargo and 1 oil tanker

Airports

Total:	19
Usable:	15
With permanent-surface runways:	14
With runways over 3,659 m:	1
With runways 2,440-3,659 m:	13
With runways 1,220-2,439 m:	0

Tourism [34]

	1987	1988	1989	1990	1991
Visitors	1,898	2,391	2,278[38]	2,633[39]	2,228
Tourists	477	608	639	572	436
Cruise passengers	2	2	2	4	
Tourism receipts	579	626	546	511	317
Tourism expenditures	444	480	419	336	281
Fare receipts	259	293	205	277	187
Fare expenditures	192	261	180	262	127

Tourists are in thousands, money in million U.S. dollars.

Manufacturing Sector

GDP and Manufacturing Summary [35]

	1980	1985	1989	1990	% change 1980-1990	% change 1989-1990
GDP (million 1980 $)	3,303	4,147	4,213	4,148	25.6	-1.5
GDP per capita (1980 $)	1,130	1,217	1,086	1,035	-8.4	-4.7
Manufacturing as % of GDP (current prices)	9.9	11.3	9.7[e]	14.6[e]	47.5	50.5
Gross output (million $)	917	1,997	2,041[e]	1,846	101.3	-9.6
Value added (million $)	406	581	827[e]	583	43.6	-29.5
Value added (million 1980 $)	363	497	566	584	60.9	3.2
Industrial production index	100	155	154	193	93.0	25.3
Employment (thousands)	25	42	45[e]	44	76.0	-2.2

Note: GDP stands for Gross Domestic Product. 'e' stands for estimated value.

Profitability and Productivity

	1980	1985	1989	1990	% change 1980-1990	% change 1989-1990
Intermediate input (%)	56	71	59[e]	68	21.4	15.3
Wages, salaries, and supplements (%)	12	9	9[e]	8	-33.3	-11.1
Gross operating surplus (%)	32	20	31[e]	24	-25.0	-22.6
Gross output per worker ($)	26,455	38,009	45,221[e]	32,641	23.4	-27.8
Value added per worker ($)	11,819	11,193	18,329[e]	10,303	-12.8	-43.8
Average wage (incl. benefits) ($)	4,418	4,326	4,202[e]	3,175	-28.1	-24.4

Profitability is in percent of gross output. Productivity is in U.S. $. 'e' stands for estimated value.

Profitability - 1990

Inputs - 68.0%
Wages - 8.0%
Surplus - 24.0%

The graphic shows percent of gross output.

Value Added in Manufacturing

	1980 $ mil.	1980 %	1985 $ mil.	1985 %	1989 $ mil.	1989 %	1990 $ mil.	1990 %	% change 1980-1990	% change 1989-1990
311 Food products	24	5.9	48	8.3	53[e]	6.4	58	9.9	141.7	9.4
313 Beverages	20	4.9	27	4.6	31[e]	3.7	28	4.8	40.0	-9.7
314 Tobacco products	50	12.3	92	15.8	100[e]	12.1	75	12.9	50.0	-25.0
321 Textiles	10	2.5	14	2.4	13[e]	1.6	20	3.4	100.0	53.8
322 Wearing apparel	8	2.0	10	1.7	12[e]	1.5	13	2.2	62.5	8.3
323 Leather and fur products	2	0.5	2	0.3	3[e]	0.4	4	0.7	100.0	33.3
324 Footwear	8	2.0	8	1.4	5[e]	0.6	3	0.5	-62.5	-40.0
331 Wood and wood products	7	1.7	7	1.2	7[e]	0.8	4	0.7	-42.9	-42.9
332 Furniture and fixtures	11	2.7	11	1.9	12[e]	1.5	14	2.4	27.3	16.7
341 Paper and paper products	9	2.2	9	1.5	20[e]	2.4	20	3.4	122.2	0.0
342 Printing and publishing	7	1.7	11	1.9	14[e]	1.7	12	2.1	71.4	-14.3
351 Industrial chemicals	10	2.5	14	2.4	42[e]	5.1	44	7.5	340.0	4.8
352 Other chemical products	20	4.9	28	4.8	45[e]	5.4	42	7.2	110.0	-6.7
353 Petroleum refineries	53	13.1	87	15.0	213[e]	25.8	55	9.4	3.8	-74.2
354 Miscellaneous petroleum and coal products	NA	0.0	NA	0.0	NA	0.0	NA	0.0	NA	NA
355 Rubber products	NA	0.0	NA	0.0	1[e]	0.1	1	0.2	NA	0.0
356 Plastic products	12	3.0	13	2.2	18[e]	2.2	17	2.9	41.7	-5.6
361 Pottery, china and earthenware	2	0.5	3	0.5	5[e]	0.6	3	0.5	50.0	-40.0
362 Glass and glass products	2	0.5	3	0.5	4[e]	0.5	3	0.5	50.0	-25.0
369 Other non-metal mineral products	98	24.1	123	21.2	128[e]	15.5	85	14.6	-10.0	-33.6
371 Iron and steel	11[e]	2.7	8	1.4	20[e]	2.4	24	4.1	118.2	20.0
372 Non-ferrous metals	5[e]	1.2	4	0.7	9[e]	1.1	9	1.5	80.0	0.0
381 Metal products	27[e]	6.7	31	5.3	22[e]	2.7	23	3.9	-14.8	4.5
382 Non-electrical machinery	2[e]	0.5	4	0.7	8[e]	1.0	9	1.5	350.0	12.5
383 Electrical machinery	2	0.5	2	0.3	5[e]	0.6	11	1.9	450.0	120.0
384 Transport equipment	NA	0.0	1	0.2	2[e]	0.2	1	0.2	NA	-50.0
385 Professional and scientific equipment	NA	0.0	NA	0.0	1[e]	0.1	2	0.3	NA	100.0
390 Other manufacturing industries	7	1.7	23	4.0	33[e]	4.0	2	0.3	-71.4	-93.9

Note: The industry codes shown are International Standard Industry codes (ISIC). Percentages are percent of total Value Added. 'e' stands for estimated value

For sources, notes, and explanations, see Annotated Source Appendix, page 1035.

475

Finance, Economics, and Trade

Economic Indicators [36]

Millions of Jordanian Dinars (JD) unless otherwise stated[95].

	1989	1990	1991[96]
Real GDP[97]	2,047	2,035	2,051
Real GDP Growth Rate (%)	-1.7	-5.7	3.0
Money Supply (MI)	1,327	1,433	1,576
Savings Rate (% GNP)	8.0	NA	NA
Investment Rate (% GNP)	24.0	NA	NA
CPI (1986 equals 100)	133.8	155.4	170
WPI (1979 equals 100)	204.4	233.8	250
External Public Debt[98]	9,412	9,139	8,236

Balance of Payments Summary [37]

Values in millions of dollars.

	1987	1988	1989	1990	1991
Exports of goods (f.o.b.)	933.1	1,007.4	1,109.4	1,063.8	1,129.5
Imports of goods (f.o.b.)	-2,400.1	-2,418.7	-1,882.5	-2,300.7	-2,224.5
Trade balance	-1,467.0	-1,411.3	-773.1	-1,236.9	-1,095.0
Services - debits	-1,576.9	-1,695.2	-1,298.9	-1,549.6	-1,560.5
Services - credits	1,350.0	1,461.3	1,278.2	1,514.6	1,465.5
Private transfers (net)	742.9	799.8	565.4	569.6	1,123.9
Government transfers (net)	599.1	551.7	613.2	587.6	475.7
Long term capital (net)	226.6	37.2	184.6	422.3	270.9
Short term capital (net)	245.4	360.7	-106.3	75.9	912.4
Errors and omissions	27.9	123.4	0.3	75.1	420.3
Overall balance	148.0	227.6	463.4	458.6	2,013.2

Exchange Rates [38]

Currency: **Jordanian dinars.**
Symbol: **JD.**

Data are currency units per $1.

January 1993	0.6890
1992	0.6797
1991	0.6808
1990	0.6636
1989	0.5704
1988	0.3709

Imports and Exports

Top Import Origins [39]

$2.3 billion (c.i.f., 1991 est.).

Origins	%
EC countries	NA
US	NA
Iraq	NA
Saudi Arabia	NA
Japan	NA
Turkey	NA

Top Export Destinations [40]

$1.0 billion (f.o.b., 1991 est.).

Destinations	%
India	NA
Iraq	NA
Saudi Arabia	NA
Indonesia	NA
Ethiopia	NA
UAE	NA
China	NA

Foreign Aid [41]

	U.S. $	
US commitments, including Ex-Im (FY70-89)	1.7	billion
Western (non-US) countries, ODA and OOF bilateral commitments (1970-89)	1.5	billion
OPEC bilateral aid (1979-89)	9.5	billion
Communist countries (1970-89)	44	million

Import and Export Commodities [42]

Import Commodities

Crude oil
Machinery
Transport equipment
Food
Live animals
Manufactured goods

Export Commodities

Phosphates
Fertilizers
Potash
Agricultural products
Manufactures

Kazakhstan

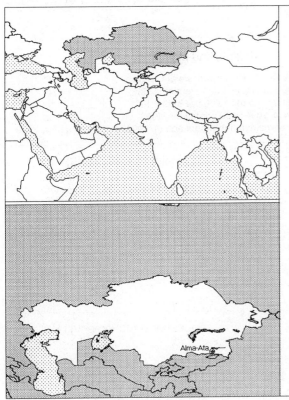

Geography [1]

Total area:
2,717,300 km2
Land area:
2,669,800 km2
Comparative area:
Slightly less than four times the size of Texas
Land boundaries:
Total 12,012 km, China 1,533 km, Kyrgyzstan 1,051 km, Russia 6,846 km, Turkmenistan 379 km, Uzbekistan 2,203 km
Coastline:
0 km
Climate:
Continental, arid and semiarid
Terrain:
Extends from the Volga to the Altai Mountains and from the plains in western Siberia to oasis and desert in Central Asia
Natural resources:
Petroleum, coal, iron, manganese, chrome, nickel, cobalt, copper, molybdenum, lead, zinc, bauxite, gold, uranium, iron
Land use:
Arable land: 15%
Permanent crops: 0%
Meadows and pastures: 57%
Forest and woodland: 4%
Other: 24%

Demographics [2]

	1960	1970	1980	1990	1991[1]	1994	2000	2010	2020
Population	9,982	13,106	14,994	16,749	NA	17,268	17,886	18,794	19,404
Population density (persons per sq. mi.)	NA	NA	NA	NA	NA	NA	NA	NA	NA
Births	NA	NA	NA	NA	NA	335	NA	NA	NA
Deaths	NA	NA	NA	NA	NA	137	NA	NA	NA
Life expectancy - males	NA	NA	NA	NA	NA	63	NA	NA	NA
Life expectancy - females	NA	NA	NA	NA	NA	73	NA	NA	NA
Birth rate (per 1,000)	NA	NA	NA	NA	NA	19	NA	NA	NA
Death rate (per 1,000)	NA	NA	NA	NA	NA	8	NA	NA	NA
Women of reproductive age (15-44 yrs.)	NA	NA	NA	4,183	NA	4,430	4,755	NA	NA
of which are currently married	NA	NA	NA	2,733	NA	2,907	3,118	NA	NA
Fertility rate	NA	NA	NA	2.8	NA	2.4	2.3	2.1	2.0

Population values are in thousands, life expectancy in years, and other items as indicated.

Health

Health Personnel [3]

Doctors per 1,000 pop., 1988-92	4.12
Nurse-to-doctor ratio, 1988-92	3.0
Hospital beds per 1,000 pop., 1985-90	13.6
Percentage of children immunized (age 1 yr. or less)	
Third dose of DPT, 1990-91	84.0
Measles, 1990-91	94.0

Health Expenditures [4]

Total health expenditure, 1990 (official exchange rate)	
Millions of dollars	2572
Millions of dollars per capita	154
Health expenditures as a percentage of GDP	
Total	4.4
Public sector	2.8
Private sector	1.7
Development assistance for health	
Total aid flows (millions of dollars)[1]	NA
Aid flows per capita (millions of dollars)	NA
Aid flows as a percentage of total health expenditure	NA

For sources, notes, and explanations, see Annotated Source Appendix, page 1035.

477

Human Factors

Health Care Ratios [5]

Population per physician, 1970	NA
Population per physician, 1990	250
Population per nursing person, 1970	NA
Population per nursing person, 1990	NA
Percent of births attended by health staff, 1985	NA

Infants and Malnutrition [6]

Percent of babies with low birth weight, 1985	NA
Infant mortality rate per 1,000 live births, 1970	NA
Infant mortality rate per 1,000 live births, 1991	32
Years of life lost per 1,000 population, 1990	19
Prevalence of malnutrition (under age 5), 1990	NA

Ethnic Division [7]

Kazakh (Qazaq)	41.9%
Russian	37.0%
Ukrainian	5.2%
German	4.7%
Uzbek	2.1%
Tatar	2.0%
Other	7.1%

Religion [8]

Muslim	47%
Russian Orthodox	15%
Protestant	2%
Other	36%

Major Languages [9]

Kazakh (Qazaq; official language), Russian (language of interethnic communication).

Education

Public Education Expenditures [10]

Educational Attainment [11]

Literacy Rate [12]

	1989[b]	1990[b]	1991[b]
Illiterate population +15 years	NA	NA	NA
Illiteracy rate - total pop. (%)	2.5	NA	NA
Illiteracy rate - males (%)	0.9	NA	NA
Illiteracy rate - females (%)	3.0	NA	NA

Libraries [13]

Daily Newspapers [14]

Cinema [15]

Science and Technology

Scientific/Technical Forces [16]

R&D Expenditures [17]

U.S. Patents Issued [18]

For sources, notes, and explanations, see Annotated Source Appendix, page 1035.

Government and Law

Organization of Government [19]

Long-form name:
Republic of Kazakhstan
Type:
republic
Independence:
16 December 1991 (from the Soviet Union)
Constitution:
adopted 18 January 1993
Legal system:
based on civil law system
National holiday:
Independence Day, 16 December
Executive branch:
president, cabinet of ministers, prime minister
Legislative branch:
unicameral Supreme Soviet
Judicial branch:
Supreme Court

Political Parties [20]

Supreme Council	% of seats
Socialist Party	94.4
Other	5.6

Government Budget [21]

For 1991.
Revenues	NA
Expenditures	NA
Capital expenditures	1.76

Defense Summary [22]

Branches: Army, Navy, National Guard, Security Forces (internal and border troops)

Manpower Availability: Males age 15-49 4,349,509; fit for military service 3,499,718; reach military age (18) annually 154,727 (1993 est.)

Defense Expenditures: 69,326 million rubles (forecast for 1993); note - conversion of the military budget into US dollars using the current exchange rate could produce misleading results

Crime [23]

Human Rights [24]

	SSTS	FL	FAPRO	PPCG	APROBC	TPW	PCPTW	STPEP	PHRFF	PRW	ASST	AFL
Observes						P	P					
	EAFRD	CPR	ESCR	SR	ACHR	MAAE	PVIAC	PVNAC	EAFDAW	TCIDTP	RC	
Observes							P	P				

P = Party; S = Signatory; see Appendix for meaning of abbreviations.

Labor Force

Total Labor Force [25]

7.563 million

Labor Force by Occupation [26]

Industry and construction	32%
Agriculture and forestry	23
Other	45

Date of data: 1990

Unemployment Rate [27]

0.4% includes only officially registered unemployed; also large numbers of underemployed workers

Production Sectors

Energy Resource Summary [28]

Energy Resources: Petroleum, coal, uranium. **Electricity**: 19,135,000 kW capacity; 81,300 million kWh produced, 4,739 kWh per capita (1992).
Pipelines: Crude oil 2,850 km, refined products 1,500 km, natural gas 3,480 km (1992).

Telecommunications [30]

- Telephone service is poor, with only about 6 telephones for each 100 persons
- Of the approximately 1 million telephones, Almaty (Alma-Ata) has 184,000
- International traffic with other former USSR republics and China carried by landline and microwave, and with other countries by satellite and through 8 international telecommunications circuits at the Moscow international gateway switch
- Satellite earth stations - INTELSAT and Orbita (TV receive only)
- New satellite ground station established at Almaty with Turkish financial help (December 1992) with 2500 channel band width

Top Agricultural Products [32]

	1990	1991
Sunflowerseeds	141.0	110.0
Grains	28.5	11.9
Potatoes	2.3	2.2
Sugar beets	1.1	0.7
Cotton	0.3	0.3

Values shown are 1,000 metric tons.

Top Mining Products [33]

Detailed information is not available. A summary of mineral resources available follows. **Mineral Resources**: Petroleum, coal, iron, manganese, chrome, nickel, cobalt, copper, molybdenum, lead, zinc, bauxite, gold, uranium, iron.

Transportation [31]

Railroads. 14,460 km (all 1.520-meter gauge); does not include industrial lines (1990)

Highways. 189,000 km total; 108,100 km hard surfaced (paved or gravel), 80,900 km earth (1990)

Airports

Total:	365
Useable:	152
With permanent-surface runways:	49
With runways over 3,659 m:	8
With runways 2,440-3,659 m:	38
With runways 1,220-2,439 m:	71

Tourism [34]

Finance, Economics, and Trade

Industrial Summary [35]

Industrial Production: Growth rate - 15% (1992 est.); accounts for 30% of net material product. **Industries**: Extractive industries (oil, coal, iron ore, manganese, chromite, lead, zinc, copper, titanium, bauxite, gold, silver, phosphates, sulfur), iron and steel, nonferrous metal, tractors and other agricultural machinery, electric motors, construction materials.

Economic Indicators [36]

National product: GDP not available. **National product real growth rate**: - 15% (1992 est.). **National product per capita**: not available. **Inflation rate (consumer prices)**: 28% per month (first quarter 1993). **External debt**: $2.6 billion (1991 est.).

Balance of Payments Summary [37]

Exchange Rates [38]

Currency: **rubles.**
Symbol: **R.**

Subject to wide fluctuations. Data are currency units per $1.

| December 24, 1992 | 415 |

Imports and Exports

Top Import Origins [39]

$500 million from outside the successor states of the former USSR (1992).

Origins	%
Russia	NA
Former Soviet republics	NA
China	NA

Top Export Destinations [40]

$1.5 billion to outside the successor states of the former USSR (1992).

Destinations	%
Russia	NA
Ukraine	NA
Uzbekistan	NA

Foreign Aid [41]

	U.S. $
Recipient of limited foreign aid (1992)	NA

Import and Export Commodities [42]

Import Commodities	Export Commodities
Machinery & parts	Oil
Industrial materials	Ferrous and nonferrous metals
	Chemicals
	Grain
	Wool
	Meat

Kenya

Geography [1]

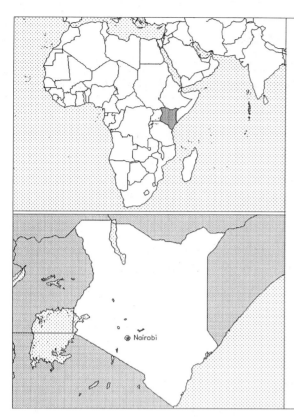

Total area:
582,650 km2
Land area:
569,250 km2
Comparative area:
Slightly more than twice the size of Nevada
Land boundaries:
Total 3,446 km, Ethiopia 830 km, Somalia 682 km, Sudan 232 km, Tanzania 769 km, Uganda 933 km
Coastline:
536 km
Climate:
Varies from tropical along coast to arid in interior
Terrain:
Low plains rise to central highlands bisected by Great Rift Valley; fertile plateau in west
Natural resources:
Gold, limestone, soda ash, salt barytes, rubies, fluorspar, garnets, wildlife
Land use:
Arable land: 3%
Permanent crops: 1%
Meadows and pastures: 7%
Forest and woodland: 4%
Other: 85%

Demographics [2]

	1960	1970	1980	1990	1991[1]	1994	2000	2010	2020
Population	8,157	11,272	16,681	24,229	25,242	28,241	37,990	37,240	44,240
Population density (persons per sq. mi.)	37	51	76	111	115	NA	156	207	261
Births	NA	NA	NA	NA	1,124	1,199	NA	NA	NA
Deaths	NA	NA	NA	NA	213	332	NA	NA	NA
Life expectancy - males	NA	NA	NA	NA	60	51	NA	NA	NA
Life expectancy - females	NA	NA	NA	NA	64	55	NA	NA	NA
Birth rate (per 1,000)	NA	NA	NA	NA	45	42	NA	NA	NA
Death rate (per 1,000)	NA	NA	NA	NA	8	12	NA	NA	NA
Women of reproductive age (15-44 yrs.)	NA	NA	NA	5,264	NA	6,299	7,477	NA	NA
of which are currently married	NA	NA	NA	3,691	NA	4,422	5,263	NA	NA
Fertility rate	NA	NA	NA	6.5	6.37	5.9	5.0	3.6	2.8

Population values are in thousands, life expectancy in years, and other items as indicated.

Health

Health Personnel [3]

Doctors per 1,000 pop., 1988-92	0.14
Nurse-to-doctor ratio, 1988-92	3.2
Hospital beds per 1,000 pop., 1985-90	1.7
Percentage of children immunized (age 1 yr. or less)	
Third dose of DPT, 1990-91	36.0
Measles, 1990-91	36.0

Health Expenditures [4]

Total health expenditure, 1990 (official exchange rate)	
Millions of dollars	375
Millions of dollars per capita	16
Health expenditures as a percentage of GDP	
Total	4.3
Public sector	2.7
Private sector	1.6
Development assistance for health	
Total aid flows (millions of dollars)[1]	84
Aid flows per capita (millions of dollars)	3.5
Aid flows as a percentage of total health expenditure	22.3

For sources, notes, and explanations, see Annotated Source Appendix, page 1035.

Human Factors

Health Care Ratios [5]

Population per physician, 1970	8000
Population per physician, 1990	10130
Population per nursing person, 1970	2520
Population per nursing person, 1990	NA
Percent of births attended by health staff, 1985	NA

Infants and Malnutrition [6]

Percent of babies with low birth weight, 1985	13
Infant mortality rate per 1,000 live births, 1970	102
Infant mortality rate per 1,000 live births, 1991	67
Years of life lost per 1,000 population, 1990	45
Prevalence of malnutrition (under age 5), 1990	NA

Ethnic Division [7]

Kikuyu	21%
Luhya	14%
Luo	13%
Kalenjin	11%
Kamba	11%
Kisii	6%
Meru	6%
Asian, European, Arab	1%

Religion [8]

Roman Catholic	28%
Protestant (including Anglican)	26%
Indigenous beliefs	18%
Muslim	6%

Major Languages [9]

English (official), Swahili (official), numerous indigenous languages.

Education

Public Education Expenditures [10]

Million Shillings	1980	1985	1987	1988	1989	1990
Total education expenditure	3,526	6,171	8,935	9,300	NA	12,955
as percent of GNP	6.8	6.4	7.1	6.4	NA	6.8
as percent of total govt. expend.	18.1	NA	22.7	27.0	NA	16.7
Current education expenditure	3,247	5,789	8,271	8,926	NA	11,000
as percent of GNP	6.2	6.0	6.6	6.2	NA	5.8
as percent of current govt. expend.	23.6	NA	25.3	31.5	NA	19.0
Capital expenditure	279	382	664	374	NA	1,955

Educational Attainment [11]

Literacy Rate [12]

In thousands and percent	1985[a]	1991[a]	2000[a]
Illiterate population +15 years	3,473	3,728	4,360
Illiteracy rate - total pop. (%)	35.0	31.0	23.8
Illiteracy rate - males (%)	22.9	20.2	15.4
Illiteracy rate - females (%)	46.8	41.5	32.0

Libraries [13]

	Admin. Units	Svc. Pts.	Vols. (000)	Shelving (meters)	Vols. Added	Reg. Users
National (1989)	1	22	603	NA	10,260	178,978
Non-specialized	NA	NA	NA	NA	NA	NA
Public	NA	NA	NA	NA	NA	NA
Higher ed.	NA	NA	NA	NA	NA	NA
School	NA	NA	NA	NA	NA	NA

Daily Newspapers [14]

	1975	1980	1985	1990
Number of papers	3	3	4	5
Circ. (000)	134	216	283	350[e]

Cinema [15]

Science and Technology

Scientific/Technical Forces [16]

Potential scientists/engineers	16,241
Number female	NA
Potential technicians	17,741
Number female	NA
Total	33,982

R&D Expenditures [17]

U.S. Patents Issued [18]

Values show patents issued to citizens of the country by the U.S. Patents Office.

	1990	1991	1992
Number of patents	0	1	0

For sources, notes, and explanations, see Annotated Source Appendix, page 1035.

Government and Law

Organization of Government [19]

Long-form name:
Republic of Kenya
Type:
republic
Independence:
12 December 1963 (from UK)
Constitution:
12 December 1963, amended as a
republic 1964; reissued with amendments
1979, 1983, 1986, 1988, 1991, and 1992
Legal system:
based on English common law, tribal law,
and Islamic law; judicial review in High
Court; accepts compulsory ICJ
jurisdiction, with reservations;
constitutional amendment of 1982 making
Kenya a de jure one-party state repealed
in 1991
National holiday:
Independence Day, 12 December (1963)
Executive branch:
president, vice president, Cabinet
Legislative branch:
unicameral National Assembly (Bunge)
Judicial branch:
Court of Appeal, High Court

Political Parties [20]

National Assembly	% of seats
Kenya African National Union (KANU)	53.2
Forum for Restoration (FORD-Kenya)	16.5
FORD-Asili	16.5
Democratic Party of Kenya (DP)	12.2
Smaller parties	1.6
Members nominated by President	6.4

Government Expenditures [21]

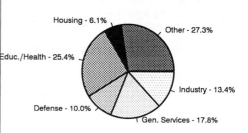

Housing - 6.1%
Other - 27.3%
Educ./Health - 25.4%
Industry - 13.4%
Defense - 10.0%
Gen. Services - 17.8%

% distribution for 1990. Expend. in FY90: 2.80 ($ bil.)

Military Expenditures and Arms Transfers [22]

	1985	1986	1987	1988	1989
Military expenditures					
Current dollars (mil.)	129	138	178	249	210[e]
1989 constant dollars (mil.)	147	153	191	259	210[e]
Armed forces (000)	19	20	21	20	20
Gross national product (GNP)					
Current dollars (mil.)	5,570	6,122	6,681	7,279	7,931
1989 constant dollars (mil.)	6,341	6,794	7,185	7,577	7,931
Central government expenditures (CGE)					
1989 constant dollars (mil.)	1,737	1,756	2,145	2,141	2,226
People (mil.)	20.2	21.0	21.8	22.6	23.5
Military expenditure as % of GNP	2.3	2.2	2.7	3.4	2.7
Military expenditure as % of CGE	8.4	8.7	8.9	12.1	9.5
Military expenditure per capita	7	7	9	11	9
Armed forces per 1,000 people	0.9	1.0	0.9	0.9	0.8
GNP per capita	314	324	330	335	338
Arms imports[6]					
Current dollars (mil.)	5	10	10	160	10
1989 constant dollars (mil.)	6	11	11	167	10
Arms exports[6]					
Current dollars (mil.)	0	0	0	0	0
1989 constant dollars (mil.)	0	0	0	0	0
Total imports[7]					
Current dollars (mil.)	1,436	1,613	1,755	1,975	2,150
1989 constant dollars	1,635	1,790	1,888	2,056	2,150
Total exports[7]					
Current dollars (mil.)	958	1,200	961	1,071	970
1989 constant dollars	1,091	1,332	1,034	1,115	970
Arms as percent of total imports[8]	0.3	0.6	0.6	8.1	0.5
Arms as percent of total exports[8]	0	0	0	0	0

Crime [23]

Crime volume
Cases known to police	87,400
Attempts (percent)	NA
Percent cases solved	48,175
Crimes per 100,000 persons	364.16
Persons responsible for offenses	
Total number offenders	61,175
Percent female	6,316
Percent juvenile[33]	3,909
Percent foreigners	NA

Human Rights [24]

	SSTS	FL	FAPRO	PPCG	APROBC	TPW	PCPTW	STPEP	PHRFF	PRW	ASST	AFL
Observes		P			P	P	P					P
	EAFRD	CPR	ESCR	SR	ACHR	MAAE	PVIAC	PVNAC	EAFDAW	TCIDTP	RC	
Observes		P	P	P		P			P		P	

P = Party; S = Signatory; see Appendix for meaning of abbreviations.

Labor Force

Total Labor Force [25]

9.2 million (includes unemployed) Total
employed is 1,370,000 (14.8% of the labor
force)

Labor Force by Occupation [26]

Services	54.8%
Industry	26.2
Agriculture	19.0
Date of data: 1989	

Unemployment Rate [27]

For sources, notes, and explanations, see Annotated Source Appendix, page 1035.

Production Sectors

Commercial Energy Production and Consumption

Production [28]

Hydro - 100.0%

Consumption [29]

Petroleum - 75.0%

Hydro - 25.0%

Data are shown in quadrillion (10^{15}) BTUs and percent for 1991

Crude oil[1]	0.00
Natural gas liquids	0.00
Dry natural gas	0.00
Coal[2]	0.00
Net hydroelectric power[3]	0.03
Net nuclear power[3]	0.00
Total	0.03

Petroleum	0.09
Dry natural gas	0.00
Coal[2]	(s)
Net hydroelectric power[3]	0.03
Net nuclear power[3]	0.00
Total	0.12

Telecommunications [30]

- In top group of African systems
- Consists primarily of radio relay links
- Over 260,000 telephones
- Broadcast stations - 16 AM
- 4 FM, 6 TV
- Satellite earth stations - 1 Atlantic Ocean INTELSAT and 1 Indian Ocean INTELSAT

Transportation [31]

Railroads. 2,040 km 1.000-meter gauge

Highways. 64,590 km total; 7,000 km paved, 4,150 km gravel, remainder improved earth

Merchant Marine. 1 oil tanker ship (1,000 GRT or over) totaling 3,727 GRT/5,558 DWT

Airports

Total:	247
Usable:	208
With permanent-surface runways:	18
With runways over 3,659 m:	2
With runways 2,440-3,659 m:	3
With runways 1,220-2,439 m:	43

Top Agricultural Products [32]

	90-91[19]	91-92[19]
Corn	2,630	2,800
Sugar[20]	432	426
Pineapples[20]	241	263
Wheat	185	210
Tea[20]	197	200
Coffee	90	99

Values shown are 1,000 metric tons.

Top Mining Products [33]

Metric tons unless otherwise specified	M.t.
Soda ash	245,000
Limestone (000 tons)	20
Flurospar, acid grade	100,000
Shale	115,000
Cement, hydraulic (000 tons)	1,500
Salt, crude, rock	102,000

Tourism [34]

	1987	1988	1989	1990	1991
Tourists[4,34]	661	695	735	814	822
Tourism receipts	354	376	400	443	424
Tourism expenditures	24	23	27	38	24
Fare receipts	87	100	93	113	108
Fare expenditures	13	16	24	26	46

Tourists are in thousands, money in million U.S. dollars.

For sources, notes, and explanations, see Annotated Source Appendix, page 1035.

485

Manufacturing Sector

GDP and Manufacturing Summary [35]

	1980	1985	1989	1990	% change 1980-1990	% change 1989-1990
GDP (million 1980 $)	7,088	8,185	9,924	10,801	52.4	8.8
GDP per capita (1980 $)	426	417	428	450	5.6	5.1
Manufacturing as % of GDP (current prices)	12.9	11.4	10.5[e]	11.1	-14.0	5.7
Gross output (million $)	3,656	4,301	6,830[e]	7,767	112.4	13.7
Value added (million $)	744	670	892[e]	921	23.8	3.3
Value added (million 1980 $)	796	960	1,245[e]	1,268	59.3	1.8
Industrial production index	100	111	139	150	50.0	7.9
Employment (thousands)	143[e]	163	189[e]	190	32.9	0.5

Note: GDP stands for Gross Domestic Product. 'e' stands for estimated value.

Profitability and Productivity

	1980	1985	1989	1990	% change 1980-1990	% change 1989-1990
Intermediate input (%)	80	84	87[e]	88	10.0	1.1
Wages, salaries, and supplements (%)	9	7	6[e]	5[e]	-44.4	-16.7
Gross operating surplus (%)	11	9	7[e]	7[e]	-36.4	0.0
Gross output per worker ($)	22,102[e]	22,384[e]	36,059[e]	40,907[e]	85.1	13.4
Value added per worker ($)	4,594	3,491[e]	4,711[e]	4,850[e]	5.6	3.0
Average wage (incl. benefits) ($)	2,269[e]	1,795	2,066[e]	2,046[e]	-9.8	-1.0

Profitability is in percent of gross output. Productivity is in U.S. $. 'e' stands for estimated value.

Profitability - 1990

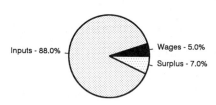

Inputs - 88.0%
Wages - 5.0%
Surplus - 7.0%

The graphic shows percent of gross output.

Value Added in Manufacturing

	1980 $ mil.	1980 %	1985 $ mil.	1985 %	1989 $ mil.	1989 %	1990 $ mil.	1990 %	% change 1980-1990	% change 1989-1990
311 Food products	177	23.8	185	27.6	261[e]	29.3	252	27.4	42.4	-3.4
313 Beverages	65	8.7	72	10.7	70[e]	7.8	90	9.8	38.5	28.6
314 Tobacco products	10	1.3	13	1.9	23[e]	2.6	12	1.3	20.0	-47.8
321 Textiles	59	7.9	40	6.0	55[e]	6.2	55	6.0	-6.8	0.0
322 Wearing apparel	17	2.3	19	2.8	25[e]	2.8	16	1.7	-5.9	-36.0
323 Leather and fur products	6	0.8	3	0.4	5[e]	0.6	4	0.4	-33.3	-20.0
324 Footwear	9	1.2	6	0.9	10[e]	1.1	13	1.4	44.4	30.0
331 Wood and wood products	20	2.7	17	2.5	20[e]	2.2	17	1.8	-15.0	-15.0
332 Furniture and fixtures	9	1.2	8	1.2	12[e]	1.3	11	1.2	22.2	-8.3
341 Paper and paper products	34	4.6	23	3.4	36[e]	4.0	42	4.6	23.5	16.7
342 Printing and publishing	22	3.0	19	2.8	27[e]	3.0	27	2.9	22.7	0.0
351 Industrial chemicals	25	3.4	16	2.4	21[e]	2.4	17	1.8	-32.0	-19.0
352 Other chemical products	39	5.2	50	7.5	57[e]	6.4	67	7.3	71.8	17.5
353 Petroleum refineries	15	2.0	6	0.9	12[e]	1.3	7	0.8	-53.3	-41.7
354 Miscellaneous petroleum and coal products	NA	0.0	NA	0.0	NA	0.0	NA	0.0	NA	NA
355 Rubber products	25	3.4	27	4.0	40[e]	4.5	33	3.6	32.0	-17.5
356 Plastic products	14	1.9	13	1.9	18[e]	2.0	24	2.6	71.4	33.3
361 Pottery, china and earthenware	1	0.1	NA	0.0	1[e]	0.1	1	0.1	0.0	0.0
362 Glass and glass products	3	0.4	4	0.6	5[e]	0.6	5	0.5	66.7	0.0
369 Other non-metal mineral products	20	2.7	17	2.5	31[e]	3.5	42	4.6	110.0	35.5
371 Iron and steel	12a[e]	0.0	6a[e]	0.0	8b[e]	0.0	12a	0.0	NA	NA
372 Non-ferrous metals	a	0.0	a	0.0	-b[e]	0.0	a	0.0	NA	NA
381 Metal products	44	5.9	31	4.6	45[e]	5.0	64	6.9	45.5	42.2
382 Non-electrical machinery	6	0.8	4	0.6	6[e]	0.7	5	0.5	-16.7	-16.7
383 Electrical machinery	40	5.4	36	5.4	41[e]	4.6	44	4.8	10.0	7.3
384 Transport equipment	64	8.6	43	6.4	52[e]	5.8	39	4.2	-39.1	-25.0
385 Professional and scientific equipment	1	0.1	1	0.1	1[e]	0.1	2	0.2	100.0	100.0
390 Other manufacturing industries	6	0.8	8	1.2	14[e]	1.6	18	2.0	200.0	28.6

Note: The industry codes shown are International Standard Industry codes (ISIC). Percentages are percent of total Value Added. 'e' stands for estimated value

For sources, notes, and explanations, see Annotated Source Appendix, page 1035.

Finance, Economics, and Trade

Economic Indicators [36]

Millions of U.S. Dollars unless otherwise indicated.

	1989	1990	1991[99]
GDP at 1982 prices	4,589	4,322	3,755
Real GDP Growth Rate (%)	5.0	4.5	3.5
Real Per Capita GDP (1982 prices)	169	153	148
Money Supply (MI)	1,316	1,459	1,470
Commercial interest rate			
(maximum, %)	18.0	19.0	22.0
Savings Rate (minimum, %)	12.5	13.5	14.5
Investment Rate (percent)	NA	NA	NA
Inflation Rate [100]	10.5	12.6	22.0
WPI	NA	NA	NA

Balance of Payments Summary [37]

Values in millions of dollars.

	1987	1988	1989	1990	1991
Exports of goods (f.o.b.)	908.7	1,017.5	926.1	1,010.5	1,053.7
Imports of goods (f.o.b.)	-1,622.6	-1,802.2	-1,963.4	-2,008.7	-1,712.9
Trade balance	-713.9	-784.7	-1,037.3	-998.2	-659.2
Services - debits	-824.6	-895.5	-933.3	-1,101.3	-1,069.9
Services - credits	-85.7	-90.0	-99.5	-111.7	-59.0
Private transfers (net)	72.0	89.0	101.5	167.8	144.4
Government transfers (net)	142.6	257.1	281.1	206.9	204.5
Long term capital (net)	263.3	331.6	601.8	202.9	136.8
Short term capital (net)	98.5	50.7	32.0	136.7	-40.2
Errors and omissions	107.9	34.7	67.7	70.0	83.2
Overall balance	-24.3	-42.8	122.0	-92.4	-51.1

Exchange Rates [38]

Currency: **Kenyan shillings.**
Symbol: **KSh.**

Data are currency units per $1.

January 1993	36.227
1992	32.217
1991	27.508
1990	22.915
1989	20.572
1988	17.747

Imports and Exports

Top Import Origins [39]

$2.05 billion (f.o.b., 1992 est.). Data are for 1988.

Origins	%
EC	45
Asia	11
Middle East	12
US	5

Top Export Destinations [40]

$1.0 billion (f.o.b., 1992 est.). Data are for 1990.

Destinations	%
EC	44
Africa	25
Asia	5
US	5
Middle East	4

Foreign Aid [41]

	U.S. $	
US commitments, including Ex-Im (FY70-89)	839	million
Western (non-US) countries, ODA and OOF bilateral commitments (1970-89)	7,490	million
OPEC bilateral aid (1979-89)	74	million
Communist countries (1970-89)	83	million

Import and Export Commodities [42]

Import Commodities

Machinery, transp. equip. 29%
Petroleum, petroleum products 15%
Iron and steel 7%
Raw materials
Food and consumer goods

Export Commodities

Tea 25%
Coffee 18%
Petroleum products 11%

Korea, North

Geography [1]

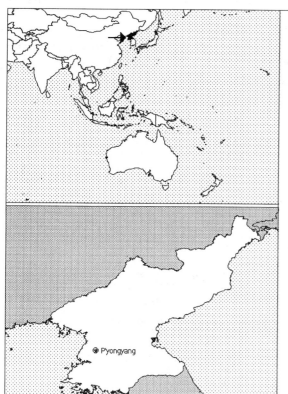

Total area:
120,540 km2
Land area:
120,410 km2
Comparative area:
Slightly smaller than Mississippi
Land boundaries:
Total 1,673 km, China 1,416 km, South Korea 238 km, Russia 19 km
Coastline:
2,495 km
Climate:
Temperate with rainfall concentrated in summer
Terrain:
Mostly hills and mountains separated by deep, narrow valleys; coastal plains wide in west, discontinuous in east
Natural resources:
Coal, lead, tungsten, zinc, graphite, magnesite, iron ore, copper, gold, pyrites, salt, fluorspar, hydropower
Land use:
Arable land:
18%
Permanent crops:
1%
Meadows and pastures:
0%
Forest and woodland:
74%
Other:
7%

Demographics [2]

	1960	1970	1980	1990	1991[1]	1994	2000	2010	2020
Population	10,568	14,388	17,999	21,412	21,815	23,067	25,491	28,491	30,969
Population density (persons per sq. mi.)	227	309	387	461	469	NA	548	613	666
Births	NA	NA	NA	NA	530	548	NA	NA	NA
Deaths	NA	NA	NA	NA	121	127	NA	NA	NA
Life expectancy - males	NA	NA	NA	NA	66	67	NA	NA	NA
Life expectancy - females	NA	NA	NA	NA	72	73	NA	NA	NA
Birth rate (per 1,000)	NA	NA	NA	NA	24	24	NA	NA	NA
Death rate (per 1,000)	NA	NA	NA	NA	6	6	NA	NA	NA
Women of reproductive age (15-44 yrs.)	NA	NA	NA	NA	NA	NA	NA	NA	NA
of which are currently married	NA	NA	NA	NA	NA	NA	NA	NA	NA
Fertility rate	NA	NA	NA	2.5	2.47	2.4	2.2	2.0	1.9

Population values are in thousands, life expectancy in years, and other items as indicated.

Health

Health Personnel [3]

Doctors per 1,000 pop., 1988-92	2.72
Nurse-to-doctor ratio, 1988-92	NA
Hospital beds per 1,000 pop., 1985-90	13.5
Percentage of children immunized (age 1 yr. or less)	
Third dose of DPT, 1990-91	90.0
Measles, 1990-91	96.0

Health Expenditures [4]

Total health expenditure, 1990 (official exchange rate)	
Millions of dollars	NA
Millions of dollars per capita	NA
Health expenditures as a percentage of GDP	
Total	NA
Public sector	NA
Private sector	NA
Development assistance for health	
Total aid flows (millions of dollars)[1]	NA
Aid flows per capita (millions of dollars)	NA
Aid flows as a percentage of total health expenditure	NA

For sources, notes, and explanations, see Annotated Source Appendix, page 1035.

Human Factors	
Health Care Ratios [5]	Infants and Malnutrition [6]

Ethnic Division [7]	Religion [8]	Major Languages [9]
Racially homogeneous.	Buddhism and Confucianism, some Christianity and syncretic Chondogyo.	Korean.

Education

Public Education Expenditures [10]	Educational Attainment [11]

Literacy Rate [12]	Libraries [13]

Daily Newspapers [14]

	1975	1980	1985	1990
Number of papers	11	11	11	11
Circ. (000)	3,500[e]	4,000[e]	4,500[e]	5,000[e]

Cinema [15]

Science and Technology

Scientific/Technical Forces [16]

R&D Expenditures [17]

	Won (000) 1988
Total expenditure[12,22]	2,347,000
Capital expenditure	925,000
Current expenditure	1,422,000
Percent current	60.6

U.S. Patents Issued [18]

Values show patents issued to citizens of the country by the U.S. Patents Office.

	1990	1991	1992
Number of patents	0	2	0

For sources, notes, and explanations, see Annotated Source Appendix, page 1035.

Government and Law

Organization of Government [19]

Long-form name:
Democratic People's Republic of Korea
Type:
Communist state; Stalinist dictatorship
Independence:
9 September 1948
Constitution:
adopted 1948, completely revised 27 December 1972, revised again in April 1992
Legal system:
based on German civil law system with Japanese influences and Communist legal theory; no judicial review of legislative acts; has not accepted compulsory ICJ jurisdiction
National holiday:
DPRK Foundation Day, 9 September (1948)
Executive branch:
president, two vice presidents, premier, ten vice premiers, State Administration Council (cabinet)
Legislative branch:
unicameral Supreme People's Assembly (Ch'oego Inmin Hoeui)
Judicial branch:
Central Court

Crime [23]

Elections [20]

Supreme People's Assembly. Last held on 7-9 April 1993 (next to be held NA); results - percent of vote by party NA; seats - (687 total) the KWP approves a single list of candidates who are elected without opposition; minor parties hold a few seats.

Government Budget [21]

For 1992.

Revenues	18.5
Expenditures	18.4
Capital expenditures	NA

Military Expenditures and Arms Transfers [22]

	1985	1986	1987	1988	1989
Military expenditures					
Current dollars (mil.)[e]	5,260	5,440	5,640	5,840	6,000
1989 constant dollars (mil.)[e]	5,988	6,037	6,066	6,079	6,000
Armed forces (000)	784	838	838	842	1,040
Gross national product (GNP)					
Current dollars (mil.)[e]	26,300	27,200	28,200	29,200	30,000
1989 constant dollars (mil.)[e]	29,940	30,180	30,330	30,400	30,000
Central government expenditures (CGE)					
1989 constant dollars (mil.)	NA	NA	NA	NA	NA
People (mil.)	19.6	19.9	20.3	20.6	21.0
Military expenditure as % of GNP	20	20	20	20	20
Military expenditure as % of CGE	NA	NA	NA	NA	NA
Military expenditure per capita	305	303	299	294	285
Armed forces per 1,000 people	40.0	42.0	41.3	40.8	49.5
GNP per capita	1,527	1,513	1,495	1,472	1,427
Arms imports[6]					
Current dollars (mil.)	380	420	420	1,000	525
1989 constant dollars (mil.)	433	466	452	1,041	525
Arms exports[6]					
Current dollars (mil.)	350	250	400	470	400
1989 constant dollars (mil.)	398	277	430	489	400
Total imports[7]					
Current dollars (mil.)	1,720	2,000	NA	3,100	NA
1989 constant dollars	1,958	2,219	NA	3,227	NA
Total exports[7]					
Current dollars (mil.)	1,380	1,700	NA	2,400	NA
1989 constant dollars	1,571	1,887	NA	2,498	NA
Arms as percent of total imports[8]	22.1	21.0	NA	32.3	NA
Arms as percent of total exports[8]	25.4	14.7	NA	19.6	NA

Human Rights [24]

	SSTS	FL	FAPRO	PPCG	APROBC	TPW	PCPTW	STPEP	PHRFF	PRW	ASST	AFL
Observes				P		P	P					
	EAFRD	CPR	ESCR	SR	ACHR	MAAE	PVIAC	PVNAC	EAFDAW	TCIDTP	RC	
Observes	P	P	P	P			P		P	P	P	

P = Party; S = Signatory; see Appendix for meaning of abbreviations.

Labor Force

Total Labor Force [25]

9.615 million

Labor Force by Occupation [26]

Agricultural	36%
Nonagricultural	64

Unemployment Rate [27]

For sources, notes, and explanations, see Annotated Source Appendix, page 1035.

Production Sectors

Energy Resource Summary [28]

Energy Resources: Coal, hydropower. **Electricity**: 7,300,000 kW capacity; 26,000 million kWh produced, 1,160 kWh per capita (1992).
Pipelines: Crude oil 37 km.

Telecommunications [30]

- Broadcast stations - 18 AM, no FM, 11 TV
- 300,000 TV sets (1989)
- 3,500,000 radio receivers
- 1 Indian Ocean INTELSAT earth station

Top Agricultural Products [32]

Agriculture accounts for about 25% of GNP and 36% of work force; principal crops—rice, corn, potatoes, soybeans, pulses; livestock and livestock products—cattle, hogs, pork, eggs; not self-sufficient in grain; fish catch estimated at 1.7 million metric tons in 1987.

Top Mining Products [33]

Metric tons unless otherwise specified	M.t.
Iron ore and concentrate, gross weight (000 tons)	10,000
Lead, mine output, Pb content	80,000
Cement, hydraulic (000 tons)	16,000
Chemical fertilizers	NA
Coal (000 tons)	90,000

Transportation [31]

Railroads. 4,915 km total; 4,250 km 1.435-meter standard gauge, 665 km 0.762-meter narrow gauge; 159 km double track; 3,084 km electrified; government owned (1989)

Highways. about 30,000 km (1991); 92.5% gravel, crushed stone, or earth surface; 7.5% paved

Merchant Marine. 80 ships (1,000 GRT and over) totaling 675,666 GRT/1,057,815 DWT; includes 1 passenger, 1 short-sea passenger, 2 passenger-cargo, 67 cargo, 2 oil tanker, 5 bulk, 1 combination bulk, 1 container

Airports

Total:	55
Usable :	55 (est.)
With permanent-surface runways:	about 30
With runways over 3,659 m:	fewer than 5
With runways 2,440-3,659 m:	20
With runways 1,220-2,439 m:	30

Tourism [34]

For sources, notes, and explanations, see Annotated Source Appendix, page 1035.

491

Finance, Economics, and Trade

GDP and Manufacturing Summary [35]

	1980	1985	1990	1991	1992
Gross Domestic Product					
Millions of 1980 dollars	12,730	20,368	26,618	25,234	23,972
Growth rate in percent	9.89	9.59	5.60	-5.20	-5.00
Manufacturing Value Added					
Millions of 1980 dollars	NA	NA	NA	NA	NA
Growth rate in percent	NA	NA	NA	NA	NA
Manufacturing share in percent of current prices	NA	NA	NA	NA	

Economic Indicators [36]

National product: GNP—purchasing power equivalent—$22 billion (1992 est.). **National product real growth rate**: - 10% to - 15% (1992 est.). **National product per capita**: $1,000 (1992 est.). **Inflation rate (consumer prices)**: NA%. **External debt**: $8 billion (1992 est.).

Balance of Payments Summary [37]

Exchange Rates [38]

Currency: **North Korean won.**
Symbol: **Wn.**

Data are currency units per $1.

May 1992	2.13
September 1991	2.14
January 1990	2.10
December 1989	2.30
December 1988	2.13
March 1987	0.94

Imports and Exports

Top Import Origins [39]

$1.9 billion (f.o.b., 1992 est.).

Origins	%
China	NA
Russia	NA
Japan	NA
Hong Kong	NA
Germany	NA
Singapore	NA

Top Export Destinations [40]

$1.3 billion (f.o.b., 1992 est.).

Destinations	%
China	NA
Japan	NA
Russia	NA
South Korea	NA
Germany	NA
Hong Kong	NA
Mexico	NA

Foreign Aid [41]

	U.S. $	
Communist countries (per year in the 1980s)	1.4	billion

Import and Export Commodities [42]

Import Commodities

Petroleum
Grain
Coking coal
Machinery & equipment
Consumer goods

Export Commodities

Minerals
Metallurgical products
Agric./fishery products
Manufactures
Armaments

Korea, South

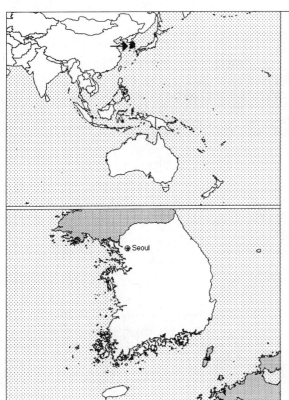

Geography [1]

Total area:
98,480 km2
Land area:
98,190 km2
Comparative area:
Slightly larger than Indiana
Land boundaries:
Total 238 km, North Korea 238 km
Coastline:
2,413 km
Climate:
Temperate, with rainfall heavier in summer than winter
Terrain:
Mostly hills and mountains; wide coastal plains in west and south
Natural resources:
Coal, tungsten, graphite, molybdenum, lead, hydropower
Land use:
Arable land:
21%
Permanent crops:
1%
Meadows and pastures:
1%
Forest and woodland:
67%
Other:
10%

Demographics [2]

	1960	1970	1980	1990	1991[1]	1994	2000	2010	2020
Population	24,784	32,241	38,124	43,237	43,134	45,083	47,861	51,677	54,014
Population density (persons per sq. mi.)	654	850	1,006	1,129	1,138	NA	1,212	1,268	1,283
Births	NA	NA	NA	NA	640	708	NA	NA	NA
Deaths	NA	NA	NA	NA	266	278	NA	NA	NA
Life expectancy - males	NA	NA	NA	NA	67	67	NA	NA	NA
Life expectancy - females	NA	NA	NA	NA	73	74	NA	NA	NA
Birth rate (per 1,000)	NA	NA	NA	NA	15	16	NA	NA	NA
Death rate (per 1,000)	NA	NA	NA	NA	6	6	NA	NA	NA
Women of reproductive age (15-44 yrs.)	NA	NA	NA	12,167	NA	12,872	13,816	NA	NA
of which are currently married	NA	NA	NA	7,293	NA	8,075	9,098	NA	NA
Fertility rate	NA	NA	NA	1.6	1.58	1.7	1.7	1.7	1.7

Population values are in thousands, life expectancy in years, and other items as indicated.

Health

Health Personnel [3]

Doctors per 1,000 pop., 1988-92	0.73
Nurse-to-doctor ratio, 1988-92	1.0
Hospital beds per 1,000 pop., 1985-90	3.0
Percentage of children immunized (age 1 yr. or less)	
Third dose of DPT, 1990-91	74.0
Measles, 1990-91	93.0

Health Expenditures [4]

Total health expenditure, 1990 (official exchange rate)	
Millions of dollars	16130
Millions of dollars per capita	377
Health expenditures as a percentage of GDP	
Total	6.6
Public sector	2.7
Private sector	3.9
Development assistance for health	
Total aid flows (millions of dollars)[1]	32
Aid flows per capita (millions of dollars)	NA
Aid flows as a percentage of total health expenditure	0.2

For sources, notes, and explanations, see Annotated Source Appendix, page 1035.

493

Human Factors

Health Care Ratios [5]

Population per physician, 1970	2220
Population per physician, 1990	1370
Population per nursing person, 1970	1190
Population per nursing person, 1990	NA
Percent of births attended by health staff, 1985	65

Infants and Malnutrition [6]

Percent of babies with low birth weight, 1985	9
Infant mortality rate per 1,000 live births, 1970	51
Infant mortality rate per 1,000 live births, 1991	16
Years of life lost per 1,000 population, 1990	10
Prevalence of malnutrition (under age 5), 1990	NA

Ethnic Division [7]

Homogeneous (except for about 20,000 Chinese).

Religion [8]

Christianity	48.6%
Buddhism	47.4%
Confucianism	3.0%
Pervasive folk religion	0.2%

Major Languages [9]

Korean, English widely taught in high school.

Education

Public Education Expenditures [10]

Billion Won	1980	1985	1987	1988	1989	1990
Total education expenditure	1,374.7	3,530.1	NA	4,047.2	5,139.4	6,159.1
as percent of GNP	3.7	4.5	NA	3.2	3.6	3.6
as percent of total govt. expend.	23.7	28.2	NA	23.2	23.3	22.4
Current education expenditure	1,159.0	2,811.9	NA	3,488.1	4,348.1	5,495.2
as percent of GNP	3.2	3.6	NA	2.8	3.1	3.2
as percent of current govt. expend.	NA	28.3	NA	26.2	24.7	24.3
Capital expenditure	215.8	718.2	NA	559.0	791.2	663.9

Educational Attainment [11]

Literacy Rate [12]

In thousands and percent	1985[a]	1991[a]	2000[a]
Illiterate population +15 years	1,524	1,185	702
Illiteracy rate - total pop. (%)	5.3	3.7	1.9
Illiteracy rate - males (%)	1.7	0.9	0.3
Illiteracy rate - females (%)	8.9	6.5	3.5

Libraries [13]

	Admin. Units	Svc. Pts.	Vols. (000)	Shelving (meters)	Vols. Added	Reg. Users
National (1991)	1	2	2,451	NA	NA	2.1mil
Non-specialized (1989)	1	1	686	NA	48,709	108,049
Public (1991)	266	266	6,419	NA	NA	15 mil
Higher ed. (1990)	305	NA	26,554	NA	NA	97 mil
School (1990)	6,468	NA	27,675	NA	1.5mil	27 mil

Daily Newspapers [14]

	1975	1980	1985	1990
Number of papers	36	30	35	39
Circ. (000)	6,010	8,000	10,000[e]	12,000[e]

Cinema [15]

Data for 1989.

Cinema seats per 1,000	NA
Annual attendance per person	1.3
Gross box office receipts (mil. Won)	125,588

Science and Technology

Scientific/Technical Forces [16]

Potential scientists/engineers	94,171[e]
Number female	NA
Potential technicians[5]	1,931,468[e]
Number female	NA
Total[5]	2,025,639[e]

R&D Expenditures [17]

U.S. Patents Issued [18]

Values show patents issued to citizens of the country by the U.S. Patents Office.

	1990	1991	1992
Number of patents	290	446	586

For sources, notes, and explanations, see Annotated Source Appendix, page 1035.

Government and Law

Organization of Government [19]

Long-form name:
Republic of Korea
Type:
republic
Independence:
15 August 1948
Constitution:
25 February 1988
Legal system:
combines elements of continental
European civil law systems, Anglo-
American law, and Chinese classical
thought
National holiday:
Independence Day, 15 August (1948)
Executive branch:
president, prime minister, two deputy
prime ministers, State Council (cabinet)
Legislative branch:
unicameral National Assembly (Kuk Hoe)
Judicial branch:
Supreme Court

Political Parties [20]

National Assembly	% of votes
Democratic Liberal Party	38.5
Democratic Party	29.2
Unification National Party (UPP)	17.3
Other	15.0

Government Expenditures [21]

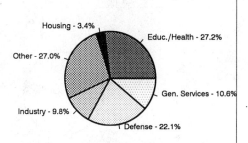

Housing - 3.4%
Educ./Health - 27.2%
Other - 27.0%
Gen. Services - 10.6%
Industry - 9.8%
Defense - 22.1%

% distribution for 1992. Expend. in 1993: 48.4 ($ bil.)

Crime [23]

Crime volume	
Cases known to police	1,147,752
Attempts (percent)	NA
Percent cases solved	91.4
Crimes per 100,000 persons	2,637.3
Persons responsible for offenses	
Total number offenders	1,146,745
Percent female	9.6
Percent juvenile[18]	7.3
Percent foreigners	NA

Military Expenditures and Arms Transfers [22]s0b4T

	1985	1986	1987	1988	1989
Military expenditures					
Current dollars (mil.)	6,109	6,745	7,120	8,030	9,100
1989 constant dollars (mil.)	6,955	7,485	7,658	8,359	9,100
Armed forces (000)	600	604	604	626	647
Gross national product (GNP)					
Current dollars (mil.)	120,600	139,700	162,500	188,400	210,100
1989 constant dollars (mil.)	137,200	155,000	174,800	196,100	210,100
Central government expenditures (CGE)					
1989 constant dollars (mil.)	26,130	27,270	30,080	33,130	38,240
People (mil.)	41.1	41.5	42.0	42.4	42.7
Military expenditure as % of GNP	5.1	4.8	4.4	4.3	4.3
Military expenditure as % of CGE	26.6	27.5	25.5	25.2	23.8
Military expenditure per capita	169	180	182	197	213
Armed forces per 1,000 people	14.6	14.5	14.4	14.8	15.1
GNP per capita	3,343	3,731	4,163	4,630	4,920
Arms imports[6]					
Current dollars (mil.)	430	550	625	675	370
1989 constant dollars (mil.)	490	610	672	703	370
Arms exports[6]					
Current dollars (mil.)	170	130	70	70	40
1989 constant dollars (mil.)	194	144	75	73	40
Total imports[7]					
Current dollars (mil.)	31,140	31,580	41,020	51,810	61,300
1989 constant dollars	35,450	35,050	44,120	53,930	61,300
Total exports[7]					
Current dollars (mil.)	30,280	34,710	47,280	60,700	62,330
1989 constant dollars	34,470	38,520	50,850	63,190	62,330
Arms as percent of total imports[8]	1.4	1.7	1.5	1.3	0.6
Arms as percent of total exports[8]	0.6	0.4	0.1	0.1	0.1

Human Rights [24]

	SSTS	FL	FAPRO	PPCG	APROBC	TPW	PCPTW	STPEP	PHRFF	PRW	ASST	AFL
Observes				P		P	P	P		P		
	EAFRD	CPR	ESCR	SR	ACHR	MAAE	PVIAC	PVNAC	EAFDAW	TCIDTP	RC	
Observes	P							P	P		P	

P = Party; S = Signatory; see Appendix for meaning of abbreviations.

Labor Force

Total Labor Force [25]

19 million

Labor Force by Occupation [26]

Services and other	52%
Mining and manufacturing	27
Agriculture, fishing, forestry	21
Date of data: 1991	

Unemployment Rate [27]

2.4% (1992 est.)

For sources, notes, and explanations, see Annotated Source Appendix, page 1035.

495

Production Sectors

Commercial Energy Production and Consumption

Production [28]

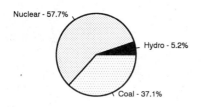

Nuclear - 57.7%
Hydro - 5.2%
Coal - 37.1%

Consumption [29]

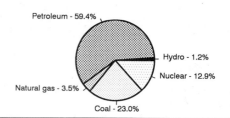

Petroleum - 59.4%
Hydro - 1.2%
Nuclear - 12.9%
Natural gas - 3.5%
Coal - 23.0%

Data are shown in quadrillion (10^{15}) BTUs and percent for 1991

Crude oil[1]	0.00
Natural gas liquids	0.00
Dry natural gas	0.00
Coal[2]	0.36
Net hydroelectric power[3]	0.05
Net nuclear power[3]	0.56
Total	0.97

Petroleum	2.58
Dry natural gas	0.15
Coal[2]	1.00
Net hydroelectric power[3]	0.05
Net nuclear power[3]	0.56
Total	4.34

Telecommunications [30]

- Excellent domestic and international services
- 13,276,449 telephone subscribers
- Broadcast stations - 79 AM, 46 FM, 256 TV (57 of 1 kW or greater)
- Satellite earth stations - 2 Pacific Ocean INTELSAT and 1 Indian Ocean INTELSAT

Transportation [31]

Railroads. 3,091 km total (1991); 3,044 km 1.435 meter standard gauge, 47 km 0.610-meter narrow gauge, 847 km double track; 525 km electrified, government owned

Highways. 63,201 km total (1991); 1,551 expressways, 12,190 km national highway, 49,460 km provincial and local roads

Merchant Marine. 431 ships (1,000 GRT or over) totaling 6,689,227 GRT/11,016,014 DWT; includes 2 short-sea passenger, 138 cargo, 61 container, 11 refrigerated cargo, 9 vehicle carrier, 45 oil tanker, 12 chemical tanker, 13 liquefied gas, 2 combination ore/oil, 135 bulk, 2 combination bulk, 1 multifunction large-load carrier

Airports

Total:	103
Usable:	93
With permanent-surface runways:	59
With runways over 3,659 m:	0
With runways 2,440-3,659 m:	22
With runways 1,220-2,439 m:	18

Top Agricultural Products [32]

	1990	1991
Rice	5,606	5,384
Chinese cabbage	3,241	2,550
Radish	1,686	1,400
Apples	629	540
Green onions	407	530
Barley	417	340

Values shown are 1,000 metric tons.

Top Mining Products [33]

Preliminary metric tons unless otherwise specified M.t.

Anthracite coal (000 tons)	14,850
Lead, mine output, Pb content	12,633
Zinc, mine output, Zn content	22,039
Tungsten, mine output, W content	780
Diatomaceous earth	91,126
Feldspar	247,969
Graphite	76,791
Mica (all grades)	5,127
Pyrophyllite	573,208
Talc	170,563

Tourism [34]

	1987	1988	1989	1990	1991
Visitors	1,875	2,340	2,728	2,959	3,196
Cruise passengers		160	301	370	481
Tourism receipts	2,299	3,265	3,556	3,559	3,426
Tourism expenditures	704	1,354	2,602	3,166	3,784
Fare receipts	661	707	818	849	933
Fare expenditures	142	182	231	352	490

Tourists are in thousands, money in million U.S. dollars.

Manufacturing Sector

GDP and Manufacturing Summary [35]

	1980	1985	1989	1990	% change 1980-1990	% change 1989-1990
GDP (million 1980 $)	62,626	93,782	132,085	151,849	142.5	15.0
GDP per capita (1980 $)	1,643	2,298	3,114	3,545	115.8	13.8
Manufacturing as % of GDP (current prices)	29.6	30.3	31.3	28.8	-2.7	-8.0
Gross output (million $)	59,725	88,541	219,760	250,507	319.4	14.0
Value added (million $)	19,520	30,731	78,922	100,210	413.4	27.0
Value added (million 1980 $)	18,600	31,629	44,457	56,673	204.7	27.5
Industrial production index	100	171	304	332	232.0	9.2
Employment (thousands)	2,015	2,395	3,137[e]	2,958	46.8	-5.7

Note: GDP stands for Gross Domestic Product. 'e' stands for estimated value.

Profitability and Productivity

	1980	1985	1989	1990	% change 1980-1990	% change 1989-1990
Intermediate input (%)	67	65	64	60	-10.4	-6.3
Wages, salaries, and supplements (%)	10	9	10[e]	11	10.0	10.0
Gross operating surplus (%)	23	25	26[e]	29	26.1	11.5
Gross output per worker ($)	29,206	36,314	70,063[e]	82,955	184.0	18.4
Value added per worker ($)	9,545	12,604	25,162[e]	33,185	247.7	31.9
Average wage (incl. benefits) ($)	2,837	3,476	7,221[e]	9,353	229.7	29.5

Profitability is in percent of gross output. Productivity is in U.S. $. 'e' stands for estimated value.

Profitability - 1990

Inputs - 60.0%
Wages - 11.0%
Surplus - 29.0%

The graphic shows percent of gross output.

Value Added in Manufacturing

	1980 $ mil.	1980 %	1985 $ mil.	1985 %	1989 $ mil.	1989 %	1990 $ mil.	1990 %	% change 1980-1990	% change 1989-1990
311 Food products	1,526	7.8	2,048	6.7	4,693[e]	5.9	6,047	6.0	296.3	28.9
313 Beverages	571	2.9	764	2.5	1,679[e]	2.1	1,889	1.9	230.8	12.5
314 Tobacco products	1,143	5.9	1,442	4.7	2,736[e]	3.5	2,793	2.8	144.4	2.1
321 Textiles	2,649	13.6	3,295	10.7	7,004[e]	8.9	6,833	6.8	157.9	-2.4
322 Wearing apparel	905	4.6	1,293	4.2	3,045[e]	3.9	3,401	3.4	275.8	11.7
323 Leather and fur products	138	0.7	270	0.9	807[e]	1.0	1,144	1.1	729.0	41.8
324 Footwear	112	0.6	211	0.7	498[e]	0.6	593	0.6	429.5	19.1
331 Wood and wood products	239	1.2	262	0.9	601[e]	0.8	876	0.9	266.5	45.8
332 Furniture and fixtures	100	0.5	203	0.7	627[e]	0.8	972	1.0	872.0	55.0
341 Paper and paper products	426	2.2	682	2.2	1,967[e]	2.5	2,122	2.1	398.1	7.9
342 Printing and publishing	440	2.3	732	2.4	1,852[e]	2.3	2,531	2.5	475.2	36.7
351 Industrial chemicals	998	5.1	1,275	4.1	2,862[e]	3.6	4,182	4.2	319.0	46.1
352 Other chemical products	1,016	5.2	1,422	4.6	3,615[e]	4.6	4,925	4.9	384.7	36.2
353 Petroleum refineries	757	3.9	1,079	3.5	2,064[e]	2.6	2,865	2.9	278.5	38.8
354 Miscellaneous petroleum and coal products	211	1.1	291	0.9	480[e]	0.6	517	0.5	145.0	7.7
355 Rubber products	657	3.4	910	3.0	2,364[e]	3.0	3,063	3.1	366.2	29.6
356 Plastic products	359	1.8	709	2.3	2,144[e]	2.7	2,734	2.7	661.6	27.5
361 Pottery, china and earthenware	89	0.5	107	0.3	245[e]	0.3	274	0.3	207.9	11.8
362 Glass and glass products	198	1.0	307	1.0	666[e]	0.8	992	1.0	401.0	48.9
369 Other non-metal mineral products	838	4.3	1,064	3.5	2,513[e]	3.2	3,698	3.7	341.3	47.2
371 Iron and steel	1,256	6.4	2,040	6.6	5,297[e]	6.7	6,187	6.2	392.6	16.8
372 Non-ferrous metals	265	1.4	334	1.1	1,270[e]	1.6	1,201	1.2	353.2	-5.4
381 Metal products	635	3.3	1,237	4.0	3,707[e]	4.7	5,144	5.1	710.1	38.8
382 Non-electrical machinery	672	3.4	1,453	4.7	4,995[e]	6.3	7,004	7.0	942.3	40.2
383 Electrical machinery	1,587	8.1	3,621	11.8	12,257[e]	15.5	15,066	15.0	849.3	22.9
384 Transport equipment	1,152	5.9	2,791	9.1	6,429[e]	8.1	10,242	10.2	789.1	59.3
385 Professional and scientific equipment	214	1.1	290	0.9	932[e]	1.2	1,144	1.1	434.6	22.7
390 Other manufacturing industries	367	1.9	598	1.9	1,572	2.0	1,769	1.8	382.0	12.5

Note: The industry codes shown are International Standard Industry codes (ISIC). Percentages are percent of total Value Added. 'e' stands for estimated value

For sources, notes, and explanations, see Annotated Source Appendix, page 1035.

497

Finance, Economics, and Trade

Economic Indicators [36]

Billions of Korean Won unless otherwise indicated[101].

	1989	1990	1991[e]
Real GNP (constant 1985)	119,577	130,373	141,455
Real GNP Growth Rate (%)	6.8	9.0	8.7[102]
Real Per Capita GNP			
(000s won, 1985 prices)	2,828	3,051	3,274
Money Supply (M1)[37]	14,329	15,905	18,609
Commercial Bank Rate (%)	12.5	12.5	12.5
Savings Rate (%)	35.3	35.3	36.0
Investment Rate (%)	33.5	37.1	38.0
CPI (1985 = 100)	119.9	130.2	143.2
WPI (1985 = 100)	103.2	107.5	114.5
External Debt Outstanding[37]	19,980	22,710	29,250

Balance of Payments Summary [37]

Values in millions of dollars.

	1987	1988	1989	1990	1991
Exports of goods (f.o.b.)	46,244.0	59,648.0	61,408.0	63,123.0	69,580.9
Imports of goods (f.o.b.)	-38,585.0	-48,203.0	-56,811.0	-65,127.0	-76,561.0
Trade balance	7,659.0	11,445.0	4,597.0	-2,004.0	-6,980.0
Services - debits	-9,034.0	-9,984.0	-12,432.0	-14,712.0	-17,124.0
Services - credits	10,011.0	11,252.0	12,643.0	14,269.0	15,531.0
Private transfers (net)	1,199.0	1,404.0	200.0	266.0	20.0
Government transfers (net)	19.0	44.0	48.0	9.0	-173.0
Long term capital (net)	-8,472.0	-3,407.0	-3,904.0	-659.0	6,069.0
Short term capital (net)	-462.0	-847.0	1,278.0	3,628.0	756.0
Errors and omissions	1,184.0	-591.0	690.0	-2,005.0	753.0
Overall balance	2,104.0	9,316.0	3,120.0	-1,208.0	-1,148.0

Exchange Rates [38]

Currency: **South Korean won.**
Symbol: **W.**

Data are currency units per $1.

January 1993	791.99
1992	780.65
1991	733.35
1990	707.76
1989	671.46
1988	731.47

Imports and Exports

Top Import Origins [39]

$81.7 billion (c.i.f., 1992).

Origins	%
Japan	24
US	22

Top Export Destinations [40]

$76.8 billion (f.o.b., 1992).

Destinations	%
US	24
Japan	15

Foreign Aid [41]

	U.S. $	
US commitments, including Ex-Im (FY70-89)	3.9	billion
Non-US countries (1970-89)	3.0	billion

Import and Export Commodities [42]

Import Commodities	Export Commodities
Machinery	Textiles
Electronics and electronic equipment	Clothing
Oil	Electronic and electrical equipment
Steel	Footwear
Transport equipment	Machinery
Textiles	Steel
Organic chemicals	Automobiles
Grains	Ships
	Fish

Kuwait

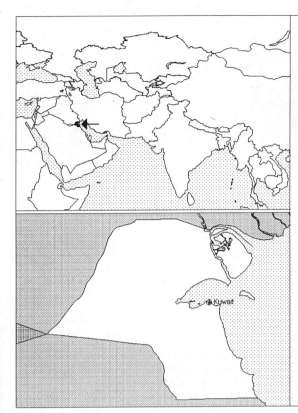

Geography [1]

Total area:
17,820 km2
Land area:
17,820 km2
Comparative area:
Slightly smaller than New Jersey
Land boundaries:
Total 464 km, Iraq 242 km, Saudi Arabia 222 km
Coastline:
499 km
Climate:
Dry desert; intensely hot summers; short, cool winters
Terrain:
Flat to slightly undulating desert plain
Natural resources:
Petroleum, fish, shrimp, natural gas
Land use:
Arable land:
0%
Permanent crops:
0%
Meadows and pastures:
8%
Forest and woodland:
0%
Other:
92%

Demographics [2]

	1960	1970	1980	1990	1991[1]	1994	2000	2010	2020
Population	292	748	1,370	2,117	2,204	1,819	2,494	3,220	4,091
Population density (persons per sq. mi.)	42	109	199	309	320	NA	418	528	663
Births	NA	NA	NA	NA	64	54	NA	NA	NA
Deaths	NA	NA	NA	NA	5	4	NA	NA	NA
Life expectancy - males	NA	NA	NA	NA	72	73	NA	NA	NA
Life expectancy - females	NA	NA	NA	NA	76	77	NA	NA	NA
Birth rate (per 1,000)	NA	NA	NA	NA	29	29	NA	NA	NA
Death rate (per 1,000)	NA	NA	NA	NA	2	2	NA	NA	NA
Women of reproductive age (15-44 yrs.)	NA	NA	NA	504	NA	422	590	NA	NA
of which are currently married	NA	NA	NA	338	NA	280	392	NA	NA
Fertility rate	NA	NA	NA	3.5	3.71	4.0	3.7	3.8	3.8

Population values are in thousands, life expectancy in years, and other items as indicated.

Health

Health Personnel [3]

Health Expenditures [4]

For sources, notes, and explanations, see Annotated Source Appendix, page 1035.

499

Human Factors

Health Care Ratios [5]	Infants and Malnutrition [6]

Ethnic Division [7]

Kuwaiti	45%
Other Arab	35%
South Asian	9%
Iranian	4%
Other	7%

Religion [8]

Muslim	85%
Shi'a	30%
Sunni	45%
Other Muslim	10%
Christian, Hindu, Parsi, and other	15%

Major Languages [9]

Arabic (official), English widely spoken.

Education

Public Education Expenditures [10]

Million Dinar	1980	1985	1986	1988[1]	1989[1]	1990
Total education expenditure	219	NA	360	305	318	NA
as percent of GNP	2.4	NA	5.0	4.0	3.4	NA
as percent of total govt. expend.	8.1	NA	11.9	10.2	9.6	NA
Current education expenditure	204	NA	344	NA	NA	NA
as percent of GNP	2.3	NA	4.8	NA	NA	NA
as percent of current govt. expend.	11.2	NA	NA	NA	NA	NA
Capital expenditure	15	NA	16	NA	NA	NA

Educational Attainment [11]

Age group	25+
Total population	565,330
Highest level attained (%)	
No schooling	50.3
First level	
Incompleted	7.2
Completed	NA
Entered second level	
S-1	10.7
S-2	19.0
Post secondary	12.7

Literacy Rate [12]

In thousands and percent	1985[a]	1991[a]	2000[a]
Illiterate population +15 years	302	346	414
Illiteracy rate - total pop. (%)	29.4	37.0	22.8
Illiteracy rate - males (%)	24.7	22.9	19.6
Illiteracy rate - females (%)	36.8	33.3	27.5

Libraries [13]

	Admin. Units	Svc. Pts.	Vols. (000)	Shelving (meters)	Vols. Added	Reg. Users
National (1986)	1	4	93	NA	43,508	29,662
Non-specialized	NA	NA	NA	NA	NA	NA
Public (1987)	1	22	737*	NA	NA	585,206
Higher ed. (1988)[20]	1	14	453	NA	18,181	NA
School (1990)	587	NA	3,117	NA	155,875	820,105

Daily Newspapers [14]

	1975	1980	1985	1990
Number of papers	8	8	8	9
Circ. (000)	180	305	380	450[e]

Cinema [15]

Data for 1990.

Cinema seats per 1,000	8.1
Annual attendance per person	NA
Gross box office receipts (mil. Dinar)	NA

Science and Technology

Scientific/Technical Forces [16]

Potential scientists/engineers	60,398
Number female	NA
Potential technicians	70,983
Number female	NA
Total	131,381

R&D Expenditures [17]

	Dinar[17] (000) 1984
Total expenditure	71,163
Capital expenditure	8,147
Current expenditure	63,016
Percent current	88.6

U.S. Patents Issued [18]

Values show patents issued to citizens of the country by the U.S. Patents Office.

	1990	1991	1992
Number of patents	2	1	1

For sources, notes, and explanations, see Annotated Source Appendix, page 1035.

Government and Law

Organization of Government [19]

Long-form name:
State of Kuwait
Type:
nominal constitutional monarchy
Independence:
19 June 1961 (from UK)
Constitution:
16 November 1962 (some provisions
suspended since 29 August 1962)
Legal system:
civil law system with Islamic law
significant in personal matters; has not
accepted compulsory ICJ jurisdiction
National holiday:
National Day, 25 February
Executive branch:
amir, prime minister, deputy prime
minister, Council of Ministers (cabinet)
Legislative branch:
unicameral National Assembly (Majlis al
'umma) dissolved 3 July 1986; elections
for new Assembly held 5 October 1992
Judicial branch:
High Court of Appeal

Crime [23]

Elections [20]

National Assembly. No political parties;
assembly dissolved 3 July 1986; new
elections were held on 5 October 1992
with a second election in the 14th and
16th constituencies scheduled for 15
February 1993.

Government Expenditures [21]

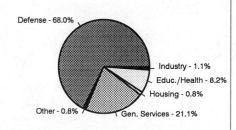

Defense - 68.0%

Industry - 1.1%
Educ./Health - 8.2%
Housing - 0.8%
Gen. Services - 21.1%
Other - 0.8%

% distribution for 1991. Expend. in FY88: 10.5 ($ bil.)

Military Expenditures and Arms Transfers [22]

	1985	1986	1987	1988	1989
Military expenditures					
Current dollars (mil.)	1,525	1,300	1,263	1,273	1,964
1989 constant dollars (mil.)	1,736	1,443	1,358	1,325	1,964
Armed forces (000)	16	18	20	15	20
Gross national product (GNP)					
Current dollars (mil.)	24,870	24,430	26,070	25,750	31,880
1989 constant dollars (mil.)[e]	28,310	27,110	28,040	26,800	31,880
Central government expenditures (CGE)					
1989 constant dollars (mil.)	11,920	11,240	9,687	9,560	9,875
People (mil.)	1.7	1.8	1.9	2.0	2.0
Military expenditure as % of GNP	6.1	5.3	4.8	4.9	6.2
Military expenditure as % of CGE	14.6	12.8	14.0	13.9	19.9
Military expenditure per capita	1,014	805	724	676	962
Armed forces per 1,000 people	9.3	10.0	10.7	7.7	9.9
GNP per capita	16,530	15,120	14,940	13,680	15,610
Arms imports[6]					
Current dollars (mil.)	350	140	160	210	490
1989 constant dollars (mil.)	398	155	172	219	490
Arms exports[6]					
Current dollars (mil.)	0	0	20	0	0
1989 constant dollars (mil.)	0	0	22	0	0
Total imports[7]					
Current dollars (mil.)	6,005	5,717	5,493	6,144	6,296
1989 constant dollars	6,836	6,344	5,908	6,396	6,296
Total exports[7]					
Current dollars (mil.)	10,490	7,383	8,264	7,661	11,480
1989 constant dollars	11,940	8,193	8,888	7,975	11,480
Arms as percent of total imports[8]	5.8	2.4	2.9	3.4	7.8
Arms as percent of total exports[8]	0	0	0.2	0	0

Human Rights [24]

	SSTS	FL	FAPRO	PPCG	APROBC	TPW	PCPTW	STPEP	PHRFF	PRW	ASST	AFL
Observes	P	P	P			P	P	P			P	P
		EAFRD	CPR	ESCR	SR	ACHR	MAAE	PVIAC	PVNAC	EAFDAW	TCIDTP	RC
Observes								P	P			P

P = Party; S = Signatory; see Appendix for meaning of abbreviations.

Labor Force

Total Labor Force [25]

566,000 (1986)

Labor Force by Occupation [26]

Services	45.0%
Construction	20.0
Trade	12.0
Manufacturing	8.6
Finance and real estate	2.6
Agriculture	1.9
Power and water	1.7
Mining and quarrying	1.4

Unemployment Rate [27]

NEGL% (1992 est.)

For sources, notes, and explanations, see Annotated Source Appendix, page 1035.

501

Production Sectors

Commercial Energy Production and Consumption

Production [28]

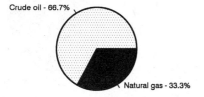

Crude oil - 66.7%

Natural gas - 33.3%

Consumption [29]

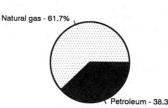

Natural gas - 61.7%

Petroleum - 38.3%

Data are shown in quadrillion (10^{15}) BTUs and percent for 1991

Crude oil[1]	0.40
Natural gas liquids	0.00
Dry natural gas	0.20
Coal[2]	0.00
Net hydroelectric power[3]	0.00
Net nuclear power[3]	0.00
Total	0.61

Petroleum	0.18
Dry natural gas	0.29
Coal[2]	0.00
Net hydroelectric power[3]	0.00
Net nuclear power[3]	0.00
Total	0.47

Telecommunications [30]

- Civil network suffered extensive damage as a result of Desert Storm and reconstruction is still under way with some restored international and domestic capabilities
- Broadcast stations - 3 AM, 0 FM, 3 TV
- Satellite earth stations - destroyed during Persian Gulf War and not rebuilt yet
- Temporary mobile satellite ground stations provide international telecommunications
- Coaxial cable and microwave radio relay to Saudi Arabia
- Service to Iraq is nonoperational

Top Agricultural Products [32]

Practically none; dependent on imports for food; about 75% of potable water must be distilled or imported.

Top Mining Products [33]

Metric tons unless otherwise specified	M.t.
Petroleum, crude (000 barrels)	68,225
Cement (000 tons)	300
Quicklime	5,000
Sulfur, elemental, petroleum byproduct	30,000

Transportation [31]

Railroads. none

Highways. 3,900 km total; 3,000 km bituminous; 900 km earth, sand, light gravel

Merchant Marine. 42 ships (1,000 GRT or over), totaling 1,996,052 GRT/3,373,088 DWT; includes 7 cargo, 4 livestock carrier, 24 oil tanker, 4 liquefied gas, 3 container

Airports

Total:	7
Usable:	4
With permanent-surface runways:	4
With runways over 3,659 m:	0
With runways 2,440-3,659 m:	4
With runways 1,220-2,439 m:	0

Tourism [34]

	1987	1988	1989	1990	1991
Visitors	1,130	1,256	1,938		3,196
Tourists	78[40]	80[40]	89[40]	50[9]	
Cruise passengers					481
Tourism receipts	75	108	123	80[9]	3,426
Tourism expenditures	2,257	2,358	2,318		3,784
Fare receipts	158	183	214		933
Fare expenditures	140	147	133		490

Tourists are in thousands, money in million U.S. dollars.

For sources, notes, and explanations, see Annotated Source Appendix, page 1035.

Manufacturing Sector

GDP and Manufacturing Summary [35]

	1980	1985	1989	1990	% change 1980-1990	% change 1989-1990
GDP (million 1980 $)	28,722	22,346	23,810	16,800	-41.5	-29.4
GDP per capita (1980 $)	20,889	12,984	12,044	8,207	-60.7	-31.9
Manufacturing as % of GDP (current prices)	5.6	5.9	22.0	12.8[e]	128.6	-41.8
Gross output (million $)	6,218	7,445	6,295[e]	5,233[e]	-15.8	-16.9
Value added (million $)	1,752	1,280	2,359[e]	2,213[e]	26.3	-6.2
Value added (million 1980 $)	1,581	1,689	2,639[e]	1,355	-14.3	-48.7
Industrial production index	100	142	176	125	25.0	-29.0
Employment (thousands)	43	46	50[e]	51[e]	18.6	2.0

Note: GDP stands for Gross Domestic Product. 'e' stands for estimated value.

Profitability and Productivity

	1980	1985	1989	1990	% change 1980-1990	% change 1989-1990
Intermediate input (%)	72	83	63[e]	58[e]	-19.4	-7.9
Wages, salaries, and supplements (%)	7[e]	8	12[e]	17[e]	142.9	41.7
Gross operating surplus (%)	21[e]	9	26[e]	25[e]	19.0	-3.8
Gross output per worker ($)	144,786	151,758	124,743[e]	91,050[e]	-37.1	-27.0
Value added per worker ($)	40,801	26,095	46,750[e]	38,511[e]	-5.6	-17.6
Average wage (incl. benefits) ($)	9,771[e]	13,015	14,573[e]	17,277[e]	76.8	18.6

Profitability is in percent of gross output. Productivity is in U.S. $. 'e' stands for estimated value.

Profitability - 1990

Wages - 17.0%
Inputs - 58.0%
Surplus - 25.0%

The graphic shows percent of gross output.

Value Added in Manufacturing

	1980 $ mil.	1980 %	1985 $ mil.	1985 %	1989 $ mil.	1989 %	1990 $ mil.	1990 %	% change 1980-1990	% change 1989-1990
311 Food products	96	5.5	101	7.9	146[e]	6.2	162[e]	7.3	68.8	11.0
313 Beverages	20	1.1	31	2.4	38[e]	1.6	38[e]	1.7	90.0	0.0
314 Tobacco products	NA	0.0	NA	0.0	NA	0.0	NA	0.0	NA	NA
321 Textiles	7	0.4	8	0.6	12[e]	0.5	20[e]	0.9	185.7	66.7
322 Wearing apparel	84	4.8	75	5.9	87[e]	3.7	123[e]	5.6	46.4	41.4
323 Leather and fur products	NA	0.0	NA	0.0	NA	0.0	NA	0.0	NA	NA
324 Footwear	NA	0.0	NA	0.0	NA	0.0	NA	0.0	NA	NA
331 Wood and wood products	40	2.3	14	1.1	10[e]	0.4	16[e]	0.7	-60.0	60.0
332 Furniture and fixtures	41	2.3	31	2.4	27[e]	1.1	36[e]	1.6	-12.2	33.3
341 Paper and paper products	5	0.3	12	0.9	22[e]	0.9	28[e]	1.3	460.0	27.3
342 Printing and publishing	40	2.3	52	4.1	47[e]	2.0	57[e]	2.6	42.5	21.3
351 Industrial chemicals	118	6.7	56	4.4	64[e]	2.7	72[e]	3.3	-39.0	12.5
352 Other chemical products	13	0.7	16	1.3	18[e]	0.8	30[e]	1.4	130.8	66.7
353 Petroleum refineries	915	52.2	561	43.8	1,505[e]	63.8	1,199[e]	54.2	31.0	-20.3
354 Miscellaneous petroleum and coal products	1	0.1	1	0.1	1[e]	0.0	2[e]	0.1	100.0	100.0
355 Rubber products	5	0.3	5	0.4	5[e]	0.2	7[e]	0.3	40.0	40.0
356 Plastic products	24	1.4	24	1.9	27[e]	1.1	47[e]	2.1	95.8	74.1
361 Pottery, china and earthenware	2	0.1	NA	0.0	4[e]	0.2	1[e]	0.0	-50.0	-75.0
362 Glass and glass products	2	0.1	4[e]	0.3	3[e]	0.1	8[e]	0.4	300.0	166.7
369 Other non-metal mineral products	143	8.2	115	9.0	146[e]	6.2	103[e]	4.7	-28.0	-29.5
371 Iron and steel	7	0.4	14	1.1	16[e]	0.7	16[e]	0.7	128.6	0.0
372 Non-ferrous metals	NA	0.0	NA	0.0	3[e]	0.1	NA	0.0	NA	NA
381 Metal products	99	5.7	88	6.9	98[e]	4.2	132[e]	6.0	33.3	34.7
382 Non-electrical machinery	10	0.6	30	2.3	28[e]	1.2	41[e]	1.9	310.0	46.4
383 Electrical machinery	22	1.3	15	1.2	21[e]	0.9	24[e]	1.1	9.1	14.3
384 Transport equipment	45	2.6	16[e]	1.3	19[e]	0.8	16[e]	0.7	-64.4	-15.8
385 Professional and scientific equipment	5	0.3	5[e]	0.4	7[e]	0.3	1[e]	0.0	-80.0	-85.7
390 Other manufacturing industries	7	0.4	5	0.4	4[e]	0.2	9[e]	0.4	28.6	125.0

Note: The industry codes shown are International Standard Industry codes (ISIC). Percentages are percent of total Value Added. 'e' stands for estimated value

For sources, notes, and explanations, see Annotated Source Appendix, page 1035.

503

Finance, Economics, and Trade

Economic Indicators [36]

Millions of Kuwaiti Dinars (KD) unless otherwise indicated.

	1989	1990	1991[e]
Nominal GDP	6,497	4,500	2,500
Nominal GDP Growth Rate	16.3	-30.8	-44.5
GNP per capita ($US)	14,800	12,200	9,500
Money Supply	860.1	904.3	NA
Interest Rate (Overnight)	8.33	8.46	NA
Savings Rate (% of GDP)	39.0	NA	NA
Investment Rate (% of GDP)	13.0	NA	NA
CPI	106.6	NA	NA
WPI	118.2	NA	NA
External Public Debt	NA	NA	NA

Balance of Payments Summary [37]

Values in millions of dollars.

	1985	1986	1987	1988	1989
Exports of goods (f.o.b.)	10,374.0	7,216.0	8,221.0	7,709.0	11,396.0
Imports of goods (f.o.b.)	-5,327.0	-5,007.0	-4,773.0	-5,447.0	-5,525.0
Trade balance	5,047.0	2,209.0	3,448.0	2,262.0	5,871.0
Services - debits	-4,741.0	-4,463.0	-4,543.0	-4,698.0	-4,953.0
Services - credits	6,417.0	9,164.0	6,897.0	8,784.0	10,164.0
Private transfers (net)	-1,044.0	-1,084.0	-1,102.0	-1,179.0	-1,283.0
Government transfers (net)	-529.0	-182.0	-158.0	-140.0	-211.0
Long term capital (net)	-712.0	-1,975.0	-240.0	-620.0	-943.0
Short term capital (net)	-1,623.0	-5,530.0	-4,568.0	-6,150.0	-6,753.0
Errors and omissions	-2,271.0	1,778.0	-1,581.0	-254.0	-638.0
Overall balance	544.0	-83.0	-1,847.0	-1,995.0	1,254.0

Exchange Rates [38]

Currency: **Kuwaiti dinars.**
Symbol: **KD.**

Data are currency units per $1.

January 1993	0.3044
1992	0.2934
1991	0.2843
1990	0.2915
1989	0.2937
1988	0.2790

Imports and Exports

Top Import Origins [39]

$4.7 billion (f.o.b., 1991 est.).

Origins	%
US	35
Japan	12
UK	9
Canada	9

Top Export Destinations [40]

$750 million (f.o.b., 1991 est.).

Destinations	%
France	16
Italy	15
Japan	12
UK	11

Foreign Aid [41]

	U.S. $	
Donor - pledged in bilateral aid to less developed countries (1979-89)	18.3	billion

Import and Export Commodities [42]

Import Commodities	Export Commodities
Food	Oil
Construction materials	
Vehicles and parts	
Clothing	

Kyrgyzstan

Geography [1]

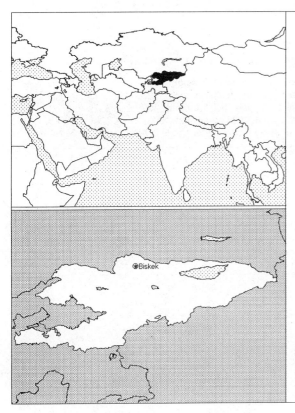

Total area:
198,500 km2
Land area:
191,300 km2
Comparative area:
Slightly smaller than South Dakota
Land boundaries:
Total 3,878 km, China 858 km, Kazakhstan 1,051 km, Tajikistan 870 km, Uzbekistan 1,099 km
Coastline:
0 km (landlocked)
Climate:
Dry continental to polar in high Tien Shan; subtropical in south (Fergana Valley)
Terrain:
Peaks of Tien Shan rise to 7,000 meters, and associated valleys and basins encompass entire nation
Natural resources:
Small amounts of coal, natural gas, oil, nepheline, rare earth metals, mercury, bismuth, gold, lead, zinc, hydroelectric power
Land use:
Arable land: NA%
Permanent crops: NA%
Meadows and pastures: NA%
Forest and woodland: NA%
Other: NA%

Demographics [2]

	1960	1970	1980	1990	1991[1]	1994	2000	2010	2020
Population	2,171	2,964	3,623	4,392	NA	4,698	5,119	5,810	6,490
Population density (persons per sq. mi.)	NA	NA	NA	NA	NA	NA	NA	NA	NA
Births	NA	NA	NA	NA	NA	124	NA	NA	NA
Deaths	NA	NA	NA	NA	NA	35	NA	NA	NA
Life expectancy - males	NA	NA	NA	NA	NA	64	NA	NA	NA
Life expectancy - females	NA	NA	NA	NA	NA	72	NA	NA	NA
Birth rate (per 1,000)	NA	NA	NA	NA	NA	26	NA	NA	NA
Death rate (per 1,000)	NA	NA	NA	NA	NA	7	NA	NA	NA
Women of reproductive age (15-44 yrs.)	NA	NA	NA	1,026	NA	1,133	1,302	NA	NA
of which are currently married	NA	NA	NA	675	NA	753	864	NA	NA
Fertility rate	NA	NA	NA	3.8	NA	3.4	3.1	2.7	2.4

Population values are in thousands, life expectancy in years, and other items as indicated.

Health

Health Personnel [3]

Doctors per 1,000 pop., 1988-92	3.67
Nurse-to-doctor ratio, 1988-92	2.8
Hospital beds per 1,000 pop., 1985-90	12.0
Percentage of children immunized (age 1 yr. or less)	
Third dose of DPT, 1990-91	78.0
Measles, 1990-91	94.0

Health Expenditures [4]

Total health expenditure, 1990 (official exchange rate)	
Millions of dollars	517
Millions of dollars per capita	118
Health expenditures as a percentage of GDP	
Total	5.0
Public sector	3.3
Private sector	1.6
Development assistance for health	
Total aid flows (millions of dollars)[1]	NA
Aid flows per capita (millions of dollars)	NA
Aid flows as a percentage of total health expenditure	NA

For sources, notes, and explanations, see Annotated Source Appendix, page 1035.

505

Human Factors

Health Care Ratios [5]

Population per physician, 1970	NA
Population per physician, 1990	280
Population per nursing person, 1970	NA
Population per nursing person, 1990	NA
Percent of births attended by health staff, 1985	NA

Infants and Malnutrition [6]

Percent of babies with low birth weight, 1985	NA
Infant mortality rate per 1,000 live births, 1970	NA
Infant mortality rate per 1,000 live births, 1991	40
Years of life lost per 1,000 population, 1990	20
Prevalence of malnutrition (under age 5), 1990	NA

Ethnic Division [7]

Kirghiz	52.4%
Russian	21.5%
Uzbek	12.9%
Ukrainian	2.5%
German	2.4%
Other	8.3%

Religion [8]

Muslim	70%
Russian Orthodox	NA

Major Languages [9]

Kirghiz (Kyrgyz)—official language, Russian.

Education

Public Education Expenditures [10]

Educational Attainment [11]

Literacy Rate [12]

	1989[b]	1990[b]	1991[b]
Illiterate population +15 years	NA	NA	NA
Illiteracy rate - total pop. (%)	3.0	NA	NA
Illiteracy rate - males (%)	1.4	NA	NA
Illiteracy rate - females (%)	4.5	NA	NA

Libraries [13]

Daily Newspapers [14]

Cinema [15]

Science and Technology

Scientific/Technical Forces [16]

R&D Expenditures [17]

U.S. Patents Issued [18]

For sources, notes, and explanations, see Annotated Source Appendix, page 1035.

Government and Law

Organization of Government [19]

Long-form name:
Republic of Kyrgyzstan
Type:
republic
Independence:
31 August 1991 (from Soviet Union)
Constitution:
adopted 5 May 1993
Legal system:
based on civil law system
National holiday:
National Day, 2 December
Executive branch:
president, Cabinet of Ministers, prime minister
Legislative branch:
unicameral Zhogorku Keneshom
Judicial branch:
Supreme Court

Political Parties [20]

Zhogorku Keneshom	% of votes
Commnunists	90.0
Other	10.0

Government Budget [21]

For FY88.

Revenues	NA
Expenditures	NA
Capital expenditures	NA

Defense Summary [22]

Branches: National Guard, Security Forces (internal and border troops), Civil Defense

Manpower Availability: Males age 15-49 1,093,694; fit for military service 890,961 (1993 est.)

Defense Expenditures: No information available.

Crime [23]

Human Rights [24]

	SSTS	FL	FAPRO	PPCG	APROBC	TPW	PCPTW	STPEP	PHRFF	PRW	ASST	AFL
Observes						P	P					
	EAFRD	CPR	ESCR	SR	ACHR	MAAE	PVIAC	PVNAC	EAFDAW	TCIDTP	RC
Observes	P						P	P	P		P

P = Party; S = Signatory; see Appendix for meaning of abbreviations.

Labor Force

Total Labor Force [25]

1.748 million

Labor Force by Occupation [26]

Agriculture and forestry	33%
Industry and construction	28
Other	39

Date of data: 1990

Unemployment Rate [27]

0.1% includes officially registered unemployed; also large numbers of underemployed workers

Production Sectors

Energy Resource Summary [28]

Energy Resources: Small amounts of coal, natural gas, oil, hydroelectric power. **Electricity**: 4,100,000 kW capacity; 11,800 million kWh produced, 2,551 kWh per capita (1992). **Pipelines**: Natural gas 200 km.

Telecommunications [30]

- Poorly developed
- 56 telephones per 1000 persons (December 1990)
- Connections with other CIS countries by landline or microwave and with other countries by leased connections with Moscow international gateway switch
- Satellite earth stations - Orbita and INTELSAT (TV receive only)
- New intelsat earth station provide TV receive-only capability for Turkish broadcasts

Top Agricultural Products [32]

	1990	1991
Grains	1,500	1,300
Vegetables	500	-
Potatoes	400	-
Grapes	100	-
Fruits and berries	100	-
Cotton	81	-

Values shown are 1,000 metric tons.

Top Mining Products [33]

Detailed information is not available. A summary of mineral resources available follows. **Mineral Resources**: Small amounts of coal, natural gas, oil, nepheline, rare earth metals, mercury, bismuth, gold, lead, zinc.

Transportation [31]

Railroads. 370 km; does not include industrial lines (1990)

Highways. 30,300 km total; 22,600 km paved or graveled, 7,700 km earth (1990)

Airports

Total:	52
Useable:	27
With permanent-surface runways:	12
With runways over 3,659 m:	1
With runways 2,440-3,659 m:	4
With runways 1,220-2,439 m:	13

Tourism [34]

Finance, Economics, and Trade

Industrial Summary [35]

Industrial Production: Growth rate NA% (1992). **Industries**: Small machinery, textiles, food-processing industries, cement, shoes, sawn logs, refrigerators, furniture, electric motors, gold, and rare earth metals.

Economic Indicators [36]

National product: GDP not available. **National product real growth rate**: - 25% (1992 est.). **National product per capita**: not available. **Inflation rate (consumer prices)**: 29% per month (first quarter 1993). **External debt**: $650 million (1991).

Balance of Payments Summary [37]

Exchange Rates [38]

Currency: **rubles.**
Symbol: **R.**

Subject to wide fluctuations. Data are currency units per $1.

December 24, 1992	415

Imports and Exports

Top Import Origins [39]

Amount not available.

Origins	%
Other CIS republics	NA

Top Export Destinations [40]

Amount not available.

Destinations	%
Russia	70
Ukraine	NA
Uzbekistan	NA
Kazakhstan	NA
Others	NA

Foreign Aid [41]

	U.S. $	
Official and commitments by foreign donors	300	million

Import and Export Commodities [42]

Import Commodities	Export Commodities
Lumber	Wool
Industrial products	Chemicals
Ferrous metals	Cotton
Fuel	Ferrous and nonferrous metals
Machinery	Shoes
Textiles	Machinery
Footwear	Tobacco

For sources, notes, and explanations, see Annotated Source Appendix, page 1035.

509

Laos

Geography [1]

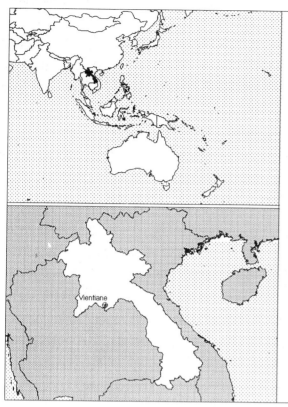

Total area:
  236,800 km2
Land area:
  230,800 km2
Comparative area:
  Slightly larger than Utah
Land boundaries:
  Total 5,083 km, Burma 235 km, Cambodia 541 km, China 423 km, Thailand 1,754 km, Vietnam 2,130 km
Coastline:
  0 km (landlocked)
Climate:
  Tropical monsoon; rainy season (May to November); dry season (December to April)
Terrain:
  Mostly rugged mountains; some plains and plateaus
Natural resources:
  Timber, hydropower, gypsum, tin, gold, gemstones
Land use:
  Arable land:
  4%
  Permanent crops:
  0%
  Meadows and pastures:
  3%
  Forest and woodland:
  58%
  Other:
  35%

Demographics [2]

	1960	1970	1980	1990	1991[1]	1994	2000	2010	2020
Population	2,309	2,845	3,293	4,191	4,113	4,702	5,557	7,168	8,923
Population density (persons per sq. mi.)	26	32	37	45	46	NA	56	67	78
Births	NA	NA	NA	NA	151	203	NA	NA	NA
Deaths	NA	NA	NA	NA	61	69	NA	NA	NA
Life expectancy - males	NA	NA	NA	NA	49	50	NA	NA	NA
Life expectancy - females	NA	NA	NA	NA	52	53	NA	NA	NA
Birth rate (per 1,000)	NA	NA	NA	NA	37	43	NA	NA	NA
Death rate (per 1,000)	NA	NA	NA	NA	15	15	NA	NA	NA
Women of reproductive age (15-44 yrs.)	NA	NA	NA	NA	NA	NA	NA	NA	NA
of which are currently married	NA	NA	NA	NA	NA	NA	NA	NA	NA
Fertility rate	NA	NA	NA	6.4	5.01	6.1	5.4	4.2	3.2

Population values are in thousands, life expectancy in years, and other items as indicated.

Health

Health Personnel [3]

Doctors per 1,000 pop., 1988-92	0.23
Nurse-to-doctor ratio, 1988-92	5.9
Hospital beds per 1,000 pop., 1985-90	2.5
Percentage of children immunized (age 1 yr. or less)	
Third dose of DPT, 1990-91	22.0
Measles, 1990-91	47.0

Health Expenditures [4]

Total health expenditure, 1990 (official exchange rate)	
Millions of dollars	22
Millions of dollars per capita	5
Health expenditures as a percentage of GDP	
Total	2.5
Public sector	1.0
Private sector	1.5
Development assistance for health	
Total aid flows (millions of dollars)[1]	5
Aid flows per capita (millions of dollars)	1.2
Aid flows as a percentage of total health expenditure	22.7

For sources, notes, and explanations, see Annotated Source Appendix, page 1035.

Human Factors

Health Care Ratios [5]

Population per physician, 1970	15160
Population per physician, 1990	4380
Population per nursing person, 1970	1390
Population per nursing person, 1990	490
Percent of births attended by health staff, 1985	NA

Infants and Malnutrition [6]

Percent of babies with low birth weight, 1985	39
Infant mortality rate per 1,000 live births, 1970	146
Infant mortality rate per 1,000 live births, 1991	100
Years of life lost per 1,000 population, 1990	93
Prevalence of malnutrition (under age 5), 1990	NA

Ethnic Division [7]

Lao	50%
Phoutheung (Kha)	15%
Tribal Thai	20%
Other	15%

Religion [8]

Buddhist	85%
Animist and other	15%

Major Languages [9]

Lao (official), French, English.

Education

Public Education Expenditures [10]

Million Kip	1980	1985	1986	1988	1989	1990
Total education expenditure	24	463	780	2,782	NA	NA
as percent of GNP	NA	0.4	0.4	1.1	NA	NA
as percent of total govt. expend.	1.3	4.5	6.6	NA	NA	NA
Current education expenditure	NA	NA	NA	2,097	NA	NA
as percent of GNP	NA	NA	NA	0.9	NA	NA
as percent of current govt. expend.	NA	NA	NA	NA	NA	NA
Capital expenditure	NA	NA	NA	685	NA	NA

Educational Attainment [11]

Literacy Rate [12]

	1970[b,4]	1980[b,4]	1985[b,4]
Illiterate population +15 years	NA	NA	NA
Illiteracy rate - total pop. (%)	NA	NA	NA
Illiteracy rate - males (%)	NA	NA	21.5
Illiteracy rate - females (%)	NA	NA	42.1

Libraries [13]

	Admin. Units	Svc. Pts.	Vols. (000)	Shelving (meters)	Vols. Added	Reg. Users
National	NA	NA	NA	NA	NA	NA
Non-specialized (1986)	1	2	145	1,328	1,500	800
Public	NA	NA	NA	NA	NA	NA
Higher ed. (1987)	5	NA	121	3,870	765	8,370
School (1987)	22	NA	33	880	800	11,000

Daily Newspapers [14]

	1975	1980	1985	1990
Number of papers	8	3	3	3
Circ. (000)	15[e]	14[e]	13[e]	14[e]

Cinema [15]

Data for 1991.

Cinema seats per 1,000	1.5
Annual attendance per person	0.2
Gross box office receipts (mil. Kip)	245

Science and Technology

Scientific/Technical Forces [16]

R&D Expenditures [17]

U.S. Patents Issued [18]

Government and Law

Organization of Government [19]

Long-form name:
Lao People's Democratic Republic
Type:
Communist state
Independence:
19 July 1949 (from France)
Constitution:
promulgated August 1991
Legal system:
based on civil law system; has not accepted compulsory ICJ jurisdiction
National holiday:
National Day, 2 December (1975) (proclamation of the Lao People's Democratic Republic)
Executive branch:
president, prime minister and two deputy prime ministers, Council of Ministers (cabinet)
Legislative branch:
National Assembly
Judicial branch:
Supreme People's Court

Crime [23]

Elections [20]

Third National Assembly. Last held on 20 December 1992 (next to be held NA); results - percent of vote by party NA; seats - (85 total) number of seats by party NA.

Government Budget [21]

For 1990 est.
Revenues	83.0
Expenditures	188.5
Capital expenditures	94.0

Military Expenditures and Arms Transfers [22]

	1985	1986	1987	1988	1989
Military expenditures					
Current dollars (mil.)	55[e]	NA	NA	NA	NA
1989 constant dollars (mil.)	63[e]	NA	NA	NA	NA
Armed forces (000)	46	48	50	56	56
Gross national product (GNP)					
Current dollars (mil.)	NA	NA	NA	NA	NA
1989 constant dollars (mil.)	NA	NA	NA	NA	NA
Central government expenditures (CGE)					
1989 constant dollars (mil.)	NA	NA	NA	302	NA
People (mil.)	3.6	3.7	3.8	3.8	3.9
Military expenditure as % of GNP	NA	NA	NA	NA	NA
Military expenditure as % of CGE	NA	NA	NA	NA	NA
Military expenditure per capita	17	NA	NA	NA	NA
Armed forces per 1,000 people	12.8	12.9	13.3	14.4	14.1
GNP per capita	NA	NA	NA	NA	NA
Arms imports[6]					
Current dollars (mil.)	100	100	140	150	100
1989 constant dollars (mil.)	114	111	151	156	100
Arms exports[6]					
Current dollars (mil.)	0	0	0	0	0
1989 constant dollars (mil.)	0	0	0	0	0
Total imports[7]					
Current dollars (mil.)	NA	205	219	NA	219
1989 constant dollars	NA	227	235	NA	219
Total exports[7]					
Current dollars (mil.)	NA	56	49	NA	58
1989 constant dollars	NA	62	52	NA	58
Arms as percent of total imports[8]	NA	48.8	64	NA	45.7
Arms as percent of total exports[8]	NA	0	0	NA	0

Human Rights [24]

	SSTS	FL	FAPRO	PPCG	APROBC	TPW	PCPTW	STPEP	PHRFF	PRW	ASST	AFL
Observes		P		P		P	P	P		P	P	
	EAFRD	CPR	ESCR	SR	ACHR	MAAE	PVIAC	PVNAC	EAFDAW	TCIDTP	RC	
Observes		P		P		P	P	P			P	P

P = Party; S = Signatory; see Appendix for meaning of abbreviations.

Labor Force

Total Labor Force [25]

1-1.5 million

Labor Force by Occupation [26]

Agriculture   85-90%

Unemployment Rate [27]

21% (1989 est.)

Production Sectors

Energy Resource Summary [28]

Energy Resources: Hydropower. **Electricity**: 226,000 kW capacity; 990 million kWh produced, 220 kWh per capita (1992). **Pipelines**: Petroleum products 136 km.

Telecommunications [30]

- Service to general public practically non-existant
- Radio communications network provides generally erratic service to government users
- 7,390 telephones (1986)
- Broadcast stations - 10 AM, no FM, 1 TV
- 1 satellite earth station

Top Agricultural Products [32]

Agriculture accounts for 60% of GDP and employs most of the work force; subsistence farming predominates; normally self-sufficient in nondrought years; principal crops—rice (80% of cultivated land), sweet potatoes, vegetables, corn, coffee, sugarcane, cotton; livestock— buffaloes, hogs, cattle, poultry.

Top Mining Products [33]

Detailed information is not available. A summary of mineral resources available follows. **Mineral Resources**: Gypsum, tin, gold, gemstones.

Transportation [31]

Railroads. none

Highways. about 27,527 km total; 1,856 km bituminous or bituminous treated; 7,451 km gravel, crushed stone, or improved earth; 18,220 km unimproved earth and often impassable during rainy season mid-May to mid-September

Airports

Total:	54
Usable:	41
With permanent-surface runways:	8
With runways over 3,659 m:	0
With runways 2,440-3,659 m:	1
With runways 1,220-2,439 m:	15

Tourism [34]

Finance, Economics, and Trade

GDP and Manufacturing Summary [35]

	1980	1985	1990	1991	1992
Gross Domestic Product					
Millions of 1980 dollars	462	661	903	942	1,011[e]
Growth rate in percent	1.70	9.83	9.10	4.30	7.30[e]
Manufacturing Value Added					
Millions of 1980 dollars	23	29	34	38	41[e]
Growth rate in percent	7.94	1.99	10.10	11.92	8.00[e]
Manufacturing share in percent of current prices	NA	NA	NA	NA	NA

Economic Indicators [36]

National product: GDP—exchange rate conversion—$900 million (1991). **National product real growth rate**: 4% (1991). **National product per capita**: $200 (1991). **Inflation rate (consumer prices)**: 10% (1991). **External debt**: $1.1 billion (1990 est.).

Balance of Payments Summary [37]

Values in millions of dollars.

	1987	1988	1989	1990	1991
Exports of goods (f.o.b.)	64.3	57.8	63.6	74.1	77.9
Imports of goods (f.o.b.)	-216.2	-188.0	-193.8	-185.4	-192.8
Trade balance	-151.9	-130.2	-130.2	-111.3	-114.9
Services - debits	-19.0	-17.6	-43.2	-35.6	-61.1
Services - credits	26.2	23.2	30.6	38.7	34.6
Private transfers (net)	3.5	6.7	8.3	10.9	20.4
Government transfers (net)	27.0	25.5	19.0	23.4	69.2
Direct investments	-	-	-	-	-
Errors and omissions	-5.3	-5.5	1.6	-3.0	-6.7
Overall balance	-8.1	-1.2	12.1	26.6	-16.9

Exchange Rates [38]

Currency: **new kips.**
Symbol: **NK.**

Data are currency units per $1.

May 1992	710
December 1991	710
September 1990	700
1989	576
1988	385
1987	200

Imports and Exports

Top Import Origins [39]

$238 million (c.i.f., 1990 est.).

Origins	%
Thailand	NA
USSR	NA
Japan	NA
France	NA
Vietnam	NA
China	NA

Top Export Destinations [40]

$72 million (f.o.b., 1990 est.).

Destinations	%
Thailand	NA
Malaysia	NA
Vietnam	NA
USSR	NA
US	NA
China	NA

Foreign Aid [41]

	U.S. $	
US commitments, including Ex-Im (FY70-79)	276	million
Western (non-US) countries, ODA and OOF bilateral commitments (1970-89)	605	million
Communist countries (1970-89)	995	million

Import and Export Commodities [42]

Import Commodities	**Export Commodities**
Food	Electricity
Fuel oil	Wood products
Consumer goods	Coffee
Manufactures	Tin

514

For sources, notes, and explanations, see Annotated Source Appendix, page 1035.

Latvia

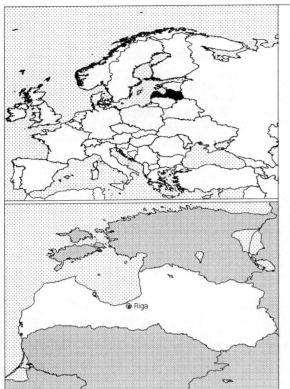

Geography [1]

Total area:
64,100 km2
Land area:
64,100 km2
Comparative area:
Slightly larger than West Virginia
Land boundaries:
Total 1,078 km, Belarus 141 km, Estonia 267 km, Lithuania 453 km, Russia 217 km
Coastline:
531 km
Climate:
Maritime; wet, moderate winters
Terrain:
Low plain
Natural resources:
Minimal; amber, peat, limestone, dolomite
Land use:
Arable land:
27%
Permanent crops:
0%
Meadows and pastures:
13%
Forest and woodland:
39%
Other:
21%

Demographics [2]

	1960	1970	1980	1990	1991	1994	2000	2010	2020
Population	2,115	2,361	2,525	2,693	NA	2,749	2,833	3,009	3,194
Population density (persons per sq. mi.)	NA	NA	NA	NA	NA	NA	NA	NA	NA
Births	NA	NA	NA	NA	NA	38	NA	NA	NA
Deaths	NA	NA	NA	NA	NA	35	NA	NA	NA
Life expectancy - males	NA	NA	NA	NA	NA	64	NA	NA	NA
Life expectancy - females	NA	NA	NA	NA	NA	75	NA	NA	NA
Birth rate (per 1,000)	NA	NA	NA	NA	NA	14	NA	NA	NA
Death rate (per 1,000)	NA	NA	NA	NA	NA	13	NA	NA	NA
Women of reproductive age (15-44 yrs.)	NA	NA	NA	654	NA	663	696	NA	NA
of which are currently married	NA	NA	NA	413	NA	419	434	NA	NA
Fertility rate	NA	NA	NA	2.1	NA	2.0	1.9	1.8	1.8

Population values are in thousands, life expectancy in years, and other items as indicated.

Health

Health Personnel [3]

Health Expenditures [4]

For sources, notes, and explanations, see Annotated Source Appendix, page 1035.

515

Human Factors

Health Care Ratios [5]

Population per physician, 1970	NA
Population per physician, 1990	200
Population per nursing person, 1970	NA
Population per nursing person, 1990	NA
Percent of births attended by health staff, 1985	NA

Infants and Malnutrition [6]

Percent of babies with low birth weight, 1985	NA
Infant mortality rate per 1,000 live births, 1970	23
Infant mortality rate per 1,000 live births, 1991	16
Years of life lost per 1,000 population, 1990	NA
Prevalence of malnutrition (under age 5), 1990	NA

Ethnic Division [7]

Latvian	51.8%
Russian	33.8%
Belarusian	4.5%
Ukrainian	3.4%
Polish	2.3%
Other	4.2%

Religion [8]

Lutheran, Roman Catholic, Russian Orthodox.

Major Languages [9]

Latvian (official)	NA
Lithuanian	NA
Russian	NA
Other	NA

Education

Public Education Expenditures [10]

Educational Attainment [11]

Age group	25+
Total population	1,725,639
Highest level attained (%)	
No schooling	0.6
First level	
Incompleted	18.5
Completed	21.2
Entered second level	
S-1	46.3
S-2	NA
Post secondary	13.4

Literacy Rate [12]

	1980[b]	1989[b]	1990[b]
Illiterate population +15 years	NA	11,476	NA
Illiteracy rate - total pop. (%)	NA	0.5	NA
Illiteracy rate - males (%)	NA	0.2	NA
Illiteracy rate - females (%)	NA	0.8	NA

Libraries [13]

Daily Newspapers [14]

Cinema [15]

Science and Technology

Scientific/Technical Forces [16]

R&D Expenditures [17]

U.S. Patents Issued [18]

Government and Law

Organization of Government [19]

Long-form name:
Republic of Latvia
Type:
republic
Independence:
6 September 1991 (from Soviet Union)
Constitution:
adopted NA May 1922, considering
rewriting constitution
Legal system:
based on civil law system
National holiday:
Independence Day, 18 November (1918)
Executive branch:
Chairman of Supreme Council
(president), prime minister, cabinet
Legislative branch:
unicameral Supreme Council
Judicial branch:
Supreme Court

Political Parties [20]

Supreme Council	% of seats
Latvian Communist Party	25.2
Latvian Democratic Workers Party	13.2
Social Democratic Party of Latvia	1.7
Green Party of Latvia	3.0
Latvian Farmers Union	3.0
Latvian Popular Front	53.8

Government Budget [21]

For 1990 est.
Revenues   NA
Expenditures   NA
 Capital expenditures   NA

Defense Summary [22]

Branches: Ground Forces, Navy, Air Force, Security Forces (internal and border troops), Border Guard, Home Guard (Zemessardze)

Manpower Availability: Males age 15-49 648,273; fit for military service 511,297; reach military age (18) annually 18,767 (1993 est.)

Defense Expenditures: 176 million rubles, 3-5% of GDP; note - conversion of the military budget into US$ using the current exchange rate could produce misleading results

Crime [23]

Human Rights [24]

	SSTS	FL	FAPRO	PPCG	APROBC	TPW	PCPTW	STPEP	PHRFF	PRW	ASST	AFL
Observes	2		P	P	P	P	P			P	P	P
		EAFRD	CPR	ESCR	SR	ACHR	MAAE	PVIAC	PVNAC	EAFDAW	TCIDTP	RC
Observes			P	P				P	P	P	P	P

P = Party; S = Signatory; see Appendix for meaning of abbreviations.

Labor Force

Total Labor Force [25]

1.407 million

Labor Force by Occupation [26]

Industry and construction	41%
Agriculture and forestry	16
Other	43
Date of data: 1990	

Unemployment Rate [27]

3.6% (March 1993); but large numbers of underemployed workers

Production Sectors

Energy Resource Summary [28]

Energy Resources: None. **Electricity**: 2,140,000 kW capacity; 5,800 million kWh produced, 2,125 kWh per capita (1992). **Pipelines**: Crude oil 750 km, refined products 780 km, natural gas 560 km (1992).

Telecommunications [30]

- NMT-450 analog cellular network is operational covering Riga, Ventspils, Daugavpils, Rezekne, and Valmiera
- Broadcast stations - NA
- International traffic carried by leased connection to the Moscow international gateway switch and through new independent international automatic telephone exchange in Riga and the Finnish cellular net

Transportation [31]

Railroads. 2,400 km; does not include industrial lines (1990)

Highways. 59,500 km total; 33,000 km hard surfaced 26,500 km earth (1990)

Merchant Marine. 96 ships (1,000 GRT or over) totaling 905,006 GRT/1,178,844 DWT; includes 14 cargo, 27 refrigerated cargo, 2 container, 9 roll-on/roll-off, 44 oil tanker

Airports

Total:	50
Useable:	15
With permanent-surface runways:	11
With runways over 3,659 m:	0
With runways 2,440-3,659 m:	7
With runways 1,220-2,439 m:	7

Top Agricultural Products [32]

Employs 16% of labor force; principally dairy farming and livestock feeding; products—meat, milk, eggs, grain, sugar beets, potatoes, vegetables; fishing and fish packing.

Top Mining Products [33]

Detailed information is not available. A summary of mineral resources available follows. **Mineral Resources**: Minimal; amber, peat, limestone, dolomite.

Tourism [34]

Finance, Economics, and Trade

Industrial Summary [35]

Industrial Production: Growth rate - 35% (1992 est.). **Industries**: Employs 33% of labor force; highly diversified; dependent on imports for energy, raw materials, and intermediate products; produces buses, vans, street and railroad cars, synthetic fibers, agricultural machinery, fertilizers, washing machines, radios, electronics, pharmaceuticals, processed foods, textiles.

Economic Indicators [36]

National product: GDP not available. **National product real growth rate**: - 30% (1992). **National product per capita**: not available. **Inflation rate (consumer prices)**: 2% per month (first quarter 1993). **External debt**: $650 million (1991 est.).

Balance of Payments Summary [37]

Exchange Rates [38]

Currency: **lats.**
Symbol: **L.**

Data are currency units per $1.

March 1993   1.32

Imports and Exports

Top Import Origins [39]

Amount not available.

Origins	%
NA	NA

Top Export Destinations [40]

Amount not available.

Destinations	%
NA	NA

Foreign Aid [41]

Import and Export Commodities [42]

Import Commodities

No information available.

Export Commodities

For sources, notes, and explanations, see Annotated Source Appendix, page 1035.

519

Lebanon

Geography [1]

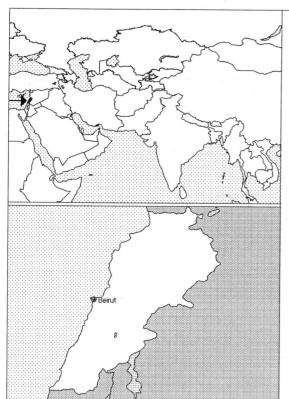

Total area:
10,400 km2

Land area:
10,230 km2

Comparative area:
About 0.8 times the size of Connecticut

Land boundaries:
Total 454 km, Israel 79 km, Syria 375 km

Coastline:
225 km

Climate:
Mediterranean; mild to cool, wet winters with hot, dry summers; Lebanon Mountians experience heavy winter snows

Terrain:
Narrow coastal plain; Al Biqa' (Bekaa Valley) separates Lebanon and Anti-Lebanon Mountains

Natural resources:
Limestone, iron ore, salt, water-surplus state in a water-deficit region

Land use:
Arable land:
21%
Permanent crops:
9%
Meadows and pastures:
1%
Forest and woodland:
8%
Other:
61%

Demographics [2]

	1960	1970	1980	1990	1991[1]	1994	2000	2010	2020
Population	1,786	2,383	3,137	3,367	3,385	3,620	4,115	4,973	5,748
Population density (persons per sq. mi.)	452	603	794	845	857	NA	1,027	1,249	1,457
Births	NA	NA	NA	NA	94	101	NA	NA	NA
Deaths	NA	NA	NA	NA	23	24	NA	NA	NA
Life expectancy - males	NA	NA	NA	NA	66	67	NA	NA	NA
Life expectancy - females	NA	NA	NA	NA	71	72	NA	NA	NA
Birth rate (per 1,000)	NA	NA	NA	NA	28	28	NA	NA	NA
Death rate (per 1,000)	NA	NA	NA	NA	7	7	NA	NA	NA
Women of reproductive age (15-44 yrs.)	NA	NA	NA	817	NA	940	1,112	NA	NA
of which are currently married	NA	NA	NA	466	NA	538	676	NA	NA
Fertility rate	NA	NA	NA	3.7	3.61	3.4	3.0	2.5	2.3

Population values are in thousands, life expectancy in years, and other items as indicated.

Health

Health Personnel [3]

Health Expenditures [4]

For sources, notes, and explanations, see Annotated Source Appendix, page 1035.

Human Factors

Health Care Ratios [5]	Infants and Malnutrition [6]

Ethnic Division [7]

Arab	95%
Armenian	4%
Other	1%

Religion [8]

Islam 70% (5 legally recognized Islamic groups - Alawite or Nusayri, Druze, Isma'ilite, Shi'a, Sunni), Christian 30% (11 legally recognized Christian groups - 4 Orthodox Christian, 6 Catholic, 1 Protestant), Judaism.

Major Languages [9]

Arabic (official)	NA
French (official)	NA
Armenian	NA
English	NA

Education

Public Education Expenditures [10]

Million Lebanese Pounds	1980	1985	1986	1987	1988	1989
Total education expenditure	511	1,639	2,163	3,192	5,976	NA
as percent of GNP	NA	NA	NA	NA	NA	NA
as percent of total govt. expend.	13.2	14.4	12.1	11.7	8.5	NA
Current education expenditure	NA	NA	NA	NA	NA	NA
as percent of GNP	NA	NA	NA	NA	NA	NA
as percent of current govt. expend.	NA	NA	NA	NA	NA	NA
Capital expenditure	NA	NA	NA	NA	NA	NA

Educational Attainment [11]

Literacy Rate [12]

In thousands and percent	1985[a]	1991[a]	2000[a]
Illiterate population + 15 years	386	382	347
Illiteracy rate - total pop. (%)	23.2	19.9	14.5
Illiteracy rate - males (%)	14.1	12.2	8.9
Illiteracy rate - females (%)	31.2	26.9	19.7

Libraries [13]

Daily Newspapers [14]

	1975	1980	1985	1990
Number of papers	33	14	13	14[e]
Circ. (000)	300[e]	290[e]	300[e]	320[e]

Cinema [15]

Science and Technology

Scientific/Technical Forces [16]

R&D Expenditures [17]

	Pound[23] (000) 1980
Total expenditure	22,000
Capital expenditure	NA
Current expenditure	NA
Percent current	NA

U.S. Patents Issued [18]

Values show patents issued to citizens of the country by the U.S. Patents Office.

	1990	1991	1992
Number of patents	1	1	1

For sources, notes, and explanations, see Annotated Source Appendix, page 1035.

Government and Law

Organization of Government [19]

Long-form name:
Republic of Lebanon
Type:
republic
Independence:
22 November 1943 (from League of
Nations mandate under French
administration)
Constitution:
26 May 1926 (amended)
Legal system:
mixture of Ottoman law, canon law,
Napoleonic code, and civil law; no judicial
review of legislative acts; has not
accepted compulsory ICJ jurisdiction
National holiday:
Independence Day, 22 November (1943)
Executive branch:
president, prime minister, Cabinet; note -
by custom, the president is a Maronite
Christian, the prime minister is a Sunni
Muslim, and the speaker of the legislature
is a Shi'a Muslim
Legislative branch:
unicameral National Assembly (Arabic -
Majlis Alnuwab, French - Assemblee
Nationale)
Judicial branch:
four Courts of Cassation

Crime [23]

Elections [20]

National Assembly. Lebanon's first
legislative election in 20 years was held
in the summer of 1992; the National
Assembly is composed of 128 deputies,
one-half Christian and one-half Muslim;
its mandate expires in 1996. Political
party activity is organized along largely
sectarian lines; numerous political
groupings exist, consisting of individual
political figures and followers motivated
by religious, clan, and economic
considerations.

Government Budget [21]

For 1991 est.
Revenues	0.533
Expenditures	1.300
Capital expenditures	NA

Military Expenditures and Arms Transfers [22]

	1985	1986	1987	1988	1989
Military expenditures					
Current dollars (mil.)	NA	NA	NA	NA	NA
1989 constant dollars (mil.)	NA	NA	NA	NA	NA
Armed forces (000)	21	36	37	20	18
Gross national product (GNP)					
Current dollars (mil.)	2,594[e]	2,697[e]	1,780[e]	NA	NA
1989 constant dollars (mil.)	2,953[e]	2,993[e]	1,915[e]	NA	NA
Central government expenditures (CGE)					
1989 constant dollars (mil.)	NA	NA	1,273[e]	NA	NA
People (mil.)	3.2	3.2	3.2	3.3	3.3
Military expenditure as % of GNP	NA	NA	NA	NA	NA
Military expenditure as % of CGE	NA	NA	NA	NA	NA
Military expenditure per capita	NA	NA	NA	NA	NA
Armed forces per 1,000 people	6.6	11.2	11.4	6.1	5.5
GNP per capita	921	929	591	NA	NA
Arms imports[6]					
Current dollars (mil.)	40	10	10	10	10
1989 constant dollars (mil.)	46	11	11	10	10
Arms exports[6]					
Current dollars (mil.)	0	0	0	0	0
1989 constant dollars (mil.)	0	0	0	0	0
Total imports[7]					
Current dollars (mil.)	2,203	2,203	1,880	2,457	2,500
1989 constant dollars	2,508	2,445	2,022	2,558	2,500
Total exports[7]					
Current dollars (mil.)	482	500	591	709	502
1989 constant dollars	549	555	636	738	502
Arms as percent of total imports[8]	1.8	0.5	0.5	0.4	0.4
Arms as percent of total exports[8]	0	0	0	0	0

Human Rights [24]

	SSTS	FL	FAPRO	PPCG	APROBC	TPW	PCPTW	STPEP	PHRFF	PRW	ASST	AFL
Observes	2	P		P	P	P	P			P		P
	EAFRD	CPR	ESCR	SR	ACHR	MAAE	PVIAC	PVNAC	EAFDAW	TCIDTP	RC
Observes	P	P	P								P

P=Party; S=Signatory; see Appendix for meaning of abbreviations.

Labor Force

Total Labor Force [25]

650,000

Labor Force by Occupation [26]

Industry, commerce, and services	79%
Agriculture	11
Government	10

Date of data: 1985

Unemployment Rate [27]

35% (1991 est.)

For sources, notes, and explanations, see Annotated Source Appendix, page 1035.

Production Sectors

Energy Resource Summary [28]

Energy Resources: None. **Electricity**: 1,300,000 kW capacity; 3,413 million kWh produced, 990 kWh per capita (1992). **Pipelines**: Crude oil 72 km (none in operation).

Telecommunications [30]

- Telecommunications system severely damaged by civil war
- Rebuilding still underway
- 325,000 telephones (95 telephones per 1,000 persons)
- Domestic traffic carried primarily by microwave radio relay and a small amount of cable
- International traffic by satellite - 1 Indian Ocean INTELSAT earth station and 1 Atlantic Ocean INTELSAT earth station (erratic operations), coaxial cable to Syria
- Microwave radio relay to Syria but inoperable beyond Syria to Jordan, 3 submarine coaxial cables
- Broadcast stations - 5 AM, 3 FM, 13 TV (numerous AM and FM stations are operated sporadically by various factions)

Top Agricultural Products [32]

Agriculture accounts for about one-third of GDP; principal products—citrus fruits, vegetables, potatoes, olives, tobacco, hemp (hashish), sheep, goats; not self-sufficient in grain.

Top Mining Products [33]

Estimated metric tons unless otherwise specified   M.t.

Salt (000 tons)	3
Cement, hydraulic (000 tons)	900
Gypsum	2,000
Lime	10,000

Transportation [31]

Railroads. system in disrepair, considered inoperable

Highways. 7,300 km total; 6,200 km paved, 450 km gravel and crushed stone, 650 km improved earth

Merchant Marine. 63 ships (1,000 GRT or over) totaling 270,505 GRT/403,328 DWT; includes 39 cargo, 1 refrigerated cargo, 2 vehicle carrier, 3 roll-on/roll-off, 1 container, 9 livestock carrier, 2 chemical tanker, 1 specialized tanker, 4 bulk, 1 combination bulk

Airports

Total:	9
Usable:	8
With permanent-surface runways:	6
With runways over 3,659 m:	0
With runways 2,440-3,659 m:	3
With runways 1,220-2,439 m:	2

Tourism [34]

Finance, Economics, and Trade

Industrial Summary [35]

Industrial Production: Growth rate NA%. **Industries**: Banking, food processing, textiles, cement, oil refining, chemicals, jewelry, some metal fabricating.

Economic Indicators [36]

National product: GDP—exchange rate conversion—$4.8 billion (1991 est.). **National product real growth rate**: NA%. **National product per capita**: $1,400 (1991 est.). **Inflation rate (consumer prices)**: 100% (1992 est.). **External debt**: $400 million (1992 est.).

Balance of Payments Summary [37]

Exchange Rates [38]

Currency: **Lebanese pounds.**
Symbol:　**#L.**

Data are currency units per $1.

April 1993	1,742.00
1992	1,712.80
1991	928.23
1990	695.09
1989	496.69
1988	409.23

Imports and Exports

Top Import Origins [39]

$3.7 billion (c.i.f., 1991).

Origins	%
Italy	14
France	12
US	6
Turkey	5
Saudi Arabia	3

Top Export Destinations [40]

$490 million (f.o.b., 1991).

Destinations	%
Saudi Arabia	21.0
Switzerland	9.5
Jordan	6.0
Kuwait	12.0
US	5.5

Foreign Aid [41]

	U.S. $	
US commitments, including Ex-Im (FY70-88)	356	million
Western (non-US) countries, ODA and OOF bilateral commitments (1970-89)	664	million
OPEC bilateral aid (1979-89)	962	million
Communist countries (1970-89)	9	million

Import and Export Commodities [42]

Import Commodities	Export Commodities
Consumer goods	Agricultural products
Machinery & transport equipment	Chemicals
Petroleum products	Textiles
	Precious metals
	Semiprecious metals
	Jewelry
	Metals and metal products

Lesotho

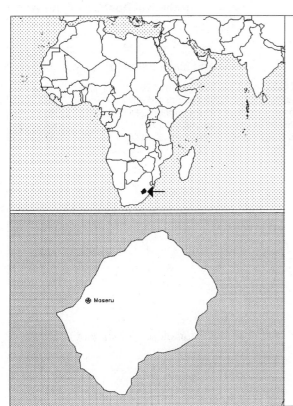

Geography [1]

Total area:
 30,350 km2
Land area:
 30,350 km2
Comparative area:
 Slightly larger than Maryland
Land boundaries:
 Total 909 km, South Africa 909 km
Coastline:
 0 km (landlocked)
Climate:
 Temperate; cool to cold, dry winters; hot, wet summers
Terrain:
 Mostly highland with some plateaus, hills, and mountains
Natural resources:
 Some diamonds and other minerals, water, agricultural and grazing land
Land use:
 Arable land:
 10%
 Permanent crops:
 0%
 Meadows and pastures:
 66%
 Forest and woodland:
 0%
 Other:
 24%

Demographics [2]

	1960	1970	1980	1990	1991[1]	1994	2000	2010	2020
Population	859	1,067	1,347	1,755	1,801	1,944	2,242	2,771	3,314
Population density (persons per sq. mi.)	73	91	115	150	154	NA	191	237	285
Births	NA	NA	NA	NA	65	66	NA	NA	NA
Deaths	NA	NA	NA	NA	18	18	NA	NA	NA
Life expectancy - males	NA	NA	NA	NA	59	60	NA	NA	NA
Life expectancy - females	NA	NA	NA	NA	63	64	NA	NA	NA
Birth rate (per 1,000)	NA	NA	NA	NA	36	34	NA	NA	NA
Death rate (per 1,000)	NA	NA	NA	NA	10	9	NA	NA	NA
Women of reproductive age (15-44 yrs.)	NA	NA	NA	411	NA	466	559	NA	NA
of which are currently married	NA	NA	NA	279	NA	315	377	NA	NA
Fertility rate	NA	NA	NA	4.9	4.78	4.5	3.9	3.1	2.6

Population values are in thousands, life expectancy in years, and other items as indicated.

Health

Health Personnel [3]

Health Expenditures [4]

For sources, notes, and explanations, see Annotated Source Appendix, page 1035.

525

Human Factors

Health Care Ratios [5]

Population per physician, 1970	30400
Population per physician, 1990	NA
Population per nursing person, 1970	3860
Population per nursing person, 1990	NA
Percent of births attended by health staff, 1985	28

Infants and Malnutrition [6]

Percent of babies with low birth weight, 1985	10
Infant mortality rate per 1,000 live births, 1970	134
Infant mortality rate per 1,000 live births, 1991	81
Years of life lost per 1,000 population, 1990	NA
Prevalence of malnutrition (under age 5), 1990	27

Ethnic Division [7]

Sotho	99.7%
Europeans	1,600
Asians	800

Religion [8]

Christian	80%
Indigenous beliefs	20%

Major Languages [9]

Sesotho (southern Sotho)	NA
English (official)	NA
Zulu	NA
Xhosa	NA

Education

Public Education Expenditures [10]

Million Maloti	1980	1985	1987	1988	1989	1990
Total education expenditure	25	NA	NA	66	83	99
as percent of GNP	5.1	NA	NA	3.8	3.7	3.8
as percent of total govt. expend.	14.8	NA	NA	13.8	14.0	12.2
Current education expenditure	20	NA	NA	61	72	81
as percent of GNP	4.1	NA	NA	3.5	3.2	3.1
as percent of current govt. expend.	19.0	NA	NA	18.9	16.9	17.4
Capital expenditure	5	NA	NA	6	11	18

Educational Attainment [11]

Literacy Rate [12]

Libraries [13]

	Admin. Units	Svc. Pts.	Vols. (000)	Shelving (meters)	Vols. Added	Reg. Users
National	NA	NA	NA	NA	NA	NA
Non-specialized	NA	NA	NA	NA	NA	NA
Public (1989)	1	3	24	672	NA	607
Higher ed.	NA	NA	NA	NA	NA	NA
School	NA	NA	NA	NA	NA	NA

Daily Newspapers [14]

	1975	1980	1985	1990
Number of papers	1	3	4	4
Circ. (000)	NA	44	47	20

Cinema [15]

Science and Technology

Scientific/Technical Forces [16]

R&D Expenditures [17]

U.S. Patents Issued [18]

526

For sources, notes, and explanations, see Annotated Source Appendix, page 1035.

Government and Law

Organization of Government [19]

Long-form name:
Kingdom of Lesotho
Type:
constitutional monarchy
Independence:
4 October 1966 (from UK)
Constitution:
4 October 1966, suspended January 1970
Legal system:
based on English common law and
Roman-Dutch law; judicial review of
legislative acts in High Court and Court of
Appeal; has not accepted compulsory ICJ
jurisdiction
National holiday:
Independence Day, 4 October (1966)
Executive branch:
monarch, chairman of the Military
Council, Military Council, Council of
Ministers (cabinet)
Legislative branch:
none - the bicameral Parliament was
dissolved following the military coup in
January 1986
Judicial branch:
High Court, Court of Appeal

Elections [20]

National Assembly. Dissolved following
the military coup in January 1986;
military has pledged elections will take
place in March 1993.

Government Expenditures [21]

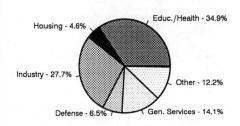

% distribution for 1991. Expend. in FY93: 399 ($ mil.)

Crime [23]

Military Expenditures and Arms Transfers [22]

	1985	1986	1987	1988	1989
Military expenditures					
Current dollars (mil.)	15	18	15	NA	NA
1989 constant dollars (mil.)	17	19	16	NA	NA
Armed forces (000)	2	2	2	2	2
Gross national product (GNP)					
Current dollars (mil.)	642	653	681	752	830
1989 constant dollars (mil.)	730	725	733	783	830
Central government expenditures (CGE)					
1989 constant dollars (mil.)	186	NA	NA	147	193
People (mil.)	1.5	1.6	1.6	1.7	1.7
Military expenditure as % of GNP	2.4	2.7	2.2	NA	NA
Military expenditure as % of CGE	9.4	NA	NA	NA	NA
Military expenditure per capita	11	12	10	NA	NA
Armed forces per 1,000 people	1.0	1.3	1.2	1.2	1.2
GNP per capita	476	459	452	471	485
Arms imports[6]					
Current dollars (mil.)	0	0	0	0	0
1989 constant dollars (mil.)	0	0	0	0	0
Arms exports[6]					
Current dollars (mil.)	0	0	0	0	0
1989 constant dollars (mil.)	0	0	0	0	0
Total imports[7]					
Current dollars (mil.)	377	393	518	523	594
1989 constant dollars	429	436	557	544	594
Total exports[7]					
Current dollars (mil.)	21	25	36	66	66
1989 constant dollars	24	28	39	69	66
Arms as percent of total imports[8]	0	0	0	0	0
Arms as percent of total exports[8]	0	0	0	0	0

Human Rights [24]

	SSTS	FL	FAPRO	PPCG	APROBC	TPW	PCPTW	STPEP	PHRFF	PRW	ASST	AFL
Observes	P	P	P	P	·P	P	P			P	P	
	EAFRD	CPR	ESCR	SR	ACHR	MAAE	PVIAC	PVNAC	EAFDAW	TCIDTP	RC	
Observes		P	P	P	P				S		P	

P = Party; S = Signatory; see Appendix for meaning of abbreviations.

Labor Force

Total Labor Force [25]

689,000 economically active

Labor Force by Occupation [26]

Subsistence agriculture	86.2%
Other	13.8

Unemployment Rate [27]

at least 55% among adult males (1991 est.)

For sources, notes, and explanations, see Annotated Source Appendix, page 1035.

527

Production Sectors

Energy Resource Summary [28]

Energy Resources: None. **Electricity**: Power supplied by South Africa.

Telecommunications [30]

- Rudimentary system consisting of a few landlines, a small microwave system, and minor radio communications stations
- 5,920 telephones
- Broadcast stations - 3 AM, 2 FM, 1 TV
- 1 Atlantic Ocean INTELSAT earth station

Transportation [31]

Railroads. 2.6 km; owned, operated by, and included in the statistics of South Africa

Highways. 7,215 km total; 572 km paved; 2,337 km crushed stone, gravel, or stabilized soil; 1,806 km improved earth, 2,500 km unimproved earth

Airports

Total:	28
Usable:	28
With permanent-surface runways:	3
With runways over 3,659 m:	0
With runways 2,440-3,659 m:	1
With runways 1,220-2,439 m:	2

Top Agricultural Products [32]

Agriculture accounts for 19% of GDP (1990 est.) and employs 60-70% of all households; exceedingly primitive, mostly subsistence farming and livestock; principal crops corn, wheat, pulses, sorghum, barley.

Top Mining Products [33]

Detailed information is not available. A summary of mineral resources available follows. **Mineral Resources**: Some diamonds and other minerals.

Tourism [34]

	1987	1988	1989	1990	1991
Visitors	220	165	216	242	357
Tourists	135	110	169	171	
Excursionists	85	55	47	71	
Tourism receipts	11	12	13	17	18
Tourism expenditures	9	11	10	12	11
Fare receipts		3	4	4	3
Fare expenditures	6	7	7	8	7

Tourists are in thousands, money in million U.S. dollars.

Finance, Economics, and Trade

GDP and Manufacturing Summary [35]

	1980	1985	1990	1991	1992
Gross Domestic Product					
Millions of 1980 dollars	368	395	552	581	601[e]
Growth rate in percent	8.35	3.49	3.96	5.28	3.40[e]
Manufacturing Value Added					
Millions of 1980 dollars	21	37	65	NA	NA
Growth rate in percent	16.00	4.36	-3.95	NA	NA
Manufacturing share in percent of current prices	6.3	10.4	13.0	15.5	NA

Economic Indicators [36]

National product: GDP—exchange rate conversion—$620 million (1991 est.). **National product real growth rate:** 5.3% (1991 est.); GNP 2.2% (1991 est.). **National product per capita**: $340 (1991 est.); GNP $570 (1991 est.). **Inflation rate (consumer prices)**: 17.9% (1991). **External debt**: $358 million (for public sector) (December 1990/91 est.).

Balance of Payments Summary [37]

Values in millions of dollars.

	1987	1988	1989	1990	1991
Exports of goods (f.o.b.)	46.5	63.7	66.4	59.5	67.2
Imports of goods (f.o.b.)	-451.5	-559.4	-592.6	-672.6	-786.9
Trade balance	-405.0	-495.7	-526.2	-613.1	-719.7
Services - debits	-75.7	-84.0	-91.2	-103.3	-243.2
Services - credits	-9.2	-8.5	-12.6	-12.7	-150.5
Private transfers (net)	0.3	4.4	4.2	4.8	2.7
Government transfers (net)	111.8	133.7	210.7	281.1	505.9
Long term capital (net)	40.3	59.1	45.8	61.4	42.8
Short term capital (net)	-37.0	-67.1	-66.0	-106.3	-67.6
Errors and omissions	-26.0	26.5	1.9	-2.8	4.1
Overall balance	1.0	-6.1	-7.9	17.4	42.7

Exchange Rates [38]

Currency: **maloti.**
Symbol: **M.**

The Basotho loti is at par with the South African rand. Data are currency units per $1.

May 1993	3.1576
1992	2.8497
1991	2.7563
1990	2.5863
1989	2.6166
1988	2.2611

Imports and Exports

Top Import Origins [39]

$805 million (c.i.f., 1991). Data are for 1989.

Origins	%
South Africa	95
EC	2

Top Export Destinations [40]

$57 million (f.o.b., 1991). Data are for 1989.

Destinations	%
South Africa	53
EC	30
North and South America	13

Foreign Aid [41]

	U.S. $	
US commitments, including Ex-Im (FY70-89)	268	million
US (1992)	10.3	million
US (1993 est.)	10.1	million
Western (non-US) countries, ODA and OOF bilateral commitments (1970-89)	819	million
OPEC bilateral aid (1979-89)	4	million
Communist countries (1970-89)	14	million

Import and Export Commodities [42]

Import Commodities	**Export Commodities**
Mainly corn	Wool
Building materials	Mohair
Clothing	Wheat
Vehicles	Cattle
Machinery	Peas
Medicines	Beans
Petroleum	Corn
	Hides
	Skins
	Baskets

Liberia

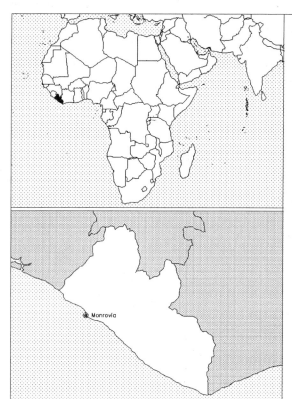

Geography [1]

Total area:
111,370 km2
Land area:
96,320 km2
Comparative area:
Slightly larger than Tennessee
Land boundaries:
Total 1,585 km, Guinea 563 km, Cote d'Ivoire 716 km, Sierra Leone 306 km
Coastline:
579 km
Climate:
Tropical; hot, humid; dry winters with hot days and cool to cold nights; wet, cloudy summers with frequent heavy showers
Terrain:
Mostly flat to rolling coastal plains rising to rolling plateau and low mountains in northeast
Natural resources:
Iron ore, timber, diamonds, gold
Land use:
Arable land:
1%
Permanent crops:
3%
Meadows and pastures:
2%
Forest and woodland:
39%
Other:
55%

Demographics [2]

	1960	1970	1980	1990	1991[1]	1994	2000	2010	2020
Population	1,055	1,397	1,900	2,311	2,730	2,973	3,620	4,903	6,449
Population density (persons per sq. mi.)	28	38	51	71	73	NA	99	134	176
Births	NA	NA	NA	NA	122	129	NA	NA	NA
Deaths	NA	NA	NA	NA	36	37	NA	NA	NA
Life expectancy - males	NA	NA	NA	NA	54	55	NA	NA	NA
Life expectancy - females	NA	NA	NA	NA	59	60	NA	NA	NA
Birth rate (per 1,000)	NA	NA	NA	NA	45	43	NA	NA	NA
Death rate (per 1,000)	NA	NA	NA	NA	13	12	NA	NA	NA
Women of reproductive age (15-44 yrs.)	NA	NA	NA	507	NA	648	793	NA	NA
of which are currently married	NA	NA	NA	350	NA	448	547	NA	NA
Fertility rate	NA	NA	NA	6.6	6.55	6.4	6.0	5.2	4.4

Population values are in thousands, life expectancy in years, and other items as indicated.

Health

Health Personnel [3]

Health Expenditures [4]

For sources, notes, and explanations, see Annotated Source Appendix, page 1035.

Human Factors

Health Care Ratios [5]	Infants and Malnutrition [6]

Ethnic Division [7]

Indigenous African tribes 95% (including Kpelle, Bassa, Gio, Kru, Grebo, Mano, Krahn, Gola, Gbandi, Loma, Kissi, Vai, and Bella), Americo-Liberians 5% (descendants of repatriated slaves).

Religion [8]

Traditional	70%
Muslim	20%
Christian	10%

Major Languages [9]

English 20% (official), Niger-Congo language group. About 20 local languages come from this group.

Education

Public Education Expenditures [10]

Million Liberian Dollars	1980	1985	1987	1988	1989	1990
Total education expenditure	62	NA	NA	NA	NA	NA
as percent of GNP	5.7	NA	NA	NA	NA	NA
as percent of total govt. expend.	24.3	NA	NA	NA	NA	NA
Current education expenditure	53	NA	NA	NA	NA	NA
as percent of GNP	4.9	NA	NA	NA	NA	NA
as percent of current govt. expend.	27.0	NA	NA	NA	NA	NA
Capital expenditure	9	NA	NA	NA	NA	NA

Educational Attainment [11]

Literacy Rate [12]

In thousands and percent	1985[a]	1991[a]	2000[a]
Illiterate population +15 years	811	839	862
Illiteracy rate - total pop. (%)	67.7	60.5	45.3
Illiteracy rate - males (%)	57.3	50.2	36.1
Illiteracy rate - females (%)	78.6	71.2	54.7

Libraries [13]

Daily Newspapers [14]

	1975	1980	1985	1990
Number of papers	3	3	5	8
Circ. (000)	13	11	28[e]	35[e]

Cinema [15]

Science and Technology

Scientific/Technical Forces [16]	R&D Expenditures [17]	U.S. Patents Issued [18]

For sources, notes, and explanations, see Annotated Source Appendix, page 1035.

531

Government and Law

Organization of Government [19]

Long-form name:
Republic of Liberia
Type:
republic
Independence:
26 July 1847
Constitution:
6 January 1986
Legal system:
dual system of statutory law based on Anglo-American common law for the modern sector and customary law based on unwritten tribal practices for indigenous sector
National holiday:
Independence Day, 26 July (1847)
Executive branch:
president, vice president, Cabinet
Legislative branch:
bicameral National Assembly consists of an upper house or Senate and a lower house or House of Representatives
Judicial branch:
People's Supreme Court

Political Parties [20]

House of Representatives	% of seats
National Democratic Party of Liberia (NDPL)	79.7
Liberian Action Party (LAP)	12.5
Unity Party (UP)	4.7
United People's Party (UPP)	3.1

Government Expenditures [21]

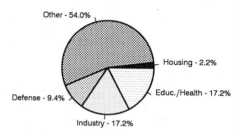

Other - 54.0%
Housing - 2.2%
Educ./Health - 17.2%
Industry - 17.2%
Defense - 9.4%

% distribution for 1988. Expend. in 1989: 435.4 ($ mil.)

Crime [23]

Military Expenditures and Arms Transfers [22]

	1985	1986	1987	1988	1989
Military expenditures					
Current dollars (mil.)	29	37[e]	41[e]	41[e]	58[e]
1989 constant dollars (mil.)	33	41[e]	44[e]	43[e]	58[e]
Armed forces (000)	6	6	6	7	7
Gross national product (GNP)					
Current dollars (mil.)	1,064	1,064	1,087	1,057[e]	1,200[e]
1989 constant dollars (mil.)	1,212	1,181	1,169	1,100[e]	1,200[e]
Central government expenditures (CGE)					
1989 constant dollars (mil.)	372	346	311	248	435
People (mil.)	2.2	2.3	2.4	2.5	2.6
Military expenditure as % of GNP	2.7	3.5	3.8	3.9	4.8
Military expenditure as % of CGE	8.8	11.8	14.2	17.3	13.3
Military expenditure per capita	15	18	18	17	23
Armed forces per 1,000 people	2.7	2.6	2.5	2.8	2.7
GNP per capita	542	511	490	446	470
Arms imports[6]					
Current dollars (mil.)	10	10	l0	10	0
1989 constant dollars (mil.)	11	11	11	10	0
Arms exports[6]					
Current dollars (mil.)	0	0	0	0	0
1989 constant dollars (mil.)	0	0	0	0	0
Total imports[7]					
Current dollars (mil.)	284	267	308	NA	335
1989 constant dollars	323	296	331	NA	335
Total exports[7]					
Current dollars (mil.)	436	404	382	NA	550
1989 constant dollars	496	448	411	NA	550
Arms as percent of total imports[8]	3.5	3.7	3.2	NA	0
Arms as percent of total exports[8]	0	0	0	NA	0

Human Rights [24]

	SSTS	FL	FAPRO	PPCG	APROBC	TPW	PCPTW	STPEP	PHRFF	PRW	ASST	AFL
Observes	2	P	P	P	P	P	P	S		S	S	P
	EAFRD	CPR	ESCR	SR	ACHR	MAAE	PVIAC	PVNAC	EAFDAW	TCIDTP	RC
Observes	P	S	S	P			P	P			P

P = Party; S = Signatory; see Appendix for meaning of abbreviations.

Labor Force

Total Labor Force [25]

510,000 including 220,000 in the monetary economy

Labor Force by Occupation [26]

Agriculture	70.5%
Services	10.8
Industry and commerce	4.5
Other	14.2

Unemployment Rate [27]

43% urban (1988)

For sources, notes, and explanations, see Annotated Source Appendix, page 1035.

Production Sectors

Energy Resource Summary [28]

Energy Resources: None. **Electricity**: 410,000 kW capacity; 750 million kWh produced, 275 kWh per capita (1991).

Telecommunications [30]

- Telephone and telegraph service via radio relay network
- Main center is Monrovia
- Broadcast stations - 3 AM, 4 FM, 5 TV
- 1 Atlantic Ocean INTELSAT earth station
- Most telecommunications services inoperable due to insurgency movement

Top Agricultural Products [32]

Agriculture accounts for about 40% of GDP (including fishing and forestry); principal products—rubber, timber, coffee, cocoa, rice, cassava, palm oil, sugarcane, bananas, sheep, goats; not self-sufficient in food, imports 25% of rice consumption.

Top Mining Products [33]

Metric tons unless otherwise specified	M.t.
Diamonds, gem and industrial (carats)	100,000[45]
Gold (kilograms)	600[45]
Iron ore (000 tons)	1,100[1]

Transportation [31]

Railroads. 480 km total; 328 km 1.435-meter standard gauge, 152 km 1.067-meter narrow gauge; all lines single track; rail systems owned and operated by foreign steel and financial interests in conjunction with Liberian Government

Highways. 10,087 km total; 603 km bituminous treated, 2,848 km all weather, 4,313 km dry weather; there are also 2,323 km of private, laterite-surfaced roads open to public use, owned by rubber and timber companies

Merchant Marine. 1,618 ships (1,000 GRT or over) totaling 57,769,476 DWT/ 101,391,576 DWT; includes 20 passenger, 1 short-sea passenger, 132 cargo, 56 refrigerated cargo, 21 roll-on/roll-off, 58 vehicle carrier, 97 container, 3 barge carrier, 499 oil tanker, 108 chemical, 68 combination ore/oil, 62 liquefied gas, 6 specialized tanker, 456 bulk, 31 combination bulk; note - a flag of convenience registry; all ships are foreign owned; the top 4 owning flags are US 16%, Japan 14%, Norway 11%, and Hong Kong 9%

Airports

Total:	59
Usable:	41
With permanent-surface runways:	2
With runways over 3,659 m:	0
With runways 2,440-3,659 m:	1
With runways 1,220-2,439 m:	4

Tourism [34]

Finance, Economics, and Trade

GDP and Manufacturing Summary [35]

	1980	1985	1990	1991	1992
Gross Domestic Product					
Millions of 1980 dollars	917	843	872	887	876[e]
Growth rate in percent	-6.29	-2.02	-1.99	1.70	-1.24[e]
Manufacturing Value Added					
Millions of 1980 dollars	77	75	80	83	83[e]
Growth rate in percent	-21.21	-1.61	-2.98	3.40	0.29[e]
Manufacturing share in percent of current prices	9.5	6.6	6.9[e]	NA	NA

Economic Indicators [36]

National product: GDP—exchange rate conversion—$988 million (1988). **National product real growth rate**: 1.5% (1988). **National product per capita**: $400 (1988). **Inflation rate (consumer prices)**: 12% (1989). **External debt**: $1.6 billion (December 1990 est.).

Balance of Payments Summary [37]

Values in millions of dollars.

	1975	1980	1985	1986	1987
Exports of goods (f.o.b.)	394.4	600.4	430.4	407.9	374.9
Imports of goods (f.o.b.)	-290.4	-478.0	-263.8	-258.8	-311.7
Trade balance	104.0	122.4	166.6	149.1	63.2
Services - debits	-45.7	-96.7	-211.2	-263.8	-262.5
Services - credits	NA	NA	-2.6	-0.7	NA
Private transfers (net)	-21.7	-29.0	-28.0	-25.4	-21.4
Government transfers (net)	27.8	36.2	91.6	96.4	45.4
Long term capital (net)	88.4	70.5	-106.5	-207.5	-188.9
Short term capital (net)	-2.7	8.2	133.3	263.3	275.7
Errors and omissions	-158.9	-175.0	-108.7	-73.8	30.3
Overall balance	0.5	-50.3	-24.6	-2.7	-0.5

Exchange Rates [38]

Currency: **Liberian dollars.**
Symbol: **L$.**

Data are currency units per $1.

Fixed rate since 1940	1.00
Unofficial parallel rate (Jan 1992)	7.00

Imports and Exports

Top Import Origins [39]

$394 million (c.i.f., 1989 est.).

Origins	%
US	NA
EC	NA
Japan	NA
China	NA
Netherlands	NA
ECOWAS	NA

Top Export Destinations [40]

$505 million (f.o.b., 1989 est.).

Destinations	%
US	NA
EC	NA
Netherlands	NA

Foreign Aid [41]

	U.S. $	
US commitments, including Ex-Im (FY70-89)	665	million
Western (non-US) countries, ODA and OOF bilateral commitments (1970-89)	870	million
OPEC bilateral aid (1979-89)	25	million
Communist countries (1970-89)	77	million

Import and Export Commodities [42]

Import Commodities

Rice
Mineral fuels
Chemicals
Machinery
Transportation equipment
Other foodstuffs

Export Commodities

Iron ore 61%
Rubber 20%
Timber 11%
Coffee

Libya

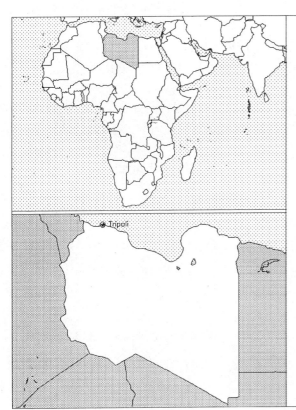

Geography [1]

Total area:
1,759,540 km2
Land area:
1,759,540 km2
Comparative area:
Slightly larger than Alaska
Land boundaries:
Total 4,383 km, Algeria 982 km, Chad 1,055 km, Egypt 1,150 km, Niger 354 km, Sudan 383 km, Tunisia 459 km
Coastline:
1,770 km
Climate:
Mediterranean along coast; dry, extreme desert interior
Terrain:
Mostly barren, flat to undulating plains, plateaus, depressions
Natural resources:
Petroleum, natural gas, gypsum
Land use:
Arable land:
2%
Permanent crops:
0%
Meadows and pastures:
8%
Forest and woodland:
0%
Other:
90%

Demographics [2]

	1960	1970	1980	1990	1991[1]	1994	2000	2010	2020
Population	1,338	2,056	3,119	4,355	4,353	5,057	6,294	8,913	12,391
Population density (persons per sq. mi.)	2	3	4	6	6	NA	8	10	13
Births	NA	NA	NA	NA	158	229	NA	NA	NA
Deaths	NA	NA	NA	NA	27	41	NA	NA	NA
Life expectancy - males	NA	NA	NA	NA	66	62	NA	NA	NA
Life expectancy - females	NA	NA	NA	NA	71	66	NA	NA	NA
Birth rate (per 1,000)	NA	NA	NA	NA	36	45	NA	NA	NA
Death rate (per 1,000)	NA	NA	NA	NA	6	8	NA	NA	NA
Women of reproductive age (15-44 yrs.)	NA	NA	NA	NA	NA	NA	NA	NA	NA
of which are currently married	NA	NA	NA	NA	NA	NA	NA	NA	NA
Fertility rate	NA	NA	NA	6.6	5.05	6.4	6.0	5.3	4.6

Population values are in thousands, life expectancy in years, and other items as indicated.

Health

Health Personnel [3]

Doctors per 1,000 pop., 1988-92	1.04
Nurse-to-doctor ratio, 1988-92	2.9
Hospital beds per 1,000 pop., 1985-90	4.1
Percentage of children immunized (age 1 yr. or less)	
Third dose of DPT, 1990-91	62.0
Measles, 1990-91	59.0

Health Expenditures [4]

Total health expenditure, 1990 (official exchange rate)	
Millions of dollars	NA
Millions of dollars per capita	NA
Health expenditures as a percentage of GDP	
Total	NA
Public sector	NA
Private sector	NA
Development assistance for health	
Total aid flows (millions of dollars)[1]	NA
Aid flows per capita (millions of dollars)	NA
Aid flows as a percentage of total health expenditure	NA

For sources, notes, and explanations, see Annotated Source Appendix, page 1035.

535

Human Factors

Health Care Ratios [5]	Infants and Malnutrition [6]

Ethnic Division [7]

Berber and Arab 97%, Greeks, Maltese, Italians, Egyptians, Pakistanis, Turks, Indians, Tunisians.

Religion [8]

Sunni Muslim   97%

Major Languages [9]

Arabic, Italian, English, all are widely understood in the major cities.

Education

Public Education Expenditures [10]

Million Dinar	1980	1985	1986	1988	1989	1990
Total education expenditure	356	575	636	NA	NA	NA
as percent of GNP	3.4	7.1	9.6	NA	NA	NA
as percent of total govt. expend.	NA	19.8	20.8	NA	NA	NA
Current education expenditure	224	457	506	NA	NA	NA
as percent of GNP	2.1	5.7	7.7	NA	NA	NA
as percent of current govt. expend.	NA	38.1	37.1	NA	NA	NA
Capital expenditure	132	117	130	NA	NA	NA

Educational Attainment [11]

Literacy Rate [12]

In thousands and percent	1985[a]	1991[a]	2000[a]
Illiterate population +15 years	883	890	848
Illiteracy rate - total pop. (%)	43.5	36.2	24.0
Illiteracy rate - males (%)	29.9	24.6	16.0
Illiteracy rate - females (%)	59.7	49.6	32.9

Libraries [13]

Daily Newspapers [14]

	1975	1980	1985	1990
Number of papers	2	3	3	3
Circ. (000)	41	55[e]	65[e]	70[e]

Cinema [15]

Science and Technology

Scientific/Technical Forces [16]

Potential scientists/engineers[16]	43,737
Number female	1,142
Potential technicians[16]	9,020
Number female	439
Total	52,757

R&D Expenditures [17]

	Dinar (000) 1980
Total expenditure	22,875
Capital expenditure	NA
Current expenditure	NA
Percent current	NA

U.S. Patents Issued [18]

For sources, notes, and explanations, see Annotated Source Appendix, page 1035.

Government and Law

Organization of Government [19]

Long-form name:
Socialist People's Libyan Arab Jamahiriya
Type:
Jamahiriya (a state of the masses) in theory, governed by the populace through local councils; in fact, a military dictatorship
Independence:
24 December 1951 (from Italy)
Constitution:
11 December 1969, amended 2 March 1977
Legal system:
based on Italian civil law system and Islamic law; separate religious courts; no constitutional provision for judicial review of legislative acts; has not accepted compulsory ICJ jurisdiction
National holiday:
Revolution Day, 1 September (1969)
Executive branch:
revolutionary leader, chairman of the General People's Committee (premier), General People's Committee (cabinet)
Legislative branch:
unicameral General People's Congress
Judicial branch:
Supreme Court

Elections [20]

None; no political parties; national elections are indirect through a hierarchy of peoples' committees.

Government Budget [21]

For 1989 est.

Revenues	8.1
Expenditures	9.8
Capital expenditures	3.1

Crime [23]

Crime volume	
Cases known to police	35,234
Attempts (percent)	NA
Percent cases solved	NA
Crimes per 100,000 persons	1,006.7
Persons responsible for offenses	
Total number offenders	38,043
Percent female	5.1
Percent juvenile[34]	3.5
Percent foreigners	5

Military Expenditures and Arms Transfers [22]

	1985	1986	1987	1988	1989
Military expenditures					
Current dollars (mil.)	NA	NA	2,774	NA	3,309
1989 constant dollars (mil.)	NA	NA	2,984	NA	3,309
Armed forces (000)	91	91[e]	91[e]	86	86
Gross national product (GNP)					
Current dollars (mil.)	27,000	22,080	23,180	20,490	22,200
1989 constant dollars (mil.)	30,740	24,500	24,930	21,330	22,200
Central government expenditures (CGE)					
1989 constant dollars (mil.)	NA	11,960[e]	NA	NA	11,330
People (mil.)	3.7	3.7	3.8	4.0	4.1
Military expenditure as % of GNP	NA	NA	12.0	NA	14.9
Military expenditure as % of CGE	NA	NA	NA	NA	29.2
Military expenditure per capita	NA	NA	776	NA	808
Armed forces per 1,000 people	24.7	24.4	23.7	21.7	21.0
GNP per capita	8,344	6,575	6,482	5,375	5,423
Arms imports[6]					
Current dollars (mil.)	1,600	1,200	625	600	975
1989 constant dollars (mil.)	1,821	1,332	672	625	975
Arms exports[6]					
Current dollars (mil.)	90	70	70	60	40
1989 constant dollars (mil.)	102	78	75	62	40
Total imports[7]					
Current dollars (mil.)	5,422	4,511	4,877	5,911	6,200
1989 constant dollars	6,172	5,006	5,245	6,153	6,200
Total exports[7]					
Current dollars (mil.)	10,930	7,720	8,052	6,000	6,100
1989 constant dollars	12,440	8,567	8,660	6,246	6,100
Arms as percent of total imports[8]	29.5	26.6	12.8	10.2	15.7
Arms as percent of total exports[8]	0.8	0.9	0.9	1.0	0.7

Human Rights [24]

	SSTS	FL	FAPRO	PPCG	APROBC	TPW	PCPTW	STPEP	PHRFF	PRW	ASST	AFL
Observes	P	P		P	P	P	P	P		P	P	P
	EAFRD	CPR	ESCR	SR	ACHR	MAAE	PVIAC	PVNAC	EAFDAW	TCIDTP	RC
Observes		P	P	P			P	P	P	P	P

P = Party; S = Signatory; see Appendix for meaning of abbreviations.

Labor Force

Total Labor Force [25]

1 million includes about 280,000 resident foreigners

Labor Force by Occupation [26]

Industry	31%
Services	27
Government	24
Agriculture	18

Unemployment Rate [27]

Production Sectors

Commercial Energy Production and Consumption

Production [28]

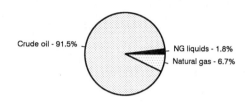

Crude oil - 91.5%

NG liquids - 1.8%

Natural gas - 6.7%

Consumption [29]

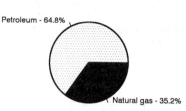

Petroleum - 64.8%

Natural gas - 35.2%

Data are shown in quadrillion (10^{15}) BTUs and percent for 1991

Crude oil[1]	3.13
Natural gas liquids	0.06
Dry natural gas	0.23
Coal[2]	0.00
Net hydroelectric power[3]	0.00
Net nuclear power[3]	0.00
Total	3.42

Petroleum	0.35
Dry natural gas	0.19
Coal[2]	(s)
Net hydroelectric power[3]	0.00
Net nuclear power[3]	0.00
Total	0.53

Telecommunications [30]

- Modern telecommunications system using radio relay, coaxial cable, tropospheric scatter, and domestic satellite stations
- 370,000 telephones
- Broadcast stations - 17 AM, 3 FM, 12 TV
- Satellite earth stations - 1 Atlantic Ocean INTELSAT, 1 Indian Ocean INTELSAT, and 14 domestic
- Submarine cables to France and Italy
- Radio relay to Tunisia and Egypt
- Tropospheric scatter to Greece
- Planned ARABSAT and Intersputnik satellite stations

Transportation [31]

Railroads. Libya has had no railroad in operation since 1965, all previous systems having been dismantled; current plans are to construct a standard gauge (1.435 m) line from the Tunisian frontier to Tripoli and Misratah, then inland to Sabha, center of a mineral rich area, but there has been no progress; other plans made jointly with Egypt would establish a rail line from As Sallum, Egypt to Tobruk with completion set for mid-1994, progress unknown

Highways. 19,300 km total; 10,800 km bituminous/bituminous treated, 8,500 km crushed stone or earth

Merchant Marine. 32 ships (1,000 GRT or over) totaling 694,883 GRT/1,215,494 DWT; includes 4 short-sea passenger, 11 cargo, 4 roll-on/roll-off, 10 oil tanker, 1 chemical tanker, 2 liquefied gas

Airports

Total:	138
Usable:	124
With permanent-surface runways:	56
With runways over 3,659 m:	9
With runways 2,440-3,659 m:	27
With runways 1,220-2,439 m:	47

Top Agricultural Products [32]

5% of GNP; cash crops—wheat, barley, olives, dates, citrus fruits, peanuts; 75% of food is imported.

Top Mining Products [33]

Estimated metric tons unless otherwise specified   M.t.

Petroleum, crude (000 barrels)	541,295[1]
Gas, natural (mil. cubic meters)	13,600
Salt (000 tons)	12,000
Gypsum (000 tons)	180
Cement, hydraulic (000 tons)	2,700
Steel, crude (000 tons)	500

Tourism [34]

For sources, notes, and explanations, see Annotated Source Appendix, page 1035.

Manufacturing Sector

GDP and Manufacturing Summary [35]

	1980	1985	1989	1990	% change 1980-1990	% change 1989-1990
GDP (million 1980 $)	35,727	29,777	25,127	31,908	-10.7	27.0
GDP per capita (1980 $)	11,737	7,865	5,730	7,014	-40.2	22.4
Manufacturing as % of GDP (current prices)	1.9	4.5	8.5[e]	8.4[e]	342.1	-1.2
Gross output (million $)	1,177	1,953[e]	3,312[e]	3,830[e]	225.4	15.6
Value added (million $)	358	638[e]	930[e]	1,211[e]	238.3	30.2
Value added (million 1980 $)	649	1,262	1,486[e]	1,617	149.2	8.8
Industrial production index	100	140	135[e]	178	78.0	31.9
Employment (thousands)	18	22[e]	24[e]	28[e]	55.6	16.7

Note: GDP stands for Gross Domestic Product. 'e' stands for estimated value.

Profitability and Productivity

	1980	1985	1989	1990	% change 1980-1990	% change 1989-1990
Intermediate input (%)	70	67[e]	72[e]	68[e]	-2.9	-5.6
Wages, salaries, and supplements (%)	13	12[e]	13[e]	12[e]	-7.7	-7.7
Gross operating surplus (%)	17	20[e]	15[e]	20[e]	17.6	33.3
Gross output per worker ($)	63,982[e]	84,676[e]	138,218[e]	133,325[e]	108.4	-3.5
Value added per worker ($)	19,492[e]	28,750[e]	38,817[e]	45,851[e]	135.2	18.1
Average wage (incl. benefits) ($)	8,326[e]	10,746[e]	17,776[e]	16,121[e]	93.6	-9.3

Profitability is in percent of gross output. Productivity is in U.S. $. 'e' stands for estimated value.

Profitability - 1990

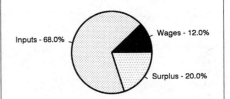

Inputs - 68.0%
Wages - 12.0%
Surplus - 20.0%

The graphic shows percent of gross output.

Value Added in Manufacturing

	1980 $ mil.	1980 %	1985 $ mil.	1985 %	1989 $ mil.	1989 %	1990 $ mil.	1990 %	% change 1980-1990	% change 1989-1990
311 Food products	35	9.8	42[e]	6.6	69[e]	7.4	67[e]	5.5	91.4	-2.9
313 Beverages	17	4.7	20[e]	3.1	30[e]	3.2	34[e]	2.8	100.0	13.3
314 Tobacco products	55	15.4	83[e]	13.0	75[e]	8.1	131[e]	10.8	138.2	74.7
321 Textiles	14	3.9	22[e]	3.4	30[e]	3.2	33[e]	2.7	135.7	10.0
322 Wearing apparel	5[e]	1.4	6[e]	0.9	5[e]	0.5	9[e]	0.7	80.0	80.0
323 Leather and fur products	7	2.0	16[e]	2.5	26[e]	2.8	33[e]	2.7	371.4	26.9
324 Footwear	14	3.9	28[e]	4.4	40[e]	4.3	53[e]	4.4	278.6	32.5
331 Wood and wood products	3[e]	0.8	6[e]	0.9	4[e]	0.4	11[e]	0.9	266.7	175.0
332 Furniture and fixtures	2[e]	0.6	4[e]	0.6	5[e]	0.5	9[e]	0.7	350.0	80.0
341 Paper and paper products	3	0.8	3[e]	0.5	6[e]	0.6	5[e]	0.4	66.7	-16.7
342 Printing and publishing	NA	0.0	1[e]	0.2	6[e]	0.6	3[e]	0.2	NA	-50.0
351 Industrial chemicals	35	9.8	46[e]	7.2	102[e]	11.0	87[e]	7.2	148.6	-14.7
352 Other chemical products	21	5.9	38[e]	6.0	19[e]	2.0	70[e]	5.8	233.3	268.4
353 Petroleum refineries	81	22.6	179[e]	28.1	288[e]	31.0	374[e]	30.9	361.7	29.9
354 Miscellaneous petroleum and coal products	NA	0.0	NA	0.0	NA	0.0	NA	0.0	NA	NA
355 Rubber products	NA	0.0	1[e]	0.2	NA	0.0	1[e]	0.1	NA	NA
356 Plastic products	2	0.6	5[e]	0.8	5[e]	0.5	9[e]	0.7	350.0	80.0
361 Pottery, china and earthenware	1	0.3	1[e]	0.2	1[e]	0.1	1[e]	0.1	0.0	0.0
362 Glass and glass products	NA	0.0	NA	0.0	NA	0.0	NA	0.0	NA	NA
369 Other non-metal mineral products	51	14.2	110[e]	17.2	202[e]	21.7	222[e]	18.3	335.3	9.9
371 Iron and steel	NA	0.0	NA	0.0	NA	0.0	NA	0.0	NA	NA
372 Non-ferrous metals	NA	0.0	NA	0.0	NA	0.0	NA	0.0	NA	NA
381 Metal products	3	0.8	8[e]	1.3	3[e]	0.3	21[e]	1.7	600.0	600.0
382 Non-electrical machinery	NA	0.0	NA	0.0	NA	0.0	NA	0.0	NA	NA
383 Electrical machinery	NA	0.0	NA	0.0	NA	0.0	NA	0.0	NA	NA
384 Transport equipment	NA	0.0	NA	0.0	NA	0.0	NA	0.0	NA	NA
385 Professional and scientific equipment	NA	0.0	NA	0.0	NA	0.0	NA	0.0	NA	NA
390 Other manufacturing industries	9	2.5	19[e]	3.0	15[e]	1.6	39[e]	3.2	333.3	160.0

Note: The industry codes shown are International Standard Industry codes (ISIC). Percentages are percent of total Value Added. 'e' stands for estimated value

Finance, Economics, and Trade

Economic Indicators [36]

National product: GDP—exchange rate conversion—$26.1 billion (1992 est.). **National product real growth rate**: 0.2% (1992 est.). **National product per capita**: $5,800 (1992 est.). **Inflation rate (consumer prices)**: 7% (1991 est.). **External debt**: $3.5 billion excluding military debt (1991 est.).

Balance of Payments Summary [37]

Exchange Rates [38]

Currency: **Libyan dinars.**
Symbol: **LD.**

Data are currency units per $1.

January 1993	0.2998
1992	0.3013
1991	0.2684
1990	0.2699
1989	0.2922
1988	0.2853

Imports and Exports

Top Import Origins [39]

$8.66 billion (f.o.b., 1992).

Origins	%
Italy	NA
Former USSR	NA
Germany	NA
UK	NA
Japan	NA
Korea	NA

Top Export Destinations [40]

$9.71 billion (f.o.b., 1992).

Destinations	%
Italy	NA
Former USSR	NA
Germany	NA
Spain	NA
France	NA
Belgium/Luxembourg	NA
Turkey	NA

Foreign Aid [41]

Note: no longer a recipient.

	U.S. $	
Western (non-US) countries, ODA and OOF bilateral commitments (1970-87)	242	million

Import and Export Commodities [42]

Import Commodities
Machinery
Transport equipment
Food
Manufactured goods

Export Commodities
Crude oil
Refined petroleum products
Natural gas

Liechtenstein

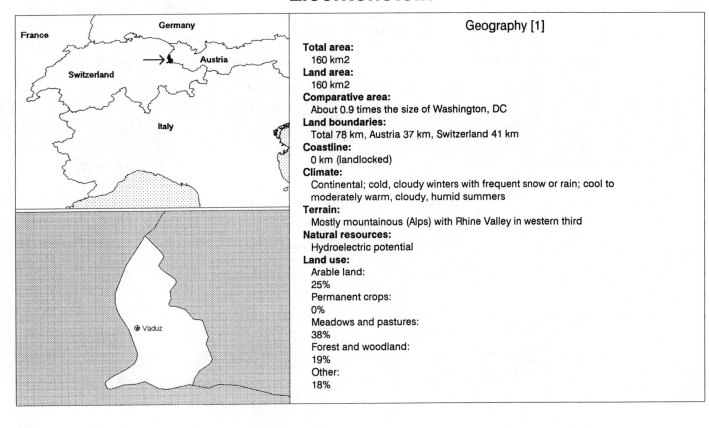

Geography [1]

Total area:
160 km2
Land area:
160 km2
Comparative area:
About 0.9 times the size of Washington, DC
Land boundaries:
Total 78 km, Austria 37 km, Switzerland 41 km
Coastline:
0 km (landlocked)
Climate:
Continental; cold, cloudy winters with frequent snow or rain; cool to moderately warm, cloudy, humid summers
Terrain:
Mostly mountainous (Alps) with Rhine Valley in western third
Natural resources:
Hydroelectric potential
Land use:
Arable land:
25%
Permanent crops:
0%
Meadows and pastures:
38%
Forest and woodland:
19%
Other:
18%

Demographics [2]

	1960	1970	1980	1990	1991[1]	1994	2000	2010	2020
Population	16	21	25	29	28	30	32	34	36
Population density (persons per sq. mi.)	265	340	404	456	459	NA	481	494	496
Births	NA	NA	NA	NA	(Z)	(Z)	NA	NA	NA
Deaths	NA	NA	NA	NA	(Z)	(Z)	NA	NA	NA
Life expectancy - males	NA	NA	NA	NA	73	74	NA	NA	NA
Life expectancy - females	NA	NA	NA	NA	81	81	NA	NA	NA
Birth rate (per 1,000)	NA	NA	NA	NA	13	13	NA	NA	NA
Death rate (per 1,000)	NA	NA	NA	NA	7	7	NA	NA	NA
Women of reproductive age (15-44 yrs.)	NA	NA	NA	8	NA	9	9	NA	NA
of which are currently married	NA	NA	NA	5	NA	5	5	NA	NA
Fertility rate	NA	NA	NA	1.4	1.52	1.5	1.5	1.5	1.6

Population values are in thousands, life expectancy in years, and other items as indicated.

Health

Health Personnel [3]

Health Expenditures [4]

For sources, notes, and explanations, see Annotated Source Appendix, page 1035.

541

Human Factors

Health Care Ratios [5]	Infants and Malnutrition [6]

Ethnic Division [7]

Alemannic	95%
Italian other	5%

Religion [8]

Roman Catholic	87.3%
Protestant	8.3%
Unknown	1.6%
Other	2.8%

Major Languages [9]

German (official), Alemannic dialect.

Education

Public Education Expenditures [10]	Educational Attainment [11]

Literacy Rate [12]

	1970[b]	1980[b]	1981[b]
Illiterate population +15 years	NA	NA	68
Illiteracy rate - total pop. (%)	NA	NA	0.3[e]
Illiteracy rate - males (%)	NA	NA	0.3[e]
Illiteracy rate - females (%)	NA	NA	0.3[e]

Libraries [13]

	Admin. Units	Svc. Pts.	Vols. (000)	Shelving (meters)	Vols. Added	Reg. Users
National (1989)	1	1	130	NA	4,000	13,000
Non-specialized	NA	NA	NA	NA	NA	NA
Public (1989)	3	3	23	NA	1,060	1,552
Higher ed.	NA	NA	NA	NA	NA	NA
School (1990)	8	NA	35	NA	1,454	2,509

Daily Newspapers [14]

	1975	1980	1985	1990
Number of papers	2	2	2	2
Circ. (000)	12	14	14	9

Cinema [15]

Science and Technology

Scientific/Technical Forces [16]	R&D Expenditures [17]	U.S. Patents Issued [18]

U.S. Patents Issued [18]

Values show patents issued to citizens of the country by the U.S. Patents Office.

	1990	1991	1992
Number of patents	16	12	16

542

For sources, notes, and explanations, see Annotated Source Appendix, page 1035.

Government and Law

Organization of Government [19]

Long-form name:
Principality of Liechtenstein
Type:
hereditary constitutional monarchy
Independence:
23 January 1719 (Imperial Principality of
Liechtenstein established)
Constitution:
5 October 1921
Legal system:
local civil and penal codes; accepts
compulsory ICJ jurisdiction, with
reservations
National holiday:
Assumption Day, 15 August
Executive branch:
reigning prince, hereditary prince, head of
government, deputy head of government
Legislative branch:
unicameral Diet (Landtag)
Judicial branch:
Supreme Court (Oberster Gerichtshof) for
criminal cases, Superior Court
(Obergericht) for civil cases

Political Parties [20]

Diet	% of seats
Progressive Citizens' Party (FBP)	48.0
Fatherland Union (VU)	44.0
Free Electoral List (FL)	8.0

Government Budget [21]

For 1990.

Revenues	259
Expenditures	292
Capital expenditures	NA

Defense Summary [22]

Note: Defense is responsibility of Switzerland

Crime [23]

Human Rights [24]

	SSTS	FL	FAPRO	PPCG	APROBC	TPW	PCPTW	STPEP	PHRFF	PRW	ASST	AFL
Observes						P	P		P			

	EAFRD	CPR	ESCR	SR	ACHR	MAAE	PVIAC	PVNAC	EAFDAW	TCIDTP	RC
Observes				P				P	P		P	S

P = Party; S = Signatory; see Appendix for meaning of abbreviations.

Labor Force

Total Labor Force [25]

19,905 of which 11,933 are foreigners
6,885 commute from Austria and
Switzerland to work each day

Labor Force by Occupation [26]

Industry, trade, and building	53.2%
Services	45
Agriculture, fishing, forestry, and horticulture	1.8

Date of data: 1990

Unemployment Rate [27]

1.5% (1990)

Production Sectors

Energy Resource Summary [28]

Energy Resources: Hydroelectric potential. **Electricity**: 23,000 kW capacity; 150 million kWh produced, 5,230 kWh per capita (1992).

Telecommunications [30]

- Limited, but sufficient automatic telephone system
- 25,400 telephones
- Linked to Swiss networks by cable and radio relay for international telephone, radio, and TV services

Transportation [31]

Railroads. 18.5 km 1.435-meter standard gauge, electrified; owned, operated, and included in statistics of Austrian Federal Railways

Highways. 130.66 km main roads, 192.27 km byroads

Airports

Airports	
Total:	0
Usable:	0
With permanent-surface runways:	0
With runways over 3,659 m:	0
With runways 2,440-3,659 m:	0
With runways 1,220-2,439 m:	0

Top Agricultural Products [32]

Livestock, vegetables, corn, wheat, potatoes, grapes.

Top Mining Products [33]

Detailed information is not available. A summary of mineral resources available follows. **Mineral Resources**: None.

Tourism [34]

	1987	1988	1989	1990	1991
Tourists[10]	75	72	77	78	71

Tourists are in thousands, money in million U.S. dollars.

544

For sources, notes, and explanations, see Annotated Source Appendix, page 1035.

Finance, Economics, and Trade

Industrial Summary [35]

Industrial Production: Growth rate NA%. **Industries**: Electronics, metal manufacturing, textiles, ceramics, pharmaceuticals, food products, precision instruments, tourism.

Economic Indicators [36]

National product: GDP—purchasing power equivalent—$630 million (1990 est.). **National product real growth rate**: NA%. **National product per capita**: $22,300 (1990 est.). **Inflation rate (consumer prices)**: 5.4% (1990). **External debt**: not available.

Balance of Payments Summary [37]

Exchange Rates [38]

Currency: **Swiss francs, franken, or franchi.**
Symbol: **SwF.**

Data are currency units per $1.

January 1993	1.4781
1992	1.4062
1991	1.4340
1990	1.3892
1989	1.6359
1988	1.4633

Imports and Exports

Top Import Origins [39]

Amount not available.

Origins	%
NA	NA

Top Export Destinations [40]

$1.6 billion. Data are for 1990.

Destinations	%
EFTA countries	20.9
EC countries	42.7
Other	36.4

Foreign Aid [41]

	U.S. $
	NA

No aid provided or received.

Import and Export Commodities [42]

Import Commodities	Export Commodities
Machinery	Small specialty machinery
Metal goods	Dental products
Textiles	Stamps
Foodstuffs	Hardware
Motor vehicles	Pottery

For sources, notes, and explanations, see Annotated Source Appendix, page 1035.

545

Lithuania

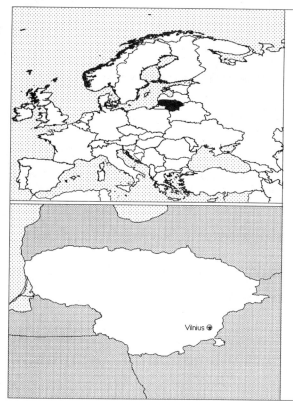

Geography [1]

Total area:
65,200 km2
Land area:
65,200 km2
Comparative area:
Slightly larger than West Virginia
Land boundaries:
Total 1,273 km, Belarus 502 km, Latvia 453 km, Poland 91 km, Russia (Kaliningrad) 227 km
Coastline:
108 km
Climate:
Maritime; wet, moderate winters
Terrain:
Lowland, many scattered small lakes, fertile soil
Natural resources:
Peat
Land use:
Arable land:
49.1%
Permanent crops:
0%
Meadows and pastures:
22.2%
Forest and woodland:
16.3%
Other:
12.4%

Demographics [2]

	1960	1970	1980	1990	1991	1994	2000	2010	2020
Population	2,765	3,138	3,436	3,727	NA	3,848	4,007	4,263	4,505
Population density (persons per sq. mi.)	NA	NA	NA	NA	NA	NA	NA	NA	NA
Births	NA	NA	NA	NA	NA	57	NA	NA	NA
Deaths	NA	NA	NA	NA	NA	42	NA	NA	NA
Life expectancy - males	NA	NA	NA	NA	NA	67	NA	NA	NA
Life expectancy - females	NA	NA	NA	NA	NA	76	NA	NA	NA
Birth rate (per 1,000)	NA	NA	NA	NA	NA	15	NA	NA	NA
Death rate (per 1,000)	NA	NA	NA	NA	NA	11	NA	NA	NA
Women of reproductive age (15-44 yrs.)	NA	NA	NA	928	NA	946	990	NA	NA
of which are currently married	NA	NA	NA	603	NA	620	647	NA	NA
Fertility rate	NA	NA	NA	2.1	NA	2.0	1.9	1.8	1.8

Population values are in thousands, life expectancy in years, and other items as indicated.

Health

Health Personnel [3]

Doctors per 1,000 pop., 1988-92	NA
Nurse-to-doctor ratio, 1988-92	NA
Hospital beds per 1,000 pop., 1985-90	NA
Percentage of children immunized (age 1 yr. or less)	
Third dose of DPT, 1990-91	80.0
Measles, 1990-91	92.0

Health Expenditures [4]

Total health expenditure, 1990 (official exchange rate)	
Millions of dollars	594
Millions of dollars per capita	159
Health expenditures as a percentage of GDP	
Total	3.6
Public sector	2.6
Private sector	1.0
Development assistance for health	
Total aid flows (millions of dollars)[1]	NA
Aid flows per capita (millions of dollars)	NA
Aid flows as a percentage of total health expenditure	NA

546

For sources, notes, and explanations, see Annotated Source Appendix, page 1035.

Human Factors

Health Care Ratios [5]

Population per physician, 1970	NA
Population per physician, 1990	220
Population per nursing person, 1970	NA
Population per nursing person, 1990	NA
Percent of births attended by health staff, 1985	NA

Infants and Malnutrition [6]

Percent of babies with low birth weight, 1985	NA
Infant mortality rate per 1,000 live births, 1970	NA
Infant mortality rate per 1,000 live births, 1991	14
Years of life lost per 1,000 population, 1990	19
Prevalence of malnutrition (under age 5), 1990	NA

Ethnic Division [7]

Lithuanian	80.1%
Russian	8.6%
Polish	7.7%
Belarusian	1.5%
Other	2.1%

Religion [8]

Roman Catholic, Lutheran, other.

Major Languages [9]

Lithuanian (official), Polish, Russian.

Education

Public Education Expenditures [10]

Educational Attainment [11]

Age group	25+
Total population	2,282,191
Highest level attained (%)	
No schooling	9.1
First level	
Incompleted	NA
Completed	21.3
Entered second level	
S-1	NA
S-2	57.0
Post secondary	12.6

Literacy Rate [12]

	1980[b]	1989[b]	1990[b]
Illiterate population +15 years	NA	44,308	NA
Illiteracy rate - total pop. (%)	NA	1.6	NA
Illiteracy rate - males (%)	NA	0.8	NA
Illiteracy rate - females (%)	NA	2.2	NA

Libraries [13]

Daily Newspapers [14]

Cinema [15]

Science and Technology

Scientific/Technical Forces [16]

R&D Expenditures [17]

U.S. Patents Issued [18]

For sources, notes, and explanations, see Annotated Source Appendix, page 1035.

Government and Law

Organization of Government [19]

Long-form name:
Republic of Lithuania
Type:
republic
Independence:
6 September 1991 (from Soviet Union)
Constitution:
adopted 25 October 1992
Legal system:
based on civil law system; no judicial
review of legislative acts
National holiday:
Independence Day, 16 February
Executive branch:
president, prime minister, cabinet
Legislative branch:
unicameral Seimas (parliament)
Judicial branch:
Supreme Court, Court of Appeals

Political Parties [20]

Seimas (parliament)	% of votes
Democratic Labor Party	51.0
Other	49.0

Government Budget [21]

For 1992 est.

Revenues	258.5
Expenditures	270.2
Capital expenditures	NA

Defense Summary [22]

Branches: Ground Forces, Navy, Air Force, Security Forces (internal and border troops), National Guard (Skat)

Manpower Availability: Males age 15-49 933,245; fit for military service 739,400; reach military age (18) annually 27,056 (1993 est.)

Defense Expenditures: Exchange rate conversion - $NA, 5.5% of GDP (1993 est.)

Crime [23]

Human Rights [24]

	SSTS	FL	FAPRO	PPCG	APROBC	TPW	PCPTW	STPEP	PHRFF	PRW	ASST	AFL
Observes												
	EAFRD	CPR	ESCR	SR	ACHR	MAAE	PVIAC	PVNAC	EAFDAW	TCIDTP	RC
Observes		P	P								P

P = Party; S = Signatory; see Appendix for meaning of abbreviations.

Labor Force

Total Labor Force [25]

1.836 million

Labor Force by Occupation [26]

Industry and construction	42%
Agriculture and forestry	18
Other	40

Date of data: 1990

Unemployment Rate [27]

1% (February 1993); but large numbers of underemployed workers

Production Sectors

Energy Resource Summary [28]

Energy Resources: None. **Electricity**: 5,925,000 kW capacity; 25,000 million kWh produced, 6,600 kWh per capita (1992). **Pipelines**: Crude oil 105 km, natural gas 760 km (1992).

Telecommunications [30]

- Better developed than in most other former USSR republics
- Operational NMT-450 analog cellular network in Vilnius
- Fiber optic cable installed beween Vilnius and Kaunas
- 224 telephones per 1000 persons
- Broadcast stations - 13 AM, 26 FM, 1 SW, 1 LW, 3 TV
- Landlines or microwave to former USSR republics
- Leased connection to the Moscow international switch for traffic with other countries
- Satellite earth stations - (8 channels to Norway)
- New international digital telephone exchange in Kaunas for direct access to 13 countries via satellite link out of Copenhagen, Denmark

Top Agricultural Products [32]

Employs around 20% of labor force; sugar, grain, potatoes, sugarbeets, vegetables, meat, milk, dairy products, eggs, fish; most developed are the livestock and dairy branches, which depend on imported grain; net exporter of meat, milk, and eggs.

Top Mining Products [33]

Detailed information is not available. A summary of mineral resources available follows. **Mineral Resources**: Peat.

Transportation [31]

Railroads. 2,100 km; does not include industrial lines (1990)

Highways. 44,200 km total 35,500 km hard surfaced, 8,700 km earth (1990)

Merchant Marine. 46 ships (1,000 GRT or over) totaling 282,633 GRT/332,447 DWT; includes 31 cargo, 3 railcar carrier, 1 roll-on/roll-off, 11 combination bulk

Airports

Total:	96
Useable:	19
With permanent-surface runways:	12
With runways over 3,659 m:	0
With runways 2,440-3,659 m:	5
With runways 1,220-2,439 m:	11

Tourism [34]

Finance, Economics, and Trade

Industrial Summary [35]

Industrial Production: Growth rate - 50% (1992 est.). **Industries:** Employs 25% of the labor force; shares in the total production of the former USSR are: metal-cutting machine tools 6.6%; electric motors 4.6%; television sets 6.2%; refrigerators and freezers 5.4%; other branches: petroleum refining, shipbuilding (small ships), furniture making, textiles, food processing, fertilizers, agricultural machinery, optical equipment, electronic components, computers, and amber.

Economic Indicators [36]

National product: GDP not available. **National product real growth rate:** - 30% (1992 est.). **National product per capita:** not available. **Inflation rate (consumer prices):** 10%- 20% per month (first quarter 1993). **External debt:** $650 million (1991 est.).

Balance of Payments Summary [37]

Exchange Rates [38]

Imports and Exports

Top Import Origins [39]

Amount not available.

Origins	%
Russia	62
Belarus	18
Former Soviet republics	10
West	10

Top Export Destinations [40]

Amount not available.

Destinations	%
Russia	40
Ukraine	16
Other former Soviet republics	32
West	12

Foreign Aid [41]

	U.S. $	
US commitments, including Ex-Im (1992)	10	million
Western (non-US) countries, ODA and OOF bilateral commitments (1970-86)	NA	million
Communist countries (1971-86)	NA	million

Import and Export Commodities [42]

Import Commodities

Oil 24%
Machinery 14%
Chemicals 8%
Grain NA

Export Commodities

Electronics 18%
Petroleum products 5%
Food 10%
Chemicals 6%

Luxembourg

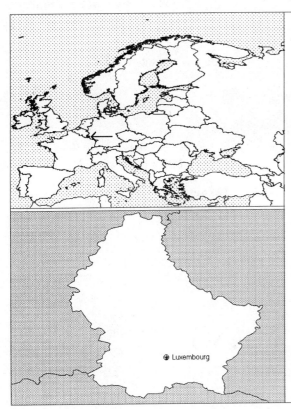

Geography [1]

Total area:
2,586 km2
Land area:
2,586 km2
Comparative area:
Slightly smaller than Rhode Island
Land boundaries:
Total 359 km, Belgium 148 km, France 73 km, Germany 138 km
Coastline:
0 km (landlocked)
Climate:
Modified continental with mild winters, cool summers
Terrain:
Mostly gently rolling uplands with broad, shallow valleys; uplands to slightly mountainous in the north; steep slope down to Moselle floodplain in the southeast
Natural resources:
Iron ore (no longer exploited)
Land use:
Arable land:
24%
Permanent crops:
1%
Meadows and pastures:
20%
Forest and woodland:
21%
Other:
34%

Demographics [2]

	1960	1970	1980	1990	1991[1]	1994	2000	2010	2020
Population	314	339	364	382	388	402	415	428	436
Population density (persons per sq. mi.)	315	340	365	385	389	NA	411	410	402
Births	NA	NA	NA	NA	5	5	NA	NA	NA
Deaths	NA	NA	NA	NA	4	4	NA	NA	NA
Life expectancy - males	NA	NA	NA	NA	73	73	NA	NA	NA
Life expectancy - females	NA	NA	NA	NA	80	81	NA	NA	NA
Birth rate (per 1,000)	NA	NA	NA	NA	12	13	NA	NA	NA
Death rate (per 1,000)	NA	NA	NA	NA	10	9	NA	NA	NA
Women of reproductive age (15-44 yrs.)	NA	NA	NA	97	NA	103	102	NA	NA
of which are currently married	NA	NA	NA	65	NA	70	70	NA	NA
Fertility rate	NA	NA	NA	1.6	1.55	1.6	1.7	1.7	1.7

Population values are in thousands, life expectancy in years, and other items as indicated.

Health

Health Personnel [3]

Health Expenditures [4]

For sources, notes, and explanations, see Annotated Source Appendix, page 1035.

551

Human Factors

Health Care Ratios [5]	Infants and Malnutrition [6]

Ethnic Division [7]

Celtic base (with French and German blend), Portuguese, Italian, and European (guest and worker residents).

Religion [8]

Roman Catholic	97%
Protestant and Jewish	3%

Major Languages [9]

Luxembourgisch	NA
German	NA
French	NA
English	NA

Education

Public Education Expenditures [10]

Million Francs	1980	1986	1987	1988	1989	1990
Total education expenditure	9,792	12,269	13,689	14,898	16,363	NA
as percent of GNP	6.1	4.0	4.5	4.5	4.4	NA
as percent of total govt. expend.	14.9	15.7	17.1	17.7	NA	NA
Current education expenditure	9,305	10,712	11,965	12,817	13,440	NA
as percent of GNP	5.8	3.5	3.9	3.8	3.6	NA
as percent of current govt. expend.	19.8	15.4	16.6	16.9	NA	NA
Capital expenditure	486	1,557	1,724	2,081	2,923	NA

Educational Attainment [11]

Literacy Rate [12]

Libraries [13]

	Admin. Units	Svc. Pts.	Vols. (000)	Shelving (meters)	Vols. Added	Reg. Users
National (1986)	1	1	685	22,000	8,000	21,500
Non-specialized	NA	NA	NA	NA	NA	NA
Public	NA	NA	NA	NA	NA	NA
Higher ed.	NA	NA	NA	NA	NA	NA
School	NA	NA	NA	NA	NA	NA

Daily Newspapers [14]

	1975	1980	1985	1990
Number of papers	7	5	4	5
Circ. (000)	130	135	140	143

Cinema [15]

Data for 1989.

Cinema seats per 1,000	8.1
Annual attendance per person	1.4
Gross box office receipts (mil. Franc)	80

Science and Technology

Scientific/Technical Forces [16]

R&D Expenditures [17]

U.S. Patents Issued [18]

Values show patents issued to citizens of the country by the U.S. Patents Office.

	1990	1991	1992
Number of patents	27	42	36

Government and Law

Organization of Government [19]

Long-form name:
Grand Duchy of Luxembourg
Type:
constitutional monarchy
Independence:
1839
Constitution:
17 October 1868, occasional revisions
Legal system:
based on civil law system; accepts
compulsory ICJ jurisdiction
National holiday:
National Day, 23 June (1921) (public
celebration of the Grand Duke's birthday)
Executive branch:
grand duke, prime minister, vice prime
minister, Council of Ministers (cabinet)
Legislative branch:
unicameral Chamber of Deputies
(Chambre des Deputes); note - the
Council of State (Conseil d'Etat) is an
advisory body whose views are
considered by the Chamber of Deputies
Judicial branch:
Superior Court of Justice (Cour
Superieure de Justice)

Political Parties [20]

Chamber of Deputies	% of votes
Christian Social Party (CSV)	31.7
Socialist Workers Party (LSAP)	27.2
Liberal (DP)	16.2
Greens	8.4
PAC	7.3
Communist (KPL)	5.1
Other	4.1

Government Expenditures [21]

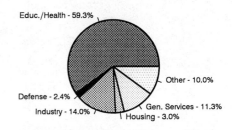

Educ./Health - 59.3%
Other - 10.0%
Defense - 2.4%
Gen. Services - 11.3%
Industry - 14.0%
Housing - 3.0%

% distribution for 1987. Expend. in 1992: 3.5 ($ bil.)

Crime [23]

Crime volume	
Cases known to police	34,657
Attempts (percent)	NA
Percent cases solved	7,107
Crimes per 100,000 persons	6,628.22
Persons responsible for offenses	
Total number offenders	9,312
Percent female	13.3
Percent juvenile[35]	8.6
Percent foreigners	43.4

Military Expenditures and Arms Transfers [22]

	1985	1986	1987	1988	1989
Military expenditures					
Current dollars (mil.)	54	58	67	79	76
1989 constant dollars (mil.)	62	64	72	82	76
Armed forces (000)	1	1	1	1	1
Gross national product (GNP)					
Current dollars (mil.)	6,938	7,243	7,486	8,121	8,879
1989 constant dollars (mil.)	7,898	8,038	8,051	8,454	8,879
Central government expenditures (CGE)					
1989 constant dollars (mil.)	2,630	2,725	2,844	3,004	NA
People (mil.)	0.4	0.4	0.4	0.4	0.4
Military expenditure as % of GNP	0.8	0.8	0.9	1.0	0.9
Military expenditure as % of CGE	2.4	2.4	2.5	2.7	NA
Military expenditure per capita	169	174	195	218	200
Armed forces per 1,000 people	2.7	2.7	2.7	2.7	2.6
GNP per capita	21,540	21,820	21,710	22,570	23,410
Arms imports[6]					
Current dollars (mil.)	0	5	10	10	10
1989 constant dollars (mil.)	0	6	11	10	10
Arms exports[6]					
Current dollars (mil.)	0	0	0	0	5
1989 constant dollars (mil.)	0	0	0	0	5
Total imports[7]					
Current dollars (mil.)	2,983	NA	NA	NA	NA
1989 constant dollars	3,396	NA	NA	NA	NA
Total exports[7]					
Current dollars (mil.)	2,830	NA	NA	NA	NA
1989 constant dollars	3,222	NA	NA	NA	NA
Arms as percent of total imports[8]	0	NA	NA	NA	NA
Arms as percent of total exports[8]	0	NA	NA	NA	NA

Human Rights [24]

	SSTS	FL	FAPRO	PPCG	APROBC	TPW	PCPTW	STPEP	PHRFF	PRW	ASST	AFL
Observes		P	P	P	P	P	P	P	P	P	P	P
	EAFRD	CPR	ESCR	SR	ACHR	MAAE	PVIAC	PVNAC	EAFDAW	TCIDTP	RC	
Observes	P	P	P	P	P	P	P	P	P	P	S	

P = Party; S = Signatory; see Appendix for meaning of abbreviations.

Labor Force

Total Labor Force [25]

177,300 One-third of labor force is foreign
workers, mostly from Portugal, Italy,
France, Belgium, and Germany

Labor Force by Occupation [26]

Services	65%
Industry	31.6
Agriculture	3.4
Date of data: 1988	

Unemployment Rate [27]

1.4% (1991)

For sources, notes, and explanations, see Annotated Source Appendix, page 1035.

553

Production Sectors

Commercial Energy Production and Consumption

Production [28]

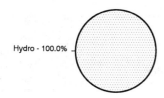

Hydro - 100.0%

Consumption [29]

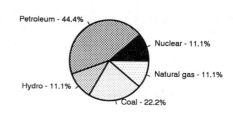

Petroleum - 44.4%
Nuclear - 11.1%
Natural gas - 11.1%
Coal - 22.2%
Hydro - 11.1%

Data are shown in quadrillion (10^{15}) BTUs and percent for 1991

Crude oil[1]	0.00
Natural gas liquids	0.00
Dry natural gas	0.00
Coal[2]	0.00
Net hydroelectric power[3]	0.01
Net nuclear power[3]	0.00
Total	0.01

Petroleum	0.08
Dry natural gas	0.02
Coal[2]	0.04
Net hydroelectric power[3]	0.02
Net nuclear power[3]	0.02
Total	0.17

Telecommunications [30]

- Highly developed, completely automated and efficient system, mainly buried cables
- 230,000 telephones
- Broadcast stations - 2 AM, 3 FM, 3 TV
- 3 channels leased on TAT-6 coaxial submarine cable
- 1 direct-broadcast satellite earth station
- Nationwide mobile phone system

Transportation [31]

Railroads. Luxembourg National Railways (CFL) operates 272 km 1.435-meter standard gauge; 178 km double track; 178 km electrified

Highways. 5,108 km total; 4,995 km paved, 57 km gravel, 56 km earth; about 80 km limited access divided highway

Merchant Marine. 53 ships (1,000 GRT or over) totaling 1,570,466 GRT/2,614,154 DWT; includes 2 cargo, 5 container, 5 roll-on/roll-off, 6 oil tanker, 4 chemical tanker, 3 combination ore/oil, 8 liquefied gas, 2 passenger, 8 bulk, 6 combination bulk, 4 refrigerated cargo

Airports

Total:	2
Usable:	2
With permanent-surface runways:	1
With runways over 3,659 m:	1
With runways 2,440-3,659 m:	0
With runways 1,220-2,439 m:	1

Top Agricultural Products [32]

Agriculture accounts for less than 3% of GDP (including forestry); principal products—barley, oats, potatoes, wheat, fruits, wine grapes; cattle raising widespread.

Top Mining Products [33]

Metric tons unless otherwise specified	M.t.
Dolomite (000 tons)	4,034[1]
Limestone (000 tons)	33,255[1]
Sand	16,707[1]
Gravel	4,192[1]
Gypsum	NA
Slate	NA

Tourism [34]

	1987	1988	1989	1990	1991
Tourists	645[41]	760[42]	875[42]	820[42]	861
Tourism receipts[43]	201	238	286		

Tourists are in thousands, money in million U.S. dollars.

Manufacturing Sector

GDP and Manufacturing Summary [35]

	1980	1985	1989	1990	% change 1980-1990	% change 1989-1990
GDP (million 1980 $)	4,546	5,146	5,995[e]	6,356	39.8	6.0
GDP per capita (1980 $)	12,454	14,021	16,114[e]	17,039	36.8	5.7
Manufacturing as % of GDP (current prices)	27.6	26.2	28.4	26.7	-3.3	-6.0
Gross output (million $)	3,269	2,806	4,893	5,854	79.1	19.6
Value added (million $)	1,168	933	1,796[e]	2,208	89.0	22.9
Value added (million 1980 $)	1,293	1,507	1,885	1,871	44.7	-0.7
Industrial production index	100	118	141	141	41.0	0.0
Employment (thousands)	38	35	33	36	-5.3	9.1

Note: GDP stands for Gross Domestic Product. 'e' stands for estimated value.

Profitability and Productivity

	1980	1985	1989	1990	% change 1980-1990	% change 1989-1990
Intermediate input (%)	64	67	63[e]	62[e]	-3.1	-1.6
Wages, salaries, and supplements (%)	27[e]	20[e]	16[e]	21[e]	-22.2	31.3
Gross operating surplus (%)	9[e]	14[e]	21[e]	17[e]	88.9	-19.0
Gross output per worker ($)	84,018	7,088	147,533	156,281[e]	86.0	5.9
Value added per worker ($)	30,461	26,420	54,144[e]	66,306[e]	117.7	22.5
Average wage (incl. benefits) ($)	23,529[e]	15,888[e]	23,564[e]	34,210[e]	45.4	45.2

Profitability is in percent of gross output. Productivity is in U.S. $. 'e' stands for estimated value.

Profitability - 1990

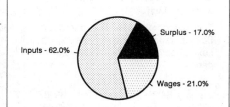

Inputs - 62.0%
Surplus - 17.0%
Wages - 21.0%

The graphic shows percent of gross output.

Value Added in Manufacturing

	1980 $ mil.	1980 %	1985 $ mil.	1985 %	1989 $ mil.	1989 %	1990 $ mil.	1990 %	% change 1980-1990	% change 1989-1990
311 Food products	31	2.7	20	2.1	37	2.1	47	2.1	51.6	27.0
313 Beverages	32[e]	2.7	24[e]	2.6	43[e]	2.4	54[e]	2.4	68.8	25.6
314 Tobacco products	10[e]	0.9	7[e]	0.8	9[e]	0.5	13[e]	0.6	30.0	44.4
321 Textiles	24	2.1	14	1.5	28[e]	1.6	59w	0.0	NA	NA
322 Wearing apparel	5	0.4	3	0.3	6[e]	0.3	13w	0.0	NA	NA
323 Leather and fur products	NA	0.0	NA	0.0	NA	0.0	w	0.0	NA	NA
324 Footwear	NA	0.0	NA	0.0	NA	0.0	w	0.0	NA	NA
331 Wood and wood products	2[e]	0.2	1[e]	0.1	1[e]	0.1	2[e]	0.1	0.0	100.0
332 Furniture and fixtures	2[e]	0.2	1[e]	0.1	2[e]	0.1	3[e]	0.1	50.0	50.0
341 Paper and paper products	14[e]	1.2	12[e]	1.3	24[e]	1.3	31[e]	1.4	121.4	29.2
342 Printing and publishing	18[e]	1.5	14[e]	1.5	27[e]	1.5	38[e]	1.7	111.1	40.7
351 Industrial chemicals	31[e]	2.7	31[e]	3.3	242[e]	13.5	258[e]	11.7	732.3	6.6
352 Other chemical products	3	0.3	7	0.8	105[e]	5.8	131[e]	5.9	4,266.7	24.8
353 Petroleum refineries	NA	0.0	NA	0.0	NA	0.0	NA	0.0	NA	NA
354 Miscellaneous petroleum and coal products	1[e]	0.1	1[e]	0.1	3[e]	0.2	3[e]	0.1	200.0	0.0
355 Rubber products	127[e]	10.9	113[e]	12.1	144[e]	8.0	185[e]	8.4	45.7	28.5
356 Plastic products	13[e]	1.1	14[e]	1.5	41[e]	2.3	45[e]	2.0	246.2	9.8
361 Pottery, china and earthenware	9[e]	0.8	6[e]	0.6	25[e]	1.4	16[e]	0.7	77.8	-36.0
362 Glass and glass products	16[e]	1.4	14[e]	1.5	46[e]	2.6	50[e]	2.3	212.5	8.7
369 Other non-metal mineral products	45[e]	3.9	40[e]	4.3	98[e]	5.5	153[e]	6.9	240.0	56.1
371 Iron and steel	592	50.7	415	44.5	540	30.1	643	29.1	8.6	19.1
372 Non-ferrous metals	32	2.7	34	3.6	59	3.3	62	2.8	93.8	5.1
381 Metal products	24	2.1	78	8.4	172	9.6	213	9.6	787.5	23.8
382 Non-electrical machinery	98	8.4	69	7.4	100	5.6	135	6.1	37.8	35.0
383 Electrical machinery	19	1.6	7	0.8	23[e]	1.3	31[e]	1.4	63.2	34.8
384 Transport equipment	7	0.6	4	0.4	9	0.5	13[e]	0.6	85.7	44.4
385 Professional and scientific equipment	10	0.9	4	0.4	9[e]	0.5	11[e]	0.5	10.0	22.2
390 Other manufacturing industries	NA	0.0	NA	0.0	NA	0.0	1[e]	0.0	NA	NA

Note: The industry codes shown are International Standard Industry codes (ISIC). Percentages are percent of total Value Added. 'e' stands for estimated value

For sources, notes, and explanations, see Annotated Source Appendix, page 1035.

555

Finance, Economics, and Trade

Economic Indicators [36]

National product: GDP—purchasing power equivalent—$8.5 billion (1992). **National product real growth rate**: 2.5% (1992). **National product per capita**: $21,700 (1992). **Inflation rate (consumer prices)**: 3.6% (1992). **External debt**: $131.6 million (1989 est.).

Balance of Payments Summary [37]

Exchange Rates [38]

Currency: **Luxembourg francs.**
Symbol: **LuxF.**

The Luxembourg franc is at par with the Belgian franc, which circulates freely in Luxembourg. Data are currency units per $1.

January 1993	33.256
1992	32.150
1991	34.148
1990	33.418
1989	39.404
1988	36.768

Imports and Exports

Top Import Origins [39]

$8.3 billion (c.i.f., 1991 est.).

Origins	%
Belgium	37
FRG	31
France	12
US	2

Top Export Destinations [40]

$6.4 billion (f.o.b., 1991 est.).

Destinations	%
EC	76
US	5

Foreign Aid [41]

	U.S. $
No aid provided or received.	NA

Import and Export Commodities [42]

Import Commodities	Export Commodities
Minerals	Finished steel products
Metals	Chemicals
Foodstuffs	Rubber products
Quality consumer goods	Glass
	Aluminum
	Other industrial products

Macedonia

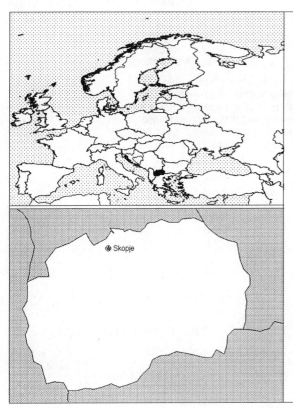

Geography [1]

Total area:
25,333 km2
Land area:
24,856 km2
Comparative area:
Slightly larger than Vermont
Land boundaries:
Total 748 km, Albania 151 km, Bulgaria 148 km, Greece 228 km, Serbia and Montenegro 221 km (all with Serbia)
Coastline:
0 km (landlocked)
Climate:
Hot, dry summers and autumns and relatively cold winters with heavy snowfall
Terrain:
Mountainous territory covered with deep basins and valleys; there are three large lakes, each divided by a frontier line
Natural resources:
Chromium, lead, zinc, manganese, tungsten, nickel, low-grade iron ore, asbestos, sulphur, timber
Land use:
Arable land: 5%
Permanent crops: 5%
Meadows and pastures: 20%
Forest and woodland: 30%
Other: 40%

Demographics [2]

	1960[2]	1970[2]	1980[2]	1990[2]	1991[1,2]	1994[2]	2000[2]	2010[2]	2020[2]
Population	1,392	1,629	1,893	2,132	NA	2,214	2,324	2,478	2,578
Population density (persons per sq. mi.)	NA	NA	NA	NA	NA	NA	NA	NA	NA
Births	NA	NA	NA	NA	NA	35	NA	NA	NA
Deaths	NA	NA	NA	NA	NA	15	NA	NA	NA
Life expectancy - males	NA	NA	NA	NA	NA	72	NA	NA	NA
Life expectancy - females	NA	NA	NA	NA	NA	76	NA	NA	NA
Birth rate (per 1,000)	NA	NA	NA	NA	NA	11	NA	NA	NA
Death rate (per 1,000)	NA	NA	NA	NA	NA	11	NA	NA	NA
Women of reproductive age (15-44 yrs.)	NA	NA	NA	548	NA	575	606	NA	NA
of which are currently married	NA	NA	NA	374	NA	394	419	NA	NA
Fertility rate	NA	NA	NA	2.1	NA	2.0	1.8	1.8	1.8

Population values are in thousands, life expectancy in years, and other items as indicated.

Health

Health Personnel [3]

Doctors per 1,000 pop., 1988-92	2.63 [2]
Nurse-to-doctor ratio, 1988-92	1.9 [2]
Hospital beds per 1,000 pop., 1985-90	6.0 [2]
Percentage of children immunized (age 1 yr. or less)	
Third dose of DPT, 1990-91	79.0 [2]
Measles, 1990-91	75.0 [2]

Health Expenditures [4]

For sources, notes, and explanations, see Annotated Source Appendix, page 1035.

557

Human Factors

Health Care Ratios [5]

Population per physician, 1970	1000 [2]
Population per physician, 1990	530 [2]
Population per nursing person, 1970	420 [2]
Population per nursing person, 1990	110 [2]
Percent of births attended by health staff, 1985	NA

Infants and Malnutrition [6]

Percent of babies with low birth weight, 1985	7
Infant mortality rate per 1,000 live births, 1970	56
Infant mortality rate per 1,000 live births, 1991	21
Years of life lost per 1,000 population, 1990	16
Prevalence of malnutrition (under age 5), 1990	NA

Ethnic Division [7]

Macedonian	67%
Albanian	21%
Turkish	4%
Serb	2%
Other	6%

Religion [8]

Eastern Orthodox	59%
Muslim	26%
Catholic	4%
Protestant	1%
Other	10%

Major Languages [9]

Macedonian	70%
Albanian	21%
Turkish	3%
Serbo-Croatian	3%
Other	3%

Education

Public Education Expenditures [10]

Million Dinar	1980[22]	1985[22]	1987[22]	1988[22]	1989[22]	1990[22]
Total education expenditure	8	42	202	544	9,555	60,318
as percent of GNP	4.7	3.4	4.2	3.6	4.3	6.1
as percent of total govt. expend.	32.5	NA	NA	NA	NA	NA
Current education expenditure	7	39	185	506	9,002	55,911
as percent of GNP	4.0	3.1	3.9	3.3	4.1	5.7
as percent of current govt. expend.	NA	NA	NA	NA	NA	NA
Capital expenditure	1	3	17	38	553	4,407

Educational Attainment [11]

Age group	25+ [1]
Total population	13,083,762 [1]
Highest level attained (%)	
No schooling	15.8 [1]
First level	
Incompleted	53.9 [1]
Completed	NA [1]
Entered second level	
S-1	23.4 [1]
S-2	NA [1]
Post secondary	6.8 [1]

Literacy Rate [12]

In thousands and percent	1985[a,1]	1991[a,1]	2000[a,1]
Illiterate population +15 years	1,614	1,342	942
Illiteracy rate - total pop. (%)	9.2	7.3	4.7
Illiteracy rate - males (%)	3.5	2.6	1.3
Illiteracy rate - females (%)	14.6	11.9	7.9

Libraries [13]

	Admin. Units[a]	Svc. Pts.[a]	Vols. (000)[a]	Shelving (meters)[a]	Vols. Added[a]	Reg. Users[a]
National (1989)	8	8	12,316	303,555	305,462	163,169
Non-specialized (1989)	20	20	3,488	60,443	72,503	49,262
Public (1989)	808	1,937	30,238	552,866	1.3mil	19.5mil
Higher ed. (1989)	409	421	14,462	319,329	469,142	529,549
School (1989)	7,784	7,784	38,430	NA	1,680	NA

Daily Newspapers [14]

	1975[b]	1980[b]	1985[b]	1990[b]
Number of papers	26	27	27	34
Circ. (000)	1,896	2,649	2,451	2,281

Cinema [15]

Data for 1989.

Cinema seats per 1,000	16.9[a]
Annual attendance per person	1.9[a]
Gross box office receipts (mil. Dinar)	643,490[a]

Science and Technology

Scientific/Technical Forces [16]

Potential scientists/engineers	563,312
Number female	NA
Potential technicians	432,380
Number female	NA
Total	995,692

R&D Expenditures [17]

	Dinar (000) 1989
Total expenditure[a,4,12]	2,152,032
Capital expenditure	815,082
Current expenditure	1,336,950
Percent current	62.1

U.S. Patents Issued [18]

For sources, notes, and explanations, see Annotated Source Appendix, page 1035.

Government and Law

Organization of Government [19]

Long-form name:
Republic of Macedonia
Type:
emerging democracy
Independence:
20 November 1991 (from Yugoslavia)
Constitution:
adopted 17 November 1991, effective 20 November 1991
Legal system:
based on civil law system; judicial review of legislative acts
National holiday:
NA
Executive branch:
president, Council of Ministers, prime minister
Legislative branch:
unicameral Assembly (Sobranje)
Judicial branch:
Constitutional Court, Judicial Court of the Republic

Political Parties [20]

Assembly	% of seats
Revolutionary Org. (VMRO-DPMNE)	30.8
Social-Democratic League (SDSM)	25.8
Party Democratic Prosperity (PDPM)	20.8
Alliance of Reform Forces (SRSM)	14.2
Party of Yugoslavs in Macedonia (SJM)	0.8
Socialist Party of Macedonia (SPM)	4.2
Others	3.3

Government Budget [21]

For 1989.

Revenues	NA
Expenditures	NA
Capital expenditures	NA

Defense Summary [22]

Branches: Army, Navy, Air and Air Defense Force, Police Force

Manpower Availability: Males age 15-49 597,024; fit for military service 484,701; reach military age (19) annually 18,979 (1993 est.)

Defense Expenditures: 7 billion denars (1993 est.); note - conversion of the military budget into US dollars using the current exchange rate could produce misleading results

Crime [23]

Human Rights [24]

Labor Force

Total Labor Force [25]

507,324

Labor Force by Occupation [26]

Agriculture	8%
Manufacturing and mining	40

Date of data: 1990

Unemployment Rate [27]

20% (1991 est.)

For sources, notes, and explanations, see Annotated Source Appendix, page 1035.

559

Production Sectors

Energy Resource Summary [28]

Energy Resources: None. **Electricity**: 1,600,000 kw capacity; 6,300 million kWh produced, 2,900 kWh per capita (1992). **Pipelines**: None.

Telecommunications [30]

- 125,000 telephones
- Broadcast stations - 6 AM, 2 FM, 5 (2 relays) TV
- 370,000 radios, 325,000 TV
- Satellite communications ground stations - none

Transportation [31]

Railroads. NA

Highways. 10,591 km total (1991); 5,091 km paved, 1,404 km gravel, 4,096 km earth

Airports

Total:	17
Useable:	17
With permanent-surface runways:	9
With runways over 3,659 m:	0
With runways 2,440-3,659 m:	2
With runways 1,220-2,439 m:	2

Top Agricultural Products [32]

Provides 12% of GDP and meets the basic need for food; principal crops are rice, tobacco, wheat, corn, and millet; also grown are cotton, sesame, mulberry leaves, citrus fruit, and vegetables; Macedonia is one of the seven legal cultivators of the opium poppy for the world pharmaceutical industry, including some exports to the US; agricultural production is highly labor intensive.

Top Mining Products [33]

Metric tons unless otherwise specified	M.t.
Magnesite, crude	175,000
Bauxite (000 tons)	2,850
Copper, Cu content of concentrate	140,000
Gypsum, crude	400,000
Pumice, volcanic tuff	380,000
Cement, hydraulic (000 tons)	7,100

Tourism [34]

	1987	1988	1989	1990	1991
Visitors[74]	26,151	29,635	34,118	39,573	
Tourists[6,74]	8,907	9,018	8,644	7,880	1,459
Tourism receipts[74]	1,668	2,024	2,230	2,774	468
Tourism expenditures[74]	90	109	131	149	103
Fare receipts[74]	340	430	475	620	105

Tourists are in thousands, money in million U.S. dollars.

Manufacturing Sector

GDP and Manufacturing Summary [35]

	1980	1985	1989	1990	% change 1980-1990	% change 1989-1990
GDP (million 1980 $)	69,958	71,058	72,234	66,371	-5.1	-8.1
GDP per capita (1980 $)	3,136	3,073	3,050	2,786	-11.2	-8.7
Manufacturing as % of GDP (current prices)	30.6	37.2	39.5	42.0	37.3	6.3
Gross output (million $)	72,629	57,020	65,078	62,136[e]	-14.4	-4.5
Value added (million $)	21,750	17,171	30,245	27,660[e]	27.2	-8.5
Value added (million 1980 $)	19,526	22,283	24,021	21,703	11.1	-9.6
Industrial production index	100	116	120	108	8.0	-10.0
Employment (thousands)	2,106	2,467	2,658	2,537[e]	20.5	-4.6

Note: GDP stands for Gross Domestic Product. 'e' stands for estimated value.

Profitability and Productivity

	1980	1985	1989	1990	% change 1980-1990	% change 1989-1990
Intermediate input (%)	70	70	54	55[e]	-21.4	1.9
Wages, salaries, and supplements (%)	14	12	12[e]	18[e]	28.6	50.0
Gross operating surplus (%)	15	18	34[e]	26[e]	73.3	-23.5
Gross output per worker ($)	34,487	23,113	24,484	24,248[e]	-29.7	-1.0
Value added per worker ($)	10,328	6,960	11,379	10,796[e]	4.5	-5.1
Average wage (incl. benefits) ($)	4,991	2,703	2,986[e]	4,488[e]	-10.1	50.3

Profitability is in percent of gross output. Productivity is in U.S. $. 'e' stands for estimated value.

Profitability - 1990

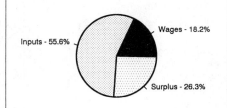

Wages - 18.2%
Inputs - 55.6%
Surplus - 26.3%

The graphic shows percent of gross output.

Value Added in Manufacturing

	1980 $ mil.	1980 %	1985 $ mil.	1985 %	1989 $ mil.	1989 %	1990 $ mil.	1990 %	% change 1980-1990	% change 1989-1990
311 Food products	1,897	8.7	1,458	8.5	3,916	12.9	3,484[e]	12.6	83.7	-11.0
313 Beverages	459	2.1	353	2.1	663	2.2	589[e]	2.1	28.3	-11.2
314 Tobacco products	184	0.8	221	1.3	344	1.1	308[e]	1.1	67.4	-10.5
321 Textiles	1,759	8.1	1,428	8.3	2,881	9.5	2,663[e]	9.6	51.4	-7.6
322 Wearing apparel	903	4.2	718	4.2	1,593	5.3	1,427[e]	5.2	58.0	-10.4
323 Leather and fur products	226	1.0	231	1.3	383	1.3	340[e]	1.2	50.4	-11.2
324 Footwear	482	2.2	503	2.9	1,022	3.4	899[e]	3.3	86.5	-12.0
331 Wood and wood products	977	4.5	530	3.1	794	2.6	706[e]	2.6	-27.7	-11.1
332 Furniture and fixtures	730	3.4	438	2.6	1,030	3.4	1,065[e]	3.9	45.9	3.4
341 Paper and paper products	529	2.4	394	2.3	759	2.5	674[e]	2.4	27.4	-11.2
342 Printing and publishing	876	4.0	462	2.7	761	2.5	678[e]	2.5	-22.6	-10.9
351 Industrial chemicals	694	3.2	631	3.7	1,107	3.7	992[e]	3.6	42.9	-10.4
352 Other chemical products	681	3.1	525	3.1	1,419	4.7	1,315[e]	4.8	93.1	-7.3
353 Petroleum refineries	454	2.1	415	2.4	260	0.9	233[e]	0.8	-48.7	-10.4
354 Miscellaneous petroleum and coal products	101	0.5	101	0.6	104	0.3	91[e]	0.3	-9.9	-12.5
355 Rubber products	276	1.3	269	1.6	479	1.6	456[e]	1.6	65.2	-4.8
356 Plastic products	413	1.9	258	1.5	397	1.3	350[e]	1.3	-15.3	-11.8
361 Pottery, china and earthenware	128	0.6	72	0.4	162	0.5	144[e]	0.5	12.5	-11.1
362 Glass and glass products	163	0.7	113	0.7	224	0.7	204[e]	0.7	25.2	-8.9
369 Other non-metal mineral products	906	4.2	513	3.0	683	2.3	604[e]	2.2	-33.3	-11.6
371 Iron and steel	1,221	5.6	1,000	5.8	1,343	4.4	1,171[e]	4.2	-4.1	-12.8
372 Non-ferrous metals	480	2.2	509	3.0	944	3.1	927[e]	3.4	93.1	-1.8
381 Metal products	2,105	9.7	1,577	9.2	1,293	4.3	1,130[e]	4.1	-46.3	-12.6
382 Non-electrical machinery	1,828	8.4	1,463	8.5	2,372	7.8	2,378[e]	8.6	30.1	0.3
383 Electrical machinery	1,600	7.4	1,544	9.0	2,640	8.7	2,334[e]	8.4	45.9	-11.6
384 Transport equipment	1,441	6.6	1,263	7.4	2,389	7.9	2,241[e]	8.1	55.5	-6.2
385 Professional and scientific equipment	101	0.5	93	0.5	154	0.5	146[e]	0.5	44.6	-5.2
390 Other manufacturing industries	134	0.6	88	0.5	128	0.4	114[e]	0.4	-14.9	-10.9

Note: The industry codes shown are International Standard Industry codes (ISIC). Percentages are percent of total Value Added. 'e' stands for estimated value

For sources, notes, and explanations, see Annotated Source Appendix, page 1035.

561

Finance, Economics, and Trade

Economic Indicators [36]

Yugoslav Dinars or U.S. Dollars as indicated[12].

	1989	1990	1991[e]
GSP[13]			
(billions current dinars)	221.9	910.9	1,490
Real GSP growth rate (%)	0.6	-7.6	-20
Money supply (M1)			
(yr-end mil.)	51,216	127,241	NA
Commercial interest			
rates (avg.)	4,354	45	80
Investment rate (%)	19.4	18.9	NA
CPI	1,356	688	310
WPI	1,406	534	320
External public debt	18,569	17,791	15,760

Balance of Payments Summary [37]

Exchange Rates [38]

Currency: **denar.**
Symbol: **MDr.**

Data are currency units per $1.

January 1991	240

Imports and Exports

Top Import Origins [39]

$1,112 million (1990).

Origins	%
Former Yugoslav republics	NA
Greece	NA
Albania	NA
Germany	NA
Bulgaria	NA

Top Export Destinations [40]

$578 million (1990).

Destinations	%
Principally Serbia and	
Montenegro and the	
other former Yugos	NA
Germany	NA
Greece	NA
Albania	NA

Foreign Aid [41]

	U.S. $	
US (for humanitarian and technical assistance)	10	million
EC promised an economic aid package (in ECU)	100	million

Import and Export Commodities [42]

Import Commodities

Fuels and lubricants 19%
Manufactured goods 18%
Machinery, transp. equip. 15%
Food and live animals 14%
Chemicals 11.4%
Raw materials 10%
Misc. manufactured articles 8.0%
Beverages and tobacco 3.5%

Export Commodities

Manufactured goods 40%
Machinery, transp. equip. 14%
Manufactured articles 23%
Raw materials 7.6%
Rice, live animals 5.7%
Beverages & tobacco 4.5%
Chemicals 4.7%

Madagascar

Geography [1]

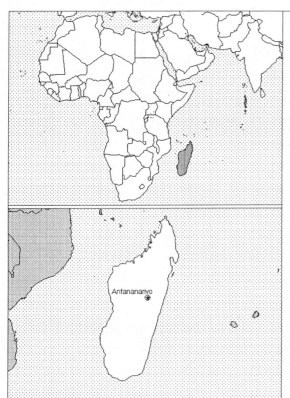

Total area:
 587,040 km2
Land area:
 581,540 km2
Comparative area:
 Slightly less than twice the size of Arizona
Land boundaries:
 0 km
Coastline:
 4,828 km
Climate:
 Tropical along coast, temperate inland, arid in south
Terrain:
 Narrow coastal plain, high plateau and mountains in center
Natural resources:
 Graphite, chromite, coal, bauxite, salt, quartz, tar sands, semiprecious stones, mica, fish
Land use:
 Arable land:
 4%
 Permanent crops:
 1%
 Meadows and pastures:
 58%
 Forest and woodland:
 26%
 Other:
 11%

Demographics [2]

	1960	1970	1980	1990	1991[1]	1994	2000	2010	2020
Population	5,482	6,766	8,700	11,811	12,185	13,428	16,232	22,064	29,362
Population density (persons per sq. mi.)	24	30	39	53	54	NA	72	98	130
Births	NA	NA	NA	NA	568	607	NA	NA	NA
Deaths	NA	NA	NA	NA	177	179	NA	NA	NA
Life expectancy - males	NA	NA	NA	NA	51	52	NA	NA	NA
Life expectancy - females	NA	NA	NA	NA	54	56	NA	NA	NA
Birth rate (per 1,000)	NA	NA	NA	NA	47	45	NA	NA	NA
Death rate (per 1,000)	NA	NA	NA	NA	15	13	NA	NA	NA
Women of reproductive age (15-44 yrs.)	NA	NA	NA	2,605	NA	2,967	3,628	NA	NA
of which are currently married	NA	NA	NA	1,726	NA	1,963	2,391	NA	NA
Fertility rate	NA	NA	NA	6.9	6.88	6.7	6.3	5.3	4.4

Population values are in thousands, life expectancy in years, and other items as indicated.

Health

Health Personnel [3]

Doctors per 1,000 pop., 1988-92	0.12
Nurse-to-doctor ratio, 1988-92	3.5
Hospital beds per 1,000 pop., 1985-90	0.9
Percentage of children immunized (age 1 yr. or less)	
Third dose of DPT, 1990-91	46.0
Measles, 1990-91	33.0

Health Expenditures [4]

Total health expenditure, 1990 (official exchange rate)	
Millions of dollars	79
Millions of dollars per capita	7
Health expenditures as a percentage of GDP	
Total	2.6
Public sector	1.3
Private sector	1.3
Development assistance for health	
Total aid flows (millions of dollars)[1]	17
Aid flows per capita (millions of dollars)	1.5
Aid flows as a percentage of total health expenditure	21.5

For sources, notes, and explanations, see Annotated Source Appendix, page 1035.

563

Human Factors

Health Care Ratios [5]

Population per physician, 1970	10120
Population per physician, 1990	8130
Population per nursing person, 1970	240
Population per nursing person, 1990	NA
Percent of births attended by health staff, 1985	62

Infants and Malnutrition [6]

Percent of babies with low birth weight, 1985	10
Infant mortality rate per 1,000 live births, 1970	181
Infant mortality rate per 1,000 live births, 1991	114
Years of life lost per 1,000 population, 1990	63
Prevalence of malnutrition (under age 5), 1990	53

Ethnic Division [7]

Malayo-Indonesian (Merina and related Betsileo), Cotiers (mixed African, Malayo-Indonesian, and Arab ancestry—Betsimisaraka, Tsimihety, Antaisaka, Sakalava), French, Indian, Creole, Comoran.

Religion [8]

Indigenous beliefs	52%
Christian	41%
Muslim	7%

Major Languages [9]

French (official), Malagasy (official).

Education

Public Education Expenditures [10]

Million Francs	1980	1985	1987[2]	1988[2]	1989[2]	1990[2]
Total education expenditure	36,896	52,182	48,906	59,842	62,111	67,038
as percent of GNP	4.4	2.9	1.9	1.9	1.7	1.5
as percent of total govt. expend.	NA	NA	NA	NA	NA	NA
Current education expenditure	31,548	49,806	47,750	58,542	61,071	64,964
as percent of GNP	3.7	2.8	1.9	1.8	1.7	1.5
as percent of current govt. expend.	NA	NA	NA	NA	NA	NA
Capital expenditure	5,348	2,377	1,156	1,300	1,040	2,074

Educational Attainment [11]

Literacy Rate [12]

In thousands and percent	1985[a]	1991[a]	2000[a]
Illiterate population +15 years	1,309	1,305	1,303
Illiteracy rate - total pop. (%)	23.1	19.8	14.5
Illiteracy rate - males (%)	14.2	12.3	9.0
Illiteracy rate - females (%)	31.6	27.1	19.7

Libraries [13]

Daily Newspapers [14]

	1975	1980	1985	1990
Number of papers	12[e]	6	7	5
Circ. (000)	55[e]	55	67	50

Cinema [15]

Data for 1991.

Cinema seats per 1,000	NA
Annual attendance per person	0.0
Gross box office receipts (mil. Franc)	209

Science and Technology

Scientific/Technical Forces [16]

R&D Expenditures [17]

	Franc (000) 1988
Total expenditure	14,371,515
Capital expenditure	13,378,000
Current expenditure	993,515
Percent current	6.9

U.S. Patents Issued [18]

564

For sources, notes, and explanations, see Annotated Source Appendix, page 1035.

Government and Law

Organization of Government [19]

Long-form name:
Republic of Madagascar
Type:
republic
Independence:
26 June 1960 (from France)
Constitution:
12 September 1992
Legal system:
based on French civil law system and traditional Malagasy law; has not accepted compulsory ICJ jurisdiction
National holiday:
Independence Day, 26 June (1960)
Executive branch:
president, prime minister, Council of Ministers
Legislative branch:
unicameral Popular National Assembly (Assemblee Nationale Populaire); note - the National Assembly has suspended its operations during 1992 and early 1993 in preparation for new legislative elections. In its place, an interim High Authority of State and a Social and Economic Recovery Council have been established
Judicial branch:
Supreme Court, High Constitutional Court

Political Parties [20]

Popular National Assembly	% of votes
Advance Guard (AREMA)	88.2
Militants for Proletarian Regime (MFM)	5.1
Congress Party (AKFM)	3.7
Movement for National Unity (VONJY)	2.2
Other	0.8

Government Expenditures [21]

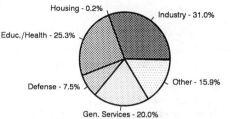

Housing - 0.2%
Industry - 31.0%
Educ./Health - 25.3%
Other - 15.9%
Defense - 7.5%
Gen. Services - 20.0%

% distribution for 1991. Expend. in 1991: 265 ($ mil.)

Crime [23]

Military Expenditures and Arms Transfers [22]

	1985	1986	1987	1988	1989
Military expenditures					
Current dollars (mil.)	36[e]	37[e]	37[e]	32[e]	35[e]
1989 constant dollars (mil.)	41[e]	41[e]	40[e]	33[e]	35[e]
Armed forces (000)	27	26	26	21	21
Gross national product (GNP)					
Current dollars (mil.)	1,891	1,970	2,016	2,164	2,340
1989 constant dollars (mil.)	2,153	2,186	2,168	2,253	2,340
Central government expenditures (CGE)					
1989 constant dollars (mil.)	517[e]	502[e]	NA	168	122
People (mil.)	10.1	10.4	10.7	11.1	11.4
Military expenditure as % of GNP	1.9	1.9	1.8	1.5	1.5
Military expenditure as % of CGE	8.0	8.2	NA	19.8	28.9
Military expenditure per capita	4	4	4	3	3
Armed forces per 1,000 people	2.7	2.5	2.4	1.9	1.8
GNP per capita	213	210	202	204	205
Arms imports[6]					
Current dollars (mil.)	30	20	30	10	30
1989 constant dollars (mil.)	34	22	32	10	30
Arms exports[6]					
Current dollars (mil.)	0	0	0	0	0
1989 constant dollars (mil.)	0	0	0	0	0
Total imports[7]					
Current dollars (mil.)	402	353	302	360	NA
1989 constant dollars	458	392	325	375	NA
Total exports[7]					
Current dollars (mil.)	274	313	331	278	319
1989 constant dollars	312	347	356	289	319
Arms as percent of total imports[8]	7.5	5.7	9.9	2.8	NA
Arms as percent of total exports[8]	0	0	0	0	0

Human Rights [24]

	SSTS	FL	FAPRO	PPCG	APROBC	TPW	PCPTW	STPEP	PHRFF	PRW	ASST	AFL
Observes	P	P	P			P	P			P	P	
	EAFRD	CPR	ESCR	SR	ACHR	MAAE	PVIAC	PVNAC	EAFDAW	TCIDTP	RC
Observes		P	P	P				P	P	P		P

P = Party; S = Signatory; see Appendix for meaning of abbreviations.

Labor Force

Total Labor Force [25]

4.9 million 90% nonsalaried family workers engaged in subsistence agriculture; 175,000 wage earners

Labor Force by Occupation [26]

Agriculture	26%
Domestic service	17
Industry	15
Commerce	14
Construction	11
Services	9
Transportation	6
Other	2

Unemployment Rate [27]

For sources, notes, and explanations, see Annotated Source Appendix, page 1035.

565

Production Sectors

Energy Resource Summary [28]

Energy Resources: Coal. **Electricity**: 125,000 kW capacity; 450 million kWh produced, 35 kWh per capita (1991).

Telecommunications [30]

- Above average system includes open-wire lines, coaxial cables, radio relay, and troposcatter links
- Submarine cable to Bahrain
- Satellite earth stations - 1 Indian Ocean INTELSAT and broadcast stations - 17 AM, 3 FM, 1 (36 repeaters) TV

Transportation [31]

Railroads. 1,020 km 1.000-meter gauge

Highways. 40,000 km total; 4,694 km paved, 811 km crushed stone, gravel, or stabilized soil, 34,495 km improved and unimproved earth (est.)

Merchant Marine. 11 ships (1,000 GRT or over) totaling 35,359 GRT/48,772 DWT; includes 6 cargo, 2 roll-on/roll-off cargo, 1 oil tanker, 1 chemical tanker, 1 liquefied gas

Airports

Total:	146
Usable:	103
With permanent-surface runways:	30
With runways over 3,659 m:	0
With runways 2,440-3,659 m:	3
With runways 1,220-2,439 m:	36

Top Agricultural Products [32]

Agriculture accounts for 31% of GDP; cash crops—coffee, vanilla, sugarcane, cloves, cocoa; food crops—rice, cassava, beans, bananas, peanuts; cattle raising widespread; almost self-sufficient in rice.

Top Mining Products [33]

Estimated metric tons unless otherwise specified   M.t.

Chromium, chromite concentrate	63,000[1]
Graphite, all grades	14,079
Cement, hydraulic	60,000
Quartz, Piezoelectric (kilograms)	66,200
Stone, labradorite (kilograms)	35,010
Quartz, crystal	32,000
Salt, marine	30,000

Tourism [34]

	1987	1988	1989	1990	1991
Visitors	33	57	67	87	
Tourists	28	35	39	53	35
Cruise passengers	2				
Tourism receipts	10	23	29	43	26
Tourism expenditures	26	30	38	40	32
Fare receipts	25	30	29	38	32
Fare expenditures	13	15	15	20	17

Tourists are in thousands, money in million U.S. dollars.

Manufacturing Sector

GDP and Manufacturing Summary [35]

	1980	1985	1989	1990	% change 1980-1990	% change 1989-1990
GDP (million 1980 $)	3,265	3,086	3,364	3,514	7.6	4.5
GDP per capita (1980 $)	372	301	289	293	-21.2	1.4
Manufacturing as % of GDP (current prices)	11.9[e]	10.2[e]	8.1[e]	12.4	4.2	53.1
Gross output (million $)	569	328	276[e]	353[e]	-38.0	27.9
Value added (million $)	221	132	103[e]	147[e]	-33.5	42.7
Value added (million 1980 $)	365	244	480[e]	282	-22.7	-41.3
Industrial production index	100	81	98[e]	101	1.0	3.1
Employment (thousands)	41	47	45[e]	46[e]	12.2	2.2

Note: GDP stands for Gross Domestic Product. 'e' stands for estimated value.

Profitability and Productivity

	1980	1985	1989	1990	% change 1980-1990	% change 1989-1990
Intermediate input (%)	61	60	63[e]	58[e]	-4.9	-7.9
Wages, salaries, and supplements (%)	15	16	13[e]	13[e]	-13.3	0.0
Gross operating surplus (%)	24	25	25[e]	28[e]	16.7	12.0
Gross output per worker ($)	14,005	6,891	6,170[e]	7,672[e]	-45.2	24.3
Value added per worker ($)	5,439	2,798	2,310[e]	3,200[e]	-41.2	38.5
Average wage (incl. benefits) ($)	2,083	1,099	786[e]	1,010[e]	-51.5	28.5

Profitability is in percent of gross output. Productivity is in U.S. $. 'e' stands for estimated value.

Profitability - 1990

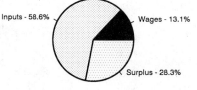

Inputs - 58.6%
Wages - 13.1%
Surplus - 28.3%

The graphic shows percent of gross output.

Value Added in Manufacturing

	1980 $ mil.	1980 %	1985 $ mil.	1985 %	1989 $ mil.	1989 %	1990 $ mil.	1990 %	% change 1980-1990	% change 1989-1990
311 Food products	23	10.4	45	34.1	18[e]	17.5	22[e]	15.0	-4.3	22.2
313 Beverages	34	15.4	16	12.1	16[e]	15.5	16[e]	10.9	-52.9	0.0
314 Tobacco products	3	1.4	3	2.3	1[e]	1.0	1[e]	0.7	-66.7	0.0
321 Textiles	67	30.3	16	12.1	33[e]	32.0	60[e]	40.8	-10.4	81.8
322 Wearing apparel	19	8.6	6	4.5	2[e]	1.9	4[e]	2.7	-78.9	100.0
323 Leather and fur products	3	1.4	1	0.8	1[e]	1.0	1[e]	0.7	-66.7	0.0
324 Footwear	8	3.6	5	3.8	3[e]	2.9	3[e]	2.0	-62.5	0.0
331 Wood and wood products	2	0.9	1	0.8	NA	0.0	NA	0.0	NA	NA
332 Furniture and fixtures	2	0.9	1	0.8	NA	0.0	NA	0.0	NA	NA
341 Paper and paper products	4	1.8	3	2.3	4[e]	3.9	5[e]	3.4	25.0	25.0
342 Printing and publishing	6	2.7	2	1.5	1[e]	1.0	1[e]	0.7	-83.3	0.0
351 Industrial chemicals	1	0.5	1	0.8	NA	0.0	NA	0.0	NA	NA
352 Other chemical products	10	4.5	11	8.3	8[e]	7.8	9[e]	6.1	-10.0	12.5
353 Petroleum refineries	11[e]	5.0	7[e]	5.3	6[e]	5.8	9[e]	6.1	-18.2	50.0
354 Miscellaneous petroleum and coal products	NA	0.0	NA	0.0	NA	0.0	NA	0.0	NA	NA
355 Rubber products	1	0.5	1	0.8	NA	0.0	1[e]	0.7	0.0	NA
356 Plastic products	3	1.4	2	1.5	1[e]	1.0	1[e]	0.7	-66.7	0.0
361 Pottery, china and earthenware	NA	0.0	NA	0.0	NA	0.0	NA	0.0	NA	NA
362 Glass and glass products	2	0.9	NA	0.0	NA	0.0	1[e]	0.7	-50.0	NA
369 Other non-metal mineral products	2[e]	0.9	1	0.8	2[e]	1.9	3[e]	2.0	50.0	50.0
371 Iron and steel	NA	0.0	NA	0.0	NA	0.0	NA	0.0	NA	NA
372 Non-ferrous metals	NA	0.0	NA	0.0	NA	0.0	NA	0.0	NA	NA
381 Metal products	9	4.1	5	3.8	3[e]	2.9	4[e]	2.7	-55.6	33.3
382 Non-electrical machinery	NA	0.0	NA	0.0	NA	0.0	NA	0.0	NA	NA
383 Electrical machinery	3	1.4	3	2.3	2[e]	1.9	2[e]	1.4	-33.3	0.0
384 Transport equipment	7	3.2	2[e]	1.5	1[e]	1.0	1[e]	0.7	-85.7	0.0
385 Professional and scientific equipment	NA	0.0	NA	0.0	NA	0.0	NA	0.0	NA	NA
390 Other manufacturing industries	2	0.9	1	0.8	NA	0.0	NA	0.0	NA	NA

Note: The industry codes shown are International Standard Industry codes (ISIC). Percentages are percent of total Value Added. 'e' stands for estimated value

Finance, Economics, and Trade

Economic Indicators [36]

National product: GDP—exchange rate conversion—$2.5 billion (1992 est.). **National product real growth rate**: 1% (1992 est.). **National product per capita**: $200 (1992 est.). **Inflation rate (consumer prices)**: 20% (1992 est.). **External debt**: $4.4 billion (1991).

Balance of Payments Summary [37]

Values in millions of dollars.

	1987	1988	1989	1990	1991
Exports of goods (f.o.b.)	327.0	284.0	321.0	320.0	344.0
Imports of goods (f.o.b.)	-315.0	-319.0	-320.0	-506.0	-445.0
Trade balance	12.0	-35.0	1.0	-186.0	-101.0
Services - debits	-410.0	-443.0	-437.0	-443.0	-417.0
Services - credits	-1.0	-1.0	-1.0	-1.0	-2.0
Private transfers (net)	34.0	38.0	42.0	43.0	53.0
Government transfers (net)	120.0	158.0	161.0	188.0	127.0
Long term capital (net)	240.0	209.0	222.0	134.0	47.0
Short term capital (net)	-41.0	-36.0	-64.0	-14.0	149.0
Errors and omissions	-11.0	53.0	-42.0	-54.0	12.0
Overall balance	49.0	76.0	33.0	-133.0	16.0

Exchange Rates [38]

Currency: **Malagasy francs.**
Symbol: **FMG.**

Data are currency units per $1.

December 1992	1,910.2
1992	1,867.9
1991	1,835.4
December 1990	1,454.6
1989	1,603.4
1988	1,407.1
1987	1,069.2

Imports and Exports

Top Import Origins [39]

$350 million (f.o.b., 1992 est.).

Origins	%
France	NA
Germany	NA
UK	NA
Other EC	NA
US	NA

Top Export Destinations [40]

$312 million (f.o.b., 1991 est.).

Destinations	%
France	NA
Japan	NA
Italy	NA
Germany	NA
US	NA

Foreign Aid [41]

	U.S. $	
US commitments, including Ex-Im (FY70-89)	136	million
Western (non-US) countries, ODA and OOF bilateral commitments (1970-89)	3,125	million
Communist countries (1970-89)	491	million

Import and Export Commodities [42]

Import Commodities

Parts 30%
Capital goods 28%
Petroleum 15%
Consumer goods 14%
Food 13%

Export Commodities

Coffee 45%
Vanilla 20%
Cloves 11%
Sugar
Petroleum products

Malawi

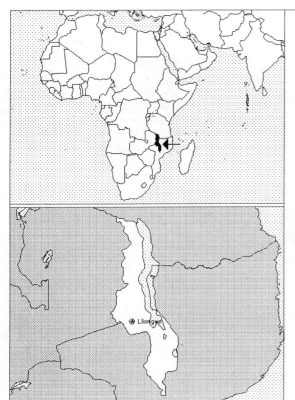

Geography [1]

Total area:
118,480 km2
Land area:
94,080 km2
Comparative area:
Slightly larger than Pennsylvania
Land boundaries:
Total 2,881 km, Mozambique 1,569 km, Tanzania 475 km, Zambia 837 km
Coastline:
0 km (landlocked)
Climate:
Tropical; rainy season (November to May); dry season (May to November)
Terrain:
Narrow elongated plateau with rolling plains, rounded hills, some mountains
Natural resources:
Limestone, unexploited deposits of uranium, coal, and bauxite
Land use:
Arable land:
25%
Permanent crops:
0%
Meadows and pastures:
20%
Forest and woodland:
50%
Other:
5%

Demographics [2]

	1960	1970	1980	1990	1991[1]	1994	2000	2010	2020
Population	3,450	4,449	6,128	9,289	9,438	9,732	11,045	13,233	16,697
Population density (persons per sq. mi.)	95	122	166	253	260	NA	327	453	612
Births	NA	NA	NA	NA	492	491	NA	NA	NA
Deaths	NA	NA	NA	NA	168	226	NA	NA	NA
Life expectancy - males	NA	NA	NA	NA	48	39	NA	NA	NA
Life expectancy - females	NA	NA	NA	NA	47	41	NA	NA	NA
Birth rate (per 1,000)	NA	NA	NA	NA	52	50	NA	NA	NA
Death rate (per 1,000)	NA	NA	NA	NA	18	23	NA	NA	NA
Women of reproductive age (15-44 yrs.)	NA	NA	NA	2,086	NA	2,142	2,393	NA	NA
of which are currently married	NA	NA	NA	1,602	NA	1,643	1,821	NA	NA
Fertility rate	NA	NA	NA	7.7	7.64	7.4	6.9	6.0	4.9

Population values are in thousands, life expectancy in years, and other items as indicated.

Health

Health Personnel [3]

Doctors per 1,000 pop., 1988-92	0.02
Nurse-to-doctor ratio, 1988-92	2.8
Hospital beds per 1,000 pop., 1985-90	1.6
Percentage of children immunized (age 1 yr. or less)	
Third dose of DPT, 1990-91	81.0
Measles, 1990-91	78.0

Health Expenditures [4]

Total health expenditure, 1990 (official exchange rate)	
Millions of dollars	93
Millions of dollars per capita	11
Health expenditures as a percentage of GDP	
Total	5.0
Public sector	2.9
Private sector	2.1
Development assistance for health	
Total aid flows (millions of dollars)[1]	22
Aid flows per capita (millions of dollars)	2.5
Aid flows as a percentage of total health expenditure	23.3

For sources, notes, and explanations, see Annotated Source Appendix, page 1035.

569

Human Factors

Health Care Ratios [5]

Population per physician, 1970	76580
Population per physician, 1990	45740
Population per nursing person, 1970	5330
Population per nursing person, 1990	1800
Percent of births attended by health staff, 1985	59

Infants and Malnutrition [6]

Percent of babies with low birth weight, 1985	10
Infant mortality rate per 1,000 live births, 1970	193
Infant mortality rate per 1,000 live births, 1991	143
Years of life lost per 1,000 population, 1990	110
Prevalence of malnutrition (under age 5), 1990	60

Ethnic Division [7]

Chewa, Nyanja, Tumbuko, Yao, Lomwe, Sena, Tonga, Ngoni, Ngonde, Asian, European.

Religion [8]

Protestant	55%
Roman Catholic	20%
Muslim	20%
Traditional indigenous beliefs	NA

Major Languages [9]

English (official), Chichewa (official), other languages important regionally.

Education

Public Education Expenditures [10]

Million Kwacha	1980	1985	1987	1988	1989	1990
Total education expenditure	31	64	NA	114	146	165
as percent of GNP	3.4	3.5	NA	3.5	3.4	3.4
as percent of total govt. expend.	8.4	9.6	NA	9.6	10.4	10.3
Current education expenditure	24	46	NA	83	106	117
as percent of GNP	2.6	2.5	NA	2.5	2.5	2.4
as percent of current govt. expend.	NA	NA	NA	9.8	9.6	9.8
Capital expenditure	8	18	NA	31	39	48

Educational Attainment [11]

Literacy Rate [12]

	1987[b]	1988[b]	1990[b]
Illiterate population +15 years	2,216,880	NA	NA
Illiteracy rate - total pop. (%)	51.5	NA	NA
Illiteracy rate - males (%)	34.7	NA	NA
Illiteracy rate - females (%)	66.5	NA	NA

Libraries [13]

	Admin. Units	Svc. Pts.	Vols. (000)	Shelving (meters)	Vols. Added	Reg. Users
National	NA	NA	NA	NA	NA	NA
Non-specialized	NA	NA	NA	NA	NA	NA
Public (1986)[5]	1	6	305	13,750	35,000	54,965
Higher ed. (1988)	5	5	310	NA	7,362	4,669
School (1987)	1	75	2	NA	500	283

Daily Newspapers [14]

	1975	1980	1985	1990
Number of papers	2	2[e]	1	1
Circ. (000)	18	20[e]	15	25

Cinema [15]

Science and Technology

Scientific/Technical Forces [16]

R&D Expenditures [17]

U.S. Patents Issued [18]

Government and Law

Organization of Government [19]

Long-form name:
Republic of Malawi
Type:
one-party republic
Independence:
6 July 1964 (from UK)
Constitution:
6 July 1964; republished as amended January 1974
Legal system:
based on English common law and customary law; judicial review of legislative acts in the Supreme Court of Appeal; has not accepted compulsory ICJ jurisdiction
National holiday:
Independence Day, 6 July (1964)
Executive branch:
president, Cabinet
Legislative branch:
unicameral National Assembly
Judicial branch:
High Court, Supreme Court of Appeal

Political Parties [20]

National Assembly	% of seats
Malawi Congress Party (MCP)	100.0

Government Expenditures [21]

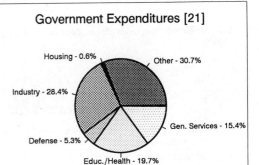

Housing - 0.6%
Other - 30.7%
Industry - 28.4%
Gen. Services - 15.4%
Defense - 5.3%
Educ./Health - 19.7%

% distribution for 1988. Expend. in FY91 est.: 510 ($ mil.)

Military Expenditures and Arms Transfers [22]

	1985	1986	1987	1988	1989
Military expenditures					
Current dollars (mil.)	24	30	24	36	35
1989 constant dollars (mil.)	27	33	26	38	35
Armed forces (000)	6	7	7	7	7
Gross national product (GNP)					
Current dollars (mil.)	1,223	1,257	1,330	1,421	1,559
1989 constant dollars (mil.)	1,392	1,395	1,431	1,479	1,559
Central government expenditures (CGE)					
1989 constant dollars (mil.)	467	502	465	434	428
People (mil.)	7.0	7.3	7.8	8.3	8.8
Military expenditure as % of GNP	2.0	2.4	1.8	2.6	2.3
Military expenditure as % of CGE	5.8	6.6	5.6	8.7	8.2
Military expenditure per capita	4	5	3	5	4
Armed forces per 1,000 people	0.9	1.0	0.9	0.8	0.8
GNP per capita	198	191	184	178	177
Arms imports[6]					
Current dollars (mil.)	5	5	0	5	5
1989 constant dollars (mil.)	6	6	0	5	5
Arms exports[6]					
Current dollars (mil.)	0	0	0	0	0
1989 constant dollars (mil.)	0	0	0	0	0
Total imports[7]					
Current dollars (mil.)	285	260	295	406	505
1989 constant dollars	324	289	317	423	505
Total exports[7]					
Current dollars (mil.)	249	248	277	289	268
1989 constant dollars	283	275	298	301	268
Arms as percent of total imports[8]	1.8	1.9	0	1.2	1.0
Arms as percent of total exports[8]	0	0	0	0	0

Crime [23]

Crime volume	
Cases known to police	80,095
Attempts (percent)	NA
Percent cases solved	39.07
Crimes per 100,000 persons	1,001.18
Persons responsible for offenses	
Total number offenders	30,470
Percent female	3.65
Percent juvenile[36]	1.47
Percent foreigners	NA

Human Rights [24]

	SSTS	FL	FAPRO	PPCG	APROBC	TPW	PCPTW	STPEP	PHRFF	PRW	ASST	AFL
Observes	P		S		P	P	P	P		P	P	
		EAFRD	CPR	ESCR	SR	ACHR	MAAE	PVIAC	PVNAC	EAFDAW	TCIDTP	RC
Observes			P					P	P	P		P

P = Party; S = Signatory; see Appendix for meaning of abbreviations.

Labor Force

Total Labor Force [25]

428,000 wage earners

Labor Force by Occupation [26]

Agriculture	43%
Manufacturing	16
Personal services	15
Commerce	9
Construction	7
Miscellaneous services	4
Other permanently employed	6
Date of data: 1986	

Unemployment Rate [27]

Production Sectors

Energy Resource Summary [28]

Energy Resources: Unexploited deposits of uranium and coal. **Electricity**: 190,000 kW capacity; 620 million kWh produced, 65 kWh per capita (1992).

Telecommunications [30]

- Fair system of open-wire lines, radio relay links, and radio communications stations
- 42,250 telephones
- Broadcast stations - 10 AM, 17 FM, no TV
- Satellite earth stations - 1 Indian Ocean INTELSAT and 1 Atlantic Ocean INTELSAT

Transportation [31]

Railroads. 789 km 1.067-meter gauge

Highways. 13,135 km total; 2,364 km paved; 251 km crushed stone, gravel, or stabilized soil; 10,520 km earth and improved earth

Airports

Total:	47
Usable:	41
With permanent-surface runways:	5
With runways over 3,659 m:	0
With runways 2,440-3,659 m:	1
With runways 1,220-2,439 m:	10

Top Agricultural Products [32]

Agriculture accounts for 40% of GDP; cash crops—tobacco, sugarcane, cotton, tea, and corn; subsistence crops—potatoes, cassava, sorghum, pulses; livestock - cattle, goats.

Top Mining Products [33]

Estimated metric tons unless otherwise specified   M.t.

Limestone	175,000
Dolomite	2,500
Cement, hydraulic	120,000
Coal	45,000
Gemstones, ruby and sapphire (grams)	1,000
Lime	4,000

Tourism [34]

	1987	1988	1989	1990	1991
Visitors[34]	76	99	117	130	132
Tourism receipts	9	10	13	11	
Tourism expenditures	6				
Fare receipts	12	11			
Fare expenditures	8	10			

Tourists are in thousands, money in million U.S. dollars.

Manufacturing Sector

GDP and Manufacturing Summary [35]

	1980	1985	1989	1990	% change 1980-1990	% change 1989-1990
GDP (million 1980 $)	1,245	1,403	1,496	1,627	30.7	8.8
GDP per capita (1980 $)	201	191	177	186	-7.5	5.1
Manufacturing as % of GDP (current prices)	12.6	12.6	15.6	13.5	7.1	-13.5
Gross output (million $)	340	330	458^e	586^e	72.4	27.9
Value added (million $)	123	90	110^e	133^e	8.1	20.9
Value added (million 1980 $)	149	155	204	196	31.5	-3.9
Industrial production index	100	116	140	146	46.0	4.3
Employment (thousands)	39	31	41^e	46^e	17.9	12.2

Note: GDP stands for Gross Domestic Product. 'e' stands for estimated value.

Profitability and Productivity

	1980	1985	1989	1990	% change 1980-1990	% change 1989-1990
Intermediate input (%)	64	73	76^e	77^e	20.3	1.3
Wages, salaries, and supplements (%)	12	10	10^e	10^e	-16.7	0.0
Gross operating surplus (%)	24	18	14^e	13^e	-45.8	-7.1
Gross output per worker ($)	8,783	10,745	11,097^e	12,767^e	45.4	15.0
Value added per worker ($)	3,174	2,923	2,671^e	3,041^e	-4.2	13.9
Average wage (incl. benefits) ($)	1,046	1,035	1,116^e	1,244^e	18.9	11.5

Profitability is in percent of gross output. Productivity is in U.S. $. 'e' stands for estimated value.

Profitability - 1990

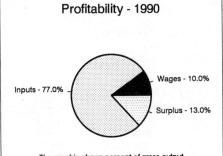

Inputs - 77.0%
Wages - 10.0%
Surplus - 13.0%

The graphic shows percent of gross output.

Value Added in Manufacturing

	1980 $ mil.	1980 %	1985 $ mil.	1985 %	1989 $ mil.	1989 %	1990 $ mil.	1990 %	% change 1980-1990	% change 1989-1990
311 Food products	54	43.9	14	15.6	18^e	16.4	26^e	19.5	-51.9	44.4
313 Beverages	8	6.5	7	7.8	9^e	8.2	12^e	9.0	50.0	33.3
314 Tobacco products	9	7.3	5	5.6	7^e	6.4	8^e	6.0	-11.1	14.3
321 Textiles	12	9.8	14	15.6	15^e	13.6	18^e	13.5	50.0	20.0
322 Wearing apparel	2	1.6	1	1.1	1^e	0.9	1^e	0.8	-50.0	0.0
323 Leather and fur products	NA	0.0	NA	0.0	NA	0.0	NA	0.0	NA	NA
324 Footwear	1^e	0.8	3	3.3	4^e	3.6	4^e	3.0	300.0	0.0
331 Wood and wood products	2	1.6	2	2.2	1^e	0.9	2^e	1.5	0.0	100.0
332 Furniture and fixtures	1	0.8	1	1.1	1^e	0.9	1^e	0.8	0.0	0.0
341 Paper and paper products	2	1.6	2	2.2	NA	0.0	1^e	0.8	-50.0	NA
342 Printing and publishing	8	6.5	6	6.7	7^e	6.4	9^e	6.8	12.5	28.6
351 Industrial chemicals	2	1.6	8	8.9	6^e	5.5	5^e	3.8	150.0	-16.7
352 Other chemical products	5	4.1	14	15.6	24^e	21.8	23^e	17.3	360.0	-4.2
353 Petroleum refineries	NA	0.0	NA	0.0	NA	0.0	NA	0.0	NA	NA
354 Miscellaneous petroleum and coal products	NA	0.0	NA	0.0	NA	0.0	NA	0.0	NA	NA
355 Rubber products	1	0.8	1	1.1	NA	0.0	NA	0.0	NA	NA
356 Plastic products	2	1.6	2	2.2	4^e	3.6	5^e	3.8	150.0	25.0
361 Pottery, china and earthenware	NA	0.0	NA	0.0	NA	0.0	NA	0.0	NA	NA
362 Glass and glass products	NA	0.0	NA	0.0	NA	0.0	NA	0.0	NA	NA
369 Other non-metal mineral products	3	2.4	1	1.1	7^e	6.4	8^e	6.0	166.7	14.3
371 Iron and steel	NA	0.0	NA	0.0	NA	0.0	NA	0.0	NA	NA
372 Non-ferrous metals	NA	0.0	NA	0.0	NA	0.0	NA	0.0	NA	NA
381 Metal products	6	4.9	6	6.7	3^e	2.7	5^e	3.8	-16.7	66.7
382 Non-electrical machinery	NA	0.0	1	1.1	2^e	1.8	3^e	2.3	NA	50.0
383 Electrical machinery	5	4.1	1	1.1	1^e	0.9	1^e	0.8	-80.0	0.0
384 Transport equipment	1^e	0.8	1	1.1	1^e	0.9	1^e	0.8	0.0	0.0
385 Professional and scientific equipment	NA	0.0	NA	0.0	NA	0.0	NA	0.0	NA	NA
390 Other manufacturing industries	NA	0.0	NA	0.0	NA	0.0	NA	0.0	NA	NA

Note: The industry codes shown are International Standard Industry codes (ISIC). Percentages are percent of total Value Added. 'e' stands for estimated value

For sources, notes, and explanations, see Annotated Source Appendix, page 1035.

573

Finance, Economics, and Trade

Economic Indicators [36]

National product: GDP—exchange rate conversion—$1.9 billion (1992 est.). **National product real growth rate**: -7.7% (1992 est.). **National product per capita**: $200 (1992 est.). **Inflation rate (consumer prices)**: 21% (1992 est.). **External debt**: $1.8 billion (December 1991 est.).

Balance of Payments Summary [37]

Values in millions of dollars.

	1980	1985	1986	1987	1988
Exports of goods (f.o.b.)	280.8	245.5	248.4	278.5	297.0
Imports of goods (f.o.b.)	-308.0	-176.7	-154.1	-177.6	-253.0
Trade balance	-27.2	68.8	94.3	100.9	44.0
Services - debits	-329.9	266.8	-254.3	-243.8	-230.3
Services - credits	-9.2	-12.9	-6.4	-2.8	-9.2
Private transfers (net)	13.3	11.1	13.1	13.8	15.1
Government transfers (net)	50.2	25.2	29.7	30.3	80.5
Long term capital (net)	181.6	5.8	37.6	92.5	171.9
Short term capital (net)	-29.7	-2.6	13.8	6.6	6.7
Errors and omissions	86.0	102.8	40.7	24.2	-18.0
Overall balance	-21.7	-18.0	2.7	68.2	107.7

Exchange Rates [38]

Currency: **Malawian kwacha.**
Symbol: **MK.**

Data are currency units per $1.

November 1992	4.3418
1991	2.8033
1990	2.7289
1989	2.7595
1988	2.5613
1987	2.2087

Imports and Exports

Top Import Origins [39]

$660 million (c.i.f., 1991 est.).

Origins	%
South Africa	NA
Japan	NA
US	NA
UK	NA
Zimbabwe	NA

Top Export Destinations [40]

$400 million (f.o.b., 1991 est.).

Destinations	%
US	NA
UK	NA
Zambia	NA
South Africa	NA
Germany	NA

Foreign Aid [41]

	U.S. $	
US commitments, including Ex-Im (FY70-89)	215	million
Western (non-US) countries, ODA and OOF bilateral commitments (1970-89)	2,150	million

Import and Export Commodities [42]

Import Commodities	**Export Commodities**
Food	Tobacco
Petroleum products	Tea
Semimanufactures	Sugar
Consumer goods	Coffee
Transportation equipment	Peanuts
	Wood products

Malaysia

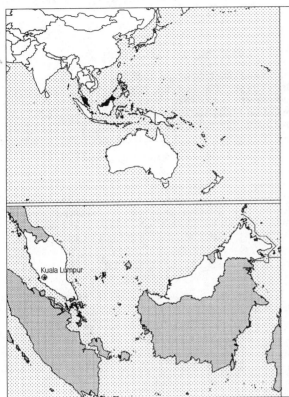

Geography [1]

Total area:
329,750 km2
Land area:
328,550 km2
Comparative area:
Slightly larger than New Mexico
Land boundaries:
Total 2,669 km, Brunei 381 km, Indonesia 1,782 km, Thailand 506 km
Coastline:
4,675 km (Peninsular Malaysia 2,068 km, East Malaysia 2,607 km)
Climate:
Tropical; annual southwest (April to October) and northeast (October to February) monsoons
Terrain:
Coastal plains rising to hills and mountains
Natural resources:
Tin, petroleum, timber, copper, iron ore, natural gas, bauxite
Land use:
Arable land:
3%
Permanent crops:
10%
Meadows and pastures:
0%
Forest and woodland:
63%
Other:
24%

Demographics [2]

	1960	1970	1980	1990	1991[1]	1994	2000	2010	2020
Population	8,428	10,910	13,764	17,556	17,982	19,283	21,953	26,589	31,681
Population density (persons per sq. mi.)	66	86	109	138	142	NA	173	209	249
Births	NA	NA	NA	NA	536	549	NA	NA	NA
Deaths	NA	NA	NA	NA	108	109	NA	NA	NA
Life expectancy - males	NA	NA	NA	NA	65	66	NA	NA	NA
Life expectancy - females	NA	NA	NA	NA	71	72	NA	NA	NA
Birth rate (per 1,000)	NA	NA	NA	NA	30	28	NA	NA	NA
Death rate (per 1,000)	NA	NA	NA	NA	6	6	NA	NA	NA
Women of reproductive age (15-44 yrs.)	NA	NA	NA	4,518	NA	4,943	5,640	NA	NA
of which are currently married	NA	NA	NA	2,845	NA	3,189	3,638	NA	NA
Fertility rate	NA	NA	NA	3.7	3.61	3.5	3.3	3.0	2.8

Population values are in thousands, life expectancy in years, and other items as indicated.

Health

Health Personnel [3]

Doctors per 1,000 pop., 1988-92	0.37
Nurse-to-doctor ratio, 1988-92	3.9
Hospital beds per 1,000 pop., 1985-90	2.4
Percentage of children immunized (age 1 yr. or less)	
Third dose of DPT, 1990-91	90.0
Measles, 1990-91	79.0

Health Expenditures [4]

Total health expenditure, 1990 (official exchange rate)	
Millions of dollars	1259
Millions of dollars per capita	67
Health expenditures as a percentage of GDP	
Total	3.0
Public sector	1.3
Private sector	1.7
Development assistance for health	
Total aid flows (millions of dollars)[1]	3
Aid flows per capita (millions of dollars)	0.1
Aid flows as a percentage of total health expenditure	0.2

For sources, notes, and explanations, see Annotated Source Appendix, page 1035.

575

Human Factors

Health Care Ratios [5]

Population per physician, 1970	4310
Population per physician, 1990	2700
Population per nursing person, 1970	1270
Population per nursing person, 1990	380
Percent of births attended by health staff, 1985	82

Infants and Malnutrition [6]

Percent of babies with low birth weight, 1985	9
Infant mortality rate per 1,000 live births, 1970	45
Infant mortality rate per 1,000 live births, 1991	15
Years of life lost per 1,000 population, 1990	15
Prevalence of malnutrition (under age 5), 1990	24

Ethnic Division [7]

Malay indigenous	59%
Chinese	32%
Indian	9%

Religion [8]

Penninsular Malaysia: Muslim (Malays), Buddhist (Chinese), Hindu (Indians); *Sabah*: Muslim 38%, Christian 17%, other 15%; *Sarawak*: Tribal religion 35%, Buddhist and Confucianist 24%, Muslim 20%, Christian 16%, other 5%.

Major Languages [9]

Penninsular Malaysia: Malay (official, English, Chinese dialects, Tamil; *State of Sabah*: English Malay, numerous tribal dialects, Chinese (Mandarin and Hakka dialects predominate); *State of Sarawak*: English Malay, Mandarin, numerous tribal languages.

Education

Public Education Expenditures [10]

Million Ringgitt	1980	1985	1987	1988[1]	1989[1]	1990[1]
Total education expenditure	3,104	4,754	NA	5,203	5,482	6,033
as percent of GNP	6.0	6.6	NA	6.1	5.7	5.5
as percent of total govt. expend.	14.7	16.3	NA	18.5	18.2	18.3
Current education expenditure	2,575	4,062	NA	4,037	4,335	4,664
as percent of GNP	5.0	5.6	NA	4.7	4.5	4.3
as percent of current govt. expend.	18.4	NA	NA	NA	NA	19.3
Capital expenditure	529	692	NA	1,166	1,147	1,369

Educational Attainment [11]

Literacy Rate [12]

In thousands and percent	1985[a]	1991[a]	2000[a]
Illiterate population +15 years	2,500	2,391	2,116
Illiteracy rate - total pop. (%)	26.0	21.6	14.9
Illiteracy rate - males (%)	16.8	13.5	8.9
Illiteracy rate - females (%)	35.0	29.6	20.8

Libraries [13]

	Admin. Units	Svc. Pts.	Vols. (000)	Shelving (meters)	Vols. Added	Reg. Users
National (1989)	1	9	703	NA	60,653	177,132
Non-specialized	NA	NA	NA	NA	NA	NA
Public (1989)[21]	13	144	4,766	NA	613,614	1 mil.
Higher ed.	NA	NA	NA	NA	NA	NA
School	NA	NA	NA	NA	NA	NA

Daily Newspapers [14]

	1975	1980	1985	1990
Number of papers	31	40	32	45
Circ. (000)	1,038	810[e]	1,500[e]	2,500[e]

Cinema [15]

Science and Technology

Scientific/Technical Forces [16]

Potential scientists/engineers	26,000
Number female	3,562
Potential technicians	NA
Number female	NA
Total	NA

R&D Expenditures [17]

	Ringgit[18] (000) 1989
Total expenditure	97,200
Capital expenditure	NA
Current expenditure	NA
Percent current	NA

U.S. Patents Issued [18]

Values show patents issued to citizens of the country by the U.S. Patents Office.

	1990	1991	1992
Number of patents	6	13	11

Government and Law

Organization of Government [19]

Long-form name:
none
Type:
constitutional monarchy
Independence:
31 August 1957 (from UK)
Constitution:
31 August 1957, amended 16 September 1963
Legal system:
based on English common law; judicial review of legislative acts in the Supreme Court at request of supreme head of the federation; has not accepted compulsory ICJ jurisdiction
National holiday:
National Day, 31 August (1957)
Executive branch:
paramount ruler, deputy paramount ruler, prime minister, deputy prime minister, Cabinet
Legislative branch:
bicameral Parliament (Parlimen) consists of an upper house or Senate (Dewan Negara) and a lower house or House of Representatives (Dewan Rakyat)
Judicial branch:
Supreme Court

Crime [23]

Political Parties [20]

House of Representatives	% of votes
National Front	52.0
Other	48.0

Government Expenditures [21]

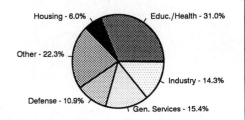

Housing - 6.0%
Educ./Health - 31.0%
Other - 22.3%
Industry - 14.3%
Defense - 10.9%
Gen. Services - 15.4%

% distribution for 1992. Expend. in 1992 est.: 18.0 ($ bil.)

Military Expenditures and Arms Transfers [22]

	1985	1986	1987	1988	1989
Military expenditures					
Current dollars (mil.)	910[e]	1,065[e]	1,242	832	1,039
1989 constant dollars (mil.)	1,036[e]	1,182[e]	1,335	866	1,039
Armed forces (000)	106[e]	106[e]	106[e]	108	110
Gross national product (GNP)					
Current dollars (mil.)	24,100	25,370	27,590	31,200	35,970
1989 constant dollars (mil.)	27,440	28,150	29,670	32,480	35,970
Central government expenditures (CGE)					
1989 constant dollars (mil.)	9,678	11,830	10,140	9,273	10,430
People (mil.)	15.5	15.9	16.3	16.7	17.1
Military expenditure as % of GNP	3.8	4.2	4.5	2.7	2.9
Military expenditure as % of CGE	10.7	10.0	13.2	9.3	10.0
Military expenditure per capita	67	74	82	52	61
Armed forces per 1,000 people	6.8	6.6	6.5	6.5	6.4
GNP per capita	1,765	1,766	1,817	1,942	2,099
Arms imports[6]					
Current dollars (mil.)	470	60	70	40	70
1989 constant dollars (mil.)	535	67	75	42	70
Arms exports[6]					
Current dollars (mil.)	0	0	0	0	0
1989 constant dollars (mil.)	0	0	0	0	0
Total imports[7]					
Current dollars (mil.)	12,300	10,820	12,700	16,550	22,500
1989 constant dollars	14,000	12,010	13,660	17,230	22,500
Total exports[7]					
Current dollars (mil.)	15,440	13,750	17,940	21,110	25,050
1989 constant dollars	17,580	15,260	19,290	21,980	25,050
Arms as percent of total imports[8]	3.8	0.6	0.6	0.2	0.3
Arms as percent of total exports[8]	0	0	0	0	0

Human Rights [24]

	SSTS	FL	FAPRO	PPCG	APROBC	TPW	PCPTW	STPEP	PHRFF	PRW	ASST	AFL
Observes		P			P	P	P				P	P
	EAFRD	CPR	ESCR	SR	ACHR	MAAE	PVIAC	PVNAC	EAFDAW	TCIDTP		RC
Observes		P			P	P	P					P

P = Party; S = Signatory; see Appendix for meaning of abbreviations.

Labor Force

Total Labor Force [25]

7.258 million (1991 est.)

Labor Force by Occupation [26]

Unemployment Rate [27]

4.1% (1992 est.)

Production Sectors

Commercial Energy Production and Consumption

Production [28]

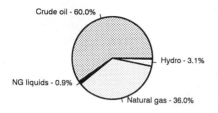

Crude oil - 60.0%
Hydro - 3.1%
NG liquids - 0.9%
Natural gas - 36.0%

Consumption [29]

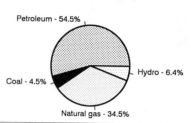

Petroleum - 54.5%
Hydro - 6.4%
Coal - 4.5%
Natural gas - 34.5%

Data are shown in quadrillion (10^{15}) BTUs and percent for 1991

Crude oil[1]	1.35
Natural gas liquids	0.02
Dry natural gas	0.81
Coal[2]	(s)
Net hydroelectric power[3]	0.07
Net nuclear power[3]	0.00
Total	2.25

Petroleum	0.60
Dry natural gas	0.38
Coal[2]	0.05
Net hydroelectric power[3]	0.07
Net nuclear power[3]	(s)
Total	1.12

Telecommunications [30]

- Good intercity service provided on Peninsular Malaysia mainly by microwave radio relay
- Adequate intercity microwave radio relay network between Sabah and Sarawak via Brunei
- International service good
- Good coverage by radio and television broadcasts
- 994,860 telephones (1984)
- Broadcast stations - 28 AM, 3 FM, 33 TV
- Submarine cables extend to India and Sarawak
- SEACOM submarine cable links to Hong Kong and Singapore
- Satellite earth stations - 1 Indian Ocean INTELSAT, 1 Pacific Ocean INTELSAT, and 2 domestic

Transportation [31]

Merchant Marine. 184 ships (1,000 GRT or over) totaling 1,869,817 GRT/2,786,765 DWT; includes 1 passenger-cargo, 2 short-sea passenger, 71 cargo, 28 container, 2 vehicle carrier, 2 roll-on/roll-off, 1 livestock carrier, 38 oil tanker, 6 chemical tanker, 6 liquefied gas, 27 bulk

Airports

Total:	111
Usable:	102
With permanent-surface runways:	32
With runways over 3,659 m:	1
With runways 2,440-3,659 m:	7
With runways 1,220-2,439 m:	18

Top Agricultural Products [32]

	1990[21]	1991[21]
Palm oil	6,412	6,035
Rubber	1,291	1,250
Palm Kernel oil	850	769
Banana[22]	505	510
Cassava[22]	410	415
Cocoa beans	240	235

Values shown are 1,000 metric tons.

Top Mining Products [33]

Preliminary metric tons unless otherwise specified M.t.

Bauxite, gross weight (000 tons)	376
Clay, kaolin	186,699
Copper, mine output, Cu content (Sabah)	25,581
Petroleum (000 42-gal. barrels)	307,693
Ilmenite concentrate, gross weight	336,347
Natural gas, gross (mil. cubic meters)[9]	22,900[e]
Iron ore and concentrate (000 tons)	375
Rare earths: monazite, gross weight	1,981
Tin, mine output, Sn content	20,710

Tourism [34]

	1987	1988	1989	1990	1991
Visitors[44]			10,540	15,957	12,042
Tourists[44]	3,359	3,624	4,846	7,446	5,847
Tourism receipts	690	745	1,038	1,667	1,530
Tourism expenditures	1,234	1,263	1,377	1,492	1,503
Fare receipts	323	357	413	447	596
Fare expenditures	190	206	224	243	281

Tourists are in thousands, money in million U.S. dollars.

578

For sources, notes, and explanations, see Annotated Source Appendix, page 1035.

Manufacturing Sector

GDP and Manufacturing Summary [35]

	1980	1985	1989	1990	% change 1980-1990	% change 1989-1990
GDP (million 1980 $)	24,487	31,408	39,659	43,541	77.8	9.8
GDP per capita (1980 $)	1,779	2,003	2,273	2,430	36.6	6.9
Manufacturing as % of GDP (current prices)	21.2	19.9	22.7^e	25.5^e	20.3	12.3
Gross output (million $)	13,400^e	18,359	27,511^e	35,421	164.3	28.8
Value added (million $)	3,727^e	4,879	6,769^e	9,068	143.3	34.0
Value added (million 1980 $)	5,054	6,511	10,403	12,327	143.9	18.5
Industrial production index	100	123	202	249	149.0	23.3
Employment (thousands)	465^e	473	620^e	831	78.7	34.0

Note: GDP stands for Gross Domestic Product. 'e' stands for estimated value.

Profitability and Productivity

	1980	1985	1989	1990	% change 1980-1990	% change 1989-1990
Intermediate input (%)	72^e	73	75^e	74	2.8	-1.3
Wages, salaries, and supplements (%)	8^e	9	7^e	8^e	0.0	14.3
Gross operating surplus (%)	20^e	18	18^e	18^e	-10.0	0.0
Gross output per worker ($)	28,564^e	38,561	44,382^e	42,489	48.8	-4.3
Value added per worker ($)	8,231^e	10,248	10,920^e	10,878	32.2	-0.4
Average wage (incl. benefits) ($)	2,253^e	3,377	2,899^e	3,226^e	43.2	11.3

Profitability is in percent of gross output. Productivity is in U.S. $. 'e' stands for estimated value.

Profitability - 1990

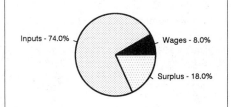

Inputs - 74.0%   Wages - 8.0%   Surplus - 18.0%

The graphic shows percent of gross output.

Value Added in Manufacturing

	1980 $ mil.	1980 %	1985 $ mil.	1985 %	1989 $ mil.	1989 %	1990 $ mil.	1990 %	% change 1980-1990	% change 1989-1990
311 Food products	686^e	18.4	703	14.4	965^e	14.3	865	9.5	26.1	-10.4
313 Beverages	108^e	2.9	122	2.5	135^e	2.0	201	2.2	86.1	48.9
314 Tobacco products	97^e	2.6	205	4.2	160^e	2.4	127	1.4	30.9	-20.6
321 Textiles	192^e	5.2	133	2.7	250^e	3.7	297	3.3	54.7	18.8
322 Wearing apparel	69^e	1.9	100	2.0	199^e	2.9	280	3.1	305.8	40.7
323 Leather and fur products	3^e	0.1	2	0.0	3^e	0.0	6	0.1	100.0	100.0
324 Footwear	11^e	0.3	5	0.1	4^e	0.1	4	0.0	-63.6	0.0
331 Wood and wood products	405^e	10.9	263	5.4	422^e	6.2	584	6.4	44.2	38.4
332 Furniture and fixtures	35^e	0.9	40	0.8	49^e	0.7	70	0.8	100.0	42.9
341 Paper and paper products	35^e	0.9	55	1.1	105^e	1.6	155	1.7	342.9	47.6
342 Printing and publishing	147^e	3.9	197	4.0	183^e	2.7	266	2.9	81.0	45.4
351 Industrial chemicals	81^e	2.2	616	12.6	777^e	11.5	748	8.2	823.5	-3.7
352 Other chemical products	120^e	3.2	153	3.1	198^e	2.9	232	2.6	93.3	17.2
353 Petroleum refineries	118^e	3.2	137	2.8	112^e	1.7	199	2.2	68.6	77.7
354 Miscellaneous petroleum and coal products	2^e	0.1	21	0.4	21^e	0.3	32	0.4	1,500.0	52.4
355 Rubber products	301^e	8.1	250	5.1	510^e	7.5	528	5.8	75.4	3.5
356 Plastic products	70^e	1.9	92	1.9	144^e	2.1	261	2.9	272.9	81.3
361 Pottery, china and earthenware	10^e	0.3	13	0.3	29^e	0.4	36	0.4	260.0	24.1
362 Glass and glass products	24^e	0.6	23	0.5	44^e	0.7	73	0.8	204.2	65.9
369 Other non-metal mineral products	172^e	4.6	297	6.1	348^e	5.1	441	4.9	156.4	26.7
371 Iron and steel	80^e	2.1	153	3.1	182^e	2.7	287	3.2	258.8	57.7
372 Non-ferrous metals	40^e	1.1	35	0.7	48^e	0.7	63	0.7	57.5	31.3
381 Metal products	142^e	3.8	147	3.0	185^e	2.7	316	3.5	122.5	70.8
382 Non-electrical machinery	119^e	3.2	99	2.0	175^e	2.6	349	3.8	193.3	99.4
383 Electrical machinery	456^e	12.2	738	15.1	1,153^e	17.0	1,945	21.4	326.5	68.7
384 Transport equipment	156^e	4.2	211	4.3	249^e	3.7	494	5.4	216.7	98.4
385 Professional and scientific equipment	26^e	0.7	30	0.6	60^e	0.9	97	1.1	273.1	61.7
390 Other manufacturing industries	23^e	0.6	39	0.8	57^e	0.8	111	1.2	382.6	94.7

Note: The industry codes shown are International Standard Industry codes (ISIC). Percentages are percent of total Value Added. 'e' stands for estimated value

Finance, Economics, and Trade

Economic Indicators [36]

In Millions of Malaysian Ringgits (M$) Unless noted.

	1989	1990	1991[e]
GDP (1978 Prices)	72,134	79,155	85,923
Percent Change	8.8	9.8	8.6
Per Cap GDP (M$)			
(1978 Prices)	4,144	4,457	4,727
Money Supply (M1)	21,249	24,240	25,020[103]
Nat Savings/GNP (%)	30.7	30.9	30.0
Investment/GNP (%)	30.3	35.2	36.5
Inflation (CPI)	2.8	3.1	4.5
Foreign Debt	42,140	41,577	40,874

Balance of Payments Summary [37]

Values in millions of dollars.

	1987	1988	1989	1990	1991
Exports of goods (f.o.b.)	17,754.0	20,852.0	24,667.0	28,877.0	33,882.0
Imports of goods (f.o.b.)	-11,918.0	-15,306.0	-20,754.0	-26,968.0	-34,049.0
Trade balance	5,836.0	5,546.0	3,913.0	1,909.0	-167.0
Services - debits	-6,567.0	-7,484.0	-8,390.0	-9,472.0	-10,521.0
Services - credits	3,229.0	3,597.0	4,185.0	5,878.0	6,048.0
Private transfers (net)	69.0	70.0	-17.0	3.0	24.0
Government transfers (net)	69.0	81.0	97.0	51.0	87.0
Long term capital (net)	-548.0	-1,224.0	991.0	2,020.0	4,884.0
Short term capital (net)	-989.0	-1,113.0	552.0	232.0	616.0
Errors and omissions	20.0	97.0	-101.0	1,333.0	270.0
Overall balance	1,119.0	-430.0	1,230.0	1,954.0	1,241.0

Exchange Rates [38]

Currency: **ringgits.**
Symbol: **M$.**

Data are currency units per $1.

January 1993	2.6238
1992	2.5475
1991	2.7501
1990	1.7048
1989	2.7088
1988	2.6188

Imports and Exports

Top Import Origins [39]

$39.1 billion (f.o.b., 1992).

Origins	%
Japan	26.0
US	15.8
Singapore	15.7
Taiwan	5.6
Germany	4.2

Top Export Destinations [40]

$39.8 billion (f.o.b., 1992).

Destinations	%
Singapore	23.0
US	18.6
Japan	13.2
UK	4.0
Germany	4.0

Foreign Aid [41]

	U.S. $	
US commitments, including Ex-Im (FY70-84)	170	million
Western (non-US) countries, ODA and OOF bilateral commitments (1970-89)	4.7	million
OPEC bilateral aid (1979-89)	42	million

Import and Export Commodities [42]

Import Commodities

Food
Consumer goods
Petroleum products
Chemicals
Capital equipment

Export Commodities

Electronic equipment
Palm oil
Petroleum & petr. products
Wood and wood products
Rubber
Textiles

Maldives

Geography [1]

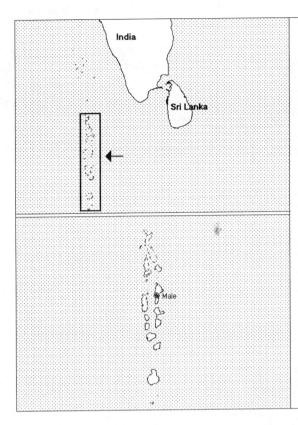

Total area:
300 km2
Land area:
300 km2
Comparative area:
Slightly more than 1.5 times the size of Washington, DC
Land boundaries:
0 km
Coastline:
644 km
Climate:
Tropical; hot, humid; dry, northeast monsoon (November to March); rainy,
southwest monsoon (June to August)
Terrain:
Flat with elevations only as high as 2.5 meters
Natural resources:
Fish
Land use:
Arable land:
10%
Permanent crops:
0%
Meadows and pastures:
3%
Forest and woodland:
3%
Other:
84%

Demographics [2]

	1960	1970	1980	1990	1991[1]	1994	2000	2010	2020
Population	92	115	154	218	226	252	310	423	554
Population density (persons per sq. mi.)	795	988	1,327	1,879	1,950	NA	2,689	3,679	4,851
Births	NA	NA	NA	NA	10	11	NA	NA	NA
Deaths	NA	NA	NA	NA	2	2	NA	NA	NA
Life expectancy - males	NA	NA	NA	NA	61	63	NA	NA	NA
Life expectancy - females	NA	NA	NA	NA	65	66	NA	NA	NA
Birth rate (per 1,000)	NA	NA	NA	NA	46	44	NA	NA	NA
Death rate (per 1,000)	NA	NA	NA	NA	9	7	NA	NA	NA
Women of reproductive age (15-44 yrs.)	NA	NA	NA	47	NA	53	67	NA	NA
of which are currently married	NA	NA	NA	35	NA	40	50	NA	NA
Fertility rate	NA	NA	NA	6.6	6.55	6.3	5.6	4.4	3.4

Population values are in thousands, life expectancy in years, and other items as indicated.

Health

Health Personnel [3]	Health Expenditures [4]

For sources, notes, and explanations, see Annotated Source Appendix, page 1035.

581

Human Factors

Health Care Ratios [5]	Infants and Malnutrition [6]

Ethnic Division [7]

Sinhalese	NA
Dravidian	NA
Arab	NA
African	NA

Religion [8]

Sunni Muslim.

Major Languages [9]

Divehi (dialect of Sinhala; script derived from Arabic), English spoken by most government officials.

Education

Public Education Expenditures [10]

Million Rufiyaa	1985	1986	1987	1988	1989	1990
Total education expenditure	NA	24	33	40	101	79
as percent of GNP	NA	NA	NA	NA	NA	NA
as percent of total govt. expend.	NA	7.2	8.5	10.0	16.0	10.0
Current education expenditure	NA	20	30	NA	NA	NA
as percent of GNP	NA	NA	NA	NA	NA	NA
as percent of current govt. expend.	NA	NA	NA	NA	NA	NA
Capital expenditure	NA	4	4	NA	NA	NA

Educational Attainment [11]

Literacy Rate [12]

	1977[b]	1985[b]	1990[b]
Illiterate population +15 years	13,814	8,568	NA
Illiteracy rate - total pop. (%)	17.6	8.7	NA
Illiteracy rate - males (%)	17.5	8.8	NA
Illiteracy rate - females (%)	17.7	8.5	NA

Libraries [13]

	Admin. Units	Svc. Pts.	Vols. (000)	Shelving (meters)	Vols. Added	Reg. Users
National (1986)	1	2	NA	337	720	703
Non-specialized	NA	NA	NA	NA	NA	NA
Public	NA	NA	NA	NA	NA	NA
Higher ed.	NA	NA	NA	NA	NA	NA
School	NA	NA	NA	NA	NA	NA

Daily Newspapers [14]

	1975	1980	1985	1990
Number of papers	1	2	2	2[e]
Circ. (000)	1[e]	1	2	2[e]

Cinema [15]

Science and Technology

Scientific/Technical Forces [16]

R&D Expenditures [17]

	Rufiyaa (000) 1986
Total expenditure	-
Capital expenditure	-
Current expenditure	-
Percent current	-

U.S. Patents Issued [18]

Government and Law

Organization of Government [19]

Long-form name:
Republic of Maldives
Type:
republic
Independence:
26 July 1965 (from UK)
Constitution:
4 June 1964
Legal system:
based on Islamic law with admixtures of English common law primarily in commercial matters; has not accepted compulsory ICJ jurisdiction
National holiday:
Independence Day, 26 July (1965)
Executive branch:
president, Cabinet
Legislative branch:
unicameral Citizens' Council (Majlis)
Judicial branch:
High Court

Elections [20]

Citizens' Council. No organized political parties; last held on 7 December 1989 (next to be held 7 December 1994); results - percent of vote NA; seats - (48 total, 40 elected).

Government Expenditures [21]

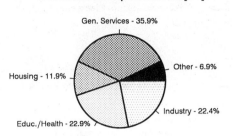

Gen. Services - 35.9%
Other - 6.9%
Housing - 11.9%
Industry - 22.4%
Educ./Health - 22.9%

% distribution for 1990. Expend. in 1991 est.: 83 ($ mil.)

Defense Summary [22]

Branches: National Security Service (paramilitary police force)

Manpower Availability: Males age 15-49 53,730; fit for military service 30,014 (1993 est.)

Defense Expenditures: No information available.

Crime [23]

Crime volume	
Cases known to police	5,018
Attempts (percent)	NA
Percent cases solved	99.38
Crimes per 100,000 persons	2,353.49
Persons responsible for offenses	
Total number offenders	1,885
Percent female	11.56
Percent juvenile[37]	12.41
Percent foreigners	93

Human Rights [24]

	SSTS	FL	FAPRO	PPCG	APROBC	TPW	PCPTW	STPEP	PHRFF	PRW	ASST	AFL
Observes				P		P	P					
	EAFRD	CPR	ESCR	SR	ACHR	MAAE	PVIAC	PVNAC	EAFDAW	TCIDTP	RC	
Observes	P						P	P			P	

P = Party; S = Signatory; see Appendix for meaning of abbreviations.

Labor Force

Total Labor Force [25]

66,000 (est.)

Labor Force by Occupation [26]

Fishing industry   25%

Unemployment Rate [27]

NEGL%

For sources, notes, and explanations, see Annotated Source Appendix, page 1035.

583

Production Sectors

Energy Resource Summary [28]

Energy Resources: None. **Electricity**: 5,000 kW capacity; 11 million kWh produced, 50 kWh per capita (1990).

Telecommunications [30]

- Minimal domestic and international facilities
- 2,804 telephones
- Broadcast stations - 2 AM, 1 FM, 1 TV
- 1 Indian Ocean INTELSAT earth station

Transportation [31]

Highways. Male has 9.6 km of coral highways within the city

Merchant Marine. 14 ships (1,000 GRT or over) totaling 38,848 GRT/58,496 DWT; includes 12 cargo, 1 container, 1 oil tanker

Airports

Total:	2
Useable:	2
With permanent-surface runways:	2
With runways over 3,659 m:	0
With runways 2,440-3,659 m:	2
With runways 1,220-2,439 m:	0

Top Agricultural Products [32]

Agriculture accounts for almost 25% of GDP (including fishing); fishing more important than farming; limited production of coconuts, corn, sweet potatoes; most staple foods must be imported; fish catch of 67,000 tons (1990 est.).

Top Mining Products [33]

Detailed information is not available. A summary of mineral resources available follows. **Mineral Resources**: None.

Tourism [34]

	1987	1988	1989	1990	1991
Tourists[16]	131	156	158	195	196
Tourism receipts	44	50	66	82	87
Tourism expenditures	6	8	10	15	18
Fare expenditures	2	3	4	5	6

Tourists are in thousands, money in million U.S. dollars.

Finance, Economics, and Trade

Industrial Summary [35]

Industrial Production: Growth rate 24.0% (1990); accounts for 6% of GDP. **Industries**: Fishing and fish processing, tourism, shipping, boat building, some coconut processing, garments, woven mats, coir (rope), handicrafts.

Economic Indicators [36]

National product: GDP—exchange rate conversion— $140 million (1991 est.). **National product real growth rate**: 4.7% (1991 est.). **National product per capita**: $620 (1991 est.). **Inflation rate (consumer prices)**: 11.5% (1991 est.). **External debt**: $90 million (1991).

Balance of Payments Summary [37]

Values in millions of dollars.

	1987	1988	1989	1990	1991
Exports of goods (f.o.b.)	34.9	44.6	51.3	58.1	59.2
Imports of goods (f.o.b.)	-71.8	-84.3	-107.6	-117.0	-141.8
Trade balance	-36.9	-39.7	-56.3	-58.9	-82.6
Services - debits	-39.5	-41.5	-41.3	-53.1	-53.2
Services - credits	60.8	73.0	87.7	111.6	119.4
Private transfers (net)	-2.4	-5.0	-5.1	-7.4	-16.6
Government transfers (net)	9.7	11.5	18.3	11.2	18.0
Long term capital (net)	-0.3	-2.4	1.1	5.1	10.4
Short term capital (net)	-9.2	-2.2	6.9	-2.0	-11.2
Errors and omissions	19.0	19.9	-8.6	-5.3	13.8
Overall balance	1.2	13.6	2.7	1.2	-2.0

Exchange Rates [38]

Currency: **rufiyaa.**
Symbol: **Rf.**

Data are currency units per $1.

January 1993	10.5060
1992	10.5690
1991	10.2530
1990	9.5090
1989	9.0408
1988	8.7846

Imports and Exports

Top Import Origins [39]

$150.9 million (c.i.f., 1991).

Origins	%
Singapore	NA
Germany	NA
Sri Lanka	NA
India	NA

Top Export Destinations [40]

$53.7 million (f.o.b., 1991).

Destinations	%
US	NA
UK	NA
Sri Lanka	NA

Foreign Aid [41]

	U.S. $	
US commitments, including Ex-Im (FY70-88)	28	million
Western (non-US) countries, ODA and OOF bilateral commitments (1970-89)	125	million
OPEC bilateral aid (1979-89)	14	million

Import and Export Commodities [42]

Import Commodities	**Export Commodities**
Consumer goods	Fish
Intermediate and capital goods	Clothing
Petroleum products	

Mali

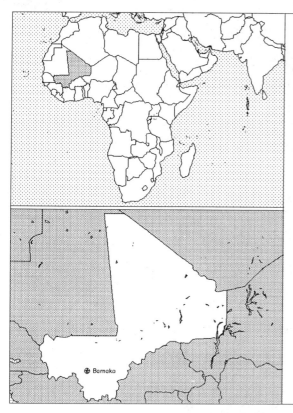

⊕ Bamako

Geography [1]

Total area:
1.24 million km2
Land area:
1.22 million km2
Comparative area:
Slightly less than twice the size of Texas
Land boundaries:
Total 7,243 km, Algeria 1,376 km, Burkina 1,000 km, Guinea 858 km, Cote D'Ivoire 532 km, Mauritania 2,237 km, Niger 821 km, Senegal 419 km
Coastline:
0 km (landlocked)
Climate:
Subtropical to arid; hot and dry February to June; rainy, humid, and mild June to November; cool and dry November to February
Terrain:
Mostly flat to rolling northern plains covered by sand; savanna in south, rugged hills in northeast
Natural resources:
Gold, phosphates, kaolin, salt, limestone, uranium, bauxite, iron ore, manganese, tin, and copper deposits are known but not exploited
Land use:
Arable land: 2%
Permanent crops: 0%
Meadows and pastures: 25%
Forest and woodland: 7%
Other: 66%

Demographics [2]

	1960	1970	1980	1990	1991[1]	1994	2000	2010	2020
Population	4,486	5,525	6,693	8,234	8,339	9,113	10,911	14,966	20,427
Population density (persons per sq. mi.)	10	12	14	17	18	NA	23	31	41
Births	NA	NA	NA	NA	423	472	NA	NA	NA
Deaths	NA	NA	NA	NA	172	186	NA	NA	NA
Life expectancy - males	NA	NA	NA	NA	45	44	NA	NA	NA
Life expectancy - females	NA	NA	NA	NA	47	48	NA	NA	NA
Birth rate (per 1,000)	NA	NA	NA	NA	51	52	NA	NA	NA
Death rate (per 1,000)	NA	NA	NA	NA	21	20	NA	NA	NA
Women of reproductive age (15-44 yrs.)	NA	NA	NA	1,826	NA	2,022	2,420	NA	NA
of which are currently married	NA	NA	NA	1,433	NA	1,575	1,880	NA	NA
Fertility rate	NA	NA	NA	7.3	7.01	7.3	6.9	6.1	5.2

Population values are in thousands, life expectancy in years, and other items as indicated.

Health

Health Personnel [3]

Doctors per 1,000 pop., 1988-92	0.05
Nurse-to-doctor ratio, 1988-92	2.5
Hospital beds per 1,000 pop., 1985-90	NA
Percentage of children immunized (age 1 yr. or less)	
Third dose of DPT, 1990-91	35.0
Measles, 1990-91	40.0

Health Expenditures [4]

Total health expenditure, 1990 (official exchange rate)	
Millions of dollars	130
Millions of dollars per capita	15
Health expenditures as a percentage of GDP	
Total	5.2
Public sector	2.8
Private sector	2.4
Development assistance for health	
Total aid flows (millions of dollars)[1]	36
Aid flows per capita (millions of dollars)	4.3
Aid flows as a percentage of total health expenditure	27.7

For sources, notes, and explanations, see Annotated Source Appendix, page 1035.

Human Factors

Health Care Ratios [5]

Population per physician, 1970	44090
Population per physician, 1990	19450
Population per nursing person, 1970	2590
Population per nursing person, 1990	1890
Percent of births attended by health staff, 1985	27

Infants and Malnutrition [6]

Percent of babies with low birth weight, 1985	17
Infant mortality rate per 1,000 live births, 1970	204
Infant mortality rate per 1,000 live births, 1991	161
Years of life lost per 1,000 population, 1990	108
Prevalence of malnutrition (under age 5), 1990	31

Ethnic Division [7]

Mande	50%
Peul	17%
Voltaic	12%
Songhai	6%
Tuareg Moor	10%
Other	5%

Religion [8]

Muslim	90%
Indigenous beliefs	9%
Christian	1%

Major Languages [9]

French (official), Bambara 80%, numerous African languages.

Education

Public Education Expenditures [10]

Million Francs C.F.A.	1980	1985	1986	1987	1988	1989
Total education expenditure	12,903	17,184	15,702	18,693	NA	NA
as percent of GNP	3.8	3.4	2.7	3.2	NA	NA
as percent of total govt. expend.	30.8	NA	22.4	17.3	NA	NA
Current education expenditure	12,752	17,048	15,538	18,285	NA	NA
as percent of GNP	3.7	3.3	2.7	3.1	NA	NA
as percent of current govt. expend.	32.1	NA	23.8	18.5	NA	NA
Capital expenditure	151	136	164	408	NA	NA

Educational Attainment [11]

Literacy Rate [12]

In thousands and percent	1985[a]	1991[a]	2000[a]
Illiterate population + 15 years	3,357	3,398	3,235
Illiteracy rate - total pop. (%)	77.3	68.0	48.0
Illiteracy rate - males (%)	69.0	59.2	40.1
Illiteracy rate - females (%)	84.6	76.1	55.4

Libraries [13]

Daily Newspapers [14]

	1975	1980	1985	1990
Number of papers	1	2	2	2
Circ. (000)	3	4[e]	10[e]	10[e]

Cinema [15]

Science and Technology

Scientific/Technical Forces [16]

R&D Expenditures [17]

U.S. Patents Issued [18]

For sources, notes, and explanations, see Annotated Source Appendix, page 1035.

Government and Law

Organization of Government [19]

Long-form name:
Republic of Mali
Type:
republic
Independence:
22 September 1960 (from France)
Constitution:
new constitution adopted in constitutional referendum in January 1992
Legal system:
based on French civil law system and customary law; judicial review of legislative acts in Constitutional Section of Court of State; has not accepted compulsory ICJ jurisdiction
National holiday:
Anniverary of the Proclamation of the Republic, 22 September (1960)
Executive branch:
Transition Committee for the Salvation of the People (CTSP) composed of 25 members, predominantly civilian
Legislative branch:
unicameral National Assembly
Judicial branch:
Supreme Court (Cour Supreme)

Crime [23]

Political Parties [20]

National Assembly	% of seats
Alliance for Democracy (Adema)	65.5
Nat. Com. Dem. Initiative (CNID)	7.8
Sudanese Union (US/RAD)	6.9
Popular Movement for Dev.	5.2
Rally for Democracy (RDP)	3.4
Union for Democracy (UDD)	3.4
Rally for Dem. & Labor (RDT)	2.6
Union Democratic Forces (UFDP)	2.6
Party for Dem. & Progress (PDP)	1.7
Malian Union (UMDD)	0.9

Government Expenditures [21]

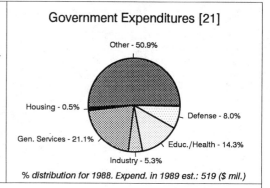

Other - 50.9%
Housing - 0.5%
Defense - 8.0%
Gen. Services - 21.1%
Educ./Health - 14.3%
Industry - 5.3%

% distribution for 1988. Expend. in 1989 est.: 519 ($ mil.)

Military Expenditures and Arms Transfers [22]

	1985	1986	1987	1988	1989
Military expenditures					
Current dollars (mil.)	38	41	40	42	41
1989 constant dollars (mil.)	44	45	43	43	41
Armed forces (000)	8	8	8	8	8
Gross national product (GNP)					
Current dollars (mil.)	1,334	1,626	1,697	1,797	2,048
1989 constant dollars (mil.)	1,519	1,804	1,826	1,870	2,048
Central government expenditures (CGE)					
1989 constant dollars (mil.)	539	555	513	550	NA
People (mil.)	7.3	7.5	7.6	7.8	8.0
Military expenditure as % of GNP	2.9	2.5	2.4	2.3	2.0
Military expenditure as % of CGE	8.1	8.1	8.4	7.9	NA
Military expenditure per capita	6	6	6	6	5
Armed forces per 1,000 people	1.1	1.1	1.1	1.0	0.9
GNP per capita	208	242	240	240	257
Arms imports[6]					
Current dollars (mil.)	10	0	40	70	10
1989 constant dollars (mil.)	11	0	43	73	10
Arms exports[6]					
Current dollars (mil.)	0	0	0	0	10
1989 constant dollars (mil.)	0	0	0	0	10
Total imports[7]					
Current dollars (mil.)	299	444	374	513	NA
1989 constant dollars	340	493	402	534	NA
Total exports[7]					
Current dollars (mil.)	124	212	179	251	NA
1989 constant dollars	141	235	193	261	NA
Arms as percent of total imports[8]	3.3	0	10.7	13.6	NA
Arms as percent of total exports[8]	0	0	0	0	NA

Human Rights [24]

	SSTS	FL	FAPRO	PPCG	APROBC	TPW	PCPTW	STPEP	PHRFF	PRW	ASST	AFL
Observes	P	P	P	P	P	P	P	P		P	P	P
	EAFRD	CPR	ESCR	SR	ACHR	MAAE	PVIAC	PVNAC	EAFDAW	TCIDTP	RC
Observes	P	P	P	P			P	P	P		P

P = Party; S = Signatory; see Appendix for meaning of abbreviations.

Labor Force

Total Labor Force [25]

2.666 million (1986 est.)

Labor Force by Occupation [26]

Agriculture	80%
Services	19
Industry and commerce	1

Date of data: 1981

Unemployment Rate [27]

588

For sources, notes, and explanations, see Annotated Source Appendix, page 1035.

Production Sectors

Energy Resource Summary [28]

Energy Resources: Uranium. **Electricity**: 260,000 kW capacity; 750 million kWh produced, 90 kWh per capita (1991).

Telecommunications [30]

- Domestic system poor but improving
- Provides only minimal service with radio relay, wire, and radio communications stations
- Expansion of radio relay in progress
- 11,000 telephones
- Broadcast stations - 2 AM, 2 FM, 2 TV
- Satellite earth stations - 1 Atlantic Ocean INTELSAT and 1 Indian Ocean INTELSAT

Top Agricultural Products [32]

Agriculture accounts for 50% of GDP; most production based on small subsistence farms; cotton and livestock products account for over 70% of exports; other crops—millet, rice, corn, vegetables, peanuts; livestock—cattle, sheep, goats.

Top Mining Products [33]

Estimated metric tons unless otherwise specified	M.t.
Cement, hydraulic	20,000
Gold, mine output, Au content (kilograms)	5,500
Phosphate rock	10,000
Salt	5,000
Marble	160
Gypsum	700
Silver (kilograms)	160[46]

Transportation [31]

Railroads. 642 km 1.000-meter gauge; linked to Senegal's rail system through Kayes

Highways. about 15,700 km total; 1,670 km paved, 3,670 km gravel and improved earth, 10,360 km unimproved earth

Airports

Total:	34
Usable:	27
With permanent-surface runways:	8
With runways over 3,659 m:	0
With runways 2,440-3,659 m:	5
With runways 1,220-2,439 m:	10

Tourism [34]

	1987	1988	1989	1990	1991
Tourists[10,30]	34	36	32	44	38
Tourism receipts	37	38	28	24	12
Tourism expenditures	57	58	56	62	60
Fare receipts	11	11	10	25	25

Tourists are in thousands, money in million U.S. dollars.

For sources, notes, and explanations, see Annotated Source Appendix, page 1035.

589

Finance, Economics, and Trade

GDP and Manufacturing Summary [35]

	1980	1985	1990	1991	1992
Gross Domestic Product					
Millions of 1980 dollars	1,629	1,666	2,252	2,248	2,255[e]
Growth rate in percent	4.01	-0.11	2.45	-0.15	0.30[e]
Manufacturing Value Added					
Millions of 1980 dollars	71	105	121	121	125[e]
Growth rate in percent	1.58	-0.47	5.15	0.27	3.34[e]
Manufacturing share in percent of current prices	4.3	7.3	12.2	NA	NA

Economic Indicators [36]

National product: GDP—exchange rate conversion—$2.3 billion (1991 est.). **National product real growth rate**: -0.2% (1991 est.). **National product per capita**: $265 (1991 est.). **Inflation rate (consumer prices)**: 1.4% (1991 est.). **External debt**: $2.6 billion (1991 est.).

Balance of Payments Summary [37]

Values in millions of dollars.

	1987	1988	1989	1990	1991
Exports of goods (f.o.b.)	255.9	251.5	269.3	337.9	354.5
Imports of goods (f.o.b.)	-335.4	-359.1	-338.8	-432.4	-447.1
Trade balance	-79.5	-107.6	-69.5	-94.5	-92.6
Services - debits	-355.1	-372.1	-357.7	-445.7	-426.6
Services - credits	85.6	88.0	75.5	99.5	89.7
Private transfers (net)	36.6	45.3	51.4	66.9	70.0
Government transfers (net)	215.6	256.5	218.5	240.1	322.5
Long term capital (net)	104.1	211.8	263.6	180.3	151.4
Short term capital (net)	4.5	-56.0	-87.7	0.9	-5.0
Errors and omissions	1.5	4.3	-5.6	-5.9	10.9
Overall balance	13.3	70.2	88.5	41.6	120.3

Exchange Rates [38]

Currency: **Communaute Financiere Africaine francs.**
Symbol: **CFAF.**

Data are currency units per $1.

January 1993	274.06
1992	264.69
1991	282.11
1990	272.26
1989	319.01
1988	297.85

Imports and Exports

Top Import Origins [39]

$390 million (f.o.b., 1991 est.).

Origins	%
Mostly franc zone and Western Europe	NA

Top Export Destinations [40]

$320 million (f.o.b., 1991 est.).

Destinations	%
Mostly franc zone and Western Europe	NA

Foreign Aid [41]

	U.S. $	
US commitments, including Ex-Im (FY70-89)	349	million
Western (non-US) countries, ODA and OOF bilateral commitments (1970-89)	3,020	million
OPEC bilateral aid (1979-89)	92	million
Communist countries (1970-89)	190	million

Import and Export Commodities [42]

Import Commodities	Export Commodities
Textiles	Livestock
Vehicles	Peanuts
Petroleum products	Dried fish
Machinery	Cotton
Sugar	Skins
Cereals	

For sources, notes, and explanations, see Annotated Source Appendix, page 1035.

Malta

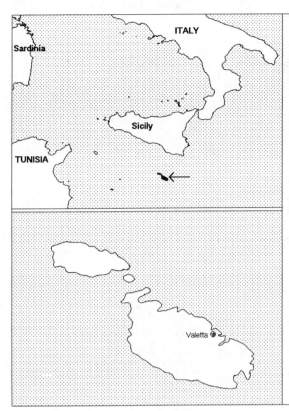

Geography [1]

Total area:
320 km2
Land area:
320 km2
Comparative area:
Slightly less than twice the size of Washington, DC
Land boundaries:
0 km
Coastline:
140 km
Climate:
Mediterranean with mild, rainy winters and hot, dry summers
Terrain:
Mostly low, rocky, flat to dissected plains; many coastal cliffs
Natural resources:
Limestone, salt
Land use:
Arable land:
38%
Permanent crops:
3%
Meadows and pastures:
0%
Forest and woodland:
0%
Other:
59%

Demographics [2]

	1960	1970	1980	1990	1991[1]	1994	2000	2010	2020
Population	329	326	364	354	356	367	382	404	420
Population density (persons per sq. mi.)	2,649	2,626	2,935	2,851	2,874	NA	3,042	3,173	3,253
Births	NA	NA	NA	NA	5	5	NA	NA	NA
Deaths	NA	NA	NA	NA	3	3	NA	NA	NA
Life expectancy - males	NA	NA	NA	NA	74	75	NA	NA	NA
Life expectancy - females	NA	NA	NA	NA	79	79	NA	NA	NA
Birth rate (per 1,000)	NA	NA	NA	NA	15	14	NA	NA	NA
Death rate (per 1,000)	NA	NA	NA	NA	8	7	NA	NA	NA
Women of reproductive age (15-44 yrs.)	NA	NA	NA	NA	NA	NA	NA	NA	NA
of which are currently married	NA	NA	NA	NA	NA	NA	NA	NA	NA
Fertility rate	NA	NA	NA	2.0	2.0	1.9	1.8	1.8	1.8

Population values are in thousands, life expectancy in years, and other items as indicated.

Health

Health Personnel [3]

Health Expenditures [4]

For sources, notes, and explanations, see Annotated Source Appendix, page 1035.

591

Human Factors

Health Care Ratios [5]	Infants and Malnutrition [6]

Ethnic Division [7]		Religion [8]		Major Languages [9]
Arab	NA	Roman Catholic	98%	Maltese (official), English (official).
Sicilian	NA			
Norman	NA			
Spanish	NA			
Italian	NA			
English	NA			

Education

Public Education Expenditures [10]

Million Lira	1980	1985	1987	1988	1989	1990
Total education expenditure	13	17	20	23	27	32
as percent of GNP	3.0	3.4	3.4	3.6	3.8	4.0
as percent of total govt. expend.	7.8	7.7	7.4	8.3	8.4	8.3
Current education expenditure	12	17	19	22	26	30
as percent of GNP	2.9	3.3	3.3	3.5	3.7	3.8
as percent of current govt. expend.	9.7	9.1	9.2	9.9	10.7	10.9
Capital expend. (mil. Lira)	NA	NA	NA	1	1	2

Educational Attainment [11]

Literacy Rate [12]

	1980[b]	1985[b]	1990[b]
Illiterate population +15 years	NA	33,740	NA
Illiteracy rate - total pop. (%)	NA	14.3	NA
Illiteracy rate - males (%)	NA	14.8	NA
Illiteracy rate - females (%)	NA	13.9	NA

Libraries [13]

	Admin. Units	Svc. Pts.	Vols. (000)	Shelving (meters)	Vols. Added	Reg. Users
National (1990)	1	NA	364	NA	NA	6,000
Non-specialized	NA	NA	NA	NA	NA	NA
Public (1990)	1	2	237	NA	NA	102,065
Higher ed. (1987)	1	3	420	8,043	7,548	2,400[e]
School (1987)	51	51	127	3,900	2,196	22,648

Daily Newspapers [14]

	1975	1980	1985	1990
Number of papers	6	5	4	3
Circ. (000)	55[e]	60[e]	56[e]	54

Cinema [15]

Data for 1989.

Cinema seats per 1,000	19.9
Annual attendance per person	0.9
Gross box office receipts (mil. Lira)	NA

Science and Technology

Scientific/Technical Forces [16]

R&D Expenditures [17]

	Lire[30] (000) 1988
Total expenditure	10
Capital expenditure	1
Current expenditure	9
Percent current	90.0

U.S. Patents Issued [18]

Values show patents issued to citizens of the country by the U.S. Patents Office.

	1990	1991	1992
Number of patents	0	0	1

Government and Law

Organization of Government [19]

Long-form name:
Republic of Malta
Type:
parliamentary democracy
Independence:
21 September 1964 (from UK)
Constitution:
26 April 1974, effective 2 June 1974
Legal system:
based on English common law and
Roman civil law; has accepted
compulsory ICJ jurisdiction, with
reservations
National holiday:
Independence Day, 21 September
Executive branch:
president, prime minister, deputy prime
minister, Cabinet
Legislative branch:
unicameral House of Representatives
Judicial branch:
Constitutional Court, Court of Appeal

Political Parties [20]

House of Representatives	% of votes
Nationalist Party (NP)	51.8
Malta Labor Party (MLP)	46.5

Government Expenditures [21]

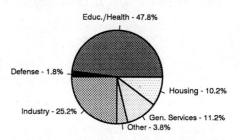

Educ./Health - 47.8%
Defense - 1.8%
Housing - 10.2%
Industry - 25.2%
Gen. Services - 11.2%
Other - 3.8%

% distribution for 1991. Expend. in 1992 est.: 1.100 ($ bil.)

Military Expenditures and Arms Transfers [22]

	1985	1986	1987	1988	1989
Military expenditures					
Current dollars (mil.)	16	17	20	19	22
1989 constant dollars (mil.)	18	19	22	20	22
Armed forces (000)	1	1	1	1	2
Gross national product (GNP)					
Current dollars (mil.)	1,442	1,514	1,626	1,813	1,981
1989 constant dollars (mil.)	1,642	1,680	1,748	1,887	1,981
Central government expenditures (CGE)					
1989 constant dollars (mil.)	690	651	680	696	897[e]
People (mil.)	0.4	0.3	0.3	0.3	0.4
Military expenditure as % of GNP	1.1	1.1	1.3	1.1	1.1
Military expenditure as % of CGE	2.6	2.8	3.2	2.9	2.4
Military expenditure per capita	49	54	64	58	62
Armed forces per 1,000 people	2.8	2.3	2.3	3.5	4.3
GNP per capita	4,529	4,914	5,080	5,438	5,656
Arms imports[6]					
Current dollars (mil.)	0	0	5	0	0
1989 constant dollars (mil.)	0	0	5	0	0
Arms exports[6]					
Current dollars (mil.)	0	0	0	0	0
1989 constant dollars (mil.)	0	0	0	0	0
Total imports[7]					
Current dollars (mil.)	759	887	1,139	1,353	1,477
1989 constant dollars	864	984	1,225	1,408	1,477
Total exports[7]					
Current dollars (mil.)	400	497	605	714	844
1989 constant dollars	455	552	651	743	844
Arms as percent of total imports[8]	0	0	4	0	0
Arms as percent of total exports[8]	0	0	0	0	0

Crime [23]

Crime volume	
Cases known to police	9,924
Attempts (percent)	5.64
Percent cases solved	7.34
Crimes per 100,000 persons	2,802.60
Persons responsible for offenses	
Total number offenders	NA
Percent female	NA
Percent juvenile	NA
Percent foreigners	NA

Human Rights [24]

	SSTS	FL	FAPRO	PPCG	APROBC	TPW	PCPTW	STPEP	PHRFF	PRW	ASST	AFL
Observes	P	P	P		P	P	P		P	P	P	P
		EAFRD	CPR	ESCR	SR	ACHR	MAAE	PVIAC	PVNAC	EAFDAW	TCIDTP	RC
Observes		P	P	P	P		P	P	P	P	P	P

P = Party; S = Signatory; see Appendix for meaning of abbreviations.

Labor Force

Total Labor Force [25]

127,200

Labor Force by Occupation [26]

Government (excluding job corps)	37%
Services	26
Manufacturing	22
Training programs	9
Construction	4
Agriculture	2

Date of data: 1990

Unemployment Rate [27]

3.6% (1992)

For sources, notes, and explanations, see Annotated Source Appendix, page 1035.

593

Production Sectors

Energy Resource Summary [28]

Energy Resources: None. **Electricity**: 328,000 kW capacity; 1,110 million kWh produced, 3,000 kWh per capita (1992).

Telecommunications [30]

- Automatic system satisfies normal requirements
- 153,000 telephones
- Excellent service by broadcast stations - 8 AM, 4 FM, and 2 TV
- Submarine cable and microwave radio relay between islands
- International service by 1 submarine cable and 1 Atlantic Ocean INTELSAT earth station

Top Agricultural Products [32]

Agriculture accounts for 3% of GDP and 2.5% of the work force (1992); overall, 20% self-sufficient; main products—potatoes, cauliflower, grapes, wheat, barley, tomatoes, citrus, cut flowers, green peppers, hogs, poultry, eggs; generally adequate supplies of vegetables, poultry, milk, pork products; seasonal or periodic shortages in grain, animal fodder, fruits, other basic foodstuffs.

Top Mining Products [33]

Metric tons unless otherwise specified	M.t.
Lime (cubic meters)	5,500
Limestone (000 cubic meters)	600
Salt (metric tons)	100

Transportation [31]

Highways. 1,291 km total; 1,179 km paved (asphalt), 77 km crushed stone or gravel, 35 km improved and unimproved earth

Merchant Marine. 789 ships (1,000 GRT or over) totaling 11,059,874 GRT/18,758,969 DWT; includes 6 passenger, 17 short-sea passenger, 272 cargo, 26 container, 2 passenger-cargo, 20 roll-on/roll-off, 2 vehicle carrier, 3 barge carrier, 17 refrigerated cargo, 19 chemical tanker, 15 combination ore/oil, 3 specialized tanker, 3 liquefied gas, 131 oil tanker, 223 bulk, 26 combination bulk, 3 multifunction large load carrier, 1 railcar carrier; note - a flag of convenience registry; China owns 2 ships, Russia owns 52 ships, Cuba owns 10, Vietnam owns 6, Croatia owns 37, Romania owns 3

Airports

Total:	1
Useable:	1
With permanent-surface runways:	1
With runways over 3,659 m:	0
With runways 2,440-3,659 m:	1
With runways 1,220-2,439 m:	0

Tourism [34]

	1987	1988	1989	1990	1991
Visitors[34]	788	841	866	917	935
Tourists[34]	746	784	828	872	895
Cruise passengers[34]	42	57	38	45	40
Tourism receipts	327	382	372	496	574
Tourism expenditures	102	120	107	134	140
Fare receipts	55	58	57	68	
Fare expenditures	16	15	11	11	

Tourists are in thousands, money in million U.S. dollars.

Manufacturing Sector

GDP and Manufacturing Summary [35]

	1980	1985	1989	1990	% change 1980-1990	% change 1989-1990
GDP (million 1980 $)	1,120	1,218	1,534	1,630	45.5	6.3
GDP per capita (1980 $)	3,068	3,530	4,370	4,617	50.5	5.7
Manufacturing as % of GDP (current prices)	33.1	29.5	24.7[e]	27.0	-18.4	9.3
Gross output (million $)	706	650	1,140[e]	1,422[e]	101.4	24.7
Value added (million $)	302	265	452[e]	554[e]	83.4	22.6
Value added (million 1980 $)	330	315	376[e]	386	17.0	2.7
Industrial production index	100	111	147[e]	169	69.0	15.0
Employment (thousands)	29	26	29[e]	28[e]	-3.4	-3.4

Note: GDP stands for Gross Domestic Product. 'e' stands for estimated value.

Profitability and Productivity

	1980	1985	1989	1990	% change 1980-1990	% change 1989-1990
Intermediate input (%)	57	59	60[e]	62[e]	8.8	3.3
Wages, salaries, and supplements (%)	23[e]	22	19[e]	18[e]	-21.7	-5.3
Gross operating surplus (%)	20[e]	19	20[e]	20[e]	0.0	0.0
Gross output per worker ($)	23,265	24,271	39,103[e]	49,778[e]	114.0	27.3
Value added per worker ($)	9,945	9,914	15,498[e]	19,302[e]	94.1	24.5
Average wage (incl. benefits) ($)	5,652[e]	5,561	7,608[e]	9,281[e]	64.2	22.0

Profitability is in percent of gross output. Productivity is in U.S. $. 'e' stands for estimated value.

Profitability - 1990

Wages - 18.0%
Inputs - 62.0%
Surplus - 20.0%

The graphic shows percent of gross output.

Value Added in Manufacturing

	1980 $ mil.	1980 %	1985 $ mil.	1985 %	1989 $ mil.	1989 %	1990 $ mil.	1990 %	% change 1980-1990	% change 1989-1990
311 Food products	20	6.6	25	9.4	50[e]	11.1	56[e]	10.1	180.0	12.0
313 Beverages	20	6.6	22	8.3	42[e]	9.3	54[e]	9.7	170.0	28.6
314 Tobacco products	8	2.6	8	3.0	7[e]	1.5	8[e]	1.4	0.0	14.3
321 Textiles	17	5.6	8	3.0	11[e]	2.4	14[e]	2.5	-17.6	27.3
322 Wearing apparel	88	29.1	65	24.5	97[e]	21.5	90[e]	16.2	2.3	-7.2
323 Leather and fur products	4	1.3	1	0.4	1[e]	0.2	1[e]	0.2	-75.0	0.0
324 Footwear	8	2.6	9	3.4	15[e]	3.3	12[e]	2.2	50.0	-20.0
331 Wood and wood products	2	0.7	1	0.4	2[e]	0.4	2[e]	0.4	0.0	0.0
332 Furniture and fixtures	14	4.6	9	3.4	17[e]	3.8	25[e]	4.5	78.6	47.1
341 Paper and paper products	2	0.7	3	1.1	4[e]	0.9	6[e]	1.1	200.0	50.0
342 Printing and publishing	22	7.3	17	6.4	24[e]	5.3	31[e]	5.6	40.9	29.2
351 Industrial chemicals	1	0.3	2	0.8	3[e]	0.7	3[e]	0.5	200.0	0.0
352 Other chemical products	5	1.7	6	2.3	9[e]	2.0	14[e]	2.5	180.0	55.6
353 Petroleum refineries	NA	0.0	NA	0.0	NA	0.0	NA	0.0	NA	NA
354 Miscellaneous petroleum and coal products	NA	0.0	NA	0.0	NA	0.0	NA	0.0	NA	NA
355 Rubber products	10	3.3	7	2.6	16[e]	3.5	17[e]	3.1	70.0	6.3
356 Plastic products	6	2.0	4	1.5	8[e]	1.8	10[e]	1.8	66.7	25.0
361 Pottery, china and earthenware	1	0.3	NA	0.0	1[e]	0.2	1[e]	0.2	0.0	0.0
362 Glass and glass products	2	0.7	1	0.4	2[e]	0.4	1[e]	0.2	-50.0	-50.0
369 Other non-metal mineral products	6	2.0	7	2.6	9[e]	2.0	11[e]	2.0	83.3	22.2
371 Iron and steel	NA	0.0	NA	0.0	NA	0.0	NA	0.0	NA	NA
372 Non-ferrous metals	NA	0.0	NA	0.0	NA	0.0	NA	0.0	NA	NA
381 Metal products	14	4.6	10	3.8	17[e]	3.8	24[e]	4.3	71.4	41.2
382 Non-electrical machinery	5	1.7	8	3.0	11[e]	2.4	12[e]	2.2	140.0	9.1
383 Electrical machinery	22	7.3	31	11.7	50[e]	11.1	98[e]	17.7	345.5	96.0
384 Transport equipment	6	2.0	3	1.1	19[e]	4.2	23[e]	4.2	283.3	21.1
385 Professional and scientific equipment	12	4.0	12	4.5	17[e]	3.8	22[e]	4.0	83.3	29.4
390 Other manufacturing industries	8	2.6	5	1.9	21[e]	4.6	21[e]	3.8	162.5	0.0

Note: The industry codes shown are International Standard Industry codes (ISIC). Percentages are percent of total Value Added. 'e' stands for estimated value

For sources, notes, and explanations, see Annotated Source Appendix, page 1035.

595

Finance, Economics, and Trade

Economic Indicators [36]

National product: GDP—exchange rate conversion—$2.7 billion (1991 est.). **National product real growth rate**: 5.9% (1991). **National product per capita**: $7,600 (1991 est.). **Inflation rate (consumer prices)**: 2.9% (1991). **External debt**: $127 million (1990 est.).

Balance of Payments Summary [37]

Values in millions of dollars.

	1986	1987	1988	1989	1990
Exports of goods (f.o.b.)	521.6	631.3	758.3	866.3	1,154.2
Imports of goods (f.o.b.)	-786.5	-1,024.6	-122.3	-1,327.7	-1,753.0
Trade balance	-264.9	-393.3	-464.0	-461.4	-598.8
Services - debits	-324.4	-384.7	-465.8	-506.2	-609.5
Services - credits	529.3	718.4	831.5	860.0	1,065.2
Private transfers (net)	48.1	63.1	97.6	54.8	32.4
Government transfers (net)	23.2	23.4	68.6	49.6	55.1
Long term capital (net)	41.2	-18.5	-75.9	-20.1	32.1
Short term capital (net)	-53.6	17.1	116.4	-14.1	-59.8
Errors and omissions	2.6	-28.8	-50.4	64.7	24.2
Overall balance	1.5	-3.3	58.0	27.3	-59.1

Exchange Rates [38]

Currency: **Maltese liri.**
Symbol: **LM.**

Data are currency units per $1.

January 1993	0.3687
1992	0.3178
1991	0.3226
1990	0.3172
1989	0.3483
1988	0.3306

Imports and Exports

Top Import Origins [39]

$2.1 billion (f.o.b., 1991).

Origins	
Italy	30
UK	16
Germany	13
US	4

Top Export Destinations [40]

$1.2 billion (f.o.b., 1991).

Destinations	%
Italy	30
Germany	22
UK	11

Foreign Aid [41]

	U.S. $	
US commitments, including Ex-Im (FY70-81)	172	million
Western (non-US) countries, ODA and OOF bilateral commitments (1970-89)	336	million
OPEC bilateral aid (1979-89)	76	million
Communist countries (1970-88)	48	million

Import and Export Commodities [42]

Import Commodities	Export Commodities
Food	Clothing
Petroleum	Textiles
Machinery	Footwear
Semimanufactured goods	Ships

For sources, notes, and explanations, see Annotated Source Appendix, page 1035.

Marshall Islands

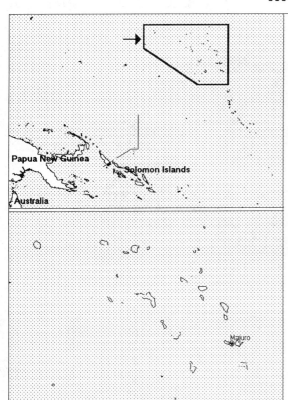

Geography [1]

Total area:
181.3 km2
Land area:
181.3 km2
Comparative area:
Slightly larger than Washington, DC
Land boundaries:
0 km
Coastline:
370.4 km
Climate:
Wet season May to November; hot and humid; islands border typhoon belt
Terrain:
Low coral limestone and sand islands
Natural resources:
Phosphate deposits, marine products, deep seabed minerals
Land use:
Arable land:
0%
Permanent crops:
60%
Meadows and pastures:
0%
Forest and woodland:
0%
Other:
40%

Demographics [2]

	1960	1970	1980	1990	1991[1]	1994	2000	2010	2020
Population	15	22	31	46	48	54	68	100	144
Population density (persons per sq. mi.)	216	310	438	661	687	NA	972	1,425	2,032
Births	NA	NA	NA	NA	2	3	NA	NA	NA
Deaths	NA	NA	NA	NA	(Z)	(Z)	NA	NA	NA
Life expectancy - males	NA	NA	NA	NA	61	62	NA	NA	NA
Life expectancy - females	NA	NA	NA	NA	64	65	NA	NA	NA
Birth rate (per 1,000)	NA	NA	NA	NA	47	46	NA	NA	NA
Death rate (per 1,000)	NA	NA	NA	NA	8	8	NA	NA	NA
Women of reproductive age (15-44 yrs.)	NA	NA	NA	NA	NA	NA	NA	NA	NA
of which are currently married	NA	NA	NA	NA	NA	NA	NA	NA	NA
Fertility rate	NA	NA	NA	7.1	7.08	6.9	6.6	6.0	5.3

Population values are in thousands, life expectancy in years, and other items as indicated.

Health

Health Personnel [3]

Health Expenditures [4]

For sources, notes, and explanations, see Annotated Source Appendix, page 1035.

597

Human Factors	
Health Care Ratios [5]	Infants and Malnutrition [6]

Ethnic Division [7]	Religion [8]	Major Languages [9]
Micronesian.	Christian (mostly Protestant).	English (universally spoken and is the official language), two major Marshallese dialects from the Malayo-Polynesian family, Japanese.

Education

Public Education Expenditures [10]	Educational Attainment [11]

Literacy Rate [12]	Libraries [13]

Daily Newspapers [14]	Cinema [15]

Science and Technology

Scientific/Technical Forces [16]	R&D Expenditures [17]	U.S. Patents Issued [18]

　　　　　　　　　　　For sources, notes, and explanations, see Annotated Source Appendix, page 1035.

Government and Law

Organization of Government [19]

Long-form name:
Republic of the Marshall Islands
Type:
constitutional government in free
association with the US; the Compact of
Free Association entered into force 21
October 1986
Independence:
21 October 1986 (from the US-
administered UN trusteeship)
Constitution:
1 May 1979
Legal system:
based on adapted Trust Territory laws,
acts of the legislature, municipal,
common, and customary laws
National holiday:
Proclamation of the Republic of the
Marshall Islands, 1 May (1979)
Executive branch:
president, Cabinet
Legislative branch:
unicameral Nitijela (parliament)
Judicial branch:
Supreme Court

Elections [20]

Parliament. No formal parties; last held
18 November 1991 (next to be held
November 1995); results - percent of
vote NA; seats - (33 total).

Government Budget [21]

For 1987 est.
Revenues	55
Expenditures	NA
Capital expenditures	NA

Defense Summary [22]

Note: Defense is the responsibility of the United States.

Crime [23]

Human Rights [24]

	SSTS	FL	FAPRO	PPCG	APROBC	TPW	PCPTW	STPEP	PHRFF	PRW	ASST	AFL
Observes												
	EAFRD	CPR	ESCR	SR	ACHR	MAAE	PVIAC	PVNAC	EAFDAW	TCIDTP	RC	
Observes												

P = Party; S = Signatory; see Appendix for meaning of abbreviations.

Labor Force

Total Labor Force [25]

4,800 (1986)

Labor Force by Occupation [26]

Unemployment Rate [27]

For sources, notes, and explanations, see Annotated Source Appendix, page 1035.

599

Production Sectors

Energy Resource Summary [28]

Energy Resources: None. **Electricity**: 42,000 kW capacity; 80 million kWh produced, 1,840 kWh per capita (1990).

Telecommunications [30]

- Telephone network - 570 lines (Majuro) and 186 (Ebeye)
- Telex services
- Islands interconnected by shortwave radio (used mostly for government purposes)
- Broadcast stations - 1 AM, 2 FM, 1 TV, 1 shortwave
- 2 Pacific Ocean INTELSAT earth stations
- US Government satellite communications system on Kwajalein

Transportation [31]

Highways. paved roads on major islands (Majuro, Kwajalein), otherwise stone-, coral-, or laterite-surfaced roads and tracks

Merchant Marine. 29 ships (1,000 GRT or over) totaling 1,786,070 GRT/3,498,895 DWT; includes 2 cargo, 1 container, 9 oil tanker, 15 bulk carrier, 2 combination ore/oil; note - a flag of convenience registry

Airports

Total:	16
Usable:	16
With permanent-surface runways:	4
With runways over 3,659m:	0
With runways 2,440-3,659 m:	0
With runways 1,220-2,439 m:	8

Top Agricultural Products [32]

Coconuts, cacao, taro, breadfruit, fruits, pigs, chickens.

Top Mining Products [33]

Detailed information is not available. A summary of mineral resources available follows. **Mineral Resources**: Phosphate deposits, marine products, deep seabed minerals.

Tourism [34]

Finance, Economics, and Trade

Industrial Summary [35]

Industrial Production: Growth rate NA%. **Industries**: Copra, fish, tourism; craft items from shell, wood, and pearls; offshore banking (embryonic).

Economic Indicators [36]

National product: GDP—exchange rate conversion—$63 million (1989 est.). **National product real growth rate**: NA%. **National product per capita**: $1,500 (1989 est.). **Inflation rate (consumer prices)**: NA%. **External debt**: not available.

Balance of Payments Summary [37]

Exchange Rates [38]

US currency is used.

Imports and Exports

Top Import Origins [39]

$29.2 million (c.i.f., 1985).

Origins	%
NA	NA

Top Export Destinations [40]

$2.5 million (f.o.b., 1985).

Destinations	%
NA	NA

Foreign Aid [41]

	U.S. $	
US aid (under the Compact of Free Association) est. per year	40	million

Import and Export Commodities [42]

Import Commodities	Export Commodities
Foodstuffs	Copra
Beverages	Copra oil
Building materials	Agricultural products
	Handicrafts

Mauritania

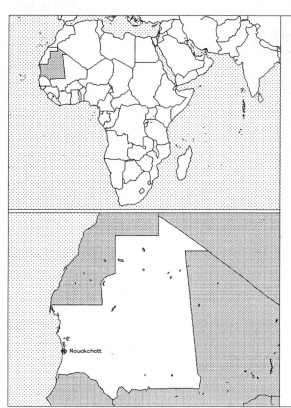

Geography [1]

Total area:
1,030,700 km2
Land area:
1,030,400 km2
Comparative area:
Slightly larger than three times the size of New Mexico
Land boundaries:
Total 5,074 km, Algeria 463 km, Mali 2,237 km, Senegal 813 km, Western Sahara 1,561 km
Coastline:
754 km
Climate:
Desert; constantly hot, dry, dusty
Terrain:
Mostly barren, flat plains of the Sahara; some central hills
Natural resources:
Iron ore, gypsum, fish, copper, phosphate
Land use:
Arable land:
1%
Permanent crops:
0%
Meadows and pastures:
38%
Forest and woodland:
5%
Other:
56%

Demographics [2]

	1960	1970	1980	1990	1991[1]	1994	2000	2010	2020
Population	1,057	1,227	1,456	1,935	1,996	2,193	2,653	3,630	4,859
Population density (persons per sq. mi.)	3	3	4	5	5	NA	7	9	12
Births	NA	NA	NA	NA	97	104	NA	NA	NA
Deaths	NA	NA	NA	NA	35	35	NA	NA	NA
Life expectancy - males	NA	NA	NA	NA	44	45	NA	NA	NA
Life expectancy - females	NA	NA	NA	NA	50	51	NA	NA	NA
Birth rate (per 1,000)	NA	NA	NA	NA	49	48	NA	NA	NA
Death rate (per 1,000)	NA	NA	NA	NA	18	16	NA	NA	NA
Women of reproductive age (15-44 yrs.)	NA	NA	NA	427	NA	485	594	NA	NA
of which are currently married	NA	NA	NA	268	NA	304	372	NA	NA
Fertility rate	NA	NA	NA	7.3	7.19	7.0	6.5	5.6	4.5

Population values are in thousands, life expectancy in years, and other items as indicated.

Health

Health Personnel [3]

Health Expenditures [4]

Human Factors

Health Care Ratios [5]

Population per physician, 1970	17960
Population per physician, 1990	NA
Population per nursing person, 1970	3740
Population per nursing person, 1990	NA
Percent of births attended by health staff, 1985	23

Infants and Malnutrition [6]

Percent of babies with low birth weight, 1985	10
Infant mortality rate per 1,000 live births, 1970	165
Infant mortality rate per 1,000 live births, 1991	119
Years of life lost per 1,000 population, 1990	NA
Prevalence of malnutrition (under age 5), 1990	30

Ethnic Division [7]

Mixed Maur/black	40%
Maur	30%
Black	30%

Religion [8]

Muslim	100%

Major Languages [9]

Hasaniya Arabic (official)	NA
Pular	NA
Soninke	NA
Wolof (official)	NA

Education

Public Education Expenditures [10]

Million Ouguiya	1980	1985	1986	1987	1988	1989
Total education expenditure	NA	NA	NA	NA	NA	NA
as percent of GNP	NA	NA	NA	NA	NA	NA
as percent of total govt. expend.	NA	NA	NA	NA	NA	NA
Current education expenditure	1,546	3,973	3,470	3,120	3,188	NA
as percent of GNP	5.0	7.7	6.0	5.1	4.7	NA
as percent of current govt. expend.	NA	33.2	26.8	22.7	22.0	NA
Capital expenditure	NA	NA	NA	NA	NA	NA

Educational Attainment [11]

Age group	6+
Total population	1,446,867
Highest level attained (%)	
No schooling	54.3
First level	
Incompleted	39.7
Completed	NA
Entered second level	
S-1	5.2
S-2	NA
Post secondary	0.8

Literacy Rate [12]

In thousands and percent	1985[a]	1991[a]	2000[a]
Illiterate population +15 years	715	740	785
Illiteracy rate - total pop. (%)	72.5	66.0	53.1
Illiteracy rate - males (%)	60.2	52.9	39.7
Illiteracy rate - females (%)	84.2	78.6	65.9

Libraries [13]

Daily Newspapers [14]

	1975	1980	1985	1990
Number of papers	-	-	-	1
Circ. (000)	-	-	-	1[e]

Cinema [15]

Science and Technology

Scientific/Technical Forces [16]

R&D Expenditures [17]

U.S. Patents Issued [18]

Values show patents issued to citizens of the country by the U.S. Patents Office.

	1990	1991	1992
Number of patents	0	0	1

For sources, notes, and explanations, see Annotated Source Appendix, page 1035.

603

Government and Law

Organization of Government [19]

Long-form name:
  Islamic Republic of Mauritania
Type:
  republic
Independence:
  28 November 1960 (from France)
Constitution:
  12 July 1991
Legal system:
  three-tier system: Islamic (Shari'a) courts, special courts, state security courts (in the process of being eliminated)
National holiday:
  Independence Day, 28 November (1960)
Executive branch:
  president
Legislative branch:
  bicameral legislature consists of an upper house or Senate (Majlis al-Shuyukh) and a lower house or National Assembly (Majlis al-Watani)
Judicial branch:
  Supreme Court (Cour Supreme)

Crime [23]

Elections [20]

National Assembly. Last held 6 and 13 March 1992 (next to be held March 1997). Legalized by constitution passed 12 July 1991, however, politics continue to be tribally based.

Government Budget [21]

For 1989 est.

Revenues	280
Expenditures	346
Capital expenditures	61

Military Expenditures and Arms Transfers [22]

	1985	1986	1987	1988	1989
Military expenditures					
Current dollars (mil.)	52[e]	48	35	NA	40
1989 constant dollars (mil.)	59[e]	53	37	NA	40
Armed forces (000)	16	16	16	14	16
Gross national product (GNP)					
Current dollars (mil.)	801	860	780	847	943
1989 constant dollars (mil.)	911	955	839	882	943
Central government expenditures (CGE)					
1989 constant dollars (mil.)	237[e]	NA	NA	NA	301
People (mil.)	1.7	1.7	1.8	1.8	1.9
Military expenditure as % of GNP	6.5	5.5	4.4	NA	4.3
Military expenditure as % of CGE	25.0	NA	NA	NA	13.4
Military expenditure per capita	36	31	21	NA	22
Armed forces per 1,000 people	9.6	9.3	9.1	7.4	8.6
GNP per capita	547	557	475	485	503
Arms imports[6]					
Current dollars (mil.)	0	5	0	10	20
1989 constant dollars (mil.)	0	6	0	10	20
Arms exports[6]					
Current dollars (mil.)	0	0	0	0	0
1989 constant dollars (mil.)	0	0	0	0	0
Total imports[7]					
Current dollars (mil.)	234	221	235	240	222
1989 constant dollars	266	245	253	250	222
Total exports[7]					
Current dollars (mil.)	374	360	428	354	238
1989 constant dollars	426	399	460	369	238
Arms as percent of total imports[8]	0	2.3	0	4.2	9.0
Arms as percent of total exports[8]	0	0	0	0	0

Human Rights [24]

	SSTS	FL	FAPRO	PPCG	APROBC	TPW	PCPTW	STPEP	PHRFF	PRW	ASST	AFL
Observes	P	P	P			P	P	P		P	P	
	EAFRD	CPR	ESCR		SR	ACHR	MAAE	PVIAC	PVNAC	EAFDAW	TCIDTP	RC
Observes		P				P		P	P			P

P = Party; S = Signatory; see Appendix for meaning of abbreviations.

Labor Force

Total Labor Force [25]

465,000 (1981 est.); 45,000 wage earners (1980)

Labor Force by Occupation [26]

Agriculture	47%
Services	29
Industry and commerce	14
Government	10

Unemployment Rate [27]

20% (1991 est.)

604

For sources, notes, and explanations, see Annotated Source Appendix, page 1035.

Production Sectors

Energy Resource Summary [28]

Energy Resources: None. **Electricity**: 190,000 kW capacity; 135 million kWh produced, 70 kWh per capita (1991).

Telecommunications [30]

- Poor system of cable and open-wire lines, minor microwave radio relay links, and radio communications stations (improvements being made)
- Broadcast stations - 2 AM, no FM, 1 TV
- Satellite earth stations - 1 Atlantic Ocean INTELSAT and 2 ARABSAT, with six planned

Transportation [31]

Railroads. 690 km 1.435-meter (standard) gauge, single track, owned and operated by government mining company

Highways. 7,525 km total; 1,685 km paved; 1,040 km gravel, crushed stone, or otherwise improved; 4,800 km unimproved roads, trails, tracks

Merchant Marine. 1 cargo ship (1,000 GRT or over) totaling 1,290 GRT/1,840 DWT

Airports

Total:	29
Usable:	29
With permanent-surface runways:	9
With runways over 3,659 m:	1
With runways 2,440-3,659 m:	5
With runways 1,220-2,439 m:	16

Top Agricultural Products [32]

Agriculture accounts for 50% of GDP (including fishing); largely subsistence farming and nomadic cattle and sheep herding except in Senegal river valley; crops—dates, millet, sorghum, root crops; fish products number-one export; large food deficit in years of drought.

Top Mining Products [33]

Estimated metric tons unless otherwise specified	M.t.
Cement	90,000[47]
Gypsum	2,839[1]
Iron ore, Fe content (000 tons)	6,500[1]
Salt	5,500

Tourism [34]

For sources, notes, and explanations, see Annotated Source Appendix, page 1035.

605

Finance, Economics, and Trade

GDP and Manufacturing Summary [35]

	1980	1985	1990	1991	1992
Gross Domestic Product					
Millions of 1980 dollars	829	874	1,023	1,018	1,032[e]
Growth rate in percent	3.93	3.11	4.00	-0.50	1.39[e]
Manufacturing Value Added					
Millions of 1980 dollars	43	66	89	95	101[e]
Growth rate in percent	-1.39	7.80	5.30	6.76	6.28[e]
Manufacturing share in percent of current prices	5.6	12.8	12.9	12.0	NA

Economic Indicators [36]

National product: GDP—exchange rate conversion—$1.1 billion (1991 est.). **National product real growth rate**: 3% (1991 est.). **National product per capita**: $555 (1991 est.). **Inflation rate (consumer prices)**: 6.2% (1991 est.). **External debt**: $1.9 billion (1990).

Balance of Payments Summary [37]

Values in millions of dollars.

	1985	1986	1987	1988	1989
Exports of goods (f.o.b.)	371.5	418.8	402.4	437.6	447.9
Imports of goods (f.o.b.)	-333.9	-401.2	-359.2	-348.9	-349.3
Trade balance	37.6	17.6	43.2	88.7	98.6
Services - debits	-297.9	-329.7	-306.3	-306.7	-251.7
Services - credits	31.0	26.0	37.0	39.4	39.2
Private transfers (net)	-20.8	-23.0	-20.6	-22.1	-25.0
Government transfers (net)	134.0	114.9	99.4	104.8	120.3
Long term capital (net)	98.0	171.8	98.7	76.9	42.9
Short term capital (net)	-3.1	12.7	12.2	23.0	-2.4
Errors and omissions	-5.6	-5.7	-101.5	-16.0	-3.6
Overall balance	-26.8	-15.4	-137.9	-12.0	18.3

Exchange Rates [38]

Currency: **ouguiya.**
Symbol: **UM.**

Data are currency units per $1.

February 1993	116.990
1992	87.082
1991	81.946
1990	80.609
1989	83.051
1988	75.261

Imports and Exports

Top Import Origins [39]

$385 million (c.i.f., 1990).

Origins	%
EC	60
Algeria	15
China	6
US	3

Top Export Destinations [40]

$447 million (f.o.b., 1990).

Destinations	%
EC	43
Japan	27
USSR	11
Cote d'Ivoire	3

Foreign Aid [41]

	U.S. $	
US commitments, including Ex-Im (FY70-89)	168	million
Western (non-US) countries, ODA and OOF bilateral commitments (1970-89)	1.3	billion
OPEC bilateral aid (1979-89)	490	million
Communist countries (1970-89)	277	million
Arab Development Bank (1991)	20	million

Import and Export Commodities [42]

Import Commodities	**Export Commodities**
Foodstuffs	Iron ore
Consumer goods	Processed fish
Petroleum products	Gum arabic & gypsum
Capital goods	Cattle

606

For sources, notes, and explanations, see Annotated Source Appendix, page 1035.

Mauritius

Geography [1]

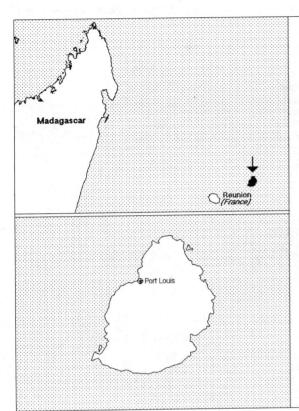

Total area:
1,860 km2
Land area:
1,850 km2
Comparative area:
Slightly less than 10.5 times the size of Washington, DC
Land boundaries:
0 km
Coastline:
177 km
Climate:
Tropical modified by southeast trade winds; warm, dry winter (May to November); hot, wet, humid summer (November to May)
Terrain:
Small coastal plain rising to discontinuous mountains encircling central plateau
Natural resources:
Arable land, fish
Land use:
Arable land:
54%
Permanent crops:
4%
Meadows and pastures:
4%
Forest and woodland:
31%
Other:
7%

Demographics [2]

	1960	1970	1980	1990	1991[1]	1994	2000	2010	2020
Population	663	830	964	1,074	1,081	1,117	1,194	1,322	1,428
Population density (persons per sq. mi.)	929	1,162	1,350	1,501	1,514	NA	1,636	1,778	1,885
Births	NA	NA	NA	NA	20	22	NA	NA	NA
Deaths	NA	NA	NA	NA	7	7	NA	NA	NA
Life expectancy - males	NA	NA	NA	NA	66	67	NA	NA	NA
Life expectancy - females	NA	NA	NA	NA	74	75	NA	NA	NA
Birth rate (per 1,000)	NA	NA	NA	NA	19	19	NA	NA	NA
Death rate (per 1,000)	NA	NA	NA	NA	6	6	NA	NA	NA
Women of reproductive age (15-44 yrs.)	NA	NA	NA	296	NA	314	337	NA	NA
of which are currently married	NA	NA	NA	179	NA	192	208	NA	NA
Fertility rate	NA	NA	NA	2.3	2.04	2.2	2.1	2.0	1.9

Population values are in thousands, life expectancy in years, and other items as indicated.

Health

Health Personnel [3]

Health Expenditures [4]

For sources, notes, and explanations, see Annotated Source Appendix, page 1035.

607

Human Factors

Health Care Ratios [5]

Population per physician, 1970	4190
Population per physician, 1990	1180
Population per nursing person, 1970	610
Population per nursing person, 1990	NA
Percent of births attended by health staff, 1985	90

Infants and Malnutrition [6]

Percent of babies with low birth weight, 1985	9
Infant mortality rate per 1,000 live births, 1970	60
Infant mortality rate per 1,000 live births, 1991	19
Years of life lost per 1,000 population, 1990	NA
Prevalence of malnutrition (under age 5), 1990	24

Ethnic Division [7]

Indo-Mauritian	68%
Creole	27%
Sino-Mauritian	3%
Franco-Mauritian	2%

Religion [8]

Hindu	52.0%
Christian	28.3%
Roman Catholic	26.0%
Protestant	2.3%
Muslim	16.6%
Other	3.1%

Major Languages [9]

English (official)	NA
Creole	NA
French	NA
Hindi	NA
Urdu	NA
Hakka	NA
Bojpoori	NA

Education

Public Education Expenditures [10]

Million Rupee	1980	1985	1987	1988	1989	1990
Total education expenditure	454	598	788	1,047	1,099	1,384
as percent of GNP	5.3	3.8	3.4	3.8	3.5	3.7
as percent of total govt. expend.	11.6	9.8	10.0	10.4	10.5	11.8
Current education expenditure	408	555	734	992	1,058	1,287
as percent of GNP	4.7	3.5	3.2	3.6	3.3	3.4
as percent of current govt. expend.	15.5	12.4	12.6	13.0	12.9	14.0
Capital expenditure	46	43	54	55	41	97

Educational Attainment [11]

Age group	25+
Total population	540,244
Highest level attained (%)	
No schooling	18.3
First level	
Incompleted	42.6
Completed	6.1
Entered second level	
S-1	18.0
S-2	13.1
Post secondary	1.9

Literacy Rate [12]

	1977[b]	1980[b]	1990[b]
Illiterate population +15 years	NA	NA	149,383
Illiteracy rate - total pop. (%)	NA	NA	20.1
Illiteracy rate - males (%)	NA	NA	14.8
Illiteracy rate - females (%)	NA	NA	25.3

Libraries [13]

	Admin. Units	Svc. Pts.	Vols. (000)	Shelving (meters)	Vols. Added	Reg. Users
National (1989)	1	1	36	390	561	-
Non-specialized	NA	NA	NA	NA	NA	NA
Public (1986)	2	2	17	260	1,953	8,147
Higher ed.	NA	NA	NA	NA	NA	NA
School (1990)[6]	26	26	161	NA	7,555	14,345

Daily Newspapers [14]

	1975	1980	1985	1990
Number of papers	12	10	7	7
Circ. (000)	82	80	70[e]	80

Cinema [15]

Data for 1991.

Cinema seats per 1,000	12.0
Annual attendance per person	0.6
Gross box office receipts (mil. Rupee)	15

Science and Technology

Scientific/Technical Forces [16]

Potential scientists/engineers	7,256
Number female	1,732
Potential technicians	10,251
Number female	3,335
Total	17,507

R&D Expenditures [17]

	Rupee (000) 1989
Total expenditure	104,300
Capital expenditure	50,000
Current expenditure	54,300
Percent current	52.1

U.S. Patents Issued [18]

Values show patents issued to citizens of the country by the U.S. Patents Office.

	1990	1991	1992
Number of patents	0	0	1

Government and Law

Organization of Government [19]

Long-form name:
Republic of Mauritius
Type:
parliamentary democracy
Independence:
12 March 1968 (from UK)
Constitution:
12 March 1968
Legal system:
based on French civil law system with
elements of English common law in
certain areas
National holiday:
Independence Day, 12 March (1968)
Executive branch:
president, vice president, prime minister,
deputy prime minister, Council of
Ministers (cabinet)
Legislative branch:
unicameral Legislative Assembly
Judicial branch:
Supreme Court

Political Parties [20]

Legislative Assembly	% of votes
Militant Socialist Movement	53.0
Mauritian Labor Party	38.0

Government Expenditures [21]

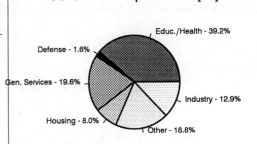

Educ./Health - 39.2%
Defense - 1.6%
Gen. Services - 19.6%
Industry - 12.9%
Housing - 8.0%
Other - 18.8%

% distribution for 1993. Expend. in FY90: 607 ($ mil.)

Crime [23]

Crime volume	
Cases known to police	29,182
Attempts (percent)	0.14
Percent cases solved	51.75
Crimes per 100,000 persons	2,770.26
Persons responsible for offenses	
Total number offenders	15.780
Percent female	7.92
Percent juvenile[38]	9.53
Percent foreigners	0.46

Military Expenditures and Arms Transfers [22]

	1985	1986	1987	1988	1989
Military expenditures					
Current dollars (mil.)	3	3	3	4	5
1989 constant dollars (mil.)	3	3	3	4	5
Armed forces (000)	1	1	1	1	1
Gross national product (GNP)					
Current dollars (mil.)	1,274	1,436	1,669	1,861	2,029
1989 constant dollars (mil.)	1,450	1,593	1,795	1,938	2,029
Central government expenditures (CGE)					
1989 constant dollars (mil.)	395	395	427	498	535
People (mil.)	1.0	1.0	1.0	1.1	1.1
Military expenditure as % of GNP	0.2	0.2	0.2	0.2	0.2
Military expenditure as % of CGE	0.8	0.8	0.8	0.8	0.9
Military expenditure per capita	3	3	3	4	4
Armed forces per 1,000 people	1.0	1.0	1.0	1.0	0.9
GNP per capita	1,419	1,543	1,722	1,841	1,910
Arms imports[6]					
Current dollars (mil.)	0	0	0	0	5
1989 constant dollars (mil.)	0	0	0	0	5
Arms exports[6]					
Current dollars (mil.)	0	0	0	0	0
1989 constant dollars (mil.)	0	0	0	0	0
Total imports[7]					
Current dollars (mil.)	523	676	993	1,261	1,324
1989 constant dollars	595	750	1,068	1,313	1,324
Total exports[7]					
Current dollars (mil.)	436	662	884	994	986
1989 constant dollars	496	735	951	1,035	986
Arms as percent of total imports[8]	0	0	0	0	0.4
Arms as percent of total exports[8]	0	0	0	0	0

Human Rights [24]

	SSTS	FL	FAPRO	PPCG	APROBC	TPW	PCPTW	STPEP	PHRFF	PRW	ASST	AFL
Observes	P	P			P	P	P	P		P	P	P
	EAFRD	CPR	ESCR	SR	ACHR	MAAE	PVIAC	PVNAC	EAFDAW	TCIDTP		RC
Observes		P	P	P		P	P	P	P			P

P = Party; S = Signatory; see Appendix for meaning of abbreviations.

Labor Force

Total Labor Force [25]

335,000

Labor Force by Occupation [26]

Government services	29%
Agriculture and fishing	27
Manufacturing	22
Other	22

Unemployment Rate [27]

2.4% (1991 est.)

Production Sectors

Energy Resource Summary [28]

Energy Resources: None. **Electricity**: 235,000 kW capacity; 630 million kWh produced, 570 kWh per capita (1992).

Telecommunications [30]

- Small system with good service utilizing primarily microwave radio relay
- New microwave link to Reunion
- High-frequency radio links to several countries
- Over 48,000 telephones
- Broadcast stations - 2 AM, no FM, 4 TV
- 1 Indian Ocean INTELSAT earth station

Transportation [31]

Highways. 1,800 km total; 1,640 km paved, 160 km earth

Merchant Marine. 7 ships (1,000 GRT or over) totaling 103,328 GRT/163,142 DWT; includes 3 cargo, 1 liquefied gas, 3 bulk

Airports

Total:	5
Usable:	4
With permanent-surface runways:	2
With runways over 3,659 m:	0
With runways 2,440-3,659 m:	1
With runways 1,220-2,439 m:	0

Top Agricultural Products [32]

Agriculture accounts for 10% of GDP; about 90% of cultivated land in sugarcane; other products—tea, corn, potatoes, bananas, pulses, cattle, goats, fish; net food importer, especially rice and fish.

Top Mining Products [33]

Metric tons unless otherwise specified	M.t.
Lime	7,000
Salt	6,000
Sand, coral	300,000
Basalt	1,000,000

Tourism [34]

	1987	1988	1989	1990	1991
Visitors	220	256			
Tourists	208	239	263	292	301
Cruise passengers	12	14			
Tourism receipts	138	172	183	264	262
Tourism expenditures	51	64	79	94	110
Fare receipts	73	90	98	141	153
Fare expenditures	16	17	16	17	20

Tourists are in thousands, money in million U.S. dollars.

For sources, notes, and explanations, see Annotated Source Appendix, page 1035.

Manufacturing Sector

GDP and Manufacturing Summary [35]

	1980	1985	1989	1990	% change 1980-1990	% change 1989-1990
GDP (million 1980 $)	1,132	1,421	1,931	2,055	81.5	6.4
GDP per capita (1980 $)	1,170	1,392	1,805	1,898	62.2	5.2
Manufacturing as % of GDP (current prices)	15.0	20.3	19.7[e]	23.8	58.7	20.8
Gross output (million $)	633	729	1,451	1,730	173.3	19.2
Value added (million $)	136	172	398	480	252.9	20.6
Value added (million 1980 $)	147	219	359[e]	368	150.3	2.5
Industrial production index	100	138[e]	177[e]	176[e]	76.0	-0.6
Employment (thousands)	43	75	115	115	167.4	0.0

Note: GDP stands for Gross Domestic Product. 'e' stands for estimated value.

Profitability and Productivity

	1980	1985	1989	1990	% change 1980-1990	% change 1989-1990
Intermediate input (%)	79	76	73	72	-8.9	-1.4
Wages, salaries, and supplements (%)	11	11	12	13	18.2	8.3
Gross operating surplus (%)	10	13	15	15	50.0	0.0
Gross output per worker ($)	14,745	9,771	12,622	15,077	2.3	19.5
Value added per worker ($)	3,163	2,307	3,458	4,180	32.2	20.9
Average wage (incl. benefits) ($)	1,654	1,063	1,566	1,904	15.1	21.6

Profitability is in percent of gross output. Productivity is in U.S. $. 'e' stands for estimated value.

Profitability - 1990

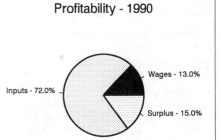

Wages - 13.0%
Inputs - 72.0%
Surplus - 15.0%

The graphic shows percent of gross output.

Value Added in Manufacturing

	1980 $ mil.	1980 %	1985 $ mil.	1985 %	1989 $ mil.	1989 %	1990 $ mil.	1990 %	% change 1980-1990	% change 1989-1990
311 Food products	36	26.5	43	25.0	67	16.8	80	16.7	122.2	19.4
313 Beverages	10	7.4	7	4.1	21	5.3	24	5.0	140.0	14.3
314 Tobacco products	2	1.5	4	2.3	6	1.5	8	1.7	300.0	33.3
321 Textiles	9	6.6	10	5.8	22	5.5	28	5.8	211.1	27.3
322 Wearing apparel	28	20.6	68	39.5	172	43.2	204	42.5	628.6	18.6
323 Leather and fur products	1	0.7	1	0.6	5	1.3	5	1.0	400.0	0.0
324 Footwear	2	1.5	2	1.2	3	0.8	3	0.6	50.0	0.0
331 Wood and wood products	1	0.7	1	0.6	2	0.5	5	1.0	400.0	150.0
332 Furniture and fixtures	2	1.5	1	0.6	3	0.8	4	0.8	100.0	33.3
341 Paper and paper products	1	0.7	2	1.2	2	0.5	3	0.6	200.0	50.0
342 Printing and publishing	5	3.7	4	2.3	8	2.0	12	2.5	140.0	50.0
351 Industrial chemicals	3	2.2	3	1.7	10	2.5	12	2.5	300.0	20.0
352 Other chemical products	4	2.9	4	2.3	8	2.0	10	2.1	150.0	25.0
353 Petroleum refineries	NA	0.0	NA	0.0	NA	0.0	NA	0.0	NA	NA
354 Miscellaneous petroleum and coal products	NA	0.0	NA	0.0	NA	0.0	NA	0.0	NA	NA
355 Rubber products	1	0.7	1	0.6	1	0.3	2	0.4	100.0	100.0
356 Plastic products	1	0.7	2	1.2	5	1.3	7	1.5	600.0	40.0
361 Pottery, china and earthenware	NA	0.0	NA	0.0	NA	0.0	NA	0.0	NA	NA
362 Glass and glass products	NA	0.0	NA	0.0	NA	0.0	NA	0.0	NA	NA
369 Other non-metal mineral products	6	4.4	4	2.3	8	2.0	11	2.3	83.3	37.5
371 Iron and steel	3	2.2	2	1.2	5	1.3	4	0.8	33.3	-20.0
372 Non-ferrous metals	NA	0.0	NA	0.0	NA	0.0	NA	0.0	NA	NA
381 Metal products	5	3.7	3	1.7	10	2.5	14	2.9	180.0	40.0
382 Non-electrical machinery	3	2.2	1	0.6	4	1.0	4	0.8	33.3	0.0
383 Electrical machinery	3	2.2	2	1.2	5	1.3	5	1.0	66.7	0.0
384 Transport equipment	2	1.5	1	0.6	4	1.0	4	0.8	100.0	0.0
385 Professional and scientific equipment	2	1.5	3	1.7	13	3.3	13	2.7	550.0	0.0
390 Other manufacturing industries	4	2.9	5	2.9	13	3.3	16	3.3	300.0	23.1

Note: The industry codes shown are International Standard Industry codes (ISIC). Percentages are percent of total Value Added. 'e' stands for estimated value

Finance, Economics, and Trade

Economic Indicators [36]

National product: GDP—exchange rate conversion—$2.5 billion (FY91 est.). **National product real growth rate**: 6.1% (FY91 est.). **National product per capita**: $2,300 (FY91 est.). **Inflation rate (consumer prices)**: 7% (FY91). **External debt**: $869 million (1991 est.).

Balance of Payments Summary [37]

Values in millions of dollars.

	1987	1988	1989	1990	1991
Exports of goods (f.o.b.)	891.4	997.9	993.1	1,205.3	1,194.5
Imports of goods (f.o.b.)	-907.6	-1,165.7	-1,204.2	-1,474.9	-1,436.2
Trade balance	-16.2	-167.8	-211.1	-269.6	-241.7
Services - debits	-319.0	-392.9	-423.3	-519.7	-552.6
Services - credits	332.4	404.2	454.4	572.6	651.2
Private transfers (net)	40.6	71.5	68.2	82.3	79.5
Government transfers (net)	25.4	21.5	7.3	14.5	2.2
Long term capital (net)	81.3	148.6	107.5	144.2	78.3
Short term capital (net)	-21.8	-21.1	-56.9	-5.6	-39.8
Errors and omissions	96.3	121.6	198.9	213.2	213.7
Overall balance	219.0	185.5	145.0	231.9	190.8

Exchange Rates [38]

Currency: **Mauritian rupees.**
Symbol: **MauRs.**

Data are currency units per $1.

January 1993	16.982
1992	15.563
1991	15.652
1990	14.839
1989	15.250
1988	13.438

Imports and Exports

Top Import Origins [39]

$1.6 billion (f.o.b., 1990).

Origins	%
EC	NA
US	NA
South Africa	NA
Japan	NA

Top Export Destinations [40]

$1.2 billion (f.o.b., 1990) EC and US have preferential treatment.

Destinations	%
EC	77
US	15

Foreign Aid [41]

	U.S. $	
US commitments, including Ex-Im (FY70-89)	76	million
Western (non-US) countries (1970-89)	709	million
Communist countries (1970-89)	54	million

Import and Export Commodities [42]

Import Commodities

Manufactured goods 50%
Capital equipment 17%
Foodstuffs 13%
Petroleum products 8%
Chemicals 7%

Export Commodities

Textiles 44%
Sugar 40%
Light manufactures 10%

            For sources, notes, and explanations, see Annotated Source Appendix, page 1035.

Mexico

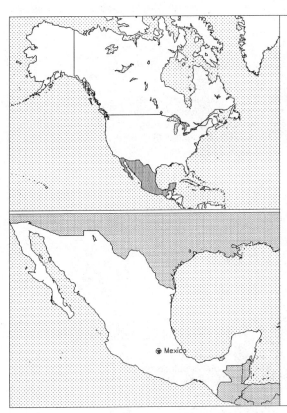

Geography [1]

Total area:
1,972,550 km2
Land area:
1,923,040 km2
Comparative area:
Slightly less than three times the size of Texas
Land boundaries:
Total 4,538 km, Belize 250 km, Guatemala 962 km, US 3,326 km
Coastline:
9,330 km
Climate:
Varies from tropical to desert
Terrain:
High, rugged mountains, low coastal plains, high plateaus, and desert
Natural resources:
Petroleum, silver, copper, gold, lead, zinc, natural gas, timber
Land use:
Arable land:
12%
Permanent crops:
1%
Meadows and pastures:
39%
Forest and woodland:
24%
Other:
24%

Demographics [2]

	1960	1970	1980	1990	1991[1]	1994	2000	2010	2020
Population	38,579	52,236	68,686	85,121	90,007	92,202	102,912	120,115	136,096
Population density (persons per sq. mi.)	52	71	94	119	121	NA	146	174	199
Births	NA	NA	NA	NA	2,580	2,505	NA	NA	NA
Deaths	NA	NA	NA	NA	438	436	NA	NA	NA
Life expectancy - males	NA	NA	NA	NA	68	69	NA	NA	NA
Life expectancy - females	NA	NA	NA	NA	76	77	NA	NA	NA
Birth rate (per 1,000)	NA	NA	NA	NA	29	27	NA	NA	NA
Death rate (per 1,000)	NA	NA	NA	NA	5	5	NA	NA	NA
Women of reproductive age (15-44 yrs.)	NA	NA	NA	21,559	NA	23,980	27,450	NA	NA
of which are currently married	NA	NA	NA	12,857	NA	14,553	17,141	NA	NA
Fertility rate	NA	NA	NA	3.5	3.36	3.2	2.8	2.4	2.2

Population values are in thousands, life expectancy in years, and other items as indicated.

Health

Health Personnel [3]

Doctors per 1,000 pop., 1988-92	0.54
Nurse-to-doctor ratio, 1988-92	0.8
Hospital beds per 1,000 pop., 1985-90	1.3
Percentage of children immunized (age 1 yr. or less)	
Third dose of DPT, 1990-91	64.0
Measles, 1990-91	78.0

Health Expenditures [4]

Total health expenditure, 1990 (official exchange rate)	
Millions of dollars	7648
Millions of dollars per capita	89
Health expenditures as a percentage of GDP	
Total	3.2
Public sector	1.6
Private sector	1.6
Development assistance for health	
Total aid flows (millions of dollars)[1]	65
Aid flows per capita (millions of dollars)	0.8
Aid flows as a percentage of total health expenditure	0.9

For sources, notes, and explanations, see Annotated Source Appendix, page 1035.

613

Human Factors

Health Care Ratios [5]

Population per physician, 1970	1480
Population per physician, 1990	NA
Population per nursing person, 1970	1610
Population per nursing person, 1990	NA
Percent of births attended by health staff, 1985	NA

Infants and Malnutrition [6]

Percent of babies with low birth weight, 1985	15
Infant mortality rate per 1,000 live births, 1970	72
Infant mortality rate per 1,000 live births, 1991	36
Years of life lost per 1,000 population, 1990	17
Prevalence of malnutrition (under age 5), 1990	14

Ethnic Division [7]

Mestizo	60%
Amerindian	30%
Caucasian	9%
Other	1%

Religion [8]

Nominally Roman Catholic	89%
Protestant	6%

Major Languages [9]

Spanish, various Mayan dialects.

Education

Public Education Expenditures [10]

Billion Pesos	1980	1985	1987	1988	1989	1990
Total education expenditure	204.3	1,767.3	NA	12,581.8	17,630.4	26,641.0
as percent of GNP	4.7	3.9	NA	3.3	3.5	4.1
as percent of total govt. expend.	NA	NA	NA	NA	NA	NA
Current education expenditure[10]	129.8	1,229.3	NA	9,584.9	12,330.5	16,617.5
as percent of GNP	3.0	2.7	NA	2.5	2.5	2.6
as percent of current govt. expend.	NA	NA	NA	NA	NA	NA
Capital expenditure[10]	10.2	99.9	NA	527.5	667.7	1,045.0

Educational Attainment [11]

Age group	25+
Total population	31,188,180
Highest level attained (%)	
No schooling	18.8
First level	
Incompleted	28.6
Completed	19.9
Entered second level	
S-1	12.7
S-2	10.7
Post secondary	9.2

Literacy Rate [12]

In thousands and percent	1985[a]	1991[a]	2000[a]
Illiterate population +15 years	7,175	7,066	6,488
Illiteracy rate - total pop. (%)	15.3	12.7	9.0
Illiteracy rate - males (%)	12.5	10.5	7.5
Illiteracy rate - females (%)	18.0	14.9	10.5

Libraries [13]

	Admin. Units	Svc. Pts.	Vols. (000)	Shelving (meters)	Vols. Added	Reg. Users
National (1989)	1	NA	1,500	NA	17,169	110,313
Non-specialized	NA	NA	NA	NA	NA	NA
Public (1989)	2,269	2,269	9,875	NA	953,554	39 mil
Higher ed. (1987)	770	770	8,347	NA	385,750	31 mil
School (1990)	3,546	3,546	9,844	NA	533,898	4.4mil

Daily Newspapers [14]

	1975	1980	1985	1990
Number of papers	216	317	332	285[e]
Circ. (000)	5,499	8,322	9,964	11,237[e]

Cinema [15]

Science and Technology

Scientific/Technical Forces [16]

R&D Expenditures [17]

	Peso[12] (000) 1989
Total expenditure	1,050,283
Capital expenditure	NA
Current expenditure	NA
Percent current	NA

U.S. Patents Issued [18]

Values show patents issued to citizens of the country by the U.S. Patents Office.

	1990	1991	1992
Number of patents	34	41	45

For sources, notes, and explanations, see Annotated Source Appendix, page 1035.

Government and Law

Organization of Government [19]

Long-form name:
United Mexican States
Type:
federal republic operating under a centralized government
Independence:
16 September 1810 (from Spain)
Constitution:
5 February 1917
Legal system:
mixture of US constitutional theory and civil law system; judicial review of legislative acts; accepts compulsory ICJ jurisdiction, with reservations
National holiday:
Independence Day, 16 September (1810)
Executive branch:
president, Cabinet
Legislative branch:
bicameral National Congress (Congreso de la Union) consists of an upper chamber or Senate (Camara de Senadores) and a lower chamber or Chamber of Deputies (Camara de Diputados)
Judicial branch:
Supreme Court of Justice (Corte Suprema de Justicia)

Crime [23]

Political Parties [20]

Chamber of Deputies	% of votes
Institutional Revolutionary Party (PRI)	53.0
National Action Party (PAN)	20.0
Cardenist Front (PFCRN)	10.0
Popular Socialist Party (PPS)	6.0
Authentic Party of the Mexican Rev. (PARM)	7.0
PMS (now part of PRD)	4.0

Government Expenditures [21]

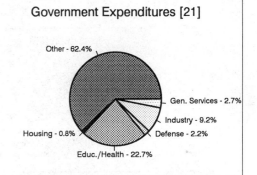

Other - 62.4%
Gen. Services - 2.7%
Industry - 9.2%
Defense - 2.2%
Educ./Health - 22.7%
Housing - 0.8%

Military Expenditures and Arms Transfers [22]

	1985	1986	1987	1988	1989
Military expenditures					
Current dollars (mil.)	1,049	967	937	962	875
1989 constant dollars (mil.)	1,194	1,073	1,008	1,002	875
Armed forces (000)	140	141[e]	141[e]	154	154
Gross national product (GNP)					
Current dollars (mil.)	159,200	155,400	164,700	174,100	186,700
1989 constant dollars (mil.)	181,200	172,500	177,100	181,300	186,700
Central government expenditures (CGE)					
1989 constant dollars (mil.)	47,440	53,200	57,890	52,010	38,420
People (mil.)	78.5	80.3	82.2	84.1	86.0
Military expenditure as % of GNP	0.7	0.6	0.6	0.6	0.5
Military expenditure as % of CGE	2.5	2.0	1.7	1.9	2.3
Military expenditure per capita	15	13	12	12	10
Armed forces per 1,000 people	1.8	1.8	1.7	1.8	1.8
GNP per capita	2,310	2,148	2,155	2,155	2,170
Arms imports[6]					
Current dollars (mil.)	35	100	240	60	20
1989 constant dollars (mil.)	34	111	258	62	20
Arms exports[6]					
Current dollars (mil.)	10	10	10	10	10
1989 constant dollars (mil.)	11	11	11	10	10
Total imports[7]					
Current dollars (mil.)	13,990	12,000	12,730	19,590	24,440
1989 constant dollars	15,930	13,310	13,690	20,390	24,440
Total exports[7]					
Current dollars (mil.)	22,110	16,350	20,890	20,760	23,050
1989 constant dollars	25,170	18,140	22,460	21,620	23,050
Arms as percent of total imports[8]	0.2	0.8	1.9	0.3	0.1
Arms as percent of total exports[8]	0	0.1	0	0	0

Human Rights [24]

	SSTS	FL	FAPRO	PPCG	APROBC	TPW	PCPTW	STPEP	PHRFF	PRW	ASST	AFL
Observes	P	P	P	P		P	P	P		P	P	P
		EAFRD	CPR	ESCR	SR	ACHR	MAAE	PVIAC	PVNAC	EAFDAW	TCIDTP	RC
Observes		P	P	P		P		P		P	P	P

P = Party; S = Signatory; see Appendix for meaning of abbreviations.

Labor Force

Total Labor Force [25]

26.2 million (1990)

Labor Force by Occupation [26]

Services
Agriculture, forestry, hunting, and fishing
Commerce
Manufacturing
Construction
Transportation
Mining and quarrying

Unemployment Rate [27]

14%-17% (1991 est.)

Production Sectors

Commercial Energy Production and Consumption

Production [28]

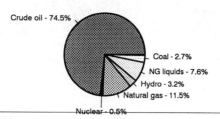

Crude oil - 74.5%
Coal - 2.7%
NG liquids - 7.6%
Hydro - 3.2%
Natural gas - 11.5%
Nuclear - 0.5%

Consumption [29]

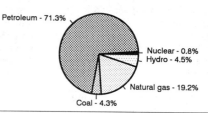

Petroleum - 71.3%
Nuclear - 0.8%
Hydro - 4.5%
Natural gas - 19.2%
Coal - 4.3%

Data are shown in quadrillion (10^{15}) BTUs and percent for 1991

Crude oil[1]	5.87
Natural gas liquids	0.60
Dry natural gas	0.91
Coal[2]	0.21
Net hydroelectric power[3]	0.25
Net nuclear power[3]	0.04
Total	7.88

Petroleum	3.67
Dry natural gas	0.99
Coal[2]	0.22
Net hydroelectric power[3]	0.23
Net nuclear power[3]	0.04
Total	5.17

Telecommunications [30]

- Highly developed system with extensive microwave radio relay links
- Privatized in December 1990
- Connected into Central America Microwave System
- 6,410,000 telephones
- Broadcast stations - 679 AM, no FM, 238 TV, 22 shortwave
- 120 domestic satellite terminals
- Earth stations - 4 Atlantic Ocean INTELSAT and 1 Pacific Ocean INTELSAT

Transportation [31]

Railroads. 24,500 km total

Highways. 212,000 km total; 65,000 km paved, 30,000 km semipaved or cobblestone, 62,000 km rural roads (improved earth) or roads under construction, 55,000 km unimproved earth roads

Merchant Marine. 58 ships (1,000 GRT or over) totaling 858,162 GRT/1,278,488 DWT; includes 4 short-sea passenger, 2 cargo, 2 refrigerated cargo, 2 roll-on/roll-off, 31 oil tanker, 4 chemical tanker, 7 liquefied gas, 1 bulk, 5 container

Airports

Total:	1,841
Usable:	1,478
With permanent-surface runways:	200
With runways over 3,659 m:	3
With runways 2,440-3,659 m:	35
With runways 1,220-2,439 m:	273

Top Agricultural Products [32]

	90-91	91-92
Corn	14,100	14,500
Wheat	3,900	3,700
Oranges	2,300	2,050
Tomatoes	1,800	1,620
Dry beans	1,300	1,000
Coffee	273	291

Values shown are 1,000 metric tons.

Top Mining Products [33]

Preliminary metric tons unless otherwise specified M.t.

Silver, mine output, Ag content (kilograms)	2,223647
Celestite	62,180
Antimony, mine output, Sb content[27]	2,753
White arsenic[28]	4,922
Bismuth[29]	651
Cadmium, mine output, Cd content	1,797
Fluorspar (000 tons)	370
Graphite, natural (amorphous and crystalline)	37,258
Mercury, mine output, Hg content	1,176

Tourism [34]

	1987	1988	1989	1990	1991
Visitors	63,182	71,960	74,301	82,409	82,755
Tourists[1]		14,142	14,964	17,176	16,560
Excursionists[45]		56,668	58,120	64,034	64,566
Tourism receipts[46]		2,902	3,388	3,934	4,355
Tourism expenditures[47]		1,323	1,750	2,171	2,146
Fare receipts	429	386	356	440	441
Fare expenditures	232	299	388	476	516

Tourists are in thousands, money in million U.S. dollars.

Manufacturing Sector

GDP and Manufacturing Summary [35]

	1980	1985	1989	1990	% change 1980-1990	% change 1989-1990
GDP (million 1980 $)	194,766	214,370	217,876	229,010	17.6	5.1
GDP per capita (1980 $)	2,766	2,701	2,512	2,582	-6.7	2.8
Manufacturing as % of GDP (current prices)	21.9	23.1	24.6	23.0	5.0	-6.5
Gross output (million $)	102,047	106,972[e]	126,238[e]	132,792[e]	30.1	5.2
Value added (million $)	43,048	46,373[e]	54,706[e]	57,482[e]	33.5	5.1
Value added (million 1980 $)	43,200	45,924	49,050	52,416	21.3	6.9
Industrial production index	100	103	104	105	5.0	1.0
Employment (thousands)	2,417	2,314[e]	2,005[e]	2,145[e]	-11.3	7.0

Note: GDP stands for Gross Domestic Product. 'e' stands for estimated value.

Profitability and Productivity

	1980	1985	1989	1990	% change 1980-1990	% change 1989-1990
Intermediate input (%)	58	57[e]	57[e]	57[e]	-1.7	0.0
Wages, salaries, and supplements (%)	14	9[e]	6[e]	9[e]	-35.7	50.0
Gross operating surplus (%)	28	34[e]	37[e]	35[e]	25.0	-5.4
Gross output per worker ($)	42,221	46,227[e]	62,970[e]	61,903[e]	46.6	-1.7
Value added per worker ($)	17,811	20,040[e]	27,289[e]	26,796[e]	50.4	-1.8
Average wage (incl. benefits) ($)	5,846	4,192[e]	3,925[e]	5,373[e]	-8.1	36.9

Profitability is in percent of gross output. Productivity is in U.S. $. 'e' stands for estimated value.

Profitability - 1990

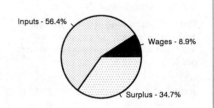

Inputs - 56.4%

Wages - 8.9%

Surplus - 34.7%

The graphic shows percent of gross output.

Value Added in Manufacturing

	1980 $ mil.	1980 %	1985 $ mil.	1985 %	1989 $ mil.	1989 %	1990 $ mil.	1990 %	% change 1980-1990	% change 1989-1990
311 Food products	6,989	16.2	7,015[e]	15.1	7,084[e]	12.9	8,661[e]	15.1	23.9	22.3
313 Beverages	2,723	6.3	2,589[e]	5.6	2,922[e]	5.3	3,299[e]	5.7	21.2	12.9
314 Tobacco products	623	1.4	740[e]	1.6	892[e]	1.6	793[e]	1.4	27.3	-11.1
321 Textiles	3,133	7.3	3,099[e]	6.7	3,208[e]	5.9	3,075[e]	5.3	-1.9	-4.1
322 Wearing apparel	1,277	3.0	1,094[e]	2.4	987[e]	1.8	1,198[e]	2.1	-6.2	21.4
323 Leather and fur products	366	0.9	397[e]	0.9	543[e]	1.0	347[e]	0.6	-5.2	-36.1
324 Footwear	845	2.0	658[e]	1.4	552[e]	1.0	575[e]	1.0	-32.0	4.2
331 Wood and wood products	919	2.1	786[e]	1.7	975[e]	1.8	845[e]	1.5	-8.1	-13.3
332 Furniture and fixtures	784	1.8	498[e]	1.1	366[e]	0.7	565[e]	1.0	-27.9	54.4
341 Paper and paper products	1,189	2.8	1,180[e]	2.5	1,319[e]	2.4	1,660[e]	2.9	39.6	25.9
342 Printing and publishing	1,050	2.4	1,250[e]	2.7	1,447[e]	2.6	1,654[e]	2.9	57.5	14.3
351 Industrial chemicals	2,235	5.2	2,982[e]	6.4	4,652[e]	8.5	3,801[e]	6.6	70.1	-18.3
352 Other chemical products	2,235	5.2	2,562[e]	5.5	3,115[e]	5.7	4,124[e]	7.2	84.5	32.4
353 Petroleum refineries	1,917	4.5	4,341[e]	9.4	7,265[e]	13.3	5,533[e]	9.6	188.6	-23.8
354 Miscellaneous petroleum and coal products	222	0.5	529[e]	1.1	705[e]	1.3	679[e]	1.2	205.9	-3.7
355 Rubber products	767	1.8	1,164[e]	2.5	1,465[e]	2.7	1,201[e]	2.1	56.6	-18.0
356 Plastic products	754	1.8	767[e]	1.7	822[e]	1.5	1,074[e]	1.9	42.4	30.7
361 Pottery, china and earthenware	383	0.9	420[e]	0.9	490[e]	0.9	398[e]	0.7	3.9	-18.8
362 Glass and glass products	566	1.3	529[e]	1.1	689[e]	1.3	709[e]	1.2	25.3	2.9
369 Other non-metal mineral products	1,464	3.4	1,113[e]	2.4	1,192[e]	2.2	1,044[e]	1.8	-28.7	-12.4
371 Iron and steel	2,070	4.8	2,227[e]	4.8	3,389[e]	6.2	2,713[e]	4.7	31.1	-19.9
372 Non-ferrous metals	562	1.3	506[e]	1.1	634[e]	1.2	597[e]	1.0	6.2	-5.8
381 Metal products	1,961	4.6	1,849[e]	4.0	1,913[e]	3.5	2,384[e]	4.1	21.6	24.6
382 Non-electrical machinery	2,074	4.8	1,643[e]	3.5	1,600[e]	2.9	2,030[e]	3.5	-2.1	26.9
383 Electrical machinery	1,900	4.4	1,635[e]	3.5	1,768[e]	3.2	1,907[e]	3.3	0.4	7.9
384 Transport equipment	2,980	6.9	3,621[e]	7.8	3,044[e]	5.6	4,915[e]	8.6	64.9	61.5
385 Professional and scientific equipment	305	0.7	381[e]	0.8	514[e]	0.9	674[e]	1.2	121.0	31.1
390 Other manufacturing industries	754	1.8	798[e]	1.7	1,153[e]	2.1	1,024[e]	1.8	35.8	-11.2

Note: The industry codes shown are International Standard Industry codes (ISIC). Percentages are percent of total Value Added. 'e' stands for estimated value

Finance, Economics, and Trade

Economic Indicators [36]

	1989	1990	1991[e]
GDP (bil. current US$)	208.5	238.2	280.0
Per Capita GDP (Curr US$)	2,623.0	2,937.0	3,386.0
Real GDP Growth Rate	3.1	3.9	4.5
Money Supply (M1 growth rate)	40.6	62.6	66.0
Commercial Interest Rates			
(Annual Perc. Comm. Paper)	45.9	36.1	23.0
Savings Rate (M4 as % of GDP)	36.0	39.0	41.0
Investment Rate (Perc. GDP)	17.4	18.9	19.0
CPI (Dec.-Dec. Growth Rate)	19.7	29.9	17.0
WPI (Dec.-Dec. Growth Rate)	15.6	29.2	14.0
External Public Debt	76.1	77.8	78.5

Balance of Payments Summary [37]

Values in millions of dollars.

	1987	1988	1989	1990	1991
Exports of goods (f.o.b.)	20,655.0	20,566.0	2,276.5	26,838.0	27,121.0
Imports of goods (f.o.b.)	-12,222.0	-18,898.0	-23,410.0	-31,271.0	-38,184.0
Trade balance	8,433.0	1,668.0	-645.0	-4,433.0	-11,063.0
Services - debits	-14,357.0	-16,119.0	-18,592.0	-20,896.0	-20,877.0
Services - credits	9,244.0	11,441.0	13,204.0	14,749.0	16,417.0
Private transfers (net)	384.0	397.0	1,922.0	2,167.0	2,055.0
Government transfers (net)	264.0	170.0	153.0	1,296.0	186.0
Long term capital (net)	4,044.0	-677.0	2,298.0	5,681.0	17,724.0
Short term capital (net)	-5,047.0	-678.0	-936.0	2,849.0	2,679.0
Errors and omissions	2,605.0	-2,840.0	2,775.0	890.0	871.0
Overall balance	5,570.0	-6,638.0	179.0	2,303.0	7,992.0

Exchange Rates [38]

Currency: **Mexican pesos.**
Symbol: **Mex$.**

The new pesos replaced the old pesos on 1 January 1993; 1 new pesos = 1,000 old pesos. Data are currency units per $1.

January 1993	3.100.0
November 1992	3,198.0
1991	3,018.4
1990	2,812.6
1989	2,461.3
1988	2,273.1

Imports and Exports

Top Import Origins [39]

$48.1 billion (c.i.f., 1992 est.). Data are for 1992.

Origins	%
US	74
Japan	11
EC	6

Top Export Destinations [40]

$27.5 billion (f.o.b., 1992 est.). Data are for 1992 est.

Destinations	%
US	74
Japan	8
EC	4

Foreign Aid [41]

	U.S. $	
US commitments, including Ex-Im (FY70-89)	3.1	billion
Western (non-US) countries, ODA and OOF bilateral commitments (1970-89)	7.7	billion
Communist countries (1970-89)	110	million

Import and Export Commodities [42]

Import Commodities	**Export Commodities**
Metal-working machines	Crude oil
Steel mill products	Oil products
Agricultural machinery	Coffee
Electrical equipment	Shrimp
Car parts for assembly	Engines
Repair parts for motor vehicles	Motor vehicles
Aircraft	Cotton
And aircraft parts	Consumer electronics

Micronesia, Federated States of

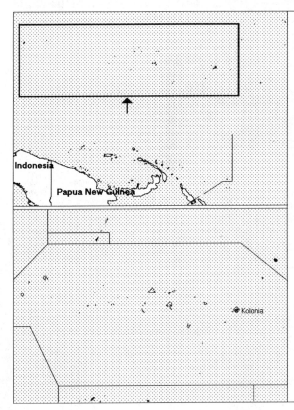

Geography [1]

Total area:
702 km2
Land area:
702 km2
Comparative area:
Slightly less than four times the size of Washington, DC
Land boundaries:
0 km
Coastline:
6,112 km
Climate:
Tropical; heavy year-round rainfall, especially in the eastern islands;
located on southern edge of the typhoon belt with occasional severe damage
Terrain:
Islands vary geologically from high mountainous islands to low, coral
atolls; volcanic outcroppings on Pohnpei, Kosrae, and Truk
Natural resources:
Forests, marine products, deep-seabed minerals
Land use:
Arable land:
NA%
Permanent crops:
NA%
Meadows and pastures:
NA%
Forest and woodland:
NA%
Other:
NA%

Demographics [2]

	1960	1970	1980	1990	1991[1]	1994	2000	2010	2020
Population	42	57	77	109	108	120	133	141	143
Population density (persons per sq. mi.)	154	210	284	387	397	NA	462	480	483
Births	NA	NA	NA	NA	4	3	NA	NA	NA
Deaths	NA	NA	NA	NA	1	1	NA	NA	NA
Life expectancy - males	NA	NA	NA	NA	NA	66	NA	NA	NA
Life expectancy - females	NA	NA	NA	NA	NA	70	NA	NA	NA
Birth rate (per 1,000)	NA	NA	NA	NA	34	28	NA	NA	NA
Death rate (per 1,000)	NA	NA	NA	NA	5	6	NA	NA	NA
Women of reproductive age (15-44 yrs.)	NA	NA	NA	NA	NA	NA	NA	NA	NA
of which are currently married	NA	NA	NA	NA	NA	NA	NA	NA	NA
Fertility rate	NA	NA	NA	4.2	4.97	4.0	3.8	NA	NA

Population values are in thousands, life expectancy in years, and other items as indicated.

Health

Health Personnel [3]

Health Expenditures [4]

For sources, notes, and explanations, see Annotated Source Appendix, page 1035.

619

Human Factors	
Health Care Ratios [5]	Infants and Malnutrition [6]

Ethnic Division [7]	Religion [8]	Major Languages [9]	
Nine ethnic Micronesian and Polynesian groups.	Christian (divided between Roman Catholic and Protestant; other churches include Assembly of God, Jehovah's Witnesses, Seventh-Day Adventist, Latter-Day Saints, and the Baha'i Faith).	English (official common language)	NA
		Trukese	NA
		Pohnpeian	NA
		Yapese	NA
		Kosrean	NA

Education

Public Education Expenditures [10]	Educational Attainment [11]
Literacy Rate [12]	Libraries [13]
Daily Newspapers [14]	Cinema [15]

Science and Technology

Scientific/Technical Forces [16]	R&D Expenditures [17]	U.S. Patents Issued [18]

Government and Law

Organization of Government [19]

Long-form name:
Federated States of Micronesia
Type:
constitutional government in free
association with the US; the Compact of
Free Association entered into force 3
November 1986
Independence:
3 November 1986 (from the US-
administered UN Trusteeship)
Constitution:
10 May 1979
Legal system:
based on adapted Trust Territory laws,
acts of the legislature, municipal,
common, and customary laws
National holiday:
Proclamation of the Federated States of
Micronesia, 10 May (1979)
Executive branch:
president, vice president, Cabinet
Legislative branch:
unicameral Congress
Judicial branch:
Supreme Court

Elections [20]

Congress. No formal parties; last held
on 5 March 1991 (next to be held March
1993); results—percent of vote NA;
seats—(14 total).

Government Budget [21]

For 1988.

Revenues	165
Expenditures	115
Capital expenditures	20

Defense Summary [22]

Note: Defense is the responsibility of the United States.

Crime [23]

Human Rights [24]

Labor Force

Total Labor Force [25]

Labor Force by Occupation [26]

Two-thirds are government employees

Unemployment Rate [27]

For sources, notes, and explanations, see Annotated Source Appendix, page 1035.

621

Production Sectors

Energy Resource Summary [28]

Energy Resources: None. **Electricity**: 18,000 kW capacity; 40 million kWh produced, 380 kWh per capita (1990).

Telecommunications [30]

- Telephone network - 960 telephone lines total at Kolonia and Truk
- Islands interconnected by shortwave radio (used mostly for government purposes)
- 16,000 radio receivers, 1,125 TV sets (est. 1987)
- Broadcast stations - 5 AM, 1 FM, 6 TV, 1 shortwave
- 4 Pacific Ocean INTELSAT earth stations

Transportation [31]

Highways. 39 km of paved roads on major islands; also 187 km stone-, coral-, or laterite-surfaced roads

Airports

Total:	6
Usable:	5
With permanent-surface runways:	4
With runways over 3,659 m:	0
With runways 2,440-3,659 m:	0
With runways 1,220-2,439 m:	4

Top Agricultural Products [32]

Mainly a subsistence economy; black pepper; tropical fruits and vegetables, coconuts, cassava, sweet potatoes, pigs, chickens.

Top Mining Products [33]

Detailed information is not available. A summary of mineral resources available follows. **Mineral Resources**: None.

Tourism [34]

Finance, Economics, and Trade

Industrial Summary [35]

Industrial Production: Growth rate NA%. **Industries**: Tourism, construction, fish processing, craft items from shell, wood, and pearls.

Economic Indicators [36]

National product: GNP—purchasing power equivalent—$150 million (1989 est.). **National product real growth rate**: NA%. **National product per capita**: $1,500 (1989 est.). **Inflation rate (consumer prices)**: NA%. **External debt**: not available.

Balance of Payments Summary [37]

Exchange Rates [38]

US currency is used.

Imports and Exports

Top Import Origins [39]

$67.7 million (c.i.f., 1988).

Origins	%
NA	NA

Top Export Destinations [40]

$2.3 million (f.o.b., 1988).

Destinations	%
NA	NA

Foreign Aid [41]

	U.S. $	
US grant aid (under the Compact of Free Association) - 1986-2001	1.3	billion

Import and Export Commodities [42]

Import Commodities	**Export Commodities**
No import data available.	Copra

For sources, notes, and explanations, see Annotated Source Appendix, page 1035.

623

Moldova

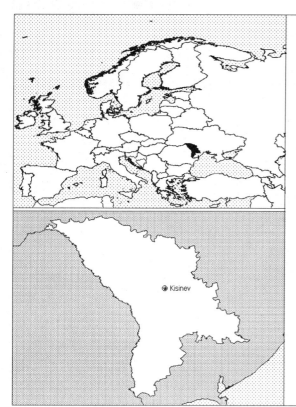

Geography [1]

Total area:
33,700 km2
Land area:
33,700 km2
Comparative area:
Slightly more than twice the size of Hawaii
Land boundaries:
Total 1,389 km, Romania 450 km, Ukraine 939 km
Coastline:
0 km (landlocked)
Climate:
Mild winters, warm summers
Terrain:
Rolling steppe, gradual slope south to Black Sea
Natural resources:
Lignite, phosphorites, gypsum
Land use:
Arable land:
50%
Permanent crops:
0%
Meadows and pastures:
9%
Forest and woodland:
0%
Other:
41%

Demographics [2]

	1960	1970	1980	1990	1991	1994	2000	2010	2020
Population	2,999	3,595	3,996	4,389	NA	4,473	4,565	4,738	4,880
Population density (persons per sq. mi.)	NA	NA	NA	NA	NA	NA	NA	NA	NA
Births	NA	NA	NA	NA	NA	72	NA	NA	NA
Deaths	NA	NA	NA	NA	NA	45	NA	NA	NA
Life expectancy - males	NA	NA	NA	NA	NA	65	NA	NA	NA
Life expectancy - females	NA	NA	NA	NA	NA	72	NA	NA	NA
Birth rate (per 1,000)	NA	NA	NA	NA	NA	16	NA	NA	NA
Death rate (per 1,000)	NA	NA	NA	NA	NA	10	NA	NA	NA
Women of reproductive age (15-44 yrs.)	NA	NA	NA	1,105	NA	1,142	1,208	NA	NA
of which are currently married	NA	NA	NA	774	NA	796	831	NA	NA
Fertility rate	NA	NA	NA	2.5	NA	2.2	2.1	1.9	1.8

Population values are in thousands, life expectancy in years, and other items as indicated.

Health

Health Personnel [3]

Doctors per 1,000 pop., 1988-92	4.00
Nurse-to-doctor ratio, 1988-92	3.0
Hospital beds per 1,000 pop., 1985-90	7.8
Percentage of children immunized (age 1 yr. or less)	
Third dose of DPT, 1990-91	87.0
Measles, 1990-91	95.0

Health Expenditures [4]

Total health expenditure, 1990 (official exchange rate)	
Millions of dollars	623
Millions of dollars per capita	143
Health expenditures as a percentage of GDP	
Total	3.9
Public sector	2.9
Private sector	1.0
Development assistance for health	
Total aid flows (millions of dollars)[1]	NA
Aid flows per capita (millions of dollars)	NA
Aid flows as a percentage of total health expenditure	NA

For sources, notes, and explanations, see Annotated Source Appendix, page 1035.

Human Factors

Health Care Ratios [5]

Population per physician, 1970	NA
Population per physician, 1990	250
Population per nursing person, 1970	NA
Population per nursing person, 1990	NA
Percent of births attended by health staff, 1985	NA

Infants and Malnutrition [6]

Percent of babies with low birth weight, 1985	NA
Infant mortality rate per 1,000 live births, 1970	NA
Infant mortality rate per 1,000 live births, 1991	23
Years of life lost per 1,000 population, 1990	19
Prevalence of malnutrition (under age 5), 1990	NA

Ethnic Division [7]

Moldovan/Romanian	64.5%
Ukrainian	13.8%
Russian	13.0%
Gagauz	3.5%
Jewish	1.5%
Bulgarian	2.0%
Other	1.7%

Religion [8]

Eastern Orthodox	98.5%
Jewish	1.5%
Baptist (only about 1,000 members)	NA

Major Languages [9]

Moldovan (official); note - virtually the same as the Romanian language, Russian.

Education

Public Education Expenditures [10]

Educational Attainment [11]

Literacy Rate [12]

	1980[b]	1989[b]	1990[b]
Illiterate population +15 years	NA	NA	NA
Illiteracy rate - total pop. (%)	NA	3.6	NA
Illiteracy rate - males (%)	NA	1.4	NA
Illiteracy rate - females (%)	NA	5.6	NA

Libraries [13]

Daily Newspapers [14]

Cinema [15]

Science and Technology

Scientific/Technical Forces [16]

R&D Expenditures [17]

U.S. Patents Issued [18]

For sources, notes, and explanations, see Annotated Source Appendix, page 1035.

625

Government and Law

Organization of Government [19]

Long-form name:
Republic of Moldova
Type:
republic
Independence:
27 August 1991 (from Soviet Union)
Constitution:
as of mid-1993 the new constitution had not been adopted; old constitution (adopted NA 1979) is still in effect but has been heavily amended during the past few years
Legal system:
based on civil law system; no judicial review of legislative acts; does not accept compulsory ICJ jurisdiction but accepts many UN and CSCE documents
National holiday:
Independence Day, 27 August 1991
Executive branch:
president, prime minister, Cabinet of Ministers
Legislative branch:
unicameral Parliament
Judicial branch:
Supreme Court

Elections [20]

Parliament. last held 25 February 1990 (next to be held NA 1995); results - percent of vote by party NA; seats - (350 total) Christian Democratic Popular Front 50; Club of Independent Deputies 25; Agrarian Club 90; Social Democrats 60-70; Russian Conciliation Club 50; 60-70 seats belong to Dniester region deputies who usually boycott Moldovan legislative proceedings; the remaining seats filled by independents; note - until May 1991 was called Supreme Soviet.

Government Budget [21]

For 1988.
Revenues	NA
Expenditures	NA
Capital expenditures	NA

Defense Summary [22]

Branches: Ground Forces, Air and Air Defence Force, Security Forces (internal and border troops)

Manpower Availability: Males age 15-49 1,082,562; fit for military service 859,948; reach military age (18) annually 35,769 (1993 est.)

Defense Expenditures: No information available.

Crime [23]

Human Rights [24]

Labor Force

Total Labor Force [25]

2.095 million

Labor Force by Occupation [26]

Agriculture	34.4%
Industry	20.1
Other	45.5

Date of data: 1985 figures

Unemployment Rate [27]

0.7% (includes only officially registered unemployed; also large numbers of underemployed workers)

Production Sectors

Energy Resource Summary [28]

Energy Resources: Lignite. **Electricity**: 3,115,000 kW capacity; 11,100 million kWh produced, 2,491 kWh per capita (1992). **Pipelines**: Natural gas 310 km (1992).

Telecommunications [30]

- Poorly supplied with telephones (as of 1991, 494,000 telephones total, with a density of 111 lines per 1000 persons)
- 215,000 unsatisfied applications for telephone installations (31 January 1990)
- Connected to Ukraine by landline and to countries beyond the former USSR through the international gateway switch in Moscow

Transportation [31]

Railroads. 1,150 km; does not include industrial lines (1990)

Highways. 20,000 km total; 13,900 km hard-surfaced, 6,100 km earth (1990)

Airports

Total:	26
Useable:	15
With permanent-surface runways:	6
With runways over 3,659 m:	0
With runways 2,440-3,659 m:	5
With runways 1,220-2,439 m:	8

Top Agricultural Products [32]

	1990	1991
Grains	2,500	3,200
Sugar beets	2,400	2,300
Vegetables	1,177	1,153
Grapes	940	780
Fruits and berries	901	646
Potatoes	295	300

Values shown are 1,000 metric tons.

Top Mining Products [33]

Detailed information is not available. A summary of mineral resources available follows. **Mineral Resources**: Lignite, phosphorites, gypsum.

Tourism [34]

Finance, Economics, and Trade

Industrial Summary [35]

Industrial Production: Growth rate - 22% (1992). **Industries**: Key products (with share of total former Soviet output in parentheses where known): agricultural machinery, foundry equipment, refrigerators and freezers (2.7%), washing machines (5.0%), hosiery (2.0%), refined sugar (3.1%), vegetable oil (3.7%), canned food (8.6%), shoes, textiles.

Economic Indicators [36]

National product: GDP not available. **National product real growth rate**: - 26% (1992). **National product per capita**: not available. **Inflation rate (consumer prices)**: 27% per month (first quarter 1993). **External debt**: $100 million (1993 est.).

Balance of Payments Summary [37]

Exchange Rates [38]

Currency: **rubles.**
Symbol: **R.**

Subject to wide fluctuations. Data are currency units per $1.

December 24, 1992	415

Imports and Exports

Top Import Origins [39]

$100 million from outside the successor states of the former USSR (1992).

Origins	%
Russia	NA
Ukraine	NA
Uzbekistan	NA
Romania	NA

Top Export Destinations [40]

$100 million to outside the successor states of the former USSR (1992).

Destinations	%
Russia	NA
Kazakhstan	NA
Ukraine	NA
Romania	NA

Foreign Aid [41]

	U.S. $	
IMF credit (1992)	18.5	million
EC agricultural credit (1992)	30	million
US commitments for grain (1992)	10	million
World Bank credit	31	million

Import and Export Commodities [42]

Import Commodities	Export Commodities
Oil	Foodstuffs
Gas	Wine
Coal	Tobacco
Steel machinery	Textiles and footwear
Foodstuffs	Machinery
Automobiles	Chemicals
Other consumer durables	

For sources, notes, and explanations, see Annotated Source Appendix, page 1035.

Monaco

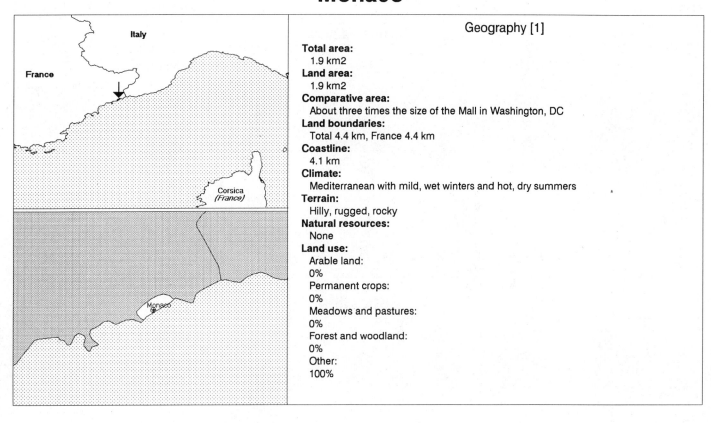

Geography [1]

Total area:
1.9 km2
Land area:
1.9 km2
Comparative area:
About three times the size of the Mall in Washington, DC
Land boundaries:
Total 4.4 km, France 4.4 km
Coastline:
4.1 km
Climate:
Mediterranean with mild, wet winters and hot, dry summers
Terrain:
Hilly, rugged, rocky
Natural resources:
None
Land use:
Arable land:
0%
Permanent crops:
0%
Meadows and pastures:
0%
Forest and woodland:
0%
Other:
100%

Demographics [2]

	1960	1970	1980	1990	1991[1]	1994	2000	2010	2020
Population	21	24	27	30	30	31	32	33	34
Population density (persons per sq. mi.)	20,757	23,752	26,567	29,453	29,712	NA	31,743	33,859	35,975
Births	NA	NA	NA	NA	(Z)	(Z)	NA	NA	NA
Deaths	NA	NA	NA	NA	(Z)	(Z)	NA	NA	NA
Life expectancy - males	NA	NA	NA	NA	72	74	NA	NA	NA
Life expectancy - females	NA	NA	NA	NA	80	82	NA	NA	NA
Birth rate (per 1,000)	NA	NA	NA	NA	7	11	NA	NA	NA
Death rate (per 1,000)	NA	NA	NA	NA	7	12	NA	NA	NA
Women of reproductive age (15-44 yrs.)	NA	NA	NA	NA	NA	NA	NA	NA	NA
of which are currently married	NA	NA	NA	NA	NA	NA	NA	NA	NA
Fertility rate	NA	NA	NA	1.7	1.12	1.7	1.7	1.7	1.7

Population values are in thousands, life expectancy in years, and other items as indicated.

Health

Health Personnel [3]

Health Expenditures [4]

For sources, notes, and explanations, see Annotated Source Appendix, page 1035.

629

Human Factors	
Health Care Ratios [5]	Infants and Malnutrition [6]

Ethnic Division [7]		Religion [8]		Major Languages [9]	
French	47%	Roman Catholic	95%	French (official)	NA
Monegasque	16%			English	NA
Italian	16%			Italian	NA
Other	21%			Monegasque	NA

Education

Public Education Expenditures [10]	Educational Attainment [11]

Literacy Rate [12]	Libraries [13]

Daily Newspapers [14]

	1975	1980	1985	1990
Number of papers	2	2	2	1
Circ. (000)	11	10	10[e]	8

Cinema [15]

Science and Technology

Scientific/Technical Forces [16]	R&D Expenditures [17]	U.S. Patents Issued [18]

Government and Law

Organization of Government [19]

Long-form name:
Principality of Monaco
Type:
constitutional monarchy
Independence:
1419 (rule by the House of Grimaldi)
Constitution:
17 December 1962
Legal system:
based on French law; has not accepted
compulsory ICJ jurisdiction
National holiday:
National Day, 19 November
Executive branch:
prince, minister of state, Council of
Government (cabinet)
Legislative branch:
unicameral National Council (Conseil
National)
Judicial branch:
Supreme Tribunal (Tribunal Supreme)

Political Parties [20]

National Council	% of seats
National and Democratic Union (UND)	100.0

Government Budget [21]

For 1991.

Revenues	424
Expenditures	376
Capital expenditures	NA

Defense Summary [22]

Note: Defense is the responsibility of France

Crime [23]

Crime volume

Cases known to police	1,383
Attempts (percent)	6.87
Percent cases solved	44.69
Crimes per 100,000 persons	4,614.31

Persons responsible for offenses

Total number offenders	705
Percent female	12.91
Percent juvenile[39]	12.91
Percent foreigners	94.47

Human Rights [24]

Labor Force

Total Labor Force [25]

Labor Force by Occupation [26]

Unemployment Rate [27]

NEGL%

For sources, notes, and explanations, see Annotated Source Appendix, page 1035.

631

Production Sectors

Energy Resource Summary [28]

Energy Resources: None **Electricity**: 10,000 kW standby capacity (1992); power imported from France.

Telecommunications [30]

- Served by cable into the French communications system
- Automatic telephone system
- 38,200 telephones
- Broadcast stations - 3 AM, 4 FM, 5 TV
- No communication satellite earth stations

Top Agricultural Products [32]

NA.

Top Mining Products [33]

Detailed information is not available. A summary of mineral resources available follows. **Mineral Resources**: None.

Transportation [31]

Railroads. 1.6 km 1.435-meter gauge

Highways. none; city streets

Merchant Marine. 1 oil tanker (1,000 GRT or over) totaling 3,268 GRT/4,959 DWT

Airports

Total:	1
Usable:	1
With permanent-surface runways:	0
With runways over 3,659 m:	0
With runways 2,440-3,659 m:	0
With runways 1,220-2,439 m:	0

Tourism [34]

	1987	1988	1989	1990	1991
Tourists[10]	214	232	245	245	239

Tourists are in thousands, money in million U.S. dollars.

Finance, Economics, and Trade

Industrial Summary [35]

Industrial Production: Growth rate NA% growth rate - 15% (1992 est.). **Industries**: Copper, processing of animal products, building materials, food and beverage, mining (particularly coal).

Economic Indicators [36]

National product: GDP—exchange rate conversion—$475 million (1991 est.). **National product real growth rate**: NA%. **National product per capita**: $16,000 (1991 est.). **Inflation rate (consumer prices)**: NA%. **External debt**: not available.

Balance of Payments Summary [37]

Exchange Rates [38]

Currency: **French francs.**
Symbol: **F.**

Data are currency units per $1.

January 1993	5.4812
1992	5.2938
1991	5.6421
1990	5.4453
1989	6.3801
1988	5.9569

Imports and Exports

Top Import Origins [39]

Full customs integration with France, which collects and rebates Monacan trade duties; also participates in EC market system through customs union with France.

Origins	%
No details available.	NA

Top Export Destinations [40]

Full customs integration with France, which collects and rebates Monacan trade duties; also participates in EC market system through customs union with France.

Destination	%
No details available.	NA

Foreign Aid [41]

	U.S. $
No aid provided or received.	NA

Import and Export Commodities [42]

Import Commodities	Export Commodities
No information available.	

For sources, notes, and explanations, see Annotated Source Appendix, page 1035.

633

Mongolia

Geography [1]

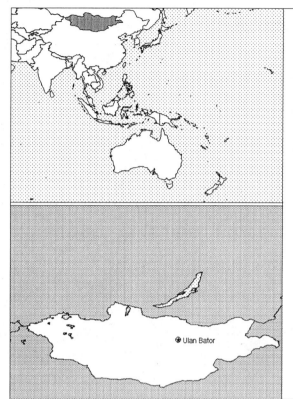

Total area:
  1.565 million km2
Land area:
  1.565 million km2
Comparative area:
  Slightly larger than Alaska
Land boundaries:
  Total 8,114 km, China 4,673 km, Russia 3,441 km
Coastline:
  0 km (landlocked)
Climate:
  Desert; continental (large daily and seasonal temperature ranges)
Terrain:
  Vast semidesert and desert plains; mountains in west and southwest; Gobi desert in southeast
Natural resources:
  Oil, coal, copper, molybdenum, tungsten, phosphates, tin, nickel, zinc, wolfram, fluorspar, gold
Land use:
  Arable land:
  1%
  Permanent crops:
  0%
  Meadows and pastures:
  79%
  Forest and woodland:
  10%
  Other:
  10%

Demographics [2]

	1960	1970	1980	1990	1991[1]	1994	2000	2010	2020
Population	955	1,248	1,662	2,186	2,247	2,430	2,826	3,545	4,309
Population density (persons per sq. mi.)	2	2	3	4	4	NA	5	6	7
Births	NA	NA	NA	NA	77	80	NA	NA	NA
Deaths	NA	NA	NA	NA	17	17	NA	NA	NA
Life expectancy - males	NA	NA	NA	NA	63	64	NA	NA	NA
Life expectancy - females	NA	NA	NA	NA	67	69	NA	NA	NA
Birth rate (per 1,000)	NA	NA	NA	NA	34	33	NA	NA	NA
Death rate (per 1,000)	NA	NA	NA	NA	8	7	NA	NA	NA
Women of reproductive age (15-44 yrs.)	NA	NA	NA	NA	NA	NA	NA	NA	NA
of which are currently married	NA	NA	NA	NA	NA	NA	NA	NA	NA
Fertility rate	NA	NA	NA	4.6	4.58	4.3	3.9	3.3	2.8

Population values are in thousands, life expectancy in years, and other items as indicated.

Health

Health Personnel [3]

Health Expenditures [4]

For sources, notes, and explanations, see Annotated Source Appendix, page 1035.

Human Factors

Health Care Ratios [5]	Infants and Malnutrition [6]

Ethnic Division [7]

Mongol	90%
Kazakh	4%
Chinese	2%
Russian	2%
Other	2%

Religion [8]

Predominantly Tibetan Buddhist, Muslim 4%.

Major Languages [9]

Khalkha Mongol	90%
Turkic	NA
Russian	NA
Chinese	NA

Education

Public Education Expenditures [10]	Educational Attainment [11]

Literacy Rate [12]

Libraries [13]

	Admin. Units	Svc. Pts.	Vols. (000)	Shelving (meters)	Vols. Added	Reg. Users
National	NA	NA	NA	NA	NA	NA
Non-specialized	NA	NA	NA	NA	NA	NA
Public	NA	NA	NA	NA	NA	NA
Higher ed. (1990)[22]	9	NA	1,581	NA	1,100	20,100
School (1990)	530	NA	3,220	NA	NA	450,600

Daily Newspapers [14]

	1975	1980	1985	1990
Number of papers	1	2	2	1
Circ. (000)	112	177	177	162

Cinema [15]

Data for 1989.

Cinema seats per 1,000	NA
Annual attendance per person	9.4
Gross box office receipts (mil. Tugrik)	NA

Science and Technology

Scientific/Technical Forces [16]	R&D Expenditures [17]	U.S. Patents Issued [18]

Government and Law

Organization of Government [19]

Long-form name:
none
Type:
republic
Independence:
13 March 1921 (from China)
Constitution:
adopted 13 January 1992
Legal system:
blend of Russian, Chinese, and Turkish systems of law; no constitutional provision for judicial review of legislative acts; has not accepted compulsory ICJ jurisdiction
National holiday:
National Day, 11 July (1921)
Executive branch:
president, vice president, prime minister, first deputy prime minister, cabinet
Legislative branch:
unicameral State Great Hural
Judicial branch:
Supreme Court serves as appeals court for people's and provincial courts, but to date rarely overturns verdicts of lower courts

Political Parties [20]

State Great Hural	% of votes
Mongolian People's Revolutionary Party (MPRP)	56.9
Other	43.1

Government Budget [21]

For 1991.
Deficit of   67

Crime [23]

Military Expenditures and Arms Transfers [22]

	1985	1986	1987	1988	1989
Military expenditures					
Current dollars (mil.)[e]	192	249	274	281	259
1989 constant dollars (mil.)[e]	219	276	295	293	259
Armed forces (000)	38	40	32	NA	33
Gross national product (GNP)					
Current dollars (mil.)	NA	NA	NA	NA	NA
1989 constant dollars (mil.)	NA	NA	NA	NA	NA
Central government expenditures (CGE)					
1989 constant dollars (mil.)	NA	NA	NA	NA	NA
People (mil.)	1.9	2.0	2.0	2.1	2.1
Military expenditure as % of GNP	NA	NA	NA	NA	NA
Military expenditure as % of CGE	NA	NA	NA	NA	NA
Military expenditure per capita	115	141	146	141	122
Armed forces per 1,000 people	19.9	20.4	15.9	NA	15.5
GNP per capita	NA	NA	NA	NA	NA
Arms imports[6]					
Current dollars (mil.)	5	10	0	0	0
1989 constant dollars (mil.)	6	11	0	0	0
Arms exports[6]					
Current dollars (mil.)	0	0	0	0	0
1989 constant dollars (mil.)	0	0	0	0	0
Total imports[7]					
Current dollars (mil.)	1,000	NA	NA	NA	NA
1989 constant dollars	1,138	NA	NA	NA	NA
Total exports[7]					
Current dollars (mil.)	388	NA	NA	NA	NA
1989 constant dollars	442	NA	NA	NA	NA
Arms as percent of total imports[8]	0.5	NA	NA	NA	NA
Arms as percent of total exports[8]	0	NA	NA	NA	NA

Human Rights [24]

	SSTS	FL	FAPRO	PPCG	APROBC	TPW	PCPTW	STPEP	PHRFF	PRW	ASST	AFL
Observes	P		P	P	P	P	P			P	P	
	EAFRD	CPR	ESCR	SR	ACHR	MAAE	PVIAC	PVNAC	EAFDAW	TCIDTP	RC	
Observes	P	P	P				S	S	P		P	

P = Party; S = Signatory; see Appendix for meaning of abbreviations.

Labor Force

Total Labor Force [25]

Labor Force by Occupation [26]

Primarily herding/agricultural

Unemployment Rate [27]

Primarily herding/agricultural 15% (1991 est.)

Production Sectors

Energy Resource Summary [28]

Energy Resources: Oil, coal. **Electricity**: 1,248,000 kW capacity; 3,740 million kWh produced, 1,622 kWh per capita (1992).

Telecommunications [30]

- 63,000 telephones (1989)
- Broadcast stations - 12 AM, 1 FM, 1 TV (with 18 provincial repeaters)
- Repeat of Russian TV
- 120,000 TVs
- 220,000 radios
- At least 1 earth station

Top Agricultural Products [32]

Agriculture accounts for about 20% of GDP and provides livelihood for about 50% of the population; livestock raising predominates (primarily sheep and goats, but also cattle, camels, and horses); crops—wheat, barley, potatoes, forage.

Top Mining Products [33]

Metric tons unless otherwise specified	M.t.
Coal (000 tons)	7,000
Copper, mine output, Cu content	90,300
Fluorspar	370
Molybdenum, mine output, Mo content	1,130
Gold, mine output, Ag content (kilograms)	800
Gypsum (000 tons)	29
Lime, hydrated and quicklime (000 tons)	100
Silver, mine output, Ag content (kilograms)	15,500
Tin, mine output, Sn content	250
Tungsten, mine output, W content	300

Transportation [31]

Railroads. 1,750 km 1.524-meter broad gauge (1988)

Highways. 46,700 km total; 1,000 km hard surface; 45,700 km other surfaces (1988)

Airports

Total:	81
Usable:	31
With permanent-surface runways:	11
With runways over 3,659 m:	fewer than 5
With runways 2,440-3,659 m:	fewer than 20
With runways 1,220-2,439 m:	12

Tourism [34]

	1987	1988	1989	1990	1991
Tourists	186	240	237	147	

Tourists are in thousands, money in million U.S. dollars.

Finance, Economics, and Trade

GDP and Manufacturing Summary [35]

	1980	1985	1990	1991	1992
Gross Domestic Product					
Millions of 1980 dollars	1,389	1,911	2,374	2,070	1,913
Growth rate in percent	3.43	5.51	4.47	-1.80	-7.60
Manufacturing Value Added					
Millions of 1980 dollars	347	516	575	500	429[e]
Growth rate in percent	8.03	3.07	2.19	-13.20	-14.20[e]
Manufacturing share in percent of current prices	29.3	32.6	35.0	NA	NA

Economic Indicators [36]

National product: GDP—exchange rate conversion—$1.8 billion (1992 est.). **National product real growth rate**: -15% (1992 est.). **National product per capita**: $800 (1992 est.). **Inflation rate (consumer prices)**: 325% (1992 est.). **External debt**: $16.8 billion (yearend 1990); 98.6% with USSR.

Balance of Payments Summary [37]

Exchange Rates [38]

Currency: **tughriks.**
Symbol: **Tug.**

Data are currency units per $1.

1992	40.00
1991	7.10
1990	5.63
1989	3.00

Imports and Exports

Top Import Origins [39]

$501 million (f.o.b., 1991 est.).

Origins	%
USSR	75
Austria	5
China	5

Top Export Destinations [40]

$347 million (f.o.b., 1991 est.).

Destinations	%
USSR	75
China	10
Japan	4

Foreign Aid [41]

	U.S. $	
USSR and CEMA countries - trade credits (est.)	300	million
USSR and other CEMA countries - grant aid	34	million
UNDP (1990)	7.4	million
Western donor coutries - grants and technical assistance	170	million
Including from World Bank	30	million
Including from IMF	30	million
Donor countries projected in 1992	200	million

Import and Export Commodities [42]

Import Commodities	**Export Commodities**
Machinery & equipment	Copper
Fuels	Livestock
Food products	Animal products
Industrial consumer goods	Cashmere
Chemicals	Wool
Building materials	Hides
Sugar	Fluorspar
Tea	Other nonferrous metals

Morocco

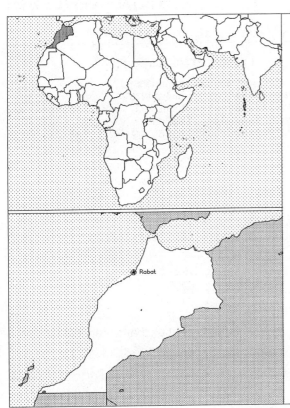

Geography [1]

Total area:
446,550 km2
Land area:
446,300 km2
Comparative area:
Slightly larger than California
Land boundaries:
Total 2,002 km, Algeria 1,559 km, Western Sahara 443 km
Coastline:
1,835 km
Climate:
Mediterranean, becoming more extreme in the interior
Terrain:
Mostly mountains with rich coastal plains
Natural resources:
Phosphates, iron ore, manganese, lead, zinc, fish, salt
Land use:
Arable land:
18%
Permanent crops:
1%
Meadows and pastures:
28%
Forest and woodland:
12%
Other:
41%

Demographics [2]

	1960	1970	1980	1990	1991[1]	1994	2000	2010	2020
Population	12,423	15,909	20,457	26,164	26,182	28,559	32,189	38,112	43,701
Population density (persons per sq. mi.)	72	92	119	149	152	NA	182	217	251
Births	NA	NA	NA	NA	787	816	NA	NA	NA
Deaths	NA	NA	NA	NA	207	179	NA	NA	NA
Life expectancy - males	NA	NA	NA	NA	63	66	NA	NA	NA
Life expectancy - females	NA	NA	NA	NA	66	70	NA	NA	NA
Birth rate (per 1,000)	NA	NA	NA	NA	30	29	NA	NA	NA
Death rate (per 1,000)	NA	NA	NA	NA	8	6	NA	NA	NA
Women of reproductive age (15-44 yrs.)	NA	NA	NA	6,270	NA	7,056	8,431	NA	NA
of which are currently married	NA	NA	NA	3,934	NA	4,475	5,411	NA	NA
Fertility rate	NA	NA	NA	4.4	3.84	3.8	3.1	2.5	2.2

Population values are in thousands, life expectancy in years, and other items as indicated.

Health

Health Personnel [3]

Doctors per 1,000 pop., 1988-92	0.21
Nurse-to-doctor ratio, 1988-92	4.5
Hospital beds per 1,000 pop., 1985-90	1.2
Percentage of children immunized (age 1 yr. or less)	
Third dose of DPT, 1990-91	79.0
Measles, 1990-91	76.0

Health Expenditures [4]

Total health expenditure, 1990 (official exchange rate)	
Millions of dollars	661
Millions of dollars per capita	26
Health expenditures as a percentage of GDP	
Total	2.6
Public sector	0.9
Private sector	1.6
Development assistance for health	
Total aid flows (millions of dollars)[1]	20
Aid flows per capita (millions of dollars)	0.8
Aid flows as a percentage of total health expenditure	3.0

For sources, notes, and explanations, see Annotated Source Appendix, page 1035.

639

Human Factors

Health Care Ratios [5]

Population per physician, 1970	13090
Population per physician, 1990	4840
Population per nursing person, 1970	NA
Population per nursing person, 1990	1050
Percent of births attended by health staff, 1985	NA

Infants and Malnutrition [6]

Percent of babies with low birth weight, 1985	9
Infant mortality rate per 1,000 live births, 1970	128
Infant mortality rate per 1,000 live births, 1991	57
Years of life lost per 1,000 population, 1990	43
Prevalence of malnutrition (under age 5), 1990	12

Ethnic Division [7]

Arab-Berber	99.1%
Other	0.7%
Jewish	0.2%

Religion [8]

Muslim	98.7%
Christian	1.1%
Jewish	0.2%

Major Languages [9]

Arabic (official), Berber dialects, French often the language of business, government, and diplomacy.

Education

Public Education Expenditures [10]

Million Dirham	1980[1]	1985[1]	1987[1]	1988[1]	1989[1]	1990[1]
Total education expenditure	4,367	7,697	NA	9,660	10,626	11,220
as percent of GNP	6.1	6.3	NA	5.6	5.8	5.5
as percent of total govt. expend.	18.5	22.9	NA	23.5	24.8	26.1
Current education expenditure	3,529	6,079	NA	7,248	9,170	10,187
as percent of GNP	4.9	5.0	NA	4.2	5.0	4.9
as percent of current govt. expend.	23.3	28.6	NA	28.8	31.8	33.6
Capital expenditure	838	1,618	2,412	1,456	1,033	

Educational Attainment [11]

Literacy Rate [12]

In thousands and percent	1985[a]	1991[a]	2000[a]
Illiterate population +15 years	7,454	7,526	7,303
Illiteracy rate - total pop. (%)	58.3	50.5	36.5
Illiteracy rate - males (%)	45.7	38.7	27.0
Illiteracy rate - females (%)	70.5	62.0	45.7

Libraries [13]

Daily Newspapers [14]

	1975	1980	1985	1990
Number of papers	7	11	14	13
Circ. (000)	250[e]	270[e]	320[e]	320[e]

Cinema [15]

Data for 1989.

Cinema seats per 1,000	6.8
Annual attendance per person	1.2
Gross box office receipts (mil. Dirham)	143

Science and Technology

Scientific/Technical Forces [16]

R&D Expenditures [17]

U.S. Patents Issued [18]

Values show patents issued to citizens of the country by the U.S. Patents Office.

	1990	1991	1992
Number of patents	0	1	0

640

For sources, notes, and explanations, see Annotated Source Appendix, page 1035.

Government and Law

Organization of Government [19]

Long-form name:
Kingdom of Morocco
Type:
constitutional monarchy
Independence:
2 March 1956 (from France)
Constitution:
10 March 1972, revised in September 1992
Legal system:
based on Islamic law and French and Spanish civil law system; judicial review of legislative acts in Constitutional Chamber of Supreme Court
National holiday:
National Day, 3 March (1961) (anniversary of King Hassan II's accession to the throne)
Executive branch:
monarch, prime minister, Council of Ministers (cabinet)
Legislative branch:
unicameral Chamber of Representatives (Majlis Nawab)
Judicial branch:
Supreme Court

Political Parties [20]

Chamber of Representatives	% of seats
Constitutional Union (UC)	27.1
National Assembly (RNI)	19.9
Popular Movement (MP)	15.4
Istiqlal (Istiqlal)	13.4
Socialist Union (USFP)	11.8
National Democratic (PND)	7.8
Other	4.6

Government Expenditures [21]

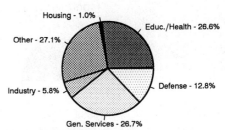

Housing - 1.0%
Educ./Health - 26.6%
Other - 27.1%
Defense - 12.8%
Industry - 5.8%
Gen. Services - 26.7%

% distribution for 1990. Expend. in 1992: 7.7 ($ bil.)

Crime [23]

Military Expenditures and Arms Transfers [22]

	1985	1986	1987	1988	1989
Military expenditures					
Current dollars (mil.)	NA	1,078[e]	1,102[e]	1,102[e]	1,203[e]
1989 constant dollars (mil.)	NA	1,196[e]	1,185[e]	1,147[e]	1,203[e]
Armed forces (000)	165	185[e]	200	195	195
Gross national product (GNP)					
Current dollars (mil.)	15,770	17,840	17,900	20,250	21,740
1989 constant dollars (mil.)	17,950	19,800	19,260	21,080	21,740
Central government expenditures (CGE)					
1989 constant dollars (mil.)	5,844	5,989	5,615	NA	NA
People (mil.)	23.0	23.5	24.0	24.5	25.1
Military expenditure as % of GNP	NA	6.0	6.2	5.4	5.5
Military expenditure as % of CGE	NA	20.0	21.1	NA	NA
Military expenditure per capita	NA	51	49	47	48
Armed forces per 1,000 people	7.2	7.9	8.3	7.9	7.8
GNP per capita	780	842	802	859	866
Arms imports[6]					
Current dollars (mil.)	110	90	410	130	40
1989 constant dollars (mil.)	125	100	441	135	40
Arms exports[6]					
Current dollars (mil.)	0	0	0	0	0
1989 constant dollars (mil.)	0	0	0	0	0
Total imports[7]					
Current dollars (mil.)	3,849	3,803	4,230	4,773	5,457
1989 constant dollars	4,382	4,220	4,550	4,969	5,457
Total exports[7]					
Current dollars (mil.)	2,165	2,454	2,826	3,603	3,307
1989 constant dollars	2,465	2,723	3,040	3,751	3,307
Arms as percent of total imports[8]	2.9	2.4	9.7	2.7	0.7
Arms as percent of total exports[8]	0	0	0	0	0

Human Rights [24]

	SSTS	FL	FAPRO	PPCG	APROBC	TPW	PCPTW	STPEP	PHRFF	PRW	ASST	AFL
Observes	P	P		P	P	P	P	P		P	P	P
	EAFRD	CPR	ESCR	SR	ACHR	MAAE	PVIAC	PVNAC	EAFDAW	TCIDTP	RC	
Observes	P	P	P	P			S	S		S	S	

P = Party; S = Signatory; see Appendix for meaning of abbreviations.

Labor Force

Total Labor Force [25]

7.4 million

Labor Force by Occupation [26]

Agriculture	50%
Services	26
Industry	15
Other	9

Date of data: 1985

Unemployment Rate [27]

19% (1992 est.)

For sources, notes, and explanations, see Annotated Source Appendix, page 1035.

641

Production Sectors

Commercial Energy Production and Consumption

Production [28]

Coal - 50.0%

Hydro - 50.0%

Consumption [29]

Petroleum - 81.2%

Hydro - 3.1%

Coal - 15.6%

Data are shown in quadrillion (10^{15}) BTUs and percent for 1991

Crude oil[1]	(s)	Petroleum	0.26
Natural gas liquids	0.00	Dry natural gas	(s)
Dry natural gas	(s)	Coal[2]	0.05
Coal[2]	0.01	Net hydroelectric power[3]	0.01
Net hydroelectric power[3]	0.01	Net nuclear power[3]	0.00
Net nuclear power[3]	0.00	Total	0.32
Total	0.03		

Telecommunications [30]

- Good system composed of wire lines, cables, and microwave radio relay links
- Principal centers are Casablanca and Rabat
- Secondary centers are Fes, Marrakech, Oujda, Tangier, and Tetouan
- 280,000 telephones (10.5 telephones per 1,000 persons)
- Broadcast stations - 20 AM, 7 FM, 26 TV and 26 repeaters
- 5 submarine cables
- Satellite earth stations - 2 Atlantic Ocean INTELSAT and 1 ARABSAT
- Microwave radio relay to Gibraltar, Spain, and Western Sahara
- Coaxial cable and microwave to Algeria
- Microwave radio relay network linking Syria, Jordan, Egypt, Libya, Tunisia, Algeria, and Morocco

Transportation [31]

Railroads. 1,893 km 1.435-meter standard gauge (246 km double track, 974 km electrified)

Highways. 59,198 km total; 27,740 km paved, 31,458 km gravel, crushed stone, improved earth, and unimproved earth

Merchant Marine. 50 ships (1,000 GRT or over) totaling 305,758 GRT/484,825 DWT; 10 cargo, 2 container, 11 refrigerated cargo, 6 roll-on/roll-off, 4 oil tanker, 11 chemical tanker, 4 bulk, 2 short-sea passenger

Airports

Total:	73
Usable:	65
With permanent-surface runways:	26
With runways over 3,659 m:	2
With runways 2,440-3,659 m:	13
With runways 1,220-2,439 m:	26

Top Agricultural Products [32]

	89-90	90-91
Wheat	3,614	4,939
Barley	2,138,138	3,253
Sugar beets	2,978	3,000
Citrus	1,468	1,126
Sugar cane	1,015	1,050
Olives	600	390

Values shown are 1,000 metric tons.

Top Mining Products [33]

Metric tons unless otherwise specified	M.t.
Phosphate rock (000 tons)	130,000[1]
Barite	433,000[1]
Copper, Cu content of concentrates and matte	14,000[1]
Flurospar, acid grade	75,000[1]
Manganese ore	59,000[1]
Salt, rock	130,000[1]

Tourism [34]

	1987	1988	1989	1990	1991
Visitors	2,342	2,919	3,566	4,138	4,211
Tourists[1]	2,248	2,841	3,468	4,024	4,162
Cruise passengers	94	78	97	114	49
Tourism receipts	936	1,110	1,146	1,259	1,052
Tourism expenditures	132	164	153	184	190
Fare receipts	49	40	31	38	74
Fare expenditures	18	31	29	31	33

Tourists are in thousands, money in million U.S. dollars.

For sources, notes, and explanations, see Annotated Source Appendix, page 1035.

Manufacturing Sector

GDP and Manufacturing Summary [35]

	1980	1985	1989	1990	% change 1980-1990	% change 1989-1990
GDP (million 1980 $)	18,997	22,336	25,545	27,085	42.6	6.0
GDP per capita (1980 $)	980	1,014	1,046	1,080	10.2	3.3
Manufacturing as % of GDP (current prices)	17.8	19.4	17.8[e]	19.4	9.0	9.0
Gross output (million $)	5,958	5,052	10,273	10,737[e]	80.2	4.5
Value added (million $)	1,485	1,374[e]	1,601[e]	3,179[e]	114.1	98.6
Value added (million 1980 $)	3,197	3,974	4,193	4,626	44.7	10.3
Industrial production index	100	104	128	108	8.0	-15.6
Employment (thousands)	176	227	330[e]	315[e]	79.0	-4.5

Note: GDP stands for Gross Domestic Product. 'e' stands for estimated value.

Profitability and Productivity

	1980	1985	1989	1990	% change 1980-1990	% change 1989-1990
Intermediate input (%)	77	74[e]	84[e]	72[e]	-6.5	-14.3
Wages, salaries, and supplements (%)	13	11	9[e]	9[e]	-30.8	0.0
Gross operating surplus (%)	10	15[e]	7[e]	18[e]	80.0	157.1
Gross output per worker ($)	33,920	22,306	31,149[e]	34,060[e]	0.4	9.3
Value added per worker ($)	7,801	5,696[e]	4,854[e]	9,442[e]	21.0	94.5
Average wage (incl. benefits) ($)	4,363	2,434	2,707[e]	3,182[e]	-27.1	17.5

Profitability is in percent of gross output. Productivity is in U.S. $. 'e' stands for estimated value.

Profitability - 1990

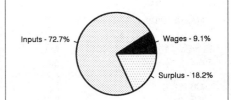

Inputs - 72.7%
Wages - 9.1%
Surplus - 18.2%

The graphic shows percent of gross output.

Value Added in Manufacturing

	1980 $ mil.	1980 %	1985 $ mil.	1985 %	1989 $ mil.	1989 %	1990 $ mil.	1990 %	% change 1980-1990	% change 1989-1990
311 Food products	130	8.8	110	8.0	308[e]	19.2	144	4.5	10.8	-53.2
313 Beverages	62	4.2	107	7.8	62[e]	3.9	278	8.7	348.4	348.4
314 Tobacco products	38	2.6	136	9.9	45[e]	2.8	354	11.1	831.6	686.7
321 Textiles	202	13.6	172	12.5	223[e]	13.9	315	9.9	55.9	41.3
322 Wearing apparel	32	2.2	45	3.3	44[e]	2.7	228	7.2	612.5	418.2
323 Leather and fur products	15	1.0	13	0.9	17[e]	1.1	29	0.9	93.3	70.6
324 Footwear	24	1.6	18	1.3	29[e]	1.8	39	1.2	62.5	34.5
331 Wood and wood products	30	2.0	25	1.8	35[e]	2.2	49	1.5	63.3	40.0
332 Furniture and fixtures	19	1.3	16	1.2	5[e]	0.3	30	0.9	57.9	500.0
341 Paper and paper products	64	4.3	64	4.7	58[e]	3.6	151	4.7	135.9	160.3
342 Printing and publishing	26	1.8	19	1.4	34[e]	2.1	43	1.4	65.4	26.5
351 Industrial chemicals	127	8.6	95	6.9	155[e]	9.7	230	7.2	81.1	48.4
352 Other chemical products	97	6.5	71	5.2	100[e]	6.2	173	5.4	78.4	73.0
353 Petroleum refineries	114[e]	7.7	84[e]	6.1	NA	0.0	213[e]	6.7	86.8	NA
354 Miscellaneous petroleum and coal products	NA	0.0	NA	0.0	NA	0.0	NA	0.0	NA	NA
355 Rubber products	34	2.3	28	2.0	25[e]	1.6	52	1.6	52.9	108.0
356 Plastic products	20	1.3	26	1.9	14[e]	0.9	48	1.5	140.0	242.9
361 Pottery, china and earthenware	6	0.4	3[e]	0.2	3[e]	0.2	8[e]	0.3	33.3	166.7
362 Glass and glass products	10	0.7	5[e]	0.4	3[e]	0.2	13[e]	0.4	30.0	333.3
369 Other non-metal mineral products	154	10.4	106[e]	7.7	178[e]	11.1	274[e]	8.6	77.9	53.9
371 Iron and steel	7	0.5	4[e]	0.3	9[e]	0.6	9[e]	0.3	28.6	0.0
372 Non-ferrous metals	8	0.5	6[e]	0.4	5[e]	0.3	9[e]	0.3	12.5	80.0
381 Metal products	110	7.4	96	7.0	120[e]	7.5	166	5.2	50.9	38.3
382 Non-electrical machinery	30	2.0	19[e]	1.4	24[e]	1.5	48[e]	1.5	60.0	100.0
383 Electrical machinery	61	4.1	56	4.1	62[e]	3.9	131[e]	4.1	114.8	111.3
384 Transport equipment	62	4.2	49	3.6	41[e]	2.6	140	4.4	125.8	241.5
385 Professional and scientific equipment	1	0.1	1[e]	0.1	1[e]	0.1	1[e]	0.0	0.0	0.0
390 Other manufacturing industries	2	0.1	1	0.1	1[e]	0.1	3[e]	0.1	50.0	200.0

Note: The industry codes shown are International Standard Industry codes (ISIC). Percentages are percent of total Value Added. 'e' stands for estimated value

For sources, notes, and explanations, see Annotated Source Appendix, page 1035.

643

Finance, Economics, and Trade

Economic Indicators [36]

Billions of Dirhams (DH) unless otherwise noted.

	1989	1990	1991[P]
GDP (current dirhams)	191.5	207.8	NA
Real GDP growth (%)	1.5	2.6	3.5-4.0
Money - M1 (% change)[104]	11.5	19.3	NA
Savings rate (% GDP)	19.4	23.0	NA
Investment rate (% GDP)	23.0	24.1	NA
CPI (% change)[104]	3.8	4.6	8.5
WPI (% change)[104]	4.8	6.7	7.4
External pub. debt ($ bil.)	20.7	23.0	20.2

Balance of Payments Summary [37]

Values in millions of dollars.

	1987	1988	1989	1990	1991
Exports of goods (f.o.b.)	2,781.0	3,608.0	3,312.0	4,210.0	4,277.0
Imports of goods (f.o.b.)	-3,850.0	-4,360.0	-4,992.0	-6,282.0	-6,253.0
Trade balance	-1,069.0	-752.0	-1,680.0	-2,072.0	-1,976.0
Services - debits	-1,938.0	-2,185.0	-2,431.0	-2,572.0	-2,688.0
Services - credits	1,411.0	1,798.0	1,700.0	2,111.0	1,975.0
Private transfers (net)	1,579.0	1,303.0	1,356.0	2,012.0	2,013.0
Government transfers (net)	191.0	303.0	265.0	320.0	280.0
Long term capital (net)	428.0	494.0	600.0	1,750.0	1,327.0
Short term capital (net)	-269.0	-724.0	211.0	175.0	130.0
Errors and omissions	38.0	22.0	-11.0	9.0	88.0
Overall balance	371.0	259.0	10.0	1,733.0	1,149.0

Exchange Rates [38]

Currency: **Moroccan dirhams.**
Symbol: **DH.**

Data are currency units per $1.

February 1993	9.207
1992	8.538
1991	8.707
1990	8.242
1989	8.488
1988	8.209

Imports and Exports

Top Import Origins [39]

$7.6 billion (f.o.b., 1992 est.).

Origins	%
EC	53
US	11
Canada	4
Iraq	3
Former USSR	3
Japan	2

Top Export Destinations [40]

$4.7 billion (f.o.b., 1992 est.).

Destinations	%
EC	58
India	7
Japan	5
Former USSR	3
US	2

Foreign Aid [41]

	U.S. $	
US commitments, including Ex-Im (FY70-89)	1.3	billion
US commitments, including Ex-Im (1992)	123.6	million
Western (non-US) countries, ODA and OOF bilateral commitments (1970-89)	7.5	billion
OPEC bilateral aid (1979-89)	4.8	billion
Communist countries (1970-89)	2.5	billion
Debt canceled by Saudi Arabia (1991)	2.8	billion
IMF standby agreement (value)	13	million
World Bank (1991)	450	million

Import and Export Commodities [42]

Import Commodities

Capital goods 24%
Semiprocessed goods 22%
Raw materials 16%
Fuel and lubricants 16%
Food and beverages 13%
Consumer goods 9%

Export Commodities

Food & beverages 30%
Semiprocessed goods 23%
Consumer goods 21%
Phosphates 17%

Mozambique

Geography [1]

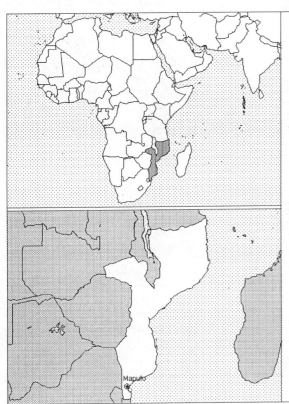

Total area:
801,590 km2
Land area:
784,090 km2
Comparative area:
Slightly less than twice the size of California
Land boundaries:
Total 4,571 km, Malawi 1,569 km, South Africa 491 km, Swaziland 105 km, Tanzania 756 km, Zambia 419 km, Zimbabwe 1,231 km
Coastline:
2,470 km
Climate:
Tropical to subtropical
Terrain:
Mostly coastal lowlands, uplands in center, high plateaus in northwest, mountains in west
Natural resources:
Coal, titanium
Land use:
Arable land:
4%
Permanent crops:
0%
Meadows and pastures:
56%
Forest and woodland:
20%
Other:
20%

Demographics [2]

	1960	1970	1980	1990	1991[1]	1994	2000	2010	2020
Population	7,472	9,304	12,103	14,438	15,113	17,346	20,868	27,381	35,240
Population density (persons per sq. mi.)	25	31	40	48	50	NA	69	91	117
Births	NA	NA	NA	NA	697	780	NA	NA	NA
Deaths	NA	NA	NA	NA	261	283	NA	NA	NA
Life expectancy - males	NA	NA	NA	NA	46	47	NA	NA	NA
Life expectancy - females	NA	NA	NA	NA	49	50	NA	NA	NA
Birth rate (per 1,000)	NA	NA	NA	NA	46	45	NA	NA	NA
Death rate (per 1,000)	NA	NA	NA	NA	17	16	NA	NA	NA
Women of reproductive age (15-44 yrs.)	NA	NA	NA	3,390	NA	4,056	4,877	NA	NA
of which are currently married	NA	NA	NA	2,544	NA	3,052	3,673	NA	NA
Fertility rate	NA	NA	NA	6.5	6.43	6.3	5.8	5.0	4.1

Population values are in thousands, life expectancy in years, and other items as indicated.

Health

Health Personnel [3]

Doctors per 1,000 pop., 1988-92	0.02
Nurse-to-doctor ratio, 1988-92	13.1
Hospital beds per 1,000 pop., 1985-90	0.9
Percentage of children immunized (age 1 yr. or less)	
Third dose of DPT, 1990-91	19.0
Measles, 1990-91	23.0

Health Expenditures [4]

Total health expenditure, 1990 (official exchange rate)	
Millions of dollars	85
Millions of dollars per capita	5
Health expenditures as a percentage of GDP	
Total	5.9
Public sector	4.4
Private sector	1.5
Development assistance for health	
Total aid flows (millions of dollars)[1]	45
Aid flows per capita (millions of dollars)	2.9
Aid flows as a percentage of total health expenditure	52.9

For sources, notes, and explanations, see Annotated Source Appendix, page 1035.

645

Human Factors

Health Care Ratios [5]

Population per physician, 1970	18860
Population per physician, 1990	NA
Population per nursing person, 1970	4280
Population per nursing person, 1990	NA
Percent of births attended by health staff, 1985	28

Infants and Malnutrition [6]

Percent of babies with low birth weight, 1985	15
Infant mortality rate per 1,000 live births, 1970	171
Infant mortality rate per 1,000 live births, 1991	149
Years of life lost per 1,000 population, 1990	141
Prevalence of malnutrition (under age 5), 1990	NA

Ethnic Division [7]

Indigenous tribal groups, Europeans 10,000, Euro-Africans 35,000, Indians 15,000.

Religion [8]

Indigenous beliefs	60%
Christian	30%
Muslim	10%

Major Languages [9]

Portuguese (official), indigenous dialects.

Education

Public Education Expenditures [10]

Million Metical	1980[4]	1985[4]	1987[4]	1988[4]	1989[4]	1990[4]
Total education expenditure	2,900	4,400	10,700	25,200	44,571	72,264
as percent of GNP	3.8	3.1	2.9	4.4	5.5	6.3
as percent of total govt. expend.	12.1	10.6	5.6	8.1	9.3	12.0
Current education expenditure	2,500	4,100	9,400	17,000	30,371	46,064
as percent of GNP	3.2	2.9	2.6	3.0	3.8	4.0
as percent of current govt. expend.	17.7	12.3	9.9	11.3	12.3	17.5
Capital expenditure	400	300	1,300	8,200	14,200	26,200

Educational Attainment [11]

Age group	25+
Total population	4,242,819
Highest level attained (%)	
No schooling	81.0[6]
First level	
Incompleted	18.1
Completed	NA
Entered second level	
S-1	0.8
S-2	NA
Post secondary	0.1

Literacy Rate [12]

In thousands and percent	1985[a]	1991[a]	2000[a]
Illiterate population +15 years	5,593	5,880	6,377
Illiteracy rate - total pop. (%)	72.4	67.1	55.4
Illiteracy rate - males (%)	60.6	54.9	43.4
Illiteracy rate - females (%)	83.6	78.7	66.9

Libraries [13]

Daily Newspapers [14]

	1975	1980	1985	1990
Number of papers	2	2	2	2
Circ. (000)	42	54	81	81

Cinema [15]

Science and Technology

Scientific/Technical Forces [16]	R&D Expenditures [17]	U.S. Patents Issued [18]

For sources, notes, and explanations, see Annotated Source Appendix, page 1035.

Government and Law

Organization of Government [19]

Long-form name:
Republic of Mozambique
Type:
republic
Independence:
25 June 1975 (from Portugal)
Constitution:
30 November 1990
Legal system:
based on Portuguese civil law system and customary law
National holiday:
Independence Day, 25 June (1975)
Executive branch:
president, prime minister, Cabinet
Legislative branch:
unicameral Assembly of the Republic (Assembleia da Republica)
Judicial branch:
Supreme Court

Elections [20]

The government plans multiparty elections as early as 1993; 14 parties, including the Liberal Democratic Party of Mozambique (PALMO), the Mozambique National Union (UNAMO), the Mozambique National Movement (MONAMO), and the Mozambique National Resistance (RENAMO, Alfonso Dhlakama, president), have already emerged; Draft electoral law provides for periodic, direct presidential and Assembly elections.

Government Budget [21]

For 1992 est.	
Revenues	252
Expenditures	607
Capital expenditures	NA

Crime [23]

Military Expenditures and Arms Transfers [22]

	1985	1986	1987	1988	1989
Military expenditures					
Current dollars (mil.)	68	NA	74	92	107
1989 constant dollars (mil.)	77	NA	79	96	107
Armed forces (000)	35	65	65	65	65
Gross national product (GNP)					
Current dollars (mil.)	919	941	918	961	1,106
1989 constant dollars (mil.)	1,046	1,045	988	1,000	1,106
Central government expenditures (CGE)					
1989 constant dollars (mil.)	204[e]	NA	230	236	NA
People (mil.)	13.8	14.1	14.1	14.0	14.2
Military expenditure as % of GNP	7.4	NA	8.0	9.6	9.7
Military expenditure as % of CGE	38.0	NA	34.6	40.7	NA
Military expenditure per capita	6	NA	6	7	8
Armed forces per 1,000 people	2.5	4.6	4.6	4.6	4.6
GNP per capita	76	74	70	71	78
Arms imports[6]					
Current dollars (mil.)	270	170	120	160	120
1989 constant dollars (mil.)	307	189	129	167	120
Arms exports[6]					
Current dollars (mil.)	0	0	0	0	0
1989 constant dollars (mil.)	0	0	0	0	0
Total imports[7]					
Current dollars (mil.)	NA	480	647	764	NA
1989 constant dollars	NA	533	696	795	NA
Total exports[7]					
Current dollars (mil.)	NA	80	86	100	NA
1989 constant dollars	NA	89	92	104	NA
Arms as percent of total imports[8]	NA	35.4	18.5	20.9	NA
Arms as percent of total exports[8]	NA	0	0	0	NA

Human Rights [24]

	SSTS	FL	FAPRO	PPCG	APROBC	TPW	PCPTW	STPEP	PHRFF	PRW	ASST	AFL
Observes				P		P	P					P
	EAFRD	CPR	ESCR	SR	ACHR	MAAE	PVIAC	PVNAC	EAFDAW	TCIDTP		RC
Observes	P			P			P					S

P = Party; S = Signatory; see Appendix for meaning of abbreviations.

Labor Force

Total Labor Force [25]

No data available.

Labor Force by Occupation [26]

Agriculture	90%

Unemployment Rate [27]

50% (1989 est.)

For sources, notes, and explanations, see Annotated Source Appendix, page 1035.

647

Production Sectors

Energy Resource Summary [28]

Energy Resources: Coal. **Electricity**: 2,270,000 kW capacity; 1,745 million kWh produced, 115 kWh per capita (1991). **Pipelines**: Crude oil (not operating) 306 km; petroleum products 289 km.

Telecommunications [30]

- Fair system of troposcatter, open-wire lines, and radio relay
- Broadcast stations - 29 AM, 4 FM, 1 TV
- Earth stations - 2 Atlantic Ocean INTELSAT and 3 domestic Indian Ocean INTELSAT

Top Agricultural Products [32]

Agriculture accounts for 50% of GDP and about 90% of exports; cash crops—cotton, cashew nuts, sugarcane, tea, shrimp; other crops—cassava, corn, rice, tropical fruits; not self-sufficient in food.

Top Mining Products [33]

Estimated metric tons unless otherwise specified	M.t.
Salt, marine	40,000
Bauxite	7,690
Clay, bentonite	682
Coal, bituminous	50,832
Gemstones, cut stones, all types (carats)	12,906
Gold (kilograms)	394

Transportation [31]

Railroads. 3,288 km total; 3,140 km 1.067-meter gauge; 148 km 0.762-meter narrow gauge; Malawi-Nacala, Malawi-Beira, and Zimbabwe-Maputo lines are subject to closure because of insurgency

Highways. 26,498 km total; 4,593 km paved; 829 km gravel, crushed stone, stabilized soil; 21,076 km unimproved earth

Merchant Marine. 4 cargo ships (1,000 GRT or over) totaling 5,686 GRT/9,742 DWT

Airports

Total:	194
Usable:	131
With permanent-surface runways:	25
With runways over 3,659 m:	1
With runways 2,440-3,659 m:	4
With runways 1,220-2,439 m:	26

Tourism [34]

Finance, Economics, and Trade

GDP and Manufacturing Summary [35]

	1980	1985	1990	1991	1992
Gross Domestic Product					
Millions of 1980 dollars	2,407	1,914	2,308	2,319	2,319[e]
Growth rate in percent	2.46	-8.82	3.10	0.50	0.00[e]
Manufacturing Value Added					
Millions of 1980 dollars	759	334	434	451	450[e]
Growth rate in percent	3.25	-11.98	4.40	3.90	-0.25[e]
Manufacturing share in percent of current prices	33.1[e]	14.9[e]	NA	NA	NA

Economic Indicators [36]

National product: GDP—exchange rate conversion—$1.75 billion (1992 est.). **National product real growth rate**: 0.3% (1992 est.). **National product per capita**: $115 (1992 est.). **Inflation rate (consumer prices)**: 50% (1992 est.). **External debt**: $5.4 billion (1991 est.).

Balance of Payments Summary [37]

Values in millions of dollars.

	1987	1988	1989	1990	1991
Exports of goods (f.o.b.)	97.0	103.0	104.8	126.4	162.3
Imports of goods (f.o.b.)	-577.8	-662.0	-726.9	-789.7	-808.8
Trade balance	-480.8	-559.0	-622.1	-663.3	-646.5
Services - debits	-324.2	-307.6	-364.8	-371.2	-417.4
Services - credits	137.0	156.6	16.7	173.4	202.8
Private transfers (net)	-25.0	52.7	57.5	72.1	78.0
Government transfers (net)	304.2	378.4	408.6	470.7	538.3
Long term capital (net)	1,014.3	270.5	328.1	696.3	410.7
Short term capital (net)	-607.6	-	14.4	-377.5	-119.3
Errors and omissions	24.6	8.3	-7.2	-9.4	-75.7
Overall balance	42.5	-0.1	-18.8	-8.9	-29.1

Exchange Rates [38]

Currency: **meticais.**
Symbol: **Mt.**

Data are currency units per $1.

January 1993	2,740.15
1992	2,433.34
1991	1,434.47
1990	929.00
1989	800.00
1988	528.60

Imports and Exports

Top Import Origins [39]

$899 million (c.i.f., 1991 est.).

Origins	%
US	NA
Western Europe	NA
USSR	NA

Top Export Destinations [40]

$162 million (f.o.b., 1991 est.).

Destinations	%
US	NA
Western Europe	NA
Germany	NA
Japan	NA

Foreign Aid [41]

	U.S. $	
US commitments, including Ex-Im (FY70-89)	350	million
Western (non-US) countries, ODA and OOF bilateral commitments (1970-89)	4.4	billion
OPEC bilateral aid (1979-89)	37	million
Communist countries (1970-89)	890	million

Import and Export Commodities [42]

Import Commodities	**Export Commodities**
Food	Shrimp 48%
Clothing	Cashews 21%
Farm equipment	Sugar 10%
Petroleum	Copra 3%
	Citrus 3%

Namibia

Geography [1]

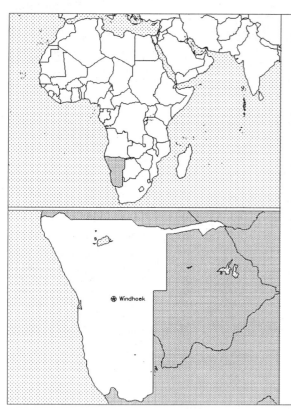

Total area:
824,290 km2
Land area:
823,290 km2
Comparative area:
Slightly more than half the size of Alaska
Land boundaries:
Total 3,935 km, Angola 1,376 km, Botswana 1,360 km, South Africa 966 km, Zambia 233 km
Coastline:
1,489 km
Climate:
Desert; hot, dry; rainfall sparse and erratic
Terrain:
Mostly high plateau; Namib Desert along coast; Kalahari Desert in east
Natural resources:
Diamonds, copper, uranium, gold, lead, tin, lithium, cadmium, zinc, salt, vanadium, natural gas, fish; suspected deposits of oil, natural gas, coal, iron ore
Land use:
Arable land: 1%
Permanent crops: 0%
Meadows and pastures: 64%
Forest and woodland: 22%
Other: 13%

Demographics [2]

	1960	1970	1980	1990	1991[1]	1994	2000	2010	2020
Population	591	765	967	1,387	1,521	1,596	1,957	2,705	3,638
Population density (persons per sq. mi.)	2	2	3	5	5	NA	7	9	12
Births	NA	NA	NA	NA	69	69	NA	NA	NA
Deaths	NA	NA	NA	NA	15	14	NA	NA	NA
Life expectancy - males	NA	NA	NA	NA	58	59	NA	NA	NA
Life expectancy - females	NA	NA	NA	NA	63	64	NA	NA	NA
Birth rate (per 1,000)	NA	NA	NA	NA	45	43	NA	NA	NA
Death rate (per 1,000)	NA	NA	NA	NA	10	9	NA	NA	NA
Women of reproductive age (15-44 yrs.)	NA	NA	NA	NA	NA	NA	NA	NA	NA
of which are currently married	NA	NA	NA	NA	NA	NA	NA	NA	NA
Fertility rate	NA	NA	NA	6.6	6.58	6.4	6.0	5.1	4.1

Population values are in thousands, life expectancy in years, and other items as indicated.

Health

Health Personnel [3]

Health Expenditures [4]

Human Factors

Health Care Ratios [5]	
Population per physician, 1970	NA
Population per physician, 1990	4620
Population per nursing person, 1970	NA
Population per nursing person, 1990	NA
Percent of births attended by health staff, 1985	NA

Infants and Malnutrition [6]	
Percent of babies with low birth weight, 1985	NA
Infant mortality rate per 1,000 live births, 1970	118
Infant mortality rate per 1,000 live births, 1991	72
Years of life lost per 1,000 population, 1990	NA
Prevalence of malnutrition (under age 5), 1990	NA

Ethnic Division [7]

Black	86%
White	6.6%
Mixed	7.4%

Religion [8]

Christian.

Major Languages [9]

English 7% (official), Afrikaans common language of most of the population and about 60% of the white population, German 32%, indigenous languages.

Education

Public Education Expenditures [10]

Educational Attainment [11]

Literacy Rate [12]

Libraries [13]

Daily Newspapers [14]

	1975	1980	1985	1990
Number of papers	3	4	3	6
Circ. (000)	16[e]	27	21	220

Cinema [15]

Science and Technology

Scientific/Technical Forces [16]

R&D Expenditures [17]

U.S. Patents Issued [18]

Government and Law

Organization of Government [19]

Long-form name:
Republic of Namibia
Type:
republic
Independence:
21 March 1990 (from South African mandate)
Constitution:
ratified 9 February 1990
Legal system:
based on Roman-Dutch law and 1990 constitution
National holiday:
Independence Day, 21 March (1990)
Executive branch:
president, Cabinet
Legislative branch:
bicameral legislature consists of an upper house or National Council and a lower house or National Assembly
Judicial branch:
Supreme Court

Political Parties [20]

National Assembly	% of seats
SW Africa People's Org. (SWAPO)	56.9
DTA of Namibia (DTA)	29.2
United Democratic (UDF)	5.6
Action Christian (ACN)	4.2
Namibia National Front (NNF)	1.4
Federal Convention (FCN)	1.4
National Patriotic Front (NPF)	1.4

Government Expenditures [21]

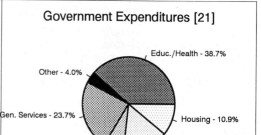

Educ./Health - 38.7%
Other - 4.0%
Gen. Services - 23.7%
Housing - 10.9%
Defense - 6.5%
Industry - 16.2%

% distribution for 1991. Expend. in FY 92: 1112 ($ mil.)

Defense Summary [22]

Branches: National Defense Force (Army), Police

Manpower Availability: Males age 15-49 324,599; fit for military service 192,381 (1993 est.)

Defense Expenditures: Exchange rate conversion - $66 million, 3.4% of GDP (FY92)

Crime [23]

Human Rights [24]

	SSTS	FL	FAPRO	PPCG	APROBC	TPW	PCPTW	STPEP	PHRFF	PRW	ASST	AFL
Observes						P	P					
	EAFRD	CPR	ESCR	SR	ACHR	MAAE	PVIAC	PVNAC	EAFDAW	TCIDTP	RC	
Observes	P						P	P	P		P	

P = Party; S = Signatory; see Appendix for meaning of abbreviations.

Labor Force

Total Labor Force [25]

500,000

Labor Force by Occupation [26]

Agriculture	60%
Industry and commerce	19
Services	8
Government	7
Mining	6

Date of data: 1981 est.

Unemployment Rate [27]

25-35% (1992)

652

For sources, notes, and explanations, see Annotated Source Appendix, page 1035.

Production Sectors

Energy Resource Summary [28]

Energy Resources: Uranium, natural gas; suspected deposits of oil, natural gas, coal. **Electricity**: 490,000 kW capacity; 1,290 million kWh produced, 850 kWh per capita (1991).

Telecommunications [30]

- Good urban, fair rural services
- Radio relay connects major towns, wires extend to other population centers
- 62,800 telephones
- Broadcast stations - 4 AM, 40 FM, 3 TV

Transportation [31]

Railroads. 2,341 km 1.067-meter gauge, single track

Highways. 54,500 km; 4,079 km paved, 2,540 km gravel, 47,881 km earth roads and tracks

Airports

Total:	137
Usable:	112
With permanent-surface runways:	21
With runways over 3,659 m:	1
With runways 2,440-3,659 m:	4
With runways 1,220-2,439 m:	62

Top Agricultural Products [32]

Agriculture accounts for 15% of GDP; mostly subsistence farming; livestock raising major source of cash income; crops—millet, sorghum, peanuts; fish catch potential of over 1 million metric tons not being fulfilled, 1988 catch reaching only 384,000 metric tons; not self-sufficient in food.

Top Mining Products [33]

Estimated metric tons unless otherwise specified	M.t.
Diamonds, gem & industrial (000 carats)	1,194
Uranium, content of concentrate	3,185
Copper, mine output, Cu content of concentrate	30,000
Silver, mine output, Ag content of concent. (kg)	89,000
Lead, mine output, Pb content of concentrate	11,800
Zinc, mine output, Zn content of concentrate	33,133

Tourism [34]

For sources, notes, and explanations, see Annotated Source Appendix, page 1035.

653

Finance, Economics, and Trade

GDP and Manufacturing Summary [35]

	1980	1985	1990	1991	1992
Gross Domestic Product					
Millions of 1980 dollars	2,007	1,871	2,156	2,219	2,241[e]
Growth rate in percent	0.18	0.00	6.00	2.90	1.00[e]
Manufacturing Value Added					
Millions of 1980 dollars	79	83	92	NA	NA
Growth rate in percent	-14.65	-3.54	5.91	NA	NA
Manufacturing share in percent of current prices	4.0	4.3[e]	4.3[e]	NA	NA

Economic Indicators [36]

National product: GDP—exchange rate conversion—$2 billion (1992 est.). **National product real growth rate**: 2% (1992 est.). **National product per capita**: $1,300 (1992 est.). **Inflation rate (consumer prices)**: 10% (1992) in urban area. **External debt**: about $220 million (1992 est.).

Balance of Payments Summary [37]

Exchange Rates [38]

Currency: **South African rand.**
Symbol: **R.**

Data are currency units per $1.

May 1993	3.1576
1992	2.8497
1991	2.7653
1990	2.5863
1989	2.6166
1988	2.2611

Imports and Exports

Top Import Origins [39]

$1.238 billion (f.o.b., 1991).

Origins	%
South Africa	NA
Germany	NA
US	NA
Switzerland	NA

Top Export Destinations [40]

$1.184 billion (f.o.b., 1991).

Destinations	%
Switzerland	NA
South Africa	NA
Germany	NA
Japan	NA

Foreign Aid [41]

	U.S. $	
Western (non-US) countries, ODA and OOF bilateral commitments (1970-87)	47.2	million

Import and Export Commodities [42]

Import Commodities	Export Commodities
Foodstuffs	Diamonds
Petroleum products and fuel	Copper
Machinery & equipment	Gold
	Zinc
	Lead
	Uranium
	Cattle
	Processed fish
	Karakul skins

For sources, notes, and explanations, see Annotated Source Appendix, page 1035.

Nepal

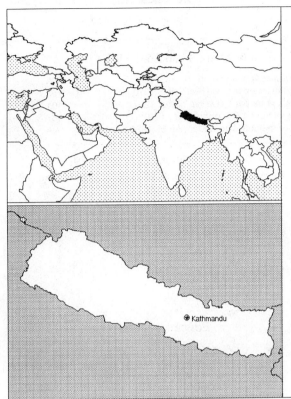

Geography [1]

Total area:
140,800 km2
Land area:
136,800 km2
Comparative area:
Slightly larger than Arkansas
Land boundaries:
Total 2,926 km, China 1,236 km, India 1,690 km
Coastline:
0 km (landlocked)
Climate:
Varies from cool summers and severe winters in north to subtropical summers and mild winters in south
Terrain:
Terai or flat river plain of the Ganges in south, central hill region, rugged Himalayas in north
Natural resources:
Quartz, water, timber, hydroelectric potential, scenic beauty, small deposits of lignite, copper, cobalt, iron ore
Land use:
Arable land: 17%
Permanent crops: 0%
Meadows and pastures: 13%
Forest and woodland: 33%
Other: 37%

Demographics [2]

	1960	1970	1980	1990	1991[1]	1994	2000	2010	2020
Population	10,035	11,919	15,001	19,104	19,612	21,042	24,364	30,783	37,767
Population density (persons per sq. mi.)	190	226	284	362	371	NA	461	580	710
Births	NA	NA	NA	NA	758	792	NA	NA	NA
Deaths	NA	NA	NA	NA	288	279	NA	NA	NA
Life expectancy - males	NA	NA	NA	NA	51	52	NA	NA	NA
Life expectancy - females	NA	NA	NA	NA	50	53	NA	NA	NA
Birth rate (per 1,000)	NA	NA	NA	NA	39	38	NA	NA	NA
Death rate (per 1,000)	NA	NA	NA	NA	15	13	NA	NA	NA
Women of reproductive age (15-44 yrs.)	NA	NA	NA	4,293	NA	4,817	5,750	NA	NA
of which are currently married	NA	NA	NA	3,518	NA	3,925	4,693	NA	NA
Fertility rate	NA	NA	NA	5.6	5.52	5.2	4.7	3.8	3.1

Population values are in thousands, life expectancy in years, and other items as indicated.

Health

Health Personnel [3]

Doctors per 1,000 pop., 1988-92	0.06
Nurse-to-doctor ratio, 1988-92	2.7
Hospital beds per 1,000 pop., 1985-90	0.3
Percentage of children immunized (age 1 yr. or less)	
Third dose of DPT, 1990-91	74.0
Measles, 1990-91	63.0

Health Expenditures [4]

Total health expenditure, 1990 (official exchange rate)	
Millions of dollars	141
Millions of dollars per capita	7
Health expenditures as a percentage of GDP	
Total	4.5
Public sector	2.2
Private sector	2.3
Development assistance for health	
Total aid flows (millions of dollars)[1]	33
Aid flows per capita (millions of dollars)	1.8
Aid flows as a percentage of total health expenditure	23.6

For sources, notes, and explanations, see Annotated Source Appendix, page 1035.

655

Human Factors

Health Care Ratios [5]

Population per physician, 1970	51360
Population per physician, 1990	16830
Population per nursing person, 1970	17700
Population per nursing person, 1990	2760
Percent of births attended by health staff, 1985	10

Infants and Malnutrition [6]

Percent of babies with low birth weight, 1985	NA
Infant mortality rate per 1,000 live births, 1970	157
Infant mortality rate per 1,000 live births, 1991	101
Years of life lost per 1,000 population, 1990	67
Prevalence of malnutrition (under age 5), 1990	NA

Ethnic Division [7]

Newars, Indians, Tibetans, Gurungs, Magars, Tamangs, Bhotias, Rais, Limbus, Sherpas.

Religion [8]

Hindu	90%
Buddhist	5%
Muslim	3%
Other	2%

Major Languages [9]

Nepali (official), 20 languages divided into numerous dialects.

Education

Public Education Expenditures [10]

Million Rupee	1985	1986	1987	1988	1989	1990
Total education expenditure	1,241	NA	NA	NA	NA	NA
as percent of GNP	2.8	NA	NA	NA	NA	NA
as percent of total govt. expend.	10.8	NA	NA	NA	NA	NA
Current education expenditure	NA	NA	NA	NA	NA	NA
as percent of GNP	NA	NA	NA	NA	NA	NA
as percent of current govt. expend.	NA	NA	NA	NA	NA	NA
Capital expenditure	NA	NA	NA	NA	NA	NA

Educational Attainment [11]

Age group	25+
Total population	1,012,465
Highest level attained (%)	
No schooling	41.2
First level	
Incompleted	29.4
Completed	NA
Entered second level	
S-1	22.7
S-2	NA
Post secondary	6.8

Literacy Rate [12]

In thousands and percent	1985[a]	1991[a]	2000[a]
Illiterate population +15 years	7,575	8,229	9,695
Illiteracy rate - total pop. (%)	77.6	74.4	66.7
Illiteracy rate - males (%)	66.2	62.4	54.0
Illiteracy rate - females (%)	89.3	86.8	80.0

Libraries [13]

	Admin. Units	Svc. Pts.	Vols. (000)	Shelving (meters)	Vols. Added	Reg. Users
National (1987)	1	1	71	NA	1,234	NA
Non-specialized	NA	NA	NA	NA	NA	NA
Public	NA	NA	NA	NA	NA	NA
Higher ed.	NA	NA	NA	NA	NA	NA
School	NA	NA	NA	NA	NA	NA

Daily Newspapers [14]

	1975	1980	1985	1990
Number of papers	29	28	28	28
Circ. (000)	110[e]	120[e]	130[e]	150[e]

Cinema [15]

Science and Technology

Scientific/Technical Forces [16]

Potential scientists/engineers	3,668
Number female	NA
Potential technicians	7,336[e]
Number female	NA
Total	11,004[e]

R&D Expenditures [17]

U.S. Patents Issued [18]

Government and Law

Organization of Government [19]

Long-form name:
Kingdom of Nepal
Type:
parliamentary democracy as of 12 May 1991
Independence:
1768 (unified by Prithvi Narayan Shah)
Constitution:
9 November 1990
Legal system:
based on Hindu legal concepts and English common law; has not accepted compulsory ICJ jurisdiction
National holiday:
Birthday of His Majesty the King, 28 December (1945)
Executive branch:
monarch, prime minister, Council of Ministers
Legislative branch:
bicameral Parliament consists of an upper house or National Council and a lower house or House of Representatives
Judicial branch:
Supreme Court (Sarbochha Adalat)

Political Parties [20]

House of Representatives	% of votes
Nepali Congress	38.0
Communist Party	28.0
National Democratic/Chand	6.0
United People's Front	5.0
National Democratic/Thapa	5.0
Terai Rights Sadbhavana	4.0
Rohit	2.0
CPN (Democratic)	1.0
Independents	4.0
Other	7.0

Government Expenditures [21]

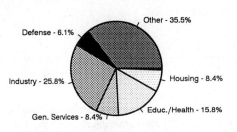

% distribution for 1990. Expend. in FY92 est.: 672.0 ($ mil.)

Crime [23]

Crime volume	
Cases known to police	5,350
Attempts (percent)	NA
Percent cases solved	82.04
Crimes per 100,000 persons	29.05
Persons responsible for offenses	
Total number offenders	8,874
Percent female	3.97
Percent juvenile[40]	0.03
Percent foreigners	2.78

Military Expenditures and Arms Transfers [22]

	1985	1986	1987	1988	1989
Military expenditures					
Current dollars (mil.)	23	26	28	29	33
1989 constant dollars (mil.)	26	29	30	31	33
Armed forces (000)	25	25	30	35	35
Gross national product (GNP)					
Current dollars (mil.)	2,050	2,192	2,335	2,656	2,811
1989 constant dollars (mil.)	2,334	2,432	2,512	2,765	2,811
Central government expenditures (CGE)					
1989 constant dollars (mil.)	422	443	467	538	535
People (mil.)	17.0	17.4	17.8	18.2	18.7
Military expenditure as % of GNP	1.1	1.2	1.2	1.1	1.2
Military expenditure as % of CGE	6.2	6.5	6.5	5.7	6.2
Military expenditure per capita	2	2	2	2	2
Armed forces per 1,000 people	1.5	1.4	1.7	1.9	1.9
GNP per capita	138	140	141	152	150
Arms imports[6]					
Current dollars (mil.)	5	0	0	20	0
1989 constant dollars (mil.)	6	0	0	21	0
Arms exports[6]					
Current dollars (mil.)	0	0	0	0	0
1989 constant dollars (mil.)	0	0	0	0	0
Total imports[7]					
Current dollars (mil.)	453	459	570	681	580
1989 constant dollars	516	509	613	709	580
Total exports[7]					
Current dollars (mil.)	160	142	151	190	158
1989 constant dollars	182	158	162	198	158
Arms as percent of total imports[8]	1.1	0	0	2.9	0
Arms as percent of total exports[8]	0	0	0	0	0

Human Rights [24]

	SSTS	FL	FAPRO	PPCG	APROBC	TPW	PCPTW	STPEP	PHRFF	PRW	ASST	AFL
Observes	P			P		P	P			P	P	
	EAFRD	CPR	ESCR	SR	ACHR	MAAE	PVIAC	PVNAC	EAFDAW	TCIDTP	RC	
Observes		P	P	P					P	P	P	

P = Party; S = Signatory; see Appendix for meaning of abbreviations.

Labor Force

Total Labor Force [25]

8.5 million (1991 est.)

Labor Force by Occupation [26]

Agriculture	93%
Services	5
Industry	2

Unemployment Rate [27]

5% (1987); underemployment estimated at 25-40%

Production Sectors

Energy Resource Summary [28]

Energy Resources: Hydroelectric potential, small deposits of lignite. **Electricity**: 300,000 kW capacity; 1,000 million kWh produced, 50 kWh per capita (1992).

Telecommunications [30]

- Poor telephone and telegraph service
- Fair radio communication and broadcast service
- International radio communication service is poor
- 50,000 telephones (1990)
- Broadcast stations - 88 AM, no FM, 1 TV
- 1 Indian Ocean INTELSAT earth station

Transportation [31]

Railroads. 52 km (1990), all 0.762-meter narrow gauge; all in Terai close to Indian border; 10 km from Raxaul to Birganj is government owned

Highways. 7,080 km total (1990); 2,898 km paved, 1,660 km gravel or crushed stone; also 2,522 km of seasonally motorable tracks

Airports

Total:	37
Usable:	37
With permanent-surface runways:	5
With runways over 3,659 m:	0
With runways 2,440-3,659 m:	1
With runways 1,220-2,439 m:	8

Top Agricultural Products [32]

Agriculture accounts for 60% of GDP and 90% of work force; farm products—rice, corn, wheat, sugarcane, root crops, milk, buffalo meat; not self-sufficient in food, particularly in drought years.

Top Mining Products [33]

Preliminary metric tons unless otherwise specified M.t.

Limestone	221,920
Cement, hydraulic	135,897
Clay for cement manufacture	8,850
Magnesite, crude	25,000[e]
Talc	3,500
Copper, gross weight	22
Lignite	10,150

Tourism [34]

	1987	1988	1989	1990	1991
Tourists[48]	248	266	240	255	293
Tourism receipts	82	94	106	109	126
Tourism expenditures	35	44	48	45	38
Fare receipts	10	6	2	6	22
Fare expenditures	16	12	4	19	31

Tourists are in thousands, money in million U.S. dollars.

658

For sources, notes, and explanations, see Annotated Source Appendix, page 1035.

Finance, Economics, and Trade

GDP and Manufacturing Summary [35]

	1980	1985	1990	1991	1992
Gross Domestic Product					
Millions of 1980 dollars	1,946	2,471	3,097	3,269	3,371
Growth rate in percent	-2.32	6.15	3.60	5.55	3.13
Manufacturing Value Added					
Millions of 1980 dollars	78	101	105	107	108[e]
Growth rate in percent	-8.24	-9.86	2.46	1.16	1.52[e]
Manufacturing share in percent of current prices	4.3	4.8	5.1	4.7[e]	NA

Economic Indicators [36]

National product: GDP—exchange rate conversion—$3.4 billion (FY92). **National product real growth rate**: 3.1% (FY92). **National product per capita**: $170 (FY92). **Inflation rate (consumer prices)**: 14% (November 1992). **External debt**: $2 billion (FY92 est.).

Balance of Payments Summary [37]

Values in millions of dollars.

	1987	1988	1989	1990	1991
Exports of goods (f.o.b.)	162.2	193.8	156.2	217.9	274.5
Imports of goods (f.o.b.)	-512.4	-664.9	-571.4	-666.6	-756.9
Trade balance	-350.2	-471.1	-415.2	-448.7	-482.4
Services - debits	-137.9	-165.3	-154.6	-178.6	-200.2
Services - credits	224.5	240.0	226.5	229.5	266.9
Private transfers (net)	67.2	60.1	52.0	60.4	53.7
Government transfers (net)	73.1	64.8	48.0	48.2	57.5
Long term capital (net)	127.1	212.9	213.3	178.7	223.6
South term capital (net)	63.6	39.7	-18.9	125.8	233.5
Errors and omissions	-3.6	12.5	4.8	4.9	10.7
Overall balance	63.8	-6.4	-44.1	20.2	163.3

Exchange Rates [38]

Currency: **Nepalese rupees.**
Symbol: **NRs.**

Data are currency units per $1.

January 1993	43.200
1992	42.742
1991	37.255
1990	29.370
1989	27.189
1988	23.289

Imports and Exports

Top Import Origins [39]

$751 million (c.i.f., FY92 est.).

Origins	%
India	NA
Singapore	NA
Japan	NA
Germany	NA

Top Export Destinations [40]

$313 million (f.o.b., FY92 est.). but does not include unrecorded border trade with India.

Destinations	%
US	NA
Germany	NA
India	NA
UK	NA

Foreign Aid [41]

	U.S. $	
US commitments, including Ex-Im (FY70-89)	304	million
Western (non-US) countries, ODA and OOF bilateral commitments (1980-89)	2,230	million
OPEC bilateral aid (1979-89)	30	million
Communist countries (1970-89)	286	million

Import and Export Commodities [42]

Import Commodities

Petroleum products 20%
Fertilizer 11%
Machinery 10%

Export Commodities

Carpets
Clothing
Leather goods
Jute goods
Grain

For sources, notes, and explanations, see Annotated Source Appendix, page 1035.

659

Netherlands

Geography [1]

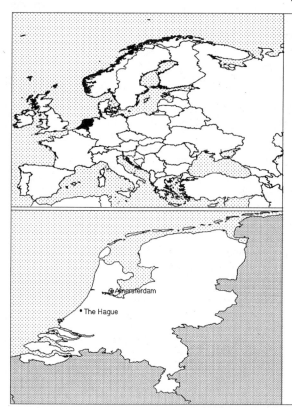

Total area:
37,330 km2
Land area:
33,920 km2
Comparative area:
Slightly less than twice the size of New Jersey
Land boundaries:
Total 1,027 km, Belgium 450 km, Germany 577 km
Coastline:
451 km
Climate:
Temperate; marine; cool summers and mild winters
Terrain:
Mostly coastal lowland and reclaimed land (polders); some hills in southeast
Natural resources:
Natural gas, petroleum, fertile soil
Land use:
Arable land:
26%
Permanent crops:
1%
Meadows and pastures:
32%
Forest and woodland:
9%
Other:
32%

Demographics [2]

	1960	1970	1980	1990	1991[1]	1994	2000	2010	2020
Population	11,486	13,032	14,144	14,952	15,022	15,368	15,801	16,140	16,222
Population density (persons per sq. mi.)	877	995	1,079	1,140	1,146	NA	1,194	1,207	1,198
Births	NA	NA	NA	NA	192	194	NA	NA	NA
Deaths	NA	NA	NA	NA	126	131	NA	NA	NA
Life expectancy - males	NA	NA	NA	NA	74	75	NA	NA	NA
Life expectancy - females	NA	NA	NA	NA	81	81	NA	NA	NA
Birth rate (per 1,000)	NA	NA	NA	NA	13	13	NA	NA	NA
Death rate (per 1,000)	NA	NA	NA	NA	8	9	NA	NA	NA
Women of reproductive age (15-44 yrs.)	NA	NA	NA	3,967	NA	4,005	3,882	NA	NA
of which are currently married	NA	NA	NA	2,215	NA	2,338	2,335	NA	NA
Fertility rate	NA	NA	NA	1.6	1.57	1.6	1.5	1.5	1.6

Population values are in thousands, life expectancy in years, and other items as indicated.

Health

Health Personnel [3]

Doctors per 1,000 pop., 1988-92	2.43
Nurse-to-doctor ratio, 1988-92	3.4
Hospital beds per 1,000 pop., 1985-90	5.9
Percentage of children immunized (age 1 yr. or less)	
Third dose of DPT, 1990-91	97.0
Measles, 1990-91	94.0

Health Expenditures [4]

Total health expenditure, 1990 (official exchange rate)	
Millions of dollars	22423
Millions of dollars per capita	1500
Health expenditures as a percentage of GDP	
Total	7.9
Public sector	5.7
Private sector	2.2
Development assistance for health	
Total aid flows (millions of dollars)[1]	NA
Aid flows per capita (millions of dollars)	NA
Aid flows as a percentage of total health expenditure	NA

For sources, notes, and explanations, see Annotated Source Appendix, page 1035.

Human Factors

Health Care Ratios [5]

Population per physician, 1970	800
Population per physician, 1990	410
Population per nursing person, 1970	300
Population per nursing person, 1990	NA
Percent of births attended by health staff, 1985	NA

Infants and Malnutrition [6]

Percent of babies with low birth weight, 1985	4
Infant mortality rate per 1,000 live births, 1970	13
Infant mortality rate per 1,000 live births, 1991	7
Years of life lost per 1,000 population, 1990	10
Prevalence of malnutrition (under age 5), 1990	NA

Ethnic Division [7]

Dutch	96%
Moroccans, Turks, other	4%

Religion [8]

Roman Catholic	36%
Protestant	27%
Other	6%
Unaffiliated	31%

Major Languages [9]

Dutch.

Education

Public Education Expenditures [10]

Million Guilder	1980	1985	1987	1988	1989	1990
Total education expenditure	26,606	28,298	31,616	30,618	30,522	32,232
as percent of GNP	7.9	6.8	7.4	6.8	6.4	6.3
as percent of total govt. expend.	NA	NA	NA	NA	NA	NA
Current education expenditure	23,079	24,796	28,013	27,907	28,272	29,005
as percent of GNP	6.9	5.9	6.5	6.2	5.9	5.7
as percent of current govt. expend.	NA	NA	NA	NA	NA	NA
Capital expenditure	3,527	3,502	3,603	2,711	2,250	3,227

Educational Attainment [11]

Literacy Rate [12]

Libraries [13]

	Admin. Units	Svc. Pts.	Vols. (000)	Shelving (meters)	Vols. Added	Reg. Users
National (1989)	1	NA	2,294	59,100	112,000	NA
Non-specialized (1989)	3	NA	995	NA	30,900	NA
Public (1989)[36]	586	1,170	40,776	NA	3.8mil	4.2mil
Higher ed. (1989)	477	NA	23,370	682,000	518,000	NA
School	NA	NA	NA	NA	NA	NA

Daily Newspapers [14]

	1975	1980	1985	1990[1]
Number of papers	84	84	88	86
Circ. (000)	4,194	4,612	4,496	4,592

Cinema [15]

Data for 1991.	
Cinema seats per 1,000	6.6
Annual attendance per person	1.0
Gross box office receipts (mil. Guilder)	182

Science and Technology

Scientific/Technical Forces [16]

Potential scientists/engineers	472,000
Number female	125,000
Potential technicians	927,000
Number female	417,000
Total	1,399,000

R&D Expenditures [17]

	Guilder[22] (000) 1988
Total expenditure	10,163,000
Capital expenditure	1,431,000
Current expenditure	8,732,000
Percent current	85.9

U.S. Patents Issued [18]

Values show patents issued to citizens of the country by the U.S. Patents Office.

	1990	1991	1992
Number of patents	1,044	1,075	974

Government and Law

Organization of Government [19]

Long-form name:
Kingdom of the Netherlands
Type:
constitutional monarchy
Independence:
1579 (from Spain)
Constitution:
17 February 1983
Legal system:
civil law system incorporating French penal theory; judicial review in the Supreme Court of legislation of lower order rather than Acts of the States General; accepts compulsory ICJ jurisdiction, with reservations
National holiday:
Queen's Day, 30 April (1938)
Executive branch:
monarch, prime minister, vice prime minister, Cabinet, Cabinet of Ministers
Legislative branch:
bicameral legislature (Staten Generaal) consists of an upper chamber or First Chamber (Eerste Kamer) and a lower chamber or Second Chamber (Tweede Kamer)
Judicial branch:
Supreme Court (De Hoge Raad)

Crime [23]

Crime volume	
Cases known to police	1,141,272
Attempts (percent)	NA
Percent cases solved	23
Crimes per 100,000 persons	7,708.57
Persons responsible for offenses	
Total number offenders	250,673
Percent female	10
Percent juvenile[42]	15
Percent foreigners	NA

Political Parties [20]

Second Chamber	% of votes
Christian Democratic (CDA)	35.3
Labor (PvdA)	31.9
Liberal (VVD)	14.6
Democrats '66 (D'66)	7.9
Other	10.3

Government Expenditures [21]

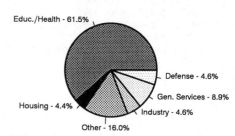

Educ./Health - 61.5%
Defense - 4.6%
Gen. Services - 8.9%
Housing - 4.4%
Industry - 4.6%
Other - 16.0%

% distribution for 1992. Expend. in 1992 est.: 122.1 ($ bil.)

Military Expenditures and Arms Transfers [22]

	1985	1986	1987	1988	1989
Military expenditures					
Current dollars (mil.)	5,433	5,675	5,949	6,070	6,399
1989 constant dollars (mil.)	6,185	6,298	6,398	6,319	6,399
Armed forces (000)	103	106	106	107	106
Gross national product (GNP)					
Current dollars (mil.)	176,400	185,100	193,200	205,000	222,500
1989 constant dollars (mil.)	200,800	205,400	207,800	213,400	222,500
Central government expenditures (CGE)					
1989 constant dollars (mil.)	114,000	111,400	116,500	118,600	119,600
People (mil.)	14.5	14.6	14.7	14.8	14.8
Military expenditure as % of GNP	3.1	3.1	3.1	3.0	2.9
Military expenditure as % of CGE	5.4	5.7	5.5	5.3	5.4
Military expenditure per capita	427	432	436	428	431
Armed forces per 1,000 people	7.1	7.3	7.2	7.2	7.1
GNP per capita	13,860	14,100	14,170	14,460	14,980
Arms imports[6]					
Current dollars (mil.)	525	500	600	410	480
1989 constant dollars (mil.)	598	555	645	427	480
Arms exports[6]					
Current dollars (mil.)	160	120	675	725	140
1989 constant dollars (mil.)	182	133	726	755	140
Total imports[7]					
Current dollars (mil.)	65,200	75,540	91,310	99,470	104,200
1989 constant dollars	74,220	83,820	98,210	103,600	104,200
Total exports[7]					
Current dollars (mil.)	67,910	80,360	92,570	103,200	115,600
1989 constant dollars	77,310	89,170	99,570	107,400	115,600
Arms as percent of total imports[8]	0.8	0.7	0.7	0.4	0.5
Arms as percent of total exports[8]	0.2	0.1	0.7	0.7	0.1

Human Rights [24]

	SSTS	FL	FAPRO	PPCG	APROBC	TPW	PCPTW	STPEP	PHRFF	PRW	ASST	AFL
Observes	P	P	P	P		P	P		P	P	P	P
	EAFRD	CPR	ESCR	SR	ACHR	MAAE	PVIAC	PVNAC	EAFDAW	TCIDTP	RC
Observes		P	P	P	P		P	P	P	P	P	S

P = Party; S = Signatory; see Appendix for meaning of abbreviations.

Labor Force

Total Labor Force [25]

5.3 million

Labor Force by Occupation [26]

Services	50.1%
Manufacturing and construction	28.2
Government	15.9
Agriculture	5.8

Date of data: 1986

Unemployment Rate [27]

5.3% (1992 est.)

662

For sources, notes, and explanations, see Annotated Source Appendix, page 1035.

Production Sectors

Commercial Energy Production and Consumption

Production [28]

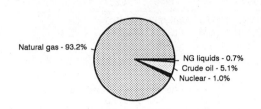

Natural gas - 93.2%
NG liquids - 0.7%
Crude oil - 5.1%
Nuclear - 1.0%

Consumption [29]

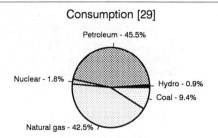

Petroleum - 45.5%
Nuclear - 1.8%
Hydro - 0.9%
Coal - 9.4%
Natural gas - 42.5%

Data are shown in quadrillion (10^{15}) BTUs and percent for 1991

Crude oil[1]	0.15
Natural gas liquids	0.02
Dry natural gas	2.74
Coal[2]	0.00
Net hydroelectric power[3]	(s)
Net nuclear power[3]	0.03
Total	2.95

Petroleum	1.55
Dry natural gas	1.45
Coal[2]	0.32
Net hydroelectric power[3]	0.03
Net nuclear power[3]	0.06
Total	3.41

Telecommunications [30]

- Highly developed, well maintained, and integrated
- Extensive redundant system of multiconductor cables, supplemented by microwave radio relay microwave links
- 9,418,000 telephones
- Broadcast stations - 3 (3 relays) AM, 12 (39 repeaters) FM, 8 (7 repeaters) TV
- 5 submarine cables
- 1 communication satellite earth station operating in INTELSAT (1 Indian Ocean and 2 Atlantic Ocean antenna) and EUTELSAT systems
- Nationwide mobile phone system

Transportation [31]

Railroads. 2,828 km 1.435-meter standard gauge operated by Netherlands Railways (NS) (includes 1,957 km electrified and 1,800 km double track)

Highways. 108,360 km total; 92,525 km paved (including 2,185 km of limited access, divided highways); 15,835 km gravel, crushed stone

Merchant Marine. 344 ships (1,000 GRT or over) totaling 2,762,000 GRT/3,675,649 DWT; includes 3 short-sea passenger, 193 cargo, 30 refrigerated cargo, 26 container, 13 roll-on/roll-off, 1 livestock carrier, 11 multifunction large-load carrier, 23 oil tanker, 22 chemical tanker, 10 liquefied gas, 2 specialized tanker, 6 bulk, 4 combination bulk; note - many Dutch-owned ships are also registered on the captive Netherlands Antilles register

Airports

Total:	28
Usable:	28
With permanent-surface runways:	20
With runways over 3,659 m:	0
With runways 2,440-3,659 m:	11
With runways 1,220-2,439 m:	6

Top Agricultural Products [32]

	1990[1]	1991[1]
Sugar beets	8,623.4	7,189.2
Potatoes	7,036.2	6,949.2
Corn for silage	2,362.9	2,336.7
Wheat	1,075.9	944.1
Grain (coarse)	282.2	305.1

Values shown are 1,000 metric tons.

Top Mining Products [33]

Metric tons unless otherwise specified	M.t.
Sand, industrial	25,000
Aluminum, primary	263,900[1]
Zinc, primary	201,300
Cement, hydraulic (000 tons)	3,255
Salt	3,400
Natural gas (mil. cubic meters)	73,000

Tourism [34]

	1987	1988	1989	1990	1991
Tourists[6]	4,922[9]	4,876	5,206	5,795	5,842
Tourism receipts	2,695	2,899	3,052	3,615	4,074
Tourism expenditures	6,408	6,750	6,481	7,340	7,886
Fare receipts	1,275	1,383	1,388	1,626	1,881
Fare expenditures	958	1,099	1,155	1,285	1,445

Tourists are in thousands, money in million U.S. dollars.

Manufacturing Sector

GDP and Manufacturing Summary [35]

	1980	1985	1989	1990	% change 1980-1990	% change 1989-1990
GDP (million 1980 $)	169,386	178,038	196,641	202,650	19.6	3.1
GDP per capita (1980 $)	11,976	12,291	13,238	13,557	13.2	2.4
Manufacturing as % of GDP (current prices)	18.9	18.9	20.2	22.2	17.5	9.9
Gross output (million $)	109,618	85,085	134,303	169,683	54.8	26.3
Value added (million $)	29,080	21,919	43,865	51,305	76.4	17.0
Value added (million 1980 $)	30,365	33,253	37,293	38,778	27.7	4.0
Industrial production index	100	106	118	120	20.0	1.7
Employment (thousands)	944	847	888	915	-3.1	3.0

Note: GDP stands for Gross Domestic Product. 'e' stands for estimated value.

Profitability and Productivity

	1980	1985	1989	1990	% change 1980-1990	% change 1989-1990
Intermediate input (%)	73	74	67	70	-4.1	4.5
Wages, salaries, and supplements (%)	20[e]	17[e]	14[e]	19[e]	-5.0	35.7
Gross operating surplus (%)	7[e]	9[e]	18[e]	11[e]	57.1	-38.9
Gross output per worker ($)	108,671	94,039	151,242	185,446	70.6	22.6
Value added per worker ($)	29,285	24,824	49,398	56,071	91.5	13.5
Average wage (incl. benefits) ($)	23,135[e]	16,940[e]	21,459[e]	35,414	53.1	65.0

Profitability is in percent of gross output. Productivity is in U.S. $. 'e' stands for estimated value.

Profitability - 1990

Inputs - 70.0%
Surplus - 11.0%
Wages - 19.0%

The graphic shows percent of gross output.

Value Added in Manufacturing

	1980 $ mil.	1980 %	1985 $ mil.	1985 %	1989 $ mil.	1989 %	1990 $ mil.	1990 %	% change 1980-1990	% change 1989-1990
311 Food products	4,562	15.7	3,388	15.5	6,627	15.1	6,476	12.6	42.0	-2.3
313 Beverages	654	2.2	458	2.1	961	2.2	851	1.7	30.1	-11.4
314 Tobacco products	282	1.0	238	1.1	605	1.4	952	1.9	237.6	57.4
321 Textiles	734	2.5	485	2.2	925	2.1	1,075	2.1	46.5	16.2
322 Wearing apparel	372	1.3	190	0.9	290	0.7	578	1.1	55.4	99.3
323 Leather and fur products	68	0.2	46	0.2	95	0.2	82	0.2	20.6	-13.7
324 Footwear	118	0.4	69	0.3	102	0.2	83	0.2	-29.7	-18.6
331 Wood and wood products	594	2.0	308	1.4	565[e]	1.3	727[e]	1.4	22.4	28.7
332 Furniture and fixtures	418	1.4	216	1.0	491[e]	1.1	690[e]	1.3	65.1	40.5
341 Paper and paper products	805	2.8	647	3.0	1,282	2.9	1,711	3.3	112.5	33.5
342 Printing and publishing	2,480	8.5	1,771	8.1	3,434	7.8	4,346	8.5	75.2	26.6
351 Industrial chemicals	2,263	7.8	2,163	9.9	4,396[e]	10.0	6,365[e]	12.4	181.3	44.8
352 Other chemical products	913	3.1	802	3.7	1,566[e]	3.6	2,668[e]	5.2	192.2	70.4
353 Petroleum refineries	533	1.8	515	2.3	1,617[e]	3.7	887	1.7	66.4	-45.1
354 Miscellaneous petroleum and coal products	101	0.3	54	0.2	206[e]	0.5	198[e]	0.4	96.0	-3.9
355 Rubber products	156	0.5	122	0.6	214[e]	0.5	270[e]	0.5	73.1	26.2
356 Plastic products	472	1.6	413	1.9	1,024[e]	2.3	1,381[e]	2.7	192.6	34.9
361 Pottery, china and earthenware	134	0.5	81	0.4	13[e]	0.0	199[e]	0.4	48.5	1,430.8
362 Glass and glass products	245	0.8	148	0.7	237[e]	0.5	379[e]	0.7	54.7	59.9
369 Other non-metal mineral products	893	3.1	539	2.5	1,512[e]	3.4	1,361[e]	2.7	52.4	-10.0
371 Iron and steel	882	3.0	784	3.6	1,005[e]	2.3	1,636[e]	3.2	85.5	62.8
372 Non-ferrous metals	371	1.3	330	1.5	946[e]	2.2	687[e]	1.3	85.2	-27.4
381 Metal products	2,455	8.4	1,780	8.1	3,781	8.6	4,246	8.3	73.0	12.3
382 Non-electrical machinery	2,369	8.1	1,774	8.1	3,700	8.4	4,001	7.8	68.9	8.1
383 Electrical machinery	3,687	12.7	2,864	13.1	4,953	11.3	5,355	10.4	45.2	8.1
384 Transport equipment	1,927	6.6	1,244	5.7	2,418	5.5	2,929	5.7	52.0	21.1
385 Professional and scientific equipment	237	0.8	198	0.9	585[e]	1.3	447	0.9	88.6	-23.6
390 Other manufacturing industries	356	1.2	296	1.4	316[e]	0.7	725[e]	1.4	103.7	129.4

Note: The industry codes shown are International Standard Industry codes (ISIC). Percentages are percent of total Value Added. 'e' stands for estimated value

Finance, Economics, and Trade

Economic Indicators [36]

Millions of Guilders unless otherwise noted[105].

	1989	1990	1991
Real GNP (1980 prices)	386,730	399,000	407,250[e]
Real GNP growth rate	4.25	3.25	2.1[e]
Total Money Supply (M1)			
(end of period)	119,025	125,612	125,000
CPI (1985 = 100)	101.2	103.7	107.1[e]
WPI (1985 = 100)	92.7	92.1	92.5[e]
External Public Debt	0	0	0

Balance of Payments Summary [37]

Values in millions of dollars.

	1987	1988	1989	1990	1991
Exports of goods (f.o.b.)	86,158	97,442	101,317	122,071	122,636
Imports of goods (f.o.b.)	-80,991	-88,966	-93,135	-111,673	-110,665
Trade balance	5,167	8,476	8,182	10,398	11,971
Services - debits	-36,304	-41,179	-45,332	-55,471	-57,262
Services - credits	37,350	41,628	49,212	57,460	58,637
Private transfers (net)	-1,050	-909	-944	-1,138	-1,163
Government transfers (net)	-1,233	-1,159	-1,309	-2,084	-3,190
Long term capital (net)	-2,793	1,492	4,713	-6,092	-6,463
Short term capital (net)	578	-489	1,208	-1,010	539
Errors and omissions	-404	-3,964	-3,881	-4,253	-5,080
Overall balance	2,862	1,627	451	277	75

Exchange Rates [38]

Currency: **Netherlands guilders, gulden, or florins**

Symbol: **f.**

Data are currency units per $1.

January 1993	1.8167
1992	1.7585
1991	1.8697
1990	1.8209
1989	2.1207
1988	1.9766

Imports and Exports

Top Import Origins [39]

$117.7 billion (f.o.b., 1992). Data are for 1991.

Origins	%
EC	64
Belgium-Luxembourg	14
UK	8
US	8

Top Export Destinations [40]

$128.5 billion (f.o.b., 1992). Data are for 1991.

Destinations	%
EC	77
Belgium-Luxembourg	15
UK	10
US	4

Foreign Aid [41]

	U.S. $	
Donor - ODA and OOF commitments (1970-89)	19.4	billion

Import and Export Commodities [42]

Import Commodities	**Export Commodities**
Raw materials	Agricultural products
Semifinished products	Foods and tobacco
Consumer goods	Natural gas
Transportation equipment	Chemicals
Crude oil	Metal products
Food products	Textiles
	Clothing

For sources, notes, and explanations, see Annotated Source Appendix, page 1035.

665

New Zealand

Geography [1]

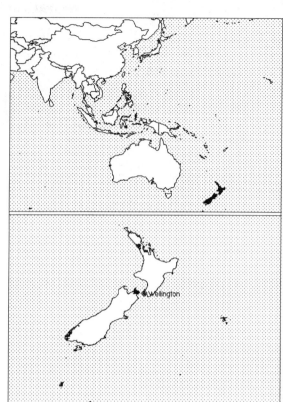

Total area:
268,680 km2
Land area:
268,670 km2
Comparative area:
About the size of Colorado
Land boundaries:
0 km
Coastline:
15,134 km
Climate:
Temperate with sharp regional contrasts
Terrain:
Predominately mountainous with some large coastal plains
Natural resources:
Natural gas, iron ore, sand, coal, timber, hydropower, gold, limestone
Land use:
Arable land:
2%
Permanent crops:
0%
Meadows and pastures:
53%
Forest and woodland:
38%
Other:
7%

Demographics [2]

	1960	1970	1980	1990	1991[1]	1994	2000	2010	2020
Population	2,372	2,811	3,113	3,300	3,309	3,389	3,476	3,543	3,586
Population density (persons per sq. mi.)	23	27	30	32	32	NA	33	33	33
Births	NA	NA	NA	NA	51	53	NA	NA	NA
Deaths	NA	NA	NA	NA	27	27	NA	NA	NA
Life expectancy - males	NA	NA	NA	NA	72	73	NA	NA	NA
Life expectancy - females	NA	NA	NA	NA	79	80	NA	NA	NA
Birth rate (per 1,000)	NA	NA	NA	NA	15	16	NA	NA	NA
Death rate (per 1,000)	NA	NA	NA	NA	8	8	NA	NA	NA
Women of reproductive age (15-44 yrs.)	NA	NA	NA	861	NA	876	864	NA	NA
of which are currently married	NA	NA	NA	497	NA	519	525	NA	NA
Fertility rate	NA	NA	NA	2.2	1.95	2.0	1.8	1.8	1.8

Population values are in thousands, life expectancy in years, and other items as indicated.

Health

Health Personnel [3]

Doctors per 1,000 pop., 1988-92	1.74
Nurse-to-doctor ratio, 1988-92	0.1
Hospital beds per 1,000 pop., 1985-90	6.6
Percentage of children immunized (age 1 yr. or less)	
Third dose of DPT, 1990-91	81.0
Measles, 1990-91	82.0

Health Expenditures [4]

Total health expenditure, 1990 (official exchange rate)	
Millions of dollars	3150
Millions of dollars per capita	925
Health expenditures as a percentage of GDP	
Total	7.2
Public sector	5.9
Private sector	1.3
Development assistance for health	
Total aid flows (millions of dollars)[1]	NA
Aid flows per capita (millions of dollars)	NA
Aid flows as a percentage of total health expenditure	NA

For sources, notes, and explanations, see Annotated Source Appendix, page 1035.

Human Factors

Health Care Ratios [5]

Population per physician, 1970	870
Population per physician, 1990	NA
Population per nursing person, 1970	150
Population per nursing person, 1990	NA
Percent of births attended by health staff, 1985	99

Infants and Malnutrition [6]

Percent of babies with low birth weight, 1985	5
Infant mortality rate per 1,000 live births, 1970	17
Infant mortality rate per 1,000 live births, 1991	9
Years of life lost per 1,000 population, 1990	11
Prevalence of malnutrition (under age 5), 1990	NA

Ethnic Division [7]

European	88%
Maori	8.9%
Pacific Islander	2.9%
Other	0.2%

Religion [8]

Anglican	24%
Presbyterian	18%
Roman Catholic	15%
Methodist	5%
Baptist	2%
Other Protestant	3%
Unspecified or none	9%

Major Languages [9]

English (official), Maori.

Education

Public Education Expenditures [10]

	1980	1985	1987	1988	1989	1990
Million New Zealand Dollars				d		
Total education expenditure	1,302	2,028	3,179	3,638	NA	4,451
as percent of GNP	5.8	4.7	5.4	5.7	NA	6.4
as percent of total govt. expend.	23.1	18.4	NA	NA	NA	NA
Current education expenditure	1,171	1,849	2,912	3,302	NA	4,252
as percent of GNP	5.2	4.3	5.0	5.2	NA	6.1
as percent of current govt. expend.	27.9	25.5	NA	NA	NA	NA
Capital expenditure	131	179	267	336	NA	199

Educational Attainment [11]

Age group	25+
Total population	1,213,566
Highest level attained (%)	
No schooling	-
First level	
Incompleted	60.4
Completed	NA
Entered second level	
S-1	18.3
S-2	12.8
Post secondary	8.5

Literacy Rate [12]

Libraries [13]

	Admin. Units	Svc. Pts.	Vols. (000)	Shelving (meters)	Vols. Added	Reg. Users
National	NA	NA	NA	NA	NA	NA
Non-specialized	NA	NA	NA	NA	NA	NA
Public	NA	NA	NA	NA	NA	NA
Higher ed. (1990)	7	33	5,910	NA	181,848	NA
School	NA	NA	NA	NA	NA	NA

Daily Newspapers [14]

	1975	1980	1985	1990
Number of papers	40	32	33	35
Circ. (000)	900[e]	1,059	1,075	1,100[e]

Cinema [15]

Science and Technology

Scientific/Technical Forces [16]

Potential scientists/engineers	139,200
Number female	54,523
Potential technicians	716,691
Number female	286,458
Total	855,891

R&D Expenditures [17]

U.S. Patents Issued [18]

Values show patents issued to citizens of the country by the U.S. Patents Office.

	1990	1991	1992
Number of patents	66	50	60

For sources, notes, and explanations, see Annotated Source Appendix, page 1035.

667

Government and Law

Organization of Government [19]

Long-form name:
none
Type:
parliamentary democracy
Independence:
26 September 1907 (from UK)
Constitution:
no formal, written constitution; consists of various documents, including certain acts of the UK and New Zealand Parliaments; Constitution Act 1986 was to have come into force 1 January 1987, but has not been enacted
Legal system:
based on English law, with special land legislation and land courts for Maoris; accepts compulsory ICJ jurisdiction, with reservations
National holiday:
Waitangi Day, 6 February (1840) (Treaty of Waitangi established British sovereignty)
Executive branch:
British monarch, governor general, prime minister, deputy prime minister, Cabinet
Legislative branch:
unicameral House of Representatives (commonly called Parliament)
Judicial branch:
High Court, Court of Appeal

Crime [23]

Political Parties [20]

House of Representatives	% of votes
National Party (NP)	49.0
New Zealand Labor Party (NZLP)	35.0
Green Party	7.0
NewLabor Party (NLP)	5.0

Government Expenditures [21]

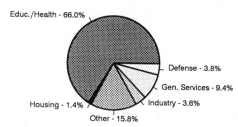

Educ./Health - 66.0%
Defense - 3.8%
Gen. Services - 9.4%
Industry - 3.6%
Housing - 1.4%
Other - 15.8%

% distribution for 1993. Expend. in 1992: 15.2 ($ bil.)

Military Expenditures and Arms Transfers [22]

	1985	1986	1987	1988	1989
Military expenditures					
Current dollars (mil.)	694	770	768	838	847
1989 constant dollars (mil.)	790	854	827	872	847
Armed forces (000)	13	12	13	13	12
Gross national product (GNP)					
Current dollars (mil.)	33,940	34,300	35,220	37,190	39,090
1989 constant dollars (mil.)	38,640	38,060	37,880	38,710	39,090
Central government expenditures (CGE)					
1989 constant dollars (mil.)	17,520	18,100	17,870	16,310	15,150[e]
People (mil.)	3.2	3.2	3.3	3.3	3.3
Military expenditure as % of GNP	2.0	2.2	2.2	2.3	2.2
Military expenditure as % of CGE	4.5	4.7	4.6	5.3	5.6
Military expenditure per capita	243	263	254	267	258
Armed forces per 1,000 people	4.0	3.8	3.9	3.9	3.8
GNP per capita	11,900	11,720	11,630	11,840	11,910
Arms imports[6]					
Current dollars (mil.)	80	50	40	90	50
1989 constant dollars (mil.)	91	55	43	94	50
Arms exports[6]					
Current dollars (mil.)	0	0	0	5	0
1989 constant dollars (mil.)	0	0	0	5	0
Total imports[7]					
Current dollars (mil.)	5,992	6,063	7,276	7,342	8,787
1989 constant dollars	6,821	6,728	7,826	7,643	8,787
Total exports[7]					
Current dollars (mil.)	5,720	5,880	7,195	8,784	8,885
1989 constant dollars	6,512	6,525	7,739	9,144	8,885
Arms as percent of total imports[8]	1.3	0.8	0.5	1.2	0.6
Arms as percent of total exports[8]	0	0	0	0.1	0

Human Rights [24]

	SSTS	FL	FAPRO	PPCG	APROBC	TPW	PCPTW	STPEP	PHRFF	PRW	ASST	AFL
Observes	P	P		P		P	P			P	P	P
	EAFRD	CPR	ESCR	SR	ACHR	MAAE	PVIAC	PVNAC	EAFDAW	TCIDTP	RC	
Observes	P	P	P	P			P	P	P	P	S	

P = Party; S = Signatory; see Appendix for meaning of abbreviations.

Labor Force

Total Labor Force [25]

1,603,500 (June 1991)

Labor Force by Occupation [26]

Services	67.4%
Manufacturing	19.8
Primary production	9.3

Date of data: 1987

Unemployment Rate [27]

10.1% (September 1992)

Production Sectors

Commercial Energy Production and Consumption

Production [28]

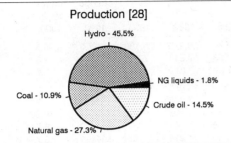

Hydro - 45.5%
NG liquids - 1.8%
Crude oil - 14.5%
Natural gas - 27.3%
Coal - 10.9%

Consumption [29]

Hydro - 37.9%
Coal - 7.6%
Natural gas - 22.7%
Petroleum - 31.8%

Data are shown in quadrillion (10^{15}) BTUs and percent for 1991

Crude oil[1]	0.08
Natural gas liquids	0.01
Dry natural gas	0.15
Coal[2]	0.06
Net hydroelectric power[3]	0.25
Net nuclear power[3]	0.00
Total	0.54

Petroleum	0.21
Dry natural gas	0.15
Coal[2]	0.05
Net hydroelectric power[3]	0.25
Net nuclear power[3]	0.00
Total	0.66

Telecommunications [30]

- Excellent international and domestic systems
- 2,110,000 telephones
- Broadcast stations - 64 AM, 2 FM, 14 TV
- Submarine cables extend to Australia and Fiji
- 2 Pacific Ocean INTELSAT earth stations

Transportation [31]

Railroads. 4,716 km total; all 1.067-meter gauge; 274 km double track; 113 km electrified; over 99% government owned

Highways. 92,648 km total; 49,547 km paved, 43,101 km gravel or crushed stone

Merchant Marine. 18 ships (1,000 GRT or over) totaling 182,206 GRT/246,446 DWT; includes 2 cargo, 5 roll-on/roll-off, 1 railcar carrier, 4 oil tanker, 1 liquefied gas, 5 bulk

Airports

Total:	120
Usable:	120
With permanent-surface runways:	33
With runways over 3,659 m:	1
With runways 2,440-3,659 m:	2
With runways 1,220-2,439 m:	42

Top Agricultural Products [32]

	1990	1991
Roundwood (000 cu. m.)	13,105	13,500
Apples	404	424
Barley	378	420
Kiwifruit	250	209
Wheat	185	198
Corn	161	195

Values shown are 1,000 metric tons.

Top Mining Products [33]

Estimated metric tons unless otherwise specified   M.t.

Clays	91,500
Coal	2,550
Sand, gravel for roads, aggregate (000 tons)	15,000
Dolomite	14,000
Limestone and marl (000 tons)	3,150
Marble	NA
Gold, mine output, Au content (kilograms)	6,611
Titaniferous magnetite sand (000 tons)	2,300

Tourism [34]

	1987	1988	1989	1990	1991
Tourists[1]	844	865	901	976	963
Tourism receipts	935	1,014	1,005	1,019	1,021
Tourism expenditures	852	1,253	1,252	1,335	1,408
Fare receipts	319	310	384	448	439
Fare expenditures	298	180	255	285	355

Tourists are in thousands, money in million U.S. dollars.

Manufacturing Sector

GDP and Manufacturing Summary [35]

	1980	1985	1989	1990	% change 1980-1990	% change 1989-1990
GDP (million 1980 $)	22,344	25,670	26,270	26,138	17.0	-0.5
GDP per capita (1980 $)	7,178	7,903	7,811	7,706	7.4	-1.3
Manufacturing as % of GDP (current prices)	21.8	21.8	18.9[e]	18.4[e]	-15.6	-2.6
Gross output (million $)	14,790	15,399	23,059[e]	23,293[e]	57.5	1.0
Value added (million $)	4,756	4,657	7,065[e]	7,159[e]	50.5	1.3
Value added (million 1980 $)	4,948	5,885	5,795[e]	5,400	9.1	-6.8
Industrial production index	100	118	107[e]	127	27.0	18.7
Employment (thousands)	285[e]	280	214	222	-22.1	3.7

Note: GDP stands for Gross Domestic Product. 'e' stands for estimated value.

Profitability and Productivity

	1980	1985	1989	1990	% change 1980-1990	% change 1989-1990
Intermediate input (%)	68	70	69[e]	69[e]	1.5	0.0
Wages, salaries, and supplements (%)	22[e]	18[e]	15[e]	17[e]	-22.7	13.3
Gross operating surplus (%)	10[e]	12[e]	15[e]	14[e]	40.0	-6.7
Gross output per worker ($)	51,964[e]	50,964	107,586[e]	97,456[e]	87.5	-9.4
Value added per worker ($)	16,711[e]	15,414	32,962[e]	29,963[e]	79.3	-9.1
Average wage (incl. benefits) ($)	11,354[e]	10,127[e]	16,535[e]	17,676[e]	55.7	6.9

Profitability is in percent of gross output. Productivity is in U.S. $. 'e' stands for estimated value.

Profitability - 1990

Inputs - 69.0%
Surplus - 14.0%
Wages - 17.0%

The graphic shows percent of gross output.

Value Added in Manufacturing

	1980 $ mil.	1980 %	1985 $ mil.	1985 %	1989 $ mil.	1989 %	1990 $ mil.	1990 %	% change 1980-1990	% change 1989-1990
311 Food products	1,098	23.1	1,082	23.2	1,677[e]	23.7	1,530[e]	21.4	39.3	-8.8
313 Beverages	110	2.3	93	2.0	208[e]	2.9	191[e]	2.7	73.6	-8.2
314 Tobacco products	30	0.6	19	0.4	55[e]	0.8	50[e]	0.7	66.7	-9.1
321 Textiles	222	4.7	193	4.1	237[e]	3.4	225[e]	3.1	1.4	-5.1
322 Wearing apparel	185	3.9	170	3.7	237[e]	3.4	217[e]	3.0	17.3	-8.4
323 Leather and fur products	45	0.9	46	1.0	67[e]	0.9	56[e]	0.8	24.4	-16.4
324 Footwear	55	1.2	46	1.0	45[e]	0.6	42[e]	0.6	-23.6	-6.7
331 Wood and wood products	253	5.3	257	5.5	327[e]	4.6	335[e]	4.7	32.4	2.4
332 Furniture and fixtures	92	1.9	95	2.0	128[e]	1.8	124[e]	1.7	34.8	-3.1
341 Paper and paper products	266	5.6	276	5.9	464[e]	6.6	491[e]	6.9	84.6	5.8
342 Printing and publishing	294	6.2	326	7.0	546[e]	7.7	546[e]	7.6	85.7	0.0
351 Industrial chemicals	140	2.9	134	2.9	237[e]	3.4	227[e]	3.2	62.1	-4.2
352 Other chemical products	155	3.3	142	3.0	202[e]	2.9	208[e]	2.9	34.2	3.0
353 Petroleum refineries	26	0.5	-1	0.0	236[e]	3.3	376[e]	5.3	1,346.2	59.3
354 Miscellaneous petroleum and coal products	9	0.2	7	0.2	12[e]	0.2	13[e]	0.2	44.4	8.3
355 Rubber products	96	2.0	70	1.5	72[e]	1.0	66[e]	0.9	-31.3	-8.3
356 Plastic products	110	2.3	138	3.0	205[e]	2.9	230[e]	3.2	109.1	12.2
361 Pottery, china and earthenware	13	0.3	11	0.2	16[e]	0.2	17[e]	0.2	30.8	6.3
362 Glass and glass products	44	0.9	41	0.9	61[e]	0.9	61[e]	0.9	38.6	0.0
369 Other non-metal mineral products	114	2.4	127	2.7	172[e]	2.4	177[e]	2.5	55.3	2.9
371 Iron and steel	93	2.0	71	1.5	85[e]	1.2	81[e]	1.1	-12.9	-4.7
372 Non-ferrous metals	82	1.7	102	2.2	201[e]	2.8	205[e]	2.9	150.0	2.0
381 Metal products	371	7.8	404	8.7	496[e]	7.0	534[e]	7.5	43.9	7.7
382 Non-electrical machinery	235	4.9	264	5.7	393[e]	5.6	350[e]	4.9	48.9	-10.9
383 Electrical machinery	239	5.0	200	4.3	249[e]	3.5	298[e]	4.2	24.7	19.7
384 Transport equipment	318	6.7	274	5.9	319[e]	4.5	392[e]	5.5	23.3	22.9
385 Professional and scientific equipment	14	0.3	20	0.4	24[e]	0.3	23[e]	0.3	64.3	-4.2
390 Other manufacturing industries	45	0.9	48	1.0	92[e]	1.3	93[e]	1.3	106.7	1.1

Note: The industry codes shown are International Standard Industry codes (ISIC). Percentages are percent of total Value Added. 'e' stands for estimated value

For sources, notes, and explanations, see Annotated Source Appendix, page 1035.

Finance, Economics, and Trade

Economic Indicators [36]

Millions of NZ Dollars unless otherwise noted[106].

	1989	1990	1991
Real GDP (constant 1983 prices)	35,264	35,288	35,621
Real GDP Growth Rate (%)	1.2	0.1	0.9
Money supply (M1)	8,188	8,837	8,812
Personal savings rate	2.0	-0.8	-2.9
Investment rate (GFCF as % of GDP)	26.1	29.2	29.2
CPI (December 1988 equals 1000)	995	1,059	1,117
External public debt	16,777	20,104	20,198

Balance of Payments Summary [37]

Values in millions of dollars.

	1987	1988	1989	1990	1991
Exports of goods (f.o.b.)	7,245	8,831	8,846	9,283	9,545
Imports of goods (f.o.b.)	-6,663	-6,667	-7,873	-8,294	-7,503
Trade balance	582	2,164	973	989	2,042
Services - debits	-5,452	-6,108	-5,906	-5,999	-5,864
Services - credits	2,819	3,092	3,048	3,106	3,125
Private transfers (net)	178	310	491	650	726
Government transfers (net)	-40	-43	-27	-44	-49
Long term capital (net)	936	-1,574	-588	339	-819
Short term capital (net)	-210	804	1,123	2,639	643
Errors and omissions	799	622	1,124	-666	-1,124
Overall balance	-388	-733	238	1,014	-1,320

Exchange Rates [38]

Currency: **New Zealand dollars.**
Symbol: **NZ$.**

Data are currency units per $1.

January 1993	1.9486
1992	1.8584
1991	1.7265
1990	1.6750
1989	1.6711
1988	1.5244

Imports and Exports

Top Import Origins [39]

$3.99 billion (f.o.b., FY92).

Origins	%
Australia	19.7
Japan	16.9
EC	16.9
US	15.3
Taiwan	3.0

Top Export Destinations [40]

$3.65 billion (f.o.b., FY92).

Destinations	%
EC	18.3
Japan	17.9
Australia	17.5
US	13.5
China	3.6
South Korea	3.1

Foreign Aid [41]

	U.S. $	
Donor - ODA and OOF commitments (1970-89)	526	million

Import and Export Commodities [42]

Import Commodities

Petroleum
Consumer goods
Motor vehicles
Industrial equipment

Export Commodities

Wool
Lamb
Mutton
Beef
Fruit
Fish
Cheese
Manufactures
Chemicals
Forestry products

For sources, notes, and explanations, see Annotated Source Appendix, page 1035.

671

Nicaragua

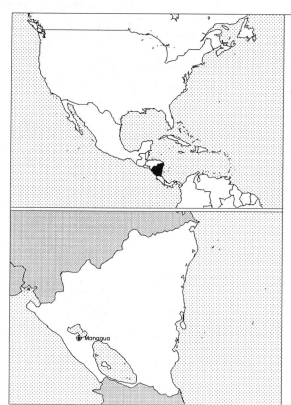

Geography [1]

Total area:
129,494 km2
Land area:
120,254 km2
Comparative area:
Slightly larger than New York State
Land boundaries:
Total 1,231 km, Costa Rica 309 km, Honduras 922 km
Coastline:
910 km
Climate:
Tropical in lowlands, cooler in highlands
Terrain:
Extensive Atlantic coastal plains rising to central interior mountains;
narrow Pacific coastal plain interrupted by volcanoes
Natural resources:
Gold, silver, copper, tungsten, lead, zinc, timber, fish
Land use:
Arable land:
9%
Permanent crops:
1%
Meadows and pastures:
43%
Forest and woodland:
35%
Other:
12%

Demographics [2]

	1960	1970	1980	1990	1991[1]	1994	2000	2010	2020
Population	1,493	2,053	2,776	3,617	3,752	4,097	4,759	5,864	6,945
Population density (persons per sq. mi.)	32	44	60	78	81	NA	102	126	151
Births	NA	NA	NA	NA	139	142	NA	NA	NA
Deaths	NA	NA	NA	NA	28	27	NA	NA	NA
Life expectancy - males	NA	NA	NA	NA	60	61	NA	NA	NA
Life expectancy - females	NA	NA	NA	NA	65	67	NA	NA	NA
Birth rate (per 1,000)	NA	NA	NA	NA	37	35	NA	NA	NA
Death rate (per 1,000)	NA	NA	NA	NA	7	7	NA	NA	NA
Women of reproductive age (15-44 yrs.)	NA	NA	NA	833	NA	980	1,218	NA	NA
of which are currently married	NA	NA	NA	496	NA	586	733	NA	NA
Fertility rate	NA	NA	NA	5.0	4.72	4.3	3.5	2.6	2.2

Population values are in thousands, life expectancy in years, and other items as indicated.

Health

Health Personnel [3]

Doctors per 1,000 pop., 1988-92	0.60
Nurse-to-doctor ratio, 1988-92	0.5
Hospital beds per 1,000 pop., 1985-90	1.8
Percentage of children immunized (age 1 yr. or less)	
Third dose of DPT, 1990-91	71.0
Measles, 1990-91	54.0

Health Expenditures [4]

Total health expenditure, 1990 (official exchange rate)	
Millions of dollars	133
Millions of dollars per capita	35
Health expenditures as a percentage of GDP	
Total	8.6
Public sector	6.7
Private sector	1.9
Development assistance for health	
Total aid flows (millions of dollars)[1]	27
Aid flows per capita (millions of dollars)	6.6
Aid flows as a percentage of total health expenditure	20.0

For sources, notes, and explanations, see Annotated Source Appendix, page 1035.

Human Factors

Health Care Ratios [5]

Population per physician, 1970	2150
Population per physician, 1990	1450
Population per nursing person, 1970	NA
Population per nursing person, 1990	NA
Percent of births attended by health staff, 1985	NA

Infants and Malnutrition [6]

Percent of babies with low birth weight, 1985	15
Infant mortality rate per 1,000 live births, 1970	106
Infant mortality rate per 1,000 live births, 1991	56
Years of life lost per 1,000 population, 1990	45
Prevalence of malnutrition (under age 5), 1990	NA

Ethnic Division [7]

Mestizo	69%
White	17%
Black	9%
Indian	5%

Religion [8]

Roman Catholic	95%
Protestant	5%

Major Languages [9]

Spanish (official).

Education

Public Education Expenditures [10]

Million Cordoba	1980	1985[1]	1986[1]	1987[1]	1988[1]	1989[1]
Total education expenditure	0.662	6.409	27	144	10,717	399,895
as percent of GNP	3.4	6.1	6.8	5.8	3.4	2.6
as percent of total govt. expend.	10.4	10.2	12.6	12.0	NA	NA
Current education expenditure	0.580	6.196	26	140	10,559	383,109
as percent of GNP	3.0	5.9	6.7	5.6	3.4	2.5
as percent of current govt. expend.	NA	12.2	15.5	13.1	NA	NA
Capital expenditure	0.082	0.213	1	4	158	16,786

Educational Attainment [11]

Literacy Rate [12]

	1971[b]	1980[b]	1990[b]
Illiterate population +15 years	410,755	NA	NA
Illiteracy rate - total pop. (%)[5]	42.5	NA	NA
Illiteracy rate - males (%)	42.0	NA	NA
Illiteracy rate - females (%)	42.9	NA	NA

Libraries [13]

	Admin. Units	Svc. Pts.	Vols. (000)	Shelving (meters)	Vols. Added	Reg. Users
National	NA	NA	NA	NA	NA	NA
Non-specialized	NA	NA	NA	NA	NA	NA
Public	NA	NA	NA	NA	NA	NA
Higher ed. (1987)	13	18	281	NA	8,252	6,962[17]
School (1987)	412	412	595	6,834	NA	71,948

Daily Newspapers [14]

	1975	1980	1985	1990
Number of papers	7	3	3	6
Circ. (000)	100[e]	136	160[e]	250[e]

Cinema [15]

Science and Technology

Scientific/Technical Forces [16]

R&D Expenditures [17]

	Cordoba[4] (000) 1987
Total expenditure	NA
Capital expenditure	NA
Current expenditure	988,970
Percent current	NA

U.S. Patents Issued [18]

For sources, notes, and explanations, see Annotated Source Appendix, page 1035.

673

Government and Law

Organization of Government [19]

Long-form name:
Republic of Nicaragua
Type:
republic
Independence:
15 September 1821 (from Spain)
Constitution:
January 1987
Legal system:
civil law system; Supreme Court may
review administrative acts
National holiday:
Independence Day, 15 September (1821)
Executive branch:
president, vice president, Cabinet
Legislative branch:
unicameral National Assembly (Asamblea
Nacional)
Judicial branch:
Supreme Court (Corte Suprema)

Political Parties [20]

National Assembly	% of votes
National Opposition Union	53.9
Sandinista National Liberation	40.8
Social Christian	1.6
Revolutionary Unity	1.0

Government Budget [21]

For 1991.
Revenues	347
Expenditures	499
Capital expenditures	NA

Crime [23]

Military Expenditures and Arms Transfers [22]

	1985	1986	1987	1988	1989
Military expenditures					
Current dollars (mil.)	192[e]	NA	NA	NA	NA
1989 constant dollars (mil.)	218[e]	NA	NA	NA	NA
Armed forces (000)	74	75	80	74	74
Gross national product (GNP)					
Current dollars (mil.)	1,114	1,032	1,182	1,094	1,106
1989 constant dollars (mil.)	1,268	1,145	1,272	1,139	1,106
Central government expenditures (CGE)					
1989 constant dollars (mil.)	834	664	NA	594	254
People (mil.)	3.3	3.4	3.4	3.5	3.6
Military expenditure as % of GNP	17.2	NA	NA	NA	NA
Military expenditure as % of CGE	26.2	NA	NA	NA	NA
Military expenditure per capita	66	NA	NA	NA	NA
Armed forces per 1,000 people	22.5	22.2	23.2	21.0	20.4
GNP per capita	385	340	369	323	305
Arms imports[6]					
Current dollars (mil.)	280	600	500	550	430
1989 constant dollars (mil.)	319	666	538	573	430
Arms exports[6]					
Current dollars (mil.)	0	0	0	0	0
1989 constant dollars (mil.)	0	0	0	0	0
Total imports[7]					
Current dollars (mil.)	964	857	923	800	550
1989 constant dollars	1,097	951	993	833	550
Total exports[7]					
Current dollars (mil.)	302	274	300	236	250
1989 constant dollars	344	274	323	245	250
Arms as percent of total imports[8]	29.0	70.0	54.2	68.8	78.2
Arms as percent of total exports[8]	0	0	0	0	0

Human Rights [24]

	SSTS	FL	FAPRO	PPCG	APROBC	TPW	PCPTW	STPEP	PHRFF	PRW	ASST	AFL
Observes	P	P	P	P	P	P	P			P	P	P
	EAFRD	CPR	ESCR	SR	ACHR	MAAE	PVIAC	PVNAC	EAFDAW	TCIDTP	RC	
Observes		P	P	P	P	P	P	S	S	P	S	P

P = Party; S = Signatory; see Appendix for meaning of abbreviations.

Labor Force

Total Labor Force [25]

1.086 million

Labor Force by Occupation [26]

Service	43%
Agriculture	44
Industry	13
Date of data: 1986

Unemployment Rate [27]

13% (1991)

For sources, notes, and explanations, see Annotated Source Appendix, page 1035.

Production Sectors

Energy Resource Summary [28]

Energy Resources: None. **Electricity**: 434,000 kW capacity; 1,118 million kWh produced, 290 kWh per capita (1992). **Pipelines**: Crude oil 56 km.

Telecommunications [30]

- Low-capacity radio relay and wire system being expanded
- Connection into Central American Microwave System
- 60,000 telephones
- Broadcast stations - 45 AM, no FM, 7 TV, 3 shortwave
- Earth stations - 1 Intersputnik and 1 Atlantic Ocean INTELSAT

Transportation [31]

Railroads. 373 km 1.067-meter narrow gauge, government owned; majority of system not operating; 3 km 1.435-meter gauge line at Puerto Cabezas (does not connect with mainline)

Highways. 25,930 km total; 4,000 km paved, 2,170 km gravel or crushed stone, 5,425 km earth or graded earth, 14,335 km unimproved; Pan-American highway 368.5 km

Merchant Marine. 2 cargo ships (1,000 GRT or over) totaling 2,161 GRT/2,500 DWT

Airports

Total:	226
Usable:	151
With permanent-surface runways:	11
With runways over 3,659 m:	0
With runways 2,440-3,659 m:	2
With runways 1,220-2,439 m:	12

Top Agricultural Products [32]

	1990[23]	1991[23]
Sugar	240	270
Corn	221	250
Bananas	95	117
Rice	74	98
Sorghum	63	72
Coffee	27	35

Values shown are 1,000 metric tons.

Top Mining Products [33]

Preliminary metric tons unless otherwise specified M.t.

Bentonite	5,070
Cement	239,300
Gold, mine output, Au content (kilograms)	1,154
Gypsum and anhydrite, crude	16,200
Lime	2,120[e1]
Petroleum refinery products (000 42-gal. barrels)	4,543[e1]
Salt, marine	15,000[e]
Sand and gravel (000 tons)	1,170[e1]
Silver, mine output, Ag content (kilograms)	1,543

Tourism [34]

	1987	1988	1989	1990	1991
Visitors	99	109	118	140	
Tourists	49	55	77	106	146
Excursionists			22	22	
Tourism receipts	9	5	4	12	17
Tourism expenditures	6	2	1	15	28

Tourists are in thousands, money in million U.S. dollars.

For sources, notes, and explanations, see Annotated Source Appendix, page 1035.

675

Manufacturing Sector

GDP and Manufacturing Summary [35]

	1980	1985	1989	1990	% change 1980-1990	% change 1989-1990
GDP (million 1980 $)	1,489	1,537	1,389	1,289	-13.4	-7.2
GDP per capita (1980 $)	537	470	371	333	-38.0	-10.2
Manufacturing as % of GDP (current prices)	24.4	27.6	17.0[e]	12.7	-48.0	-25.3
Gross output (million $)	612	1,587	NA	2,733[e]	346.6	NA
Value added (million $)	242	982	NA	1,781[e]	636.0	NA
Value added (million 1980 $)	351	366	270	232	-33.9	-14.1
Industrial production index	100	116	96	121	21.0	26.0
Employment (thousands)	34	39	45[e]	46[e]	35.3	2.2

Note: GDP stands for Gross Domestic Product. 'e' stands for estimated value.

Profitability and Productivity

	1980	1985	1989	1990	% change 1980-1990	% change 1989-1990
Intermediate input (%)	60	38	NA	35[e]	-41.7	NA
Wages, salaries, and supplements (%)	12	10	NA	11[e]	-8.3	NA
Gross operating surplus (%)	28	52	NA	54[e]	92.9	NA
Gross output per worker ($)	18,017	38,009	NA	55,171[e]	206.2	NA
Value added per worker ($)	7,131	23,515	NA	35,959[e]	404.3	NA
Average wage (incl. benefits) ($)	2,078	4,152	NA	6,439[e]	209.9	NA

Profitability is in percent of gross output. Productivity is in U.S. $. 'e' stands for estimated value.

Profitability - 1990

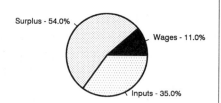

Surplus - 54.0%
Wages - 11.0%
Inputs - 35.0%

The graphic shows percent of gross output.

Value Added in Manufacturing

	1980 $ mil.	1980 %	1985 $ mil.	1985 %	1989 $ mil.	1989 %	1990 $ mil.	1990 %	% change 1980-1990	% change 1989-1990
311 Food products	52	21.5	268	27.3	NA	NA	451[e]	25.3	767.3	NA
313 Beverages	48	19.8	227	23.1	NA	NA	474[e]	26.6	887.5	NA
314 Tobacco products	28	11.6	64	6.5	NA	NA	134[e]	7.5	378.6	NA
321 Textiles	9	3.7	70	7.1	NA	NA	117[e]	6.6	1,200.0	NA
322 Wearing apparel	4	1.7	23	2.3	NA	NA	40[e]	2.2	900.0	NA
323 Leather and fur products	2	0.8	6	0.6	NA	NA	14[e]	0.8	600.0	NA
324 Footwear	4	1.7	27	2.7	NA	NA	48[e]	2.7	1,100.0	NA
331 Wood and wood products	3	1.2	10	1.0	NA	NA	16[e]	0.9	433.3	NA
332 Furniture and fixtures	1	0.4	4	0.4	NA	NA	5[e]	0.3	400.0	NA
341 Paper and paper products	1	0.4	3	0.3	NA	NA	3[e]	0.2	200.0	NA
342 Printing and publishing	4	1.7	22	2.2	NA	NA	41[e]	2.3	925.0	NA
351 Industrial chemicals	11	4.5	23	2.3	NA	NA	36[e]	2.0	227.3	NA
352 Other chemical products	14	5.8	56	5.7	NA	NA	122[e]	6.9	771.4	NA
353 Petroleum refineries	35	14.5	78	7.9	NA	NA	116[e]	6.5	231.4	NA
354 Miscellaneous petroleum and coal products	NA	0.0	1	0.1	NA	NA	2[e]	0.1	NA	NA
355 Rubber products	1	0.4	6	0.6	NA	NA	12[e]	0.7	1,100.0	NA
356 Plastic products	4	1.7	20	2.0	NA	NA	31[e]	1.7	675.0	NA
361 Pottery, china and earthenware	NA	0.0	2	0.2	NA	NA	NA	0.0	NA	NA
362 Glass and glass products	NA	0.0	1	0.1	NA	NA	2[e]	0.1	NA	NA
369 Other non-metal mineral products	7	2.9	17	1.7	NA	NA	27[e]	1.5	285.7	NA
371 Iron and steel	NA	0.0	1	0.1	NA	NA	2[e]	0.1	NA	NA
372 Non-ferrous metals	NA	0.0	NA	0.0	NA	NA	NA	0.0	NA	NA
381 Metal products	9	3.7	40	4.1	NA	NA	70[e]	3.9	677.8	NA
382 Non-electrical machinery	NA	0.0	3	0.3	NA	NA	4[e]	0.2	NA	NA
383 Electrical machinery	1	0.4	5	0.5	NA	NA	9[e]	0.5	800.0	NA
384 Transport equipment	1	0.4	3	0.3	NA	NA	5[e]	0.3	400.0	NA
385 Professional and scientific equipment	1	0.4	NA	0.0	NA	NA	NA	0.0	NA	NA
390 Other manufacturing industries	NA	0.0	2	0.2	NA	NA	3[e]	0.2	NA	NA

Note: The industry codes shown are International Standard Industry codes (ISIC). Percentages are percent of total Value Added. 'e' stands for estimated value

For sources, notes, and explanations, see Annotated Source Appendix, page 1035.

Finance, Economics, and Trade

Economic Indicators [36]

Monetary Units in U.S. Dollars[107].

	1989	1990[p]	1991[e]
GDP (millions US$)[108]	1,323	1,339	1,421
Percent Growth in Real GDP[109]	-2.8	-4.4	1.0
GDP per capita (US$)	353	347	NA
Money supply (US$ million)	46.6	68.5	78.1
Commercial interest rate	NA	18.0	18.0
Savings rate	-22.2	-26.8	-13.2
Investment rate[110]	26.1	24.7	NA
CPI (% change)[111]	4,770	7,485	2,741
External public debt	9,597	10,585	9,996

Balance of Payments Summary [37]

Values in millions of dollars.

	1987	1988	1989	1990	1991
Exports of goods (f.o.b.)	295.1	235.7	318.7	332.4	268.1
Imports of goods (f.o.b.)	-734.4	-718.3	-547.3	-569.7	-688.0
Trade balance	-439.3	-482.6	-228.6	-237.3	-419.9
Services - debits	-405.8	-402.3	-330.8	-341.1	-509.2
Services - credits	30.9	39.5	28.8	71.6	79.9
Private transfers (net)	-	-	-	-	-
Government transfers (net)	135.4	130.0	168.9	201.6	844.4
Long term capital (net)	80.4	203.2	-98.6	-167.9	189.7
Short term capital (net)	697.0	564.7	593.6	615.0	-183.5
Errors and omissions	-78.9	51.9	-69.2	-181.2	84.7
Overall balance	19.9	104.4	64.1	-39.3	86.1

Exchange Rates [38]

Currency: **cordobas.**
Symbol: **C$.**

New gold cordoba issued in 1992. Data are currency units per $1.

10 January 1993	6
March 1992	25,000,000
1991	21,354,000
1989	15,655
1988	270
1987	103

Imports and Exports

Top Import Origins [39]

$720 million (c.i.f., 1992 est.). Data are for 1990 est.

Origins	%
Latin America	30
US	25
EC	20
USSR and Eastern Europe	10
Other	15

Top Export Destinations [40]

$280 million (f.o.b., 1992 est.).

Destinations	%
OECD	75
USSR and Eastern Europe	15
Other	10

Foreign Aid [41]

	U.S. $	
US commitments, including Ex-Im (FY70-89)	294	million
Western (non-US) countries, ODA and OOF bilateral commitments (1970-89)	1,381	million
Communist countries (1970-89)	3.5	billion

Import and Export Commodities [42]

Import Commodities	**Export Commodities**
Petroleum	Coffee
Food	Cotton
Chemicals	Sugar
Machinery	Bananas
Clothing	Seafood
	Meat
	Chemicals

Niger

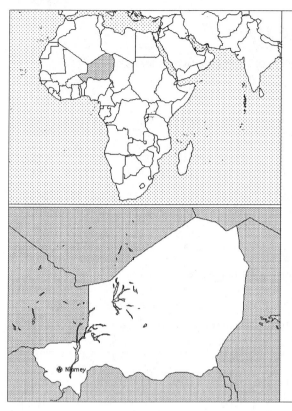

Geography [1]

Total area:
1.267 million km2
Land area:
1,266,700 km2
Comparative area:
Slightly less than twice the size of Texas
Land boundaries:
Total 5,697 km, Algeria 956 km, Benin 266 km, Burkina 628 km, Chad 1,175 km, Libya 354 km, Mali 821 km, Nigeria 1,497 km
Coastline:
0 km (landlocked)
Climate:
Desert; mostly hot, dry, dusty; tropical in extreme south
Terrain:
Predominately desert plains and sand dunes; flat to rolling plains in south; hills in north
Natural resources:
Uranium, coal, iron ore, tin, phosphates
Land use:
Arable land:
3%
Permanent crops:
0%
Meadows and pastures:
7%
Forest and woodland:
2%
Other:
88%

Demographics [2]

	1960	1970	1980	1990	1991[1]	1994	2000	2010	2020
Population	3,105	4,100	5,563	7,521	8,154	8,635	10,651	14,652	20,166
Population density (persons per sq. mi.)	6	8	11	16	17	NA	23	31	42
Births	NA	NA	NA	NA	409	494	NA	NA	NA
Deaths	NA	NA	NA	NA	130	189	NA	NA	NA
Life expectancy - males	NA	NA	NA	NA	49	43	NA	NA	NA
Life expectancy - females	NA	NA	NA	NA	53	46	NA	NA	NA
Birth rate (per 1,000)	NA	NA	NA	NA	50	57	NA	NA	NA
Death rate (per 1,000)	NA	NA	NA	NA	16	22	NA	NA	NA
Women of reproductive age (15-44 yrs.)	NA	NA	NA	NA	NA	NA	NA	NA	NA
of which are currently married	NA	NA	NA	NA	NA	NA	NA	NA	NA
Fertility rate	NA	NA	NA	7.4	7.03	7.4	7.0	6.2	5.3

Population values are in thousands, life expectancy in years, and other items as indicated.

Health

Health Personnel [3]

Doctors per 1,000 pop., 1988-92	0.03
Nurse-to-doctor ratio, 1988-92	11.3
Hospital beds per 1,000 pop., 1985-90	NA
Percentage of children immunized (age 1 yr. or less)	
Third dose of DPT, 1990-91	18.0
Measles, 1990-91	24.0

Health Expenditures [4]

Total health expenditure, 1990 (official exchange rate)	
Millions of dollars	126
Millions of dollars per capita	16
Health expenditures as a percentage of GDP	
Total	5.0
Public sector	3.4
Private sector	1.6
Development assistance for health	
Total aid flows (millions of dollars)[1]	43
Aid flows per capita (millions of dollars)	5.6
Aid flows as a percentage of total health expenditure	34.0

For sources, notes, and explanations, see Annotated Source Appendix, page 1035.

Human Factors

Health Care Ratios [5]

Population per physician, 1970	60090
Population per physician, 1990	34850
Population per nursing person, 1970	5610
Population per nursing person, 1990	650
Percent of births attended by health staff, 1985	47

Infants and Malnutrition [6]

Percent of babies with low birth weight, 1985	20
Infant mortality rate per 1,000 live births, 1970	170
Infant mortality rate per 1,000 live births, 1991	126
Years of life lost per 1,000 population, 1990	121
Prevalence of malnutrition (under age 5), 1990	49

Ethnic Division [7]

Hausa	56.0%
Djerma	22.0%
Fula	8.5%
Tuareg	8.0%
Beri Beri (Kanouri)	4.3%
Arab	NA
Toubou Gourmantche	1.2%
French expatriates	4,000

Religion [8]

Muslim	80%
Indigenous beliefs and Christians	20%

Major Languages [9]

French (official), Hausa, Djerma.

Education

Public Education Expenditures [10]

Million Francs C.F.A.	1980	1985	1987	1988	1989[2]	1990
Total education expenditure	16,533	NA	NA	NA	19,873	NA
as percent of GNP	3.1	NA	NA	NA	3.1	NA
as percent of total govt. expend.	22.9	NA	NA	NA	9.0	NA
Current education expenditure	7,763	NA	NA	NA	15,545	NA
as percent of GNP	1.5	NA	NA	NA	2.5	NA
as percent of current govt. expend.	16.8	NA	NA	NA	13.6	NA
Capital expenditure	8,770	NA	NA	NA	4,328	

Educational Attainment [11]

Literacy Rate [12]

In thousands and percent	1985[a]	1991[a]	2000[a]
Illiterate population + 15 years	2,558	2,683	2,945
Illiteracy rate - total pop. (%)	78.5	71.6	57.7
Illiteracy rate - males (%)	67.9	59.6	44.5
Illiteracy rate - females (%)	88.7	83.2	70.6

Libraries [13]

	Admin. Units	Svc. Pts.	Vols. (000)	Shelving (meters)	Vols. Added	Reg. Users
National	NA	NA	NA	NA	NA	NA
Non-specialized	NA	NA	NA	NA	NA	NA
Public	NA	NA	NA	NA	NA	NA
Higher ed. (1987)	1	1	21	NA	638	1,392
School	NA	NA	NA	NA	NA	NA

Daily Newspapers [14]

	1975	1980	1985	1990
Number of papers	2	1	1	
Circ. (000)	4	3	3	3

Cinema [15]

Science and Technology

Scientific/Technical Forces [16]

R&D Expenditures [17]

U.S. Patents Issued [18]

For sources, notes, and explanations, see Annotated Source Appendix, page 1035.

679

Government and Law

Organization of Government [19]

Long-form name:
Republic of Niger
Type:
transition government as of November 1991, appointed by national reform conference; scheduled to turn over power to democratically elected government in March 1993
Independence:
3 August 1960 (from France)
Constitution:
December 1989 constitution revised November 1991 by National Democratic Reform Conference
Legal system:
based on French civil law system and customary law; has not accepted compulsory ICJ jurisdiction
National holiday:
Republic Day, 18 December (1958)
Executive branch:
president (ceremonial), prime minister, Cabinet
Legislative branch:
unicameral National Assembly
Judicial branch:
State Court (Cour d'Etat), Court of Appeal (Cour d'Apel)

Crime [23]

Elections [20]

National Assembly. Last held 10 December 1989 (next to be held NA); results - MNSD was the only party; seats—(150 total) MNSD 150 (indirectly elected); note—Niger held a national conference from July to November 1991 to decide upon a transitional government and an agenda for multiparty elections.

Government Budget [21]

For 1991 est.

Revenues	193
Expenditures	355
Capital expenditures	106

Military Expenditures and Arms Transfers [22]

	1985	1986	1987	1988	1989
Military expenditures					
Current dollars (mil.)	14[e]	15[e]	NA	19	27
1989 constant dollars (mil.)	16[e]	17[e]	NA	19	27
Armed forces (000)	5[e]	4	5	4	4
Gross national product (GNP)					
Current dollars (mil.)	1,689	1,852	1,818	1,970	1,987
1989 constant dollars (mil.)	1,926	2,056	1,955	2,050	1,987
Central government expenditures (CGE)					
1989 constant dollars (mil.)	321[e]	292[e]	NA	463	NA
People (mil.)	6.6	6.9	7.1	7.4	7.6
Military expenditure as % of GNP	0.8	0.8	NA	1.0	1.3
Military expenditure as % of CGE	5.0	5.7	NA	4.2	NA
Military expenditure per capita	2	2	NA	3	3
Armed forces per 1,000 people	0.8	0.6	0.7	0.6	0.5
GNP per capita	290	299	275	279	261
Arms imports[6]					
Current dollars (mil.)	0	20	5	5	5
1989 constant dollars (mil.)	0	22	5	5	5
Arms exports[6]					
Current dollars (mil.)	0	0	0	0	0
1989 constant dollars (mil.)	0	0	0	0	0
Total imports[7]					
Current dollars (mil.)	345	NA	NA	441	NA
1989 constant dollars	393	NA	NA	459	NA
Total exports[7]					
Current dollars (mil.)	209	243	NA	371	NA
1989 constant dollars	238	270	NA	386	NA
Arms as percent of total imports[8]	0	NA	NA	1.1	NA
Arms as percent of total exports[8]	0	0	NA	0	NA

Human Rights [24]

	SSTS	FL	FAPRO	PPCG	APROBC	TPW	PCPTW	STPEP	PHRFF	PRW	ASST	AFL
Observes	P	P	P		P	P	P	P		P	P	P
	EAFRD	CPR	ESCR	SR	ACHR	MAAE	PVIAC	PVNAC	EAFDAW	TCIDTP	RC
Observes	P	P	P	P		P	P	P			P

P=Party; S=Signatory; see Appendix for meaning of abbreviations.

Labor Force

Total Labor Force [25]

2.5 million wage earners (1982)

Labor Force by Occupation [26]

Agriculture	90%
Industry and commerce	6
Government	4

Unemployment Rate [27]

Production Sectors

Energy Resource Summary [28]

Energy Resources: Uranium, coal. **Electricity**: 105,000 kW capacity; 230 million kWh produced, 30 kWh per capita (1991).

Telecommunications [30]

- Small system of wire, radiocommunications, and radio relay links concentrated in southwestern area
- 14,260 telephones
- Broadcast stations - 15 AM, 5 FM, 18 TV
- Satellite earth stations - 1 Atlantic Ocean INTELSAT, 1 Indian Ocean INTELSAT, and 3 domestic, with 1 planned

Transportation [31]

Highways. 39,970 km total; 3,170 km bituminous, 10,330 km gravel and laterite, 3,470 km earthen, 23,000 km tracks

Airports

Total:	28
Usable:	26
With permanent-surface runways:	9
With runways over 3,659 m:	0
With runways 2,440-3,659 m:	2
With runways 1,220-2,439 m:	13

Top Agricultural Products [32]

Agriculture accounts for roughly 40% of GDP and 90% of labor force; cash crops—cowpeas, cotton, peanuts; food crops—millet, sorghum, cassava, rice; livestock—cattle, sheep, goats; self-sufficient in food except in drought years.

Top Mining Products [33]

Estimated metric tons unless otherwise specified   M.t.

Coal	156,542
Cement, hydraulic	20,109
Uranium, content of concentrate	3,330[1]
Salt	3,000[1]
Gypsum	1,000[1]
Tin, mine output, Sn content	20

Tourism [34]

	1987	1988	1989	1990	1991
Tourists[16]	31	33	24	21	16
Tourism receipts	9	11	14	17	16
Tourism expenditures	34	35	32	44	40
Fare receipts	30	13	15	18	18
Fare expenditures	34	33	32	40	43

Tourists are in thousands, money in million U.S. dollars.

Finance, Economics, and Trade

GDP and Manufacturing Summary [35]

	1980	1985	1990	1991	1992
Gross Domestic Product					
Millions of 1980 dollars	2,538	2,473	2,639	2,675	2,718[e]
Growth rate in percent	4.90	5.70	2.80	1.35	1.60[e]
Manufacturing Value Added					
Millions of 1980 dollars	94	100	98	99	101[e]
Growth rate in percent	4.68	8.25	1.70	1.73	1.27[e]
Manufacturing share in percent of current prices	3.8	7.4	8.9[e]	NA	NA

Economic Indicators [36]

National product: GDP—exchange rate conversion—$2.3 billion (1991 est.). **National product real growth rate**: 1.9% (1991 est.). **National product per capita**: $290 (1991 est.). **Inflation rate (consumer prices)**: 1.3% (1991 est.). **External debt**: $1.2 billion (December 1991 est.).

Balance of Payments Summary [37]

Values in millions of dollars.

	1987	1988	1989	1990	1991
Exports of goods (f.o.b.)	411.9	369.0	311.0	303.4	283.9
Imports of goods (f.o.b.)	-409.6	-392.5	-368.6	-337.5	-273.3
Trade balance	2.3	-23.5	-57.6	-34.1	10.6
Services - debits	-245.2	-218.6	-202.5	-256.4	-195.7
Services - credits	60.6	50.0	57.07	71.3	58.5
Private transfers (net)	-49.9	-34.9	-40.4	-48.8	-37.9
Government transfers (net)	142.1	143.7	132.0	184.4	160.6
Long term capital (net)	102.5	95.7	95.9	105.8	5.3
Short term capital (net)	20.4	-21.9	-	-6.6	34.4
Errors and omissions	-15.4	43.4	-4.1	-25.2	-40.4
Overall balance	17.4	33.9	-19.0	-9.6	-4.6

Exchange Rates [38]

Currency: **Communaute Financiere Africaine francs.**
Symbol: **CFAF.**

Data are currency units per $1.

January 1993	274.06
1992	264.69
1991	282.11
1990	272.26
1989	319.01
1988	297.85

Imports and Exports

Top Import Origins [39]

$346 million (c.i.f., 1991).

Origins	%
Germany	26
Cote d'Ivoire	11
France	5
Italy	4
Nigeria	2

Top Export Destinations [40]

$294 million (f.o.b., 1991).

Destinations	%
France	77
Nigeria	8
Cote d'Ivoire	NA
Italy	NA

Foreign Aid [41]

	U.S. $	
US commitments, including Ex-Im (FY70-89)	380	million
Western (non-US) countries, ODA and OOF bilateral commitments (1970-89)	3,165	million
OPEC bilateral aid (1979-89)	504	million
Communist countries (1970-89)	61	million

Import and Export Commodities [42]

Import Commodities	**Export Commodities**
Primary materials	Uranium ore 60%
Machinery	Livestock products 20%
Vehicles and parts	Cowpeas
Electronic equipment	Onions
Cereals	
Petroleum products	
Pharmaceuticals	
Chemical products	
Foodstuffs	

682

For sources, notes, and explanations, see Annotated Source Appendix, page 1035.

Nigeria

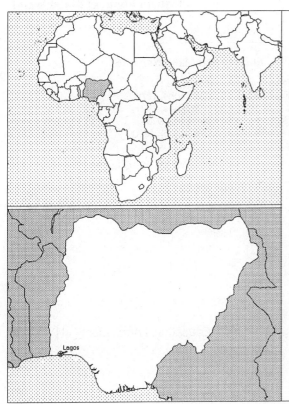

Geography [1]

Total area:
923,770 km2
Land area:
910,770 km2
Comparative area:
Slightly more than twice the size of California
Land boundaries:
Total 4,047 km, Benin 773 km, Cameroon 1,690 km, Chad 87 km, Niger 1,497 km
Coastline:
853 km
Climate:
Varies; equatorial in south, tropical in center, arid in north
Terrain:
Southern lowlands merge into central hills and plateaus; mountains in southeast, plains in north
Natural resources:
Petroleum, tin, columbite, iron ore, coal, limestone, lead, zinc, natural gas
Land use:
Arable land:
31%
Permanent crops:
3%
Meadows and pastures:
23%
Forest and woodland:
15%
Other:
28%

Demographics [2]

	1960	1970	1980	1990	1991[1]	1994	2000	2010	2020
Population	39,230	49,309	65,699	86,551	122,471	98,091	118,620	161,969	215,893
Population density (persons per sq. mi.)	145	190	256	338	348	NA	457	606	777
Births	NA	NA	NA	NA	5,641	4,269	NA	NA	NA
Deaths	NA	NA	NA	NA	1,993	1,219	NA	NA	NA
Life expectancy - males	NA	NA	NA	NA	48	54	NA	NA	NA
Life expectancy - females	NA	NA	NA	NA	50	57	NA	NA	NA
Birth rate (per 1,000)	NA	NA	NA	NA	46	44	NA	NA	NA
Death rate (per 1,000)	NA	NA	NA	NA	16	12	NA	NA	NA
Women of reproductive age (15-44 yrs.)	NA	NA	NA	18,957	NA	21,539	26,276	NA	NA
of which are currently married	NA	NA	NA	14,871	NA	16,871	20,505	NA	NA
Fertility rate	NA	NA	NA	6.6	6.52	6.4	6.0	5.1	4.2

Population values are in thousands, life expectancy in years, and other items as indicated.

Health

Health Personnel [3]

Doctors per 1,000 pop., 1988-92	0.15
Nurse-to-doctor ratio, 1988-92	6.0
Hospital beds per 1,000 pop., 1985-90	1.4
Percentage of children immunized (age 1 yr. or less)	
Third dose of DPT, 1990-91	65.0
Measles, 1990-91	70.0

Health Expenditures [4]

Total health expenditure, 1990 (official exchange rate)	
Millions of dollars	906
Millions of dollars per capita	9
Health expenditures as a percentage of GDP	
Total	2.7
Public sector	1.2
Private sector	1.6
Development assistance for health	
Total aid flows (millions of dollars)[1]	58
Aid flows per capita (millions of dollars)	0.6
Aid flows as a percentage of total health expenditure	6.4

For sources, notes, and explanations, see Annotated Source Appendix, page 1035.

Human Factors

Health Care Ratios [5]

Population per physician, 1970	19830
Population per physician, 1990	NA
Population per nursing person, 1970	4240
Population per nursing person, 1990	NA
Percent of births attended by health staff, 1985	NA

Infants and Malnutrition [6]

Percent of babies with low birth weight, 1985	25
Infant mortality rate per 1,000 live births, 1970	139
Infant mortality rate per 1,000 live births, 1991	85
Years of life lost per 1,000 population, 1990	98
Prevalence of malnutrition (under age 5), 1990	NA

Ethnic Division [7]

No information provided.

Religion [8]

Muslim	50%
Christian	40%
Indigenous beliefs	10%

Major Languages [9]

English (official)	NA
Hausa	NA
Yoruba	NA
Ibo	NA
Fulani	NA

Education

Public Education Expenditures [10]

Million Naira	1985[5]	1986[5]	1987	1988	1989	1990
Total education expenditure	814	1,170	NA	NA	NA	NA
as percent of GNP	1.2	1.7	NA	NA	NA	NA
as percent of total govt. expend.	8.7	12.0	NA	NA	NA	NA
Current education expenditure	699	728	NA	NA	NA	NA
as percent of GNP	1.0	1.1	NA	NA	NA	NA
as percent of current govt. expend.	19.7	19.0	NA	NA	NA	NA
Capital expenditure	115	442	NA	NA	NA	NA

Educational Attainment [11]

Literacy Rate [12]

In thousands and percent	1985[a]	1991[a]	2000[a]
Illiterate population +15 years	28,224	28,723	28,448
Illiteracy rate - total pop. (%)	57.3	49.3	34.4
Illiteracy rate - males (%)	45.2	37.7	24.9
Illiteracy rate - females (%)	68.9	60.5	43.6

Libraries [13]

	Admin. Units	Svc. Pts.	Vols. (000)	Shelving (meters)	Vols. Added	Reg. Users
National (1989)	1	15	558	17,120	6,713	29,906
Non-specialized	NA	NA	NA	NA	NA	NA
Public (1989)	12	92	1,108	15,361	38,877	46,728
Higher ed. (1987)	63	144	3,842	NA	105,057	139,938
School (1987)	213	16,714	1,088	-	16,100	13,763

Daily Newspapers [14]

	1975	1980	1985	1990
Number of papers	12	16	19	31
Circ. (000)	650[e]	1,100[e]	1,400[e]	1,700[e]

Cinema [15]

Science and Technology

Scientific/Technical Forces [16]

Potential scientists/engineers	22,050
Number female	NA
Potential technicians	79,550
Number female	NA
Total	101,600

R&D Expenditures [17]

	Naira[6] (000) 1987
Total expenditure	86,270
Capital expenditure	16,655
Current expenditure	69,615
Percent current	80.7

U.S. Patents Issued [18]

Values show patents issued to citizens of the country by the U.S. Patents Office.

	1990	1991	1992
Number of patents	0	1	2

For sources, notes, and explanations, see Annotated Source Appendix, page 1035.

Government and Law

Organization of Government [19]

Long-form name:
Federal Republic of Nigeria
Type:
military government since 31 December 1983; plans to turn over power to elected civilians in August 1993
Independence:
1 October 1960 (from UK)
Constitution:
1 October 1979, amended 9 February 1984, revised 1989
Legal system:
based on English common law, Islamic law, and tribal law
National holiday:
Independence Day, 1 October (1960)
Executive branch:
president, vice-president, cabinet
Legislative branch:
bicameral National Assembly consists of an upper house or Senate and a lower house or House of Representatives
Judicial branch:
Supreme Court, Federal Court of Appeal

Political Parties [20]

House of Representatives	% of seats
Social Democratic (SDP)	53.7
National Republican (NRC)	46.3

Government Budget [21]

For 1992 est.
Revenues	9.0
Expenditures	10.8
Capital expenditures	NA

Crime [23]

Military Expenditures and Arms Transfers [22]

	1985	1986	1987	1988	1989
Military expenditures					
Current dollars (mil.)	294	247	184	201	130
1989 constant dollars (mil.)	335	274	198	210	130
Armed forces (000)	134	138	138	107	107
Gross national product (GNP)					
Current dollars (mil.)	22,350	23,960	23,540	25,210	27,520
1989 constant dollars (mil.)	25,450	26,590	25,320	26,240	27,520
Central government expenditures (CGE)					
1989 constant dollars (mil.)	3,552	5,087	7,400	8,221	5,460
People (mil.)	102.8	105.4	108.6	111.9	115.3
Military expenditure as % of GNP	1.3	1.0	0.8	0.8	0.5
Military expenditure as % of CGE	9.4	5.4	2.7	2.6	2.4
Military expenditure per capita	3	3	2	2	1
Armed forces per 1,000 people	1.3	1.3	1.3	1.0	0.9
GNP per capita	248	252	233	235	239
Arms imports[6]					
Current dollars (mil.)	310	190	170	40	5
1989 constant dollars (mil.)	353	211	183	42	5
Arms exports[6]					
Current dollars (mil.)	0	5	0	0	20
1989 constant dollars (mil.)	0	6	0	0	20
Total imports[7]					
Current dollars (mil.)	8,877	3,405	3,909	5,700	6,200
1989 constant dollars	10,110	3,779	4,204	5,934	6,200
Total exports[7]					
Current dollars (mil.)	12,550	5,157	7,365	6,876	7,871
1989 constant dollars	14,280	5,723	7,921	7,158	7,871
Arms as percent of total imports[8]	3.5	5.6	4.3	0.7	0.1
Arms as percent of total exports[8]	0	0.1	0	0	0.3

Human Rights [24]

	SSTS	FL	FAPRO	PPCG	APROBC	TPW	PCPTW	STPEP	PHRFF	PRW	ASST	AFL
Observes	P	P	P		P	P	P			P	P	P
	EAFRD	CPR	ESCR	SR	ACHR	MAAE	PVIAC	PVNAC	EAFDAW	TCIDTP	RC
Observes		P		P			P	P	P	S	P

P = Party; S = Signatory; see Appendix for meaning of abbreviations.

Labor Force

Total Labor Force [25]

42.844 million

Labor Force by Occupation [26]

Agriculture	54%
Industry, commerce, and services	19
Government	15

Unemployment Rate [27]

28% (1992 est.)

Production Sectors

Commercial Energy Production and Consumption

Production [28]

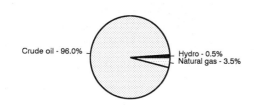

Crude oil - 96.0%
Hydro - 0.5%
Natural gas - 3.5%

Consumption [29]

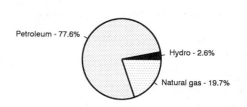

Petroleum - 77.6%
Hydro - 2.6%
Natural gas - 19.7%

Data are shown in quadrillion (10^{15}) BTUs and percent for 1991

Crude oil[1]	4.06		Petroleum	0.59
Natural gas liquids	0.00		Dry natural gas	0.15
Dry natural gas	0.15		Coal[2]	(s)
Coal[2]	(s)		Net hydroelectric power[3]	0.02
Net hydroelectric power[3]	0.02		Net nuclear power[3]	0.00
Net nuclear power[3]	0.00		Total	0.77
Total	4.24			

Telecommunications [30]

- Above-average system limited by poor maintenance
- Major expansion in progress
- Radio relay microwave and cable routes
- Broadcast stations - 35 AM, 17 FM, 28 TV
- Satellite earth stations - 2 Atlantic Ocean INTELSAT, 1 Indian Ocean INTELSAT, 20 domestic station
- 1 coaxial submarine cable

Transportation [31]

Railroads. 3,505 km 1.067-meter gauge

Highways. 107,990 km total 30,019 km paved (mostly bituminous-surface treatment); 25,411 km laterite, gravel, crushed stone, improved earth; 52,560 km unimproved

Merchant Marine. 28 ships (1,000 GRT or over) totaling 418,046 GRT/664,949 DWT; includes 17 cargo, 1 refrigerated cargo, 1 roll-on/roll-off, 7 oil tanker, 1 chemical tanker, 1 bulk

Airports

Total:	76
Usable:	63
With permanent-surface runways:	34
With runways over 3,659 m:	1
With runways 2,440-3,659 m:	15
With runways 1,220-2,439 m:	23

Top Agricultural Products [32]

	90-91[9,24]	91-92[9,24]
Cassava	24,000	26,000
Yams	13,000	13,000
Sorghum	2,800	3,300
Millet	2,300	2,600
Corn	1,520	1,850
Vegetable oil	755	795

Values shown are 1,000 metric tons.

Top Mining Products [33]

Estimated metric tons unless otherwise specified   M.t.

Petroleum, crude (000 barrels)	689,800
Gas, natural (mil. cubic meters)	31,286
Shale (000 tons)	70
Cement, hydraulic (000 tons)	3,500
Steel, crude	137
Marble	1,600

Tourism [34]

	1987	1988	1989	1990	1991
Visitors	329	264			
Tourists[9]	177	143	161	190	
Tourism receipts	77	53	21	25	39
Tourism expenditures	55	41	416	576	839
Fare receipts	11	11	1	5	19
Fare expenditures	69	68	123	34	44

Tourists are in thousands, money in million U.S. dollars.

Manufacturing Sector

GDP and Manufacturing Summary [35]

	1980	1985	1989	1990	% change 1980-1990	% change 1989-1990
GDP (million 1980 $)	23,795	21,341	24,009	27,976	17.6	16.5
GDP per capita (1980 $)	303	232	229	325[e]	7.3	41.9
Manufacturing as % of GDP (current prices)	4.7	6.1[e]	7.6[e]	NA	NA	-100.0
Gross output (million $)	4,740	3,454	4,294[e]	5,797[e]	22.3	35.0
Value added (million $)	2,422	1,667	2,283[e]	3,606[e]	48.9	58.0
Value added (million 1980 $)	1,161	1,078	903	1,344	15.8	48.8
Industrial production index	100	85	83	90[e]	-10.0	8.4
Employment (thousands)	432	330	363[e]	418[e]	-3.2	15.2

Note: GDP stands for Gross Domestic Product. 'e' stands for estimated value.

Profitability and Productivity

	1980	1985	1989	1990	% change 1980-1990	% change 1989-1990
Intermediate input (%)	49	52	47[e]	38[e]	-22.4	-19.1
Wages, salaries, and supplements (%)	11[e]	10[e]	10[e]	10[e]	-9.1	0.0
Gross operating surplus (%)	40[e]	38[e]	43[e]	52[e]	30.0	20.9
Gross output per worker ($)	10,273	10,005	11,819[e]	13,472[e]	31.1	14.0
Value added per worker ($)	5,211	4,844	6,283[e]	8,657[e]	66.1	37.8
Average wage (incl. benefits) ($)	1,226[e]	1,037[e]	1,202[e]	1,422[e]	16.0	18.3

Profitability is in percent of gross output. Productivity is in U.S. $. 'e' stands for estimated value.

Profitability - 1990

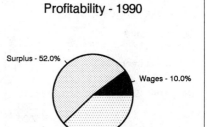

Surplus - 52.0%
Wages - 10.0%
Inputs - 38.0%

The graphic shows percent of gross output.

Value Added in Manufacturing

	1980 $ mil.	1980 %	1985 $ mil.	1985 %	1989 $ mil.	1989 %	1990 $ mil.	1990 %	% change 1980-1990	% change 1989-1990
311 Food products	149	6.2	251	15.1	288[e]	12.6	505[e]	14.0	238.9	75.3
313 Beverages	267	11.0	173[e]	10.4	453[e]	19.8	432[e]	12.0	61.8	-4.6
314 Tobacco products	96	4.0	32[e]	1.9	38[e]	1.7	71[e]	2.0	-26.0	86.8
321 Textiles	231	9.5	233	14.0	370[e]	16.2	449[e]	12.5	94.4	21.4
322 Wearing apparel	3	0.1	1	0.1	5[e]	0.2	2[e]	0.1	-33.3	-60.0
323 Leather and fur products	12	0.5	23	1.4	31[e]	1.4	47[e]	1.3	291.7	51.6
324 Footwear	12	0.5	28	1.7	22[e]	1.0	61[e]	1.7	408.3	177.3
331 Wood and wood products	88	3.6	14	0.8	6[e]	0.3	26[e]	0.7	-70.5	333.3
332 Furniture and fixtures	56	2.3	14	0.8	20[e]	0.9	33[e]	0.9	-41.1	65.0
341 Paper and paper products	38	1.6	51	3.1	62[e]	2.7	115[e]	3.2	202.6	85.5
342 Printing and publishing	75	3.1	45	2.7	60[e]	2.6	107[e]	3.0	42.7	78.3
351 Industrial chemicals	30	1.2	9	0.5	14[e]	0.6	19[e]	0.5	-36.7	35.7
352 Other chemical products	265	10.9	213	12.8	179[e]	7.8	461[e]	12.8	74.0	157.5
353 Petroleum refineries	75[e]	3.1	-7[e]	-0.4	70[e]	3.1	40[e]	1.1	-46.7	-42.9
354 Miscellaneous petroleum and coal products	3[e]	0.1	NA	0.0	7[e]	0.3	2[e]	0.1	-33.3	-71.4
355 Rubber products	26	1.1	31	1.9	35[e]	1.5	67[e]	1.9	157.7	91.4
356 Plastic products	98	4.0	49	2.9	30[e]	1.3	107[e]	3.0	9.2	256.7
361 Pottery, china and earthenware	NA	0.0	2	0.1	NA	0.0	2[e]	0.1	NA	NA
362 Glass and glass products	24	1.0	7	0.4	14[e]	0.6	18[e]	0.5	-25.0	28.6
369 Other non-metal mineral products	87	3.6	106	6.4	136[e]	6.0	228[e]	6.3	162.1	67.6
371 Iron and steel	3	0.1	17	1.0	38[e]	1.7	25[e]	0.7	733.3	-34.2
372 Non-ferrous metals	33	1.4	27	1.6	57[e]	2.5	73[e]	2.0	121.2	28.1
381 Metal products	140	5.8	92	5.5	176[e]	7.7	201[e]	5.6	43.6	14.2
382 Non-electrical machinery	23	0.9	19	1.1	28[e]	1.2	42[e]	1.2	82.6	50.0
383 Electrical machinery	46	1.9	36	2.2	52[e]	2.3	78[e]	2.2	69.6	50.0
384 Transport equipment	526	21.7	193	11.6	82[e]	3.6	386[e]	10.7	-26.6	370.7
385 Professional and scientific equipment	NA	0.0	NA	0.0	NA	0.0	1[e]	0.0	NA	NA
390 Other manufacturing industries	13	0.5	6	0.4	10[e]	0.4	10[e]	0.3	-23.1	0.0

Note: The industry codes shown are International Standard Industry codes (ISIC). Percentages are percent of total Value Added. 'e' stands for estimated value

Finance, Economics, and Trade

Economic Indicators [36]

In Naira (N) or U.S. Dollars as indicated[112].

	1989	1990[P]	1991[P]
GDP (current prices) (N bil.)	234.7	285.0	355.0
Real GDP growth rate (%)	6.3	5.1	4.0
GDP per capita ($)	270	300	300
Money supply (M1) (N bil.)	25.7	37.2	49.0
Gross domestic savings (% of GDP)	23.4	29.6	21.0
Gross domestic investment (% of GDP)	13.9	14.6	16.0
Consumer prices (% chg.)	50.5	7.5	15.0
Foreign Debt (year end)	3l,164	33,958	35,000

Balance of Payments Summary [37]

Values in millions of dollars.

	1987	1988	1989	1990	1991
Exports of goods (f.o.b.)	7,545.0	6,897.0	7,870.0	13,585.0	12,254.0
Imports of goods (f.o.b.)	-4,097.0	-4,271.0	-3,692.0	-4,932.0	-7,813.0
Trade balance	3,448.0	2,626.0	4,178.0	8,653.0	4,441.0
Services - debits	-3,762.0	-3,212.0	-3,918.0	-4,926.0	-5,080.0
Services - credits	270.0	404.0	704.0	1,176.0	1,097.0
Private transfers (net)	-19.0	-33.0	-19.0	1.0	12.0
Government transfers (net)	-5.0	21.0	145.0	84.0	732.0
Long term capital (net)	-1,788.0	-2,426.0	-1,170.0	-1,248.0	-2,437.0
Short term capital (net)	2,210.0	2,398.0	1,374.0	-1,497.0	1,967.0
Errors and omissions	-30.60	-215.0	-110.0	235.0	-92.0
Overall balance	48.0	-437.0	1,184.0	2,478.0	640.0

Exchange Rates [38]

Currency: **naira.**
Symbol: **N.**

Data are currency units per $1.

December 1992	19.6610
1992	17.2980
1991	9.9090
1990	8.0380
1989	7.3647
1988	4.5370
1987	4.0160

Imports and Exports

Top Import Origins [39]

$7.8 billion (c.i.f., 1991).

Origins	%
EC countries	70
US	16

Top Export Destinations [40]

$12.7 billion (f.o.b., 1991).

Destinations	%
EC countries	43
US	41

Foreign Aid [41]

	U.S. $	
US commitments, including Ex-Im (FY70-89)	705	million
Western (non-US) countries, ODA and OOF bilateral commitments (1970-89)	3.0	billion
Communist countries (1970-89)	2.2	billion

Import and Export Commodities [42]

Import Commodities

Consumer goods
Capital equipment
Chemicals
Raw materials

Export Commodities

Oil 95%
Cocoa
Rubber

Norway

Geography [1]

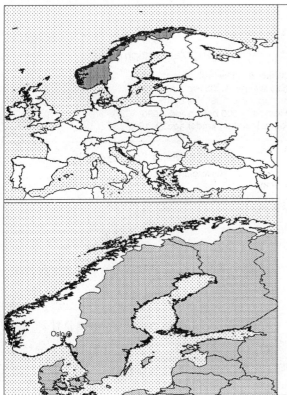

Total area:
  324,220 km2
Land area:
  307,860 km2
Comparative area:
  Slightly larger than New Mexico
Land boundaries:
  Total 2,515 km, Finland 729 km, Sweden 1,619 km, Russia 167 km
Coastline:
  21,925 km (includes mainland 3,419 km, large islands 2,413 km, long fjords, numerous small islands, and minor indentations 16,093 km)
Climate:
  Temperate along coast, modified by North Atlantic Current; colder interior; rainy year-round on west coast
Terrain:
  Glaciated; mostly high plateaus and rugged mountains broken by fertile valleys; small, scattered plains; coastline deeply indented by fjords; arctic tundra in north
Natural resources:
  Petroleum, copper, natural gas, pyrites, nickel, iron ore, zinc, lead, fish, timber, hydropower
Land use:
  Arable land: 3%
  Permanent crops: 0%
  Meadows and pastures: 0%
  Forest and woodland: 27%
  Other: 70%

Demographics [2]

	1960	1970	1980	1990	1991[1]	1994	2000	2010	2020
Population	3,581	3,877	4,086	4,242	4,273	4,315	4,387	4,424	4,446
Population density (persons per sq. mi.)	30	33	34	36	36	NA	37	38	38
Births	NA	NA	NA	NA	58	57	NA	NA	NA
Deaths	NA	NA	NA	NA	46	45	NA	NA	NA
Life expectancy - males	NA	NA	NA	NA	74	74	NA	NA	NA
Life expectancy - females	NA	NA	NA	NA	81	81	NA	NA	NA
Birth rate (per 1,000)	NA	NA	NA	NA	14	13	NA	NA	NA
Death rate (per 1,000)	NA	NA	NA	NA	11	10	NA	NA	NA
Women of reproductive age (15-44 yrs.)	NA	NA	NA	1,056	NA	1,068	1,044	NA	NA
of which are currently married	NA	NA	NA	586	NA	613	617	NA	NA
Fertility rate	NA	NA	NA	2.0	1.83	1.8	1.5	1.5	1.6

Population values are in thousands, life expectancy in years, and other items as indicated.

Health

Health Personnel [3]

Doctors per 1,000 pop., 1988-92	2.43
Nurse-to-doctor ratio, 1988-92	4.4
Hospital beds per 1,000 pop., 1985-90	4.8
Percentage of children immunized (age 1 yr. or less)	
Third dose of DPT, 1990-91	89.0
Measles, 1990-91	90.0

Health Expenditures [4]

Total health expenditure, 1990 (official exchange rate)	
Millions of dollars	7782
Millions of dollars per capita	1835
Health expenditures as a percentage of GDP	
Total	7.4
Public sector	7.0
Private sector	0.3
Development assistance for health	
Total aid flows (millions of dollars)[1]	NA
Aid flows per capita (millions of dollars)	NA
Aid flows as a percentage of total health expenditure	NA

For sources, notes, and explanations, see Annotated Source Appendix, page 1035.

Human Factors

Health Care Ratios [5]

Population per physician, 1970	720
Population per physician, 1990	NA
Population per nursing person, 1970	160
Population per nursing person, 1990	NA
Percent of births attended by health staff, 1985	100

Infants and Malnutrition [6]

Percent of babies with low birth weight, 1985	4
Infant mortality rate per 1,000 live births, 1970	13
Infant mortality rate per 1,000 live births, 1991	8
Years of life lost per 1,000 population, 1990	10
Prevalence of malnutrition (under age 5), 1990	NA

Ethnic Division [7]

Germanic (Nordic, Alpine, Baltic), Lapps 20,000.

Religion [8]

Evangelical Lutheran (state church)	87.8%
Other Protestant and Roman Catholic	3.8%
None	3.2%
Unknown	5.2%

Major Languages [9]

Norwegian (official).

Education

Public Education Expenditures [10]

Million Krone	1980[7]	1985[7]	1987[7]	1988[7]	1989[7]	1990[7]
Total education expenditure	19,731	31,680	NA	42,239	47,044	51,119
as percent of GNP	7.2	6.5	NA	7.4	7.8	7.9
as percent of total govt. expend.	13.7	14.6	NA	14.1	14.3	14.6
Current education expenditure	16,448	27,979	NA	37,140	40,768	44,109
as percent of GNP	6.0	5.7	NA	6.5	6.8	6.8
as percent of current govt. expend.	14.4	15.3	NA	14.6	14.7	14.5
Capital expenditure	3,283	3,701	NA	5,099	6,276	7,010

Educational Attainment [11]

Age group	25+
Total population	2,574,641
Highest level attained (%)	
No schooling	1.8[9]
First level	
Incompleted	0.0
Completed	NA
Entered second level	
S-1	60.0
S-2	26.2
Post secondary	11.9

Literacy Rate [12]

Libraries [13]

	Admin. Units	Svc. Pts.	Vols. (000)	Shelving (meters)	Vols. Added	Reg. Users
National (1990)	2	NA	1,988	NA	NA	NA
Non-specialized (1989)	20	NA	1,054	NA	55,692	NA
Public (1989)	446	1,290	18,551	NA	826,438	NA
Higher ed. (1990)	99	210	8,718	261,326	277,698	89,292
School (1990)[37]	3,383	3,383	6,858	NA	307,861	464,557

Daily Newspapers [14]

	1975	1980	1985	1990
Number of papers	80	85	82	85
Circ. (000)	1,657	1,892	2,120	2,588

Cinema [15]

Data for 1991.

Cinema seats per 1,000	24.8
Annual attendance per person	2.5
Gross box office receipts (mil. Krone)	382

Science and Technology

Scientific/Technical Forces [16]

Potential scientists/engineers	120,780
Number female	22,630
Potential technicians	NA
Number female	NA
Total	NA

R&D Expenditures [17]

	Kroner (000) 1989
Total expenditure	12,000,000[e]
Capital expenditure	1,300,000[e]
Current expenditure	10,700,000[e]
Percent current	89.2

U.S. Patents Issued [18]

Values show patents issued to citizens of the country by the U.S. Patents Office.

	1990	1991	1992
Number of patents	119	119	121

Government and Law

Organization of Government [19]

Long-form name:
Kingdom of Norway
Type:
constitutional monarchy
Independence:
26 October 1905 (from Sweden)
Constitution:
17 May 1814, modified in 1884
Legal system:
mixture of customary law, civil law
system, and common law traditions;
Supreme Court renders advisory opinions
to legislature when asked; accepts
compulsory ICJ jurisdiction, with
reservations
National holiday:
Constitution Day, 17 May (1814)
Executive branch:
monarch, prime minister, State Council
(cabinet)
Legislative branch:
unicameral Parliament (Storting) with an
Upper Chamber (Lagting) and a Lower
Chamber (Odelsting)
Judicial branch:
Supreme Court (Hoyesterett)

Crime [23]

Political Parties [20]

Storting	% of votes
Labor	34.3
Conservative	22.2
Progress	13.0
Socialist Left	10.1
Christian People's	8.5
Center Party	6.6
Finnmark List	0.3
Other	5.0

Government Expenditures [21]

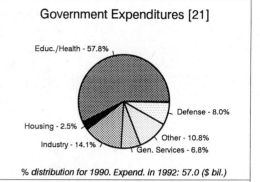

% distribution for 1990. Expend. in 1992: 57.0 ($ bil.)

Military Expenditures and Arms Transfers [22]

	1985	1986	1987	1988	1989
Military expenditures					
Current dollars (mil.)	2,152	2,328	2,622	2,680	2,925
1989 constant dollars (mil.)	2,450	2,583	2,820	2,790	2,925
Armed forces (000)	36	38	38	40	43
Gross national product (GNP)					
Current dollars (mil.)	68,350	73,260	78,310	81,240	88,510
1989 constant dollars (mil.)	77,810	81,290	84,230	84,570	88,510
Central government expenditures (CGE)					
1989 constant dollars (mil.)	32,700	36,830	39,060	40,330	42,160
People (mil.)	4.2	4.2	4.2	4.2	4.2
Military expenditure as % of GNP	3.1	3.2	3.3	3.3	3.3
Military expenditure as % of CGE	7.5	7.0	7.2	6.9	6.9
Military expenditure per capita	590	620	674	663	691
Armed forces per 1,000 people	8.7	9.1	9.1	9.5	10.2
GNP per capita	18,740	19,510	20,120	20,090	20,920
Arms imports[6]					
Current dollars (mil.)	170	200	260	280	340
1989 constant dollars (mil.)	194	222	280	291	340
Arms exports[6]					
Current dollars (mil.)	30	30	40	50	30
1989 constant dollars (mil.)	34	33	43	52	30
Total imports[7]					
Current dollars (mil.)	15,560	20,300	22,640	23,220	23,670
1989 constant dollars	17,710	22,530	24,350	24,170	23,670
Total exports[7]					
Current dollars (mil.)	19,980	18,090	21,490	22,440	27,060
1989 constant dollars	22,750	20,080	23,110	23,360	27,060
Arms as percent of total imports[8]	1.1	1.0	1.1	1.2	1.4
Arms as percent of total exports[8]	0.2	0.2	0.2	0.2	0.1

Human Rights [24]

	SSTS	FL	FAPRO	PPCG	APROBC	TPW	PCPTW	STPEP	PHRFF	PRW	ASST	AFL
Observes	P	P	P	P	P	P	P	P	P	P	P	P
	EAFRD	CPR	ESCR	SR	ACHR	MAAE	PVIAC	PVNAC	EAFDAW	TCIDTP	RC	
Observes		P	P	P	P		P	P	P	P	P	

P = Party; S = Signatory; see Appendix for meaning of abbreviations.

Labor Force

Total Labor Force [25]

2.004 million (1992)

Labor Force by Occupation [26]

Services	39.1%
Commerce	17.6
Mining, oil, and manufacturing	16.0
Banking and financial services	7.6
Transportation and communications	7.8
Construction	6.1
Agriculture, forestry, and fishing	5.5

Date of data: 1989

Unemployment Rate [27]

5.9% (excluding people in job-training
programs) (1992)

For sources, notes, and explanations, see Annotated Source Appendix, page 1035.

Production Sectors

Commercial Energy Production and Consumption

Production [28]

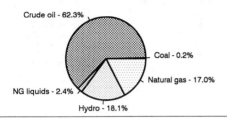

- Crude oil - 62.3%
- Coal - 0.2%
- Natural gas - 17.0%
- NG liquids - 2.4%
- Hydro - 18.1%

Consumption [29]

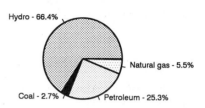

- Hydro - 66.4%
- Natural gas - 5.5%
- Coal - 2.7%
- Petroleum - 25.3%

Data are shown in quadrillion (10^{15}) BTUs and percent for 1991

Crude oil[1]	3.88
Natural gas liquids	0.15
Dry natural gas	1.06
Coal[2]	0.01
Net hydroelectric power[3]	1.13
Net nuclear power[3]	0.00
Total	6.23

Petroleum	0.37
Dry natural gas	0.08
Coal[2]	0.04
Net hydroelectric power[3]	0.97
Net nuclear power[3]	(s)
Total	1.46

Telecommunications [30]

- High-quality domestic and international telephone, telegraph, and telex services
- 2 buried coaxial cable systems
- 3,102,000 telephones
- Broadcast stations - 46 AM, 350 private and 143 government FM, 54 (2,100 repeaters) TV
- 4 coaxial submarine cables
- 3 communications satellite earth stations operating in the EUTELSAT, INTELSAT (1 Atlantic Ocean), MARISAT, and domestic systems

Top Agricultural Products [32]

	89-90	90-91
Grains	1,165	1,541
Potatoes	500	509
Root crops	151	179
Vegetables	114	115
Fruits[25]	43	26
Berries[25]	21	15

Values shown are 1,000 metric tons.

Top Mining Products [33]

Estimated metric tons unless otherwise specified	M.t.
Iron ore and concentrate (000 tons)	2,209
Cement, hydraulic (000 tons)	1,147
Steel, crude (000 tons)	438
Aluminum, primary	832,558
Nepheline syenite (000 tons)	250
Titanium, ilmenite concentrate (000 tons)	800

Transportation [31]

Railroads. 4,223 km 1.435-meter standard gauge; Norwegian State Railways (NSB) operates 4,219 km (2,450 km electrified and 96 km double track); 4 km other

Highways. 79,540 km total; 38,580 km paved; 40,960 km gravel, crushed stone, and earth

Merchant Marine. 829 ships (1,000 GRT or over) totaling 22,312,412 GRT/38,532,109 DWT; includes 13 passenger, 20 short-sea passenger, 106 cargo, 2 passenger-cargo, 19 refrigerated cargo, 15 container, 49 roll-on/roll-off, 23 vehicle carrier, 1 railcar carrier, 174 oil tanker, 91 chemical tanker, 82 liquefied gas, 25 combination ore/oil, 201 bulk, 8 combination bulk; note - the government has created a captive register, the Norwegian International Ship Register (NIS), as a subset of the Norwegian register; ships on the NIS enjoy many benefits of flags of convenience and do not have to be crewed by Norwegians; the majority of ships (777) under the Norwegian flag are now registered with the NIS

Airports

Total:	103
Usable:	102
With permanent-surface runways:	63
With runways over 3,659 m:	0
With runways 2,440-3,659 m:	12
With runways 1,220-2,439 m:	16

Tourism [34]

	1987	1988	1989	1990	1991
Tourists[49]	1,782	1,704	1,867	1,955	2,114
Tourism receipts	1,255	1,466	1,336	1,517	1,574
Tourism expenditures	3,067	3,442	2,855	3,413	3,207
Fare receipts	581	580	520	633	585

Tourists are in thousands, money in million U.S. dollars.

For sources, notes, and explanations, see Annotated Source Appendix, page 1035.

Manufacturing Sector

GDP and Manufacturing Summary [35]

	1980	1985	1989	1990	% change 1980-1990	% change 1989-1990
GDP (million 1980 $)	57,713	68,041	71,442	73,638	27.6	3.1
GDP per capita (1980 $)	14,125	16,383	17,006	17,479	23.7	2.8
Manufacturing as % of GDP (current prices)	17.3	15.3	14.5	14.6	-15.6	0.7
Gross output (million $)	31,936	28,185	44,190	50,107	56.9	13.4
Value added (million $)	9,339	7,948	12,753	14,015	50.1	9.9
Value added (million 1980 $)	9,240	9,889	9,982	9,520	3.0	-4.6
Industrial production index	100	109	107	113	13.0	5.6
Employment (thousands)	354	312	276	271	-23.4	-1.8

Note: GDP stands for Gross Domestic Product. 'e' stands for estimated value.

Profitability and Productivity

	1980	1985	1989	1990	% change 1980-1990	% change 1989-1990
Intermediate input (%)	71	72	71	72	1.4	1.4
Wages, salaries, and supplements (%)	21	20	16	19	-9.5	18.8
Gross operating surplus (%)	8	8	13	9	12.5	-30.8
Gross output per worker ($)	89,656	89,848	160,283	184,350	105.6	15.0
Value added per worker ($)	26,217	25,335	46,256	51,564	96.7	11.5
Average wage (incl. benefits) ($)	19,129	17,851	25,143	35,540	85.8	41.4

Profitability is in percent of gross output. Productivity is in U.S. $. 'e' stands for estimated value.

Profitability - 1990

Inputs - 72.0% Surplus - 9.0% Wages - 19.0%

The graphic shows percent of gross output.

Value Added in Manufacturing

	1980 $ mil.	1980 %	1985 $ mil.	1985 %	1989 $ mil.	1989 %	1990 $ mil.	1990 %	% change 1980-1990	% change 1989-1990
311 Food products	908	9.7	922	11.6	1,592	12.5	1,819	13.0	100.3	14.3
313 Beverages	292	3.1	297	3.7	556	4.4	660	4.7	126.0	18.7
314 Tobacco products	168	1.8	220	2.8	400	3.1	478	3.4	184.5	19.5
321 Textiles	213	2.3	126	1.6	159	1.2	191	1.4	-10.3	20.1
322 Wearing apparel	101	1.1	59	0.7	49	0.4	58	0.4	-42.6	18.4
323 Leather and fur products	18	0.2	9	0.1	13	0.1	16	0.1	-11.1	23.1
324 Footwear	27	0.3	9	0.1	10	0.1	11	0.1	-59.3	10.0
331 Wood and wood products	587	6.3	365	4.6	545	4.3	619	4.4	5.5	13.6
332 Furniture and fixtures	196	2.1	164	2.1	206	1.6	236	1.7	20.4	14.6
341 Paper and paper products	452	4.8	400	5.0	729	5.7	787	5.6	74.1	8.0
342 Printing and publishing	668	7.2	717	9.0	1,212	9.5	1,381	9.9	106.7	13.9
351 Industrial chemicals	452	4.8	422	5.3	713	5.6	811	5.8	79.4	13.7
352 Other chemical products	227	2.4	184	2.3	353	2.8	393	2.8	73.1	11.3
353 Petroleum refineries	103	1.1	24	0.3	97	0.8	195	1.4	89.3	101.0
354 Miscellaneous petroleum and coal products	53	0.6	58	0.7	58	0.5	63	0.4	18.9	8.6
355 Rubber products	51	0.5	38	0.5	51	0.4	58	0.4	13.7	13.7
356 Plastic products	170	1.8	147	1.8	226	1.8	278	2.0	63.5	23.0
361 Pottery, china and earthenware	26	0.3	17	0.2	25	0.2	27	0.2	3.8	8.0
362 Glass and glass products	55	0.6	50	0.6	67	0.5	77	0.5	40.0	14.9
369 Other non-metal mineral products	281	3.0	215	2.7	314	2.5	361	2.6	28.5	15.0
371 Iron and steel	385	4.1	276	3.5	494	3.9	347	2.5	-9.9	-29.8
372 Non-ferrous metals	743	8.0	550	6.9	1,192	9.3	826	5.9	11.2	-30.7
381 Metal products	595	6.4	465	5.9	688	5.4	784	5.6	31.8	14.0
382 Non-electrical machinery	933	10.0	1,079	13.6	1,359	10.7	1,590	11.3	70.4	17.0
383 Electrical machinery	547	5.9	498	6.3	702	5.5	751	5.4	37.3	7.0
384 Transport equipment	1,000	10.7	555	7.0	805	6.3	1,028	7.3	2.8	27.7
385 Professional and scientific equipment	32	0.3	38	0.5	71	0.6	82	0.6	156.3	15.5
390 Other manufacturing industries	59	0.6	42	0.5	68	0.5	89	0.6	50.8	30.9

Note: The industry codes shown are International Standard Industry codes (ISIC). Percentages are percent of total Value Added. 'e' stands for estimated value

For sources, notes, and explanations, see Annotated Source Appendix, page 1035.

693

Finance, Economics, and Trade

Economic Indicators [36]

Mil. Norwegian Krone (NOK) unless otherwise noted.

	1989	1990	1991[e]
Real GDP (1989 Prices)	622,992	633,919	651,669
Real GDP Growth (%)	0.4	1.8	2.8
Real GDP Per Capita (1989 Prices NOK)	147,279	156,237	163,286
Money Supply (M2; EOP)	445,001	460,365	497,194
Loan Int. Rate (EOP)[113]	12.28	11.36	10.50
Savings Rate[114]	11.3	11.9	12.0
Investment Rate[115]	27.6	18.9	19.0
CPI (1979 = 100)	222.1	231.2	239.3
WPI (1981 = 100)	152.2	157.8	163.6
Net External Debt[116]	130,400	91,600	73,000

Balance of Payments Summary [37]

Values in millions of dollars.

	1987	1988	1989	1990	1991
Exports of goods (f.o.b.)	21,191	23,075	27,171	34,315	34,186
Imports of goods (f.o.b.)	-21,951	-23,284	-23,401	-26,545	-25,468
Trade balance	-760	-209	3,770	770	8,718
Services - debits	-13,949	-15,544	-16,619	-19,080	-19,317
Services - credits	11,585	12,994	14,195	16,626	17,023
Private transfers (net)	-194	-168	-222	-222	-299
Government transfers (net)	-788	-962	-910	-1,208	-1,185
Long term capital (net)	-13	4,655	3,021	-1,511	-2,682
Short term capital (net)	5,246	245	-966	1,030	-4,760
Errors and omissions	-1,349	-1,149	-1,305	-2,990	-247
Overall balance	-221	-138	964	415	-2,749

Exchange Rates [38]

Currency: **Norwegian kroner.**
Symbol: **NKr.**

Data are currency units per $1.

January 1993	6.8774
1992	6.2145
1991	6.4829
1990	6.2597
1989	6.9045
1988	6.5170

Imports and Exports

Top Import Origins [39]

$26.8 billion (c.i.f., 1992).

Origins	%
EC	48.7
Nordic countries	26.8
Developing countries	9.3
US	8.6
Japan	6.3

Top Export Destinations [40]

$35.3 billion (f.o.b., 1992).

Destinations	%
EC	67.0
Nordic countries	18.2
Developing countries	7.9
US	5.1
Japan	1.6

Foreign Aid [41]

	U.S. $	
Donor - ODA and OOF commitments (1970-89)	4.4	billion

Import and Export Commodities [42]

Import Commodities

Machinery
Fuels and lubricants
Transportation equipment
Chemicals
Foodstuffs
Clothing
Ships

Export Commodities

Oil & petr. products 37.8%
Metals and products 10.7%
Natural gas 7.3%
Fish 6.6%
Chemicals 6.3%
Ships 5.4%

For sources, notes, and explanations, see Annotated Source Appendix, page 1035.

Oman

Geography [1]

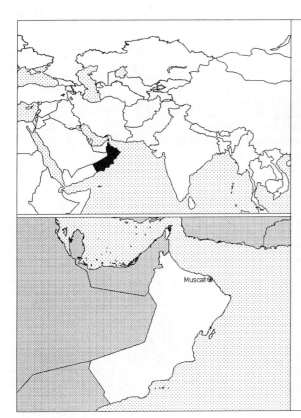

Total area:
212,460 km2
Land area:
212,460 km2
Comparative area:
Slightly smaller than Kansas
Land boundaries:
Total 1,374 km, Saudi Arabia 676 km, UAE 410 km, Yemen 288 km
Coastline:
2,092 km
Climate:
Dry desert; hot, humid along coast; hot, dry interior; strong southwest summer monsoon (May to September) in far south
Terrain:
Vast central desert plain, rugged mountains in north and south
Natural resources:
Petroleum, copper, asbestos, some marble, limestone, chromium, gypsum, natural gas
Land use:
Arable land:
Less than 2%
Permanent crops:
0%
Meadows and pastures:
5%
Forest and woodland:
0%
Other:
93%

Demographics [2]

	1960	1970	1980	1990	1991[1]	1994	2000	2010	2020
Population	505	654	984	1,481	1,534	1,701	2,098	2,991	4,175
Population density (persons per sq. mi.)	6	8	12	18	19	NA	26	36	51
Births	NA	NA	NA	NA	63	69	NA	NA	NA
Deaths	NA	NA	NA	NA	10	10	NA	NA	NA
Life expectancy - males	NA	NA	NA	NA	65	66	NA	NA	NA
Life expectancy - females	NA	NA	NA	NA	68	70	NA	NA	NA
Birth rate (per 1,000)	NA	NA	NA	NA	41	40	NA	NA	NA
Death rate (per 1,000)	NA	NA	NA	NA	6	6	NA	NA	NA
Women of reproductive age (15-44 yrs.)	NA	NA	NA	NA	NA	NA	NA	NA	NA
of which are currently married	NA	NA	NA	NA	NA	NA	NA	NA	NA
Fertility rate	NA	NA	NA	6.7	6.71	6.5	6.2	5.6	4.9

Population values are in thousands, life expectancy in years, and other items as indicated.

Health

Health Personnel [3]

Health Expenditures [4]

For sources, notes, and explanations, see Annotated Source Appendix, page 1035.

695

Human Factors

Health Care Ratios [5]

Population per physician, 1970	8380
Population per physician, 1990	1060
Population per nursing person, 1970	3420
Population per nursing person, 1990	400
Percent of births attended by health staff, 1985	60

Infants and Malnutrition [6]

Percent of babies with low birth weight, 1985	14
Infant mortality rate per 1,000 live births, 1970	159
Infant mortality rate per 1,000 live births, 1991	31
Years of life lost per 1,000 population, 1990	NA
Prevalence of malnutrition (under age 5), 1990	NA

Ethnic Division [7]

Arab, Balochi, Zanzibari, South Asian (Indian, Pakistani, Bangladeshi).

Religion [8]

Ibadhi Muslim 75%, Sunni Muslim, Shi'a Muslim, Hindu.

Major Languages [9]

Arabic (official)	NA
English	NA
Balochi	NA
Urdu	NA
Indian dialects	NA

Education

Public Education Expenditures [10]

Million Rial	1980	1985	1987	1988	1989	1990
Total education expenditure	38	123	NA	110	109	128
as percent of GNP	2.1	4.0	NA	4.4	3.8	3.5
as percent of total govt. expend.	NA	NA	NA	14.9	12.4	11.1
Current education expenditure	31	77	NA	99	102	117
as percent of GNP	1.7	2.5	NA	3.9	3.6	3.2
as percent of current govt. expend.	NA	NA	NA	NA	18.8	18.8
Capital expenditure	7	46	NA	12	6	10

Educational Attainment [11]

Literacy Rate [12]

Libraries [13]

	Admin. Units	Svc. Pts.	Vols. (000)	Shelving (meters)	Vols. Added	Reg. Users
National	NA	NA	NA	NA	NA	NA
Non-specialized	NA	NA	NA	NA	NA	NA
Public	NA	NA	NA	NA	NA	NA
Higher ed. (1991)	2	6	65	NA	11,900	6,516
School (1987)	130	130	132	NA	-	243,840

Daily Newspapers [14]

	1975	1980	1985	1990
Number of papers	-	-	3	4
Circ. (000)	-	-	51	62

Cinema [15]

Science and Technology

Scientific/Technical Forces [16]

R&D Expenditures [17]

U.S. Patents Issued [18]

Government and Law

Organization of Government [19]

Long-form name:
Sultanate of Oman
Type:
absolute monarchy with residual UK influence
Independence:
1650 (expulsion of the Portuguese)
Constitution:
none
Legal system:
based on English common law and Islamic law; ultimate appeal to the sultan; has not accepted compulsory ICJ jurisdiction
National holiday:
National Day, 18 November
Executive branch:
sultan, Cabinet
Legislative branch:
unicameral National Assembly
Judicial branch:
none; traditional Islamic judges and a nascent civil court system

Crime [23]

Elections [20]

No parties; elections scheduled for October 1992.

Government Expenditures [21]

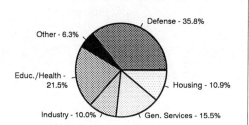

Defense - 35.8%
Other - 6.3%
Educ./Health - 21.5%
Housing - 10.9%
Industry - 10.0%
Gen. Services - 15.5%

% distribution for 1992. Expend. in 1991: 4.8 ($ bil.)

Military Expenditures and Arms Transfers [22]

	1985	1986	1987	1988	1989
Military expenditures					
Current dollars (mil.)	1,937	1,731	1,518	1,350	1,552
1989 constant dollars (mil.)	2,205	1,920	1,632	1,405	1,552
Armed forces (000)	25	26[e]	27[e]	27	29
Gross national product (GNP)					
Current dollars (mil.)	7,950	6,397	7,294	7,178	7,635
1989 constant dollars (mil.)	9,051	7,099	7,845	7,472	7,635
Central government expenditures (CGE)					
1989 constant dollars (mil.)	5,216	4,604	3,764	3,725	3,752
People (mil.)	1.2	1.3	1.3	1.4	1.4
Military expenditure as % of GNP	24.4	27.1	20.8	18.8	20.3
Military expenditure as % of CGE	42.3	41.7	43.4	37.7	41.4
Military expenditure per capita	1,780	1,494	1,225	1,018	1,085
Armed forces per 1,000 people	20.2	20.2	20.3	19.3	20.3
GNP per capita	7,305	5,522	5,889	5,414	5,340
Arms imports[6]					
Current dollars (mil.)	140	110	110	30	60
1989 constant dollars (mil.)	159	122	118	31	60
Arms exports[6]					
Current dollars (mil.)	0	5	0	0	0
1989 constant dollars (mil.)	0	6	0	0	0
Total imports[7]					
Current dollars (mil.)	3,153	2,402	1,822	2,202	2,257
1989 constant dollars	3,589	2,666	1,960	2,292	2,257
Total exports[7]					
Current dollars (mil.)	4,705	2,516	3,198	2,625	NA
1989 constant dollars	5,356	2,792	3,440	2,733	NA
Arms as percent of total imports[8]	4.4	4.6	6.0	1.4	2.7
Arms as percent of total exports[8]	0	0.2	0	0	NA

Human Rights [24]

	SSTS	FL	FAPRO	PPCG	APROBC	TPW	PCPTW	STPEP	PHRFF	PRW	ASST	AFL
Observes						P	P					

	EAFRD	CPR	ESCR	SR	ACHR	MAAE	PVIAC	PVNAC	EAFDAW	TCIDTP	RC
Observes							P	P			

P = Party; S = Signatory; see Appendix for meaning of abbreviations.

Labor Force

Total Labor Force [25]

430,000

Labor Force by Occupation [26]

Agriculture   40%
Date of data: est.

Unemployment Rate [27]

For sources, notes, and explanations, see Annotated Source Appendix, page 1035.

697

Production Sectors

Commercial Energy Production and Consumption

Production [28]

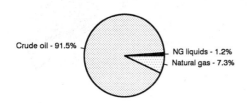

Crude oil - 91.5%
NG liquids - 1.2%
Natural gas - 7.3%

Consumption [29]

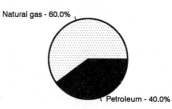

Natural gas - 60.0%
Petroleum - 40.0%

Data are shown in quadrillion (10^{15}) BTUs and percent for 1991

Crude oil[1]	1.51
Natural gas liquids	0.02
Dry natural gas	0.12
Coal[2]	0.00
Net hydroelectric power[3]	0.00
Net nuclear power[3]	0.00
Total	1.64

Petroleum	0.08
Dry natural gas	0.12
Coal[2]	0.00
Net hydroelectric power[3]	0.00
Net nuclear power[3]	0.00
Total	0.20

Telecommunications [30]

- Modern system consisting of open-wire, microwave, and radio communications stations
- Limited coaxial cable
- 50,000 telephones
- Broadcast stations - 2 AM, 3 FM, 7 TV
- Satellite earth stations - 2 Indian Ocean INTELSAT, 1 ARABSAT, and 8 domestic

Transportation [31]

Highways. 26,000 km total; 6,000 km paved, 20,000 km motorable track

Merchant Marine. 1 passenger ship (1,000 GRT or over) totaling 4,442 GRT/1,320 DWT

Airports

Total:	138
Usable:	130
With permanent-surface runways:	6
With runways over 3,659 m:	1
With runways 2,440-3,659 m:	9
With runways 1,220-2,439 m:	74

Top Agricultural Products [32]

Agriculture accounts for 6% of GDP and 40% of the labor force (including fishing); less than 2% of land cultivated; largely subsistence farming (dates, limes, bananas, alfalfa, vegetables, camels, cattle); not self-sufficient in food; annual fish catch averages 100,000 metric tons.

Top Mining Products [33]

Detailed information is not available. A summary of mineral resources available follows. **Mineral Resources**: Petroleum, copper, asbestos, some marble, limestone, chromium, gypsum, natural gas.

Tourism [34]

	1987	1988	1989	1990	1991
Tourists[10]	112	126	136	149	161
Tourism receipts[50]	47	49	56	69	63
Tourism expenditures	47	47	47	47	

Tourists are in thousands, money in million U.S. dollars.

Finance, Economics, and Trade

GDP and Manufacturing Summary [35]

	1980	1985	1990	1991	1992
Gross Domestic Product					
Millions of 1980 dollars	5,896	11,850	14,054	14,757	15,790[e]
Growth rate in percent	6.03	13.76	9.08	5.00	7.00[e]
Manufacturing Value Added					
Millions of 1980 dollars	45	240	339	NA	NA
Growth rate in percent	19.05	20.39	8.34	NA	NA
Manufacturing share in percent of current prices	0.8	2.4	3.7	4.2	NA

Economic Indicators [36]

Millions of Rials Omani (RO) unless otherwise noted.

	1989	1990	1991
Real GDP (purchaser's value)	2,289.0	2,496.6	NA
Real GDP growth rate (%)	3	9	NA
Money Supply (M1)	344.1	390.1	393.9[117]
Commercial interest - loans[118]	11.25	11.25	11.25
Savings rate	NA	NA	NA
Investment rate	NA	NA	NA
CPI (1988 = 100)	103.3	115	118.7[119]
WPI	NA	NA	NA
External public debt	NA	NA	NA

Balance of Payments Summary [37]

Values in millions of dollars.

	1986	1987	1988	1989	1990
Exports of goods (f.o.b.)	2,861.0	3,805.0	3,342.0	4,047.0	5,488.0
Imports of goods (f.o.b.)	-2,309.0	-1,769.0	-2,107.0	-2,130.0	-2,519.0
Trade balance	552.0	2,036.0	1,235.0	1,917.0	2,969.0
Services - debits	-1,357.0	-1,091.0	-1,095.0	-1,151.0	-1,338.0
Services - credits	610.0	533.0	270.0	333.0	367.0
Private transfers (net)	-846.0	-702.0	-762.0	-791.0	-845.0
Government transfers (net)	-	8.0	42.0	16.0	-57.0
Long term capital (net)	687.0	-100.0	273.0	162.0	-249.0
Short term capital (net)	328.0	-90.0	-52.0	-139.0	-261.0
Errors and omissions	-588.0	-486.0	-379.0	-64.0	-282.0
Overall balance	-614.0	108.0	-468.0	283.0	304.0

Exchange Rates [38]

Currency: **Omani rials.**
Symbol: **RO.**

Data are currency units per $1.

Fixed rate since 1986	0.3845

Imports and Exports

Top Import Origins [39]

$3.0 billion (f.o.b, 1991).

Origins	%
Japan	20
UAE	19
UK	19
US	7

Top Export Destinations [40]

$4.9 billion (f.o.b., 1991).

Destinations	%
UAE	30
Japan	27
South Korea	10
Singapore	5

Foreign Aid [41]

	U.S. $	
US commitments, including Ex-Im (FY70-89)	137	million
Western (non-US) countries, ODA and OOF bilateral commitments (1970-89)	148	million
OPEC bilateral aid (1979-89)	797	million

Import and Export Commodities [42]

Import Commodities	**Export Commodities**
Machinery	Petroleum 87%
Transportation equipment	Reexports
Manufactured goods	Fish
Food	Processed copper
Livestock	Textiles
Lubricants	

For sources, notes, and explanations, see Annotated Source Appendix, page 1035.

699

Pakistan

Geography [1]

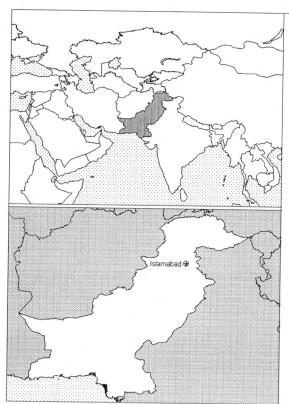

Total area:
803,940 km2
Land area:
778,720 km2
Comparative area:
Slightly less than twice the size of California
Land boundaries:
Total 6,774 km, Afghanistan 2,430 km, China 523 km, India 2,912 km, Iran 909 km
Coastline:
1,046 km
Climate:
Mostly hot, dry desert; temperate in northwest; arctic in north
Terrain:
Flat Indus plain in east; mountains in north and northwest; Balochistan plateau in west
Natural resources:
Land, extensive natural gas reserves, limited petroleum, poor quality coal, iron ore, copper, salt, limestone
Land use:
Arable land: 26%
Permanent crops: 0%
Meadows and pastures: 6%
Forest and woodland: 4%
Other: 64%

Demographics [2]

	1960	1970	1980	1990	1991[1]	1994	2000	2010	2020
Population	50,387	65,706	85,219	114,842	117,490	128,856	148,540	195,108	251,330
Population density (persons per sq. mi.)	168	219	283	381	391	NA	496	649	836
Births	NA	NA	NA	NA	5,085	5,440	NA	NA	NA
Deaths	NA	NA	NA	NA	1,530	1,595	NA	NA	NA
Life expectancy - males	NA	NA	NA	NA	56	57	NA	NA	NA
Life expectancy - females	NA	NA	NA	NA	57	58	NA	NA	NA
Birth rate (per 1,000)	NA	NA	NA	NA	43	42	NA	NA	NA
Death rate (per 1,000)	NA	NA	NA	NA	13	12	NA	NA	NA
Women of reproductive age (15-44 yrs.)	NA	NA	NA	24,861	NA	27,995	32,797	NA	NA
of which are currently married	NA	NA	NA	18,414	NA	20,723	24,270	NA	NA
Fertility rate	NA	NA	NA	6.7	6.64	6.4	5.9	5.0	4.1

Population values are in thousands, life expectancy in years, and other items as indicated.

Health

Health Personnel [3]

Doctors per 1,000 pop., 1988-92	0.34
Nurse-to-doctor ratio, 1988-92	0.8
Hospital beds per 1,000 pop., 1985-90	0.6
Percentage of children immunized (age 1 yr. or less)	
Third dose of DPT, 1990-91	81.0
Measles, 1990-91	77.0

Health Expenditures [4]

Total health expenditure, 1990 (official exchange rate)	
Millions of dollars	1394
Millions of dollars per capita	12
Health expenditures as a percentage of GDP	
Total	3.4
Public sector	1.8
Private sector	1.6
Development assistance for health	
Total aid flows (millions of dollars)[1]	76
Aid flows per capita (millions of dollars)	0.7
Aid flows as a percentage of total health expenditure	5.4

For sources, notes, and explanations, see Annotated Source Appendix, page 1035.

Human Factors

Health Care Ratios [5]

Population per physician, 1970	4310
Population per physician, 1990	2940
Population per nursing person, 1970	6600
Population per nursing person, 1990	5040
Percent of births attended by health staff, 1985	24

Infants and Malnutrition [6]

Percent of babies with low birth weight, 1985	25
Infant mortality rate per 1,000 live births, 1970	142
Infant mortality rate per 1,000 live births, 1991	97
Years of life lost per 1,000 population, 1990	61
Prevalence of malnutrition (under age 5), 1990	57

Ethnic Division [7]

Punjabi, Sindhi, Pashtun (Pathan), Baloch, Muhajir (immigrants from India and their descendents).

Religion [8]

Muslim	97%
Sunni	77%
Shi'a	20%
Christian, Hindu, and other	3%

Major Languages [9]

Urdu (official), English (official; spoken by Pakistani elite, most government ministries).

Punjabi	64%
Sindhi	12%
Pashtu	8%
Urdu	7%
Balochi other	9%

Education

Public Education Expenditures [10]

Million Rupee	1980	1985	1986	1987	1988	1989
Total education expenditure	4,619	12,645	15,320	18,962	NA	NA
as percent of GNP	2.0	2.5	2.8	3.2	NA	NA
as percent of total govt. expend.	5.0	NA	NA	NA	NA	NA
Current education expenditure	3,379	9,390	11,210	14,280	NA	NA
as percent of GNP	1.5	1.9	2.1	2.4	NA	NA
as percent of current govt. expend.	5.2	NA	NA	NA	NA	NA
Capital expenditure	1,240	3,255	4,110	4,682	NA	NA

Educational Attainment [11]

Age group	25+
Total population	30,707,279
Highest level attained (%)	
No schooling	78.9[10]
First level	
Incompleted	8.7
Completed	NA
Entered second level	
S-1	10.5
S-2	NA
Post secondary	1.9

Literacy Rate [12]

In thousands and percent	1985[a]	1991[a]	2000[a]
Illiterate population +15 years	39,411	43,459	51,902
Illiteracy rate - total pop. (%)	69.0	65.2	56.4
Illiteracy rate - males (%)	56.9	52.7	43.8
Illiteracy rate - females (%)	82.3	78.9	70.2

Libraries [13]

Daily Newspapers [14]

	1975	1980	1985	1990
Number of papers	102	106	118	237
Circ. (000)	1,900[e]	2,650[e]	3,670	1,817

Cinema [15]

Data for 1989.

Cinema seats per 1,000	NA
Annual attendance per person	NA
Gross box office receipts (mil. Rupee)	NA

Science and Technology

Scientific/Technical Forces [16]

Potential scientists/engineers	287,000
Number female	NA
Potential technicians	210,000
Number female	100,000
Total	497,000

R&D Expenditures [17]

	Rupee (000) 1987
Total expenditure[4,24]	5,582,081
Capital expenditure	NA
Current expenditure	NA
Percent current	NA

U.S. Patents Issued [18]

Values show patents issued to citizens of the country by the U.S. Patents Office.

	1990	1991	1992
Number of patents	0	0	1

For sources, notes, and explanations, see Annotated Source Appendix, page 1035.

Government and Law

Organization of Government [19]

Long-form name:
Islamic Republic of Pakistan
Type:
republic
Independence:
14 August 1947 (from UK)
Constitution:
10 April 1973, suspended 5 July 1977, restored with amendments, 30 December 1985
Legal system:
based on English common law with provisions to accommodate Pakistan's stature as an Islamic state; accepts compulsory ICJ jurisdiction, with reservations
National holiday:
Pakistan Day, 23 March (1956) (proclamation of the republic)
Executive branch:
president, prime minister, Cabinet
Legislative branch:
bicameral Parliament (Majlis-e-Shoora) consists of an upper house or Senate and a lower house or National Assembly
Judicial branch:
Supreme Court, Federal Islamic (Shari'at) Court

Crime [23]

Elections [20]

National Assembly. Last held on 24 October 1990 (next to be held by October 1995); results - percent of vote by party NA; seats - (217 total) number of seats by party NA; note - President Ghulam Ishaq Khan dismissed the National Assembly on 18 April 1993; it was reestablished, however, on 26 May 1993 by the Supreme Court, which ruled the dismissal order unconstitutional.

Government Budget [21]

For FY93 est.
Revenues	9.4
Expenditures	10.9
Capital expenditures	3.1

Military Expenditures and Arms Transfers [22]

	1985	1986	1987	1988	1989
Military expenditures					
Current dollars (mil.)	1,748	1,937	2,098	2,309	2,488
1989 constant dollars (mil.)	1,990	2,149	2,257	2,404	2,488
Armed forces (000)	644	573	572	484	520
Gross national product (GNP)					
Current dollars (mil.)	25,570	27,580	30,240	33,680	36,810
1989 constant dollars (mil.)	29,100	30,610	32,520	35,060	36,810
Central government expenditures (CGE)					
1989 constant dollars (mil.)	7,071	8,569	8,766	8,876	10,130
People (mil.)	99.3	102.3	105.4	108.5	111.7
Military expenditure as % of GNP	6.8	7.0	6.9	6.9	6.8
Military expenditure as % of CGE	28.1	25.1	25.7	27.1	24.5
Military expenditure per capita	20	21	21	22	22
Armed forces per 1,000 people	6.5	5.6	5.4	4.5	4.7
GNP per capita	293	299	309	323	330
Arms imports[6]					
Current dollars (mil.)	470	310	320	420	460
1989 constant dollars (mil.)	535	344	344	437	460
Arms exports[6]					
Current dollars (mil.)	40	5	5	10	20
1989 constant dollars (mil.)	46	6	5	10	20
Total imports[7]					
Current dollars (mil.)	5,890	5,374	5,822	6,590	7,143
1989 constant dollars	6,705	5,964	6,262	6,860	7,143
Total exports[7]					
Current dollars (mil.)	2,740	3,384	4,172	4,522	4,709
1989 constant dollars	3,119	3,755	4,487	4,707	4,709
Arms as percent of total imports[8]	8.0	5.8	5.5	6.4	6.4
Arms as percent of total exports[8]	1.5	0.1	0.1	0.2	0.4

Human Rights [24]

	SSTS	FL	FAPRO	PPCG	APROBC	TPW	PCPTW	STPEP	PHRFF	PRW	ASST	AFL
Observes	P	P	P	P	P	P	P	P		P	P	P
	EAFRD	CPR	ESCR	SR	ACHR	MAAE	PVIAC	PVNAC	EAFDAW	TCIDTP		RC
Observes		P					S	S				P

P = Party; S = Signatory; see Appendix for meaning of abbreviations.

Labor Force

Total Labor Force [25]

28.9 million

Labor Force by Occupation [26]

Agriculture	54%
Mining and manufacturing	13
Services	33

Extensive export of labor
Date of data: 1987 est.

Unemployment Rate [27]

10% (FY91 est.)

Production Sectors

Commercial Energy Production and Consumption

Production [28]

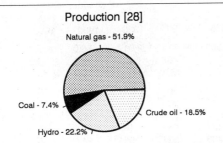

Natural gas - 51.9%
Coal - 7.4%
Hydro - 22.2%
Crude oil - 18.5%

Consumption [29]

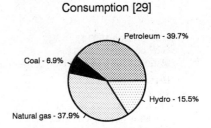

Petroleum - 39.7%
Coal - 6.9%
Hydro - 15.5%
Natural gas - 37.9%

Data are shown in quadrillion (10^{15}) BTUs and percent for 1991

Crude oil[1]	0.15
Natural gas liquids	(s)
Dry natural gas	0.42
Coal[2]	0.06
Net hydroelectric power[3]	0.18
Net nuclear power[3]	(s)
Total	0.82

Petroleum	0.46
Dry natural gas	0.44
Coal[2]	0.08
Net hydroelectric power[3]	0.18
Net nuclear power[3]	(s)
Total	1.16

Telecommunications [30]

- The domestic telephone system is poor, adequate only for government and business use
- About 7 telephones per 1,000 persons
- The system for international traffic is better and employs both microwave radio relay and satellites
- Satellite ground stations - 1 Atlantic Ocean INTELSAT and 2 Indian Ocean INTELSAT
- Broadcast stations - 19 AM, 8 FM, 29 TV

Transportation [31]

Railroads. 8,773 km total; 7,718 km broad gauge, 445 km 1-meter gauge, and 610 km less than 1-meter gauge; 1,037 km broad-gauge double track; 286 km electrified; all government owned (1985)

Highways. 101,315 km total (1987); 40,155 km paved, 23,000 km gravel, 29,000 km improved earth, and 9,160 km unimproved earth or sand tracks (1985)

Merchant Marine. 29 ships (1,000 GRT or over) totaling 350,916 GRT/530,855 DWT; includes 3 passenger-cargo, 24 cargo, 1 oil tanker, 1 bulk

Airports

Total:	111
Usable:	104
With permanent-surface runways:	75
With runways over 3,659 m:	1
With runways 2,440-3,659 m:	31
With runways 1,220-2,439 m:	42

Top Agricultural Products [32]

	90-91[26]	91-92[26]
Sugar cane	35,939	37,414
Wheat	14,300	14,300
Cottonseed	3,274	3,572
Rice	3,265	3,250
Cotton	1,637	1,786
Corn	1,185	1,200

Values shown are 1,000 metric tons.

Top Mining Products [33]

Preliminary metric tons unless otherwise specified M.t.

Petroleum, crude (000 42-gal. barrels)	23,027
Refinery products (000 42-gal. barrels)	45,600[e]
Natural gas, gross production (mil. cubic feet)	518,483
Liquified natural gas (000 42-gal. barrels)	80
Coal, all grades (000 tons)	3,040

Tourism [34]

	1987	1988	1989	1990	1991
Tourists	425	460	495	424	438
Tourism receipts	178	147	169	146	163
Tourism expenditures	248	338	337	440	555
Fare receipts	159	57	260	468	470
Fare expenditures	121	126	132	116	169

Tourists are in thousands, money in million U.S. dollars.

For sources, notes, and explanations, see Annotated Source Appendix, page 1035.

703

Manufacturing Sector

GDP and Manufacturing Summary [35]

	1980	1985	1989	1990	% change 1980-1990	% change 1989-1990
GDP (million 1980 $)	28,418	38,555	48,319	52,017	83.0	7.7
GDP per capita (1980 $)	333	373	407	423	27.0	3.9
Manufacturing as % of GDP (current prices)	14.6	15.9	15.3[e]	17.4	19.2	13.7
Gross output (million $)	7,144	10,132	12,052[e]	13,354[e]	86.9	10.8
Value added (million $)	2,423	3,174	3,907[e]	4,299[e]	77.4	10.0
Value added (million 1980 $)	4,294	6,557	7,958	9,010	109.8	13.2
Industrial production index	100	133	153	160	60.0	4.6
Employment (thousands)	452	493	534[e]	561[e]	24.1	5.1

Note: GDP stands for Gross Domestic Product. 'e' stands for estimated value.

Profitability and Productivity

	1980	1985	1989	1990	% change 1980-1990	% change 1989-1990
Intermediate input (%)	66	69	68[e]	68[e]	3.0	0.0
Wages, salaries, and supplements (%)	7	6	7[e]	7[e]	0.0	0.0
Gross operating surplus (%)	27	25	25[e]	25[e]	-7.4	0.0
Gross output per worker ($)	14,606	20,486	22,587[e]	23,742[e]	62.5	5.1
Value added per worker ($)	4,953	6,418	7,321[e]	7,644[e]	54.3	4.4
Average wage (incl. benefits) ($)	1,122	1,324	1,582[e]	1,689[e]	50.5	6.8

Profitability is in percent of gross output. Productivity is in U.S. $. 'e' stands for estimated value.

Profitability - 1990

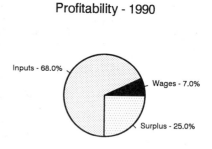

Inputs - 68.0%
Wages - 7.0%
Surplus - 25.0%

The graphic shows percent of gross output.

Value Added in Manufacturing

	1980 $ mil.	1980 %	1985 $ mil.	1985 %	1989 $ mil.	1989 %	1990 $ mil.	1990 %	% change 1980-1990	% change 1989-1990
311 Food products	431	17.8	580	18.3	711[e]	18.2	690[e]	16.1	60.1	-3.0
313 Beverages	45	1.9	74	2.3	89[e]	2.3	79[e]	1.8	75.6	-11.2
314 Tobacco products	300	12.4	372	11.7	415[e]	10.6	440[e]	10.2	46.7	6.0
321 Textiles	483	19.9	562	17.7	704[e]	18.0	792[e]	18.4	64.0	12.5
322 Wearing apparel	7	0.3	18	0.6	47[e]	1.2	74[e]	1.7	957.1	57.4
323 Leather and fur products	41	1.7	35	1.1	70[e]	1.8	26[e]	0.6	-36.6	-62.9
324 Footwear	4	0.2	3	0.1	6[e]	0.2	30[e]	0.7	650.0	400.0
331 Wood and wood products	4	0.2	9	0.3	12[e]	0.3	14[e]	0.3	250.0	16.7
332 Furniture and fixtures	3	0.1	6	0.2	3[e]	0.1	5[e]	0.1	66.7	66.7
341 Paper and paper products	29	1.2	33	1.0	46[e]	1.2	46[e]	1.1	58.6	0.0
342 Printing and publishing	26	1.1	36	1.1	34[e]	0.9	39[e]	0.9	50.0	14.7
351 Industrial chemicals	127	5.2	281	8.9	347[e]	8.9	281[e]	6.5	121.3	-19.0
352 Other chemical products	156	6.4	230	7.2	316[e]	8.1	308[e]	7.2	97.4	-2.5
353 Petroleum refineries	158	6.5	108	3.4	214[e]	5.5	263[e]	6.1	66.5	22.9
354 Miscellaneous petroleum and coal products	9	0.4	17	0.5	16[e]	0.4	32[e]	0.7	255.6	100.0
355 Rubber products	28	1.2	41	1.3	51[e]	1.3	40[e]	0.9	42.9	-21.6
356 Plastic products	12	0.5	21	0.7	24[e]	0.6	21[e]	0.5	75.0	-12.5
361 Pottery, china and earthenware	5	0.2	8	0.3	10[e]	0.3	15[e]	0.3	200.0	50.0
362 Glass and glass products	11	0.5	17	0.5	22[e]	0.6	31[e]	0.7	181.8	40.9
369 Other non-metal mineral products	171	7.1	199	6.3	260[e]	6.7	338[e]	7.9	97.7	30.0
371 Iron and steel	99	4.1	216	6.8	156[e]	4.0	292[e]	6.8	194.9	87.2
372 Non-ferrous metals	1	0.0	1	0.0	1[e]	0.0	1[e]	0.0	0.0	0.0
381 Metal products	38	1.6	33	1.0	30[e]	0.8	42[e]	1.0	10.5	40.0
382 Non-electrical machinery	43	1.8	80	2.5	89[e]	2.3	79[e]	1.8	83.7	-11.2
383 Electrical machinery	78	3.2	98	3.1	119[e]	3.0	138[e]	3.2	76.9	16.0
384 Transport equipment	97	4.0	83	2.6	97[e]	2.5	155[e]	3.6	59.8	59.8
385 Professional and scientific equipment	6	0.2	6	0.2	5[e]	0.1	11[e]	0.3	83.3	120.0
390 Other manufacturing industries	11	0.5	11	0.3	12[e]	0.3	16[e]	0.4	45.5	33.3

Note: The industry codes shown are International Standard Industry codes (ISIC). Percentages are percent of total Value Added. 'e' stands for estimated value

For sources, notes, and explanations, see Annotated Source Appendix, page 1035.

Finance, Economics, and Trade

Economic Indicators [36]

Millions of Pakistani Rupees unless otherwise indicated.

	FY89	FY90	FY91[120]
GDP (Nominal)	769,745	862,452	1,016,728
Real GDP Growth Rate	5.0	5.3	6.5
Real GNP Per Capita (US$)	389	385	388
Money Supply (M1)	201,401	233,833	279,325
Comm. Interest Rate (%)[121]	11.2	11.0	11.0
Saving Rate (% of GNP)	12.4	13.4	13.8
Investment Rate (% of GNP)	17.2	17.6	17.8
CPI (annual % change)	9.1	6.0	12.6
WPI (annual % change)	9.7	7.3	13.4
External Public Debt	14,190	15,094	15,961

Balance of Payments Summary [37]

Values in millions of dollars.

	1987	1988	1989	1990	1991
Exports of goods (f.o.b.)	3,938.0	4,405.0	4,796.0	5,380.0	6,352.0
Imports of goods (f.o.b.)	-6,254.0	-7,097.0	-7,366.0	-8,094.0	-8,599.0
Trade balance	-2,316.0	-2,692.0	-2,570.0	-2,714.0	-2,247.0
Services - debits	-2,187.0	-2,393.0	-2,808.0	-3,238.0	-3,551.0
Services - credits	1,083.0	946.0	1,323.0	1,518.0	1,585.0
Private transfers (net)	2,440.0	2,101.0	2,207.0	2,276.0	1,797.0
Government transfers (net)	434.0	631.0	524.0	513.0	627.0
Long term capital (net)	411.0	1,245.0	1,488.0	1,384.0	1,485.0
Short term capital (net)	216.0	416.0	-106.0	62.0	333.0
Errors and omissions	17.0	23.0	-210.0	-103.0	-46.0
Overall balance	98.0	277.0	-152.0	-302.0	-17.0

Exchange Rates [38]

Currency: **Pakistani rupees.**
Symbol: **PRs.**

Data are currency units per $1.

January 1993	25.904
1992	25.083
1991	23.801
1990	21.707
1989	20.541
1988	18.003

Imports and Exports

Top Import Origins [39]

$9.1 billion (f.o.b., FY92). Data are for FY91.

Origins	%
EC	29
Japan	13
US	12

Top Export Destinations [40]

$6.8 billion (f.o.b., FY92) Data are for FY91.

Destinations	%
EC	35
US	11
Japan	8

Foreign Aid [41]

Note: Figures include Bangladesh prior to 1972.

	U.S. $	
US commitments, including Ex-Im (FY70-89)	4.5	billion
Western (non-US) countries, ODA and OOF bilateral commitments (1980-89)	9.1	billion
OPEC bilateral aid (1979-89)	2.3	billion
Communist countries (1970-89)	3.2	billion

Import and Export Commodities [42]

Import Commodities	Export Commodities
Petroleum	Cotton
Petroleum products	Textiles
Machinery	Clothing
Transportation equipment	Rice
Vegetable oils	
Animal fats	
Chemicals	

For sources, notes, and explanations, see Annotated Source Appendix, page 1035.

705

Panama

Geography [1]

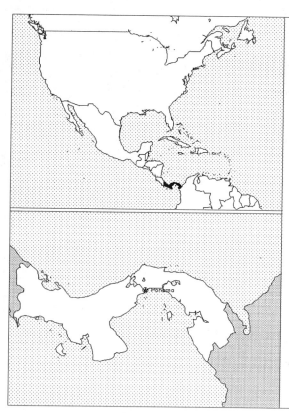

Total area:
78,200 km2
Land area:
75,990 km2
Comparative area:
Slightly smaller than South Carolina
Land boundaries:
Total 555 km, Colombia 225 km, Costa Rica 330 km
Coastline:
2,490 km
Climate:
Tropical; hot, humid, cloudy; prolonged rainy season (May to January), short dry season (January to May)
Terrain:
Interior mostly steep, rugged mountains and dissected, upland plains; coastal areas largely plains and rolling hills
Natural resources:
Copper, mahogany forests, shrimp
Land use:
Arable land:
6%
Permanent crops:
2%
Meadows and pastures:
15%
Forest and woodland:
54%
Other:
23%

Demographics [2]

	1960	1970	1980	1990	1991[1]	1994	2000	2010	2020
Population	1,148	1,531	1,956	2,427	2,476	2,630	2,934	3,422	3,886
Population density (persons per sq. mi.)	39	52	67	83	84	NA	100	117	133
Births	NA	NA	NA	NA	64	65	NA	NA	NA
Deaths	NA	NA	NA	NA	12	13	NA	NA	NA
Life expectancy - males	NA	NA	NA	NA	72	72	NA	NA	NA
Life expectancy - females	NA	NA	NA	NA	76	78	NA	NA	NA
Birth rate (per 1,000)	NA	NA	NA	NA	26	25	NA	NA	NA
Death rate (per 1,000)	NA	NA	NA	NA	5	5	NA	NA	NA
Women of reproductive age (15-44 yrs.)	NA	NA	NA	617	NA	680	768	NA	NA
of which are currently married	NA	NA	NA	350	NA	391	451	NA	NA
Fertility rate	NA	NA	NA	3.1	3.02	2.9	2.6	2.3	2.2

Population values are in thousands, life expectancy in years, and other items as indicated.

Health

Health Personnel [3]

Health Expenditures [4]

Human Factors

Health Care Ratios [5]

Population per physician, 1970	1660
Population per physician, 1990	840
Population per nursing person, 1970	1560
Population per nursing person, 1990	NA
Percent of births attended by health staff, 1985	83

Infants and Malnutrition [6]

Percent of babies with low birth weight, 1985	8
Infant mortality rate per 1,000 live births, 1970	47
Infant mortality rate per 1,000 live births, 1991	21
Years of life lost per 1,000 population, 1990	NA
Prevalence of malnutrition (under age 5), 1990	25

Ethnic Division [7]

Mestizo	70%
West Indian	14%
White	10%
Indian	6%

Religion [8]

Roman Catholic	85%
Protestant	15%

Major Languages [9]

Spanish (official), English 14%.

Education

Public Education Expenditures [10]

Million Balboa	1980	1985	1987	1988	1989	1990
Total education expenditure	166	237	NA	249	251	248
as percent of GNP	4.9	5.2	NA	6.0	6.1	5.5
as percent of total govt. expend.	19.0	18.7	NA	NA	NA	NA
Current education expenditure	156	231	NA	244	246	241
as percent of GNP	4.6	5.1	NA	5.9	5.9	5.3
as percent of current govt. expend.	19.8	19.9	NA	NA	NA	NA
Capital expenditure	10	6	NA	5	5	7

Educational Attainment [11]

Age group	25+
Total population	1,035,339
Highest level attained (%)	
No schooling	12.9
First level	
Incompleted	20.0
Completed	21.6
Entered second level	
S-1	17.2
S-2	11.5
Post secondary	16.8

Literacy Rate [12]

In thousands and percent	1985[a]	1991[a]	2000[a]
Illiterate population +15 years	185	187	180
Illiteracy rate - total pop. (%)	13.6	11.9	9.1
Illiteracy rate - males (%)	13.5	11.9	9.2
Illiteracy rate - females (%)	13.8	11.8	9.0

Libraries [13]

Daily Newspapers [14]

	1975	1980	1985	1990
Number of papers	6	5	7	8
Circ. (000)	131	110[e]	245	170[e]

Cinema [15]

Science and Technology

Scientific/Technical Forces [16]

R&D Expenditures [17]

	Balboa[13] (000) 1986
Total expenditure	173
Capital expenditure	-
Current expenditure	173
Percent current	100.0

U.S. Patents Issued [18]

Values show patents issued to citizens of the country by the U.S. Patents Office.

	1990	1991	1992
Number of patents	1	2	0

Government and Law

Organization of Government [19]

Long-form name:
Republic of Panama
Type:
centralized republic
Independence:
3 November 1903 (from Colombia;
became independent from Spain 28
November 1821)
Constitution:
11 October 1972; major reforms adopted
April 1983
Legal system:
based on civil law system; judicial review
of legislative acts in the Supreme Court of
Justice; accepts compulsory ICJ
jurisdiction, with reservations
National holiday:
Independence Day, 3 November (1903)
Executive branch:
president, two vice presidents, Cabinet
Legislative branch:
unicameral Legislative Assembly
(Asamblea Legislativa)
Judicial branch:
Supreme Court of Justice (Corte Suprema
de Justicia), 5 superior courts, 3 courts of
appeal

Crime [23]

Political Parties [20]

Legislative Assembly	% of seats
Nationalist Republican	22.4
Arnulfista Party	11.9
Authentic Liberal	6.0
Christian Democratic	41.8
Democratic Revolutionary	14.9
Agrarian Labor	1.5
Liberal	1.5

Government Expenditures [21]

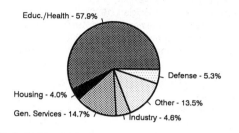

Educ./Health - 57.9%
Defense - 5.3%
Housing - 4.0%
Other - 13.5%
Gen. Services - 14.7%
Industry - 4.6%

% distribution for 1990. Expend. in 1992 est.: 1.900 ($ bil.)

Military Expenditures and Arms Transfers [22]

	1985	1986	1987	1988	1989
Military expenditures					
Current dollars (mil.)	94[e]	98[e]	100	101	141
1989 constant dollars (mil.)	107[e]	109[e]	107	105	141
Armed forces (000)	12	12	12	11	14
Gross national product (GNP)					
Current dollars (mil.)	4,266	4,552	4,804	4,049	4,134
1989 constant dollars (mil.)	4,856	5,051	5,167	4,215	4,134
Central government expenditures (CGE)					
1989 constant dollars (mil.)	1,773	1,932	1,938	1,445	NA
People (mil.)	2.2	2.2	2.3	2.3	2.4
Military expenditure as % of GNP	2.2	2.2	2.1	2.5	3.4
Military expenditure as % of CGE	6.0	5.7	5.6	7.3	NA
Military expenditure per capita	49	49	47	45	59
Armed forces per 1,000 people	5.5	5.4	5.4	4.7	5.9
GNP per capita	2,227	2,267	2,270	1,813	1,741
Arms imports[6]					
Current dollars (mil.)	10	10	20	10	5
1989 constant dollars (mil.)	11	11	22	10	5
Arms exports[6]					
Current dollars (mil.)	0	0	0	0	0
1989 constant dollars (mil.)	0	0	0	0	0
Total imports[7]					
Current dollars (mil.)	1,392	1,229	1,306	751	986
1989 constant dollars	1,585	1,364	1,405	782	986
Total exports[7]					
Current dollars (mil.)	333	341	348	280	297
1989 constant dollars	379	378	374	291	297
Arms as percent of total imports[8]	0.7	0.8	1.5	1.3	0.5
Arms as percent of total exports[8]	0	0	0	0	0

Human Rights [24]

	SSTS	FL	FAPRO	PPCG	APROBC	TPW	PCPTW	STPEP	PHRFF	PRW	ASST	AFL
Observes		P	P	P	P	P	P					P
	EAFRD	CPR	ESCR	SR	ACHR	MAAE	PVIAC	PVNAC	EAFDAW	TCIDTP	RC	
Observes		P	P	P	P	P		S	S	S	P	P

P = Party; S = Signatory; see Appendix for meaning of abbreviations.

Labor Force

Total Labor Force [25]

921,000 (1992 est.)

Labor Force by Occupation [26]

Government and community services	31.8%
Agriculture, hunting, and fishing	26.8
Commerce, restaurants, and hotels	16.4
Manufacturing and mining	9.4
Construction	3.2
Transportation and communications	6.2
Finance, insurance, and real estate	4.3

Unemployment Rate [27]

15% (1992 est.)

Production Sectors

Commercial Energy Production and Consumption

Production [28]

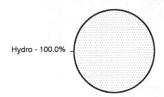

Hydro - 100.0%

Consumption [29]

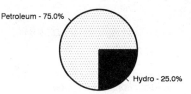

Petroleum - 75.0%

Hydro - 25.0%

Data are shown in quadrillion (10^{15}) BTUs and percent for 1991

Crude oil[1]	0.00
Natural gas liquids	0.00
Dry natural gas	0.00
Coal[2]	0.00
Net hydroelectric power[3]	0.02
Net nuclear power[3]	0.00
Total	0.02

Petroleum	0.06
Dry natural gas	0.00
Coal[2]	(s)
Net hydroelectric power[3]	0.02
Net nuclear power[3]	(s)
Total	0.08

Telecommunications [30]

- Domestic and international facilities well developed
- Connection into Central American Microwave System
- 220,000 telephones
- Broadcast stations - 91 AM, no FM, 23 TV
- 1 coaxial submarine cable
- Satellite ground stations - 2 Atlantic Ocean INTELSAT

Top Agricultural Products [32]

	1989	1990
Bananas	1,253.6	1,264.7
Rice, paddy	207.2	216.1
Fish	206.1	142.5
Sugar cane	110.7	112.7
Corn	90.3	93.8
Fish meal	33.5	22.9

Values shown are 1,000 metric tons.

Top Mining Products [33]

Preliminary metric tons unless otherwise specified M.t.

Cement	275,000[e]
Clay (for cement and other products)	462,294
Limestone (for cement and other uses)	337,823
Salt, marine	5,000
Sand and gravel (000 tons)	1,941
Gold (kilograms)[32]	194
Petroleum refinery products (000 42-gal. barrels)	8,000

Transportation [31]

Railroads. 238 km total; 78 km 1.524-meter gauge, 160 km 0.914-meter gauge

Highways. 8,530 km total; 2,745 km paved, 3,270 km gravel or crushed stone, 2,515 km improved and unimproved earth

Merchant Marine. 3,244 ships (1,000 GRT or over) totaling 51,353,963 GRT/82,138,537 DWT; includes 22 passenger, 26 short-sea passenger, 3 passenger-cargo, 1,091 cargo, 246 refrigerated cargo, 196 container, 63 roll-on/roll-off cargo, 121 vehicle carrier, 9 livestock carrier, 5 multifunction large-load carrier, 403 oil tanker, 180 chemical tanker, 26 combination ore/oil, 121 liquefied gas, 9 specialized tanker, 688 bulk, 34 combination bulk, 1 barge carrier; note - all but 5 are foreign owned and operated; the top 4 foreign owners are Japan 36%, Greece 8%, Hong Kong 8%, and Taiwan 5%; (China owns at least 131 ships, Vietnam 3, Croatia 3, Cuba 4, Cyprus 6, and Russia 16)

Airports

Total:	112
Usable:	104
With permanent-surface runways:	39
With runways over 3,659 m:	0
With runways 2,440-3,659 m:	2
With runways 1,220-2,439 m:	15

Tourism [34]

	1987	1988	1989	1990	1991
Visitors[51]	422	290	268	278	350
Tourists	271	199	192	214	279
Excursionists	111	78	76	63	71
Tourism receipts	188	168	157	167	196
Tourism expenditures	90	88	86	99	108
Fare receipts	26	17	17	16	11
Fare expenditures	57	39	42	54	59

Tourists are in thousands, money in million U.S. dollars.

For sources, notes, and explanations, see Annotated Source Appendix, page 1035.

709

Manufacturing Sector

GDP and Manufacturing Summary [35]

	1980	1985	1989	1990	% change 1980-1990	% change 1989-1990
GDP (million 1980 $)	3,559	4,094	3,623	3,809	7.0	5.1
GDP per capita (1980 $)	1,818	1,877	1,529	1,574	-13.4	2.9
Manufacturing as % of GDP (current prices)	9.8	8.3	7.7	7.3[e]	-25.5	-5.2
Gross output (million $)	1,473[e]	1,765	1,353[e]	1,504[e]	2.1	11.2
Value added (million $)	477	585	486[e]	547[e]	14.7	12.6
Value added (million 1980 $)	356	351	302	339	-4.8	12.3
Industrial production index	100	106	90	101	1.0	12.2
Employment (thousands)	31[e]	36	31[e]	36[e]	16.1	16.1

Note: GDP stands for Gross Domestic Product. 'e' stands for estimated value.

Profitability and Productivity

	1980	1985	1989	1990	% change 1980-1990	% change 1989-1990
Intermediate input (%)	68[e]	67	64[e]	64[e]	-5.9	0.0
Wages, salaries, and supplements (%)	9[e]	13[e]	13[e]	14[e]	55.6	7.7
Gross operating surplus (%)	23[e]	20[e]	23[e]	22[e]	-4.3	-4.3
Gross output per worker ($)	46,756[e]	48,648	43,455[e]	41,614[e]	-11.0	-4.2
Value added per worker ($)	15,159[e]	16,134	15,610[e]	15,650[e]	3.2	0.3
Average wage (incl. benefits) ($)	4,238[e]	6,270[e]	5,479[e]	5,868[e]	38.5	7.1

Profitability is in percent of gross output. Productivity is in U.S. $. 'e' stands for estimated value.

Profitability - 1990

Inputs - 64.0%
Wages - 14.0%
Surplus - 22.0%

The graphic shows percent of gross output.

Value Added in Manufacturing

	1980 $ mil.	1980 %	1985 $ mil.	1985 %	1989 $ mil.	1989 %	1990 $ mil.	1990 %	% change 1980-1990	% change 1989-1990
311 Food products	155	32.5	179	30.6	159[e]	32.7	195[e]	35.6	25.8	22.6
313 Beverages	52	10.9	63	10.8	64[e]	13.2	74[e]	13.5	42.3	15.6
314 Tobacco products	26	5.5	31	5.3	25[e]	5.1	28[e]	5.1	7.7	12.0
321 Textiles	4	0.8	3	0.5	4[e]	0.8	5[e]	0.9	25.0	25.0
322 Wearing apparel	31	6.5	27	4.6	20[e]	4.1	15[e]	2.7	-51.6	-25.0
323 Leather and fur products	4	0.8	4	0.7	2[e]	0.4	2[e]	0.4	-50.0	0.0
324 Footwear	7	1.5	9	1.5	6[e]	1.2	6[e]	1.1	-14.3	0.0
331 Wood and wood products	8	1.7	8	1.4	3[e]	0.6	2[e]	0.4	-75.0	-33.3
332 Furniture and fixtures	8	1.7	11	1.9	4[e]	0.8	5[e]	0.9	-37.5	25.0
341 Paper and paper products	20	4.2	33	5.6	22[e]	4.5	22[e]	4.0	10.0	0.0
342 Printing and publishing	22	4.6	30	5.1	23[e]	4.7	23[e]	4.2	4.5	0.0
351 Industrial chemicals	4	0.8	9	1.5	6[e]	1.2	8[e]	1.5	100.0	33.3
352 Other chemical products	26	5.5	42	7.2	36[e]	7.4	39[e]	7.1	50.0	8.3
353 Petroleum refineries	27	5.7	25	4.3	37[e]	7.6	40[e]	7.3	48.1	8.1
354 Miscellaneous petroleum and coal products	NA	0.0	2	0.3	3[e]	0.6	3[e]	0.5	NA	0.0
355 Rubber products	2	0.4	2	0.3	2[e]	0.4	2[e]	0.4	0.0	0.0
356 Plastic products	12	2.5	21	3.6	19[e]	3.9	19[e]	3.5	58.3	0.0
361 Pottery, china and earthenware	NA	0.0	NA	0.0	NA	0.0	NA	0.0	NA	NA
362 Glass and glass products	1[e]	0.2	7	1.2	5[e]	1.0	6[e]	1.1	500.0	20.0
369 Other non-metal mineral products	31	6.5	27	4.6	14[e]	2.9	17[e]	3.1	-45.2	21.4
371 Iron and steel	5	1.0	4	0.7	1[e]	0.2	1[e]	0.2	-80.0	0.0
372 Non-ferrous metals	2	0.4	3	0.5	2[e]	0.4	2[e]	0.4	0.0	0.0
381 Metal products	19	4.0	21	3.6	12[e]	2.5	14[e]	2.6	-26.3	16.7
382 Non-electrical machinery	1	0.2	1	0.2	1[e]	0.2	1[e]	0.2	0.0	0.0
383 Electrical machinery	3	0.6	4	0.7	2[e]	0.4	3[e]	0.5	0.0	50.0
384 Transport equipment	4	0.8	13	2.2	8[e]	1.6	8[e]	1.5	100.0	0.0
385 Professional and scientific equipment	1	0.2	3	0.5	2[e]	0.4	3[e]	0.5	200.0	50.0
390 Other manufacturing industries	2	0.4	2	0.3	3[e]	0.6	3[e]	0.5	50.0	0.0

Note: The industry codes shown are International Standard Industry codes (ISIC). Percentages are percent of total Value Added. 'e' stands for estimated value

710

For sources, notes, and explanations, see Annotated Source Appendix, page 1035.

Finance, Economics, and Trade

Economic Indicators [36]

Millions of U.S. Dollars unless otherwise noted.

	1989	1990	1991[e]
Real GDP (1970 prices)	1,786.0	1,868.0	1,952.0
GDP growth (%)	-0.4	4.6	4.5
Real GDP/capita (1970 prices)	754	773	791
Money and quasi-money	1,600.0	2,184.0	2,556.0
Savings Rate (% GDP)	4.4	9.3	10.9
Investment rate (% GDP)	2.8	16.6	15.2
Consumer prices (%, annual avg.)	-0.1	0.5	2.0
Wholesale prices (%, annual avg.)	2.5	9.5	9.0
External public debt[122]	5,039	5,358	5,099

Balance of Payments Summary [37]

Values in millions of dollars.

	1987	1988	1989	1990	1991
Exports of goods (f.o.b.)	2,491.7	2,346.5	2,680.8	3,318.1	4,153.6
Imports of goods (f.o.b.)	-3,058.3	-2,531.4	-3,084.2	-3,803.6	-4,981.0
Trade balance	-566.6	-184.9	-403.4	-485.5	-827.4
Services - debits	-2,527.0	-1,179.7	-1,319.7	-1,505.6	-1,480.9
Services - credits	3,235.1	2,059.1	2,100.7	2,159.1	2,227.4
Private transfers (net)	-50.8	-40.0	-35.8	-35.0	-23.8
Government transfers (net)	113.7	111.5	105.9	117.5	239.2
Long term capital (net)	3.7	54.7	-7.9	-147.0	140.1
Short term capital (net)	313.7	403.2	9.7	-351.3	-805.1
Errors and omissions	-519.4	-1,237.3	-393.8	583.3	718.0
Overall balance	2.4	-13.4	55.7	335.5	187.5

Exchange Rates [38]

Currency: **balboas.**
Symbol: **B.**

Data are currency units per $1.

Fixed rate	1.000

Imports and Exports

Top Import Origins [39]

$2.0 billion (f.o.b., 1992 est.).

Origins	%
US	36
Japan	NA
EC	NA
Central America	NA
Caribbean	NA
Mexico	NA
Venezuela 1992 est.	NA

Top Export Destinations [40]

$486 million (f.o.b., 1992 est.).

Destinations	%
US	38
Central America and Caribbean	NA
EC	NA

Foreign Aid [41]

	U.S. $	
US commitments, including Ex-Im (FY70-89)	516	million
Western (non-US) countries, ODA and OOF bilateral commitments (1970-89)	582	million
Communist countries (1970-89)	4	million

Import and Export Commodities [42]

Import Commodities

Capital goods 21%
Crude oil 11%
Foodstuffs 9%
Consumer goods
Chemicals

Export Commodities

Bananas 43%
Shrimp 11%
Sugar 4%
Clothing 5%
Coffee 2%

For sources, notes, and explanations, see Annotated Source Appendix, page 1035.

711

Papua New Guinea

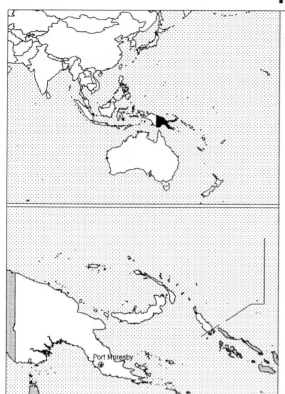

Geography [1]

Total area:
461,690 km2
Land area:
451,710 km2
Comparative area:
Slightly larger than California
Land boundaries:
Total 820 km, Indonesia 820 km
Coastline:
5,152 km
Climate:
Tropical; northwest monsoon (December to March), southeast monsoon (May to October); slight seasonal temperature variation
Terrain:
Mostly mountains with coastal lowlands and rolling foothills
Natural resources:
Gold, copper, silver, natural gas, timber, oil potential
Land use:
Arable land:
0%
Permanent crops:
1%
Meadows and pastures:
0%
Forest and woodland:
71%
Other:
28%

Demographics [2]

	1960	1970	1980	1990	1991[1]	1994	2000	2010	2020
Population	1,747	2,288	2,991	3,823	3,913	4,197	4,812	5,925	7,044
Population density (persons per sq. mi.)	10	13	17	22	22	NA	28	34	41
Births	NA	NA	NA	NA	133	141	NA	NA	NA
Deaths	NA	NA	NA	NA	42	44	NA	NA	NA
Life expectancy - males	NA	NA	NA	NA	55	56	NA	NA	NA
Life expectancy - females	NA	NA	NA	NA	56	57	NA	NA	NA
Birth rate (per 1,000)	NA	NA	NA	NA	34	34	NA	NA	NA
Death rate (per 1,000)	NA	NA	NA	NA	11	10	NA	NA	NA
Women of reproductive age (15-44 yrs.)	NA	NA	NA	NA	NA	NA	NA	NA	NA
of which are currently married	NA	NA	NA	NA	NA	NA	NA	NA	NA
Fertility rate	NA	NA	NA	5.1	4.91	4.7	4.1	3.3	2.7

Population values are in thousands, life expectancy in years, and other items as indicated.

Health

Health Personnel [3]

Doctors per 1,000 pop., 1988-92	0.08
Nurse-to-doctor ratio, 1988-92	8.1
Hospital beds per 1,000 pop., 1985-90	3.4
Percentage of children immunized (age 1 yr. or less)	
Third dose of DPT, 1990-91	64.0
Measles, 1990-91	63.0

Health Expenditures [4]

Total health expenditure, 1990 (official exchange rate)	
Millions of dollars	142
Millions of dollars per capita	36
Health expenditures as a percentage of GDP	
Total	4.4
Public sector	2.8
Private sector	1.6
Development assistance for health	
Total aid flows (millions of dollars)[1]	7
Aid flows per capita (millions of dollars)	1.8
Aid flows as a percentage of total health expenditure	4.9

For sources, notes, and explanations, see Annotated Source Appendix, page 1035.

Human Factors

Health Care Ratios [5]

Population per physician, 1970	11640
Population per physician, 1990	12870
Population per nursing person, 1970	1710
Population per nursing person, 1990	1180
Percent of births attended by health staff, 1985	34

Infants and Malnutrition [6]

Percent of babies with low birth weight, 1985	25
Infant mortality rate per 1,000 live births, 1970	112
Infant mortality rate per 1,000 live births, 1991	55
Years of life lost per 1,000 population, 1990	79
Prevalence of malnutrition (under age 5), 1990	NA

Ethnic Division [7]

Melanesian, Papuan, Negrito, Micronesian, Polynesian.

Religion [8]

Roman Catholic	22%
Lutheran	16%
Presbyterian/Methodist/ London Missionary Society	8%
Anglican	5%
Evangelical Alliance	4%
Other Protestant sects	11%
Indigenous beliefs	34%

Major Languages [9]

English spoken by 1-2%, pidgin English widespread, Motu spoken in Papua region.

Education

Public Education Expenditures [10]

Educational Attainment [11]

Age group	25+
Total population	1,135,783
Highest level attained (%)	
No schooling	82.6
First level	
Incompleted	8.2
Completed	5.0
Entered second level	
S-1	3.9
S-2	0.3
Post secondary	NA

Literacy Rate [12]

In thousands and percent	1985[a]	1991[a]	2000[a]
Illiterate population + 15 years	1,093	1,119	1,134
Illiteracy rate - total pop. (%)	53.3	48.0	38.1
Illiteracy rate - males (%)	39.8	35.1	26.7
Illiteracy rate - females (%)	68.0	62.2	50.3

Libraries [13]

	Admin. Units	Svc. Pts.	Vols. (000)	Shelving (meters)	Vols. Added	Reg. Users
National (1989)	1	4	60	4,000	2,408	3,600
Non-specialized	NA	NA	NA	NA	NA	NA
Public (1989)	19	NA	151	NA	6,000	46,095
Higher ed.	NA	NA	NA	NA	NA	NA
School	NA	NA	NA	NA	NA	NA

Daily Newspapers [14]

	1975	1980	1985	1990
Number of papers	1	1	2	2
Circ. (000)	20[e]	27	45	49

Cinema [15]

Science and Technology

Scientific/Technical Forces [16]

R&D Expenditures [17]

U.S. Patents Issued [18]

For sources, notes, and explanations, see Annotated Source Appendix, page 1035.

713

Government and Law

Organization of Government [19]

Long-form name:
Independent State of Papua New Guinea
Type:
parliamentary democracy
Independence:
16 September 1975 (from UN trusteeship under Australian administration)
Constitution:
16 September 1975
Legal system:
based on English common law
National holiday:
Independence Day, 16 September (1975)
Executive branch:
British monarch, governor general, prime minister, deputy prime minister, National Executive Council (cabinet)
Legislative branch:
unicameral National Parliament (sometimes referred to as the House of Assembly)
Judicial branch:
Supreme Court

Political Parties [20]

National Parliament	% of seats
Pangu Party	22.0
People's Democratic Movement (PDM)	15.6
People's Progress Party (PPP)	9.2
People's Action Party (PAP)	9.2
Independents	27.5
Others	16.5

Government Expenditures [21]

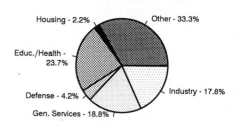

Housing - 2.2%
Other - 33.3%
Educ./Health - 23.7%
Defense - 4.2%
Industry - 17.8%
Gen. Services - 18.8%

% distribution for 1992. Expend. in 1993 est.: 1.49 ($ bil.)

Defense Summary [22]

Branches: Papua New Guinea Defense Force (including Army, Navy, Air Force)

Manpower Availability: Males age 15-49 1,046,929; fit for military service 582,685 (1993 est.)

Defense Expenditures: Exchange rate conversion - $55 million, 1.8% of GDP (1993 est.)

Crime [23]

Crime volume	
Cases known to police	27,556
Attempts (percent)	NA
Percent cases solved	34.84
Crimes per 100,000 persons	750.60
Persons responsible for offenses	
Total number offenders	11,313
Percent female	5.12
Percent juvenile[41]	12.89
Percent foreigners	0.15

Human Rights [24]

	SSTS	FL	FAPRO	PPCG	APROBC	TPW	PCPTW	STPEP	PHRFF	PRW	ASST	AFL
Observes	P	P		P	P	P	P			P		P
	EAFRD	CPR	ESCR	SR	ACHR	MAAE	PVIAC	PVNAC	EAFDAW	TCIDTP	RC
Observes		P		P						S	S

P = Party; S = Signatory; see Appendix for meaning of abbreviations.

Labor Force

Total Labor Force [25]

Labor Force by Occupation [26]

Unemployment Rate [27]

Production Sectors

Energy Resource Summary [28]

Energy Resources: Natural gas, oil potential. **Electricity**: 400,000 kW capacity; 1,600 million kWh produced, 400 kWh per capita (1992).

Telecommunications [30]

- Services are adequate and being improved
- Facilities provide radiobroadcast, radiotelephone and telegraph, coastal radio, aeronautical radio, and international radiocommunication services
- Submarine cables extend to Australia and Guam
- More than 70,000 telephones (1987)
- Broadcast stations - 31 AM, 2 FM, 2 TV (1987)
- 1 Pacific Ocean INTELSAT earth station

Top Agricultural Products [32]

One-third of GDP; livelihood for 85% of population; fertile soils and favorable climate permits cultivating a wide variety of crops; cash crops—coffee, cocoa, coconuts, palm kernels; other products—tea, rubber, sweet potatoes, fruit, vegetables, poultry, pork; net importer of food for urban centers.

Top Mining Products [33]

Preliminary metric tons unless otherwise specified M.t.

Copper, mine output, Cu content	204,459
Gold, mine output, Au content (kilograms)	60,780
Silver, mine output, Ag content (kilograms)	124,880

Transportation [31]

Railroads. none

Highways. 19,200 km total; 640 km paved, 10,960 km gravel, crushed stone, or stabilized-soil surface, 7,600 km unimproved earth

Merchant Marine. 11 ships (1,000 GRT or over) totaling 20,523 GRT/24,774 DWT; includes 2 cargo, 1 roll-on/roll-off cargo, 5 combination ore/oil, 2 bulk, 1 container

Airports

Total:	504
Usable:	457
With permanent-surface runways:	18
With runways over 3,659 m:	0
With runways 2,440-3,659 m:	1
With runways 1,220-2,439 m:	39

Tourism [34]

	1987	1988	1989	1990	1991
Tourists	35	41	49	41	37
Tourism receipts	21	25	40	41[52]	47
Tourism expenditures	32	48	42		
Fare receipts	15	12	20		
Fare expenditures	54	55			

Tourists are in thousands, money in million U.S. dollars.

Finance, Economics, and Trade

GDP and Manufacturing Summary [35]

	1980	1985	1990	1991	1992
Gross Domestic Product					
Millions of 1980 dollars	2,549	2,739	2,841	3,112	3,391
Growth rate in percent	-1.91	4.31	-3.71	9.55	8.96
Manufacturing Value Added					
Millions of 1980 dollars	242	268	239	243	279[e]
Growth rate in percent	-3.02	2.80	-14.54	1.50	15.06[e]
Manufacturing share in percent of current prices	10.5	11.0	12.2	9.7	9.7[e]

Economic Indicators [36]

National product: GDP—exchange rate conversion—$3.4 billion (1992). **National product real growth rate**: 8.5% (1992). **National product per capita**: $850 (1992). **Inflation rate (consumer prices)**: 4.5% (1992-93). **External debt**: $2.2 billion (April 1991).

Balance of Payments Summary [37]

Values in millions of dollars.

	1985	1986	1987	1988	1989
Exports of goods (f.o.b.)	926.3	1,030.7	1,243.9	1,475.3	1,318.5
Imports of goods (f.o.b.)	-874.8	-928.8	-1,129.6	-1,384.5	-1,341.3
Trade balance	51.5	101.9	114.3	90.8	-22.8
Services - debits	-450.3	-505.2	-583.9	-761.3	-674.0
Services - credits	119.6	164.0	155.2	240.9	254.9
Private transfers (net)	-88.8	-78.7	-104.5	-124.6	-130.7
Government transfers (net)	218.4	212.7	204.3	217.7	217.3
Long term capital (net)	120.7	143.8	167.8	210.5	252.0
Short term capital (net)	3.4	-9.3	9.3	34.6	13.0
Errors and omissions	24.3	-26.0	39.5	37.9	31.6
Overall balance	-1.2	3.2	2.0	-53.5	-58.7

Exchange Rates [38]

Currency: **kina.**
Symbol: **K.**

Data are currency units per $1.

January 1993	1.0065
1992	1.0367
1991	1.0504
1990	1.0467
1989	1.1685
1988	1.1538

Imports and Exports

Top Import Origins [39]

$1.6 billion (c.i.f., 1990).

Origins	%
Australia	NA
Singapore	NA
Japan	NA
US	NA
New Zealand	NA
UK	NA

Top Export Destinations [40]

$1.3 billion (f.o.b., 1990).

Destinations	%
FRG	NA
Japan	NA
Australia	NA
UK	NA
Spain	NA
US	NA

Foreign Aid [41]

	U.S. $	
US commitments, including Ex-Im (FY70-89)	40.6	million
Western (non-US) countries, ODA and OOF bilateral commitments (1970-89)	6.5	billion
OPEC bilateral aid (1979-89)	17	million

Import and Export Commodities [42]

Import Commodities	**Export Commodities**
Machinery & transport equipment	Gold
Food	Copper ore
Fuels	Coffee
Chemicals	Logs
Consumer goods	Palm oil
	Cocoa
	Lobster

Paraguay

Geography [1]

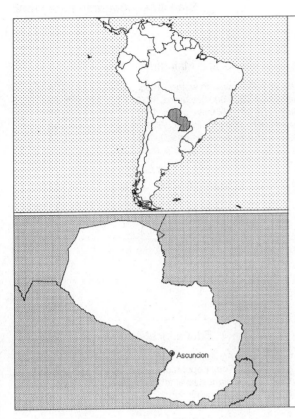

Total area:
406,750 km2
Land area:
397,300 km2
Comparative area:
Slightly smaller than California
Land boundaries:
Total 3,920 km, Argentina 1,880 km, Bolivia 750 km, Brazil 1,290 km
Coastline:
0 km (landlocked)
Climate:
Varies from temperate in east to semiarid in far west
Terrain:
Grassy plains and wooded hills east of Rio Paraguay; Gran Chaco region west of Rio Paraguay mostly low, marshy plain near the river, and dry forest and thorny scrub elsewhere
Natural resources:
Hydropower, timber, iron ore, manganese, limestone
Land use:
Arable land:
20%
Permanent crops:
1%
Meadows and pastures:
39%
Forest and woodland:
35%
Other:
5%

Demographics [2]

	1960	1970	1980	1990	1991[1]	1994	2000	2010	2020
Population	1,910	2,477	3,379	4,651	4,799	5,214	6,104	7,730	9,474
Population density (persons per sq. mi.)	12	16	22	30	31	NA	39	48	57
Births	NA	NA	NA	NA	167	167	NA	NA	NA
Deaths	NA	NA	NA	NA	29	23	NA	NA	NA
Life expectancy - males	NA	NA	NA	NA	67	72	NA	NA	NA
Life expectancy - females	NA	NA	NA	NA	72	75	NA	NA	NA
Birth rate (per 1,000)	NA	NA	NA	NA	35	32	NA	NA	NA
Death rate (per 1,000)	NA	NA	NA	NA	6	4	NA	NA	NA
Women of reproductive age (15-44 yrs.)	NA	NA	NA	1,095	NA	1,237	1,498	NA	NA
of which are currently married	NA	NA	NA	641	NA	729	874	NA	NA
Fertility rate	NA	NA	NA	4.6	4.69	4.3	3.9	3.2	2.8

Population values are in thousands, life expectancy in years, and other items as indicated.

Health

Health Personnel [3]

Doctors per 1,000 pop., 1988-92	0.62
Nurse-to-doctor ratio, 1988-92	1.7
Hospital beds per 1,000 pop., 1985-90	1.0
Percentage of children immunized (age 1 yr. or less)	
Third dose of DPT, 1990-91	79.0
Measles, 1990-91	74.0

Health Expenditures [4]

Total health expenditure, 1990 (official exchange rate)	
Millions of dollars	160
Millions of dollars per capita	37
Health expenditures as a percentage of GDP	
Total	2.8
Public sector	1.2
Private sector	1.6
Development assistance for health	
Total aid flows (millions of dollars)[1]	10
Aid flows per capita (millions of dollars)	2.4
Aid flows as a percentage of total health expenditure	6.4

For sources, notes, and explanations, see Annotated Source Appendix, page 1035.

717

Human Factors

Health Care Ratios [5]

Population per physician, 1970	2300
Population per physician, 1990	NA
Population per nursing person, 1970	2210
Population per nursing person, 1990	NA
Percent of births attended by health staff, 1985	22

Infants and Malnutrition [6]

Percent of babies with low birth weight, 1985	6
Infant mortality rate per 1,000 live births, 1970	57
Infant mortality rate per 1,000 live births, 1991	35
Years of life lost per 1,000 population, 1990	22
Prevalence of malnutrition (under age 5), 1990	4

Ethnic Division [7]

Mestizo	95%
White Indian	5%

Religion [8]

Roman Catholic 90%, Mennonite and other Protestant denominations.

Major Languages [9]

Spanish (official), Guarani.

Education

Public Education Expenditures [10]

Million Guarani	1980	1985	1987	1988[1]	1989[1]	1990[1]
Total education expenditure	8,793	20,662	NA	31,937	56,512	74,387
as percent of GNP	1.5	1.5	NA	1.0	1.2	1.2
as percent of total govt. expend.	16.4	16.7	NA	12.7	11.4	9.1
Current education expenditure	6,267	16,822	NA	29,382	54,252	71,134
as percent of GNP	1.1	1.2	NA	0.9	1.2	1.1
as percent of current govt. expend.	NA	18.8	NA	NA	NA	NA
Capital expenditure	1,158	3,840	NA	2,555	2,260	3,253

Educational Attainment [11]

Age group	25+
Total population	1,141,841
Highest level attained (%)	
No schooling	14.2[9]
First level	
Incompleted	51.0
Completed	15.4
Entered second level	
S-1	16.0
S-2	NA
Post secoondary	3.4

Literacy Rate [12]

In thousands and percent	1985[a]	1991[a]	2000[a]
Illiterate population +15 years	255	252	239
Illiteracy rate - total pop. (%)	11.7	9.9	7.0
Illiteracy rate - males (%)	9.1	7.9	5.8
Illiteracy rate - females (%)	14.2	11.9	8.2

Libraries [13]

	Admin. Units	Svc. Pts.	Vols. (000)	Shelving (meters)	Vols. Added	Reg. Users
National (1990)	1	NA	NA	NA	NA	NA
Non-specialized	NA	NA	NA	NA	NA	NA
Public (1990)	28	NA	NA	NA	NA	NA
Higher ed. (1990)	26	NA	NA	NA	NA	NA
School (1990)	145	NA	NA	NA	NA	NA

Daily Newspapers [14]

	1975	1980	1985	1990
Number of papers	8	5	6	5
Circ. (000)	140[e]	160[e]	170[e]	165[e]

Cinema [15]

Science and Technology

Scientific/Technical Forces [16]

R&D Expenditures [17]

U.S. Patents Issued [18]

Values show patents issued to citizens of the country by the U.S. Patents Office.

	1990	1991	1992
Number of patents	1	0	0

Government and Law

Organization of Government [19]

Long-form name:
Republic of Paraguay
Type:
republic
Independence:
14 May 1811 (from Spain)
Constitution:
25 August 1967; Constituent Assembly
rewrote the Constitution that was
promulgated on 20 June 1992
Legal system:
based on Argentine codes, Roman law,
and French codes; judicial review of
legislative acts in Supreme Court of
Justice; does not accept compulsory ICJ
jurisdiction
National holiday:
Independence Days, 14-15 May (1811)
Executive branch:
president, Council of Ministers (cabinet),
Council of State
Legislative branch:
bicameral Congress (Congreso) consists
of an upper chamber or Chamber of
Senators (Camara de Senadores) and a
lower chamber or Chamber of Deputies
(Camara de Diputados)
Judicial branch:
Supreme Court of Justice

Political Parties [20]

Chamber of Deputies	% of seats
Colorado Party	66.7
Authentic Radical Liberal Party	26.4
Feberista Revolutionary Party	2.8
Christian Democratic Party	1.4
Other	2.8

Government Budget [21]

For 1991.
Revenues	1.200
Expenditures	1.200
Capital expenditures	0.487

Crime [23]

Crime volume	
Cases known to police	7,332
Attempts (percent)	100
Percent cases solved	67.52
Crimes per 100,000 persons	461.37
Persons responsible for offenses	
Total number offenders	6,466
Percent female	9.17
Percent juvenile	12.62
Percent foreigners	10.33

Military Expenditures and Arms Transfers [22]

	1985	1986	1987	1988	1989
Military expenditures					
Current dollars (mil.)	37[e]	NA	37[e]	55[e]	61
1989 constant dollars (mil.)	43[e]	NA	40[e]	57[e]	61
Armed forces (000)	14	16	16	16	16
Gross national product (GNP)					
Current dollars (mil.)	3,180	3,222	3,449	3,811	4,323
1989 constant dollars (mil.)	3,620	3,575	3,709	3,967	4,323
Central government expenditures (CGE)					
1989 constant dollars (mil.)	357	312	362	368	NA
People (mil.)	4.0	4.1	4.3	4.4	4.5
Military expenditure as % of GNP	1.2	NA	1.1	1.4	1.4
Military expenditure as % of CGE	11.9	NA	11.0	15.6	NA
Military expenditure per capita	11	NA	9	13	13
Armed forces per 1,000 people	3.5	3.9	3.8	3.6	3.5
GNP per capita	907	868	872	905	956
Arms imports[6]					
Current dollars (mil.)	20	10	0	30	0
1989 constant dollars (mil.)	23	11	0	31	0
Arms exports[6]					
Current dollars (mil.)	0	0	0	0	0
1989 constant dollars (mil.)	0	0	0	0	0
Total imports[7]					
Current dollars (mil.)	502	578	595	574	798
1989 constant dollars	571	641	640	598	798
Total exports[7]					
Current dollars (mil.)	304	234	353	510	691
1989 constant dollars	346	260	380	531	691
Arms as percent of total imports[8]	4.0	1.7	0	5.2	0
Arms as percent of total exports[8]	0	0	0	0	0

Human Rights [24]

	SSTS	FL	FAPRO	PPCG	APROBC	TPW	PCPTW	STPEP	PHRFF	PRW	ASST	AFL
Observes		P	P	S	P	P	P			P		P
	EAFRD	CPR	ESCR	SR	ACHR	MAAE	PVIAC	PVNAC	EAFDAW	TCIDTP	RC	
Observes		P		P	P				P	P	P	

P = Party; S = Signatory; see Appendix for meaning of abbreviations.

Labor Force

Total Labor Force [25]

1.641 million (1992 est.)

Labor Force by Occupation [26]

Agriculture
Industry and commerce
Services
Government
Date of data: 1986

Unemployment Rate [27]

10% (1992 est.)

Production Sectors

Energy Resource Summary [28]

Energy Resources: Hydropower. **Electricity**: 5,257,000 kW capacity; 16,200 million kWh produced, 3,280 kWh per capita (1992).

Telecommunications [30]

- Principal center in Asuncion
- Fair intercity microwave net
- 78,300 telephones
- Broadcast stations - 40 AM, no FM, 5 TV, 7 shortwave
- 1 Atlantic Ocean INTELSAT earth station

Top Agricultural Products [32]

	1990	1991
Sugar cane	2,255	2,200
Soybeans	1,575	1,300
Corn	1,000	980
Wheat	375	300
Cotton (lint)	225	260
Tung	163	165

Values shown are 1,000 metric tons.

Top Mining Products [33]

Estimated metric tons unless otherwise specified	M.t.
Clay, except kaolin (000 metric tons)	1,900
Sand, incl. glass sand (000 metric tons)	2,000
Gypsum	4,500
Kaolin	74,000
Limestone, for cement and lime (000 metric tons)	600
Pigments, mineral (natural ochre)	330
Iron and steel	123,000
Stone, dimension, crushed, broken (000 m.t.)	3,420
Talc, soapstone, pyrophylite	200

Transportation [31]

Railroads. 970 km total; 440 km 1.435-meter standard gauge, 60 km 1.000-meter gauge, 470 km various narrow gauge (privately owned)

Highways. 21,960 km total; 1,788 km paved, 474 km gravel, and 19,698 km earth

Merchant Marine. 13 ships (1,000 GRT or over) totaling 16,747 GRT/19,865 DWT; includes 11 cargo, 2 oil tanker; note - 1 naval cargo ship is sometimes used commercially

Airports

Total:	862
Usable:	719
With permanent-surface runways:	7
With runways over 3,659 m:	0
With runways 2,440-3,659 m:	4
With runways 1,220-2,439 m:	64

Tourism [34]

	1987	1988	1989	1990	1991
Tourists[53]	303	284	279	280	361
Tourism receipts	121	114	112	112	145
Tourism expenditures	51	59	75	58	122
Fare receipts	2	16	17	22	24
Fare expenditures	13	17	18	21	22

Tourists are in thousands, money in million U.S. dollars.

Manufacturing Sector

GDP and Manufacturing Summary [35]

	1980	1985	1989	1990	% change 1980-1990	% change 1989-1990
GDP (million 1980 $)	3,844	4,302	5,052	5,227	36.0	3.5
GDP per capita (1980 $)	1,222	1,164	1,215	1,220	-0.2	0.4
Manufacturing as % of GDP (current prices)	16.5	15.2	17.0	17.3	4.8	1.8
Gross output (million $)	1,312	1,395	1,534[e]	1,408	7.3	-8.2
Value added (million $)	575	622[e]	490[e]	633[e]	10.1	29.2
Value added (million 1980 $)	633	669	765	784	23.9	2.5
Industrial production index	100	113	124	125	25.0	0.8
Employment (thousands)	146[e]	128[e]	92[e]	153[e]	4.8	66.3

Note: GDP stands for Gross Domestic Product. 'e' stands for estimated value.

Profitability and Productivity

	1980	1985	1989	1990	% change 1980-1990	% change 1989-1990
Intermediate input (%)	NA	NA	NA	NA	NA	NA
Wages, salaries, and supplements (%)	NA	NA	NA	NA	NA	NA
Gross operating surplus (%)	NA	NA	NA	NA	NA	NA
Gross output per worker ($)	8,962[e]	10,740[e]	16,719[e]	9,102[e]	1.6	-45.6
Value added per worker ($)	4,061[e]	4,824[e]	5,340[e]	4,140[e]	1.9	-22.5
Average wage (incl. benefits) ($)	NA	NA	NA	NA	NA	NA

Profitability is in percent of gross output. Productivity is in U.S. $. 'e' stands for estimated value.

Profitability - 1990

Value Added in Manufacturing

	1980 $ mil.	1980 %	1985 $ mil.	1985 %	1989 $ mil.	1989 %	1990 $ mil.	1990 %	% change 1980-1990	% change 1989-1990
311 Food products	170	29.6	232	37.3	132[e]	26.9	224	35.4	31.8	69.7
313 Beverages	43	7.5	58	9.3	40[e]	8.2	55	8.7	27.9	37.5
314 Tobacco products	6	1.0	9	1.4	7[e]	1.4	9	1.4	50.0	28.6
321 Textiles	44	7.7	54	8.7	29[e]	5.9	41	6.5	-6.8	41.4
322 Wearing apparel	2	0.3	3	0.5	2[e]	0.4	3	0.5	50.0	50.0
323 Leather and fur products	7	1.2	11	1.8	19[e]	3.9	14	2.2	100.0	-26.3
324 Footwear	18	3.1	18	2.9	23[e]	4.7	18	2.8	0.0	-21.7
331 Wood and wood products	95	16.5	87	14.0	75[e]	15.3	92	14.5	-3.2	22.7
332 Furniture and fixtures	6	1.0	10	1.6	9[e]	1.8	9	1.4	50.0	0.0
341 Paper and paper products	NA	0.0	1[e]	0.2	2[e]	0.4	1[e]	0.2	NA	-50.0
342 Printing and publishing	24	4.2	27[e]	4.3	23[e]	4.7	32[e]	5.1	33.3	39.1
351 Industrial chemicals	4	0.7	4[e]	0.6	7[e]	1.4	2[e]	0.3	-50.0	-71.4
352 Other chemical products	10	1.7	8[e]	1.3	5[e]	1.0	6[e]	0.9	-40.0	20.0
353 Petroleum refineries	94	16.3	45	7.2	62[e]	12.7	62	9.8	-34.0	0.0
354 Miscellaneous petroleum and coal products	NA	0.0	NA	0.0	NA	0.0	NA	0.0	NA	NA
355 Rubber products	NA	0.0	NA	0.0	NA	0.0	NA	0.0	NA	NA
356 Plastic products	6	1.0	10[e]	1.6	9[e]	1.8	12[e]	1.9	100.0	33.3
361 Pottery, china and earthenware	NA	0.0	NA	0.0	NA	0.0	NA	0.0	NA	NA
362 Glass and glass products	1	0.2	2[e]	0.3	3[e]	0.6	2[e]	0.3	100.0	-33.3
369 Other non-metal mineral products	26	4.5	18	2.9	21[e]	4.3	26	4.1	0.0	23.8
371 Iron and steel	NA	0.0	NA	0.0	NA	0.0	NA	0.0	NA	NA
372 Non-ferrous metals	1	0.2	2[e]	0.3	2[e]	0.4	2[e]	0.3	100.0	0.0
381 Metal products	9	1.6	12[e]	1.9	9[e]	1.8	12[e]	1.9	33.3	33.3
382 Non-electrical machinery	1	0.2	1[e]	0.2	1[e]	0.2	1[e]	0.2	0.0	0.0
383 Electrical machinery	NA	0.0	NA	0.0	NA	0.0	NA	0.0	NA	NA
384 Transport equipment	5	0.9	6[e]	1.0	6[e]	1.2	6[e]	0.9	20.0	0.0
385 Professional and scientific equipment	1	0.2	1[e]	0.2	1[e]	0.2	NA	0.0	NA	NA
390 Other manufacturing industries	2	0.3	3[e]	0.5	2[e]	0.4	3[e]	0.5	50.0	50.0

Note: The industry codes shown are International Standard Industry codes (ISIC). Percentages are percent of total Value Added. 'e' stands for estimated value

Finance, Economics, and Trade

Economic Indicators [36]

	1989	1990	1991[e,p]
Real GDP (mil. 1982 $)	6,614	6,818	7,022
Real GDP growth rate	5.8	3.1	3.0
Money Supply (bil. Guaranies)	609	808.2	1,006
Commercial Banks (Lending Rates %)	28.0	36.0	37.0
Savings Rate (as a % of GDP)	17.1	16.9	18.3
Investment rate (as a % of GDP)	23.8	23.0	NA
CPI	28.5	44.1	15.0
WPI	26.1	67.2	5.0
External Public Debt (mil. $)	2,490	2,433	1,690

Balance of Payments Summary [37]

Values in millions of dollars.

	1987	1988	1989	1990	1991
Exports of goods (f.o.b.)	597.4	871.0	1,242.1	1,376.0	1,268.1
Imports of goods (f.o.b.)	-918.7	-1,030.1	-1,015.9	-1,473.3	-1,680.4
Trade balance	-321.3	-159.1	226.2	-97.3	-412.3
Services - debits	-417.8	-438.8	-415.9	-558.2	-612.4
Services - credits	222.4	352.6	483.6	604.3	646.9
Private transfers (net)	2.0	2.0	1.6	7.2	2.3
Government transfers (net)	25.0	33.2	22.3	48.4	52.7
Long term capital (net)	-63.8	-98.3	29.7	-87.8	52.0
Short term capital (net)	266.9	-40.9	-49.5	126.7	111.6
Errors and omissions	338.3	198.3	-152.8	206.0	458.1
Overall balance	51.7	-151.0	145.2	249.3	298.9

Exchange Rates [38]

Currency: **guaranies.**
Symbol: **G.**

Data are currency units per $1.

January 1993	1,637.60
1992	1,500.30
March 1992	447.50
1991	1,325.20
1990	1,229.80
1989	1,056.20
Fixed rate 1986-February 1989	550.00

Imports and Exports

Top Import Origins [39]

$1.33 billion (c.i.f., 1992).

Origins	%
Brazil	30
EC	20
US	18
Argentina	8
Japan	7

Top Export Destinations [40]

$719 million (f.o.b., 1992).

Destinations	%
EC	37
Brazil	25
Argentina	10
Chile	6
US	6

Foreign Aid [41]

	U.S. $	
US commitments, including Ex-Im (FY70-89)	172	million
Western (non-US) countries, ODA and OOF bilateral commitments (1970-89)	1.1	billion

Import and Export Commodities [42]

Import Commodities	Export Commodities
Capital goods 35%	Cotton
Consumer goods 20%	Soybean
Fuels and lubricants 19%	Timber
Raw materials 16%	Vegetable oils
Foodstuffs	Coffee
Beverages	Tung oil
And tobacco 10%	Meat products

For sources, notes, and explanations, see Annotated Source Appendix, page 1035.

Peru

Geography [1]

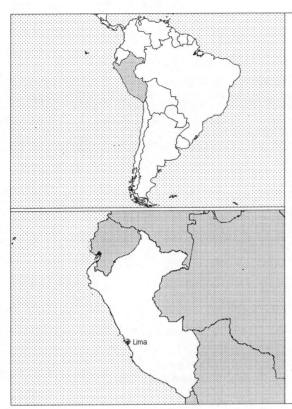

Total area:
1,285,220 km2
Land area:
1.28 million km2
Comparative area:
Slightly smaller than Alaska
Land boundaries:
Total 6,940 km, Bolivia 900 km, Brazil 1,560 km, Chile 160 km, Colombia 2,900 km, Ecuador 1,420 km
Coastline:
2,414 km
Climate:
Varies from tropical in east to dry desert in west
Terrain:
Western coastal plain (costa), high and rugged Andes in center (sierra), eastern lowland jungle of Amazon Basin (selva)
Natural resources:
Copper, silver, gold, petroleum, timber, fish, iron ore, coal, phosphate, potash
Land use:
Arable land: 3%
Permanent crops: 0%
Meadows and pastures: 21%
Forest and woodland: 55%
Other: 21%

Demographics [2]

	1960	1970	1980	1990	1991[1]	1994	2000	2010	2020
Population	9,931	13,193	17,295	21,879	22,362	23,651	26,258	30,483	34,340
Population density (persons per sq. mi.)	20	27	35	44	45	NA	53	62	71
Births	NA	NA	NA	NA	625	604	NA	NA	NA
Deaths	NA	NA	NA	NA	168	166	NA	NA	NA
Life expectancy - males	NA	NA	NA	NA	62	63	NA	NA	NA
Life expectancy - females	NA	NA	NA	NA	67	68	NA	NA	NA
Birth rate (per 1,000)	NA	NA	NA	NA	28	26	NA	NA	NA
Death rate (per 1,000)	NA	NA	NA	NA	8	7	NA	NA	NA
Women of reproductive age (15-44 yrs.)	NA	NA	NA	5,419	NA	6,068	7,063	NA	NA
of which are currently married	NA	NA	NA	3,187	NA	3,605	4,276	NA	NA
Fertility rate	NA	NA	NA	3.6	3.51	3.1	2.6	2.2	2.1

Population values are in thousands, life expectancy in years, and other items as indicated.

Health

Health Personnel [3]

Doctors per 1,000 pop., 1988-92	1.03
Nurse-to-doctor ratio, 1988-92	0.9
Hospital beds per 1,000 pop., 1985-90	1.5
Percentage of children immunized (age 1 yr. or less)	
Third dose of DPT, 1990-91	71.0
Measles, 1990-91	59.0

Health Expenditures [4]

Total health expenditure, 1990 (official exchange rate)	
Millions of dollars	1065
Millions of dollars per capita	49
Health expenditures as a percentage of GDP	
Total	3.2
Public sector	1.9
Private sector	1.3
Development assistance for health	
Total aid flows (millions of dollars)[1]	29
Aid flows per capita (millions of dollars)	1.4
Aid flows as a percentage of total health expenditure	2.7

For sources, notes, and explanations, see Annotated Source Appendix, page 1035.

723

Human Factors

Health Care Ratios [5]

Population per physician, 1970	1920
Population per physician, 1990	NA
Population per nursing person, 1970	NA
Population per nursing person, 1990	NA
Percent of births attended by health staff, 1985	55

Infants and Malnutrition [6]

Percent of babies with low birth weight, 1985	9
Infant mortality rate per 1,000 live births, 1970	108
Infant mortality rate per 1,000 live births, 1991	53
Years of life lost per 1,000 population, 1990	32
Prevalence of malnutrition (under age 5), 1990	13

Ethnic Division [7]

Indian	45%
Mestizo	37%
White	15%
Black,	
Japanese,	
Chinese,	
other	3%

Religion [8]

Roman Catholic.

Major Languages [9]

Spanish (official), Quechua (official), Aymara.

Education

Public Education Expenditures [10]

Million Inti	1980	1985	1987	1988	1989	1990
Total education expenditure	176	5,042	24,853	NA	NA	NA
as percent of GNP	3.1	2.9	3.5	NA	NA	NA
as percent of total govt. expend.	15.2	15.7	NA	NA	NA	NA
Current education expenditure	166	4,855	23,507	NA	NA	NA
as percent of GNP	2.9	2.7	3.4	NA	NA	NA
as percent of current govt. expend.	18.5	17.9	NA	NA	NA	NA
Capital expenditure	10	188	1,346	NA	NA	NA

Educational Attainment [11]

Age group	25+
Total population	6,526,328
Highest level attained (%)	
No schooling	24.0[9]
First level	
Incompleted	27.3
Completed	17.2
Entered second level	
S-1	10.7
S-2	10.7
Post secondary	10.1

Literacy Rate [12]

In thousands and percent	1985[a]	1991[a]	2000[a]
Illiterate population +15 years	2,111	2,025	1,800
Illiteracy rate - total pop. (%)	18.0	14.9	10.0
Illiteracy rate - males (%)	10.5	8.5	5.5
Illiteracy rate - females (%)	25.5	21.3	14.5

Libraries [13]

	Admin. Units	Svc. Pts.	Vols. (000)	Shelving (meters)	Vols. Added	Reg. Users
National (1986)	1	NA	4,016	NA	68,788	26,310
Non-specialized	NA	NA	NA	NA	NA	NA
Public (1986)	NA	687	5,802	NA	4,000	NA
Higher ed.	NA	NA	NA	NA	NA	NA
School (1987)	143	143	18	NA	7,272	NA

Daily Newspapers [14]

	1975	1980	1985	1990
Number of papers	49	66	55	54
Circ. (000)	1,300[e]	2,937	2,700[e]	2,800[e]

Cinema [15]

Science and Technology

Scientific/Technical Forces [16]

Potential scientists/engineers	291,812
Number female	NA
Potential technicians	165,673
Number female	92,807
Total	457,485

R&D Expenditures [17]

	Sol[17] (000) 1984
Total expenditure	159,024,000
Capital expenditure	NA
Current expenditure	NA
Percent current	NA

U.S. Patents Issued [18]

Values show patents issued to citizens of the country by the U.S. Patents Office.

	1990	1991	1992
Number of patents	5	2	5

Government and Law

Organization of Government [19]

Long-form name:
Republic of Peru
Type:
republic
Independence:
28 July 1821 (from Spain)
Constitution:
28 July 1980 (often referred to as the 1979
Constitution because the Constituent
Assembly met in 1979, but the
Constitution actually took effect the
following year); suspended 5 April 1992;
being revised or replaced
Legal system:
based on civil law system; has not
accepted compulsory ICJ jurisdiction
National holiday:
Independence Day, 28 July (1821)
Executive branch:
president, prime minister, Council of
Ministers (cabinet)
Legislative branch:
unicameral Democratic Constituent
Congress (CCD)
Judicial branch:
Supreme Court of Justice (Corte Suprema
de Justicia)

Political Parties [20]

Democratic Constituent Congress	% of seats
New Majority/Change 90	55.0
Popular Christian Party	10.0
Independent Moralization Front	8.8
Renewal	7.5
Movement of the Democratic Left	5.0
Democratic Coordinator	5.0
Others	8.8

Government Budget [21]

For 1992 est.

Revenues	2.000
Expenditures	2.700
Capital expenditures	0.300

Crime [23]

Crime volume	
Cases known to police	102,210
Attempts (percent)	NA
Percent cases solved	40
Crimes per 100,000 persons	474.28
Persons responsible for offenses	
Total number offenders	88,922
Percent female	6
Percent juvenile[43]	2
Percent foreigners	1

Military Expenditures and Arms Transfers [22]

	1985	1986	1987	1988	1989
Military expenditures					
Current dollars (mil.)	2,356[e]	2,801[e]	2,450[e]	NA	NA
1989 constant dollars (mil.)	2,682[e]	3,109[e]	2,635[e]	NA	NA
Armed forces (000)	128	127	127	111	110
Gross national product (GNP)					
Current dollars (mil.)	36,430	41,840	46,840	43,280	40,820
1989 constant dollars (mil.)	41,470	46,430	50,380	45,050	40,820
Central government expenditures (CGE)					
1989 constant dollars (mil.)	8,448	9,118	9,545	6,395	5,554
People (mil.)	19.6	20.1	20.5	21.0	21.4
Military expenditure as % of GNP	6.5	6.7	5.2	NA	NA
Military expenditure as % of CGE	31.7	34.1	27.6	NA	NA
Military expenditure per capita	137	155	128	NA	NA
Armed forces per 1,000 people	6.5	6.3	6.2	5.3	5.1
GNP per capita	2,113	2,312	2,453	2,146	1,903
Arms imports[6]					
Current dollars (mil.)	60	140	440	30	180
1989 constant dollars (mil.)	68	155	473	31	180
Arms exports[6]					
Current dollars (mil.)	0	0	0	0	0
1989 constant dollars (mil.)	0	0	0	0	0
Total imports[7]					
Current dollars (mil.)	1,835	2,909	3,562	3,080	2,398
1989 constant dollars	2,089	3,228	3,831	3,206	2,398
Total exports[7]					
Current dollars (mil.)	2,979	2,531	2,661	2,695	3,522
1989 constant dollars	3,391	2,809	2,862	2,805	3,522
Arms as percent of total imports[8]	3.3	4.8	12.4	1.0	7.5
Arms as percent of total exports[8]	0	0	0	0	0

Human Rights [24]

	SSTS	FL	FAPRO	PPCG	APROBC	TPW	PCPTW	STPEP	PHRFF	PRW	ASST	AFL
Observes		P	P	P	P	P	P			P	S	P
	EAFRD	CPR	ESCR	SR	ACHR	MAAE	PVIAC	PVNAC	EAFDAW	TCIDTP	RC
Observes		P	P	P	P	P		P	P	P	P	P

P = Party; S = Signatory; see Appendix for meaning of abbreviations.

Labor Force

Total Labor Force [25]

8 million (1992)

Labor Force by Occupation [26]

Government and other services	44%
Agriculture	37
Industry	19

Date of data: 1988 est.

Unemployment Rate [27]

15% (1992 est.); underemployment 70%
(1992 est.)

For sources, notes, and explanations, see Annotated Source Appendix, page 1035.

Production Sectors

Commercial Energy Production and Consumption

Production [28]

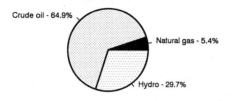

Crude oil - 64.9%
Natural gas - 5.4%
Hydro - 29.7%

Consumption [29]

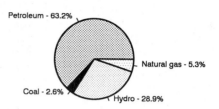

Petroleum - 63.2%
Natural gas - 5.3%
Coal - 2.6%
Hydro - 28.9%

Data are shown in quadrillion (10^{15}) BTUs and percent for 1991

Crude oil[1]	0.24
Natural gas liquids	(s)
Dry natural gas	0.02
Coal[2]	(s)
Net hydroelectric power[3]	0.11
Net nuclear power[3]	0.00
Total	0.38

Petroleum	0.24
Dry natural gas	0.02
Coal[2]	0.01
Net hydroelectric power[3]	0.11
Net nuclear power[3]	0.00
Total	0.38

Telecommunications [30]

- Fairly adequate for most requirements
- Nationwide microwave system
- 544,000 telephones
- Broadcast stations - 273 AM, no FM, 140 TV, 144 shortwave
- Satellite earth stations - 2 Atlantic Ocean INTELSAT, 12 domestic

Transportation [31]

Railroads. 1,801 km total; 1,501 km 1.435-meter gauge, 300 km 0.914-meter gauge

Highways. 69,942 km total; 7,459 km paved, 13,538 km improved, 48,945 km unimproved earth

Merchant Marine. 21 ships (1,000 GRT or over) totaling 194,473 GRT/307,845 DWT; includes 13 cargo, 1 refrigerated cargo, 1 roll-on/roll-off cargo, 2 oil tanker, 4 bulk; note - in addition, 6 naval tankers and 1 naval cargo are sometimes used commercially

Airports

Total:	228
Usable:	199
With permanent-surface runways:	37
With runways over 3,659 m:	2
With runways 2,440-3,659 m:	23
With runways 1,220-2,439 m:	46

Top Agricultural Products [32]

	1991	1992[4]
Potatoes	450	1,200
Rice (milled)	545	510
Sugar cane	550	500
Corn for feed	450	500
Cotton (raw)	176	158
Barley	80	100

Values shown are 1,000 metric tons.

Top Mining Products [33]

Preliminary metric tons unless otherwise specified M.t.

Arsenic, white[35]	661
Copper, mine output, Cu content	381,991
Lead, mine output, Pb content	199,811
Molybdenum, mine output, Mo content	3,045
Silver, mine output, Ag content	1,769
Tellurium, metal (kilograms)	13,355
Zinc, mine output, Zn content	627,824

Tourism [34]

	1987	1988	1989	1990	1991
Tourists	330	359	334	317	232
Tourism receipts	341	448	402	398	277
Tourism expenditures	335	344	417	770	881
Fare receipts	43	45	48	57	57
Fare expenditures	107	108	98	81	88

Tourists are in thousands, money in million U.S. dollars.

Manufacturing Sector

GDP and Manufacturing Summary [35]

	1980	1985	1989	1990	% change 1980-1990	% change 1989-1990
GDP (million 1980 $)	20,579	20,167	19,500	18,418	-10.5	-5.5
GDP per capita (1980 $)	1,190	1,039	924	854	-28.2	-7.6
Manufacturing as % of GDP (current prices)	20.4	24.1	20.9	27.1	32.8	29.7
Gross output (million $)	12,977	9,573	22,625[e]	14,267[e]	9.9	-36.9
Value added (million $)	4,984	3,918	8,614[e]	7,265[e]	45.8	-15.7
Value added (million 1980 $)	4,159	3,741	3,635	3,386	-18.6	-6.9
Industrial production index	100	86	80	76	-24.0	-5.0
Employment (thousands)	273	263	300[e]	295[e]	8.1	-1.7

Note: GDP stands for Gross Domestic Product. 'e' stands for estimated value.

Profitability and Productivity

	1980	1985	1989	1990	% change 1980-1990	% change 1989-1990
Intermediate input (%)	62	59	62[e]	49[e]	-21.0	-21.0
Wages, salaries, and supplements (%)	7[e]	6	7[e]	10[e]	42.9	42.9
Gross operating surplus (%)	32[e]	35	31[e]	41[e]	28.1	32.3
Gross output per worker ($)	47,484	36,350	75,312[e]	48,193[e]	1.5	-36.0
Value added per worker ($)	18,238	14,877	28,673[e]	24,575[e]	34.7	-14.3
Average wage (incl. benefits) ($)	3,176[e]	2,154	5,281[e]	4,619[e]	45.4	-12.5

Profitability is in percent of gross output. Productivity is in U.S. $. 'e' stands for estimated value.

Profitability - 1990

Inputs - 49.0%
Wages - 10.0%
Surplus - 41.0%

The graphic shows percent of gross output.

Value Added in Manufacturing

	1980 $ mil.	1980 %	1985 $ mil.	1985 %	1989 $ mil.	1989 %	1990 $ mil.	1990 %	% change 1980-1990	% change 1989-1990
311 Food products	767	15.4	402	10.3	1,027[e]	11.9	127[e]	1.7	-83.4	-87.6
313 Beverages	379	7.6	303	7.7	1,061[e]	12.3	605[e]	8.3	59.6	-43.0
314 Tobacco products	84	1.7	61	1.6	154[e]	1.8	95[e]	1.3	13.1	-38.3
321 Textiles	466	9.3	352	9.0	820[e]	9.5	704[e]	9.7	51.1	-14.1
322 Wearing apparel	65	1.3	52	1.3	195[e]	2.3	181[e]	2.5	178.5	-7.2
323 Leather and fur products	56	1.1	20	0.5	51[e]	0.6	37[e]	0.5	-33.9	-27.5
324 Footwear	41	0.8	20	0.5	68[e]	0.8	41[e]	0.6	0.0	-39.7
331 Wood and wood products	81	1.6	32	0.8	91[e]	1.1	59[e]	0.8	-27.2	-35.2
332 Furniture and fixtures	40	0.8	19	0.5	65[e]	0.8	49[e]	0.7	22.5	-24.6
341 Paper and paper products	156	3.1	77	2.0	210[e]	2.4	209[e]	2.9	34.0	-0.5
342 Printing and publishing	100	2.0	80	2.0	265[e]	3.1	228[e]	3.1	128.0	-14.0
351 Industrial chemicals	215	4.3	158	4.0	248[e]	2.9	313[e]	4.3	45.6	26.2
352 Other chemical products	289	5.8	193	4.9	471[e]	5.5	440[e]	6.1	52.2	-6.6
353 Petroleum refineries	192	3.9	1,154	29.5	1,479[e]	17.2	1,032[e]	14.2	437.5	-30.2
354 Miscellaneous petroleum and coal products	6	0.1	1	0.0	2[e]	0.0	1[e]	0.0	-83.3	-50.0
355 Rubber products	62	1.2	52	1.3	109[e]	1.3	142[e]	2.0	129.0	30.3
356 Plastic products	89	1.8	90	2.3	311[e]	3.6	233[e]	3.2	161.8	-25.1
361 Pottery, china and earthenware	15	0.3	8	0.2	21[e]	0.2	7[e]	0.1	-53.3	-66.7
362 Glass and glass products	47	0.9	15	0.4	69[e]	0.8	55[e]	0.8	17.0	-20.3
369 Other non-metal mineral products	129	2.6	113	2.9	274[e]	3.2	197[e]	2.7	52.7	-28.1
371 Iron and steel	192	3.9	123	3.1	111[e]	1.3	285[e]	3.9	48.4	156.8
372 Non-ferrous metals	604	12.1	172	4.4	238[e]	2.8	210[e]	2.9	-65.2	-11.8
381 Metal products	188	3.8	113	2.9	268[e]	3.1	262[e]	3.6	39.4	-2.2
382 Non-electrical machinery	156	3.1	58	1.5	178[e]	2.1	210[e]	2.9	34.6	18.0
383 Electrical machinery	211	4.2	111	2.8	421[e]	4.9	258[e]	3.6	22.3	-38.7
384 Transport equipment	278	5.6	106	2.7	267[e]	3.1	278[e]	3.8	0.0	4.1
385 Professional and scientific equipment	14	0.3	10	0.3	38[e]	0.4	30[e]	0.4	114.3	-21.1
390 Other manufacturing industries	58	1.2	25	0.6	104[e]	1.2	74[e]	1.0	27.6	-28.8

Note: The industry codes shown are International Standard Industry codes (ISIC). Percentages are percent of total Value Added. 'e' stands for estimated value

For sources, notes, and explanations, see Annotated Source Appendix, page 1035.

727

Finance, Economics, and Trade

Economic Indicators [36]

Millions of U.S. Dollars unless otherwise noted[123].

	1989	1990	1991[e]
Real GDP (1979 US$)	15,268	14,520	14,927
Real GDP Growth Rate	-11.6	-4.9	2.8
Per Capita GDP (1979 US$)	701	650	652
Money Supply (M1) (000 soles)	7,739	392,769	706,400
Consumer Prices (%. Chg.)	2,775	7,649	132
Wholesale Prices (%. Chg.)	1,918	6,534	130
External Public Debt[124]	15,796	16,301	15,857

Balance of Payments Summary [37]

Values in millions of dollars.

	1987	1988	1989	1990	1991
Exports of goods (f.o.b.)	2,661.0	2,691.0	3,488.0	3,231.0	3,329.0
Imports of goods (f.o.b.)	-3,182.0	-2,790.0	-2,291.0	-2,891.0	-3,494.0
Trade balance	-521.0	-99.0	1,197.0	340.0	-165.0
Services - debits	-2,138.0	-2,187.0	-2,133.0	-2,531.0	-2,986.0
Services - credits	998.0	1,038.0	1,062.0	995.0	964.0
Private transfers (net)	-	-	-	-	-
Government transfers (net)	180.0	157.0	236.0	247.0	316.0
Long term capital (net)	-1,226.0	-1,271.0	-855.0	-970.0	-268.0
Short term capital (net)	2,093.0	2,486.0	1,302.0	1,517.0	1,793.0
Errors and omissions	-54.0	-1,140	-214.0	687.0	1,722.0
Overall balance	-668.0	10.0	595.0	285.0	1,376.0

Exchange Rates [38]

Currency: **nuevo sol.**
Symbol: **S/.**

Data are currency units per $1.

January 1993	1.690
1992	1.245
1991	0.772
1990	0.187
1989	2.666
1988	0.129

Imports and Exports

Top Import Origins [39]

$4.1 billion (f.o.b., 1992). Data are for 1991.

Origins	%
US	32
Latin America	22
EC	17
Switzerland	6
Japan	3

Top Export Destinations [40]

$3.5 billion (f.o.b., 1992). Data are for 1991.

Destinations	%
EC	28
US	22
Japan	13
Latin America	12
Former USSR	2

Foreign Aid [41]

	U.S. $	
US commitments, including Ex-Im (FY70-89)	1.7	billion
Western (non-US) countries, ODA and OOF bilateral commitments (1970-89)	4.3	billion
Communist countries (1970-89)	577	million

Import and Export Commodities [42]

Import Commodities

Foodstuffs
Machinery
Transport equipment
Iron and steel semimanufactures
Chemicals
Pharmaceuticals

Export Commodities

Copper
Fishmeal
Zinc
Crude oil & byproducts
Lead
Refined silver
Coffee
Cotton

Philippines

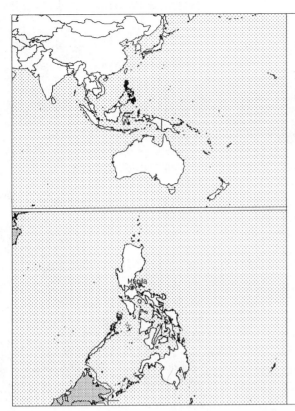

Geography [1]

Total area:
300,000 km2
Land area:
298,170 km2
Comparative area:
Slightly larger than Arizona
Land boundaries:
0 km
Coastline:
36,289 km
Climate:
Tropical marine; northeast monsoon (November to April); southwest monsoon (May to October)
Terrain:
Mostly mountains with narrow to extensive coastal lowlands
Natural resources:
Timber, petroleum, nickel, cobalt, silver, gold, salt, copper
Land use:
Arable land:
26%
Permanent crops:
11%
Meadows and pastures:
4%
Forest and woodland:
40%
Other:
19%

Demographics [2]

	1960	1970	1980	1990	1991[1]	1994	2000	2010	2020
Population	28,557	38,680	50,864	64,405	65,759	69,809	77,747	90,316	101,530
Population density (persons per sq. mi.)	248	336	442	559	571	NA	675	784	881
Births	NA	NA	NA	NA	1,908	1,909	NA	NA	NA
Deaths	NA	NA	NA	NA	474	484	NA	NA	NA
Life expectancy - males	NA	NA	NA	NA	62	63	NA	NA	NA
Life expectancy - females	NA	NA	NA	NA	67	68	NA	NA	NA
Birth rate (per 1,000)	NA	NA	NA	NA	29	27	NA	NA	NA
Death rate (per 1,000)	NA	NA	NA	NA	7	7	NA	NA	NA
Women of reproductive age (15-44 yrs.)	NA	NA	NA	16,269	NA	18,163	21,039	NA	NA
of which are currently married	NA	NA	NA	10,182	NA	11,466	13,488	NA	NA
Fertility rate	NA	NA	NA	3.7	3.64	3.4	2.9	2.4	2.2

Population values are in thousands, life expectancy in years, and other items as indicated.

Health

Health Personnel [3]

Doctors per 1,000 pop., 1988-92	0.12
Nurse-to-doctor ratio, 1988-92	3.1
Hospital beds per 1,000 pop., 1985-90	1.3
Percentage of children immunized (age 1 yr. or less)	
Third dose of DPT, 1990-91	88.0
Measles, 1990-91	85.0

Health Expenditures [4]

Total health expenditure, 1990 (official exchange rate)	
Millions of dollars	883
Millions of dollars per capita	14
Health expenditures as a percentage of GDP	
Total	2.0
Public sector	1.0
Private sector	1.0
Development assistance for health	
Total aid flows (millions of dollars)[1]	69
Aid flows per capita (millions of dollars)	1.1
Aid flows as a percentage of total health expenditure	7.8

For sources, notes, and explanations, see Annotated Source Appendix, page 1035.

729

Human Factors

Health Care Ratios [5]

Population per physician, 1970	9270
Population per physician, 1990	8120
Population per nursing person, 1970	2690
Population per nursing person, 1990	NA
Percent of births attended by health staff, 1985	NA

Infants and Malnutrition [6]

Percent of babies with low birth weight, 1985	18
Infant mortality rate per 1,000 live births, 1970	66
Infant mortality rate per 1,000 live births, 1991	41
Years of life lost per 1,000 population, 1990	27
Prevalence of malnutrition (under age 5), 1990	19

Ethnic Division [7]

Christian Malay	91.5%
Muslim Malay	4.0%
Chinese	1.5%
Other	3.0%

Religion [8]

Roman Catholic	83%
Protestant	9%
Muslim	5%
Buddhist and other	3%

Major Languages [9]

Pilipino (official; based on Tagalog), English (official).

Education

Public Education Expenditures [10]

Million Pesos	1980	1985	1987	1988	1989	1990
Total education expenditure	4,191	7,524	NA	21,953	26,470	31,067
as percent of GNP	1.7	1.4	NA	2.8	2.9	2.9
as percent of total govt. expend.	9.1	7.4	NA	12.7	11.5	10.1
Current education expenditure	4,023	7,026	NA	19,376	24,621	28,713
as percent of GNP	1.7	1.3	NA	2.4	2.7	2.7
as percent of current govt. expend.	13.0	10.0	NA	12.8	12.5	11.1
Capital expenditure	168	498	NA	2,577	1,849	2,354

Educational Attainment [11]

Age group	25+
Total population	17,865,290
Highest level attained (%)	
No schooling	11.7
First level	
Incompleted	31.3
Completed	22.8
Entered second level	
S-1	18.9
S-2	NA
Post secondary	15.2

Literacy Rate [12]

In thousands and percent	1985[a]	1991[a]	2000[a]
Illiterate population +15 years	3,993	3,852	3,561
Illiteracy rate - total pop. (%)	12.3	10.3	7.2
Illiteracy rate - males (%)	11.8	10.0	7.3
Illiteracy rate - females (%)	12.7	10.5	7.1

Libraries [13]

	Admin. Units	Svc. Pts.	Vols. (000)	Shelving (meters)	Vols. Added	Reg. Users
National (1989)	1	1	33	712	6,728	216,663
Non-specialized (1986)	1	1	NA	31	NA	232
Public (1989)	1	517	5,756	NA	NA	1.6mil
Higher ed.	NA	NA	NA	NA	NA	NA
School	NA	NA	NA	NA	NA	NA

Daily Newspapers [14]

	1975	1980	1985	1990
Number of papers	15	22	15	47
Circ. (000)	850	2,000	2,170	3,400[e]

Cinema [15]

Science and Technology

Scientific/Technical Forces [16]

Potential scientists/engineers	1,770,762
Number female	1,006,402
Potential technicians	NA
Number female	NA
Total	NA

R&D Expenditures [17]

	Peso (000) 1984
Total expenditure	613,410
Capital expenditure	98,610
Current expenditure	514,800
Percent current	83.9

U.S. Patents Issued [18]

Values show patents issued to citizens of the country by the U.S. Patents Office.

	1990	1991	1992
Number of patents	8	8	7

Government and Law

Organization of Government [19]

Long-form name:
Republic of the Philippines
Type:
republic
Independence:
4 July 1946 (from US)
Constitution:
2 February 1987, effective 11 February 1987
Legal system:
based on Spanish and Anglo-American law; accepts compulsory ICJ jurisdiction, with reservations
National holiday:
Independence Day, 12 June (1898) (from Spain)
Executive branch:
president, vice president, Cabinet
Legislative branch:
bicameral Congress (Kongreso) consists of an upper house or Senate (Senado) and a lower house or House of Representatives (Kapulungan Ng Mga Kinatawan)
Judicial branch:
Supreme Court

Political Parties [20]

House of Representatives	% of votes
Democratic Filipino Struggle (LDP)	43.5
People Power (NUCD)	25.0
Nationalist People's Coalition (NPC)	23.5
Liberal	5.0
New Society Movement (KBL)	3.0

Government Expenditures [21]

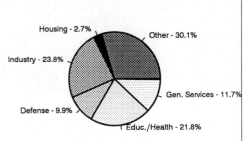

Housing - 2.7%
Other - 30.1%
Industry - 23.8%
Gen. Services - 11.7%
Defense - 9.9%
Educ./Health - 21.8%

% distribution for 1992. Expend. in 1992 est.: 12.0 ($ bil.)

Military Expenditures and Arms Transfers [22]

	1985	1986	1987	1988	1989
Military expenditures					
Current dollars (mil.)	405	627	648	675	960
1989 constant dollars (mil.)	461	695	697	702	960
Armed forces (000)	157	161	161	105	106
Gross national product (GNP)					
Current dollars (mil.)	31,810	33,240	36,330	40,010	43,960
1989 constant dollars (mil.)	36,220	36,890	39,080	41,650	43,960
Central government expenditures (CGE)					
1989 constant dollars (mil.)	4,854	6,595	6,664	6,890	7,912
People (mil.)	57.6	59.0	60.3	61.7	63.0
Military expenditure as % of GNP	1.3	1.9	1.8	1.7	2.2
Military expenditure as % of CGE	9.5	10.5	10.5	10.2	12.1
Military expenditure per capita	8	12	12	11	15
Armed forces per 1,000 people	2.7	2.7	2.7	1.7	1.7
GNP per capita	629	626	648	675	697
Arms imports[6]					
Current dollars (mil.)	30	40	50	60	70
1989 constant dollars (mil.)	34	44	54	62	70
Arms exports[6]					
Current dollars (mil.)	10	10	0	0	0
1989 constant dollars (mil.)	11	11	0	0	0
Total imports[7]					
Current dollars (mil.)	5,459	5,394	7,144	8,721	11,170
1989 constant dollars	6,215	5,986	7,684	9,078	11,170
Total exports[7]					
Current dollars (mil.)	4,607	4,770	5,649	7,032	7,755
1989 constant dollars	5,245	5,293	6,076	7,320	7,755
Arms as percent of total imports[8]	0.5	0.7	0.7	0.7	0.6
Arms as percent of total exports[8]	0.2	0.2	0	0	0

Crime [23]

Crime volume	
Cases known to police	139,392
Attempts (percent)	NA
Percent cases solved	NA
Crimes per 100,000 persons	229.72
Persons responsible for offenses	
Total number offenders	NA
Percent female	NA
Percent juvenile	NA
Percent foreigners	NA

Human Rights [24]

	SSTS	FL	FAPRO	PPCG	APROBC	TPW	PCPTW	STPEP	PHRFF	PRW	ASST	AFL
Observes	P		P	P	P	P	P	P		P	P	P
		EAFRD	CPR	ESCR	SR	ACHR	MAAE	PVIAC	PVNAC	EAFDAW	TCIDTP	RC
Observes		P	P	P	P			S	P	P	P	P

P=Party; S=Signatory; see Appendix for meaning of abbreviations.

Labor Force

Total Labor Force [25]

24.12 million

Labor Force by Occupation [26]

Agriculture	46%
Industry and commerce	16
Services	18.5
Government	10
Other	9.5

Date of data: 1989

Unemployment Rate [27]

9.8% (1992 est.)

For sources, notes, and explanations, see Annotated Source Appendix, page 1035.

731

Production Sectors

Commercial Energy Production and Consumption

Production [28]

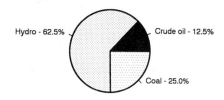

Hydro - 62.5%   Crude oil - 12.5%   Coal - 25.0%

Consumption [29]

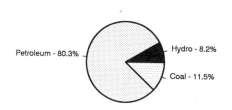

Petroleum - 80.3%   Hydro - 8.2%   Coal - 11.5%

Data are shown in quadrillion (10^{15}) BTUs and percent for 1991

Crude oil[1]	0.01
Natural gas liquids	0.00
Dry natural gas	0.00
Coal[2]	0.02
Net hydroelectric power[3]	0.05
Net nuclear power[3]	0.00
Total	0.08

Petroleum	0.49
Dry natural gas	0.00
Coal[2]	0.07
Net hydroelectric power[3]	0.05
Net nuclear power[3]	0.00
Total	0.62

Telecommunications [30]

- Good international radio and submarine cable services
- Domestic and interisland service adequate
- 872,900 telephones
- Broadcast stations - 267 AM (including 6 US), 55 FM, 33 TV (including 4 US)
- Submarine cables extended to Hong Kong, Guam, Singapore, Taiwan, and Japan
- Satellite earth stations - 1 Indian Ocean INTELSAT, 2 Pacific Ocean INTELSAT, and 11 domestic

Transportation [31]

Railroads. 378 km operable on Luzon, 34% government owned (1982)

Highways. 157,450 km total (1988); 22,400 km paved; 85,050 km gravel, crushed-stone, or stabilized-soil surface; 50,000 km unimproved earth

Merchant Marine. 562 ships (1,000 GRT or over) totaling 8,282,936 GRT/13,772,023 DWT; includes 1 passenger, 11 short-sea passenger, 13 passenger-cargo, 155 cargo, 27 refrigerated cargo, 25 vehicle carrier, 9 livestock carrier, 13 roll-on/roll-off cargo, 8 container, 38 oil tanker, 1 chemical tanker, 3 liquefied gas, 1 combination ore/oil, 249 bulk, 8 combination bulk; note - many Philippine flag ships are foreign owned and are on the register for the purpose of long-term bare-boat charter back to their original owners who are principally in Japan and Germany

Airports

Total:	270
Usable:	238
With permanent-surface runways:	73
With runways over 3,659 m:	0
With runways 2,440-3,659 m:	9
With runways 1,220-2,439 m:	57

Top Agricultural Products [32]

	1990[28]	1991[27,28]
Rice (raw)	9,883	9,300
Corn	5,102	4,850
Bananas	2,913	3,000
Copra	2,202	2,000
Sugar	1,718	1,900
Cassavas	1,854	1,860

Values shown are 1,000 metric tons.

Top Mining Products [33]

Estimated metric tons unless otherwise specified   M.t.

Gold, mine output, Au content (kilograms)	24,938[1]
Copper, mine output, Cu content	150,000
Silver, mine output, Ag content (kilograms)	38,414[1]
Chromite, gross weight	184,010
Nickel ore, mine output, Zn content	100
Manganese ore and concentrate, gross weight	14,000
Zinc, mine output, Zn content	100

Tourism [34]

	1987	1988	1989	1990	1991
Visitors	795	1,043	1,190	1,025	951
Tourists	781	1,023	1,076	893	849
Cruise passengers	14	20	10	23	5
Tourism receipts[54]	1,029	1,301	1,465	1,306	1,281
Tourism expenditures	88	76	77	111	61
Fare receipts	41	50	39	40	37
Fare expenditures	105	156	137	149	236

Tourists are in thousands, money in million U.S. dollars.

732

For sources, notes, and explanations, see Annotated Source Appendix, page 1035.

Manufacturing Sector

GDP and Manufacturing Summary [35]

	1980	1985	1989	1990	% change 1980-1990	% change 1989-1990
GDP (million 1980 $)	35,235	34,221	40,916	41,770	18.5	2.1
GDP per capita (1980 $)	729	621	672	669	-8.2	-0.4
Manufacturing as % of GDP (current prices)	25.7	25.2	24.6[e]	25.4	-1.2	3.3
Gross output (million $)	17,369	12,081	18,205[e]	21,234[e]	22.3	16.6
Value added (million $)	4,861	3,448	4,787[e]	7,215[e]	48.4	50.7
Value added (million 1980 $)	8,595	7,989	9,965	10,172	18.3	2.1
Industrial production index	100	74	389	137[e]	37.0	-64.8
Employment (thousands)	949	618	734[e]	865[e]	-8.9	17.8

Note: GDP stands for Gross Domestic Product. 'e' stands for estimated value.

Profitability and Productivity

	1980	1985	1989	1990	% change 1980-1990	% change 1989-1990
Intermediate input (%)	72	71	74[e]	66[e]	-8.3	-10.8
Wages, salaries, and supplements (%)	6	6	7[e]	8[e]	33.3	14.3
Gross operating surplus (%)	22	22	19[e]	26[e]	18.2	36.8
Gross output per worker ($)	16,263	19,386	24,787[e]	24,174[e]	48.6	-2.5
Value added per worker ($)	4,552	5,586	6,518[e]	8,221[e]	80.6	26.1
Average wage (incl. benefits) ($)	1,127	1,258	1,818[e]	1,991[e]	76.7	9.5

Profitability is in percent of gross output. Productivity is in U.S. $. 'e' stands for estimated value.

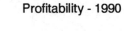

Profitability - 1990

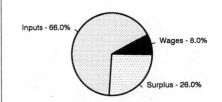

Inputs - 66.0%
Wages - 8.0%
Surplus - 26.0%

The graphic shows percent of gross output.

Value Added in Manufacturing

	1980 $ mil.	1980 %	1985 $ mil.	1985 %	1989 $ mil.	1989 %	1990 $ mil.	1990 %	% change 1980-1990	% change 1989-1990
311 Food products	969	19.9	658	19.1	911[e]	19.0	1,343[e]	18.6	38.6	47.4
313 Beverages	195	4.0	423	12.3	563[e]	11.8	913[e]	12.7	368.2	62.2
314 Tobacco products	309	6.4	209	6.1	472[e]	9.9	457[e]	6.3	47.9	-3.2
321 Textiles	395	8.1	109	3.2	209[e]	4.4	315[e]	4.4	-20.3	50.7
322 Wearing apparel	205	4.2	105	3.0	208[e]	4.3	419[e]	5.8	104.4	101.4
323 Leather and fur products	8	0.2	3	0.1	4[e]	0.1	11[e]	0.2	37.5	175.0
324 Footwear	13	0.3	9	0.3	10[e]	0.2	14[e]	0.2	7.7	40.0
331 Wood and wood products	229	4.7	86	2.5	90[e]	1.9	196[e]	2.7	-14.4	117.8
332 Furniture and fixtures	75	1.5	22	0.6	39[e]	0.8	89[e]	1.2	18.7	128.2
341 Paper and paper products	128	2.6	97	2.8	141[e]	2.9	196[e]	2.7	53.1	39.0
342 Printing and publishing	89	1.8	46	1.3	56[e]	1.2	102[e]	1.4	14.6	82.1
351 Industrial chemicals	296	6.1	101	2.9	176[e]	3.7	302[e]	4.2	2.0	71.6
352 Other chemical products	389	8.0	205	5.9	313[e]	6.5	653[e]	9.1	67.9	108.6
353 Petroleum refineries	328	6.7	715	20.7	641[e]	13.4	444[e]	6.2	35.4	-30.7
354 Miscellaneous petroleum and coal products	2	0.0	3	0.1	3[e]	0.1	6[e]	0.1	200.0	100.0
355 Rubber products	103	2.1	34	1.0	83[e]	1.7	182[e]	2.5	76.7	119.3
356 Plastic products	85	1.7	32	0.9	50[e]	1.0	123[e]	1.7	44.7	146.0
361 Pottery, china and earthenware	33	0.7	9	0.3	14[e]	0.3	26[e]	0.4	-21.2	85.7
362 Glass and glass products	42	0.9	28	0.8	44[e]	0.9	93[e]	1.3	121.4	111.4
369 Other non-metal mineral products	63	1.3	60	1.7	77[e]	1.6	118[e]	1.6	87.3	53.2
371 Iron and steel	98	2.0	164	4.8	173[e]	3.6	283[e]	3.9	188.8	63.6
372 Non-ferrous metals	35	0.7	28	0.8	13[e]	0.3	159[e]	2.2	354.3	1,123.1
381 Metal products	127	2.6	49	1.4	65[e]	1.4	98[e]	1.4	-22.8	50.8
382 Non-electrical machinery	98	2.0	31	0.9	56[e]	1.2	58[e]	0.8	-40.8	3.6
383 Electrical machinery	260	5.3	156	4.5	255[e]	5.3	410[e]	5.7	57.7	60.8
384 Transport equipment	234	4.8	35	1.0	74[e]	1.5	124[e]	1.7	-47.0	67.6
385 Professional and scientific equipment	5	0.1	5	0.1	10[e]	0.2	19[e]	0.3	280.0	90.0
390 Other manufacturing industries	49	1.0	28	0.8	37[e]	0.8	63[e]	0.9	28.6	70.3

Note: The industry codes shown are International Standard Industry codes (ISIC). Percentages are percent of total Value Added. 'e' stands for estimated value

For sources, notes, and explanations, see Annotated Source Appendix, page 1035.

733

Finance, Economics, and Trade

Economic Indicators [36]

Millions of Pesos unless otherwise noted[125].

	1989	1990	1991[P]
Real GDP (1985 pesos)	697,816	712,869	713,582
Real GDP growth rate (%)	5.9	2.2	0.1
Real per capita GNP (pesos)	11,466	11,618	11,394
Money supply (M1) (year-end)	78,530	89,012	100,000
Commercial interest rate (%)[126]	19.1	24.4	23.0[127]
Savings interest rate (%)	4.3	5.1	4.7[127]
Investment rate (%)[128]	14.6	19.1	17.8[129]
CPI (1978 = 100)	443.5	499.7	589.6
WPI, Metro Manila (1978 = 100)	551.5	607.6	706.0
External public debt	22,222	23,052	23,815

Balance of Payments Summary [37]

Values in millions of dollars.

	1987	1988	1989	1990	1991
Exports of goods (f.o.b.)	5,720.0	7,074.0	7,821.0	8,186.0	8,840.0
Imports of goods (f.o.b.)	-6,737.0	-8,159.0	-10,419.0	-12,206.0	-12,051.0
Trade balance	-1,017.0	-1,085.0	-2,598.0	-4,020.0	-3,211.0
Services - debits	-3,454.0	-3,672.0	-4,274.0	-4,231.0	-4,273.0
Services - credits	3,454.0	3,592.0	4,586.0	4,842.0	5,623.0
Private transfers (net)	376.0	500.0	473.0	357.0	473.0
Government transfers (net)	197.0	275.0	357.0	357.0	354.0
Long term capital (net)	564.0	605.0	1,424.0	1,707.0	1,890.0
Short term capital (net)	-246.0	-34.0	-70.0	350.0	1,039.0
Errors and omissions	68.0	479.0	402.0	593.0	-140.0
Overall balance	-58.0	660.0	300.0	-45.0	1,755.0

Exchange Rates [38]

Currency: **Philippine pesos.**
Symbol: **P.**

Data are currency units per $1.

April 1993	25.817
1992	25.512
1991	27.479
1990	24.311
1989	21.737
1988	21.095

Imports and Exports

Top Import Origins [39]

$14.5 billion (f.o.b., 1992).

Origins	%
US	NA
Japan	NA
Taiwan	NA
Saudi Arabia	NA

Top Export Destinations [40]

$9.8 billion (f.o.b., 1992).

Destinations	%
US	39
EC	NA
Japan	NA
ASEAN	NA

Foreign Aid [41]

	U.S. $	
US commitments, including Ex-Im (FY70-89)	3.6	billion
Western (non-US) countries, ODA and OOF bilateral commitments (1970-88)	7.9	billion
OPEC bilateral aid (1979-89)	5	million
Communist countries (1975-89)	123	million

Import and Export Commodities [42]

Import Commodities	Export Commodities
Raw materials 45%	Electronics
Capital goods 26%	Textiles
Petroleum products 18%	Coconut oil
	Copper

Poland

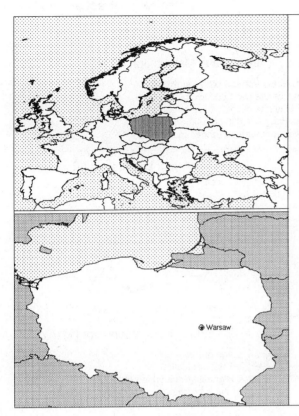

Geography [1]

Total area:
312,680 km2
Land area:
304,510 km2
Comparative area:
Slightly smaller than New Mexico
Land boundaries:
Total 3,114 km, Belarus 605 km, Czech Republic 658 km, Germany 456 km, Lithuania 91 km, Russia (Kaliningrad Oblast) 432 km, Slovakia 444 km, Ukraine 428 km
Coastline:
491 km
Climate:
Temperate with cold, cloudy, moderately severe winters with frequent precipitation; mild summers with frequent showers and thundershowers
Terrain:
Mostly flat plain; mountains along southern border
Natural resources:
Coal, sulfur, copper, natural gas, silver, lead, salt
Land use:
Arable land: 46%
Permanent crops: 1%
Meadows and pastures: 13%
Forest and woodland: 28%
Other: 12%

Demographics [2]

	1960	1970	1980	1990	1991[1]	1994	2000	2010	2020
Population	29,590	32,526	35,578	38,112	37,800	38,655	39,531	41,332	42,474
Population density (persons per sq. mi.)	252	277	303	321	322	NA	331	345	355
Births	NA	NA	NA	NA	531	520	NA	NA	NA
Deaths	NA	NA	NA	NA	353	363	NA	NA	NA
Life expectancy - males	NA	NA	NA	NA	69	69	NA	NA	NA
Life expectancy - females	NA	NA	NA	NA	77	77	NA	NA	NA
Birth rate (per 1,000)	NA	NA	NA	NA	14	13		NA	NA
Death rate (per 1,000)	NA	NA	NA	NA	9	9	NA	NA	NA
Women of reproductive age (15-44 yrs.)	NA	NA	NA	9,389	NA	9,861	10,247	NA	NA
of which are currently married	NA	NA	NA	6,485	NA	6,715	6,875	NA	NA
Fertility rate	NA	NA	NA	2.0	2.06	1.9	1.8	1.8	1.8

Population values are in thousands, life expectancy in years, and other items as indicated.

Health

Health Personnel [3]

Doctors per 1,000 pop., 1988-92	2.06
Nurse-to-doctor ratio, 1988-92	NA
Hospital beds per 1,000 pop., 1985-90	6.6
Percentage of children immunized (age 1 yr. or less)	
Third dose of DPT, 1990-91	98.0
Measles, 1990-91	94.0

Health Expenditures [4]

Total health expenditure, 1990 (official exchange rate)	
Millions of dollars	3157
Millions of dollars per capita	83
Health expenditures as a percentage of GDP	
Total	5.1
Public sector	4.1
Private sector	1.0
Development assistance for health	
Total aid flows (millions of dollars)[1]	NA
Aid flows per capita (millions of dollars)	NA
Aid flows as a percentage of total health expenditure	NA

For sources, notes, and explanations, see Annotated Source Appendix, page 1035.

735

Human Factors

Health Care Ratios [5]

Population per physician, 1970	700
Population per physician, 1990	490
Population per nursing person, 1970	250
Population per nursing person, 1990	NA
Percent of births attended by health staff, 1985	NA

Infants and Malnutrition [6]

Percent of babies with low birth weight, 1985	8
Infant mortality rate per 1,000 live births, 1970	33
Infant mortality rate per 1,000 live births, 1991	15
Years of life lost per 1,000 population, 1990	16
Prevalence of malnutrition (under age 5), 1990	NA

Ethnic Division [7]

Polish	97.6%
German	1.3%
Ukrainian	0.6%
Belarusian	0.5%

Religion [8]

Roman Catholic (about 75% practicing)	95%
Eastern Orthodox, Protestant, and other	5%

Major Languages [9]

Polish.

Education

Public Education Expenditures [10]

Billion Zloty	1980	1985	1987	1988	1989	1990
Total education expenditure	NA	497.5	NA	1,010.7	4,342.5	28,249.9
as percent of GNP	NA	4.9	NA	3.6	3.8	4.9
as percent of total govt. expend.	NA	12.2	NA	10.1	12.9	14.6
Current education expenditure	80.0	405.6	NA	NA	NA	NA
as percent of GNP	3.3	4.0	NA	NA	NA	NA
as percent of current govt. expend.	7.0	11.6	NA	NA	NA	NA
Capital expenditure	NA	92.0	NA	NA	NA	NA

Educational Attainment [11]

Age group	25+
Total population	22,986,018
Highest level attained (%)	
No schooling	1.5
First level	
Incompleted	5.6
Completed	37.2
Entered second level	
S-1	NA
S-2	47.8
Post secondary	7.9

Literacy Rate [12]

	1970[b]	1978[b]	1980[b]
Illiterate population +15 years	537,149	334,586	NA
Illiteracy rate - total pop. (%)	2.2	1.2	NA
Illiteracy rate - males (%)	1.3	0.7	NA
Illiteracy rate - females (%)	3.1	1.7	NA

Libraries [13]

	Admin. Units	Svc. Pts.	Vols. (000)	Shelving (meters)	Vols. Added	Reg. Users
National (1989)	1	1	2,176	NA	34,208	1,056
Non-specialized (1989)	124	234	13,730	NA	271,023	138,562
Public (1990)	10,269	17,565	136,641	NA	NA	7.4mil
Higher ed. (1991)	932	932	39,557	NA	1 mil.	797,884
School (1991)	21,538	NA	157,901	NA	5.4mil	7.5mil

Daily Newspapers [14]

	1975	1980	1985	1990
Number of papers	44	43	45	67
Circ. (000)	8,429	9,407	7,714	4,889

Cinema [15]

Data for 1991.

Cinema seats per 1,000	7.0
Annual attendance per person	0.5
Gross box office receipts (mil. zloty)	162,383

Science and Technology

Scientific/Technical Forces [16]

Potential scientists/engineers	1,423,000
Number female	629,000
Potential technicians[5]	4,841,000
Number female[5]	2,671,000
Total[5]	6,264,000

R&D Expenditures [17]

	Zloty[4] (000) 1989
Total expenditure	NA
Capital expenditure	NA
Current expenditure	1,226,962
Percent current	NA

U.S. Patents Issued [18]

Values show patents issued to citizens of the country by the U.S. Patents Office.

	1990	1991	1992
Number of patents	18	8	5

Government and Law

Organization of Government [19]

Long-form name:
Republic of Poland
Type:
democratic state
Independence:
11 November 1918 (independent republic proclaimed)
Constitution:
interim "small constitution" came into effect in December 1992 replacing the Communist-imposed Constitution of 22 July 1952
Legal system:
mixture of Continental (Napoleonic) civil law and Communist theory; changes being introduced as part of broader democratization process; limited judicial review; has not accepted compulsory ICJ jurisdiction
National holiday:
Constitution Day, 3 May (1791)
Executive branch:
president, prime minister, Council of Ministers (cabinet)
Legislative branch:
bicameral National Assembly: Senate and Diet
Judicial branch:
Supreme Court

Crime [23]

Political Parties [20]

Sejm	% of seats
Democratic Union	13.5
Christian National Union	10.7
Centrum	9.6
Other post-Solidarity bloc	22.8
Confederation Independent Poland	10.0
PPPP	3.5
German Minority	1.5
Other	5.0
SLD	13.0
Polish Peasants' Party	10.4

Government Budget [21]

For 1992 est.
Revenues	17.5
Expenditures	22.0
Capital expenditures	1.5

Military Expenditures and Arms Transfers [22]

	1985	1986	1987	1988	1989
Military expenditures					
Current dollars (mil.)[e]	14,670	15,230	15,450	15,350	15,480
1989 constant dollars (mil.)[e]	16,700	16,900	16,620	15,980	15,480
Armed forces (000)	439	443	441	430	350
Gross national product (GNP)					
Current dollars (mil.)	144,400	152,800	155,400	170,900	174,700
1989 constant dollars (mil.)	164,300	169,600	167,100	177,900	174,700
Central government expenditures (CGE)					
1989 constant dollars (mil.)	41,050	40,240	35,690	35,160	66,160
People (mil.)	37.2	37.5	37.7	37.8	37.8
Military expenditure as % of GNP[3]	10.2	10.0	9.9	9.0	8.9
Military expenditure as % of CGE[3]	40.7	42.0	46.6	45.5	23.4
Military expenditure per capita	449	451	441	423	410
Armed forces per 1,000 people	11.8	11.8	11.7	11.4	9.3
GNP per capita	4,417	4,528	4,437	4,709	4,625
Arms imports[6]					
Current dollars (mil.)	1,100	1,200	825	1,000	625
1989 constant dollars (mil.)	1,252	1,332	887	1,041	625
Arms exports[6]					
Current dollars (mil.)	1,300	1,500	1,300	1,200	400
1989 constant dollars (mil.)	1,480	1,665	1,398	1,249	400
Total imports[7]					
Current dollars (mil.)	17,420	22,310	22,840	26,610	24,380
1989 constant dollars	19,830	24,760	24,560	27,700	24,380
Total exports[7]					
Current dollars (mil.)	17,710	23,360	24,690	30,750	28,480
1989 constant dollars	20,160	25,920	26,560	32,010	28,480
Arms as percent of total imports[8]	6.3	5.4	3.6	3.8	2.6
Arms as percent of total exports[8]	7.3	6.4	5.3	3.9	1.4

Human Rights [24]

	SSTS	FL	FAPRO	PPCG	APROBC	TPW	PCPTW	STPEP	PHRFF	PRW	ASST	AFL
Observes	P	P	P	P	P	P	P	P	S	P	P	P
	EAFRD	CPR	ESCR	SR	ACHR	MAAE	PVIAC	PVNAC	EAFDAW	TCIDTP	RC	
Observes		P	P	P	P		P	P	P	P	P	

P = Party; S = Signatory; see Appendix for meaning of abbreviations.

Labor Force

Total Labor Force [25]

15.609 million

Labor Force by Occupation [26]

Industry and construction	34.4%
Agriculture	27.3
Trade, transport, and communications	16.1
Government and other	22.2
Date of data: 1991	

Unemployment Rate [27]

13.6% (December 1992)

For sources, notes, and explanations, see Annotated Source Appendix, page 1035.

737

Production Sectors

Commercial Energy Production and Consumption

Production [28]

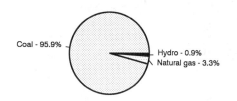

Coal - 95.9%
Hydro - 0.9%
Natural gas - 3.3%

Consumption [29]

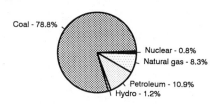

Coal - 78.8%
Nuclear - 0.8%
Natural gas - 8.3%
Petroleum - 10.9%
Hydro - 1.2%

Data are shown in quadrillion (10^{15}) BTUs and percent for 1991

Crude oil[1]	(s)	
Natural gas liquids	(s)	
Dry natural gas	0.15	
Coal[2]	4.41	
Net hydroelectric power[3]	0.04	
Net nuclear power[3]	0.00	
Total	4.61	

Petroleum	0.54	
Dry natural gas	0.41	
Coal[2]	3.90	
Net hydroelectric power[3]	0.06	
Net nuclear power[3]	0.04	
Total	4.97	

Telecommunications [30]

- Severely underdeveloped and outmoded system
- Cable, open wire and microwave
- Phone density is 10.5 phones per 100 residents (October 1990)
- 3.6 million telephone subscribers
- Exchanges are 86% automatic (1991)
- Broadcast stations - 27 AM, 27 FM, 40 (5 Soviet repeaters) TV
- 9.6 million TVs
- 1 satellite earth station using INTELSAT, EUTELSAT, INMARSAT and Intersputnik

Transportation [31]

Railroads. 26,250 km total; 23,857 km 1.435-meter gauge, 397 km 1.520-meter gauge, 1,996 km narrow gauge; 8,987 km double track; 11,510 km electrified; government owned (1991)

Highways. 360,629 km total (excluding farm, factory and forest roads); 220 km limited access expressways, 45,257 km main highways, 128,775 km regional roads, 186,377 urban or village roads (local traffic); 220,000 km are paved (including all main and regional highways) (1988)

Merchant Marine. 209 ships (1,000 GRT or over) totaling 2,747,631 GRT/3,992,053 DWT; includes 5 short-sea passenger, 76 cargo, 1 refrigerated cargo, 11 roll-on/roll-off cargo, 9 container, 1 oil tanker, 4 chemical tanker, 101 bulk, 1 passenger; Poland owns 1 ship of 6,333 DWT operating under Liberian registry

Airports

Total:	163
Usable:	163
With permanent-surface runways:	100
With runway over 3,659 m:	0
With runways 2,440-3,659 m:	51
With runways 1,220-2,439 m:	95

Top Agricultural Products [32]

	1990	1991
Potatoes	36,313	29,038
Grains	28,014	27,811
Sugar beets	16,721	11,879
Fruit	1,416	1,875
Rapeseed	1,206	1,045
Hay	29,113	NA

Values shown are 1,000 metric tons.

Top Mining Products [33]

Estimated metric tons unless otherwise specified   M.t.

Copper, mine output, Cu content	390,000
Zinc, mine output, Zn content	175,000
Silver, mine output, Ag content (000 kilograms)	899[1]
Lime, hydrated and quicklime (000 tons)	2,469[1]
Salt (000 tons)	3,840[1]
Sulfur	4,052

Tourism [34]

	1987	1988	1989	1990	1991
Visitors	4,748	6,196	8,233	18,211	36,846
Tourists					11,350
Excursionists					25,496
Tourism receipts	184	206	202	358	149
Tourism expenditures	203	251	215	423	143
Fare receipts	73	144	300	334	207
Fare expenditures	68	104	179	177	103

Tourists are in thousands, money in million U.S. dollars.

Manufacturing Sector

GDP and Manufacturing Summary [35]

	1980	1985	1989	1990	% change 1980-1990	% change 1989-1990
GDP (million 1980 $)	56,707	54,570	61,240	53,163	-6.2	-13.2
GDP per capita (1980 $)	1,594	1,467	1,603	1,382	-13.3	-13.8
Manufacturing as % of GDP (current prices)	44.0	41.2	39.1[e]	47.0	6.8	20.2
Gross output (million $)	NA	NA	NA	NA	NA	NA
Value added (million $)	22,833	24,432	31,740	23,017	0.8	-27.5
Value added (million 1980 $)	26,384	24,563	27,571	21,006	-20.4	-23.8
Industrial production index	100	98	113	84	-16.0	-25.7
Employment (thousands)	4,126	3,578	3,326	3,014	-27.0	-9.4

Note: GDP stands for Gross Domestic Product. 'e' stands for estimated value.

Profitability and Productivity

	1980	1985	1989	1990	% change 1980-1990	% change 1989-1990
Intermediate input (%)	NA	NA	NA	NA	NA	NA
Wages, salaries, and supplements (%)	NA	NA	NA	NA	NA	NA
Gross operating surplus (%)	NA	NA	NA	NA	NA	NA
Gross output per worker ($)	NA	NA	NA	NA	NA	NA
Value added per worker ($)	5,361	6,242	9,543	7,637	42.5	-20.0
Average wage (incl. benefits) ($)	1,551	1,627	1,774[e]	1,257	-19.0	-29.1

Profitability is in percent of gross output. Productivity is in U.S. $. 'e' stands for estimated value.

Profitability - 1990

Value Added in Manufacturing

	1980 $ mil.	1980 %	1985 $ mil.	1985 %	1989 $ mil.	1989 %	1990 $ mil.	1990 %	% change 1980-1990	% change 1989-1990
311 Food products	-889	-3.9	144	0.6	1,978	6.2	2,595	11.3	-391.9	31.2
313 Beverages	3,062	13.4	3,582	14.7	2,757	8.7	1,838	8.0	-40.0	-33.3
314 Tobacco products	636	2.8	74	0.3	313	1.0	379	1.6	-40.4	21.1
321 Textiles	2,795	12.2	2,444	10.0	3,009	9.5	1,222	5.3	-56.3	-59.4
322 Wearing apparel	572	2.5	801	3.3	1,033	3.3	432	1.9	-24.5	-58.2
323 Leather and fur products	122	0.5	221	0.9	295	0.9	120	0.5	-1.6	-59.3
324 Footwear	403	1.8	430	1.8	641	2.0	263	1.1	-34.7	-59.0
331 Wood and wood products	423	1.9	434	1.8	643	2.0	325	1.4	-23.2	-49.5
332 Furniture and fixtures	491	2.2	500	2.0	627	2.0	307	1.3	-37.5	-51.0
341 Paper and paper products	224	1.0	269	1.1	461	1.5	348	1.5	55.4	-24.5
342 Printing and publishing	154	0.7	208	0.9	217	0.7	166	0.7	7.8	-23.5
351 Industrial chemicals	837	3.7	734	3.0	1,164	3.7	1,056	4.6	26.2	-9.3
352 Other chemical products	961	4.2	644	2.6	853	2.7	649	2.8	-32.5	-23.9
353 Petroleum refineries	1,058	4.6	1,239	5.1	1,496	4.7	1,419	6.2	34.1	-5.1
354 Miscellaneous petroleum and coal products	54	0.2	60	0.2	85	0.3	249	1.1	361.1	192.9
355 Rubber products	317	1.4	341	1.4	330	1.0	209	0.9	-34.1	-36.7
356 Plastic products	360	1.6	296	1.2	338	1.1	274	1.2	-23.9	-18.9
361 Pottery, china and earthenware	97	0.4	146	0.6	178	0.6	107	0.5	10.3	-39.9
362 Glass and glass products	269	1.2	282	1.2	340	1.1	227	1.0	-15.6	-33.2
369 Other non-metal mineral products	335	1.5	634	2.6	687	2.2	602	2.6	79.7	-12.4
371 Iron and steel	868	3.8	1,161	4.8	2,110	6.6	1,887	8.2	117.4	-10.6
372 Non-ferrous metals	602	2.6	336	1.4	1,094	3.4	951	4.1	58.0	-13.1
381 Metal products	1,343	5.9	1,347	5.5	1,606	5.1	1,081	4.7	-19.5	-32.7
382 Non-electrical machinery	3,263	14.3	3,360	13.8	3,650	11.5	2,604	11.3	-20.2	-28.7
383 Electrical machinery	1,558	6.8	1,801	7.4	2,334	7.4	1,420	6.2	-8.9	-39.2
384 Transport equipment	2,436	10.7	2,255	9.2	2,739	8.6	1,855	8.1	-23.9	-32.3
385 Professional and scientific equipment	244	1.1	251	1.0	273	0.9	173	0.8	-29.1	-36.6
390 Other manufacturing industries	237	1.0	438	1.8	490	1.5	258	1.1	8.9	-47.3

Note: The industry codes shown are International Standard Industry codes (ISIC). Percentages are percent of total Value Added. 'e' stands for estimated value

For sources, notes, and explanations, see Annotated Source Appendix, page 1035.

739

Finance, Economics, and Trade

Economic Indicators [36]

Billions of Zlotys unless otherwise noted.

	1989	1990	1991
GDP (current zlotys)	118,319	606,700	NA
Real GDP growth rate (%)	0.2	-11.6	NA
GDP per capita (000s zlotys)	2,638	15,920	NA
Money supply (tril. zlotys at yr-end)	69.5	189.1	263.4[130]
Investment rate (% of GDP)	15.9	19.6	NA
Savings rate[131]	6.7	6.8	NA
CPI (previous year equals 100)[132]	351	685	151[130]
WPI[6]	313	722	150[130]
External public debt	40,300	49,000	NA

Balance of Payments Summary [37]

Values in millions of dollars.

	1987	1988	1989	1990	1991
Exports of goods (f.o.b.)	12,026	13,846	12,869	15,837	14,393
Imports of goods (f.o.b.)	-11,236	-12,757	-12,822	-12,248	-15,104
Trade balance	790	1,089	47	3,589	-711
Services - debits	-5,160	-5,630	-6,676	-6,836	-6,463
Services - credits	2,433	2,743	3,611	3,803	4,260
Private transfers (net)	1,558	1,691	1,521	2,206	723
Government transfers (net)	-	-	-	-7	34
Long term capital (net)	4,383	893	-1,426	6,470	-2,316
Short term capital (net)	244	-42	-176	-3,292	-1,205
Errors and omissions	160	50	-141	661	-82
Overall balance	-281	158	172	-413	1,796

Exchange Rates [38]

Currency: **zlotych.**
Symbol: **Zl.**

Data are currency units per $1.

January 1993	15,879.00
1992	13,626.00
1991	10,576.00
1990	9,500.00
1989	1,439.18
1988	430.55

Imports and Exports

Top Import Origins [39]

$12.9 billion (f.o.b., 1992 est.). Data are for 1991.

Origins	%
Germany	17.4
Former USSR	25.6
Italy	5.3
Austria	5.2

Top Export Destinations [40]

$12.8 billion (f.o.b., 1992 est.). Data are for 1991.

Destinations	%
Germany	28.0
Former USSR	11.7
UK	8.8
Switzerland	5.5

Foreign Aid [41]

	U.S. $	
Donor - bilateral aid to non-Communist less developed countries (1954-89)	2.2	billion
G-24 pledge of grants and credit	8	billion

Import and Export Commodities [42]

Import Commodities

Machinery 38%
Fuels and power 20%
Chemicals 13%
Food 10%
Light industry 6%

Export Commodities

Machinery 22%
Metals 16%
Chemicals 12%
Fuels and power 11%
Food 10%

Portugal

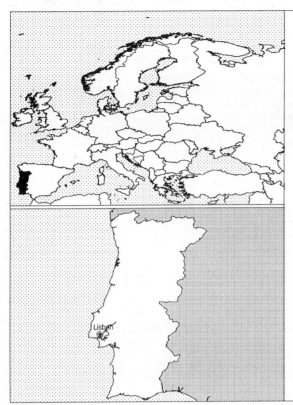

Geography [1]

Total area:
92,080 km2
Land area:
91,640 km2
Comparative area:
Slightly smaller than Indiana
Land boundaries:
Total 1,214 km, Spain 1,214 km
Coastline:
1,793 km
Climate:
Maritime temperate; cool and rainy in north, warmer and drier in south
Terrain:
Mountainous north of the Tagus, rolling plains in south
Natural resources:
Fish, forests (cork), tungsten, iron ore, uranium ore, marble
Land use:
Arable land:
32%
Permanent crops:
6%
Meadows and pastures:
6%
Forest and woodland:
40%
Other:
16%

Demographics [2]

	1960	1970	1980	1990	1991[1]	1994	2000	2010	2020
Population	9,037	9,044	9,778	10,365	10,388	10,524	10,744	10,997	11,038
Population density (persons per sq. mi.)	255	256	276	293	294	NA	301	304	302
Births	NA	NA	NA	NA	125	123	NA	NA	NA
Deaths	NA	NA	NA	NA	100	102	NA	NA	NA
Life expectancy - males	NA	NA	NA	NA	71	72	NA	NA	NA
Life expectancy - females	NA	NA	NA	NA	78	79	NA	NA	NA
Birth rate (per 1,000)	NA	NA	NA	NA	12	12	NA	NA	NA
Death rate (per 1,000)	NA	NA	NA	NA	10	10	NA	NA	NA
Women of reproductive age (15-44 yrs.)	NA	NA	NA	2,617	NA	2,729	2,785	NA	NA
of which are currently married	NA	NA	NA	1,773	NA	1,879	1,982	NA	NA
Fertility rate	NA	NA	NA	1.4	1.53	1.5	1.5	1.5	1.6

Population values are in thousands, life expectancy in years, and other items as indicated.

Health

Health Personnel [3]

Doctors per 1,000 pop., 1988-92	2.57
Nurse-to-doctor ratio, 1988-92	0.8
Hospital beds per 1,000 pop., 1985-90	4.2
Percentage of children immunized (age 1 yr. or less)	
Third dose of DPT, 1990-91	95.0
Measles, 1990-91	96.0

Health Expenditures [4]

Total health expenditure, 1990 (official exchange rate)	
Millions of dollars	3970
Millions of dollars per capita	383
Health expenditures as a percentage of GDP	
Total	7.0
Public sector	4.3
Private sector	2.7
Development assistance for health	
Total aid flows (millions of dollars)[1]	NA
Aid flows per capita (millions of dollars)	NA
Aid flows as a percentage of total health expenditure	NA

For sources, notes, and explanations, see Annotated Source Appendix, page 1035.

Human Factors

Health Care Ratios [5]

Population per physician, 1970	1110
Population per physician, 1990	490
Population per nursing person, 1970	820
Population per nursing person, 1990	NA
Percent of births attended by health staff, 1985	NA

Infants and Malnutrition [6]

Percent of babies with low birth weight, 1985	8
Infant mortality rate per 1,000 live births, 1970	56
Infant mortality rate per 1,000 live births, 1991	11
Years of life lost per 1,000 population, 1990	12
Prevalence of malnutrition (under age 5), 1990	NA

Ethnic Division [7]

Homogeneous Mediterranean stock in mainland, Azores, Madeira Islands; citizens of black African descent who immigrated to mainland during decolonization number less than 100,000.

Religion [8]

Roman Catholic	97%
Protestant denominations	1%
Other	2%

Major Languages [9]

Portuguese.

Education

Public Education Expenditures [10]

Million Escudos	1980	1985	1987[1]	1988	1989	1990
Total education expenditure	53,234	152,886	220,242	282,610	342,381	412,481
as percent of GNP	4.4	4.6	4.4	4.8	4.9	5.1
as percent of total govt. expend.	NA	NA	NA	NA	NA	NA
Current education expenditure	45,443	135,612	194,776	257,210	313,506	378,252
as percent of GNP	3.8	4.1	3.9	4.4	4.5	4.7
as percent of current govt. expend.	NA	NA	NA	NA	NA	NA
Capital expenditure	7,791	17,274	25,466	25,400	28,875	34,229

Educational Attainment [11]

Age group	25+
Total population	5,696,282
Highest level attained (%)	
No schooling	27.5
First level	
Incompleted	NA
Completed	58.5
Entered second level	
S-1	NA
S-2	10.6
Post secondary	3.5

Literacy Rate [12]

In thousands and percent	1985[a]	1991[a]	2000[a]
Illiterate population +15 years	1,429	1,215	829
Illiteracy rate - total pop. (%)	18.4	15.0	9.7
Illiteracy rate - males (%)	13.6	11.2	6.8
Illiteracy rate - females (%)	22.8	18.5	12.3

Libraries [13]

	Admin. Units	Svc. Pts.	Vols. (000)	Shelving (meters)	Vols. Added	Reg. Users
National (1990)	1	1	2,236	44,785	15,730	60
Non-specialized (1990)	11	12	1,995	34,181	77,789	5,711
Public (1990)	167	234	3,371	96,289	178,647	381,006
Higher ed. (1990)	204	264	5,080	128,438	179,164	154,883
School (1990)	675	707	2,896	65,563	156,413	175,970

Daily Newspapers [14]

	1975	1980	1985	1990
Number of papers	30	28	25	24
Circ. (000)	612[e]	480[e]	413	390

Cinema [15]

Data for 1990.

Cinema seats per 1,000	11.2
Annual attendance per person	1.0
Gross box office receipts (mil. escudo)	2,856

Science and Technology

Scientific/Technical Forces [16]

R&D Expenditures [17]

	Escudo (000) 1988
Total expenditure	29,910,800
Capital expenditure	6,390,500
Current expenditure	23,520,300
Percent current	78.6

U.S. Patents Issued [18]

Values show patents issued to citizens of the country by the U.S. Patents Office.

	1990	1991	1992
Number of patents	7	8	8

Government and Law

Organization of Government [19]

Long-form name:
Portuguese Republic
Type:
republic
Independence:
1140 (independent republic proclaimed 5 October 1910)
Constitution:
25 April 1976, revised 30 October 1982 and 1 June 1989
Legal system:
civil law system; the Constitutional Tribunal reviews the constitutionality of legislation; accepts compulsory ICJ jurisdiction, with reservations
National holiday:
Day of Portugal, 10 June
Executive branch:
president, Council of State, prime minister, deputy prime minister, Council of Ministers (cabinet)
Legislative branch:
unicameral Assembly of the Republic (Assembleia da Republica)
Judicial branch:
Supreme Tribunal of Justice (Supremo Tribunal de Justica)

Political Parties [20]

Assembly of the Republic	% of votes
Social Democratic Party (PSD)	50.4
Portuguese Socialist Party (PS)	29.3
United Democratic Coalition (CDU)	8.8
Center Democratic Party	4.4
National Solidarity Party	1.7
Democratic Renewal (PRD)	0.6
Other	4.8

Government Expenditures [21]

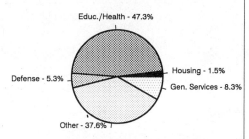

Educ./Health - 47.3%
Housing - 1.5%
Gen. Services - 8.3%
Defense - 5.3%
Other - 37.6%

% distribution for 1990. Expend. in 1991: 33.2 ($ bil.)

Military Expenditures and Arms Transfers [22]

	1985	1986	1987	1988	1989
Military expenditures					
Current dollars (mil.)	1,051	1,124	1,183	1,338	1,457
1989 constant dollars (mil.)	1,196	1,248	1,272	1,393	1,457
Armed forces (000)	102	101	105	104	104
Gross national product (GNP)					
Current dollars (mil.)	31,470	34,280	37,520	40,520	44,620
1989 constant dollars (mil.)	35,820	38,050	40,350	42,180	44,620
Central government expenditures (CGE)					
1989 constant dollars (mil.)	18,360	19,200	18,660	19,080	19,440
People (mil.)	10.2	10.2	10.2	10.3	10.3
Military expenditure as % of GNP	3.3	3.3	3.2	3.3	3.3
Military expenditure as % of CGE	6.5	6.5	6.8	7.3	7.5
Military expenditure per capita	118	122	124	135	141
Armed forces per 1,000 people	10.0	9.9	10.2	10.1	10.1
GNP per capita	3,525	3,727	3,937	4,100	4,323
Arms imports[6]					
Current dollars (mil.)	220	30	30	50	60
1989 constant dollars (mil.)	250	33	32	52	60
Arms exports[6]					
Current dollars (mil.)	210	220	60	100	40
1989 constant dollars (mil.)	239	244	65	104	40
Total imports[7]					
Current dollars (mil.)	7,652	9,650	13,970	17,870	19,070
1989 constant dollars	8,711	10,710	15,020	18,600	19,070
Total exports[7]					
Current dollars (mil.)	5,685	7,242	9,320	10,990	12,510
1989 constant dollars	6,472	8,037	10,020	11,440	12,510
Arms as percent of total imports[8]	2.9	0.3	0.2	0.3	0.3
Arms as percent of total exports[8]	3.7	3.0	0.6	0.9	0.3

Crime [23]

Crime volume	
Cases known to police	83,205
Attempts (percent)	NA
Percent cases solved	NA
Crimes per 100,000 persons	804.92
Persons responsible for offenses	
Total number offenders	NA
Percent female	NA
Percent juvenile	NA
Percent foreigners	NA

Human Rights [24]

	SSTS	FL	FAPRO	PPCG	APROBC	TPW	PCPTW	STPEP	PHRFF	PRW	ASST	AFL
Observes	P	P	P		P	P	P	P	P		P	P
	EAFRD	CPR	ESCR	SR	ACHR	MAAE	PVIAC	PVNAC	EAFDAW	TCIDTP	RC	
Observes		P	P	P	P		P	P	P	P	P	

P = Party; S = Signatory; see Appendix for meaning of abbreviations.

Labor Force

Total Labor Force [25]

4,605,700

Labor Force by Occupation [26]

Services	45%
Industry	35
Agriculture	20

Date of data: 1988

Unemployment Rate [27]

5% (1992)

For sources, notes, and explanations, see Annotated Source Appendix, page 1035.

743

Production Sectors

Commercial Energy Production and Consumption

Production [28]

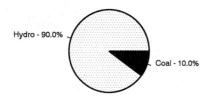

Hydro - 90.0%

Coal - 10.0%

Consumption [29]

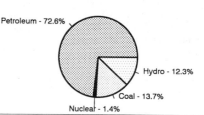

Petroleum - 72.6%

Hydro - 12.3%

Coal - 13.7%

Nuclear - 1.4%

Data are shown in quadrillion (10^{15}) BTUs and percent for 1991

Crude oil[1]	0.00
Natural gas liquids	0.00
Dry natural gas	0.00
Coal[2]	0.01
Net hydroelectric power[3]	0.09
Net nuclear power[3]	0.00
Total	0.10

Petroleum	0.53
Dry natural gas	0.00
Coal[2]	0.10
Net hydroelectric power[3]	0.09
Net nuclear power[3]	0.01
Total	0.73

Telecommunications [30]

- Generally adequate integrated network of coaxial cables, open wire and microwave radio relay
- 2,690,000 telephones
- Broadcast stations - 57 AM, 66 (22 repeaters) FM, 66 (23 repeaters) TV
- 6 submarine cables
- 3 INTELSAT earth stations (2 Atlantic Ocean, 1 Indian Ocean), EUTELSAT, domestic satellite systems (mainland and Azores)
- Tropospheric link to Azores

Top Agricultural Products [32]

	1990	1991[4]
Wine grapes (000 hl.)	10,431	10,118
Potatoes	999	939
Tomatoes for processing	712	730
Corn	643	621
Wheat	268	322
Olives (000 hl.)	294	320

Values shown are 1,000 metric tons.

Top Mining Products [33]

Metric tons unless otherwise specified	M.t.
Copper, Cu content of concentrate	164,768[1]
Tin, mine output, Sn content	10,360[1]
Salt, rock and marine	649,800
Gypsum	300,000
Lime, hydrated and quicklime	200,000
Pyrite and pyrrhotite	138,760[1]

Transportation [31]

Railroads. 3,625 km total; state-owned Portuguese Railroad Co. (CP) operates 2,858 km 1.665-meter gauge (434 km electrified and 426 km double track), 755 km 1.000-meter gauge; 12 km (1.435-meter gauge) electrified, double track, privately owned

Highways. 73,661 km total; 61,599 km surfaced (bituminous, gravel, and crushed stone), including 140 km of limited-access divided highway; 7,962 km improved earth; 4,100 km unimproved earth (motorable tracks)

Merchant Marine. 51 ships (1,000 GRT or over) totaling 634,072 GRT/1,130,515 DWT; includes 1 short-sea passenger, 21 cargo, 3 refrigerated cargo, 3 container, 1 roll-on/roll-off cargo, 13 oil tanker, 2 chemical tanker, 5 bulk, 2 liquified gas; note - Portugal has created a captive register on Madeira (MAR) for Portuguese-owned ships that will have the taxation and crewing benefits of a flag of convenience; although only one ship currently is known to fly the Portuguese flag on the MAR register, it is likely that a majority of Portuguese flag ships will transfer to this subregister in a few years

Airports

Total:	64
Usable:	62
With permanent-surface runways:	36
With runways over 3,659 m:	2
With runways 2,440-3,659 m:	10
With runways 1,220-2,439 m:	11

Tourism [34]

	1987	1988	1989	1990	1991
Visitors	16,173	16,077	16,476	18,422	19,641
Tourists[4,55]	6,102	6,624	7,116	8,020	8,657
Excursionists	9,891	9,282	9,153	10,179	10,755
Tourism receipts	2,145	2,402	2,685	3,555	3,700
Tourism expenditures	421	533	583	867	1,027
Fare receipts	130	113	106	20	156
Fare expenditures	21	28	32	42	36

Tourists are in thousands, money in million U.S. dollars.

744

For sources, notes, and explanations, see Annotated Source Appendix, page 1035.

Manufacturing Sector

GDP and Manufacturing Summary [35]

	1980	1985	1989	1990	% change 1980-1990	% change 1989-1990
GDP (million 1980 $)	25,090	26,223	31,792	32,855	30.9	3.3
GDP per capita (1980 $)	2,569	2,581	3,097	3,193	24.3	3.1
Manufacturing as % of GDP (current prices)	30.0	29.3	26.7[e]	29.6[e]	-1.3	10.9
Gross output (million $)	17,932	15,793	28,011[e]	37,287[e]	107.9	33.1
Value added (million $)	5,602	4,147	8,784[e]	11,680[e]	108.5	33.0
Value added (million 1980 $)	7,576	8,169	9,981	10,144	33.9	1.6
Industrial production index	100	96	108	111	11.0	2.8
Employment (thousands)	680	623	602	577	-15.1	-4.2

Note: GDP stands for Gross Domestic Product. 'e' stands for estimated value.

Profitability and Productivity

	1980	1985	1989	1990	% change 1980-1990	% change 1989-1990
Intermediate input (%)	69	74	69[e]	69[e]	0.0	0.0
Wages, salaries, and supplements (%)	17	14	10[e]	16[e]	-5.9	60.0
Gross operating surplus (%)	14	12	21[e]	15[e]	7.1	-28.6
Gross output per worker ($)	25,884	24,974	46,504[e]	63,758[e]	146.3	37.1
Value added per worker ($)	8,087	6,559	14,583[e]	19,989[e]	147.2	37.1
Average wage (incl. benefits) ($)	4,541	3,490	4,775[e]	10,367[e]	128.3	117.1

Profitability is in percent of gross output. Productivity is in U.S. $. 'e' stands for estimated value.

Profitability - 1990

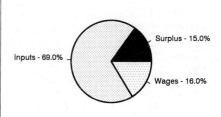

Surplus - 15.0%
Inputs - 69.0%
Wages - 16.0%

The graphic shows percent of gross output.

Value Added in Manufacturing

	1980 $ mil.	1980 %	1985 $ mil.	1985 %	1989 $ mil.	1989 %	1990 $ mil.	1990 %	% change 1980-1990	% change 1989-1990
311 Food products	544	9.7	490	11.8	1,011	11.5	1,360	11.6	150.0	34.5
313 Beverages	135	2.4	133	3.2	305	3.5	408	3.5	202.2	33.8
314 Tobacco products	64	1.1	93	2.2	170	1.9	226	1.9	253.1	32.9
321 Textiles	905	16.2	679	16.4	1,291	14.7	1,738	14.9	92.0	34.6
322 Wearing apparel	186	3.3	182	4.4	387[e]	4.4	486	4.2	161.3	25.6
323 Leather and fur products	41	0.7	41	1.0	87	1.0	91	0.8	122.0	4.6
324 Footwear	86	1.5	86	2.1	193[e]	2.2	293	2.5	240.7	51.8
331 Wood and wood products	325	5.8	150	3.6	252[e]	2.9	292[e]	2.5	-10.2	15.9
332 Furniture and fixtures	106	1.9	30	0.7	110[e]	1.3	135[e]	1.2	27.4	22.7
341 Paper and paper products	274	4.9	276	6.7	546[e]	6.2	694[e]	5.9	153.3	27.1
342 Printing and publishing	180	3.2	140	3.4	261[e]	3.0	330[e]	2.8	83.3	26.4
351 Industrial chemicals	147	2.6	215	5.2	457	5.2	601	5.1	308.8	31.5
352 Other chemical products	224	4.0	190	4.6	482	5.5	541[e]	4.6	141.5	12.2
353 Petroleum refineries	219[e]	3.9	-18[e]	-0.4	245[e]	2.8	339	2.9	54.8	38.4
354 Miscellaneous petroleum and coal products	1[e]	0.0	NA	0.0	1[e]	0.0	NA	0.0	NA	NA
355 Rubber products	58	1.0	49[e]	1.2	108	1.2	108[e]	0.9	86.2	0.0
356 Plastic products	128	2.3	92[e]	2.2	183	2.1	278[e]	2.4	117.2	51.9
361 Pottery, china and earthenware	80	1.4	67	1.6	180	2.0	250[e]	2.1	212.5	38.9
362 Glass and glass products	87	1.6	53	1.3	147	1.7	215	1.8	147.1	46.3
369 Other non-metal mineral products	295	5.3	200	4.8	511	5.8	712	6.1	141.4	39.3
371 Iron and steel	207	3.7	98	2.4	247	2.8	318	2.7	53.6	28.7
372 Non-ferrous metals	33	0.6	26	0.6	38	0.4	55	0.5	66.7	44.7
381 Metal products	323	5.8	219	5.3	386	4.4	546	4.7	69.0	41.5
382 Non-electrical machinery	170	3.0	143	3.4	233	2.7	303	2.6	78.2	30.0
383 Electrical machinery	319	5.7	263	6.3	518	5.9	809	6.9	153.6	56.2
384 Transport equipment	428	7.6	222	5.4	381	4.3	477	4.1	11.4	25.2
385 Professional and scientific equipment	15	0.3	16	0.4	32[e]	0.4	49[e]	0.4	226.7	53.1
390 Other manufacturing industries	20	0.4	11	0.3	21[e]	0.2	28[e]	0.2	40.0	33.3

Note: The industry codes shown are International Standard Industry codes (ISIC). Percentages are percent of total Value Added. 'e' stands for estimated value

For sources, notes, and explanations, see Annotated Source Appendix, page 1035.

745

Finance, Economics, and Trade

Economic Indicators [36]

Millions of U.S. Dollars, unless otherwise stated[133].

	1989	1990	1991[e]
GDP at current market prices	45,511	59,712	69,317
Real GDP growth rate (%)	5.5	4.2	3.0
Per capita GDP (US$)	4,410	5,778	6,701
Money Supply (M1)	1,828	2,352	2,947
Internal savings rate (% of GDP)	28.8	28.5	NA
Gross investment rate (% of GDP)	28.5	27.0	27.6
CPI (annual avg. % change)	12.6	13.4	12.0
WPI	NA	NA	NA
External public debt	5,889	5,193	4,575

Balance of Payments Summary [37]

Values in millions of dollars.

	1987	1988	1989	1990	1991
Exports of goods (f.o.b.)	9,266	10,874	12,720	16,311	16,231
Imports of goods (f.o.b.)	-12,847	-16,392	-17,585	-23,141	-24,079
Trade balance	-3,581	-5,518	-4,865	-6,830	-7,848
Services - debits	-3,401	-3,906	-4,152	-5,461	-5,784
Services - credits	3,646	4,037	4,630	6,603	6,941
Private transfers (net)	3,404	3,597	3,726	4,509	4,593
Government transfers (net)	368	725	814	998	1,381
Long term capital (net)	110	782	2,796	3,574	3,998
Short term capital (net)	578	-489	1,208	-1,010	539
Errors and omissions	653	1,640	497	1,160	1,893
Overall balance	1,777	868	4,654	3,543	5,713

Exchange Rates [38]

Currency: **Portuguese escudos.**
Symbol: **Esc.**

Data are currency units per $1.

January 1993	145.51
1992	135.00
1991	144.48
1990	142.55
1989	157.46
1988	143.95

Imports and Exports

Top Import Origins [39]

$26.0 billion (c.i.f., 1992 est.).

Origins	%
EC	72.0
Other developed countries	10.9
Less developed countries	12.9
US	3.4

Top Export Destinations [40]

$16.3 billion (f.o.b., 1992 est.). Data are for 1991.

Destinations	%
EC	75.4
Other developed countries	12.4
US	3.8

Foreign Aid [41]

	U.S. $	
US commitments, including Ex-Im (FY70-89)	1.8	billion
Western (non-US) countries, ODA and OOF bilateral commitments (1970-89)	1.2	billion

Import and Export Commodities [42]

Import Commodities

Machinery & transport equipment
Agricultural products
Chemicals
Petroleum
Textiles

Export Commodities

Cotton textiles
Cork and paper products
Canned fish
Wine
Timber and timber products
Resin
Machinery
Appliances

Qatar

Geography [1]

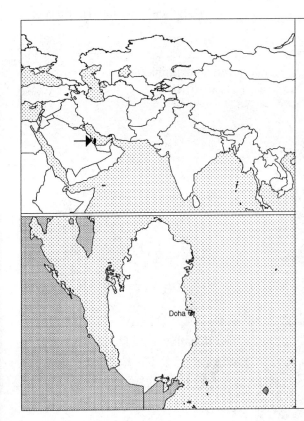

Total area:
  11,000 km2
Land area:
  11,000 km2
Comparative area:
  Slightly smaller than Connecticut
Land boundaries:
  Total 60 km, Saudi Arabia 60 km
Coastline:
  563 km
Climate:
  Desert; hot, dry; humid and sultry in summer
Terrain:
  Mostly flat and barren desert covered with loose sand and gravel
Natural resources:
  Petroleum, natural gas, fish
Land use:
  Arable land:
  0%
  Permanent crops:
  0%
  Meadows and pastures:
  5%
  Forest and woodland:
  0%
  Other:
  95%

Demographics [2]

	1960	1970	1980	1990	1991[1]	1994	2000	2010	2020
Population	45	113	231	452	518	513	572	645	713
Population density (persons per sq. mi.)	11	27	54	116	122	NA	175	225	263
Births	NA	NA	NA	NA	11	10	NA	NA	NA
Deaths	NA	NA	NA	NA	2	2	NA	NA	NA
Life expectancy - males	NA	NA	NA	NA	69	70	NA	NA	NA
Life expectancy - females	NA	NA	NA	NA	74	75	NA	NA	NA
Birth rate (per 1,000)	NA	NA	NA	NA	21	19	NA	NA	NA
Death rate (per 1,000)	NA	NA	NA	NA	3	4	NA	NA	NA
Women of reproductive age (15-44 yrs.)	NA	NA	NA	78	NA	90	108	NA	NA
of which are currently married	NA	NA	NA	56	NA	64	73	NA	NA
Fertility rate	NA	NA	NA	4.3	4.04	3.7	2.9	2.4	2.1

Population values are in thousands, life expectancy in years, and other items as indicated.

Health

Health Personnel [3]

Health Expenditures [4]

For sources, notes, and explanations, see Annotated Source Appendix, page 1035.

747

Human Factors

Health Care Ratios [5]	Infants and Malnutrition [6]

Ethnic Division [7]

Arab	40%
Pakistani	18%
Indian	18%
Iranian	10%
Other	14%

Religion [8]

Muslim	95%

Major Languages [9]

Arabic (official), English commonly used as a second language.

Education

Public Education Expenditures [10]

Million Riyal	1980	1985	1987	1988	1989	1990
Total education expenditure	792	1,012	NA	995	917	929
as percent of GNP	NA	NA	NA	NA	NA	NA
as percent of total govt. expend.	7.2	NA	NA	NA	NA	NA
Current education expenditure	598	767	NA	791	869	904
as percent of GNP	NA	NA	NA	NA	NA	NA
as percent of current govt. expend.	7.8	NA	NA	NA	NA	NA
Capital expenditure	194	246	NA	204	49	25

Educational Attainment [11]

Age group	25+
Total population	211,485
Highest level attained (%)	
No schooling	53.5
First level	
Incompleted	9.8
Completed	NA
Entered second level	
S-1	10.1
S-2	13.3
Post secondary	13.3

Literacy Rate [12]

In thousands and percent	1985[a]	1991[a]	2000[a]
Illiterate population +15 years	65	NA	NA
Illiteracy rate - total pop. (%)	24.3	NA	NA
Illiteracy rate - males (%)	23.2	NA	NA
Illiteracy rate - females (%)	27.5	NA	NA

Libraries [13]

	Admin. Units	Svc. Pts.	Vols. (000)	Shelving (meters)	Vols. Added	Reg. Users
National (1989)	1	1	165	NA	9,378	500
Non-specialized (1986)	5	6	NA	NA	NA	NA
Public (1989)	6	6	152	NA	4,180	5,224
Higher ed.	NA	NA	NA	NA	NA	NA
School (1987)	156	NA	471	NA	30,572	NA

Daily Newspapers [14]

	1975	1980	1985	1990
Number of papers	1	3	4	5
Circ. (000)	20[e]	30[e]	60	80[e]

Cinema [15]

Data for 1989.

Cinema seats per 1,000	9.6
Annual attendance per person	0.7
Gross box office receipts (mil. riyal)	NA

Science and Technology

Scientific/Technical Forces [16]

Potential scientists/engineers[17]	6,302
Number female	1,701
Potential technicians[17]	3,500
Number female	1,265
Total	9,802

R&D Expenditures [17]

	Riyal (000) 1986
Total expenditure	6,650
Capital expenditure	-
Current expenditure	6,650
Percent current	100.0

U.S. Patents Issued [18]

Government and Law

Organization of Government [19]

Long-form name:
State of Qatar
Type:
traditional monarchy
Independence:
3 September 1971 (from UK)
Constitution:
provisional constitution enacted 2 April 1970
Legal system:
discretionary system of law controlled by the amir, although civil codes are being implemented; Islamic law is significant in personal matters
National holiday:
Independence Day, 3 September (1971)
Executive branch:
amir, Council of Ministers (cabinet)
Legislative branch:
unicameral Advisory Council (Majlis al-Shura)
Judicial branch:
Court of Appeal

Elections [20]

Advisory Council. None; constitution calls for elections for part of this consultative body, but no elections have been held; seats - (30 total).

Government Budget [21]

For FY92 est.	
Revenues	2.500
Expenditures	3.000
Capital expenditures	0.440

Military Expenditures and Arms Transfers [22]

	1985	1986	1987	1988	1989
Military expenditures					
Current dollars (mil.)	NA	NA	NA	NA	NA
1989 constant dollars (mil.)	NA	NA	NA	NA	NA
Armed forces (000)	7[e]	9[e]	11	7	7
Gross national product (GNP)					
Current dollars (mil.)	6,722	5,677	5,923	6,030	6,870
1989 constant dollars (mil.)	7,652	6,300	6,371	6,277	6,870
Central government expenditures (CGE)					
1989 constant dollars (mil.)	4,895[e]	5,105[e]	3,620[e]	3,539[e]	NA
People (mil.)	0.3	0.4	0.4	0.4	0.5
Military expenditure as % of GNP	NA	NA	NA	NA	NA
Military expenditure as % of CGE	NA	NA	NA	NA	NA
Military expenditure per capita	NA	NA	NA	NA	NA
Armed forces per 1,000 people	20.1	24.0	27.2	16.1	15.1
GNP per capita	21,950	16,800	15,740	14,460	14,850
Arms imports[6]					
Current dollars (mil.)	40	80	0	30	0
1989 constant dollars (mil.)	46	89	0	31	0
Arms exports[6]					
Current dollars (mil.)	0	0	0	0	0
1989 constant dollars (mil.)	0	0	0	0	0
Total imports[7]					
Current dollars (mil.)	1,139	1,099	1,162	1,267	1,326
1989 constant dollars	1,297	1,220	1,250	1,319	1,326
Total exports[7]					
Current dollars (mil.)	4,203	2,720	2,113	2,200	NA
1989 constant dollars	4,785	3,018	2,273	2,290	NA
Arms as percent of total imports[8]	3.5	7.3	0	2.4	0
Arms as percent of total exports[8]	0	0	0	0	NA

Crime [23]

Crime volume	
Cases known to police	1,327
Attempts (percent)	2.33
Percent cases solved	72.79
Crimes per 100,000 persons	357.68
Persons responsible for offenses	
Total number offenders	1,454
Percent female	8
Percent juvenile[44]	14
Percent foreigners	57.29

Human Rights [24]

	SSTS	FL	FAPRO	PPCG	APROBC	TPW	PCPTW	STPEP	PHRFF	PRW	ASST	AFL
Observes						P	P					
	EAFRD	CPR	ESCR	SR	ACHR	MAAE	PVIAC	PVNAC	EAFDAW	TCIDTP	RC	
Observes	P						P				S	

P = Party; S = Signatory; see Appendix for meaning of abbreviations.

Labor Force

Total Labor Force [25]

104,000; 85% non-Qatari in private sector (1983)

Labor Force by Occupation [26]

Unemployment Rate [27]

Production Sectors

Commercial Energy Production and Consumption

Production [28]	Consumption [29]
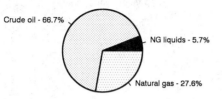	
Crude oil - 66.7%	Natural gas - 82.9%
NG liquids - 5.7%	Petroleum - 17.1%
Natural gas - 27.6%	

Data are shown in quadrillion (10^{15}) BTUs and percent for 1991

Crude oil[1]	0.82	Petroleum	0.07
Natural gas liquids	0.07	Dry natural gas	0.34
Dry natural gas	0.34	Coal[2]	0.00
Coal[2]	0.00	Net hydroelectric power[3]	0.00
Net hydroelectric power[3]	0.00	Net nuclear power[3]	0.00
Net nuclear power[3]	0.00	Total	0.41
Total	1.23		

Telecommunications [30]

- Modern system centered in Doha
- 110,000 telephones
- Tropospheric scatter to Bahrain
- Microwave radio relay to Saudi Arabia and UAE
- Submarine cable to Bahrain and UAE
- Satellite earth stations - 1 Atlantic Ocean INTELSAT, 1 Indian Ocean INTELSAT, 1 ARABSAT
- Broadcast stations - 2 AM, 3 FM, 3 TV

Transportation [31]

Highways. 1,500 km total; 1,000 km paved, 500 km gravel or natural surface (est.)

Merchant Marine. 20 ships (1,000 GRT or over) totaling 390,072 GRT/593,508 DWT; includes 13 cargo, 4 container, 2 oil tanker, 1 refrigerated cargo

Airports

Total:	4
Usable:	4
With permanent-surface runways:	1
With runways over 3,659 m:	1
With runways 2,440-3,659 m:	0
With runways 1,220-2,439 m:	2

Top Agricultural Products [32]

Farming and grazing on small scale, less than 2% of GDP; agricultural area is small and government-owned; commercial fishing increasing in importance; most food imported.

Top Mining Products [33]

Estimated metric tons unless otherwise specified   M.t.

Cement, hydraulic	336,000[1]
Limestone (000 tons)	850
Steel, crude (000 tons)	580
Sulfur	53,000
Petroleum, crude (000 barrels)	137,970[1]

Tourism [34]

	1987	1988	1989	1990	1991
Tourists[56]	101	113	110	136	143

Tourists are in thousands, money in million U.S. dollars.

750

For sources, notes, and explanations, see Annotated Source Appendix, page 1035.

Finance, Economics, and Trade

GDP and Manufacturing Summary [35]

	1980	1985	1990	1991	1992
Gross Domestic Product					
Millions of 1980 dollars	7,838	6,342	7,609	7,876	8,191[e]
Growth rate in percent	7.10	-3.91	10.50	3.50	4.00[e]
Manufacturing Value Added					
Millions of 1980 dollars	258	389	598	651	714[e]
Growth rate in percent	11.20	3.64	14.61	8.96	9.66[e]
Manufacturing share in percent of current prices	3.3	7.8	NA	NA	NA

Economic Indicators [36]

National product: GDP—exchange rate conversion—$8.1 billion (1991 est.). **National product real growth rate**: 3% (1991 est.). **National product per capita**: $17,000 (1991 est.). **Inflation rate (consumer prices)**: 3% (1990). **External debt**: $1.1 billion (December 1989 est.).

Balance of Payments Summary [37]

Exchange Rates [38]

Currency: **Qatari riyals.**
Symbol: **QR.**

Data are currency units per $1.

Fixed rate	3.6400

Imports and Exports

Top Import Origins [39]

$1.4 billion (f.o.b., 1991 est.).

Origins	%
France	13
Japan	12
UK	11
Germany	9

Top Export Destinations [40]

$3.2 billion (f.o.b., 1991).

Destinations	%
Japan	61
Brazil	6
South Korea	5
UAE	4

Foreign Aid [41]

	U.S. $	
Donor - pledge ODA to less developed countries (1979-88)	2.7	billion

Import and Export Commodities [42]

Import Commodities

Machinery & equipment
Consumer goods
Food
Chemicals

Export Commodities

Petroleum products 85%
Steel
Fertilizers

For sources, notes, and explanations, see Annotated Source Appendix, page 1035.

751

Romania

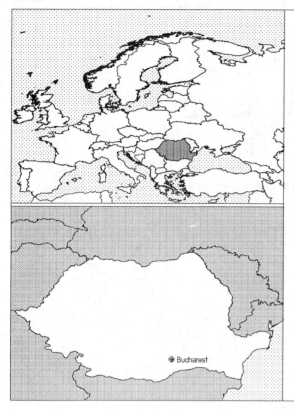

Geography [1]

Total area:
237,500 km2
Land area:
230,340 km2
Comparative area:
Slightly smaller than Oregon
Land boundaries:
Total 2,508 km, Bulgaria 608 km, Hungary 443 km, Moldova 450 km, Serbia and Montenegro 476 km (all with Serbia), Ukraine (north) 362 km, Ukraine (south) 169 km
Coastline:
225 km
Climate:
Temperate; cold, cloudy winters with frequent snow and fog; sunny summers with frequent showers and thunderstorms
Terrain:
Central Transylvanian Basin is separated from the plain of Moldavia on the east by the Carpathian Mountains and separated from the Walachian Plain on the south by the Transylvanian Alps
Natural resources:
Petroleum (reserves being exhausted), timber, natural gas, coal, iron ore, salt
Land use:
Arable land: 43%
Permanent crops: 3%
Meadows and pastures: 19%
Forest and woodland: 28%
Other: 7%

Demographics [2]

	1960	1970	1980	1990	1991[1]	1994	2000	2010	2020
Population	18,403	20,253	22,201	23,191	23,397	23,181	23,383	23,950	24,337
Population density (persons per sq. mi.)	207	228	250	262	263	NA	276	286	292
Births	NA	NA	NA	NA	372	317	NA	NA	NA
Deaths	NA	NA	NA	NA	232	232	NA	NA	NA
Life expectancy - males	NA	NA	NA	NA	69	69	NA	NA	NA
Life expectancy - females	NA	NA	NA	NA	75	75	NA	NA	NA
Birth rate (per 1,000)	NA	NA	NA	NA	16	14	NA	NA	NA
Death rate (per 1,000)	NA	NA	NA	NA	10	10	NA	NA	NA
Women of reproductive age (15-44 yrs.)	NA	NA	NA	5,586	NA	5,736	5,773	NA	NA
of which are currently married	NA	NA	NA	3,976	NA	4,090	4,222	NA	NA
Fertility rate	NA	NA	NA	1.8	2.12	1.8	1.8	1.8	1.8

Population values are in thousands, life expectancy in years, and other items as indicated.

Health

Health Personnel [3]

Doctors per 1,000 pop., 1988-92	1.79
Nurse-to-doctor ratio, 1988-92	NA
Hospital beds per 1,000 pop., 1985-90	8.9
Percentage of children immunized (age 1 yr. or less)	
Third dose of DPT, 1990-91	97.0
Measles, 1990-91	92.0

Health Expenditures [4]

Total health expenditure, 1990 (official exchange rate)	
Millions of dollars	1455
Millions of dollars per capita	63
Health expenditures as a percentage of GDP	
Total	3.9
Public sector	2.4
Private sector	1.5
Development assistance for health	
Total aid flows (millions of dollars)[1]	NA
Aid flows per capita (millions of dollars)	NA
Aid flows as a percentage of total health expenditure	NA

For sources, notes, and explanations, see Annotated Source Appendix, page 1035.

Human Factors

Health Care Ratios [5]

Population per physician, 1970	840
Population per physician, 1990	560
Population per nursing person, 1970	430
Population per nursing person, 1990	NA
Percent of births attended by health staff, 1985	99

Infants and Malnutrition [6]

Percent of babies with low birth weight, 1985	6
Infant mortality rate per 1,000 live births, 1970	49
Infant mortality rate per 1,000 live births, 1991	27
Years of life lost per 1,000 population, 1990	19
Prevalence of malnutrition (under age 5), 1990	NA

Ethnic Division [7]

Romanian	89.1%
Hungarian	8.9%
German	0.4%
Other	1.6%

Religion [8]

Romanian Orthodox	70%
Roman Catholic	
(of which 3% are Uniate)	6%
Protestant	6%
Unaffiliated	18%

Major Languages [9]

Romanian, Hungarian, German.

Education

Public Education Expenditures [10]

Million Leu	1980	1985	1987	1988	1989	1990
Total education expenditure	19,930	17,941	NA	19,797	NA	24,270
as percent of GNP	3.3	2.2	NA	2.3	NA	3.1
as percent of total govt. expend.	6.7	NA	NA	NA	NA	7.3
Current education expenditure	17,691	17,345	NA	NA	NA	23,881
as percent of GNP	2.9	2.1	NA	NA	NA	3.0
as percent of current govt. expend.	NA	NA	NA	NA	NA	9.0
Capital expenditure	2,239	596	NA	NA	NA	389

Educational Attainment [11]

Age group	25+
Total population	13,636,685
Highest level attained (%)	
No schooling	5.4
First level	
Incompleted	23.8
Completed	NA
Entered second level	
S-1	36.0
S-2	27.4
Post secondary	7.3

Literacy Rate [12]

	1980[b]	1990[b]	1992[b]
Illiterate population +15 years	NA	NA	586,334
Illiteracy rate - total pop. (%)	NA	NA	3.1
Illiteracy rate - males (%)	NA	NA	1.4
Illiteracy rate - females (%)	NA	NA	4.8

Libraries [13]

	Admin. Units	Svc. Pts.	Vols. (000)	Shelving (meters)	Vols. Added	Reg. Users
National (1988)	2	NA	14,743	NA	NA	38,000
Non-specialized	NA	NA	NA	NA	NA	NA
Public (1988)	7,227	NA	70,673	NA	NA	5.2mil
Higher ed. (1991)	46	NA	21,632	NA	388,453	284,104
School (1991)	10,246	NA	59,856	NA	2.1mil	2.8mil

Daily Newspapers [14]

	1975	1980	1985	1990
Number of papers	20	35	36	65
Circ. (000)	3,015[e]	4,024[e]	3,601[e]	NA

Cinema [15]

Data for 1990.

Cinema seats per 1,000	10.2[2]
Annual attendance per person	5.6
Gross box office receipts (mil. leu)	NA

Science and Technology

Scientific/Technical Forces [16]

R&D Expenditures [17]

	Leu (000) 1989
Total expenditure	20,866,000
Capital expenditure	2,417,000
Current expenditure	18,449,000
Percent current	88.4

U.S. Patents Issued [18]

Values show patents issued to citizens of the country by the U.S. Patents Office.

	1990	1991	1992
Number of patents	1	1	0

For sources, notes, and explanations, see Annotated Source Appendix, page 1035.

Government and Law

Organization of Government [19]

Long-form name:
none
Type:
republic
Independence:
1881 (from Turkey; republic proclaimed 30 December 1947)
Constitution:
8 December 1991
Legal system:
former mixture of civil law system and Communist legal theory that increasingly reflected Romanian traditions is being revised
National holiday:
National Day of Romania, 1 December (1990)
Executive branch:
president, prime minister, Council of Ministers (cabinet)
Legislative branch:
bicameral Parliament consists of an upper house or Senate (Senat) and a lower house or House of Deputies (Adunarea Deputatilor)
Judicial branch:
Supreme Court of Justice, Constitutional Court

Political Parties [20]

House of Deputies	% of votes
DFSN	27.5
Democratic Convention (CDR)	22.5
National Salvation Front (FSN)	11.0
Others	38.5

Government Expenditures [21]

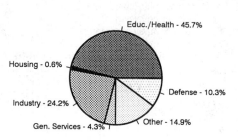

Educ./Health - 45.7%
Housing - 0.6%
Defense - 10.3%
Industry - 24.2%
Other - 14.9%
Gen. Services - 4.3%

% distribution for 1991. Expend. in 1991 est.: 20.0 ($ bil.)

Crime [23]

Crime volume	
Cases known to police	64,005
Attempts (percent)	2.9
Percent cases solved	95.5
Crimes per 100,000 persons	276.4
Persons responsible for offenses	
Total number offenders	56,282
Percent female	7.9
Percent juvenile[45]	9.7
Percent foreigners	0.3

Military Expenditures and Arms Transfers [22]

	1985	1986	1987	1988	1989
Military expenditures					
Current dollars (mil.)[e]	6,840	7,059	7,485	7,004	6,916
1989 constant dollars (mil.)[e]	7,787	7,833	8,050	7,291	6,916
Armed forces (000)	237	238	248	220	207
Gross national product (GNP)					
Current dollars (mil.)	99,330	103,600	104,600	112,300	113,400
1989 constant dollars (mil.)	113,100	114,900	112,500	117,000	113,400
Central government expenditures (CGE)					
1989 constant dollars (mil.)	39,010	41,510	37,470	39,120	40,970
People (mil.)	22.7	22.8	22.9	23.0	23.2
Military expenditure as % of GNP[3]	6.9	6.8	7.2	6.2	6.1
Military expenditure as % of CGE[3]	20.0	18.9	21.5	18.6	16.9
Military expenditure per capita	343	343	351	316	299
Armed forces per 1,000 people	10.4	10.4	10.8	9.5	8.9
GNP per capita	4,977	5,037	4,906	5,076	4,896
Arms imports[6]					
Current dollars (mil.)	40	500	120	20	20
1989 constant dollars (mil.)	46	555	129	21	20
Arms exports[6]					
Current dollars (mil.)	440	310	210	150	70
1989 constant dollars (mil.)	501	344	226	156	70
Total imports[7]					
Current dollars (mil.)	10,430	10,600	11,480	NA	NA
1989 constant dollars	11,880	11,760	12,340	NA	NA
Total exports[7]					
Current dollars (mil.)	12,170	12,540	13,180	NA	NA
1989 constant dollars	13,850	13,920	14,180	NA	NA
Arms as percent of total imports[8]	0.4	4.7	1.0	NA	NA
Arms as percent of total exports[8]	3.6	2.5	1.6	NA	NA

Human Rights [24]

	SSTS	FL	FAPRO	PPCG	APROBC	TPW	PCPTW	STPEP	PHRFF	PRW	ASST	AFL
Observes	P	P	P	P	P	P	P	P		P	P	P
	EAFRD	CPR	ESCR	SR	ACHR	MAAE	PVIAC	PVNAC	EAFDAW	TCIDTP	RC	
Observes		P	P	P		P	S	S	P	P	P	

P = Party; S = Signatory; see Appendix for meaning of abbreviations.

Labor Force

Total Labor Force [25]

10,945,700

Labor Force by Occupation [26]

Industry	38%
Agriculture	28
Other	34

Date of data: 1989

Unemployment Rate [27]

9% (January 1993)

For sources, notes, and explanations, see Annotated Source Appendix, page 1035.

Production Sectors

Commercial Energy Production and Consumption

Production [28]

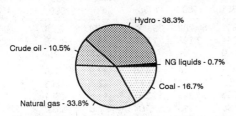

Hydro - 38.3%
Crude oil - 10.5%
NG liquids - 0.7%
Coal - 16.7%
Natural gas - 33.8%

Consumption [29]

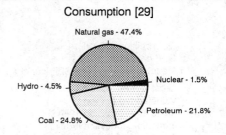

Natural gas - 47.4%
Nuclear - 1.5%
Hydro - 4.5%
Petroleum - 21.8%
Coal - 24.8%

Data are shown in quadrillion (10^{15}) BTUs and percent for 1991

Crude oil[1]	0.30
Natural gas liquids	0.02
Dry natural gas	0.97
Coal[2]	0.48
Net hydroelectric power[3]	1.10
Net nuclear power[3]	0.00
Total	1.87

Petroleum	0.58
Dry natural gas	1.26
Coal[2]	0.66
Net hydroelectric power[3]	0.12
Net nuclear power[3]	0.04
Total	2.67

Telecommunications [30]

- Poor service
- About 2.3 million telephone customers
- 89% of phone network is automatic
- Cable and open wire
- Trunk network is microwave
- Present phone density is 9.85 per 100 residents
- Roughly 3,300 villages with no service (February 1990)
- Broadcast stations - 12 AM, 5 FM, 13 TV (1990)
- 1 satellite ground station using INTELSAT

Transportation [31]

Railroads. 11,275 km total; 10,860 km 1.435-meter gauge, 370 km narrow gauge, 45 km broad gauge; 3,411 km electrified, 3,060 km double track; government owned (1987)

Highways. 72,799 km total; 35,970 km paved; 27,729 km gravel, crushed stone, and other stabilized surfaces; 9,100 km unsurfaced roads (1985)

Merchant Marine. 249 ships (1,000 GRT or over) totaling 2,882,727 GRT/4,463,879 DWT; includes 1 passenger-cargo, 170 cargo, 2 container, 1 rail-car carrier, 9 roll-on/roll-off cargo, 15 oil tanker, 51 bulk

Airports

Total:	158
Usable:	158
With permanent-surface runways:	27
With runways over 3,659 m:	0
With runways 2,440-3,659 m:	21
With runways 1,220-2,439 m:	26

Top Agricultural Products [32]

	90-91	91-92
Corn	6,810	10,498
Wheat	7,300	5,517
Barley	2,680	2,951
Vegetables	2,225	2,112
Potatoes	2,852	1,635
Grapes	954	849

Values shown are 1,000 metric tons.

Top Mining Products [33]

Estimated metric tons unless otherwise specified   M.t.

Bauxite	200,000
Copper, mine output, Cu content	27,000
Coal, all types (000 tons)	40,000
Clay, kaolin	250,000
Steel, crude (000 tons)	7,092[1]
Cement, hydraulic (000 tons)	7,300

Tourism [34]

	1987	1988	1989	1990	1991
Visitors	5,142	5,514	4,852	6,533	5,360
Tourism receipts	176	171	167	106	103
Tourism expenditures	30	33	35	103	114

Tourists are in thousands, money in million U.S. dollars.

Manufacturing Sector

GDP and Manufacturing Summary [35]

	1980	1985	1989	1990	% change 1980-1990	% change 1989-1990
GDP (million 1980 $)	34,272	44,851	50,782[e]	39,901	16.4	-21.4
GDP per capita (1980 $)	1,544	1,974	2,193[e]	1,715	11.1	-21.8
Manufacturing as % of GDP (current prices)	52.7	53.0	NA	48.2	-8.5	NA
Gross output (million $)	34,513[e]	60,429	NA	44,146	27.9	NA
Value added (million $)	14,815[e]	21,752[e]	51,996[e]	15,944	7.6	-69.3
Value added (million 1980 $)	17,333	24,038	31,943[e]	19,351	11.6	-39.4
Industrial production index	100	120	145[e]	107	7.0	-26.2
Employment (thousands)	2,884	3,197	3,630[e]	3,599	24.8	-0.9

Note: GDP stands for Gross Domestic Product. 'e' stands for estimated value.

Profitability and Productivity

	1980	1985	1989	1990	% change 1980-1990	% change 1989-1990
Intermediate input (%)	57[e]	64[e]	NA	64	12.3	NA
Wages, salaries, and supplements (%)	12[e]	12[e]	NA	14	16.7	NA
Gross operating surplus (%)	30[e]	24[e]	NA	22	-26.7	NA
Gross output per worker ($)	11,967[e]	18,902	NA	11,725	-2.0	NA
Value added per worker ($)	6,137[e]	6,804[e]	14,324[e]	4,305	-29.9	-69.9
Average wage (incl. benefits) ($)	1,487[e]	2,247[e]	121[e]	1,728	16.2	1,328.1

Profitability is in percent of gross output. Productivity is in U.S. $. 'e' stands for estimated value.

Profitability - 1990

Inputs - 64.0%
Wages - 14.0%
Surplus - 22.0%

The graphic shows percent of gross output.

Value Added in Manufacturing

	1980 $ mil.	1980 %	1985 $ mil.	1985 %	1989 $ mil.	1989 %	1990 $ mil.	1990 %	% change 1980-1990	% change 1989-1990
311 Food products	1,162[e]	7.8	1,830[e]	8.4	6,852[e]	13.2	1,345	8.4	15.7	-80.4
313 Beverages	536[e]	3.6	853[e]	3.9	1,061[e]	2.0	660	4.1	23.1	-37.8
314 Tobacco products	7[e]	0.0	15[e]	0.1	287[e]	0.6	15	0.1	114.3	-94.8
321 Textiles	1,285[e]	8.7	1,969[e]	9.1	2,680[e]	5.2	1,607	10.1	25.1	-40.0
322 Wearing apparel	726[e]	4.9	1,149[e]	5.3	1,334[e]	2.6	877	5.5	20.8	-34.3
323 Leather and fur products	202[e]	1.4	310[e]	1.4	558[e]	1.1	246	1.5	21.8	-55.9
324 Footwear	226[e]	1.5	357[e]	1.6	497[e]	1.0	275	1.7	21.7	-44.7
331 Wood and wood products	482[e]	3.3	752[e]	3.5	831[e]	1.6	577	3.6	19.7	-30.6
332 Furniture and fixtures	294[e]	2.0	458[e]	2.1	667[e]	1.3	357	2.2	21.4	-46.5
341 Paper and paper products	175[e]	1.2	262[e]	1.2	663[e]	1.3	205	1.3	17.1	-69.1
342 Printing and publishing	39[e]	0.3	66[e]	0.3	80[e]	0.2	72	0.5	84.6	-10.0
351 Industrial chemicals	353[e]	2.4	538[e]	2.5	4,941[e]	9.5	439	2.8	24.4	-91.1
352 Other chemical products	345[e]	2.3	545[e]	2.5	289[e]	0.6	393	2.5	13.9	36.0
353 Petroleum refineries	383[e]	2.6	518[e]	2.4	496[e]	1.0	288	1.8	-24.8	-41.9
354 Miscellaneous petroleum and coal products	36[e]	0.2	61[e]	0.3	35[e]	0.1	43	0.3	19.4	22.9
355 Rubber products	176[e]	1.2	273[e]	1.3	83[e]	0.2	231	1.4	31.3	178.3
356 Plastic products	146[e]	1.0	238[e]	1.1	87[e]	0.2	205	1.3	40.4	135.6
361 Pottery, china and earthenware	76[e]	0.5	114[e]	0.5	262[e]	0.5	98	0.6	28.9	-62.6
362 Glass and glass products	128[e]	0.9	202[e]	0.9	262[e]	0.5	148	0.9	15.6	-43.5
369 Other non-metal mineral products	412[e]	2.8	644[e]	3.0	2,148[e]	4.1	487	3.1	18.2	-77.3
371 Iron and steel	615[e]	4.2	970[e]	4.5	3,410[e]	6.6	693	4.3	12.7	-79.7
372 Non-ferrous metals	205[e]	1.4	278[e]	1.3	985[e]	1.9	225	1.4	9.8	-77.2
381 Metal products	678[e]	4.6	1,058[e]	4.9	3,945[e]	7.6	808	5.1	19.2	-79.5
382 Non-electrical machinery	1,844[e]	12.4	3,093[e]	14.2	11,694[e]	22.5	2,296	14.4	24.5	-80.4
383 Electrical machinery	666[e]	4.5	1,051[e]	4.8	3,910[e]	7.5	829	5.2	24.5	-78.8
384 Transport equipment	2,025[e]	13.7	1,849[e]	8.5	2,720[e]	5.2	913	5.7	-54.9	-66.4
385 Professional and scientific equipment	585[e]	3.9	695[e]	3.2	126[e]	0.2	397	2.5	-32.1	215.1
390 Other manufacturing industries	1,010[e]	6.8	1,607[e]	7.4	1,093[e]	2.1	1,216	7.6	20.4	11.3

Note: The industry codes shown are International Standard Industry codes (ISIC). Percentages are percent of total Value Added. 'e' stands for estimated value

For sources, notes, and explanations, see Annotated Source Appendix, page 1035.

Finance, Economics, and Trade

Economic Indicators [36]

Romanian Lei (RL) or U.S. Dollars as indicated.

	1989	1990	1991[134]
Real GDP (in bil. 1985 lei)[135,136]	839.7	790.8	753.1
Real GDP Growth Rate (%)[137]	-1.9	-7.9	-5.0
Money Supply (M1)[135,136]	276.9	341.3	NA
Commercial Interest Rate (%)[136]	NA	NA	12-20
Savings Rate	NA	NA	NA
Investment Rate	NA	NA	NA
CPI (Oct. 1990 = 100)[135,136]	94	137.7	360.0
WPI[135,136]	100	123.2	NA
External Public Debt (bil. $U.S)[137]	0.8	0.8	3.2

Balance of Payments Summary [37]

Values in millions of dollars.

	1987	1988	1989	1990	1991
Exports of goods (f.o.b.)	10,491	11,392	10,487	5,770	4,125
Imports of goods (f.o.b.)	-8,313	-7,642	-8,437	-9,114	-5,345
Trade balance	2,178	3,750	2,050	-3,344	-1,220
Services - debits	-1,043	-851	-551	-801	-833
Services - credits	908	1,023	1,015	785	747
Private transfers (net)	-	-	-	-	-
Government transfers (net)	-	-	-	106	122
Long term captial (net)	-1,512	-3,619	-1,707	40	294
Short term capital (net)	NA	NA	NA	NA	NA
Errors and omissions	81	16	114	147	34
Overall balance	NA	NA	NA	NA	NA

Exchange Rates [38]

Currency: **lei.**
Symbol: **L.**

Data are currency units per $1.

January 1993	470.100
1992	307.950
1991	76.390
1990	22.432
1989	14.922
1988	14.277

Imports and Exports

Top Import Origins [39]

$5.1 billion (f.o.b., 1991). Data are for 1987.

Origins	%
Communist countries	60
Non-Communist countries	40

Top Export Destinations [40]

$3.5 billion (f.o.b., 1991). Data are for 1987.

Destinations	%
USSR	27
Eastern Europe	23
EC	15
US	5
China	4

Foreign Aid [41]

	U.S. $	
Donor - bilateral aid to non-Communist less developed countries (1956-89)	4.4	billion

Import and Export Commodities [42]

Import Commodities

Fuels
Minerals
And metals 56.0%
Machinery & equipment 25.5%
Agric. and forestry products 8.6%
Consumer goods 3.4%
Other 6.5%

Export Commodities

Machinery & equipment 29.3%
Fuels
Minerals and metals 32.1%
Consumer goods 18.1%
Agric./forestry prod. 9.0%
Other 11.5%

Russia

Geography [1]

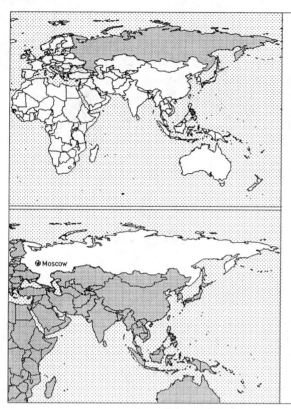

Total area:
17,075,200 km2
Land area:
16,995,800 km2
Comparative area:
Slightly more than 1.8 times the size of the US
Land boundaries:
Total 20,139 km, Azerbaijan 284 km, Belarus 959 km, China (southeast) 3,605 km, China (south) 40 km, Estonia 290 km, Finland 1,313 km, Georgia 723 km, Kazakhstan 6,846 km, North Korea 19 km, Latvia 217 km, Lithuania (Kaliningrad Oblast) 227 km, Mongolia 3,441 km, Norway 167 km, Poland (Kaliningrad Oblast) 432 km, Ukraine 1,576 km
Coastline:
37,653 km
Climate:
Ranges from steppes in the south through humid continental in much of European Russia; subarctic in Siberia to tundra climate in the polar north; winters vary from cool along Black Sea coast to frigid in Siberia; summers vary from warm in the steppes to cool along Arctic coast
Terrain:
Broad plain with low hills west of Urals; vast coniferous forest and tundra in Siberia; uplands and mountains along southern border regions
Land use:
No information available.

Demographics [2]

	1960	1970	1980	1990	1991	1994	2000	2010	2020
Population	119,632	130,245	139,045	148,124	NA	149,609	151,460	155,933	159,263
Population density (persons per sq. mi.)	NA	NA	NA	NA	NA	NA	NA	NA	NA
Births	NA	NA	NA	NA	NA	1,896	NA	NA	NA
Deaths	NA	NA	NA	NA	NA	1,697	NA	NA	NA
Life expectancy - males	NA	NA	NA	NA	NA	64	NA	NA	NA
Life expectancy - females	NA	NA	NA	NA	NA	74	NA	NA	NA
Birth rate (per 1,000)	NA	NA	NA	NA	NA	13	NA	NA	NA
Death rate (per 1,000)	NA	NA	NA	NA	NA	11	NA	NA	NA
Women of reproductive age (15-44 yrs.)	NA	NA	NA	36,026	NA	37,553	39,182	NA	NA
of which are currently married	NA	NA	NA	24,360	NA	25,229	25,887	NA	NA
Fertility rate	NA	NA	NA	2.0	NA	1.8	1.8	1.8	1.7

Population values are in thousands, life expectancy in years, and other items as indicated.

Health

Health Personnel [3]

Doctors per 1,000 pop., 1988-92	4.69
Nurse-to-doctor ratio, 1988-92	NA
Hospital beds per 1,000 pop., 1985-90	13.8
Percentage of children immunized (age 1 yr. or less)	
Third dose of DPT, 1990-91	65.0
Measles, 1990-91	83.0

Health Expenditures [4]

Total health expenditure, 1990 (official exchange rate)	
Millions of dollars	23527
Millions of dollars per capita	157
Health expenditures as a percentage of GDP	
Total	3.0
Public sector	2.0
Private sector	1.0
Development assistance for health	
Total aid flows (millions of dollars)[1]	NA
Aid flows per capita (millions of dollars)	NA
Aid flows as a percentage of total health expenditure	NA

For sources, notes, and explanations, see Annotated Source Appendix, page 1035.

Human Factors

Health Care Ratios [5]

Population per physician, 1970	NA
Population per physician, 1990	210
Population per nursing person, 1970	NA
Population per nursing person, 1990	NA
Percent of births attended by health staff, 1985	NA

Infants and Malnutrition [6]

Percent of babies with low birth weight, 1985	NA
Infant mortality rate per 1,000 live births, 1970	NA
Infant mortality rate per 1,000 live births, 1991	20
Years of life lost per 1,000 population, 1990	17
Prevalence of malnutrition (under age 5), 1990	NA

Ethnic Division [7]

Russian	81.5%
Tatar	3.8%
Ukrainian	3.0%
Chuvash	1.2%
Bashkir	0.9%
Belarusian	0.8%
Moldavian	0.7%
Other	8.1%

Religion [8]

Russian Orthodox, Muslim, other.

Major Languages [9]

Russian, other.

Education

Public Education Expenditures [10]

Million Roubles	1980[18,19]	1985[18,19]	1987[18,19]	1988[18,19]	1989[18,19]	1990[18,19]
Total education expenditure	3,205	NA	3,810	NA	3,540	5,667
as percent of GNP	NA	NA	NA	NA	NA	NA
as percent of total govt. expend.	12.7	NA	4.0	NA	NA	NA
Current education expenditures	2,886	NA	3,150	NA	3,474	5,282
as percent of GNP	NA	NA	NA	NA	NA	NA
as percent of current govt. expend.	14.4	NA	3.7	NA	NA	NA
Capital expenditures	319	NA	660	NA	66	385

Educational Attainment [11]

Age group	25+[13]
Total population	170,405,095[13]
Highest level attained (%)	
No schooling	7.9[13]
First level	
Incompleted	NA[13]
Completed	14.9[13]
Entered second level	
S-1	63.3[13]
S-2	NA[13]
Post secondary	13.9[13]

Literacy Rate [12]

	1980[b]	1989[b]	1990[b]
Illiterate population +15 years	NA	NA	NA
Illiteracy rate - total pop. (%)	NA	1.3	NA
Illiteracy rate - males (%)	NA	0.4	NA
Illiteracy rate - females (%)	NA	2.1	NA

Libraries [13]

	Admin. Units	Svc. Pts.	Vols. (000)	Shelving (meters)	Vols. Added	Reg. Users
National	NA	NA	NA	NA	NA	NA
Non-specialized	NA	NA	NA	NA	NA	NA
Public	NA	NA	NA	NA	NA	NA
Higher ed. (1990)	491	NA	323,763	NA	16 mil	4 mil.
School (1990)	64,263	NA	739,822	NA	97 mil	NA

Daily Newspapers [14]

	1975[d]	1980[d]	1985[d]	1990[d,2]
Number of papers	691	713	727	719
Circ. (000)	100,928	109,089[e]	120,027	137,764

Cinema [15]

Data for 1990.

Cinema seats per 1,000	NA[c]
Annual attendance per person	10.3[c]
Gross box office receipts (mil. Rouble)	NA[c]

Science and Technology

Scientific/Technical Forces [16]

Potential scientists/engineers	15,530,862
Number female	8,470,356
Potential technicians	20,161,646
Number female	13,146,876
Total	35,692,508

R&D Expenditures [17]

U.S. Patents Issued [18]

Values show patents issued to citizens of the country by the U.S. Patents Office.

	1990	1991	1992
Number of patents	177	182	67

Government and Law

Organization of Government [19]

Long-form name:
Russian Federation
Type:
federation
Independence:
24 August 1991 (from Soviet Union)
Constitution:
adopted in 1978; a new constitution is in the process of being drafted
Legal system:
based on civil law system; judicial review of legislative acts; does not accept compulsory ICJ jurisdiction
National holiday:
Independence Day, June 12
Executive branch:
president, vice president, Security Council, Presidential Administration, Council of Ministers, Group of Assistants, Council of Heads of Republics
Legislative branch:
unicameral Congress of People's Deputies, bicameral Supreme Soviet
Judicial branch:
Constitutional Court, Supreme Court

Elections [20]

Congress of People's Deputies. Last held March 1990 (next to be held 1995); results - percent of vote by party NA%; seats - (1,063 total) number of seats by party NA; election held before parties were formed.

Supreme Soviet. last held May 1990 (next to be held 1995); results - percent of vote by party NA%; seats - (252 total) number of seats by party NA; elected from Congress of People's Deputies.

Government Budget [21]

For 1991 est.
Revenues	NA
Expenditures	NA
Capital expenditures	NA

Crime [23]

Crime volume[57]	
Cases known to police	2,786,605
Attempts (percent)	NA
Percent cases solved	1,440,424
Crimes per 100,000 persons	971.90
Persons responsible for offenses[57]	
Total number offenders	1,383,552
Number female	191,031
Number juvenile[57]	221,350
Number foreigners	NA

Military Expenditures and Arms Transfers [22]

	1985	1986	1987	1988	1989
Military expenditures					
Current dollars (mil.)[e]	277,200	287,600	303,000	317,900	311,000
1989 constant dollars (mil.)[e]	315,600	319,200	325,900	330,900	311,000
Armed forces (000)	3,900	3,900	3,900	3,900	3,700
Gross national product (GNP)					
Current dollars (bil.)	2,145	2,275	2,393	2,526	2,664
1989 constant dollars (bil.)	2,442	2,525	2,574	2,630	2,664
Central government expenditures (CGE)					
1989 constant dollars (mil.)	630,700	680,300	709,900	712,500	680,000
People (mil.)	278.9	281.5	284.0	286.4	288.7
Military expenditure as % of GNP[3]	12.9	12.6	12.7	12.6	11.7
Military expenditure as % of CGE[3,5]	50.0	46.9	45.9	46.4	45.7
Military expenditure per capita	1,132	1,134	1,147	1,155	1,077
Armed forces per 1,000 people	14.0	13.9	13.7	13.6	12.8
GNP per capita	8,757	8,970	9,062	9,180	9,226
Arms imports[6]					
Current dollars (mil.)	1,200	1,300	1,300	1,200	900
1989 constant dollars (mil.)	1,366	1,443	1,398	1,249	900
Arms exports[6]					
Current dollars (mil.)[11]	17,100	21,300	22,600	21,600	19,600
1989 constant dollars (mil.)[11]	19,470	23,640	24,310	22,490	19,600
Total imports[7]					
Current dollars (mil.)	83,310	88,870	95,970	107,300	114,700
1989 constant dollars	94,850	98,620	103,200	111,700	114,700
Total exports[7]					
Current dollars (mil.)	87,200	97,050	107,700	110,700	109,300
1989 constant dollars	99,260	107,700	115,800	115,300	109,300
Arms as percent of total imports[8]	1.4	1.5	1.4	1.1	0.8
Arms as percent of total exports[8]	19.6	21.9	21.0	19.5	17.9

Human Rights [24]

	SSTS	FL	FAPRO	PPCG	APROBC	TPW	PCPTW	STPEP	PHRFF	PRW	ASST	AFL
Observes	P	P	P	P	P	P	P	P		P	P	

	EAFRD	CPR	ESCR	SR	ACHR	MAAE	PVIAC	PVNAC	EAFDAW	TCIDTP	RC
Observes	P	P	P			P	P	P	P	P	P

P = Party; S = Signatory; see Appendix for meaning of abbreviations.

Labor Force

Total Labor Force [25]

75 million (1993 est.)

Labor Force by Occupation [26]

Production and economic services	83.9%
Government	16.1

Unemployment Rate [27]

3%-4% of labor force (1 January 1993 est.)

Production Sectors

Commercial Energy Production and Consumption

Production [28]

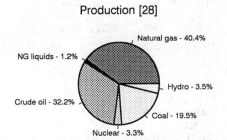

Natural gas - 40.4%
NG liquids - 1.2%
Hydro - 3.5%
Crude oil - 32.2%
Coal - 19.5%
Nuclear - 3.3%

Consumption [29]

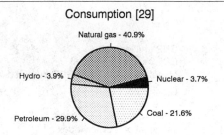

Natural gas - 40.9%
Hydro - 3.9%
Nuclear - 3.7%
Petroleum - 29.9%
Coal - 21.6%

Data are shown in quadrillion (10^{15}) BTUs and percent for 1991

Crude oil[1]	21.22
Natural gas liquids	0.80
Dry natural gas	26.64
Coal[2]	12.852
Net hydroelectric power[3]	2.29
Net nuclear power[3]	2.20
Total	66.00

Petroleum	17.08
Dry natural gas	23.35
Coal[2]	12.34
Net hydroelectric power[3]	2.24
Net nuclear power[3]	2.14
Total	57.15

Telecommunications [30]

- NMT-450 analog cellular phone networks in Moscow, St. Petersburg
- Expanding access to international E-mail service via Sprint networks
- Inadequacy of the telecommunications is a severe handicap to the economy, especially in international communications.
- Installed telephones 24.4 mil.; urban: 20.9, rural: 3.5 mil.
- International traffic handled by a system of satellites, land lines, microwave radio relay and submarine cables; a Raduga satellite will soon link Moscow and St. Petersburg with Rome
- Satellite ground stations - INTELSAT, Intersputnik, Eutelsat (Moscow), INMARSAT, Orbita
- Broadcast stations - 1,050 AM/FM/SW (reach 98.6% of population), 7,183 TV receiving sets - 54.2 mil. TV, 48.8 mil. radio receivers

Top Agricultural Products [32]

	1990	1991
Grains	116.7	89.1
Potatoes	30.8	34.1
Sugar beets	31.1	24.4
Vegetables	10.3	10.1
Sunflowerseed	3.4	2.9
Fruits and berries	2.4	1.7

Values shown are 1,000 metric tons.

Top Mining Products [33]

Estimated metric tons unless otherwise specified   M.t.

Potash ore	50,000
Salt, all types	13,000[1]
Nickel, mine output, Ni content	245
Asbestos, grades I-VII	2,100
Perlite	140
Phosphate rock, crude ore	69,000

Transportation [31]

Railroads. 158,100 km all 1.520-meter broad gauge; 86,800 km in common carrier service, of which 48,900 km are diesel traction and 37,900 km are electric traction; 71,300 km serves specific industry and is not available for common carrier use (31 December 1991)

Highways. 893,000 km total, of which 677,000 km are paved or gravelled and 216,000 km are dirt; 456,000 km are for general use and are maintained by the Russian Highway Corporation (formerly Russian Highway Ministry); the 437,000 km not in general use are the responsibility of various other organizations (formerly ministries); of the 456,000 km in general use, 265,000 km are paved, 140,000 km are gravelled, and 51,000 km are dirt; of the 437,000 km not in general use, 272,000 km are paved or gravelled and 165,000 km are dirt (31 December 1991)

Merchant Marine. 865 ships (1,000 GRT or over) totaling 8,073,954 GRT/11,138,336 DWT; includes 457 cargo, 82 container, 3 multi-function large load carrier, 2 barge carrier, 72 roll-on/roll-off, 124 oil tanker, 25 bulk cargo, 9 chemical tanker, 2 specialized tanker, 16 combination ore/oil, 5 passenger cargo, 18 short-sea passenger, 6 passenger, 28 combination bulk, 16 refrigerated cargo

Airports

Total:	2,550
Useable:	964
With permanent surface runways:	565
With runways over 3,659 m:	19
With runways 2,440-3,659 m:	275
With runways 1,220-2,439 m:	426

Tourism [34]

	1987	1988	1989	1990	1991
Visitors[73]	5,246	6,007	7,752	7,204	6,895
Tourists[73]	2,250	2,458	2,740	2,286	2,235

Tourists are in thousands, money in million U.S. dollars.

Manufacturing Sector

GDP and Manufacturing Summary [35]

	1980	1985	1989	1990	% change 1980-1990	% change 1989-1990
GDP (million 1980 $)	892,879	1,044,668	1,212,263	1,132,755	26.9	-6.6
GDP per capita (1980 $)	3,362	3,764	4,232	3,923	16.7	-7.3
Manufacturing as % of GDP (current prices)	45.7	40.0	37.5[e]	34.9	-23.6	-6.9
Gross output (million $)	834,089	867,602	1,295,101	568,264[e]	-31.9	-56.1
Value added (million $)	362,425	436,103	501,622	502,119	38.5	0.1
Value added (million 1980 $)	404,805	464,867	564,471	524,952	29.7	-7.0
Industrial production index	100	120	138	139	39.0	0.7
Employment (thousands)	31,464	32,794	31,207	30,596[e]	-2.8	-2.0

Note: GDP stands for Gross Domestic Product. 'e' stands for estimated value.

Profitability and Productivity

	1980	1985	1989	1990	% change 1980-1990	% change 1989-1990
Intermediate input (%)	NA	NA	NA	NA	NA	NA
Wages, salaries, and supplements (%)	12	11	NA	12[e]	0.0	NA
Gross operating surplus (%)	NA	NA	NA	NA	NA	NA
Gross output per worker ($)	26,509	26,456	41,500	18,515[e]	-30.2	-55.4
Value added per worker ($)	11,519	13,298	16,074	16,512[e]	43.3	2.7
Average wage (incl. benefits) ($)	3,247	3,002	4,836	2,159[e]	-33.5	-55.4

Profitability is in percent of gross output. Productivity is in U.S. $. 'e' stands for estimated value.

Profitability - 1990

Value Added in Manufacturing

	1980 $ mil.	1980 %	1985 $ mil.	1985 %	1989 $ mil.	1989 %	1990 $ mil.	1990 %	% change 1980-1990	% change 1989-1990
311 Food products	66,053	18.2	75,960	17.4	85,208	17.0	85,868	17.1	30.0	0.8
313 Beverages	10,336	2.9	9,303	2.1	8,889	1.8	8,786	1.7	-15.0	-1.2
314 Tobacco products	2,032	0.6	2,866	0.7	2,398	0.5	2,398	0.5	18.0	0.0
321 Textiles	32,553	9.0	34,506	7.9	38,086	7.6	37,435	7.5	15.0	-1.7
322 Wearing apparel	19,633	5.4	21,792	5.0	23,559	4.7	24,345	4.8	24.0	3.3
323 Leather and fur products	2,443	0.7	2,345	0.5	2,345	0.5	2,345	0.5	-4.0	0.0
324 Footwear	3,892	1.1	4,593	1.1	5,371	1.1	5,488	1.1	41.0	2.2
331 Wood and wood products	4,932	1.4	5,771	1.3	6,560	1.3	6,412	1.3	30.0	-2.3
332 Furniture and fixtures	3,457	1.0	4,459	1.0	5,427	1.1	5,669	1.1	64.0	4.5
341 Paper and paper products	2,784	0.8	3,424	0.8	3,981	0.8	4,065	0.8	46.0	2.1
342 Printing and publishing	2,613	0.7	3,214[e]	0.7	3,736	0.7	3,815[e]	0.8	46.0	2.1
351 Industrial chemicals	14,704	4.1	17,939	4.1	20,144	4.0	19,703	3.9	34.0	-2.2
352 Other chemical products	7,584	2.1	8,419	1.9	9,632	1.9	9,632	1.9	27.0	0.0
353 Petroleum refineries	5,490	1.5	6,093	1.4	6,972	1.4	6,972	1.4	27.0	0.0
354 Miscellaneous petroleum and coal products	11,003	3.0	12,214	2.8	13,974	2.8	13,974	2.8	27.0	0.0
355 Rubber products	4,154	1.1	4,861	1.1	5,401	1.1	5,276	1.1	NA	-2.3
356 Plastic products	1,546	0.4	2,273	0.5	2,969	0.6	2,969	0.6	92.0	0.0
361 Pottery, china and earthenware	2,014	0.6	2,457	0.6	2,860	0.6	3,001	0.6	49.0	4.9
362 Glass and glass products	1,204	0.3	1,517	0.3	1,878	0.4	1,914	0.4	59.0	1.9
369 Other non-metal mineral products	13,769	3.8	15,696	3.6	18,037	3.6	17,761	3.5	29.0	-1.5
371 Iron and steel	14,418	4.0	15,139	3.5	15,860	3.2	15,139	3.0	5.0	-4.5
372 Non-ferrous metals	7,716	2.1	8,333	1.9	8,256	1.6	7,793	1.6	1.0	-5.6
381 Metal products	7,130	2.0	9,625	2.2	11,693	2.3	11,764	2.3	65.0	0.6
382 Non-electrical machinery	79,367	21.9	107,146	24.6	130,162	25.9	130,956	26.1	65.0	0.6
383 Electrical machinery	9,105	2.5	12,292	2.8	14,932	3.0	15,023	3.0	65.0	0.6
384 Transport equipment	11,574	3.2	15,625	3.6	18,982	3.8	19,097	3.8	65.0	0.6
385 Professional and scientific equipment	9,711	2.7	13,110	3.0	15,927	3.2	16,024	3.2	65.0	0.6
390 Other manufacturing industries	11,210	3.1	15,133[e]	3.5	18,384	3.7	18,496[e]	3.7	65.0	0.6

Note: The industry codes shown are International Standard Industry codes (ISIC). Percentages are percent of total Value Added. 'e' stands for estimated value

For sources, notes, and explanations, see Annotated Source Appendix, page 1035.

Finance, Economics, and Trade

Economic Indicators [36]

Rubles or U.S. Dollars as noted.

	1989	1990	1991[e]
GNP (billion 1990 $)[138]	2,760	2,660	NA
Real GNP growth (%)[139]	1.5	-2.4-5.0	-13.4[140]
GNP per capita (1990 $)	9,500	9,130	NA
Money supply (billion rubles)	614.0	733.0	NA
Savings rate (% natl. income)	47.6	NA	NA
CPI (U.S. estimate)	131	149	NA

Balance of Payments Summary [37]

Exchange Rates [38]

Currency: **rubles.**
Symbol: **R.**

Subject to wide fluctuations. Data are currency units per $1.

December 24, 1994	415

Imports and Exports

Top Import Origins [39]

$35.0 billion (f.o.b., 1992).

Origins	%
Europe	NA
North America	NA
Japan	NA
Third World countries	NA
Cuba	NA

Top Export Destinations [40]

$39.2 billion (f.o.b., 1992).

Destinations	%
Europe	NA

Foreign Aid [41]

	U.S. $	
US commitments, including Ex-Im (1990-92)	9.0	billion
Other countries, ODA and OOF bilateral commitments (1988-92)	91	billion

Import and Export Commodities [42]

Import Commodities	**Export Commodities**
Machinery & equipment	Petroleum & petr. products
Chemicals	Natural gas
Consumer goods	Wood & wood products
Grain	Metals
Meat	Chemicals
Sugar	Civilian manufactures
Semifinished metal products	Military manufactures

For sources, notes, and explanations, see Annotated Source Appendix, page 1035.

763

Rwanda

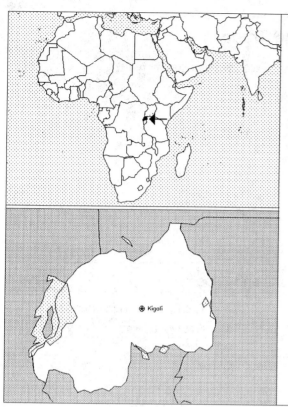

Geography [1]

Total area:
26,340 km2
Land area:
24,950 km2
Comparative area:
Slightly smaller than Maryland
Land boundaries:
Total 893 km, Burundi 290 km, Tanzania 217 km, Uganda 169 km, Zaire 217 km
Coastline:
0 km (landlocked)
Climate:
Temperate; two rainy seasons (February to April, November to January); mild in mountains with frost and snow possible
Terrain:
Mostly grassy uplands and hills; mountains in west
Natural resources:
Gold, cassiterite (tin ore), wolframite (tungsten ore), natural gas, hydropower
Land use:
Arable land:
29%
Permanent crops:
11%
Meadows and pastures:
18%
Forest and woodland:
10%
Other:
32%

Demographics [2]

	1960	1970	1980	1990	1991[1]	1994	2000	2010	2020
Population	3,083	3,813	5,170	7,415	7,903	8,374	9,715	11,755	15,006
Population density (persons per sq. mi.)	323	402	548	790	820	NA	1,147	1,639	2,278
Births	NA	NA	NA	NA	414	412	NA	NA	NA
Deaths	NA	NA	NA	NA	116	179	NA	NA	NA
Life expectancy - males	NA	NA	NA	NA	51	39	NA	NA	NA
Life expectancy - females	NA	NA	NA	NA	54	41	NA	NA	NA
Birth rate (per 1,000)	NA	NA	NA	NA	52	49	NA	NA	NA
Death rate (per 1,000)	NA	NA	NA	NA	15	21	NA	NA	NA
Women of reproductive age (15-44 yrs.)	NA	NA	NA	1,540	NA	1,745	2,070	NA	NA
of which are currently married	NA	NA	NA	1,029	NA	1,131	1,288	NA	NA
Fertility rate	NA	NA	NA	8.5	8.42	8.2	7.6	6.4	5.0

Population values are in thousands, life expectancy in years, and other items as indicated.

Health

Health Personnel [3]

Doctors per 1,000 pop., 1988-92	0.02
Nurse-to-doctor ratio, 1988-92	1.7
Hospital beds per 1,000 pop., 1985-90	1.7
Percentage of children immunized (age 1 yr. or less)	
Third dose of DPT, 1990-91	89.0
Measles, 1990-91	89.0

Health Expenditures [4]

Total health expenditure, 1990 (official exchange rate)	
Millions of dollars	74
Millions of dollars per capita	10
Health expenditures as a percentage of GDP	
Total	3.5
Public sector	1.9
Private sector	1.6
Development assistance for health	
Total aid flows (millions of dollars)[1]	29
Aid flows per capita (millions of dollars)	4.1
Aid flows as a percentage of total health expenditure	39.5

For sources, notes, and explanations, see Annotated Source Appendix, page 1035.

Human Factors

Health Care Ratios [5]

Population per physician, 1970	59600
Population per physician, 1990	72990
Population per nursing person, 1970	5610
Population per nursing person, 1990	4190
Percent of births attended by health staff, 1985	NA

Infants and Malnutrition [6]

Percent of babies with low birth weight, 1985	17
Infant mortality rate per 1,000 live births, 1970	142
Infant mortality rate per 1,000 live births, 1991	111
Years of life lost per 1,000 population, 1990	124
Prevalence of malnutrition (under age 5), 1990	33

Ethnic Division [7]

Hutu	90%
Tutsi	9%
Twa (Pygmoid)	1%

Religion [8]

Roman Catholic	65%
Protestant	9%
Muslim	1%
Indigenous beliefs and other	25%

Major Languages [9]

Kinyarwanda (official), French (official), Kiswahili used in commercial centers.

Education

Public Education Expenditures [10]

Million Francs	1980	1986	1987	1988	1989	1990
Total education expenditure	2,880	5,737	6,010	6,138	7,222	NA
as percent of GNP	2.7	3.4	3.5	3.5	4.2	NA
as percent of total govt. expend.	21.6	22.6	22.9	22.2	25.4	NA
Current education expenditure	2,439	NA	5,642	5,653	6,793	NA
as percent of GNP	2.3	NA	3.3	3.2	3.9	NA
as percent of current govt. expend.	21.5	NA	25.0	25.0	29.1	NA
Capital expenditure	442	NA	368	485	429	NA

Educational Attainment [11]

Literacy Rate [12]

In thousands and percent	1985[a]	1991[a]	2000[a]
Illiterate population +15 years	1,701	1,838	2,149
Illiteracy rate - total pop. (%)	54.6	49.8	40.9
Illiteracy rate - males (%)	40.7	36.1	28.5
Illiteracy rate - females (%)	67.9	62.9	52.8

Libraries [13]

	Admin. Units	Svc. Pts.	Vols. (000)	Shelving (meters)	Vols. Added	Reg. Users
National	NA	NA	NA	NA	NA	NA
Non-specialized	NA	NA	NA	NA	NA	NA
Public	NA	NA	NA	NA	NA	Na
Higher ed. (1987)[7]	3	4	163	NA	6,416	1,901
School	NA	NA	NA	NA	NA	NA

Daily Newspapers [14]

	1975	1980	1985	1990
Number of papers	1	1	1	1
Circ. (000)	0.2	0.3	0.3	0.5

Cinema [15]

Science and Technology

Scientific/Technical Forces [16]

R&D Expenditures [17]

	Franc (000) 1985
Total expenditure	918,560
Capital expenditure	819,280
Current expenditure	99,280
Percent current	10.8

U.S. Patents Issued [18]

For sources, notes, and explanations, see Annotated Source Appendix, page 1035.

765

Government and Law

Organization of Government [19]

Long-form name:
Republic of Rwanda
Type:
republic; presidential system
Independence:
1 July 1962 (from UN trusteeship under Belgian administration)
Constitution:
18 June 1991
Legal system:
based on German and Belgian civil law systems and customary law; judicial review of legislative acts in the Supreme Court; has not accepted compulsory ICJ jurisdiction
National holiday:
Independence Day, 1 July (1962)
Executive branch:
president, prime minister, Council of Ministers (cabinet)
Legislative branch:
unicameral National Development Council (Conseil National de Developpement)
Judicial branch:
Constitutional Court (consists of the Court of Cassation and the Council of State in joint session)

Political Parties [20]

National Development Council	% of seats
Republican National Movement (MRND)	100.0

Government Budget [21]

For 1992 est.
Revenues	350.0
Expenditures	453.7
Capital expenditures	NA

Crime [23]

Crime volume	
Cases known to police	22,891
Attempts (percent)	NA
Percent cases solved	NA
Crimes per 100,000 persons	327.01
Persons responsible for offenses	
Total number offenders	29,464
Percent female	7.2
Percent juvenile[46]	0.6
Percent foreigners	0.8

Military Expenditures and Arms Transfers [22]

	1985	1986	1987	1988	1989
Military expenditures					
Current dollars (mil.)	33[e]	NA	42	35[e]	NA
1989 constant dollars (mil.)	38[e]	NA	45	37[e]	NA
Armed forces (000)	5	5	5	5	6
Gross national product (GNP)					
Current dollars (mil.)	1,928	2,084	2,144	2,218	2,160
1989 constant dollars (mil.)	2,195	2,313	2,305	2,309	2,160
Central government expenditures (CGE)					
1989 constant dollars (mil.)	404[e]	NA	377	365	355
People (mil.)	6.3	6.6	6.8	7.1	7.3
Military expenditure as % of GNP	1.7	NA	2.0	1.6	NA
Military expenditure as % of CGE	9.4	NA	12.0	10.0	NA
Military expenditure per capita	6	NA	7	5	NA
Armed forces per 1,000 people	0.8	0.8	0.7	0.7	0.8
GNP per capita	346	352	338	327	295
Arms imports[6]					
Current dollars (mil.)	0	0	0	0	20
1989 constant dollars (mil.)	0	0	0	0	20
Arms exports[6]					
Current dollars (mil.)	0	0	0	0	0
1989 constant dollars (mil.)	0	0	0	0	0
Total imports[7]					
Current dollars (mil.)	298	349	352	370	333
1989 constant dollars	339	387	379	385	333
Total exports[7]					
Current dollars (mil.)	131	189	114	108	95
1989 constant dollars	149	210	123	112	95
Arms as percent of total imports[8]	0	0	0	0	6.0
Arms as percent of total exports[8]	0	0	0	0	0

Human Rights [24]

	SSTS	FL	FAPRO	PPCG	APROBC	TPW	PCPTW	STPEP	PHRFF	PRW	ASST	AFL
Observes		P	P	P	P	P	P					P
	EAFRD	CPR	ESCR	SR	ACHR	MAAE	PVIAC	PVNAC	EAFDAW	TCIDTP	RC
Observes	P	P	P	P		P	P	P	P		P

P = Party; S = Signatory; see Appendix for meaning of abbreviations.

Labor Force

Total Labor Force [25]

3.6 million

Labor Force by Occupation [26]

Agriculture	93%
Government and services	5
Industry and commerce	2

Unemployment Rate [27]

Production Sectors

Energy Resource Summary [28]

Energy Resources: Natural gas, hydropower. **Electricity**: 30,000 kW capacity; 130 million kWh produced, 15 kWh per capita (1991).

Telecommunications [30]

- Fair system with low-capacity radio relay system centered on Kigali
- Broadcast stations - 2 AM, 1 (7 repeaters) FM, no TV
- Satellite earth stations - 1 Indian Ocean INTELSAT and 1 SYMPHONIE

Transportation [31]

Highways. 4,885 km total; 460 km paved, 1,725 km gravel and/or improved earth, 2,700 km unimproved

Airports

Total:	8
Usable:	7
With permanent-surface runways:	3
With runways over 3,659 m:	0
With runways 2,440-3,659 m:	1
With runways 1,220-2,439 m:	2

Top Agricultural Products [32]

Agriculture accounts for almost 50% of GDP and about 90% of the labor force; cash crops - coffee, tea, pyrethrum (insecticide made from chrysanthemums); main food crops—bananas, beans, sorghum, potatoes; stock raising; self-sufficiency declining; country imports foodstuffs as farm production fails to keep up with a 3.8% annual growth in population.

Top Mining Products [33]

Estimated metric tons unless otherwise specified   M.t.

Cement	60,000
Columbite-tantalite, ore and concentrate	150[48]
Gold, mine output, Au content (kilograms)	700[49]
Gas, natural (000 cubic meters)	970
Tin, mine output, Sn content	730[50]
Tungsten, mine output, W content	175[51]

Tourism [34]

Finance, Economics, and Trade

GDP and Manufacturing Summary [35]

	1980	1985	1990	1991	1992
Gross Domestic Product					
Millions of 1980 dollars	1,163	1,347	1,365	1,324	1,073
Growth rate in percent	6.01	4.41	-1.65	-2.96	-19.00
Manufacturing Value Added					
Millions of 1980 dollars	178	202	233	215	NA
Growth rate in percent	12.30	6.96	-4.00	-8.00	NA
Manufacturing share in percent of current prices	15.8	14.2	15.2	14.8[e]	NA

Economic Indicators [36]

National product: GDP—exchange rate conversion—$2.35 billion (1992 est.). **National product real growth rate**: 1.3% (1992 est.). **National product per capita**: $290 (1992 est.). **Inflation rate (consumer prices)**: 6% (1992 est.). **External debt**: $911 million (1990 est.).

Balance of Payments Summary [37]

Values in millions of dollars.

	1987	1988	1989	1990	1991
Exports of goods (f.o.b.)	121.4	117.9	104.7	102.6	95.6
Imports of goods (f.o.b.)	-267.0	-278.6	-254.1	-227.7	-228.1
Trade balance	-145.6	-160.7	-149.4	-125.1	-132.5
Services - debits	-171.3	-165.0	-142.0	-152.0	-128.8
Services - credits	56.4	56.9	52.3	46.6	46.5
Private transfers (net)	7.5	10.5	7.9	5.8	20.9
Government transfers (net)	118.6	139.3	129.2	138.2	159.8
Long term capital (net)	110.9	87.6	59.6	47.0	79.8
Short term capital (net)	11.4	6.1	-5.7	8.7	19.3
Errors and omissions	1.5	0.4	1.9	30.3	0.2
Overall balance	-10.6	-24.9	-46.2	-0.5	65.2

Exchange Rates [38]

Currency: **Rwandan francs.**
Symbol: **RF.**

Data are currency units per $1.

January 1993	146.34
1992	133.35
1991	125.14
1990	82.60
1989	79.98
1988	76.45

Imports and Exports

Top Import Origins [39]

$259.5 million (f.o.b., 1992 est.).

Origins	%
US	NA
Belgium	NA
Germany	NA
Kenya	NA
Japan	NA

Top Export Destinations [40]

$66.6 million (f.o.b., 1992 est.).

Destinations	%
Germany	NA
Belgium	NA
Italy	NA
Uganda	NA
UK	NA
France	NA
US	NA

Foreign Aid [41]

Note - In October 1990 Rwanda launched a Structural Adjustment Program with the IMF; since September 1991, the EC has given $46 million and the US $25 million in support of this program.

	U.S. $	
US commitments, including Ex-Im (FY70-89)	128	million
Western (non-US) countries, ODA and OOF bilateral commitments (1970-89)	2.0	billion
OPEC bilateral aid (1979-89)	45	million
Communist countries (1970-89)	58	million

Import and Export Commodities [42]

Import Commodities	Export Commodities
Textiles	Coffee 85%
Foodstuffs	Tea
Machines and equipment	Tin
Capital goods	Cassiterite
Steel	Wolframite
Petroleum products	Pyrethrum
Cement and construction material	

Saint Kitts and Nevis

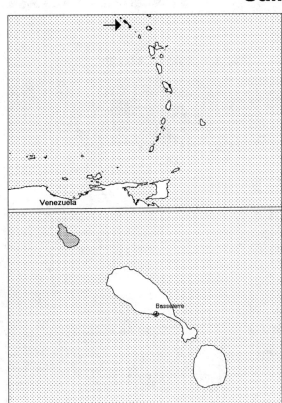

Geography [1]

Total area:
269 km2
Land area:
269 km2
Comparative area:
Slightly more than 1.5 times the size of Washington, DC
Land boundaries:
0 km
Coastline:
135 km
Climate:
Subtropical tempered by constant sea breezes; little seasonal temperature variation; rainy season (May to November)
Terrain:
Volcanic with mountainous interiors
Natural resources:
Negligible
Land use:
Arable land:
22%
Permanent crops:
17%
Meadows and pastures:
3%
Forest and woodland:
17%
Other:
41%

Demographics [2]

	1960	1970	1980	1990	1991[1]	1994	2000	2010	2020
Population	51	46	44	40	40	41	43	50	57
Population density (persons per sq. mi.)	368	332	318	289	290	NA	315	366	418
Births	NA	NA	NA	NA	1	1	NA	NA	NA
Deaths	NA	NA	NA	NA	(Z)	(Z)	NA	NA	NA
Life expectancy - males	NA	NA	NA	NA	64	63	NA	NA	NA
Life expectancy - females	NA	NA	NA	NA	71	69	NA	NA	NA
Birth rate (per 1,000)	NA	NA	NA	NA	24	24	NA	NA	NA
Death rate (per 1,000)	NA	NA	NA	NA	10	10	NA	NA	NA
Women of reproductive age (15-44 yrs.)	NA	NA	NA	9	NA	10	12	NA	NA
of which are currently married	NA	NA	NA	2	NA	2	3	NA	NA
Fertility rate	NA	NA	NA	2.8	2.62	2.6	2.4	2.1	2.0

Population values are in thousands, life expectancy in years, and other items as indicated.

Health

Health Personnel [3]

Health Expenditures [4]

For sources, notes, and explanations, see Annotated Source Appendix, page 1035.

769

Human Factors

Health Care Ratios [5]	Infants and Malnutrition [6]

Ethnic Division [7]	Religion [8]	Major Languages [9]
Black African.	Anglican, other Protestant sects, Roman Catholic.	English.

Education

Public Education Expenditures [10]

Million E. Carribbean Dollars	1980	1985	1987[6]	1988[6]	1989[6]	1990
Total education expenditure	7	12	11	10	12	NA
as percent of GNP	5.2	6.0	3.7	3.0	3.2	NA
as percent of total govt. expend.	9.4	18.5	14.6	12.5	12.0	NA
Current education expenditure	7	12	10	10	11	NA
as percent of GNP	5.2	6.0	3.7	3.0	3.1	NA
as percent of current govt. expend.	13.6	19.1	14.4	NA	12.3	NA
Capital expenditure	0.0	0.0	1	0.0	1	NA

Educational Attainment [11]

Age group	25+
Total population	16,695
Highest level attained (%)	
No schooling	1.1
First level	
Incompleted	29.6
Completed	NA
Entered second level	
S-1	67.2
S-2	NA
Post seocondary	2.1

Literacy Rate [12]

	1970[b,3]	1980[b,3]	1990[b,3]
Illiterate population +15 years	546	NA	NA
Illiteracy rate - total pop. (%)	2.4	NA	NA
Illiteracy rate - males (%)	2.4	NA	NA
Illiteracy rate - females (%)	2.3	NA	NA

Libraries [13]

Daily Newspapers [14]

Cinema [15]

Science and Technology

Scientific/Technical Forces [16]	R&D Expenditures [17]	U.S. Patents Issued [18]

     For sources, notes, and explanations, see Annotated Source Appendix, page 1035.

Government and Law

Organization of Government [19]

Long-form name:
Federation of Saint Kitts and Nevis
Type:
constitutional monarchy
Independence:
19 September 1983 (from UK)
Constitution:
19 September 1983
Legal system:
based on English common law
National holiday:
Independence Day, 19 September (1983)
Executive branch:
British monarch, governor general, prime minister, deputy prime minister, Cabinet
Legislative branch:
unicameral House of Assembly
Judicial branch:
Eastern Caribbean Supreme Court

Political Parties [20]

House of Assembly	% of seats
People's Action Movement (PAM)	42.9
Labor Party (SKNLP)	14.3
Nevis Reformation Party (NRP)	14.3
Concerned Citizens Movement (CCM)	7.1

Government Budget [21]

For 1993.
Revenues	85.7
Expenditures	85.8
Capital expenditures	42.4

Defense Summary [22]

Branches: Royal Saint Kitts and Nevis Police Force, Coast Guard

Manpower Availability: No information available.

Defense Expenditures: No information available.

Crime [23]

Human Rights [24]

	SSTS	FL	FAPRO	PPCG	APROBC	TPW	PCPTW	STPEP	PHRFF	PRW	ASST	AFL
Observes	1					P	P			1	1	
	EAFRD	CPR	ESCR	SR	ACHR	MAAE	PVIAC	PVNAC	EAFDAW	TCIDTP	RC	
Observes							P	P	P		P	

P = Party; S = Signatory; see Appendix for meaning of abbreviations.

Labor Force

Total Labor Force [25]

20,000 (1981)

Labor Force by Occupation [26]

Unemployment Rate [27]

12.2% (1990)

Production Sectors

Energy Resource Summary [28]

Energy Resources: None. **Electricity**: 15,800 kW capacity; 45 million kWh produced, 1,120 kWh per capita (1992).

Telecommunications [30]

- Good interisland VHF/UHF/SHF radio connections and international link via Antigua and Barbuda and Saint Martin
- 2,400 telephones
- Broadcast stations - 2 AM, no FM, 4 TV

Transportation [31]

Railroads. 58 km 0.760-meter gauge on Saint Kitts for sugarcane

Highways. 300 km total; 125 km paved, 125 km otherwise improved, 50 km unimproved earth

Airports

Total:	2
Usable:	2
With permanent-surface runways:	2
With runways over 3,659 m:	0
With runways 2,440-3,659 m:	1
With runways 1,220-2,439 m:	0

Top Agricultural Products [32]

Agriculture accounts for 7% of GDP; cash crop—sugarcane; subsistence crops - rice, yams, vegetables, bananas; fishing potential not fully exploited; most food imported.

Top Mining Products [33]

Detailed information is not available. A summary of mineral resources available follows. **Mineral Resources**: Negligible.

Tourism [34]

	1987	1988	1989	1990	1991
Visitors	97	124	109	110	137
Tourists[57]	66	70	72	76	84
Cruise passengers[58]	31	54	37	34	53
Tourism receipts	47	54	60	63	74
Tourism expenditures	3	3	3	4	4
Fare expenditures	3	4	3	3	5

Tourists are in thousands, money in million U.S. dollars.

Finance, Economics, and Trade

Industrial Summary [35]

Industrial Production: Growth rate 11.8% (1988 est.); accounts for 11% of GDP. **Industries**: Sugar processing, tourism, cotton, salt, copra, clothing, footwear, beverages.

Economic Indicators [36]

National product: GDP—exchange rate conversion— $142 million (1991). **National product real growth rate**: 6.8% (1991). **National product per capita**: $3,500 (1991). **Inflation rate (consumer prices)**: 4.2% (1991). **External debt**: $37.2 million (1990).

Balance of Payments Summary [37]

Values in millions of dollars.

	1987	1988	1989	1990	1991
Exports of goods (f.o.b.)	29.8	29.1	32.8	24.6	21.2
Imports of goods (f.o.b.)	-70.0	-81.6	-89.7	-103.2	-97.6
Trade balance	-40.2	-52.5	-56.9	-78.6	-76.4
Services - debits	-21.6	-25.5	-35.7	-42.3	-41.6
Services - credits	40.9	46.8	49.2	56.9	62.5
Private transfers (net)	9.2	9.9	10.5	12.5	6.4
Government transfers (net)	3.1	3.7	4.7	0.7	-0.6
Long term capital (net)	14.5	20.2	35.9	49.3	47.7
Short term capital (net)	-7.4	-2.3	6.7	-1.1	3.7
Errors and omissions	2.3	-0.2	-8.0	2.7	-1.1
Overall balance	0.8	0.1	6.4	0.1	0.6

Exchange Rates [38]

Currency: **East Caribbean dollars.**
Symbol: **EC$.**

Data are currency units per $1.

Fixed rate since 1976	2.70

Imports and Exports

Top Import Origins [39]

$103.2 million (f.o.b., 1990). Data are for 1988.

Origins	%
US	36
UK	17
Trinidad and Tobago	6
Canada	3
Japan	3
OECS	4

Top Export Destinations [40]

$24.6 million (f.o.b., 1990). Data are for 1988.

Destinations	%
US	53
UK	22
Trinidad and Tobago	5
OECS	5

Foreign Aid [41]

	U.S. $	
US commitments, including Ex-Im (FY85-88)	10.7	million
Western (non-US) countries, ODA and OOF bilateral commitments (1970-89)	67	million

Import and Export Commodities [42]

Import Commodities	Export Commodities
Foodstuffs	Sugar
Intermediate manufactures	Clothing
Machinery	Electronics
Fuels	Postage stamps

For sources, notes, and explanations, see Annotated Source Appendix, page 1035.

773

Saint Lucia

Geography [1]

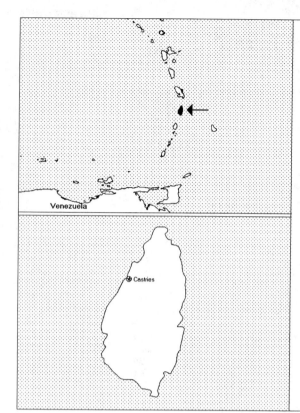

Total area:
620 km2
Land area:
610 km2
Comparative area:
Slightly less than 3.5 times the size of Washington, DC
Land boundaries:
0 km
Coastline:
158 km
Climate:
Tropical, moderated by northeast trade winds; dry season from January to April, rainy season from May to August
Terrain:
Volcanic and mountainous with some broad, fertile valleys
Natural resources:
Forests, sandy beaches, minerals (pumice), mineral springs, geothermal potential
Land use:
Arable land:
8%
Permanent crops:
20%
Meadows and pastures:
5%
Forest and woodland:
13%
Other:
54%

Demographics [2]

	1960	1970	1980	1990	1991[1]	1994	2000	2010	2020
Population	88	103	122	142	153	145	151	169	193
Population density (persons per sq. mi.)	372	432	519	635	649	NA	788	957	1,139
Births	NA	NA	NA	NA	5	3	NA	NA	NA
Deaths	NA	NA	NA	NA	1	1	NA	NA	NA
Life expectancy - males	NA	NA	NA	NA	69	67	NA	NA	NA
Life expectancy - females	NA	NA	NA	NA	74	72	NA	NA	NA
Birth rate (per 1,000)	NA	NA	NA	NA	31	23	NA	NA	NA
Death rate (per 1,000)	NA	NA	NA	NA	5	6	NA	NA	NA
Women of reproductive age (15-44 yrs.)	NA	NA	NA	35	NA	38	42	NA	NA
of which are currently married	NA	NA	NA	18	NA	20	23	NA	NA
Fertility rate	NA	NA	NA	2.8	3.46	2.5	2.2	2.0	2.0

Population values are in thousands, life expectancy in years, and other items as indicated.

Health

Health Personnel [3]

Health Expenditures [4]

For sources, notes, and explanations, see Annotated Source Appendix, page 1035.

Human Factors

Health Care Ratios [5]	Infants and Malnutrition [6]

Ethnic Division [7]

African descent	90.3%
Mixed	5.5%
East Indian	3.2%
Caucasian	0.8%

Religion [8]

Roman Catholic	90%
Protestant	7%
Anglican	3%

Major Languages [9]

English (official), French patois.

Education

Public Education Expenditures [10]

Million E. Carribbean Dollars	1980	1986	1987	1988	1989	1990
Total education expenditure	NA	38	NA	NA	NA	NA
as percent of GNP	NA	5.7	NA	NA	NA	NA
as percent of total govt. expend.	NA	NA	NA	NA	NA	NA
Current education expenditure	19	36	37	42	NA	NA
as percent of GNP	6.9	5.4	5.0	5.0	NA	NA
as percent of current govt. expend.	NA	NA	NA	NA	NA	NA
Capital expenditure	NA	2	NA	NA	NA	NA

Educational Attainment [11]

Age group	25+
Total population	39,599
Highest level attained (%)	
No schooling	17.5
First level	
Incompleted	74.5
Completed	NA
Entered second level	
S-1	6.8
S-2	NA
Post secondary	1.3

Literacy Rate [12]

	1970[b,3]	1980[b,3]	1990[b,3]
Illiterate population +15 years	9,195	NA	NA
Illiteracy rate - total pop. (%)	18.3	NA	NA
Illiteracy rate - males (%)	19.2	NA	NA
Illiteracy rate - females (%)	17.6	NA	NA

Libraries [13]

Daily Newspapers [14]

Cinema [15]

Science and Technology

Scientific/Technical Forces [16]

R&D Expenditures [17]

	Dollar (000) 1984
Total expenditure	12,150
Capital expenditure	NA
Current expenditure	NA
Percent current	NA

U.S. Patents Issued [18]

For sources, notes, and explanations, see Annotated Source Appendix, page 1035.

775

Government and Law

Organization of Government [19]

Long-form name:
none
Type:
parliamentary democracy
Independence:
22 February 1979 (from UK)
Constitution:
22 February 1979
Legal system:
based on English common law
National holiday:
Independence Day, 22 February (1979)
Executive branch:
British monarch, governor general, prime minister, Cabinet
Legislative branch:
bicameral Parliament consists of an upper house or Senate and a lower house or House of Assembly
Judicial branch:
Eastern Caribbean Supreme Court

Political Parties [20]

House of Assembly	% of seats
United Workers' Party (UWP)	64.7
Saint Lucia Labor Party (SLP)	35.3

Government Budget [21]

For FY90 est.

Revenues	131
Expenditures	149
Capital expenditures	71

Defense Summary [22]

Branches: Royal Saint Lucia Police Force, Coast Guard

Manpower Availability: No information available.

Defense Expenditures: No information available.

Crime [23]

Human Rights [24]

	SSTS	FL	FAPRO	PPCG	APROBC	TPW	PCPTW	STPEP	PHRFF	PRW	ASST	AFL
Observes	P	P	P	S	P	P	P			P	P	P
	EAFRD	CPR	ESCR	SR	ACHR	MAAE	PVIAC	PVNAC	EAFDAW	TCIDTP	RC	
Observes		P	1		1			P	P	P		S

P = Party; S = Signatory; see Appendix for meaning of abbreviations.

Labor Force

Total Labor Force [25]

43,800

Labor Force by Occupation [26]

Agriculture	43.4%
Services	38.9
Industry and commerce	17.7

Date of data: 1983 est.

Unemployment Rate [27]

16% (1988)

Production Sectors

Energy Resource Summary [28]

Energy Resources: Geothermal potential. **Electricity**: 32,500 kW capacity; 112 million kWh produced, 740 kWh per capita (1992).

Telecommunications [30]

- Fully automatic telephone system
- 9,500 telephones
- Direct microwave link with Martinique and Saint Vincent and the Grenadines
- Interisland troposcatter link to Barbados
- Broadcast stations - 4 AM, 1 FM, 1 TV (cable)

Top Agricultural Products [32]

Agriculture accounts for 12% of GDP and 43% of labor force; crops—bananas, coconuts, vegetables, citrus fruit, root crops, cocoa; imports food for the tourist industry.

Top Mining Products [33]

Detailed information is not available. A summary of mineral resources available follows. **Mineral Resources**: Pumice.

Transportation [31]

Highways. 760 km total; 500 km paved; 260 km otherwise improved

Airports

Total:	2
Usable:	2
With permanent-surface runways:	2
With runways over 3,659 m:	0
With runways 2,440-3,659 m:	1
With runways 1,220-2,439:	1

Tourism [34]

	1987	1988	1989	1990	1991
Visitors	202	214[59]	242[59]	251[2]	325[2]
Tourists	118	133	135	147	165
Excursionists	1	2	3	2	7
Cruise passengers	83	79	104	102	153
Tourism receipts	126	134	145	154	173
Tourism expenditures	15	21	21	18	17

Tourists are in thousands, money in million U.S. dollars.

Finance, Economics, and Trade

Industrial Summary [35]

Industrial Production: Growth rate 3.5% (1990 est.); accounts for 12% of GDP. **Industries**: Clothing, assembly of electronic components, beverages, corrugated boxes, tourism, lime processing, coconut processing.

Economic Indicators [36]

National product: GDP—exchange rate conversion—$250 million (1991 est.). **National product real growth rate**: 2.5% (1991 est.). **National product per capita**: $1,650 (1991 est.). **Inflation rate (consumer prices)**: 6.1% (1991). **External debt**: $65.7 million (1991 est.).

Balance of Payments Summary [37]

Values in millions of dollars.

	1987	1988	1989	1990	1991
Exports of goods (f.o.b.)	77.3	119.1	111.9	127.3	105.4
Imports of goods (f.o.b.)	-161.9	-201.0	-235.2	-244.2	-267.3
Trade balance	-84.6	-81.9	-123.3	-116.9	-161.9
Services - debits	-47.5	-59.9	-66.5	-124.8	-116.9
Services - credits	94.5	124.4	135.8	168.4	176.1
Private transfers (net)	15.0	10.8	8.7	15.2	16.7
Government transfers (net)	9.7	4.0	3.7	2.8	4.9
Long term capital (net)	27.3	28.4	34.4	57.4	87.3
Short term capital (net)	-6.8	-19.7	16.1	4.9	2.2
Errors and omissions	1.5	-3.5	-2.2	1.5	-3.4
Overall balance	9.1	2.6	6.7	8.5	5.0

Exchange Rates [38]

Currency: **East Caribbean dollars.**
Symbol: **EC$.**

Data are currency units per $1.

Fixed rate since 1976	2.70

Imports and Exports

Top Import Origins [39]

$267 million (f.o.b., 1991).

Origins	%
US	34
CARICOM	17
UK	14
Japan	7
Canada	4

Top Export Destinations [40]

$105 million (f.o.b., 1991).

Destinations	%
UK	56
US	22
CARICOM	19

Foreign Aid [41]

	U.S. $
Western (non-US) countries, ODA and OOF bilateral commitments (1970-89)	120 million

Import and Export Commodities [42]

Import Commodities
Manufactured goods 21%
Machinery, transp. equip. 21%
Food and live animals
Chemicals
Fuels

Export Commodities
Bananas 58%
Clothing
Cocoa
Vegetables
Fruits
Coconut oil

778

For sources, notes, and explanations, see Annotated Source Appendix, page 1035.

Saint Vincent and the Grenadines

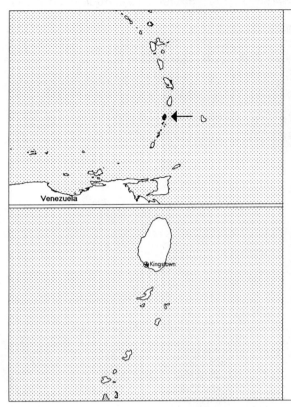

Geography [1]

Total area:
340 km2
Land area:
340 km2
Comparative area:
Slightly less than twice the size of Washington, DC
Land boundaries:
0 km
Coastline:
84 km
Climate:
Tropical; little seasonal temperature variation; rainy season (May to November)
Terrain:
Volcanic, mountainous; Soufriere volcano on the island of Saint Vincent
Natural resources:
Negligible
Land use:
Arable land:
38%
Permanent crops:
12%
Meadows and pastures:
6%
Forest and woodland:
41%
Other:
3%

Demographics [2]

	1960	1970	1980	1990	1991[1]	1994	2000	2010	2020
Population	81	88	98	112	114	115	122	136	152
Population density (persons per sq. mi.)	618	668	747	860	872	NA	1,005	1,193	1,398
Births	NA	NA	NA	NA	3	2	NA	NA	NA
Deaths	NA	NA	NA	NA	1	1	NA	NA	NA
Life expectancy - males	NA	NA	NA	NA	68	71	NA	NA	NA
Life expectancy - females	NA	NA	NA	NA	72	74	NA	NA	NA
Birth rate (per 1,000)	NA	NA	NA	NA	27	20	NA	NA	NA
Death rate (per 1,000)	NA	NA	NA	NA	6	5	NA	NA	NA
Women of reproductive age (15-44 yrs.)	NA	NA	NA	29	NA	31	36	NA	NA
of which are currently married	NA	NA	NA	14	NA	15	19	NA	NA
Fertility rate	NA	NA	NA	2.4	2.84	2.1	1.9	1.8	1.8

Population values are in thousands, life expectancy in years, and other items as indicated.

Health

Health Personnel [3]

Health Expenditures [4]

For sources, notes, and explanations, see Annotated Source Appendix, page 1035.

779

Human Factors

Health Care Ratios [5]	Infants and Malnutrition [6]

Ethnic Division [7]	Religion [8]	Major Languages [9]
Black African descent, white, East Indian, Carib Indian.	Anglican, Methodist, Roman Catholic, Seventh-Day Adventist.	English, French patois.

Education

Public Education Expenditures [10]

Million E. Carribbean Dollars	1985	1986	1987	1988	1989[6]	1990[6]
Total education expenditure	NA	19	NA	NA	26	34
as percent of GNP	NA	5.8	NA	NA	5.9	7.0
as percent of total govt. expend.	NA	11.6	NA	NA	10.5	13.8
Current education expenditure	NA	18	NA	NA	23	26
as percent of GNP	NA	5.4	NA	NA	5.2	5.2
as percent of current govt. expend.	NA	16.8	NA	NA	17.1	17.2
Capital expenditure	NA	1	NA	NA	3	8

Educational Attainment [11]

Age group	25+
Total population	32,444
Highest level attained (%)	
No schooling	2.4
First level	
Incompleted	88.0
Completed	NA
Entered second level	
S-1	8.2
S-2	NA
Post secondary	1.4

Literacy Rate [12]

	1970[b,3]	1980[b,3]	1990[b,3]
Illiterate population +15 years	1,839	NA	NA
Illiteracy rate - total pop. (%)	4.4	NA	NA
Illiteracy rate - males (%)	4.2	NA	NA
Illiteracy rate - females (%)	4.5	NA	NA

Libraries [13]

	Admin. Units	Svc. Pts.	Vols. (000)	Shelving (meters)	Vols. Added	Reg. Users
National	NA	NA	NA	NA	NA	NA
Non-specialized	NA	NA	NA	NA	NA	NA
Public (1989)	1	16	100	NA	5,000	NA
Higher ed. (1990)	2	2	12	NA	550	1,100
School (1990)	2	3	NA	NA	NA	1,400

Daily Newspapers [14]	Cinema [15]

Science and Technology

Scientific/Technical Forces [16]	R&D Expenditures [17]	U.S. Patents Issued [18]

780

For sources, notes, and explanations, see Annotated Source Appendix, page 1035.

Government and Law

Organization of Government [19]

Long-form name:
none
Type:
constitutional monarchy
Independence:
27 October 1979 (from UK)
Constitution:
27 October 1979
Legal system:
based on English common law
National holiday:
Independence Day, 27 October (1979)
Executive branch:
British monarch, governor general, prime minister, Cabinet
Legislative branch:
unicameral House of Assembly
Judicial branch:
Eastern Caribbean Supreme Court

Elections [20]

House of Assembly. last held 16 May 1989 (next to be held NA July 1994); results - percent of vote by party NA; seats - (21 total; 15 elected representatives and 6 appointed senators) New Democratic Party 15.

Government Expenditures [21]

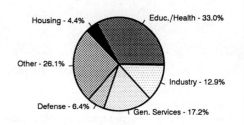

Housing - 4.4%
Educ./Health - 33.0%
Other - 26.1%
Industry - 12.9%
Defense - 6.4%
Gen. Services - 17.2%

% distribution for 1988. Expend. in FY90 est.: 67 ($ mil.)

Defense Summary [22]

Branches: Royal Saint Vincent and the Grenadines Police Force, Coast Guard
Manpower Availability: No information available.
Defense Expenditures: No information available.

Crime [23]

Crime volume	
Cases known to police	4,255
Attempts (percent)	NA
Percent cases solved	436.56
Crimes per 100,000 persons	3,976.63
Persons responsible for offenses	
Total number offenders	3,306
Percent female	71.75
Percent juvenile[47]	29.36
Percent foreigners	17

Human Rights [24]

	SSTS	FL	FAPRO	PPCG	APROBC	TPW	PCPTW	STPEP	PHRFF	PRW	ASST	AFL
Observes	P			P		P	P			P	P	
	EAFRD	CPR	ESCR	SR	ACHR	MAAE	PVIAC	PVNAC	EAFDAW	TCIDTP	RC	
Observes		P	P	P				P	P	P	P	P

P = Party; S = Signatory; see Appendix for meaning of abbreviations.

Labor Force

Total Labor Force [25]

67,000 (1984 est.)

Labor Force by Occupation [26]

Unemployment Rate [27]

35%-40% (1992 est.)

Production Sectors

Energy Resource Summary [28]

Energy Resources: None. **Electricity**: 16,600 kW capacity; 64 million kWh produced, 555 kWh per capita (1992).

Telecommunications [30]

- Islandwide fully automatic telephone system
- 6,500 telephones
- VHF/UHF interisland links from Saint Vincent to Barbados and the Grenadines
- New SHF links to Grenada and Saint Lucia
- Broadcast stations - 2 AM, no FM, 1 TV (cable)

Transportation [31]

Highways. 1,000 km total; 300 km paved; 400 km improved; 300 km unimproved (est.)

Merchant Marine. 407 ships (1,000 GRT or over) totaling 3,388,427 GRT/5,511,325 DWT; includes 3 passenger, 2 passenger-cargo, 222 cargo, 22 container, 19 roll-on/roll-off cargo, 14 refrigerated cargo, 24 oil tanker, 7 chemical tanker, 4 liquefied gas, 73 bulk, 13 combination bulk, 2 vehicle carrier, 1 livestock carrier, 1 specialized tanker; note - China owns 3 ships; a flag of convenience registry

Airports

Total:	6
Usable:	6
With permanent-surface runways:	5
With runways over 3,659 m:	0
With runways 2,440-3,659 m:	0
With runways 1,220-2,439 m:	1

Top Agricultural Products [32]

Agriculture accounts for 15% of GDP and 60% of labor force; provides bulk of exports; products—bananas, coconuts, sweet potatoes, spices; small numbers of cattle, sheep, hogs, goats; small fish catch used locally.

Top Mining Products [33]

Detailed information is not available. A summary of mineral resources available follows. **Mineral Resources**: Negligible.

Tourism [34]

	1987	1988	1989	1990	1991
Visitors	128	129	129	158	173
Tourists[60]	46	47	50	54	52
Cruise passengers	68	65	57	83	92
Tourism receipts	35	39	43	54	53
Tourism expenditures	4	4	5	7	7
Fare receipts	1	1	1	1	1
Fare expenditures	6	4	5	5	6

Tourists are in thousands, money in million U.S. dollars.

Finance, Economics, and Trade

Industrial Summary [35]

Industrial Production: Growth rate 0% (1989); accounts for 14% of GDP. **Industries**: Food processing, cement, furniture, clothing, starch.

Economic Indicators [36]

National product: GDP—exchange rate conversion—$171 million (1992 est.). **National product real growth rate**: 3% (1992 est.). **National product per capita**: $1,500 (1992 est.). **Inflation rate (consumer prices)**: 2.3% (1991 est.). **External debt**: $50.9 million (1989).

Balance of Payments Summary [37]

Values in millions of dollars.

	1987	1988	1989	1990	1991
Exports of goods (f.o.b.)	52.3	85.3	74.6	82.7	65.7
Imports of goods (f.o.b.)	-89.5	-110.0	-114.8	-122.5	-110.7
Trade balance	-37.2	-24.7	-40.1	-39.8	-45.0
Services - debits	-30.9	-35.2	-39.2	-43.8	-44.5
Services - credits	31.3	34.4	38.7	42.1	49.1
Private transfers (net)	17.8	20.0	21.7	23.6	24.6
Government transfers (net)	6.6	6.3	8.7	10.0	6.8
Long term capital (net)	13.3	15.7	10.6	11.4	12.5
Short term capital (net)	-3.9	-12.0	2.2	-10.8	6.6
Errors and omissions	-2.2	-2.2	-0.7	11.4	-13.5
Overall balance	-5.2	2.3	1.9	4.1	-3.4

Exchange Rates [38]

Currency: **East Caribbean dollars.**
Symbol: **EC$.**

Data are currency units per $1.

Fixed rate since 1976	2.70

Imports and Exports

Top Import Origins [39]

$110.7 million (f.o.b., 1991).

Origins	%
US	42
CARICOM	19
UK	15

Top Export Destinations [40]

$65.7 million (f.o.b., 1991).

Destinations	%
UK	43
CARICOM	37
US	15

Foreign Aid [41]

	U.S. $	
US commitments, including Ex-Im (FY70-87)	11	million
Western (non-US) countries, ODA and OOF bilateral commitments (1970-89)	81	million

Import and Export Commodities [42]

Import Commodities

Foodstuffs
Machinery & equipment
Chemicals and fertilizers
Minerals and fuels

Export Commodities

Bananas
Eddoes and dasheen (taro)
Arrowroot starch
Tennis racquets

San Marino

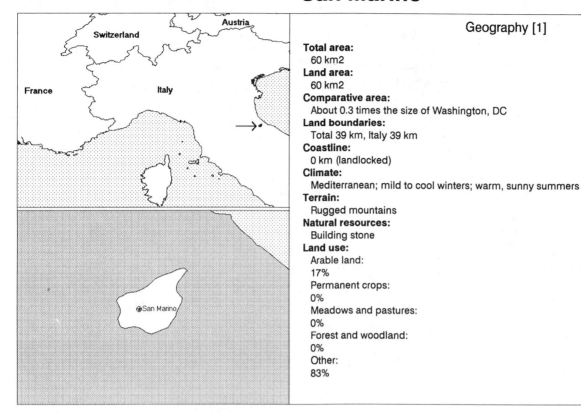

Geography [1]

Total area:
60 km2
Land area:
60 km2
Comparative area:
About 0.3 times the size of Washington, DC
Land boundaries:
Total 39 km, Italy 39 km
Coastline:
0 km (landlocked)
Climate:
Mediterranean; mild to cool winters; warm, sunny summers
Terrain:
Rugged mountains
Natural resources:
Building stone
Land use:
Arable land:
17%
Permanent crops:
0%
Meadows and pastures:
0%
Forest and woodland:
0%
Other:
83%

Demographics [2]

	1960	1970	1980	1990	1991[1]	1994	2000	2010	2020
Population	15	19	21	23	23		25	26	27
Population density (persons per sq. mi.)	670	834	932	1,005	1,011	NA	1,063	1,111	1,150
Births	NA	NA	NA	NA	(Z)	(Z)	NA	NA	NA
Deaths	NA	NA	NA	NA	(Z)	(Z)	NA	NA	NA
Life expectancy - males	NA	NA	NA	NA	74	77	NA	NA	NA
Life expectancy - females	NA	NA	NA	NA	79	85	NA	NA	NA
Birth rate (per 1,000)	NA	NA	NA	NA	8	11	NA	NA	NA
Death rate (per 1,000)	NA	NA	NA	NA	7	7	NA	NA	NA
Women of reproductive age (15-44 yrs.)	NA	NA	NA	NA	NA	NA	NA	NA	NA
of which are currently married	NA	NA	NA	NA	NA	NA	NA	NA	NA
Fertility rate	NA	NA	NA	1.6	1.26	1.5	1.5	1.5	1.6

Population values are in thousands, life expectancy in years, and other items as indicated.

Health

Health Personnel [3]

Health Expenditures [4]

For sources, notes, and explanations, see Annotated Source Appendix, page 1035.

Human Factors	
Health Care Ratios [5]	Infants and Malnutrition [6]

Ethnic Division [7]	Religion [8]	Major Languages [9]
Sammarinese, Italian.	Roman Catholic.	Italian.

Education

Public Education Expenditures [10]

Million Lira	1980	1985	1987	1988	1989	1990
Total education expenditure	7,252	NA	NA	19,140	19,922	22,048
as percent of GNP	NA	NA	NA	NA	NA	NA
as percent of total govt. expend.	7.5	NA	NA	NA	NA	NA
Current education expenditure	6,263	NA	NA	17,946	19,278	21,723
as percent of GNP	NA	NA	NA	NA	NA	NA
as percent of current govt. expend.	9.5	NA	NA	NA	NA	NA
Capital expenditure	989	NA	NA	1,194	644	325

Educational Attainment [11]

Literacy Rate [12]

	1976[b]	1980[b]	1990[b]
Illiterate population +15 years	640	NA	NA
Illiteracy rate - total pop. (%)	3.9	NA	NA
Illiteracy rate - males (%)	3.2	NA	NA
Illiteracy rate - females (%)	4.7	NA	NA

Libraries [13]

	Admin. Units	Svc. Pts.	Vols. (000)	Shelving (meters)	Vols. Added	Reg. Users
National	NA	NA	NA	NA	NA	NA
Non-specialized	NA	NA	NA	NA	NA	NA
Public	NA	NA	NA	NA	NA	NA
Higher ed.	NA	NA	NA	NA	NA	NA
School (1987)	5	17	NA	NA	NA	NA

Daily Newspapers [14]

	1975	1980	1985	1990
Number of papers	4	3	16	6
Circ. (000)	1	1	2	2[e]

Cinema [15]

Data for 1991.

Cinema seats per 1,000	87.0
Annual attendance per person	1.3
Gross box office receipts (mil. Lira)	90

Science and Technology

Scientific/Technical Forces [16]

Potential scientists/engineers	566
Number female	213
Potential technicians	2,004
Number female	883
Total	2,570

R&D Expenditures [17]

	Lire (000) 1986
Total expenditure	-
Capital expenditure	-
Current expenditure	-
Percent current	-

U.S. Patents Issued [18]

Government and Law

Organization of Government [19]

Long-form name:
Republic of San Marino

Type:
republic

Independence:
301 AD (by tradition)

Constitution:
8 October 1600; electoral law of 1926 serves some of the functions of a constitution

Legal system:
based on civil law system with Italian law influences; has not accepted compulsory ICJ jurisdiction

National holiday:
Anniversary of the Foundation of the Republic, 3 September

Executive branch:
two captains regent, Congress of State (cabinet); real executive power is wielded by the secretary of state for foreign affairs and the secretary of state for internal affairs

Legislative branch:
unicameral Great and General Council (Consiglio Grande e Generale)

Judicial branch:
Council of Twelve (Consiglio dei XII)

Political Parties [20]

Great and General Council	% of seats
Christian Democratic Party (DCS)	45.0
San Marino Communist Party (PCS)	30.0
Unitary Socialst Party (PSU)	13.3
San Marino Socialist Party (PSS)	11.7

Government Budget [21]

For 1991.

Revenues	NA
Expenditures	300
Capital expenditures	NA

Defense Summary [22]

Branches: Public security or police force

Manpower Availability: All fit men ages 16-60 constitute a militia that can serve as an army

Defense Expenditures: No information available.

Crime [23]

Human Rights [24]

	SSTS	FL	FAPRO	PPCG	APROBC	TPW	PCPTW	STPEP	PHRFF	PRW	ASST	AFL
Observes		P			P	P	P		P		P	

	EAFRD	CPR	ESCR	SR	ACHR	MAAE	PVIAC	PVNAC	EAFDAW	TCIDTP	RC
Observes		P	P				S	S			P

P = Party; S = Signatory; see Appendix for meaning of abbreviations.

Labor Force

Total Labor Force [25]

4,300 (est.)

Labor Force by Occupation [26]

Unemployment Rate [27]

3% (1991)

For sources, notes, and explanations, see Annotated Source Appendix, page 1035.

Production Sectors

Energy Resource Summary [28]

Energy Resources: None. **Electricity**: Supplied by Italy.

Telecommunications [30]

- Automatic telephone system completely integrated into Italian system
- 11,700 telephones
- Broadcast services from Italy
- Microwave and cable links into Italian networks
- No communication satellite facilities

Transportation [31]

Highways. 104 km

Airports

Total:	0
Usable:	0
With permanent-surface runways:	0
With runways 3,659 m:	0
With runways 2,440-3,659 m:	0
With runways 1,220-2,439 m:	0

Top Agricultural Products [32]

Employs 3% of labor force; products—wheat, grapes, maize, olives, meat, cheese, hides; small numbers of cattle, pigs, horses; depends on Italy for food imports.

Top Mining Products [33]

Detailed information is not available. A summary of mineral resources available follows. **Mineral Resources**: Building stone.

Tourism [34]

	1987	1988	1989	1990	1991
Visitors[61]	505	504	437	582	

Tourists are in thousands, money in million U.S. dollars.

Finance, Economics, and Trade

Industrial Summary [35]

Industrial Production: Growth rate NA%; accounts for 42% of workforce. **Industries**: Wine, olive oil, cement, leather, textile, tourism.

Economic Indicators [36]

National product: GDP—purchasing power equivalent—$465 million (1992 est.). **National product real growth rate**: NA%. **National product per capita**: $20,000 (1992 est.). **Inflation rate (consumer prices)**: 5% (1992 est.). **External debt**: not available.

Balance of Payments Summary [37]

Exchange Rates [38]

Currency: **Italian lire.**
Symbol: **Lit.**

Data are currency units per $1.

January 1993	1,482.5
1992	1,232.4
1991	1,240.6
1990	1,198.1
1989	1,372.1
1988	1,301.6

Imports and Exports

Top Import Origins [39]

Amount not available.

Origins	%
See exports	NA

Top Export Destinations [40]

Trade data are included with the statistics for Italy; commodity trade consists primarily of exchanging building stone, lime, wood, chestnuts, wheat, wine, baked goods, hides, and ceramics for a wide variety of consumer manufactures.

Destination	%
No details available.	NA

Foreign Aid [41]

Import and Export Commodities [42]

Import Commodities   **Export Commodities**

No information available.

Sao Tome and Principe

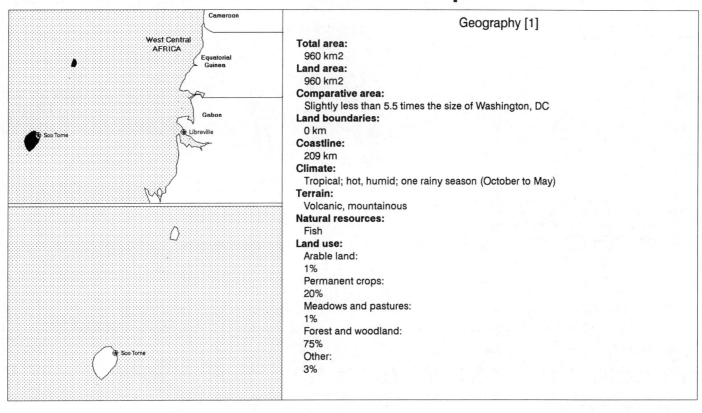

Geography [1]

Total area:
960 km2
Land area:
960 km2
Comparative area:
Slightly less than 5.5 times the size of Washington, DC
Land boundaries:
0 km
Coastline:
209 km
Climate:
Tropical; hot, humid; one rainy season (October to May)
Terrain:
Volcanic, mountainous
Natural resources:
Fish
Land use:
Arable land:
1%
Permanent crops:
20%
Meadows and pastures:
1%
Forest and woodland:
75%
Other:
3%

Demographics [2]

	1960	1970	1980	1990	1991[1]	1994	2000	2010	2020
Population	63	74	94	123	128	137	159	196	232
Population density (persons per sq. mi.)	171	198	254	336	346	NA	447	569	699
Births	NA	NA	NA	NA	5	5	NA	NA	NA
Deaths	NA	NA	NA	NA	1	1	NA	NA	NA
Life expectancy - males	NA	NA	NA	NA	64	61	NA	NA	NA
Life expectancy - females	NA	NA	NA	NA	68	65	NA	NA	NA
Birth rate (per 1,000)	NA	NA	NA	NA	38	35	NA	NA	NA
Death rate (per 1,000)	NA	NA	NA	NA	8	9	NA	NA	NA
Women of reproductive age (15-44 yrs.)	NA	NA	NA	NA	NA	NA	NA	NA	NA
of which are currently married	NA	NA	NA	NA	NA	NA	NA	NA	NA
Fertility rate	NA	NA	NA	4.9	5.27	4.5	3.9	3.0	2.5

Population values are in thousands, life expectancy in years, and other items as indicated.

Health

Health Personnel [3]

Health Expenditures [4]

For sources, notes, and explanations, see Annotated Source Appendix, page 1035.

789

Human Factors	
Health Care Ratios [5]	Infants and Malnutrition [6]

Ethnic Division [7]	Religion [8]	Major Languages [9]
Mestico, angolares (descendents of Angolan slaves), forros (descendents of freed slaves), servicais (contract laborers from Angola, Mozambique, and Cape Verde), tongas (children of servicais born on the islands), Europeans (primarily Portuguese).	Roman Catholic, Evangelical Protestant, Seventh-Day Adventist.	Portuguese (official).

Education

Public Education Expenditures [10]

Million Dobra	1985	1986	1987	1988	1989	1990
Total education expenditure	NA	100	NA	NA	NA	NA
as percent of GNP	NA	4.3	NA	NA	NA	NA
as percent of total govt. expend.	NA	18.8	NA	NA	NA	NA
Current education expenditure	NA	NA	NA	NA	NA	NA
as percent of GNP	NA	NA	NA	NA	NA	NA
as percent of current govt. expend.	NA	NA	NA	NA	NA	NA
Capital expenditure	NA	NA	NA	NA	NA	NA

Educational Attainment [11]

Age group	25+
Total population	33,308
Highest level attained (%)	
No schooling	56.6[6]
First level	
Incompleted	18.0
Completed	19.3
Entered second level	
S-1	4.6
S-2	1.3
Post secondary	0.3

Literacy Rate [12]

	1981[b]	1990[b]	1991[b]
Illiterate population +15 years	22,080	NA	NA
Illiteracy rate - total pop. (%)	42.6	NA	NA
Illiteracy rate - males (%)	26.8	NA	NA
Illiteracy rate - females (%)	57.6	NA	NA

Libraries [13]

Daily Newspapers [14]

Cinema [15]

Science and Technology

Scientific/Technical Forces [16]	R&D Expenditures [17]	U.S. Patents Issued [18]

For sources, notes, and explanations, see Annotated Source Appendix, page 1035.

Government and Law

Organization of Government [19]

Long-form name:
Democratic Republic of Sao Tome and
Principe
Type:
republic
Independence:
12 July 1975 (from Portugal)
Constitution:
5 November 1975, approved 15
December 1982
Legal system:
based on Portuguese law system and
customary law; has not accepted
compulsory ICJ jurisdiction
National holiday:
Independence Day, 12 July (1975)
Executive branch:
president, prime minister, Council of
Ministers (cabinet)
Legislative branch:
unicameral National People's Assembly
(Assembleia Popular Nacional)
Judicial branch:
Supreme Court

Political Parties [20]

National People's Assembly	% of votes
Democratic Convergence (PCD-GR)	54.4
Movement for the Liberation (MLSTP)	30.5
Democratic Opposition Coalition (CODO)	5.2
Christian Democratic Front (FDC)	1.5
Other	8.4

Government Budget [21]

For 1989.

Revenues	10.2
Expenditures	36.8
Capital expenditures	22.5

Crime [23]

Military Expenditures and Arms Transfers [22]

	1985	1986	1987	1988	1989
Military expenditures					
Current dollars (mil.)	NA	NA	NA	NA	NA
1989 constant dollars (mil.)	NA	NA	NA	NA	NA
Armed forces (000)	2	1	1	1	1
Gross national product (GNP)					
Current dollars (mil.)	37	39	40	42	45
1989 constant dollars (mil.)	42	44	43	44	45
Central government expenditures (CGE)					
1989 constant dollars (mil.)	NA	NA	12	7	NA
People (mil.)	0.1	0.1	0.1	0.1	0.1
Military expenditure as % of GNP	NA	NA	NA	NA	NA
Military expenditure as % of CGE	NA	NA	NA	NA	NA
Military expenditure per capita	NA	NA	NA	NA	NA
Armed forces per 1,000 people	18.6	9.0	8.8	8.5	8.3
GNP per capita	389	394	377	374	368
Arms imports[6]					
Current dollars (mil.)	0	20	5	0	5
1989 constant dollars (mil.)	0	22	5	0	5
Arms exports[6]					
Current dollars (mil.)	0	0	0	0	0
1989 constant dollars (mil.)	0	0	0	0	0
Total imports[7]					
Current dollars (mil.)	NA	16	12	17	NA
1989 constant dollars	NA	18	13	18	NA
Total exports[7]					
Current dollars (mil.)	NA	9	6	10	NA
1989 constant dollars	NA	10	6	10	NA
Arms as percent of total imports[8]	NA	125.0	41.7	0	NA
Arms as percent of total exports[8]	NA	0	0	0	NA

Human Rights [24]

	SSTS	FL	FAPRO	PPCG	APROBC	TPW	PCPTW	STPEP	PHRFF	PRW	ASST	AFL
Observes						P	P					

	EAFRD	CPR	ESCR	SR	ACHR	MAAE	PVIAC	PVNAC	EAFDAW	TCIDTP	RC
Observes				P							P

P = Party; S = Signatory; see Appendix for meaning of abbreviations.

Labor Force

Total Labor Force [25]

21,096 (1981); most of population
engaged in subsistence agriculture and
fishing; labor shortages on plantations
and of skilled workers; 56% of population
is of working age (1983)

Labor Force by Occupation [26]

Subsistance agriculture
Fishing

Unemployment Rate [27]

Rate is not available.

For sources, notes, and explanations, see Annotated Source Appendix, page 1035.

791

Production Sectors

Energy Resource Summary [28]

Energy Resources: None. **Electricity**: 5,000 kW capacity; 10 million kWh produced, 80 kWh per capita (1991).

Telecommunications [30]

- Minimal system
- Broadcast stations - 1 AM, 2 FM, no TV
- 1 Atlantic Ocean INTELSAT earth station

Top Agricultural Products [32]

Dominant sector of economy, primary source of exports; cash crops—cocoa (85%), coconuts, palm kernels, coffee; food products—bananas, papaya, beans, poultry, fish; not self-sufficient in food grain and meat.

Top Mining Products [33]

Detailed information is not available. A summary of mineral resources available follows. **Mineral Resources**: None.

Transportation [31]

Highways. 300 km (two-thirds are paved); roads on Principe are mostly unpaved and in need of repair

Merchant Marine. 1 cargo ship (1,000 GRT or over) totaling 1,096 GRT/1,105 DWT

Airports

Total:	2
Usable:	2
With permanent-surface runways :	2
With runways over 3,659 m:	0
With runways 2,440-3,659 m:	0
With runways 1,220-2,439 m:	2

Tourism [34]

Finance, Economics, and Trade

GDP and Manufacturing Summary [35]

	1980	1985	1990	1991	1992
Gross Domestic Product					
Millions of 1980 dollars	47	33	39	40	40[e]
Growth rate in percent	2.59	-5.01	3.81	1.51	0.80[e]
Manufacturing Value Added					
Millions of 1980 dollars	3	3	3	3	3[e]
Growth rate in percent	0.00	-8.74	3.28	1.91	1.33[e]
Manufacturing share in percent of current prices	7.3	7.2[e]	NA	NA	NA

Economic Indicators [36]

National product: GDP—exchange rate conversion—$41.4 million (1992 est.). **National product real growth rate**: 1.5% (1992 est.). **National product per capita**: $315 (1992 est.). **Inflation rate (consumer prices)**: 27% (1992 est.). **External debt**: $163.6 million (1992).

Balance of Payments Summary [37]

Values in millions of dollars.

	1986	1987	1988	1989	1990
Exports of goods (f.o.b.)	9.9	6.5	9.5	4.9	4.2
Imports of goods (f.o.b.)	-23.2	-13.6	-14.1	-13.3	-13.0
Trade balance	-13.3	-7.0	-4.6	-8.4	-8.8
Services - debits	-14.2	-9.3	-9.0	-8.8	-9.3
Services - credits	3.7	1.7	2.0	4.6	3.9
Private transfers (net)	-0.8	-0.3	-	-0.2	0.1
Government transfers (net)	5.8	1.9	0.9	1.5	2.1
Long term capital (net)	5.4	3.2	2.0	3.9	11.2
Short term capial (net)	9.8	5.8	4.4	2.8	3.0
Errors and omissions	2.2	-	-	-1.0	-2.7
Overall balance	-1.4	-4.0	-4.3	-5.7	-0.5

Exchange Rates [38]

Currency: **dobras.**
Symbol: **Db.**

Data are currency units per $1.

1992	230.000
November 1991	260.000
December 1988	122.480
1987	72.827
1986	36.993

Imports and Exports

Top Import Origins [39]

$24.5 million (f.o.b., 1991).

Origins	%
Portugal	NA
Germany	NA
Angola	NA
China	NA

Top Export Destinations [40]

$5.5 million (f.o.b., 1991 est.).

Destinations	%
Germany	NA
Netherlands	NA
China	NA

Foreign Aid [41]

	U.S. $	
US commitments, including Ex-Im (FY70-89)	8	million
Western (non-US) countries, ODA and OOF bilateral commitments (1970-89)	89	million

Import and Export Commodities [42]

Import Commodities	Export Commodities
Machinery & electrical equip. 54%	Cocoa 85%
Food products 23%	Copra
Other 23%	Coffee
	Palm oil

For sources, notes, and explanations, see Annotated Source Appendix, page 1035.

793

Saudi Arabia

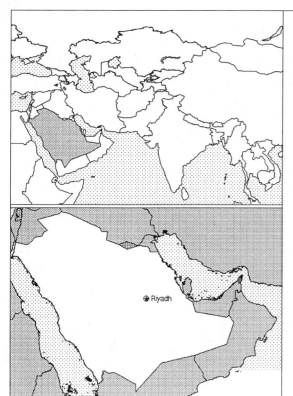

Geography [1]

Total area:
1,960,582 km2
Land area:
1,960,582 km2
Comparative area:
Slightly less than one-fourth the size of the US
Land boundaries:
Total 4,415 km, Iraq 814 km, Jordan 728 km, Kuwait 222 km, Oman 676 km, Qatar 60 km, UAE 457 km, Yemen 1,458 km
Coastline:
2,640 km
Climate:
Harsh, dry desert with great extremes of temperature
Terrain:
Mostly uninhabited, sandy desert
Natural resources:
Petroleum, natural gas, iron ore, gold, copper
Land use:
Arable land:
1%
Permanent crops:
0%
Meadows and pastures:
39%
Forest and woodland:
1%
Other:
59%

Demographics [2]

	1960	1970	1980	1990	1991[1]	1994	2000	2010	2020
Population	4,718	6,109	10,139	16,271	17,870	18,197	22,070	30,494	42,085
Population density (persons per sq. mi.)	6	7	12	21	22	NA	30	41	55
Births	NA	NA	NA	NA	654	696	NA	NA	NA
Deaths	NA	NA	NA	NA	112	106	NA	NA	NA
Life expectancy - males	NA	NA	NA	NA	65	66	NA	NA	NA
Life expectancy - females	NA	NA	NA	NA	68	70	NA	NA	NA
Birth rate (per 1,000)	NA	NA	NA	NA	37	38	NA	NA	NA
Death rate (per 1,000)	NA	NA	NA	NA	6	6	NA	NA	NA
Women of reproductive age (15-44 yrs.)	NA	NA	NA	NA	NA	NA	NA	NA	NA
of which are currently married	NA	NA	NA	NA	NA	NA	NA	NA	NA
Fertility rate	NA	NA	NA	6.8	6.71	6.7	6.4	5.9	5.2

Population values are in thousands, life expectancy in years, and other items as indicated.

Health

Health Personnel [3]

Doctors per 1,000 pop., 1988-92	1.52
Nurse-to-doctor ratio, 1988-92	1.5
Hospital beds per 1,000 pop., 1985-90	2.7
Percentage of children immunized (age 1 yr. or less)	
Third dose of DPT, 1990-91	94.0
Measles, 1990-91	90.0

Health Expenditures [4]

Total health expenditure, 1990 (official exchange rate)	
Millions of dollars	4784
Millions of dollars per capita	322
Health expenditures as a percentage of GDP	
Total	4.8
Public sector	3.1
Private sector	1.7
Development assistance for health	
Total aid flows (millions of dollars)[1]	1
Aid flows per capita (millions of dollars)	0.1
Aid flows as a percentage of total health expenditure	NA

For sources, notes, and explanations, see Annotated Source Appendix, page 1035.

Human Factors

Health Care Ratios [5]

Population per physician, 1970	7460
Population per physician, 1990	660
Population per nursing person, 1970	2070
Population per nursing person, 1990	420
Percent of births attended by health staff, 1985	78

Infants and Malnutrition [6]

Percent of babies with low birth weight, 1985	6
Infant mortality rate per 1,000 live births, 1970	119
Infant mortality rate per 1,000 live births, 1991	32
Years of life lost per 1,000 population, 1990	37
Prevalence of malnutrition (under age 5), 1990	NA

Ethnic Division [7]

Arab	90%
Afro-Asian	10%

Religion [8]

Muslim	100%

Major Languages [9]

Arabic.

Education

Public Education Expenditures [10]

Million Riyal	1980	1985	1987	1988	1989	1990
Total education expenditure	21,294	23,540	23,181	22,909	23,582	25,460
as percent of GNP	5.5	6.7	7.4	7.1	6.8	6.2
as percent of total govt. expend.	8.7	12.0	13.6	16.2	16.7	17.8
Current education expenditure	13,526	19,283	20,304	20,818	21,698	NA
as percent of GNP	3.5	5.5	6.5	6.5	6.2	NA
as percent of current govt. expend.	NA	NA	NA	NA	NA	NA
Capital expenditure	7,768	4,257	2,877	2,091	1,884	NA

Educational Attainment [11]

Literacy Rate [12]

In thousands and percent	1985	1991	2000
Illiterate population +15 years	2,689	2,897	3,293
Illiteracy rate - total pop. (%)	42.1	37.6	29.3
Illiteracy rate - males (%)	30.6	26.9	20.4
Illiteracy rate - females (%)	57.5	51.9	40.6

Libraries [13]

	Admin. Units	Svc. Pts.	Vols. (000)	Shelving (meters)	Vols. Added	Reg. Users
National	NA	NA	NA	NA	NA	NA
Non-specialized	NA	NA	NA	NA	NA	NA
Public (1986)	1	50	630	24,000	NA	NA
Higher ed.	NA	NA	NA	NA	NA	NA
School	NA	NA	NA	NA	NA	NA

Daily Newspapers [14]

	1975	1980	1985	1990
Number of papers	12	11	13	12
Circ. (000)	215[e]	350[e]	450[e]	600[e]

Cinema [15]

Science and Technology

Scientific/Technical Forces [16]

R&D Expenditures [17]

U.S. Patents Issued [18]

Values show patents issued to citizens of the country by the U.S. Patents Office.

	1990	1991	1992
Number of patents	7	5	8

For sources, notes, and explanations, see Annotated Source Appendix, page 1035.

Government and Law

Organization of Government [19]

Long-form name:
Kingdom of Saudi Arabia
Type:
monarchy
Independence:
23 September 1932 (unification)
Constitution:
none; governed according to Shari'a (Islamic law)
Legal system:
based on Islamic law, several secular codes have been introduced; commercial disputes handled by special committees; has not accepted compulsory ICJ jurisdiction
National holiday:
Unification of the Kingdom, 23 September (1932)
Executive branch:
monarch and prime minister, crown prince and deputy prime minister, Council of Ministers
Legislative branch:
none
Judicial branch:
Supreme Council of Justice

Elections [20]

No political parties allowed; no elections.

Government Budget [21]

For 1993 est.

Revenues	45.1
Expenditures	52.5
Capital expenditures	NA

Crime [23]

Crime volume (for 1990)	
Cases known to police	17,793
Attempts (percent)	NA
Percent cases solved	NA
Crimes per 100,000 persons	119.65
Persons responsible for offenses	
Total number offenders	18,089
Percent female	6.5
Percent juvenile	5.0
Percent foreigners	38.0

Military Expenditures and Arms Transfers [22]

	1985	1986	1987	1988	1989
Military expenditures					
Current dollars (mil.)	21,340	17,290	16,210[e]	13,600	14,690
1989 constant dollars (mil.)	24,290	19,190	17,430[e]	14,160	14,690
Armed forces (000)	80[e]	80[e]	80[e]	84	82
Gross national product (GNP)					
Current dollars (mil.)	94,140	82,860	83,420	85,320	91,670
1989 constant dollars (mil.)[e]	107,200	91,950	89,730	88,820	91,670
Central government expenditures (CGE)					
1989 constant dollars (mil.)	89,960	59,960	37,650[e]	39,220[e]	38,100[e]
People (mil.)	13.5	14.2	14.9	15.6	16.4
Military expenditure as % of GNP	22.7	20.9	19.4	15.9	16.0
Military expenditure as % of CGE	27.0	32.0	46.3	36.1	38.5
Military expenditure per capita	1,803	1,353	1,170	906	897
Armed forces per 1,000 people	5.9	5.6	5.4	5.3	5.0
GNP per capita	7,954	6,482	6,020	5,681	5,600
Arms imports[6]					
Current dollars (mil.)[12]	3,800	5,500	7,000	2,700	4,200
1989 constant dollars (mil.)[12]	4,326	6,103	7,529	2,811	4,200
Arms exports[6]					
Current dollars (mil.)	10	20	20	5	5
1989 constant dollars (mil.)	11	22	22	5	5
Total imports[7]					
Current dollars (mil.)	23,620	19,110	20,110	21,780	19,770
1989 constant dollars	26,890	21,210	21,630	22,680	19,770
Total exports[7]					
Current dollars (mil.)	27,480	20,180	23,200	23,740	27,740
1989 constant dollars	31,280	22,400	24,950	24,710	27,740
Arms as percent of total imports[8]	16.1	28.8	34.8	12.4	21.2
Arms as percent of total exports[8]	0	0.1	0.1	0	0

Human Rights [24]

	SSTS	FL	FAPRO	PPCG	APROBC	TPW	PCPTW	STPEP	PHRFF	PRW	ASST	AFL
Observes	P	P		P		P	P					P
		EAFRD	CPR	ESCR	SR	ACHR	MAAE	PVIAC	PVNAC	EAFDAW	TCIDTP	P / RC
Observes								P				P

P = Party; S = Signatory; see Appendix for meaning of abbreviations.

Labor Force

Total Labor Force [25]

5 million

Labor Force by Occupation [26]

Government	34%
Industry and oil	28
Services	22
Agriculture	16

Unemployment Rate [27]

6.5% (1992 est.)

For sources, notes, and explanations, see Annotated Source Appendix, page 1035.

Production Sectors

Commercial Energy Production and Consumption

Production [28]

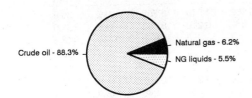

Crude oil - 88.3%
Natural gas - 6.2%
NG liquids - 5.5%

Consumption [29]

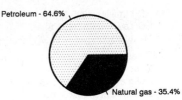

Petroleum - 64.6%
Natural gas - 35.4%

Data are shown in quadrillion (10^{15}) BTUs and percent for 1991

Crude oil[1]	17.65
Natural gas liquids	1.10
Dry natural gas	1.23
Coal[2]	0.00
Net hydroelectric power[3]	0.00
Net nuclear power[3]	0.00
Total	19.98

Petroleum	2.23
Dry natural gas	1.22
Coal[2]	0.00
Net hydroelectric power[3]	0.00
Net nuclear power[3]	0.00
Total	3.45

Telecommunications [30]

- Modern system with extensive microwave and coaxial and fiber optic cable systems
- 1,624,000 telephones
- Broadcast stations - 43 AM, 13 FM, 80 TV
- Microwave radio relay to Bahrain, Jordan, Kuwait, Qatar, UAE, Yemen, and Sudan
- Coaxial cable to Kuwait and Jordan
- Submarine cable to Djibouti, Egypt and Bahrain
- Earth stations - 3 Atlantic Ocean INTELSAT, 2 Indian Ocean INTELSAT, 1 ARABSAT, 1 INMARSAT

Transportation [31]

Railroads. 1390 km 1.435-meter standard gauge; 448 km are double tracked

Highways. 74,000 km total; 35,000 km paved, 39,000 km gravel and improved earth

Merchant Marine. 77 ships (1,000 GRT or over) totaling 860,818 GRT/1,219,345 DWT; includes 1 passenger, 6 short-sea passenger, 11 cargo, 13 roll-on/roll-off cargo, 3 container, 6 refrigerated cargo, 5 livestock carrier, 23 oil tanker, 6 chemical tanker, 1 liquefied gas, 1 specialized tanker, 1 bulk

Airports

Total:	213
Usable:	193
With permanent-surface runways:	71
With runways over 3,659 m:	14
With runways 2,440-3,659 m:	36
With runways 1,220-2,439 m:	107

Top Agricultural Products [32]

	1990	1991[29]
Wheat	3,600	4,000
Vegetables and melons	2,000	2,000
Date palm	502	505
Barley	350	375
Fruits	300	305
Sorghum	81	85

Values shown are 1,000 metric tons.

Top Mining Products [33]

Estimated metric tons unless otherwise specified   M.t.

Cement, hydraulic (000 tons)	13,000
Gypsum	375,000
Steel, crude (000 tons)	1,850
Gold, Au content of concentr. and bullion (kg)	4,300
Natural gas liquids, all forms	196,000
Petroleum, crude (mil. barrels)	2,985[21]

Tourism [34]

	1987	1988	1989	1990	1991
Visitors[62]	2,000	2,089	1,865	1,984	2,290
Tourists[63]	960	763	775	827	720
Tourism receipts	2,600	2,066	2,050		

Tourists are in thousands, money in million U.S. dollars.

Manufacturing Sector

GDP and Manufacturing Summary [35]

	1980	1985	1989	1990	% change 1980-1990	% change 1989-1990
GDP (million 1980 $)	115,962	95,862	84,548	115,756	-0.2	36.9
GDP per capita (1980 $)	12,372	8,267	6,219	8,183	-33.9	31.6
Manufacturing as % of GDP (current prices)	5.0	7.8	8.8	10.0	100.0	13.6
Gross output (million $)	10,798[e]	12,741[e]	NA	17,995[e]	66.7	NA
Value added (million $)	2,594[e]	3,286[e]	13,185[e]	5,387[e]	107.7	-59.1
Value added (million 1980 $)	5,800	9,805	9,252[e]	12,401	113.8	34.0
Industrial production index	100	175	227[e]	245	145.0	7.9
Employment (thousands)	79[e]	138[e]	NA	127[e]	60.8	NA

Note: GDP stands for Gross Domestic Product. 'e' stands for estimated value.

Profitability and Productivity

	1980	1985	1989	1990	% change 1980-1990	% change 1989-1990
Intermediate input (%)	NA	NA	NA	NA	NA	NA
Wages, salaries, and supplements (%)	NA	NA	NA	NA	NA	NA
Gross operating surplus (%)	NA	NA	NA	NA	NA	NA
Gross output per worker ($)	137,239[e]	92,275[e]	NA	141,222[e]	2.9	NA
Value added per worker ($)	NA	NA	NA	NA	NA	NA
Average wage (incl. benefits) ($)	NA	NA	NA	NA	NA	NA

Profitability is in percent of gross output. Productivity is in U.S. $. 'e' stands for estimated value.

Profitability - 1990

Value Added in Manufacturing

	1980 $ mil.	1980 %	1985 $ mil.	1985 %	1989 $ mil.	1989 %	1990 $ mil.	1990 %	% change 1980-1990	% change 1989-1990
311 Food products	306[e]	11.8	241[e]	7.3	834[e]	6.3	317[e]	5.9	3.6	-62.0
313 Beverages	15[e]	0.6	25[e]	0.8	104[e]	0.8	42[e]	0.8	180.0	-59.6
314 Tobacco products	29[e]	1.1	23[e]	0.7	175[e]	1.3	31[e]	0.6	6.9	-82.3
321 Textiles	23[e]	0.9	17[e]	0.5	170[e]	1.3	21[e]	0.4	-8.7	-87.6
322 Wearing apparel	3[e]	0.1	4[e]	0.1	201[e]	1.5	5[e]	0.1	66.7	-97.5
323 Leather and fur products	5[e]	0.2	4[e]	0.1	52[e]	0.4	5[e]	0.1	0.0	-90.4
324 Footwear	1[e]	0.0	1[e]	0.0	80[e]	0.6	1[e]	0.0	0.0	-98.8
331 Wood and wood products	11[e]	0.4	8[e]	0.2	122[e]	0.9	10[e]	0.2	-9.1	-91.8
332 Furniture and fixtures	13[e]	0.5	21[e]	0.6	72[e]	0.5	36[e]	0.7	176.9	-50.0
341 Paper and paper products	51[e]	2.0	75[e]	2.3	119[e]	0.9	110[e]	2.0	115.7	-7.6
342 Printing and publishing	58[e]	2.2	45[e]	1.4	115[e]	0.9	57[e]	1.1	-1.7	-50.4
351 Industrial chemicals	802[e]	30.9	963[e]	29.3	NA	0.0	1,908[e]	35.4	137.9	NA
352 Other chemical products	63[e]	2.4	99[e]	3.0	NA	0.0	167[e]	3.1	165.1	NA
353 Petroleum refineries	432[e]	16.7	582[e]	17.7	8,983	68.1	804	14.9	86.1	-91.0
354 Miscellaneous petroleum and coal products	25[e]	1.0	46[e]	1.4	NA	0.0	119[e]	2.2	376.0	NA
355 Rubber products	3[e]	0.1	4[e]	0.1	NA	0.0	8[e]	0.1	166.7	NA
356 Plastic products	61[e]	2.4	92[e]	2.8	809[e]	6.1	148[e]	2.7	142.6	-81.7
361 Pottery, china and earthenware	10[e]	0.4	14[e]	0.4	NA	0.0	25[e]	0.5	150.0	NA
362 Glass and glass products	12[e]	0.5	16[e]	0.5	NA	0.0	23[e]	0.4	91.7	NA
369 Other non-metal mineral products	238[e]	9.2	395[e]	12.0	692[e]	5.2	677[e]	12.6	184.5	-2.2
371 Iron and steel	175[e]	6.7	241[e]	7.3	26[e]	0.2	344[e]	6.4	96.6	1,223.1
372 Non-ferrous metals	8[e]	0.3	12[e]	0.4	9[e]	0.1	17[e]	0.3	112.5	88.9
381 Metal products	150[e]	5.8	204[e]	6.2	173[e]	1.3	271[e]	5.0	80.7	56.6
382 Non-electrical machinery	39[e]	1.5	53[e]	1.6	115[e]	0.9	61[e]	1.1	56.4	-47.0
383 Electrical machinery	35[e]	1.3	60[e]	1.8	125[e]	0.9	106[e]	2.0	202.9	-15.2
384 Transport equipment	11[e]	0.4	18[e]	0.5	148[e]	1.1	32[e]	0.6	190.9	-78.4
385 Professional and scientific equipment	1[e]	0.0	1[e]	0.0	NA	0.0	3[e]	0.1	200.0	NA
390 Other manufacturing industries	16[e]	0.6	24[e]	0.7	61[e]	0.5	38[e]	0.7	137.5	-37.7

Note: The industry codes shown are International Standard Industry codes (ISIC). Percentages are percent of total Value Added. 'e' stands for estimated value

798

For sources, notes, and explanations, see Annotated Source Appendix, page 1035.

Finance, Economics, and Trade

Economic Indicators [36]

Billions of Saudi Riyals (SR) unless otherwise indicated[141].

	1989	1990	1991[e]
Real GDP (1989 prices)	304.0	362.0	368.0
Real GDP Growth (%.)	1.0	19.0	1.5
Real Per Capita GDP (1989 prices, thousand riyals)	21.7	25.3	25.0
Money Supply (M2) (billion riyals, annual avg)	137.5	142.6	149.0
Comm. interest rates (%)	8.9	7.9	6.2
CPI (1988 = 100)	101.1	103.1	106.4
WPI (1988 = 100)	101.1	102.9	106.1
External Public Debt	0.7	0.7	5.2

Balance of Payments Summary [37]

Values in millions of dollars.

	1987	1988	1989	1990	1991
Exports of goods (f.o.b.)	23,138.0	24,315.0	28,299.0	44,296.0	48,219.0
Imports of goods (f.o.b.)	-18,283.0	-19,805.0	-19,231.0	-21,490.0	-26,123.0
Trade balance	4,855.0	4,510.0	9,068.0	22,806.0	22,096.0
Services - debits	-19,506.0	-15,650.0	-20,790.0	-23,268.0	-39,485.0
Services - credits	13,113.0	12,809.0	13,015.0	12,198.0	11,885.0
Private transfers (net)	-4,935.0	-6,510.0	-8,264.0	-11,602.0	-13,746.0
Government transfers (net)	-3,300.0	-2,499.0	-2,200.0	-4,401.0	-6,489.0
Long term capital (net)	4,975.0	2,729.0	-2,075.0	160.0	26,385.0
Short term capital (net)	7,438.0	3,092.0	7,739.0	-1,270.0	-597.0
Errors and omissions	-	-	-	-	-
Overall balance	2,640.0	-1,519.0	-3,507.0	-5,377.0	49.0

Exchange Rates [38]

Currency: **Saudi riyals.**
Symbol: **SR.**

Data are currency units per $1.

Fixed rate since late 1986	3.7450
1986	3.7033

Imports and Exports

Top Import Origins [39]

$26.1 billion (f.o.b., 1991).

Origins	%
US	21
UK	13
Japan	12
Germany	8
France	6

Top Export Destinations [40]

$48.2 billion (f.o.b., 1991).

Destinations	%
US	21
Japan	18
Singapore	6
France	6
Korea	5

Foreign Aid [41]

	U.S. $	
Donor - pledges of bilateral aid (1979-89)	64.7	billion

Import and Export Commodities [42]

Import Commodities

Food stuffs
Manufactured goods
Transportation equipment
Chemical products
Textiles

Export Commodities

Petroleum & petr. products 92%

For sources, notes, and explanations, see Annotated Source Appendix, page 1035.

799

Senegal

Geography [1]

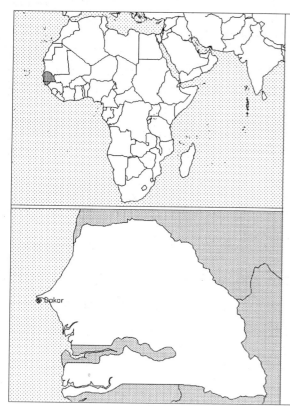

Total area:
196,190 km2
Land area:
192,000 km2
Comparative area:
Slightly smaller than South Dakota
Land boundaries:
Total 2,640 km, The Gambia 740 km, Guinea 330 km, Guinea-Bissau 338 km, Mali 419 km, Mauritania 813 km
Coastline:
531 km
Climate:
Tropical; hot, humid; rainy season (December to April) has strong southeast winds; dry season (May to November) dominated by hot, dry harmattan wind
Terrain:
Generally low, rolling, plains rising to foothills in southeast
Natural resources:
Fish, phosphates, iron ore
Land use:
Arable land:
27%
Permanent crops:
0%
Meadows and pastures:
30%
Forest and woodland:
31%
Other:
12%

Demographics [2]

	1960	1970	1980	1990	1991[1]	1994	2000	2010	2020
Population	3,270	4,318	5,731	7,715	7,953	8,731	10,533	14,318	19,127
Population density (persons per sq. mi.)	44	58	77	104	107	NA	141	190	249
Births	NA	NA	NA	NA	348	377	NA	NA	NA
Deaths	NA	NA	NA	NA	105	105	NA	NA	NA
Life expectancy - males	NA	NA	NA	NA	54	55	NA	NA	NA
Life expectancy - females	NA	NA	NA	NA	56	58	NA	NA	NA
Birth rate (per 1,000)	NA	NA	NA	NA	44	43	NA	NA	NA
Death rate (per 1,000)	NA	NA	NA	NA	13	12	NA	NA	NA
Women of reproductive age (15-44 yrs.)	NA	NA	NA	1,754	NA	2,007	2,420	NA	NA
of which are currently married	NA	NA	NA	1,345	NA	1,538	1,867	NA	NA
Fertility rate	NA	NA	NA	6.3	6.24	6.1	5.7	5.0	4.4

Population values are in thousands, life expectancy in years, and other items as indicated.

Health

Health Personnel [3]

Doctors per 1,000 pop., 1988-92	0.05
Nurse-to-doctor ratio, 1988-92	2.6
Hospital beds per 1,000 pop., 1985-90	0.8
Percentage of children immunized (age 1 yr. or less)	
Third dose of DPT, 1990-91	60.0
Measles, 1990-91	59.0

Health Expenditures [4]

Total health expenditure, 1990 (official exchange rate)	
Millions of dollars	214
Millions of dollars per capita	29
Health expenditures as a percentage of GDP	
Total	3.7
Public sector	2.3
Private sector	1.4
Development assistance for health	
Total aid flows (millions of dollars)[1]	36
Aid flows per capita (millions of dollars)	4.9
Aid flows as a percentage of total health expenditure	16.9

For sources, notes, and explanations, see Annotated Source Appendix, page 1035.

Human Factors

Health Care Ratios [5]

Population per physician, 1970	15810
Population per physician, 1990	17650
Population per nursing person, 1970	1670
Population per nursing person, 1990	NA
Percent of births attended by health staff, 1985	NA

Infants and Malnutrition [6]

Percent of babies with low birth weight, 1985	10
Infant mortality rate per 1,000 live births, 1970	135
Infant mortality rate per 1,000 live births, 1991	81
Years of life lost per 1,000 population, 1990	99
Prevalence of malnutrition (under age 5), 1990	22

Ethnic Division [7]

Wolof	36%
Fulani	17%
Serer	17%
Toucouleur	9%
Diola	9%
Mandingo	9%
European Lebanese	1%
Other	2%

Religion [8]

Muslim	92%
Indigenous beliefs	6%
Christian (mostly Roman Catholic)	2%

Major Languages [9]

French (official)	NA
Wolof	NA
Pulaar	NA
Diola	NA
Mandingo	NA

Education

Public Education Expenditures [10]

Million Francs C.F.A.	1980	1985	1986	1987	1988	1989
Total education expenditure	NA	NA	NA	NA	NA	NA
as percent of GNP	NA	NA	NA	NA	NA	NA
as percent of total govt. expend.	NA	NA	NA	NA	NA	NA
Current education expenditure	27,485	46,118	47,097	48,037	51,906	NA
as percent of GNP	4.5	4.2	3.8	3.7	3.7	NA
as percent of current govt. expend.	23.5	24.4	23.0	23.3	24.1	NA
Capital expenditure	NA	NA	NA	NA	NA	NA

Educational Attainment [11]

Literacy Rate [12]

In thousands and percent	1985[a]	1991[a]	2000[a]
Illiterate population +15 years	2,433	2,525	2,672
Illiteracy rate - total pop. (%)	67.9	61.7	49.5
Illiteracy rate - males (%)	54.6	48.1	36.3
Illiteracy rate - females (%)	80.7	74.9	62.2

Libraries [13]

	Admin. Units	Svc. Pts.	Vols. (000)	Shelving (meters)	Vols. Added	Reg. Users
National	NA	NA	NA	NA	NA	NA
Non-specialized	NA	NA	NA	NA	NA	NA
Public (1987)	10	11	15	382	1,721	959
Higher ed. (1987)[8]	1	NA	380	NA	4,134	NA
School	NA	NA	NA	NA	NA	NA

Daily Newspapers [14]

	1975	1980	1985	1990
Number of papers	1	1	3	1
Circ. (000)	25	35	53	50

Cinema [15]

Science and Technology

Scientific/Technical Forces [16]

R&D Expenditures [17]

U.S. Patents Issued [18]

Values show patents issued to citizens of the country by the U.S. Patents Office.

	1990	1991	1992
Number of patents	0	0	1

For sources, notes, and explanations, see Annotated Source Appendix, page 1035.

801

Government and Law

Organization of Government [19]

Long-form name:
Republic of Senegal
Type:
republic under multiparty democratic rule
Independence:
20 August 1960 (from France)
Constitution:
3 March 1963, last revised in 1991
Legal system:
based on French civil law system; judicial review of legislative acts in Supreme Court, which also audits the government's accounting office; has not accepted compulsory ICJ jurisdiction
National holiday:
Independence Day, 4 April (1960)
Executive branch:
president, prime minister, Council of Ministers (cabinet)
Legislative branch:
unicameral National Assembly (Assemblee Nationale)
Judicial branch:
Supreme Court (Cour Supreme)

Political Parties [20]

National Assembly	% of votes
Socialist Party (PS)	71.0
Senegalese Democratic Party (PDS)	25.0
Other	4.0

Government Budget [21]

For FY89 est.

Revenues	921
Expenditures	1,024
Capital expenditures	14

Military Expenditures and Arms Transfers [22]

	1985	1986	1987	1988	1989
Military expenditures					
Current dollars (mil.)	95[e]	99[e]	88[e]	89[e]	90[e]
1989 constant dollars (mil.)	108[e]	110[e]	94[e]	92[e]	90[e]
Armed forces (000)	18	18	18	14	15
Gross national product (GNP)					
Current dollars (mil.)	3,433	3,696	3,971	4,318	4,441
1989 constant dollars (mil.)	3,908	4,101	4,271	4,495	4,441
Central government expenditures (CGE)					
1989 constant dollars (mil.)	1,226[e]	1,476[e]	1,405[e]	1,488[e]	1,419
People (mil.)	6.6	6.8	7.0	7.3	7.5
Military expenditure as % of GNP	2.8	2.7	2.2	2.1	2.0
Military expenditure as % of CGE	8.8	7.5	6.7	6.2	6.3
Military expenditure per capita	16	16	13	13	12
Armed forces per 1,000 people	2.7	2.6	2.6	2.0	1.9
GNP per capita	589	600	606	619	593
Arms imports[6]					
Current dollars (mil.)	5	5	30	10	5
1989 constant dollars (mil.)	6	6	32	10	5
Arms exports[6]					
Current dollars (mil.)	0	0	0	0	0
1989 constant dollars (mil.)	0	0	0	0	0
Total imports[7]					
Current dollars (mil.)	826	961	1,024	1,100	1,127
1989 constant dollars	940	1,066	1,101	1,145	1,127
Total exports[7]					
Current dollars (mil.)	562	620	606	748	797
1989 constant dollars	640	688	652	779	797
Arms as percent of total imports[8]	0.6	0.5	2.9	0.9	0.4
Arms as percent of total exports[8]	0	0	0	0	0

Crime [23]

Crime volume

Cases known to police	10,061
Attempts (percent)	NA
Percent cases solved	88.07
Crimes per 100,000 persons	149.45
Persons responsible for offenses	
Total number offenders	8,129
Percent female	11.57
Percent juvenile[48]	6.70
Percent foreigners	10.99

Human Rights [24]

	SSTS	FL	FAPRO	PPCG	APROBC	TPW	PCPTW	STPEP	PHRFF	PRW	ASST	AFL
Observes	P	P	P	P	P	P	P	P		P	P	P
	EAFRD	CPR	ESCR	SR	ACHR	MAAE	PVIAC	PVNAC	EAFDAW	TCIDTP	RC
Observes		P	P	P	P			P	P	P	P	P

P = Party; S = Signatory; see Appendix for meaning of abbreviations.

Labor Force

Total Labor Force [25]

2.509 million (77% are engaged in subsistence farming; 175,000 wage earners)

Labor Force by Occupation [26]

Private sector	40%
Government and parapublic	60

Unemployment Rate [27]

For sources, notes, and explanations, see Annotated Source Appendix, page 1035.

Production Sectors

Energy Resource Summary [28]

Energy Resources: Electricity: 215,000 kW capacity; 760 million kWh produced, 100 kWh per capita (1991).

Telecommunications [30]

- Above-average urban system, using microwave and cable
- Broadcast stations - 8 AM, no FM, 1 TV
- 3 submarine cables
- 1 Atlantic Ocean INTELSAT earth station

Transportation [31]

Railroads. 1,034 km 1.000-meter gauge; all single track except 70 km double track Dakar to Thies

Highways. 14,007 km total; 3,777 km paved, 10,230 km laterite or improved earth

Merchant Marine. 1 bulk ship (1,000 GRT and over) totaling 1,995 GRT/3,775 DWT

Airports

Total:	25
Usable:	19
With permanent-surface runways:	10
With runways over 3,659 m:	0
With runways 2,440-3,659 m:	1
With runways 1,220-2,439 m:	15

Top Agricultural Products [32]

	89-90[1]	90-91[1]
Peanuts	820	679
Millet	640	514
Rice	168	156
Sorghum	127	147
Corn	131	133
Cassava	59	70

Values shown are 1,000 metric tons.

Top Mining Products [33]

Estimated metric tons unless otherwise specified   M.t.

Cement, hydraulic	503,317
Clay, attapulgite	129,403
Phosphate rock, crude	
Aluminum phosphate (000 tons)	2 9
Calcium phosphate (000 tons)	,7411
Salt	102,000

Tourism [34]

	1987	1988	1989	1990	1991
Tourists[10]	235	256	259	246	234
Cruise passengers	7	5	10	6	5
Tourism receipts	130	147	142	168	171
Tourism expenditures	55	76	72	105	103
Fare expenditures	53	52	50	78	79

Tourists are in thousands, money in million U.S. dollars.

Manufacturing Sector

GDP and Manufacturing Summary [35]

	1980	1985	1989	1990	% change 1980-1990	% change 1989-1990
GDP (million 1980 $)	2,970	3,446	3,843	4,050	36.4	5.4
GDP per capita (1980 $)	536	540	539	553	3.2	2.6
Manufacturing as % of GDP (current prices)	15.3	12.8	18.2[e]	13.8	-9.8	-24.2
Gross output (million $)	1,070	926	1,054[e]	1,475[e]	37.9	39.9
Value added (million $)	258	268	292[e]	639[e]	147.7	118.8
Value added (million 1980 $)	438	560	530	713	62.8	34.5
Industrial production index	100	94	104	115	15.0	10.6
Employment (thousands)	32	30	41[e]	26[e]	-18.8	-36.6

Note: GDP stands for Gross Domestic Product. 'e' stands for estimated value.

Profitability and Productivity

	1980	1985	1989	1990	% change 1980-1990	% change 1989-1990
Intermediate input (%)	76	71	72[e]	57[e]	-25.0	-20.8
Wages, salaries, and supplements (%)	10[e]	11[e]	15[e]	12[e]	20.0	-20.0
Gross operating surplus (%)	14[e]	18[e]	12[e]	31[e]	121.4	158.3
Gross output per worker ($)	33,812	22,546	25,468[e]	37,211[e]	10.1	46.1
Value added per worker ($)	8,164	6,528	7,069[e]	16,120[e]	97.5	128.0
Average wage (incl. benefits) ($)	3,508[e]	3,282[e]	3,941[e]	7,075[e]	101.7	79.5

Profitability is in percent of gross output. Productivity is in U.S. $. 'e' stands for estimated value.

Profitability - 1990

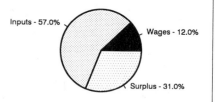

Inputs - 57.0%
Wages - 12.0%
Surplus - 31.0%

The graphic shows percent of gross output.

Value Added in Manufacturing

	1980 $ mil.	1980 %	1985 $ mil.	1985 %	1989 $ mil.	1989 %	1990 $ mil.	1990 %	% change 1980-1990	% change 1989-1990
311 Food products	106	41.1	93[e]	34.7	118[e]	40.4	232[e]	36.3	118.9	96.6
313 Beverages	11	4.3	13[e]	4.9	12[e]	4.1	19[e]	3.0	72.7	58.3
314 Tobacco products	7	2.7	10[e]	3.7	11[e]	3.8	6[e]	0.9	-14.3	-45.5
321 Textiles	33	12.8	26[e]	9.7	26[e]	8.9	10[e]	1.6	-69.7	-61.5
322 Wearing apparel	10	3.9	7[e]	2.6	10[e]	3.4	4[e]	0.6	-60.0	-60.0
323 Leather and fur products	5	1.9	1[e]	0.4	4[e]	1.4	NA	0.0	NA	NA
324 Footwear	2	0.8	1[e]	0.4	1[e]	0.3	NA	0.0	NA	NA
331 Wood and wood products	2	0.8	2[e]	0.7	1[e]	0.3	5[e]	0.8	150.0	400.0
332 Furniture and fixtures	2	0.8	3[e]	1.1	1[e]	0.3	6[e]	0.9	200.0	500.0
341 Paper and paper products	4	1.6	7[e]	2.6	3[e]	1.0	19[e]	3.0	375.0	533.3
342 Printing and publishing	6	2.3	9[e]	3.4	7[e]	2.4	26[e]	4.1	333.3	271.4
351 Industrial chemicals	16	6.2	22[e]	8.2	13[e]	4.5	66[e]	10.3	312.5	407.7
352 Other chemical products	5	1.9	9[e]	3.4	5[e]	1.7	26[e]	4.1	420.0	420.0
353 Petroleum refineries	18	7.0	8[e]	3.0	15[e]	5.1	66[e]	10.3	266.7	340.0
354 Miscellaneous petroleum and coal products	NA	0.0	NA	0.0	NA	0.0	NA	0.0	NA	NA
355 Rubber products	NA	0.0	NA	0.0	NA	0.0	NA	0.0	NA	NA
356 Plastic products	NA	0.0	NA	0.0	NA	0.0	NA	0.0	NA	NA
361 Pottery, china and earthenware	NA	0.0	NA	0.0	NA	0.0	NA	0.0	NA	NA
362 Glass and glass products	NA	0.0	NA	0.0	NA	0.0	NA	0.0	NA	NA
369 Other non-metal mineral products	12	4.7	19[e]	7.1	23[e]	7.9	51[e]	8.0	325.0	121.7
371 Iron and steel	NA	0.0	NA	0.0	NA	0.0	NA	0.0	NA	NA
372 Non-ferrous metals	NA	0.0	NA	0.0	NA	0.0	NA	0.0	NA	NA
381 Metal products	10	3.9	16[e]	6.0	23[e]	7.9	43[e]	6.7	330.0	87.0
382 Non-electrical machinery	3	1.2	5[e]	1.9	8[e]	2.7	17[e]	2.7	466.7	112.5
383 Electrical machinery	1	0.4	4[e]	1.5	2[e]	0.7	7[e]	1.1	600.0	250.0
384 Transport equipment	5	1.9	11[e]	4.1	11[e]	3.8	35[e]	5.5	600.0	218.2
385 Professional and scientific equipment	NA	0.0	NA	0.0	NA	0.0	NA	0.0	NA	NA
390 Other manufacturing industries	NA	0.0	NA	0.0	NA	0.0	NA	0.0	NA	NA

Note: The industry codes shown are International Standard Industry codes (ISIC). Percentages are percent of total Value Added. 'e' stands for estimated value

For sources, notes, and explanations, see Annotated Source Appendix, page 1035.

Finance, Economics, and Trade

Economic Indicators [36]

National product: GDP—exchange rate conversion—$5.4 billion (1991 est.). **National product real growth rate**: 1.2% (1991 est.). **National product per capita**: $780 (1991 est.). **Inflation rate (consumer prices)**: 2% (1990). **External debt**: $2.9 billion (1990).

Balance of Payments Summary [37]

Values in millions of dollars.

	1987	1988	1989	1990	1991
Exports of goods (f.o.b.)	670.9	678.6	758.6	911.6	903.2
Imports of goods (f.o.b.)	-955.8	-956.0	-998.4	-1,176.1	-1,187.1
Trade balance	-284.9	-277.4	-239.8	-264.5	-283.9
Services - debits	-708.9	-752.0	-722.5	-831.5	-830.2
Services - credits	453.2	491.3	498.0	585.8	582.8
Private transfers (net)	6.3	5.9	6.3	29.4	28.4
Government transfers (net)	228.2	271.2	259.9	355.9	370.4
Long term capital (net)	401.9	250.0	184.2	275.8	139.6
Short term capital (net)	-114.1	18.1	23.8	-116.3	-26.5
Errors and omissions	-0.2	-3.3	1.6	-16.7	10.9
Overall balance	-18.5	3.8	11.5	17.9	-8.5

Exchange Rates [38]

Currency: **Communaute Financi- ere Africaine francs.**

Symbol: **CFAF.**

Data are currency units per $1.

January 1993	274.06
1992	264.69
1991	282.11
1990	272.26
1989	319.01
1988	297.85

Imports and Exports

Top Import Origins [39]

$1.2 billion (c.i.f., 1991 est.).

Origins	%
France	NA
Other EC	NA
Cote d'Ivoire	NA
Nigeria	NA
Algeria	NA
China	NA
Japan	NA

Top Export Destinations [40]

$904 million (f.o.b., 1991 est.).

Destinations	%
France	NA
Other EC members	NA
Mali	NA
Cote d'Ivoire	NA
India	NA

Foreign Aid [41]

	U.S. $	
US commitments, including Ex-Im (FY70-89)	551	million
Western (non-US) countries, ODA and OOF bilateral commitments (1970-89)	5.23	billion
OPEC bilateral aid (1979-89)	589	million
Communist countries (1970-89)	295	million

Import and Export Commodities [42]

Import Commodities

Semimanufactures 30%
Food 27%
Durable consumer goods 17%
Petroleum 12%
Capital goods 14%

Export Commodities

Manufactures 30%
Fish products 23%
Peanuts 12%
Petroleum products 16%
Phosphates 9%

Serbia and Montenegro

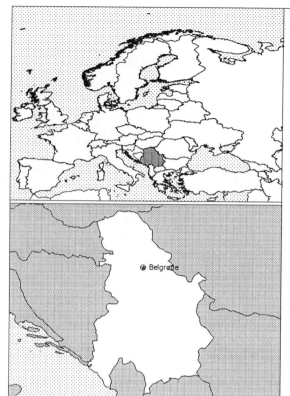

Geography [1]

Total area:
102,350 km2

Land area:
102,136 km2

Comparative area:
Slightly larger than Kentucky

Land boundaries:
Total 2,234 km, Albania 287 km (114 km with Serbia; 173 km with Motenegro), Bosnia and Herzegovina 527 km (312 km with Serbia; 215 km with Montenegro), Bulgaria 318 km, Croatia (north) 239 km, Croatia (south) 15 km, Hungary 151 km, Macedonia 221 km, Romania 476 km

Coastline:
199 km (Montenegro 199 km, Serbia 0 km)

Climate:
In the north, continental climate; central portion, continental and Mediterranean climate; to the south, Adriatic along the coast, hot, dry summers and autumns and cold winters inland

Terrain:
Extremely varied; to the north, rich fertile plains; to the east, limestone ranges and basins; to the southeast, ancient mountain and hills; to the southwest, extremely high shoreline with no islands off the coast

Land use:
Arable land: 30%
Permanent crops: 5%
Meadows and pastures: 20%
Forest and woodland: 25%
Other: 20%

Demographics [2]

	1960[2]	1970[2]	1980[2]	1990[2]	1991[1,2]	1994[2]	2000	2010	2020[2]
Population	8,050	8,910	9,841	10,528	NA	10,760	11,121	11,625	11,881
Population density (persons per sq. mi.)	NA	NA	NA	NA	NA	NA	NA	NA	NA
Births	NA	NA	NA	NA	NA	154	NA	NA	NA
Deaths	NA	NA	NA	NA	NA	94	NA	NA	NA
Life expectancy - males	NA	NA	NA	NA	NA	71	NA	NA	NA
Life expectancy - females	NA	NA	NA	NA	NA	76	NA	NA	NA
Birth rate (per 1,000)	NA	NA	NA	NA	NA	14	NA	NA	NA
Death rate (per 1,000)	NA	NA	NA	NA	NA	9	NA	NA	NA
Women of reproductive age (15-44 yrs.)	NA	NA	NA	2,538	NA	2,619	2,695	NA	NA
of which are currently married	NA	NA	NA	1,759	NA	1,812	1,874	NA	NA
Fertility rate	NA	NA	NA	2.1	NA	2.0	2.0	1.9	1.8

Population values are in thousands, life expectancy in years, and other items as indicated.

Health

Health Personnel [3]

Doctors per 1,000 pop., 1988-92	2.63 [2]
Nurse-to-doctor ratio, 1988-92	1.9 [2]
Hospital beds per 1,000 pop., 1985-90	6.0 [2]
Percentage of children immunized (age 1 yr. or less)	
Third dose of DPT, 1990-91	79.0 [2]
Measles, 1990-91	75.0 [2]

Health Expenditures [4]

Total health expenditure, 1990 (official exchange rate)	
Millions of dollars	4512
Millions of dollars per capita	205
Health expenditures as a percentage of GDP	
Total	3.0
Public sector	4.0
Private sector	1.0
Development assistance for health	
Total aid flows (millions of dollars)[1]	NA
Aid flows per capita (millions of dollars)	NA
Aid flows as a percentage of total health expenditure	NA

For sources, notes, and explanations, see Annotated Source Appendix, page 1035.

Human Factors

Health Care Ratios [5]

Population per physician, 1970	1000 [2]
Population per physician, 1990	530 [2]
Population per nursing person, 1970	420 [2]
Population per nursing person, 1990	110 [2]
Percent of births attended by health staff, 1985	NA

Infants and Malnutrition [6]

Percent of babies with low birth weight, 1985	7
Infant mortality rate per 1,000 live births, 1970	56
Infant mortality rate per 1,000 live births, 1991	21
Years of life lost per 1,000 population, 1990	16
Prevalence of malnutrition (under age 5), 1990	NA

Ethnic Division [7]

Serbs	63%
Albanians	14%
Montenegrins	6%
Hungarians	4%
Other	13%

Religion [8]

Orthodox	65%
Muslim	19%
Roman Catholic	4%
Protestant	1%
Other	11%

Major Languages [9]

Serbo-Croatian	95%
Albanian	5%

Education

Public Education Expenditures [10]

Million Dinar	1980[22]	1985[22]	1987[22]	1988[22]	1989[22]	1990[22]
Total education expenditure	8	42	202	544	9,555	60,318
as percent of GNP	4.7	3.4	4.2	3.6	4.3	6.1
as percent of total govt. expend.	32.5	NA	NA	NA	NA	NA
Current education expenditure	7	39	185	506	9,002	55,911
as percent of GNP	4.0	3.1	3.9	3.3	4.1	5.7
as percent of current govt. expend.	NA	NA	NA	NA	NA	NA
Capital expenditure	1	3	17	38	553	4,407

Educational Attainment [11]

Age group	25+ [1]
Total population	13,083,762 [1]
Highest level attained (%)	
No schooling	15.8 [1]
First level	
Incompleted	53.9 [1]
Completed	NA [1]
Entered second level	
S-1	23.4 [1]
S-2	NA [1]
Post secondary	6.8 [1]

Literacy Rate [12]

In thousands and percent	1985[a]	1991[a]	2000[a]
Illiterate population +15 years	1,614	1,342	942
Illiteracy rate - total pop. (%)	9.2	7.3	4.7
Illiteracy rate - males (%)	3.5	2.6	1.3
Illiteracy rate - females (%)	14.6	11.9	7.9

Libraries [13]

	Admin. Units[a]	Svc. Pts.[a]	Vols. (000)[a]	Shelving (meters)[a]	Vols. Added[a]	Reg. Users[a]
National (1989)	8	8	12,316	303,555	305,462	163,169
Non-specialized (1989)	20	20	3,488	60,443	72,503	49,262
Public (1989)	808	1,937	30,238	552,866	1.3mil	19.5mil
Higher ed. (1989)	409	421	14,462	319,329	469,142	529,549
School (1989)	7,784	7,784	38,430	NA	1,680	NA

Daily Newspapers [14]

	1975[b]	1980[b]	1985[b]	1990[b]
Number of papers	26	27	27	34
Circ. (000)	1,896	2,649	2,451	2,281

Cinema [15]

Data for 1989.

Cinema seats per 1,000	16.9 [a]
Annual attendance per person	1.9 [a]
Gross box office receipts (mil. Dinar)	643,490 [a]

Science and Technology

Scientific/Technical Forces [16]

Potential scientists/engineers	563,312
Number female	NA
Potential technicians	432,380
Number female	NA
Total	995,692

R&D Expenditures [17]

	Dinar (000) 1989
Total expenditure[a,4,12]	2,152,032
Capital expenditure	815,082
Current expenditure	1,336,950
Percent current	62.1

U.S. Patents Issued [18]

Values show patents issued to citizens of the country by the U.S. Patents Office.

	1990	1991	1992
Number of patents	24	25	18

Government and Law

Organization of Government [19]

Long-form name:
none
Type:
republic
Independence:
11 April 1992 (from Yugoslavia)
Constitution:
27 April 1992
Legal system:
based on civil law system
National holiday:
NA
Executive branch:
president, vice president, prime minister, deputy prime minister, cabinet
Legislative branch:
bicameral Federal Assembly consists of an upper house or Chamber of Republics and a lower house or Chamber of Deputies
Judicial branch:
Savezni Sud (Federal Court), Constitutional Court

Political Parties [20]

Chamber of Citizens	% of seats
Serbian Socialist Party (SPS)	52.9
Serbian Radical Party (SRS)	23.9
Dem. Party of Socialists (DSSCG)	16.7
League of Communists (SK-PJ)	1.4
Democratic Community (DZVM)	1.4
Independents	1.4
Vacant	2.2

Government Budget [21]

For FY89 est.

Revenues	NA
Expenditures	NA
Capital expenditures	NA

Crime [23]

Military Expenditures and Arms Transfers [22]

	1985	1986	1987	1988	1989
Military expenditures					
Current dollars (mil.)	1,884	2,161	2,287	2,528	2,126
1989 constant dollars (mil.)	2,144	2,398	2,460	2,631	2,126
Armed forces (000)	258	234	234	229	225
Gross national product (GNP)					
Current dollars (mil.)	51,150	54,740	55,610	55,890	58,640
1989 constant dollars (mil.)	58,230	60,750	59,810	58,180	58,640
Central government expenditures (CGE)					
1989 constant dollars (mil.)	3,910	4,000	4,465	5,406	3,981
People (mil.)	23.1	23.3	23.4	23.6	23.7
Military expenditure as % of GNP	3.7	3.9	4.1	4.5	3.6
Military expenditure as % of CGE	54.8	60.0	55.1	48.7	53.4
Military expenditure per capita	93	103	105	112	90
Armed forces per 1,000 people	11.2	10.0	10.0	9.7	9.5
GNP per capita	2,518	2,610	2,554	2,469	2,474
Arms imports[6]					
Current dollars (mil.)	30	30	625	40	120
1989 constant dollars (mil.)	34	33	672	42	120
Arms exports[6]					
Current dollars (mil.)	420	360	320	230	150
1989 constant dollars (mil.)	478	399	344	239	150
Total imports[7]					
Current dollars (mil.)	12,210	11,750	12,630	13,170	14,830
1989 constant dollars	13,900	13,040	13,590	13,710	14,830
Total exports[7]					
Current dollars (mil.)	10,700	10,350	11,440	12,660	13,460
1989 constant dollars	12,180	11,490	12,310	13,180	13,460
Arms as percent of total imports[8]	0.2	0.3	4.9	0.3	0.8
Arms as percent of total exports[8]	3.9	3.5	2.8	1.8	1.1

Human Rights [24]

Labor Force

Total Labor Force [25]

2,640,909

Labor Force by Occupation [26]

Industry, mining	40%
Agriculture	5

Date of data: 1990

Unemployment Rate [27]

25%-40% (1991 est.)

For sources, notes, and explanations, see Annotated Source Appendix, page 1035.

Production Sectors

Commercial Energy Production and Consumption

Production [28]

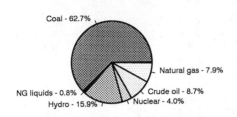

Coal - 62.7%
Natural gas - 7.9%
NG liquids - 0.8%
Crude oil - 8.7%
Hydro - 15.9%
Nuclear - 4.0%

Consumption [29]

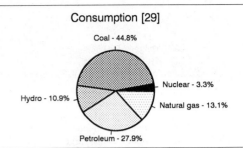

Coal - 44.8%
Nuclear - 3.3%
Hydro - 10.9%
Natural gas - 13.1%
Petroleum - 27.9%

Data are shown in quadrillion (10^{15}) BTUs and percent for 1991

Crude oil[1]	0.11[a]
Natural gas liquids	0.01[a]
Dry natural gas	0.10[a]
Coal[2]	0.79[a]
Net hydroelectric power[3]	0.20[a]
Net nuclear power[3]	0.05[a]
Total	1.26[a]

Petroleum	0.51[a]
Dry natural gas	0.24[a]
Coal[2]	0.82[a]
Net hydroelectric power[3]	0.20[a]
Net nuclear power[3]	0.06[a]
Total	1.83[a]

Telecommunications [30]

- 700,000 telephones
- Broadcast stations - 26 AM, 9 FM, 18 TV
- 2,015,000 radios
- 1,000,000 TVs
- Satellite ground stations - 1 Atlantic Ocean INTELSAT

Transportation [31]

Railroads. NA

Highways. 46,019 km total (1990); 26,949 km paved, 10,373 km gravel, 8,697 km earth

Airports

Total:	48
Useable:	48
With permanent-surface runways:	16
With runways over 3,659 m:	0
With runways 2,440-3,659 m:	6
With runways 1,220-2,439 m:	9

Top Agricultural Products [32]

The fertile plains of Vojvodina produce 80% of the cereal production of the former Yugoslavia and most of the cotton, oilseeds, and chicory; Vojvodina also produces fodder crops; Serbia produces fruit, grapes, and cereals; livestock production and dairy farming prosper; Kosovo produces fruits, vegetables, tobacco, and some cereals; Kosovo and Montenegro support sheep and goat husbandry; Montenegro also grows olives, citrus, grapes, and rice.

Top Mining Products [33]

Metric tons unless otherwise specified	M.t.
Magnesite, crude	175,000
Bauxite (000 tons)	2,850
Copper, Cu content of concentrate	140,000
Gypsum, crude	400,000
Pumice, volcanic tuff	380,000
Cement, hydraulic (000 tons)	7,100

Tourism [34]

	1987	1988	1989	1990	1991
Visitors[74]	26,151	29,635	34,118	39,573	
Tourists[6,74]	8,907	9,018	8,644	7,880	1,459
Tourism receipts[74]	1,668	2,024	2,230	2,774	468
Tourism expenditures[74]	90	109	131	149	103
Fare receipts[74]	340	430	475	620	105

Tourists are in thousands, money in million U.S. dollars.

Manufacturing Sector

GDP and Manufacturing Summary [35]

	1980	1985	1989	1990	% change 1980-1990	% change 1989-1990
GDP (million 1980 $)	69,958	71,058	72,234	66,371	-5.1	-8.1
GDP per capita (1980 $)	3,136	3,073	3,050	2,786	-11.2	-8.7
Manufacturing as % of GDP (current prices)	30.6	37.2	39.5	42.0	37.3	6.3
Gross output (million $)	72,629	57,020	65,078	62,136[e]	-14.4	-4.5
Value added (million $)	21,750	17,171	30,245	27,660[e]	27.2	-8.5
Value added (million 1980 $)	19,526	22,283	24,021	21,703	11.1	-9.6
Industrial production index	100	116	120	108	8.0	-10.0
Employment (thousands)	2,106	2,467	2,658	2,537[e]	20.5	-4.6

Note: GDP stands for Gross Domestic Product. 'e' stands for estimated value.

Profitability and Productivity

	1980	1985	1989	1990	% change 1980-1990	% change 1989-1990
Intermediate input (%)	70	70	54	55[e]	-21.4	1.9
Wages, salaries, and supplements (%)	14	12	12[e]	18[e]	28.6	50.0
Gross operating surplus (%)	15	18	34[e]	26[e]	73.3	-23.5
Gross output per worker ($)	34,487	23,113	24,484	24,248[e]	-29.7	-1.0
Value added per worker ($)	10,328	6,960	11,379	10,796[e]	4.5	-5.1
Average wage (incl. benefits) ($)	4,991	2,703	2,986[e]	4,488[e]	-10.1	50.3

Profitability is in percent of gross output. Productivity is in U.S. $. 'e' stands for estimated value.

Profitability - 1990

Wages - 18.2%
Inputs - 55.6%
Surplus - 26.3%

The graphic shows percent of gross output.

Value Added in Manufacturing

	1980 $ mil.	1980 %	1985 $ mil.	1985 %	1989 $ mil.	1989 %	1990 $ mil.	1990 %	% change 1980-1990	% change 1989-1990
311 Food products	1,897	8.7	1,458	8.5	3,916	12.9	3,484[e]	12.6	83.7	-11.0
313 Beverages	459	2.1	353	2.1	663	2.2	589[e]	2.1	28.3	-11.2
314 Tobacco products	184	0.8	221	1.3	344	1.1	308[e]	1.1	67.4	-10.5
321 Textiles	1,759	8.1	1,428	8.3	2,881	9.5	2,663[e]	9.6	51.4	-7.6
322 Wearing apparel	903	4.2	718	4.2	1,593	5.3	1,427[e]	5.2	58.0	-10.4
323 Leather and fur products	226	1.0	231	1.3	383	1.3	340[e]	1.2	50.4	-11.2
324 Footwear	482	2.2	503	2.9	1,022	3.4	899[e]	3.3	86.5	-12.0
331 Wood and wood products	977	4.5	530	3.1	794	2.6	706[e]	2.6	-27.7	-11.1
332 Furniture and fixtures	730	3.4	438	2.6	1,030	3.4	1,065[e]	3.9	45.9	3.4
341 Paper and paper products	529	2.4	394	2.3	759	2.5	674[e]	2.4	27.4	-11.2
342 Printing and publishing	876	4.0	462	2.7	761	2.5	678[e]	2.5	-22.6	-10.9
351 Industrial chemicals	694	3.2	631	3.7	1,107	3.7	992[e]	3.6	42.9	-10.4
352 Other chemical products	681	3.1	525	3.1	1,419	4.7	1,315[e]	4.8	93.1	-7.3
353 Petroleum refineries	454	2.1	415	2.4	260	0.9	233[e]	0.8	-48.7	-10.4
354 Miscellaneous petroleum and coal products	101	0.5	101	0.6	104	0.3	91[e]	0.3	-9.9	-12.5
355 Rubber products	276	1.3	269	1.6	479	1.6	456[e]	1.6	65.2	-4.8
356 Plastic products	413	1.9	258	1.5	397	1.3	350[e]	1.3	-15.3	-11.8
361 Pottery, china and earthenware	128	0.6	72	0.4	162	0.5	144[e]	0.5	12.5	-11.1
362 Glass and glass products	163	0.7	113	0.7	224	0.7	204[e]	0.7	25.2	-8.9
369 Other non-metal mineral products	906	4.2	513	3.0	683	2.3	604[e]	2.2	-33.3	-11.6
371 Iron and steel	1,221	5.6	1,000	5.8	1,343	4.4	1,171[e]	4.2	-4.1	-12.8
372 Non-ferrous metals	480	2.2	509	3.0	944	3.1	927[e]	3.4	93.1	-1.8
381 Metal products	2,105	9.7	1,577	9.2	1,293	4.3	1,130[e]	4.1	-46.3	-12.6
382 Non-electrical machinery	1,828	8.4	1,463	8.5	2,372	7.8	2,378[e]	8.6	30.1	0.3
383 Electrical machinery	1,600	7.4	1,544	9.0	2,640	8.7	2,334[e]	8.4	45.9	-11.6
384 Transport equipment	1,441	6.6	1,263	7.4	2,389	7.9	2,241[e]	8.1	55.5	-6.2
385 Professional and scientific equipment	101	0.5	93	0.5	154	0.5	146[e]	0.5	44.6	-5.2
390 Other manufacturing industries	134	0.6	88	0.5	128	0.4	114[e]	0.4	-14.9	-10.9

Note: The industry codes shown are International Standard Industry codes (ISIC). Percentages are percent of total Value Added. 'e' stands for estimated value

Finance, Economics, and Trade

Economic Indicators [36]

Yugoslav Dinars or U.S. Dollars as indicated[12].

	1989	1990	1991[e]
GSP[13]			
(billions current dinars)	221.9	910.9	1,490
Real GSP growth rate (%)	0.6	-7.6	-20
Money supply (M1)			
(yr-end mil.)	51,216	127,241	NA
Commercial interest			
rates (avg.)	4,354	45	80
Investment rate (%)	19.4	18.9	NA
CPI	1,356	688	310
WPI	1,406	534	320
External public debt	18,569	17,791	15,760

Balance of Payments Summary [37]

Exchange Rates [38]

Currency: **Yugoslav New Dinars.**
Symbol: **YD.**

Data are currency units per $1.

December 1991	28.230
1990	15.162
1989	15.528
1988	0.701
1987	0.176

Imports and Exports

Top Import Origins [39]

$6.4 billion (c.i.f., 1990). Prior to the imposition of sanctions by the UN Security Council the trade partners were principally the other former Yugoslav republics; the successor states of the former USSR, EC countries (mainly Italy and Germany), East European countries, US.

Origins	%
No details available	NA

Top Export Destinations [40]

$4.4 billion (f.o.b., 1990). Data relate to period before imposition of sanctions.

Destinations	%
Germany	NA
Other EC	NA
Successor states of the	
former USSR	NA
East European countries	NA
US	NA

Foreign Aid [41]

Import and Export Commodities [42]

Import Commodities

Machinery, transp. equip. 26%
Fuels and lubricants 18%
Manufactured goods 16%
Chemicals 12.5%
Food and live animals 11%
Misc. manufactured items 8%
Raw materials (coking coal) 7%
Beverages
Tobacco
Edible oils 1.5%

Export Commodities

Machinery & transp. equip. 29%
Manufactured goods 28.5%
Manufactured articles 13.5%
Chemicals 11%
Food and live animals 9%
Raw materials 6%
Fuels and lubricants 2%
Beverages & tobacco 1%

Seychelles

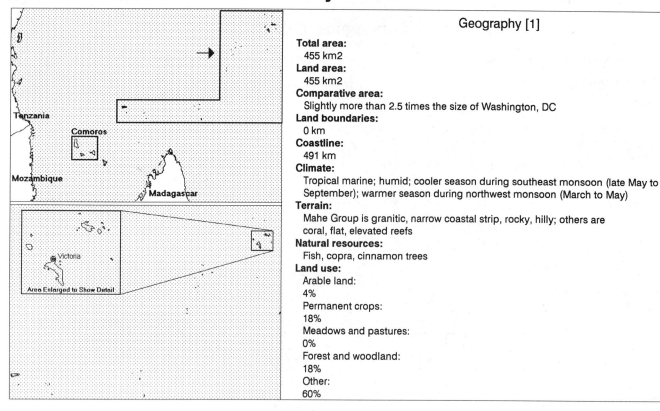

Geography [1]

Total area:
455 km2
Land area:
455 km2
Comparative area:
Slightly more than 2.5 times the size of Washington, DC
Land boundaries:
0 km
Coastline:
491 km
Climate:
Tropical marine; humid; cooler season during southeast monsoon (late May to September); warmer season during northwest monsoon (March to May)
Terrain:
Mahe Group is granitic, narrow coastal strip, rocky, hilly; others are coral, flat, elevated reefs
Natural resources:
Fish, copra, cinnamon trees
Land use:
Arable land:
4%
Permanent crops:
18%
Meadows and pastures:
0%
Forest and woodland:
18%
Other:
60%

Demographics [2]

	1960	1970	1980	1990	1991[1]	1994	2000	2010	2020
Population	42	54	65	70	69	72	75	81	86
Population density (persons per sq. mi.)	237	303	359	388	392	NA	420	456	494
Births	NA	NA	NA	NA	2	2	NA	NA	NA
Deaths	NA	NA	NA	NA	(Z)	1	NA	NA	NA
Life expectancy - males	NA	NA	NA	NA	65	66	NA	NA	NA
Life expectancy - females	NA	NA	NA	NA	75	73	NA	NA	NA
Birth rate (per 1,000)	NA	NA	NA	NA	23	22	NA	NA	NA
Death rate (per 1,000)	NA	NA	NA	NA	7	7	NA	NA	NA
Women of reproductive age (15-44 yrs.)	NA	NA	NA	18	NA	20	22	NA	NA
of which are currently married	NA	NA	NA	7	NA	8	10	NA	NA
Fertility rate	NA	NA	NA	2.5	2.50	2.2	2.0	1.9	1.9

Population values are in thousands, life expectancy in years, and other items as indicated.

Health

Health Personnel [3]

Health Expenditures [4]

For sources, notes, and explanations, see Annotated Source Appendix, page 1035.

Human Factors

Health Care Ratios [5]	Infants and Malnutrition [6]

Ethnic Division [7]	Religion [8]		Major Languages [9]
Seychellois (mixture of Asians, Africans, Europeans).	Roman Catholic Anglican Other	90% 8% 2%	English (official), French (official), Creole.

Education

Public Education Expenditures [10]

Million Rupee	1980	1985	1986	1988	1989	1990
Total education expenditure	52	125	126	128	143	153
as percent of GNP	5.8	10.7	10.2	9.0	8.9	8.5
as percent of total govt. expend.	14.4	21.3	16.0	15.2	15.2	11.9
Current education expenditure	50	120	121	126	133	153
as percent of GNP	5.5	10.3	9.8	8.8	8.3	8.5
as percent of current govt. expend.	14.0	21.9	20.7	18.4	16.9	NA
Capital expenditure	2	4	5	3	10	-

Educational Attainment [11]

Literacy Rate [12]

	1971[b]	1980[b]	1990[b]
Illiterate population +15 years	12,494	NA	NA
Illiteracy rate - total pop. (%)	42.3	NA	NA
Illiteracy rate - males (%)	44.4	NA	NA
Illiteracy rate - females (%)	40.2	NA	NA

Libraries [13]

	Admin. Units	Svc. Pts.	Vols. (000)	Shelving (meters)	Vols. Added	Reg. Users
National	NA	NA	NA	NA	NA	NA
Non-specialized	NA	NA	NA	NA	NA	NA
Public (1989)	1	4	42	NA	1,285[9]	16,664
Higher ed. (1990)	1	5	26	510	2,270	1,610[e]
School (1990)	24	24	NA	1,083	8,106[9]	7,940

Daily Newspapers [14]

	1975	1980	1985	1990
Number of papers	2	1	1	1
Circ. (000)	4	3	3	3

Cinema [15]

Science and Technology

Scientific/Technical Forces [16]

Potential scientists/engineers[18]	900
Number female	NA
Potential technicians	NA
Number female	NA
Total	NA

R&D Expenditures [17]

	Rupee[4] (000) 1983
Total expenditure	12,854
Capital expenditure	6,771
Current expenditure	6,083
Percent current	47.3

U.S. Patents Issued [18]

For sources, notes, and explanations, see Annotated Source Appendix, page 1035.

Government and Law

Organization of Government [19]

Long-form name:
Republic of Seychelles
Type:
republic
Independence:
29 June 1976 (from UK)
Constitution:
5 June 1979
Legal system:
based on English common law, French civil law, and customary law
National holiday:
Liberation Day, 5 June (1977) (anniversary of coup)
Executive branch:
president, Council of Ministers
Legislative branch:
unicameral People's Assembly (Assemblee du Peuple)
Judicial branch:
Court of Appeal, Supreme Court

Political Parties [20]

People's Assembly	% of seats
Seychelles People's Progressive Front (SPPF)	92.0
Appointed	8.0

Government Budget [21]

For 1989.

Revenues	180
Expenditures	202
Capital expenditures	32

Defense Summary [22]

Branches: Army, National Guard, Marines, Coast Guard, Presidential Protection Unit, Police Force

Manpower Availability: Males age 15-49 18,982; fit for military service 9,710 (1993 est.)

Defense Expenditures: Exchange rate conversion - $12 million, 4% of GDP (1990 est.)

Crime [23]

Crime volume	
Cases known to police	3,088
Attempts (percent)	NA
Percent cases solved	NA
Crimes per 100,000 persons	4,583.09
Persons responsible for offenses	
Total number offenders	428
Percent female	NA
Percent juvenile	NA
Percent foreigners	NA

Human Rights [24]

	SSTS	FL	FAPRO	PPCG	APROBC	TPW	PCPTW	STPEP	PHRFF	PRW	ASST	AFL
Observes	2	P	P	P		P	P			1	P	P
	EAFRD	CPR	ESCR	SR	ACHR	MAAE	PVIAC	PVNAC	EAFDAW	TCIDTP	RC	
Observes	P	P	P	P			P	P	P	P	P	

P = Party; S = Signatory; see Appendix for meaning of abbreviations.

Labor Force

Total Labor Force [25]

27,700 (1985)

Labor Force by Occupation [26]

Industry and commerce	31%
Services	21
Government	20
Agriculture, forestry, and fishing	12
Other	16

Date of data: 1985

Unemployment Rate [27]

9% (1987)

Production Sectors

Energy Resource Summary [28]

Energy Resources: None. **Electricity**: 30,000 kW capacity; 80 million kWh produced, 1,160 kWh per capita (1991).

Telecommunications [30]

- Direct radio communications with adjacent islands and African coastal countries
- 13,000 telephones
- Broadcast stations - 2 AM, no FM, 2 TV
- 1 Indian Ocean INTELSAT earth station
- USAF tracking station

Top Agricultural Products [32]

Agriculture accounts for 7% of GDP, mostly subsistence farming; cash crops—coconuts, cinnamon, vanilla; other products—sweet potatoes, cassava, bananas; broiler chickens; large share of food needs imported; expansion of tuna fishing under way.

Top Mining Products [33]

Detailed information is not available. A summary of mineral resources available follows. **Mineral Resources**: None.

Transportation [31]

Highways. 260 km total; 160 km paved, 100 km crushed stone or earth

Merchant Marine. 1 refrigerated cargo totaling 1,827 GRT/2,170 DWT

Airports

Total:	14
Usable:	14
With permanent-surface runways:	8
With runways over 3,659 m:	0
With runways 2,440-3,659 m:	1
With runways 1,220-2,439 m:	1

Tourism [34]

	1987	1988	1989	1990	1991
Visitors	77	80	88	112	98
Tourists	72	77	86[16]	104[57]	90
Cruise passengers	5	3	2	8	8
Tourism receipts	67	81	91	120	99
Tourism expenditures	12	13	17	20	12
Fare receipts	18	16	21	19	33
Fare expenditures	5	6	6	6	6

Tourists are in thousands, money in million U.S. dollars.

For sources, notes, and explanations, see Annotated Source Appendix, page 1035.

815

Finance, Economics, and Trade

GDP and Manufacturing Summary [35]

	1980	1985	1990	1991	1992
Gross Domestic Product					
Millions of 1980 dollars	147	158	197	201	205[e]
Growth rate in percent	-2.55	10.26	6.62	2.50	1.91[e]
Manufacturing Value Added					
Millions of 1980 dollars	11	11	17	19	20[e]
Growth rate in percent	18.21	8.44	11.04	8.21	7.89[e]
Manufacturing share in percent of current prices	8.0	10.6	9.8	NA	NA

Economic Indicators [36]

National product: GDP—exchange rate conversion— $350 million (1991 est.). **National product real growth rate**: - 4.5% (1991 est.). **National product per capita**: $5,200 (1991 est.). **Inflation rate (consumer prices)**: 1.8% (1990 est.). **External debt**: $189 million (1991 est.).

Balance of Payments Summary [37]

Values in millions of dollars.

	1987	1988	1989	1990	1991
Exports of goods (f.o.b.)	8.1	17.3	14.5	28.1	18.4
Imports of goods (f.o.b.)	-96.3	-135.0	-139.6	-158.4	-146.5
Trade balance	-88.2	-117.7	-125.1	-130.3	-128.1
Services - debits	-101.7	-102.1	-114.1	-130.1	-131.4
Services - credits	147.6	167.5	189.9	233.2	238.5
Private transfers (net)	-2.3	-4.9	-5.2	-2.5	-2.5
Government transfers (net)	23.5	28.9	31.7	29.3	27.6
Long term capital (net)	17.8	20.9	28.3	15.1	20.5
Short term capital (net)	1.1	-0.9	4.6	-2.7	-4.2
Errors and omissions	6.2	4.2	-6.5	-7.9	-9.9
Overall balance	4.0	-4.2	3.6	4.1	10.5

Exchange Rates [38]

Currency: **Seychelles rupees.**
Symbol: **SRe.**

Data are currency units per $1.

January 1993	5.2545
1992	5.1220
1991	5.2893
1990	5.3369
1989	5.6457
1988	5.3836

Imports and Exports

Top Import Origins [39]

$186 million (f.o.b., 1990 est.). Data are for 1987.

Origins	%
UK	20
France	14
South Africa	13
Yemen	13
Singapore	8
Japan	6

Top Export Destinations [40]

$40 million (f.o.b., 1990 est.). Data are for 1987.

Destinations	%
France	63
Pakistan	12
Reunion	10
UK	7

Foreign Aid [41]

	U.S. $	
US commitments, including Ex-Im (FY78-89)	26	million
Western (non-US) countries, ODA and OOF bilateral commitments (1978-89)	315	million
OPEC bilateral aid (1979-89)	5	million
Communist countries (1970-89)	60	million

Import and Export Commodities [42]

Import Commodities

Manufactured goods
Food
Tobacco
Beverages
Machinery & transport equipment
Petroleum products

Export Commodities

Fish
Copra
Cinnamon bark
Petroleum products (reexports)

Sierra Leone

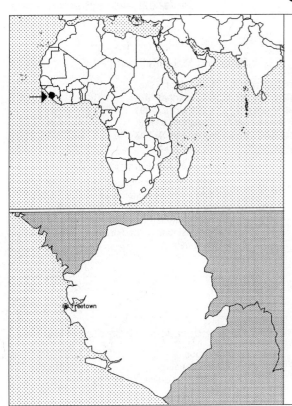

Geography [1]

Total area:
71,740 km2
Land area:
71,620 km2
Comparative area:
Slightly smaller than South Carolina
Land boundaries:
Total 958 km, Guinea 652 km, Liberia 306 km
Coastline:
402 km
Climate:
Tropical; hot, humid; summer rainy season (May to December); winter dry season (December to April)
Terrain:
Coastal belt of mangrove swamps, wooded hill country, upland plateau, mountains in east
Natural resources:
Diamonds, titanium ore, bauxite, iron ore, gold, chromite
Land use:
Arable land:
25%
Permanent crops:
2%
Meadows and pastures:
31%
Forest and woodland:
29%
Other:
13%

Demographics [2]

	1960	1970	1980	1990	1991[1]	1994	2000	2010	2020
Population	2,396	2,789	3,315	4,230	4,275	4,630	5,421	7,041	9,036
Population density (persons per sq. mi.)	87	101	120	151	155	NA	195	252	323
Births	NA	NA	NA	NA	197	209	NA	NA	NA
Deaths	NA	NA	NA	NA	87	87	NA	NA	NA
Life expectancy - males	NA	NA	NA	NA	42	44	NA	NA	NA
Life expectancy - females	NA	NA	NA	NA	48	49	NA	NA	NA
Birth rate (per 1,000)	NA	NA	NA	NA	46	45	NA	NA	NA
Death rate (per 1,000)	NA	NA	NA	NA	20	19	NA	NA	NA
Women of reproductive age (15-44 yrs.)	NA	NA	NA	NA	NA	NA	NA	NA	NA
of which are currently married	NA	NA	NA	NA	NA	NA	NA	NA	NA
Fertility rate	NA	NA	NA	6.2	6.13	6.0	5.6	4.8	4.0

Population values are in thousands, life expectancy in years, and other items as indicated.

Health

Health Personnel [3]

Doctors per 1,000 pop., 1988-92	0.07
Nurse-to-doctor ratio, 1988-92	5.0
Hospital beds per 1,000 pop., 1985-90	1.0
Percentage of children immunized (age 1 yr. or less)	
Third dose of DPT, 1990-91	75.0
Measles, 1990-91	74.0

Health Expenditures [4]

Total health expenditure, 1990 (official exchange rate)	
Millions of dollars	22
Millions of dollars per capita	5
Health expenditures as a percentage of GDP	
Total	2.4
Public sector	1.7
Private sector	0.8
Development assistance for health	
Total aid flows (millions of dollars)[1]	7
Aid flows per capita (millions of dollars)	1.7
Aid flows as a percentage of total health expenditure	33.0

For sources, notes, and explanations, see Annotated Source Appendix, page 1035.

817

Human Factors

Health Care Ratios [5]

Population per physician, 1970	17830
Population per physician, 1990	NA
Population per nursing person, 1970	2700
Population per nursing person, 1990	NA
Percent of births attended by health staff, 1985	25

Infants and Malnutrition [6]

Percent of babies with low birth weight, 1985	14
Infant mortality rate per 1,000 live births, 1970	197
Infant mortality rate per 1,000 live births, 1991	145
Years of life lost per 1,000 population, 1990	188
Prevalence of malnutrition (under age 5), 1990	NA

Ethnic Division [7]

13 native African tribes 99% (Temne 30%, Mende 30%, other 39%), Creole, European, Lebanese, and Asian 1%.

Religion [8]

Muslim	30%
Indigenous beliefs	30%
Christian	10%
Other or none	30%

Major Languages [9]

English (official; regular use limited to literate minority), Mende principal vernacular in the south, Temne principal vernacular in the north, Krio the language of the re-settled ex-slave population of the Freetown area and is lingua franca.

Education

Public Education Expenditures [10]

Million Leone	1980	1985	1986	1987	1988	1989
Total education expenditure	43	112	146	432	529	604
as percent of GNP	3.8	2.4	2.0	2.2	1.9	1.4
as percent of total govt. expend.	11.8	12.4	NA	NA	NA	NA
Current education expenditure	41	106	134	417	499	577
as percent of GNP	3.6	2.3	1.8	2.1	1.8	1.3
as percent of current govt. expend.	14.5	15.5	NA	NA	NA	NA
Capital expenditure	2	6	12	15	30	27

Educational Attainment [11]

Age group	5+
Total population	1,315,897
Highest level attained (%)	
No schooling	64.1
First level	
Incompleted	18.7
Completed	1.8
Entered second level	
S-1	9.7
S-2	3.8
Post secondary	1.5

Literacy Rate [12]

In thousands and percent	1985[a]	1991[a]	2000[a]
Illiterate population +15 years	1,783	1,830	1,909
Illiteracy rate - total pop. (%)	86.7	79.3	64.2
Illiteracy rate - males (%)	79.2	69.3	51.5
Illiteracy rate - females (%)	93.8	88.7	76.2

Libraries [13]

Daily Newspapers [14]

	1975	1980	1985	1990
Number of papers	2	1	1	1
Circ. (000)	30	10	10	10

Cinema [15]

Science and Technology

Scientific/Technical Forces [16]

R&D Expenditures [17]

U.S. Patents Issued [18]

Government and Law

Organization of Government [19]

Long-form name:
Republic of Sierra Leone
Type:
military government
Independence:
27 April 1961 (from UK)
Constitution:
1 October 1991; amended September 1991
Legal system:
based on English law and customary laws indigenous to local tribes; has not accepted compulsory ICJ jurisdiction
National holiday:
Republic Day, 27 April (1961)
Executive branch:
National Provisional Ruling Council
Legislative branch:
unicameral House of Representatives (suspended after coup of 29 April 1992)
Judicial branch:
Supreme Court (suspended after coup of 29 April 1992)

Elections [20]

Suspended after 29 April 1992 coup; Chairman Strasser promises multi-party elections sometime within three years.

Government Budget [21]

For FY92 est.

Revenues	68
Expenditures	118
Capital expenditures	28

Crime [23]

Military Expenditures and Arms Transfers [22]

	1985	1986	1987	1988	1989
Military expenditures					
Current dollars (mil.)	5[e]	NA	6[e]	5[e]	NA
1989 constant dollars (mil.)	5[e]	NA	6[e]	5[e]	NA
Armed forces (000)	4	4	6	4	4
Gross national product (GNP)					
Current dollars (mil.)	609	545	650	672	686
1989 constant dollars (mil.)	694	605	699	699	686
Central government expenditures (CGE)					
1989 constant dollars (mil.)	104	NA	NA	69	NA
People (mil.)	3.7	3.8	3.9	4.0	4.1
Military expenditure as % of GNP	0.7	NA	0.9	0.7	NA
Military expenditure as % of CGE	5.0	NA	NA	7.0	NA
Military expenditure per capita	1	NA	2	1	NA
Armed forces per 1,000 people	1.1	1.1	1.6	1.0	1.0
GNP per capita	188	160	181	177	169
Arms imports[6]					
Current dollars (mil.)	0	0	10	0	0
1989 constant dollars (mil.)	0	0	11	0	0
Arms exports[6]					
Current dollars (mil.)	0	0	0	0	0
1989 constant dollars (mil.)	0	0	0	0	0
Total imports[7]					
Current dollars (mil.)	151	132	137	156	183
1989 constant dollars	172	146	147	162	183
Total exports[7]					
Current dollars (mil.)	130	144	130	106	138
1989 constant dollars	148	160	140	110	1,338
Arms as percent of total imports[8]	0	0	7.3	0	0
Arms as percent of total exports[8]	0	0	0	0	0

Human Rights [24]

	SSTS	FL	FAPRO	PPCG	APROBC	TPW	PCPTW	STPEP	PHRFF	PRW	ASST	AFL
Observes	P	P	P		P	P	P			P	P	P
	EAFRD	CPR	ESCR	SR	ACHR	MAAE	PVIAC	PVNAC	EAFDAW	TCIDTP	RC	
Observes		P			P		P	P	P	S	P	

P = Party; S = Signatory; see Appendix for meaning of abbreviations.

Labor Force

Total Labor Force [25]

1.369 million (1981 est.)

Labor Force by Occupation [26]

Agriculture	65%
Industry	19
Services	16

Date of data: 1981 est.

Unemployment Rate [27]

For sources, notes, and explanations, see Annotated Source Appendix, page 1035.

Production Sectors

Energy Resource Summary [28]

Energy Resources: None. **Electricity**: 85,000 kW capacity; 185 million kWh produced, 45 kWh per capita (1991).

Telecommunications [30]

- Marginal telephone and telegraph service
- National microwave radio relay system unserviceable at present
- 23,650 telephones
- Broadcast stations - 1 AM, 1 FM, 1 TV
- 1 Atlantic Ocean INTELSAT earth station

Top Agricultural Products [32]

Agriculture accounts for over 30% of GDP and two-thirds of the labor force; largely subsistence farming; cash crops—coffee, cocoa, palm kernels; harvests of food staple rice meets 80% of domestic needs; annual fish catch averages 53,000 metric tons.

Top Mining Products [33]

Detailed information is not available. A summary of mineral resources available follows. **Mineral Resources**: Diamonds, titanium ore, bauxite, iron ore, gold, chromite.

Transportation [31]

Railroads. 84 km 1.067-meter narrow-gauge mineral line is used on a limited basis because the mine at Marampa is closed

Highways. 7,400 km total; 1,150 km paved, 490 km laterite (some gravel), 5,760 km improved earth

Merchant Marine. 1 cargo ship totaling 5,592 GRT/9,107 DWT

Airports

Total:	11
Usable:	7
With permanent-surface runways:	4
With runways over 3,659 m:	0
With runways 2,440-3,659 m:	1
With runways 1,220-2,439 m:	3

Tourism [34]

	1987	1988	1989	1990	1991
Tourists	88	75	86	103	118
Tourism receipts	14	15	17	19	
Tourism expenditures	17	7	3		
Fare expenditures	1		1		

Tourists are in thousands, money in million U.S. dollars.

For sources, notes, and explanations, see Annotated Source Appendix, page 1035.

Finance, Economics, and Trade

GDP and Manufacturing Summary [35]

	1980	1985	1990	1991	1992
Gross Domestic Product					
Millions of 1980 dollars	758	828	888	913	946[e]
Growth rate in percent	3.00	8.53	3.01	2.82	3.54[e]
Manufacturing Value Added					
Millions of 1980 dollars	55	54	44	44	44[e]
Growth rate in percent	-5.57	-13.93	-3.96	0.00	0.81[e]
Manufacturing share in percent of current prices	7.5	4.8	9.7	NA	NA

Economic Indicators [36]

National product: GDP—exchange rate conversion—$1.4 billion (FY92 est.). **National product real growth rate**: -1% (FY92 est.). **National product per capita**: $330 (FY92 est.). **Inflation rate (consumer prices)**: 5% (1992). **External debt**: $633 million (FY92 est.).

Balance of Payments Summary [37]

Values in millions of dollars.

	1985	1986	1987	1988	1989
Exports of goods (f.o.b.)	131.9	126.0	138.9	104.5	139.5
Imports of goods (f.o.b.)	-141.2	-111.4	-114.8	-138.2	-160.3
Trade balance	-9.3	14.6	24.1	-33.7	-20.8
Services - debits	-34.3	93.3	-105.2	-29.9	-43.6
Services - credits	28.2	26.7	44.1	52.1	38.5
Private transfers (net)	2.3	1.5	-	0.3	0.1
Government transfers (net)	16.9	4.6	6.8	8.5	7.2
Long term capital (net)	-78.4	-303.4	44.0	-63.4	-38.5
Short term capital (net)	87.3	144.2	10.4	135.5	59.5
Errors and omissions	-9.3	42.0	-21.9	-62.5	19.9
Overall balance	3.4	23.5	2.3	6.8	22.3

Exchange Rates [38]

Currency: **leones.**
Symbol: **Le.**

Data are currency units per $1.

January 1993	552.4300
1992	499.4400
1991	295.3400
1990	144.9275
1989	58.1395
1988	31.2500

Imports and Exports

Top Import Origins [39]

$62 million (c.i.f., FY92 est.).

Origins	%
US	NA
EC countries	NA
Japan	NA
China	NA
Nigeria	NA

Top Export Destinations [40]

$75 million (f.o.b., FY92 est.).

Destinations	%
US	NA
UK	NA
Belgium	NA
Germany	NA
Other Western Europe	NA

Foreign Aid [41]

	U.S. $	
US commitments, including Ex-Im (FY70-89)	161	million
Western (non-US) countries, ODA and OOF bilateral commitments (1970-89)	848	million
OPEC bilateral aid (1979-89)	18	million
Communist countries (1970-89)	101	million

Import and Export Commodities [42]

Import Commodities	Export Commodities
Capital goods 40%	Rutile 50%
Food 32%	Bauxite 17%
Petroleum 12%	Cocoa 11%
Consumer goods 7%	Diamonds 3%
Light industrial goods	Coffee 3%

For sources, notes, and explanations, see Annotated Source Appendix, page 1035.

821

Singapore

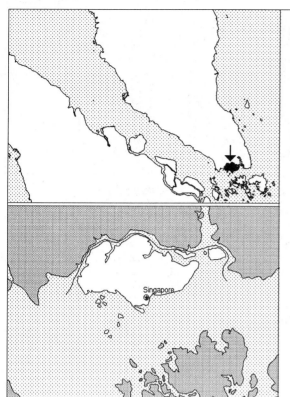

Geography [1]

Total area:
632.6 km2
Land area:
622.6 km2
Comparative area:
Slightly less than 3.5 times the size of Washington, DC
Land boundaries:
0 km
Coastline:
193 km
Climate:
Tropical; hot, humid, rainy; no pronounced rainy or dry seasons; thunderstorms occur on 40% of all days (67% of days in April)
Terrain:
Lowland; gently undulating central plateau contains water catchment area and nature preserve
Natural resources:
Fish, deepwater ports
Land use:
Arable land:
4%
Permanent crops:
7%
Meadows and pastures:
0%
Forest and woodland:
5%
Other:
84%

Demographics [2]

	1960	1970	1980	1990	1991[1]	1994	2000	2010	2020
Population	1,646	2,075	2,414	2,720	2,756	2,859	3,025	3,206	3,335
Population density (persons per sq. mi.)	6,832	8,608	10,016	11,290	11,437	NA	12,535	13,414	14,113
Births	NA	NA	NA	NA	49	47	NA	NA	NA
Deaths	NA	NA	NA	NA	14	15	NA	NA	NA
Life expectancy - males	NA	NA	NA	NA	72	73	NA	NA	NA
Life expectancy - females	NA	NA	NA	NA	77	79	NA	NA	NA
Birth rate (per 1,000)	NA	NA	NA	NA	18	17	NA	NA	NA
Death rate (per 1,000)	NA	NA	NA	NA	5	5	NA	NA	NA
Women of reproductive age (15-44 yrs.)	NA	NA	NA	794	NA	816	814	NA	NA
of which are currently married	NA	NA	NA	457	NA	493	501	NA	NA
Fertility rate	NA	NA	NA	1.9	1.97	1.9	1.8	1.8	1.8

Population values are in thousands, life expectancy in years, and other items as indicated.

Health

Health Personnel [3]

Doctors per 1,000 pop., 1988-92	1.09
Nurse-to-doctor ratio, 1988-92	3.8
Hospital beds per 1,000 pop., 1985-90	3.3
Percentage of children immunized (age 1 yr. or less)	
Third dose of DPT, 1990-91	91.0
Measles, 1990-91	92.0

Health Expenditures [4]

Total health expenditure, 1990 (official exchange rate)	
Millions of dollars	658
Millions of dollars per capita	219
Health expenditures as a percentage of GDP	
Total	1.9
Public sector	1.1
Private sector	0.8
Development assistance for health	
Total aid flows (millions of dollars)[1]	1
Aid flows per capita (millions of dollars)	0.2
Aid flows as a percentage of total health expenditure	0.1

For sources, notes, and explanations, see Annotated Source Appendix, page 1035.

Human Factors

Health Care Ratios [5]

Population per physician, 1970	1370
Population per physician, 1990	820
Population per nursing person, 1970	250
Population per nursing person, 1990	NA
Percent of births attended by health staff, 1985	100

Infants and Malnutrition [6]

Percent of babies with low birth weight, 1985	7
Infant mortality rate per 1,000 live births, 1970	20
Infant mortality rate per 1,000 live births, 1991	6
Years of life lost per 1,000 population, 1990	9
Prevalence of malnutrition (under age 5), 1990	NA

Ethnic Division [7]

Chinese	76.4%
Malay	14.9%
Indian	6.4%
Other	2.3%

Religion [8]

Buddhist (Chinese), Atheist (Chinese), Muslim (Malays), Christian, Hindu, Sikh, Taoist, Confucianist.

Major Languages [9]

Chinese (official), Malay (official and national), Tamil (official), English (official).

Education

Public Education Expenditures [10]

Million Singapore Dollars	1980	1985	1986	1987	1988	1989
Total education expenditure	686	1,776	1,639	1,664	1,718	NA
as percent of GNP	2.8	4.4	4.1	3.9	3.4	NA
as percent of total govt. expend.	7.3	NA	NA	11.5	NA	NA
Current education expenditure	587	1,388	1,277	1,368	1,523	NA
as percent of GNP	2.4	3.4	3.2	3.2	3.0	NA
as percent of current govt. expend.	10.3	NA	NA	15.3	NA	NA
Capital expenditure	99	387	361	295	196	NA

Educational Attainment [11]

Age group	25+
Total population	1,596,600
Highest level attained (%)	
No schooling	NA
First level	
Incompleted	NA
Completed	64.0
Entered second level	
S-1	23.2
S-2	8.1
Post secondary	4.7

Literacy Rate [12]

	1970[b]	1980[b]	1990[b]
Illiterate population + 15 years	394,543	300,994	NA
Illiteracy rate - total pop. (%)	31.1	17.1	NA
Illiteracy rate - males (%)	17.0	8.4	NA
Illiteracy rate - females (%)	45.7	26.0	NA

Libraries [13]

	Admin. Units	Svc. Pts.	Vols. (000)	Shelving (meters)	Vols. Added	Reg. Users
National (1989)[23a]	1	14	2,319	NA	269,766	615,180
Non-specialized	NA	NA	NA	NA	NA	NA
Public	NA	NA	NA	NA	NA	NA
Higher ed. (1990)[23b]	5	12	2,354	14,865	137,181	77,934
School (1990)	366	NA	4,640	NA	NA	440,042

Daily Newspapers [14]

	1975	1980	1985	1990
Number of papers	10	12	10	8
Circ. (000)	449	690	924	763

Cinema [15]

Science and Technology

Scientific/Technical Forces [16]

Potential scientists/engineers	38,259
Number female	10,246
Potential technicians[5]	25,920
Number female	4,769
Total[5]	64,179

R&D Expenditures [17]

	Dollar[25] (000) 1987
Total expenditure	374,744
Capital expenditure	151,021
Current expenditure	223,723
Percent current	59.7

U.S. Patents Issued [18]

Values show patents issued to citizens of the country by the U.S. Patents Office.

	1990	1991	1992
Number of patents	16	25	35

For sources, notes, and explanations, see Annotated Source Appendix, page 1035.

Government and Law

Organization of Government [19]

Long-form name:
Republic of Singapore
Type:
republic within Commonwealth
Independence:
9 August 1965 (from Malaysia)
Constitution:
3 June 1959, amended 1965; based on preindependence State of Singapore Constitution
Legal system:
based on English common law; has not accepted compulsory ICJ jurisdiction
National holiday:
National Day, 9 August (1965)
Executive branch:
president, prime minister, two deputy prime ministers, Cabinet
Legislative branch:
unicameral Parliament
Judicial branch:
Supreme Court

Political Parties [20]

Parliament	% of seats
People's Action Party	95.1
Singapore Democratic Party	3.7
Workers' Party	1.2

Government Expenditures [21]

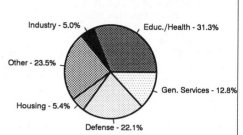

Industry - 5.0%
Educ./Health - 31.3%
Other - 23.5%
Gen. Services - 12.8%
Housing - 5.4%
Defense - 22.1%

% distribution for 1991. Expend. in 1993: 9.4 ($ bil.)

Military Expenditures and Arms Transfers [22]

	1985	1986	1987	1988	1989
Military expenditures					
Current dollars (mil.)	1,125	1,093	1,137	1,343	1,475
1989 constant dollars (mil.)	1,281	1,213	1,223	1,398	1,475
Armed forces (000)	59	56	55	56	56
Gross national product (GNP)					
Current dollars (mil.)	19,080	19,940	22,020	25,430	28,890
1989 constant dollars (mil.)	21,720	22,120	23,680	26,480	28,890
Central government expenditures (CGE)					
1989 constant dollars (mil.)	7,511	8,048	8,207	5,770	7,807
People (mil.)	2.6	2.6	2.6	2.6	2.7
Military expenditure as % of GNP	5.9	5.5	5.2	5.3	5.1
Military expenditure as % of CGE	17.0	15.1	14.9	24.2	18.9
Military expenditure per capita	501	469	468	528	550
Armed forces per 1,000 people	23.1	21.5	21.1	21.0	20.7
GNP per capita	8,491	8,559	9,063	10,000	10,760
Arms imports[6]					
Current dollars (mil.)	120	280	260	320	120
1989 constant dollars (mil.)	137	311	280	333	120
Arms exports[6]					
Current dollars (mil.)	40	60	40	50	70
1989 constant dollars (mil.)	46	67	43	52	70
Total imports[7]					
Current dollars (mil.)	26,280	25,510	32,560	43,860	49,670
1989 constant dollars	29,920	28,310	35,020	45,660	49,670
Total exports[7]					
Current dollars (mil.)	22,810	22,490	28,690	39,310	44,660
1989 constant dollars	25,970	24,960	30,850	40,920	44,660
Arms as percent of total imports[8]	0.5	1.1	0.8	0.7	0.2
Arms as percent of total exports[8]	0.2	0.3	0.1	0.1	0.2

Crime [23]

Crime volume	
Cases known to police	45,251
Attempts (percent)	NA
Percent cases solved	23.8
Crimes per 100,000 persons	1,507
Persons responsible for offenses	
Total number offenders	12,317
Percent female	NA
Percent juvenile[49]	1,205
Percent foreigners	NA

Human Rights [24]

	SSTS	FL	FAPRO	PPCG	APROBC	TPW	PCPTW	STPEP	PHRFF	PRW	ASST	AFL
Observes		P				P	P	P	P			P
	EAFRD	CPR	ESCR	SR	ACHR	MAAE	PVIAC	PVNAC	EAFDAW	TCIDTP	RC	
Observes		P				P	P	P	P			

P = Party; S = Signatory; see Appendix for meaning of abbreviations.

Labor Force

Total Labor Force [25]

1,485,800

Labor Force by Occupation [26]

Financial, business, and other services	30.2%
Manufacturing	28.4
Commerce	22.0
Construction	9.0
Other	10.4

Date of data: 1990

Unemployment Rate [27]

2.7% (June 1992)

For sources, notes, and explanations, see Annotated Source Appendix, page 1035.

Production Sectors

Energy Resource Summary [28]

Energy Resources: None. **Electricity**: 4,860,000 kW capacity; 18,000 million kWh produced, 6,420 kWh per capita (1992).

Telecommunications [30]

- Good domestic facilities
- Good international service
- Good radio and television broadcast coverage
- 1,110,000 telephones
- Broadcast stations - 13 AM, 4 FM, 2 TV
- Submarine cables extend to Malaysia (Sabah and peninsular Malaysia), Indonesia, and the Philippine
- Satellite earth stations - 1 Indian Ocean INTELSAT and 1 Pacific Ocean INTELSAT

Top Agricultural Products [32]

	1990	1991[4]
Vegetables	9	10
Fruits	1	1

Values shown are 1,000 metric tons.

Top Mining Products [33]

Detailed information is not available. A summary of mineral resources available follows. **Mineral Resources**: None.

Transportation [31]

Railroads. 38 km of 1.000-meter gauge

Highways. 2,644 km total (1985)

Merchant Marine. 492 ships (1,000 GRT or over) totaling 9,763,511 GRT/15,816,384 DWT; includes 1 passenger-cargo, 125 cargo, 72 container, 7 roll-on/roll-off cargo, 4 refrigerated cargo, 18 vehicle carrier, 1 livestock carrier, 165 oil tanker, 8 chemical tanker, 7 combination ore/oil, 2 specialized tanker, 5 liquefied gas, 74 bulk, 3 combination bulk; note - many Singapore flag ships are foreign owned

Airports

Total:	10
Usable:	10
With permanent-surface runways:	10
With runways over 3,659 m:	2
With runways 2,440-3,659 m:	4
With runways 1,220-2,439 m:	3

Tourism [34]

	1987	1988	1989	1990	1991
Visitors[64]	3,695	4,201	4,843	5,331	5,432
Tourists[64]	3,373	3,833	4,397	4,842	4,913
Excursionists[65]	306	353	433	481	502
Cruise passengers[64]	16	15	13	8	17
Tourism receipts	2,088	2,622	3,307	4,719	5,020
Tourism expenditures	795	930	1,334	1,821	2,019

Tourists are in thousands, money in million U.S. dollars.

For sources, notes, and explanations, see Annotated Source Appendix, page 1035.

825

Manufacturing Sector

GDP and Manufacturing Summary [35]

	1980	1985	1989	1990	% change 1980-1990	% change 1989-1990
GDP (million 1980 $)	11,719	15,821	21,264	23,179	97.8	9.0
GDP per capita (1980 $)	4,853	6,182	7,902	8,506	75.3	7.6
Manufacturing as % of GDP (current prices)	28.0	22.0	28.4[e]	26.3	-6.1	-7.4
Gross output (million $)	15,278	17,570	33,255	39,414	158.0	18.5
Value added (million $)	4,004	4,868	10,279	11,923	197.8	16.0
Value added (million 1980 $)	3,415	3,689	6,070	6,662	95.1	9.8
Industrial production index	100	103	159	165	65.0	3.8
Employment (thousands)	285	252	348	350	22.8	0.6

Note: GDP stands for Gross Domestic Product. 'e' stands for estimated value.

Profitability and Productivity

	1980	1985	1989	1990	% change 1980-1990	% change 1989-1990
Intermediate input (%)	74	72	69	70	-5.4	1.4
Wages, salaries, and supplements (%)	8	11	9	10	25.0	11.1
Gross operating surplus (%)	18	17	22	21	16.7	-4.5
Gross output per worker ($)	53,196	69,162	95,460	112,015	110.6	17.3
Value added per worker ($)	13,942	19,161	29,507	33,885	143.0	14.8
Average wage (incl. benefits) ($)	4,170	7,316	8,931	10,790	158.8	20.8

Profitability is in percent of gross output. Productivity is in U.S. $. 'e' stands for estimated value.

Profitability - 1990

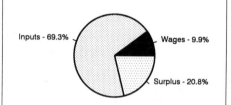

Inputs - 69.3%
Wages - 9.9%
Surplus - 20.8%

The graphic shows percent of gross output.

Value Added in Manufacturing

	1980 $ mil.	1980 %	1985 $ mil.	1985 %	1989 $ mil.	1989 %	1990 $ mil.	1990 %	% change 1980-1990	% change 1989-1990
311 Food products	121	3.0	180	3.7	308	3.0	322	2.7	166.1	4.5
313 Beverages	52	1.3	76	1.6	127	1.2	139	1.2	167.3	9.4
314 Tobacco products	25	0.6	35	0.7	50	0.5	64	0.5	156.0	28.0
321 Textiles	70	1.7	29	0.6	64	0.6	72	0.6	2.9	12.5
322 Wearing apparel	127	3.2	157	3.2	302	2.9	294	2.5	131.5	-2.6
323 Leather and fur products	7	0.2	5	0.1	15	0.1	11	0.1	57.1	-26.7
324 Footwear	9	0.2	6	0.1	10	0.1	9	0.1	0.0	-10.0
331 Wood and wood products	84	2.1	43	0.9	59	0.6	55	0.5	-34.5	-6.8
332 Furniture and fixtures	40	1.0	60	1.2	96	0.9	89	0.7	122.5	-7.3
341 Paper and paper products	45	1.1	82	1.7	153	1.5	189	1.6	320.0	23.5
342 Printing and publishing	128	3.2	232	4.8	423	4.1	515	4.3	302.3	21.7
351 Industrial chemicals	52	1.3	163	3.3	607	5.9	584	4.9	1,023.1	-3.8
352 Other chemical products	143	3.6	244	5.0	493	4.8	600	5.0	319.6	21.7
353 Petroleum refineries	636[e]	15.9	362[e]	7.4	377[e]	3.7	843[e]	7.1	32.5	123.6
354 Miscellaneous petroleum and coal products	50[e]	1.2	35[e]	0.7	263[e]	2.6	74[e]	0.6	48.0	-71.9
355 Rubber products	44	1.1	21	0.4	42	0.4	39	0.3	-11.4	-7.1
356 Plastic products	84	2.1	105	2.2	254	2.5	297	2.5	253.6	16.9
361 Pottery, china and earthenware	1[e]	0.0	NA	0.0	1	0.0	2[e]	0.0	100.0	100.0
362 Glass and glass products	10[e]	0.2	5[e]	0.1	11	0.1	31[e]	0.3	210.0	181.8
369 Other non-metal mineral products	82	2.0	140	2.9	119	1.2	148	1.2	80.5	24.4
371 Iron and steel	62	1.5	49	1.0	97	0.9	97	0.8	56.5	0.0
372 Non-ferrous metals	9	0.2	14	0.3	30	0.3	41	0.3	355.6	36.7
381 Metal products	206	5.1	310	6.4	655	6.4	730	6.1	254.4	11.5
382 Non-electrical machinery	319	8.0	359	7.4	610	5.9	699	5.9	119.1	14.6
383 Electrical machinery	950	23.7	1,527	31.4	4,054	39.4	4,741	39.8	399.1	16.9
384 Transport equipment	500	12.5	470	9.7	742	7.2	890	7.5	78.0	19.9
385 Professional and scientific equipment	80	2.0	89	1.8	176	1.7	200	1.7	150.0	13.6
390 Other manufacturing industries	69	1.7	70	1.4	144	1.4	147	1.2	113.0	2.1

Note: The industry codes shown are International Standard Industry codes (ISIC). Percentages are percent of total Value Added. 'e' stands for estimated value

826

For sources, notes, and explanations, see Annotated Source Appendix, page 1035.

Finance, Economics, and Trade

Economic Indicators [36]

Millions of Singapore Dollars Unless stated Otherwise.

	1989	1990	1991[e]
Real GDP (1985 Market Prices)	52,679	57,016	61,292
Growth Rate	9.2	8.3	7.0
Per capita GNP (US$)	9,963	11,769	13,770
Money Supply (M1) (end of period)	13,745	15,261	16,000
Savings Rate (% of GNP)	43.3	44.6	45.0
Investment Rate (% of GNP)	34.5	37.9	38.0
CPI (Base June 82-May 83)	102.8	106.3	110.0
WPI (Base 1985)	91.9	93.5	88.0
External Public Debt (Yr-end)	138.7	69.5	50.0

Balance of Payments Summary [37]

Values in millions of dollars.

	1987	1988	1989	1990	1991
Exports of goods (f.o.b.)	27,464.0	37,993.0	43,239.0	50,684.0	56,819.0
Imports of goods (f.o.b.)	-29,910.0	-40,338.0	-45,687.0	-55,803.0	-60,948.0
Trade balance	-2,446.0	-2,345.0	-2,448.0	-5,119.0	-4,129.0
Services - debits	-7,825.0	-9,629.0	-10,970.0	-13,985.0	-15,392.0
Services - credits	10,348.0	13,162.0	16,225.0	21,677.0	24,209.0
Private transfers (net)	-170.0	-209.0	-254.0	-270.0	-339.0
Government transfers (net)	-64.0	-90.0	-125.0	-134.0	-141.0
Long term capital (net)	2,566.0	3,040.0	2,646.0	3,287.0	3,129.0
Short term captal (net)	-2,096.0	-2,052.0	-1,019.0	3,154.0	-21.0
Errors and omissions	782.0	-218.0	-1,318.0	-3,180.0	-3,122.0
Overall balance	1,095.0	1,659.0	2,737.0	5,430.0	4,194.0

Exchange Rates [38]

Currency: **Singapore dollars.**
Symbol: **S$.**

Data are currency units per $1.

January 1993	1.6531
1992	1.6290
1991	1.7276
1990	1.8125
1989	1.9503
1988	2.0124

Imports and Exports

Top Import Origins [39]

$66.4 billion (f.o.b., 1992).

Origins	%
Japan	21
US	16
Malaysia	14
Taiwan	4

Top Export Destinations [40]

$61.5 billion (f.o.b., 1992).

Destinations	%
US	21
Malaysia	13
Hong Kong	8
Japan	7
Thailand	6

Foreign Aid [41]

	U.S. $	
US commitments, including Ex-Im (FY70-83)	590	million
Western (non-US) countries, ODA and OOF bilateral commitments (1970-89)	1.0	billion

Import and Export Commodities [42]

Import Commodities	**Export Commodities**
Aircraft	Computer equipment
Petroleum	Rubber and rubber products
Chemicals	Petroleum products
Foodstuffs	Telecommunications equipment

Slovakia

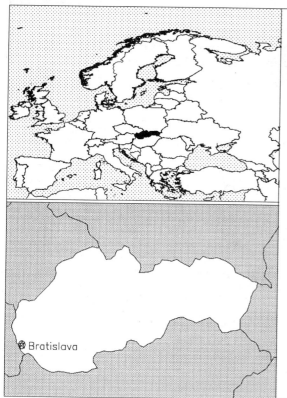

Geography [1]

Total area:
48,845 km2
Land area:
48,800 km2
Comparative area:
About twice the size of New Hampshire
Land boundaries:
Total 1,355 km, Austria 91 km, Czech Republic 215 km, Hungary 515 km, Poland 444 km, Ukraine 90 km
Coastline:
0 km (landlocked)
Climate:
Temperate; cool summers; cold, cloudy, humid winters
Terrain:
Rugged mountains in the central and northern part and lowlands in the south
Natural resources:
Brown coal and lignite; small amounts of iron ore, copper and manganese ore; salt; gas
Land use:
Arable land:
NA%
Permanent crops:
NA%
Meadows and pastures:
NA%
Forest and woodland:
NA%
Other:
NA%

Demographics [2]

	1960	1970	1980	1990	1991	1994	2000	2010	2020
Population	3,994	4,524	4,966	5,298	NA	5,404	5,585	5,883	6,078
Population density (persons per sq. mi.)	NA	NA	NA	NA	NA	NA	NA	NA	NA
Births	NA	NA	NA	NA	NA	79	NA	NA	NA
Deaths	NA	NA	NA	NA	NA	50	NA	NA	NA
Life expectancy - males	NA	NA	NA	NA	NA	69	NA	NA	NA
Life expectancy - females	NA	NA	NA	NA	NA	77	NA	NA	NA
Birth rate (per 1,000)	NA	NA	NA	NA	NA	15	NA	NA	NA
Death rate (per 1,000)	NA	NA	NA	NA	NA	9	NA	NA	NA
Women of reproductive age (15-44 yrs.)	NA	NA	NA	1,328	NA	1,396	1,453	NA	NA
of which are currently married	NA	NA	NA	889	NA	923	973	NA	NA
Fertility rate	NA	NA	NA	2.1	NA	2.0	1.8	1.8	1.8

Population values are in thousands, life expectancy in years, and other items as indicated.

Health

Health Personnel [3]

Doctors per 1,000 pop., 1988-92	3.23 [3]
Nurse-to-doctor ratio, 1988-92	2.4 [3]
Hospital beds per 1,000 pop., 1985-90	7.9 [3]
Percentage of children immunized (age 1 yr. or less)	
Third dose of DPT, 1990-91	99.0 [3]
Measles, 1990-91	98.0 [3]

Health Expenditures [4]

Total health expenditure, 1990 (official exchange rate)	
Millions of dollars	2711
Millions of dollars per capita	173
Health expenditures as a percentage of GDP	
Total	5.9
Public sector	5.0
Private sector	0.9
Development assistance for health	
Total aid flows (millions of dollars)[1]	NA
Aid flows per capita (millions of dollars)	NA
Aid flows as a percentage of total health expenditure	NA

For sources, notes, and explanations, see Annotated Source Appendix, page 1035.

Human Factors

Health Care Ratios [5]

Population per physician, 1970	470 [3]
Population per physician, 1990	310 [3]
Population per nursing person, 1970	170 [3]
Population per nursing person, 1990	NA [3]
Percent of births attended by health staff, 1985	100 [3]

Infants and Malnutrition [6]

Percent of babies with low birth weight, 1985	6
Infant mortality rate per 1,000 live births, 1970	22
Infant mortality rate per 1,000 live births, 1991	11
Years of life lost per 1,000 population, 1990	16
Prevalence of malnutrition (under age 5), 1990	NA

Ethnic Division [7]

Slovak 85.6%, Hungarian 10.8%, Gypsy 1.5% (the 1992 census figures underreport the Gypsy/Romany community, which could reach 500,000 or more), Czech 1.1%, Ruthenian 15,000, Ukrainian 13,000, Moravian 6,000, German 5,000, Polish 3,000.

Religion [8]

Roman Catholic	60.3%
Atheist	9.7%
Protestant	8.4%
Orthodox	4.1%
Other	17.5%

Major Languages [9]

Slovak (official), Hungarian.

Education

Public Education Expenditures [10]

Million Koruna	1980[1,15]	1985[1,15]	1987[1,15]	1988[1,15]	1989[1,15]	1990[1,15]
Total education expenditure	23,181	28,201	30,549	32,254	NA	37,323
as percent of GNP	4.0	4.2	4.3	4.4	NA	4.6
as percent of total govt. expend.	NA	7.9	8.0	8.0	NA	8.2
Current education expenditure	21,802	26,959	29,247	31,005	NA	35,482
as percent of GNP	3.8	4.0	4.1	4.2	NA	4.4
as percent of current govt. expend.	NA	8.6	8.7	8.7	NA	8.7
Capital expenditure	1,379	1,242	1,302	1,249	NA	1,841

Educational Attainment [11]

Age group	25+ [4]
Total population	9,274,694 [4]
Highest level attained (%)	
No schooling	0.4 [4]
First level	
Incompleted	47.6 [4]
Completed	NA [4]
Entered second level	
S-1	45.9 [4]
S-2	NA [4]
Post secondary	6.0 [4]

Literacy Rate [12]

Libraries [13]

	Admin. Units	Svc. Pts.	Vols. (000)	Shelving (meters)	Vols. Added	Reg. Users
National (1989)	19	NA	26,962	NA	1.2mil	455,748
Non-specialized	NA	NA	NA	NA	NA	NA
Public (1989)	8,398	11,454	58,627	NA	3.1mil	2.8mil
Higher ed. (1990)	1,590	1,590	12,720	NA	366,500	269,591
School	NA	NA	NA	NA	NA	NA

Daily Newspapers [14]

	1975	1980	1985	1990
Number of papers	29	30	30	48
Circ. (000)	4,436	4,798	5,124	7,943

Cinema [15]

Data for 1990.

Cinema seats per 1,000	51.2 [b]
Annual attendance per person	3.2 [b]
Gross box office receipts (mil. Koruna)	386 [b]

Science and Technology

Scientific/Technical Forces [16]

Potential scientists/engineers	542,706
Number female	191,256
Potential technicians	NA
Number female	NA
Total	NA

R&D Expenditures [17]

	Koruny (000) 1989
Total expenditure[b,27]	24,721,000
Capital expenditure	2,621,000
Current expenditure	22,100,000
Percent current	89.4

U.S. Patents Issued [18]

Values show patents issued to citizens of the country by the U.S. Patents Office.

	1990	1991	1992
Number of patents	39	27	18

For sources, notes, and explanations, see Annotated Source Appendix, page 1035.

829

Government and Law

Organization of Government [19]

Long form name:
Slovak Republic
Type:
parliamentary democracy
Independence:
1 January 1993 (from Czechoslovakia)
Constitution:
ratified 3 September 1992; fully effective 1 January 1993
Legal system:
civil law system based on Austro-Hungarian codes; has not accepted compulsory ICJ jurisdiction; legal code modified to comply with the obligations of Conference on Security and Cooperation in Europe (CSCE) and to expunge Marxist-Leninist legal theory
National holiday:
Slovak National Uprising, August 29 (1944)
Executive branch:
president, prime minister, Cabinet
Legislative branch:
unicameral National Council (Narodni Rada)
Judicial branch:
Supreme Court

Political Parties [20]

National Council	% of votes
Movement for a Dem. Slovakia	37.0
Party of the Democratic Left	15.0
Christian Democratic Movement	9.0
Slovak National Party	8.0
Hungarian Christ. Dem. Movement/Coexistence	7.0

Government Budget [21]

For 1993.

Revenues	NA
Expenditures	NA
Capital expenditures	NA

Defense Summary [22]

Branches: Army, Air and Air Defense Forces, Civil Defense, Railroad Units

Manpower Availability: Males age 15-49 1,407,908; fit for military service 1,082,790; reach military age (18) annually 47,973 (1993 est.)

Defense Expenditures: 8.2 billion koruny (1993 est.); note - conversion of defense expenditures into US dollars using the current exchange rate could produce misleading results

Crime [23]

Crime volume
Cases known to police	286,724
Attempts (percent)	11,381
Percent cases solved	124,633
Crimes per 100,000 persons	1,911.49

Persons responsible for offenses
Total number offenders	107,032
Percent female	9,699
Percent juvenile[55]	10,775
Percent foreigners	2,133

Human Rights [24]

Labor Force

Total Labor Force [25]

2.484 million

Labor Force by Occupation [26]

Industry	33.2%
Agriculture	12.2
Construction	10.3
Communication and other	44.3

Date of data: 1990

Unemployment Rate [27]

11.3% (1992 est.)

For sources, notes, and explanations, see Annotated Source Appendix, page 1035.

Production Sectors

Commercial Energy Production and Consumption

Production [28]	Consumption [29]
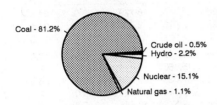	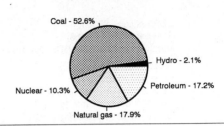
Coal - 81.2%	Coal - 52.6%
Crude oil - 0.5%	Hydro - 2.1%
Hydro - 2.2%	Petroleum - 17.2%
Nuclear - 15.1%	Nuclear - 10.3%
Natural gas - 1.1%	Natural gas - 17.9%

Data are shown in quadrillion (10^{15}) BTUs and percent for 1991

Crude oil[1]	0.01[b]	Petroleum	0.50[b]
Natural gas liquids	(s)[b]	Dry natural gas	0.52[b]
Dry natural gas	0.02[b]	Coal[2]	1.53[b]
Coal[2]	1.51[b]	Net hydroelectric power[3]	0.06[b]
Net hydroelectric power[3]	0.04[b]	Net nuclear power[3]	0.30[b]
Net nuclear power[3]	0.28[b]	Total	2.92[b]
Total	1.85[b]		

Telecommunications [30]

Transportation [31]

Railroads. 3,669 km total (1990)

Highways. 17,650 km total (1990)

Merchant Marine. the former Czechoslovakia had 22 ships (1,000 GRT or over) totaling 290,185 GRT/437,291 DWT; includes 13 cargo, 9 bulk; may be shared with the Czech Republic

Airports

Total:	34
Usable:	34
With permanent-surface runways:	9
With runways over 3,659 m:	0
With runways 2,440-3,659 m:	1
With runways 1,220-2,439 m:	5

Top Agricultural Products [32]

Largely self-sufficient in food production; diversified crop and livestock production, including grains, potatoes, sugar beets, hops, fruit, hogs, cattle, and poultry; exporter of forest products.

Top Mining Products [33]

Estimated metric tons unless otherwise specified   M.t.

Gypsum, crude	624,000
Magnesite, crude	328,000
Clay, kaolin	705,000
Zinc, mine output, Zn content of concentrate	533,000
Iron ore, Fe content (000 tons)	460
Limestone (000 tons)	7,442

Tourism [34]

	1987	1988	1989	1990	1991
Visitors[21]	21,756	24,593	29,683	46,607	64,801
Tourists[21]	6,126	14,028	8,036		
Excursionists[21]	15,630	10,565	21,647		
Tourism receipts	493	608	581	470	825
Tourism expenditures	409	399	431	636[22]	393

Tourists are in thousands, money in million U.S. dollars.

For sources, notes, and explanations, see Annotated Source Appendix, page 1035.

831

Manufacturing Sector

GDP and Manufacturing Summary [35]

	1980	1985	1989	1990	% change 1980-1990	% change 1989-1990
GDP (million 1980 $)	40,327	43,826	62,060	46,773	16.0	-24.6
GDP per capita (1980 $)	2,634	2,827	3,970	2,986	13.4	-24.8
Manufacturing as % of GDP (current prices)	45.6	42.3	47.9[e]	40.1	-12.1	-16.3
Gross output (million $)	41,415	45,108	53,480	44,915	8.5	-16.0
Value added (million $)	17,194	13,083	15,456	12,471	-27.5	-19.3
Value added (million 1980 $)	22,261	24,404	35,035	26,006	16.8	-25.8
Industrial production index	100	121	133	130	30.0	-2.3
Employment (thousands)	2,518	2,588	2,572	2,448	-2.8	-4.8

Note: GDP stands for Gross Domestic Product. 'e' stands for estimated value.

Profitability and Productivity

	1980	1985	1989	1990	% change 1980-1990	% change 1989-1990
Intermediate input (%)	58	71	71	72	24.1	1.4
Wages, salaries, and supplements (%)	15	13	12	13	-13.3	8.3
Gross operating surplus (%)	27	16	17	15	-44.4	-11.8
Gross output per worker ($)	16,448	17,430	20,793	18,348	11.6	-11.8
Value added per worker ($)	6,828	5,055	6,010	5,094	-25.4	-15.2
Average wage (incl. benefits) ($)	2,438	2,264	2,522	2,396	-1.7	-5.0

Profitability is in percent of gross output. Productivity is in U.S. $. 'e' stands for estimated value.

Profitability - 1990

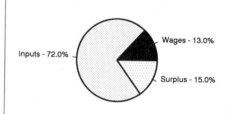

Inputs - 72.0%
Wages - 13.0%
Surplus - 15.0%

The graphic shows percent of gross output.

Value Added in Manufacturing

	1980 $ mil.	1980 %	1985 $ mil.	1985 %	1989 $ mil.	1989 %	1990 $ mil.	1990 %	% change 1980-1990	% change 1989-1990
311 Food products	1,257	7.3	911	7.0	1,126	7.3	916	7.3	-27.1	-18.7
313 Beverages	285	1.7	209	1.6	272	1.8	258	2.1	-9.5	-5.1
314 Tobacco products	33	0.2	23	0.2	27	0.2	24	0.2	-27.3	-11.1
321 Textiles	1,100	6.4	848	6.5	1,046	6.8	790	6.3	-28.2	-24.5
322 Wearing apparel	271	1.6	236	1.8	281	1.8	223	1.8	-17.7	-20.6
323 Leather and fur products	94	0.5	69	0.5	94	0.6	66	0.5	-29.8	-29.8
324 Footwear	299	1.7	244	1.9	306	2.0	256	2.1	-14.4	-16.3
331 Wood and wood products	387	2.3	259	2.0	318	2.1	289	2.3	-25.3	-9.1
332 Furniture and fixtures	210	1.2	162	1.2	183	1.2	154	1.2	-26.7	-15.8
341 Paper and paper products	391	2.3	287	2.2	391	2.5	255	2.0	-34.8	-34.8
342 Printing and publishing	136	0.8	103	0.8	141	0.9	127	1.0	-6.6	-9.9
351 Industrial chemicals	1,262	7.3	862	6.6	875	5.7	698	5.6	-44.7	-20.2
352 Other chemical products	178	1.0	130	1.0	162	1.0	177	1.4	-0.6	9.3
353 Petroleum refineries	497	2.9	390	3.0	429	2.8	316	2.5	-36.4	-26.3
354 Miscellaneous petroleum and coal products	120	0.7	74	0.6	92	0.6	209	1.7	74.2	127.2
355 Rubber products	214	1.2	158	1.2	203	1.3	131	1.1	-38.8	-35.5
356 Plastic products	50	0.3	34	0.3	32	0.2	49	0.4	-2.0	53.1
361 Pottery, china and earthenware	45	0.3	39	0.3	40	0.3	46	0.4	2.2	15.0
362 Glass and glass products	422	2.5	263	2.0	361	2.3	298	2.4	-29.4	-17.5
369 Other non-metal mineral products	773	4.5	488	3.7	574	3.7	411	3.3	-46.8	-28.4
371 Iron and steel	1,753	10.2	1,312	10.0	1,910	12.4	1,271	10.2	-27.5	-33.5
372 Non-ferrous metals	327	1.9	214	1.6	264	1.7	236	1.9	-27.8	-10.6
381 Metal products	792	4.6	590	4.5	710	4.6	602	4.8	-24.0	-15.2
382 Non-electrical machinery	3,452	20.1	2,827	21.6	2,866	18.5	2,597	20.8	-24.8	-9.4
383 Electrical machinery	853	5.0	828	6.3	1,025	6.6	894	7.2	4.8	-12.8
384 Transport equipment	1,677	9.8	1,315	10.1	1,441	9.3	903	7.2	-46.2	-37.3
385 Professional and scientific equipment	94	0.5	67	0.5	94	0.6	84	0.7	-10.6	-10.6
390 Other manufacturing industries	223	1.3	140	1.1	194	1.3	192	1.5	-13.9	-1.0

Note: The industry codes shown are International Standard Industry codes (ISIC). Percentages are percent of total Value Added. 'e' stands for estimated value

832

For sources, notes, and explanations, see Annotated Source Appendix, page 1035.

Finance, Economics, and Trade

Economic Indicators [36]

In Crowns or U.S. Dollars as indicated[31].

	1989	1990	1991
Real GDP (Bil. Crowns)	736.6	730.5	NA
Real GDP Growth Rate	1.4	-0.3	NA
Money Supply (MI) (Bil. Crowns, est.)	311.1	291.2	346.5
Commercial Interest Rate (Jan) (%)	5.20	6.16	15.76
Savings Rate (%)	3.5	-0.2	7.3
Investment Rate (%)	2.6	5.7	NA
CPI (1989 equals 100)	100.0	110.0	160.7[32]
WPI (1985 equals 100)	99.4	103.8	165.7[33]
External Public Debt (Bil. $US)	7.92	8.1	9.27[34]

Balance of Payments Summary [37]

Exchange Rates [38]

Currency: **koruny.**
Symbol: **Kcs.**

Data are currency units per $1.

December 1992	28.59
1992	28.26
1991	29.53
1990	17.95
1989	15.05
1988	14.36

Imports and Exports

Top Import Origins [39]

$3.6 billion (f.o.b., 1992).

Origins	%
Czech Republic	NA
CIS republics	NA
Germany	NA
Austria	NA
Poland	NA
Switzerland	NA
Hungary	NA
UK	NA
Italy	NA

Top Export Destinations [40]

$3.6 billion (f.o.b., 1992).

Destinations	%
Czech Republic	NA
CIS republics	NA
Germany	NA
Poland	NA
Austria	NA
Hungary	NA
Italy	NA
France	NA
US	NA
UK	NA

Foreign Aid [41]

	U.S. $	
The former Czechoslovakia was a donor - in bilateral aid to non-Communist less developed countries	4.2	billion

Import and Export Commodities [42]

Import Commodities

Machinery & transport equipment
Fuels and lubricants
Manufactured goods
Raw materials
Chemicals
Agricultural products

Export Commodities

Machinery & transport equip.
Chemicals
Fuels
Minerals
Metals
Agricultural products

For sources, notes, and explanations, see Annotated Source Appendix, page 1035.

833

Slovenia

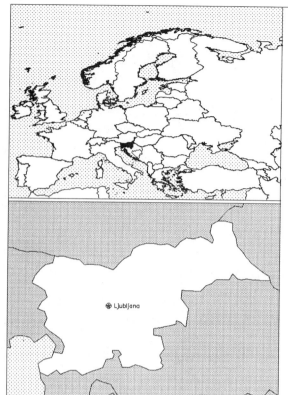

Geography [1]

Total area:
20,296 km2
Land area:
20,296 km2
Comparative area:
Slightly larger than New Jersey
Land boundaries:
Total 999 km, Austria 262 km, Croatia 455 km, Italy 199 km, Hungary 83 km
Coastline:
32 km
Climate:
Mediterranean climate on the coast, continental climate with mild to hot summers and cold winters in the plateaus and valleys to the east
Terrain:
A short coastal strip on the Adriatic, an alpine mountain region adjacent to Italy, mixed mountain and valleys with numerous rivers to the east
Natural resources:
Lignite coal, lead, zinc, mercury, uranium, silver
Land use:
Arable land:
10%
Permanent crops:
2%
Meadows and pastures:
20%
Forest and woodland:
45%
Other:
23%

Demographics [2]

	1960	1970	1980	1990	1991	1994	2000	2010	2020
Population	1,580	1,718	1,885	1,954	NA	1,972	1,998	2,025	2,008
Population density (persons per sq. mi.)	NA	NA	NA	NA	NA	NA	NA	NA	NA
Births	NA	NA	NA	NA	NA	23	NA	NA	NA
Deaths	NA	NA	NA	NA	NA	19	NA	NA	NA
Life expectancy - males	NA	NA	NA	NA	NA	70	NA	NA	NA
Life expectancy - females	NA	NA	NA	NA	NA	78	NA	NA	NA
Birth rate (per 1,000)	NA	NA	NA	NA	NA	12	NA	NA	NA
Death rate (per 1,000)	NA	NA	NA	NA	NA	10	NA	NA	NA
Women of reproductive age (15-44 yrs.)	NA	NA	NA	489	NA	496	501	NA	NA
of which are currently married	NA	NA	NA	344	NA	348	354	NA	NA
Fertility rate	NA	NA	NA	1.7	NA	1.7	1.6	1.6	1.6

Population values are in thousands, life expectancy in years, and other items as indicated.

Health

Health Personnel [3]

Health Expenditures [4]

For sources, notes, and explanations, see Annotated Source Appendix, page 1035.

Human Factors

Health Care Ratios [5]	Infants and Malnutrition [6]

Ethnic Division [7]

Slovene	91%
Croat	3%
Serb	2%
Muslim	1%
Other	3%

Religion [8]

Roman Catholic (including 2% Uniate)	96%
Muslim	1%
Other	3%

Major Languages [9]

Slovenian	91%
Serbo-Croatian	7%
Other	2%

Education

Public Education Expenditures [10]

Million Dinar	1980[22]	1985[22]	1987[22]	1988[22]	1989[22]	1990[22]
Total education expenditure	8	42	202	544	9,555	60,318
as percent of GNP	4.7	3.4	4.2	3.6	4.3	6.1
as percent of total govt. expend.	32.5	NA	NA	NA	NA	NA
Current education expenditure	7	39	185	506	9,002	55,911
as percent of GNP	4.0	3.1	3.9	3.3	4.1	5.7
as percent of current govt. expend.	NA	NA	NA	NA	NA	NA
Capital expenditure	1	3	17	38	553	4,407

Educational Attainment [11]

Literacy Rate [12]

In thousands and percent	1985[a,1]	1991[a,1]	2000[a,1]
Illiterate population +15 years	1,614	1,342	942
Illiteracy rate - total pop. (%)	9.2	7.3	4.7
Illiteracy rate - males (%)	3.5	2.6	1.3
Illiteracy rate - females (%)	14.6	11.9	7.9

Libraries [13]

	Admin. Units	Svc. Pts.	Vols. (000)	Shelving (meters)	Vols. Added	Reg. Users
National	NA	NA	NA	NA	NA	NA
Non-specialized	NA	NA	NA	NA	NA	NA
Public (1991)	60	916	5,428	NA	NA	4.4mil
Higher ed.	NA	NA	NA	NA	NA	NA
School	NA	NA	NA	NA	NA	NA

Daily Newspapers [14]

	1975[b]	1980[b]	1985[b]	1990[b]
Number of papers	26	27	27	34
Circ. (000)	1,896	2,649	2,451	2,281

Cinema [15]

Data for 1989.

Cinema seats per 1,000	16.9[a]
Annual attendance per person	1.9[a]
Gross box office receipts (mil. Dinar)	643,490[a]

Science and Technology

Scientific/Technical Forces [16]

Potential scientists/engineers	563,312
Number female	NA
Potential technicians	432,380
Number female	NA
Total	995,692

R&D Expenditures [17]

	Dinar (000) 1989
Total expenditure[a,4,12]	2,152,032
Capital expenditure	815,082
Current expenditure	1,336,950
Percent current	62.1

U.S. Patents Issued [18]

Government and Law

Organization of Government [19]

Long-form name:
Republic of Slovenia
Type:
emerging democracy
Independence:
25 June 1991 (from Yugoslavia)
Constitution:
adopted 23 December 1991, effective 23
December 1991
Legal system:
based on civil law system
National holiday:
Statehood Day, 25 June
Executive branch:
president, prime minister, deputy prime
ministers, cabinet
Legislative branch:
bicameral National Assembly; consists of
the State Assembly and the State Council;
note - State Council will become
operational after next election
Judicial branch:
Supreme Court, Constitutional Court

Political Parties [20]

State Assembly	% of seats
Liberal Democratic (LDS)	24.4
Slovene Chris. Democratic (SKD)	16.7
United List	15.6
Slovene National Party	13.3
SN	11.1
Democratic Party	6.7
Greens of Slovenia (ZS)	5.6
Social-Democratic Party (SDSS)	4.4
Hungarian minority	1.1
Italian minority	1.1

Government Budget [21]

For 1993.
Revenues	NA
Expenditures	NA
Capital expenditures	NA

Defense Summary [22]

Branches: Slovene Defense Forces

Manpower Availability: Males age 15-49 512,186; fit for military service 410,594; reach military age (19) annually 14,970 (1993 est.)

Defense Expenditures: 13.5 billion tolars, 4.5% of GDP (1993); note - conversion of the military budget into US dollars using the current exchange rate could produce misleading results

Crime [23]

Human Rights [24]

	SSTS	FL	FAPRO	PPCG	APROBC	TPW	PCPTW	STPEP	PHRFF	PRW	ASST	AFL
Observes			P	P	P	P	P			P	P	
	EAFRD	CPR	ESCR	SR	ACHR	MAAE	PVIAC	PVNAC	EAFDAW	TCIDTP	RC
Observes	P	P	P	P		P	P	P	P		P

P = Party; S = Signatory; see Appendix for meaning of abbreviations.

Labor Force

Total Labor Force [25]

786,036

Labor Force by Occupation [26]

Agriculture	2%
Manufacturing and mining	46

Unemployment Rate [27]

10% (April 1992)

For sources, notes, and explanations, see Annotated Source Appendix, page 1035.

Production Sectors

Energy Resource Summary [28]

Energy Resources: Lignite coal, uranium. **Electricity**: 2,900,000 kW capacity; 10,000 million kWh produced, 5,090 kWh per capita (1992).
Pipelines: Crude oil 290 km, natural gas 305 km.

Telecommunications [30]

- 130,000 telephones
- Broadcast stations - 6 AM, 5 FM, 7 TV
- 370,000 radios
- 330,000 TVs

Top Agricultural Products [32]

Dominated by stock breeding (sheep and cattle) and dairy farming; main crops - potatoes, hops, hemp, flax; an export surplus in these commodities; Slovenia must import many other agricultural products and has a negative overall trade balance in this sector.

Top Mining Products [33]

Metric tons unless otherwise specified	M.t.
Magnesite, crude	175,000
Bauxite (000 tons)	2,850
Copper, Cu content of concentrate	140,000
Gypsum, crude	400,000
Pumice, volcanic tuff	380,000
Cement, hydraulic (000 tons)	7,100

Transportation [31]

Railroads. 1,200 km, 1.435 m gauge (1991)

Highways. 14,553 km total; 10,525 km paved, 4,028 km gravel

Merchant Marine. 22 ships (1,000 GRT or over) totaling 348,784 GRT/596,740 DWT; includes 15 bulk, 7 cargo; all under the flag of Saint Vincent and the Grenadines except for 1 bulk under Liberian flag

Airports

Total:	13
Useable:	13
With permanent-surface runways:	5
With runways over 3,659 m:	0
With runways 2,440-3,659 m:	2
With runways 1,220-2,439 m:	4

Tourism [34]

	1987	1988	1989	1990	1991
Visitors[74]	26,151	29,635	34,118	39,573	
Tourists[6,74]	8,907	9,018	8,644	7,880	1,459
Tourism receipts[74]	1,668	2,024	2,230	2,774	468
Tourism expenditures[74]	90	109	131	149	103
Fare receipts[74]	340	430	475	620	105

Tourists are in thousands, money in million U.S. dollars.

Manufacturing Sector

GDP and Manufacturing Summary [35]

	1980	1985	1989	1990	% change 1980-1990	% change 1989-1990
GDP (million 1980 $)	69,958	71,058	72,234	66,371	-5.1	-8.1
GDP per capita (1980 $)	3,136	3,073	3,050	2,786	-11.2	-8.7
Manufacturing as % of GDP (current prices)	30.6	37.2	39.5	42.0	37.3	6.3
Gross output (million $)	72,629	57,020	65,078	62,136[e]	-14.4	-4.5
Value added (million $)	21,750	17,171	30,245	27,660[e]	27.2	-8.5
Value added (million 1980 $)	19,526	22,283	24,021	21,703	11.1	-9.6
Industrial production index	100	116	120	108	8.0	-10.0
Employment (thousands)	2,106	2,467	2,658	2,537[e]	20.5	-4.6

Note: GDP stands for Gross Domestic Product. 'e' stands for estimated value.

Profitability and Productivity

	1980	1985	1989	1990	% change 1980-1990	% change 1989-1990
Intermediate input (%)	70	70	54	55[e]	-21.4	1.9
Wages, salaries, and supplements (%)	14	12	12[e]	18[e]	28.6	50.0
Gross operating surplus (%)	15	18	34[e]	26[e]	73.3	-23.5
Gross output per worker ($)	34,487	23,113	24,484	24,248[e]	-29.7	-1.0
Value added per worker ($)	10,328	6,960	11,379	10,796[e]	4.5	-5.1
Average wage (incl. benefits) ($)	4,991	2,703	2,986[e]	4,488[e]	-10.1	50.3

Profitability is in percent of gross output. Productivity is in U.S. $. 'e' stands for estimated value.

Profitability - 1990

Inputs - 55.6%
Wages - 18.2%
Surplus - 26.3%

The graphic shows percent of gross output.

Value Added in Manufacturing

	1980 $ mil.	1980 %	1985 $ mil.	1985 %	1989 $ mil.	1989 %	1990 $ mil.	1990 %	% change 1980-1990	% change 1989-1990
311 Food products	1,897	8.7	1,458	8.5	3,916	12.9	3,484[e]	12.6	83.7	-11.0
313 Beverages	459	2.1	353	2.1	663	2.2	589[e]	2.1	28.3	-11.2
314 Tobacco products	184	0.8	221	1.3	344	1.1	308[e]	1.1	67.4	-10.5
321 Textiles	1,759	8.1	1,428	8.3	2,881	9.5	2,663[e]	9.6	51.4	-7.6
322 Wearing apparel	903	4.2	718	4.2	1,593	5.3	1,427[e]	5.2	58.0	-10.4
323 Leather and fur products	226	1.0	231	1.3	383	1.3	340[e]	1.2	50.4	-11.2
324 Footwear	482	2.2	503	2.9	1,022	3.4	899[e]	3.3	86.5	-12.0
331 Wood and wood products	977	4.5	530	3.1	794	2.6	706[e]	2.6	-27.7	-11.1
332 Furniture and fixtures	730	3.4	438	2.6	1,030	3.4	1,065[e]	3.9	45.9	3.4
341 Paper and paper products	529	2.4	394	2.3	759	2.5	674[e]	2.4	27.4	-11.2
342 Printing and publishing	876	4.0	462	2.7	761	2.5	678[e]	2.5	-22.6	-10.9
351 Industrial chemicals	694	3.2	631	3.7	1,107	3.7	992[e]	3.6	42.9	-10.4
352 Other chemical products	681	3.1	525	3.1	1,419	4.7	1,315[e]	4.8	93.1	-7.3
353 Petroleum refineries	454	2.1	415	2.4	260	0.9	233[e]	0.8	-48.7	-10.4
354 Miscellaneous petroleum and coal products	101	0.5	101	0.6	104	0.3	91[e]	0.3	-9.9	-12.5
355 Rubber products	276	1.3	269	1.6	479	1.6	456[e]	1.6	65.2	-4.8
356 Plastic products	413	1.9	258	1.5	397	1.3	350[e]	1.3	-15.3	-11.8
361 Pottery, china and earthenware	128	0.6	72	0.4	162	0.5	144[e]	0.5	12.5	-11.1
362 Glass and glass products	163	0.7	113	0.7	224	0.7	204[e]	0.7	25.2	-8.9
369 Other non-metal mineral products	906	4.2	513	3.0	683	2.3	604[e]	2.2	-33.3	-11.6
371 Iron and steel	1,221	5.6	1,000	5.8	1,343	4.4	1,171[e]	4.2	-4.1	-12.8
372 Non-ferrous metals	480	2.2	509	3.0	944	3.1	927[e]	3.4	93.1	-1.8
381 Metal products	2,105	9.7	1,577	9.2	1,293	4.3	1,130[e]	4.1	-46.3	-12.6
382 Non-electrical machinery	1,828	8.4	1,463	8.5	2,372	7.8	2,378[e]	8.6	30.1	0.3
383 Electrical machinery	1,600	7.4	1,544	9.0	2,640	8.7	2,334[e]	8.4	45.9	-11.6
384 Transport equipment	1,441	6.6	1,263	7.4	2,389	7.9	2,241[e]	8.1	55.5	-6.2
385 Professional and scientific equipment	101	0.5	93	0.5	154	0.5	146[e]	0.5	44.6	-5.2
390 Other manufacturing industries	134	0.6	88	0.5	128	0.4	114[e]	0.4	-14.9	-10.9

Note: The industry codes shown are International Standard Industry codes (ISIC). Percentages are percent of total Value Added. 'e' stands for estimated value

For sources, notes, and explanations, see Annotated Source Appendix, page 1035.

Finance, Economics, and Trade

Economic Indicators [36]

Yugoslav Dinars or U.S. Dollars as indicated[12].

	1989	1990	1991[e]
GSP[13]			
(billions current dinars)	221.9	910.9	1,490
Real GSP growth rate (%)	0.6	-7.6	-20
Money supply (M1)			
(yr-end mil.)	51,216	127,241	NA
Commercial interest			
rates (avg.)	4,354	45	80
Investment rate (%)	19.4	18.9	NA
CPI	1,356	688	310
WPI	1,406	534	320
External public debt	18,569	17,791	15,760

Balance of Payments Summary [37]

Exchange Rates [38]

Currency: **tolars.**
Symbol: **SIT.**

Data are currency units per $1.

June 1993	112
January 1992	28

Imports and Exports

Top Import Origins [39]

$4.679 billion (c.i.f., 1990) Principally the other former Yugoslav republics, Germany, successor states of the former USSR, US, Hungary, Italy, Austria.

Origins	%
No details available.	NA

Top Export Destinations [40]

$4.12 billion (f.o.b., 1990).

Destinations	%
Principally the other former Yugoslav republics	NA
Austria	NA
And Italy	NA

Foreign Aid [41]

Import and Export Commodities [42]

Import Commodities

Machinery, transp. equip. 35%
Other manufactured goods 26.7%
Chemicals 14.5%
Raw materials 9.4%
Fuels and lubricants 7%
Food and live animals 6%

Export Commodities

Machinery, transp. equip. 38%
Other manufactured goods 44%
Chemicals 9%
Food and live animals 4.6%
Raw materials 3%
Beverages & tobacco <1%

For sources, notes, and explanations, see Annotated Source Appendix, page 1035.

839

Solomon Islands

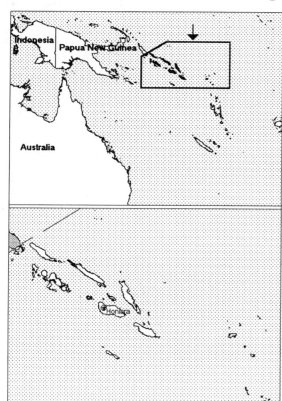

Geography [1]

Total area:
28,450 km2
Land area:
27,540 km2
Comparative area:
Slightly larger than Maryland
Land boundaries:
0 km
Coastline:
5,313 km
Climate:
Tropical monsoon; few extremes of temperature and weather
Terrain:
Mostly rugged mountains with some low coral atolls
Natural resources:
Fish, forests, gold, bauxite, phosphates
Land use:
Arable land:
1%
Permanent crops:
1%
Meadows and pastures:
1%
Forest and woodland:
93%
Other:
4%

Demographics [2]

	1960	1970	1980	1990	1991[1]	1994	2000	2010	2020
Population	126	163	233	336	347	386	470	620	767
Population density (persons per sq. mi.)	12	15	22	32	33	NA	44	58	73
Births	NA	NA	NA	NA	14	15	NA	NA	NA
Deaths	NA	NA	NA	NA	2	2	NA	NA	NA
Life expectancy - males	NA	NA	NA	NA	67	68	NA	NA	NA
Life expectancy - females	NA	NA	NA	NA	72	73	NA	NA	NA
Birth rate (per 1,000)	NA	NA	NA	NA	40	39	NA	NA	NA
Death rate (per 1,000)	NA	NA	NA	NA	5	5	NA	NA	NA
Women of reproductive age (15-44 yrs.)	NA	NA	NA	71	NA	84	107	NA	NA
of which are currently married	NA	NA	NA	46	NA	54	70	NA	NA
Fertility rate	NA	NA	NA	6.3	6.16	5.7	4.8	3.4	2.6

Population values are in thousands, life expectancy in years, and other items as indicated.

Health

Health Personnel [3]

Health Expenditures [4]

For sources, notes, and explanations, see Annotated Source Appendix, page 1035.

Human Factors

Health Care Ratios [5]	Infants and Malnutrition [6]

Ethnic Division [7]

Melanesian	93.0%
Polynesian	4.0%
Micronesian	1.5%
European	0.8%
Chinese	0.3%
Other	0.4%

Religion [8]

Anglican	34%
Roman Catholic	19%
Baptist	17%
United (Methodist/Presbyterian)	11%
Seventh-Day Adventist	10%
Other Protestant	5%

Major Languages [9]

Melanesian pidgin in much of the country is lingua franca, English spoken by 1-2% of population.

Education

Public Education Expenditures [10]	Educational Attainment [11]
Literacy Rate [12]	Libraries [13]
Daily Newspapers [14]	Cinema [15]

Science and Technology

Scientific/Technical Forces [16]	R&D Expenditures [17]	U.S. Patents Issued [18]

For sources, notes, and explanations, see Annotated Source Appendix, page 1035.

Government and Law

Organization of Government [19]

Long-form name:
none
Type:
parliamentary democracy
Independence:
7 July 1978 (from UK)
Constitution:
7 July 1978
Legal system:
common law
National holiday:
Independence Day, 7 July (1978)
Executive branch:
British monarch, governor general, prime minister, Cabinet
Legislative branch:
unicameral National Parliament
Judicial branch:
High Court

Political Parties [20]

National Parliament	% of seats
People's Alliance Party (PAP)	34.2
United Party (UP)	15.8
Nationalist Front (NFP)	10.5
Solomon Islands Liberal (SILP)	10.5
Labor Party (LP)	5.3
Independents	23.7

Government Budget [21]

For 1991 est.	
Revenues	48
Expenditures	107
Capital expenditures	45

Defense Summary [22]

Branches: Police Force

Manpower Availability: No information available.

Defense Expenditures: No information available.

Crime [23]

Human Rights [24]

	SSTS	FL	FAPRO	PPCG	APROBC	TPW	PCPTW	STPEP	PHRFF	PRW	ASST	AFL
Observes	P	P				P	P			P	P	
	EAFRD	CPR	ESCR	SR	ACHR	MAAE	PVIAC	PVNAC	EAFDAW	TCIDTP	RC	
Observes	1		1				P	P	S	P	P	

P = Party; S = Signatory; see Appendix for meaning of abbreviations.

Labor Force

Total Labor Force [25]

23,448 economically active

Labor Force by Occupation [26]

Agriculture, forestry, and fishing
Services
Construction, manufacturing, and mining
Commerce, transport, and finance
Date of data: 1984

Unemployment Rate [27]

For sources, notes, and explanations, see Annotated Source Appendix, page 1035.

Production Sectors

Energy Resource Summary.[28]

Energy Resources: None. **Electricity**: 21,000 kW capacity; 39 million kWh produced, 115 kWh per capita (1990).

Telecommunications [30]

- 3,000 telephones
- Broadcast stations - 4 AM, no FM, no TV
- 1 Pacific Ocean INTELSAT earth station

Transportation [31]

Highways. about 2,100 km total (1982); 30 km paved, 290 km gravel, 980 km earth, 800 private logging and plantation roads of varied construction

Airports

Total:	30
Usable:	29
With permanent-surface runways:	2
With runways over 3,659 m:	0
With runways 2,440-3,659 m:	0
With runways 1,220-2,439 m:	3

Top Agricultural Products [32]

Including fishing and forestry, accounts for about 70% of GDP; mostly subsistence farming; cash crops—cocoa, beans, coconuts, palm kernels, timber; other products—rice, potatoes, vegetables, fruit, cattle, pigs; not self-sufficient in food grains; 90% of the total fish catch of 44,500 metric tons was exported (1988).

Top Mining Products [33]

Detailed information is not available. A summary of mineral resources available follows. **Mineral Resources:** Gold, bauxite, phosphates.

Tourism [34]

	1987	1988	1989	1990	1991
Visitors	19	16	13	2	38
Tourists	13	11	10	9	11
Cruise passengers	6	5	3	3	26
Tourism receipts	6	5	5	4	5
Tourism expenditures	6	9	9	11	13
Fare receipts	1	1	1	1	3
Fare expenditures	4	5	5	5	4

Tourists are in thousands, money in million U.S. dollars.

For sources, notes, and explanations, see Annotated Source Appendix, page 1035.

843

Finance, Economics, and Trade

Industrial Summary [35]

Industrial Production: Growth rate 0% (1987); accounts for 5% of GDP. **Industries**: Copra, fish (tuna).

Economic Indicators [36]

National product: GDP—exchange rate conversion—$200 million (1990 est.). **National product real growth rate**: 6% (1990 est.). **National product per capita**: $600 (1990 est.). **Inflation rate (consumer prices)**: 14.3% (1991). **External debt**: $128 million (1988 est.).

Balance of Payments Summary [37]

Values in millions of dollars.

	1987	1988	1989	1990	1991
Exports of goods (f.o.b.)	63.2	81.9	74.7	70.4	83.4
Imports of goods (f.o.b.)	-69.4	-105.1	-97.6	-77.2	-91.7
Trade balance	-6.2	-23.2	-22.8	-6.8	-8.3
Services - debits	-63.8	-74.4	-87.7	-82.7	-98.4
Services - credits	25.9	29.8	33.9	30.7	36.0
Private transfers (net)	-0.1	-1.4	1.2	1.4	-2.0
Government transfers (net)	31.9	46.1	37.3	33.5	35.3
Long term capital (net)	18.1	23.6	20.6	22.9	29.0
Short term capital (net)	-10.0	16.2	0.3	0.4	4.7
Errors and omissions	8.5	-10.3	5.6	-4.6	0.4
Overall balance	4.2	6.4	-11.6	-5.1	-3.3

Exchange Rates [38]

Currency: **Solomon Islands dollars.**
Symbol: **SI$.**

Data are currency units per $1.

January 1993	3.1211
1992	2.9281
1991	2.7148
1990	2.5288
1989	2.2932
1988	2.0825

Imports and Exports

Top Import Origins [39]

$87.1 million (c.i.f., 1991 est.). Data are for 1985.

Origins	%
Japan	36
US	23
Singapore	9
UK	9
NZ	9
Australia	4
Hong Kong	4
China	3

Top Export Destinations [40]

$74.2 million (f.o.b., 1991 est.). Data are for 1985.

Destinations	%
Japan	51
UK	12
Thailand	9
Netherlands	8
Australia	2
US	2

Foreign Aid [41]

	U.S. $	
Western (non-US) countries, ODA and OOF bilateral commitments (1980-89)	250	million

Import and Export Commodities [42]

Import Commodities	**Export Commodities**
Plant and machinery 30%	Fish 46%
Fuel 19%	Timber 31%
Food 16%	Copra 5%
	Palm oil 5%

Somalia

Geography [1]

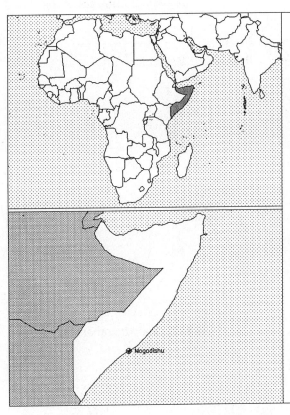

Total area:
637,660 km2
Land area:
627,340 km2
Comparative area:
Slightly smaller than Texas
Land boundaries:
Total 2,366 km, Djibouti 58 km, Ethiopia 1,626 km, Kenya 682 km
Coastline:
3,025 km
Climate:
Desert; northeast monsoon (December to February), cooler southwest monsoon (May to October); irregular rainfall; hot, humid periods (tangambili) between monsoons
Terrain:
Mostly flat to undulating plateau rising to hills in north
Natural resources:
Uranium and largely unexploited reserves of iron ore, tin, gypsum, bauxite, copper, salt
Land use:
Arable land: 2%
Permanent crops: 0%
Meadows and pastures: 46%
Forest and woodland: 14%
Other: 38%

Demographics [2]

	1960	1970	1980	1990	1991[1]	1994	2000	2010	2020
Population	2,956	3,667	5,799	6,667	6,709	6,667	9,176	12,588	16,832
Population density (persons per sq. mi.)	12	15	24	27	28	NA	39	53	71
Births	NA	NA	NA	NA	307	306	NA	NA	NA
Deaths	NA	NA	NA	NA	89	90	NA	NA	NA
Life expectancy - males	NA	NA	NA	NA	56	54	NA	NA	NA
Life expectancy - females	NA	NA	NA	NA	56	55	NA	NA	NA
Birth rate (per 1,000)	NA	NA	NA	NA	46	46	NA	NA	NA
Death rate (per 1,000)	NA	NA	NA	NA	13	14	NA	NA	NA
Women of reproductive age (15-44 yrs.)	NA	NA	NA	NA	NA	NA	NA	NA	NA
of which are currently married	NA	NA	NA	NA	NA	NA	NA	NA	NA
Fertility rate	NA	NA	NA	7.3	7.19	7.3	6.5	5.4	4.2

Population values are in thousands, life expectancy in years, and other items as indicated.

Health

Health Personnel [3]

Doctors per 1,000 pop., 1988-92	0.07
Nurse-to-doctor ratio, 1988-92	7.1
Hospital beds per 1,000 pop., 1985-90	0.8
Percentage of children immunized (age 1 yr. or less)	
Third dose of DPT, 1990-91	18.0
Measles, 1990-91	30.0

Health Expenditures [4]

Total health expenditure, 1990 (official exchange rate)	
Millions of dollars	60
Millions of dollars per capita	8
Health expenditures as a percentage of GDP	
Total	1.5
Public sector	0.9
Private sector	0.6
Development assistance for health	
Total aid flows (millions of dollars)[1]	27
Aid flows per capita (millions of dollars)	3.5
Aid flows as a percentage of total health expenditure	45.6

For sources, notes, and explanations, see Annotated Source Appendix, page 1035.

845

Human Factors

Health Care Ratios [5]	Infants and Malnutrition [6]

Ethnic Division [7]

Somali	85%
Bantu	NA
Arabs	30,000
Europeans	3,000
Asians	800

Religion [8]

Sunni Muslim.

Major Languages [9]

Somali (official)	NA
Arabic	NA
Italian	NA
English	NA

Education

Public Education Expenditures [10]

Million Shilling	1980[6]	1985[6]	1986[6]	1988[6]	1989[6]	1990[6]
Total education expenditure	169	371	434	NA	NA	NA
as percent of GNP	1.0	0.5	0.4	NA	NA	NA
as percent of total govt. expend.	8.7	4.1	2.8	NA	NA	NA
Current education expenditure	154	274	290	NA	NA	NA
as percent of GNP	0.9	0.3	0.3	NA	NA	NA
as percent of current govt. expend.	NA	NA	NA	NA	NA	NA
Capital expenditure	15	97	143	NA	NA	NA

Educational Attainment [11]

Literacy Rate [12]

In thousands and percent	1985[a]	1991[a]	2000[a]
Illiterate population +15 years	2,877	3,003	3,235
Illiteracy rate - total pop. (%)	83.1	75.9	61.3
Illiteracy rate - males (%)	73.3	63.9	47.9
Illiteracy rate - females (%)	91.2	86.0	73.5

Libraries [13]

Daily Newspapers [14]

	1975	1980	1985	1990
Number of papers	1	2	2	1
Circ. (000)	3[e]	5[e]	7[e]	9[e]

Cinema [15]

Science and Technology

Scientific/Technical Forces [16]	R&D Expenditures [17]	U.S. Patents Issued [18]

For sources, notes, and explanations, see Annotated Source Appendix, page 1035.

Government and Law

Organization of Government [19]

Long-form name:
none
Type:
none
Independence:
1 July 1960 (from a merger of British Somaliland, which became independent from the UK on 26 June 1960, and Italian Somaliland, which became independent from the Italian-administered UN trusteeship on 1 July 1960, to form the Somali Republic)
Constitution:
25 August 1979, presidential approval 23 September 1979
Legal system:
NA
National holiday:
NA
Executive branch:
president, two vice presidents, prime minister, Council of Ministers (cabinet)
Legislative branch:
unicameral People's Assembly (Golaha Shacbiga); non-functioning
Judicial branch:
Supreme Court (non-functioning)

Crime [23]

Elections [20]

People's Assembly. Last held 31 December 1984 (next to be held NA); results - SRSP (Somali Revolutionary Socialist Party) was the only party; seats—(177 total, 171 elected) SRSP 171; note—the United Somali Congress (USC) ousted the regime of Maj. Gen. Mohamed Siad Barre on 27 January 1991; the provisional government has promised that a democratically elected government will be established.

Government Budget [21]

For 1991 est.

Revenues	NA
Expenditures	NA
Capital expenditures	NA

Military Expenditures and Arms Transfers [22]

	1985	1986	1987	1988	1989
Military expenditures					
Current dollars (mil.)	NA	31[e]	NA	NA	NA
1989 constant dollars (mil.)	NA	35[e]	NA	NA	NA
Armed forces (000)	43	50	50	47	47
Gross national product (GNP)					
Current dollars (mil.)	934	985	1,095	1,111	1,173
1989 constant dollars (mil.)	1,063	1,093	1,178	1,156	1,173
Central government expenditures (CGE)					
1989 constant dollars (mil.)	NA	116[e]	NA	NA	NA
People (mil.)	6.5	6.8	7.0	7.0	6.8
Military expenditure as % of GNP	NA	3.2	NA	NA	NA
Military expenditure as % of CGE	NA	30.0	NA	NA	NA
Military expenditure per capita	NA	5	NA	NA	NA
Armed forces per 1,000 people	6.6	7.4	7.1	6.7	6.9
GNP per capita	164	161	168	164	172
Arms imports[6]					
Current dollars (mil.)	60	20	20	30	30
1989 constant dollars (mil.)	68	22	22	31	30
Arms exports[6]					
Current dollars (mil.)	0	0	0	0	0
1989 constant dollars (mil.)	0	0	0	0	0
Total imports[7]					
Current dollars (mil.)	112	279	132	354	NA
1989 constant dollars	128	310	142	369	NA
Total exports[7]					
Current dollars (mil.)	91	85	104	58	NA
1989 constant dollars	104	94	112	60	NA
Arms as percent of total imports[8]	53.6	7.2	15.2	8.5	NA
Arms as percent of total exports[8]	0	0	0	0	NA

Human Rights [24]

	SSTS	FL	FAPRO	PPCG	APROBC	TPW	PCPTW	STPEP	PHRFF	PRW	ASST	AFL
Observes		P				P	P					P
	EAFRD	CPR	ESCR	SR	ACHR	MAAE	PVIAC	PVNAC	EAFDAW	TCIDTP	RC	
Observes	P	P	P	P						P		

P = Party; S = Signatory; see Appendix for meaning of abbreviations.

Labor Force

Total Labor Force [25]

2.2 million (very few are skilled laborers)

Labor Force by Occupation [26]

Pastoral nomad	70%
Other	30

Unemployment Rate [27]

For sources, notes, and explanations, see Annotated Source Appendix, page 1035.

847

Production Sectors

Energy Resource Summary [28]

Energy Resources: Uranium. **Electricity**: Former public power capacity of 75,000 kW is completely shut down by the destruction of the civil war; UN, relief organizations, and foreign military units in Somalia use their own portable power systems. **Pipelines**: Crude oil 15 km.

Telecommunications [30]

- The public telecommunications system was completely destroyed or dismantled by the civil war factions
- All relief organizations depend on their own private systems (1993)

Top Agricultural Products [32]

Dominant sector, led by livestock raising (cattle, sheep, goats); crops—bananas, sorghum, corn, mangoes, sugarcane; not self-sufficient in food; distribution of food disrupted by civil strife; fishing potential largely unexploited.

Top Mining Products [33]

Estimated metric tons unless otherwise specified M.t.

	M.t.
Cement, hydraulic	10,000
Gypsum	1,000
Limestone	17,000[52]
Salt, marine	500
Sepiolite, (meerschaum)	4

Transportation [31]

Highways. 22,500 km total; including 2,700 km paved, 3,000 km gravel, and 16,800 km improved earth or stabilized soil (1992)

Merchant Marine. 3 ships (1,000 GRT or over) totaling 6,913 GRT/ 8,718 DWT; includes 2 cargo, 1 refrigerated cargo

Airports

Total:	69
Usable:	48
With permanent-surface runways:	8
With runways over 3,659 m:	2
With runways 2,440-3,659 m:	6
With runways 1,220-2,439 m:	20

Tourism [34]

Finance, Economics, and Trade

GDP and Manufacturing Summary [35]

	1980	1985	1990	1991	1992
Gross Domestic Product					
Millions of 1980 dollars	2,755	3,658	4,075	3,260	3,032[e]
Growth rate in percent	-2.25	7.93	-1.60	-20.00	-7.00[e]
Manufacturing Value Added					
Millions of 1980 dollars	123	103	137	NA	NA
Growth rate in percent	9.17	7.55	0.00	NA	NA
Manufacturing share in percent of current prices	4.7	4.9	4.3[e]	NA	NA

Economic Indicators [36]

National product: not available. **National product real growth rate**: NA%. **National product per capita**: not available. **Inflation rate (consumer prices)**: NA%. **External debt**: $1.9 billion (1989).

Balance of Payments Summary [37]

Values in millions of dollars.

	1985	1986	1987	1988	1989
Exports of goods (f.o.b.)	90.6	94.7	94.0	58.4	67.7
Imports of goods (f.o.b.)	-330.7	-342.1	-358.5	-216.0	-346.3
Trade balance	-240.1	-247.4	-264.5	-157.6	-278.6
Services - debits	-123.4	-188.5	-179.7	-164.6	-206.4
Services - credits	37.0	-	-	-	-
Private transfers (net)	19.4	5.3	-13.1	6.4	-2.9
Government transfers (net)	204.3	304.9	343.3	217.3	331.2
Long term capital (net)	76.3	33.3	77.2	-33.5	-26.0
Short term capital (net)	18.2	100.0	11.6	108.0	163.1
Errors and omissions	15.5	19.2	40.7	21.1	-0.8
Overall balance	7.2	26.8	15.5	-2.9	-20.4

Exchange Rates [38]

Currency: **Somali shillings.**
Symbol: **So. Sh..**

Data are currency units per $1.

December 1992	4,200.00
December 1990	3,800.00
1989	490.70
1988	170.45
1987	105.18
1986	72.00

Imports and Exports

Top Import Origins [39]

Amount not available.

Origins	%
US	13
Italy	NA
FRG	NA
Kenya	NA
UK	NA
Saudi Arabia 1986	NA

Top Export Destinations [40]

Amount not available.

Destinations	%
Saudi Arabia	NA
Italy	NA
FRG (1986)	NA

Foreign Aid [41]

	U.S. $	
US commitments, including Ex-Im (FY70-89)	639	million
Western (non-US) countries, ODA and OOF bilateral commitments (1970-89)	3.8	billion
OPEC bilateral aid (1979-89)	1.1	billion
Communist countries (1970-89)	336	million

Import and Export Commodities [42]

Import Commodities	**Export Commodities**
Petroleum products	Bananas
Foodstuffs	Livestock
Construction materials	Fish
	Hides
	Skins

South Africa

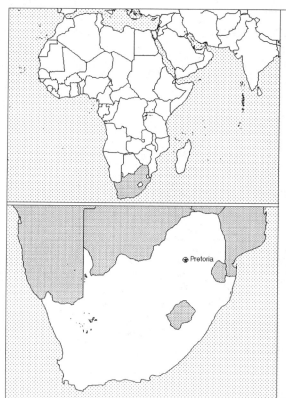

Geography [1]

Total area:
1,221,040 km2
Land area:
1,221,040 km2
Comparative area:
Slightly less than twice the size of Texas
Land boundaries:
Total 4,973 km, Botswana 1,840 km, Lesotho 909 km, Mozambique 491 km, Namibia 1,078 km, Swaziland 430 km, Zimbabwe 225 km
Coastline:
2,881 km
Climate:
Mostly semiarid; subtropical along coast; sunny days, cool nights
Terrain:
Vast interior plateau rimmed by rugged hills and narrow coastal plain
Natural resources:
Gold, chromium, antimony, coal, iron ore, manganese, nickel, phosphates, tin, uranium, gem diamonds, platinum, copper, vanadium, salt, natural gas
Land use:
Arable land:
10%
Permanent crops:
1%
Meadows and pastures:
65%
Forest and woodland:
3%
Other:
21%

Demographics [2]

	1960	1970	1980	1990	1991[1]	1994	2000	2010	2020
Population	17,258	22,562	30,270	39,535	40,601	43,931	51,334	65,850	82,502
Population density (persons per sq. mi.)	37	48	64	84	86	NA	109	140	176
Births	NA	NA	NA	NA	1,388	1,475	NA	NA	NA
Deaths	NA	NA	NA	NA	318	331	NA	NA	NA
Life expectancy - males	NA	NA	NA	NA	61	62	NA	NA	NA
Life expectancy - females	NA	NA	NA	NA	67	68	NA	NA	NA
Birth rate (per 1,000)	NA	NA	NA	NA	34	34	NA	NA	NA
Death rate (per 1,000)	NA	NA	NA	NA	8	8	NA	NA	NA
Women of reproductive age (15-44 yrs.)	NA	NA	NA	9,648	NA	10,707	12,536	NA	NA
of which are currently married	NA	NA	NA	4,641	NA	5,214	6,140	NA	NA
Fertility rate	NA	NA	NA	4.5	4.45	4.4	4.2	3.9	3.4

Population values are in thousands, life expectancy in years, and other items as indicated.

Health

Health Personnel [3]

Doctors per 1,000 pop., 1988-92	0.61
Nurse-to-doctor ratio, 1988-92	4.5
Hospital beds per 1,000 pop., 1985-90	4.1
Percentage of children immunized (age 1 yr. or less)	
Third dose of DPT, 1990-91	67.0
Measles, 1990-91	63.0

Health Expenditures [4]

Total health expenditure, 1990 (official exchange rate)	
Millions of dollars	5,671
Millions of dollars per capita	158
Health expenditures as a percentage of GDP	
Total	5.6
Public sector	3.2
Private sector	2.4
Development assistance for health	
Total aid flows (millions of dollars)[1]	2
Aid flows per capita (millions of dollars)	NA
Aid flows as a percentage of total health expenditure	NA

For sources, notes, and explanations, see Annotated Source Appendix, page 1035.

Human Factors

Health Care Ratios [5]

Population per physician, 1970	NA
Population per physician, 1990	1750
Population per nursing person, 1970	300
Population per nursing person, 1990	NA
Percent of births attended by health staff, 1985	NA

Infants and Malnutrition [6]

Percent of babies with low birth weight, 1985	12
Infant mortality rate per 1,000 live births, 1970	79
Infant mortality rate per 1,000 live births, 1991	54
Years of life lost per 1,000 population, 1990	40
Prevalence of malnutrition (under age 5), 1990	NA

Ethnic Division [7]

Black	75.2%
White	13.6%
Colored	8.6%
Indian	2.6%

Religion [8]

Christian (most whites and Coloreds and about 60% of blacks), Hindu (60% of Indians), Muslim 20%.

Major Languages [9]

Afrikaans (official)	NA
English (official)	NA
Zulu	NA
Xhosa	NA
North Sotho	NA
South Sotho	NA
Tswana	NA
Many other vernacular languages	NA

Education

Public Education Expenditures [10]

Educational Attainment [11]

Age group	25+[11]
Total population	10,388,428[11]
Highest level attained (%)	
No schooling	24.8[11]
First level	
Incompleted	41.6[11]
Completed	4.8[11]
Entered second level	
S-1	22.1[11]
S-2	4.4[11]
Post secondary	2.3[11]

Literacy Rate [12]

	1980[b,6]	1985[b,6]	1990[b,6]
Illiterate population +15 years	3,711,776	NA	NA
Illiteracy rate - total pop. (%)	23.8	NA	NA
Illiteracy rate - males (%)	22.5	NA	NA
Illiteracy rate - females (%)	25.2	NA	NA

Libraries [13]

Daily Newspapers [14]

	1975	1980	1985	1990
Number of papers	20[e]	24[e]	24	22
Circ. (000)	1,000[e]	1,400[e]	1,440	1,340

Cinema [15]

Science and Technology

Scientific/Technical Forces [16]

R&D Expenditures [17]

U.S. Patents Issued [18]

Values show patents issued to citizens of the country by the U.S. Patents Office.

	1990	1991	1992
Number of patents	123	111	101

For sources, notes, and explanations, see Annotated Source Appendix, page 1035.

851

Government and Law

Organization of Government [19]

Long-form name:
Republic of South Africa
Type:
republic
Independence:
31 May 1910 (from UK)
Constitution:
3 September 1984
Legal system:
based on Roman-Dutch law and English
common law; accepts compulsory ICJ
jurisdiction, with reservations
National holiday:
Republic Day, 31 May (1910)
Executive branch:
state president, Executive Council
(cabinet), Ministers' Councils (from the
three houses of Parliament)
Legislative branch:
tricameral Parliament (Parlement)
consists of the House of Assembly
(Volksraad; whites), House of
Representatives (Raad van
Verteenwoordigers; Coloreds), and House
of Delegates (Raad van Afgevaardigdes;
Indians)
Judicial branch:
Supreme Court

Crime [23]

Elections [20]

Tentative agreement to hold national
election open to all races for a 400-seat
constitutient assembly on 27 April 1994.

Government Budget [21]

For FY93 est.
Revenues	28
Expenditures	36
Capital expenditures	3

Military Expenditures and Arms Transfers [22]

	1985	1986	1987	1988	1989
Military expenditures					
Current dollars (mil.)	2,612	2,771	3,323[e]	3,555	3,786
1989 constant dollars (mil.)	2,974	3,074	3,574[e]	3,701	3,786
Armed forces (000)	95	90	102	100	100
Gross national product (GNP)					
Current dollars (mil.)	70,500	72,560	77,110	83,610	86,750
1989 constant dollars (mil.)	80,260	80,530	82,930	87,040	86,750
Central government expenditures (CGE)					
1989 constant dollars (mil.)	25,670	26,700	28,520	24,640	28,130
People (mil.)	34.6	35.6	36.5	37.5	38.5
Military expenditure as % of GNP	3.7	3.8	4.3	4.3	4.4
Military expenditure as % of CGE	11.6	11.5	12.5	15.0	13.5
Military expenditure per capita	86	86	98	99	98
Armed forces per 1,000 people	2.7	2.5	2.8	2.7	2.6
GNP per capita	2,317	2,264	2,271	2,321	2,253
Arms imports[6]					
Current dollars (mil.)	20	20	40	10	100
1989 constant dollars (mil.)	23	22	43	10	100
Arms exports[6]					
Current dollars (mil.)	90	10	80	60	0
1989 constant dollars (mil.)	102	11	86	62	0
Total imports[7]					
Current dollars (mil.)	11,470	12,990	15,330	18,760	18,450
1989 constant dollars	13,060	14,410	16,490	19,530	18,450
Total exports[7]					
Current dollars (mil.)	16,520	18,450	23,540	21,550	22,220
1989 constant dollars	18,810	20,480	25,320	22,430	22,220
Arms as percent of total imports[8]	0.2	0.2	0.3	0.1	0.5
Arms as percent of total exports[8]	0.5	0.1	0.3	0.3	0

Human Rights [24]

	SSTS	FL	FAPRO	PPCG	APROBC	TPW	PCPTW	STPEP	PHRFF	PRW	ASST	AFL
Observes	2					P	P	P				
	EAFRD	CPR	ESCR	SR	ACHR	MAAE	PVIAC	PVNAC	EAFDAW	TCIDTP	RC	
Observes	2					P	P	P				

P = Party; S = Signatory; see Appendix for meaning of abbreviations.

Labor Force

Total Labor Force [25]

13.4 million economically active (1990)

Labor Force by Occupation [26]

Services	55%
Agriculture	10
Industry	20
Mining	9
Other	6

Unemployment Rate [27]

45% (well over 50% in some homeland areas)
(1992 est.)

For sources, notes, and explanations, see Annotated Source Appendix, page 1035.

Production Sectors

Commercial Energy Production and Consumption

Production [28]

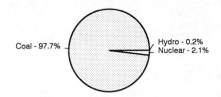

Coal - 97.7%
Hydro - 0.2%
Nuclear - 2.1%

Consumption [29]

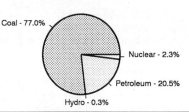

Coal - 77.0%
Nuclear - 2.3%
Petroleum - 20.5%
Hydro - 0.3%

Data are shown in quadrillion (10^{15}) BTUs and percent for 1991

Crude oil[1]	0.00
Natural gas liquids	0.00
Dry natural gas	0.00
Coal[2]	4.20
Net hydroelectric power[3]	0.01
Net nuclear power[3]	0.09
Total	4.30

Petroleum	0.81
Dry natural gas	0.00
Coal[2]	3.05
Net hydroelectric power[3]	0.01
Net nuclear power[3]	0.09
Total	3.96

Telecommunications [30]

- The system is the best developed, most modern, and has the highest capacity in Africa
- It consists of carrier-equipped open-wire lines, coaxial cables, radio relay links, fiber optic cable, and radiocommunication stations
- Key centers are Bloemfontein, Cape Town, Durban, Johannesburg, Port Elizabeth, and Pretoria
- Over 4,500,000 telephones
- Broadcast stations - 14 AM, 286 FM, 67 TV
- 1 submarine cable
- Satellite earth stations - 1 Indian Ocean INTELSAT and 2 Atlantic Ocean INTELSAT

Transportation [31]

Railroads. 20,638 km route distance total; 20,324 km of 1.067-meter gauge trackage (counts double and multiple tracking as single track); 314 km of 610 mm gauge; substantial electrification of 1.067 meter gauge

Highways. 188,309 km total; 54,013 km paved, 134,296 km crushed stone, gravel, or improved earth

Merchant Marine. 5 ships (1,000 GRT or over) totaling 213,708 GRT/201,043 DWT; includes 4 container, 1 vehicle carrier

Airports

Total:	899
Usable:	713
With permanent-surface runways:	136
With runways over 3,659 m:	5
With runways 2,440-3,659 m:	10
With runways 1,220-2,439 m:	221

Top Agricultural Products [32]

	89-90[9]	90-91[9]
Sugar cane	18,636	18,198
Corn	8,900	8,200
Wheat	1,702	2,238
Potatoes	1,269	1,383
Deciduous fruits	915	943
Citrus	763	880

Values shown are 1,000 metric tons.

Top Mining Products [33]

Detailed information is not available. A summary of mineral resources available follows. **Mineral Resources**: Gold, chromium, antimony, coal, iron ore, manganese, nickel, phosphates, tin, uranium, gem diamonds, platinum, copper, vanadium, salt, natural gas.

Tourism [34]

	1987	1988	1989	1990	1991
Tourists[4]	703	805	930	1,029	1,710[66]
Tourism receipts	589	673	709	1,029	1,046
Tourism expenditures	838	958	936	1,065	1,047
Fare receipts	207	211	245	340	305
Fare expenditures	218	216	290	393	443

Tourists are in thousands, money in million U.S. dollars.

For sources, notes, and explanations, see Annotated Source Appendix, page 1035.

853

Manufacturing Sector

GDP and Manufacturing Summary [35]

	1980	1985	1989	1990	% change 1980-1990	% change 1989-1990
GDP (million 1980 $)	77,542	82,915	90,496	89,579	15.5	-1.0
GDP per capita (1980 $)	2,743	2,626	2,622	2,539	-7.4	-3.2
Manufacturing as % of GDP (current prices)	22.6	22.4	21.8[e]	24.7	9.3	13.3
Gross output (million $)	53,686[e]	36,059	57,822	63,685	18.6	10.1
Value added (million $)	17,866	12,584	20,673	22,787	27.5	10.2
Value added (million 1980 $)	16,625	15,765	17,747	16,970	2.1	-4.4
Industrial production index	100	100	103	105	5.0	1.9
Employment (thousands)	1,392	1,423	1,459	1,461	5.0	0.1

Note: GDP stands for Gross Domestic Product. 'e' stands for estimated value.

Profitability and Productivity

	1980	1985	1989	1990	% change 1980-1990	% change 1989-1990
Intermediate input (%)	67[e]	65	64	64	-4.5	0.0
Wages, salaries, and supplements (%)	16[e]	18[e]	17	18[e]	12.5	5.9
Gross operating surplus (%)	17[e]	17[e]	19	18[e]	5.9	-5.3
Gross output per worker ($)	38,568[e]	25,024	39,631	43,590	13.0	10.0
Value added per worker ($)	12,835	8,769	14,169	15,597	21.5	10.1
Average wage (incl. benefits) ($)	6,118	4,505[e]	6,636	7,783[e]	27.2	17.3

Profitability is in percent of gross output. Productivity is in U.S. $. 'e' stands for estimated value.

Profitability - 1990

Surplus - 18.0%
Inputs - 64.0%
Wages - 18.0%

The graphic shows percent of gross output.

Value Added in Manufacturing

	1980 $ mil.	1980 %	1985 $ mil.	1985 %	1989 $ mil.	1989 %	1990 $ mil.	1990 %	% change 1980-1990	% change 1989-1990
311 Food products	1,626	9.1	1,277	10.1	1,757	8.5	2,220	9.7	36.5	26.4
313 Beverages	458	2.6	418	3.3	883	4.3	1,055	4.6	130.3	19.5
314 Tobacco products	111	0.6	108	0.9	118	0.6	83	0.4	-25.2	-29.7
321 Textiles	886	5.0	408	3.2	742	3.6	851	3.7	-4.0	14.7
322 Wearing apparel	477	2.7	334	2.7	559	2.7	701	3.1	47.0	25.4
323 Leather and fur products	40	0.2	44	0.3	91	0.4	75	0.3	87.5	-17.6
324 Footwear	152	0.9	113	0.9	272	1.3	316	1.4	107.9	16.2
331 Wood and wood products	213	1.2	190	1.5	272	1.3	469	2.1	120.2	72.4
332 Furniture and fixtures	219	1.2	138	1.1	199	1.0	307	1.3	40.2	54.3
341 Paper and paper products	591	3.3	471	3.7	1,325	6.4	1,208	5.3	104.4	-8.8
342 Printing and publishing	549	3.1	392	3.1	601	2.9	763	3.3	39.0	27.0
351 Industrial chemicals	1,006	5.6	717	5.7	1,206	5.8	932	4.1	-7.4	-22.7
352 Other chemical products	639	3.6	1,047	8.3	1,054	5.1	1,255	5.5	96.4	19.1
353 Petroleum refineries	634	3.5	1,038	8.2	1,045	5.1	1,244	5.5	96.2	19.0
354 Miscellaneous petroleum and coal products	111	0.6	182	1.4	182	0.9	217	1.0	95.5	19.2
355 Rubber products	297	1.7	157	1.2	276	1.3	401	1.8	35.0	45.3
356 Plastic products	355	2.0	225	1.8	396	1.9	560	2.5	57.7	41.4
361 Pottery, china and earthenware	28	0.2	24	0.2	33	0.2	42	0.2	50.0	27.3
362 Glass and glass products	154	0.9	102	0.8	227	1.1	292	1.3	89.6	28.6
369 Other non-metal mineral products	754	4.2	481	3.8	779	3.8	794	3.5	5.3	1.9
371 Iron and steel	2,135	12.0	986	7.8	2,068	10.0	2,343	10.3	9.7	13.3
372 Non-ferrous metals	555	3.1	418	3.3	746	3.6	642	2.8	15.7	-13.9
381 Metal products	1,576	8.8	860	6.8	1,401	6.8	1,697	7.4	7.7	21.1
382 Non-electrical machinery	1,351	7.6	805	6.4	1,168	5.6	1,432	6.3	6.0	22.6
383 Electrical machinery	1,229	6.9	607	4.8	866	4.2	970	4.3	-21.1	12.0
384 Transport equipment	1,258	7.0	741	5.9	1,755	8.5	1,311	5.8	4.2	-25.3
385 Professional and scientific equipment	49	0.3	54	0.4	148	0.7	160	0.7	226.5	8.1
390 Other manufacturing industries	415	2.3	246	2.0	501	2.4	448	2.0	8.0	-10.6

Note: The industry codes shown are International Standard Industry codes (ISIC). Percentages are percent of total Value Added. 'e' stands for estimated value

854

For sources, notes, and explanations, see Annotated Source Appendix, page 1035.

Finance, Economics, and Trade

Economic Indicators [36]

Billions of Rand Except Where noted.

	1989	1990	1991[e]
Real GDP (1985 prices)	121.1	119.9	119.3
Real GDP growth rate (%)	2.1	-0.9	-0.5
GDP per capita (est., 1985 Rand)[142]	3,264	3,155	3,066
Money supply (M1)	45.8	53.0	NA
Gross savings/GDP (%)	22.5	21.5	19.5
Gross domestic fixed investment/GDP (%)	19.7	19.6	19.0
CPI (year-end % change)	14.7	14.4	15.0
External public debt	7.9	6.8	NA

Balance of Payments Summary [37]

Exchange Rates [38]

Currency: **rand.**
Symbol: **R.**

Data are currency units per $1.

May 1993	3.1576
1992	2.8497
1991	2.7563
1990	2.5863
1989	2.6166
1988	2.2611

Imports and Exports

Top Import Origins [39]

$18.2 billion (f.o.b., 1992).

Origins	%
Germany	NA
Japan	NA
UK	NA
US	NA
Italy	NA

Top Export Destinations [40]

$23.5 billion (f.o.b., 1992).

Destinations	%
Italy	NA
Japan	NA
US	NA
Germany	NA
UK	NA
Other EC countries	NA
Hong Kong	NA

Foreign Aid [41]

Import and Export Commodities [42]

Import Commodities

Machinery 32%
Transport equipment 15%
Chemicals 11%
Oil
Textiles
Scientific instruments

Export Commodities

Gold 27%
Minerals and metals 20-25%
Food 5%
Chemicals 3%

Spain

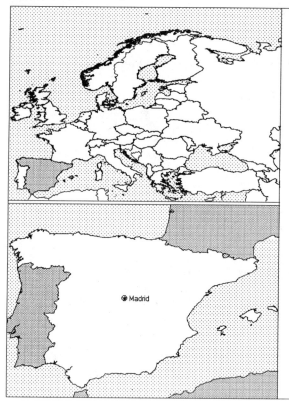

Geography [1]

Total area:
504,750 km2

Land area:
499,400 km2

Comparative area:
Slightly more than twice the size of Oregon

Land boundaries:
Total 1,903.2 km, Andorra 65 km, France 623 km, Gibraltar 1.2 km, Portugal 1,214 km

Coastline:
4,964 km

Climate:
Temperate; clear, hot summers in interior, more moderate and cloudy along coast; cloudy, cold winters in interior, partly cloudy and cool along coast

Terrain:
Large, flat to dissected plateau surrounded by rugged hills; Pyrenees in north

Natural resources:
Coal, lignite, iron ore, uranium, mercury, pyrites, fluorspar, gypsum, zinc, lead, tungsten, copper, kaolin, potash, hydropower

Land use:
Arable land: 31%
Permanent crops: 10%
Meadows and pastures: 21%
Forest and woodland: 31%
Other: 7%

Demographics [2]

	1960	1970	1980	1990	1991[1]	1994	2000	2010	2020
Population	30,641	33,876	37,488	38,964	39,385	39,303	39,972	40,682	40,241
Population density (persons per sq. mi.)	159	176	194	204	204	NA	210	213	210
Births	NA	NA	NA	NA	442	434	NA	NA	NA
Deaths	NA	NA	NA	NA	326	347	NA	NA	NA
Life expectancy - males	NA	NA	NA	NA	75	74	NA	NA	NA
Life expectancy - females	NA	NA	NA	NA	82	81	NA	NA	NA
Birth rate (per 1,000)	NA	NA	NA	NA	11	11	NA	NA	NA
Death rate (per 1,000)	NA	NA	NA	NA	8	9	NA	NA	NA
Women of reproductive age (15-44 yrs.)	NA	NA	NA	9,757	NA	10,108	10,137	NA	NA
of which are currently married	NA	NA	NA	5,521	NA	5,865	6,244	NA	NA
Fertility rate	NA	NA	NA	1.3	1.45	1.4	1.5	1.5	1.6

Population values are in thousands, life expectancy in years, and other items as indicated.

Health

Health Personnel [3]

Doctors per 1,000 pop., 1988-92	3.60
Nurse-to-doctor ratio, 1988-92	1.1
Hospital beds per 1,000 pop., 1985-90	4.8
Percentage of children immunized (age 1 yr. or less)	
Third dose of DPT, 1990-91	73.0
Measles, 1990-91	84.0

Health Expenditures [4]

Total health expenditure, 1990 (official exchange rate)	
Millions of dollars	32375
Millions of dollars per capita	831
Health expenditures as a percentage of GDP	
Total	6.6
Public sector	5.2
Private sector	1.4
Development assistance for health	
Total aid flows (millions of dollars)[1]	NA
Aid flows per capita (millions of dollars)	NA
Aid flows as a percentage of total health expenditure	NA

For sources, notes, and explanations, see Annotated Source Appendix, page 1035.

Human Factors

Health Care Ratios [5]

Population per physician, 1970	750
Population per physician, 1990	280
Population per nursing person, 1970	NA
Population per nursing person, 1990	NA
Percent of births attended by health staff, 1985	96

Infants and Malnutrition [6]

Percent of babies with low birth weight, 1985	NA
Infant mortality rate per 1,000 live births, 1970	28
Infant mortality rate per 1,000 live births, 1991	8
Years of life lost per 1,000 population, 1990	10
Prevalence of malnutrition (under age 5), 1990	NA

Ethnic Division [7]

Composite of Mediterranean and Nordic types.

Religion [8]

Roman Catholic	99%
Other sects	1%

Major Languages [9]

Castilian Spanish	NA
Catalan	17%
Galician	7%
Basque	2%

Education

Public Education Expenditures [10]

Billion Peseta	1985	1986	1987	1988	1989	1990
Total education expenditure	917.8	1,010.4	1,334.0	1,578.9	1,880.6	NA
as percent of GNP	3.3	3.2	3.7	4.0	4.2	NA
as percent of total govt. expend.	14.1	13.3	8.8	9.6	9.7	NA
Current education expenditure	821.1	914.0	1,207.9	1,402.4	1,624.5	NA
as percent of GNP	2.9	2.9	3.4	3.5	3.6	NA
as percent of current govt. expend.	NA	NA	NA	NA	NA	NA
Capital expenditure	96.0	96.4	126.1	176.4	256.1	NA

Educational Attainment [11]

Age group	25+
Total population	23,164,032
Highest level attained (%)	
No schooling	5.2[10]
First level	
Incompleted	40.3
Completed	29.9
Entered second level	
S-1	8.9
S-2	8.7
Post secondary	7.0

Literacy Rate [12]

In thousands and percent	1985[a]	1991[a]	2000[a]
Illiterate population +15 years	1,697	1,440	1,063
Illiteracy rate - total pop. (%)	5.7	4.6	3.2
Illiteracy rate - males (%)	3.1	2.6	1.9
Illiteracy rate - females (%)	8.0	6.6	4.4

Libraries [13]

	Admin. Units	Svc. Pts.	Vols. (000)	Shelving (meters)	Vols. Added	Reg. Users
National (1988)	1	7	3,421	NA	137,265	NA
Non-specialized	NA	NA	NA	NA	NA	NA
Public (1988)	2,982	3,206	23,274	NA	1.7mil	8.2mil
Higher ed. (1990)	567	1,028	16,050	642,005	781,080	912,359[38]
School	NA	NA	NA	NA	NA	NA

Daily Newspapers [14]

	1975	1980	1985	1990[1]
Number of papers	115	111	102	102
Circ. (000)	3,491	3,487	3,078	3,200

Cinema [15]

Data for 1991.

Cinema seats per 1,000	NA
Annual attendance per person	2.0
Gross box office receipts (mil. Peseta)	30,956

Science and Technology

Scientific/Technical Forces [16]

Potential scientists/engineers[12]	1,697,800
Number female	776,400
Potential technicians	NA
Number female	NA
Total	1,697,800

R&D Expenditures [17]

	Peseta[31] (000) 1987
Total expenditure	221,154,466
Capital expenditure	50,148,290
Current expenditure	169,059,260
Percent current	77.1

U.S. Patents Issued [18]

Values show patents issued to citizens of the country by the U.S. Patents Office.

	1990	1991	1992
Number of patents	148	185	149

For sources, notes, and explanations, see Annotated Source Appendix, page 1035.

857

Government and Law

Organization of Government [19]

Long-form name:
Kingdom of Spain
Type:
parliamentary monarchy
Independence:
1492 (expulsion of the Moors and unification)
Constitution:
6 December 1978, effective 29 December 1978
Legal system:
civil law system, with regional applications; does not accept compulsory ICJ jurisdiction
National holiday:
National Day, 12 October
Executive branch:
monarch, president of the government (prime minister), deputy prime minister, Council of Ministers (cabinet), Council of State
Legislative branch:
bicameral The General Courts or National Assembly (Las Cortes Generales) consists of an upper house or Senate (Senado) and a lower house or Congress of Deputies (Congreso de los Diputados)
Judicial branch:
Supreme Court (Tribunal Supremo)

Political Parties [20]

Congress of Deputies	% of votes
Spanish Socialist Workers' Party	39.6
Popular Party	25.8
Social Democratic Center	9.0
United Left	9.0
Convergence and Unity	5.0
Basque Nationalist Party	1.2
Basque Popular Unity	1.0
Andalusian Party	1.0
Other	8.4

Government Expenditures [21]

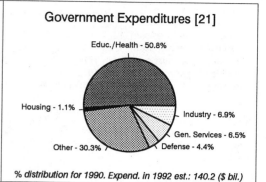

Educ./Health - 50.8%
Housing - 1.1%
Industry - 6.9%
Gen. Services - 6.5%
Defense - 4.4%
Other - 30.3%

% distribution for 1990. Expend. in 1992 est.: 140.2 ($ bil.)

Crime [23]

Crime volume	
Cases known to police	1,054,551
Attempts (percent)	2.97
Percent cases solved	24.82
Crimes per 100,000 persons	2,688.96
Persons responsible for offenses	
Total number offenders	239,167
Percent female	NA
Percent juvenile[21]	1.80
Percent foreigners	NA

Military Expenditures and Arms Transfers [22]

	1985	1986	1987	1988	1989
Military expenditures					
Current dollars (mil.)	6,639	6,505	7,554	7,244	7,775
1989 constant dollars (mil.)	7,558	7,219	8,125	7,541	7,775
Armed forces (000)	314	314	314	304	277
Gross national product (GNP)					
Current dollars (mil.)	271,000	287,800	313,800	340,200	370,700
1989 constant dollars (mil.)	308,500	319,300	337,500	354,200	370,700
Central government expenditures (CGE)					
1989 constant dollars (mil.)	108,800	110,000	118,500	103,800[e]	135,500[e]
People (mil.)	38.5	38.7	38.8	39.0	39.1
Military expenditure as % of GNP	2.4	2.3	2.4	2.1	2.1
Military expenditure as % of CGE	6.9	6.6	6.9	7.3	5.7
Military expenditure per capita	196	187	209	193	199
Armed forces per 1,000 people	8.2	8.1	8.1	7.8	7.1
GNP per capita	8,014	8,260	8,692	9,084	9,471
Arms imports[6]					
Current dollars (mil.)	120	330	1,000	1,000	750
1989 constant dollars (mil.)	137	366	1,076	1,041	750
Arms exports[6]					
Current dollars (mil.)	525	170	460	280	130
1989 constant dollars (mil.)	598	189	495	291	130
Total imports[7]					
Current dollars (mil.)	29,960	35,060	49,110	60,530	71,470
1989 constant dollars	34,110	38,900	52,820	63,010	71,470
Total exports[7]					
Current dollars (mil.)	24,250	27,210	34,190	40,340	44,490
1989 constant dollars	27,600	30,190	36,780	41,990	44,490
Arms as percent of total imports[8]	0.4	0.9	2.0	1.7	1.0
Arms as percent of total exports[8]	2.2	0.6	1.3	0.7	0.3

Human Rights [24]

	SSTS	FL	FAPRO	PPCG	APROBC	TPW	PCPTW	STPEP	PHRFF	PRW	ASST	AFL
Observes	P	P	P	P	P	P	P	P	P	P	P	P
	EAFRD	CPR	ESCR	SR	ACHR	MAAE	PVIAC	PVNAC	EAFDAW	TCIDTP	RC	
Observes		P	P	P		P	P	P	P	P	P	

P = Party; S = Signatory; see Appendix for meaning of abbreviations.

Labor Force

Total Labor Force [25]

14.621 million

Labor Force by Occupation [26]

Services	53%
Industry	24
Agriculture	14
Construction	9

Date of data: 1988

Unemployment Rate [27]

19% (year-end 1992)

858

For sources, notes, and explanations, see Annotated Source Appendix, page 1035.

Production Sectors

Commercial Energy Production and Consumption

Production [28]

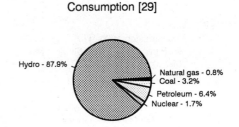

Coal - 42.7%
NG liquids - 1.2%
Natural gas - 2.9%
Nuclear - 33.3%
Hydro - 17.0%
Crude oil - 2.9%

Consumption [29]

Hydro - 87.9%
Natural gas - 0.8%
Coal - 3.2%
Petroleum - 6.4%
Nuclear - 1.7%

Data are shown in quadrillion (10^{15}) BTUs and percent for 1991

Crude oil[1]	0.05
Natural gas liquids	0.02
Dry natural gas	0.05
Coal[2]	0.73
Net hydroelectric power[3]	0.29
Net nuclear power[3]	0.57
Total	1.70

Petroleum	2.12
Dry natural gas	0.25
Coal[2]	1.07
Net hydroelectric power[3]	29
Net nuclear power[3]	0.57
Total	4.30

Telecommunications [30]

- Generally adequate, modern facilities
- 15,350,464 telephones
- Broadcast stations - 190 AM, 406 (134 repeaters) FM, 100 (1,297 repeaters) TV
- 22 coaxial submarine cables
- 2 communications satellite earth stations operating in INTELSAT (Atlantic Ocean and Indian Ocean)
- MARECS, INMARSAT, and EUTELSAT systems
- Tropospheric links

Top Agricultural Products [32]

	1990	1991
Wine (000 hl.)	42,597	32,569
Vegetables and melons	15,578	13,935
Alfalfa	13,477	13,991
Barley	9,414	8,500
Wheat	4,700	4,900
Oranges	2,565	2,490

Values shown are 1,000 metric tons.

Top Mining Products [33]

Metric tons unless otherwise specified	M.t.
Lead, mine output, Pb content	50,000
Zinc, mine output, Zn content	261,000
Pyrite, including cuprous (000 tons)	1,600
Sepiolite	500,000
Silver, mine output, Ag content	400,000
Iron ore and concentrate (000 tons)	3,920

Transportation [31]

Railroads. 15,430 km total; Spanish National Railways (RENFE) operates 12,691 km (all 1.668-meter gauge, 6,184 km electrified, and 2,295 km double track); FEVE (government-owned narrow-gauge railways) operates 1,821 km (predominantly 1.000-meter gauge, 441 km electrified); privately owned railways operate 918 km (predominantly 1.000-meter gauge, 512 km electrified, and 56 km double track)

Highways. 150,839 km total; 82,513 km national (includes 2,433 km limited-access divided highway, 63,042 km bituminous treated, 17,038 km intermediate bituminous, concrete, or stone block) and 68,326 km provincial or local roads (bituminous treated, intermediate bituminous, or stone block)

Merchant Marine. 242 ships (1,000 GRT or over) totaling 2,394,175 GRT/4,262,868 DWT; includes 2 passenger, 8 short-sea passenger, 71 cargo, 12 refrigerated cargo, 12 container, 32 roll-on/roll-off cargo, 4 vehicle carrier, 41 oil tanker, 14 chemical tanker, 7 liquefied gas, 3 specialized tanker, 36 bulk

Airports

Total:	105
Usable:	99
With permanent-surface runways:	60
With runways over 3,659 m:	4
With runways 2,440-3,659 m:	22
With runways 1,220-2,439 m:	26

Tourism [34]

	1987	1988	1989	1990	1991
Visitors[1]	50,545	54,178	54,058	52,044	53,495
Tourists	32,900	35,000	35,350	34,300	35,347
Excursionists	16,750	18,311	17,727	16,899	17,322
Tourism receipts	14,760	16,686	16,174	18,593	19,004
Tourism expenditures	1,938	2,440	3,080	4,254	4,530
Fare receipts	1,558	1,733	1,717	1,852	1,596
Fare expenditures	403	587	604	792	752

Tourists are in thousands, money in million U.S. dollars.

For sources, notes, and explanations, see Annotated Source Appendix, page 1035.

859

Manufacturing Sector

GDP and Manufacturing Summary [35]

	1980	1985	1989	1990	% change 1980-1990	% change 1989-1990
GDP (million 1980 $)	212,115	227,090	273,044	282,764	33.3	3.6
GDP per capita (1980 $)	5,650	5,883	6,988	7,216	27.7	3.3
Manufacturing as % of GDP (current prices)	26.7	25.4	25.8	25.4	-4.9	-1.6
Gross output (million $)	149,786	104,581	199,859	259,413	73.2	29.8
Value added (million $)	51,944	33,140	72,070	87,329	68.1	21.2
Value added (million 1980 $)	59,751	61,143	74,522	74,001	23.8	-0.7
Industrial production index	100	99	121	121	21.0	0.0
Employment (thousands)	2,383	1,792	1,809[e]	1,907	-20.0	5.4

Note: GDP stands for Gross Domestic Product. 'e' stands for estimated value.

Profitability and Productivity

	1980	1985	1989	1990	% change 1980-1990	% change 1989-1990
Intermediate input (%)	65	68	64	66	1.5	3.1
Wages, salaries, and supplements (%)	20	17	14[e]	18[e]	-10.0	28.6
Gross operating surplus (%)	14	15	22[e]	16[e]	14.3	-27.3
Gross output per worker ($)	59,041	53,991	110,477[e]	126,977	115.1	14.9
Value added per worker ($)	20,475	17,109	39,838[e]	42,746	108.8	7.3
Average wage (incl. benefits) ($)	12,852	9,700	15,307[e]	24,115[e]	87.6	57.5

Profitability is in percent of gross output. Productivity is in U.S. $. 'e' stands for estimated value.

Profitability - 1990

Inputs - 66.0%
Surplus - 16.0%
Wages - 18.0%

The graphic shows percent of gross output.

Value Added in Manufacturing

	1980 $ mil.	1980 %	1985 $ mil.	1985 %	1989 $ mil.	1989 %	1990 $ mil.	1990 %	% change 1980-1990	% change 1989-1990
311 Food products	5,665	10.9	4,193	12.7	8,816[e]	12.2	10,773	12.3	90.2	22.2
313 Beverages	1,932	3.7	1,576	4.8	3,122[e]	4.3	3,663	4.2	89.6	17.3
314 Tobacco products	649	1.2	471	1.4	691	1.0	911	1.0	40.4	31.8
321 Textiles	3,289	6.3	1,613	4.9	2,900	4.0	3,314	3.8	0.8	14.3
322 Wearing apparel	1,502	2.9	753	2.3	1,500	2.1	2,242	2.6	49.3	49.5
323 Leather and fur products	375	0.7	269	0.8	492	0.7	614	0.7	63.7	24.8
324 Footwear	810	1.6	415	1.3	458	0.6	781	0.9	-3.6	70.5
331 Wood and wood products	1,258	2.4	707	2.1	1,582[e]	2.2	2,163	2.5	71.9	36.7
332 Furniture and fixtures	1,262	2.4	617	1.9	1,068[e]	1.5	1,534	1.8	21.6	43.6
341 Paper and paper products	1,278	2.5	947	2.9	1,899	2.6	2,101	2.4	64.4	10.6
342 Printing and publishing	1,506	2.9	1,198	3.6	3,016	4.2	4,403	5.0	192.4	46.0
351 Industrial chemicals	2,006	3.9	1,737	5.2	3,371[e]	4.7	3,427	3.9	70.8	1.7
352 Other chemical products	2,506	4.8	1,922	5.8	4,011[e]	5.6	5,609	6.4	123.8	39.8
353 Petroleum refineries	1,409	2.7	969	2.9	2,642	3.7	1,348	1.5	-4.3	-49.0
354 Miscellaneous petroleum and coal products	229	0.4	191	0.6	562	0.8	384	0.4	67.7	-31.7
355 Rubber products	955	1.8	597	1.8	1,316[e]	1.8	1,490	1.7	56.0	13.2
356 Plastic products	1,098	2.1	814	2.5	1,719[e]	2.4	2,452	2.8	123.3	42.6
361 Pottery, china and earthenware	346	0.7	174	0.5	351	0.5	432	0.5	24.9	23.1
362 Glass and glass products	640	1.2	442	1.3	893	1.2	1,128	1.3	76.3	26.3
369 Other non-metal mineral products	2,522	4.9	1,617	4.9	3,719	5.2	4,797	5.5	90.2	29.0
371 Iron and steel	3,255	6.3	1,756	5.3	3,497	4.9	3,762	4.3	15.6	7.6
372 Non-ferrous metals	948	1.8	616	1.9	1,136	1.6	1,275	1.5	34.5	12.2
381 Metal products	3,720	7.2	2,044	6.2	4,620	6.4	5,437	6.2	46.2	17.7
382 Non-electrical machinery	3,595	6.9	2,225	6.7	4,912	6.8	5,745	6.6	59.8	17.0
383 Electrical machinery	3,669	7.1	2,064	6.2	4,478	6.2	5,978	6.8	62.9	33.5
384 Transport equipment	4,743	9.1	2,776	8.4	8,333	11.6	10,320	11.8	117.6	23.8
385 Professional and scientific equipment	205	0.4	122	0.4	333	0.5	375	0.4	82.9	12.6
390 Other manufacturing industries	573	1.1	316	1.0	632[e]	0.9	870	1.0	51.8	37.7

Note: The industry codes shown are International Standard Industry codes (ISIC). Percentages are percent of total Value Added. 'e' stands for estimated value

For sources, notes, and explanations, see Annotated Source Appendix, page 1035.

Finance, Economics, and Trade

Economic Indicators [36]

Millions of Pesetas (Ptas) unless otherwise stated.

	1989	1990	1991[e]
Real GDP (bil. ptas, 1985 prices)	33,919	35,160	36,039
Real GDP growth (%)	4.8	3.7	2.5
GDP per capita (US$)	9,706	12,492	13,523
M1 (bil. ptas)	11,155	14,114	16,000
Commercial Interest rate (%)	14.1	14.6	12.5
Savings rate (% of GDP)	22.1	22.2	22.5
Investment rate (% of GDP)	25.3	25.7	26.0
CPI (1983 equals 100)	158.3	168.6	179.6
WPI (1974 equals 100)	460.5	471.1	480.5
External debt (mil. US$)	34,764	44,973	45,000

Balance of Payments Summary [37]

Values in millions of dollars.

	1987	1988	1989	1990	1991
Exports of goods (f.o.b.)	33,561	39,652	43,301	53,888	59,960
Imports of goods (f.o.b.)	-46,547	-57,650	-67,797	-83,454	-89,987
Trade balance	-12,986	-17,998	-24,496	-29,566	-30,027
Services - debits	-13,749	-17,729	-19,919	-26,006	-30,191
Services - credits	23,867	27,440	28,875	34,496	38,208
Private transfers (net)	2,277	3,018	3,163	3,053	2,199
Government transfers (net)	358	1,485	1,444	1,204	3,856
Long term capital (net)	9,291	9,610	16,879	19,223	33,592
Short term capital (net)	4,939	5,005	1,463	9,079	2,573
Errors and omissions	-1,291	-2,414	-2,693	-4,521	-6,062
Overall balance	12,706	8,417	4,716	6,962	14,148

Exchange Rates [38]

Currency: **pesetas.**
Symbol: **Ptas.**

Data are currency units per $1.

January 1993	114.59
1992	102.38
1991	103.91
1990	101.93
1989	118.38
1988	116.49

Imports and Exports

Top Import Origins [39]

$100 billion (c.i.f., 1992 est.). Data are for 1991.

Origins	%
EC	60.0
US	8.0
Other developed countries	11.5
Middle East	2.6

Top Export Destinations [40]

$62 billion (f.o.b., 1992 est.). Data are for 1991.

Destinations	%
EC	71.0
US	4.9
Other developed countries	7.9

Foreign Aid [41]

Note: Not currently a recipient.

	U.S. $	
US commitments, including Ex-Im (FY70-87)	1.9	billion
Western (non-US) countries, ODA and OOF bilateral commitments (1970-79)	545.0	million

Import and Export Commodities [42]

Import Commodities
Machinery
Transport equipment
Fuels
Semifinished goods
Foodstuffs
Consumer goods
Chemicals

Export Commodities
Cars and trucks
Components
Foodstuffs
Machinery

Sri Lanka

Geography [1]

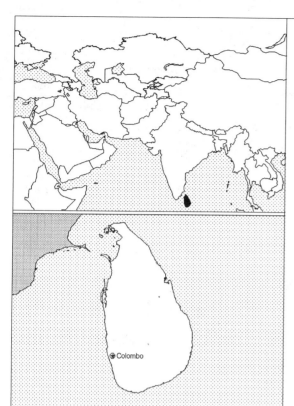

Total area:
65,610 km2
Land area:
64,740 km2
Comparative area:
Slightly larger than West Virginia
Land boundaries:
0 km
Coastline:
1,340 km
Climate:
Tropical monsoon; northeast monsoon (December to March); southwest monsoon (June to October)
Terrain:
Mostly low, flat to rolling plain; mountains in south-central interior
Natural resources:
Limestone, graphite, mineral sands, gems, phosphates, clay
Land use:
Arable land:
16%
Permanent crops:
17%
Meadows and pastures:
7%
Forest and woodland:
37%
Other:
23%

Demographics [2]

	1960	1970	1980	1990	1991[1]	1994	2000	2010	2020
Population	9,879	12,532	14,900	17,208	17,424	18,033	19,146	20,972	22,463
Population density (persons per sq. mi.)	395	501	596	688	697	NA	772	858	931
Births	NA	NA	NA	NA	347	328	NA	NA	NA
Deaths	NA	NA	NA	NA	103	105	NA	NA	NA
Life expectancy - males	NA	NA	NA	NA	69	69	NA	NA	NA
Life expectancy - females	NA	NA	NA	NA	74	74	NA	NA	NA
Birth rate (per 1,000)	NA	NA	NA	NA	20	18	NA	NA	NA
Death rate (per 1,000)	NA	NA	NA	NA	6	6	NA	NA	NA
Women of reproductive age (15-44 yrs.)	NA	NA	NA	4,660	NA	4,991	5,466	NA	NA
of which are currently married	NA	NA	NA	2,833	NA	3,089	3,426	NA	NA
Fertility rate	NA	NA	NA	2.3	2.26	2.1	1.9	1.8	1.8

Population values are in thousands, life expectancy in years, and other items as indicated.

Health

Health Personnel [3]

Doctors per 1,000 pop., 1988-92	0.14
Nurse-to-doctor ratio, 1988-92	5.1
Hospital beds per 1,000 pop., 1985-90	2.8
Percentage of children immunized (age 1 yr. or less)	
Third dose of DPT, 1990-91	86.0
Measles, 1990-91	79.0

Health Expenditures [4]

Total health expenditure, 1990 (official exchange rate)	
Millions of dollars	305
Millions of dollars per capita	18
Health expenditures as a percentage of GDP	
Total	3.7
Public sector	1.8
Private sector	1.9
Development assistance for health	
Total aid flows (millions of dollars)[1]	26
Aid flows per capita (millions of dollars)	1.5
Aid flows as a percentage of total health expenditure	7.4

For sources, notes, and explanations, see Annotated Source Appendix, page 1035.

Human Factors

Health Care Ratios [5]

Population per physician, 1970	5900
Population per physician, 1990	NA
Population per nursing person, 1970	1280
Population per nursing person, 1990	NA
Percent of births attended by health staff, 1985	87

Infants and Malnutrition [6]

Percent of babies with low birth weight, 1985	28
Infant mortality rate per 1,000 live births, 1970	53
Infant mortality rate per 1,000 live births, 1991	18
Years of life lost per 1,000 population, 1990	14
Prevalence of malnutrition (under age 5), 1990	45

Ethnic Division [7]

Sinhalese	74%
Tamil	18%
Moor	7%
Burgher, Malay, Vedda	1%

Religion [8]

Buddhist	69%
Hindu	15%
Christian	8%
Muslim	8%

Major Languages [9]

Sinhala (official national language)	74%
Tamil (national language)	18%
Other	8%

Education

Public Education Expenditures [10]

Million Rupee	1980	1985	1987	1988	1989	1990
Total education expenditure	1,799	4,183	5,202	6,274	NA	8,621
as percent of GNP	2.7	2.6	2.7	2.8	NA	2.7
as percent of total govt. expend.	7.7	6.9	7.8	7.3	NA	8.1
Current education expenditure	1,535	3,530	4,141	5,204	NA	7,024
as percent of GNP	2.3	2.2	2.1	2.4	NA	2.2
as percent of current govt. expend.	13.5	11.4	10.0	12.5	NA	10.7
Capital expenditure	264	653	1,061	1,070	NA	1,597

Educational Attainment [11]

Age group	25+
Total population	6,490,502
Highest level attained (%)	
No schooling	15.9
First level	
Incompleted	48.9
Completed	NA
Entered second level	
S-1	34.1
S-2	NA
Post secondary	1.1

Literacy Rate [12]

In thousands and percent	1985[a]	1991[a]	2000[a]
Illiterate population +15 years	1,373	1,347	1,199
Illiteracy rate - total pop. (%)	13.3	11.6	8.5
Illiteracy rate - males (%)	7.6	6.6	4.8
Illiteracy rate - females (%)	19.1	16.5	12.1

Libraries [13]

	Admin. Units	Svc. Pts.	Vols. (000)	Shelving (meters)	Vols. Added	Reg. Users
National (1989)	1	1	133	NA	-	61
Non-specialized	NA	NA	NA	NA	NA	NA
Public (1989)	15	154	481	3,030	10,500	98,006
Higher ed.	NA	NA	NA	NA	NA	NA
School	NA	NA	NA	NA	NA	NA

Daily Newspapers [14]

	1975	1980	1985	1990
Number of papers	15	21	17	18
Circ. (000)	480	450	390	550[e]

Cinema [15]

Science and Technology

Scientific/Technical Forces [16]

Potential scientists/engineers	21,533
Number female	NA
Potential technicians	NA
Number female	NA
Total	NA

R&D Expenditures [17]

	Rupee (000) 1984
Total expenditure	256,799
Capital expenditure	82,464
Current expenditure	174,335
Percent current	67.9

U.S. Patents Issued [18]

Values show patents issued to citizens of the country by the U.S. Patents Office.

	1990	1991	1992
Number of patents	0	2	2

For sources, notes, and explanations, see Annotated Source Appendix, page 1035.

Government and Law

Organization of Government [19]

Long-form name:
Democratic Socialist Republic of Sri Lanka

Type:
republic

Independence:
4 February 1948 (from UK)

Constitution:
31 August 1978

Legal system:
a highly complex mixture of English common law, Roman-Dutch, Muslim, Sinhalese, and customary law; has not accepted compulsory ICJ jurisdiction

National holiday:
Independence and National Day, 4 February (1948)

Executive branch:
president, prime minister, Cabinet

Legislative branch:
unicameral Parliament

Judicial branch:
Supreme Court

Crime [23]

Crime volume	
Cases known to police	52,917
Attempts (percent)	NA
Percent cases solved	54.6
Crimes per 100,000 persons	309.4
Persons responsible for offenses	
Total number offenders	43,278
Percent female	3.7
Percent juvenile[50]	0.8
Percent foreigners	0.03

Political Parties [20]

Parliament	% of votes
United National Party (UNP)	51.0
Sri Lanka Freedom Party (SLFP)	32.0
Sri Lanka Muslim Congress (SLMC)	4.0
Tamil United Liberation Front (TULF)	3.0
United Socialist Alliance (USA)	3.0
Eelam Revolutionary Org. (EROS)	3.0
People's United Front (MEP)	1.0
Other	3.0

Government Expenditures [21]

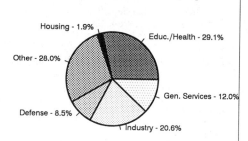

Housing - 1.9%
Educ./Health - 29.1%
Other - 28.0%
Gen. Services - 12.0%
Defense - 8.5%
Industry - 20.6%

% distribution for 1992. Expend. in 1992: 3.700 ($ bil.)

Military Expenditures and Arms Transfers [22]

	1985	1986	1987	1988	1989
Military expenditures					
Current dollars (mil.)	157	144	189	300	223
1989 constant dollars (mil.)	179	159	203	313	223
Armed forces (000)	21	25[e]	30[e]	47	47
Gross national product (GNP)					
Current dollars (mil.)	5,517	5,903	6,161	6,519	6,939
1989 constant dollars (mil.)	6,280	6,551	6,626	6,786	6,939
Central government expenditures (CGE)					
1989 constant dollars (mil.)	2,145	2,170	2,161	2,345	2,194
People (mil.)	16.0	16.2	16.5	16.7	17.0
Military expenditure as % of GNP	2.9	2.4	3.1	4.6	3.2
Military expenditure as % of CGE	8.4	7.4	9.4	13.3	10.2
Military expenditure per capita	11	10	12	19	13
Armed forces per 1,000 people	1.3	1.5	1.8	2.8	2.8
GNP per capita	392	403	402	406	409
Arms imports[6]					
Current dollars (mil.)	40	10	50	20	10
1989 constant dollars (mil.)	46	11	54	21	10
Arms exports[6]					
Current dollars (mil.)	0	0	0	0	0
1989 constant dollars (mil.)	0	0	0	0	0
Total imports[7]					
Current dollars (mil.)	1,843	1,857	2,058	2,262	2,188
1989 constant dollars	2,098	2,061	2,213	2,355	2,188
Total exports[7]					
Current dollars (mil.)	1,293	1,215	1,368	1,479	1,545
1989 constant dollars	1,472	1,348	1,471	1,540	1,545
Arms as percent of total imports[8]	2.2	0.5	2.4	0.9	0.5
Arms as percent of total exports[8]	0	0	0	0	0

Human Rights [24]

	SSTS	FL	FAPRO	PPCG	APROBC	TPW	PCPTW	STPEP	PHRFF	PRW	ASST	AFL
Observes	P	P		P	P	P	P	P			P	
	EAFRD	CPR	ESCR	SR	ACHR	MAAE	PVIAC	PVNAC	EAFDAW	TCIDTP	RC	
Observes		P	P	P					P		P	

P = Party; S = Signatory; see Appendix for meaning of abbreviations.

Labor Force

Total Labor Force [25]

6.6 million

Labor Force by Occupation [26]

Agriculture	45.9%
Mining and manufacturing	13.3
Trade and transport	12.4
Services and other	28.4

Date of data: 1985 est.

Unemployment Rate [27]

15% (1991 est.)

For sources, notes, and explanations, see Annotated Source Appendix, page 1035.

Production Sectors

Energy Resource Summary [28]

Energy Resources: None. **Electricity**: 1,300,000 kW capacity; 3,600 million kWh produced, 200 kWh per capita (1992). **Pipelines**: Crude oil and petroleum products 62 km (1987).

Telecommunications [30]

- Very inadequate domestic service, good international service
- 114,000 telephones (1982)
- Broadcast stations - 12 AM, 5 FM, 5 TV
- Submarine cables extend to Indonesia and Djibouti
- 2 Indian Ocean INTELSAT earth stations

Top Agricultural Products [32]

	1989	1990
Rice	1,403	1,726
Cassava	420	384
Tea	207	233
Sugar cane	132	120
Rubber	111	114
Green chilies	68	100

Values shown are 1,000 metric tons.

Top Mining Products [33]

Preliminary metric tons unless otherwise specified M.t.

Gemstones, precious/semiprecious, other than diamond (val. $000)	57,000[e]
Ilmenite	60,861
Rutile	3,085
Monazite concentrate, gross weight	200[e]
Clay	107,737
Feldspar, crude and ground	9,908
Graphite, all grades	6,381
Quartz, massive	978
Salt	52,888

Transportation [31]

Railroads. 1,948 km total (1990); all 1.868-meter broad gauge; 102 km double track; no electrification; government owned

Highways. 75,749 km total (1990); 27,637 km paved (mostly bituminous treated), 32,887 km crushed stone or gravel, 14,739 km improved earth or unimproved earth; several thousand km of mostly unmotorable tracks (1988 est.)

Merchant Marine. 27 ships (1,000 GRT or over) totaling 276,074 GRT/443,266 DWT; includes 12 cargo, 6 refrigerated cargo, 3 container, 3 oil tanker, 3 bulk

Airports

Total:	14
Usable:	13
With permanent-surface runways:	12
With runways over 3,659 m:	0
With runways 2,440-3,659 m:	1
With runways 1,220-2,439 m:	8

Tourism [34]

	1987	1988	1989	1990	1991
Visitors	185	189	189	302	321
Tourists[4]	183	183	185	298	318
Excursionists	2	6	4	4	3
Tourism receipts	82	77	76	132	156
Tourism expenditures	63	68	68	79	92
Fare receipts	34	36	43	96	108
Fare expenditures	44	42	47	51	46

Tourists are in thousands, money in million U.S. dollars.

For sources, notes, and explanations, see Annotated Source Appendix, page 1035.

865

Manufacturing Sector

GDP and Manufacturing Summary [35]

	1980	1985	1989	1990	% change 1980-1990	% change 1989-1990
GDP (million 1980 $)	4,133	5,303	5,903	6,257	51.4	6.0
GDP per capita (1980 $)	279	329	347	363	30.1	4.6
Manufacturing as % of GDP (current prices)	19.0	17.5	16.4e	14.4	-24.2	-12.2
Gross output (million $)	1,129	1,815	1,923e	2,218e	96.5	15.3
Value added (million $)	307	628e	809e	929e	202.6	14.8
Value added (million 1980 $)	751	936	1,143e	1,188	58.2	3.9
Industrial production index	100	109	132	138	38.0	4.5
Employment (thousands)	163	211	225e	251e	54.0	11.6

Note: GDP stands for Gross Domestic Product. 'e' stands for estimated value.

Profitability and Productivity

	1980	1985	1989	1990	% change 1980-1990	% change 1989-1990
Intermediate input (%)	73	65e	58e	58e	-20.5	0.0
Wages, salaries, and supplements (%)	7	6	7e	7e	0.0	0.0
Gross operating surplus (%)	20	28e	35e	35e	75.0	0.0
Gross output per worker ($)	6,934	8,599	8,535e	8,830e	27.3	3.5
Value added per worker ($)	1,887	2,973e	3,590e	3,703e	96.2	3.1
Average wage (incl. benefits) ($)	486	529	599e	642e	32.1	7.2

Profitability is in percent of gross output. Productivity is in U.S. $. 'e' stands for estimated value.

Profitability - 1990

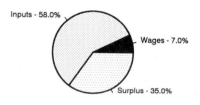

Inputs - 58.0%
Wages - 7.0%
Surplus - 35.0%

The graphic shows percent of gross output.

Value Added in Manufacturing

	1980 $ mil.	1980 %	1985 $ mil.	1985 %	1989 $ mil.	1989 %	1990 $ mil.	1990 %	% change 1980-1990	% change 1989-1990
311 Food products	28	9.1	180e	28.7	169e	20.9	250e	26.9	792.9	47.9
313 Beverages	8	2.6	34e	5.4	73e	9.0	75e	8.1	837.5	2.7
314 Tobacco products	63	20.5	151e	24.0	177e	21.9	150e	16.1	138.1	-15.3
321 Textiles	27	8.8	49e	7.8	68e	8.4	80e	8.6	196.3	17.6
322 Wearing apparel	12	3.9	33e	5.3	84e	10.4	101e	10.9	741.7	20.2
323 Leather and fur products	1	0.3	2e	0.3	2e	0.2	1e	0.1	0.0	-50.0
324 Footwear	2	0.7	4e	0.6	12e	1.5	22e	2.4	1,000.0	83.3
331 Wood and wood products	5	1.6	8e	1.3	9e	1.1	7e	0.8	40.0	-22.2
332 Furniture and fixtures	1	0.3	2e	0.3	1e	0.1	1e	0.1	0.0	0.0
341 Paper and paper products	8	2.6	12e	1.9	18e	2.2	23e	2.5	187.5	27.8
342 Printing and publishing	4	1.3	8e	1.3	13e	1.6	11e	1.2	175.0	-15.4
351 Industrial chemicals	6	2.0	4e	0.6	3e	0.4	3e	0.3	-50.0	0.0
352 Other chemical products	12	3.9	18e	2.9	26e	3.2	25e	2.7	108.3	-3.8
353 Petroleum refineries	55	17.9	23e	3.7	10e	1.2	11e	1.2	-80.0	10.0
354 Miscellaneous petroleum and coal products	NA	0.0	NA	0.0	NA	0.0	NA	0.0	NA	NA
355 Rubber products	14	4.6	30e	4.8	32e	4.0	39e	4.2	178.6	21.9
356 Plastic products	4	1.3	4e	0.6	9e	1.1	5e	0.5	25.0	-44.4
361 Pottery, china and earthenware	4	1.3	6e	1.0	14e	1.7	17e	1.8	325.0	21.4
362 Glass and glass products	2	0.7	2e	0.3	3e	0.4	4e	0.4	100.0	33.3
369 Other non-metal mineral products	21	6.8	28e	4.5	32e	4.0	34e	3.7	61.9	6.3
371 Iron and steel	3	1.0	2e	0.3	5e	0.6	5e	0.5	66.7	0.0
372 Non-ferrous metals	2	0.7	1e	0.2	1e	0.1	2e	0.2	0.0	100.0
381 Metal products	7	2.3	9e	1.4	11e	1.4	14e	1.5	100.0	27.3
382 Non-electrical machinery	4	1.3	6e	1.0	3e	0.4	15e	1.6	275.0	400.0
383 Electrical machinery	10	3.3	5e	0.8	6e	0.7	11e	1.2	10.0	83.3
384 Transport equipment	4	1.3	2e	0.3	6e	0.7	10e	1.1	150.0	66.7
385 Professional and scientific equipment	1	0.3	NA	0.0	NA	0.0	NA	0.0	NA	NA
390 Other manufacturing industries	1	0.3	5e	0.8	22e	2.7	14e	1.5	1,300.0	-36.4

Note: The industry codes shown are International Standard Industry codes (ISIC). Percentages are percent of total Value Added. 'e' stands for estimated value

Finance, Economics, and Trade

Economic Indicators [36]

National product: GDP—exchange rate conversion—$7.75 billion (1992 est.). **National product real growth rate**: 4.5% (1992 est.). **National product per capita**: $440 (1992 est.). **Inflation rate (consumer prices)**: 10% (1992). **External debt**: $5.7 billion (1991 est.).

Balance of Payments Summary [37]

Values in millions of dollars.

	1987	1988	1989	1990	1991
Exports of goods (f.o.b.)	1,393.9	1,477.1	1,505.1	1,853.0	2,009.1
Imports of goods (f.o.b.)	-1,866.0	-2,017.5	-2,055.1	-2,325.6	-2,483.0
Trade balance	-472.1	-540.4	-550.0	-472.6	-473.9
Services - debits	-744.0	-787.9	-787.1	-898.9	-1,052.8
Services - credits	397.5	407.9	404.2	532.6	653.8
Private transfers (net)	312.8	320.0	330.7	362.4	401.3
Government transfers (net)	179.9	206.1	188.6	178.1	203.9
Long term capital (net)	268.1	269.2	184.7	405.7	591.7
Short term capital (net)	126.7	-13.3	392.3	72.4	109.2
Errors and omissions	-122.5	37.4	-115.0	-115.1	-113.4
Overall balance	-53.6	-101.0	48.4	64.6	319.8

Exchange Rates [38]

Currency: **Sri Lankan rupees.**
Symbol: **SLRes.**

Data are currency units per $1.

January 1993	46.342
1992	43.687
1991	41.372
1990	40.063
1989	36.047
1988	31.807

Imports and Exports

Top Import Origins [39]

$3.1 billion (c.i.f., 1991).

Origins	%
Japan	NA
Iran	NA
US	5.7
India	NA
Taiwan	NA
Singapore	NA
Germany	NA
UK	NA

Top Export Destinations [40]

$2.0 billion (f.o.b., 1991).

Destinations	%
US	27.4
Germany	NA
Japan	NA
UK	NA
Belgium	NA
Taiwan	NA
Hong Kong	NA
China	NA

Foreign Aid [41]

	U.S. $	
US commitments, including Ex-Im (FY70-89)	1.0	billion
Western (non-US) countries, ODA and OOF bilateral commitments (1980-89)	5.1	billion
OPEC bilateral aid (1979-89)	169	million
Communist countries (1970-89)	369	million

Import and Export Commodities [42]

Import Commodities

Food and beverages
Textiles
Petroleum and petroleum products
Machinery & equipment

Export Commodities

Textiles and garments
Teas
Petroleum products
Coconuts
Rubber
Other agricultural products
Gems and jewelry
Marine products
Graphite

Sudan

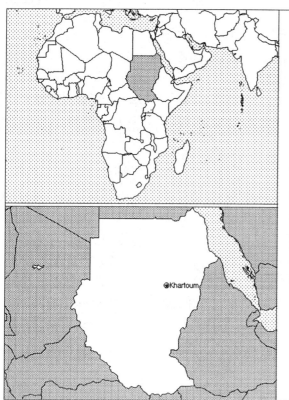

Geography [1]

Total area:
2,505,810 km2
Land area:
2.376 million km2
Comparative area:
Slightly more than one-quarter the size of the US
Land boundaries:
Total 7,697 km, Central African Republic 1,165 km, Chad 1,360 km, Egypt 1,273 km, Ethiopia 2,221 km, Kenya 232 km, Libya 383 km, Uganda 435 km, Zaire 628 km
Coastline:
853 km
Climate:
Tropical in south; arid desert in north; rainy season (April to October)
Terrain:
Generally flat, featureless plain; mountains in east and west
Natural resources:
Small reserves of petroleum, iron ore, copper, chromium ore, zinc, tungsten, mica, silver
Land use:
Arable land: 5%
Permanent crops: 0%
Meadows and pastures: 24%
Forest and woodland: 20%
Other: 51%

Demographics [2]

	1960	1970	1980	1990	1991[1]	1994	2000	2010	2020
Population	10,589	13,788	19,064	26,542	27,220	29,420	35,236	46,167	58,090
Population density (persons per sq. mi.)	12	15	21	29	30	NA	39	51	65
Births	NA	NA	NA	NA	1,199	1,234	NA	NA	NA
Deaths	NA	NA	NA	NA	358	356	NA	NA	NA
Life expectancy - males	NA	NA	NA	NA	52	53	NA	NA	NA
Life expectancy - females	NA	NA	NA	NA	54	55	NA	NA	NA
Birth rate (per 1,000)	NA	NA	NA	NA	44	42	NA	NA	NA
Death rate (per 1,000)	NA	NA	NA	NA	13	12	NA	NA	NA
Women of reproductive age (15-44 yrs.)	NA	NA	NA	6,038	NA	6,633	8,097	NA	NA
of which are currently married	NA	NA	NA	4,431	NA	4,860	5,915	NA	NA
Fertility rate	NA	NA	NA	6.5	6.37	6.1	5.5	4.4	3.4

Population values are in thousands, life expectancy in years, and other items as indicated.

Health

Health Personnel [3]

Doctors per 1,000 pop., 1988-92	0.09
Nurse-to-doctor ratio, 1988-92	2.7
Hospital beds per 1,000 pop., 1985-90	0.9
Percentage of children immunized (age 1 yr. or less)	
Third dose of DPT, 1990-91	63.0
Measles, 1990-91	58.0

Health Expenditures [4]

Total health expenditure, 1990 (official exchange rate)	
Millions of dollars	300
Millions of dollars per capita	12
Health expenditures as a percentage of GDP	
Total	3.3
Public sector	0.5
Private sector	2.8
Development assistance for health	
Total aid flows (millions of dollars)[1]	39
Aid flows per capita (millions of dollars)	1.5
Aid flows as a percentage of total health expenditure	13.0

For sources, notes, and explanations, see Annotated Source Appendix, page 1035.

Human Factors

Health Care Ratios [5]

Population per physician, 1970	14520
Population per physician, 1990	NA
Population per nursing person, 1970	990
Population per nursing person, 1990	NA
Percent of births attended by health staff, 1985	20

Infants and Malnutrition [6]

Percent of babies with low birth weight, 1985	15
Infant mortality rate per 1,000 live births, 1970	149
Infant mortality rate per 1,000 live births, 1991	101
Years of life lost per 1,000 population, 1990	84
Prevalence of malnutrition (under age 5), 1990	55

Ethnic Division [7]

Black	52%
Arab	39%
Beja	6%
Foreigners	2%
Other	1%

Religion [8]

Sunni Muslim	70%
Indigenous beliefs	25%
Christian	5%

Major Languages [9]

Arabic (official), Nubian, Ta Bedawie, diverse dialects of Nilotic, Nilo-Hamitic, Sudanic languages, English.

Education

Public Education Expenditures [10]

Million Sudanese Pounds	1980	1985[6]	1987	1988	1989	1990
Total education expenditure	187	NA	NA	NA	NA	NA
as percent of GNP	4.8	NA	NA	NA	NA	NA
as percent of total govt. expend.	9.1	NA	NA	NA	NA	NA
Current education expenditure	172	580	NA	NA	NA	NA
as percent of GNP	4.4	4.0	NA	NA	NA	NA
as percent of current govt. expend.	12.6	15.0	NA	NA	NA	NA
Capital expenditure	15	NA	NA	NA	NA	NA

Educational Attainment [11]

Literacy Rate [12]

In thousands and percent	1985[a]	1991[a]	2000[a]
Illiterate population + 15 years	9,040	10,061	12,541
Illiteracy rate - total pop. (%)	75.6	72.9	66.9
Illiteracy rate - males (%)	60.7	57.3	50.4
Illiteracy rate - females (%)	90.3	88.3	83.3

Libraries [13]

	Admin. Units	Svc. Pts.	Vols. (000)	Shelving (meters)	Vols. Added	Reg. Users
National	NA	NA	NA	NA	NA	NA
Non-specialized	NA	NA	NA	NA	NA	NA
Public	NA	NA	NA	NA	NA	NA
Higher ed. (1987)[10]	1	5	500	NA	10,000	20,000
School	NA	NA	NA	NA	NA	NA

Daily Newspapers [14]

	1975	1980	1985	1990
Number of papers	4	6	5	5
Circ. (000)	30[e]	105	250[e]	610[e]

Cinema [15]

Science and Technology

Scientific/Technical Forces [16]

R&D Expenditures [17]

U.S. Patents Issued [18]

For sources, notes, and explanations, see Annotated Source Appendix, page 1035.

869

Government and Law

Organization of Government [19]

Long-form name:
Republic of the Sudan
Type:
military civilian government suspended
and martial law imposed after 30 June
1989 coup
Independence:
1 January 1956 (from Egypt and UK)
Constitution:
12 April 1973, suspended 6 April 1985
(coup); 10 October 1985 interim
constitution, suspended 30 June 1989
(coup)
Legal system:
English common law and Islamic law; as
of 20 January 1991, six northern provinces
under Islamic law; accepts compulsory
ICJ jurisdiction, with reservations
National holiday:
Independence Day, 1 January (1956)
Executive branch:
executive and legislative authority vested
in a Revolutionary Command Council
(RCC)
Legislative branch:
appointed 300-member assembly
Judicial branch:
Supreme Court, Special Revolutionary
Courts

Crime [23]

Elections [20]

Political parties banned following 30
June 1989 coup; no elections.

Government Budget [21]

For FY91 est.

Revenues	1.300
Expenditures	2.100
Capital expenditures	0.505

Military Expenditures and Arms Transfers [22]

	1985	1986	1987	1988	1989
Military expenditures					
Current dollars (mil.)[e]	220	201	313	286	339
1989 constant dollars (mil.)[e]	250	223	337	298	339
Armed forces (000)	65	59	59[e]	65	65
Gross national product (GNP)					
Current dollars (mil.)	11,700	13,270	13,950	13,640	15,640
1989 constant dollars (mil.)	13,310	14,720	15,000	14,200	15,640
Central government expenditures (CGE)					
1989 constant dollars (mil.)	NA	NA	NA	2,550[e]	NA
People (mil.)	23.5	24.2	24.8	25.2	25.7
Military expenditure as % of GNP	1.9	1.5	2.2	2.1	2.2
Military expenditure as % of CGE	NA	NA	NA	11.7	NA
Military expenditure per capita	11	9	14	12	13
Armed forces per 1,000 people	2.8	2.4	2.4	2.6	2.5
GNP per capita	567	607	606	563	608
Arms imports[6]					
Current dollars (mil.)	40	50	80	90	70
1989 constant dollars (mil.)	46	55	86	94	70
Arms exports[6]					
Current dollars (mil.)	0	5	0	0	0
1989 constant dollars (mil.)	0	6	0	0	0
Total imports[7]					
Current dollars (mil.)	771	961	871	1,060	1,200
1989 constant dollars	878	1,066	937	1,103	1,200
Total exports[7]					
Current dollars (mil.)	374	333	504	509	671
1989 constant dollars	426	370	542	530	671
Arms as percent of total imports[8]	5.2	5.2	9.2	8.5	5.8
Arms as percent of total exports[8]	0	1.5	0	0	0

Human Rights [24]

	SSTS	FL	FAPRO	PPCG	APROBC	TPW	PCPTW	STPEP	PHRFF	PRW	ASST	AFL
Observes	P	P			P	P	P				P	P
	EAFRD	CPR	ESCR	SR	ACHR	MAAE	PVIAC	PVNAC	EAFDAW	TCIDTP	RC	
Observes		P	P	P	P					S	P	

P = Party; S = Signatory; see Appendix for meaning of abbreviations.

Labor Force

Total Labor Force [25]

6.5 million

Labor Force by Occupation [26]

Agriculture	80%
Industry and commerce	10
Government	6

Unemployment Rate [27]

30% (FY92 est.)

For sources, notes, and explanations, see Annotated Source Appendix, page 1035.

Production Sectors

Energy Resource Summary [28]

Energy Resources: Small reserves of petroleum. **Electricity**: 610,000 kW capacity; 905 million kWh produced, 40 kWh per capita (1991).
Pipelines: Refined products 815 km.

Telecommunications [30]

- Large, well-equipped system by African standards, but barely adequate and poorly maintained by modern standards
- Consists of microwave radio relay, cable, radio communications, troposcatter, and a domestic satellite system with 14 stations
- Broadcast stations - 11 AM, 3 TV
- Satellite earth stations for international traffic - 1 Atlantic Ocean INTELSAT and 1 ARABSAT

Transportation [31]

Railroads. 5,516 km total; 4,800 km 1.067-meter gauge, 716 km 1.6096-meter-gauge plantation line

Highways. 20,703 km total; 2,000 km bituminous treated, 4,000 km gravel, 2,304 km improved earth, 12,399 km unimproved earth and track

Merchant Marine. 5 ships (1,000 GRT or over) totaling 42,277 GRT/ 59,588 DWT; includes 3 cargo, 2 roll-on/roll-off

Airports

Total:	68
Usable:	56
With permanent-surface runways:	10
With runways over 3,659 m:	0
With runways 2,440-3,659 m:	6
With runways 1,220-2,439 m:	30

Top Agricultural Products [32]

Agriculture accounts for 35% of GDP and 80% of labor force; water shortages; two-thirds of land area suitable for raising crops and livestock; major products—cotton, oilseeds, sorghum, millet, wheat, gum arabic, sheep; marginally self-sufficient in most foods.

Top Mining Products [33]

Estimated metric tons unless otherwise specified   M.t.

Cement, hydraulic	170,000
Chromium, content of mine output	10,000
Gold, mine output, Au content (kilograms)	50
Gypsum and anhydrite, crude	7,000
Salt	75,000

Tourism [34]

	1987	1988	1989	1990	1991
Tourists	52	37	23	33[57]	16
Tourism receipts	14	29	45	21	8
Tourism expenditures	35	99	144	51	12
Fare receipts			36	13	9
Fare expenditures			11	7	11

Tourists are in thousands, money in million U.S. dollars.

Finance, Economics, and Trade

GDP and Manufacturing Summary [35]

	1980	1985	1990	1991	1992
Gross Domestic Product					
Millions of 1980 dollars	7,807	7,688	7,355	7,403	7,551[e]
Growth rate in percent	-3.41	-2.90	-8.04	0.65	2.00[e]
Manufacturing Value Added					
Millions of 1980 dollars	673	744	816	869	878[e]
Growth rate in percent	-7.69	-0.26	4.99	6.45	1.05[e]
Manufacturing share in percent of current prices	8.9	8.8	9.0	9.5	NA

Economic Indicators [36]

National product: GDP—exchange rate conversion—$5.2 billion (FY92 est.). **National product real growth rate**: 9% (FY92 est.). **National product per capita**: $184 (FY92 est.). **Inflation rate (consumer prices)**: 150% (FY92 est.). **External debt**: $15 billion (June 1992 est.).

Balance of Payments Summary [37]

Values in millions of dollars.

	1987	1988	1989	1990	1991
Exports of goods (f.o.b.)	265.7	427.0	544.4	326.5	302.5
Imports of goods (f.o.b.)	-696.6	-948.5	-1,051.0	-648.8	-1,138.2
Trade balance	-430.9	-521.5	-506.6	-322.3	-835.7
Services - debits	-321.2	-341.3	-495.7	-373.0	-329.5
Services - credits	192.2	171.6	279.7	184.9	79.7
Private transfers (net)	133.7	216.3	412.4	59.8	45.2
Government transfers (net)	197.6	117.0	161.3	81.4	82.5
Long term capital (net)	-235.8	63.1	114.9	102.7	486.3
Short term capital (net)	614.8	292.1	198.3	252.5	369.8
Errors and omissions	-185.9	7.5	-160.3	10.9	97.9
Overall balance	-35.5	4.8	4.0	-3.1	-3.8

Exchange Rates [38]

Currency: **Sudanese pounds.**
Symbol: **#Sd.**

Data are currency units per $1.

January 1993	124.0000
January 1993 free market rate	155.000
March 1992	90.1000
1991	5.4288
Fixed rate since 1987	4.5004
1987	2.8121

Imports and Exports

Top Import Origins [39]

$1.3 billion (c.i.f., FY92 est.). Data are for FY88.

Origins	%
Western Europe	32
Africa and Asia	15
US	13
Eastern Europe	3

Top Export Destinations [40]

$315 million (f.o.b., FY92 est.). Data are for FY88.

Destinations	%
Western Europe	46
Saudi Arabia	14
Eastern Europe	9
Japan	9
US	3

Foreign Aid [41]

	U.S. $	
US commitments, including Ex-Im (FY70-89)	1.5	billion
Western (non-US) countries, ODA and OOF bilateral commitments (1970-89)	5.1	billion
OPEC bilateral aid (1979-89)	3.1	billion
Communist countries (1970-89)	588	million

Import and Export Commodities [42]

Import Commodities

Foodstuffs
Petroleum products
Manufactured goods
Machinery & equipment
Medicines and chemicals
Textiles

Export Commodities

Cotton 52%
Sesame
Gum arabic
Peanuts

For sources, notes, and explanations, see Annotated Source Appendix, page 1035.

Suriname

Geography [1]

Total area:
163,270 km2
Land area:
161,470 km2
Comparative area:
Slightly larger than Georgia
Land boundaries:
Total 1,707 km, Brazil 597 km, French Guiana 510 km, Guyana 600 km
Coastline:
386 km
Climate:
Tropical; moderated by trade winds
Terrain:
Mostly rolling hills; narrow coastal plain with swamps
Natural resources:
Timber, hydropower potential, fish, shrimp, bauxite, iron ore, and small amounts of nickel, copper, platinum, gold
Land use:
Arable land:
0%
Permanent crops:
0%
Meadows and pastures:
0%
Forest and woodland:
97%
Other:
3%

Demographics [2]

	1960	1970	1980	1990	1991[1]	1994	2000	2010	2020
Population	285	373	355	398	402	423	465	534	598
Population density (persons per sq. mi.)	5	6	6	6	6	NA	7	9	10
Births	NA	NA	NA	NA	11	11	NA	NA	NA
Deaths	NA	NA	NA	NA	3	3	NA	NA	NA
Life expectancy - males	NA	NA	NA	NA	66	67	NA	NA	NA
Life expectancy - females	NA	NA	NA	NA	71	72	NA	NA	NA
Birth rate (per 1,000)	NA	NA	NA	NA	26	25	NA	NA	NA
Death rate (per 1,000)	NA	NA	NA	NA	6	6	NA	NA	NA
Women of reproductive age (15-44 yrs.)	NA	NA	NA	NA	NA	NA	NA	NA	NA
of which are currently married	NA	NA	NA	NA	NA	NA	NA	NA	NA
Fertility rate	NA	NA	NA	3.0	2.88	2.8	2.5	2.2	2.0

Population values are in thousands, life expectancy in years, and other items as indicated.

Health

Health Personnel [3]

Health Expenditures [4]

For sources, notes, and explanations, see Annotated Source Appendix, page 1035.

873

Human Factors

Health Care Ratios [5]	Infants and Malnutrition [6]

Ethnic Division [7]

Hindustani (East Indian)	37.0%
Creole (black mixed)	31.0%
Javanese	15.3%
Bush black	10.3%
Amerindian	2.6%
Chinese	1.7%
Europeans	1.0%
Other	1.1%

Religion [8]

Hindu	27.4%
Muslim	19.6%
Roman Catholic	22.8%
Protestant	25.2%
Indigenous beliefs	5.0%

Major Languages [9]

Dutch (official), English widely spoken, Sranan Tongo (Surinamese, sometimes called Taki-Taki) is native language of Creoles and much of the younger population and is lingua franca among others, Hindi Suriname Hindustani (a variant of Bhoqpuri), Javanese.

Education

Public Education Expenditures [10]

Million Guilder	1980	1985	1986	1988	1989	1990
Total education expenditure	105	158	182	209	224	250
as percent of GNP	6.7	9.4	10.2	9.1	8.3	8.3
as percent of total govt. expend.	22.5	NA	22.8	NA	NA	NA
Current education expenditure	105	NA	182	208	223	249
as percent of GNP	6.7	NA	10.2	9.1	8.3	8.2
as percent of current govt. expend.	22.8	NA	NA	NA	NA	NA
Capital expenditure	-	NA	-	1	1	1

Educational Attainment [11]

Literacy Rate [12]

In thousands and percent	1985[a]	1991[a]	2000[a]
Illiterate population +15 years	17	13	7
Illiteracy rate - total pop. (%)	7.3	5.1	2.1
Illiteracy rate - males (%)	6.9	4.9	2.3
Illiteracy rate - females (%)	7.6	5.3	2.0

Libraries [13]

Daily Newspapers [14]

	1975	1980	1985	1990
Number of papers	7	4	5	2
Circ. (000)	35[e]	45[e]	55[e]	40

Cinema [15]

Science and Technology

Scientific/Technical Forces [16]	R&D Expenditures [17]	U.S. Patents Issued [18]

Government and Law

Organization of Government [19]

Long-form name:
Republic of Suriname
Type:
republic
Independence:
25 November 1975 (from Netherlands)
Constitution:
ratified 30 September 1987
Legal system:
NA
National holiday:
Independence Day, 25 November (1975)
Executive branch:
president, vice president and prime minister, Cabinet of Ministers, Council of State; note - Commander in Chief of the National Army maintains significant power
Legislative branch:
unicameral National Assembly (Assemblee Nationale)
Judicial branch:
Supreme Court

Political Parties [20]

National Assembly	% of seats
The New Front (NF)	58.8
National Democratic Party (NDP)	19.6
DA'91	17.6
Independent	3.9

Government Budget [21]

For 1989 est.

Revenues	466
Expenditures	716
Capital expenditures	123

Military Expenditures and Arms Transfers [22]

	1985	1986	1987	1988	1989
Military expenditures					
Current dollars (mil.)	32[e]	31[e]	NA	38	39
1989 constant dollars (mil.)	36[e]	35[e]	NA	40	39
Armed forces (000)	2	3[e]	4	4	4
Gross national product (GNP)					
Current dollars (mil.)	1,184	1,218	1,114	1,195	1,319
1989 constant dollars (mil.)	1,348	1,352	1,198	1,244	1,319
Central government expenditures (CGE)					
1989 constant dollars (mil.)	674	713	614	550	543
People (mil.)	0.4	0.4	0.4	0.4	0.4
Military expenditure as % of GNP	2.7	2.6	NA	3.2	3.0
Military expenditure as % of CGE	5.4	4.9	NA	7.2	7.2
Military expenditure per capita	96	91	NA	103	100
Armed forces per 1,000 people	5.3	7.9	9.7	10.4	10.2
GNP per capita	3,591	3,564	3,134	3,221	3,371
Arms imports[6]					
Current dollars (mil.)	10	10	0	0	0
1989 constant dollars (mil.)	11	11	0	0	0
Arms exports[6]					
Current dollars (mil.)	0	0	0	0	0
1989 constant dollars (mil.)	0	0	0	0	0
Total imports[7]					
Current dollars (mil.)	299	244	294	365	NA
1989 constant dollars	340	271	316	380	NA
Total exports[7]					
Current dollars (mil.)	329	241	301	425	NA
1989 constant dollars	375	267	324	442	NA
Arms as percent of total imports[8]	3.3	4.1	0	0	NA
Arms as percent of total exports[8]	0	0	0	0	NA

Crime [23]

Human Rights [24]

	SSTS	FL	FAPRO	PPCG	APROBC	TPW	PCPTW	STPEP	PHRFF	PRW	ASST	AFL
Observes	P	P	P			P	P			P	P	P
	EAFRD	CPR	ESCR	SR	ACHR	MAAE	PVIAC	PVNAC	EAFDAW	TCIDTP	RC	
Observes		P	P	P	P		P	P				S

P = Party; S = Signatory; see Appendix for meaning of abbreviations.

Labor Force

Total Labor Force [25]

104,000 (1984)

Labor Force by Occupation [26]

Unemployment Rate [27]

16.5% (1990)

For sources, notes, and explanations, see Annotated Source Appendix, page 1035.

875

Production Sectors

Energy Resource Summary [28]

Energy Resources: Hydropower potential. **Electricity**: 458,000 kW capacity; 2,018 million kWh produced, 4,920 kWh per capita (1992).

Telecommunications [30]

- International facilities good
- Domestic microwave system
- 27,500 telephones
- Broadcast stations - 5 AM, 14 FM, 6 TV, 1 shortwave
- 2 Atlantic Ocean INTELSAT earth stations

Top Agricultural Products [32]

Agriculture accounts for 10.4% of GDP and 25% of export earnings; paddy rice planted on 85% of arable land and represents 60% of total farm output; other products—bananas, palm kernels, coconuts, plantains, peanuts, beef, chicken; shrimp and forestry products of increasing importance; self-sufficient in most foods.

Top Mining Products [33]

Thousand metric tons unless otherwise specified   M.t.

Bauxite, gross weight	3,198[1]
Alumina	1,510[1]
Aluminum metal, primary (exports)	29[1]
Cement, hydraulic	50[e]
Clay, common	16[e]
Gold, mine output, Au content (kilograms)	30[e]
Petroleum, crude (000 42-gal. barrels)	1,500
Gravel	35[e]
Sand, common	160[e]
Stone, crushed and broken	50[e]

Transportation [31]

Railroads. 166 km total; 86 km 1.000-meter gauge, government owned, and 80 km 1.435-meter standard gauge; all single track

Highways. 8,300 km total; 500 km paved; 5,400 km bauxite gravel, crushed stone, or improved earth; 2,400 km sand or clay

Merchant Marine. 3 ships (1,000 GRT or over) totaling 6,472 GRT/ 8,914 DWT; includes 2 cargo, 1 container

Airports

Total:	46
Usable:	39
With permanent-surface runways:	6
With runways over 3,659 m:	0
With runways 2,440-3,659 m:	1
With runways 1,220-2,439 m:	3

Tourism [34]

	1987	1988	1989	1990	1991
Tourists[16]	27	21	21	28	30
Tourism receipts	11	8	8	11	11
Tourism expenditures	8	10	10	12	
Fare receipts	2	2	2	2	
Fare expenditures	2	2	6	6	

Tourists are in thousands, money in million U.S. dollars.

Finance, Economics, and Trade

GDP and Manufacturing Summary [35]

	1980	1985	1990	1991	1992
Gross Domestic Product					
Millions of 1980 dollars	891	879	921	895	908[e]
Growth rate in percent	-8.57	2.02	-1.66	-2.81	1.43[e]
Manufacturing Value Added					
Millions of 1980 dollars	140	112	95	88	89[e]
Growth rate in percent	-10.52	6.45	-9.58	-6.95	1.00[e]
Manufacturing share in percent of current prices	17.6	12.5	10.2	9.9[e]	NA

Economic Indicators [36]

National product: GDP—exchange rate conversion—$1.35 billion (1991 est.). **National product real growth rate**: - 2.5% (1991 est.). **National product per capita**: $3,300 (1991 est.). **Inflation rate (consumer prices)**: 26% (1991). **External debt**: $138 million (1990 est.).

Balance of Payments Summary [37]

Values in millions of dollars.

	1986	1987	1988	1989	1990
Exports of goods (f.o.b.)	337.1	338.8	358.4	549.2	465.9
Imports of goods (f.o.b.)	-304.1	-274.3	-239.4	-330.9	-374.4
Trade balance	33.0	64.5	119.0	218.3	91.5
Services - debits	-81.6	-73.8	-86.1	-98.0	-106.7
Services - credits	27.3	82.1	24.0	24.6	23.0
Private transfers (net)	-1.9	-0.4	-4.6	-5.6	-7.5
Government transfers (net)	0.9	2.4	10.3	23.5	31.7
Long term capital (net)	-15.1	-59.2	-90.4	-151.9	-35.7
Short term capital (net)	16.1	8.6	-71.5	-188.5	-21.7
Errors and omissions	-18.7	-33.6	-1.8	9.9	-6.7
Overall balance	-40.0	-9.4	-101.1	-167.7	-32.1

Exchange Rates [38]

Currency: **Surinamese guilders, gulden, or florins**
Symbol: **Sf.**

Data are currency units per $1.

Fixed rate until	
October 1992	1.7850
January 1992	25.0400

Imports and Exports

Top Import Origins [39]

$514 million (f.o.b., 1992 est.). Data are for 1989.

Origins	%
US	41
Netherlands	24
Trinidad and Tobago	9
Brazil	4

Top Export Destinations [40]

$417 million (f.o.b., 1992 est.). Data are for 1989.

Destinations	%
Norway	36
Netherlands	28
US	11
Japan	7
Brazil	5
UK	5

Foreign Aid [41]

	U.S. $	
US commitments, including Ex-Im (FY70-83)	2.5	million
Western (non-US) countries, ODA and OOF bilateral commitments (1970-89)	1.5	billion

Import and Export Commodities [42]

Import Commodities	**Export Commodities**
Capital equipment	Alumina
Petroleum	Aluminum
Foodstuffs	Shrimp and fish
Cotton	Rice
Consumer goods	Bananas

For sources, notes, and explanations, see Annotated Source Appendix, page 1035.

877

Swaziland

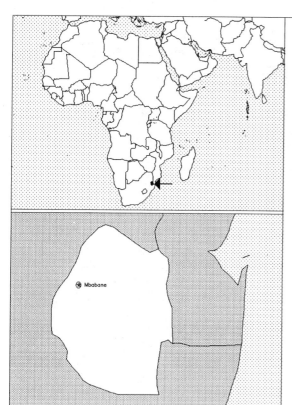

Geography [1]

Total area:
  17,360 km2
Land area:
  17,200 km2
Comparative area:
  Slightly smaller than New Jersey
Land boundaries:
  Total 535 km, Mozambique 105 km, South Africa 430 km
Coastline:
  0 km (landlocked)
Climate:
  Varies from tropical to near temperate
Terrain:
  Mostly mountains and hills; some moderately sloping plains
Natural resources:
  Asbestos, coal, clay, cassiterite, hydropower, forests, small gold and diamond deposits, quarry stone, and talc
Land use:
  Arable land:
  8%
  Permanent crops:
  0%
  Meadows and pastures:
  67%
  Forest and woodland:
  6%
  Other:
  19%

Demographics [2]

	1960	1970	1980	1990	1991[1]	1994	2000	2010	2020
Population	352	455	607	853	859	936	1,137	1,566	2,128
Population density (persons per sq. mi.)	53	69	91	126	129	NA	169	233	316
Births	NA	NA	NA	NA	38	40	NA	NA	NA
Deaths	NA	NA	NA	NA	10	10	NA	NA	NA
Life expectancy - males	NA	NA	NA	NA	51	52	NA	NA	NA
Life expectancy - females	NA	NA	NA	NA	59	61	NA	NA	NA
Birth rate (per 1,000)	NA	NA	NA	NA	44	43	NA	NA	NA
Death rate (per 1,000)	NA	NA	NA	NA	12	11	NA	NA	NA
Women of reproductive age (15-44 yrs.)	NA	NA	NA	204	NA	223	272	NA	NA
of which are currently married	NA	NA	NA	74	NA	80	99	NA	NA
Fertility rate	NA	NA	NA	6.2	6.21	6.1	5.9	5.4	4.9

Population values are in thousands, life expectancy in years, and other items as indicated.

Health

Health Personnel [3]

Health Expenditures [4]

For sources, notes, and explanations, see Annotated Source Appendix, page 1035.

Human Factors	
Health Care Ratios [5]	Infants and Malnutrition [6]

Ethnic Division [7]		Religion [8]		Major Languages [9]
African	97%	Christian	60%	English (official; government business
European	3%	Indigenous beliefs	40%	conducted in English), siSwati (official).

Education

Public Education Expenditures [10]

Million Lilangeni	1980	1985	1986	1987	1988	1989
Total education expenditure	26	52	60	66	90	101
as percent of GNP	6.1	6.0	5.6	5.6	5.5	6.4
as percent of total govt. expend.	NA	20.3	20.6	20.6	22.8	22.5
Current education expenditure	20	44	57	63	72	88
as percent of GNP	4.7	5.0	5.3	5.3	4.4	5.6
as percent of current govt. expend.	23.1	25.9	26.1	23.3	23.7	25.1
Capital expenditure	6	8	3	3	18	13

Educational Attainment [11]

Literacy Rate [12]

	1976[b]	1986[b]	1990[b]
Illiterate population + 15 years	115,036	116,464	NA
Illiteracy rate - total pop. (%)	44.8	32.7	NA
Illiteracy rate - males (%)	42.7	30.3	NA
Illiteracy rate - females (%)	46.5	34.8	NA

Libraries [13]

	Admin. Units	Svc. Pts.	Vols. (000)	Shelving (meters)	Vols. Added	Reg. Users
National (1989)	1	NA	3	32	20	NA
Non-specialized	NA	NA	NA	NA	NA	NA
Public	NA	NA	NA	NA	NA	NA
Higher ed.	NA	NA	NA	NA	NA	NA
School	NA	NA	NA	NA	NA	NA

Daily Newspapers [14]

	1975	1980	1985	1990
Number of papers	1	1	2	3
Circ. (000)	5	9	10	10[e]

Cinema [15]

Science and Technology

Scientific/Technical Forces [16]	R&D Expenditures [17]	U.S. Patents Issued [18]

For sources, notes, and explanations, see Annotated Source Appendix, page 1035.

Government and Law

Organization of Government [19]

Long-form name:
Kingdom of Swaziland

Type:
monarchy independent member of Commonwealth

Independence:
6 September 1968 (from UK)

Constitution:
none; constitution of 6 September 1968 was suspended on 12 April 1973; a new constitution was promulgated 13 October 1978, but has not been formally presented to the people

Legal system:
based on South African Roman-Dutch law in statutory courts, Swazi traditional law and custom in traditional courts; has not accepted compulsory ICJ jurisdiction

National holiday:
Somhlolo (Independence) Day, 6 September (1968)

Executive branch:
monarch, prime minister, Cabinet

Legislative branch:
bicameral Parliament is advisory and consists of an upper house or Senate and a lower house or House of Assembly

Judicial branch:
High Court, Court of Appeal

Elections [20]

Political parties banned by the Constitution promulgated on 13 October 1978. Direct legislative elections rescheduled for June 1993.

Government Expenditures [21]

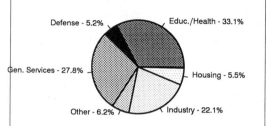

Defense - 5.2%
Educ./Health - 33.1%
Gen. Services - 27.8%
Housing - 5.5%
Other - 6.2%
Industry - 22.1%

% distribution for 1989. Expend. in FY94 est.: 410 ($ mil.)

Military Expenditures and Arms Transfers [22]

	1985	1986	1987	1988	1989
Military expenditures					
Current dollars (mil.)	8	8	7	8	11
1989 constant dollars (mil.)	9	9	8	8	11
Armed forces (000)	3	3	3	3	4
Gross national product (GNP)					
Current dollars (mil.)	471	501	511	594	644
1989 constant dollars (mil.)	536	556	550	619	644
Central government expenditures (CGE)					
1989 constant dollars (mil.)	177	171	152	158	197
People (mil.)	0.7	0.7	0.8	0.8	0.8
Military expenditure as % of GNP	1.7	1.6	1.4	1.4	1.7
Military expenditure as % of CGE	5.2	5.4	5.0	5.3	5.5
Military expenditure per capita	13	13	10	11	13
Armed forces per 1,000 people	4.3	4.1	3.9	3.8	4.9
GNP per capita	761	764	721	779	790
Arms imports[6]					
Current dollars (mil.)	0	0	0	0	0
1989 constant dollars (mil.)	0	0	0	0	0
Arms exports[6]					
Current dollars (mil.)	0	0	0	0	0
1989 constant dollars (mil.)	0	0	0	0	0
Total imports[7]					
Current dollars (mil.)	323	352	431	516	581
1989 constant dollars	368	391	464	538	581
Total exports[7]					
Current dollars (mil.)	176	267	406	453	440
1989 constant dollars	200	296	437	472	440
Arms as percent of total imports[8]	0	0	0	0	0
Arms as percent of total exports[8]	0	0	0	0	0

Crime [23]

Crime volume	
Cases known to police	34,868
Attempts (percent)	NA
Percent cases solved	76.98
Crimes per 100,000 persons	4,310.01
Persons responsible for offenses	
Total number offenders	16,591
Percent female	16.46
Percent juvenile[53]	11
Percent foreigners	NA

Human Rights [24]

	SSTS	FL	FAPRO	PPCG	APROBC	TPW	PCPTW	STPEP	PHRFF	PRW	ASST	AFL
Observes	1	P	P		P	P	P			P	1	P
	EAFRD	CPR	ESCR	SR	ACHR	MAAE	PVIAC	PVNAC	EAFDAW	TCIDTP	RC	
Observes		P		P						S		S

P = Party; S = Signatory; see Appendix for meaning of abbreviations.

Labor Force

Total Labor Force [25]

195,000 (over 60,000 engaged in subsistence agriculture; about 92,000 wage earners - many only intermittently)

Labor Force by Occupation [26]

Agriculture and forestry	36%
Community and social service	20
Manufacturing	14
Construction	9
Other	21

Unemployment Rate [27]

For sources, notes, and explanations, see Annotated Source Appendix, page 1035.

Production Sectors

Energy Resource Summary [28]

Energy Resources: Hydropower. **Electricity**: 60,000 kW capacity; 155 million kWh produced, 180 kWh per capita (1991).

Telecommunications [30]

- System consists of carrier-equipped open-wire lines and low-capacity microwave links
- 17,000 telephones
- Broadcast stations - 7 AM, 6 FM, 10 TV
- 1 Atlantic Ocean INTELSAT earth station

Transportation [31]

Railroads. 297 km (plus 71 km disused), 1.067-meter gauge, single track

Highways. 2,853 km total; 510 km paved, 1,230 km crushed stone, gravel, or stabilized soil, and 1,113 km improved earth

Airports

Total:	23
Usable:	21
With permanent-surfaced runways:	1
With runways over 3,659 m:	0
With runways 2,440-3,659 m:	1
With runways 1,220-2,439 m:	1

Top Agricultural Products [32]

Agriculture accounts for 23% of GDP and over 60% of labor force; mostly subsistence agriculture; cash crops— sugarcane, cotton, maize, tobacco, rice, citrus fruit, pineapples; other crops and livestock—corn, sorghum, peanuts, cattle, goats, sheep; not self-sufficient in grain.

Top Mining Products [33]

Metric tons unless otherwise specified	M.t.
Asbestos, chrysotile fiber	13,888
Coal, anthracite	122,502
Diamonds (carats)	57,420
Stone, quarry product (cubic meters)	128,759

Tourism [34]

	1987	1988	1989	1990	1991
Tourists[67]	194	196	248	294	279
Tourism receipts	20	22	25	25	26
Tourism expenditures	14	18	10	14	20
Fare receipts	3	4	4	5	6
Fare expenditures	3	4	4	5	5

Tourists are in thousands, money in million U.S. dollars.

Finance, Economics, and Trade

GDP and Manufacturing Summary [35]

	1980	1985	1990	1991	1992
Gross Domestic Product					
Millions of 1980 dollars	543	637	836	856	891[e]
Growth rate in percent	3.28	2.35	4.60	2.50	4.05[e]
Manufacturing Value Added					
Millions of 1980 dollars	102	119	171	179	189[e]
Growth rate in percent	11.17	-1.28	7.70	4.42	5.98[e]
Manufacturing share in percent of current prices	21.3	15.3	30.9[e]	NA	NA

Economic Indicators [36]

National product: GDP—exchange rate conversion—$700 million (1991 est.). **National product real growth rate**: 2.5% (1991 est.). **National product per capita**: $800 (1991 est.). **Inflation rate (consumer prices)**: 13% (1991 est.). **External debt**: $290 million (1990).

Balance of Payments Summary [37]

Values in millions of dollars.

	1987	1988	1989	1990	1991
Exports of goods (f.o.b.)	423.5	466.2	493.8	564.2	564.9
Imports of goods (f.o.b.)	-369.4	-441.0	-515.4	-603.1	-604.6
Trade balance	54.1	25.2	-21.6	-38.9	-39.7
Services - debits	-171.5	-194.6	-275.6	-276.9	-262.6
Services - credits	148.3	179.6	216.5	241.8	226.8
Private transfers (net)	1.7	3.9	1.4	2.1	-2.0
Government transfers (net)	48.0	68.6	82.1	86.8	86.6
Long term capital (net)	22.7	29.9	66.5	16.5	7.4
Short term capital (net)	-39.9	-86.7	-65.0	-16.5	-5.8
Errors and omissions	-45.1	-9.0	52.6	-1.0	12.0
Overall balance	18.3	16.9	56.9	13.9	22.7

Exchange Rates [38]

Currency: **emalangeni.**
Symbol: **E.**

The Swazi emalangeni is at par with the South African rand. Data are currency units per $1.

May 1993	3.1576
1992	2.8497
1991	2.7563
1990	2.5863
1989	2.6166
1988	2.2611

Imports and Exports

Top Import Origins [39]

$730 million (c.i.f., 1991).

Origins	%
South Africa	75
Japan	NA
Belgium	NA
UK	NA

Top Export Destinations [40]

$575 million (f.o.b., 1991).

Destinations	%
South Africa	50
EC countries	NA
Canada	NA

Foreign Aid [41]

	U.S. $	
US commitments, including Ex-Im (FY70-89)	142	million
Western (non-US) countries, ODA and OOF bilateral commitments (1970-89)	518	million

Import and Export Commodities [42]

Import Commodities

Motor vehicles
Machinery
Transport equipment
Petroleum products
Foodstuffs
Chemicals

Export Commodities

Soft drink concentrates
Sugar
Wood pulp
Citrus
Canned fruit

882

For sources, notes, and explanations, see Annotated Source Appendix, page 1035.

Sweden

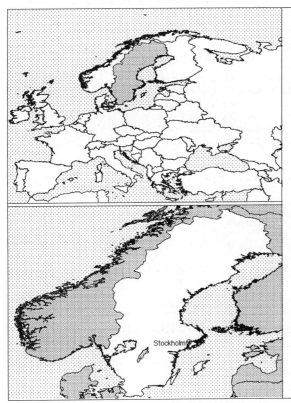

Geography [1]

Total area:
449,964 km2
Land area:
410,928 km2
Comparative area:
Slightly smaller than California
Land boundaries:
Total 2,205 km, Finland 586 km, Norway 1,619 km
Coastline:
3,218 km
Climate:
Temperate in south with cold, cloudy winters and cool, partly cloudy summers; subarctic in north
Terrain:
Mostly flat or gently rolling lowlands; mountains in west
Natural resources:
Zinc, iron ore, lead, copper, silver, timber, uranium, hydropower potential
Land use:
Arable land:
7%
Permanent crops:
0%
Meadows and pastures:
2%
Forest and woodland:
64%
Other:
27%

Demographics [2]

	1960	1970	1980	1990	1991[1]	1994	2000	2010	2020
Population	7,480	8,043	8,310	8,559	8,564	8,778	8,994	9,228	9,469
Population density (persons per sq. mi.)	47	51	52	54	54	NA	55	55	54
Births	NA	NA	NA	NA	112	119	NA	NA	NA
Deaths	NA	NA	NA	NA	97	96	NA	NA	NA
Life expectancy - males	NA	NA	NA	NA	75	75	NA	NA	NA
Life expectancy - females	NA	NA	NA	NA	81	81	NA	NA	NA
Birth rate (per 1,000)	NA	NA	NA	NA	13	14	NA	NA	NA
Death rate (per 1,000)	NA	NA	NA	NA	11	11	NA	NA	NA
Women of reproductive age (15-44 yrs.)	NA	NA	NA	2,048	NA	2,052	1,985	NA	NA
of which are currently married	NA	NA	NA	911	NA	929	910	NA	NA
Fertility rate	NA	NA	NA	2.1	1.92	2.0	1.8	1.8	1.8

Population values are in thousands, life expectancy in years, and other items as indicated.

Health

Health Personnel [3]

Doctors per 1,000 pop., 1988-92	2.73
Nurse-to-doctor ratio, 1988-92	3.4
Hospital beds per 1,000 pop., 1985-90	6.2
Percentage of children immunized (age 1 yr. or less)	
Third dose of DPT, 1990-91	99.0
Measles, 1990-91	95.0

Health Expenditures [4]

Total health expenditure, 1990 (official exchange rate)	
Millions of dollars	20055
Millions of dollars per capita	2343
Health expenditures as a percentage of GDP	
Total	8.8
Public sector	7.9
Private sector	0.9
Development assistance for health	
Total aid flows (millions of dollars)[1]	NA
Aid flows per capita (millions of dollars)	NA
Aid flows as a percentage of total health expenditure	NA

For sources, notes, and explanations, see Annotated Source Appendix, page 1035.

883

Human Factors

Health Care Ratios [5]

Population per physician, 1970	730
Population per physician, 1990	370
Population per nursing person, 1970	140
Population per nursing person, 1990	NA
Percent of births attended by health staff, 1985	100

Infants and Malnutrition [6]

Percent of babies with low birth weight, 1985	4
Infant mortality rate per 1,000 live births, 1970	11
Infant mortality rate per 1,000 live births, 1991	6
Years of life lost per 1,000 population, 1990	11
Prevalence of malnutrition (under age 5), 1990	NA

Ethnic Division [7]

White, Lapp, foreign born or first-generation immigrants 12% (Finns, Yugoslavs, Danes, Norwegians, Greeks, Turks).

Religion [8]

Evangelical Lutheran	94.0%
Roman Catholic	1.5%
Pentecostal	1.0%
Other	3.5%

Major Languages [9]

Swedish.

Education

Public Education Expenditures [10]

Million Krona	1980	1985	1987	1988	1989	1990
Total education expenditure	47,322	65,001	73,428	73,353	87,290	101,363
as percent of GNP	9.0	7.7	7.3	6.7	7.3	7.7
as percent of total govt. expend.	14.1	12.6	12.8	12.3	13.1	13.8
Current education expenditure	40,886	57,703	66,180	67,564	79,777	93,083
as percent of GNP	7.8	6.8	6.6	6.2	6.7	7.1
as percent of current govt. expend.	NA	NA	NA	NA	NA	NA
Capital expenditure	6,437	7,298	7,248	5,789	7,513	8,280

Educational Attainment [11]

Literacy Rate [12]

Libraries [13]

	Admin. Units	Svc. Pts.	Vols. (000)	Shelving (meters)	Vols. Added	Reg. Users
National (1990)	1	NA	3,677[e]	91,915	NA	NA
Non-specialized (1989)	26	128	18,035	452,390	365,000	NA
Public (1991)	375	NA	47,203	NA	NA	NA
Higher ed. (1990)	26	137	18,275	461,426	305,000	NA
School (1989)	5,226	NA	30,600	NA	NA	NA

Daily Newspapers [14]

	1975	1980	1985	1990
Number of papers	112	114	115	107
Circ. (000)	4,413	4,386	4,389	4,499

Cinema [15]

Data for 1991.

Cinema seats per 1,000	25.3
Annual attendance per person	1.8
Gross box office receipts (mil. Krona)	847

Science and Technology

Scientific/Technical Forces [16]

R&D Expenditures [17]

	Krona[32] (000) 1987
Total expenditure	30,554,000
Capital expenditure	2,668,000
Current expenditure	27,886,000
Percent current	91.3

U.S. Patents Issued [18]

Values show patents issued to citizens of the country by the U.S. Patents Office.

	1990	1991	1992
Number of patents	885	831	728

Government and Law

Organization of Government [19]

Long-form name:
Kingdom of Sweden
Type:
constitutional monarchy
Independence:
6 June 1809 (constitutional monarchy
established)
Constitution:
1 January 1975
Legal system:
civil law system influenced by customary
law; accepts compulsory ICJ jurisdiction,
with reservations
National holiday:
Day of the Swedish Flag, 6 June
Executive branch:
monarch, prime minister, Cabinet
Legislative branch:
unicameral parliament (Riksdag)
Judicial branch:
Supreme Court (Hogsta Domstolen)

Political Parties [20]

Riksdag	% of votes
Social Democratic Party	37.6
Moderate Party (conservative)	21.9
Liberal People's Party	9.1
Center Party	8.5
Christian Democrats	7.1
New Democracy	6.7
Left Party (Communist)	4.5
Green Party	3.4
Other	1.2

Government Expenditures [21]

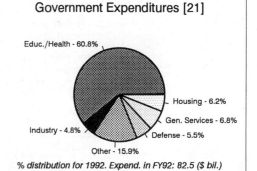

Educ./Health - 60.8%
Housing - 6.2%
Gen. Services - 6.8%
Defense - 5.5%
Other - 15.9%
Industry - 4.8%

% distribution for 1992. Expend. in FY92: 82.5 ($ bil.)

Crime [23]

Crime volume	
Cases known to police	1,218,820
Attempts (percent)	NA
Percent cases solved	30
Crimes per 100,000 persons	14,187.76
Persons responsible for offenses	
Total number offenders	320,773
Percent female	14.5
Percent juvenile[51]	12.4
Percent foreigners	NA

Military Expenditures and Arms Transfers [22]

	1985	1986	1987	1988	1989
Military expenditures					
Current dollars (mil.)	4,390	4,447	4,641	4,875	4,872
1989 constant dollars (mil.)	4,998	4,935	4,991	5,075	4,872
Armed forces (000)	69[e]	66	66	65	62
Gross national product (GNP)					
Current dollars (mil.)	147,600	155,300	165,800	174,800	185,800
1989 constant dollars (mil.)	168,000	172,300	178,300	182,000	185,800
Central government expenditures (CGE)					
1989 constant dollars (mil.)	79,240	75,190	74,500	74,020	74,600
People (mil.)	8.4	8.4	8.4	8.4	8.5
Military expenditure as % of GNP	3.0	2.9	2.8	2.8	2.6
Military expenditure as % of CGE	6.3	6.6	6.7	6.9	6.5
Military expenditure per capita	598	589	594	601	574
Armed forces per 1,000 people	8.3	7.8	7.8	7.7	7.4
GNP per capita	20,110	20,580	21,220	21,550	21,900
Arms imports[6]					
Current dollars (mil.)	80	80	90	210	70
1989 constant dollars (mil.)	91	89	97	219	70
Arms exports[6]					
Current dollars (mil.)	210	250	290	450	575
1989 constant dollars (mil.)	239	277	312	468	575
Total imports[7]					
Current dollars (mil.)	28,550	32,690	40,710	45,630	48,970
1989 constant dollars	32,500	36,280	43,780	47,500	48,970
Total exports[7]					
Current dollars (mil.)	30,460	37,260	44,510	49,750	51,550
1989 constant dollars	34,680	41,350	47,870	51,790	51,550
Arms as percent of total imports[8]	0.3	0.2	0.2	0.5	0.1
Arms as percent of total exports[8]	0.7	0.7	0.7	0.9	1.1

Human Rights [24]

	SSTS	FL	FAPRO	PPCG	APROBC	TPW	PCPTW	STPEP	PHRFF	PRW	ASST	AFL
Observes	P	P	P	P	P	P	P		P	P	P	P
		EAFRD	CPR	ESCR	SR	ACHR	MAAE	PVIAC	PVNAC	EAFDAW	TCIDTP	RC
Observes		P	P	P	P		P	P	P	P	P	P

P = Party; S = Signatory; see Appendix for meaning of abbreviations.

Labor Force

Total Labor Force [25]

4.552 million

Labor Force by Occupation [26]

Community, social and personal services
Mining and manufacturing
Commerce, hotels, and restaurants
Banking, insurance
Communications
Construction
Agriculture, fishing, and forestry
Date of data: 1991

Unemployment Rate [27]

5.3% (1992)

Production Sectors

Commercial Energy Production and Consumption

Production [28]

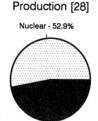

Nuclear - 52.9%

Hydro - 47.1%

Consumption [29]

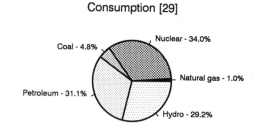

Coal - 4.8%

Nuclear - 34.0%

Natural gas - 1.0%

Petroleum - 31.1%

Hydro - 29.2%

Data are shown in quadrillion (10^{15}) BTUs and percent for 1991

Crude oil[1]	0.00
Natural gas liquids	0.00
Dry natural gas	0.00
Coal[2]	(s)
Net hydroelectric power[3]	0.65
Net nuclear power[3]	0.73
Total	1.38

Petroleum	0.65
Dry natural gas	0.02
Coal[2]	0.10
Net hydroelectric power[3]	0.61
Net nuclear power[3]	0.71
Total	2.09

Telecommunications [30]

- Excellent domestic and international facilities
- 8,200,000 telephones
- Mainly coaxial and multiconductor cables carry long-distance network
- Parallel microwave network carries primarily radio, TV and some telephone channels
- Automatic system
- Broadcast stations - 5 AM, 360 (mostly repeaters) FM, 880 (mostly repeaters) TV
- 5 submarine coaxial cables
- Satellite earth stations - 1 Atlantic Ocean INTELSAT and 1 EUTELSAT

Transportation [31]

Railroads. 12,000 km total; Swedish State Railways (SJ) - 10,819 km 1.435-meter standard gauge, 6,955 km electrified and 1,152 km double track; 182 km 0.891-meter gauge; 117 km rail ferry service; privately-owned railways - 511 km 1.435-meter standard gauge (332 km electrified) and 371 km 0.891-meter gauge (all electrified)

Highways. 97,400 km total; 51,899 km paved, 20,659 km gravel, 24,842 km unimproved earth

Merchant Marine. 179 ships (1,000 GRT or over) totaling 2,473,769 GRT/3,227,366 DWT; includes 10 short-sea passenger, 29 cargo, 3 container, 43 roll-on/roll-off cargo, 13 vehicle carrier, 2 railcar carrier, 32 oil tanker, 27 chemical tanker, 4 specialized tanker, 2 liquefied gas, 2 combination ore/oil, 10 bulk, 1 combination bulk, 1 refrigerated cargo

Airports

Total:	253
Usable:	250
With permanent-surface runways:	139
With runways over 3,659 m:	0
With runways 2,440-3,659 m:	12
With runways 1,220-2,439 m:	94

Top Agricultural Products [32]

	1990	1991
Barley	2,122	1,869
Wheat	2,243	1,525
Sugar beets	2,776	1,430
Oats	1,584	1,412
Potatoes	1,180	1,200
Rapeseed	367	332

Values shown are 1,000 metric tons.

Top Mining Products [33]

Metric tons unless otherwise specified	M.t.
Iron ore and concentrate (000 tons)	19,328[1]
Copper, mine output, Cu content	81,650[1]
Lead, mine output, Pb content	91,127[1]
Zinc, mine output, Zn content	161,170[1]
Silver, mine output, Ag content (kilograms)	239,321[1]
Steel, crude (000 tons)	4,248[1]

Tourism [34]

For sources, notes, and explanations, see Annotated Source Appendix, page 1035.

Manufacturing Sector

GDP and Manufacturing Summary [35]

	1980	1985	1989	1990	% change 1980-1990	% change 1989-1990
GDP (million 1980 $)	124,883	136,691	147,391	151,406	21.2	2.7
GDP per capita (1980 $)	15,026	16,368	17,492	17,928	19.3	2.5
Manufacturing as % of GDP (current prices)	23.0	23.7	21.0	21.6	-6.1	2.9
Gross output (million $)	73,194	59,391	103,106	115,465	57.8	12.0
Value added (million $)	30,905	24,486	45,863	51,429	66.4	12.1
Value added (million 1980 $)	26,351	29,786	31,329	31,454	19.4	0.4
Industrial production index	100	109	117	115	15.0	-1.7
Employment (thousands)	853	769	769	719	-15.7	-6.5

Note: GDP stands for Gross Domestic Product. 'e' stands for estimated value.

Profitability and Productivity

	1980	1985	1989	1990	% change 1980-1990	% change 1989-1990
Intermediate input (%)	58	59	56	55	-5.2	-1.8
Wages, salaries, and supplements (%)	18	15	15[e]	16	-11.1	6.7
Gross operating surplus (%)	24	26	29[e]	29	20.8	0.0
Gross output per worker ($)	85,747	77,211	134,067	160,524	87.2	19.7
Value added per worker ($)	36,206	31,833	59,635	71,499	97.5	19.9
Average wage (incl. benefits) ($)	15,835	11,676	20,139[e]	24,885	57.2	23.6

Profitability is in percent of gross output. Productivity is in U.S. $. 'e' stands for estimated value.

Profitability - 1990

Inputs - 55.0% Wages - 16.0% Surplus - 29.0%

The graphic shows percent of gross output.

Value Added in Manufacturing

	1980 $ mil.	1980 %	1985 $ mil.	1985 %	1989 $ mil.	1989 %	1990 $ mil.	1990 %	% change 1980-1990	% change 1989-1990
311 Food products	2,719	8.8	2,107	8.6	3,627	7.9	4,249	8.3	56.3	17.1
313 Beverages	338	1.1	250	1.0	531	1.2	743	1.4	119.8	39.9
314 Tobacco products	104	0.3	108	0.4	194	0.4	257	0.5	147.1	32.5
321 Textiles	534	1.7	379	1.5	586	1.3	618	1.2	15.7	5.5
322 Wearing apparel	274	0.9	157	0.6	215	0.5	199	0.4	-27.4	-7.4
323 Leather and fur products	54	0.2	40	0.2	41	0.1	52	0.1	-3.7	26.8
324 Footwear	61	0.2	24	0.1	29	0.1	27	0.1	-55.7	-6.9
331 Wood and wood products	2,102	6.8	1,154	4.7	2,308[e]	5.0	3,046	5.9	44.9	32.0
332 Furniture and fixtures	452	1.5	285	1.2	498[e]	1.1	551	1.1	21.9	10.6
341 Paper and paper products	2,596	8.4	2,230	9.1	4,509	9.8	4,524	8.8	74.3	0.3
342 Printing and publishing	1,842	6.0	1,517	6.2	2,671	5.8	3,158	6.1	71.4	18.2
351 Industrial chemicals	986	3.2	841	3.4	1,899[e]	4.1	1,983	3.9	101.1	4.4
352 Other chemical products	1,246	4.0	1,090	4.5	2,100[e]	4.6	2,546	5.0	104.3	21.2
353 Petroleum refineries	359	1.2	396	1.6	538	1.2	1,325	2.6	269.1	146.3
354 Miscellaneous petroleum and coal products	137	0.4	122	0.5	208[e]	0.5	218	0.4	59.1	4.8
355 Rubber products	314	1.0	225	0.9	367[e]	0.8	387	0.8	23.2	5.4
356 Plastic products	402	1.3	334	1.4	650[e]	1.4	786	1.5	95.5	20.9
361 Pottery, china and earthenware	87	0.3	71	0.3	101	0.2	123	0.2	41.4	21.8
362 Glass and glass products	175	0.6	124	0.5	241	0.5	294	0.6	68.0	22.0
369 Other non-metal mineral products	801	2.6	510	2.1	1,029	2.2	1,129	2.2	40.9	9.7
371 Iron and steel	1,650	5.3	1,185	4.8	2,027	4.4	2,097	4.1	27.1	3.5
372 Non-ferrous metals	390	1.3	331	1.4	709	1.5	640	1.2	64.1	-9.7
381 Metal products	2,598	8.4	2,048	8.4	3,829	8.3	4,448	8.6	71.2	16.2
382 Non-electrical machinery	3,936	12.7	3,185	13.0	5,360	11.7	6,226	12.1	58.2	16.2
383 Electrical machinery	2,570	8.3	2,132	8.7	3,564	7.8	4,023	7.8	56.5	12.9
384 Transport equipment	3,652	11.8	3,153	12.9	7,053	15.4	6,457	12.6	76.8	-8.5
385 Professional and scientific equipment	371	1.2	400	1.6	805	1.8	1,166	2.3	214.3	44.8
390 Other manufacturing industries	154	0.5	87	0.4	175[e]	0.4	157	0.3	1.9	-10.3

Note: The industry codes shown are International Standard Industry codes (ISIC). Percentages are percent of total Value Added. 'e' stands for estimated value

For sources, notes, and explanations, see Annotated Source Appendix, page 1035.

887

Finance, Economics, and Trade

Economic Indicators [36]

Billions Swedish Kronor (SEK) unless otherwise noted[143].

	1989	1990	1991[e]
Real GDP (1985 prices)	951.0	953.6	947.5
Real GDP growth rate (%)	2.1	0.3	-0.6
Money Supply (M3)[144]	674.2	790.8	756.7
Comm'l interest rates[145]	13.75	14.45	12.20
Savings rate (%)[146]	-4.6	0.0	2.2
Investment rate (%)[147]	21.2	20.4	19.0
Consumer prices (% chg)[148]	6.4	10.4	9.2
External public debt[149]	94.9	77.5	70.0

Balance of Payments Summary [37]

Values in millions of dollars.

	1987	1988	1989	1990	1991
Exports of goods (f.o.b.)	44,011	49,369	51,071	56,835	54,538
Imports of goods (f.o.b.)	-39,528	-44,487	-47,056	-53,433	-48,560
Trade balance	4,483	4,882	4,015	3,402	5,978
Services - debits	-16,520	-19,435	-23,084	-31,485	-33,064
Services - credits	13,291	15,477	17,848	23,467	25,845
Private transfers (net)	-379	-472	-709	-645	-395
Government transfers (net)	-1,016	-1,140	-1,338	-1,645	-1,608
Long term capital (net)	254	-928	-7,681	-21,600	8,564
Short term capital (net)	660	3,425	17,826	36,414	-17,572
Errors and omissions	108	-1,142	-5,654	-564	12,422
Overall balance	881	667	1,223	7,344	170

Exchange Rates [38]

Currency: **Swedish kronor.**
Symbol: **SKr.**

Data are currency units per $1.

December 1992	6.8812
1992	5.8238
1991	6.0475
1990	5.9188
1989	6.4469
1988	6.1272

Imports and Exports

Top Import Origins [39]

$51.7 billion (c.i.f., 1992).

Origins	%
EC	53.6
UK	6.3
Denmark	7.5
France	4.9
EFTA Norway	6.6
Finland	6
US	8.4
Central and Eastern Europe	3

Top Export Destinations [40]

$56 billion (f.o.b., 1992).

Destinations	%
EC	55.8
UK	9.7
Denmark	7.2
France	5.8
EFTA	17.4
Finland	5.1
US	8.2
Central and Eastern Europe	2.5

Foreign Aid [41]

	U.S. $	
Donor - ODA and OOF commitments (1970-89)	10.3	billion

Import and Export Commodities [42]

Import Commodities	Export Commodities
Machinery	Machinery
Petroleum and petroleum products	Motor vehicles
Chemicals	Paper products
Motor vehicles	Pulp and wood
Foodstuffs	Iron and steel products
Iron and steel	Chemicals
Clothing	Petroleum & petr. products

888

For sources, notes, and explanations, see Annotated Source Appendix, page 1035.

Switzerland

Geography [1]

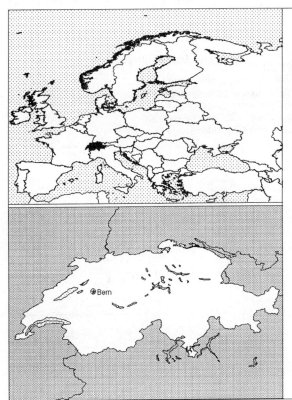

Total area:
41,290 km2
Land area:
39,770 km2
Comparative area:
Slightly more than twice the size of New Jersey
Land boundaries:
Total 1,852 km, Austria 164 km, France 573 km, Italy 740 km, Liechtenstein 41 km, Germany 334 km
Coastline:
0 km (landlocked)
Climate:
Temperate, but varies with altitude; cold, cloudy, rainy/snowy winters; cool to warm, cloudy, humid summers with occasional showers
Terrain:
Mostly mountains (Alps in south, Jura in northwest) with a central plateau of rolling hills, plains, and large lakes
Natural resources:
Hydropower potential, timber, salt
Land use:
Arable land: 10%
Permanent crops: 1%
Meadows and pastures: 40%
Forest and woodland: 26%
Other: 23%

Demographics [2]

	1960	1970	1980	1990	1991[1]	1994	2000	2010	2020
Population	5,362	6,267	6,385	6,779	6,784	7,040	7,268	7,519	7,696
Population density (persons per sq. mi.)	349	408	416	439	442	NA	457	455	446
Births	NA	NA	NA	NA	82	86	NA	NA	NA
Deaths	NA	NA	NA	NA	61	65	NA	NA	NA
Life expectancy - males	NA	NA	NA	NA	75	75	NA	NA	NA
Life expectancy - females	NA	NA	NA	NA	83	82	NA	NA	NA
Birth rate (per 1,000)	NA	NA	NA	NA	12	12	NA	NA	NA
Death rate (per 1,000)	NA	NA	NA	NA	9	9	NA	NA	NA
Women of reproductive age (15-44 yrs.)	NA	NA	NA	1,734	NA	1,776	1,754	NA	NA
of which are currently married	NA	NA	NA	1,052	NA	1,105	1,105	NA	NA
Fertility rate	NA	NA	NA	1.6	1.57	1.6	1.6	1.6	1.6

Population values are in thousands, life expectancy in years, and other items as indicated.

Health

Health Personnel [3]

Doctors per 1,000 pop., 1988-92	1.59
Nurse-to-doctor ratio, 1988-92	2.6
Hospital beds per 1,000 pop., 1985-90	11.0
Percentage of children immunized (age 1 yr. or less)	
Third dose of DPT, 1990-91	90.0
Measles, 1990-91	90.0

Health Expenditures [4]

Total health expenditure, 1990 (official exchange rate)	
Millions of dollars	16916
Millions of dollars per capita	2520
Health expenditures as a percentage of GDP	
Total	7.5
Public sector	5.1
Private sector	2.4
Development assistance for health	
Total aid flows (millions of dollars)[1]	NA
Aid flows per capita (millions of dollars)	NA
Aid flows as a percentage of total health expenditure	NA

For sources, notes, and explanations, see Annotated Source Appendix, page 1035.

889

Human Factors

Health Care Ratios [5]

Population per physician, 1970	700
Population per physician, 1990	630
Population per nursing person, 1970	NA
Population per nursing person, 1990	NA
Percent of births attended by health staff, 1985	NA

Infants and Malnutrition [6]

Percent of babies with low birth weight, 1985	5
Infant mortality rate per 1,000 live births, 1970	15
Infant mortality rate per 1,000 live births, 1991	7
Years of life lost per 1,000 population, 1990	10
Prevalence of malnutrition (under age 5), 1990	NA

Ethnic Division [7]

No detailed information provided.

Religion [8]

Roman Catholic	47.6%
Protestant	44.3%
Other	8.1%

Major Languages [9]

German	65%
French	18%
Italian	12%
Romansch	1%
Other	4%

Education

Public Education Expenditures [10]

Million Swiss Francs	1980[7]	1985[7]	1987[7]	1988[7]	1989[7]	1990[7,17]
Total education expenditure	8,873	11,696	12,754	13,757	14,560	16,215
as percent of GNP	5.0	4.8	4.8	4.9	4.8	5.0
as percent of total govt. expend.	18.8	18.6	18.9	18.8	18.7	NA
Current education expenditure	7,937	10,638	11,555	12,311	12,941	14,395
as percent of GNP	4.5	4.4	4.3	4.4	4.2	4.4
as percent of current govt. expend.	20.0	19.9	20.0	20.0	19.6	NA
Capital expenditure	936	1,058	1,199	1,446	1,619	1,820

Educational Attainment [11]

Literacy Rate [12]

Libraries [13]

	Admin. Units	Svc. Pts.	Vols. (000)	Shelving (meters)	Vols. Added	Reg. Users
National (1990)	1	1	2,922	NA	56,212	7,603
Non-specialized (1989)	33	33	7,721	216,176	247,908	940,667
Public	NA	NA	NA	NA	NA	NA
Higher ed. (1990)[39]	9	NA	13,519	339,908	293,603	136,780
School	NA	NA	NA	NA	NA	NA

Daily Newspapers [14]

	1975	1980	1985	1990
Number of papers	95	89	97	94
Circ. (000)	2,574	2,483	3,213	3,063

Cinema [15]

Data for 1991.

Cinema seats per 1,000	14.7
Annual attendance per person	2.3
Gross box office receipts (mil. Franc)	NA

Science and Technology

Scientific/Technical Forces [16]

Potential scientists/engineers	348,167
Number female	61,729
Potential technicians	NA
Number female	NA
Total	NA

R&D Expenditures [17]

	Franc[33] (000) 1986
Total expenditure	7,015,000
Capital expenditure	NA
Current expenditure	NA
Percent current	NA

U.S. Patents Issued [18]

Values show patents issued to citizens of the country by the U.S. Patents Office.

	1990	1991	1992
Number of patents	1,347	1,448	1,293

Government and Law

Organization of Government [19]

Long-form name:
Swiss Confederation
Type:
federal republic
Independence:
1 August 1291
Constitution:
29 May 1874
Legal system:
civil law system influenced by customary
law; judicial review of legislative acts,
except with respect to federal decrees of
general obligatory character; accepts
compulsory ICJ jurisdiction, with
reservations
National holiday:
Anniversary of the Founding of the Swiss
Confederation, 1 August (1291)
Executive branch:
president, vice president, Federal Council
(German - Bundesrat, French - Conseil
Federal, Italian - Consiglio Federale)
Legislative branch:
bicameral Federal Assembly consists of
an upper council or Council of States and
a lower council or National Council
Judicial branch:
Federal Supreme Court

Political Parties [20]

National Council	% of seats
Free Democratic Party (FDP)	22.0
Social Democratic Party (SPS)	21.0
Christian Democratic (CVP)	18.5
Swiss People's Party (SVP)	12.5
Green Party (GPS)	7.0
Liberal Party (LPS)	5.0
Automobile Party (AP)	4.0
Alliance of Independents (LdU)	3.0
Swiss Democratic Party (SD)	2.5
Other	4.5

Government Expenditures [21]

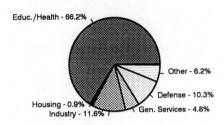

Educ./Health - 66.2%
Other - 6.2%
Defense - 10.3%
Gen. Services - 4.8%
Industry - 11.6%
Housing - 0.9%

% distribution for 1984. Expend. in 1990: 23.8 ($ bil.)

Military Expenditures and Arms Transfers [22]

	1985	1986	1987	1988	1989
Military expenditures					
Current dollars (mil.)	3,487[e]	3,512[e]	3,357[e]	3,468[e]	3,806[e]
1989 constant dollars (mil.)	3,970[e]	3,898[e]	3,610[e]	3,610[e]	3,806[e]
Armed forces (000)	23[e]	21	20	23	17
Gross national product (GNP)					
Current dollars (mil.)	146,400	153,100	160,800	172,300	184,300
1989 constant dollars (mil.)	166,700	169,800	172,900	179,400	184,300
Central government expenditures (CGE)					
1989 constant dollars (mil.)	33,490[e]	35,350[e]	35,350[e]	37,760[e]	NA
People (mil.)	6.5	6.6	6.6	6.7	6.7
Military expenditure as % of GNP	2.4	2.3	2.1	2.0	2.1
Military expenditure as % of CGE	11.9	11.0	10.2	9.6	NA
Military expenditure per capita	608	593	545	542	568
Armed forces per 1,000 people	3.5	3.2	3.0	3.5	2.5
GNP per capita	25,510	25,840	26,120	26,940	27,510
Arms imports[6]					
Current dollars (mil.)	750	550	260	330	300
1989 constant dollars (mil.)	854	610	280	344	300
Arms exports[6]					
Current dollars (mil.)	330	290	400	210	60
1989 constant dollars (mil.)	376	322	430	219	60
Total imports[7]					
Current dollars (mil.)	30,700	41,050	50,650	56,360	58,190
1989 constant dollars	34,940	45,550	54,480	58,670	58,190
Total exports[7]					
Current dollars (mil.)	27,430	37,450	45,510	50,700	51,520
1989 constant dollars	27,430	41,560	48,950	52,780	51,520
Arms as percent of total imports[8]	2.4	1.3	0.5	0.6	0.5
Arms as percent of total exports[8]	1.2	0.8	0.9	0.4	0.1

Crime [23]

Crime volume	
Cases known to police	356,129
Attempts (percent)	NA
Percent cases solved	NA
Crimes per 100,000 persons	5,275.59
Persons responsible for offenses	
Total number offenders	75,869
Percent female	16.7
Percent juvenile[52]	16.2
Percent foreigners	37.9

Human Rights [24]

	SSTS	FL	FAPRO	PPCG	APROBC	TPW	PCPTW	STPEP	PHRFF	PRW	ASST	AFL
Observes	P	P	P			P	P		P		P	P
		EAFRD	CPR	ESCR	SR	ACHR	MAAE	PVIAC	PVNAC	EAFDAW	TCIDTP	RC
Observes			P	P	P			P	P		P	S

P = Party; S = Signatory; see Appendix for meaning of abbreviations.

Labor Force

Total Labor Force [25]

3.31 million (904,095 foreign workers,
mostly Italian)

Labor Force by Occupation [26]

Services	50%
Industry and crafts	33
Government	10
Agriculture and forestry	6
Other	1

Date of data: 1989

Unemployment Rate [27]

3% (1992 est.)

For sources, notes, and explanations, see Annotated Source Appendix, page 1035.

891

Production Sectors

Commercial Energy Production and Consumption

Production [28]

Hydro - 58.2%

Nuclear - 41.8%

Consumption [29]

Petroleum - 50.9%

Coal - 0.9%

Natural gas - 7.1%

Nuclear - 18.8%

Hydro - 22.3%

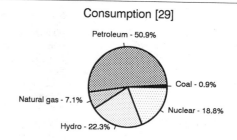

Data are shown in quadrillion (10^{15}) BTUs and percent for 1991

Crude oil[1]	0.00
Natural gas liquids	0.00
Dry natural gas	0.00
Coal[2]	0.00
Net hydroelectric power[3]	0.32
Net nuclear power[3]	0.23
Total	0.56

Petroleum	0.57
Dry natural gas	0.08
Coal[2]	0.01
Net hydroelectric power[3]	0.25
Net nuclear power[3]	0.21
Total	1.11

Telecommunications [30]

- Excellent domestic, international, and broadcast services
- 5,890,000 telephones
- Extensive cable and microwave networks
- Broadcast stations - 7 AM, 265 FM, 18 (1,322 repeaters) TV
- Communications satellite earth station operating in the INTELSAT (Atlantic Ocean and Indian Ocean) system

Transportation [31]

Railroads. 4,418 km total; 3,073 km are government owned and 1,345 km are nongovernment owned; the government network consists of 2,999 km 1.435-meter standard gauge and 74 km 1.000-meter narrow gauge track; 1,432 km double track, 99% electrified; the nongovernment network consists of 510 km 1.435-meter standard gauge, and 835 km 1.000-meter gauge, 100% electrified

Highways. 62,145 km total (all paved); 18,620 km are canton, 1,057 km are national highways (740 km autobahn), 42,468 km are communal roads

Merchant Marine. 23 ships (1,000 GRT or over) totaling 308,725 GRT/548,244 DWT; includes 5 cargo, 2 roll-on/roll-off cargo, 5 chemical tanker, 2 specialized tanker, 8 bulk, 1 oil tanker

Airports

Total:	66
Usable:	65
With permanent-surface runways:	42
With runways over 3,659 m:	2
With runways 2,440-3,659 m:	5
With runways 1,220-2,439 m:	18

Top Agricultural Products [32]

	1989	1990
Wine grapes (000 hl.)	1,747	1,334
Sugar beets	889	975
potatoes	890	857
Grains, feed	745	690
Grains, bread	642	592
Apples	198	311

Values shown are 1,000 metric tons.

Top Mining Products [33]

Estimated metric tons unless otherwise specified   M.t.

Aluminum, smelter, primary (tons)	71,000
Cement, hydraulic	5,000
Salt	250
Gypsum	200
Lime	25
Steel, crude	1,000

Tourism [34]

	1987	1988	1989	1990	1991
Visitors[9]	117,900	112,200	122,900	129,200	137,000
Tourists[9]	11,600	11,700	12,600	13,200	12,600
Excursionists[69]	106,300	100,500	110,300	116,000	124,400
Tourism receipts	5,345	5,720	5,543	6,789	7,064
Tourism expenditures	4,339	5,019	4,907	5,817	5,682
Fare receipts	1,312	1,398	1,395	1,630	1,702
Fare expenditures	846	937	862	1,110	1,115

Tourists are in thousands, money in million U.S. dollars.

For sources, notes, and explanations, see Annotated Source Appendix, page 1035.

Manufacturing Sector

GDP and Manufacturing Summary [35]

	1980	1985	1989	1990	% change 1980-1990	% change 1989-1990
GDP (million 1980 $)	101,629	108,881	121,786	124,863	22.9	2.5
GDP per capita (1980 $)	16,081	16,826	18,489	18,873	17.4	2.1
Manufacturing as % of GDP (current prices)	27.4	25.0	24.2[e]	26.2[e]	-4.4	8.3
Gross output (million $)	NA	NA	NA	NA	NA	NA
Value added (million $)	27,450	23,504	44,792	58,051	111.5	29.6
Value added (million 1980 $)	28,371	29,222	32,068	34,896	23.0	8.8
Industrial production index	100	99	112	111	11.0	-0.9
Employment (thousands)	686	656	676[e]	677	-1.3	0.1

Note: GDP stands for Gross Domestic Product. 'e' stands for estimated value.

Profitability and Productivity

	1980	1985	1989	1990	% change 1980-1990	% change 1989-1990
Intermediate input (%)	NA	NA	NA	NA	NA	NA
Wages, salaries, and supplements (%)	NA	NA	NA	NA	NA	NA
Gross operating surplus (%)	NA	NA	NA	NA	NA	NA
Gross output per worker ($)	NA	NA	NA	NA	NA	NA
Value added per worker ($)	40,026	35,808	66,225[e]	85,691	114.1	29.4
Average wage (incl. benefits) ($)	NA	NA	NA	NA	NA	NA

Profitability is in percent of gross output. Productivity is in U.S. $. 'e' stands for estimated value.

Profitability - 1990

Value Added in Manufacturing

	1980 $ mil.	1980 %	1985 $ mil.	1985 %	1989 $ mil.	1989 %	1990 $ mil.	1990 %	% change 1980-1990	% change 1989-1990
311 Food products	2,905	10.6	2,571[e]	10.9	4,729[e]	10.6	5,874[e]	10.1	102.2	24.2
313 Beverages	499	1.8	464[e]	2.0	794[e]	1.8	1,130[e]	1.9	126.5	42.3
314 Tobacco products	292	1.1	161[e]	0.7	246[e]	0.5	293[e]	0.5	0.3	19.1
321 Textiles	972	3.5	878	3.7	1,440	3.2	1,722	3.0	77.2	19.6
322 Wearing apparel	864	3.1	633	2.7	888	2.0	1,140	2.0	31.9	28.4
323 Leather and fur products	124	0.5	61[e]	0.3	85[e]	0.2	95[e]	0.2	-23.4	11.8
324 Footwear	324	1.2	255	1.1	326	0.7	372	0.6	14.8	14.1
331 Wood and wood products	1,079	3.9	873[e]	3.7	1,698[e]	3.8	2,245[e]	3.9	108.1	32.2
332 Furniture and fixtures	707	2.6	572[e]	2.4	1,113[e]	2.5	1,472[e]	2.5	108.2	32.3
341 Paper and paper products	624	2.3	558	2.4	1,129	2.5	1,405	2.4	125.2	24.4
342 Printing and publishing	1,471	5.4	1,703	7.2	3,361	7.5	4,222	7.3	187.0	25.6
351 Industrial chemicals	1,530	5.6	1,627[e]	6.9	3,328[e]	7.4	4,267[e]	7.4	178.9	28.2
352 Other chemical products	1,332	4.9	1,363[e]	5.8	3,621[e]	8.1	4,405[e]	7.6	230.7	21.7
353 Petroleum refineries	585	2.1	522[e]	2.2	981[e]	2.2	1,138[e]	2.0	94.5	16.0
354 Miscellaneous petroleum and coal products	95	0.3	85[e]	0.4	182[e]	0.4	186[e]	0.3	95.8	2.2
355 Rubber products	226	0.8	240[e]	1.0	402[e]	0.9	768[e]	1.3	239.8	91.0
356 Plastic products	625	2.3	665[e]	2.8	1,156[e]	2.6	2,131[e]	3.7	241.0	84.3
361 Pottery, china and earthenware	137	0.5	148[e]	0.6	223[e]	0.5	308[e]	0.5	124.8	38.1
362 Glass and glass products	187	0.7	203[e]	0.9	306[e]	0.7	422[e]	0.7	125.7	37.9
369 Other non-metal mineral products	651	2.4	433[e]	1.8	770[e]	1.7	861[e]	1.5	32.3	11.8
371 Iron and steel	455	1.7	468[e]	2.0	940[e]	2.1	1,163[e]	2.0	155.6	23.7
372 Non-ferrous metals	584	2.1	428[e]	1.8	798[e]	1.8	1,007[e]	1.7	72.4	26.2
381 Metal products	1,922	7.0	1,545[e]	6.6	2,933[e]	6.5	3,909[e]	6.7	103.4	33.3
382 Non-electrical machinery	3,777	13.8	3,037[e]	12.9	5,764[e]	12.9	7,683[e]	13.2	103.4	33.3
383 Electrical machinery	2,860	10.4	2,300[e]	9.8	4,365[e]	9.7	5,819[e]	10.0	103.5	33.3
384 Transport equipment	508	1.9	409[e]	1.7	775[e]	1.7	1,034[e]	1.8	103.5	33.4
385 Professional and scientific equipment	1,977	7.2	1,217[e]	5.2	2,277[e]	5.1	2,787[e]	4.8	41.0	22.4
390 Other manufacturing industries	138	0.5	85	0.4	159	0.4	193	0.3	39.9	21.4

Note: The industry codes shown are International Standard Industry codes (ISIC). Percentages are percent of total Value Added. 'e' stands for estimated value

Finance, Economics, and Trade

Economic Indicators [36]

Millions of Swiss Francs unless otherwise noted[150].

	1989	1990	1991
GDP at current prices	289,800	316,715	366,000
GNP per capita (SF)	45,399	46,576	248,700
M1 money supply (% change)	-5.9	-5.0	NA
Savings rates	3.45	4.55	5.06[151]
CPI (% change)	3.2	5.4	5.7[152]
WPI (% change)	4.3	1.5	-0.2[152]

Balance of Payments Summary [37]

Values in millions of dollars.

	1987	1988	1989	1990	1991
Exports of goods (f.o.b.)	55,219	62,725	65,366	77,488	73,745
Imports of goods (f.o.b.)	-60,647	-67,301	-69,690	-83,878	-77,553
Trade balance	-5,428	-4,576	-4,324	-6,390	-3,808
Services - debits	-19,266	-20,754	-24,155	-31,563	-30,212
Services - credits	32,484	35,887	38,204	47,226	46,600
Private transfers (net)	-1,550	-1,711	-1,665	-2,183	-2,273
Government transfers (net)	45	-2	-18	-146	-460
Long term captial (net)	-4,396	-18,766	-11,215	-7,153	-17,219
Short term capital (net)	-1,078	4,821	17,943	10,071	4,602
Errors and omissions	4,983	3,472	995	5,686	2,900
Overall balance	3,205	-2,426	1,419	1,173	970

Exchange Rates [38]

Currency: **Swiss francs, franken, or franchi.**
Symbol: **SwF.**

Data are currency units per $1.

January 1993	1.4781
1992	1.4062
1991	1.4340
1990	1.3892
1989	1.6359
1988	1.4633

Imports and Exports

Top Import Origins [39]

$68.5 billion (c.i.f., 1991 est.).

Origins	%
Western Europe	78
Other	7
US	6

Top Export Destinations [40]

$62.2 billion (f.o.b., 1991 est.).

Destinations	%
Western Europe	64
Other	8
US	9
Japan	4

Foreign Aid [41]

	U.S. $	
Donor - ODA and OOF commitments (1970-89)	3.5	billion

Import and Export Commodities [42]

Import Commodities

Agricultural products
Machinery & transport equipment
Chemicals
Textiles
Construction materials

Export Commodities

Machinery & equipment
Precision instruments
Metal products
Foodstuffs
Textiles and clothing

Syria

Geography [1]

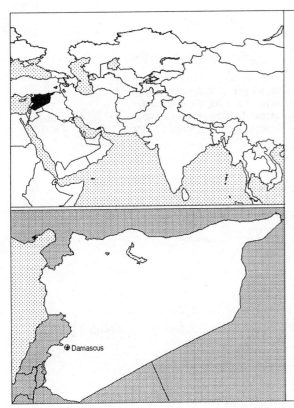

Total area:
185,180 km2
Land area:
184,050 km2
Comparative area:
Slightly larger than North Dakota
Land boundaries:
Total 2,253 km, Iraq 605 km, Israel 76 km, Jordan 375 km, Lebanon 375 km, Turkey 822 km
Coastline:
193 km
Climate:
Mostly desert; hot, dry, sunny summers (June to August) and mild, rainy winters (December to February) along coast
Terrain:
Primarily semiarid and desert plateau; narrow coastal plain; mountains in west
Natural resources:
Petroleum, phosphates, chrome and manganese ores, asphalt, iron ore, rock salt, marble, gypsum
Land use:
Arable land: 28%
Permanent crops: 3%
Meadows and pastures: 46%
Forest and woodland: 3%
Other: 20%

Demographics [2]

	1960	1970	1980	1990	1991[1]	1994	2000	2010	2020
Population	4,533	6,258	8,692	12,762	12,966	14,887	18,519	25,768	34,309
Population density (persons per sq. mi.)	64	88	122	176	182	NA	256	365	503
Births	NA	NA	NA	NA	563	650	NA	NA	NA
Deaths	NA	NA	NA	NA	71	93	NA	NA	NA
Life expectancy - males	NA	NA	NA	NA	68	65	NA	NA	NA
Life expectancy - females	NA	NA	NA	NA	71	68	NA	NA	NA
Birth rate (per 1,000)	NA	NA	NA	NA	43	44	NA	NA	NA
Death rate (per 1,000)	NA	NA	NA	NA	5	6	NA	NA	NA
Women of reproductive age (15-44 yrs.)	NA	NA	NA	2,660	NA	3,138	4,026	NA	NA
of which are currently married	NA	NA	NA	1,762	NA	2,101	2,712	NA	NA
Fertility rate	NA	NA	NA	7.1	6.66	6.7	6.0	4.9	3.8

Population values are in thousands, life expectancy in years, and other items as indicated.

Health

Health Personnel [3]

Doctors per 1,000 pop., 1988-92	0.85
Nurse-to-doctor ratio, 1988-92	1.2
Hospital beds per 1,000 pop., 1985-90	1.1
Percentage of children immunized (age 1 yr. or less)	
Third dose of DPT, 1990-91	89.0
Measles, 1990-91	84.0

Health Expenditures [4]

Total health expenditure, 1990 (official exchange rate)	
Millions of dollars	283
Millions of dollars per capita	23
Health expenditures as a percentage of GDP	
Total	2.1
Public sector	0.4
Private sector	1.6
Development assistance for health	
Total aid flows (millions of dollars)[1]	20
Aid flows per capita (millions of dollars)	1.6
Aid flows as a percentage of total health expenditure	7.1

For sources, notes, and explanations, see Annotated Source Appendix, page 1035.

895

Human Factors

Health Care Ratios [5]

Population per physician, 1970	3860
Population per physician, 1990	1160
Population per nursing person, 1970	1790
Population per nursing person, 1990	870
Percent of births attended by health staff, 1985	37

Infants and Malnutrition [6]

Percent of babies with low birth weight, 1985	9
Infant mortality rate per 1,000 live births, 1970	96
Infant mortality rate per 1,000 live births, 1991	37
Years of life lost per 1,000 population, 1990	25
Prevalence of malnutrition (under age 5), 1990	NA

Ethnic Division [7]

Arab	90.3%
Kurds, Armenians, other	9.7%

Religion [8]

Sunni Muslim	74%
Alawite, Druze, and other Muslim sects	16%
Christian (various sects)	10%
Jewish	

Major Languages [9]

French widely understood.

Arabic (official)	NA
Kurdish	NA
Armenian	NA
Aramaic	NA
Circassian	NA

Education

Public Education Expenditures [10]

Million Syrian Pounds	1980	1985	1987	1988	1989	1990
Total education expenditure	2,347	5,060	NA	6,727	8,439	10,720
as percent of GNP	4.6	6.1	NA	3.8	4.4	4.5
as percent of total govt. expend.	8.1	11.8	NA	13.1	15.0	17.3
Current education expenditure[6]	1,272	2,799	NA	4,552	5,748	NA
as percent of GNP	2.5	3.4	NA	2.6	3.0	NA
as percent of current govt. expend.	NA	NA	NA	NA	NA	NA
Capital expenditure[6]	457	846	NA	522	939	NA

Educational Attainment [11]

Literacy Rate [12]

In thousands and percent	1985[a]	1991[a]	2000[a]
Illiterate population + 15 years	2,218	2,304	2,435
Illiteracy rate - total pop. (%)	40.9	35.5	25.5
Illiteracy rate - males (%)	25.8	21.7	15.3
Illiteracy rate - females (%)	56.5	49.2	35.9

Libraries [13]

	Admin. Units	Svc. Pts.	Vols. (000)	Shelving (meters)	Vols. Added	Reg. Users
National (1989)	1	NA	127	NA	13,418	NA
Non-specialized	NA	NA	NA	NA	NA	NA
Public	NA	NA	NA	NA	NA	NA
Higher ed. (1987)[24]	1	21	84	NA	NA	NA
School	NA	NA	NA	NA	NA	NA

Daily Newspapers [14]

	1975	1980	1985	1990
Number of papers	6	7	7	10
Circ. (000)	77[e]	114[e]	163[e]	280[e]

Cinema [15]

Data for 1989.

Cinema seats per 1,000	3.4
Annual attendance per person	0.6
Gross box office receipts (mil. Pound)	NA

Science and Technology

Scientific/Technical Forces [16]	R&D Expenditures [17]	U.S. Patents Issued [18]

Government and Law

Organization of Government [19]

Long-form name:
Syrian Arab Republic
Type:
republic under leftwing military regime since March 1963
Independence:
17 April 1946 (from League of Nations mandate under French administration)
Constitution:
13 March 1973
Legal system:
based on Islamic law and civil law system; special religious courts; has not accepted compulsory ICJ jurisdiction
National holiday:
National Day, 17 April (1946)
Executive branch:
president, three vice presidents, prime minister, three deputy prime ministers, Council of Ministers (cabinet)
Legislative branch:
unicameral People's Council (Majlis al-Chaab)
Judicial branch:
Supreme Constitutional Court, High Judicial Council, Court of Cassation, State Security Courts

Political Parties [20]

People's Council	% of votes
Arab Socialist Resurrectionist	53.6
Arab Socialist Union (ASU)	3.2
Communist Party (SCP)	3.2
Arab Socialist Unionist Movement	2.8
Arab Socialist Party (ASP)	2.0
Democratic Socialist Union	1.6
Independents	33.6

Government Expenditures [21]

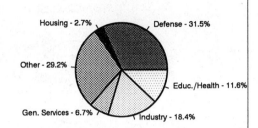

Housing - 2.7%
Defense - 31.5%
Other - 29.2%
Educ./Health - 11.6%
Gen. Services - 6.7%
Industry - 18.4%

% distribution for 1990. Expend. in 1991 est.: 7.5 ($ bil.)

Crime [23]

Crime volume	
Cases known to police	8,892
Attempts (percent)	100
Percent cases solved	90
Crimes per 100,000 persons	73.40
Persons responsible for offenses	
Total number offenders	3,410
Percent female	1,161
Percent juvenile[54]	2,249
Percent foreigners	NA

Military Expenditures and Arms Transfers [22]

	1985	1986	1987	1988	1989
Military expenditures					
Current dollars (mil.)[4]	3,627[e]	2,911[e]	1,801[e]	1,801[e]	2,234
1989 constant dollars (mil.)[4]	4,129[e]	3,230[e]	1,937[e]	1,875[e]	2,234
Armed forces (000)	402	400	400	400	400
Gross national product (GNP)					
Current dollars (mil.)	16,670	16,210	16,680	19,460	19,320
1989 constant dollars (mil.)	18,980	17,980	17,940	20,260	19,320
Central government expenditures (CGE)					
1989 constant dollars (mil.)	9,831[e]	6,735	4,793	4,376	3,200
People (mil.)	10.4	10.7	11.2	11.6	12.0
Military expenditure as % of GNP	21.8	18.0	10.8	9.3	11.6
Military expenditure as % of CGE	42.0	48.0	40.4	42.8	69.8
Military expenditure per capita	399	301	174	162	186
Armed forces per 1,000 people	38.8	37.2	35.9	34.5	33.3
GNP per capita	1,832	1,673	1,608	1,750	1,608
Arms imports[6]					
Current dollars (mil.)	1,600	1,200	2,000	1,300	1,000
1989 constant dollars (mil.)	1,821	1,332	2,151	1,353	1,000
Arms exports[6]					
Current dollars (mil.)	0	20	0	0	0
1989 constant dollars (mil.)	0	22	0	0	0
Total imports[7]					
Current dollars (mil.)	3,967	2,728	7,112	2,231	2,097
1989 constant dollars	4,516	3,027	7,649	2,322	2,097
Total exports[7]					
Current dollars (mil.)	1,640	1,325	3,871	1,345	3,006
1989 constant dollars	1,867	1,470	4,163	1,400	3,006
Arms as percent of total imports[8]	40.3	44.0	28.1	58.3	47.7
Arms as percent of total exports[8]	0	1.5	0	0	0

Human Rights [24]

	SSTS	FL	FAPRO	PPCG	APROBC	TPW	PCPTW	STPEP	PHRFF	PRW	ASST	AFL
Observes	P	P	P	P	P	P	P	P			P	P
	EAFRD	CPR	ESCR	SR	ACHR	MAAE	PVIAC	PVNAC	EAFDAW	TCIDTP	RC	
Observes		P	P	P				P				S

P=Party; S=Signatory; see Appendix for meaning of abbreviations.

Labor Force

Total Labor Force [25]

2.951 million (1989)

Labor Force by Occupation [26]

Miscellaneous and government services
Agriculture
Industry and construction

Unemployment Rate [27]

5.7% (1989)

For sources, notes, and explanations, see Annotated Source Appendix, page 1035.

897

Production Sectors

Commercial Energy Production and Consumption

Production [28]

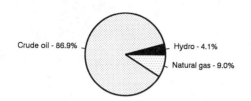

Crude oil - 86.9%
Hydro - 4.1%
Natural gas - 9.0%

Consumption [29]

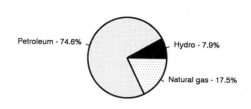

Petroleum - 74.6%
Hydro - 7.9%
Natural gas - 17.5%

Data are shown in quadrillion (10^{15}) BTUs and percent for 1991

Crude oil[1]	1.06
Natural gas liquids	(s)
Dry natural gas	0.11
Coal[2]	0.00
Net hydroelectric power[3]	0.05
Net nuclear power[3]	0.00
Total	1.23

Petroleum	0.47
Dry natural gas	0.11
Coal[2]	0.00
Net hydroelectric power[3]	0.05
Net nuclear power[3]	0.00
Total	0.63

Telecommunications [30]

- Fair system currently undergoing significant improvement and digital upgrades, including fiber optic technology
- 512,600 telephones (37 telephones per 1,000 persons)
- Broadcast stations - 9 AM, 1 FM, 17 TV
- Satellite earth stations - 1 Indian Ocean INTELSAT and 1 Intersputnik
- 1 submarine cable
- Coaxial cable and microwave radio relay to Iraq, Jordan, Lebanon, and Turkey

Transportation [31]

Railroads. 1,998 km total; 1,766 km standard gauge, 232 km 1.050-meter (narrow) gauge

Highways. 29,000 km total; 670 km expressways; 5,000 km main or national roads; 23,330 km secondary or regional roads (not including municipal roads); 22,680 km of the total is paved (1988)

Merchant Marine. 41 ships (1,000 GRT or over) totaling 117,247 GRT/183,607 DWT; includes 36 cargo, 2 vehicle carrier, 3 bulk

Airports

Total:	104
Usable:	100
With permanent-surface runways:	24
With runways over 3,659 m:	0
With runways 2,440-3,659 m:	21
With runways 1,220-2,439 m:	3

Top Agricultural Products [32]

	1990	1991
Barley	500	926
Sugar beets	582	600
Cotton	441	540
Citrus	363	430
Olives	461	250
Apples	205	225

Values shown are 1,000 metric tons.

Top Mining Products [33]

Metric tons unless otherwise specified   M.t.

Cement, hydraulic (000 tons)	3,500
Gypsum	175,000
Phosphate rock (000 tons)	1,359[1]
Salt	127,000
Marble (cubic meters)	18,000
Sand and gravel (000 tons)	8,000

Tourism [34]

	1987	1988	1989	1990	1991
Visitors[4]	1,218	1,274	1,363	1,442	1,570
Tourists[10]	493	421	411	562	622
Excursionists	725	853	621	880	948
Tourism receipts	204	332	290	270	300
Tourism expenditures	131	257	190	200	210

Tourists are in thousands, money in million U.S. dollars.

Manufacturing Sector

GDP and Manufacturing Summary [35]

	1980	1985	1989	1990	% change 1980-1990	% change 1989-1990
GDP (million 1980 $)	10,593	12,231	11,019	13,598	28.4	23.4
GDP per capita (1980 $)	1,204	1,169	912	1,085	-9.9	19.0
Manufacturing as % of GDP (current prices)	3.6	7.7	6.0[e]	NA	NA	-100.0
Gross output (million $)	3,362	5,914	6,349[e]	9,058	169.4	42.7
Value added (million $)	1,256	1,435	1,461[e]	1,833	45.9	25.5
Value added (million 1980 $)	377	529	NA	449	19.1	NA
Industrial production index	100	147	114[e]	95	-5.0	-16.7
Employment (thousands)	195	182[e]	130	125	-35.9	-3.8

Note: GDP stands for Gross Domestic Product. 'e' stands for estimated value.

Profitability and Productivity

	1980	1985	1989	1990	% change 1980-1990	% change 1989-1990
Intermediate input (%)	63	76	77[e]	80	27.0	3.9
Wages, salaries, and supplements (%)	10[e]	8[e]	6[e]	5	-50.0	-16.7
Gross operating surplus (%)	27[e]	16[e]	17[e]	15	-44.4	-11.8
Gross output per worker ($)	17,278	32,511[e]	48,863[e]	72,252	318.2	47.9
Value added per worker ($)	6,452	7,892[e]	11,243[e]	14,617	126.5	30.0
Average wage (incl. benefits) ($)	1,778[e]	2,738[e]	3,043	3,843	116.1	26.3

Profitability is in percent of gross output. Productivity is in U.S. $. 'e' stands for estimated value.

Profitability - 1990

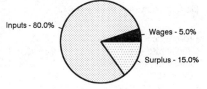

Inputs - 80.0%
Wages - 5.0%
Surplus - 15.0%

The graphic shows percent of gross output.

Value Added in Manufacturing

	1980 $ mil.	1980 %	1985 $ mil.	1985 %	1989 $ mil.	1989 %	1990 $ mil.	1990 %	% change 1980-1990	% change 1989-1990
311 Food products	214	17.0	235	16.4	235[e]	16.1	325	17.7	51.9	38.3
313 Beverages	37	2.9	42	2.9	40[e]	2.7	58	3.2	56.8	45.0
314 Tobacco products	146	11.6	163	11.4	168[e]	11.5	225	12.3	54.1	33.9
321 Textiles	273	21.7	154	10.7	326[e]	22.3	369	20.1	35.2	13.2
322 Wearing apparel	14	1.1	9	0.6	18[e]	1.2	21	1.1	50.0	16.7
323 Leather and fur products	26	2.1	19	1.3	41[e]	2.8	45	2.5	73.1	9.8
324 Footwear	43	3.4	28	2.0	61[e]	4.2	67	3.7	55.8	9.8
331 Wood and wood products	29	2.3	27	1.9	18[e]	1.2	21	1.1	-27.6	16.7
332 Furniture and fixtures	74	5.9	69	4.8	45[e]	3.1	55	3.0	-25.7	22.2
341 Paper and paper products	6	0.5	8	0.6	4[e]	0.3	8	0.4	33.3	100.0
342 Printing and publishing	14	1.1	16	1.1	9[e]	0.6	18	1.0	28.6	100.0
351 Industrial chemicals	3	0.2	7	0.5	6[e]	0.4	7	0.4	133.3	16.7
352 Other chemical products	31	2.5	73	5.1	64[e]	4.4	75	4.1	141.9	17.2
353 Petroleum refineries	100	8.0	112	7.8	104[e]	7.1	115	6.3	15.0	10.6
354 Miscellaneous petroleum and coal products	4	0.3	4	0.3	4[e]	0.3	4	0.2	0.0	0.0
355 Rubber products	15	1.2	16	1.1	13[e]	0.9	16	0.9	6.7	23.1
356 Plastic products	13	1.0	14	1.0	12[e]	0.8	14	0.8	7.7	16.7
361 Pottery, china and earthenware	7	0.6	13	0.9	10[e]	0.7	10	0.5	42.9	0.0
362 Glass and glass products	13	1.0	24	1.7	15[e]	1.0	18	1.0	38.5	20.0
369 Other non-metal mineral products	72	5.7	135	9.4	98[e]	6.7	103	5.6	43.1	5.1
371 Iron and steel	NA	0.0	NA	0.0	NA	0.0	NA	0.0	NA	NA
372 Non-ferrous metals	13	1.0	28	2.0	10[e]	0.7	20	1.1	53.8	100.0
381 Metal products	53	4.2	100	7.0	66[e]	4.5	97	5.3	83.0	47.0
382 Non-electrical machinery	18	1.4	42	2.9	28[e]	1.9	41	2.2	127.8	46.4
383 Electrical machinery	16	1.3	62	4.3	41[e]	2.8	60	3.3	275.0	46.3
384 Transport equipment	3	0.2	11	0.8	8[e]	0.5	11	0.6	266.7	37.5
385 Professional and scientific equipment	NA	0.0	NA	0.0	NA	0.0	NA	0.0	NA	NA
390 Other manufacturing industries	19	1.5	23	1.6	20[e]	1.4	26	1.4	36.8	30.0

Note: The industry codes shown are International Standard Industry codes (ISIC). Percentages are percent of total Value Added. 'e' stands for estimated value

Finance, Economics, and Trade

Economic Indicators [36]

Millions of Syrian pounds (SP) unless otherwise noted[153].

	1989	1990	1991[e]
GDP (1985 prices)	87,609	99,596	110,000
GDP per capita (SP at 1985 prices)	7,476	8,220	NA
Money supply (M1) current prices	116,500	147,500	177,000
CPI Damascus (1985 = 100)	325	360	432
External (est. nonmilitary public debt in billion US$)[154]	3.3	3.3	NA

Balance of Payments Summary [37]

Values in millions of dollars.

	1986	1987	1988	1989	1990
Exports of goods (f.o.b.)	1,037.0	1,357.0	1,348.0	3,013.0	4,221.0
Imports of goods (f.o.b.)	-2,363.0	-2,226.0	-1,986.0	-1,821.0	-2,062.0
Trade balance	-1,326.0	-869.0	-638.0	1,192.0	2,159.0
Services - debits	-836.0	-1,148.0	-1,097.0	-1,475.0	-1,650.0
Services - credits	576.0	625.0	689.0	836.0	864.0
Private transfers (net)	323.0	334.0	360.0	395.0	375.0
Government transfers (net)	759.0	760.0	536.0	223.0	80.0
Long term capital (net)	144.0	207.0	297.0	-472.0	-795.0
Short term capital (net)	447.0	193.0	-212.0	18.0	37.0
Errors and omissions	-26.0	-23.0	34.0	71.0	-348.0
Overall balance	61.0	79.0	-31.0	788.0	722.0

Exchange Rates [38]

Currency: **Syrian pounds.**
Symbol: **#S.**

Data are currency units per $1.

Promotional rate since 1991	22.0000
Official rate since 1991	22.0000
Official parallel rate since 1991	42.0000
Fixed rate 1987-90	11.2250

Imports and Exports

Top Import Origins [39]

$2.7 billion (f.o.b., 1992 est.). Data are for 1990.

Origins	%
EC	42
USSR and Eastern Europe	13
Other Europe	13
US/Canada	11
Arab countries	6

Top Export Destinations [40]

$3.5 billion (f.o.b., 1992 est.). Data are for 1990.

Destinations	%
USSR and Eastern Europe	44
EC	34
Arab countries	17
US/Canada	1

Foreign Aid [41]

	U.S. $	
US commitments, including Ex-Im (FY70-81)	538	million
Western (non-US) ODA and OOF bilateral commitments (1970-89)	1.23	billion
OPEC bilateral aid (1979-89)	12.3	billion
Former Communist countries (1970-89)	3.3	billion

Import and Export Commodities [42]

Import Commodities

Foodstuffs and beverages 21%
Machinery 15%
Metal and metal products 15%
Textiles 7%
Petroleum products

Export Commodities

Petroleum 45%
Farm products 11%
Textiles
Phosphates 5%

Taiwan

Geography [1]

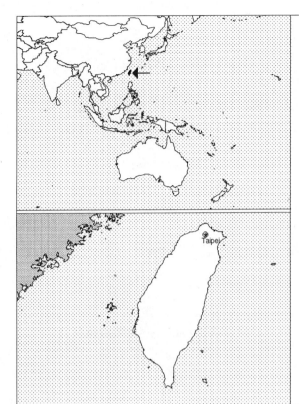

Total area:
  35,980 km2
Land area:
  32,260 km2
Comparative area:
  Slightly larger than Maryland and Delaware combined
Land boundaries:
  0 km
Coastline:
  1,448 km
Climate:
  Tropical; marine; rainy season during southwest monsoon (June to August); cloudiness is persistent and extensive all year
Terrain:
  Eastern two-thirds mostly rugged mountains; flat to gently rolling plains in west
Natural resources:
  Small deposits of coal, natural gas, limestone, marble, and asbestos
Land use:
  Arable land:
  24%
  Permanent crops:
  1%
  Meadows and pastures:
  5%
  Forest and woodland:
  55%
  Other:
  15%

Demographics [2]

	1960	1970	1980	1990	1991[1]	1994	2000	2010	2020
Population	11,209	14,598	17,848	20,436	20,659	21,299	22,448	24,092	25,122
Population density (persons per sq. mi.)	900	1,172	1,433	1,641	1,659	NA	1,802	1,932	2,012
Births	NA	NA	NA	NA	340	332	NA	NA	NA
Deaths	NA	NA	NA	NA	111	120	NA	NA	NA
Life expectancy - males	NA	NA	NA	NA	NA	72	NA	NA	NA
Life expectancy - females	NA	NA	NA	NA	78	79	NA	NA	NA
Birth rate (per 1,000)	NA	NA	NA	NA	16	16	NA	NA	NA
Death rate (per 1,000)	NA	NA	NA	NA	5	6	NA	NA	NA
Women of reproductive age (15-44 yrs.)	NA	NA	NA	5,536	NA	5,592	6,379	NA	NA
of which are currently married	NA	NA	NA	3,764	NA	4,107	4,484	NA	NA
Fertility rate	NA	NA	NA	1.8	1.81	1.8	1.8	1.8	1.8

Population values are in thousands, life expectancy in years, and other items as indicated.

Health

Health Personnel [3]

Health Expenditures [4]

For sources, notes, and explanations, see Annotated Source Appendix, page 1035.

901

Human Factors

Health Care Ratios [5]	Infants and Malnutrition [6]

Ethnic Division [7]

Taiwanese	84%
Mainand Chinese	14%
Aborigine	2%

Religion [8]

Mixture of Buddhist, Confucian, and Taoist 93%, Christian 4.5%, other 2.5%.

Major Languages [9]

Madarin Chinese (official), Taiwanese (Min), Hakka dialects.

Education

Public Education Expenditures [10]	Educational Attainment [11]
Literacy Rate [12]	Libraries [13]
Daily Newspapers [14]	Cinema [15]

Science and Technology

Scientific/Technical Forces [16]	R&D Expenditures [17]	U.S. Patents Issued [18]

U.S. Patents Issued [18]

Values show patents issued to citizens of the country by the U.S. Patents Office.

	1990	1991	1992
Number of patents	861	1,094	1,252

Government and Law

Organization of Government [19]

Long-form name:
none
Type:
multiparty democratic regime; opposition political parties legalized in March, 1989
Constitution:
25 December 1947, presently undergoing revision
Legal system:
based on civil law system; accepts compulsory ICJ jurisdiction, with reservations
National holiday:
National Day, 10 October (1911) (Anniversary of the Revolution)
Executive branch:
president, vice president, premier of the Executive Yuan, vice premier of the Executive Yuan, Executive Yuan
Legislative branch:
unicameral Legislative Yuan and unicameral National Assembly
Judicial branch:
Judicial Yuan

Political Parties [20]

Legislative Yuan	% of votes
Kuomintang (KMT)	60.0
Democratic Progressive (DPP)	31.0
Independents	9.0

Government Budget [21]

For FY91 est.

Revenues	30.3
Expenditures	30.1
Capital expenditures	NA

Crime [23]

Military Expenditures and Arms Transfers [22]

	1985	1986	1987	1988	1989
Military expenditures					
Current dollars (mil.)	6,377[e]	6,790[e]	5,498[e]	6,607[e]	8,060[e]
1989 constant dollars (mil.)	7,260[e]	7,535[e]	5,913[e]	6,877[e]	8,060[e]
Armed forces (000)	440	390	365	390	379
Gross national product (GNP)					
Current dollars (mil.)	90,620	104,600	120,800	134,600	150,200
1989 constant dollars (mil.)	103,200	116,100	129,900	140,100	150,200
Central government expenditures (CGE)					
1989 constant dollars (mil.)	14,520[e]	15,070[e]	14,240[e]	21,940	26,560
People (mil.)	19.4	19.6	19.9	20.1	20.3
Military expenditure as % of GNP	7.0	6.5	4.6	4.9	5.4
Military expenditure as % of CGE	50.0	50.0	41.5	31.3	30.3
Military expenditure per capita	374	384	298	342	397
Armed forces per 1,000 people	22.7	19.9	18.4	19.4	18.6
GNP per capita	5,317	5,911	6,537	6,969	7,390
Arms imports[6]					
Current dollars (mil.)	575	390	1,300	1,200	430
1989 constant dollars (mil.)	655	433	1,398	1,249	430
Arms exports[6]					
Current dollars (mil.)	5	5	10	20	10
1989 constant dollars (mil.)	6	6	11	21	10
Total imports[7]					
Current dollars (mil.)	20,120	24,230	34,800	49,760	52,530
1989 constant dollars	22,910	26,890	37,430	51,800	52,530
Total exports[7]					
Current dollars (mil.)	30,700	39,750	53,820	60,500	66,200
1989 constant dollars	34,940	44,120	57,890	62,980	66,200
Arms as percent of total imports[8]	2.9	1.6	3.7	2.4	0.8
Arms as percent of total exports[8]	0	0	0	0	0

Human Rights [24]

	SSTS	FL	FAPRO	PPCG	APROBC	TPW	PCPTW	STPEP	PHRFF	PRW	ASST	AFL
Observes	P			P						P	P	
	EAFRD	CPR	ESCR	SR	ACHR	MAAE	PVIAC	PVNAC	EAFDAW	TCIDTP	RC	
Observes		P	S	S	P					P	P	

P = Party; S = Signatory; see Appendix for meaning of abbreviations.

Labor Force

Total Labor Force [25]

7.9 million

Labor Force by Occupation [26]

Industry and commerce	53%
Services	22
Agriculture	15.6
Civil administration	7

Date of data: 1989

Unemployment Rate [27]

1.6% (1992 est.)

For sources, notes, and explanations, see Annotated Source Appendix, page 1035.

Production Sectors

Commercial Energy Production and Consumption

Production [28]

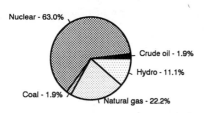

Nuclear - 63.0%
Crude oil - 1.9%
Hydro - 11.1%
Coal - 1.9%
Natural gas - 22.2%

Consumption [29]

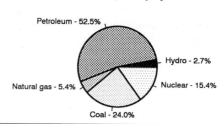

Petroleum - 52.5%
Hydro - 2.7%
Nuclear - 15.4%
Natural gas - 5.4%
Coal - 24.0%

Data are shown in quadrillion (10^{15}) BTUs and percent for 1991

Crude oil[1]	0.01
Natural gas liquids	(s)
Dry natural gas	0.12
Coal[2]	0.01
Net hydroelectric power[3]	0.06
Net nuclear power[3]	0.34
Total	0.55

Petroleum	1.16
Dry natural gas	0.12
Coal[2]	0.53
Net hydroelectric power[3]	0.06
Net nuclear power[3]	0.34
Total	2.22

Telecommunications [30]

- Best developed system in Asia outside of Japan
- 7,800,000 telephones
- Extensive microwave radio relay links on east and west coasts
- Broadcast stations - 91 AM, 23 FM, 15 TV (13 repeaters)
- 8,620,000 radios
- 6,386,000 TVs (5,680,000 color, 706,000 monochrome)
- Satellite earth stations - 1 Pacific Ocean INTELSAT and 1 Indian Ocean INTELSAT
- Submarine cable links to Japan (Okinawa), the Philippines, Guam, Singapore, Hong Kong, Indonesia, Australia, Middle East, and Western Europe

Transportation [31]

Railroads. about 4,600 km total track with 1,075 km common carrier lines and 3,525 km industrial lines; common carrier lines consist of the 1.067-meter gauge 708 km West Line and the 367 km East Line; a 98.25 km South Link Line connection was completed in late 1991; common carrier lines owned by the government and operated by the Railway Administration under Ministry of Communications; industrial lines owned and operated by government enterprises

Highways. 20,041 km total; 17,095 km bituminous or concrete pavement, 2,371 km crushed stone or gravel, 575 km graded earth

Merchant Marine. 223 ships (1,000 GRT or over) totaling 6,761,609 GRT/9,375,677 DWT; includes 1 passenger-cargo, 43 cargo, 11 refrigerated cargo, 85 container, 19 oil tanker, 2 combination ore/oil, 1 specialized tanker, 57 bulk, 1 roll-on/roll-off, 2 combination bulk, 1 chemical tanker

Airports

Total:	40
Usable:	38
With permanent-surface runways:	36
With runways over 3,659 m:	3
With runways 2,440-3,659 m:	16
With runways 1,220-2,439 m:	7

Top Agricultural Products [32]

	1990	1991
Vegetables	2,713	2,847
Fruits	2,357	2,313
Rice	1,807	1,828
Sugar[30]	514	464
Corn	394	383
Tobacco	19	21

Values shown are 1,000 metric tons.

Top Mining Products [33]

Preliminary metric tons unless otherwise specified M.t.

Natural gas, gross (mil. cubic meters)	776[e]
Coal, bituminous (000 tons)	403
Dolomite (000 tons)	363
Marble	11,352
Limestone	15,352
Clay	172,467
Feldspar	1,339
Sodium compounds	228,920
Talc	18,518

Tourism [34]

	1987	1988	1989	1990	1991
Tourists	1,761	1,935	2,004	1,934	1,855
Tourism receipts	1,619	2,289	2,698	1,740	2,018
Tourism expenditures				4,984	5,678

Tourists are in thousands, money in million U.S. dollars.

Manufacturing Sector

GDP and Manufacturing Summary [35]

	1980	1985	1989	1990	% change 1980-1990	% change 1989-1990
GDP (million 1980 $)	41,384	57,275	82,933	86,947	110.1	4.8
GDP per capita (1980 $)	2,324	2,974	4,125	4,277	84.0	3.7
Manufacturing as % of GDP (current prices)	36.2	36.9	35.6	32.9	-9.1	-7.6
Gross output (million $)	55,343	69,206	132,678	143,687	159.6	8.3
Value added (million $)	14,907	23,557	48,995	55,424	271.8	13.1
Value added (million 1980 $)	14,907	21,734	30,544	30,484	104.5	-0.2
Industrial production index	100	138	191	188	88.0	-1.6
Employment (thousands)	1,997	2,459	2,453	2,260	13.2	-7.9

Note: GDP stands for Gross Domestic Product. 'e' stands for estimated value.

Profitability and Productivity

	1980	1985	1989	1990	% change 1980-1990	% change 1989-1990
Intermediate input (%)	73	68	63	61	-16.4	-3.2
Wages, salaries, and supplements (%)	10	14	15	16	60.0	6.7
Gross operating surplus (%)	17	20	22	23	35.3	4.5
Gross output per worker ($)	27,719	28,144	54,097	63,575	129.4	17.5
Value added per worker ($)	7,466	9,580	19,977	24,523	228.5	22.8
Average wage (incl. benefits) ($)	2,678	3,862	8,323	10,168	279.7	22.2

Profitability is in percent of gross output. Productivity is in U.S. $. 'e' stands for estimated value.

Profitability - 1990

Inputs - 61.0%
Wages - 16.0%
Surplus - 23.0%

The graphic shows percent of gross output.

Value Added in Manufacturing

	1980 $ mil.	1980 %	1985 $ mil.	1985 %	1989 $ mil.	1989 %	1990 $ mil.	1990 %	% change 1980-1990	% change 1989-1990
311 Food products	1,464	9.8	2,444	10.4	4,320	8.8	5,239	9.5	257.9	21.3
313 Beverages	204	1.4	312	1.3	717	1.5	668	1.2	227.5	-6.8
314 Tobacco products	170	1.1	223	0.9	237	0.5	408	0.7	140.0	72.2
321 Textiles	1,885	12.6	2,687	11.4	4,198	8.6	4,680	8.4	148.3	11.5
322 Wearing apparel	337	2.3	720	3.1	1,021	2.1	1,139	2.1	238.0	11.6
323 Leather and fur products	176	1.2	431	1.8	797	1.6	889	1.6	405.1	11.5
324 Footwear	46	0.3	119	0.5	212	0.4	236	0.4	413.0	11.3
331 Wood and wood products	316	2.1	394	1.7	695	1.4	547	1.0	73.1	-21.3
332 Furniture and fixtures	119	0.8	146	0.6	258	0.5	325	0.6	173.1	26.0
341 Paper and paper products	424	2.8	647	2.7	1,662	3.4	1,950	3.5	359.9	17.3
342 Printing and publishing	263	1.8	294	1.2	733	1.5	860	1.6	227.0	17.3
351 Industrial chemicals	718	4.8	1,121	4.8	2,557	5.2	2,435	4.4	239.1	-4.8
352 Other chemical products	502	3.4	890	3.8	2,326	4.7	2,245	4.1	347.2	-3.5
353 Petroleum refineries	834	5.6	1,340	5.7	2,142	4.4	2,768	5.0	231.9	29.2
354 Miscellaneous petroleum and coal products	19	0.1	23	0.1	196	0.4	38	0.1	100.0	-80.6
355 Rubber products	198	1.3	349	1.5	691	1.4	776	1.4	291.9	12.3
356 Plastic products	870	5.8	1,535	6.5	3,605	7.4	3,736	6.7	329.4	3.6
361 Pottery, china and earthenware	76	0.5	156	0.7	504	1.0	576	1.0	657.9	14.3
362 Glass and glass products	64	0.4	73	0.3	255	0.5	291	0.5	354.7	14.1
369 Other non-metal mineral products	542	3.6	656	2.8	1,128	2.3	1,290	2.3	138.0	14.4
371 Iron and steel	828	5.6	1,242	5.3	3,074	6.3	3,392	6.1	309.7	10.3
372 Non-ferrous metals	139	0.9	146	0.6	349	0.7	385	0.7	177.0	10.3
381 Metal products	584	3.9	1,115	4.7	2,346	4.8	3,052	5.5	422.6	30.1
382 Non-electrical machinery	524	3.5	827	3.5	1,659	3.4	1,973	3.6	276.5	18.9
383 Electrical machinery	1,890	12.7	2,882	12.2	6,027	12.3	7,247	13.1	283.4	20.2
384 Transport equipment	686	4.6	1,182	5.0	2,411	4.9	2,959	5.3	331.3	22.7
385 Professional and scientific equipment	254	1.7	388	1.6	518	1.1	1,144	2.1	350.4	120.8
390 Other manufacturing industries	774	5.2	1,216	5.2	4,357	8.9	4,176	7.5	439.5	-4.2

Note: The industry codes shown are International Standard Industry codes (ISIC). Percentages are percent of total Value Added. 'e' stands for estimated value

Finance, Economics, and Trade

Economic Indicators [36]

In Billions of New Taiwan Dollars (NTD) Unless noted.

	1989	1990	1991[e]
Real GDP (at 1986 prices)	3,703	3,884	4,166[155]
Real GDP Growth (%)	7.6	4.9	7.3[155]
Money Supply (M1)	2,069	1,932	2,280[155]
Comm'l Interest Rate (%)	10-12	10-12	9.3-11.5[156]
Savings Rate (%)	30.82	29.15	29.92[155]
Investment Rate (%)	21.55	21.90	21.79[155]
CPI (1986 = 100)	106.30	110.69	114.59[155]
WPI (1986 = 100)	94.88	94.30	95.12[155]
External Public Debt	1,404	1,121	800

Balance of Payments Summary [37]

Exchange Rates [38]

Currency: **New Taiwan dollars.**
Symbol: **T$.**

Data are currency units per $1.

1991	25.748
1990	27.108
1989	26.407
1988	28.589
1987	31.845

Imports and Exports

Top Import Origins [39]

$72.1 billion (c.i.f., 1992 est.).

Origins	%
Japan	30.3
US	21.9
EC countries	17.1

Top Export Destinations [40]

$82.4 billion (f.o.b., 1992 est.).

Destinations	%
US	29.1
Hong Kong	18.7
EC countries	17.1

Foreign Aid [41]

	U.S. $	
US, including Ex-Im (FY46-82)	4.6	billion
Western (non-US) countries, ODA and OOF bilateral commitments (1970-89)	500	million

Import and Export Commodities [42]

Import Commodities

Machinery & equipment 15.8%
Chemicals 10.0%
Crude oil 4.2%
Foodstuffs 2.1%

Export Commodities

Electrical machinery 18.5%
Textiles 14.7%
Machinery & equip. 17.7%
Footwear 4.5%
Foodstuffs 1.1%
Plywood & wood prod. 1.1%

Tajikistan

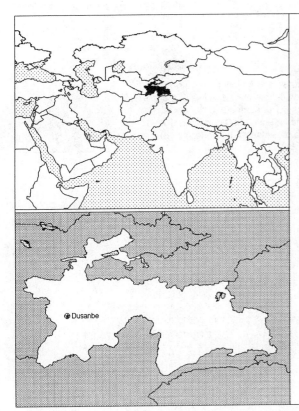

Geography [1]

Total area:
143,100 km2
Land area:
142,700 km2
Comparative area:
Slightly smaller than Wisconsin
Land boundaries:
Total 3,651 km, Afghanistan 1,206 km, China 414 km, Kyrgyzstan 870 km, Uzbekistan 1,161 km
Coastline:
0 km (landlocked)
Climate:
Midlatitude; semiarid to polar in Pamir Mountains
Terrain:
Pamir and Altay Mountains dominate landscape; western Fergana Valley in north, Kafirnigan and Vakhsh Valleys in south or southwest
Natural resources:
Significant hydropower potential, petroleum, uranium, mercury, brown coal, lead, zinc, antimony, tungsten
Land use:
Arable land: 6%
Permanent crops: 0%
Meadows and pastures: 23%
Forest and woodland: 0%
Other: 71%

Demographics [2]

	1960	1970	1980	1990	1991	1994	2000	2010	2020
Population	2,081	2,939	3,969	5,346	NA	5,995	6,956	8,619	10,429
Population density (persons per sq. mi.)	NA	NA	NA	NA	NA	NA	NA	NA	NA
Births	NA	NA	NA	NA	NA	209	NA	NA	NA
Deaths	NA	NA	NA	NA	NA	40	NA	NA	NA
Life expectancy - males	NA	NA	NA	NA	NA	66	NA	NA	NA
Life expectancy - females	NA	NA	NA	NA	NA	72	NA	NA	NA
Birth rate (per 1,000)	NA	NA	NA	NA	NA	35	NA	NA	NA
Death rate (per 1,000)	NA	NA	NA	NA	NA	7	NA	NA	NA
Women of reproductive age (15-44 yrs.)	NA	NA	NA	1,193	NA	1,357	1,654	NA	NA
of which are currently married	NA	NA	NA	810	NA	936	1,136	NA	NA
Fertility rate	NA	NA	NA	5.3	NA	4.6	4.1	3.4	2.8

Population values are in thousands, life expectancy in years, and other items as indicated.

Health

Health Personnel [3]

Doctors per 1,000 pop., 1988-92	2.71
Nurse-to-doctor ratio, 1988-92	2.8
Hospital beds per 1,000 pop., 1985-90	10.6
Percentage of children immunized (age 1 yr. or less)	
Third dose of DPT, 1990-91	89.0
Measles, 1990-91	89.0

Health Expenditures [4]

Total health expenditure, 1990 (official exchange rate)	
Millions of dollars	532
Millions of dollars per capita	100
Health expenditures as a percentage of GDP	
Total	6.0
Public sector	4.4
Private sector	1.6
Development assistance for health	
Total aid flows (millions of dollars)[1]	NA
Aid flows per capita (millions of dollars)	NA
Aid flows as a percentage of total health expenditure	NA

For sources, notes, and explanations, see Annotated Source Appendix, page 1035.

Human Factors

Health Care Ratios [5]

Population per physician, 1970	NA
Population per physician, 1990	350
Population per nursing person, 1970	NA
Population per nursing person, 1990	NA
Percent of births attended by health staff, 1985	NA

Infants and Malnutrition [6]

Percent of babies with low birth weight, 1985	NA
Infant mortality rate per 1,000 live births, 1970	NA
Infant mortality rate per 1,000 live births, 1991	50
Years of life lost per 1,000 population, 1990	24
Prevalence of malnutrition (under age 5), 1990	NA

Ethnic Division [7]

Tajik	64.9%
Uzbek	25.0%
Russian	3.5%
Other	6.6%

Religion [8]

Sunni Muslim	80%
Shi'a Muslim	5%

Major Languages [9]

Tajik (official).

Education

Public Education Expenditures [10]

Educational Attainment [11]

Literacy Rate [12]

	1970[b]	1989[b]	1990[b]
Illiterate population +15 years	NA	NA	NA
Illiteracy rate - total pop. (%)	NA	2.3	NA
Illiteracy rate - males (%)	NA	1.2	NA
Illiteracy rate - females (%)	NA	3.4	NA

Libraries [13]

Daily Newspapers [14]

Cinema [15]

Science and Technology

Scientific/Technical Forces [16]

R&D Expenditures [17]

U.S. Patents Issued [18]

Government and Law

Organization of Government [19]

Long-form name:
Republic of Tajikistan
Type:
republic
Independence:
9 September 1991 (from Soviet Union)
Constitution:
as of mid-1993, a new constitution had
not been formally approved
Legal system:
based on civil law system; no judicial
review of legislative acts
National holiday:
NA
Executive branch:
president, prime minister, cabinet
Legislative branch:
unicameral Assembly (Majlis)
Judicial branch:
NA

Elections [20]

Supreme Soviet. last held 25 February
1990 (next to be held NA); results -
Communist Party 99%, other 1%;
seats—(230 total) Communist Party 227,
other 3. *Note*: In May 1992, the Supreme
Soviet was replaced by the transitional
80-member Assembly (Majlis) and in
November 1992 Emomili Rakhmanov,
chairman of the Assembly, became
Chief of State.

Government Budget [21]

For 1991 est.	
Revenues	NA
Expenditures	NA
Capital expenditures	NA

Defense Summary [22]

Branches: Army (being formed), National Guard, Security Forces (internal and border troops)

Manpower Availability: Males age 15-49 1,313,676; fit for military service 1,079,935; reach
military age (18) annually 56,862 (1993 est.)

Defense Expenditures: No information available.

Crime [23]

Human Rights [24]

Labor Force

Total Labor Force [25]

1.938 million

Labor Force by Occupation [26]

Agriculture and forestry	43%
Industry and construction	22
Other	35

Date of data: 1990

Unemployment Rate [27]

0.4% includes only officially registered
unemployed; also large numbers of
underemployed workers

Production Sectors

Energy Resource Summary [28]

Energy Resources: Significant hydropower potential, petroleum, uranium. **Electricity**: 4,585,000 kW capacity; 16,800 million kWh produced, 2,879 kWh per capita (1992). **Pipelines**: Natural gas 400 km (1992).

Telecommunications [30]

- Poorly developed and not well maintained
- Many towns are not reached by the national network
- Telephone density in urban locations is about 100 per 1000 persons
- Linked by cable and microwave to other CIS republics, and by leased connections to the Moscow international gateway switch
- Satellite earth stations - 1 orbita and 2 INTELSAT (TV receive-only
- The second INTELSAT earth station provides TV receive-only service from Turkey)

Transportation [31]

Railroads. 480 km; does not include industrial lines (1990)

Highways. 29,900 km total (1990); 21,400 km hard surfaced, 8,500 km earth

Airports

Total:	58
Useable:	30
With permanent-surface runways:	12
With runways over 3,659 m:	0
With runways 2,440-3,659 m:	4
With runways 1,220-2,439 m:	13

Top Agricultural Products [32]

	1990	1991
Cotton	842	800
Vegetables	528	517
Grains	300	300
Potatoes	207	200
Grapes	1,189	157

Values shown are 1,000 metric tons.

Top Mining Products [33]

Detailed information is not available. A summary of mineral resources available follows. **Mineral Resources**: Petroleum, uranium, mercury, brown coal, lead, zinc, antimony, tungsten.

Tourism [34]

Finance, Economics, and Trade

Industrial Summary [35]

Industrial Production: Growth rate - 25% (1992 est.). **Industries**: Aluminum, zinc, lead, chemicals and fertilizers, cement, vegetable oil, metal-cutting machine tools, refrigerators and freezers.

Economic Indicators [36]

National product: GDP not available. **National product real growth rate**: - 34% (1992 est.). **National product per capita**: not available. **Inflation rate (consumer prices)**: 35% per month (first quarter 1993). **External debt**: $650 million (end of 1991 est.).

Balance of Payments Summary [37]

Exchange Rates [38]

Currency: **rubles.**
Symbol: **R.**

Subject to wide fluctuations. Data are currency units per $1.

December 24, 1994	415

Imports and Exports

Top Import Origins [39]

$100 million imports from outside the successor states of the former USSR (1992).

Origins	%
No details available.	NA

Top Export Destinations [40]

$100 million to outside successor states of the former USSR (1992).

Destinations	%
Russia	NA
Kazakhstan	NA
Ukraine	NA
Uzbekistan	NA

Foreign Aid [41]

	U.S. $	
Offical and commitments by foreign donors (1992)	700	million

Import and Export Commodities [42]

Import Commodities	Export Commodities
Chemicals	Aluminum
Machinery & transport equipment	Cotton
Textiles	Fruits
Foodstuffs	Vegetable oil
	Textiles

Tanzania

Geography [1]

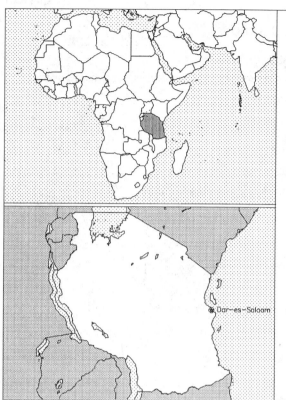

Total area:
945,090 km2
Land area:
886,040 km2
Comparative area:
Slightly larger than twice the size of California
Land boundaries:
Total 3,402 km, Burundi 451 km, Kenya 769 km, Malawi 475 km, Mozambique 756 km, Rwanda 217 km, Uganda 396 km, Zambia 338 km
Coastline:
1,424 km
Climate:
Varies from tropical along coast to temperate in highlands
Terrain:
Plains along coast; central plateau; highlands in north, south
Natural resources:
hydropower potential, tin, phosphates, iron ore, coal, diamonds, gemstones, Gold, natural gas, nickel
Land use:
Arable land:
5%
Permanent crops:
1%
Meadows and pastures:
40%
Forest and woodland:
47%
Other:
7%

Demographics [2]

	1960	1970	1980	1990	1991[1]	1994	2000	2010	2020
Population	10,876	14,038	18,695	25,155	26,869	27,986	32,254	38,651	48,526
Population density (persons per sq. mi.)	32	41	55	76	NA	79	107	148	201
Births	NA	NA	NA	NA	1,331	1,273	NA	NA	NA
Deaths	NA	NA	NA	NA	408	543	NA	NA	NA
Life expectancy - males	NA	NA	NA	NA	50	42	NA	NA	NA
Life expectancy - females	NA	NA	NA	NA	55	45	NA	NA	NA
Birth rate (per 1,000)	NA	NA	NA	NA	50	45	NA	NA	NA
Death rate (per 1,000)	NA	NA	NA	NA	15	19	NA	NA	NA
Women of reproductive age (15-44 yrs.)	NA	NA	NA	5,659	NA	6,354	7,304	NA	NA
of which are currently married	NA	NA	NA	4,097	NA	4,579	5,250	NA	NA
Fertility rate	NA	NA	NA	6.4	7.03	6.2	5.9	5.2	4.4

Population values are in thousands, life expectancy in years, and other items as indicated.

Health

Health Personnel [3]

Doctors per 1,000 pop., 1988-92	0.03
Nurse-to-doctor ratio, 1988-92	7.3
Hospital beds per 1,000 pop., 1985-90	1.1
Percentage of children immunized (age 1 yr. or less)	
Third dose of DPT, 1990-91	79.0
Measles, 1990-91	75.0

Health Expenditures [4]

Total health expenditure, 1990 (official exchange rate)	
Millions of dollars	109
Millions of dollars per capita	4
Health expenditures as a percentage of GDP	
Total	4.7
Public sector	3.2
Private sector	1.5
Development assistance for health	
Total aid flows (millions of dollars)[1]	53
Aid flows per capita (millions of dollars)	2.1
Aid flows as a percentage of total health expenditure	48.3

For sources, notes, and explanations, see Annotated Source Appendix, page 1035.

Human Factors

Health Care Ratios [5]

Population per physician, 1970	22600
Population per physician, 1990	24880
Population per nursing person, 1970	3310
Population per nursing person, 1990	5470
Percent of births attended by health staff, 1985	74

Infants and Malnutrition [6]

Percent of babies with low birth weight, 1985	14
Infant mortality rate per 1,000 live births, 1970	132
Infant mortality rate per 1,000 live births, 1991	115
Years of life lost per 1,000 population, 1990	112
Prevalence of malnutrition (under age 5), 1990	20

Ethnic Division [7]

No detailed information provided.

Religion [8]

Mainland	
Christian	40%
Muslim	33%
Indigenous beliefs	25%
Zanzibar	
Muslim	100%

Major Languages [9]

Swahili (official; widely understood and generally used for communication between ethnic groups and is used in primary education), English (official; primary language of commerce, administration, and higher education).

Education

Public Education Expenditures [10]

Million Shilling	1985	1986	1987	1988	1989	1990
Total education expenditure	4,234	NA	7,675	11,694	13,997	23,426
as percent of GNP	3.5	NA	3.6	3.8	3.7	5.8
as percent of total govt. expend.	14.0	NA	9.9	9.3	14.0	11.4
Current education expenditure	3,643	NA	6,811	10,566	13,069	20,599
as percent of GNP	3.0	NA	3.2	3.4	3.5	5.1
as percent of current govt. expend.	15.6	NA	11.3	11.1	17.0	NA
Capital expenditure	591	NA	864	1,128	928	2,827

Educational Attainment [11]

Literacy Rate [12]

Libraries [13]

Daily Newspapers [14]

	1975	1980	1985	1990
Number of papers	3	3	2	3
Circ. (000)	70	208	101	200[e]

Cinema [15]

Data for 1991.

Cinema seats per 1,000	0.5
Annual attendance per person	0.1[1]
Gross box office receipts (mil. Shilling)	148

Science and Technology

Scientific/Technical Forces [16]

R&D Expenditures [17]

U.S. Patents Issued [18]

Values show patents issued to citizens of the country by the U.S. Patents Office.

	1990	1991	1992
Number of patents	0	0	1

For sources, notes, and explanations, see Annotated Source Appendix, page 1035.

913

Government and Law

Organization of Government [19]

Long-form name:
United Republic of Tanzania

Type:
republic

Independence:
26 April 1964; Tanganyika became independent 9 December 1961 (from UN trusteeship); Zanzibar on 19 December 1963 (from UK); Tanganyika united with Zanzibar 26 April 1964

Constitution:
15 March 1984 (Zanzibar has its own constitution but remains subject to provisions of the union constitution)

Legal system:
based on English common law; limited judicial review; has not accepted compulsory ICJ jurisdiction

National holiday:
Union Day, 26 April (1964)

Executive branch:
president, first vp and prime minister of the union, second vp and president of Zanzibar, cabinet

Legislative branch:
unicameral National Assembly (Bunge)

Judicial branch:
Court of Appeal, High Court

Elections [20]

National Assembly. Last held 28 October 1990 (next to be held NA October 1995); results - CCM (Chama Chr Mapinduzi) was the only party; seats - (241 total, 168 elected) CCM 168.

Government Budget [21]

For FY90.

Revenues	495
Expenditures	631
Capital expenditures	118

Military Expenditures and Arms Transfers [22]

	1985	1986	1987	1988	1989
Military expenditures					
Current dollars (mil.)	64	NA	78	93	110
1989 constant dollars (mil.)	73	NA	84	97	110
Armed forces (000)	43	40	40[e]	40	40
Gross national product (GNP)					
Current dollars (mil.)	2,078	2,217	2,281	2,461	2,642
1989 constant dollars (mil.)	2,366	2,460	2,453	2,562	2,642
Central government expenditures (CGE)					
1989 constant dollars (mil.)	567	499	571	664	1,095
People (mil.)	21.9	22.6	23.4	24.3	25.1
Military expenditure as % of GNP	3.1	NA	3.4	3.8	4.1
Military expenditure as % of CGE	12.8	NA	14.8	14.6	10.0
Military expenditure per capita	3	NA	4	4	4
Armed forces per 1,000 people	2.0	1.8	1.7	1.6	1.6
GNP per capita	108	109	105	106	105
Arms imports[6]					
Current dollars (mil.)	50	30	110	70	40
1989 constant dollars (mil.)	57	33	118	73	40
Arms exports[6]					
Current dollars (mil.)	0	0	0	0	0
1989 constant dollars (mil.)	0	0	0	0	0
Total imports[7]					
Current dollars (mil.)	1,017	868	923	800	1,300
1989 constant dollars	1,158	963	993	833	1,300
Total exports[7]					
Current dollars (mil.)	255	343	288	276	NA
1989 constant dollars	290	381	310	287	NA
Arms as percent of total imports[8]	4.9	3.5	11.9	8.8	3.1
Arms as percent of total exports[8]	0	0	0	0	NA

Crime [23]

Crime volume	
Cases known to police	289,972
Attempts (percent)	NA
Percent cases solved	32.16
Crimes per 100,000 persons	1,249.88
Persons responsible for offenses	
Total number offenders	27,442
Percent female	NA
Percent juvenile	NA
Percent foreigners	NA

Human Rights [24]

	SSTS		FL	FAPRO	PPCG	APROBC	TPW	PCPTW	STPEP	PHRFF	PRW	ASST	AFL
Observes	P	P		P	P	P	P			P	P		
	EAFRD		CPR		ESCR		SR	ACHR	MAAE	PVIAC	PVNAC	EAFDAW	TCIDTP	RC
Observes			S	P	S					P	P	P		P

P=Party; S=Signatory; see Appendix for meaning of abbreviations.

Labor Force

Total Labor Force [25]

732,200 wage earners

Labor Force by Occupation [26]

Agriculture	90%
Industry and commerce	10

Date of data: 1986 est.

Unemployment Rate [27]

For sources, notes, and explanations, see Annotated Source Appendix, page 1035.

Production Sectors

Energy Resource Summary [28]

Energy Resources: Hydropower potential, natural gas. **Electricity**: 405,000 kW capacity; 600 million kWh produced, 20 kWh per capita (1991).
Pipelines: Crude oil 982 km.

Telecommunications [30]

- Fair system operating below capacity
- Open wire, radio relay, and troposcatter
- 103,800 telephones
- Broadcast stations - 12 AM, 4 FM, 2 TV
- 1 Indian Ocean and 1 Atlantic Ocean INTELSAT earth station

Top Agricultural Products [32]

Agriculture accounts for over 58% of GDP; topography and climatic conditions limit cultivated crops to only 5% of land area; cash crops—coffee, sisal, tea, cotton, pyrethrum (insecticide made from chrysanthemums), cashews, tobacco, cloves (Zanzibar); food crops—corn, wheat, cassava, bananas, fruits, vegetables; small numbers of cattle, sheep, and goats; not self-sufficient in food grain production.

Top Mining Products [33]

Metric tons unless otherwise specified	M.t.
Coal, bituminous	33,213
Diamonds (carats)	99,763[53]
Gemstones, precious and semiprecious (kilograms)	59,630[54]
Apatite	22,419
Salt, all types	64,419
Tin, mine output, Sn content	6

Transportation [31]

Railroads. 3,555 km total; 960 km 1.067-meter gauge (including the 962 km Tazara Railroad); 2,595 km 1.000-meter gauge, including 6.4 km double track; 115 km of 1.000-meter gauge planned by end of decade

Highways. 81,900 km total, 3,600 km paved; 5,600 km gravel or crushed stone; 72,700 km improved and unimproved earth

Merchant Marine. 6 ships (1,000 GRT or over) totaling 19,185 GRT/22,916 DWT; includes 2 passenger-cargo, 2 cargo, 1 roll-on/roll-off cargo, 1 oil tanker

Airports

Total:	103
Usable:	92
With permanent-surface runways:	12
With runways over 3,659 m:	0
With runways 2,440-3,659 m:	4
With runways 1,220-2,439 m:	40

Tourism [34]

	1987	1988	1989	1990	1991
Visitors	131	130	138	153	187
Tourism receipts	31	40	60	65	95
Tourism expenditures	23	23	25	19	

Tourists are in thousands, money in million U.S. dollars.

For sources, notes, and explanations, see Annotated Source Appendix, page 1035.

915

Manufacturing Sector

GDP and Manufacturing Summary [35]

	1980	1985	1989	1990	% change 1980-1990	% change 1989-1990
GDP (million 1980 $)	5,138	5,327	6,210	6,450	25.5	3.9
GDP per capita (1980 $)	272	234	236	236	-13.2	0.0
Manufacturing as % of GDP (current prices)	10.7	6.1	3.5[e]	3.9	-63.6	11.4
Gross output (million $)	1,266	1,145	464[e]	396[e]	-68.7	-14.7
Value added (million $)	361	278	111[e]	87[e]	-75.9	-21.6
Value added (million 1980 $)	500	387	428[e]	482	-3.6	12.6
Industrial production index	100	81	101[e]	104	4.0	3.0
Employment (thousands)	101	94	110[e]	123[e]	21.8	11.8

Note: GDP stands for Gross Domestic Product. 'e' stands for estimated value.

Profitability and Productivity

	1980	1985	1989	1990	% change 1980-1990	% change 1989-1990
Intermediate input (%)	71	76	76[e]	78[e]	9.9	2.6
Wages, salaries, and supplements (%)	9	9	9[e]	6[e]	-33.3	-33.3
Gross operating surplus (%)	19	16	15[e]	16[e]	-15.8	6.7
Gross output per worker ($)	12,457	12,141	4,206[e]	3,209[e]	-74.2	-23.7
Value added per worker ($)	3,555	2,952	1,008[e]	707[e]	-80.1	-29.9
Average wage (incl. benefits) ($)	1,174	1,042	358[e]	203[e]	-82.7	-43.3

Profitability is in percent of gross output. Productivity is in U.S. $. 'e' stands for estimated value.

Profitability - 1990

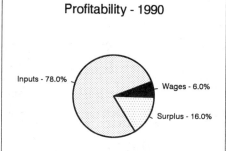

Inputs - 78.0%
Wages - 6.0%
Surplus - 16.0%

The graphic shows percent of gross output.

Value Added in Manufacturing

	1980 $ mil.	1980 %	1985 $ mil.	1985 %	1989 $ mil.	1989 %	1990 $ mil.	1990 %	% change 1980-1990	% change 1989-1990
311 Food products	58	16.1	58	20.9	23[e]	20.7	11[e]	12.6	-81.0	-52.2
313 Beverages	14	3.9	21	7.6	9[e]	8.1	5[e]	5.7	-64.3	-44.4
314 Tobacco products	12	3.3	16	5.8	7[e]	6.3	9[e]	10.3	-25.0	28.6
321 Textiles	95	26.3	43	15.5	17[e]	15.3	15[e]	17.2	-84.2	-11.8
322 Wearing apparel	10	2.8	4	1.4	1[e]	0.9	1[e]	1.1	-90.0	0.0
323 Leather and fur products	7	1.9	4	1.4	2[e]	1.8	1[e]	1.1	-85.7	-50.0
324 Footwear	8	2.2	6	2.2	3[e]	2.7	1[e]	1.1	-87.5	-66.7
331 Wood and wood products	7	1.9	6	2.2	2[e]	1.8	2[e]	2.3	-71.4	0.0
332 Furniture and fixtures	6	1.7	3	1.1	1[e]	0.9	1[e]	1.1	-83.3	0.0
341 Paper and paper products	8	2.2	7	2.5	4[e]	3.6	3[e]	3.4	-62.5	-25.0
342 Printing and publishing	14	3.9	12	4.3	5[e]	4.5	2[e]	2.3	-85.7	-60.0
351 Industrial chemicals	11	3.0	9	3.2	4[e]	3.6	12[e]	13.8	9.1	200.0
352 Other chemical products	10	2.8	7	2.5	2[e]	1.8	2[e]	2.3	-80.0	0.0
353 Petroleum refineries	15	4.2	10	3.6	3[e]	2.7	3[e]	3.4	-80.0	0.0
354 Miscellaneous petroleum and coal products	NA	0.0	NA	0.0	NA	0.0	NA	0.0	NA	NA
355 Rubber products	11	3.0	11	4.0	5[e]	4.5	1[e]	1.1	-90.9	-80.0
356 Plastic products	8	2.2	2	0.7	NA	0.0	1[e]	1.1	-87.5	NA
361 Pottery, china and earthenware	NA	0.0	NA	0.0	NA	0.0	NA	0.0	NA	NA
362 Glass and glass products	NA	0.0	NA	0.0	NA	0.0	NA	0.0	NA	NA
369 Other non-metal mineral products	11	3.0	4	1.4	1[e]	0.9	4[e]	4.6	-63.6	300.0
371 Iron and steel	2[e]	0.6	6[e]	2.2	3[e]	2.7	2[e]	2.3	0.0	-33.3
372 Non-ferrous metals	4[e]	1.1	4[e]	1.4	2[e]	1.8	1[e]	1.1	-75.0	-50.0
381 Metal products	20	5.5	15	5.4	5[e]	4.5	4[e]	4.6	-80.0	-20.0
382 Non-electrical machinery	3	0.8	4	1.4	1[e]	0.9	1[e]	1.1	-66.7	0.0
383 Electrical machinery	6	1.7	6	2.2	3[e]	2.7	1[e]	1.1	-83.3	-66.7
384 Transport equipment	19	5.3	19	6.8	8[e]	7.2	5[e]	5.7	-73.7	-37.5
385 Professional and scientific equipment	NA	0.0	NA	0.0	NA	0.0	NA	0.0	NA	NA
390 Other manufacturing industries	2	0.6	2	0.7	NA	0.0	NA	0.0	NA	NA

Note: The industry codes shown are International Standard Industry codes (ISIC). Percentages are percent of total Value Added. 'e' stands for estimated value

Finance, Economics, and Trade

Economic Indicators [36]

National product: GDP—exchange rate conversion—$7.2 billion (1992 est.). **National product real growth rate**: 4.5% (1992 est.). **National product per capita**: $260 (1992 est.). **Inflation rate (consumer prices)**: 22% (1992 est.). **External debt**: $6.44 billion (1992).

Balance of Payments Summary [37]

Values in millions of dollars.

	1986	1987	1988	1989	1990
Exports of goods (f.o.b.)	335.9	346.8	386.5	415.1	407.8
Imports of goods (f.o.b.)	-913.3	-1,000.5	-1,033.0	-1,070.1	-1,186.3
Trade balance	-577.4	-653.7	-646.5	-655.0	-778.5
Services - debits	-327.9	-420.2	-471.1	-479.2	-481.1
Services - credits	110.2	101.6	120.6	123.2	140.1
Private transfers (net)	250.6	230.0	231.9	182.4	164.5
Government transfers (net)	223.5	477.0	389.3	469.8	529.0
Long term capital (net)	1,300.0	151.5	261.0	411.8	424.2
Short term capital (net)	-897.6	129.0	169.0	47.8	9.5
Errors and omissions	-43.5	-74.2	-42.5	-114.9	133.1
Overall balance	37.9	-59.0	11.7	-14.1	140.8

Exchange Rates [38]

Currency: **Tanzanian shillings.**
Symbol: **TSh.**

Data are currency units per $1.

November 1992	325.00
1991	219.16
1990	195.06
1989	143.38
1988	99.29
1987	64.26

Imports and Exports

Top Import Origins [39]

$1.43 billion (c.i.f., 1991).

Origins	%
Germany	NA
UK	NA
US	NA
Japan	NA
Italy	NA
Denmark	NA

Top Export Destinations [40]

$422 million (f.o.b., 1991).

Destinations	%
FRG	NA
UK	NA
Japan	NA
Netherlands	NA
Kenya	NA
Hong Kong	NA
US	NA

Foreign Aid [41]

	U.S. $	
US commitments, including Ex-Im (FY70-89)	400	million
Western (non-US) countries, ODA and OOF bilateral commitments (1970-89)	9.8	billion
OPEC bilateral aid (1979-89)	44	million
Communist countries (1970-89)	614	million

Import and Export Commodities [42]

Import Commodities

Manufactured goods
Machinery & transport equipment
Cotton piece goods
Crude oil
Foodstuffs

Export Commodities

Coffee
Cotton
Tobacco
Tea
Cashew nuts
Sisal

Thailand

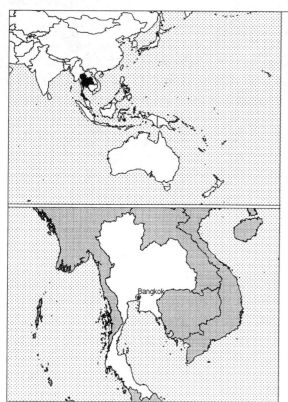

Geography [1]

Total area:
514,000 km2
Land area:
511,770 km2
Comparative area:
Slightly more than twice the size of Wyoming
Land boundaries:
Total 4,863 km, Burma 1,800 km, Cambodia 803 km, Laos 1,754 km, Malaysia 506 km
Coastline:
3,219 km
Climate:
Tropical; rainy, warm, cloudy southwest monsoon (mid-May to September); dry, cool northeast monsoon (November to mid-March); southern isthmus always hot and humid
Terrain:
Central plain; eastern plateau (Khorat); mountains elsewhere
Natural resources:
Tin, rubber, natural gas, tungsten, tantalum, timber, lead, fish, gypsum, lignite, fluorite
Land use:
Arable land: 34%
Permanent crops: 4%
Meadows and pastures: 1%
Forest and woodland: 30%
Other: 31%

Demographics [2]

	1960	1970	1980	1990	1991[1]	1994	2000	2010	2020
Population	27,513	37,091	47,026	56,220	56,814	59,510	63,620	64,181	62,941
Population density (persons per sq. mi.)	139	188	239	283	288	NA	323	358	385
Births	NA	NA	NA	NA	1,152	1,156	NA	NA	NA
Deaths	NA	NA	NA	NA	340	381	NA	NA	NA
Life expectancy - males	NA	NA	NA	NA	66	65	NA	NA	NA
Life expectancy - females	NA	NA	NA	NA	71	72	NA	NA	NA
Birth rate (per 1,000)	NA	NA	NA	NA	20	19	NA	NA	NA
Death rate (per 1,000)	NA	NA	NA	NA	6	6	NA	NA	NA
Women of reproductive age (15-44 yrs.)	NA	NA	NA	15,169	NA	16,474	18,086	NA	NA
of which are currently married	NA	NA	NA	9,438	NA	10,570	11,915	NA	NA
Fertility rate	NA	NA	NA	2.4	2.21	2.1	1.9	1.8	1.8

Population values are in thousands, life expectancy in years, and other items as indicated.

Health

Health Personnel [3]

Doctors per 1,000 pop., 1988-92	0.20
Nurse-to-doctor ratio, 1988-92	5.5
Hospital beds per 1,000 pop., 1985-90	1.6
Percentage of children immunized (age 1 yr. or less)	
Third dose of DPT, 1990-91	69.0
Measles, 1990-91	60.0

Health Expenditures [4]

Total health expenditure, 1990 (official exchange rate)	
Millions of dollars	4061
Millions of dollars per capita	73
Health expenditures as a percentage of GDP	
Total	5.0
Public sector	1.1
Private sector	3.9
Development assistance for health	
Total aid flows (millions of dollars)[1]	36
Aid flows per capita (millions of dollars)	0.7
Aid flows as a percentage of total health expenditure	0.9

For sources, notes, and explanations, see Annotated Source Appendix, page 1035.

Human Factors

Health Care Ratios [5]

Population per physician, 1970	8290
Population per physician, 1990	5000
Population per nursing person, 1970	1170
Population per nursing person, 1990	550
Percent of births attended by health staff, 1985	33

Infants and Malnutrition [6]

Percent of babies with low birth weight, 1985	12
Infant mortality rate per 1,000 live births, 1970	73
Infant mortality rate per 1,000 live births, 1991	27
Years of life lost per 1,000 population, 1990	22
Prevalence of malnutrition (under age 5), 1990	26

Ethnic Division [7]

Thai	75%
Chinese	14%
Other	11%

Religion [8]

Buddhism	95.0%
Muslim	3.8%
Christianity	0.5%
Hinduism	0.1%
Other	0.6%

Major Languages [9]

Thai, English the secondary language of the elite, ethnic and regional dialects.

Education

Public Education Expenditures [10]

Million Baht	1980	1985	1986	1987	1988	1990
Total education expenditure	22,489	39,367	41,214	43,668	47,283	77,420
as percent of GNP	3.4	3.9	3.8	3.5	3.2	3.8
as percent of total govt. expend.	20.6	18.5	19.4	17.9	16.6	20.0
Current education expenditure	15,867	33,830	35,858	38,019	40,916	64,702
as percent of GNP	2.4	3.4	3.3	3.1	2.8	3.1
as percent of current govt. expend.	19.1	19.2	20.4	17.9	16.7	21.0
Capital expenditure	6,622	5,537	5,356	5,649	6,367	12,718

Educational Attainment [11]

Age group	25+
Total population	17,491,470
Highest level attained (%)	
No schooling	20.5
First level	
Incompleted	67.3
Completed	2.4
Entered second level	
S-1	4.5
S-2	2.3
Post secondary	2.9

Literacy Rate [12]

In thousands and percent	1985[a]	1991[a]	2000[a]
Illiterate population +15 years	3,049	2,627	1,871
Illiteracy rate - total pop. (%)	9.3	7.0	4.0
Illiteracy rate - males (%)	5.3	3.9	2.2
Illiteracy rate - females (%)	13.3	10.1	5.8

Libraries [13]

	Admin. Units	Svc. Pts.	Vols. (000)	Shelving (meters)	Vols. Added	Reg. Users
National (1989)	1	24	1,390	5,900	158,845	1.5mil
Non-specialized	NA	NA	NA	NA	NA	NA
Public	NA	NA	NA	NA	NA	NA
Higher ed.	NA	NA	NA	NA	NA	NA
School	NA	NA	NA	NA	NA	NA

Daily Newspapers [14]

	1975	1980	1985	1990
Number of papers	22	27	32	34
Circ. (000)	2,540	2,680	4,350	4,000[e]

Cinema [15]

Data for 1989.

Cinema seats per 1,000	0.1
Annual attendance per person	NA
Gross box office receipts (mil. Baht)	NA

Science and Technology

Scientific/Technical Forces [16]

R&D Expenditures [17]

	Baht (000) 1987
Total expenditure	2,664,380
Capital expenditure	543,680
Current expenditure	2,120,700
Percent current	79.6

U.S. Patents Issued [18]

Values show patents issued to citizens of the country by the U.S. Patents Office.

	1990	1991	1992
Number of patents	3	2	2

For sources, notes, and explanations, see Annotated Source Appendix, page 1035.

919

Government and Law

Organization of Government [19]

Long-form name:
Kingdom of Thailand

Type:
constitutional monarchy

Independence:
1238 (traditional founding date; never colonized)

Constitution:
22 December 1978; new constitution approved 7 December 1991; amended 10 June 1992

Legal system:
based on civil law system, with influences of common law; has not accepted compulsory ICJ jurisdiction; martial law since 23 February 1991 coup

National holiday:
Birthday of His Majesty the King, 5 December (1927)

Executive branch:
monarch, prime minister, four deputy prime ministers, Council of Ministers (cabinet), Privy Council

Legislative branch:
bicameral National Assembly: Senate, House of Representatives

Judicial branch:
Supreme Court (Sarndika)

Political Parties [20]

House of Representatives	% of seats
Democrat Party (DP)	21.9
Thai Nation Pary (TNP)	21.4
National Development Party (NDP)	16.7
NAP	14.2
Phalang Tham	13.1
Social Action Party (SAP)	6.1
Liberal Democratic Party (LDP)	2.2
Solidarity Party (SP)	2.2
Mass Party (Mass Party)	1.1
Other	1.1

Government Expenditures [21]

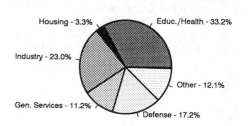

Housing - 3.3%
Educ./Health - 33.2%
Industry - 23.0%
Other - 12.1%
Gen. Services - 11.2%
Defense - 17.2%

% distribution for 1992. Expend. in FY93 est.: 22.40 ($ bil.)

Crime [23]

Crime volume	
Cases known to police	809,811
Attempts (percent)	NA
Percent cases solved	95.9
Crimes per 100,000 persons	1,449
Persons responsible for offenses	
Total number offenders	NA
Percent female	NA
Percent juvenile	NA
Percent foreigners	NA

Military Expenditures and Arms Transfers [22]

	1985	1986	1987	1988	1989
Military expenditures					
Current dollars (mil.)	1,814	1,730	1,730	1,741	1,843
1989 constant dollars (mil.)	2,065	1,920	1,860	1,813	1,843
Armed forces (000)	270	275[e]	275[e]	273	273
Gross national product (GNP)					
Current dollars (mil.)	41,280	44,100	49,400	58,670	68,770
1989 constant dollars (mil.)	46,990	48,940	53,130	61,080	68,770
Central government expenditures (CGE)					
1989 constant dollars (mil.)	10,470	10,300	10,160	9,973	10,410
People (mil.)	51.7	52.6	53.5	54.3	55.2
Military expenditure as % of GNP	4.4	3.9	3.5	3.0	2.7
Military expenditure as % of CGE	19.7	18.6	18.3	18.2	17.7
Military expenditure per capita	40	36	35	33	33
Armed forces per 1,000 people	5.2	5.2	5.1	5.0	4.9
GNP per capita	908	930	993	1,124	1,246
Arms imports[6]					
Current dollars (mil.)	150	140	390	525	240
1989 constant dollars (mil.)	171	155	419	547	240
Arms exports[6]					
Current dollars (mil.)	0	0	0	0	0
1989 constant dollars (mil.)	0	0	0	0	0
Total imports[7]					
Current dollars (mil.)	9,242	9,178	12,920	20,280	25,090
1989 constant dollars	10,520	10,180	13,900	21,120	25,090
Total exports[7]					
Current dollars (mil.)	7,120	8,794	11,650	15,830	20,060
1989 constant dollars	8,105	9,759	12,530	16,480	20,060
Arms as percent of total imports[8]	1.6	1.5	3.0	2.6	1.0
Arms as percent of total exports[8]	0	0	0	0	0

Human Rights [24]

	SSTS	FL	FAPRO	PPCG	APROBC	TPW	PCPTW	STPEP	PHRFF	PRW	ASST	AFL
Observes	P					P	P			P		P
	EAFRD	CPR	ESCR	SR	ACHR	MAAE	PVIAC	PVNAC	EAFDAW	TCIDTP		RC
Observes									P			P

P = Party; S = Signatory; see Appendix for meaning of abbreviations.

Labor Force

Total Labor Force [25]

30.87 million

Labor Force by Occupation [26]

Agriculture	62%
Industry	13
Commerce	11
Services (including government)	14

Date of data: 1989 est.

Unemployment Rate [27]

4.7% (1992 est.)

Production Sectors

Commercial Energy Production and Consumption

Production [28]

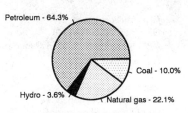

Natural gas - 48.4%
NG liquids - 6.3%
Crude oil - 15.6%
Coal - 21.9%
Hydro - 7.8%

Consumption [29]

Petroleum - 64.3%
Coal - 10.0%
Natural gas - 22.1%
Hydro - 3.6%

Data are shown in quadrillion (10^{15}) BTUs and percent for 1991

Crude oil[1]	0.10		Petroleum	0.90
Natural gas liquids	0.04		Dry natural gas	0.31
Dry natural gas	0.31		Coal[2]	0.14
Coal[2]	0.14		Net hydroelectric power[3]	0.05
Net hydroelectric power[3]	0.05		Net nuclear power[3]	(s)
Net nuclear power[3]	0.00		Total	1.40
Total	0.63			

Telecommunications [30]

- Service to general public inadequate
- Bulk of service to government activities provided by multichannel cable and microwave radio relay network
- 739,500 telephones (1987)
- Broadcast stations - over 200 AM, 100 FM, and 11 TV in government-controlled networks
- Satellite earth stations - 1 Indian Ocean INTELSAT and 1 Pacific Ocean INTELSAT
- Domestic satellite system being developed

Transportation [31]

Railroads. 3,940 km 1.000-meter gauge, 99 km double track

Highways. 77,697 km total; 35,855 km paved (including 88 km expressways), 14,092 km gravel or other stabilization, 27,750 km mostly dirt and other (1988)

Merchant Marine. 169 ships (1,000 GRT or over) totaling 752,055 GRT/1,166,136 DWT; includes 1 short-sea passenger, 91 cargo, 12 container, 40 oil tanker, 9 liquefied gas, 2 chemical tanker, 5 bulk, 6 refrigerated cargo, 2 combination bulk, 1 passenger

Airports

Total:	106
Usable:	95
With permanent-surface runways:	51
With runways over 3,659 m:	1
With runways 2,440-3,659 m:	14
With runways 1,220-2,439 m:	28

Top Agricultural Products [32]

	1990[31]	1991[31]
Rice	17,500	20,000
Cassava	8,260	8,208
Corn	3,800	3,700
Sugar	3,502	3,700
Pineapples	1,510	1,500
Soybeans	530	460

Values shown are 1,000 metric tons.

Top Mining Products [33]

Preliminary metric tons unless otherwise specified M.t.

Kaolin, marketable	915,578
Lignite (000 tons)	14,689
Limestone for cement mfg. only (000 tons)	19,517
Tin, mine output, Sn content	14,937
Zinc, mine ouput, Zn content	87,000

Tourism [34]

	1987	1988	1989	1990	1991
Tourists[4]	3,483	4,231	4,810	5,299	5,087
Cruise passengers	19	12	8	12	
Tourism receipts	1,947	3,120	3,753	4,326	3,923
Tourism expenditures	381	602	750	854	1,266
Fare receipts	227	401	416	663	655
Fare expenditures	115	161	196	252	227

Tourists are in thousands, money in million U.S. dollars.

Manufacturing Sector

GDP and Manufacturing Summary [35]

	1980	1985	1989	1990	% change 1980-1990	% change 1989-1990
GDP (million 1980 $)	32,160	42,323	60,195	67,828	110.9	12.7
GDP per capita (1980 $)	688	820	1,096	1,216	76.7	10.9
Manufacturing as % of GDP (current prices)	21.3	22.1	23.2[e]	26.1	22.5	12.5
Gross output (million $)	26,353[e]	29,388	50,688[e]	62,726	138.0	23.7
Value added (million $)	9,028[e]	10,078	16,677[e]	22,670	151.1	35.9
Value added (million 1980 $)	6,834	8,567	13,828	16,411	140.1	18.7
Industrial production index	100	117	184	211	111.0	14.7
Employment (thousands)	1,591[e]	1,860[e]	1,413[e]	2,520[e]	58.4	78.3

Note: GDP stands for Gross Domestic Product. 'e' stands for estimated value.

Profitability and Productivity

	1980	1985	1989	1990	% change 1980-1990	% change 1989-1990
Intermediate input (%)	66[e]	66	67[e]	64	-3.0	-4.5
Wages, salaries, and supplements (%)	7[e]	9	8[e]	9[e]	28.6	12.5
Gross operating surplus (%)	27[e]	25	25[e]	27[e]	0.0	8.0
Gross output per worker ($)	16,565[e]	15,799	35,880[e]	24,879[e]	50.2	-30.7
Value added per worker ($)	5,675[e]	5,418	11,805[e]	9,990[e]	76.0	-15.4
Average wage (incl. benefits) ($)	1,235[e]	1,411[e]	2,709[e]	2,286[e]	85.1	-15.6

Profitability is in percent of gross output. Productivity is in U.S. $. 'e' stands for estimated value.

Profitability - 1990

Inputs - 64.0%
Wages - 9.0%
Surplus - 27.0%

The graphic shows percent of gross output.

Value Added in Manufacturing

	1980 $ mil.	1980 %	1985 $ mil.	1985 %	1989 $ mil.	1989 %	1990 $ mil.	1990 %	% change 1980-1990	% change 1989-1990
311 Food products	2,039[e]	22.6	2,274	22.6	2,677[e]	16.1	3,033	13.4	48.7	13.3
313 Beverages	682[e]	7.6	786	7.8	1,192[e]	7.1	1,671	7.4	145.0	40.2
314 Tobacco products	375[e]	4.2	470	4.7	569[e]	3.4	728	3.2	94.1	27.9
321 Textiles	1,118[e]	12.4	1,044	10.4	1,770[e]	10.6	2,642	11.7	136.3	49.3
322 Wearing apparel	591[e]	6.5	1,025	10.2	1,163[e]	7.0	2,882	12.7	387.6	147.8
323 Leather and fur products	38[e]	0.4	85	0.8	38[e]	0.2	889[e]	3.9	2,239.5	2,239.5
324 Footwear	47[e]	0.5	54	0.5	105[e]	0.6	276[e]	1.2	487.2	162.9
331 Wood and wood products	244[e]	2.7	180	1.8	309[e]	1.9	98	0.4	-59.8	-68.3
332 Furniture and fixtures	132[e]	1.5	173	1.7	184[e]	1.1	339	1.5	156.8	84.2
341 Paper and paper products	213[e]	2.4	120	1.2	497[e]	3.0	207[e]	0.9	-2.8	-58.4
342 Printing and publishing	110[e]	1.2	161	1.6	122[e]	0.7	310[e]	1.4	181.8	154.1
351 Industrial chemicals	94[e]	1.0	63	0.6	426[e]	2.6	122	0.5	29.8	-71.4
352 Other chemical products	245[e]	2.7	238	2.4	948[e]	5.7	567	2.5	131.4	-40.2
353 Petroleum refineries	537[e]	5.9	683	6.8	1,134[e]	6.8	1,791	7.9	233.5	57.9
354 Miscellaneous petroleum and coal products	27[e]	0.3	21	0.2	70[e]	0.4	23	0.1	-14.8	-67.1
355 Rubber products	221[e]	2.4	147	1.5	548[e]	3.3	352	1.6	59.3	-35.8
356 Plastic products	102[e]	1.1	103	1.0	224[e]	1.3	327	1.4	220.6	46.0
361 Pottery, china and earthenware	35[e]	0.4	48	0.5	66[e]	0.4	78	0.3	122.9	18.2
362 Glass and glass products	64[e]	0.7	54	0.5	193[e]	1.2	141	0.6	120.3	-26.9
369 Other non-metal mineral products	267[e]	3.0	424	4.2	422[e]	2.5	678	3.0	153.9	60.7
371 Iron and steel	316[e]	3.5	236	2.3	266[e]	1.6	412[e]	1.8	30.4	54.9
372 Non-ferrous metals	118[e]	1.3	74	0.7	118[e]	0.7	97[e]	0.4	-17.8	-17.8
381 Metal products	226[e]	2.5	208	2.1	332[e]	2.0	481	2.1	112.8	44.9
382 Non-electrical machinery	168[e]	1.9	243	2.4	259[e]	1.6	809	3.6	381.5	212.4
383 Electrical machinery	340[e]	3.8	355	3.5	487[e]	2.9	1,473	6.5	333.2	202.5
384 Transport equipment	338[e]	3.7	337	3.3	1,249[e]	7.5	1,245	5.5	268.3	-0.3
385 Professional and scientific equipment	26[e]	0.3	56	0.6	51[e]	0.3	122[e]	0.5	369.2	139.2
390 Other manufacturing industries	314[e]	3.5	414	4.1	1,257[e]	7.5	877	3.9	179.3	-30.2

Note: The industry codes shown are International Standard Industry codes (ISIC). Percentages are percent of total Value Added. 'e' stands for estimated value

For sources, notes, and explanations, see Annotated Source Appendix, page 1035.

Finance, Economics, and Trade

Economic Indicators [36]

	1989	1990	1991[157]
GDP at Current Prices (mil. $US)	69,100	81,54	794,000
Real GDP growth rate (%)	12.0	10.0	8.0
Per Capita GDP (U.S.$)	1,230	1,437	1,630
Money Supply (M1 - bil. baht)	174.7	195.4	215.0
National Savings to GDP (%)	30.2	30.4	32.5
Domestic Investment to GDP (%)	31.5	36.8	35.5
Wholesale Prices (1976=100)	201.9	209.6	NA
Consumer Prices (1986=100)	112.1	118.8	125.5
External Public Debt	11,660	11,252	NA

Balance of Payments Summary [37]

Values in millions of dollars.

	1987	1988	1989	1990	1991
Exports of goods (f.o.b.)	11,595.0	15,781.0	19,834.0	22,811.0	28,232.0
Imports of goods (f.o.b.)	-12,019.0	-17,856.0	-22,750.0	-29,561.0	-34,218.0
Trade balance	-424.0	-2,075.0	-2,916.0	-6,750.0	-5,986.0
Services - debits	-4,334.0	-5,760.0	-6,874.0	-9,222.0	-11,366.0
Services - credits	4,168.0	5,944.0	7,046.0	8,478.0	9,526.0
Private transfers (net)	100.0	47.0	47.0	26.0	216.0
Government transfers (net)	125.0	189.0	199.0	187.0	45.0
Long term capital (net)	600.0	1,343.0	4,271.0	3,602.0	5,094.0
Short term capital (net)	414.0	2,582.0	2,380.0	5,698.0	7,090.0
Errors and omissions	248.0	411.0	928.0	1,419.0	424.0
Overall balance	897.0	2,681.0	5,081.0	3,438.0	5,043.0

Exchange Rates [38]

Currency: **baht.**
Symbol: **B.**

Data are currency units per $1.

April 1993	25.280
1992	25.400
1991	25.517
1990	25.585
1989	25.702
1988	25.294

Imports and Exports

Top Import Origins [39]

$41.5 billion (c.i.f., 1992).

Origins	%
Japan	29.3
US	11.4
Singapore	7.6
Taiwan	5.5
Germany	5.4
South Korea	4.6
Malaysia	4.2
China	3.3
Hong Kong	3.3
UK 1992 est.	NA

Top Export Destinations [40]

$32.9 billion (f.o.b., 1992).

Destinations	%
US	21.6
Japan	18.0
Singapore	8.7
Hong Kong	4.8
Germany	4.4
Netherlands	4.2
UK	3.4
Malaysia	NA
France	NA
China 1992 est.	NA

Foreign Aid [41]

	U.S. $	
US commitments, including Ex-Im (FY70-89)	870	million
Western (non-US) countries, ODA and OOF bilateral commitments (1970-89)	8.6	billion
OPEC bilateral aid (1979-89)	19	million

Import and Export Commodities [42]

Import Commodities

Capital goods 41.4%
Parts & raw materials 32.8%
Consumer goods 10.4%
Oil 8.2%

Export Commodities

Machinery 76.9%
Agricultural products 14.9%
Fisheries products 5.9%

For sources, notes, and explanations, see Annotated Source Appendix, page 1035.

923

Togo

Geography [1]

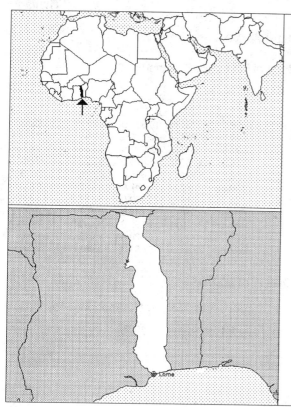

Total area:
56,790 km2
Land area:
54,390 km2
Comparative area:
Slightly smaller than West Virginia
Land boundaries:
Total 1,647 km, Benin 644 km, Burkina 126 km, Ghana 877 km
Coastline:
56 km
Climate:
Tropical; hot, humid in south; semiarid in north
Terrain:
Gently rolling savanna in north; central hills; southern plateau; low coastal plain with extensive lagoons and marshes
Natural resources:
Phosphates, limestone, marble
Land use:
Arable land:
25%
Permanent crops:
1%
Meadows and pastures:
4%
Forest and woodland:
28%
Other:
42%

Demographics [2]

	1960	1970	1980	1990	1991[1]	1994	2000	2010	2020
Population	1,456	1,964	2,596	3,680	3,811	4,255	5,263	7,401	10,146
Population density (persons per sq. mi.)	69	94	124	175	181	NA	335	467	628
Births	NA	NA	NA	NA	188	NA	335	467	628
Deaths	NA	NA	NA	NA	50	201	NA	NA	NA
Life expectancy - males	NA	NA	NA	NA	54	55	NA	NA	NA
Life expectancy - females	NA	NA	NA	NA	58	59	NA	NA	NA
Birth rate (per 1,000)	NA	NA	NA	NA	49	47	NA	NA	NA
Death rate (per 1,000)	NA	NA	NA	NA	13	11	NA	NA	NA
Women of reproductive age (15-44 yrs.)	NA	NA	NA	832	NA	951	1,177	NA	NA
of which are currently married	NA	NA	NA	653	NA	744	916	NA	NA
Fertility rate	NA	NA	NA	7.2	7.10	6.9	6.5	5.6	4.6

Population values are in thousands, life expectancy in years, and other items as indicated.

Health

Health Personnel [3]

Doctors per 1,000 pop., 1988-92	0.08
Nurse-to-doctor ratio, 1988-92	6.2
Hospital beds per 1,000 pop., 1985-90	1.6
Percentage of children immunized (age 1 yr. or less)	
Third dose of DPT, 1990-91	73.0
Measles, 1990-91	61.0

Health Expenditures [4]

Total health expenditure, 1990 (official exchange rate)	
Millions of dollars	67
Millions of dollars per capita	18
Health expenditures as a percentage of GDP	
Total	4.1
Public sector	2.5
Private sector	1.6
Development assistance for health	
Total aid flows (millions of dollars)[1]	14
Aid flows per capita (millions of dollars)	3.9
Aid flows as a percentage of total health expenditure	21.0

For sources, notes, and explanations, see Annotated Source Appendix, page 1035.

Human Factors

Health Care Ratios [5]

Population per physician, 1970	28860
Population per physician, 1990	NA
Population per nursing person, 1970	1590
Population per nursing person, 1990	NA
Percent of births attended by health staff, 1985	NA

Infants and Malnutrition [6]

Percent of babies with low birth weight, 1985	20
Infant mortality rate per 1,000 live births, 1970	134
Infant mortality rate per 1,000 live births, 1991	87
Years of life lost per 1,000 population, 1990	79
Prevalence of malnutrition (under age 5), 1990	14

Ethnic Division [7]

37 tribes; largest and most important are Ewe, Mina, and Kabye, European and Syrian-Lebanese under 1%.

Religion [8]

Indigenous beliefs	70%
Christian	20%
Muslim	10%

Major Languages [9]

French (official and the language of commerce), Ewe (one of the two major African languages in the south), Mina (one of the two major African languages in the south), Dagomba (one of the two major African languages in the north), Kabye (one of the two major African languages in the north).

Education

Public Education Expenditures [10]

Million Francs C.F.A.	1980	1985	1987	1988	1989	1990
Total education expenditure	13,049	15,880	17,710	20,400	22,801	24,420
as percent of GNP	5.6	5.0	4.9	5.2	5.5	5.7
as percent of total govt. expend.	19.4	19.4	19.7	21.2	24.7	NA
Current education expenditure	12,575	15,028	16,533	19,144	21,384	22,720
as percent of GNP	5.4	4.7	4.6	4.9	5.2	5.3
as percent of current govt. expend.	21.0	19.2	NA	22.2	NA	NA
Capital expenditure	474	852	1,177	1,256	1,417	1,700

Educational Attainment [11]

Age group	25+
Total population	1,084,488
Highest level attained (%)	
No schooling	76.5
First level	
Incompleted	13.5
Completed	NA
Entered second level	
S-1	8.7
S-2	NA
Post secondary	1.3

Literacy Rate [12]

In thousands and percent	1985[a]	1991[a]	2000[a]
Illiterate population +15 years	1,015	1,070	1,173
Illiteracy rate - total pop. (%)	62.1	56.7	45.6
Illiteracy rate - males (%)	48.6	43.6	33.9
Illiteracy rate - females (%)	74.9	69.3	56.8

Libraries [13]

	Admin. Units	Svc. Pts.	Vols. (000)	Shelving (meters)	Vols. Added	Reg. Users
National (1989)	1	NA	14	NA	504	-
Non-specialized	NA	NA	NA	NA	NA	NA
Public (1989)	1	26	63	600	2,000	7,706
Higher ed.	NA	NA	NA	NA	NA	NA
School	NA	NA	NA	NA	NA	NA

Daily Newspapers [14]

	1975	1980	1985	1990
Number of papers	1	3	2	1
Circ. (000)	7	16[e]	11[e]	10

Cinema [15]

Science and Technology

Scientific/Technical Forces [16]

R&D Expenditures [17]

U.S. Patents Issued [18]

Government and Law

Organization of Government [19]

Long-form name:
Republic of Togo
Type:
republic under transition to multiparty democratic rule
Independence:
27 April 1960 (from UN trusteeship under French administration)
Constitution:
1980 constitution nullified during national reform conference; transition constitution adopted 24 August 1991; multiparty draft constitution adopted by public referendum September 1992
Legal system:
French-based court system
National holiday:
Independence Day, 27 April (1960)
Executive branch:
president, prime minister, Council of Ministers (cabinet)
Legislative branch:
National Assembly dissolved during national reform conference; High Council for the Republic (HCR) formed to act as legislature during transition; legislative elections scheduled to be held in 1993
Judicial branch:
Court of Appeal, Supreme Court

Crime [23]

Elections [20]

National Assembly. Last held 4 March 1990; dissolved during national reform conference (next to be held 1993); results - RPT was the only party; seats - (77 total) RPT 77; interim legislative High Council of the Republic (HCR) in place since August 1991. Rally of the Togolese People (RPT) led by President Eyadema was the only party until the formation of multiple parties was legalized 12 April 1991; transition regime in place since August 1991.

Government Budget [21]

For 1991 est.
Revenues	284.8
Expenditures	407.0
Capital expenditures	NA

Military Expenditures and Arms Transfers [22]

	1985	1986	1987	1988	1989
Military expenditures					
Current dollars (mil.)	26	31	40	NA	43[e]
1989 constant dollars (mil.)	29	35	43	NA	43[e]
Armed forces (000)	7	7[e]	8	6	6
Gross national product (GNP)					
Current dollars (mil.)	992	1,056	1,103	1,200	1,292
1989 constant dollars (mil.)	1,130	1,171	1,186	1,249	1,292
Central government expenditures (CGE)					
1989 constant dollars (mil.)	425	461	388	280[e]	265[e]
People (mil.)	3.1	3.2	3.3	3.4	3.5
Military expenditure as % of GNP	2.6	3.0	3.6	NA	3.3
Military expenditure as % of CGE	6.9	7.6	11.1	NA	16.3
Military expenditure per capita	10	11	13	NA	12
Armed forces per 1,000 people	2.3	2.2	2.4	1.7	1.7
GNP per capita	368	368	360	365	365
Arms imports[6]					
Current dollars (mil.)	0	10	0	0	5
1989 constant dollars (mil.)	0	11	0	0	5
Arms exports[6]					
Current dollars (mil.)	0	0	0	0	0
1989 constant dollars (mil.)	0	0	0	0	0
Total imports[7]					
Current dollars (mil.)	288	312	424	487	NA
1989 constant dollars	328	346	456	507	NA
Total exports[7]					
Current dollars (mil.)	190	204	244	242	NA
1989 constant dollars	216	226	262	252	NA
Arms as percent of total imports[8]	0	3.2	0	0	NA
Arms as percent of total exports[8]	0	0	0	0	NA

Human Rights [24]

	SSTS	FL	FAPRO	PPCG	APROBC	TPW	PCPTW	STPEP	PHRFF	PRW	ASST	AFL
Observes	P	P	P	P	P	P	P	P			P	
	EAFRD	CPR	ESCR	SR	ACHR	MAAE	PVIAC	PVNAC	EAFDAW	TCIDTP	RC	
Observes		P	P	P	P		P	P	P	P	P	P

P = Party; S = Signatory; see Appendix for meaning of abbreviations.

Labor Force

Total Labor Force [25]

Labor Force by Occupation [26]

Agriculture	78%
Industry	22

Unemployment Rate [27]

2% (1987)

Production Sectors

Energy Resource Summary [28]

Energy Resources: None. **Electricity**: 179,000 kW capacity; 209 million kWh produced, 60 kWh per capita (1990).

Telecommunications [30]

- Fair system based on network of radio relay routes supplemented by open wire lines
- Broadcast stations - 2 AM, no FM, 3 (2 relays) TV
- Satellite earth stations - 1 Atlantic Ocean INTELSAT and 1 SYMPHONIE

Transportation [31]

Railroads. 570 km 1.000-meter gauge, single track

Highways. 6,462 km total; 1,762 km paved; 4,700 km unimproved roads

Merchant Marine. 2 roll-on/roll-off ships (1,000 GRT or over) totaling 11,118 GRT/20,529 DWT

Airports

Total:	9
Usable:	9
With permanent-surface runways:	2
With runways over 3,659 m:	0
With runways 2,440-3,659 m:	2
With runways 1,220-2,439 m:	0

Top Agricultural Products [32]

Agriculture accounts for 33% of GDP; cash crops—coffee, cocoa, cotton; food crops - yams, cassava, corn, beans, rice, millet, sorghum; livestock production not significant; annual fish catch, 10,000-14,000 tons.

Top Mining Products [33]

Estimated metric tons unless otherwise specified   M.t.

Cement	388,000[55]
Phosphate rock (000 tons)	1,076
Marble	
Blocks (square meters)	250
Crushed	600

Tourism [34]

	1987	1988	1989	1990	1991
Tourists[10]	98	104	115	103	
Tourism receipts	35	44	41	50	50
Tourism expenditures	29	32	37	46	48
Fare receipts		1	5	6	7
Fare expenditures	22	21	27	34	38

Tourists are in thousands, money in million U.S. dollars.

Manufacturing Sector

GDP and Manufacturing Summary [35]

	1980	1985	1989	1990	% change 1980-1990	% change 1989-1990
GDP (million 1980 $)	1,131	1,056	NA	1,317	16.4	NA
GDP per capita (1980 $)	432	349	NA	373	-13.7	NA
Manufacturing as % of GDP (current prices)	7.0	6.6	NA	9.4	34.3	NA
Gross output (million $)	155e	104	NA	254	63.9	NA
Value added (million $)	52e	43e	NA	102e	96.2	NA
Value added (million 1980 $)	79	78	NA	91	15.2	NA
Industrial production index	100	92	NA	114	14.0	NA
Employment (thousands)	5e	5e	NA	6e	20.0	NA

Note: GDP stands for Gross Domestic Product. 'e' stands for estimated value.

Profitability and Productivity

	1980	1985	1989	1990	% change 1980-1990	% change 1989-1990
Intermediate input (%)	61e	71e	NA	79e	29.5	NA
Wages, salaries, and supplements (%)	14e	12e	NA	13e	-7.1	NA
Gross operating surplus (%)	25e	18e	NA	8e	-68.0	NA
Gross output per worker ($)	28,454e	23,960e	NA	53,412e	87.7	NA
Value added per worker ($)	9,532e	8,800e	NA	19,640e	106.0	NA
Average wage (incl. benefits) ($)	3,447e	2,559e	NA	5,207e	51.1	NA

Profitability is in percent of gross output. Productivity is in U.S. $. 'e' stands for estimated value.

Profitability - 1990

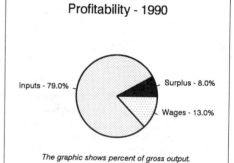

Inputs - 79.0%
Surplus - 8.0%
Wages - 13.0%

The graphic shows percent of gross output.

Value Added in Manufacturing

	1980 $ mil.	1980 %	1985 $ mil.	1985 %	1989 $ mil.	1989 %	1990 $ mil.	1990 %	% change 1980-1990	% change 1989-1990
311 Food products	4	7.7	11e	25.6	NA	NA	11e	10.8	175.0	NA
313 Beverages	16	30.8	15e	34.9	NA	NA	46e	45.1	187.5	NA
314 Tobacco products	NA	0.0	NA	0.0	NA	NA	NA	0.0	NA	NA
321 Textiles	8e	15.4	5e	11.6	NA	NA	12e	11.8	50.0	NA
322 Wearing apparel	NA	0.0	-	0.0	NA	NA	-	0.0	NA	NA
323 Leather and fur products	-	0.0	-	0.0	NA	NA	-	0.0	NA	NA
324 Footwear	6	11.5	2e	4.7	NA	NA	4e	3.9	-33.3	NA
331 Wood and wood products	1	1.9	-	0.0	NA	NA	1e	1.0	0.0	NA
332 Furniture and fixtures	NA	0.0	-	0.0	NA	NA	1e	1.0	NA	NA
341 Paper and paper products	-	0.0	-	0.0	NA	NA	-	0.0	NA	NA
342 Printing and publishing	3	5.8	1e	2.3	NA	NA	3e	2.9	0.0	NA
351 Industrial chemicals	3	5.8	1e	2.3	NA	NA	4e	3.9	33.3	NA
352 Other chemical products	1e	1.9	-	0.0	NA	NA	1e	1.0	0.0	NA
353 Petroleum refineries	NA	0.0	NA	0.0	NA	NA	NA	0.0	NA	NA
354 Miscellaneous petroleum and coal products	NA	0.0	NA	0.0	NA	NA	NA	0.0	NA	NA
355 Rubber products	NA	0.0	NA	0.0	NA	NA	NA	0.0	NA	NA
356 Plastic products	NA	0.0	NA	0.0	NA	NA	NA	0.0	NA	NA
361 Pottery, china and earthenware	NA	0.0	3	0.0	NA	NA	9w	0.0	NA	NA
362 Glass and glass products	1e	1.9	2e	4.7	NA	NA	7e	6.9	600.0	NA
369 Other non-metal mineral products	6	11.5	-	0.0	NA	NA	-	0.0	NA	NA
371 Iron and steel	2	3.8	1	0.0	NA	NA	3	0.0	NA	NA
372 Non-ferrous metals	NA	0.0	-	0.0	NA	NA	-	0.0	NA	NA
381 Metal products	1	1.9	-	0.0	NA	NA	-	0.0	NA	NA
382 Non-electrical machinery	NA	0.0	-	0.0	NA	NA	-	0.0	NA	NA
383 Electrical machinery	NA	0.0	-	0.0	NA	NA	-	0.0	NA	NA
384 Transport equipment	NA	0.0	-	0.0	NA	NA	-	0.0	NA	NA
385 Professional and scientific equipment	NA	0.0	-	0.0	NA	NA	-	0.0	NA	NA
390 Other manufacturing industries	NA	0.0	-	0.0	NA	NA	NA	0.0	NA	NA

Note: The industry codes shown are International Standard Industry codes (ISIC). Percentages are percent of total Value Added. 'e' stands for estimated value

For sources, notes, and explanations, see Annotated Source Appendix, page 1035.

Finance, Economics, and Trade

Economic Indicators [36]

National product: GDP—exchange rate conversion—$1.5 billion (1991 est.). **National product real growth rate**: 0% (1991 est.). **National product per capita**: $400 (1991 est.). **Inflation rate (consumer prices)**: 0.5% (1991 est.). **External debt**: $1.3 billion (1991).

Balance of Payments Summary [37]

Values in millions of dollars.

	1987	1988	1989	1990	1991
Exports of goods (f.o.b.)	397.5	435.0	468.0	502.1	511.9
Imports of goods (f.o.b.)	-437.1	-504.5	-503.4	-616.7	-583.1
Trade balance	-39.6	-69.5	-35.4	-114.6	-71.2
Services - debits	-262.7	-271.1	-272.4	-315.1	-311.9
Services - credits	137.7	133.7	163.6	197.6	195.7
Private transfers (net)	7.0	2.6	11.9	13.6	17.7
Government transfers (net)	97.1	124.2	78.7	108.7	86.5
Long term capital (net)	-26.6	170.5	47.6	47.9	57.4
Short term capital (net)	68.6	-94.8	46.1	87.4	48.9
Errors and omissions	-15.7	-34.7	-0.8	-1.3	-1.1
Overall balance	-34.2	-39.1	39.3	24.2	22.0

Exchange Rates [38]

Currency: **Communaute Financiere Africaine francs.**

Symbol: **CFAF.**

Data are currency units per $1.

January 1993	274.06
1992	264.69
1991	282.11
1990	272.26
1989	319.01
1988	297.85

Imports and Exports

Top Import Origins [39]

$583 million (f.o.b., 1991 est.). Data are for 1990.

Origins	%
EC	57
Africa	17
US	5
Japan	4

Top Export Destinations [40]

$512 million (f.o.b., 1991 est.). Data are for 1990.

Destinations	%
EC	40
Africa	16
US	1

Foreign Aid [41]

	U.S. $	
US commitments, including Ex-Im (FY70-90)	142	million
Western (non-US) countries, ODA and OOF bilateral commitments (1970-90)	2	billion
OPEC bilateral aid (1979-89)	35	million
Communist countries (1970-89)	51	million

Import and Export Commodities [42]

Import Commodities

Machinery & equipment
Consumer goods
Food
Chemical products

Export Commodities

Phosphates
Cotton
Cocoa
Coffee

For sources, notes, and explanations, see Annotated Source Appendix, page 1035.

929

Trinidad and Tobago

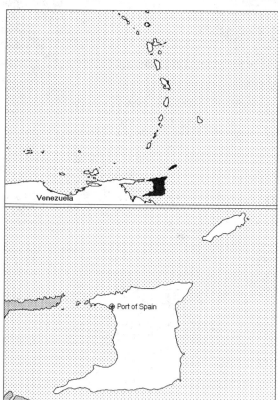

Geography [1]

Total area:
  5,130 km2
Land area:
  5,130 km2
Comparative area:
  Slightly smaller than Delaware
Land boundaries:
  0 km
Coastline:
  362 km
Climate:
  Tropical; rainy season (June to December)
Terrain:
  Mostly plains with some hills and low mountains
Natural resources:
  Petroleum, natural gas, asphalt
Land use:
  Arable land:
  14%
  Permanent crops:
  17%
  Meadows and pastures:
  2%
  Forest and woodland:
  44%
  Other:
  23%

Demographics [2]

	1960	1970	1980	1990	1991[1]	1994	2000	2010	2020
Population	841	955	1,091	1,271	1,285	1,328	1,420	1,583	1,722
Population density (persons per sq. mi.)	425	482	550	642	649	NA	719	808	889
Births	NA	NA	NA	NA	27	26	NA	NA	NA
Deaths	NA	NA	NA	NA	8	8	NA	NA	NA
Life expectancy - males	NA	NA	NA	NA	68	68	NA	NA	NA
Life expectancy - females	NA	NA	NA	NA	73	73	NA	NA	NA
Birth rate (per 1,000)	NA	NA	NA	NA	21	20	NA	NA	NA
Death rate (per 1,000)	NA	NA	NA	NA	6	6	NA	NA	NA
Women of reproductive age (15-44 yrs.)	NA	NA	NA	330	NA	352	397	NA	NA
of which are currently married	NA	NA	NA	169	NA	183	204	NA	NA
Fertility rate	NA	NA	NA	2.5	2.44	2.3	2.1	1.9	1.8

Population values are in thousands, life expectancy in years, and other items as indicated.

Health

Health Personnel [3]

Health Expenditures [4]

For sources, notes, and explanations, see Annotated Source Appendix, page 1035.

Human Factors

Health Care Ratios [5]

Population per physician, 1970	2250
Population per physician, 1990	NA
Population per nursing person, 1970	190
Population per nursing person, 1990	NA
Percent of births attended by health staff, 1985	90

Infants and Malnutrition [6]

Percent of babies with low birth weight, 1985	NA
Infant mortality rate per 1,000 live births, 1970	44
Infant mortality rate per 1,000 live births, 1991	19
Years of life lost per 1,000 population, 1990	NA
Prevalence of malnutrition (under age 5), 1990	9

Ethnic Division [7]

Black	43%
East Indian	40%
Mixed	14%
White	1%
Chinese	1%
Other	1%

Religion [8]

Roman Catholic	32.2%
Hindu	24.3%
Anglican	14.4%
Other Protestant	14.0%
Muslim	6.0%
None or unknown	9.1%

Major Languages [9]

English (official)	NA
Hindi	NA
French	NA
Spanish	NA

Education

Public Education Expenditures [10]

Million Trinidad Dollars	1980	1985	1987	1988	1989	1990
Total education expenditure	564	1,042	1,022	838	772	788
as percent of GNP	4.0	6.2	6.3	5.2	4.7	4.1
as percent of total govt. expend.	11.5	NA	14.0	NA	12.1	11.6
Current education expenditure	431	912	960	773	701	727
as percent of GNP	3.0	5.4	5.9	4.8	4.2	3.7
as percent of current govt. expend.	17.6	NA	18.3	NA	13.4	13.5
Capital expenditure	133	130	62	65	71	61

Educational Attainment [11]

Age group	25+
Total population	408,215
Highest level attained (%)	
No schooling	1.3
First level	
Incompleted	29.4
Completed	42.6
Entered second level	
S-1	19.7
S-2	4.0
Post secondary	2.9

Literacy Rate [12]

	1970[b]	1980[b]	1990[b]
Illiterate population +15 years	41,750	34,800	NA
Illiteracy rate - total pop. (%)	7.8	5.1	NA
Illiteracy rate - males (%)	5.3	3.5	NA
Illiteracy rate - females (%)	10.3	6.6	NA

Libraries [13]

	Admin. Units	Svc. Pts.	Vols. (000)	Shelving (meters)	Vols. Added	Reg. Users
National (1986)	1	1	17	NA	NA	NA
Non-specialized	NA	NA	NA	NA	NA	NA
Public	NA	NA	NA	NA	NA	NA
Higher ed.	NA	NA	NA	NA	NA	NA
School	NA	NA	NA	NA	NA	NA

Daily Newspapers [14]

	1975	1980	1985	1990
Number of papers	3	4	4	2
Circ. (000)	135	155[e]	173	95

Cinema [15]

Science and Technology

Scientific/Technical Forces [16]

R&D Expenditures [17]

	Dollar (000) 1984
Total expenditure	143,257
Capital expenditure	33,336
Current expenditure	109,921
Percent current	76.7

U.S. Patents Issued [18]

Values show patents issued to citizens of the country by the U.S. Patents Office.

	1990	1991	1992
Number of patents	1	1	2

For sources, notes, and explanations, see Annotated Source Appendix, page 1035.

931

Government and Law

Organization of Government [19]

Long-form name:
Republic of Trinidad and Tobago
Type:
parliamentary democracy
Independence:
31 August 1962 (from UK)
Constitution:
31 August 1976
Legal system:
based on English common law; judicial review of legislative acts in the Supreme Court; has not accepted compulsory ICJ jurisdiction
National holiday:
Independence Day, 31 August (1962)
Executive branch:
president, prime minister, Cabinet
Legislative branch:
bicameral Parliament consists of an upper house or Senate and a lower house or House of Representatives
Judicial branch:
Court of Appeal, Supreme Court

Political Parties [20]

House of Representatives	% of votes
People's National Movement (PNM)	32.0
United National Congress (UNC)	13.0
National Alliance (NAR)	2.0

Government Budget [21]

For 1993 est.

Revenues	1.600
Expenditures	1.600
Capital expenditures	0.158

Crime [23]

Crime volume	
Cases known to police	65,853
Attempts (percent)	NA
Percent cases solved	28.65
Crimes per 100,000 persons	5,334.87
Persons responsible for offenses	
Total number offenders	17,262
Percent female	10.09
Percent juvenile[56]	2.91
Percent foreigners	0.30

Military Expenditures and Arms Transfers [22]

	1985	1986	1987	1988	1989
Military expenditures					
Current dollars (mil.)	NA	NA	NA	NA	59
1989 constant dollars (mil.)	NA	NA	NA	NA	59
Armed forces (000)	2	2	2	2[e]	2
Gross national product (GNP)					
Current dollars (mil.)	4,292	4,380	4,046	3,843	3,735
1989 constant dollars (mil.)	4,886	4,860	4,352	4,000	3,735
Central government expenditures (CGE)					
1989 constant dollars (mil.)	2,189	1,916	1,694	1,492	1,338
People (mil.)	1.2	1.2	1.2	1.2	1.3
Military expenditure as % of GNP	NA	NA	NA	NA	1.6
Military expenditure as % of CGE	NA	NA	NA	NA	4.4
Military expenditure per capita	NA	NA	NA	NA	47
Armed forces per 1,000 people	1.7	1.7	1.6	1.6	1.7
GNP per capita	4,108	4,015	3,543	3,218	2,972
Arms imports[6]					
Current dollars (mil.)	0	0	0	0	0
1989 constant dollars (mil.)	0	0	0	0	0
Arms exports[6]					
Current dollars (mil.)	0	0	0	0	0
1989 constant dollars (mil.)	0	0	0	0	0
Total imports[7]					
Current dollars (mil.)	1,586	1,345	1,261	1,123	1,217
1989 constant dollars	1,805	1,493	1,356	1,169	1,217
Total exports[7]					
Current dollars (mil.)	2,196	1,376	1,460	1,391	1,558
1989 constant dollars	2,500	1,527	1,570	1,448	1,558
Arms as percent of total imports[8]	0	0	0	0	0
Arms as percent of total exports[8]	0	0	0	0	0

Human Rights [24]

	SSTS	FL	FAPRO	PPCG	APROBC	TPW	PCPTW	STPEP	PHRFF	PRW	ASST	AFL
Observes	P	P	P		P	P	P			P	P	P
	EAFRD	CPR	ESCR	SR	ACHR	MAAE	PVIAC	PVNAC	EAFDAW	TCIDTP	RC
Observes	P	P	P		P				P		P

P = Party; S = Signatory; see Appendix for meaning of abbreviations.

Labor Force

Total Labor Force [25]

463,900

Labor Force by Occupation [26]

Construction and utilities	18.1%
Manufacturing, mining, and quarrying	14.8
Agriculture	10.9
Other	56.2

Date of data: 1985 est.

Unemployment Rate [27]

18.5% (1991)

Production Sectors

Commercial Energy Production and Consumption

Production [28]

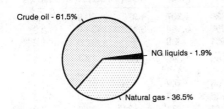

Crude oil - 61.5%
NG liquids - 1.9%
Natural gas - 36.5%

Consumption [29]

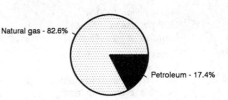

Natural gas - 82.6%
Petroleum - 17.4%

Data are shown in quadrillion (10^{15}) BTUs and percent for 1991

Crude oil[1]	0.32		Petroleum	0.04
Natural gas liquids	0.01		Dry natural gas	0.19
Dry natural gas	0.19		Coal[2]	(s)
Coal[2]	0.00		Net hydroelectric power[3]	0.00
Net hydroelectric power[3]	0.00		Net nuclear power[3]	0.00
Net nuclear power[3]	0.00		Total	0.23
Total	0.51			

Telecommunications [30]

- Excellent international service via tropospheric scatter links to Barbados and Guyana
- Good local service
- 109,000 telephones
- Broadcast stations - 2 AM, 4 FM, 5 TV
- 1 Atlantic Ocean INTELSAT earth station

Transportation [31]

Railroads. minimal agricultural railroad system near San Fernando

Highways. 8,000 km total; 4,000 km paved, 1,000 km improved earth, 3,000 km unimproved earth

Merchant Marine. 2 cargo ships (1,000 GRT or over) totaling 12,507 GRT/21,923 DWT

Airports

Total:	6
Usable:	5
With permanent-surface runways:	2
With runways over 3,659 m:	0
With runways 2,440-3,659 m:	2
With runways 1,220-2,439 m:	1

Top Agricultural Products [32]

Agriculture accounts for 3% of GDP; highly subsidized sector; major crops—cocoa, sugarcane; sugarcane acreage is being shifted into rice, citrus, coffee, vegetables; poultry sector most important source of animal protein; must import large share of food needs.

Top Mining Products [33]

Estimated metric tons unless otherwise specified M.t.

Petroleum, crude (000 42-gal. barrels)	55,000[1]
Natural gas, gross (mil. cubic meters)	7,000
Petroleum refinery products (000 42-gal. barrels)	30,200[e]
Steel, crude	444,000[e]
Limestone	600,000[e]
Nitrogen, N content of ammonia (000 tons)	1,524[1]
Cement, hydraulic	485,396

Tourism [34]

	1987	1988	1989	1990	1991
Visitors	218	197	211	227	252
Tourists[16]	202	186	194	195	220
Cruise passengers	16	11	17	32	32
Tourism receipts	94	98	85	95	101
Tourism expenditures	158	168	119	112	111
Fare receipts	67	87	88	127	132
Fare expenditures	26	25	20	26	21

Tourists are in thousands, money in million U.S. dollars.

For sources, notes, and explanations, see Annotated Source Appendix, page 1035.

Manufacturing Sector

GDP and Manufacturing Summary [35]

	1980	1985	1989	1990	% change 1980-1990	% change 1989-1990
GDP (million 1980 $)	5,486	4,828	4,207	4,376	-20.2	4.0
GDP per capita (1980 $)	5,070	4,095	3,334	3,408	-32.8	2.2
Manufacturing as % of GDP (current prices)	5.0	7.1	9.8	8.7	74.0	-11.2
Gross output (million $)	1,605[e]	1,765	1,645[e]	1,885[e]	17.4	14.6
Value added (million $)	492	387	413[e]	471[e]	-4.3	14.0
Value added (million 1980 $)	490	405	442	426	-13.1	-3.6
Industrial production index	100	86	88	104	4.0	18.2
Employment (thousands)	44	34	31[e]	31[e]	-29.5	0.0

Note: GDP stands for Gross Domestic Product. 'e' stands for estimated value.

Profitability and Productivity

	1980	1985	1989	1990	% change 1980-1990	% change 1989-1990
Intermediate input (%)	69[e]	78	75[e]	75[e]	8.7	0.0
Wages, salaries, and supplements (%)	15[e]	18[e]	16[e]	17[e]	13.3	6.3
Gross operating surplus (%)	15[e]	4[e]	9[e]	8[e]	-46.7	-11.1
Gross output per worker ($)	36,225[e]	52,667	52,877[e]	56,036[e]	54.7	6.0
Value added per worker ($)	11,099	11,544	13,267[e]	14,091[e]	27.0	6.2
Average wage (incl. benefits) ($)	5,568[e]	9,488[e]	8,718[e]	9,936[e]	78.4	14.0

Profitability is in percent of gross output. Productivity is in U.S. $. 'e' stands for estimated value.

Profitability - 1990

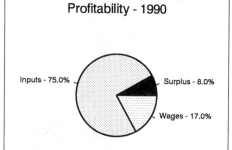

Inputs - 75.0% Surplus - 8.0% Wages - 17.0%

The graphic shows percent of gross output.

Value Added in Manufacturing

	1980 $ mil.	1980 %	1985 $ mil.	1985 %	1989 $ mil.	1989 %	1990 $ mil.	1990 %	% change 1980-1990	% change 1989-1990
311 Food products	67	13.6	95	24.5	123[e]	29.8	131[e]	27.8	95.5	6.5
313 Beverages	27	5.5	34	8.8	36[e]	8.7	40[e]	8.5	48.1	11.1
314 Tobacco products	14	2.8	35	9.0	33[e]	8.0	33[e]	7.0	135.7	0.0
321 Textiles	1	0.2	2	0.5	1[e]	0.2	2[e]	0.4	100.0	100.0
322 Wearing apparel	16	3.3	13	3.4	10[e]	2.4	13[e]	2.8	-18.8	30.0
323 Leather and fur products	NA	0.0	NA	0.0	NA	0.0	NA	0.0	NA	NA
324 Footwear	4	0.8	5	1.3	2[e]	0.5	2[e]	0.4	-50.0	0.0
331 Wood and wood products	6	1.2	4	1.0	3[e]	0.7	4[e]	0.8	-33.3	33.3
332 Furniture and fixtures	9	1.8	7	1.8	1[e]	0.2	1[e]	0.2	-88.9	0.0
341 Paper and paper products	9	1.8	14	3.6	22[e]	5.3	24[e]	5.1	166.7	9.1
342 Printing and publishing	13	2.6	19	4.9	17[e]	4.1	21[e]	4.5	61.5	23.5
351 Industrial chemicals	5	1.0	6	1.6	4[e]	1.0	4[e]	0.8	-20.0	0.0
352 Other chemical products	12	2.4	10	2.6	8[e]	1.9	12[e]	2.5	0.0	50.0
353 Petroleum refineries	190[e]	38.6	17[e]	4.4	34[e]	8.2	36[e]	7.6	-81.1	5.9
354 Miscellaneous petroleum and coal products	2[e]	0.4	NA	0.0	NA	0.0	NA	0.0	NA	NA
355 Rubber products	9	1.8	10	2.6	11[e]	2.7	11[e]	2.3	22.2	0.0
356 Plastic products	2	0.4	8	2.1	12[e]	2.9	12[e]	2.5	500.0	0.0
361 Pottery, china and earthenware	NA	0.0	NA	0.0	NA	0.0	NA	0.0	NA	NA
362 Glass and glass products	3	0.6	4	1.0	3[e]	0.7	3[e]	0.6	0.0	0.0
369 Other non-metal mineral products	23	4.7	31	8.0	24[e]	5.8	26[e]	5.5	13.0	8.3
371 Iron and steel	NA	0.0	NA	0.0	NA	0.0	NA	0.0	NA	NA
372 Non-ferrous metals	NA	0.0	NA	0.0	NA	0.0	NA	0.0	NA	NA
381 Metal products	26	5.3	11	2.8	17[e]	4.1	20[e]	4.2	-23.1	17.6
382 Non-electrical machinery	13	2.6	NA	0.0	16[e]	3.9	31[e]	6.6	138.5	93.8
383 Electrical machinery	3	0.6	13	3.4	13[e]	3.1	13[e]	2.8	333.3	0.0
384 Transport equipment	28	5.7	43	11.1	17[e]	4.1	23[e]	4.9	-17.9	35.3
385 Professional and scientific equipment	NA	0.0	NA	0.0	NA	0.0	NA	0.0	NA	NA
390 Other manufacturing industries	8	1.6	6	1.6	6[e]	1.5	8[e]	1.7	0.0	33.3

Note: The industry codes shown are International Standard Industry codes (ISIC). Percentages are percent of total Value Added. 'e' stands for estimated value

For sources, notes, and explanations, see Annotated Source Appendix, page 1035.

Finance, Economics, and Trade

Economic Indicators [36]

In Millions of U.S. Dollars unless otherwise specified.

	1989	1990	1991
GDP (current market prices)	4,283.9	4,970.6	4,956.1
GDP (1985 TT $) growth (%)	-2.4	-0.5	2.7
GDP Per Capita (US$1)	3,531.1	4,049.5	3,940.1
Money Supply (M1)	404.8	461.6	545
Commercial Interest Rate	12.04	11.73	11.63[158]
Savings Rate	17.6	24.7	-
CPI (aver. quarterly; 9/82 = 100)	201.2	223.4	231.5
PPI (aver. quarterly; 10/78 = 100)	279.4	283.4	-
External Public Debt (yr-end)	1,458.7	1,536.7	1,460

Balance of Payments Summary [37]

Values in millions of dollars.

	1987	1988	1989	1990	1991
Exports of goods (f.o.b.)	1,396.9	1,453.3	1,534.6	1,935.2	1,751.3
Imports of goods (f.o.b.)	-1,057.6	-1,064.2	-1,045.2	-947.6	-1,210.3
Trade balance	339.3	389.1	489.4	987.6	541.0
Services - debits	-785.2	-776.2	-849.9	-915.6	-1,001.5
Services - credits	243.6	307.7	329.5	393.1	455.7
Private transfers (net)	-20.2	-23.0	-19.2	-21.0	-15.4
Government transfers (net)	-16.6	-6.6	-5.5	-4.4	3.5
Long term capital (net)	8.3	21.7	147.2	-245.3	-114.3
Short term capital (net)	67.7	-89.6	-62.3	-9.4	6.2
Errors and omissions	-94.8	21.1	45.4	-112.0	-32.3
Overall balance	-257.9	-155.8	74.6	73.0	-157.1

Exchange Rates [38]

Currency: **Trinidad and Tobago dollars**

Symbol: **TT$.**

Data are currency units per $1.

Fixed rate since 1989	4.25

Imports and Exports

Top Import Origins [39]

$1.7 billion (c.i.f., 1991). Data are for 1991.

Origins	%
US	39
Venezuela	14
UK	7
CARICOM	5

Top Export Destinations [40]

$2.2 billion (f.o.b., 1991).

Destinations	%
US	49
CARICOM	12

Foreign Aid [41]

	U.S. $	
US commitments, including Ex-Im (FY70-89)	373	million
Western (non-US) countries, ODA and OOF bilateral commitments (1970-89)	518	million

Import and Export Commodities [42]

Import Commodities	Export Commodities
Raw materials/intermed. goods 48%	Oil & petr. products 82%
Capital goods 29%	Steel products 9%
Consumer goods 23%	Fertilizer
	Sugar
	Cocoa
	Coffee
	Citrus

Tunisia

Geography [1]

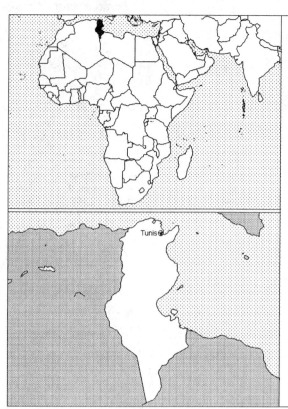

Total area:
163,610 km2
Land area:
155,360 km2
Comparative area:
Slightly larger than Georgia
Land boundaries:
Total 1,424 km, Algeria 965 km, Libya 459 km
Coastline:
1,148 km
Climate:
Temperate in north with mild, rainy winters and hot, dry summers; desert in south
Terrain:
Mountains in north; hot, dry central plain; semiarid south merges into the Sahara
Natural resources:
Petroleum, phosphates, iron ore, lead, zinc, salt
Land use:
Arable land:
20%
Permanent crops:
10%
Meadows and pastures:
19%
Forest and woodland:
4%
Other:
47%

Demographics [2]

	1960	1970	1980	1990	1991[1]	1994	2000	2010	2020
Population	4,149	5,099	6,452	8,084	8,276	8,727	9,599	10,937	12,144
Population density (persons per sq. mi.)	69	85	108	135	138	NA	162	187	210
Births	NA	NA	NA	NA	214	204	NA	NA	NA
Deaths	NA	NA	NA	NA	43	43	NA	NA	NA
Life expectancy - males	NA	NA	NA	NA	70	71	NA	NA	NA
Life expectancy - females	NA	NA	NA	NA	74	75	NA	NA	NA
Birth rate (per 1,000)	NA	NA	NA	NA	26	23	NA	NA	NA
Death rate (per 1,000)	NA	NA	NA	NA	5	5	NA	NA	NA
Women of reproductive age (15-44 yrs.)	NA	NA	NA	1,964	NA	2,199	2,578	NA	NA
of which are currently married	NA	NA	NA	1,155	NA	1,317	1,587	NA	NA
Fertility rate	NA	NA	NA	3.5	3.31	2.9	2.3	2.1	2.0

Population values are in thousands, life expectancy in years, and other items as indicated.

Health

Health Personnel [3]

Doctors per 1,000 pop., 1988-92	0.53
Nurse-to-doctor ratio, 1988-92	2.7
Hospital beds per 1,000 pop., 1985-90	2.0
Percentage of children immunized (age 1 yr. or less)	
Third dose of DPT, 1990-91	90.0
Measles, 1990-91	80.0

Health Expenditures [4]

Total health expenditure, 1990 (official exchange rate)	
Millions of dollars	614
Millions of dollars per capita	76
Health expenditures as a percentage of GDP	
Total	4.9
Public sector	3.3
Private sector	1.6
Development assistance for health	
Total aid flows (millions of dollars)[1]	18
Aid flows per capita (millions of dollars)	2.3
Aid flows as a percentage of total health expenditure	3.0

For sources, notes, and explanations, see Annotated Source Appendix, page 1035.

Human Factors

Health Care Ratios [5]

Population per physician, 1970	5930
Population per physician, 1990	1870
Population per nursing person, 1970	940
Population per nursing person, 1990	300
Percent of births attended by health staff, 1985	60

Infants and Malnutrition [6]

Percent of babies with low birth weight, 1985	7
Infant mortality rate per 1,000 live births, 1970	127
Infant mortality rate per 1,000 live births, 1991	38
Years of life lost per 1,000 population, 1990	21
Prevalence of malnutrition (under age 5), 1990	10

Ethnic Division [7]

Arab-Berber	98%
European	1%
Jewish	<1%

Religion [8]

Muslim	98%
Christian	1%
Jewish	1%

Major Languages [9]

Arabic (official and one of the languages of commerce), French (commerce).

Education

Public Education Expenditures [10]

Million Dinar	1980	1985	1987	1988	1989	1990
Total education expenditure	185	389	NA	491	NA	648
as percent of GNP	5.4	5.9	NA	5.9	NA	6.1
as percent of total govt. expend.	16.4	14.1	NA	14.1	NA	14.3
Current education expenditure	162	351	NA	449	515	569
as percent of GNP	4.7	5.3	NA	5.4	5.6	5.3
as percent of current govt. expend.	23.5	20.8	NA	19.2	18.2	17.7
Capital expenditure	23	38	NA	42	NA	79

Educational Attainment [11]

Age group	25+
Total population	2,714,100
Highest level attained (%)	
No schooling	66.3
First level	
Incompleted	18.9
Completed	NA
Entered second level	
S-1	12.0
S-2	NA
Post secondary	2.8

Literacy Rate [12]

In thousands and percent	1985[a]	1991[a]	2000[a]
Illiterate population +15 years	1,858	1,762	1,497
Illiteracy rate - total pop. (%)	42.4	34.7	22.5
Illiteracy rate - males (%)	32.2	25.8	16.0
Illiteracy rate - females (%)	52.7	43.7	29.1

Libraries [13]

	Admin. Units	Svc. Pts.	Vols. (000)	Shelving (meters)	Vols. Added	Reg. Users
National	NA	NA	NA	NA	NA	NA
Non-specialized	NA	NA	NA	NA	NA	NA
Public (1989)	252	340	NA	NA	84,109	NA
Higher ed.	NA	NA	NA	NA	NA	NA
School	NA	NA	NA	NA	NA	NA

Daily Newspapers [14]

	1975	1980	1985	1990
Number of papers	4	5	6	6
Circ. (000)	190	272	280[e]	300[e]

Cinema [15]

Science and Technology

Scientific/Technical Forces [16]

R&D Expenditures [17]

U.S. Patents Issued [18]

Values show patents issued to citizens of the country by the U.S. Patents Office.

	1990	1991	1992
Number of patents	0	1	0

For sources, notes, and explanations, see Annotated Source Appendix, page 1035.

Government and Law

Organization of Government [19]

Long-form name:
Republic of Tunisia
Type:
republic
Independence:
20 March 1956 (from France)
Constitution:
1 June 1959
Legal system:
based on French civil law system and
Islamic law; some judicial review of
legislative acts in the Supreme Court in
joint session
National holiday:
National Day, 20 March (1956)
Executive branch:
president, prime minister, Cabinet
Legislative branch:
unicameral Chamber of Deputies (Majlis
al-Nuwaab)
Judicial branch:
Court of Cassation (Cour de Cassation)

Political Parties [20]

Chamber of Deputies	% of votes
Constitutional Democratic	
Rally Party	80.7
Independents/Islamists	13.7
Movement of Democratic Socialists	3.2
Other	2.4

Government Expenditures [21]

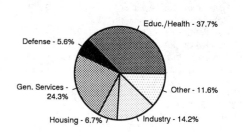

Educ./Health - 37.7%
Defense - 5.6%
Gen. Services - 24.3%
Housing - 6.7%
Industry - 14.2%
Other - 11.6%

% distribution for 1991. Expend. in 1993 est.: 5.5 ($ bil.)

Crime [23]

Military Expenditures and Arms Transfers [22]

	1985	1986	1987	1988	1989
Military expenditures					
Current dollars (mil.)	277[e]	282[e]	269[e]	238	273
1989 constant dollars (mil.)	315[e]	313[e]	289[e]	247	273
Armed forces (000)	38[e]	38[e]	38[e]	40	40
Gross national product (GNP)					
Current dollars (mil.)	7,724	7,801	8,459	8,873	9,616
1989 constant dollars (mil.)	8,793	8,657	9,098	9,237	9,616
Central government expenditures (CGE)					
1989 constant dollars (mil.)	3,573	3,754	3,482	3,484	3,669
People (mil.)	7.2	7.4	7.6	7.8	7.9
Military expenditure as % of GNP	3.6	3.6	3.2	2.7	2.8
Military expenditure as % of CGE	8.8	8.3	8.3	7.1	7.4
Military expenditure per capita	44	42	38	32	34
Armed forces per 1,000 people	5.3	5.1	5.0	5.2	5.0
GNP per capita	1,218	1,169	1,199	1,190	1,212
Arms imports[6]					
Current dollars (mil.)	300	90	50	20	20
1989 constant dollars (mil.)	342	100	54	21	20
Arms exports[6]					
Current dollars (mil.)	0	0	0	0	0
1989 constant dollars (mil.)	0	0	0	0	0
Total imports[7]					
Current dollars (mil.)	2,757	2,890	3,039	3,689	4,374
1989 constant dollars	3,139	3,207	3,269	3,840	4,374
Total exports[7]					
Current dollars (mil.)	1,738	1,759	2,139	2,395	2,930
1989 constant dollars	1,979	1,952	2,301	2,493	2,930
Arms as percent of total imports[8]	10.9	3.1	1.6	0.5	0.5
Arms as percent of total exports[8]	0	0	0	0	0

Human Rights [24]

	SSTS	FL	FAPRO	PPCG	APROBC	TPW	PCPTW	STPEP	PHRFF	PRW	ASST	AFL
Observes	P	P	P	P	P	P	P			P	P	P
	EAFRD	CPR	ESCR	SR	ACHR	MAAE	PVIAC	PVNAC	EAFDAW	TCIDTP	RC	
Observes		P	P	P	P			P	P	P	P	

P = Party; S = Signatory; see Appendix for meaning of abbreviations.

Labor Force

Total Labor Force [25]

2.25 million

Labor Force by Occupation [26]

Agriculture   32%

Unemployment Rate [27]

15.7% (1992)

For sources, notes, and explanations, see Annotated Source Appendix, page 1035.

Production Sectors

Commercial Energy Production and Consumption

Production [28]

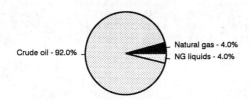

Crude oil - 92.0%

Natural gas - 4.0%
NG liquids - 4.0%

Consumption [29]

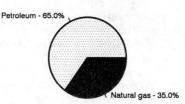

Petroleum - 65.0%

Natural gas - 35.0%

Data are shown in quadrillion (10^{15}) BTUs and percent for 1991

Crude oil[1]	0.23
Natural gas liquids	0.01
Dry natural gas	0.01
Coal[2]	0.00
Net hydroelectric power[3]	(s)
Net nuclear power[3]	0.00
Total	0.25

Petroleum	0.13
Dry natural gas	0.07
Coal[2]	(s)
Net hydroelectric power[3]	(s)
Net nuclear power[3]	0.00
Total	0.21

Telecommunications [30]

- The system is above the African average
- Facilities consist of open-wire lines, coaxial cable, and microwave radio relay
- Key centers are Sfax, Sousse, Bizerte, and Tunis
- 233,000 telephones (28 telephones per 1,000 persons)
- Broadcast stations - 7 AM, 8 FM, 19 TV
- 5 submarine cables
- Satellite earth stations - 1 Atlantic Ocean INTELSAT and 1 ARABSAT with back-up control station
- Coaxial cable and microwave radio relay to Algeria and Libya

Transportation [31]

Railroads. 2,115 km total; 465 km 1.435-meter (standard) gauge; 1,650 km 1.000-meter gauge

Highways. 17,700 km total; 9,100 km bituminous; 8,600 km improved and unimproved earth

Merchant Marine. 22 ships (1,000 GRT or over) totaling 161,661 GRT/221,959 DWT; includes 1 short-sea passenger, 4 cargo, 2 roll-on/roll-off cargo, 2 oil tanker, 6 chemical tanker, 1 liquefied gas, 6 bulk

Airports

Total:	29
Usable:	26
With permanent-surface runways:	13
With runways over 3,659 m:	0
With runways 2,440-3,659 m:	7
With runways 1,220-2,439 m:	7

Top Agricultural Products [32]

	1990	1991
Grains	1,633	2,551
Vegetables	1,807	1,948
Olives for oil	650	825
Oranges	123	117
Dates	75	81
Almonds	52	55

Values shown are 1,000 metric tons.

Top Mining Products [33]

Metric tons unless otherwise specified	M.t.
Petroleum, crude (000 barrels)	38,690[1]
Phosphate rock (000 tons)	6,400[1]
Barite	322,366
Flourspar	37,580
Zinc, mine output, Zn content	9,353
Cement, hydraulic (000 tons)	3,300

Tourism [34]

	1987	1988	1989	1990	1991
Visitors	1,933	3,515	3,272	3,249	3,268
Tourists[4]	1,875	3,468	3,222	3,204	3,224
Cruise passengers	58	47	50	45	44
Tourism receipts	672	1,234	933	953	685
Tourism expenditures	95	119	134	179	129
Fare receipts	167	180	188	212	202
Fare expenditures	74	77	87	109	95

Tourists are in thousands, money in million U.S. dollars.

For sources, notes, and explanations, see Annotated Source Appendix, page 1035.

939

Manufacturing Sector

GDP and Manufacturing Summary [35]

	1980	1985	1989	1990	% change 1980-1990	% change 1989-1990
GDP (million 1980 $)	8,742	10,733	11,699	12,646	44.7	8.1
GDP per capita (1980 $)	1,369	1,478	1,463	1,544	12.8	5.5
Manufacturing as % of GDP (current prices)	13.6	13.5	16.4	17.0	25.0	3.7
Gross output (million $)	3,579	3,449[e]	5,039[e]	5,547[e]	55.0	10.1
Value added (million $)	939	949[e]	1,469[e]	1,612[e]	71.7	9.7
Value added (million 1980 $)	1,030	1,443	1,744[e]	1,982	92.4	13.6
Industrial production index	100	126	128	143	43.0	11.7
Employment (thousands)	125	165[e]	195[e]	213[e]	70.4	9.2

Note: GDP stands for Gross Domestic Product. 'e' stands for estimated value.

Profitability and Productivity

	1980	1985	1989	1990	% change 1980-1990	% change 1989-1990
Intermediate input (%)	74	72[e]	71[e]	71[e]	-4.1	0.0
Wages, salaries, and supplements (%)	12	13[e]	15[e]	15[e]	25.0	0.0
Gross operating surplus (%)	14	14[e]	15[e]	14[e]	0.0	-6.7
Gross output per worker ($)	28,669	20,841[e]	25,899[e]	26,013[e]	-9.3	0.4
Value added per worker ($)	7,525	5,853[e]	7,550[e]	7,905[e]	5.0	4.7
Average wage (incl. benefits) ($)	3,499	2,811[e]	3,784[e]	3,834[e]	9.6	1.3

Profitability is in percent of gross output. Productivity is in U.S. $. 'e' stands for estimated value.

Profitability - 1990

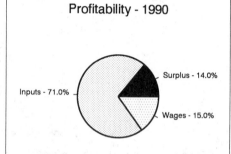

Surplus - 14.0%
Inputs - 71.0%
Wages - 15.0%

The graphic shows percent of gross output.

Value Added in Manufacturing

	1980 $ mil.	1980 %	1985 $ mil.	1985 %	1989 $ mil.	1989 %	1990 $ mil.	1990 %	% change 1980-1990	% change 1989-1990
311 Food products	96	10.2	78[e]	8.2	112[e]	7.6	120[e]	7.4	25.0	7.1
313 Beverages	49	5.2	54[e]	5.7	87[e]	5.9	93[e]	5.8	89.8	6.9
314 Tobacco products	22	2.3	22[e]	2.3	37[e]	2.5	36[e]	2.2	63.6	-2.7
321 Textiles	55	5.9	60[e]	6.3	104[e]	7.1	83[e]	5.1	50.9	-20.2
322 Wearing apparel	92	9.8	87[e]	9.2	171[e]	11.6	122[e]	7.6	32.6	-28.7
323 Leather and fur products	6	0.6	6[e]	0.6	8[e]	0.5	10[e]	0.6	66.7	25.0
324 Footwear	21	2.2	21[e]	2.2	33[e]	2.2	38[e]	2.4	81.0	15.2
331 Wood and wood products	12	1.3	12[e]	1.3	15[e]	1.0	21[e]	1.3	75.0	40.0
332 Furniture and fixtures	13	1.4	12[e]	1.3	20[e]	1.4	16[e]	1.0	23.1	-20.0
341 Paper and paper products	24	2.6	21[e]	2.2	30[e]	2.0	34[e]	2.1	41.7	13.3
342 Printing and publishing	17	1.8	16[e]	1.7	20[e]	1.4	26[e]	1.6	52.9	30.0
351 Industrial chemicals	42[e]	4.5	26[e]	2.7	29[e]	2.0	43[e]	2.7	2.4	48.3
352 Other chemical products	96[e]	10.2	77[e]	8.1	122[e]	8.3	140[e]	8.7	45.8	14.8
353 Petroleum refineries	13	1.4	10[e]	1.1	13[e]	0.9	14[e]	0.9	7.7	7.7
354 Miscellaneous petroleum and coal products	NA	0.0	NA	0.0	NA	0.0	NA	0.0	NA	NA
355 Rubber products	8	0.9	11[e]	1.2	10[e]	0.7	20[e]	1.2	150.0	100.0
356 Plastic products	18	1.9	22[e]	2.3	30[e]	2.0	35[e]	2.2	94.4	16.7
361 Pottery, china and earthenware	11	1.2	9[e]	0.9	12[e]	0.8	14[e]	0.9	27.3	16.7
362 Glass and glass products	7	0.7	6[e]	0.6	9[e]	0.6	11[e]	0.7	57.1	22.2
369 Other non-metal mineral products	156	16.6	149[e]	15.7	280[e]	19.1	246[e]	15.3	57.7	-12.1
371 Iron and steel	45	4.8	81[e]	8.5	111[e]	7.6	144[e]	8.9	220.0	29.7
372 Non-ferrous metals	8	0.9	7[e]	0.7	6[e]	0.4	12[e]	0.7	50.0	100.0
381 Metal products	53	5.6	77[e]	8.1	120[e]	8.2	151[e]	9.4	184.9	25.8
382 Non-electrical machinery	2	0.2	2[e]	0.2	2[e]	0.1	4[e]	0.2	100.0	100.0
383 Electrical machinery	35	3.7	38[e]	4.0	52[e]	3.5	85[e]	5.3	142.9	63.5
384 Transport equipment	30	3.2	38[e]	4.0	26[e]	1.8	85[e]	5.3	183.3	226.9
385 Professional and scientific equipment	1	0.1	1[e]	0.1	2[e]	0.1	2[e]	0.1	100.0	0.0
390 Other manufacturing industries	5	0.5	6[e]	0.6	7[e]	0.5	9[e]	0.6	80.0	28.6

Note: The industry codes shown are International Standard Industry codes (ISIC). Percentages are percent of total Value Added. 'e' stands for estimated value

Finance, Economics, and Trade

Economic Indicators [36]

All figures are in Millions of Dinar.

	1989[r]	1990[r]	1991[r]
Real GDP (1989 base year)	8,924	9,504	9,770
Real GDP growth rate	3.5	6.5	2.8
Money Supply (M1)	2,494	2,643	2,807
Commercial Interest rates (%.)	< = 14	NA	NA
Savings rate (%.)	Avg. 8	NA	NA
Investment rate (%.)	< = 14	NA	NA
CPI	154.9	165.0	178.2
WPI	NA	NA	NA
External Public Debt	5,350	5,730	6,682

Balance of Payments Summary [37]

Values in millions of dollars.

	1987	1988	1989	1990	1991
Exports of goods (f.o.b.)	2,101.0	2,399.0	2,931.0	3,515.0	4,021.0
Imports of goods (f.o.b.)	-2,829.0	-3,496.0	-4,139.0	-5,193.0	-4,895.0
Trade balance	-728.0	-1,097.0	-1,208.0	-1,678.0	-874.0
Services - debits	-1,119.0	-1,256.0	-1,280.0	-1,452.0	-1,482.0
Services - credits	1,267.0	1,905.0	1,631.0	1,787.0	1,456.0
Private transfers (net)	481.0	548.0	485.0	594.0	578.0
Government transfers (net)	37.0	115.0	213.0	225.0	131.0
Long term capital (net)	129.0	100.0	119.0	71.0	414.0
Short term capital (net)	8.0	98.0	179.0	665.0	-25.0
Errors and omissions	48.0	27.0	-73.0	-336.0	-253.0
Overall balance	123.0	440.0	66.0	-124.0	-55.0

Exchange Rates [38]

Currency: **Tunisian dinars.**
Symbol: **TD.**

Data are currency units per $1.

February 1993	0.9931
1992	0.8844
1991	0.9246
1990	0.8783
1989	0.9493
1988	0.8578

Imports and Exports

Top Import Origins [39]

$6.1 billion (c.i.f., 1992).

Origins	%
EC countries	67
US	6
Canada	NA
Japan	NA
Switzerland	NA
Turkey	NA
Algeria	NA

Top Export Destinations [40]

$3.7 billion (f.o.b., 1992).

Destinations	%
EC countries	74
Middle East	11
US	2
Turkey	NA
Former USSR republics	NA

Foreign Aid [41]

	U.S. $	
US commitments, including Ex-Im (FY70-89)	730	million
Western (non-US) countries, ODA and OOF bilateral commitments (1970-89)	5.2	billion
OPEC bilateral aid (1979-89)	684	million
Communist countries (1970-89)	410	million

Import and Export Commodities [42]

Import Commodities

Industrial goods and equipment 57%
Hydrocarbons 13%
Food 12%
Consumer goods

Export Commodities

Hydrocarbons
Agricultural products
Phosphates and chemicals

For sources, notes, and explanations, see Annotated Source Appendix, page 1035.

941

Turkey

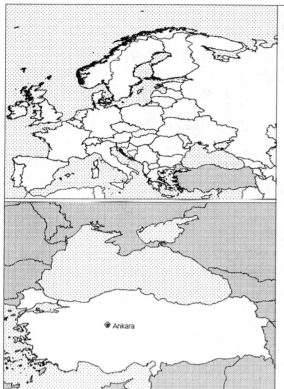

Geography [1]

Total area:
780,580 km2
Land area:
770,760 km2
Comparative area:
Slightly larger than Texas
Land boundaries:
Total 2,627 km, Armenia 268 km, Azerbaijan 9 km, Bulgaria 240 km, Georgia 252 km, Greece 206 km, Iran 499 km, Iraq 331 km, Syria 822 km
Coastline:
7,200 km
Climate:
Temperate; hot, dry summers with mild, wet winters; harsher in interior
Terrain:
Mostly mountains; narrow coastal plain; high central plateau (Anatolia)
Natural resources:
Antimony, coal, chromium, mercury, copper, borate, sulphur, iron ore
Land use:
Arable land:
30%
Permanent crops:
4%
Meadows and pastures:
12%
Forest and woodland:
26%
Other:
28%

Demographics [2]

	1960	1970	1980	1990	1991[1]	1994	2000	2010	2020
Population	28,217	35,758	45,121	57,130	58,581	62,154	69,624	81,790	93,362
Population density (persons per sq. mi.)	95	120	152	192	197	NA	236	281	324
Births	NA	NA	NA	NA	1,663	1,615	NA	NA	NA
Deaths	NA	NA	NA	NA	363	360	NA	NA	NA
Life expectancy - males	NA	NA	NA	NA	68	69	NA	NA	NA
Life expectancy - females	NA	NA	NA	NA	72	73	NA	NA	NA
Birth rate (per 1,000)	NA	NA	NA	NA	28	26	NA	NA	NA
Death rate (per 1,000)	NA	NA	NA	NA	6	6	NA	NA	NA
Women of reproductive age (15-44 yrs.)	NA	NA	NA	13,795	NA	15,381	17,892	NA	NA
of which are currently married	NA	NA	NA	10,140	NA	11,403	13,425	NA	NA
Fertility rate	NA	NA	NA	3.6	3.56	3.2	2.8	2.4	2.2

Population values are in thousands, life expectancy in years, and other items as indicated.

Health

Health Personnel [3]

Doctors per 1,000 pop., 1988-92	0.74
Nurse-to-doctor ratio, 1988-92	1.5
Hospital beds per 1,000 pop., 1985-90	2.1
Percentage of children immunized (age 1 yr. or less)	
Third dose of DPT, 1990-91	72.0
Measles, 1990-91	66.0

Health Expenditures [4]

Total health expenditure, 1990 (official exchange rate)	
Millions of dollars	4281
Millions of dollars per capita	76
Health expenditures as a percentage of GDP	
Total	4.0
Public sector	1.5
Private sector	2.5
Development assistance for health	
Total aid flows (millions of dollars)[1]	23
Aid flows per capita (millions of dollars)	0.4
Aid flows as a percentage of total health expenditure	0.5

For sources, notes, and explanations, see Annotated Source Appendix, page 1035.

Human Factors

Health Care Ratios [5]

Population per physician, 1970	2230
Population per physician, 1990	1260
Population per nursing person, 1970	1010
Population per nursing person, 1990	NA
Percent of births attended by health staff, 1985	78

Infants and Malnutrition [6]

Percent of babies with low birth weight, 1985	7
Infant mortality rate per 1,000 live births, 1970	147
Infant mortality rate per 1,000 live births, 1991	58
Years of life lost per 1,000 population, 1990	31
Prevalence of malnutrition (under age 5), 1990	NA

Ethnic Division [7]

Turkish	80%
Kurdish	20%

Religion [8]

Muslim (mostly Sunni)	99.8%
Other (Christian and Jews)	0.2%

Major Languages [9]

Turkish (official), Kurdish, Arabic.

Education

Public Education Expenditures [10]

Billion Turkish Lira	1980	1985	1987	1988[6]	1989[6]	1990[6]
Total education expenditure	117.7	627.1	NA	1,797.4	2,967.1	8,506.5
as percent of GNP	2.8	2.3	NA	1.8	1.8	3.1
as percent of total govt. expend.	10.5	NA	NA	NA	NA	13.3
Current education expenditure	98.6	523.1	NA	1,470.3	2,418.1	7,580.8
as percent of GNP	2.3	1.9	NA	1.5	1.5	2.7
as percent of current govt. expend.	NA	NA	NA	NA	NA	NA
Capital expenditure	19.2	104.0	NA	327.1	549.0	NA

Educational Attainment [11]

Age group	25+
Total population	18,277,340
Highest level attained (%)	
No schooling	52.4
First level	
Incompleted	35.3
Completed	NA
Entered second level	
S-1	8.7
S-2	NA
Post secondary	3.6

Literacy Rate [12]

In thousands and percent	1985[a]	1991[a]	2000[a]
Illiterate population +15 years	7,689	7,046	7,459
Illiteracy rate - total pop. (%)	24.0	19.3	16.4
Illiteracy rate - males (%)	12.4	10.3	8.5
Illiteracy rate - females (%)	35.7	28.9	24.5

Libraries [13]

	Admin. Units	Svc. Pts.	Vols. (000)	Shelving (meters)	Vols. Added	Reg. Users
National (1989)	1	1	1,016	NA	26,237	31,250
Non-specialized	NA	NA	NA	NA	NA	NA
Public (1989)	NA	854	7,339	NA	374,445	719,417
Higher ed. (1987)[25]	139	241	3,462	117,515	189,632	342,175
School	NA	NA	NA	NA	NA	NA

Daily Newspapers [14]

	1975	1980	1985	1990
Number of papers	500[e]	400[e]	366	399
Circ. (000)	2,000[e]	2,500[e]	3,020	4,000

Cinema [15]

Data for 1991. Note that the first week's gross for *Jurrasic Park* in Turkey was $410,956.

Cinema seats per 1,000	3.1
Annual attendance per person	0.3
Gross box office receipts (mil. Lira)	NA

Science and Technology

Scientific/Technical Forces [16]

Potential scientists/engineers[22]	708,000
Number female[22]	163,000
Potential technicians[23]	830,000
Number female[23]	285,000
Total	1,538,000

R&D Expenditures [17]

	Lira (000) 1985
Total expenditure	192,465,000
Capital expenditure	491,150,000
Current expenditure	143,315,000
Percent current	74.5

U.S. Patents Issued [18]

Values show patents issued to citizens of the country by the U.S. Patents Office.

	1990	1991	1992
Number of patents	3	1	4

Government and Law

Organization of Government [19]

Long-form name:
Republic of Turkey
Type:
republican parliamentary democracy
Independence:
29 October 1923 (successor state to the Ottoman Empire)
Constitution:
7 November 1982
Legal system:
derived from various continental legal systems; accepts compulsory ICJ jurisdiction, with reservations
National holiday:
Anniversary of the Declaration of the Republic, 29 October (1923)
Executive branch:
president, Presidential Council, prime minister, deputy prime minister, Cabinet
Legislative branch:
unicameral Grand National Assembly (Buyuk Millet Meclisi)
Judicial branch:
Court of Cassation

Political Parties [20]

Grand National Assembly	% of votes
Correct Way Party (DYP)	27.0
Motherland Party (ANAP)	24.0
Social Democratic Populist (SHP)	20.8
Refah Party (RP)	16.9
Democratic Left (DSP)	10.8
Socialist Unity (SBP)	0.4
Independent	0.1

Government Expenditures [21]

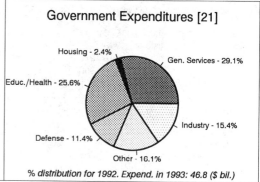

Housing - 2.4%
Gen. Services - 29.1%
Educ./Health - 25.6%
Industry - 15.4%
Defense - 11.4%
Other - 16.1%

% distribution for 1992. Expend. in 1993: 46.8 ($ bil.)

Military Expenditures and Arms Transfers [22]

	1985	1986	1987	1988	1989
Military expenditures					
Current dollars (mil.)	2,552	3,010	2,980	2,826	3,150
1989 constant dollars (mil.)	2,906	3,340	3,205	2,942	3,150
Armed forces (000)	814	860	879	847	780
Gross national product (GNP)					
Current dollars (mil.)	55,550	61,500	68,290	73,050	77,280
1989 constant dollars (mil.)	63,240	68,250	73,450	76,050	77,280
Central government expenditures (CGE)					
1989 constant dollars (mil.)	16,210	14,860	16,550	16,650	18,320
People (mil.)	51.0	52.2	53.5	54.7	56.0
Military expenditure as % of GNP	4.6	4.9	4.4	3.9	4.1
Military expenditure as % of CGE	17.9	22.5	19.4	17.7	17.2
Military expenditure per capita	57	64	60	54	56
Armed forces per 1,000 people	16.0	16.5	16.4	15.5	13.9
GNP per capita	1,241	1,308	1,374	1,390	1,380
Arms imports[6]					
Current dollars (mil.)	450	625	950	975	1,100
1989 constant dollars (mil.)	512	694	1,022	1,015	1,100
Arms exports[6]					
Current dollars (mil.)	120	0	10	10	0
1989 constant dollars (mil.)	137	0	11	10	0
Total imports[7]					
Current dollars (mil.)	11,270	11,020	14,160	14,330	15,760
1989 constant dollars	12,840	12,230	15,230	14,920	15,760
Total exports[7]					
Current dollars (mil.)	7,958	7,466	10,190	11,660	11,630
1989 constant dollars	9,059	8,285	10,960	12,140	11,630
Arms as percent of total imports[8]	4.0	5.7	6.7	6.8	7.0
Arms as percent of total exports[8]	1.5	0	0.1	0.1	0

Crime [23]

Crime volume	
Cases known to police	76,535
Attempts (percent)	na
Percent cases solved	78
Crimes per 100,000 persons	134.27
Persons responsible for offenses	
Total number offenders	103,933
Percent female	5.2
Percent juvenile	na
Percent foreigners	na

Human Rights [24]

	SSTS	FL	FAPRO	PPCG	APROBC	TPW	PCPTW	STPEP	PHRFF	PRW	ASST	AFL
Observes	P	S		P	P	P	P		P	P	P	P
	EAFRD	CPR	ESCR	SR	ACHR	MAAE	PVIAC	PVNAC	EAFDAW	TCIDTP	RC
Observes		S		P					P	P	S

P=Party; S=Signatory; see Appendix for meaning of abbreviations.

Labor Force

Total Labor Force [25]

20.7 million

Labor Force by Occupation [26]

Agriculture	50%
Services	35
Industry	15

Unemployment Rate [27]

11.1% (1992 est.)

For sources, notes, and explanations, see Annotated Source Appendix, page 1035.

Production Sectors

Commercial Energy Production and Consumption

Production [28]

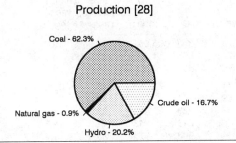

Coal - 62.3%
Crude oil - 16.7%
Natural gas - 0.9%
Hydro - 20.2%

Consumption [29]

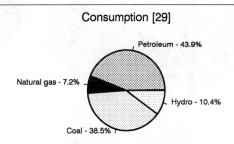

Petroleum - 43.9%
Natural gas - 7.2%
Hydro - 10.4%
Coal - 38.5%

Data are shown in quadrillion (10^{15}) BTUs and percent for 1991

Crude oil[1]	0.19
Natural gas liquids	0.00
Dry natural gas	0.01
Coal[2]	0.71
Net hydroelectric power[3]	0.23
Net nuclear power[3]	0.00
Total	1.14

Petroleum	0.97
Dry natural gas	0.16
Coal[2]	0.85
Net hydroelectric power[3]	0.23
Net nuclear power[3]	(s)
Total	2.21

Telecommunications [30]

- Fair domestic and international systems
- Trunk radio relay microwave network
- Limited open wire network
- 3,400,000 telephones
- Broadcast stations - 15 AM
- 94 FM
- 357 TV
- 1 satellite ground station operating in the INTELSAT (2 Atlantic Ocean antennas) and EUTELSAT systems
- 1 submarine cable

Transportation [31]

Railroads. 8,429 km 1.435-meter gauge (including 795 km electrified)

Highways. 320,611 km total; 138 km limited access expressways, 31,062 km national (main) roads, 27,853 km regional (secondary) roads, 261,558 km local and municipal roads; 45,526 km of hard surfaced roads (of which about 27,000 km are paved and about 18,500 km are surfaced with gravel or crushed stone) (1988 est.)

Merchant Marine. 353 ships (1,000 GRT or over) totaling 3,825,274 GRT/6,628,207 DWT; includes 7 short-sea passenger, 1 passenger-cargo, 189 cargo, 1 container, 6 roll-on/roll-off cargo, 2 refrigerated cargo, 1 livestock carrier, 39 oil tanker, 10 chemical tanker, 3 liquefied gas, 9 combination ore/oil, 2 specialized tanker, 80 bulk, 3 combination bulk

Airports

Total:	110
Usable:	102
With permanent-surface runways:	65
With runways over 3,659 m:	3
With runways 2,440-3,659 m:	32
With runways 1,220-2,439 m:	26

Top Agricultural Products [32]

	1990	1991
Wheat	15.5	16.0
Barley	6.6	6.8
Potatoes	4.3	4.5
Grapes	3.5	3.6
Corn	2.1	2.2
Apples	1.9	2.0

Values shown are 1,000 metric tons.

Top Mining Products [33]

Metric tons unless otherwise specified	M.t.
Boron concentrates	1,100,000
Emery	23,000
Celestite	65,000
Perlite	165,000[1]
Pumice	1,538,000[1,22]
Trona (000 tons)	385

Tourism [34]

	1987	1988	1989	1990	1991
Visitors	2,856	4,173	4,459	5,389	5,518
Tourists	2,468	3,715	3,921	4,799	5,158
Excursionists[70]	388	458	538	590	360
Tourism receipts	1,721	2,355	2,557	3,308	2,654
Tourism expenditures	448	358	565	520	592
Fare receipts	260	320	265	326	288

Tourists are in thousands, money in million U.S. dollars.

For sources, notes, and explanations, see Annotated Source Appendix, page 1035.

945

Manufacturing Sector

GDP and Manufacturing Summary [35]

	1980	1985	1989	1990	% change 1980-1990	% change 1989-1990
GDP (million 1980 $)	56,918	71,874	87,353	96,659	69.8	10.7
GDP per capita (1980 $)	1,281	1,428	1,595	1,729	35.0	8.4
Manufacturing as % of GDP (current prices)	22.6	25.6	26.8	26.8[e]	18.6	0.0
Gross output (million $)	29,413	32,471	56,991[e]	73,064	148.4	28.2
Value added (million $)	10,837	10,449	22,211[e]	28,958	167.2	30.4
Value added (million 1980 $)	12,770	18,457	22,928	25,450	99.3	11.0
Industrial production index	100	174	218	243	143.0	11.5
Employment (thousands)	787	844	962	975	23.9	1.4

Note: GDP stands for Gross Domestic Product. 'e' stands for estimated value.

Profitability and Productivity

	1980	1985	1989	1990	% change 1980-1990	% change 1989-1990
Intermediate input (%)	63	68	61[e]	60	-4.8	-1.6
Wages, salaries, and supplements (%)	16	10	6[e]	12[e]	-25.0	100.0
Gross operating surplus (%)	20	23	33[e]	28[e]	40.0	-15.2
Gross output per worker ($)	36,960	38,378	59,255[e]	74,807	102.4	26.2
Value added per worker ($)	13,617	12,350	23,094[e]	29,649	117.7	28.4
Average wage (incl. benefits) ($)	6,142	3,716	3,359[e]	9,013[e]	46.7	168.3

Profitability is in percent of gross output. Productivity is in U.S. $. 'e' stands for estimated value.

Profitability - 1990

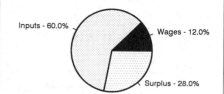

Inputs - 60.0%   Wages - 12.0%   Surplus - 28.0%

The graphic shows percent of gross output.

Value Added in Manufacturing

	1980 $ mil.	1980 %	1985 $ mil.	1985 %	1989 $ mil.	1989 %	1990 $ mil.	1990 %	% change 1980-1990	% change 1989-1990
311 Food products	1,185	10.9	973	9.3	1,882[e]	8.5	2,541	8.8	114.4	35.0
313 Beverages	335	3.1	331	3.2	581[e]	2.6	893	3.1	166.6	53.7
314 Tobacco products	467	4.3	877	8.4	1,213[e]	5.5	1,168	4.0	150.1	-3.7
321 Textiles	1,535	14.2	1,289	12.3	2,582[e]	11.6	3,222	11.1	109.9	24.8
322 Wearing apparel	60	0.6	146	1.4	665[e]	3.0	947	3.3	1,478.3	42.4
323 Leather and fur products	25	0.2	37	0.4	38[e]	0.2	60	0.2	140.0	57.9
324 Footwear	33	0.3	22	0.2	39[e]	0.2	69	0.2	109.1	76.9
331 Wood and wood products	118	1.1	64	0.6	149[e]	0.7	187	0.6	58.5	25.5
332 Furniture and fixtures	16	0.1	55	0.5	63[e]	0.3	81	0.3	406.3	28.6
341 Paper and paper products	205	1.9	241	2.3	376[e]	1.7	559	1.9	172.7	48.7
342 Printing and publishing	97	0.9	133	1.3	271[e]	1.2	434	1.5	347.4	60.1
351 Industrial chemicals	719	6.6	457	4.4	2,029[e]	9.1	1,517	5.2	111.0	-25.2
352 Other chemical products	387	3.6	394	3.8	835[e]	3.8	1,449	5.0	274.4	73.5
353 Petroleum refineries	1,352	12.5	1,514	14.5	2,706[e]	12.2	4,525	15.6	234.7	67.2
354 Miscellaneous petroleum and coal products	222	2.0	152	1.5	294[e]	1.3	458	1.6	106.3	55.8
355 Rubber products	201	1.9	151	1.4	393[e]	1.8	452	1.6	124.9	15.0
356 Plastic products	125	1.2	76	0.7	217[e]	1.0	328	1.1	162.4	51.2
361 Pottery, china and earthenware	93	0.9	102	1.0	276[e]	1.2	467	1.6	402.2	69.2
362 Glass and glass products	110	1.0	167	1.6	416[e]	1.9	531	1.8	382.7	27.6
369 Other non-metal mineral products	535	4.9	428	4.1	1,039[e]	4.7	1,365	4.7	155.1	31.4
371 Iron and steel	783	7.2	734	7.0	1,759[e]	7.9	1,403	4.8	79.2	-20.2
372 Non-ferrous metals	292	2.7	181	1.7	491[e]	2.2	580	2.0	98.6	18.1
381 Metal products	395	3.6	344	3.3	709[e]	3.2	904	3.1	128.9	27.5
382 Non-electrical machinery	506	4.7	456	4.4	1,001[e]	4.5	1,423	4.9	181.2	42.2
383 Electrical machinery	463	4.3	531	5.1	1,002[e]	4.5	1,482	5.1	220.1	47.9
384 Transport equipment	541	5.0	534	5.1	1,100[e]	5.0	1,743	6.0	222.2	58.5
385 Professional and scientific equipment	8	0.1	9	0.1	43[e]	0.2	87	0.3	987.5	102.3
390 Other manufacturing industries	28	0.3	49	0.5	42[e]	0.2	84	0.3	200.0	100.0

Note: The industry codes shown are International Standard Industry codes (ISIC). Percentages are percent of total Value Added. 'e' stands for estimated value

Finance, Economics, and Trade

Economic Indicators [36]

Billions of Turkish Lira (TL) unless otherwise noted.

	1989	1990	1991
Real GNP (87 prod.'s val.)	78,469	86,208	NA
Real GNP growth rate (%)	0.9	9.9	2.2[P]
Per capita GNP (US$)	2,005	2,667	NA
Money supply (M1)[159]	18,232	27,354	39,430
Savings rate			
(dom svgs/GNP)[160]	17.8	15.6[161]	NA
Investment (fixed) rate[160]	16.3	15.5	NA
CPI (% chg)[162]	64.3	60.4	71.1
WPI (% chg)[162]	62.3	48.6	59.2
External public debt	41,751	49,035	45,598[163]

Balance of Payments Summary [37]

Values in millions of dollars.

	1987	1988	1989	1990	1991
Exports of goods (f.o.b.)	10,322.0	11,929.0	11,780.0	13,026.0	13,672.0
Imports of goods (f.o.b.)	-13,551.0	-13,706.0	-15,999.0	-22,581.0	-20,998.0
Trade balance	-3,229.0	-1,777.0	-4,219.0	-9,555.0	-7,326.0
Services - debits	-4,162.0	-4,812.0	-5,476.0	-6,496.0	-6,816.0
Services - credits	4,195.0	6,026.0	7,098.0	8,933.0	9,315.0
Private transfers (net)	2,066.0	1,827.0	3,135.0	3,349.0	2,854.0
Government transfers (net)	324.0	332.0	423.0	1,144.0	2,245.0
Long term capital (net)	1,841.0	1,323.0	1,364.0	1,037.0	623.0
Short term capital (net)	50.0	-2,281.0	-528.0	3,000.0	-3,020.0
Errors and omissions	-505.0	515.0	915.0	-469.0	926.0
Overall balance	580.0	1,153.0	2,712.0	943.0	-1,199.0

Exchange Rates [38]

Currency: **Turkish liras.**
Symbol: **TL.**

Data are currency units per $1.

January 1993	8,814.3
1992	6,872.4
1991	4,171.8
1990	2,608.6
1989	2,121.7
1988	1,422.3

Imports and Exports

Top Import Origins [39]

$21.1 billion (c.i.f., 1991).

Origins	%
EC countries	44
US	12
Former USSR	5

Top Export Destinations [40]

$13.7 billion (f.o.b., 1991).

Destinations	%
EC countries	51
US	7
Iran	5
Former USSR	5

Foreign Aid [41]

	U.S. $	
US commitments, including Ex-Im (FY70-89)	2.3	billion
Western (non-US) countries, ODA and OOF bilateral commitments (1970-89)	10.1	billion
OPEC bilateral aid (1979-89)	665	million
Communist countries (1970-89)	4.5	billion
Aid for Persian Gulf war (1991)	4.1	billion
Aid pledged for Turkish Defense Fund	2.5	billion

Import and Export Commodities [42]

Import Commodities

Manufactured goods 61%
Foodstuffs 8%
Fuels 21%

Export Commodities

Manufactured goods 69%
Foodstuffs 22%
Fuels 2%

For sources, notes, and explanations, see Annotated Source Appendix, page 1035.

947

Turkmenistan

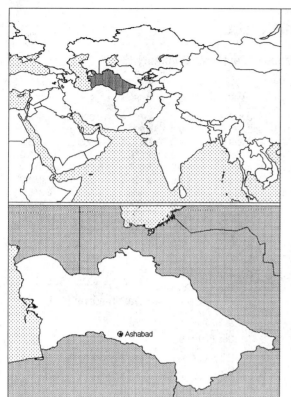

Geography [1]

Total area:
488,100 km2
Land area:
488,100 km2
Comparative area:
Slightly larger than California
Land boundaries:
Total 3,736 km, Afghanistan 744 km, Iran 992 km, Kazakhstan 379 km, Uzbekistan 1,621 km
Coastline:
0 km
Climate:
Subtropical desert
Terrain:
Flat-to-rolling sandy desert with dunes; borders Caspian Sea in west
Natural resources:
Petroleum, natural gas, coal, sulphur, salt
Land use:
Arable land:
3%
Permanent crops:
0%
Meadows and pastures:
69%
Forest and woodland:
0%
Other:
28%

Demographics [2]

	1960	1970	1980	1990	1991	1994	2000	2010	2020
Population	1,585	2,181	2,875	3,660	NA	3,995	4,474	5,277	6,116
Population density (persons per sq. mi.)	NA	NA	NA	NA	NA	NA	NA	NA	NA
Births	NA	NA	NA	NA	NA	122	NA	NA	NA
Deaths	NA	NA	NA	NA	NA	30	NA	NA	NA
Life expectancy - males	NA	NA	NA	NA	NA	62	NA	NA	NA
Life expectancy - females	NA	NA	NA	NA	NA	69	NA	NA	NA
Birth rate (per 1,000)	NA	NA	NA	NA	NA	30	NA	NA	NA
Death rate (per 1,000)	NA	NA	NA	NA	NA	7	NA	NA	NA
Women of reproductive age (15-44 yrs.)	NA	NA	NA	872	NA	981	1,156	NA	NA
of which are currently married	NA	NA	NA	547	NA	625	741	NA	NA
Fertility rate	NA	NA	NA	4.4	NA	3.8	3.4	2.9	2.5

Population values are in thousands, life expectancy in years, and other items as indicated.

Health

Health Personnel [3]

Doctors per 1,000 pop., 1988-92	3.57
Nurse-to-doctor ratio, 1988-92	2.8
Hospital beds per 1,000 pop., 1985-90	11.3
Percentage of children immunized (age 1 yr. or less)	
Third dose of DPT, 1990-91	78.0
Measles, 1990-91	68.0

Health Expenditures [4]

Total health expenditure, 1990 (official exchange rate)	
Millions of dollars	459
Millions of dollars per capita	125
Health expenditures as a percentage of GDP	
Total	5.0
Public sector	3.3
Private sector	1.7
Development assistance for health	
Total aid flows (millions of dollars)[1]	2.0
Aid flows per capita (millions of dollars)	0.5
Aid flows as a percentage of total health expenditure	0.4

For sources, notes, and explanations, see Annotated Source Appendix, page 1035.

Human Factors

Health Care Ratios [5]	
Population per physician, 1970	NA
Population per physician, 1990	290
Population per nursing person, 1970	NA
Population per nursing person, 1990	NA
Percent of births attended by health staff, 1985	NA

Infants and Malnutrition [6]	
Percent of babies with low birth weight, 1985	NA
Infant mortality rate per 1,000 live births, 1970	NA
Infant mortality rate per 1,000 live births, 1991	56
Years of life lost per 1,000 population, 1990	29
Prevalence of malnutrition (under age 5), 1990	NA

Ethnic Division [7]	
Turkmen	73.3%
Russian	9.8%
Uzbek	9.0%
Kazakhs	2.0%
Other	5.9%

Religion [8]	
Muslim	87%
Eastern Orthodox	11%
Unknown	2%

Major Languages [9]	
Turkmen	72%
Russian	12%
Uzbek	9%
Other	7%

Education

Public Education Expenditures [10]

Educational Attainment [11]

Literacy Rate [12]

	1989[b]	1990[b]	1991[b]
Illiterate population +15 years	NA	NA	NA
Illiteracy rate - total pop. (%)	2.3	NA	NA
Illiteracy rate - males (%)	1.2	NA	NA
Illiteracy rate - females (%)	3.4	NA	NA

Libraries [13]

Daily Newspapers [14]

Cinema [15]

Science and Technology

Scientific/Technical Forces [16]

R&D Expenditures [17]

U.S. Patents Issued [18]

For sources, notes, and explanations, see Annotated Source Appendix, page 1035.

Government and Law

Organization of Government [19]

Long-form name:
Republic of Turkmenistan
Type:
republic
Independence:
27 October 1991 (from the Soviet Union)
Constitution:
adopted 18 May 1992
Legal system:
based on civil law system
National holiday:
Independence Day, 27 October (1991)
Executive branch:
president, prime minister, nine deputy prime ministers, Council of Ministers
Legislative branch:
under 1992 constitution there are two parliamentary bodies, a unicameral People's Council (Halk Maslahaty - having more than 100 members and meeting infrequently) and a 50-member unicameral Assembly (Majlis)
Judicial branch:
Supreme Court

Elections [20]

Majlis. Last held 7 January 1990 (next to be held NA 1995); results - percent of vote by party NA; seats - (175 total) elections not officially by party, but Communist Party members won nearly 90% of seats; note - seats to be reduced to 50 at next election.

Government Budget [21]

For 1993.
Revenues	NA
Expenditures	NA
Capital expenditures	NA

Defense Summary [22]

Branches: National Guard, Republic Security Forces (internal and border troops), Joint Command Turkmenistan/Russia (Ground, Navy or Caspian Sea Flotilla, Air, and Air Defense)

Manpower Availability: Males age 15-49 933,285; fit for military service 765,824; reach military age (18) annually 39,254 (1993 est.)

Defense Expenditures: No information available.

Crime [23]

Human Rights [24]

	SSTS	FL	FAPRO	PPCG	APROBC	TPW	PCPTW	STPEP	PHRFF	PRW	ASST	AFL
Observes						P	P					
	EAFRD	CPR	ESCR	SR	ACHR	MAAE	PVIAC	PVNAC	EAFDAW	TCIDTP	RC	
Observes							P	P				

P = Party; S = Signatory; see Appendix for meaning of abbreviations.

Labor Force

Total Labor Force [25]

1.542 million

Labor Force by Occupation [26]

Agriculture and forestry	42%
Industry and construction	21
Other	37

Date of data: 1990

Unemployment Rate [27]

15%-20% (1992 est.)

Production Sectors

Energy Resource Summary [28]

Energy Resources: Petroleum, natural gas, coal. **Electricity**: 2,920,000 kW capacity; 13,100 million kWh produced, 3,079 kWh per capita (1992). **Pipelines**: Crude oil 250 km, natural gas 4,400 km.

Telecommunications [30]

- Poorly developed
- Only 65 telephones per 1000 persons (1991)
- Linked by cable and microwave to other CIS republics and to other countries by leased connections to the Moscow international gateway switch
- A new direct telephone link from Ashgabat (Ashkhabad) to Iran has been established
- Satellite earth stations - 1 Orbita and 1 INTELSAT for TV receive-only service
- A newly installed satellite earth station provides TV receiver-only capability for Turkish broadcasts

Transportation [31]

Railroads. 2,120 km; does not include industrial lines (1990)

Highways. 23,000 km total; 18,300 km hard surfaced, 4,700 km earth (1990)

Airports

Total:	7
Useable:	7
With permanent-surface runways:	4
With runways over 3,659 m:	0
With runways 2,440-3,659 m:	0
With runways 1,220-2,439 m:	4

Top Agricultural Products [32]

	1990	1991
Grains	400	500
Grapes	169	140
Cotton	1,457	-
Vegetables	411	-
Rice	42	-
Fruits and berries	47	-

Values shown are 1,000 metric tons.

Top Mining Products [33]

Detailed information is not available. A summary of mineral resources available follows. **Mineral Resources**: Petroleum, natural gas, coal, sulphur, salt.

Tourism [34]

Finance, Economics, and Trade

Industrial Summary [35]

Industrial Production: Growth rate - 17% (1992 est.). **Industries**: Oil and gas, petrochemicals, fertilizers, food processing, textiles.

Economic Indicators [36]

National product: GDP not available. **National product real growth rate**: - 10% (1992 est.). **National product per capita**: not available. **Inflation rate (consumer prices)**: 53% per month (first quarter 1993). **External debt**: $650 million (end 1991 est.).

Balance of Payments Summary [37]

Exchange Rates [38]

Currency: **rubles.**
Symbol: **R.**

Subject to wide fluctuations. Data are currency units per $1.

December 24, 1994	415

Imports and Exports

Top Import Origins [39]

$100 million from outside the successor states of the former USSR (1992).

Origins	%
No details available	NA

Top Export Destinations [40]

$100 million to outside the successor states of the former USSR (1992).

Destinations	%
Russia	NA
Ukraine	NA
Uzbekistan	NA

Foreign Aid [41]

	U.S. $	
Offical aid commitments by foreign donors (1992)	280	million

Import and Export Commodities [42]

Import Commodities	**Export Commodities**
Machinery & parts	Natural gas
Plastics and rubber	Oil
Consumer durables	Chemicals
Textiles	Cotton
	Textiles
	Carpets

Uganda

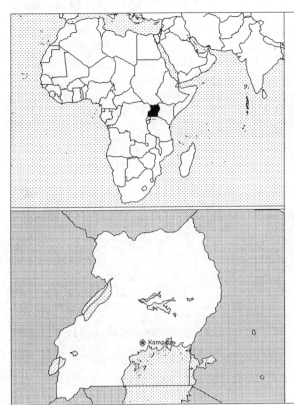

Geography [1]

Total area:
236,040 km2
Land area:
199,710 km2
Comparative area:
Slightly smaller than Oregon
Land boundaries:
Total 2,698 km, Kenya 933 km, Rwanda 169 km, Sudan 435 km, Tanzania 396 km, Zaire 765 km
Coastline:
0 km (landlocked)
Climate:
Tropical; generally rainy with two dry seasons (December to February, June to August); semiarid in northeast
Terrain:
Mostly plateau with rim of mountains
Natural resources:
Copper, cobalt, limestone, salt
Land use:
Arable land:
23%
Permanent crops:
9%
Meadows and pastures:
25%
Forest and woodland:
30%
Other:
13%

Demographics [2]

	1960	1970	1980	1990	1991[1]	1994	2000	2010	2020
Population	7,286	9,806	12,765	17,745	18,690	19,859	22,748	26,997	34,106
Population density (persons per sq. mi.)	94	127	166	234	242	NA	335	467	628
Births	NA	NA	NA	NA	960	979	NA	NA	NA
Deaths	NA	NA	NA	NA	275	48	NA	NA	NA
Life expectancy - males	NA	NA	NA	NA	50	37	NA	NA	NA
Life expectancy - females	NA	NA	NA	NA	52	38	NA	NA	NA
Birth rate (per 1,000)	NA	NA	NA	NA	51	49	NA	NA	NA
Death rate (per 1,000)	NA	NA	NA	NA	15	24	NA	NA	NA
Women of reproductive age (15-44 yrs.)	NA	NA	NA	3,859	NA	4,264	4,839	NA	NA
of which are currently married	NA	NA	NA	2,805	NA	3,087	3,476	NA	NA
Fertility rate	NA	NA	NA	7.4	7.29	7.1	6.7	5.8	4.9

Population values are in thousands, life expectancy in years, and other items as indicated.

Health

Health Personnel [3]

Doctors per 1,000 pop., 1988-92	0.04
Nurse-to-doctor ratio, 1988-92	8.4
Hospital beds per 1,000 pop., 1985-90	0.8
Percentage of children immunized (age 1 yr. or less)	
Third dose of DPT, 1990-91	77.0
Measles, 1990-91	74.0

Health Expenditures [4]

Total health expenditure, 1990 (official exchange rate)	
Millions of dollars	95
Millions of dollars per capita	6
Health expenditures as a percentage of GDP	
Total	3.4
Public sector	1.6
Private sector	1.8
Development assistance for health	
Total aid flows (millions of dollars)[1]	46
Aid flows per capita (millions of dollars)	2.8
Aid flows as a percentage of total health expenditure	48.4

For sources, notes, and explanations, see Annotated Source Appendix, page 1035.

953

Human Factors

Health Care Ratios [5]

Population per physician, 1970	9210
Population per physician, 1990	NA
Population per nursing person, 1970	NA
Population per nursing person, 1990	NA
Percent of births attended by health staff, 1985	NA

Infants and Malnutrition [6]

Percent of babies with low birth weight, 1985	10
Infant mortality rate per 1,000 live births, 1970	109
Infant mortality rate per 1,000 live births, 1991	118
Years of life lost per 1,000 population, 1990	107
Prevalence of malnutrition (under age 5), 1990	45

Ethnic Division [7]

African	99%
European	NA
Asian	NA
Arab	1%

Religion [8]

Roman Catholic	33%
Protestant	33%
Muslim	16%
Indigenous beliefs	18%

Major Languages [9]

English (official)	NA
Luganda	NA
Swahili	NA
Bantu languages	NA
Nilotic languages	NA

Education

Public Education Expenditures [10]

Million Shilling	1980[1]	1985[1]	1987[1]	1988[1]	1989[1]	1990[1]
Total education expenditure	15	NA	6,349	NA	NA	NA
as percent of GNP	1.2	NA	2.9	NA	NA	NA
as percent of total govt. expend.	11.3	NA	22.5	NA	NA	NA
Current education expenditure	14	NA	6,188	NA	NA	NA
as percent of GNP	1.1	NA	2.8	NA	NA	NA
as percent of current govt. expend.	12.8	NA	30.6	NA	NA	NA
Capital expenditure	2	NA	161	NA	NA	NA

Educational Attainment [11]

Literacy Rate [12]

In thousands and percent	1985[a]	1991[a]	2000[a]
Illiterate population +15 years	4,600	4,908	5,545
Illiteracy rate - total pop. (%)	57.2	51.7	41.1
Illiteracy rate - males (%)	42.9	37.8	28.8
Illiteracy rate - females (%)	71.0	65.1	53.0

Libraries [13]

	Admin. Units	Svc. Pts.	Vols. (000)	Shelving (meters)	Vols. Added	Reg. Users
National (1986)	1	2	15	NA	1,100	450
Non-specialized	NA	NA	NA	NA	NA	NA
Public (1986)	1	18	NA	NA	7,362	6,273
Higher ed.	NA	NA	NA	NA	NA	NA
School	NA	NA	NA	NA	NA	NA

Daily Newspapers [14]

	1975	1980	1985	1990
Number of papers	3[e]	1	1	2
Circ. (000)	46[e]	25	25	30

Cinema [15]

Science and Technology

Scientific/Technical Forces [16]

R&D Expenditures [17]

U.S. Patents Issued [18]

Government and Law

Organization of Government [19]

Long-form name:
Republic of Uganda
Type:
republic
Independence:
9 October 1962 (from UK)
Constitution:
8 September 1967, in process of
constitutional revision
Legal system:
government plans to restore system
based on English common law and
customary law and reinstitute a normal
judicial system; accepts compulsory ICJ
jurisdiction, with reservations
National holiday:
Independence Day, 9 October (1962)
Executive branch:
president, vice president, prime minister,
three deputy prime ministers, Cabinet
Legislative branch:
unicameral National Resistance Council
Judicial branch:
Court of Appeal, High Court

Crime [23]

Elections [20]

National Resistance Council. Last held
11-28 February 1989 (next to be held by
January 1995); results - NRM was the
only party; seats - (278 total, 210
indirectly elected) 210 members elected
without party affiliation. Only party -
National Resistance Movement (NRM).
Other parties continue to exist but are
all proscribed from conducting public
political activities.

Government Budget [21]

For FY89 est.

Revenues	365
Expenditures	545
Capital expenditures	165

Military Expenditures and Arms Transfers [22]

	1985	1986	1987	1988	1989
Military expenditures					
Current dollars (mil.)	51	80	NA	59	NA
1989 constant dollars (mil.)	58	89	NA	61	NA
Armed forces (000)	15[e]	15[e]	15[e]	25	25
Gross national product (GNP)					
Current dollars (mil.)	2,824	2,954	3,299	3,646	4,045
1989 constant dollars (mil.)	3,215	3,279	3,549	3,796	4,045
Central government expenditures (CGE)					
1989 constant dollars (mil.)	375	338	283	294	NA
People (mil.)	14.9	15.5	16.1	16.7	17.3
Military expenditure as % of GNP	1.8	2.7	NA	1.6	NA
Military expenditure as % of CGE	15.6	26.3	NA	20.9	NA
Military expenditure per capita	4	6	NA	4	NA
Armed forces per 1,000 people	1.0	1.0	0.9	1.5	1.4
GNP per capita	216	211	220	227	233
Arms imports[6]					
Current dollars (mil.)	10	20	60	90	20
1989 constant dollars (mil.)	11	22	65	94	20
Arms exports[6]					
Current dollars (mil.)	0	0	0	0	0
1989 constant dollars (mil.)	0	0	0	0	0
Total imports[7]					
Current dollars (mil.)	298	307	848	887	652
1989 constant dollars	339	341	912	923	652
Total exports[7]					
Current dollars (mil.)	387	436	319	274	273
1989 constant dollars	441	484	343	285	273
Arms as percent of total imports[8]	3.4	6.5	7.1	10.1	3.1
Arms as percent of total exports[8]	0	0	0	0	0

Human Rights [24]

	SSTS	FL	FAPRO	PPCG	APROBC	TPW	PCPTW	STPEP	PHRFF	PRW	ASST	AFL
Observes	P	P			P	P	P				P	P
		EAFRD	CPR	ESCR	SR	ACHR	MAAE	PVIAC	PVNAC	EAFDAW	TCIDTP	RC
Observes		P		P	P			P	P	P	P	P

P = Party; S = Signatory; see Appendix for meaning of abbreviations.

Labor Force

Total Labor Force [25]

4.5 million (est.)

Labor Force by Occupation [26]

Agriculture   80%

Unemployment Rate [27]

Production Sectors

Energy Resource Summary [28]

Energy Resources: None. **Electricity**: 200,000 kW capacity; 610 million kWh produced, 30 kWh per capita (1991).

Telecommunications [30]

- Fair system with microwave and radio communications stations
- Broadcast stations - 10 AM, no FM, 9 TV
- Satellite communications ground stations - 1 Atlantic Ocean INTELSAT

Transportation [31]

Railroads. 1,300 km, 1.000-meter-gauge single track

Highways. 26,200 km total; 1,970 km paved; 5,849 km crushed stone, gravel, and laterite; remainder earth roads and tracks

Merchant Marine. 3 roll-on/roll-off (1,000 GRT or over) totaling 15,091 GRT

Airports

Total:	31
Usable:	23
With permanent-surface runways:	5
With runways over 3,659 m:	1
With runways 2,440-3,659 m:	3
With runways 1,220-2,439 m:	11

Top Agricultural Products [32]

Mainly subsistence; accounts for 57% of GDP and over 80% of labor force; cash crops—coffee, tea, cotton, tobacco; food crops - cassava, potatoes, corn, millet, pulses; livestock products—beef, goat meat, milk, poultry; self-sufficient in food.

Top Mining Products [33]

Estimated metric tons unless otherwise specified   M.t.

Cement, hydraulic	50,000
Lime	2,000
Apatite	100
Salt, evaporated	5,000
Tin, mine output, Sn content	25
Tungsten, mine output, w content	4

Tourism [34]

	1987	1988	1989	1990	1991
Tourists	37	40	41	69	69
Tourism receipts	5	8	9	10	15
Tourism expenditures	13	11	10	8	18

Tourists are in thousands, money in million U.S. dollars.

Finance, Economics, and Trade

GDP and Manufacturing Summary [35]

	1980	1985	1990	1991	1992
Gross Domestic Product					
Millions of 1980 dollars	4,644	5,284	6,846	7,130	7,379
Growth rate in percent	-3.40	2.00	4.30	4.14	3.50
Manufacturing Value Added					
Millions of 1980 dollars	192	194	333	354	374[e]
Growth rate in percent	6.10	-9.80	7.50	6.32	5.74[e]
Manufacturing share in percent of current prices	4.2	2.2	4.1	NA	NA

Economic Indicators [36]

National product: GDP—exchange rate conversion—$6 billion (1992 est.). **National product real growth rate**: 4% (1992 est.). **National product per capita**: $300 (1992 est.). **Inflation rate (consumer prices)**: 41.5% (1992 est.). **External debt**: $1.9 billion (1991 est.).

Balance of Payments Summary [37]

Values in millions of dollars.

	1987	1988	1989	1990	1991
Exports of goods (f.o.b.)	333.6	266.3	277.7	177.8	169.6
Imports of goods (f.o.b.)	-475.6	-523.5	-588.3	-491.0	-467.7
Trade balance	-142.0	-257.2	-310.6	-313.2	-298.1
Services - debits	-236.2	-260.4	-260.5	-243.1	-325.2
Services - credits	NA	NA	NA	NA	30.4
Private transfers (net)	NA	NA	NA	NA	90.7
Government transfers (net)	266.2	322.4	311.6	293.0	214.3
Long term capital (net)	103.5	19.2	320.6	251.9	166.9
Short term capital (net)	17.5	26.5	-4.2	-39.8	-20.9
Errors and omissions	26.4	154.9	-38.0	9.5	115.3
Overall balance	35.4	5.4	18.9	-41.7	-26.6

Exchange Rates [38]

Currency: **Ugandan shillings.**
Symbol: **USh.**

Data are currency units per $1.

January 1993	1,217.10
1992	1.133.80
1991	734.00
1990	428.85
1989	223.10
1988	106.10

Imports and Exports

Top Import Origins [39]

$610 million (c.i.f., 1991 est.).

Origins	%
Kenya	25
UK	14
Italy	13

Top Export Destinations [40]

$170 million (f.o.b., 1991 est.).

Destinations	%
US	25
UK	18
France	11
Spain	10

Foreign Aid [41]

	U.S. $	
US commitments, including Ex-Im (1970-89)	145	million
Western (non-US) countries, ODA and OOF bilateral commitments (1970-89)	1.4	billion
OPEC bilateral aid (1979-89)	60	million
Communist countries (1970-89)	169	million

Import and Export Commodities [42]

Import Commodities

Petroleum products
Machinery
Cotton piece goods
Metals
Transportation equipment
Food

Export Commodities

Coffee 97%
Cotton
Tea

For sources, notes, and explanations, see Annotated Source Appendix, page 1035.

957

Ukraine

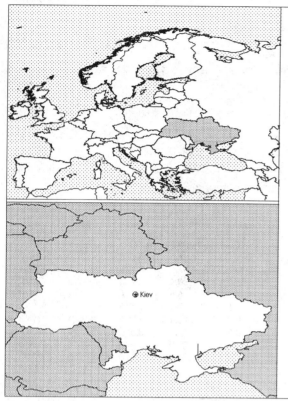

Geography [1]

Total area:
603,700 km2
Land area:
603,700 km2
Comparative area:
Slightly smaller than Texas
Land boundaries:
Total 4,558 km, Belarus 891 km, Hungary 103 km, Moldova 939 km, Poland 428 km, Romania (southwest) 169 km, Romania (west) 362 km, Russia 1,576 km, Slovakia 90 km
Coastline:
2,782 km
Climate:
Temperate continental; subtropical only on the southern Crimean coast; precipitation disproportionately distributed, highest in west and north, lesser in east and southeast; winters vary from cool along the Black Sea to cold farther inland; summers are warm across the greater part of the country, hot in the south
Terrain:
Most of Ukraine consists of fertile plains (steppes) and plateaux, mountains being found only in the west (the Carpathians), and in the Crimean Peninsula in the extreme south
Land use:
Arable land: 56%
Permanent crops: 2%
Meadows and pastures: 12%
Forest and woodland: 0%
Other: 30%

Demographics [2]

	1960	1970	1980	1990	1991	1994	2000	2010	2020
Population	42,644	47,236	50,047	51,674	NA	51,847	51,931	52,280	52,337
Population density (persons per sq. mi.)	NA	NA	NA	NA	NA	NA	NA	NA	NA
Births	NA	NA	NA	NA	NA	640	NA	NA	NA
Deaths	NA	NA	NA	NA	NA	653	NA	NA	NA
Life expectancy - males	NA	NA	NA	NA	NA	65	NA	NA	NA
Life expectancy - females	NA	NA	NA	NA	NA	75	NA	NA	NA
Birth rate (per 1,000)	NA	NA	NA	NA	NA	12	NA	NA	NA
Death rate (per 1,000)	NA	NA	NA	NA	NA	13	NA	NA	NA
Women of reproductive age (15-44 yrs.)	NA	NA	NA	12,307	NA	12,497	12,808	NA	NA
of which are currently married	NA	NA	NA	8,510	NA	8,647	8,807	NA	NA
Fertility rate	NA	NA	NA	1.9	NA	1.8	1.8	1.8	1.7

Population values are in thousands, life expectancy in years, and other items as indicated.

Health

Health Personnel [3]

Doctors per 1,000 pop., 1988-92	4.40
Nurse-to-doctor ratio, 1988-92	2.7
Hospital beds per 1,000 pop., 1985-90	13.6
Percentage of children immunized (age 1 yr. or less)	
Third dose of DPT, 1990-91	78.0
Measles, 1990-91	88.0

Health Expenditures [4]

Total health expenditure, 1990 (official exchange rate)	
Millions of dollars	6803
Millions of dollars per capita	131
Health expenditures as a percentage of GDP	
Total	3.3
Public sector	2.3
Private sector	1.0
Development assistance for health	
Total aid flows (millions of dollars)[1]	NA
Aid flows per capita (millions of dollars)	NA
Aid flows as a percentage of total health expenditure	NA

For sources, notes, and explanations, see Annotated Source Appendix, page 1035.

Human Factors

Health Care Ratios [5]

Population per physician, 1970	NA
Population per physician, 1990	230
Population per nursing person, 1970	NA
Population per nursing person, 1990	NA
Percent of births attended by health staff, 1985	NA

Infants and Malnutrition [6]

Percent of babies with low birth weight, 1985	NA
Infant mortality rate per 1,000 live births, 1970	NA
Infant mortality rate per 1,000 live births, 1991	18
Years of life lost per 1,000 population, 1990	16
Prevalence of malnutrition (under age 5), 1990	NA

Ethnic Division [7]

Ukrainian	73%
Russian	22%
Jewish	1%
Other	4%

Religion [8]

Ukrainian Orthodox - Moscow Patriarchate,
Ukrainian Orthodox - Kiev Patriarchate,
Ukrainian Autocephalous Orthodox,
Ukrainian Catholic (Uniate), Protestant,
Jewish.

Major Languages [9]

Ukrainian	NA
Russian	NA
Romanian	NA
Polish	NA

Education

Public Education Expenditures [10]

Million Roubles	1980	1985	1987	1988	1989	1990
Total education expenditure	5,927	6,721	NA	7,849	7,911	8,606
as percent of GNP	NA	NA	NA	NA	NA	NA
as percent of total govt. expend.	24.5	21.1	NA	NA	NA	NA
Current education expenditure	5,114	5,708	NA	6,398	6,245	6,903
as percent of GNP	NA	NA	NA	NA	NA	NA
as percent of current govt. expend.	NA	NA	NA	NA	NA	NA
Capital expenditure	812	1,013	NA	1,451	1,666	1,704

Educational Attainment [11]

Literacy Rate [12]

	1970[b]	1979[b]	1989[b]
Illiterate population +15 years	NA	NA	NA
Illiteracy rate - total pop. (%)	0.2	0.1	1.6
Illiteracy rate - males (%)	0.2	0.1	0.5
Illiteracy rate - females (%)	0.2	0.1	2.6

Libraries [13]

	Admin. Units	Svc. Pts.	Vols. (000)	Shelving (meters)	Vols. Added	Reg. Users
National	NA	NA	NA	NA	NA	NA
Non-specialized	NA	NA	NA	NA	NA	NA
Public (1989)	25,762	NA	419,800	NA	NA	26 mil[40]
Higher ed.	NA	NA	NA	NA	NA	NA
School (1988)	22,500	NA	299,900	NA	NA	NA

Daily Newspapers [14]

	1975	1980	1985	1990
Number of papers	NA	NA	NA	127
Circ. (000)	NA	NA	NA	13,026

Cinema [15]

Data for 1991.	
Cinema seats per 1,000	NA
Annual attendance per person	11.9
Gross box office receipts (mil. Rouble)	286

Science and Technology

Scientific/Technical Forces [16]

Potential scientists/engineers	449,782
Number female	NA
Potential technicians[25]	243,019
Number female	NA
Total	692,801

R&D Expenditures [17]

U.S. Patents Issued [18]

Government and Law

Organization of Government [19]

Long-form name:
none
Type:
republic
Independence:
1 December 1991 (from Soviet Union)
Constitution:
using 1978 pre-independence
constitution; new consitution currently
being drafted
Legal system:
based on civil law system; no judicial
review of legislative acts
National holiday:
Independence Day, 24 August (1991)
Executive branch:
president, prime minister, cabinet
Legislative branch:
unicameral Supreme Council
Judicial branch:
being organized

Elections [20]

Supreme Council. Last held 4 March
1990 (next scheduled for 1995, may be
held earlier in late 1993); results -
percent of vote by party NA; seats - (450
total) number of seats by party NA.

Government Budget [21]

For FY89 est.
Revenues	NA
Expenditures	NA
Capital expenditures	NA

Defense Summary [22]

Branches: Army, Navy, Airspace Defense Forces, Republic Security Forces (internal and
border troops), National Guard

Manpower Availability: Males age 15-49 12,070,775; fit for military service 9,521,697; reach
military age (18) annually 365,534 (1993 est.)

Defense Expenditures: 544,256 million karbovantsi (forecast for 1993); note - conversion of
the military budget into US dollars using the current exchange rate could produce misleading
results

Crime [23]

Human Rights [24]

	SSTS	FL	FAPRO	PPCG	APROBC	TPW	PCPTW	STPEP	PHRFF	PRW	ASST	AFL
Observes	P	P	P	P	P	P	P	P		P	P	
	EAFRD	CPR	ESCR	SR	ACHR	MAAE	PVIAC	PVNAC	EAFDAW	TCIDTP	RC	
Observes		P	P	P			P	P	P	P	P	P

P = Party; S = Signatory; see Appendix for meaning of abbreviations.

Labor Force

Total Labor Force [25]

25.277 million

Labor Force by Occupation [26]

Industry and construction	41%
Agriculture and forestry	19
Health, education, and culture	18
Trade and distribution	8
Transport and communication	7
Other	7

Date of data: 1990

Unemployment Rate [27]

Production Sectors

Energy Resource Summary [28]

Energy Resources: Coal, natural gas, oil. **Electricity**: 55,882,000 kW capacity; 281,000 million kWh produced, 5,410 kWh per capita (1992).
Pipelines: Crude oil 2,010 km, petroleum products 1,920 km, natural gas 7,800 km (1992).

Telecommunications [30]

- International electronic mail system established in Kiev
- 7 million phone lines (135 per 1000 persons, 650 per 1000 in Kiev)
- NMT-450 analog cellular network under construction in Kiev
- International calls via satellite, by landline to other CIS countries, through Moscow international switching center on 150 international lines
- Satellite earth stations employ INTELSAT, INMARSAT, and Intersputnik
- New international digital telephone exchange operational in Kiev for direct communication with 167 countries

Top Agricultural Products [32]

	1990	1991
Grains	51.0	38.6
Sugar beets	44.3	36.3
Potatoes	16.7	14.6
Sunflower seed	2.7	2.4
Fruits and berries	2.9	2.1
Grapes	0.8	0.7

Values shown are 1,000 metric tons.

Top Mining Products [33]

Detailed information is not available. A summary of mineral resources available follows. **Mineral Resources**: Iron ore, coal, manganese, natural gas, oil, salt, sulphur, graphite, titanium, magnesium, kaolin, nickel, mercury.

Transportation [31]

Railroads. 22,800 km; does not include industrial lines (1990)

Highways. 273,700 km total (1990); 236,400 km hard surfaced, 37,300 km earth

Merchant Marine. 394 ships (1,000 GRT or over) totaling 3,952,328 GRT/5,262,161 DWT; includes 234 cargo, 18 container, 7 barge carriers, 55 bulk cargo, 10 oil tanker, 2 chemical tanker, 1 liquefied gas, 12 passenger, 5 passenger cargo, 9 short-sea passenger, 33 roll-on/roll-off, 2 railcar carrier, 1 multi-function-large-load-carrier, 5 refrigerated cargo

Airports

Total:	694
Useable:	100
With permanent-surface runways:	111
With runways over 3,659 m:	3
With runways 2,440-3,659 m:	81
With runways 1,220-2,439 m:	78

Tourism [34]

For sources, notes, and explanations, see Annotated Source Appendix, page 1035.

961

Finance, Economics, and Trade

Industrial Summary [35]

Industrial Production: Growth rate - 9% (1992). **Industries**: Coal, electric power, ferrous and nonferrous metals, machinery and transport equipment, chemicals, food-processing (especially sugar).

Economic Indicators [36]

National product: GDP not available. **National product real growth rate**: - 13% (1992 est.). **National product per capita**: not available. **Inflation rate (consumer prices)**: 20%- 30% per month (first quarter 1993). **External debt**: $12 billion (1992 est.).

Balance of Payments Summary [37]

Exchange Rates [38]

Currency: **Ukrainian karbovantsi.**
Symbol: **UK.**

Data are currency units per $1.

April 1993	3,000

Imports and Exports

Top Import Origins [39]

$16.7 billion. Imports from outside of the successor states of the former USSR (1990).

Origins	%
No details available.	NA

Top Export Destinations [40]

$13.5 billion to outside of the successor states of the former USSR (1990).

Destinations	%
NA	NA

Foreign Aid [41]

Import and Export Commodities [42]

Import Commodities

Machinery & parts
Transportation equipment
Chemicals
Textiles

Export Commodities

Coal
Electric power
Ferrous and nonferrous metals
Chemicals
Machinery, transp. equip.
Grain
Meat

For sources, notes, and explanations, see Annotated Source Appendix, page 1035.

United Arab Emirates

Geography [1]

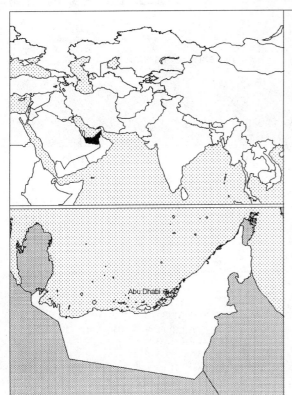

Total area:
75,581 km2
Land area:
75,581 km2
Comparative area:
Slightly smaller than Maine
Land boundaries:
Total 867 km, Oman 410 km, Saudi Arabia 457 km
Coastline:
1,318 km
Climate:
Desert; cooler in eastern mountains
Terrain:
Flat, barren coastal plain merging into rolling sand dunes of vast desert wasteland; mountains in east
Natural resources:
Petroleum, natural gas
Land use:
Arable land:
0%
Permanent crops:
0%
Meadows and pastures:
2%
Forest and woodland:
0%
Other:
98%

Demographics [2]

	1960	1970	1980	1990	1991[1]	1994	2000	2010	2020
Population	103	249	1,000	2,252	2,390	2,791	3,582	4,873	6,080
Population density (persons per sq. mi.)	3	8	31	70	74	NA	111	152	192
Births	NA	NA	NA	NA	72	77	NA	NA	NA
Deaths	NA	NA	NA	NA	7	9	NA	NA	NA
Life expectancy - males	NA	NA	NA	NA	69	70	NA	NA	NA
Life expectancy - females	NA	NA	NA	NA	74	74	NA	NA	NA
Birth rate (per 1,000)	NA	NA	NA	NA	30	28	NA	NA	NA
Death rate (per 1,000)	NA	NA	NA	NA	3	3	NA	NA	NA
Women of reproductive age (15-44 yrs.)	NA	NA	NA	433	NA	560	760	NA	NA
of which are currently married	NA	NA	NA	336	NA	422	549	NA	NA
Fertility rate	NA	NA	NA	4.9	4.85	4.6	4.2	3.5	3.0

Population values are in thousands, life expectancy in years, and other items as indicated.

Health

Health Personnel [3]

Health Expenditures [4]

For sources, notes, and explanations, see Annotated Source Appendix, page 1035.

963

Human Factors

Health Care Ratios [5]	Infants and Malnutrition [6]

Ethnic Division [7]

Emirian	19%
Other Arab	23%
South Asian	50%
Other	8%

Religion [8]

Muslim	96%
Shi'a	16%
Christian Hindu and other	4%

Major Languages [9]

Arabic (official)	NA
Persian	NA
English	NA
Hindi	NA
Urdu	NA

Education

Public Education Expenditures [10]

Million Dirham	1980	1985	1987	1988	1989	1990
Total education expenditure	1,460	1,738	NA	2,026	2,179	2,280
as percent of GNP	1.3	1.7	NA	2.2	2.1	1.9
as percent of total govt. expend.	NA	10.4	NA	14.2	14.9	14.6
Current education expenditure	1,153	1,637	NA	1,907	2,029	2,174
as percent of GNP	1.0	1.6	NA	2.1	2.0	1.8
as percent of current govt. expend.	NA	10.6	NA	13.8	14.3	14.3
Capital expenditure	307	101	NA	119	150	106

Educational Attainment [11]

Literacy Rate [12]

	1970[b]	1975[b]	1980[b]
Illiterate population +15 years	NA	186,058	NA
Illiteracy rate - total pop. (%)	NA	46.5	NA
Illiteracy rate - males (%)	NA	41.6	NA
Illiteracy rate - females (%)	NA	61.9	NA

Libraries [13]

	Admin. Units	Svc. Pts.	Vols. (000)	Shelving (meters)	Vols. Added	Reg. Users
National	NA	NA	NA	NA	NA	NA
Non-specialized	NA	NA	NA	NA	NA	NA
Public	NA	NA	NA	NA	NA	NA
Higher ed. (1990)	3	22	248	11,177	34,009	12,519
School (1990)	290	290	667	17,300	53,000	NA

Daily Newspapers [14]

Cinema [15]

Data for 1989.

Cinema seats per 1,000	19.4
Annual attendance per person	NA
Gross box office receipts (mil. Dirham)	NA

Science and Technology

Scientific/Technical Forces [16]

R&D Expenditures [17]

U.S. Patents Issued [18]

Values show patents issued to citizens of the country by the U.S. Patents Office.

	1990	1991	1992
Number of patents	1	2	1

Government and Law

Organization of Government [19]

Long-form name:
United Arab Emirates
Type:
federation with specified powers
delegated to the UAE central government
and other powers reserved to member
emirates
Independence:
2 December 1971 (from UK)
Constitution:
2 December 1971 (provisional)
Legal system:
secular codes are being introduced by the
UAE Government and in several member
emirates; Islamic law remains influential
National holiday:
National Day, 2 December (1971)
Executive branch:
president, vice president, Supreme
Council of Rulers, prime minister, deputy
prime minister, Council of Ministers
Legislative branch:
unicameral Federal National Council
(Majlis Watani Itihad)
Judicial branch:
Union Supreme Court

Crime [23]

Elections [20]

No political parties; no elections.

Government Expenditures [21]

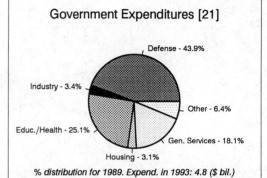

% distribution for 1989. Expend. in 1993: 4.8 ($ bil.)

Military Expenditures and Arms Transfers [22]

	1985	1986	1987	1988	1989
Military expenditures					
Current dollars (mil.)	1,901[e]	1,580[e]	1,590	1,587	1,471
1989 constant dollars (mil.)	2,165[e]	1,753[e]	1,710	1,652	1,471
Armed forces (000)	44	44[e]	44[e]	43	43
Gross national product (GNP)					
Current dollars (mil.)	28,520	22,700	24,760	24,590	27,760
1989 constant dollars (mil.)[e]	32,470	25,190	26,630	25,590	27,760
Central government expenditures (CGE)					
1989 constant dollars (mil.)	4,943	4,041	3,884	3,739	3,613
People (mil.)	1.6	1.7	1.8	2.0	2.1
Military expenditure as % of GNP	6.7	7.0	6.4	6.5	5.3
Military expenditure as % of CGE	43.8	43.4	44.0	44.2	40.7
Military expenditure per capita	1,378	1,023	926	834	695
Armed forces per 1,000 people	28.0	25.7	23.8	21.7	20.3
GNP per capita	20,680	14,700	14,420	12,910	13,110
Arms imports[6]					
Current dollars (mil.)	190	150	230	60	850
1989 constant dollars (mil.)	216	166	247	62	850
Arms exports[6]					
Current dollars (mil.)	0	0	0	0	5
1989 constant dollars (mil.)	0	0	0	0	5
Total imports[7]					
Current dollars (mil.)	6,549	6,422	7,226	8,522	10,010
1989 constant dollars	7,455	7,127	7,772	8,871	10,010
Total exports[7]					
Current dollars (mil.)	14,040	15,840	12,000	10,600	NA
1989 constant dollars	15,990	17,570	12,910	11,030	NA
Arms as percent of total imports[8]	2.9	2.3	3.2	0.7	8.5
Arms as percent of total exports[8]	0	0	0	0	NA

Human Rights [24]

	SSTS	FL	FAPRO	PPCG	APROBC	TPW	PCPTW	STPEP	PHRFF	PRW	ASST	AFL
Observes		P				P	P					
	EAFRD	CPR	ESCR	SR	ACHR	MAAE	PVIAC	PVNAC	EAFDAW	TCIDTP	RC	
Observes		P						P	P			

P = Party; S = Signatory; see Appendix for meaning of abbreviations.

Labor Force

Total Labor Force [25]

580,000 (1986 est.)

Labor Force by Occupation [26]

Industry and commerce	85%
Agriculture	5
Services	5
Government	5

Unemployment Rate [27]

NEGL% (1988)

For sources, notes, and explanations, see Annotated Source Appendix, page 1035.

965

Production Sectors

Commercial Energy Production and Consumption

Production [28]

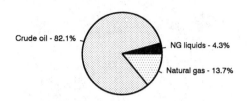

Crude oil - 82.1%
NG liquids - 4.3%
Natural gas - 13.7%

Consumption [29]

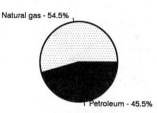

Natural gas - 54.5%
Petroleum - 45.5%

Data are shown in quadrillion (10^{15}) BTUs and percent for 1991

Crude oil[1]	5.17
Natural gas liquids	0.27
Dry natural gas	0.86
Coal[2]	0.00
Net hydroelectric power[3]	0.00
Net nuclear power[3]	0.00
Total	6.30

Petroleum	0.61
Dry natural gas	0.73
Coal[2]	0.00
Net hydroelectric power[3]	0.00
Net nuclear power[3]	0.00
Total	1.35

Telecommunications [30]

- Modern system consisting of microwave and coaxial cable
- Key centers are Abu Dhabi and Dubayy
- 386,600 telephones
- Satellite ground stations - 1 Atlantic Ocean INTELSAT, 2 Indian Ocean INTELSAT and 1 ARABSAT
- Submarine cables to Qatar, Bahrain, India, and Pakistan
- Tropospheric scatter to Bahrain
- Microwave radio relay to Saudi Arabia
- Broadcast stations - 8 AM, 3 FM, 12 TV

Transportation [31]

Highways. 2,000 km total; 1,800 km bituminous, 200 km gravel and graded earth

Merchant Marine. 56 ships (1,000 GRT or over) totaling 1,197,306 GRT/2,153,673 DWT; includes 15 cargo, 8 container, 3 roll-on/roll-off, 23 oil tanker, 4 bulk, 1 refrigerated cargo, 1 liquified gas, 1 chemical tanker

Airports

Total:	37
Usable:	34
With permanent-surface runways:	20
With runways over 3,659 m:	7
With runways 2,440-3,659 m:	5
With runways 1,220-2,439 m:	5

Top Agricultural Products [32]

	88-89[1]	89-90[1]
Green fodder	351	467
Tomatoes	26	41
Cabbage	12	41
Eggplants	15	39
Lemons	14	19
Mangoes	7	7

Values shown are 1,000 metric tons.

Top Mining Products [33]

Detailed information is not available. A summary of mineral resources available follows. **Mineral Resources**: Petroleum, natural gas.

Tourism [34]

	1987	1988	1989	1990	1991
Tourists[25]	543	598	629	633	717

Tourists are in thousands, money in million U.S. dollars.

For sources, notes, and explanations, see Annotated Source Appendix, page 1035.

Finance, Economics, and Trade

GDP and Manufacturing Summary [35]

	1980	1985	1990	1991	1992
Gross Domestic Product					
Millions of 1980 dollars	29,629	27,036	30,375	32,654	33,796
Growth rate in percent	26.42	-2.39	17.75	7.50	3.50
Manufacturing Value Added					
Millions of 1980 dollars	1,131	2,547	2,400	2,636	2,910[e]
Growth rate in percent	64.87	-2.20	5.38	9.83	10.40[e]
Manufacturing share in percent of current prices	3.7	9.0	7.2	7.5[e]	NA

Economic Indicators [36]

In Millions of U.S. Dollars unless otherwise indicated.

	1989	1990	1991[p]
GDP	27,988	34,243	33,097
GDP Growth Rate (%)	-17.4	22.3	-3.4
Money Supply (M1)	2,805	2,942	3,000
Commercial Interest Rates (%)	11.5	11.5	11.5
Savings Rate	NA	NA	NA
Investment Rate	NA	NA	NA
CPI	NA	NA	NA
WPI	NA	NA	NA
External Federal Public Debt	0	0	0

Balance of Payments Summary [37]

Exchange Rates [38]

Currency: **Emirian dirhams.**
Symbol: **Dh.**

Data are currency units per $1.

Fixed rate	3.6710

Imports and Exports

Top Import Origins [39]

$13.9 billion (f.o.b., 1991 est.).

Origins	%
Japan	15
US	10
UK	9
Germany	7
Korea	4

Top Export Destinations [40]

$21.2 billion (f.o.b., 1991 est.).

Destinations	%
Japan	39
Singapore	5
Korea	4
Iran	4
India	NA

Foreign Aid [41]

	U.S. $	
Donor - pledges in bilateral aid to less developed countries (1979-89)	9.1	billion

Import and Export Commodities [42]

Import Commodities	Export Commodities
Capital goods	Crude oil 66%
Consumer goods	Natural gas
Food	Reexports
	Dried fish
	Dates

For sources, notes, and explanations, see Annotated Source Appendix, page 1035.

967

United Kingdom

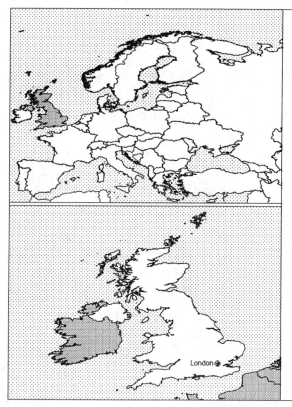

Geography [1]

Total area:
244,820 km2
Land area:
241,590 km2
Comparative area:
Slightly smaller than Oregon
Land boundaries:
Total 360 km, Ireland 360 km
Coastline:
12,429 km
Climate:
Temperate; moderated by prevailing southwest winds over the North Atlantic current; more than half of the days are overcast
Terrain:
Mostly rugged hills and low mountains; level to rolling plains in east and southeast
Natural resources:
Coal, petroleum, natural gas, tin, limestone, iron ore, salt, clay, chalk, gypsum, lead, silica
Land use:
Arable land: 29%
Permanent crops: 0%
Meadows and pastures: 48%
Forest and woodland: 9%
Other: 14%

Demographics [2]

	1960	1970	1980	1990	1991[1]	1994	2000	2010	2020
Population	52,372	55,632	56,314	57,418	57,515	58,135	58,951	59,617	60,042
Population density (persons per sq. mi.)	561	596	604	615	617	NA	630	634	637
Births	NA	NA	NA	NA	789	778	NA	NA	NA
Deaths	NA	NA	NA	NA	638	626	NA	NA	NA
Life expectancy - males	NA	NA	NA	NA	73	74	NA	NA	NA
Life expectancy - females	NA	NA	NA	NA	79	80	NA	NA	NA
Birth rate (per 1,000)	NA	NA	NA	NA	14	13	NA	NA	NA
Death rate (per 1,000)	NA	NA	NA	NA	11	11	NA	NA	NA
Women of reproductive age (15-44 yrs.)	NA	NA	NA	14,138	NA	14,107	13,814	NA	NA
of which are currently married	NA	NA	NA	9,351	NA	9,607	9,411	NA	NA
Fertility rate	NA	NA	NA	1.9	1.82	1.8	1.8	1.8	1.8

Population values are in thousands, life expectancy in years, and other items as indicated.

Health

Health Personnel [3]

Doctors per 1,000 pop., 1988-92	1.40
Nurse-to-doctor ratio, 1988-92	2.0
Hospital beds per 1,000 pop., 1985-90	6.3
Percentage of children immunized (age 1 yr. or less)	
Third dose of DPT, 1990-91	85.0
Measles, 1990-91	89.0

Health Expenditures [4]

Total health expenditure, 1990 (official exchange rate)	
Millions of dollars	59623
Millions of dollars per capita	1039
Health expenditures as a percentage of GDP	
Total	6.1
Public sector	5.2
Private sector	0.9
Development assistance for health	
Total aid flows (millions of dollars)[1]	NA
Aid flows per capita (millions of dollars)	NA
Aid flows as a percentage of total health expenditure	NA

For sources, notes, and explanations, see Annotated Source Appendix, page 1035.

Human Factors

Health Care Ratios [5]

Population per physician, 1970	810
Population per physician, 1990	NA
Population per nursing person, 1970	240
Population per nursing person, 1990	NA
Percent of births attended by health staff, 1985	98

Infants and Malnutrition [6]

Percent of babies with low birth weight, 1985	7
Infant mortality rate per 1,000 live births, 1970	19
Infant mortality rate per 1,000 live births, 1991	7
Years of life lost per 1,000 population, 1990	12
Prevalence of malnutrition (under age 5), 1990	NA

Ethnic Division [7]

English	81.5%
Scottish	9.6%
Irish	2.4%
Welsh	1.9%
Ulster	1.8%
Other	2.8%

Religion [8]

Anglican 27 million, Roman Catholic 9 million, Muslim 1 million, Presbyterian 800,000, Methodist 760,000, Sikh 400,000, Hindu 350,000, Jewish 300,000.

Major Languages [9]

English, Welsh (about 26% of the population of Wales), Scottish form of Gaelic (about 60,000 in Scotland).

Education

Public Education Expenditures [10]

Million Pound Sterling	1980	1985	1987	1988	1989	1990
Total education expenditure	12,856	17,501	20,401	22,148	24,102	26,677
as percent of GNP	5.6	4.9	4.9	4.8	4.7	4.9
as percent of total govt. expend.	13.9	NA	NA	NA	NA	NA
Current education expenditure	12,094	16,764	19,596	21,275	22,829	25,318
as percent of GNP	5.2	4.7	4.7	4.6	4.5	4.6
as percent of current govt. expend.	NA	NA	NA	NA	NA	NA
Capital expenditure	762	737	805	873	1,273	1,359

Educational Attainment [11]

Literacy Rate [12]

Libraries [13]

	Admin. Units	Svc. Pts.	Vols. (000)	Shelving (meters)	Vols. Added	Reg. Users
National (1990)	3	23	27,500	790,000	542,606	82,945[41]
Non-specialized	NA	NA	NA	NA	NA	NA
Public (1989)	165	5,270	156,700	NA	14 mil	-
Higher ed.	NA	NA	NA	NA	NA	NA
School	NA	NA	NA	NA	NA	NA

Daily Newspapers [14]

	1975	1980	1985	1990[1]
Number of papers	109	113	104	104
Circ. (000)	24,805	23,472	22,495	22,494

Cinema [15]

Data for 1991.

Cinema seats per 1,000	NA
Annual attendance per person	1.8[1]
Gross box office receipts (mil. Pound Sterling)	295

Science and Technology

Scientific/Technical Forces [16]

Potential scientists/engineers	3,038,000
Number female	1,018,000
Potential technicians	2,102,000
Number female	1,135,000
Total	5,140,000

R&D Expenditures [17]

	Pound[34] (000) 1986
Total expenditure	8,777,900
Capital expenditure	NA
Current expenditure	NA
Percent current	NA

U.S. Patents Issued [18]

Values show patents issued to citizens of the country by the U.S. Patents Office.

	1990	1991	1992
Number of patents	3,016	3,048	2,631

For sources, notes, and explanations, see Annotated Source Appendix, page 1035.

Government and Law

Organization of Government [19]

Long-form name:
United Kingdom of Great Britain and Northern Ireland
Type:
constitutional monarchy
Independence:
1 January 1801 (United Kingdom established)
Constitution:
unwritten; partly statutes, partly common law and practice
Legal system:
common law tradition with early Roman and modern continental influences; no judicial review of Acts of Parliament; accepts compulsory ICJ jurisdiction, with reservations
National holiday:
Celebration of the Birthday of the Queen (second Saturday in June)
Executive branch:
monarch, prime minister, Cabinet
Legislative branch:
bicameral Parliament consists of an upper house or House of Lords and a lower house or House of Commons
Judicial branch:
House of Lords

Political Parties [20]

House of Commons	% of votes
Conservative	41.9
Labor	34.5
Liberal Democratic	17.9
Other	5.7

Government Expenditures [21]

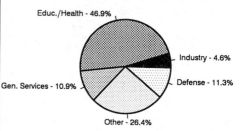

Educ./Health - 46.9%
Industry - 4.6%
Defense - 11.3%
Other - 26.4%
Gen. Services - 10.9%

% distribution for 1991. Expend. in FY92 est.: 439.3 ($ bil.)

Crime [23]

Crime volume	
Cases known to police	5,559,872
Attempts (percent)	NA
Percent cases solved	NA
Crimes per 100,000 persons	31,414.06
Persons responsible for offenses	
Total number offenders	683,310
Percent female	NA
Percent juvenile	NA
Percent foreigners	NA

Military Expenditures and Arms Transfers [22]

	1985	1986	1987	1988	1989
Military expenditures					
Current dollars (mil.)	32,560	32,780	33,320	32,710	34,630
1989 constant dollars (mil.)	37,070	36,380	35,840	34,050	34,630
Armed forces (000)	334	331	328	324	318
Gross national product (GNP)					
Current dollars (mil.)	632,700	673,000	724,400	783,300	834,400
1989 constant dollars (mil.)	720,200	746,900	779,200	815,400	834,400
Central government expenditures (CGE)					
1989 constant dollars (mil.)	292,500	292,300	289,100	282,400	285,700
People (mil.)	56.6	56.8	56.9	57.1	57.2
Military expenditure as % of GNP	5.1	4.9	4.6	4.2	4.2
Military expenditure as % of CGE	12.7	12.4	12.4	12.1	12.1
Military expenditure per capita	655	641	630	597	605
Armed forces per 1,000 people	5.9	5.8	5.8	5.7	5.6
GNP per capita	12,720	13,160	13,690	14,290	14,580
Arms imports[6]					
Current dollars (mil.)	675	675	550	650	650
1989 constant dollars (mil.)	768	749	592	677	650
Arms exports[6]					
Current dollars (mil.)	1,500	3,700	4,700	1,600	3,000
1989 constant dollars (mil.)	1,708	4,106	5,055	1,666	3,000
Total imports[7]					
Current dollars (mil.)	109,000	126,300	154,400	189,300	197,700
1989 constant dollars	124,000	140,200	166,100	197,100	197,700
Total exports[7]					
Current dollars (mil.)	101,300	107,200	131,300	145,200	152,300
1989 constant dollars	115,300	119,000	141,200	151,100	152,300
Arms as percent of total imports[8]	0.6	0.5	0.4	0.3	0.3
Arms as percent of total exports[8]	1.5	3.5	3.6	1.1	2.0

Human Rights [24]

	SSTS	FL	FAPRO	PPCG	APROBC	TPW	PCPTW	STPEP	PHRFF	PRW	ASST	AFL
Observes	P	P	P	P	P	P	P		P	P	P	P
	EAFRD	CPR	ESCR	SR	ACHR	MAAE	PVIAC	PVNAC	EAFDAW	TCIDTP	RC	
Observes		P	P	P	P			S	S	P	P	P

P = Party; S = Signatory; see Appendix for meaning of abbreviations.

Labor Force

Total Labor Force [25]

28.048 million

Labor Force by Occupation [26]

Services	62.8%
Manufacturing and construction	25.0
Government	9.1
Energy	1.9
Agriculture	1.2

Date of data: June 1992

Unemployment Rate [27]

9.8% (1992)

Production Sectors

Commercial Energy Production and Consumption

Production [28]

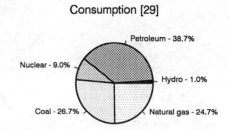

Crude oil - 41.7%
Hydro - 0.7%
Coal - 24.0%
NG liquids - 2.4%
Nuclear - 8.7%
Natural gas - 22.6%

Consumption [29]

Petroleum - 38.7%
Nuclear - 9.0%
Hydro - 1.0%
Coal - 26.7%
Natural gas - 24.7%

Data are shown in quadrillion (10^{15}) BTUs and percent for 1991

Crude oil[1]	3.81
Natural gas liquids	0.22
Dry natural gas	2.06
Coal[2]	2.19
Net hydroelectric power[3]	0.06
Net nuclear power[3]	0.79
Total	9.13

Petroleum	3.62
Dry natural gas	2.31
Coal[2]	2.50
Net hydroelectric power[3]	0.09
Net nuclear power[3]	0.84
Total	9.36

Telecommunications [30]

- Technologically advanced domestic and international system
- 30,200,000 telephones
- Equal mix of buried cables, microwave and optical-fiber systems
- Excellent countrywide broadcast systems
- Broadcast stations - 225 AM, 525 (mostly repeaters) FM, 207 (3,210 repeaters) TV
- 40 coaxial submarine cables
- 5 satellite ground stations operating in INTELSAT (7 Atlantic Ocean and 3 Indian Ocean), INMARSAT, and EUTELSAT systems
- At least 8 large international switching centers

Top Agricultural Products [32]

	90-91[1]	91-92[1]
Wheat	14.10	14.50
Barley	8.0	7.9
Potatoes	6.5	6.3
Rapeseed	1.3	1.4
Sugar (raw)	1.36	1.36
Oats	0.6	0.5

Values shown are 1,000 metric tons.

Top Mining Products [33]

Estimated metric tons unless otherwise specified   M.t.

Copper ore, Cu content of concentrate	600
Clay, kaolin	3,000
Gypsum (000 tons)	4,000
Sand and gravel, common & industrial	124,000
Clay, ball	800
Salt, rock	600,000

Transportation [31]

Railroads. UK, 16,914 km total; Great Britain's British Railways (BR) operates 16,584 km 1.435-meter (standard) gauge (including 4,545 km electrified and 12,591 km double or multiple track), several additional small standard-gauge and narrow-gauge lines are privately owned and operated; Northern Ireland Railways (NIR) operates 330 km 1.600-meter gauge (including 190 km double track)

Highways. UK, 362,982 km total; Great Britain, 339,483 km paved (including 2,573 km limited-access divided highway); Northern Ireland, 23,499 km (22,907 paved, 592 km gravel)

Merchant Marine. 204 ships (1,000 GRT or over) totaling 3,819,719 GRT/4,941,785 DWT; includes 7 passenger, 16 short-sea passenger, 37 cargo, 25 container, 14 roll-on/roll-off, 5 refrigerated cargo, 1 vehicle carrier, 65 oil tanker, 1 chemical tanker, 8 liquefied gas, 1 specialized tanker, 22 bulk, 1 combination bulk, 1 passenger cargo

Airports

Total:	496
Usable:	385
With permanent-surface runways:	249
With runways over 3,659 m:	1
With runways 2,440-3,659 m:	37
With runways 1,220-2,439 m:	134

Tourism [34]

	1987	1988	1989	1990	1991
Visitors[71]	15,566	15,799	17,338	18,021	16,664
Excursionists[71]	646	767	1,063	883	1,009
Tourism receipts	10,225	11,008	11,182	13,910	12,635
Tourism expenditures	11,939	14,624	15,111	17,614	18,850
Fare receipts	3,902	4,305	4,548	5,664	4,971
Fare expenditures	3,794	4,629	4,321	5,113	4,726

Tourists are in thousands, money in million U.S. dollars.

For sources, notes, and explanations, see Annotated Source Appendix, page 1035.

971

Manufacturing Sector

GDP and Manufacturing Summary [35]

	1980	1985	1989	1990	% change 1980-1990	% change 1989-1990
GDP (million 1980 $)	536,588	590,748	671,543	691,358	28.8	3.0
GDP per capita (1980 $)	9,493	10,398	11,719	12,038	26.8	2.7
Manufacturing as % of GDP (current prices)	25.7	22.9	20.4[e]	21.5[e]	-16.3	5.4
Gross output (million $)	400,929	306,225	523,446	576,765	43.9	10.2
Value added (million $)	163,790	124,409	231,549	253,630	54.9	9.5
Value added (million 1980 $)	124,163	128,881	152,418	152,981	23.2	0.4
Industrial production index	100	103	122	122	22.0	0.0
Employment (thousands)	6,462	4,932	4,869[e]	4,785	-26.0	-1.7

Note: GDP stands for Gross Domestic Product. 'e' stands for estimated value.

Profitability and Productivity

	1980	1985	1989	1990	% change 1980-1990	% change 1989-1990
Intermediate input (%)	59	59	56	56	-5.1	0.0
Wages, salaries, and supplements (%)	23	20	18[e]	21[e]	-8.7	16.7
Gross operating surplus (%)	17	20	27[e]	23[e]	35.3	-14.8
Gross output per worker ($)	61,483	61,368	107,497[e]	118,945	93.5	10.6
Value added per worker ($)	25,117	24,932	47,552[e]	52,306	108.2	10.0
Average wage (incl. benefits) ($)	14,579	12,528	18,998[e]	25,318[e]	73.7	33.3

Profitability is in percent of gross output. Productivity is in U.S. $. 'e' stands for estimated value.

Profitability - 1990

Wages - 21.0%
Inputs - 56.0%
Surplus - 23.0%

The graphic shows percent of gross output.

Value Added in Manufacturing

	1980 $ mil.	1980 %	1985 $ mil.	1985 %	1989 $ mil.	1989 %	1990 $ mil.	1990 %	% change 1980-1990	% change 1989-1990
311 Food products	14,744	9.0	12,179	9.8	20,696	8.9	25,025	9.9	69.7	20.9
313 Beverages	5,419	3.3	3,554	2.9	6,072	2.6	6,620	2.6	22.2	9.0
314 Tobacco products	1,814	1.1	1,479	1.2	1,885	0.8	2,364	0.9	30.3	25.4
321 Textiles	5,419	3.3	3,917	3.1	6,636	2.9	6,995	2.8	29.1	5.4
322 Wearing apparel	3,395	2.1	2,633	2.1	4,235	1.8	4,682	1.8	37.9	10.6
323 Leather and fur products	558	0.3	363	0.3	522	0.2	529	0.2	-5.2	1.3
324 Footwear	1,093	0.7	752	0.6	1,150	0.5	1,254	0.5	14.7	9.0
331 Wood and wood products	2,349	1.4	1,556	1.3	3,281[e]	1.4	3,195	1.3	36.0	-2.6
332 Furniture and fixtures	2,558	1.6	2,101	1.7	3,960[e]	1.7	4,526	1.8	76.9	14.3
341 Paper and paper products	4,860	3.0	3,800	3.1	7,434	3.2	7,988	3.1	64.4	7.5
342 Printing and publishing	9,814	6.0	8,807	7.1	17,726	7.7	19,538	7.7	99.1	10.2
351 Industrial chemicals	8,233	5.0	7,328	5.9	14,403[e]	6.2	14,094	5.6	71.2	-2.1
352 Other chemical products	7,512	4.6	6,641	5.3	13,410[e]	5.8	14,826	5.8	97.4	10.6
353 Petroleum refineries	4,512	2.8	1,712	1.4	2,707	1.2	4,398	1.7	-2.5	62.5
354 Miscellaneous petroleum and coal products	721	0.4	428	0.3	746[e]	0.3	744	0.3	3.2	-0.3
355 Rubber products	2,349	1.4	1,505	1.2	2,757[e]	1.2	2,996	1.2	27.5	8.7
356 Plastic products	3,698	2.3	3,087	2.5	6,891[e]	3.0	8,204	3.2	121.8	19.1
361 Pottery, china and earthenware	977	0.6	765	0.6	1,396	0.6	1,464	0.6	49.8	4.9
362 Glass and glass products	1,442	0.9	960	0.8	2,046	0.9	2,078	0.8	44.1	1.6
369 Other non-metal mineral products	5,698	3.5	4,215	3.4	8,845	3.8	8,979	3.5	57.6	1.5
371 Iron and steel	5,860	3.6	4,345	3.5	8,431	3.6	8,050	3.2	37.4	-4.5
372 Non-ferrous metals	2,581	1.6	1,505	1.2	2,978	1.3	2,769	1.1	7.3	-7.0
381 Metal products	10,140	6.2	7,211	5.8	12,381	5.3	14,936	5.9	47.3	20.6
382 Non-electrical machinery	21,326	13.0	15,110	12.1	27,198	11.7	29,931	11.8	40.3	10.0
383 Electrical machinery	15,209	9.3	12,399	10.0	22,600	9.8	22,245	8.8	46.3	-1.6
384 Transport equipment	17,512	10.7	12,944	10.4	25,339	10.9	28,805	11.4	64.5	13.7
385 Professional and scientific equipment	2,209	1.3	1,803	1.4	3,323	1.4	3,634	1.4	64.5	9.4
390 Other manufacturing industries	1,791	1.1	1,310	1.1	2,501[e]	1.1	2,762	1.1	54.2	10.4

Note: The industry codes shown are International Standard Industry codes (ISIC). Percentages are percent of total Value Added. 'e' stands for estimated value

Finance, Economics, and Trade

Economic Indicators [36]

Billions of Pounds Sterling (BPS) unless otherwise noted.

	1989	1990	1991[164]
Real GDP (1985 prices)[165]	352.8	355.8	350.5
Real GDP Growth (%)	2.2	0.9	-1.9
Real Per Capita GDP ('85 BPS)	7,167	7,200	7,080
Money Supply (M2)	9.2	7.9	11.0
Base Interest Rate[166]	13.9	14.8	12.3
Personal Saving Rate	7.1	9.2	10.0
Retail Inflation	7.8	9.5	5.8
Wholesale Inflation	5.1	5.9	5.8

Balance of Payments Summary [37]

Values in millions of dollars.

	1987	1988	1989	1990	1991
Exports of goods (f.o.b.)	129,847	143,078	150,696	181,729	182,577
Imports of goods (f.o.b.)	-148,866	-181,237	-191,239	-214,471	-200,868
Trade balance	-19,019	-38,159	-40,543	-32,742	-18,291
Services - debits	-107,243	-133,475	-157,455	-186,567	-183,749
Services - credits	124,248	149,183	169,867	198,662	193,005
Private transfers (net)	-205	-480	-491	-535	-531
Government transfers (net)	-5,340	-5,857	-6,964	-8,208	-1,871
Long term capital (net)	24,899	-5,319	-28,208	7,236	-2,123
Short term capital (net)	-23,669	20,355	44,037	10,386	26,184
Errors and omissions	659	12,077	3,455	9,814	1,412
Overall balance	-5,670	-1,675	-16,302	-1,954	14,036

Exchange Rates [38]

Currency: **British pounds.**
Symbol: **#.**

Data are currency units per $1.

January 1993	0.6527
1992	0.5664
1991	0.5652
1990	0.5603
1989	0.6099
1988	0.5614

Imports and Exports

Top Import Origins [39]

$210.7 billion (c.i.f., 1992).

Origins	%
Germany	14.9
France	9.3
Netherlands	8.4
US	11.6

Top Export Destinations [40]

$187.4 billion (f.o.b., 1992).

Destinations	%
EC countries	56.7
France	11.1
Netherlands	7.9
US	10.9

Foreign Aid [41]

	U.S. $	
Donor - ODA and OOF commitments (1970-89)	21.0	billion

Import and Export Commodities [42]

Import Commodities

Manufactured goods
Machinery
Semifinished goods
Foodstuffs
Consumer goods

Export Commodities

Manufactured goods
Machinery
Fuels
Chemicals
Semifinished goods
Transport equipment

For sources, notes, and explanations, see Annotated Source Appendix, page 1035.

973

United States

Geography [1]

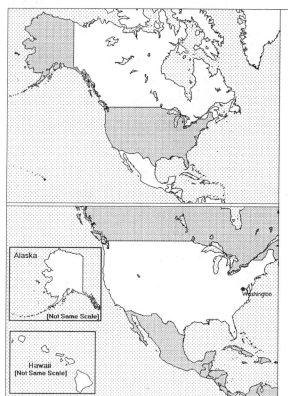

Alaska
[Not Same Scale]

Hawaii
[Not Same Scale]

Washington

Total area:
9,372,610 km2
Land area:
9,166,600 km2
Comparative area:
About half the size of Russia; about three-tenths the size of Africa; about one-half the size of South America (or slightly larger than Brazil)
Land boundaries:
Total 12,248 km, Canada 8,893 km (including 2,477 km with Alaska), Cuba 29 km (US naval base at Guantanamo), Mexico 3,326 km
Coastline:
19,924 km
Climate:
Mostly temperate, but tropical in Hawaii and Florida and arctic in Alaska.
Terrain:
Vast central plain, mountains in west, hills and low mountains in east; rugged mountains and broad river valleys in Alaska; rugged, volcanic topography in Hawaii
Land use:
Arable land: 20%
Permanent crops: 0%
Meadows and pastures: 26%
Forest and woodland: 29%
Other: 25%

Demographics [2]

	1960	1970	1980	1990	1991[1]	1994	2000	2010	2020
Population	180,671	205,052	227,726	249,924	252,502	260,327	275,327	298,621	323,113
Population density (persons per sq. mi.)	51	58	64	71	71	NA	76	80	83
Births	NA	NA	NA	NA	3,687	3,963	NA	NA	NA
Deaths	NA	NA	NA	NA	2,197	2,263	NA	NA	NA
Life expectancy - males	NA	NA	NA	NA	72	73	NA	NA	NA
Life expectancy - females	NA	NA	NA	NA	79	79	NA	NA	NA
Birth rate (per 1,000)	NA	NA	NA	NA	15	15	NA	NA	NA
Death rate (per 1,000)	NA	NA	NA	NA	9	9	NA	NA	NA
Women of reproductive age (15-44 yrs.)	NA	NA	NA	65,802	NA	67,787	69,970	NA	NA
of which are currently married	NA	NA	NA	41,700	NA	43,597	44,717	NA	NA
Fertility rate	NA	NA	NA	2.1	1.85	2.1	2.1	2.1	2.1

Population values are in thousands, life expectancy in years, and other items as indicated.

Health

Health Personnel [3]

Doctors per 1,000 pop., 1988-92	2.38
Nurse-to-doctor ratio, 1988-92	2.8
Hospital beds per 1,000 pop., 1985-90	5.3
Percentage of children immunized (age 1 yr. or less)	
Third dose of DPT, 1990-91	67.0
Measles, 1990-91	80.0

Health Expenditures [4]

Total health expenditure, 1990 (official exchange rate)	
Millions of dollars	690667
Millions of dollars per capita	2763
Health expenditures as a percentage of GDP	
Total	12.7
Public sector	5.6
Private sector	7.0
Development assistance for health	
Total aid flows (millions of dollars)[1]	NA
Aid flows per capita (millions of dollars)	NA
Aid flows as a percentage of total health expenditure	NA

For sources, notes, and explanations, see Annotated Source Appendix, page 1035.

Human Factors

Health Care Ratios [5]

Population per physician, 1970	630
Population per physician, 1990	420
Population per nursing person, 1970	160
Population per nursing person, 1990	NA
Percent of births attended by health staff, 1985	100

Infants and Malnutrition [6]

Percent of babies with low birth weight, 1985	7
Infant mortality rate per 1,000 live births, 1970	20
Infant mortality rate per 1,000 live births, 1991	9
Years of life lost per 1,000 population, 1990	11
Prevalence of malnutrition (under age 5), 1990	NA

Ethnic Division [7]

White	83.4%
Black	12.4%
Asian	3.3%
Native American	0.8%

Religion [8]

Protestant	56%
Roman Catholic	28%
Jewish	2%
Other	4%
None	10%

Major Languages [9]

English, Spanish (spoken by a sizable minority).

Education

Public Education Expenditures [10]

Million U.S. Dollars	1980[7]	1985	1987	1988	1989	1990
Total education expenditure	182,849	199,372	227,800	251,921	275,044	NA
as percent of GNP	6.7	5.0	5.1	5.2	5.3	NA
as percent of total govt. expend.	NA	15.5	11.9	12.4	12.4	NA
Current education expenditure	NA	182,875	207,053	228,864	247,590	NA
as percent of GNP	NA	4.6	4.6	4.7	4.8	NA
as percent of current govt. expend.	NA	16.3	12.0	12.6	12.4	NA
Capital expenditure	NA	16,497	20,747	23,057	27,454	NA

Educational Attainment [11]

Age group	25+
Total population	132,899,000
Highest level attained (%)	
No schooling	3.3
First level	
Incompleted	NA
Completed	64.6
Entered second level	
S-1	NA
S-2	NA
Post secondary	32.2

Literacy Rate [12]

Libraries [13]

	Admin. Units	Svc. Pts.	Vols. (000)	Shelving (meters)	Vols. Added	Reg. Users
National	NA	NA	NA	NA	NA	NA
Non-specialized	NA	NA	NA	NA	NA	NA
Public	NA	NA	NA	NA	NA	NA
Higher ed. (1988)	3,438	NA	718,503	NA	22 mil	NA
School (1988)	92,438	NA	738,706	NA	28 mil	NA

Daily Newspapers [14]

	1975	1980	1985	1990
Number of papers	1,775	1,750[e]	1,640	1,611
Circ. (000)	60,655	62,000[e]	62,700[e]	62,328

Cinema [15]

Data for 1991.

Cinema seats per 1,000	NA
Annual attendance per person	3.9
Gross box office receipts (mil. Dollar)	4,803

Science and Technology

Scientific/Technical Forces [16]

Potential scientists/engineers	5,286,400
Number female	NA
Potential technicians	NA
Number female	NA
Total	NA

R&D Expenditures [17]

	Dollar[14] (000) 1988
Total expenditure	139,255,000
Capital expenditure	4,024,000
Current expenditure	135,231,000
Percent current	97.1

U.S. Patents Issued [18]

Values show patents issued to citizens of the country by the U.S. Patents Office.

	1990	1991	1992
Number of patents	52,985	57,804	58,804

Government and Law

Organization of Government [19]

Long-form name:
United States of America
Type:
federal republic; strong democratic tradition
Independence:
4 July 1776 (from England)
Constitution:
17 September 1787, effective 4 June 1789
Legal system:
based on English common law; judicial review of legislative acts; accepts compulsory ICJ jurisdiction, with reservations
National holiday:
Independence Day, 4 July (1776)
Executive branch:
president, vice president, Cabinet
Legislative branch:
bicameral Congress consists of an upper house or Senate and a lower house or House of Representatives
Judicial branch:
Supreme Court

Political Parties [20]

House of Representatives	% of votes
Democratic Party	52.0
Republican Party	46.0
Other	2.0

Government Expenditures [21]

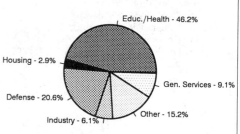

Educ./Health - 46.2%
Housing - 2.9%
Gen. Services - 9.1%
Defense - 20.6%
Other - 15.2%
Industry - 6.1%

% distribution for 1992. Expend. in FY92: 1382 ($ bil.)

Crime [23]

Crime volume	
Cases known to police	14,475,600
Attempts (percent)	NA
Percent cases solved	21.6
Crimes per 100,000 persons	5,820.3
Persons responsible for offenses	
Total number offenders	11,250,083
Percent female	18.4
Percent juvenile[22]	15.6
Percent foreigners	NA

Military Expenditures and Arms Transfers [22]

	1985	1986	1987	1988	1989
Military expenditures					
Current dollars (mil.)	265,800	280,900	288,200	293,100	304,100
1989 constant dollars (mil.)	302,600	311,700	309,900	305,100	304,100
Armed forces (000)	2,244	2,269	2,279	2,246	2,241
Gross national product (GNP)					
Current dollars (bil.)	4,015	4,232	4,516	4,874	5,201
1989 constant dollars (bil.)	4,571	4,696	4,857	5,073	5,201
Central government expenditures (CGE)					
1989 constant dollars (mil.)	1,143,000	1,149,000	1,137,000	1,164,000	1,190,000
People (mil.)	239.3	241.6	243.9	246.3	248.8
Military expenditure as % of GNP	6.6	6.6	6.4	6.0	5.8
Military expenditure as % of CGE	26.5	27.1	27.2	26.2	25.5
Military expenditure per capita	1,265	1,290	1,271	1,239	1,222
Armed forces per 1,000 people	9.4	9.4	9.3	9.1	9.0
GNP per capita	19,100	19,430	19,910	20,600	20,910
Arms imports[6]					
Current dollars (mil.)[13]	1,800	1,900	2,200	2,500	1,600
1989 constant dollars (mil.)[13]	2,049	2,108	2,366	2,602	1,600
Arms exports[6]					
Current dollars (mil.)	11,100	9,200	14,300	14,800	11,200
1989 constant dollars (mil.)	12,640	10,210	15,380	15,410	11,200
Total imports[7]					
Current dollars (mil.)	352,500	382,300	424,400	459,500	492,900
1989 constant dollars	401,200	424,200	456,500	478,400	492,900
Total exports[7]					
Current dollars (mil.)	218,800	227,200	254,100	322,400	364,000
1989 constant dollars	249,100	252,100	273,300	335,600	364,000
Arms as percent of total imports[8]	0.5	0.5	0.5	0.5	0.3
Arms as percent of total exports[8]	5.1	4.1	5.6	4.6	3.1

Human Rights [24]

	SSTS	FL	FAPRO	PPCG	APROBC	TPW	PCPTW	STPEP	PHRFF	PRW	ASST	AFL
Observes	P			P		P	P			P	P	P

	EAFRD	CPR	ESCR	SR	ACHR	MAAE	PVIAC	PVNAC	EAFDAW	TCIDTP	RC
Observes	S	P	S	P	S		S	S	S	S	P

P = Party; S = Signatory; see Appendix for meaning of abbreviations.

Labor Force

Total Labor Force [25]

128.548 million (includes armed forces and unemployed; civilian labor force 126.982 million) (1992)

Labor Force by Occupation [26]

Very diverse distribution.

Unemployment Rate [27]

7% (April 1993)

Production Sectors

Commercial Energy Production and Consumption

Production [28]

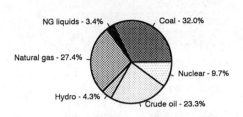

NG liquids - 3.4%
Coal - 32.0%
Natural gas - 27.4%
Nuclear - 9.7%
Hydro - 4.3%
Crude oil - 23.3%

Consumption [29]

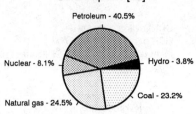

Petroleum - 40.5%
Nuclear - 8.1%
Hydro - 3.8%
Natural gas - 24.5%
Coal - 23.2%

Data are shown in quadrillion (10^{15}) BTUs and percent for 1991

Crude oil[1]	15.70
Natural gas liquids	2.31
Dry natural gas	18.49
Coal[2]	21.56
Net hydroelectric power[3]	2.88
Net nuclear power[3]	6.54
Total	67.49

Petroleum	32.85
Dry natural gas	19.84
Coal[2]	18.79
Net hydroelectric power[3]	3.08
Net nuclear power[3]	6.54
Total	81.10

Telecommunications [30]

- 126,000,000 telephone access lines
- 7,557,000 cellular phone subscribers
- Broadcast stations - 4,987 AM, 4,932 FM, 1,092 TV
- About 9,000 TV cable systems
- 530,000,000 radio sets and 193,000,000 TV sets in use
- 16 satellites and 24 ocean cable systems in use
- Satellite ground stations - 45 Atlantic Ocean INTELSAT and 16 Pacific Ocean INTELSAT (1990)

Transportation [31]

Railroads. 240,000 km of mainline routes, all standard 1.435 meter track, no government ownership (1989)

Highways. 7,599,250 km total; 6,230,000 km state-financed roads; 1,369,250 km federally-financed roads (including 71,825 km interstate limited access freeways) (1988)

Merchant Marine. 385 ships (1,000 GRT or over) totaling 12,567,000 GRT/19,511,000 DWT; includes 3 passenger-cargo, 36 cargo, 23 bulk, 169 tanker, 13 tanker tug-barge, 13 liquefied gas, 128 intermodal; in addition, there are 219 government-owned vessels

Airports

Total:	14,177
Usable:	12,417
With permanent-surface runways:	4,820
With runways over 3,659 m:	63
With runways 2,440-3,659 m:	325
With runways 1,220-2,439 m:	2,524

Top Agricultural Products [32]

Agriculture accounts for 2% of GDP and 2.8% of labor force; favorable climate and soils support a wide variety of crops and livestock production; world's second largest producer and number one exporter of grain; surplus food producer; fish catch of 4.4 million metric tons (1990).

Top Mining Products [33]

Detailed information is not available. A summary of mineral resources available follows. **Mineral Resources**: Coal, copper, lead, molybdenum, phosphates, uranium, bauxite, gold, iron, mercury, nickel, potash, silver, tungsten, zinc, petroleum, natural gas.

Tourism [34]

	1987	1988	1989	1990	1991
Tourists[72]	29,658	34,245	36,604	39,772	42,723
Tourism receipts	23,505	28,935	34,432	40,579	45,551
Tourism expenditures	29,215	33,098	34,977	38,671	39,418
Fare receipts	6,870	8,770	10,100	12,251	18,833
Fare expenditures	7,410	7,930	8,520	8,890	8,176

Tourists are in thousands, money in million U.S. dollars.

Manufacturing Sector

GDP and Manufacturing Summary [35]

	1980	1985	1989	1990	% change 1980-1990	% change 1989-1990
GDP (billion 1980 $)	2,688.5	3,095.6	3,522.4	3,575.1	33.0	1.5
GDP per capita (1980 $)	11,804	12,937	14,242	14,336	21.5	0.7
Manufacturing as % of GDP (current prices)	21.4	19.6	19.4[e]	17.8[e]	-16.8	-8.2
Gross output (billion $)	1,857.1	2,267.0	2,874.8	2,861.3	54.1	-0.5
Value added (billion $)	769.9	996.4	1,345.9	1,322.1	71.7	-1.8
Value added (billion 1980 $)	586.4	694.5	808.6	812.8	38.6	0.5
Industrial production index	100	113	131	132	32.0	0.8
Employment (thousands)	19,210	17,422	17,820	17,498	-8.9	-1.8

Note: GDP stands for Gross Domestic Product. 'e' stands for estimated value.

Profitability and Productivity

	1980	1985	1989	1990	% change 1980-1990	% change 1989-1990
Intermediate input (%)	59	56	53	54	-8.5	1.9
Wages, salaries, and supplements (%)	21	21	16[e]	21	0.0	31.3
Gross operating surplus (%)	21	22	30[e]	26	23.8	-13.3
Gross output per worker ($)	96,673	130,122	161,325	163,521	69.1	1.4
Value added per worker ($)	40,078	57,191	75,527	75,555	88.5	0.0
Average wage (incl. benefits) ($)	20,044	27,955	26,356[e]	33,573	67.5	27.4

Profitability is in percent of gross output. Productivity is in U.S. $. 'e' stands for estimated value.

Profitability - 1990

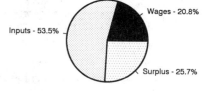

Wages - 20.8%
Inputs - 53.5%
Surplus - 25.7%

The graphic shows percent of gross output.

Value Added in Manufacturing

	1980 $ mil.	1980 %	1985 $ mil.	1985 %	1989 $ mil.	1989 %	1990 $ mil.	1990 %	% change 1980-1990	% change 1989-1990
311 Food products	63,460	8.2	87,960	8.8	116,264	8.6	119,840	9.1	88.8	3.1
313 Beverages	11,810	1.5	16,160	1.6	22,154	1.6	21,140	1.6	79.0	-4.6
314 Tobacco products	6,160	0.8	11,890	1.2	17,719	1.3	22,560	1.7	266.2	27.3
321 Textiles	23,030	3.0	26,910	2.7	36,379	2.7	34,950	2.6	51.8	-3.9
322 Wearing apparel	19,780	2.6	22,150	2.2	27,371	2.0	25,480	1.9	28.8	-6.9
323 Leather and fur products	1,850	0.2	1,570	0.2	2,483	0.2	2,210	0.2	19.5	-11.0
324 Footwear	2,950	0.4	2,470	0.2	2,414	0.2	2,320	0.2	-21.4	-3.9
331 Wood and wood products	12,970	1.7	15,390	1.5	21,758[e]	1.6	20,820	1.6	60.5	-4.3
332 Furniture and fixtures	9,840	1.3	13,250	1.3	17,382[e]	1.3	16,910	1.3	71.8	-2.7
341 Paper and paper products	29,790	3.9	40,390	4.1	58,011	4.3	57,200	4.3	92.0	-1.4
342 Printing and publishing	44,390	5.8	73,050	7.3	92,665	6.9	103,180	7.8	132.4	11.3
351 Industrial chemicals	38,920	5.1	43,360	4.4	73,209[e]	5.4	73,480	5.6	88.8	0.4
352 Other chemical products	35,530	4.6	54,280	5.4	77,988[e]	5.8	81,760	6.2	130.1	4.8
353 Petroleum refineries	23,010	3.0	13,890	1.4	21,759	1.6	22,820	1.7	-0.8	4.9
354 Miscellaneous petroleum and coal products	2,670	0.3	3,450	0.3	5,134[e]	0.4	4,390	0.3	64.4	-14.5
355 Rubber products	8,030	1.0	10,970	1.1	13,365[e]	1.0	13,430	1.0	67.2	0.5
356 Plastic products	14,540	1.9	24,740	2.5	37,445[e]	2.8	37,320	2.8	156.7	-0.3
361 Pottery, china and earthenware	1,210	0.2	1,300	0.1	1,813[e]	0.1	1,840	0.1	52.1	1.5
362 Glass and glass products	6,470	0.8	7,660	0.8	10,876[e]	0.8	10,080	0.8	55.8	-7.3
369 Other non-metal mineral products	16,300	2.1	19,880	2.0	25,507[e]	1.9	23,990	1.8	47.2	-5.9
371 Iron and steel	30,780	4.0	24,070	2.4	33,832	2.5	31,780	2.4	3.2	-6.1
372 Non-ferrous metals	14,340	1.9	11,440	1.1	20,064	1.5	17,510	1.3	22.1	-12.7
381 Metal products	53,180	6.9	61,810	6.2	73,822	5.5	70,350	5.3	32.3	-4.7
382 Non-electrical machinery	102,760	13.3	115,550	11.6	159,978	11.9	145,050	11.0	41.2	-9.3
383 Electrical machinery	74,850	9.7	111,220	11.2	115,619	8.6	112,400	8.5	50.2	-2.8
384 Transport equipment	81,280	10.6	128,230	12.9	161,828	12.0	154,020	11.7	89.5	-4.8
385 Professional and scientific equipment	27,940	3.6	40,280	4.0	80,399	6.0	76,510	5.8	173.8	-4.8
390 Other manufacturing industries	12,060	1.6	13,060	1.3	18,655[e]	1.4	18,720	1.4	55.2	0.3

Note: The industry codes shown are International Standard Industry codes (ISIC). Percentages are percent of total Value Added. 'e' stands for estimated value

978

For sources, notes, and explanations, see Annotated Source Appendix, page 1035.

Finance, Economics, and Trade

Economic Indicators [36]

National product: GDP—purchasing power equivalent—$5.951 trillion (1992). **National product real growth rate**: 2.1% (1992). **National product per capita**: $23,400 (1992). **Inflation rate (consumer prices)**: 3% (1992). **External debt**: not available.

Balance of Payments Summary [37]

Values in millions of dollars.

	1987	1988	1989	1990	1991
Exports of goods (f.o.b.)	250,280	320,340	361,670	388,710	415,960
Imports of goods (f.o.b.)	-409,770	-447,310	-477,380	-497,550	-489,400
Trade balance	-159,490	-126,970	-115,710	-108,840	-73,440
Services - debits	-167,900	-197,480	-227,670	-240,850	-227,200
Services - credits	181,520	213,130	267,790	292,170	288,960
Private transfers (net)	-1,840	-1,760	-12,320	-12,390	-12,990
Government transfers (net)	-12,490	-13,290	-13,290	-20,550	20,980
Long term capital (net)	57,520	93,480	85,580	2,000	2,940
Short term capital (net)	52,540	5,750	30,110	11,190	-20,710
Errors and omissions	-6,720	-9,130	47,460	-1,120	-1,120
Overall balance	-56,860	-36,270	-29,810	-22,580	-22,580

Exchange Rates [38]

Currency: **dollar.**
Symbol: **$.**

Data are currency units per $1.

Basis for all exchange rates   1.00

Imports and Exports

Top Import Origins [39]

$544.1 billion (c.i.f., 1992). Data are for 1989.

Origins	%
Western Europe	21.5
Japan	19.7
Canada	18.8

Top Export Destinations [40]

$442.3 billion (f.o.b., 1992). Data are for 1989.

Destinations	%
Western Europe	27.3
Canada	22.1
Japan	12.1

Foreign Aid [41]

	U.S. $	
Donor - commitments, including ODA and OOF, (FY80-89)	115.7	billion

Import and Export Commodities [42]

Import Commodities	**Export Commodities**
Crude & refined petroleum products	Capital goods
Machinery	Automobiles
Automobiles	Industrial supplies
Consumer goods	Raw materials
Industrial raw materials	Consumer goods
Food and beverages	Agricultural products

For sources, notes, and explanations, see Annotated Source Appendix, page 1035.

979

Uruguay

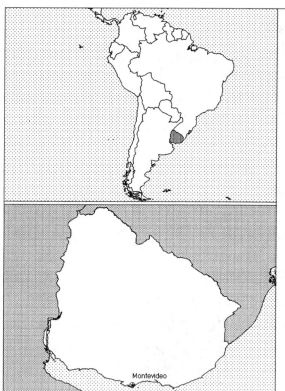

Montevideo

Geography [1]

Total area:
176,220 km2
Land area:
173,620 km2
Comparative area:
Slightly smaller than Washington State
Land boundaries:
Total 1,564 km, Argentina 579 km, Brazil 985 km
Coastline:
660 km
Climate:
Warm temperate; freezing temperatures almost unknown
Terrain:
Mostly rolling plains and low hills; fertile coastal lowland
Natural resources:
Soil, hydropower potential, minor minerals
Land use:
 Arable land:
8%
 Permanent crops:
0%
 Meadows and pastures:
78%
 Forest and woodland:
4%
 Other:
10%

Demographics [2]

	1960	1970	1980	1990	1991[1]	1994	2000	2010	2020
Population	2,531	2,824	2,920	3,106	3,121	3,199	3,344	3,594	3,822
Population density (persons per sq. mi.)	38	42	44	46	47	NA	49	52	54
Births	NA	NA	NA	NA	54	57	NA	NA	NA
Deaths	NA	NA	NA	NA	31	30	NA	NA	NA
Life expectancy - males	NA	NA	NA	NA	69	71	NA	NA	NA
Life expectancy - females	NA	NA	NA	NA	76	77	NA	NA	NA
Birth rate (per 1,000)	NA	NA	NA	NA	17	18	NA	NA	NA
Death rate (per 1,000)	NA	NA	NA	NA	10	9	NA	NA	NA
Women of reproductive age (15-44 yrs.)	NA	NA	NA	740	NA	774	811	NA	NA
of which are currently married	NA	NA	NA	454	NA	473	504	NA	NA
Fertility rate	NA	NA	NA	2.5	2.39	2.4	2.3	2.1	2.0

Population values are in thousands, life expectancy in years, and other items as indicated.

Health

Health Personnel [3]

Doctors per 1,000 pop., 1988-92	2.90
Nurse-to-doctor ratio, 1988-92	0.2
Hospital beds per 1,000 pop., 1985-90	4.6
Percentage of children immunized (age 1 yr. or less)	
Third dose of DPT, 1990-91	88.0
Measles, 1990-91	82.0

Health Expenditures [4]

Total health expenditure, 1990 (official exchange rate)	
Millions of dollars	383
Millions of dollars per capita	124
Health expenditures as a percentage of GDP	
Total	4.6
Public sector	2.5
Private sector	2.1
Development assistance for health	
Total aid flows (millions of dollars)[1]	5
Aid flows per capita (millions of dollars)	1.7
Aid flows as a percentage of total health expenditure	1.4

For sources, notes, and explanations, see Annotated Source Appendix, page 1035.

Human Factors

Health Care Ratios [5]

Population per physician, 1970	910
Population per physician, 1990	NA
Population per nursing person, 1970	NA
Population per nursing person, 1990	NA
Percent of births attended by health staff, 1985	NA

Infants and Malnutrition [6]

Percent of babies with low birth weight, 1985	8
Infant mortality rate per 1,000 live births, 1970	46
Infant mortality rate per 1,000 live births, 1991	21
Years of life lost per 1,000 population, 1990	15
Prevalence of malnutrition (under age 5), 1990	9

Ethnic Division [7]

White	88%
Mestizo	8%
Black	4%

Religion [8]

Less than half of the adult Catholic population attends church regularly.

Roman Catholic	66%
Protestant	2%
Jewish	2%
Nonprofessing or other	30%

Major Languages [9]

Spanish.

Education

Public Education Expenditures [10]

Million Pesos	1980	1985	1987	1988	1989	1990
Total education expenditure	2,035	12,565	NA	85,884	149,091	289,354
as percent of GNP	2.3	2.8	NA	3.3	3.2	3.1
as percent of total govt. expend.	10.0	9.3	NA	15.1	14.2	15.9
Current education expenditure	1,927	12,068	NA	78,989	134,277	265,660
as percent of GNP	2.2	2.7	NA	3.0	2.8	2.9
as percent of current govt. expend.	NA	9.3	NA	15.5	14.4	16.7
Capital expenditure	108	497	NA	6,895	14,814	23,694

Educational Attainment [11]

Age group	25+
Total population	1,594,747
Highest level attained (%)	
No schooling	62.4
First level	
Incompleted	NA
Completed	NA
Entered second level	
S-1	28.6
S-2	NA
Post secondary	9.0

Literacy Rate [12]

In thousands and percent	1985[a]	1991[a]	2000[a]
Illiterate population +15 years	104	88	61
Illiteracy rate - total pop. (%)	4.7	3.8	2.4
Illiteracy rate - males (%)	4.4	3.4	2.0
Illiteracy rate - females (%)	4.9	4.1	2.8

Libraries [13]

	Admin. Units	Svc. Pts.	Vols. (000)	Shelving (meters)	Vols. Added	Reg. Users
National (1986)	1	1	890	25,672	6,658	-
Non-specialized	NA	NA	NA	NA	NA	NA
Public	NA	NA	NA	NA	NA	NA
Higher ed.	NA	NA	NA	NA	NA	NA
School	NA	NA	NA	NA	NA	NA

Daily Newspapers [14]

	1975	1980	1985	1990
Number of papers	30	24	25	30
Circ. (000)	800[e]	700[e]	680[e]	720[e]

Cinema [15]

Science and Technology

Scientific/Technical Forces [16]

Potential scientists/engineers	57,650[e]
Number female	26,838[e]
Potential technicians	NA
Number female	NA
Total	NA

R&D Expenditures [17]

U.S. Patents Issued [18]

Values show patents issued to citizens of the country by the U.S. Patents Office.

	1990	1991	1992
Number of patents	3	0	2

For sources, notes, and explanations, see Annotated Source Appendix, page 1035.

Government and Law

Organization of Government [19]

Long-form name:
Oriental Republic of Uruguay
Type:
republic
Independence:
25 August 1828 (from Brazil)
Constitution:
27 November 1966, effective February 1967, suspended 27 June 1973, new constitution rejected by referendum 30 November 1980
Legal system:
based on Spanish civil law system; accepts compulsory ICJ jurisdiction
National holiday:
Independence Day, 25 August (1828)
Executive branch:
president, vice president, Council of Ministers (cabinet)
Legislative branch:
bicameral General Assembly (Asamblea General) consists of an upper chamber or Chamber of Senators (Camara de Senadores) and a lower chamber or Chamber of Representatives (Camera de Representantes)
Judicial branch:
Supreme Court

Crime [23]

Political Parties [20]

Chamber of Representatives	% of votes
Blanco	39.0
Colorado	30.0
Broad Front	22.0
New Space	8.0
Other	1.0

Government Expenditures [21]

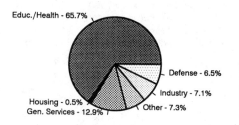

Educ./Health - 65.7%
Defense - 6.5%
Industry - 7.1%
Other - 7.3%
Gen. Services - 12.9%
Housing - 0.5%

% distribution for 1991. Expend. in 1991: 3.000 ($ bil.)

Military Expenditures and Arms Transfers [22]

	1985	1986	1987	1988	1989
Military expenditures					
Current dollars (mil.)	155	163	138	162	NA
1989 constant dollars (mil.)	176	181	148	169	NA
Armed forces (000)	30	30[e]	28	29	27
Gross national product (GNP)					
Current dollars (mil.)	5,938	6,720	7,394	7,661	8,066
1989 constant dollars (mil.)	6,760	7,457	7,953	7,975	8,066
Central government expenditures (CGE)					
1989 constant dollars (mil.)	1,660	1,799	1,926	2,072	1,500
People (mil.)	3.0	3.0	3.0	3.1	3.1
Military expenditure as % of GNP	2.6	2.4	1.9	2.1	NA
Military expenditure as % of CGE	10.6	10.1	7.7	8.1	NA
Military expenditure per capita	59	60	49	55	NA
Armed forces per 1,000 people	10.0	9.9	9.2	9.5	8.8
GNP per capita	2,247	2,464	2,612	2,603	2,617
Arms imports[6]					
Current dollars (mil.)	0	0	0	5	20
1989 constant dollars (mil.)	0	0	0	5	20
Arms exports[6]					
Current dollars (mil.)	0	0	0	0	0
1989 constant dollars (mil.)	0	0	0	0	0
Total imports[7]					
Current dollars (mil.)	708	870	1,142	1,157	1,174
1989 constant dollars	806	965	1,228	1,204	1,174
Total exports[7]					
Current dollars (mil.)	909	1,088	1,189	1,405	1,500
1989 constant dollars	1,035	1,207	1,279	1,463	1,500
Arms as percent of total imports[8]	0	0	0	0.4	1.7
Arms as percent of total exports[8]	0	0	0	0	0

Human Rights [24]

	SSTS	FL	FAPRO	PPCG	APROBC	TPW	PCPTW	STPEP	PHRFF	PRW	ASST	AFL
Observes			P	P	P	P	P			S		P
	EAFRD	CPR	ESCR	SR	ACHR	MAAE	PVIAC	PVNAC	EAFDAW	TCIDTP	RC	
Observes							P	P			S	

P = Party; S = Signatory; see Appendix for meaning of abbreviations.

Labor Force

Total Labor Force [25]

1.355 million (1991 est.)

Labor Force by Occupation [26]

Government	25%
Manufacturing	19
Agriculture	11
Commerce	12
Utilities, other	12
Other services	21

Date of data: 1988 est.

Unemployment Rate [27]

9% (1992 est.)

For sources, notes, and explanations, see Annotated Source Appendix, page 1035.

Production Sectors

Energy Resource Summary [28]

Energy Resources: Hydropower potential. **Electricity**: 2,168,000 kW capacity; 5,960 million kWh produced, 1,900 kWh per capita (1992).

Telecommunications [30]

- Most modern facilities concentrated in Montevideo
- New nationwide microwave network
- 337,000 telephones
- Broadcast stations - 99 AM, no FM, 26 TV, 9 shortwave
- 2 Atlantic Ocean INTELSAT earth stations

Transportation [31]

Railroads. 3,000 km, all 1.435-meter (standard) gauge and government owned

Highways. 49,900 km total; 6,700 km paved, 3,000 km gravel, 40,200 km earth

Merchant Marine. 4 ships (1,000 GRT or over) totaling 84,797 GRT/132,296 DWT; includes 1 cargo, 2 container, 1 oil tanker

Airports

Total:	88
Usable:	81
With permanent-surface runways:	16
With runways over 3,659 m:	0
With runways 2,440-3,659 m:	2
With runways 1,220-2,439 m:	14

Top Agricultural Products [32]

	1990	1991
Sugar cane	600	620
Rice[32]	517	610
Wheat	416	208
Barley	133	158
Corn	124	140
Sugar beets	142	120[32]

Values shown are 1,000 metric tons.

Top Mining Products [33]

Estimated metric tons unless otherwise specified   M.t.

Clays, unspecified	150,000
Dimension stone	10,000
Dolomite	19,000
Gypsum	145,000
Limestone	750,000
Marble	4,000
Quartz	300
Sand and gravel (000 metric tons)	2,000

Tourism [34]

	1987	1988	1989	1990	1991
Visitors[1]	1,047	1,036	1,240	1,267	1,510
Tourism receipts	208	203	228	262	333
Tourism expenditures	129	138	167	111	100
Fare receipts	33	37	25	28	31
Fare expenditures	44	45	46	46	45

Tourists are in thousands, money in million U.S. dollars.

For sources, notes, and explanations, see Annotated Source Appendix, page 1035.

983

Manufacturing Sector

GDP and Manufacturing Summary [35]

	1980	1985	1989	1990	% change 1980-1990	% change 1989-1990
GDP (million 1980 $)	5,970	5,208	5,778	6,205	3.9	7.4
GDP per capita (1980 $)	2,049	1,731	1,878	2,005	-2.1	6.8
Manufacturing as % of GDP (current prices)	26.0	27.1	18.7	24.7	-5.0	32.1
Gross output (million $)	3,302	3,189	4,761e	5,110e	54.8	7.3
Value added (million $)	1,286	1,344	2,106e	2,251e	75.0	6.9
Value added (million 1980 $)	1,334	1,012	1,167	1,211	-9.2	3.8
Industrial production index	100	74	89	88	-12.0	-1.1
Employment (thousands)	160	123	124e	122e	-23.8	-1.6

Note: GDP stands for Gross Domestic Product. 'e' stands for estimated value.

Profitability and Productivity

	1980	1985	1989	1990	% change 1980-1990	% change 1989-1990
Intermediate input (%)	61	58	56e	56e	-8.2	0.0
Wages, salaries, and supplements (%)	13e	9	11e	11e	-15.4	0.0
Gross operating surplus (%)	26e	33	33e	33e	26.9	0.0
Gross output per worker ($)	20,456	26,012	38,479e	41,689e	103.8	8.3
Value added per worker ($)	7,971	10,961	17,018e	18,504e	132.1	8.7
Average wage (incl. benefits) ($)	2,635e	2,448	4,263e	4,529e	71.9	6.2

Profitability is in percent of gross output. Productivity is in U.S. $. 'e' stands for estimated value.

Profitability - 1990

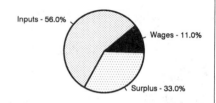

Inputs - 56.0%
Wages - 11.0%
Surplus - 33.0%

The graphic shows percent of gross output.

Value Added in Manufacturing

	1980 $ mil.	1980 %	1985 $ mil.	1985 %	1989 $ mil.	1989 %	1990 $ mil.	1990 %	% change 1980-1990	% change 1989-1990
311 Food products	165	12.8	266	19.8	379e	18.0	407e	18.1	146.7	7.4
313 Beverages	104	8.1	92	6.8	163e	7.7	178e	7.9	71.2	9.2
314 Tobacco products	90	7.0	68	5.1	101e	4.8	107e	4.8	18.9	5.9
321 Textiles	109	8.5	137	10.2	203e	9.6	247e	11.0	126.6	21.7
322 Wearing apparel	59	4.6	43	3.2	67e	3.2	65e	2.9	10.2	-3.0
323 Leather and fur products	31	2.4	76	5.7	100e	4.7	83e	3.7	167.7	-17.0
324 Footwear	18	1.4	8	0.6	19e	0.9	17e	0.8	-5.6	-10.5
331 Wood and wood products	14e	1.1	8	0.6	11e	0.5	12e	0.5	-14.3	9.1
332 Furniture and fixtures	7e	0.5	2	0.1	4e	0.2	4e	0.2	-42.9	0.0
341 Paper and paper products	30	2.3	47	3.5	75e	3.6	82e	3.6	173.3	9.3
342 Printing and publishing	37	2.9	27	2.0	43e	2.0	51e	2.3	37.8	18.6
351 Industrial chemicals	20	1.6	26	1.9	41e	1.9	49e	2.2	145.0	19.5
352 Other chemical products	75	5.8	112	8.3	173e	8.2	170e	7.6	126.7	-1.7
353 Petroleum refineries	192	14.9	194	14.4	239e	11.3	234e	10.4	21.9	-2.1
354 Miscellaneous petroleum and coal products	2	0.2	4	0.3	5e	0.2	5e	0.2	150.0	0.0
355 Rubber products	40	3.1	34	2.5	65e	3.1	72e	3.2	80.0	10.8
356 Plastic products	24	1.9	25	1.9	42e	2.0	51e	2.3	112.5	21.4
361 Pottery, china and earthenware	13	1.0	7	0.5	17e	0.8	20e	0.9	53.8	17.6
362 Glass and glass products	14	1.1	7	0.5	28e	1.3	34e	1.5	142.9	21.4
369 Other non-metal mineral products	41	3.2	24	1.8	34e	1.6	37e	1.6	-9.8	8.8
371 Iron and steel	10	0.8	14	1.0	18e	0.9	19e	0.8	90.0	5.6
372 Non-ferrous metals	3	0.2	3	0.2	4e	0.2	5e	0.2	66.7	25.0
381 Metal products	53e	4.1	32	2.4	61e	2.9	66e	2.9	24.5	8.2
382 Non-electrical machinery	16e	1.2	12	0.9	15e	0.7	15e	0.7	-6.3	0.0
383 Electrical machinery	33	2.6	31	2.3	59e	2.8	71e	3.2	115.2	20.3
384 Transport equipment	78	6.1	38	2.8	131e	6.2	142e	6.3	82.1	8.4
385 Professional and scientific equipment	1	0.1	1	0.1	2e	0.1	2e	0.1	100.0	0.0
390 Other manufacturing industries	8	0.6	6	0.4	7e	0.3	7e	0.3	-12.5	0.0

Note: The industry codes shown are International Standard Industry codes (ISIC). Percentages are percent of total Value Added. 'e' stands for estimated value

984

For sources, notes, and explanations, see Annotated Source Appendix, page 1035.

Finance, Economics, and Trade

Economic Indicators [36]

	1989	1990	1991[167]
Real GDP (mil. 1983 pesos)[168]	207,857	209,747	213,942
Real GDP growth rate (%)[168]	0.5	0.9	2.0
Real per capita GDP (US$)[169]	2,541	2,656	2,935
Money supply (M1) (nominal % increase at the end of CY)[168]	67.0	103.2	100.0
Investment rate % of GDP[170]	10.9	10.3	11.0
Consumer price inflation (%)[171]	89.2	129.0	90.0
Wholesale price inflation (%)[168]	80.7	120.7	75.0
Net external debt (yr-end mil. $US)[168]	3,245	3,120	2,650

Balance of Payments Summary [37]

Values in millions of dollars.

	1987	1988	1989	1990	1991
Exports of goods (f.o.b.)	1,182.3	1,404.5	1,599.0	1,692.9	1,604.7
Imports of goods (f.o.b.)	-1,079.9	-1,112.2	-1,136.2	-1,266.9	-1,543.7
Trade balance	102.4	292.3	462.8	426.0	61.0
Services - debits	-760.7	-777.6	-916.5	-895.1	-781.2
Services - credits	516.3	498.2	600.8	697.3	785.2
Private transfers (net)	-	-	-	-	-
Government transfers (net)	8.0	21.3	8.0	8.1	40.1
Long term capital (net)	130.1	22.1	47.2	9.6	-148.2
Short term capital (net)	218.8	222.6	-18.1	-75.5	-206.0
Errors and omissions	-117.1	-256.8	-62.4	114.9	397.2
Overall balance	97.8	22.1	121.8	285.1	148.1

Exchange Rates [38]

Currency: **new Uruguayan pesos.**
Symbol: **N$Ur.**

Data are currency units per $1.

December 1992	3,457.5
1992	3,026.9
1991	2,489.0
1990	1,594.0
1989	805.0
1988	451.0
1987	281.0

Imports and Exports

Top Import Origins [39]

$1.7 billion (f.o.b., 1992 est.). Data are for 1990.

Origins	%
Brazil	23
Argentina	17
US	10
EC	27.1

Top Export Destinations [40]

$1.7 billion (f.o.b., 1992 est.).

Destinations	%
Argentina	NA
Brazil	NA
US	NA
Germany	NA

Foreign Aid [41]

	U.S. $	
US commitments, including Ex-Im (FY70-88)	105	million
Western (non-US) countries, ODA and OOF bilateral commitments (1970-89)	420	million
Communist countries (1970-89)	69	million

Import and Export Commodities [42]

Import Commodities	**Export Commodities**
Crude oil	Hides and leather goods 17%
Fuels	Beef 10%
And lubricants	Wool 9%
Metals	Fish 7%
Machinery	Rice 4%
Transportation equipment	
Industrial chemicals	

Uzbekistan

Geography [1]

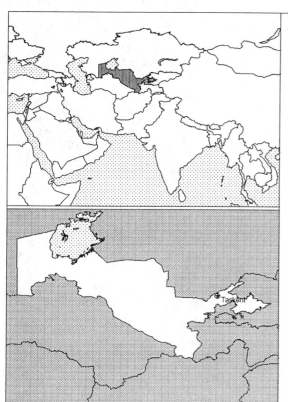

Total area:
447,400 km2
Land area:
425,400 km2
Comparative area:
Slightly larger than California
Land boundaries:
Total 6,221 km, Afghanistan 137 km, Kazakhstan 2,203 km, Kyrgyzstan 1,099 km, Tajikistan 1,161 km, Turkmenistan 1,621 km
Coastline:
0 km
Climate:
Mostly mid latitude desert; semiarid grassland in east
Terrain:
Mostly flat-to-rolling sandy desert with dunes; Fergana Valley in east surrounded by mountainous Tajikistan and Kyrgyzstan; shrinking Aral Sea in west
Natural resources:
Natural gas, petroleum, coal, gold, uranium, silver, copper, lead and zinc, tungsten, molybdenum
Land use:
Arable land: 10%
Permanent crops: 0%
Meadows and pastures: 47%
Forest and woodland: 0%
Other: 43%

Demographics [2]

	1960	1970	1980	1990	1991	1994	2000	2010	2020
Population	8,531	11,940	16,000	20,615	NA	22,609	25,467	30,380	35,4212
Population density (persons per sq. mi.)	NA	NA	NA	NA	NA	NA	NA	NA	NA
Births	NA	NA	NA	NA	NA	678	NA	NA	NA
Deaths	NA	NA	NA	NA	NA	147	NA	NA	NA
Life expectancy - males	NA	NA	NA	NA	NA	65	NA	NA	NA
Life expectancy - females	NA	NA	NA	NA	NA	72	NA	NA	NA
Birth rate (per 1,000)	NA	NA	NA	NA	NA	30	NA	NA	NA
Death rate (per 1,000)	NA	NA	NA	NA	NA	7	NA	NA	NA
Women of reproductive age (15-44 yrs.)	NA	NA	NA	4,787	NA	5,382	6,413	NA	NA
of which are currently married	NA	NA	NA	3,226	NA	3,670	4,363	NA	NA
Fertility rate	NA	NA	NA	4.3	NA	3.7	3.4	2.9	2.5

Population values are in thousands, life expectancy in years, and other items as indicated.

Health

Health Personnel [3]

Doctors per 1,000 pop., 1988-92	3.58
Nurse-to-doctor ratio, 1988-92	2.9
Hospital beds per 1,000 pop., 1985-90	12.4
Percentage of children immunized (age 1 yr. or less)	
Third dose of DPT, 1990-91	57.0
Measles, 1990-91	81.0

Health Expenditures [4]

Total health expenditure, 1990 (official exchange rate)	
Millions of dollars	2388
Millions of dollars per capita	116
Health expenditures as a percentage of GDP	
Total	5.9
Public sector	4.3
Private sector	1.6
Development assistance for health	
Total aid flows (millions of dollars)[1]	NA
Aid flows per capita (millions of dollars)	NA
Aid flows as a percentage of total health expenditure	NA

For sources, notes, and explanations, see Annotated Source Appendix, page 1035.

Human Factors

Health Care Ratios [5]

Population per physician, 1970	NA
Population per physician, 1990	280
Population per nursing person, 1970	NA
Population per nursing person, 1990	NA
Percent of births attended by health staff, 1985	NA

Infants and Malnutrition [6]

Percent of babies with low birth weight, 1985	NA
Infant mortality rate per 1,000 live births, 1970	NA
Infant mortality rate per 1,000 live births, 1991	44
Years of life lost per 1,000 population, 1990	20
Prevalence of malnutrition (under age 5), 1990	NA

Ethnic Division [7]

Uzbek	71.4%
Russian	8.3%
Tajik	4.7%
Kazakhs	4.1%
Tartars	2.4%
Karakalpaks	2.1%
Other	7.0%

Religion [8]

Muslim (mostly Sunnis)	88%
Eastern Orthodox	9%
Other	3%

Major Languages [9]

Uzbek	85%
Russian	5%
Other	10%

Education

Public Education Expenditures [10]

Educational Attainment [11]

Literacy Rate [12]

	1980[b]	1989[b]	1990[b]
Illiterate population +15 years	NA	NA	NA
Illiteracy rate - total pop. (%)	NA	2.8	NA
Illiteracy rate - males (%)	NA	1.5	NA
Illiteracy rate - females (%)	NA	4.0	NA

Libraries [13]

Daily Newspapers [14]

Cinema [15]

Science and Technology

Scientific/Technical Forces [16]

R&D Expenditures [17]

U.S. Patents Issued [18]

Government and Law

Organization of Government [19]

Long-form name:
Republic of Uzbekistan
Type:
republic
Independence:
31 August 1991 (from Soviet Union)
Constitution:
new constitution adopted 8 December 1992
Legal system:
evolution of Soviet civil law
National holiday:
Independence Day, 1 September (1991)
Executive branch:
president, prime minister, cabinet
Legislative branch:
unicameral Supreme Soviet
Judicial branch:
Supreme Court

Political Parties [20]

Supreme Soviet	% of seats
Communist	90.0
Freedom Democratic Party	2.0
Other	8.0

Government Budget [21]

For 1991.

Revenues	NA
Expenditures	NA
Capital expenditures	NA

Defense Summary [22]

Branches: Army, National Guard, Republic Security Forces (internal and border troops)

Manpower Availability: Males age 15-49 5,214,075; fit for military service 4,272,398; reach military age (18) annually 218,916 (1993 est.)

Defense Expenditures: No information available.

Crime [23]

Human Rights [24]

Labor Force

Total Labor Force [25]

7.941 million

Labor Force by Occupation [26]

Agriculture and forestry	39%
Industry and construction	24
Other	37

Date of data: 1990

Unemployment Rate [27]

0.1% includes only officially registered unemployed; there are also large numbers of underemployed workers

988

For sources, notes, and explanations, see Annotated Source Appendix, page 1035.

Production Sectors

Energy Resource Summary [28]

Energy Resources: Natural gas, petroleum, coal, uranium. **Electricity**: 11,950,000 kW capacity; 50,900 million kWh produced, 2,300 kWh per capita (1992). **Pipelines**: Crude oil 250 km, petroleum products 40 km, natural gas 810 km (1992).

Telecommunications [30]

- Poorly developed
- NMT-450 analog cellular network established in Tashkent
- 1.4 million telephone lines with 7.2 lines per 100 persons (1992)
- Linked by landline or microwave with CIS member states and by leased connection via the Moscow international gateway switch to other countries
- Satellite earth stations - Orbita and INTELSAT (TV receive only)
- New intelsat earth station provides TV receive only capability for Turkish broadcasts
- New satellite ground station also installed in Tashkent for direct linkage to Tokyo.

Top Agricultural Products [32]

	1990	1991
Cotton	5,100	4,700
Grain	1,900	1,900
Potatoes	300	300
Vegetables	2,700	-
Fruits and berries	700	-
Grapes	700	-

Values shown are 1,000 metric tons.

Top Mining Products [33]

Detailed information is not available. A summary of mineral resources available follows. **Mineral Resources**: Natural gas, petroleum, coal, gold, uranium, silver, copper, lead and zinc, tungsten, molybdenum.

Transportation [31]

Railroads. 3,460 km; does not include industrial lines (1990)

Highways. 78,400 km total; 67,000 km hard-surfaced, 11,400 km earth (1990)

Airports

Totol:	265
Useable:	74
With permanent-surface runways:	30
With runways over 3,659 m:	2
With runways 2,440-3,659 m:	20
With runways 1,220-2,439 m:	19

Tourism [34]

Finance, Economics, and Trade

Industrial Summary [35]

Industrial Production: Growth rate - 6%. **Industries**: Chemical and mineral fertilizers, vegetable oil, textiles.

Economic Indicators [36]

National product: GDP not available. **National product real growth rate**: - 10% (1992). **National product per capita**: not available. **Inflation rate (consumer prices)**: at least 17% per month (first quarter 1993). **External debt**: $2 billion (end 1991 est.).

Balance of Payments Summary [37]

Exchange Rates [38]

Currency: **rubles.**
Symbol: **R.**

Subject to wide fluctuations. Data are currency units per $1.

December 24, 1994	415

Imports and Exports

Top Import Origins [39]

$900 million from outside the successor states of the former USSR (1992).

Origins	%
Former Soviet republics	NA

Top Export Destinations [40]

$900 million to outside the successor states of the former USSR (1992).

Destinations	%
Russia	NA
Ukraine	NA
Eastern Europe	NA

Foreign Aid [41]

	U.S. $	
Official aid commitments by foreign donors (1992)	950	million

Import and Export Commodities [42]

Import Commodities

Machinery & parts
Consumer durables
Grain
Other foods

Export Commodities

Cotton
Gold
Textiles
Chemical, mineral fertilizers
Vegetable oil

990

For sources, notes, and explanations, see Annotated Source Appendix, page 1035.

Vanuatu

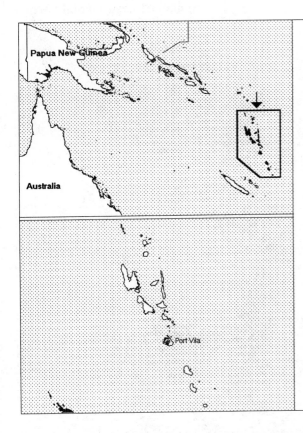

Geography [1]

Total area:
14,760 km2
Land area:
14,760 km2
Comparative area:
Slightly larger than Connecticut
Land boundaries:
0 km
Coastline:
2,528 km
Climate:
Tropical; moderated by southeast trade winds
Terrain:
Mostly mountains of volcanic origin; narrow coastal plains
Natural resources:
Manganese, hardwood forests, fish
Land use:
Arable land:
1%
Permanent crops:
5%
Meadows and pastures:
2%
Forest and woodland:
1%
Other:
91%

Demographics [2]

	1960	1970	1980	1990	1991[1]	1994	2000	2010	2020
Population	66	85	117	154	170	170	193	230	266
Population density (persons per sq. mi.)	12	15	21	29	30	NA	39	49	59
Births	NA	NA	NA	NA	6	5	NA	NA	NA
Deaths	NA	NA	NA	NA	1	2	NA	NA	NA
Life expectancy - males	NA	NA	NA	NA	67	58	NA	NA	NA
Life expectancy - females	NA	NA	NA	NA	72	61	NA	NA	NA
Birth rate (per 1,000)	NA	NA	NA	NA	36	32	NA	NA	NA
Death rate (per 1,000)	NA	NA	NA	NA	5	9	NA	NA	NA
Women of reproductive age (15-44 yrs.)	NA	NA	NA	35	NA	40	48	NA	NA
of which are currently married	NA	NA	NA	22	NA	26	31	NA	NA
Fertility rate	NA	NA	NA	5.0	5.35	4.3	3.5	2.6	2.2

Population values are in thousands, life expectancy in years, and other items as indicated.

Health

Health Personnel [3]

Health Expenditures [4]

For sources, notes, and explanations, see Annotated Source Appendix, page 1035.

991

Human Factors

Health Care Ratios [5]	Infants and Malnutrition [6]

Ethnic Division [7]

Indigenous Melanesian	94%
French	4%

Religion [8]

Presbyterian	36.7%
Anglican	15.0%,
Catholic	15.0%
Indigenous beliefs	7.6%
Seventh-Day Adventist	6.2%
Church of Christ	3.8%
Other	15.7%

Major Languages [9]

English (official), French (official), pidgin (known as Bislama or Bichelama).

Education

Public Education Expenditures [10]	Educational Attainment [11]

Literacy Rate [12]	

Libraries [13]

	Admin. Units	Svc. Pts.	Vols. (000)	Shelving (meters)	Vols. Added	Reg. Users
National (1989)[43]	1	1	50	700	1,000	200
Non-specialized	NA	NA	NA	NA	NA	NA
Public	NA	NA	NA	NA	NA	NA
Higher ed.	NA	NA	NA	NA	NA	NA
School	NA	NA	NA	NA	NA	NA

Daily Newspapers [14]	Cinema [15]

Science and Technology

Scientific/Technical Forces [16]	R&D Expenditures [17]	U.S. Patents Issued [18]

Government and Law

Organization of Government [19]

Long-form name:
Republic of Vanuatu
Type:
republic
Independence:
30 July 1980 (from France and UK)
Constitution:
30 July 1980
Legal system:
unified system being created from former dual French and British systems
National holiday:
Independence Day, 30 July (1980)
Executive branch:
president, prime minister, deputy prime minister, Council of Ministers (cabinet)
Legislative branch:
unicameral Parliament; note - the National Council of Chiefs advises on matters of custom and land
Judicial branch:
Supreme Court

Political Parties [20]

Parliament	% of seats
Union of Moderate Parties (UMP)	41.3
National United Party (NUP)	21.7
Vanuatu Party (VP)	21.7
Melanesian Progressive Party (MPP)	8.7
Tan Union Party (TUP)	2.2
Nagriamel	2.2
Friend	2.2

Government Expenditures [21]

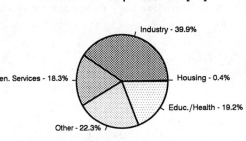

Industry - 39.9%
Gen. Services - 18.3%
Housing - 0.4%
Educ./Health - 19.2%
Other - 22.3%

% distribution for 1989. Expend. in 1989 est.: 103 ($ mil.)

Defense Summary [22]

Branches: Vanuatu Police Force (VPF), paramilitary Vanuatu Mobile Force (VMF)

Note: No military forces

Manpower Availability: No information available.

Defense Expenditures: No information available.

Crime [23]

Human Rights [24]

	SSTS	FL	FAPRO	PPCG	APROBC	TPW	PCPTW	STPEP	PHRFF	PRW	ASST	AFL
Observes						P	P				1	
	EAFRD	CPR	ESCR	SR	ACHR	MAAE	PVIAC	PVNAC	EAFDAW	TCIDTP	RC	
Observes	P	P	P	P	P	P			S	P	P	

P = Party; S = Signatory; see Appendix for meaning of abbreviations.

Labor Force

Total Labor Force [25]

Labor Force by Occupation [26]

Unemployment Rate [27]

For sources, notes, and explanations, see Annotated Source Appendix, page 1035.

993

Production Sectors

Energy Resource Summary [28]

Energy Resources: None. **Electricity**: 17,000 kW capacity; 30 million kWh produced, 180 kWh per capita (1990).

Telecommunications [30]

- Broadcast stations - 2 AM, no FM, no TV
- 3,000 telephones
- 1 Pacific Ocean INTELSAT ground station

Transportation [31]

Railroads. none

Highways. 1,027 km total; at least 240 km sealed or all-weather roads

Merchant Marine. 125 ships (1,000 GRT or over) totaling 2,121,819 GRT/3,193,942 DWT; includes 23 cargo, 16 refrigerated cargo, 6 container, 11 vehicle carrier, 1 livestock carrier, 6 oil tanker, 2 chemical tanker, 3 liquefied gas, 54 bulk, 1 combination bulk, 1 passenger, 1 short-sea passenger; note - a flag of convenience registry

Airports

Total:	31
Usable:	31
With permanent-surface runways:	2
With runways over 3,659 m:	0
With runways 2,440-3,659 m:	1
With runways 1,220-2,439 m:	2

Top Agricultural Products [32]

Agriculture accounts for 40% of GDP; export crops—coconuts, cocoa, coffee, fish; subsistence crops—taro, yams, coconuts, fruits, vegetables.

Top Mining Products [33]

Detailed information is not available. A summary of mineral resources available follows. **Mineral Resources**: Manganese.

Tourism [34]

	1987	1988	1989	1990	1991
Tourists	15	18	24	35	40
Tourism receipts	8	11	16	25	30
Tourism expenditures	1	2	1	1	1
Fare expenditures	2	1	3	4	4

Tourists are in thousands, money in million U.S. dollars.

Finance, Economics, and Trade

GDP and Manufacturing Summary [35]

	1980	1985	1990	1991	1992
Gross Domestic Product					
Millions of 1980 dollars	113	171	185	191	191
Growth rate in percent	-11.46	1.11	4.70	3.40	0.00
Manufacturing Value Added					
Millions of 1980 dollars	3	7	12	13	15[e]
Growth rate in percent	13.98	11.23	1.94	13.40	13.69[e]
Manufacturing share in percent of current prices	4.2	3.8	5.9	NA	NA

Economic Indicators [36]

National product: GDP—exchange rate conversion—$142 million (1988 est.). **National product real growth rate**: 6% (1990). **National product per capita**: $900 (1988 est.). **Inflation rate (consumer prices)**: 5% (1990). **External debt**: $30 million (1990 est.).

Balance of Payments Summary [37]

Values in millions of dollars.

	1987	1988	1989	1990	1991
Exports of goods (f.o.b.)	13.7	15.4	13.7	13.8	14.9
Imports of goods (f.o.b.)	-57.1	-57.9	-57.9	-79.6	-74.0
Trade balance	-43.4	-42.5	-44.2	-65.8	-59.1
Services - debits	-67.5	-58.8	-48.3	-57.9	-77.7
Services - credits	67.3	63.4	63.7	95.7	106.8
Private transfers (net)	5.8	9.8	4.6	6.7	-19.1
Government transfers (net)	46.4	36.2	20.7	29.4	35.5
Long term capital (net)	28.8	27.2	10.1	16.1	20.7
Short term capital (net)	-9.5	-17.2	14.2	61	-79.9
Errors and omissions	-15.1	-17.3	-13.0	-34.2	70.9
Overall balance	12.9	0.8	7.8	-3.9	-1.9

Exchange Rates [38]

Currency: **vatu.**
Symbol: **VT.**

Data are currency units per $1.

January 1993	120.77
1992	113.39
1991	111.68
1990	116.57
1989	116.04
1988	104.43

Imports and Exports

Top Import Origins [39]

$60.4 million (f.o.b., 1990 est.).

Origins	%
Australia	36
Japan	13
NZ	10
France	8
Fiji	8

Top Export Destinations [40]

$15.6 million (f.o.b., 1990 est.).

Destinations	%
Netherlands	NA
Japan	NA
France	NA
New Caledonia	NA
Belgium	NA

Foreign Aid [41]

	U.S. $	
Western (non-US) countries, ODA and OOF bilateral commitments (1970-89)	606	million

Import and Export Commodities [42]

Import Commodities	**Export Commodities**
Machines and vehicles 25%	Copra 59%
Food and beverages 23%	Cocoa 11%
Basic manufactures 18%	Meat 9%
Raw materials and fuels 11%	Fish 8%
Chemicals 6%	Timber 4%

For sources, notes, and explanations, see Annotated Source Appendix, page 1035.

995

Venezuela

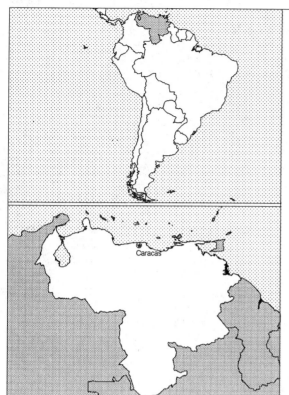

Geography [1]

Total area:
912,050 km2
Land area:
882,050 km2
Comparative area:
Slightly more than twice the size of California
Land boundaries:
Total 4,993 km, Brazil 2,200 km, Colombia 2,050 km, Guyana 743 km
Coastline:
2,800 km
Climate:
Tropical; hot, humid; more moderate in highlands
Terrain:
Andes mountains and Maracaibo lowlands in northwest; central plains (llanos); Guyana highlands in southeast
Natural resources:
Petroleum, natural gas, iron ore, gold, bauxite, other minerals, hydropower, diamonds
Land use:
Arable land:
3%
Permanent crops:
1%
Meadows and pastures:
20%
Forest and woodland:
39%
Other:
37%

Demographics [2]

	1960	1970	1980	1990	1991[1]	1994	2000	2010	2020
Population	7,502	10,604	14,452	18,776	20,189	20,562	23,196	27,407	31,312
Population density (persons per sq. mi.)	22	31	44	58	59	NA	72	87	101
Births	NA	NA	NA	NA	562	529	NA	NA	NA
Deaths	NA	NA	NA	NA	84	95	NA	NA	NA
Life expectancy - males	NA	NA	NA	NA	71	70	NA	NA	NA
Life expectancy - females	NA	NA	NA	NA	78	76	NA	NA	NA
Birth rate (per 1,000)	NA	NA	NA	NA	28	26	NA	NA	NA
Death rate (per 1,000)	NA	NA	NA	NA	4	5	NA	NA	NA
Women of reproductive age (15-44 yrs.)	NA	NA	NA	4,770	NA	5,360	6,219	NA	NA
of which are currently married	NA	NA	NA	2,669	NA	3,022	3,561	NA	NA
Fertility rate	NA	NA	NA	3.4	3.39	3.1	2.7	2.3	2.2

Population values are in thousands, life expectancy in years, and other items as indicated.

Health

Health Personnel [3]

Doctors per 1,000 pop., 1988-92	1.55
Nurse-to-doctor ratio, 1988-92	0.5
Hospital beds per 1,000 pop., 1985-90	2.9
Percentage of children immunized (age 1 yr. or less)	
Third dose of DPT, 1990-91	54.0
Measles, 1990-91	54.0

Health Expenditures [4]

Total health expenditure, 1990 (official exchange rate)	
Millions of dollars	1747
Millions of dollars per capita	89
Health expenditures as a percentage of GDP	
Total	3.6
Public sector	2.0
Private sector	1.6
Development assistance for health	
Total aid flows (millions of dollars)[1]	2.0
Aid flows per capita (millions of dollars)	0.1
Aid flows as a percentage of total health expenditure	0.1

For sources, notes, and explanations, see Annotated Source Appendix, page 1035.

Human Factors

Health Care Ratios [5]

Population per physician, 1970	1120
Population per physician, 1990	630
Population per nursing person, 1970	440
Population per nursing person, 1990	330
Percent of births attended by health staff, 1985	82

Infants and Malnutrition [6]

Percent of babies with low birth weight, 1985	9
Infant mortality rate per 1,000 live births, 1970	53
Infant mortality rate per 1,000 live births, 1991	34
Years of life lost per 1,000 population, 1990	13
Prevalence of malnutrition (under age 5), 1990	5

Ethnic Division [7]

Mestizo	67%
White	21%
Black	10%
Indian	2%

Religion [8]

Nominally Roman Catholic	96%
Protestant	2%

Major Languages [9]

Spanish (official), Indian dialects spoken by about 200,000 Amerindians in the remote interior.

Education

Public Education Expenditures [10]

Million Bolivar	1980	1985	1986	1987	1988	1989
Total education expenditure	13,162	23,068	23,481	34,499	40,463	60,782
as percent of GNP	4.4	5.1	4.9	5.1	4.8	4.3
as percent of total govt. expend.	14.7	20.3	18.9	19.6	21.1	19.0
Current education expenditure	12,524	NA	NA	34,171	38,116	NA
as percent of GNP	4.2	NA	NA	5.1	4.5	NA
as percent of current govt. expend.	24.3	NA	NA	27.5	28.5	NA
Capital expenditure	639	NA	NA	328	2,347	NA

Educational Attainment [11]

Age group	25+[12]
Total population	5,542,852[12]
Highest level attained (%)	
No schooling	23.5[12]
First level	
Incompleted	47.2[12]
Completed	NA[12]
Entered second level	
S-1	22.3[12]
S-2	NA[12]
Post secondary	7.0[12]

Literacy Rate [12]

In thousands and percent	1985[a]	1991[a]	2000[a]
Illiterate population +15 years	1,498	1,450	1,280
Illiteracy rate - total pop. (%)	14.3	11.9	7.9
Illiteracy rate - males (%)	16.2	13.3	8.7
Illiteracy rate - females (%)	12.5	10.4	7.1

Libraries [13]

	Admin. Units	Svc. Pts.	Vols. (000)	Shelving (meters)	Vols. Added	Reg. Users
National (1991)	1	NA	2,197	NA	97,713[e]	115,015
Non-specialized	NA	NA	NA	NA	NA	NA
Public (1991)	24	649	3,556	NA	NA	36,692
Higher ed. (1989)	78	168	NA	NA	NA	NA
School (1989)	443	3,033	NA	NA	NA	NA

Daily Newspapers [14]

	1975	1980	1985	1990
Number of papers	49	66	55	54
Circ. (000)	1,300[e]	2,937	2,700[e]	2,800[e]

Cinema [15]

Data for 1991.

Cinema seats per 1,000	6.2[4]
Annual attendance per person	1.0[4]
Gross box office receipts (mil. Bolivar)	983

Science and Technology

Scientific/Technical Forces [16]

Potential scientists/engineers	942,972
Number female	401,711
Potential technicians	1,159,798
Number female	457,395
Total	2,102,770

R&D Expenditures [17]

	Bolivar (000) 1985
Total expenditure[4,18]	1,411,720
Capital expenditure	NA
Current expenditure	NA
Percent current	NA

U.S. Patents Issued [18]

Values show patents issued to citizens of the country by the U.S. Patents Office.

	1990	1991	1992
Number of patents	20	25	24

For sources, notes, and explanations, see Annotated Source Appendix, page 1035.

997

Government and Law

Organization of Government [19]

Long-form name:
Republic of Venezuela
Type:
republic
Independence:
5 July 1811 (from Spain)
Constitution:
23 January 1961
Legal system:
based on Napoleonic code; judicial review of legislative acts in Cassation Court only; has not accepted compulsory ICJ jurisdiction
National holiday:
Independence Day, 5 July (1811)
Executive branch:
president, Council of Ministers (cabinet)
Legislative branch:
bicameral Congress of the Republic (Congreso de la Republica) consists of an upper chamber or Senate (Senado) and a lower chamber or Chamber of Deputies (Camara de Diputados)
Judicial branch:
Supreme Court of Justice (Corte Suprema de Justicia)

Political Parties [20]

Chamber of Deputies	% of votes
Democratic Action (AD)	43.7
Social Christian Party (COPEI)	31.4
Movement Toward Socialism (MAS)	10.3
Other	14.6

Government Budget [21]

For 1992.
Revenues	13.2
Expenditures	13.1
Capital expenditures	NA

Crime [23]

Military Expenditures and Arms Transfers [22]

	1985	1986	1987	1988	1989
Military expenditures					
Current dollars (mil.)	358	490	1,207[e]	647[e]	407[e]
1989 constant dollars (mil.)	407	544	1,298[e]	674[e]	407[e]
Armed forces (000)	71	66	69	73	75
Gross national product (GNP)					
Current dollars (mil.)	34,740	38,360	40,810	44,740	41,460
1989 constant dollars (mil.)	39,550	42,570	43,900	46,570	41,460
Central government expenditures (CGE)					
1989 constant dollars (mil.)	9,076	10,580	10,830[e]	9,629[e]	7,792[e]
People (mil.)	17.3	17.8	18.2	18.7	19.2
Military expenditure as % of GNP	1.0	1.3	3.0	1.4	1.0
Military expenditure as % of CGE	4.5	5.1	12.0	7.0	5.2
Military expenditure per capita	24	31	71	36	21
Armed forces per 1,000 people	4.1	3.7	3.8	3.9	3.9
GNP per capita	2,285	2,395	2,406	2,487	2,158
Arms imports[6]					
Current dollars (mil.)	525	100	100	130	80
1989 constant dollars (mil.)	598	111	108	135	80
Arms exports[6]					
Current dollars (mil.)	0	0	0	0	0
1989 constant dollars (mil.)	0	0	0	0	0
Total imports[7]					
Current dollars (mil.)	8,178	8,504	9,659	12,730	7,803
1989 constant dollars	9,310	9,437	10,390	13,250	7,803
Total exports[7]					
Current dollars (mil.)	12,270	8,660	10,580	10,240	13,140
1989 constant dollars	13,970	9,610	11,380	10,660	13,140
Arms as percent of total imports[8]	6.4	1.2	1.0	1.0	1.0
Arms as percent of total exports[8]	0	0	0	0	0

Human Rights [24]

	SSTS	FL	FAPRO	PPCG	APROBC	TPW	PCPTW	STPEP	PHRFF	PRW	ASST	AFL
Observes	P	P	P	P	P	P	P		P			P

	EAFRD	CPR	ESCR	SR	ACHR	MAAE	PVIAC	PVNAC	EAFDAW	TCIDTP	RC
Observes	P	P	P	P	P	P			S	P	P

P = Party; S = Signatory; see Appendix for meaning of abbreviations.

Labor Force

Total Labor Force [25]

5.8 million

Labor Force by Occupation [26]

Services	56%
Industry	28
Agriculture	16
Date of data: 1985	

Unemployment Rate [27]

8.4% (1992 est.)

For sources, notes, and explanations, see Annotated Source Appendix, page 1035.

Production Sectors

Commercial Energy Production and Consumption

Production [28]

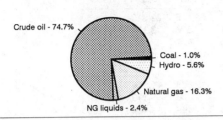

Crude oil - 74.7%
Coal - 1.0%
Hydro - 5.6%
Natural gas - 16.3%
NG liquids - 2.4%

Consumption [29]

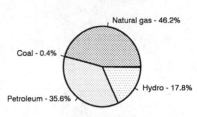

Natural gas - 46.2%
Coal - 0.4%
Hydro - 17.8%
Petroleum - 35.6%

Data are shown in quadrillion (10^{15}) BTUs and percent for 1991

Crude oil[1]	5.32
Natural gas liquids	0.17
Dry natural gas	1.16
Coal[2]	0.07
Net hydroelectric power[3]	0.40
Net nuclear power[3]	0.00
Total	7.11

Petroleum	0.80
Dry natural gas	1.04
Coal[2]	0.01
Net hydroelectric power[3]	0.40
Net nuclear power[3]	0.00
Total	2.25

Telecommunications [30]

- Modern and expanding
- 1,440,000 telephones
- Broadcast stations - 181 AM, no FM, 59 TV, 26 shortwave
- 3 submarine coaxial cables
- Satellite ground stations - 1 Atlantic Ocean INTELSAT and 3 domestic

Top Agricultural Products [32]

	1989	1990
Sugar cane	7,809	6,902
Fruits	2,421	2,480
Cereals	1,684	1,780
Roots and tubers	685	626
Vegetables	330	400
Oilseeds and textiles	326	290

Values shown are 1,000 metric tons.

Top Mining Products [33]

Preliminary metric tons unless otherwise specified M.t.

Aluminum	3,287,348
Cement, hydraulic (000 tons)	6,337
Diamonds, gem and industrial (carats)	213,557
Ferroalloys (000 tons)	85
Gold, mine output, Au content (kilograms)[37]	4,215
Iron ore and concentrate (000 tons)	21,241
Steel, crude (000 tons)	3,119
Petroleum, crude (000 42-gal. barrels)	871,762
Natural gas, gross (mil. cubic meters)	43,326
Petroleum refinery products (000 42-gal. barrels)	389,638

Transportation [31]

Railroads. 542 km total; 363 km 1.435-meter standard gauge all single track, government owned; 179 km 1.435-meter gauge, privately owned

Highways. 77,785 km total; 22,780 km paved, 24,720 km gravel, 14,450 km earth roads, and 15,835 km unimproved earth

Merchant Marine. 56 ships (1,000 GRT or over) totaling 837,375 GRT/1,344,795 DWT; includes 1 short-sea passenger, 1 passenger cargo, 19 cargo, 2 container, 4 roll-on/roll-off, 18 oil tanker, 1 chemical tanker, 2 liquefied gas, 6 bulk, 1 vehicle carrier, 1 combination bulk

Airports

Total:	360
Usable:	331
With permanent-surface runways:	133
With runways over 3,659 m:	0
With runways 2,440-3,659 m:	15
With runways 1,220-2,439 m:	87

Tourism [34]

	1987	1988	1989	1990	1991
Visitors	442	485	492	655	724
Tourists	338	373	412	525	598
Excursionists	104	112	80	130	126
Tourism receipts	416	291	389	359	365
Tourism expenditures	509	509	640	945	1,011
Fare receipts	105	160	159	153	209
Fare expenditures	63	124	92	91	133

Tourists are in thousands, money in million U.S. dollars.

For sources, notes, and explanations, see Annotated Source Appendix, page 1035.

999

Manufacturing Sector

GDP and Manufacturing Summary [35]

	1980	1985	1989	1990	% change 1980-1990	% change 1989-1990
GDP (million 1980 $)	59,213	55,446	69,778	62,315	5.2	-10.7
GDP per capita (1980 $)	3,941	3,202	3,626	3,154	-20.0	-13.0
Manufacturing as % of GDP (current prices)	18.6	22.5	20.9	20.8	11.8	-0.5
Gross output (million $)	30,213[e]	30,305	21,843	24,128	-20.1	10.5
Value added (million $)	14,461[e]	14,071	9,965	12,175	-15.8	22.2
Value added (million 1980 $)	9,596	10,507	12,868	11,314	17.9	-12.1
Industrial production index	100	101	506	113	13.0	-77.7
Employment (thousands)	426[e]	406	469	464	8.9	-1.1

Note: GDP stands for Gross Domestic Product. 'e' stands for estimated value.

Profitability and Productivity

	1980	1985	1989	1990	% change 1980-1990	% change 1989-1990
Intermediate input (%)	52[e]	54	54	50	-3.8	-7.4
Wages, salaries, and supplements (%)	15[e]	13	9	9	-40.0	0.0
Gross operating surplus (%)	33[e]	34	36	42	27.3	16.7
Gross output per worker ($)	62,664[e]	71,154	46,614	51,776	-17.4	11.1
Value added per worker ($)	30,311[e]	33,038	21,265	26,127	-13.8	22.9
Average wage (incl. benefits) ($)	10,359[e]	9,495	4,382	4,651	-55.1	6.1

Profitability is in percent of gross output. Productivity is in U.S. $. 'e' stands for estimated value.

Profitability - 1990

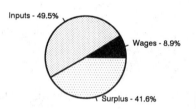

Inputs - 49.5%
Wages - 8.9%
Surplus - 41.6%

The graphic shows percent of gross output.

Value Added in Manufacturing

	1980 $ mil.	1980 %	1985 $ mil.	1985 %	1989 $ mil.	1989 %	1990 $ mil.	1990 %	% change 1980-1990	% change 1989-1990
311 Food products	1,425[e]	9.9	1,597	11.3	1,052	10.6	1,210	9.9	-15.1	15.0
313 Beverages	953[e]	6.6	836	5.9	551	5.5	583	4.8	-38.8	5.8
314 Tobacco products	409[e]	2.8	597	4.2	310	3.1	273	2.2	-33.3	-11.9
321 Textiles	430[e]	3.0	505	3.6	346	3.5	291	2.4	-32.3	-15.9
322 Wearing apparel	348[e]	2.4	359	2.6	178	1.8	160	1.3	-54.0	-10.1
323 Leather and fur products	57[e]	0.4	58	0.4	34	0.3	40	0.3	-29.8	17.6
324 Footwear	197[e]	1.4	158	1.1	86	0.9	90	0.7	-54.3	4.7
331 Wood and wood products	106[e]	0.7	80	0.6	50	0.5	36	0.3	-66.0	-28.0
332 Furniture and fixtures	188[e]	1.3	142	1.0	72	0.7	65	0.5	-65.4	-9.7
341 Paper and paper products	395[e]	2.7	357	2.5	254	2.5	277	2.3	-29.9	9.1
342 Printing and publishing	376[e]	2.6	299	2.1	213	2.1	182	1.5	-51.6	-14.6
351 Industrial chemicals	325[e]	2.2	498	3.5	418	4.2	443	3.6	36.3	6.0
352 Other chemical products	858[e]	5.9	890	6.3	607	6.1	662	5.4	-22.8	9.1
353 Petroleum refineries	4,222[e]	29.2	3,634	25.8	2,563	25.7	4,734	38.9	12.1	84.7
354 Miscellaneous petroleum and coal products	25[e]	0.2	30	0.2	17	0.2	19	0.2	-24.0	11.8
355 Rubber products	151[e]	1.0	188	1.3	101	1.0	139	1.1	-7.9	37.6
356 Plastic products	394[e]	2.7	348	2.5	240	2.4	215	1.8	-45.4	-10.4
361 Pottery, china and earthenware	60[e]	0.4	39	0.3	28	0.3	18	0.1	-70.0	-35.7
362 Glass and glass products	137[e]	0.9	132	0.9	96	1.0	109	0.9	-20.4	13.5
369 Other non-metal mineral products	489[e]	3.4	378	2.7	272	2.7	290	2.4	-40.7	6.6
371 Iron and steel	651[e]	4.5	855	6.1	557	5.6	498	4.1	-23.5	-10.6
372 Non-ferrous metals	256[e]	1.8	447	3.2	730	7.3	788	6.5	207.8	7.9
381 Metal products	652[e]	4.5	503	3.6	378	3.8	336	2.8	-48.5	-11.1
382 Non-electrical machinery	287[e]	2.0	241	1.7	219	2.2	180	1.5	-37.3	-17.8
383 Electrical machinery	345[e]	2.4	307	2.2	314	3.2	245	2.0	-29.0	-22.0
384 Transport equipment	605[e]	4.2	486	3.5	190	1.9	198	1.6	-67.3	4.2
385 Professional and scientific equipment	38[e]	0.3	26	0.2	32	0.3	37	0.3	-2.6	15.6
390 Other manufacturing industries	82[e]	0.6	81	0.6	56	0.6	56	0.5	-31.7	0.0

Note: The industry codes shown are International Standard Industry codes (ISIC). Percentages are percent of total Value Added. 'e' stands for estimated value

For sources, notes, and explanations, see Annotated Source Appendix, page 1035.

Finance, Economics, and Trade

Economic Indicators [36]

Billions of Bolivars (Bs) unless otherwise noted.

	1989	1990	1991[172]
GDP (Bs billion)	1,510.4	2,264.0	3,138.4
Real GDP growth rate	-8.6	5.3	9.2
M1 (Dec 31) (Bs billion)	171.3	241.8	315.8
Commercial interest lending rate	34.1	34.9	36.4
Savings rate (%. of GDP)	14.0	19.1	16.6
Investment rate (%. of GDP)	9.8	7.2	15.4
CPI (1984 = 100)	380.2	534.8	700.0
WPI (1984 = 100)	529.7	637.6	800.8
External public debt	26,427	27,077	29,904

Balance of Payments Summary [37]

Values in millions of dollars.

	1987	1988	1989	1990	1991
Exports of goods (f.o.b.)	10,437.0	10,082.0	12,915.0	17,444.0	14,892.0
Imports of goods (f.o.b.)	-8,870.0	-12,080.0	-7,283.0	-6,807.0	-10,101.0
Trade balance	1,567.0	-1,998.0	5,632.0	10,637.0	4,791.0
Services - debits	-5,312.0	-6,287.0	-5,979.0	-6,107.0	-6,275.0
Services - credits	2,446.0	2,623.0	2,695.0	4,032.0	3,490.0
Private transfers (net)	-73.0	-123.0	-171.0	-259.0	-310.0
Government transfers (net)	-18.0	-24.0	-16.0	-24.0	-33.0
Long term capital (net)	-1,732.0	-491.0	-1,318.0	108.0	2,697.0
Short term capital (net)	1,981.0	-1,399.0	-2,059.0	-3,229.0	521.0
Errors and omissions	-478.0	3,105.0	1,418.0	-1,742.0	-2,422.0
Overall balance	-1,628.0	-4,549.0	202.0	3,416.0	2,459.0

Exchange Rates [38]

Currency: **bolivares.**
Symbol: **Bs.**

Data are currency units per $1.

January 1993	80.18
1992	68.38
1991	56.82
1990	46.90
1989	34.68
Fixed rate 1987-88	14.50

Imports and Exports

Top Import Origins [39]

$12.4 billion (f.o.b., 1992 est.). Data are for 1989.

Origins	%
US	44
FRG	8.0
Japan	4
Italy	7
Canada	2

Top Export Destinations [40]

$14.0 billion (f.o.b., 1992 est.). Data are for 1989.

Destinations	%
US	50.7
Europe	13.7
Japan	4.0

Foreign Aid [41]

	U.S. $	
US commitments, including Ex-Im (FY70-86)	488	million
Communist countries (1970-89)	10	million

Import and Export Commodities [42]

Import Commodities	Export Commodities
Foodstuffs	Petroleum 82%
Chemicals	Bauxite and aluminum
Manufactures	Iron ore
Machinery & transport equipment	Agricultural products
	Basic manufactures

Vietnam

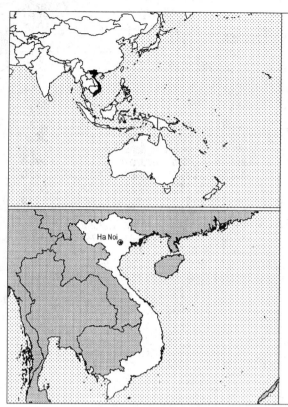

Geography [1]

Total area:
329,560 km2
Land area:
325,360 km2
Comparative area:
Slightly larger than New Mexico
Land boundaries:
Total 3,818 km, Cambodia 982 km, China 1,281 km, Laos 1,555 km
Coastline:
3,444 km (excludes islands)
Climate:
Tropical in south; monsoonal in north with hot, rainy season (mid-May to mid-September) and warm, dry season (mid-October to mid-March)
Terrain:
Low, flat delta in south and north; central highlands; hilly, mountainous in far north and northwest
Natural resources:
Phosphates, coal, manganese, bauxite, chromate, offshore oil deposits, forests
Land use:
Arable land: 22%
Permanent crops: 2%
Meadows and pastures: 1%
Forest and woodland: 40%
Other: 35%

Demographics [2]

	1960	1970	1980	1990	1991[1]	1994	2000	2010	2020
Population	31,955	42,978	54,234	67,718	67,568	73,104	80,533	91,729	102,359
Population density (persons per sq. mi.)	251	338	425	527	538	NA	635	733	820
Births	NA	NA	NA	NA	1,991	1,983	NA	NA	NA
Deaths	NA	NA	NA	NA	544	567	NA	NA	NA
Life expectancy - males	NA	NA	NA	NA	63	65	NA	NA	NA
Life expectancy - females	NA	NA	NA	NA	67	69	NA	NA	NA
Birth rate (per 1,000)	NA	NA	NA	NA	29	27	NA	NA	NA
Death rate (per 1,000)	NA	NA	NA	NA	8	8	NA	NA	NA
Women of reproductive age (15-44 yrs.)	NA	NA	NA	16,987	NA	18,789	21,746	NA	NA
of which are currently married	NA	NA	NA	10,299	NA	11,707	13,754	NA	NA
Fertility rate	NA	NA	NA	3.8	3.70	3.3	2.8	2.3	2.1

Population values are in thousands, life expectancy in years, and other items as indicated.

Health

Health Personnel [3]

Doctors per 1,000 pop., 1988-92	0.35
Nurse-to-doctor ratio, 1988-92	4.9
Hospital beds per 1,000 pop., 1985-90	3.3
Percentage of children immunized (age 1 yr. or less)	
Third dose of DPT, 1990-91	85.0
Measles, 1990-91	85.0

Health Expenditures [4]

Total health expenditure, 1990 (official exchange rate)	
Millions of dollars	157
Millions of dollars per capita	2
Health expenditures as a percentage of GDP	
Total	2.1
Public sector	1.1
Private sector	1.0
Development assistance for health	
Total aid flows (millions of dollars)[1]	25
Aid flows per capita (millions of dollars)	0.4
Aid flows as a percentage of total health expenditure	15.9

For sources, notes, and explanations, see Annotated Source Appendix, page 1035.

Human Factors

Health Care Ratios [5]	Infants and Malnutrition [6]

Ethnic Division [7]

Vietnamese	85-90%
Chinese	3%
Muong	NA
Thai	NA
Meo	NA
Khmer	NA
Man	NA
Cham	NA

Religion [8]

Buddhist, Taoist, Roman Catholic, indigenous beliefs, Islamic, Protestant.

Major Languages [9]

Vietnamese (official), French, Chinese, English, Khmer, tribal languages (Mon-Khmer and Malayo-Polynesian).

Education

Public Education Expenditures [10]	Educational Attainment [11]

Literacy Rate [12]

In thousands and percent	1985[a]	1991[a]	2000[a]
Illiterate population + 15 years	5,563	5,061	4,654
Illiteracy rate - total pop. (%)	15.6	12.4	8.7
Illiteracy rate - males (%)	10.4	8.0	5.5
Illiteracy rate - females (%)	20.3	16.4	11.7

Libraries [13]

Daily Newspapers [14]

Cinema [15]

Data for 1988.

Cinema seats per 1,000	NA
Annual attendance per person	3.8
Gross box office receipts (mil. Dong)	NA

Science and Technology

Scientific/Technical Forces [16]

R&D Expenditures [17]

	Dong[18] (000) 1985
Total expenditure	498,000
Capital expenditure	NA
Current expenditure	NA
Percent current	NA

U.S. Patents Issued [18]

For sources, notes, and explanations, see Annotated Source Appendix, page 1035.

Government and Law

Organization of Government [19]

Long-form name:
Socialist Republic of Vietnam
Type:
Communist state
Independence:
2 September 1945 (from France)
Constitution:
NA April 1992
Legal system:
based on Communist legal theory and French civil law system
National holiday:
Independence Day, 2 September (1945)
Executive branch:
president, prime minister, three deputy prime ministers
Legislative branch:
unicameral National Assembly (Quoc-Hoi)
Judicial branch:
Supreme People's Court

Elections [20]

National Assembly. Only party - Vietnam Communist Party (VCP). Elections last held 19 July 1992 (next to be held NA July 1997); results - VCP is the only party; seats - (395 total) VCP or VCP-approved 395.

Government Budget [21]

For 1990.

Revenues	1.7
Expenditures	1.9
Capital expenditures	NA

Crime [23]

Military Expenditures and Arms Transfers [22]

	1985	1986	1987	1988	1989
Military expenditures					
Current dollars (mil.)	NA	2,406[e]	NA	NA	NA
1989 constant dollars (mil.)	NA	2,670[e]	NA	NA	NA
Armed forces (000)	1,000	1,300	1,300	1,100	1,000
Gross national product (GNP)					
Current dollars (mil.)	NA	12,400[e]	12,600[e]	13,200[e]	14,200[e]
1989 constant dollars (mil.)	NA	13,760[e]	13,550[e]	13,740[e]	14,200[e]
Central government expenditures (CGE)					
1989 constant dollars (mil.)	NA	6,547[e]	4,507[e]	NA	NA
People (mil.)	59.3	60.7	62.0	63.4	64.8
Military expenditure as % of GNP	NA	19.4	NA	NA	NA
Military expenditure as % of CGE	NA	40.8	NA	NA	NA
Military expenditure per capita	NA	44	NA	NA	NA
Armed forces per 1,000 people	16.9	21.4	21.0	17.3	15.4
GNP per capita	NA	227	218	217	219
Arms imports[6]					
Current dollars (mil.)	1,500	2,100	1,900	1,500	1,300
1989 constant dollars (mil.)	1,708	2,330	2,044	1,561	1,300
Arms exports[6]					
Current dollars (mil.)	20	0	110	80	0
1989 constant dollars (mil.)	23	0	118	83	0
Total imports[7]					
Current dollars (mil.)	NA	1,590	2,190	2,500	NA
1989 constant dollars	NA	1,764	2,355	2,602	NA
Total exports[7]					
Current dollars (mil.)	NA	785	880	1,100	NA
1989 constant dollars	NA	871	946	1,145	NA
Arms as percent of total imports[8]	NA	132.1	86.8	60.0	NA
Arms as percent of total exports[8]	NA	0	12.5	7.3	NA

Human Rights [24]

	SSTS	FL	FAPRO	PPCG	APROBC	TPW	PCPTW	STPEP	PHRFF	PRW	ASST	AFL
Observes	P	P		P	P	P	P				S	
	EAFRD	CPR	ESCR	SR	ACHR	MAAE	PVIAC	PVNAC	EAFDAW	TCIDTP	RC
Observes	P	P	P					P		S		P

P = Party; S = Signatory; see Appendix for meaning of abbreviations.

Labor Force

Total Labor Force [25]

32.7 million

Labor Force by Occupation [26]

Agricultural	65%
Industrial and service	35

Date of data: 1990 est.

Unemployment Rate [27]

25% (1992 est.)

For sources, notes, and explanations, see Annotated Source Appendix, page 1035.

Production Sectors

Energy Resource Summary [28]

Energy Resources: Coal, offshore oil deposits. **Electricity**: 3,300,000 kW capacity; 9,000 million kWh produced, 130 kWh per capita (1992).
Pipelines: Petroleum products 150 km.

Telecommunications [30]

- The inadequacies of the obsolete switching equipment and cable system is a serious constraint on the business sector and on economic growth, and restricts access to the international links that Vietnam has established with most major countries
- The telephone system is not generally available for private use (25 telephones for each 10,000 persons)
- 3 satellite earth stations
- Broadcast stations - NA AM, 288 FM
- 36 (77 repeaters) TV
- About 2,500,000 TV receivers and 7,000,000 radio receivers in use (1991)

Top Agricultural Products [32]

Agriculture accounts for half of GNP; paddy rice, corn, potatoes make up 50% of farm output; commercial crops (rubber, soybeans, coffee, tea, bananas) and animal products 50%; since 1989 self-sufficient in food staple rice; fish catch of 943,100 metric tons (1989 est.).

Top Mining Products [33]

Metric tons unless otherwise specified	M.t.
Bauxite, gross weight	6,000
Chromite	3,500
Coal: anthracite (000 tons)	4,000
Gold	1,300
Iron ore	NA
Manganese	NA
Petroleum, crude (000 42-gal. barrels)	13,670
Phosphate rock, gross weight (000 tons)	274
Tin, mine output, Sn content	850
Zinc, mine output, Zn content	5,500

Transportation [31]

Railroads. 3,059 km total; 2,454 1.000-meter gauge, 151 km 1.435-meter (standard) gauge, 230 km dual gauge (three rails), and 224 km not restored to service after war damage

Highways. 85,000 km total; 9,400 km paved, 48,700 km gravel or improved earth, 26,900 km unimproved earth (est.)

Merchant Marine. 99 ships (1,000 GRT or over) totaling 460,712 GRT/739,246 DWT; includes 84 cargo, 3 refrigerated cargo, 1 roll-on/roll-off, 8 oil tanker, 3 bulk

Airports

Total:	100
Usable:	100
With permanent-surface runways:	50
With runways over 3,659 m:	0
With runways 2,440-3,659 m:	10
With runways 1,220-2,439 m:	20

Tourism [34]

Finance, Economics, and Trade

GDP and Manufacturing Summary [35]

	1980	1985	1990	1991	1992
Gross Domestic Product					
Millions of 1980 dollars	5,630	7,791	9,338	9,899	10,700
Growth rate in percent	-4.81	6.20	2.40	6.00	8.10
Manufacturing Value Added					
Millions of 1980 dollars	NA	NA	NA	NA	NA
Growth rate in percent	NA	NA	NA	NA	NA
Manufacturing share in percent of current prices	NA	NA	NA	NA	NA

Economic Indicators [36]

National product: GNP—exchange rate conversion—$16 billion (1992 est.). **National product real growth rate**: 7.4% (1992 est.). **National product per capita**: $230 (1992 est.). **Inflation rate (consumer prices)**: 15%- 20% (1992 est.). **External debt**: $16.8 billion (1990 est.).

Balance of Payments Summary [37]

Exchange Rates [38]

Currency: **new dong.**
Symbol: **D.**

1985-89 figures are end of year. Data are currency units per $1.

November 1992	10,800
July 1991	8,100
December 1990	7,280
March 1990	3,996
1988	2,047
1987	225

Imports and Exports

Top Import Origins [39]

$1.9 billion (c.i.f., 1992).

Origins	%
Japan	NA
Singapore	NA
Thailand	NA

Top Export Destinations [40]

$2.3 billion (f.o.b., 1992).

Destinations	%
Japan	NA
Singapore	NA
Thailand	NA
Hong Kong	NA
Taiwan	NA

Foreign Aid [41]

	U.S. $	
US commitments, including Ex-Im (FY70-74)	3.1	billion
Western (non-US) countries, ODA and OOF bilateral commitments (1970-89)	2.9	billion
OPEC bilateral aid (1979-89)	61	million
Communist countries (1970-89)	12.0	billion

Import and Export Commodities [42]

Import Commodities

Petroleum products
Steel products
Railroad equipment
Chemicals
Medicines
Raw cotton
Fertilizer
Grain

Export Commodities

Agric./handicraft products
Coal
Minerals
Crude oil
Ores
Seafood

Western Samoa

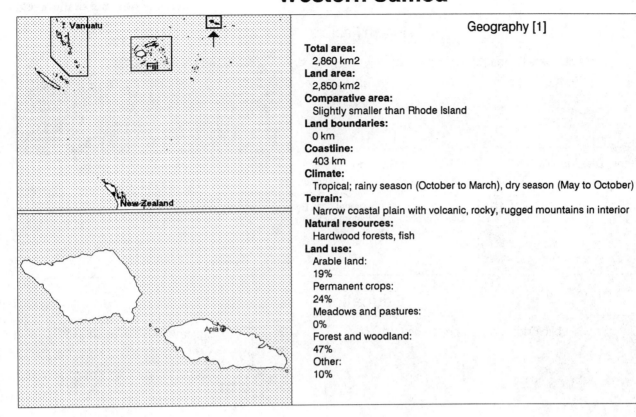

Total area:
 2,860 km2
Land area:
 2,850 km2
Comparative area:
 Slightly smaller than Rhode Island
Land boundaries:
 0 km
Coastline:
 403 km
Climate:
 Tropical; rainy season (October to March), dry season (May to October)
Terrain:
 Narrow coastal plain with volcanic, rocky, rugged mountains in interior
Natural resources:
 Hardwood forests, fish
Land use:
 Arable land:
 19%
 Permanent crops:
 24%
 Meadows and pastures:
 0%
 Forest and woodland:
 47%
 Other:
 10%

Demographics [2]

	1960	1970	1980	1990	1991[1]	1994	2000	2010	2020
Population	110	142	155	186	190	204	235	288	341
Population density (persons per sq. mi.)	100	129	141	169	173	NA	214	262	313
Births	NA	NA	NA	NA	6	7	NA	NA	NA
Deaths	NA	NA	NA	NA	1	1	NA	NA	NA
Life expectancy - males	NA	NA	NA	NA	64	66	NA	NA	NA
Life expectancy - females	NA	NA	NA	NA	69	70	NA	NA	NA
Birth rate (per 1,000)	NA	NA	NA	NA	34	32	NA	NA	NA
Death rate (per 1,000)	NA	NA	NA	NA	6	6	NA	NA	NA
Women of reproductive age (15-44 yrs.)	NA	NA	NA	43	NA	47	57	NA	NA
of which are currently married	NA	NA	NA	23	NA	27	34	NA	NA
Fertility rate	NA	NA	NA	4.7	4.51	4.2	3.5	2.7	2.3

Population values are in thousands, life expectancy in years, and other items as indicated.

Health

Health Personnel [3]

Health Expenditures [4]

For sources, notes, and explanations, see Annotated Source Appendix, page 1035.

1007

Human Factors	
Health Care Ratios [5]	Infants and Malnutrition [6]

Ethnic Division [7]		Religion [8]	Major Languages [9]
Samoan	92.6%	Christian 99.7% (about half of population associated with the London Missionary Society; includes Congregational, Roman Catholic, Methodist, Latter Day Saints, Seventh-Day Adventist).	Samoan (Polynesian), English.
Euronesians	7%		
Europeans	0.4%		

Education

Public Education Expenditures [10]	Educational Attainment [11]

Literacy Rate [12]

	1971[b]	1980[b]	1990[b]
Illiterate population +15 years	1,581	NA	NA
Illiteracy rate - total pop. (%)	2.2	NA	NA
Illiteracy rate - males (%)	2.2	NA	NA
Illiteracy rate - females (%)	2.1	NA	NA

Libraries [13]

Daily Newspapers [14]

Cinema [15]

Science and Technology

Scientific/Technical Forces [16]	R&D Expenditures [17]	U.S. Patents Issued [18]

Government and Law

Organization of Government [19]

Long-form name:
Independent State of Western Samoa
Type:
constitutional monarchy under native chief
Independence:
1 January 1962 (from UN trusteeship administered by New Zealand)
Constitution:
1 January 1962
Legal system:
based on English common law and local customs; judicial review of legislative acts with respect to fundamental rights of the citizen; has not accepted compulsory ICJ jurisdiction
National holiday:
National Day, 1 June
Executive branch:
chief, Executive Council, prime minister, Cabinet
Legislative branch:
unicameral Legislative Assembly (Fono)
Judicial branch:
Supreme Court, Court of Appeal

Political Parties [20]

Legislative Assembly	% of seats
Human Rights Protection Party (HRPP)	59.6
Samoan National Development Party (SNDP)	38.3
Independents	2.1

Government Budget [21]

For FY92.

Revenues	95.3
Expenditures	95.4
Capital expenditures	41.0

Defense Summary [22]

Branches: Department of Police and Prisons

Manpower Availability: Males age 15-49 NA; fit for military service NA

Defense Expenditures: No information available.

Crime [23]

Human Rights [24]

	SSTS	FL	FAPRO	PPCG	APROBC	TPW	PCPTW	STPEP	PHRFF	PRW	ASST	AFL
Observes						P	P					
	EAFRD	CPR	ESCR	SR	ACHR	MAAE	PVIAC	PVNAC	EAFDAW	TCIDTP	RC	
Observes	P	P	P	P			P	P	S	P	P	

P = Party; S = Signatory; see Appendix for meaning of abbreviations.

Labor Force

Total Labor Force [25]

38,000

Labor Force by Occupation [26]

Agriculture 22,000
Date of data: 1987 est.

Unemployment Rate [27]

For sources, notes, and explanations, see Annotated Source Appendix, page 1035.

1009

Production Sectors

Energy Resource Summary [28]

Energy Resources: None. **Electricity**: 29,000 kW capacity; 45 million kWh produced, 240 kWh per capita (1990).

Telecommunications [30]

- 7,500 telephones
- 70,000 radios
- Broadcast stations - 1 AM, no FM, no TV
- 1 Pacific Ocean INTELSAT ground station

Top Agricultural Products [32]

Agriculture accounts for 50% of GDP; coconuts, fruit (including bananas, taro, yams).

Top Mining Products [33]

Detailed information is not available. A summary of mineral resources available follows. **Mineral Resources**: None.

Transportation [31]

Highways. 2,042 km total; 375 km sealed; 1,667 km mostly gravel, crushed stone, or earth

Merchant Marine. 1 roll-on/roll-off ship (1,000 GRT or over) totaling 3,838 GRT/5,536 DWT

Airports

Total:	3
Usable:	3
With permanent-surface runways:	1
With runways over 3,659 m:	0
With runways 2,440-3,659 m:	1
With runways 1,220-2,439 m:	0

Tourism [34]

	1987	1988	1989	1990	1991
Visitors		93			
Tourists[16]	49	49	54	48	39
Tourism receipts	18	18	19	20	18
Tourism expenditures	2	1	2	2	2
Fare receipts	1	1	1	2	2
Fare expenditures	2	2	2	2	3

Tourists are in thousands, money in million U.S. dollars.

For sources, notes, and explanations, see Annotated Source Appendix, page 1035.

Finance, Economics, and Trade

Industrial Summary [35]

Industrial Production: Growth rate - 4% (1990 est.); accounts for 14% of GDP. **Industries**: Timber, tourism, food processing, fishing.

Economic Indicators [36]

National product: GDP—exchange rate conversion— $115 million (1990). **National product real growth rate**: -4.5% (1990 est.). **National product per capita**: $690 (1990). **Inflation rate (consumer prices)**: 15% (1990). **External debt**: $83 million (December 1990 est.).

Balance of Payments Summary [37]

Exchange Rates [38]

Currency: **tala.**
Symbol: **WS$.**

Data are currency units per $1.

January 1993	2.5681
1992	2.4655
1991	2.3975
1990	2.3095
1989	2.2686
1988	2.0790

Imports and Exports

Top Import Origins [39]

$75 million (c.i.f., 1990).

Origins	%
New Zealand	41
Australia	18
Japan	13
UK	6
US	6

Top Export Destinations [40]

$9 million (f.o.b., 1990).

Destinations	%
NZ	28
American Samoa	23
Germany	22
US	6

Foreign Aid [41]

	U.S. $	
US commitments, including Ex-Im (FY70-89)	18	million
Western (non-US) countries, ODA and OOF bilateral commitments (1970-89)	306	million
OPEC bilateral aid (1979-89)	4	million

Import and Export Commodities [42]

Import Commodities

Intermediate goods 58%
Food 17%
Capital goods 12%

Export Commodities

Coconut oil and cream 54%
Taro 12%
Copra 9%
Cocoa 3%

For sources, notes, and explanations, see Annotated Source Appendix, page 1035.

1011

Yemen

Geography [1]

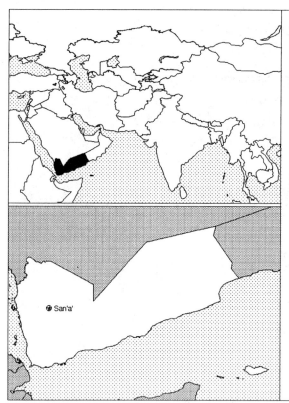

San'a'

Total area:
527,970 km2
Land area:
527,970 km2
Comparative area:
Slightly larger than twice the size of Wyoming
Land boundaries:
Total 1,746 km, Oman 288 km, Saudi Arabia 1,458 km
Coastline:
1,906 km
Climate:
Mostly desert; hot and humid along west coast; temperate in western mountains affected by seasonal monsoon; extraordinarily hot, dry, harsh desert in east
Terrain:
Narrow coastal plain backed by flat-topped hills and rugged mountains; dissected upland desert plains in center slope into the desert interior of the Arabian Peninsula
Natural resources:
Petroleum, fish, rock salt, marble, small deposits of coal, gold, lead, nickel, and copper, fertile soil in west
Land use:
Arable land: 6%
Permanent crops: 0%
Meadows and pastures: 30%
Forest and woodland: 7%
Other: 57%

Demographics [2]

	1960	1970	1980	1990	1991[1]	1994	2000	2010	2020
Population	4,783	5,782	7,324	9,746	10,063	11,105	13,603	18,985	25,907
Population density (persons per sq. mi.)	23	28	36	48	49	NA	67	93	127
Births	NA	NA	NA	NA	516	563	NA	NA	NA
Deaths	NA	NA	NA	NA	163	166	NA	NA	NA
Life expectancy - males	NA	NA	NA	NA	49	50	NA	NA	NA
Life expectancy - females	NA	NA	NA	NA	51	53	NA	NA	NA
Birth rate (per 1,000)	NA	NA	NA	NA	51	51	NA	NA	NA
Death rate (per 1,000)	NA	NA	NA	NA	16	15	NA	NA	NA
Women of reproductive age (15-44 yrs.)	NA	NA	NA	NA	NA	NA	NA	NA	NA
of which are currently married	NA	NA	NA	NA	NA	NA	NA	NA	NA
Fertility rate	NA	NA	NA	7.5	7.40	7.2	6.7	5.8	4.8

Population values are in thousands, life expectancy in years, and other items as indicated.

Health

Health Personnel [3]

Doctors per 1,000 pop., 1988-92	0.18
Nurse-to-doctor ratio, 1988-92	2.9
Hospital beds per 1,000 pop., 1985-90	0.9
Percentage of children immunized (age 1 yr. or less)	
Third dose of DPT, 1990-91	62.0
Measles, 1990-91	57.0

Health Expenditures [4]

Total health expenditure, 1990 (official exchange rate)	
Millions of dollars	217
Millions of dollars per capita	19
Health expenditures as a percentage of GDP	
Total	3.2
Public sector	1.5
Private sector	1.7
Development assistance for health	
Total aid flows (millions of dollars)[1]	25
Aid flows per capita (millions of dollars)	2.2
Aid flows as a percentage of total health expenditure	11.6

For sources, notes, and explanations, see Annotated Source Appendix, page 1035.

Human Factors

Health Care Ratios [5]

Population per physician, 1970	34790
Population per physician, 1990	NA
Population per nursing person, 1970	NA
Population per nursing person, 1990	NA
Percent of births attended by health staff, 1985	NA

Infants and Malnutrition [6]

Percent of babies with low birth weight, 1985	NA
Infant mortality rate per 1,000 live births, 1970	175
Infant mortality rate per 1,000 live births, 1991	109
Years of life lost per 1,000 population, 1990	104
Prevalence of malnutrition (under age 5), 1990	NA

Ethnic Division [7]

Predominantly Arab; Afro-Arab concentrations in coastal locations; South Asians in southern regions; small European communities in major metropolitan areas; 60,000 (est.) Somali refugees encamped near Aden.

Religion [8]

Muslim (including Sha'fi, Sunni, and Zaydi Shi'a), Jewish, Christian, Hindu.

Major Languages [9]

Arabic.

Education

Public Education Expenditures [10]

Million Dinar	1980[20]	1985[20]	1987[20]	1988[20]	1989[20]	1990[20]
Total education expenditure	17	NA	NA	NA	NA	NA
as percent of GNP	NA	NA	NA	NA	NA	NA
as percent of total govt. expend.	16.9	NA	NA	NA	NA	NA
Current education expenditure	NA	NA	NA	NA	NA	NA
as percent of GNP	NA	NA	NA	NA	NA	NA
as percent of current govt. expend.	NA	NA	NA	NA	NA	NA
Capital expenditure	NA	NA	NA	NA	NA	NA

Educational Attainment [11]

Literacy Rate [12]

In thousands and percent	1985[a,2]	1991[a,2]	2000[a,2]
Illiterate population +15 years	2,423	2,559	2,881
Illiteracy rate - total pop. (%)	67.7	61.5	49.3
Illiteracy rate - males (%)	52.9	46.7	35.5
Illiteracy rate - females (%)	79.5	73.7	61.4

Libraries [13]

Daily Newspapers [14]

Cinema [15]

Science and Technology

Scientific/Technical Forces [16]

R&D Expenditures [17]

U.S. Patents Issued [18]

Government and Law

Organization of Government [19]

Long-form name:
Republic of Yemen
Type:
republic
Independence:
22 May 1990. Established on 22 May 1990 with the merger of the Yemen Arab Republic (North) and the People's Democratic Republic of Yemen (South); North Yemen had become independent November 1918 (Ottoman Empire) and South Yemen on 30 November 1967 (UK)
Constitution:
16 April 1991
Legal system:
based on Islamic, Turkish, English law, and local custom; does not accept compulsory ICJ jurisdiction
National holiday:
Proclamation of the Republic, 22 May (1990)
Executive branch:
5-member Presidential Council, prime minister
Legislative branch:
unicameral House of Representatives
Judicial branch:
Supreme Court

Crime [23]

Elections [20]

House of Representatives. Last held NA (next to be held 27 April 1993); results - percent of vote NA; seats - (301); number of seats by party NA; note - the 301 members of the new House of Representatives come from North Yemen's Consultative Assembly (159 members), South Yemen's Supreme People's Council (111 members), and appointments by the New Presidential Council (31 members).

Government Expenditures [21]

Housing - 100.0%

1991

Military Expenditures and Arms Transfers [22]

	1985	1986	1987	1988	1989
Military expenditures					
Current dollars (mil.)	379	368	379	641	618
1989 constant dollars (mil.)	432	408	408	668	618
Armed forces (000)	28	28[e]	43	62	62
Gross national product (GNP)					
Current dollars (mil.)	4,887	5,327	5,693	6,285	6,776
1989 constant dollars (mil.)	5,564	5,911	6,123	6,543	6,776
Central government expenditures (CGE)					
1989 constant dollars (mil.)	1,532	1,487	1,865	2,223	2,076
People (mil.)	6.2	6.3	6.5	6.7	6.9
Military expenditure as % of GNP	7.8	6.9	6.7	10.2	9.1
Military expenditure as % of CGE	28.2	27.5	21.9	30.0	29.8
Military expenditure per capita	70	64	62	99	89
Armed forces per 1,000 people	4.5	4.4	6.6	9.2	8.9
GNP per capita	902	932	937	972	976
Arms imports[6]					
Current dollars (mil.)	230	280	400	430	420
1989 constant dollars (mil.)	262	311	430	448	420
Arms exports[6]					
Current dollars (mil.)	0	0	0	0	0
1989 constant dollars (mil.)	0	0	0	0	0
Total imports[7]					
Current dollars (mil.)	1,300	1,157	883	1,384	NA
1989 constant dollars	1,480	1,284	950	1,441	NA
Total exports[7]					
Current dollars (mil.)	13	8	48	NA	NA
1989 constant dollars	15	9	52	NA	NA
Arms as percent of total imports[8]	17.7	24.2	45.3	31.1	NA
Arms as percent of total exports[8]	0	0	0	NA	NA

Human Rights [24]

	SSTS	FL	FAPRO	PPCG	APROBC	TPW	PCPTW	STPEP	PHRFF	PRW	ASST	AFL
Observes	P	P	P	P	P	P	P	P		P	P	P
	EAFRD	CPR	ESCR	SR	ACHR	MAAE	PVIAC	PVNAC	EAFDAW	TCIDTP	RC	
Observes	P	P	P	P			P	P	S	P	P	

P = Party; S = Signatory; see Appendix for meaning of abbreviations.

Labor Force

Total Labor Force [25]

Labor Force by Occupation [26]

Agriculture and herding	70%
Expatriate laborers	30
(est.) agriculture	45.2
Services	21.2
Construction	13.4
Industry	10.6
Commerce and other	9.6

Date of data: 1983

Unemployment Rate [27]

30% (December 1992)

Production Sectors

Energy Resource Summary [28]

Energy Resources: Petroleum, small deposits of coal. **Electricity**: 714,000 kW capacity; 1,224 million kWh produced, 120 kWh per capita (1992). **Pipelines**: Crude oil 644 km, petroleum products 32 km.

Telecommunications [30]

- Since unification in 1990, efforts are still being made to create a national domestic civil telecommunications network
- The network consists of microwave radio relay, cable and troposcatter
- 65,000 telephones (est.)
- Broadcast stations - 4 AM, 1 FM, 10 TV
- Satellite earth stations - 2 Indian Ocean INTELSAT, 1 Atlantic Ocean INTELSAT, 1 Intersputnik, 2 ARABSAT
- Microwave radio relay to Saudi Arabia, and Djibouti

Top Agricultural Products [32]

Accounted for 26% of GDP; products—grain, fruits, vegetables, qat (mildly narcotic shrub), coffee, cotton, dairy, poultry, meat, fish; not self-sufficient in grain.

Top Mining Products [33]

Estimated metric tons unless otherwise specified   M.t.

Stone, dimension (cubic meters)	410,000
Salt	225,000
Gypsum	75,000
Cement	850,000
Petroleum, crude (000 barrels)	80,000
Gas, natural (mil. cubic meters)	50,000

Transportation [31]

Highways. 15,500 km total; 4,000 km paved, 11,500 km natural surface (est.)

Merchant Marine. 3 ships (1,000 GRT or over) totaling 4,309 GRT/ 6,568 DWT; includes 2 cargo, 1 oil tanker

Airports

Total:	45
Usable:	39
With permanent-surface runways:	10
With runways over 3,659 m:	0
With runways 2,440-3,659 m:	18
With runways 1,220-2,439 m:	11

Tourism [34]

	1987	1988	1989	1990	1991
Tourists[10]	51	60	65	52	44
Tourism receipts	48	21	26	20	21
Tourism expenditures	37	47	81		
Fare receipts	4				

Tourists are in thousands, money in million U.S. dollars.

Finance, Economics, and Trade

GDP and Manufacturing Summary [35]

	1980	1985	1990	1991	1992
Gross Domestic Product					
Millions of 1980 dollars	2,779	3,692	5,433	5,868	6,090[e]
Growth rate in percent	6.04	10.31	6.80	8.020	3.79[e]
Manufacturing Value Added					
Millions of 1980 dollars	160	295	363	398	436[e]
Growth rate in percent	7.70	1.46	2.99	9.51	9.67[e]
Manufacturing share in percent of current prices	8.5	12.1	5.2[e]	NA	NA

Economic Indicators [36]

National product: GDP—exchange rate conversion—$8 billion (1992 est.). **National product real growth rate**: NA%. **National product per capita**: $775 (1992 est.). **Inflation rate (consumer prices)**: 100% (December 1992). **External debt**: $5.75 billion (December 1989 est.).

Balance of Payments Summary [37]

Values in millions of dollars.

	1985	1986	1987	1988	1989
Exports of goods (f.o.b.)	50.8	46.5	119.1	529.2	719.8
Imports of goods (f.o.b.)	-1,702.5	-1,244.5	-1,646.3	-1,905.5	-1,836.6
Trade balance	-1,651.7	-1,198.0	-1,527.2	-1,376.3	-1,116.8
Services - debits	-483.6	-409.0	-549.0	-766.0	-841.1
Services - credits	306.8	255.9	290.8	350.6	396.7
Private transfers (net)	1,188.8	819.8	1,010.3	566.5	414.0
Government transfers (net)	121.4	230.1	193.0	126.6	151.5
Long term capital (net)	190.1	338.0	512.1	764.3	845.9
Short term capital (net)	149.9	-48.8	51.2	84.5	85.4
Errors and omissions	58.2	41.5	4.1	4.1	55.6
Overall balance	-120.1	29.5	-245.7	-245.7	-8.8

Exchange Rates [38]

Currency: **Yemeni rials.**
Symbol: **YR.**

Data are currency units per $1.

Unofficial	30-40.000
Official	12.000
North Yemeni YR	
June 1992	12.100
1991	12.000
1990	9.760
January 1989	9.760
1988	9.772
1987	10.342
South Yemenu YD	
fixed rate	0.345

Imports and Exports

Top Import Origins [39]

$2.1 billion (f.o.b., 1990 est.).

Origins	%
Japan	NA
Saudi Arabia	NA
Australia	NA
EC countries	NA
China	NA
Russia	NA
US	NA

Top Export Destinations [40]

$908 million (f.o.b., 1990 est.).

Destinations	%
US	NA
EC countries	NA
South Korea	NA
Saudi Arabia	NA

Foreign Aid [41]

	U.S. $	
US commitments, including Ex-Im (FY70-89)	389	million
Western (non-US) countries, ODA and OOF bilateral commitments (1970-89)	2.0	billion
OPEC bilateral aid (1979-89)	3.2	billion
Communist countries (1970-89)	2.4	billion

Import and Export Commodities [42]

Import Commodities	Export Commodities
Textiles	Crude oil
Consumer goods	Cotton
Petroleum products	Coffee
Sugar	Hides
Grain	Vegetables
Flour	Dried and salted fish
Other foodstuffs	
Cement	
Machinery	
Chemicals	

1016

For sources, notes, and explanations, see Annotated Source Appendix, page 1035.

Zaire

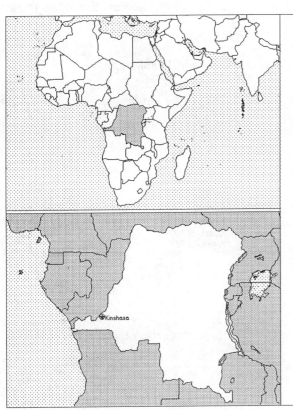

Geography [1]

Total area:
2,345,410 km2
Land area:
2,267,600 km2
Comparative area:
Slightly more than one-quarter the size of US
Land boundaries:
Total 10,271 km, Angola 2,511 km, Burundi 233 km, Central African Republic 1,577 km, Congo 2,410 km, Rwanda 217 km, Sudan 628 km, Uganda 765 km, Zambia 1,930 km
Coastline:
37 km
Climate:
Tropical; hot and humid in equatorial river basin; cooler and drier in southern highlands; cooler and wetter in eastern highlands; north of Equator: wet season April to October, dry season December to February; south of Equator: wet season November to March, dry season April to October
Terrain:
Vast central basin is a low-lying plateau; mountains in east
Land use:
Arable land: 3%
Permanent crops: 0%
Meadows and pastures: 4%
Forest and woodland: 78%
Other: 15%

Demographics [2]

	1960	1970	1980	1990	1991[1]	1994	2000	2010	2020
Population	15,860	20,934	27,954	37,903	37,832	42,684	51,413	69,079	92,860
Population density (persons per sq. mi.)	18	23	30	42	43	NA	57	77	102
Births	NA	NA	NA	NA	1,732	2,065	NA	NA	NA
Deaths	NA	NA	NA	NA	497	715	NA	NA	NA
Life expectancy - males	NA	NA	NA	NA	52	46	NA	NA	NA
Life expectancy - females	NA	NA	NA	NA	56	49	NA	NA	NA
Birth rate (per 1,000)	NA	NA	NA	NA	46	48	NA	NA	NA
Death rate (per 1,000)	NA	NA	NA	NA	13	17	NA	NA	NA
Women of reproductive age (15-44 yrs.)	NA	NA	NA	8,465	NA	9,488	11,332	NA	NA
of which are currently married	NA	NA	NA	6,466	NA	7,224	8,631	NA	NA
Fertility rate	NA	NA	NA	6.7	6.17	6.7	6.4	5.6	4.7

Population values are in thousands, life expectancy in years, and other items as indicated.

Health

Health Personnel [3]

Doctors per 1,000 pop., 1988-92	0.07
Nurse-to-doctor ratio, 1988-92	2.1
Hospital beds per 1,000 pop., 1985-90	1.6
Percentage of children immunized (age 1 yr. or less)	
Third dose of DPT, 1990-91	32.0
Measles, 1990-91	31.0

Health Expenditures [4]

Total health expenditure, 1990 (official exchange rate)	
Millions of dollars	179
Millions of dollars per capita	5
Health expenditures as a percentage of GDP	
Total	2.4
Public sector	0.8
Private sector	1.5
Development assistance for health	
Total aid flows (millions of dollars)[1]	48
Aid flows per capita (millions of dollars)	1.3
Aid flows as a percentage of total health expenditure	26.7

For sources, notes, and explanations, see Annotated Source Appendix, page 1035.

1017

Human Factors

Health Care Ratios [5]	Infants and Malnutrition [6]

Ethnic Division [7]
More than 200 African ethnic groups, the majority are Bantu; four largest tribes — Mongo, Luba, Kongo (all Bantu), and the Mangbetu-Azande (Hamitic) make up about 45% of the population.

Religion [8]
Roman Catholic	50%
Protestant	20%
Kimbanguist	10%
Muslim	10%
Other sects and traditional beliefs	10%

Major Languages [9]
French	NA
Lingala	NA
Swahili	NA
Kingwana	NA
Kikongo	NA
Tshiluba	NA

Education

Public Education Expenditures [10]

Million Zaires	1980	1985	1986	1987	1988	1989
Total education expenditure	1,015	3,291	3,874	8,238	15,006	NA
as percent of GNP	2.6	1.0	0.8	1.0	0.9	NA
as percent of total govt. expend.	24.2	7.3	6.9	8.2	6.4	NA
Current education expenditure	998	3,239	3,833	7,926	14,357	NA
as percent of GNP	2.5	1.0	0.8	1.0	0.9	NA
as percent of current govt. expend.	25.3	7.3	6.9	8.2	6.4	NA
Capital expenditure	17	52	41	312	649	NA

Educational Attainment [11]

Literacy Rate [12]

In thousands and percent	1985[a]	1991[a]	2000[a]
Illiterate population +15 years	5,641	5,466	4,919
Illiteracy rate - total pop. (%)	34.1	28.2	18.3
Illiteracy rate - males (%)	20.6	16.4	10.1
Illiteracy rate - females (%)	46.8	39.3	26.3

Libraries [13]

Daily Newspapers [14]

	1975	1980	1985	1990
Number of papers	6	5	4	5
Circ. (000)	50[e]	60[e]	50[e]	NA

Cinema [15]

Science and Technology

Scientific/Technical Forces [16]	R&D Expenditures [17]	U.S. Patents Issued [18]

 For sources, notes, and explanations, see Annotated Source Appendix, page 1035.

Government and Law

Organization of Government [19]

Long-form name:
Republic of Zaire
Type:
republic with a strong presidential system
Independence:
30 June 1960 (from Belgium)
Constitution:
24 June 1967, amended August 1974, revised 15 February 1978; amended April 1990; new constitution to be put to referendum in 1993
Legal system:
based on Belgian civil law system and tribal law; has not accepted compulsory ICJ jurisdiction
National holiday:
Anniversary of the Regime (Second Republic), 24 November (1965)
Executive branch:
president, prime minister, Executive Council (cabinet)
Legislative branch:
unicameral National Parliament; anti-Mobutu opposition claims National Parliament replaced by High Council
Judicial branch:
Supreme Court (Cour Supreme)

Crime [23]

Elections [20]

Legislative Council. Last held 6 September 1987 (next to be scheduled by High Council); results - MPR was the only party; seats - (210 total) MPR 210. Sole legal party until January 1991 - Popular Movement of the Revolution (MPR); other parties include Union for Democracy and Social Progress, Democratic Social Christian Party, Union of Federalists and Independent Republicans, Unified Lumumbast Party.

Government Budget [21]

For FY92.
Revenues	NA
Expenditures	NA
Capital expenditures	NA

Military Expenditures and Arms Transfers [22]

	1985	1986	1987	1988	1989
Military expenditures					
Current dollars (mil.)	90	174	NA	233	NA
1989 constant dollars (mil.)	103	193	NA	242	NA
Armed forces (000)	62	53	53	51	51
Gross national product (GNP)					
Current dollars (mil.)	7,535	8,089	8,345	8,944	9,152
1989 constant dollars (mil.)	8,578	8,976	8,975	9,310	9,152
Central government expenditures (CGE)					
1989 constant dollars (mil.)	1,045	1,062	NA	1,715	1,517
People (mil.)	31.1	32.1	33.2	34.3	35.4
Military expenditure as % of GNP	1.2	2.2	NA	2.6	NA
Military expenditure as % of CGE	9.8	18.2	NA	14.1	NA
Military expenditure per capita	3	6	NA	7	NA
Armed forces per 1,000 people	2.0	1.7	1.6	1.5	1.4
GNP per capita	276	280	271	272	258
Arms imports[6]					
Current dollars (mil.)	30	30	50	40	0
1989 constant dollars (mil.)	34	33	54	42	0
Arms exports[6]					
Current dollars (mil.)	0	0	0	0	0
1989 constant dollars (mil.)	0	0	0	0	0
Total imports[7]					
Current dollars (mil.)	793	871	764	765	849
1989 constant dollars	903	967	822	796	849
Total exports[7]					
Current dollars (mil.)	950	1,093	983	1,121	1,254
1989 constant dollars	1,081	1,213	1,057	1,167	1,254
Arms as percent of total imports[8]	3.8	3.4	6.5	5.2	0
Arms as percent of total exports[8]	0	0	0	0	0

Human Rights [24]

	SSTS	FL	FAPRO	PPCG	APROBC	TPW	PCPTW	STPEP	PHRFF	PRW	ASST	AFL
Observes		P		P	P	P	P			P	P	
	EAFRD	CPR	ESCR	SR	ACHR	MAAE	PVIAC	PVNAC	EAFDAW	TCIDTP	RC	
Observes	P	P	P	P			P		S		P	

P=Party; S=Signatory; see Appendix for meaning of abbreviations.

Labor Force

Total Labor Force [25]

15 million (13% of the labor force is wage earners; 51% of the population is of working age)

Labor Force by Occupation [26]

Agriculture	75%
Industry	13
Services	12

Date of data: 1985

Unemployment Rate [27]

For sources, notes, and explanations, see Annotated Source Appendix, page 1035.

1019

Production Sectors

Energy Resource Summary [28]

Energy Resources: Petroleum, uranium, coal, hydropower potential. **Electricity**: 2,580,000 kW capacity; 6,000 million kWh produced, 160 kWh per capita (1991). **Pipelines**: Petroleum products 390 km.

Telecommunications [30]

- Barely adequate wire and microwave service
- Broadcast stations - 10 AM, 4 FM, 18 TV
- Satellite earth stations - 1 Atlantic Ocean INTELSAT, 14 domestic

Top Agricultural Products [32]

Cash crops—coffee, palm oil, rubber, quinine; food crops - cassava, bananas, root crops, corn.

Transportation [31]

Railroads. 5,254 km total; 3,968 km 1.067-meter gauge (851 km electrified); 125 km 1.000-meter gauge; 136 km 0.615-meter gauge; 1,025 km 0.600-meter gauge; limited trackage in use because of civil strife

Highways. 146,500 km total; 2,800 km paved, 46,200 km gravel and improved earth; 97,500 unimproved earth

Merchant Marine. 1 passenger cargo ship (1,000 GRT or over) totaling 15,489 GRT/13,481 DWT

Airports

Total:	281
Usable:	235
With permanent-surface runways:	25
With runways over 3,659 m:	1
With runways 2,440-3,659 m:	6
With runways 1,220-2,439 m:	73

Top Mining Products [33]

Estimated metric tons unless otherwise specified   M.t.

Copper, mine output,	
Cu content of ore (000 tons)	310
Cobalt, mine output,	
Co content of ore (000 tons)	20,900
Cement, hydraulic	250,000
Diamonds, gem and industrial (000 carats)	17,814
Stone, crushed	360,000
Gold	80,000

Tourism [34]

	1987	1988	1989	1990	1991
Visitors	56	109	110		
Tourists[4,75]	36	39	51		
Excursionists	20	70	59		
Tourism receipts	18	7	6	7	
Tourism expenditures	22	16	17	16	

Tourists are in thousands, money in million U.S. dollars.

Manufacturing Sector

GDP and Manufacturing Summary [35]

	1980	1985	1989	1990	% change 1980-1990	% change 1989-1990
GDP (million 1980 $)	6,137	6,653	7,164	6,916	12.7	-3.5
GDP per capita (1980 $)	234	219	208	195	-16.7	-6.3
Manufacturing as % of GDP (current prices)	3.1	1.7	11.4[e]	2.3[e]	-25.8	-79.8
Gross output (million $)	NA	NA	NA	NA	NA	NA
Value added (million $)	170	66[e]	93[e]	96[e]	-43.5	3.2
Value added (million 1980 $)	184	188	169[e]	173	-6.0	2.4
Industrial production index	100	119	119[e]	133	33.0	11.8
Employment (thousands)	50[e]	50[e]	31[e]	50[e]	0.0	61.3

Note: GDP stands for Gross Domestic Product. 'e' stands for estimated value.

Profitability and Productivity

	1980	1985	1989	1990	% change 1980-1990	% change 1989-1990
Intermediate input (%)	NA	NA	NA	NA	NA	NA
Wages, salaries, and supplements (%)	NA	NA	NA	NA	NA	NA
Gross operating surplus (%)	NA	NA	NA	NA	NA	NA
Gross output per worker ($)	NA	NA	NA	NA	NA	NA
Value added per worker ($)	2,929[e]	1,117[e]	2,956[e]	1,616[e]	-44.8	-45.3
Average wage (incl. benefits) ($)	4,535[e]	1,589[e]	3,947[e]	2,092[e]	-53.9	-47.0

Profitability is in percent of gross output. Productivity is in U.S. $. 'e' stands for estimated value.

Profitability - 1990

Value Added in Manufacturing

	1980 $ mil.	1980 %	1985 $ mil.	1985 %	1989 $ mil.	1989 %	1990 $ mil.	1990 %	% change 1980-1990	% change 1989-1990
311 Food products	20	11.8	5[e]	7.6	11[e]	11.8	5[e]	5.2	-75.0	-54.5
313 Beverages	35	20.6	20[e]	30.3	19[e]	20.4	28[e]	29.2	-20.0	47.4
314 Tobacco products	9	5.3	7[e]	10.6	11[e]	11.8	15[e]	15.6	66.7	36.4
321 Textiles	10	5.9	2[e]	3.0	5[e]	5.4	5[e]	5.2	-50.0	0.0
322 Wearing apparel	7	4.1	1[e]	1.5	2[e]	2.2	2[e]	2.1	-71.4	0.0
323 Leather and fur products	NA	0.0	NA	0.0	1[e]	1.1	1[e]	1.0	NA	0.0
324 Footwear	8	4.7	2[e]	3.0	4[e]	4.3	4[e]	4.2	-50.0	0.0
331 Wood and wood products	4	2.4	1[e]	1.5	1[e]	1.1	2[e]	2.1	-50.0	100.0
332 Furniture and fixtures	1	0.6	NA	0.0	1[e]	1.1	NA	0.0	NA	NA
341 Paper and paper products	NA	0.0	NA	0.0	NA	0.0	NA	0.0	NA	NA
342 Printing and publishing	2	1.2	1[e]	1.5	1[e]	1.1	1[e]	1.0	-50.0	0.0
351 Industrial chemicals	12	7.1	6[e]	9.1	12[e]	12.9	8[e]	8.3	-33.3	-33.3
352 Other chemical products	NA	0.0	NA	0.0	NA	0.0	NA	0.0	NA	NA
353 Petroleum refineries	14	8.2	1[e]	1.5	1[e]	1.1	1[e]	1.0	-92.9	0.0
354 Miscellaneous petroleum and coal products	NA	0.0	NA	0.0	NA	0.0	NA	0.0	NA	NA
355 Rubber products	NA	0.0	NA	0.0	NA	0.0	NA	0.0	NA	NA
356 Plastic products	NA	0.0	NA	0.0	NA	0.0	NA	0.0	NA	NA
361 Pottery, china and earthenware	NA	0.0	NA	0.0	NA	0.0	NA	0.0	NA	NA
362 Glass and glass products	1	0.6	NA	0.0	NA	0.0	NA	0.0	NA	NA
369 Other non-metal mineral products	4	2.4	1[e]	1.5	1[e]	1.1	2[e]	2.1	-50.0	100.0
371 Iron and steel	4	2.4	1[e]	1.5	NA	0.0	2[e]	2.1	-50.0	NA
372 Non-ferrous metals	2	1.2	NA	0.0	NA	0.0	1[e]	1.0	-50.0	NA
381 Metal products	5	2.9	2[e]	3.0	3[e]	3.2	3[e]	3.1	-40.0	0.0
382 Non-electrical machinery	5	2.9	2[e]	3.0	3[e]	3.2	3[e]	3.1	-40.0	0.0
383 Electrical machinery	3	1.8	1[e]	1.5	1[e]	1.1	2[e]	2.1	-33.3	100.0
384 Transport equipment	5	2.9	3[e]	4.5	5[e]	5.4	3[e]	3.1	-40.0	-40.0
385 Professional and scientific equipment	NA	0.0	NA	0.0	NA	0.0	NA	0.0	NA	NA
390 Other manufacturing industries	15	8.8	7[e]	10.6	9[e]	9.7	9[e]	9.4	-40.0	0.0

Note: The industry codes shown are International Standard Industry codes (ISIC). Percentages are percent of total Value Added. 'e' stands for estimated value

For sources, notes, and explanations, see Annotated Source Appendix, page 1035.

1021

Finance, Economics, and Trade

Economic Indicators [36]

Millions of U.S. Dollars unless otherwise noted.

	1989	1990	1991e
GDP (1980 Prices)	5,633	5,633	NA
GDP Growth (%)	1.2	NA	NA
GDP per capita ($)	180	180	NA
Money supply (mil. zaires)	282,042	560,000	NA
Interest rate (rediscount)	55	45	55
Savings rate (% of GDP)	11	NA	NA
Investment rate (% of GDP)	12	NA	NA
CPI (% change)	101	242	1,500
WPI	NA	NA	NA
External public debt (excl. IMF)	8,843	10,008	NA

Balance of Payments Summary [37]

Values in millions of dollars.

	1986	1987	1988	1989	1990
Exports of goods (f.o.b.)	1,844.0	1,731.0	2,178.0	2,201.0	2,138.0
Imports of goods (f.o.b.)	-1,283.0	-1,376.0	-1,645.0	-1,683.0	-1,539.0
Trade balance	561.0	355.0	533.0	518.0	599.0
Services - debits	-1,286.0	-1,412.0	-1,458.0	-1,461.0	-1,549.0
Services - credits	189.0	262.0	185.0	165.0	171.0
Private transfers (net)	-62.0	-70.0	-67.0	-109.0	-81.0
Government transfers (net)	184.0	220.0	226.0	276.0	217.0
Long term capital (net)	317.0	639.0	332.0	1,150.0	122.0
Short term capital (net)	120.0	105.0	420.0	-442.0	502.0
Errors and omissions	-17.0	13.0	-134.0	113.0	105.0
Overall balance	6.0	112.0	37.0	210.0	86.0

Exchange Rates [38]

Currency: **zaire**.
Symbol: **Z**.

Data are currency units per $1.

January 1993	2,000,000
1991	15,587
1990	719
1989	381
1988	187
1987	112

Imports and Exports

Top Import Origins [39]

$1.2 billion (f.o.b., 1992 est.).

Origins	%
South Africa	NA
US	NA
Belgium	NA
France	NA
Germany	NA
Italy	NA
Japan	NA
UK	NA

Top Export Destinations [40]

$1.5 billion (f.o.b., 1992 est.).

Destinations	%
US	NA
Belgium	NA
France	NA
Germany	NA
Italy	NA
UK	NA
Japan	NA
South Africa	NA

Foreign Aid [41]

Note: except for humanitarian aid to private organizations, no US assistance was given to Zaire in 1

	U.S. $	
US commitments, including Ex-Im (FY70-89)	1.1	billion
Western (non-US) countries, ODA and OOF bilateral commitments (1970-89)	6.9	billion
OPEC bilateral aid (1979-89)	35	million
Communist countries (1970-89)	263	million

Import and Export Commodities [42]

Import Commodities	Export Commodities
Consumer goods	Copper
Foodstuffs	Coffee
Mining & other machinery	Diamonds
Transport equipment	Cobalt
Fuels	Crude oil

Zambia

Geography [1]

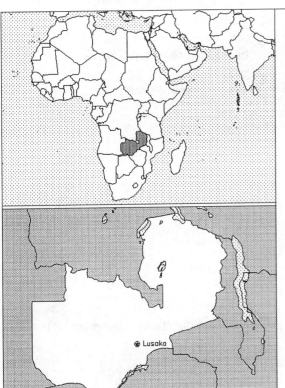

Total area:
752,610 km2
Land area:
740,720 km2
Comparative area:
Slightly larger than Texas
Land boundaries:
Total 5,664 km, Angola 1,110 km, Malawi 837 km, Mozambique 419 km, Namibia 233 km, Tanzania 338 km, Zaire 1,930 km, Zimbabwe 797 km
Coastline:
0 km (landlocked)
Climate:
Tropical; modified by altitude; rainy season (October to April)
Terrain:
Mostly high plateau with some hills and mountains
Natural resources:
Copper, cobalt, zinc, lead, coal, emeralds, gold, silver, uranium, hydropower potential
Land use:
Arable land:
7%
Permanent crops:
0%
Meadows and pastures:
47%
Forest and woodland:
27%
Other:
19%

Demographics [2]

	1960	1970	1980	1990	1991[1]	1994	2000	2010	2020
Population	3,254	4,247	5,638	8,233	8,446	9,188	10,625	12,614	15,828
Population density (persons per sq. mi.)	11	15	20	29	30	NA	40	57	77
Births	NA	NA	NA	NA	414	423	NA	NA	NA
Deaths	NA	NA	NA	NA	99	162	NA	NA	NA
Life expectancy - males	NA	NA	NA	NA	55	44	NA	NA	NA
Life expectancy - females	NA	NA	NA	NA	58	45	NA	NA	NA
Birth rate (per 1,000)	NA	NA	NA	NA	49	46	NA	NA	NA
Death rate (per 1,000)	NA	NA	NA	NA	12	18	NA	NA	NA
Women of reproductive age (15-44 yrs.)	NA	NA	NA	1,799	NA	1,999	2,298	NA	NA
of which are currently married	NA	NA	NA	1,241	NA	1,353	1,517	NA	NA
Fertility rate	NA	NA	NA	6.9	6.94	6.7	6.3	5.4	4.5

Population values are in thousands, life expectancy in years, and other items as indicated.

Health

Health Personnel [3]

Doctors per 1,000 pop., 1988-92	0.09
Nurse-to-doctor ratio, 1988-92	6.0
Hospital beds per 1,000 pop., 1985-90	NA
Percentage of children immunized (age 1 yr. or less)	
Third dose of DPT, 1990-91	79.0
Measles, 1990-91	76.0

Health Expenditures [4]

Total health expenditure, 1990 (official exchange rate)	
Millions of dollars	117
Millions of dollars per capita	14
Health expenditures as a percentage of GDP	
Total	3.2
Public sector	2.2
Private sector	1.0
Development assistance for health	
Total aid flows (millions of dollars)[1]	6
Aid flows per capita (millions of dollars)	0.7
Aid flows as a percentage of total health expenditure	4.9

For sources, notes, and explanations, see Annotated Source Appendix, page 1035.

1023

Human Factors

Health Care Ratios [5]

Population per physician, 1970	13640
Population per physician, 1990	11290
Population per nursing person, 1970	1730
Population per nursing person, 1990	600
Percent of births attended by health staff, 1985	NA

Infants and Malnutrition [6]

Percent of babies with low birth weight, 1985	14
Infant mortality rate per 1,000 live births, 1970	106
Infant mortality rate per 1,000 live births, 1991	106
Years of life lost per 1,000 population, 1990	86
Prevalence of malnutrition (under age 5), 1990	NA

Ethnic Division [7]

African	98.7%
European	1.1%
Other	0.2%

Religion [8]

Christian	50-75%
Muslim and Hindu	24-49%
Indigenous beliefs	1%

Major Languages [9]

English (official).

Education

Public Education Expenditures [10]

Million Kwacha	1980	1985	1987	1988	1989	1990
Total education expenditure	127	293	574	737	1,352	2,737
as percent of GNP	4.5	4.6	3.4	2.8	2.5	2.9
as percent of total govt. expend.	7.6	13.4	9.8	8.8	10.9	8.7
Current education expenditure	120	272	502	683	1,209	2,382
as percent of GNP	4.2	4.3	3.0	2.6	2.2	2.5
as percent of current govt. expend.	11.1	14.3	9.7	9.8	11.6	8.7
Capital expenditure	6	21	72	54	143	355

Educational Attainment [11]

Age group	25+
Total population	1,880,124
Highest level attained (%)	
No schooling	49.8
First level	
Incompleted	37.0
Completed	NA
Entered second level	
S-1	12.8
S-2	NA
Post secondary	0.4

Literacy Rate [12]

In thousands and percent	1985[a]	1991[a]	2000[a]
Illiterate population + 15 years	1,172	1,170	1,127
Illiteracy rate - total pop. (%)	32.6	27.2	18.2
Illiteracy rate - males (%)	23.3	19.2	12.6
Illiteracy rate - females (%)	41.3	34.7	23.6

Libraries [13]

	Admin. Units	Svc. Pts.	Vols. (000)	Shelving (meters)	Vols. Added	Reg. Users
National	NA	NA	NA	NA	NA	NA
Non-specialized	NA	NA	NA	NA	NA	NA
Public	NA	NA	NA	NA	NA	NA
Higher ed. (1987)[11]	1	1	10	850	1,737	750
School	NA	NA	NA	NA	NA	NA

Daily Newspapers [14]

	1975	1980	1985	1990
Number of papers	2	2	2	2
Circ. (000)	106	110	95	99

Cinema [15]

Science and Technology

Scientific/Technical Forces [16]

R&D Expenditures [17]

U.S. Patents Issued [18]

1024

For sources, notes, and explanations, see Annotated Source Appendix, page 1035.

Government and Law

Organization of Government [19]

Long-form name:
Republic of Zambia
Type:
republic
Independence:
24 October 1964 (from UK)
Constitution:
NA August 1991
Legal system:
based on English common law and
customary law; judicial review of
legislative acts in an ad hoc constitutional
council; has not accepted compulsory ICJ
jurisdiction
National holiday:
Independence Day, 24 October (1964)
Executive branch:
president, Cabinet
Legislative branch:
unicameral National Assembly
Judicial branch:
Supreme Court

Crime [23]

Political Parties [20]

National Assembly	% of seats
Movement for Multiparty Democracy (MMD)	83.3
United National Independence (UNIP)	16.7

Government Expenditures [21]

Gen. Services - 38.9%
Housing - 2.0%
Educ./Health - 17.6%
Other - 19.2%
Industry - 22.3%

% distribution for 1988. Expend. in 1991 est.: 767 ($ mil.)

Military Expenditures and Arms Transfers [22]

	1985	1986	1987	1988	1989
Military expenditures					
Current dollars (mil.)	NA	NA	NA	105[e]	65[e]
1989 constant dollars (mil.)	NA	NA	NA	109[e]	65[e]
Armed forces (000)	16	17	17	17	17
Gross national product (GNP)					
Current dollars (mil.)	3,624	3,459	3,722	4,330	4,655
1989 constant dollars (mil.)	4,126	3,839	4,003	4,507	4,655
Central government expenditures (CGE)					
1989 constant dollars (mil.)	1,721	2,148	1,668	1,381	770
People (mil.)	6.8	7.0	7.3	7.6	7.9
Military expenditure as % of GNP	NA	NA	NA	2.4	1.4
Military expenditure as % of CGE	NA	NA	NA	7.9	8.4
Military expenditure per capita	NA	NA	NA	14	8
Armed forces per 1,000 people	2.4	2.4	2.3	2.2	2.2
GNP per capita	610	545	546	593	591
Arms imports[6]					
Current dollars (mil.)	10	5	5	0	60
1989 constant dollars (mil.)	11	6	5	0	60
Arms exports[6]					
Current dollars (mil.)	0	0	0	0	0
1989 constant dollars (mil.)	0	0	0	0	0
Total imports[7]					
Current dollars (mil.)	654	648	816	835	928
1989 constant dollars	745	719	878	869	928
Total exports[7]					
Current dollars (mil.)	784	517	873	1,179	1,334
1989 constant dollars	893	574	939	1,227	1,334
Arms as percent of total imports[8]	1.5	0.8	0.6	0	6.5
Arms as percent of total exports[8]	0	0	0	0	0

Human Rights [24]

	SSTS	FL	FAPRO	PPCG	APROBC	TPW	PCPTW	STPEP	PHRFF	PRW	ASST	AFL
Observes	P	P				P	P			P	P	P
	EAFRD	CPR	ESCR	SR	ACHR	MAAE	PVIAC	PVNAC	EAFDAW	TCIDTP	RC	
Observes		P	P	P	P		P			S		P

P = Party; S = Signatory; see Appendix for meaning of abbreviations.

Labor Force

Total Labor Force [25]

2.455 million

Labor Force by Occupation [26]

Agriculture	85%
Mining, mfg., constr.	6
Transport and services	9

Unemployment Rate [27]

Production Sectors

Energy Resource Summary [28]

Energy Resources: Coal, uranium, hydropower potential. **Electricity**: 2,775,000 kW capacity; 12,000 million kWh produced, 1,400 kWh per capita (1991). **Pipelines**: Crude oil 1,724 km.

Telecommunications [30]

- Facilities are among the best in Sub-Saharan Africa
- High-capacity microwave connects most larger towns and cities
- Broadcast stations - 11 AM, 5 FM, 9 TV
- Satellite earth stations - 1 Indian Ocean INTELSAT and 1 Atlantic Ocean INTELSAT

Transportation [31]

Railroads. 1,266 km, all 1.067-meter gauge; 13 km double track

Highways. 36,370 km total; 6,500 km paved, 7,000 km crushed stone, gravel, or stabilized soil; 22,870 km improved and unimproved earth

Airports

Total:	116
Usable:	104
With permanent-surface runways:	13
With runways over 3,659 m:	1
With runways 2,440-3,659 m:	4
With runways 1,220-2,439 m:	22

Top Agricultural Products [32]

Agriculture accounts for 17% of GDP and 85% of labor force; crops—corn (food staple), sorghum, rice, peanuts, sunflower, tobacco, cotton, sugarcane, cassava; cattle, goats, beef, eggs.

Top Mining Products [33]

Metric tons unless otherwise specified	M.t.
Cobalt, mine output, Co content of ore	10,751
Copper, mine output, Cu content of ore	479,511
Cement, hydraulic	366,914
Zinc, Zn content of ore milled	19,825
Limestone (000 tons)	720[56]
Amethyst (kilograms)	168,220

Tourism [34]

	1987	1988	1989	1990	1991
Tourists	121	108	113	141	171
Tourism receipts	6	5	12	41	35
Tourism expenditures	46	49	98		91
Fare receipts	8	11	25		
Fare expenditures	23	30	100		

Tourists are in thousands, money in million U.S. dollars.

Manufacturing Sector

GDP and Manufacturing Summary [35]

	1980	1985	1989	1990	% change 1980-1990	% change 1989-1990
GDP (million 1980 $)	3,883	3,978	3,981	4,307	10.9	8.2
GDP per capita (1980 $)	677	568	489	510	-24.7	4.3
Manufacturing as % of GDP (current prices)	19.0	23.8	31.6[e]	33.2	74.7	5.1
Gross output (million $)	1,671	1,378[e]	2,668	2,610[e]	56.2	-2.2
Value added (million $)	780	575[e]	1,133	1,028[e]	31.8	-9.3
Value added (million 1980 $)	717	789	821[e]	1,098	53.1	33.7
Industrial production index	100	106	113	122	22.0	8.0
Employment (thousands)	59	62[e]	61	61[e]	3.4	0.0

Note: GDP stands for Gross Domestic Product. 'e' stands for estimated value.

Profitability and Productivity

	1980	1985	1989	1990	% change 1980-1990	% change 1989-1990
Intermediate input (%)	53	58[e]	58	61[e]	15.1	5.2
Wages, salaries, and supplements (%)	11	11[e]	11	11[e]	0.0	0.0
Gross operating surplus (%)	35	30[e]	31	29[e]	-17.1	-6.5
Gross output per worker ($)	28,232	22,254[e]	43,950	43,052[e]	52.5	-2.0
Value added per worker ($)	13,184	9,280[e]	18,661	16,966[e]	28.7	-9.1
Average wage (incl. benefits) ($)	3,245	2,542[e]	4,980	4,642[e]	43.1	-6.8

Profitability is in percent of gross output. Productivity is in U.S. $. 'e' stands for estimated value.

Profitability - 1990

Inputs - 60.4%
Wages - 10.9%
Surplus - 28.7%

The graphic shows percent of gross output.

Value Added in Manufacturing

	1980 $ mil.	1980 %	1985 $ mil.	1985 %	1989 $ mil.	1989 %	1990 $ mil.	1990 %	% change 1980-1990	% change 1989-1990
311 Food products	92	11.8	62[e]	10.8	100	8.8	87[e]	8.5	-5.4	-13.0
313 Beverages	193	24.7	104[e]	18.1	243	21.4	237[e]	23.1	22.8	-2.5
314 Tobacco products	58	7.4	39[e]	6.8	111	9.8	97[e]	9.4	67.2	-12.6
321 Textiles	51	6.5	32[e]	5.6	69	6.1	62[e]	6.0	21.6	-10.1
322 Wearing apparel	34	4.4	23[e]	4.0	47	4.1	46[e]	4.5	35.3	-2.1
323 Leather and fur products	4	0.5	3[e]	0.5	6	0.5	6[e]	0.6	50.0	0.0
324 Footwear	15	1.9	13[e]	2.3	28	2.5	29[e]	2.8	93.3	3.6
331 Wood and wood products	8	1.0	11[e]	1.9	29	2.6	28[e]	2.7	250.0	-3.4
332 Furniture and fixtures	12	1.5	10[e]	1.7	24	2.1	21[e]	2.0	75.0	-12.5
341 Paper and paper products	15	1.9	8[e]	1.4	13	1.1	11[e]	1.1	-26.7	-15.4
342 Printing and publishing	17	2.2	13[e]	2.3	24	2.1	24[e]	2.3	41.2	0.0
351 Industrial chemicals	22	2.8	26[e]	4.5	43	3.8	37[e]	3.6	68.2	-14.0
352 Other chemical products	47	6.0	51[e]	8.9	84	7.4	73[e]	7.1	55.3	-13.1
353 Petroleum refineries	9	1.2	5[e]	0.9	7	0.6	6[e]	0.6	-33.3	-14.3
354 Miscellaneous petroleum and coal products	3	0.4	2[e]	0.3	3	0.3	3[e]	0.3	0.0	0.0
355 Rubber products	20	2.6	16[e]	2.8	27	2.4	23[e]	2.2	15.0	-14.8
356 Plastic products	7	0.9	7[e]	1.2	13	1.1	12[e]	1.2	71.4	-7.7
361 Pottery, china and earthenware	1	0.1	1[e]	0.2	1	0.1	1[e]	0.1	0.0	0.0
362 Glass and glass products	3	0.4	3[e]	0.5	4	0.4	4[e]	0.4	33.3	0.0
369 Other non-metal mineral products	33	4.2	45[e]	7.8	60	5.3	55[e]	5.4	66.7	-8.3
371 Iron and steel	10	1.3	5[e]	0.9	8	0.7	7[e]	0.7	-30.0	-12.5
372 Non-ferrous metals	2	0.3	1[e]	0.2	1	0.1	1[e]	0.1	-50.0	0.0
381 Metal products	50	6.4	47[e]	8.2	99	8.7	82[e]	8.0	64.0	-17.2
382 Non-electrical machinery	18	2.3	11[e]	1.9	20	1.8	20[e]	1.9	11.1	0.0
383 Electrical machinery	26	3.3	13[e]	2.3	21	1.9	18[e]	1.8	-30.8	-14.3
384 Transport equipment	28	3.6	24[e]	4.2	45	4.0	39[e]	3.8	39.3	-13.3
385 Professional and scientific equipment	NA	0.0	NA	0.0	NA	0.0	NA	0.0	NA	NA
390 Other manufacturing industries	2	0.3	1[e]	0.2	1	0.1	1[e]	0.1	-50.0	0.0

Note: The industry codes shown are International Standard Industry codes (ISIC). Percentages are percent of total Value Added. 'e' stands for estimated value

Finance, Economics, and Trade

Economic Indicators [36]

National product: GDP—exchange rate conversion—$4.7 billion (1992 est.). **National product real growth rate**: -3% (1992 est.). **National product per capita**: $550 (1992 est.). **Inflation rate (consumer prices)**: 170% (1992 est.). **External debt**: $7.6 billion (1991).

Balance of Payments Summary [37]

Values in millions of dollars.

	1985	1986	1987	1988	1989
Exports of goods (f.o.b.)	797.0	692.0	852.0	1,189.0	1,340.0
Imports of goods (f.o.b.)	-571.0	-518.0	-585.0	-687.0	-774.0
Trade balance	226.0	174.0	267.0	502.0	566.0
Services - debits	-667.0	-553.0	-553.0	-893.0	-915.0
Services - credits	71.0	48.0	49.0	61.0	87.0
Private transfers (net)	-33.0	-42.0	-20.0	-25.0	-30.0
Government transfers (net)	6.0	22.0	8.0	59.0	109.0
Long term capital (net)	373.0	81.0	108.0	154.0	158.0
Short term capital (net)	323.0	-79.0	190.0	209.0	84.0
Errors and omissions	-145.0	258.0	-116.0	-17.0	41.0
Overall balance	154.0	-91.0	-67.0	50.0	100.0

Exchange Rates [38]

Currency: **Zambian kwacha.**
Symbol: **ZK.**

Data are currency units per $1.

August 1992	178.5714
1991	61.7284
1990	28.9855
1989	12.9032
1988	8.2237
1987	8.8889

Imports and Exports

Top Import Origins [39]

$1.2 billion (c.i.f., 1992 est.).

Origins	%
EC countries	NA
Japan	NA
Saudi Arabia	NA
South Africa	NA
US	NA

Top Export Destinations [40]

$1.0 billion (f.o.b., 1992 est.).

Destinations	%
EC countries	NA
Japan	NA
South Africa	NA
US	NA
India	NA

Foreign Aid [41]

	U.S. $	
US commitments, including Ex-Im (1970-89)	4.8	billion
Western (non-US) countries, ODA and OOF bilateral commitments (1970-89)	4.8	billion
OPEC bilateral aid (1979-89)	60	million
Communist countries (1970-89)	533	million

Import and Export Commodities [42]

Import Commodities	**Export Commodities**
Machinery	Copper
Transportation equipment	Zinc
Foodstuffs	Cobalt
Fuels	Lead
Manufactures	Tobacco

For sources, notes, and explanations, see Annotated Source Appendix, page 1035.

Zimbabwe

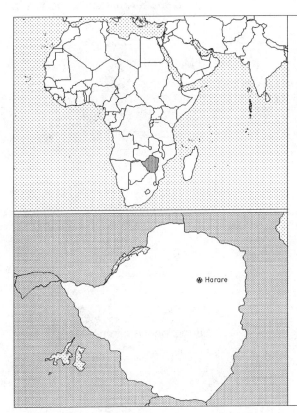

Geography [1]

Total area:
390,580 km2
Land area:
386,670 km2
Comparative area:
Slightly larger than Montana
Land boundaries:
Total 3,066 km, Botswana 813 km, Mozambique 1,231 km, South Africa 225 km, Zambia 797 km
Coastline:
0 km (landlocked)
Climate:
Tropical; moderated by altitude; rainy season (November to March)
Terrain:
Mostly high plateau with higher central plateau (high veld); mountains in east
Natural resources:
Coal, chromium ore, asbestos, gold, nickel, copper, iron ore, vanadium, lithium, tin, platinum group metals
Land use:
Arable land: 7%
Permanent crops: 0%
Meadows and pastures: 12%
Forest and woodland: 62%
Other: 19%

Demographics [2]

	1960	1970	1980	1990	1991[1]	1994	2000	2010	2020
Population	4,011	5,515	7,298	10,187	10,720	10,975	12,013	12,990	14,620
Population density (persons per sq. mi.)	27	37	49	70	72	NA	92	118	143
Births	NA	NA	NA	NA	435	409	NA	NA	NA
Deaths	NA	NA	NA	NA	89	199	NA	NA	NA
Life expectancy - males	NA	NA	NA	NA	60	40	NA	NA	NA
Life expectancy - females	NA	NA	NA	NA	54	44	NA	NA	NA
Birth rate (per 1,000)	NA	NA	NA	NA	41	37	NA	NA	NA
Death rate (per 1,000)	NA	NA	NA	NA	8	18	NA	NA	NA
Women of reproductive age (15-44 yrs.)	NA	NA	NA	2,308	NA	2,540	2,929	NA	NA
of which are currently married	NA	NA	NA	1,530	NA	1,667	1,905	NA	NA
Fertility rate	NA	NA	NA	5.8	5.60	5.1	4.1	2.9	2.3

Population values are in thousands, life expectancy in years, and other items as indicated.

Health

Health Personnel [3]

Doctors per 1,000 pop., 1988-92	0.16
Nurse-to-doctor ratio, 1988-92	6.1
Hospital beds per 1,000 pop., 1985-90	2.1
Percentage of children immunized (age 1 yr. or less)	
Third dose of DPT, 1990-91	89.0
Measles, 1990-91	87.0

Health Expenditures [4]

Total health expenditure, 1990 (official exchange rate)	
Millions of dollars	416
Millions of dollars per capita	42
Health expenditures as a percentage of GDP	
Total	6.2
Public sector	3.2
Private sector	3.0
Development assistance for health	
Total aid flows (millions of dollars)[1]	42
Aid flows per capita (millions of dollars)	4.2
Aid flows as a percentage of total health expenditure	10.0

For sources, notes, and explanations, see Annotated Source Appendix, page 1035.

1029

Human Factors

Health Care Ratios [5]

Population per physician, 1970	6300
Population per physician, 1990	7180
Population per nursing person, 1970	640
Population per nursing person, 1990	1000
Percent of births attended by health staff, 1985	69

Infants and Malnutrition [6]

Percent of babies with low birth weight, 1985	15
Infant mortality rate per 1,000 live births, 1970	96
Infant mortality rate per 1,000 live births, 1991	48
Years of life lost per 1,000 population, 1990	37
Prevalence of malnutrition (under age 5), 1990	12

Ethnic Division [7]

African	98%
Shona	71%
Ndebele	16%
Other	11%
White	1%
Mixed Asian	1%

Religion [8]

Syncretic (part Christian, part indigenous beliefs)	50%
Christian	25%
Indigenous beliefs	24%
Muslim and other	1%

Major Languages [9]

English (official), Shona, Sindebele.

Education

Public Education Expenditures [10]

Million Zimbabwe Dollars	1980	1986	1987	1988	1989[6]	1990
Total education expenditure	224	726	NA	962	1,035	1,661
as percent of GNP	6.6	9.1	NA	9.1	8.3	10.6
as percent of total govt. expend.	13.7	15.0	NA	NA	NA	NA
Current education expenditure	218	721	NA	947	1,025	NA
as percent of GNP	6.4	9.0	NA	9.0	8.2	NA
as percent of current govt. expend.	14.1	NA	NA	NA	NA	Na
Capital expenditure	6	6	NA	16	10	NA

Educational Attainment [11]

Literacy Rate [12]

In thousands and percent	1985[a]	1991[a]	2000[a]
Illiterate population + 15 years	1,683	1,776	1,900
Illiteracy rate - total pop. (%)	37.7	33.1	25.4
Illiteracy rate - males (%)	30.2	26.3	20.0
Illiteracy rate - females (%)	45.0	39.7	30.6

Libraries [13]

	Admin. Units	Svc. Pts.	Vols. (000)	Shelving (meters)	Vols. Added	Reg. Users
National (1988)	1	1	80	NA	1,545	31,144
Non-specialized	NA	NA	NA	NA	NA	NA
Public (1989)	76	83	1,038	NA	3,195	151,563
Higher ed. (1990)	25	31	764	NA	7,609	30,707
School	NA	NA	NA	NA	NA	NA

Daily Newspapers [14]

	1975	1980	1985	1990
Number of papers	3	2	3	2
Circ. (000)	116	133	203	206

Cinema [15]

Data for 1991.

Cinema seats per 1,000	1.2
Annual attendance per person	0.2
Gross box office receipts (mil. Dollar)	7.6

Science and Technology

Scientific/Technical Forces [16]

R&D Expenditures [17]

U.S. Patents Issued [18]

Values show patents issued to citizens of the country by the U.S. Patents Office.

	1990	1991	1992
Number of patents	0	1	0

For sources, notes, and explanations, see Annotated Source Appendix, page 1035.

Government and Law

Organization of Government [19]

Long-form name:
Republic of Zimbabwe
Type:
parliamentary democracy
Independence:
18 April 1980 (from UK)
Constitution:
21 December 1979
Legal system:
mixture of Roman-Dutch and English
common law
National holiday:
Independence Day, 18 April (1980)
Executive branch:
executive president, 2 vice presidents,
Cabinet
Legislative branch:
unicameral Parliament
Judicial branch:
Supreme Court

Political Parties [20]

Parliament	% of seats
African National Union (ZANU-PF)	78.0
Unity Movement (ZUM)	1.3
African National Union (ZANU-S)	0.7

Government Budget [21]

For FY91.
Revenues	2.700
Expenditures	3.300
Capital expenditures	0.330

Military Expenditures and Arms Transfers [22]

	1985	1986	1987	1988	1989
Military expenditures					
Current dollars (mil.)	266	297	359	355	386
1989 constant dollars (mil.)	303	330	386	369	386
Armed forces (000)	46	45	45	45	51
Gross national product (GNP)					
Current dollars (mil.)	4,411	4,634	4,747	5,275	5,742
1989 constant dollars (mil.)	5,021	5,142	5,105	5,491	5,742
Central government expenditures (CGE)					
1989 constant dollars (mil.)	2,101	2,110	2,389	2,467	2,580
People (mil.)	8.7	9.0	9.3	9.7	10.1
Military expenditure as % of GNP	6.0	6.4	7.6	6.7	6.7
Military expenditure as % of CGE	14.4	15.6	16.1	15.0	15.0
Military expenditure per capita	35	37	41	38	38
Armed forces per 1,000 people	5.3	5.0	4.8	4.6	5.1
GNP per capita	580	573	547	566	571
Arms imports[6]					
Current dollars (mil.)	0	110	80	0	10
1989 constant dollars (mil.)	0	122	86	0	10
Arms exports[6]					
Current dollars (mil.)	0	0	0	0	0
1989 constant dollars (mil.)	0	0	0	0	0
Total imports[7]					
Current dollars (mil.)	1,031	1,132	1,209	1,100	NA
1989 constant dollars	1,174	1,256	1,300	1,145	NA
Total exports[7]					
Current dollars (mil.)	959	1,054	1,427	1,600	NA
1989 constant dollars	1,092	1,170	1,535	1,666	NA
Arms as percent of total imports[8]	0	9.7	6.6	0	NA
Arms as percent of total exports[8]	0	0	0	0	NA

Crime [23]

Crime volume	
Cases known to police	427,550
Attempts (percent)	NA
Percent cases solved	NA
Crimes per 100,000 persons	4,275.50
Persons responsible for offenses	
Total number offenders	NA
Percent female	NA
Percent juvenile	NA
Percent foreigners	NA

Human Rights [24]

	SSTS	FL	FAPRO	PPCG	APROBC	TPW	PCPTW	STPEP	PHRFF	PRW	ASST	AFL
Observes	1			P		P	P				P	
	EAFRD	CPR	ESCR	SR	ACHR	MAAE	PVIAC	PVNAC	EAFDAW	TCIDTP	RC	
Observes	P	P	P	P					P		P	

P=Party; S=Signatory; see Appendix for meaning of abbreviations.

Labor Force

Total Labor Force [25]

3.1 million

Labor Force by Occupation [26]

Agriculture	74%
Transport and services	16
Mining, manufacturing, construction	10
Date of data: 1987	

Unemployment Rate [27]

at least 35% (1993 est.)

Production Sectors

Energy Resource Summary [28]

Energy Resources: Coal. **Electricity**: 3,650,000 kW capacity; 8,920 million kWh produced, 830 kWh per capita (1991). **Pipelines**: Petroleum products 212 km.

Telecommunications [30]

- System was once one of the best in Africa, but now suffers from poor maintenance
- Consists of microwave links, open-wire lines, and radio communications stations
- 247,000 telephones
- Broadcast stations - 8 AM, 18 FM, 8 TV
- 1 Atlantic Ocean INTELSAT earth station

Transportation [31]

Railroads. 2,745 km 1.067-meter gauge (including 42 km double track, 355 km electrified)

Highways. 85,237 km total; 15,800 km paved, 39,090 km crushed stone, gravel, stabilized soil: 23,097 km improved earth; 7,250 km unimproved earth

Airports

Total:	485
Usable:	403
With permanent-surface runways:	22
With runways over 3,659 m:	2
With runways 2,440-3,659 m:	3
With runways 1,220-2,439 m:	29

Top Agricultural Products [32]

	89-90	90-91
Corn	1,980	1,596
Sugar	502	492
Wheat	326	265
Cotton, seed	188	205
Tobacco	120	153
Peanuts	95	94

Values shown are 1,000 metric tons.

Top Mining Products [33]

Estimated metric tons unless otherwise specified   M.t.

Chromite	563,634
Asbestos	141,697
Gold (kilograms)	17,820
Nickel, mine output, Ni content of concentrate	12,371
Coal, bituminous (000 tons)	5,616
Tin, mine output, Sn content of concentrate	12,371

Tourism [34]

	1987	1988	1989	1990	1991
Visitors[1]	488	489	504	636	697
Tourists[76]	372	449	474	606	664
Excursionists	116	40	30	30	33
Tourism receipts	43	54	40	47	
Tourism expenditures	48	58			
Fare receipts	28	35			
Fare expenditures	21	29			

Tourists are in thousands, money in million U.S. dollars.

For sources, notes, and explanations, see Annotated Source Appendix, page 1035.

Manufacturing Sector

GDP and Manufacturing Summary [35]

	1980	1985	1989	1990	% change 1980-1990	% change 1989-1990
GDP (million 1980 $)	5,351	6,586	7,264	7,930	48.2	9.2
GDP per capita (1980 $)	751	794	772	817	8.8	5.8
Manufacturing as % of GDP (current prices)	24.1	22.2	20.4[e]	24.2	0.4	18.6
Gross output (million $)	3,579	3,020	4,543[e]	4,944[e]	38.1	8.8
Value added (million $)	1,480	1,278	1,867[e]	2,323[e]	57.0	24.4
Value added (million 1980 $)	1,247	1,403	1,634	1,712	37.3	4.8
Industrial production index	100	112	131	139	39.0	6.1
Employment (thousands)	161	163	186[e]	194[e]	20.5	4.3

Note: GDP stands for Gross Domestic Product. 'e' stands for estimated value.

Profitability and Productivity

	1980	1985	1989	1990	% change 1980-1990	% change 1989-1990
Intermediate input (%)	59	58	59[e]	53[e]	-10.2	-10.2
Wages, salaries, and supplements (%)	17	18	15[e]	16[e]	-5.9	6.7
Gross operating surplus (%)	24	25	26[e]	31[e]	29.2	19.2
Gross output per worker ($)	22,265	18,449	24,481[e]	25,497[e]	14.5	4.2
Value added per worker ($)	9,205	7,808	10,062[e]	11,980[e]	30.1	19.1
Average wage (incl. benefits) ($)	3,848	3,241	3,751[e]	4,169[e]	8.3	11.1

Profitability is in percent of gross output. Productivity is in U.S. $. 'e' stands for estimated value.

Profitability - 1990

Inputs - 53.0%
Wages - 16.0%
Surplus - 31.0%

The graphic shows percent of gross output.

Value Added in Manufacturing

	1980 $ mil.	1980 %	1985 $ mil.	1985 %	1989 $ mil.	1989 %	1990 $ mil.	1990 %	% change 1980-1990	% change 1989-1990
311 Food products	193	13.0	130	10.2	217[e]	11.6	253[e]	10.9	31.1	16.6
313 Beverages	92	6.2	189	14.8	281[e]	15.1	352[e]	15.2	282.6	25.3
314 Tobacco products	55	3.7	72	5.6	108[e]	5.8	104[e]	4.5	89.1	-3.7
321 Textiles	147	9.9	114	8.9	152[e]	8.1	218[e]	9.4	48.3	43.4
322 Wearing apparel	70	4.7	55	4.3	86[e]	4.6	100[e]	4.3	42.9	16.3
323 Leather and fur products	4	0.3	4	0.3	8[e]	0.4	8[e]	0.3	100.0	0.0
324 Footwear	34	2.3	42	3.3	64[e]	3.4	68[e]	2.9	100.0	6.3
331 Wood and wood products	38	2.6	17	1.3	33[e]	1.8	32[e]	1.4	-15.8	-3.0
332 Furniture and fixtures	26	1.8	15	1.2	23[e]	1.2	20[e]	0.9	-23.1	-13.0
341 Paper and paper products	30	2.0	37	2.9	58[e]	3.1	56[e]	2.4	86.7	-3.4
342 Printing and publishing	59	4.0	45	3.5	66[e]	3.5	71[e]	3.1	20.3	7.6
351 Industrial chemicals	58	3.9	67	5.2	70[e]	3.7	78[e]	3.4	34.5	11.4
352 Other chemical products	80	5.4	78	6.1	121[e]	6.5	148[e]	6.4	85.0	22.3
353 Petroleum refineries	NA	0.0	1	0.1	1[e]	0.1	1[e]	0.0	NA	0.0
354 Miscellaneous petroleum and coal products	7	0.5	8	0.6	9[e]	0.5	10[e]	0.4	42.9	11.1
355 Rubber products	30	2.0	24	1.9	35[e]	1.9	48[e]	2.1	60.0	37.1
356 Plastic products	25	1.7	37	2.9	56[e]	3.0	81[e]	3.5	224.0	44.6
361 Pottery, china and earthenware	3	0.2	2	0.2	3[e]	0.2	3[e]	0.1	0.0	0.0
362 Glass and glass products	9	0.6	5	0.4	7[e]	0.4	11[e]	0.5	22.2	57.1
369 Other non-metal mineral products	44	3.0	28	2.2	71[e]	3.8	74[e]	3.2	68.2	4.2
371 Iron and steel	194	13.1	105	8.2	110[e]	5.9	261[e]	11.2	34.5	137.3
372 Non-ferrous metals	10	0.7	9	0.7	10[e]	0.5	13[e]	0.6	30.0	30.0
381 Metal products	132	8.9	78	6.1	113[e]	6.1	124[e]	5.3	-6.1	9.7
382 Non-electrical machinery	39	2.6	22	1.7	27[e]	1.4	32[e]	1.4	-17.9	18.5
383 Electrical machinery	44	3.0	36	2.8	50[e]	2.7	63[e]	2.7	43.2	26.0
384 Transport equipment	38	2.6	48	3.8	73[e]	3.9	78[e]	3.4	105.3	6.8
385 Professional and scientific equipment	2	0.1	1	0.1	2[e]	0.1	2[e]	0.1	0.0	0.0
390 Other manufacturing industries	17	1.1	9	0.7	13[e]	0.7	11[e]	0.5	-35.3	-15.4

Note: The industry codes shown are International Standard Industry codes (ISIC). Percentages are percent of total Value Added. 'e' stands for estimated value

For sources, notes, and explanations, see Annotated Source Appendix, page 1035.

1033

Finance, Economics, and Trade

Economic Indicators [36]

National product: GDP—exchange rate conversion—$6.2 billion (1992 est.). **National product real growth rate**: -10% (1992 est.). **National product per capita**: $545 (1992 est.). **Inflation rate (consumer prices)**: 45% (1992 est.). **External debt**: $3.9 billion (March 1993 est.).

Balance of Payments Summary [37]

Values in millions of dollars.

	1980	1985	1986	1987	1988
Exports of goods (f.o.b.)	1,445.5	1,119.6	1,322.7	1,452.0	1,664.9
Imports of goods (f.o.b.)	-1,339.0	-918.9	-1,011.6	-1,071.0	-1,163.6
Trade balance	106.5	200.7	311.1	381.0	501.3
Services - debits	-561.3	-629.2	-541.5	-578.2	-644.9
Services - credits	273.9	333.3	205.4	197.6	207.8
Private transfers (net)	-120.6	-45.2	-26.6	-32.1	-13.2
Government transfers (net)	57.7	64.8	58.3	79.7	65.6
Long term capital (net)	-52.8	108.2	85.0	69.9	36.0
Short term capital (net)	22.9	12.5	33.1	-10.0	12.0
Errors and omissions	187.2	37.2	-69.4	16.6	-62.9
Overall balance	-86.5	82.3	55.4	124.5	101.7

Exchange Rates [38]

Currency: **Zimbabwean dollars.**
Symbol: **Z$.**

Data are currency units per $1.

February 1993	6.3532
1992	5.1046
1991	3.4282
1990	2.4480
1989	2.1133
1988	1.8018

Imports and Exports

Top Import Origins [39]

$1.8 billion (c.i.f., 1992 est.). Data are for 1991.

Origins	%
UK	15
Germany	9
South Africa	5
Botswana	5
US	5
Japan	5

Top Export Destinations [40]

$1.5 billion (f.o.b., 1992 est.). Data are for 1991.

Destinations	%
UK	14
Germany	11
South Africa	10
Japan	7
US	5

Foreign Aid [41]

	U.S. $	
US commitments, including Ex-Im (FY80-89)	389	million
Western (non-US) countries, ODA and OOF bilateral commitments (1970-89)	2.6	billion
OPEC bilateral aid (1979-89)	36	million
Communist countries (1970-89)	134	million

Import and Export Commodities [42]

Import Commodities

Machinery, transp. equip. 37%
Other manufactures 22%
Chemicals 16%
Fuels 15%

Export Commodities

Tobacco 20%
Other 15%
Manufactures 20%
Gold 10%
Ferrochrome 10%
Cotton 5%

ANNOTATED SOURCE APPENDIX

Table of Contents

Maps

Each national entry has both a regional map and a national map. The regional map indicates the location of the nation within its region, normally a continent or major sub-continental area. The maps of the Regions of the World, provided in the front section, show the location of all nations within a region in one map.

National and regional maps for Africa and North America are Mercator projections produced using the *CIA World Database* and Allison Software's *MAPIT* v1.3 software. The corresponding maps for all other nations are Plate Carée projections using the Defense Mapping Agency's *Digital Chart of the World*. The Regions of the World maps are all Plate Carrée projections.

Each national map is presented at the scale which produces the largest map which fits within the panel. Thus the maps are at different scales. The regional maps of the same region are at identical scales, again determined by the maximum scale which fits the format.

Due to the small size of the maps, no rivers are shown. Only those inland seas and lakes are shown that can be seen easily.

The Mercator projection is probably the most familiar to the public. It projects the earth's sphere onto a cylinder. Lines of latitude and longitude are all parallel straight lines at right angles to each other. Therefore, it is absolutely accurate only at the equator. Distortion of east to west distances increases with the distance from the equator. The Plate Carrée method projects the earth's sphere onto a cone (i.e. a conic projection) with the equator being the standard and absolutely accurate length latitude line. Lines of latitude are concentric circles, and lines of longitude are non-parallel straight lines. The lines of latitude and longitude intersect at right angles. Again, this projection is only absolutely accurate at the equator. However, distortions of distance and shape away from the equator are less than those associated with the Mercator projection. It is useful for maps that have a greater east to west than north to south expanse (as is the case in *SAW*). This projection was used where it was available to provide greater accuracy while providing a view of an area similar to the more familiar Mercator projection.

Panel 1 - Geography

Source. *The World Factbook 1993*, Central Intelligence Agency (CIA), Washington, DC: CIA, 1994.

Notes. *Total area* is the sum of all land and water areas delimited by international boundaries and/or coastlines.

Land area is the aggregate of all surfaces delimited by international boundaries and/or coastlines, excluding inland water bodies (lakes, reservoirs, rivers).

Comparative areas are based on total area equivalents. Most entities are compared with the entire U.S. or one of the 50 states. The smaller entities are compared with Washington DC (178 square km, 69 square miles) or The Mall in Washington DC (0.59 square km, 0.23 square miles, 146 acres).

Land use: Human use of the land surface is categorized as *arable land*—land cultivated for crops that are replanted after each harvest (wheat, maize, rice); *permanent crops*—land cultivated for crops that are not replanted after each harvest (citrus, coffee, rubber); *meadows and pastures*—land permanently used for herbaceous forage crops; *forest and woodland*—land under dense or open stands of trees; and *other*—any land type not specifically mentioned above (urban areas, roads, desert).

Panel 2 - Demographics

Source. U.S. Bureau of the Census, Report WP/94, *World Population Profile: 1994*. U.S. Government Printing Office, Washington, DC, 1994.

Notes. Primary Source: U.S. Bureau of the Census, International Data Base.

New estimates and projections of population and vital rates are made for each issue of the *World Population Profile* based on the latest information available. Sometimes the latest information requires making a revision to estimated data for the past as well as new projections for the future. Therefore, the user is cautioned against creating time series of population of population or vital rates from different issues of the report.

Footnotes. Z stands for less than 500 or between 0.05 and -0.05 percent. Minus sign (-) denotes a negative natural increase. NA stands for Data not available. The category "currently married women" includes women in consensual unions. Estimates are based on component projections of the female population and the percent of women who are married or in consensual unions.

1.	The data for 1991 have been included from *World Population Profile: 1991* for the sole purpose of providing the reader with an additional year of recent demographic data. Therefore, when using these data in a time-series analysis, the 1991 column should be omitted from the set. See technical note, above.

2.	The U.S. view is that the Socialist Federal Republic of Yugoslavia has dissolved and no successor state represents its continuation. Macedonia has proclaimed independent statehood, but has not been recognized as a state by the United States. Serbia and Montenegro have asserted the formation of a joint independent state, but this entity has not been recognized as a state by the U.S.

Panel 3 - Health Personnel

Source. International Bank for Reconstruction and Development/World Bank, *World Development Report 1993: Investing in Health*, New York: Oxford University Press, 1993. Used with permission.

Notes. *Doctor* is defined to include only individuals with the professional degree of medical doctor. The definition of *nurse* includes only registered nurses and registered midwives. *Hospital bed* is defined as beds in clinics and hospitals; beds in long-term care facilities and nursing homes are excluded. Data sources are the World Bank, the Organization for Economic Cooperation and Development (OECD), PAHO, and WHO.

Immunization data refer to DPT3—three completed doses of vaccine against diphtheria, pertussis (whooping cough), and tetanus—and to measles. The denominator for estimating coverage is the number of surviving infants age 1 year. The source of data is WHO's Expanded Programme on Immunization.

Footnotes. NA stands for Not Available. Each value refers to one particular (but unspecified year) within the time range.

1. DPT data refer to three doses of vaccine against diphtheria, pertussis, and tetanus [see also Technical Notes].

2. Refers to former Socialist Federal Republic of Yugoslavia because disaggregated data are not yet available.

3. Refers to former Czechoslovakia because disaggregated data are not yet available.

Panel 4 - Health Expenditures

Source. International Bank for Reconstruction and Development/World Bank, *World Development Report 1993: Investing in Health*, New York: Oxford University Press, 1993. Used with permission.

Notes. Health expenditure includes outlays for prevention, promotion, rehabilitation, and care; population activities; nutrition activities; program food aid; and emergency aid specifically for health. It does not include water and sanitation. Per capita expenditures and per capita flows are based on World Bank midyear population estimates.

Total health expenditure is expressed in official exchange rate U.S. dollars. Data on public and private health expenditure for the established market economies and Turkey are from the OECD. For other countries, information on government health expenditures is from national sources, supplemented by *Government Finance Statistics* (published by the International Monetary Fund), World Bank sector studies, and other studies. Data on parastatal expenditures (for health-related social security and social insurance programs) are from the Social

Security Division of the International Labour Office (ILO) and the World Bank. Data are drawn from Murray, Govindaraj, and Chellaraj, background paper.

Public sector expenditures include government health expenditures, parastatal expenditures, and foreign aid, making the figures comparable with those for OECD countries. *Private sector* expenditures for countries other than OECD members are based on household surveys carried out by the ILO and other sources, supplemented by information from United Nations National Income Accounts, World Bank studies, and other studies published in the scientific literature.

Estimates for countries with incomplete data were calculated in three steps. First, where data on either private or public expenditures were lacking, the missing figures were imputed from data from countries for which information was available. The imputation followed regressions relating public or private expenditure to GDP per capita. Second, for a country with no health expenditure data, it was assumed that the share of GDP spent on health care was the same as the average for the corresponding demographic region. Third, if GDP was also unknown but population was known, it was assumed that per capita health spending was the same as the regional average.

Estimates for *development assistance for health* are expressed in official exchange rate U.S. dollars. *Total aid flows* represent the sum of all health assistance for health to each country by bilateral and multilateral agencies and by international nongovernmental organizations (NGOs). Direct bilateral official development assistance (ODA) comes from the OECD countries. Sources of multilateral development assistance include United Nations agencies, development banks (including the World Bank), the European Community, and the Organization of Petroleum Exporting Countries (OPEC). Major international NGOs include the International Committee for the Red Cross (ICRC) and the International Planned Parenthood Federation (IPPF). National NGOs were not included because the available information was not separated by recipient country.

Information on ODA from bilateral and multilateral organizations was completed by data from the OECD's Development Assistance Committee (DAC) and Creditor Reporting System (CRS) and from the Advisory Committee for the Coordination of Information Systems (ACCIS). DAC has compiled annual aggregate ODA statistics, by sector, since 1960. The OECD's CRS, established in 1970, complements the DAC statistics by identifying contributions allocated by sector. The CRS data base is the most complete source of information for bilateral ODA, but its completeness varies among OECD countries and from year to year. ACCIS has kept, since 1987, a Register of Development Activities of the United Nations that lists sources of funds and executing agencies for all United Nations projects by sector.

The estimates of development assistance in this table were prepared by the Harvard Center for Population and Development Studies as a background paper for this report.

Footnotes. NA stands for Not Available.

1. Aid flows are official development assistance and include only a small portion of private flows, that is NGO assistance.

2. Refers to former Czechoslovakia because disaggregated data are not yet available.

3. Refers to former Socialist Federal Republic of Yugoslavia because disaggregated data are not yet available.

Panel 5 - Health Care Ratios

Source. International Bank for Reconstruction and Development/World Bank, *World Development Report 1993: Investing in Health*, New York: Oxford University Press, 1993. Used with permission.

Notes. The estimates of *population per physician* and *per nursing person* are derived from World Health Organization (WHO) data and are supplemented by data obtained directly by the World Bank from national sources. The data refer to a variety of years, generally no more than two years before the year specified. Nursing persons include auxiliary nurses as well as paraprofessional personnel such as traditional birth attendants. The inclusion of auxiliary and paraprofessional personnel provided more realistic estimates of available nursing care. Because definitions of doctors and nursing personnel vary—and because the data shown are for a variety of years—the data for these two indicators are not strictly comparable across countries.

Data on *births attended by health staff* show the percentage of births recorded where a recognized health service worker was in attendance. The data are from WHO, supplemented by UNICEF data. They are based on national sources, derived mostly from official community reports and hospital records; some reflect only births in hospitals and other medical institutions. Sometimes smaller private and rural hospitals are excluded, and sometimes even relatively primitive local facilities are included. The coverage is therefore not always comprehensive, and the figures should be treated with extreme caution.

Footnotes

1. Data refer to the Federal Republic of Germany before unification.

2. Data refer to former Yugoslavia.

3. Data refer to former Czechoslovakia.

Panel 6 - Infants and Malnutrition

Source. International Bank for Reconstruction and Development/World Bank, *World Development Report 1993: Investing*

in Health, New York: Oxford University Press, 1993. Used with permission.

Notes. In tables for the following countries, data refer to former Yugoslavia: Bosnia and Hercegovina, Croatia, Macedonia, Serbia and Montenegro, Slovenia.

In the tables for Russia, data refer to the former Soviet Union.

In tables for the Czech and Slovak Republics, data refer to former Czechoslovakia.

Babies with low birth weight are children born weighing less than 2,500 grams. Low birth weight is frequently associated with maternal malnutrition. It tends to raise the risk of infant mortality and lead to poor growth in infancy and childhood, thus increasing the incidence of other forms of retarded development. The figures are derived from both WHO and UNICEF sources and are based on national data. The data are not strictly comparable across countries since they are compiled from a combination of surveys and administrative records that may not have representative national coverage.

The *infant mortality rate* is the number of infants who die before reaching one year of age, per thousand live births in a given year. The data are from the U.N. publication *Mortality of Children Under Age 5: Projections, 1950-2025* as well as from the World Bank.

The years of life lost (per 1,000 population) conveys the burden of mortality in absolute terms. It is composed of the sum of the years lost to premature death per 1,000 population. Years of life lost at age x are measured by subtracting the remaining expected years of life, given a life expectancy at birth fixed at 80 years for men and 82.5 for women. This indicator depends on the effect of three variables: the age structure of the population, the overall rate of mortality, and the age structure of mortality.

Child malnutrition measures the percentage of children under five with a deficiency or an excess of nutrients that interfere with their health and genetic potential for growth. Methods of assessment vary, but the most commonly used are the following: less than 80 percent of the standard weight for age; less than minus two standard deviation from the 50th percentile of the weight for age reference population; and the Gomez scale of malnutrition. Note that for a few countries the figures are for children of three or four years of age and younger. The summary measures in this table are country data weighted by each country's share in the aggregate population.

Footnotes

1. Data refer to the Federal Republic of Germany before unification.

Panel 7 - Ethnic Division

Source. *The World Factbook 1993*, Central Intelligence Agency (CIA), Washington, DC: CIA, 1994.

Notes. These tables show the major ethnic divisions of peoples in the given country. Where available, the distribution is shown in percent.

Panel 8 - Religion

Source. *The World Factbook 1993*, Central Intelligence Agency (CIA), Washington, DC: CIA, 1994.

Notes. These tables show the major religious denominations of the peoples of the given country. Where available, a percent distribution is shown.

Panel 9 - Major Languages

Source. *The World Factbook 1993*, Central Intelligence Agency (CIA), Washington, DC: CIA, 1994.

Notes. These tables show the major language(s) spoken by inhabitants of the given country. Where available, percent distribution is shown.

Panel 10 - Public Education Expenditures

Source. United Nations Educational, Scientific, and Cultural Organization (UNESCO), *Statistical Yearbook 1993*, Paris: UNESCO, 1993. Used with permission.

Notes. These tables present total public expenditure on education distributed between current and capital.

Educational expenditure is also expressed as a percentage of the Gross national product (GNP) and of total public expenditure.

For almost all countries, data on GNP are supplied by the World Bank. Because these data are revised every year by the World Bank, the percentages of educational expenditure in relation to GNP may sometimes differ from those shown in previous editions of the *UNESCO Statistical Yearbook*.

Capital expenditures are generally defined as fixed investment in infrastructure, machinery, and equipment.

Current expenditures include expenditures such as salaries and benefits, research and development, training, goods and services, and the like.

Gross National Product (GNP) is the sum goods and services produced by a country's residents or capital regardless of location.

Footnotes. NA stands for not available. Dash (-) stands for nil.

1. Data refer to expenditure of the Ministry of Education only.

2. Data refer to expenditure of the Ministry of Primary and Secondary Education only.

3. From 1985 to 1989, expenditure of Al-Azhar is not included.

4. Data include foreign aid received for education.

5. For 1985 and 1986, data refer to expenditure of the Federal government only.

6. Expenditure on third level education is not included.

7. Data refer to public and private expenditure on education.

8. For 1990, expenditure from private sources represent 8.8% of the total amount.

9. Expenditure on education is calculated as percentage of global social product.

10. Data on current and capital expenditure refer to the ministry of Education only.

11. Expenditure on universities is not included.

12. Data refer to expenditure on education of the central government only.

13. Expenditure of the Office of Greek Education only.

14. For 1980, data on current and capital expenditure do not include public subsidies to private education. From 1985 to 1988, these data refer to total public and private expenditure on education.

15. Data refer to former Czechoslovakia.

16. Metropolitan France.

17. For 1990, expenditure from private sources represent 3.6% of the total amount.

18. Data refer to the former Union of Soviet Socialist Republics.

19. Expenditure on education is calculated as percentage of net material product.

20. Data refer to North Yemen only.

21. Data refer to West Germany only.

22. Data refer to former Yugoslavia.

Panel 11 - Educational Attainment

Source. United Nations Educational, Scientific, and Cultural Organization (UNESCO), *Statistical Yearbook 1993*, Paris: UNESCO, 1993. Used with permission.

Notes. These data show the percentage distribution of the highest level of educational attainment of the adult population. The data were derived from national censuses or sample surveys that were provided by the United Nations Statistical Office or were derived from regional or national publications.

Readers interested in data for earlier years are referred to the publication *Statistics of Educational Attainment and Illiteracy, 1970-1980*, CSR-E-44 (UNESCO, 1983).

The six levels of educational attainment are based on categories of the Standard Classification of Education (ISCED) and may be defined as follows:

No schooling. This term applies to those who have completed less than one year of schooling.

Incompleted first level. This category includes all those who completed at least one year of education at the first level but who did not complete the final grade at this level. The number of years of education included in the first level may vary depending on the country.

Completed first level. Those who completed the final grade of education at the first level (ISCED category 1) but who did not go on to second level studies are included in this group.

S-1: Entered second level, first stage. This group includes persons who completed no more than the lower stage of education at the second level.

S-2: Entered second level, second stage. This group corresponds to ISCED category 3, and includes persons who moved to the higher stage of the second level education but did not proceed to studies at the post secondary level.

Post secondary. Anyone who undertook third level studies (ISCED categories 5, 6, or 7), whether or not they completed the full course, would be counted in this group.

At the post secondary education level there is a usually a larger number of persons in the 25-34 age group than in the 15-24 age group. This is because many of the persons in the 15-24 age group are too young to have reached entrance age. For this reason the total adult age range is taken as 25+ (and not 15+) for the purposes of these data.

Data for the following countries refer to 1980: Argentina, Barbados, Czech Republic, El Salvador, Guyana, Indonesia, Mozambique, Norway, Papua New Guinea, Philippines, Saint Kitts, Saint Lucia, Saint Vincent and the Grenadines, Slovakia, Thailand, Trinidad and Tobago, Turkey, and Zambia

Data for the following countries refer to 1981: Austria, Bangladesh, Bosnia and Hercegovina, Botswana, Brunei, Canada, Croatia, Cuba, Dominica, [East] Germany, Greece, Grenada, Guatemala, India, Ireland, Italy, Macedonia, Nepal, Pakistan, Peru, Portugal, Sao Tome and Principe, Serbia and Montenegro, Sri Lanka, Togo, United States, and Venezuela

Data for the following countries refer to 1982: Chile, China, Haiti, Israel, Jamaica, and Paraguay

Data for the following countries refer to 1983: Honduras

Data for the following countries refer to 1984: Congo, Tunisia

Data for the following countries refer to 1985: Kuwait, Sierra Leone, South Africa, and Uruguay

Data for the following countries refer to 1986: Egypt, Fiji, Qatar, and Spain

Data for the following countries refer to 1987: Cyprus

Data for the following countries refer to 1988: Mauritania, Poland

Data for the following countries refer to 1989: Brazil, Estonia, Latvia, Lithuania, and Russia

Data for the following countries refer to 1990: Bahamas, Ecuador, Finland, Hungary, Japan, Mauritius, Mexico, Panama, and Singapore

Data for the following countries refer to 1991: Belize, New Zealand

Data for the following countries refer to 1992: Romania

S-1 = First stage

S-2 = Second stage

A= All ages

NA stands for Data not available.

- Stands for nil.

Unless otherwise indicated, the number of persons whose level of education is not stated have been subtracted from the total population.

Footnotes

1. Data refer to former Yugoslavia.

2. Data do not include rural population of the region north of Brazil.

3. Data based on a 10% sample of census returns.

4. Data refer to former Czechoslovakia.

5. Egypt: Second level also includes third level education not leading to a university degree.

6. Illiteracy data have been used for the category *no schooling*.

7.   Data refer to East Germany only.

8.   Based on a sample survey referring to 51,372 persons.

9.   Those persons who did not state their level of education have been included in the category *no schooling*.

10. The category *no schooling* comprises illiterates and those persons who did not state their level of education.

11. Data do not include Bothuthatswana Transkei and Veda.

12. Estimates based on a sample.

13. Data refer to former Union of Soviet Socialist Republics.

Panel 12 - Literacy Rate

Source. For years marked with footnote (a): United Nations Educational, Scientific, and Cultural Organization (UNESCO), *Compendium of Statistics on Illiteracy - 1990 Edition*, Paris: UNESCO, 1990. Used with permission.

For years marked with footnote (b): United Nations Educational, Scientific, and Cultural Organization (UNESCO), *Statistical Yearbook 1993*, Paris: UNESCO, 1993. Used with permission.

Footnotes

a.   **Technical Notes**: These data present estimates and projections of illiteracy rates in 1985, 1990 and 2000, prepared by UNESCO. The reader should keep in mind the conditional nature of these projections.

For countries providing the relevant information, the total rate of illiteracy was derived from an analysis of the rates by demographical generation. Following the rate of a cohort from one census to another gives a relatively stable curve; adjusting this statistical curve permits the estimation and projection of the illiteracy rates in a satisying manner.

For some countries, with insufficient statistical information, an estimation of the global illiteracy rate (15 years and over) was made, without taking into account the generation rates. In this way, by using all the available data for all countries, a preliminary analysis was made to decide the most significant correlation between illiteracy and several socio-economic and educational variables: infant mortality, fertility rate, and enrollment ratio in primary education. Following this procedure a certain number of rates were estimated but due to the uncertainty of some of them, not all were included in these tables.

Large decreases of the illiteracy rates are expected not only in those countries where enrollment has rapidly increased but also in countries where important literacy campaigns have been carried out or are currently underway. These campaigns can completely change the illiteracy rates of cohorts but due to insufficient information the impact of the more recent campaigns has not been reflected in these projections. Conse-

quently, it was considered preferable not to present estimates for certain countries such as Ethiopia, Nicaragua, and United Republic of Tanzania which have recently carried out important literacy campaigns.

b.   **Technical Notes**: These tables present data relating to the illiterate population from the latest census or survey held since 1970. These figures were provided by the United Nations Statistical Office or were derived from regional or national publications.

1.   Data refer to former Yugoslavia.

2.   Data refer to North Yemen.

3.   Persons with no schooling are defined as illiterates.

4.   Data are estimates made by national authorities.

5.   *De jure* population. In 1980, after the National Literacy Campaign, the Ministry of Education estimated that of the 722,431 illiterates identified in the census of October 1979, 130,372 were *analfabetos inaptos* (or 12.96% of the population 10 years and older).

6.   Data do not include Bothuthatswana, Tanskei, and Veda.

Panel 13 - Libraries

Source. United Nations Educational, Scientific, and Cultural Organization (UNESCO), *Statistical Yearbook 1993*, Paris: UNESCO, 1993. Used with permission.

Notes. These tables present selected statistics on collections, annual additions and registered users for the different categories of libraries.

From 1950 UNESCO collected library statistics every other year. The frequency of surveys was changed to three years following the *Recommendation Concerning the International Standardization of Library Statistics*, adopted by the General Conference of UNESCO at its sixteenth session in 1970. The first three surveys conducted after the adoption of the 1970 Recommendation requested data for 1971, 1974, and 1977 respectively. In order to facilitate the collection of data and improve the response rate from Member States, it was decided as from 1980 to survey not more than two categories of library at any one time. The questionnaire on libraries was therefore divided into three parts dealing respectively with (1) national and public libraries, (2) libraries of institutions of higher education and school libraries and (3) special libraries, each part being sent out in turn. Thus, the 1985, 1988 and 1991 surveys dealt with national and public libraries while those carried out in early 1983, 1986, 1989 and 1992 concentrated on libraries of institutions of higher education and school libraries. Statistics on special libraries were obtained from the 1984 and 1987 surveys. For various reasons, surveys on special libraries are no longer done.

The majority of the definitions that follow are taken from the above- mentioned Recommendation. There are, nevertheless, a few such as those concerning audio-visual and other library materials or registered users which are either not covered by the Recommendation or have undergone modifications in order to better respond to certain developments in librarianship.

Library. Irrespective of its title, any organized collection of printed books and periodicals or any other graphic or audio-visual materials, and the services of a staff to provide and facilitate the use of such materials necessary to meet the informational, research, educational or recreational needs of its users.

Libraries thus defined are counted in numbers of administrative units and service points, as follows: (a) *administrative unit*, any independent library, or group of libraries, under a single director or a single administration; (b) *service point*, any library which provides in separate quarters a service for users, whether it is an independent library or part of a larger administrative unit. Libraries are classified as follows:

National libraries. Libraries which, irrespective of their title, are responsible for acquiring and conserving copies of all significant publications produced in the country and functioning as a 'deposit' library, either by law or other arrangement, and normally compiling a national bibliography.

Libraries of institutions of higher education. Libraries primarily serving students and teachers in universities and other institutions of education at the third level.

Other major non-specialized libraries. Non-specialized libraries of a learned character which are neither libraries of institutions of higher education nor national libraries, though they may fulfill the functions of a national library for a specified geographical area.

School libraries. Those attached to all types of schools below the third level of education and serving primarily the pupils and teachers of such schools, even though they may also be open to the general public.

Special libraries. Those maintained by an association, government service, parliament, research institution (excluding university institutes), learned society, professional association, museum, business firm, industrial enterprise, chamber of commerce, etc. or other organized group, the greater part of their collections covering a specified field or subject, e.g. natural sciences, social sciences, agriculture, chemistry, medicine, economics, engineering, law, history.

Public (or popular) libraries. Those which serve the population of a community or region free of charge or for a nominal fee; they may serve the general public or special categories of users such as children, members of the armed forces, hospital patients, prisoners, workers and employees.

With respect to library holdings, acquisitions, loans, expenditure, personnel, etc., the following definitions and classifications are given:

Annual additions. All materials added to collections during the year whether by purchase, donation, exchange or any other method. Statistics cover the following *documents available to users*: a) books and bound periodicals; b) manuscripts; c) microforms; d) audio-visual documents and e) other library materials.

Volume. Any printed or manuscript work contained in one binding or portfolio.

Audio-visual materials. Non-book, non-microform library materials which require the use of special equipment to be seen and/or heard. This includes materials such as records, tapes, cassettes, motion pictures, slides, transparencies, video recordings, etc.

Other library materials. All materials other than books, bound periodicals, manuscripts, microforms and audio-visual materials. This includes materials such as maps, charts, art prints, photographs, dioramas, etc.

Registered user. A person registered with a library in order to use materials of its collection on or off the premises.

Footnotes. NA stands for not available. A dash (-) stands for nil. An e. stands for provisional or estimated figure.

a.　　Data refer to former Yugoslavia.

b.　　Data refer to former Czechoslovakia.

1.　　The figure in column 6 refers to the number of readers only.

2.　　Data refer only to libraries of institutions of higher education that are not part of a university library.

3.　　Data on libraries of institutions of higher education refer to the Institut Superieur d'Agriculture only.

4.　　Data on libraries of institutions of higher education refer only to Ecole Normale Superieure which is not a part of a university.

5.　　Data on public libraries refer to a library financed by public authorities only.

6.　　Data on school libraries refer to state school libraries only.

7.　　Data refer only to the Bibliotheque de l'universite nationale du Rwanda: campus de Butare et campus de Rumengeri.

8.　　Data on libraries of institutions of higher education refer to the main or central library of the University of Dakar.

9.　　The figures in column 5 on public and school libraries refer to the number of titles only.

10.　　Data refer to the library of the University of Kartoum only.

11. Data on libraries of institutions of higher education refer to the Copperbelt University Library only.

12. Data on libraries of higher education refer only to libraries of Barbados Community College and Samuel Jackman Prescod Polytechnic.

13. The public library also serves as a national library.

14. Data for public libraries refer only to libraries financed by public authorities and the figure in column 2 includes mobile station stops.

15. Data on libraries of institutions of higher education refer only to main or central university libraries.

16. Data on libraries of institutions of higher education refer to the Universidad San Carlos de Guatemala only.

17. The figure in column 6 on libraries of institutions of higher education refers to the number of registered users of the main or central university libraries only.

18. Data on libraries of institutions of higher education refer to the main or central library only.

19. The figure on the national library in column 3 refer to books only and in column 6 to the number of readers.

20. Data on libraries of institutions of higher education refer to the main or central university library only.

21. Data on public libraries refers only to libraries financed by public authorities.

22. Data on libraries of institutions of higher education refer to main or central university libraries only.

23a. The national library also serves as a public library.

23b. The figure in column 4 on libraries of institutions of higher education refers to 4 libraries only.

24. Data on libraries of institutions of higher education refer to the University of Damascus only.

25. Data on libraries of institutions of higher education do not include libraries which are not part of a university.

26. The figure in column 6 on public libraries refers to the number of readers.

27. Data on libraries of institutions of higher education refer to 19 main or central university libraries, to an unknown number of libraries of institutes or department and to 2 libraries of institutions of higher education which are not part of a university.

28. The figure in column 6 on national library refers to the number of readers only.

29. The figure in column 5 on national and public libraries includes manuscripts, microforms and audiovisual documents.

30. All data refer to Metropolitan France and overseas departments.

31. Data on non-specialized libraries refer only to the Bibliotheque publique d'Information (BP) de Beaubourg and the figures in column 6 refer to the number of visitors.

32. Data on libraries of institutions of higher education do not include 33 libraries that are not part of a university nor 3,549 libraries attached to university institutes or departments.

33. Data relating to public libraries include the national library and special libraries.

34. Data relating to public libraries do not include libraries financed from private sources.

35. Data relating to public libraries and libraries of institutions of higher education refer only to libraries dependent of the Ministry of Culture and Environment.

36. Data relating to public libraries do not include libraries financed from private sources.

37. Data on school libraries refer only to libraries of primary schools.

38. The figure in column 6 on libraries of institutions of higher education refers to registered borrowers only.

39. Data on libraries of institutions of higher education refer only to main or central university libraries.

40. The figure in column 6 on public libraries refers to registered persons: borrowers and readers.

41. The figures in column 6 on national libraries do not include data for The National Library of Scotland.

42. The figure in column 3 on national library includes manuscripts and microforms. Data on libraries of institutions of higher education include research libraries.

43. The national library also serves as public library.

Panel 14 - Newspapers

Source. United Nations Educational, Scientific, and Cultural Organization (UNESCO), *Statistical Yearbook 1993*, Paris: UNESCO, 1993. Used with permission.

Notes. National statistics on newspapers have been drawn up in accordance with the definitions and classifications set out in the 1985 Revised Recommendation concerning the International Standardization of Statistics on the Production and Distribution of Books, Newspapers and Periodicals. Data shown in these tables are published regularly, while data for newspapers by type may be found in the 1992 edition of *UNESCO Statistical Yearbook*. According to the 1985 Recommendation, national statistics on the press cover printed periodic publications which are published in a particular country and made available to the public. Exception are publications issued for advertising purposes, those of a transitory character, and those in which the text is not the most important part.

Newspapers are periodic publications intended for the general public and mainly designed to be a primary source of written information on current events connected with public affairs, international questions, politics, etc. A newspaper issued at least four times a week is considered to be a *daily newspaper*; those appearing three times a week or less frequently are considered to be *non-daily newspapers*.

Circulation figures show the average daily circulation. These figures include the number of copies (a) sold directly, (b) sold by subscription, and (c) mainly distributed free of charge both inside the country and abroad.

When interpreting these data, it should be noted that in some cases, definitions, classifications, and statistical methods applied by certain countries do not entirely conform to the standards recommended by UNESCO. For example, circulation data refer to the number of copies distributed as defined above. It appears, however, that some countries have reported the number of copies printed, which is usually higher than the distribution figure.

Footnotes

a.   Data refer to former Czechoslovakia.

b.   Data refer to former Yugoslavia.

c.   Data refer to West Germany.

d.   Data refer to the former Union of Soviet Socialist Republics.

1.   Data shown for 1990 refer to 1988.

2.   Data shown for 1990 refer to 1989.

Panel 15 - Cinema

Source. United Nations Educational, Scientific, and Cultural Organization (UNESCO), *Statistical Yearbook 1993*, Paris: UNESCO, 1993. Used with permission.

Notes. The statistics shown in these tables refer to fixed cinemas and mobile units regularly used for commercial exhibition of long films of 16 mm and over.

Cinema attendance is calculated from the number of tickets sold for all types of cinemas during a given year.

As a rule, figures refer only to commercial establishments. Exceptions may occur, however, in countries that include non-commercial units in their report of mobile units. Gross receipts are given in the national currency.

The term *fixed cinema* as used above refers to establishments possessing their own equipment and includes indoor cinemas (those with a permanent fixed roof over most of the seating accommodation), outdoor cinemas and drive-ins (establishments designed to enable the audience to watch a film while seated in an automobile). *Mobile units* are defined as projection units equipped and used to serve more than one site.

The capacity for fixed cinemas refers to the number of seats in the case of cinema halls and to the number of places for automobiles multiplied by a factor of 4 in the case of drive-ins.

Footnotes. A dash (-) stands for nil. NA stands for Not Available. An e. stands for estimated or provisional.

a.   Data refer to former Yugoslavia.

b.   Data refer to former Czechoslovakia.

c.   Data refer to the former Union of Soviet Socialist Republics.

1.   Receipts do not include taxes.

2.   Data on seating capacity refer to 35 mm cinemas only.

3.   Data refer to 15 non-commercial units attached to foreign cultural centers.

4.   Data on seating capacity and attendance refer only to 204 fixed cinemas and six drive-in cinemas.

Panel 16 - Scientific/Technical Forces

Source. United Nations Educational, Scientific, and Cultural Organization (UNESCO), *Statistical Yearbook 1993*, Paris: UNESCO, 1993. Used with permission.

Notes. These data present selected results of the world-wide data collection effort by UNESCO in the field of science and technology. Most of the data were obtained from replies to the annual statistical questionnaires on manpower and expenditure for research and experimental development sent to the Member States of UNESCO during recent years, completed or supplemented by data collected in the earlier surveys and from official reports and publications.

The definitions and concepts suggested for use in the *Statistical Questionnaire on Scientific Research and Experimental Development* (most recent documents UNESCO STCC/Q/9001 and UNESCO STS/Q/921) are based on the *Recommendation concerning the International Standardization of Statistics on Science and Technology* and can be found in the *Manual for Statistics on Science and Technological Activities* (document UNESCO ST-84/WS/12). They can also be found in previous editions of the *UNESCO Statistical Year-book*.

Abridged versions of the definitions set out in the above mentioned Recommendation are given below.

For an explanation of ISCED levels, please see notes to Panel 11 (above).

Type of personnel

The following three categories of scientific and technical personnel are defined according to the work they are engaged in and their qualifications:

(1) *Scientists and engineers* includes persons working in those capacities, i.e. as persons with scientific or technological training (usually completion of post secondary education) in a field of science who are engaged in professional work on R&D activities, administrators and other high-level personnel who direct the execution of R&D activities.

(2) *Technicians* include persons engaged in that capacity in R&D activities who have received vocational or technical training in any branch of knowledge or technology of a specified standard (usually at least three years after the first stage of second-level education).

(3) *Auxiliary personnel* includes persons whose work is *directly* associated with the performance of R&D activities, i.e. clerical, secretarial and administrative personnel, skilled, semi-skilled and unskilled workers in the various trades and all other auxiliary personnel. *Excludes* security, janitorial and maintenance personnel engaged in general house-keeping activities.

It should be noted that in general all personnel are considered for inclusion in the appropriate categories regardless of citizenship status or country of origin.

Scientific and technical manpower. An indication of the total numerical strength of qualified human resources is obtained either from the total stock or the number of economically active persons who possess the necessary qualifications to be scientists, engineers or technicians. Missing data on potential scientists and engineers have been estimated by UNESCO using the number of persons who have completed education at ISCED levels 6 and 7; for technicians, missing data are estimated using the number of persons who have completed education at ISCED level 5.

Total stock of qualified manpower (ST): The number of persons as described above, regardless of economic activity, age, sex, nationality or other characteristics present in the domestic territory of a country at a given reference date.

Number of economically active qualified manpower (EA): This group includes all persons of either sex, as specified above, who are engaged in, or actively seeking work in, any branch of the economy at a given reference date.

Footnotes. NA stands for Not Available. ** stands for included elsewhere in data. In 1991, due to a change of methodology

whereby persons with qualifications at ISCED level 3 have been excluded from the count of potential technicians, data relating to scientific and technical manpower for the following countries have been revised: Argentina, Australia, Austria, Brazil, Canada, Cyprus, Finland, Greece, Guatemala, Japan, Kenya, Mauritius, Nigeria, Peru, Qatar and the former Yugoslavia. The usual footnote for Bahrain, India, Israel and Nepal indicating the non-inclusion of persons having these qualifications from the total number of potential technicians has been deleted. For certain countries (namely: Brunei Darussalam, Bulgaria, Guyana, Ireland, Italy, Republic of Korea, Poland and Singapore) where the number of persons with qualifications at ISCED level 3 cannot be separated from the number of potential technicians previously reported, an appropriate footnote has been provided.

a. Data refer to former Yugoslavia.

b. Data refer to former Czechoslovakia.

c. Data refer to West Germany.

d. Data refer to the former Union of Soviet Socialist Republics.

e. Stands for values that are provisional or estimated.

1. 7,464 (Females—2,015) of the potential scientists and engineers in the first two rows and 7,066 (Females—2,313) of the technicians in the third and fourth rows are foreigners.

2. Refers to specialists in the national economy, i.e. persons having completed education at the third level for potential scientists and engineers and secondary specialized education for technicians.

3. 270 of the potential scientists and engineers in this figure are foreigners.

4. 1,495 of the potential scientists and engineers in and 919 of the technicians are foreigners.

5. Data include persons with qualifications at ISCED level 3.

6. Data refer to employed persons in the national economy.

7. Figures include persons with qualifications at ISCED level 3.

8. Data not included either for social sciences and humanities or for collective organizations.

9. Data are based on a 2.5% sample of the 1982 census returns.

10. Figure refers to persons having completed education at the third level.

11. Figure refers to those having completed post-secondary education.

12. Data for technicians are included with those for potential scientists and engineers.

13. Not including data for social sciences and humanities.

14. Data refer to the annual civilian labor force.

15. Data refer to persons in gainful employment.

16. 32,135 of the potential scientists and engineers and 5,673 of the technicians are foreigners.

17. 4,782 of the potential scientists and engineers and 2,461 of the technicians are foreigners.

18. 600 of the potential scientists and engineers are foreigners.

19. Data refer to persons aged 15 years and over with an education at the third level.

20. Data refer to persons in gainful employment aged 25 years and over with an education at the third level.

21. Data are based on a 1% sample.

22. Figure refers to persons having completed faculty and other higher education.

23. Figure refers to persons having completed vocational education.

24. Data refer to total number of persons engaged in scientific and technical activities.

25. Data include specialists having completed third level education.

26. Data not included for law, humanities and education.

27. Data refer to persons aged 25 years and over having completed education at the third level.

Panel 17 - R&D Expenditures

Source. United Nations Educational, Scientific, and Cultural Organization (UNESCO), *Statistical Yearbook 1993*, Paris: UNESCO, 1993. Used with permission.

Notes. In general, *R&D* is defined as any creative systematic activity undertaken in order to increase the stock of knowledge of man, culture, and society, and the use of this knowledge to devise new applications. It includes fundamental research (i.e. experimental or theoretical work undertaken with no immediate practical purpose in mind), applied research in such fields as agriculture, medicine, industrial chemistry, etc. (i.e. research directed primarily towards a special practical aim or objective), and experimental development work leading to new devices, products or processes.

Total domestic expenditure on R&D activities refers to all expenditure made for this purpose in the course of a reference year in institutions and installations established in the national territory, as well as installations physically situated abroad.

The total *expenditure for R&D* as defined above comprises *current expenditure*, including overheads, and *capital expenditure*.

Current expenditures usually include spending on labor, training, goods and services, and the like. Capital expenditures usually include spending on equipment and infrastructure.

Footnotes

a. Data refer to former Yugoslavia.

b. Data refer to former Czechoslovakia.

e. stands for provisional or estimated data.

1. Not including data for the productive sector (non-integrated R&D).

2. Not including data for the productive sector nor labor costs at the Ministry of Public Health.

3. Not including data for the general service sector.

4. Not including military and defense R&D.

5. Not including data for the productive sector.

6. Data relate only to 23 out of 26 national research institutes under the Federal ministry of Science and Technology.

7. Not including social sciences and humanities in the productive sector (integrated R&D).

8. Data relate to government funds only and do not include military and defense R&D.

9. Data refer to the R&D activities performed in public enterprises. Not including data for the higher education sector and foreign funds.

10. Data refer to the productive sector (integrated R&D) and the higher education sector only.

11. Data relate to the Scientific Research Council only.

12. Figures in millions.

13. Data refer to the central government only.

14. Not including data for law, humanities and education. Data do not include capital expenditure in the productive sector.

15. Not including either military and defense R&D nor private productive enterprises.

16. Data for the general service sector and for medical sciences in the higher education sector are also excluded.

17. Data refer to scientific and technological activities.

18. Data relate to government expenditure only.

19. Data relate to 2 research institutes only.

20. Data relate to the general service sector only.

21. Data refer to the civilian sector only.

22. Not including military and defense R&D nor social sciences and humanities.

23. Data refer to the Faculty of Science at the University of Lebanon only.

24. Data relate to R&D activities concentrated mainly in government- financed research establishments only; social sciences and humanities in the higher education and general service sectors are excluded. Not including military and defense R&D.

25. Not including foreign funds nor social sciences and humanities.

26. Not including data from communities and regions.

27. Of military R&D, only that part carried out in civil establishments is included.

28. Data in total expenditure figure include 1,680 million markkas in the higher education sector for which a distribution by type of expenditure is not available; this figure has been excluded from the percentage calculation.

29. Data refer to West Germany only. Data in Total figure include 644 million Deutsche marks for which a distribution by type of expenditure is not available; this figure has been excluded from the percentage calculation. Due to methodological changes, data are not strictly comparable with the previous years. Not including data for social sciences and humanities in the productive sector.

30. Data relate to the higher education sector only.

31. Data in Total figure include 1,946,916 thousand pesetas (disbursed by private non-profit organizations) for which a distribution by type of expenditure is not available; this figure has been excluded from the percentage calculation.

32. Not including data for social sciences and humanities in the productive and general service sectors.

33. Not including foreign and other funds.

34. Not including social sciences and humanities nor funds for R&D performed abroad.

35. Data relate to one research institute only.

36. Data refer only to 6 out of 11 research institutes.

37. Expenditure on Science from the national budget and other sources.

Panel 18 - U.S. Patents Issued

Source. *CASSIS (Classification and Search Support Information System) 1992.* [machine-readable datafiles]. Prepared by U.S. Office of Trademarks and Patents. Washington, DC: PTO, 1992.

Notes. These tables show the number of U.S. patents issued to residents of the given country in each year from 1990 through 1992. Countries have been included in this publication if any residents received patents in one or more of these years. Countries have *not* been included if there were no patents issued over that three year period, even if residents may have received patents before 1990.

Panel 19 - Organization of Government

Source. *The World Factbook 1993*, Central Intelligence Agency (CIA), Washington, DC: CIA, 1994.

Note. ICJ stands for International Court of Justice.

Panel 20 - Political Parties

Source. *The World Factbook 1993*, Central Intelligence Agency (CIA), Washington, DC: CIA, 1994.

Notes. Where available, political party representation is shown for the lower house of the legislative branch of government. The lower house was chosen in order to present as accurate a picture as possible of the electoral results of voting by the general public. The name of this legislative body is shown in the legend of the given table.

When election results were available, representation is shown as percent distribution of votes in the most recent election.

When election results were not available, percent distribution of seats in the lower house is shown by political party.

If there are no political parties or there is one-party rule, this information is provided in place of tabular data, with reasons for this situation when available.

Wherever possible, political party names have been represented in their English translations. Some names have been abbreviated or truncated to ensure fit in the panel.

Panel 21 - Government Budget

Sources

International Monetary Fund, *Government Finance Statistics Yearbook (Volume XVII)*, Washington, DC: IMF, 1993. Used with permission.

When data from the IMF were unavailable, data were obtained from:

The World Factbook 1993, Central Intelligence Agency (CIA), Washington, DC: CIA, 1994.

The following countries have CIA data:

Afghanistan	Albania	Algeria
Angola	Antigua and Barbuda	Armenia
Azerbaijan	The Bahamas	Bangladesh
Belgium	Benin	Bosnia and Herzegovina
Brunei	Burkina	Burma
Burundi	Cambodia	Cameroon
Cape Verde	Central African Republic	Chad
China	Colombia	Comoros
Congo	Cote d'Ivoire	Croatia
Cuba	Denmark	Djibouti
Dominica	Equatorial Guinea	France
Gabon	Georgia	Grenada
Guinea	Guyana	Haiti
Honduras	Hong Kong	Hungary
Iraq	Italy	Jamaica
Japan	Kazakhstan	Korea, North
Kyrgyzstan	Laos	Latvia
Lebanon	Libya	Liechtenstein
Lithuania	Macedonia	Marshall Islands
Mauritania	Micronesia, Federated States of	Moldova
Monaco	Mongolia	Mozambique
Nicaragua	Niger	Nigeria
Pakistan	Paraguay	Peru
Poland	Qatar	Russia
Rwanda	Saint Kitts and Nevis	Saint Lucia
San Marino	Sao Tome and Principe	Saudi Arabia
Senegal	Serbia and Montenegro	Seychelles
Sierra Leone	Slovakia	Slovenia
Solomon Islands	Somalia	South Africa
Sudan	Suriname	Taiwan
Tajikistan	Tanzania	Togo
Trinidad and Tobago	Turkmenistan	Uganda
Ukraine	Uzbekistan	Venezuela
Vietnam	Western Samoa	Zaire
Zimbabwe		

Notes to IMF Data. Information for the IMF data was obtained primarily by means of a detailed questionnaire distributed to government finance statistics correspondents, who are usually located in each country's respective ministry of finance or central bank.

Data relating to a fiscal year are presented within the calendar year containing the greatest number of months for that fiscal year. Fiscal years ending June 30 are presented within the same calendar year. For example, the fiscal year July 1, 1991 - June 30, 1992 is shown within the calendar year 1992.

Government is defined as all units that implement public policy by providing primarily nonmarket services and transferring income; these units are financed mainly by compulsory levies on other sectors.

Central government includes all units representing the territorial jurisdiction of the central authority throughout a country.

Expenditures include all nonrepayable payments by government, including both capital and current expenditures and regardless of whether goods or services were received for such expenditures.

Panel 22 - Military Expenditures

Sources

U.S. Arms Control and Disarmament Agency, *World Military Expenditures and Arms Transfers 1990*, Washington, DC: USGPO, November 1991.

When data from the ACDA were unavailable, a summary of defense information was used from:

The World Factbook 1993, Central Intelligence Agency (CIA), Washington, DC: CIA, 1994.

The following countries have CIA data:

Antigua and Barbuda	Armenia	Azerbaijan
The Bahamas	Belarus	Belize
Bhutan	Bosnia and Herzegovina	Brunei
Comoros	Croatia	Djibouti
Dominica	Estonia	Georgia
Grenada	Hong Kong	Kazakhstan
Kyrgyzstan	Latvia	Liechtenstein
Lithuania	Macedonia	Maldives
Marshall Islands	Micronesia, Federated States of	Moldova
Monaco	Namibia	Papua New Guinea
Saint Kitts and Nevis	Saint Lucia	Saint Vincent and the Grenadines
San Marino	Seychelles	Slovakia
Slovenia	Solomon Islands	Tajikistan
Turkmenistan	Ukraine	Uzbekistan
Vanuatu	Western Samoa	

Notes to ACDA Data. Most data shown are for calendar years. For some countries, however, expenditure data are available only for fiscal years which diverge from calendar years. In such cases, the fiscal year which contains the most months of a given calendar year is assigned to that year; e.g., data for fiscal year April 1977 through March 1978 would be shown under 1977. Fiscal years ending on June 30 are normally listed in the calendar year in which they end.

Military Expenditures:

For NATO countries, military expenditures are from NATO publications and are based on the NATO definition. In this definition, (a) civilian-type expenditures of the defense ministry are excluded and military-type expenditures of other ministries are included; (b) grant military assistance is included in the expenditures of the donor country; and (c) purchases of military equipment for credit are included at the time the debt is incurred, not at the time of payment.

For non-communist countries, data are generally the expenditures of the ministry of defense. When these are known to include the costs of internal security, an attempt is made to remove these expenditures. In view of the discontinuance of such data collection by the Agency for International Development, a major source of data for these countries in the past, a number of other sources are consulted, including the *Government Finance Statistics Yearbook*, issued by the International

Monetary Fund, the publications and files of other U.S. government agencies, standardized reporting of countries to the United Nations, and other international sources.

It should be recognized by users of the statistical tables that the military expenditure data are of uneven accuracy and completeness. For example, there are indications to believe that the military expenditures reported by some countries consist mainly or entirely of recurring or operating expenditures and omit all or most capital expenditures, including arms purchases, e.g, Algeria, Chile, Cuba, Ecuador, Egypt, Honduras, Iran, Iraq, Libya, and Syria.

In some of these cases (as noted in subsequent footnotes), it is believed that a better estimate of total military expenditures is obtained by adding to nominal military expenditures the value of arms imports. This method may over- or underestimate the actual expenditures in a given year due to the fact that payment for arms may not coincide in time with deliveries, which the data in lines 13 through 22 reflect. Also, in some cases arms acquisitions may be financed by, or consist of grants from other countries.

In these tables, the symbol "e" denotes rough estimates such as those described above and others made on the basis of partial or uncertain data.

For countries that have major clandestine nuclear or other military weapons development programs, such as Iraq, estimation of military expenditures is extremely difficult and especially subject to errors of underestimation.

Particular problems arise in estimating the military expenditures of communist countries due to the exceptional scarcity and ambiguity of released information. Data on the military expenditures of the [former] Soviet Union are based on Central Intelligence Agency (CIA) estimates....Estimates for the most recent year are based on the change in the index of CIA-estimated military expenditures in ruble terms, as reported in the Joint Economic Committee of Congress series, *Allocation of Resources in the Soviet Union and China*.

For Warsaw Pact countries other than the Soviet Union, the estimates of military expenditures....refer only to the officially announced state budget expenditures on national defense. These figures understate total military expenditures in view of defense outlays by non-defense agencies of the central government, local governments, and economic enterprises. Possible subsidization of military procurement may also cause understatement. The dollar estimates were derived by calculating full U.S. average rates for officers and for lower ranks. After subtraction of pay and allowances, the remainder of the official defense budgets in national currencies was converted into dollars at overall rates based on comparisons of the various countries' GNPs expressed in dollars and in national currencies. The rates are based on the purchasing power parities estimated by the International Comparison Project of the United Nations, including their latest (Phase V) versions. These conversion rates are not as specific as might be desired and, when the problems mentioned above are taken into

account, the resulting estimates must be considered subject to limitations. Another omission in all Warsaw Pact data is that the nonpersonnel component of military assistance is not covered.

Data used here for China are based on U.S. Government estimates of the yuan costs of Chinese forces, weapons, programs, and activities. Costs in yuan are here converted to dollars using the same estimated coversion rate as used for GNP (see below). Due to the exceptional difficulties in both estimating yuan costs and converting them to dollars, comparisons of Chinese military spending with other data should be treated as having a wide margin of error.

Other sources used include the *Government Finance Statistics Yearbook* issued by the International Monetary Fund, the *SIPRI Yearbook: World Armaments and Disarmament* issued by the Stockholm International Institute for Strategic Studies, and the *World Factbook*, produced annually by the Central Intelligence Agency.

Gross National Product (GNP):

GNP represents the total output of goods and services produced by residents of a country and valued at market prices. The source of the GNP data for most non-communist countries is the International Bank for Reconstruction and Development (World Bank).

For a number of countries whose GNP is dominated by oil exports (Bahrain, Kuwait, Libya, Oman, Qatar, Saudi Arabia, and the United Arab Emirates), the World Bank's estimate of deflated (or constant price) GNP in domestic currency tends to underestimate increases in the monetary value of oil exports, and thus, of GNP, resulting from oil price increases. These World Bank estimates are designed to measure real (or physical) product. An alternative estimate of constant-price GNP was therefore obtained using the implicit price deflator [the ratio of GNP in current prices to GNP in constant prices] for U.S. GNP (for lack of good national deflator). This was considered appropriate because a large share of the GNP of these countries is realized in US dollars.

GNP estimates of the Soviet Union are by the CIA, as published in its *Handbook of Economic Statistics 1990* and updated. GNP data for other Warsaw Pact countries are in updated and substantially revised version of estimates in "East European Military Expenditures, 1965-1978" by Thad P. Alton and others.

GNP estimates for China are based on World Bank estimates in yuan. These are in line with estimates of GDP in Western accounting terms made by Chinese authorities. Conversion to dollars is a highly uncertain matter, however. A survey of various recent estimates yuan purchasing power parities reports a kind of benchmark estimate for an overall yuan/dollar purchasing power parity, which serves as the basis for the conversion rate employed here. This estimate is published in a Chinese source and is based on actual Chinese trade statistics. It serves as the basis for the yuan conversion rate employed since *WMEAT 1986*.

GNP estimates for a few non-communist countries are from the CIA's *Handbook of Economic Statistics* cited above. Estimates for other communist countries are rough approximations.

Military-Expenditures-to-GNP Ratio:

It should be noted that the meaning of the ratio of military expenditures to GNP, shown in these tables, differs somewhat between most communist countries; both military expenditures and GNP are converted from the national currency and reflects national relative prices. For communist countries, however, military expenditures and GNP are converted differently. Soviet military expenditures, as already noted, are estimated in a way designed to show the cost of the Soviet armed forces in U.S. prices, e.g., as if purchased in this country. On the other hand, the Soviet GNP estimates used here are designed to show average relative size when both U.S. and Soviet GNP are valued and compared at both dollar and ruble prices. The Soviet ratio of military expenditures to GNP in ruble terms, the preferred method of comparison, is estimated to have been 15-18 percent in recent years.

For Eastern European countries, the ratios of military expenditures to GNP in dollars are about twice the ratios that would obtain in domestic currencies. However, since official military budgets in these countries probably substantially understate their actual military expenditures, the larger ratios based on dollar estimates are believed to be the better approximations of the actual ratios.

Central Government Expenditures (CGE):

These expenditures include current and capital (developmental) expenditures plus net lending to government enterprises, by central (or federal) governments. A major source is the International Monetary Fund's *Government Finance Statistics Yearbook*. The category used here is "Total Expenditures and Lending minus Repayment, Consolidated Central Government."

Other sources for these data are the International Monetary Fund monthly, *International Finance Statistics*; OECD, *Economic Surveys*; and CIA, *World Factbook (annual)*. Data for Warsaw Pact countries are from national publications and are supplied by Thad P. Alton and others. For all Warsaw Pact countries and China, conversion to dollars is at the implicit rates used for calculating dollar estimates of GNP.

For all countries, with the same exceptions as noted above for the military-expenditures-to-GNP ratio, military expenditures and central government expenditures are converted to dollars at the same rate; the ratio of the two variables in dollars thus remains the same as in national currency.

It should be noted that for the Soviet Union, China, Iran, Jordan, and possibly others, the ratio of military expenditures to central government expenditures may be overstated, inasmuch as the same estimate for military expenditures is obtained at least in part independently of nominal budget or government expenditure data, and it is possible that all estimated military expenditures do not pass through the nominal central government budget.

Arms Transfers:

Arms transfers (arms imports and exports) represent international transfer (under terms of grant, credit, barter or cash) of military equipment, usually referred to as "conventional," including weapons of war, parts thereof, ammunition, support equipment, and other commodities designed for military use. Among the items included are tactical guided missiles and rockets, military aircraft, naval vessels, armored and nonarmored military vehicles, communications and electronic equipment, artillery, infantry weapons, small arms, ammunition, other ordnance, parachutes, and uniforms. Dual use equipment, which can have application in both military and civilian sectors, is included when its primary mission is identified as military. The building of defense production facilities and licensing fees paid as royalties for the production of military equipment are included when they are contained in military transfer agreements. There have been no international transfers of purely strategic weaponry. Excluded are foodstuffs, medical equipment, petroleum products, and other supplies. Military services such as construction, training, and technical support are not included for the United States, whose services consist mainly of construction (primarily for Saudi Arabia). Data on military services of other countries, which are normally of a much smaller magnitude, are included when available.

The statistics contained in these tables are estimates of the value of goods actually delivered during the reference year, in contrast both to the value of programs, agreements, contracts, or orders which may result in future deliveries, and to payments made during the period....Both deliveries and agreements data represent arms transfers to governments and do not include the value of arms obtained by subnational groups.

Figures for U.S. arms exports are for fiscal years and are obtained from official data on arms transfers compiled by the U.S. Departments of Defense and State. They include commercial deliveries of items on the U.S. Munitions Control List, some of which may be intended for civilian rather than military use.

The arms import figures in these tables consist of data obtained from the Department of Commerce, Bureau of Economic Analysis (BEA), including (a) imports of military-type goods, as compiled by the Bureau of the Census, and (b) Department of Defense (DOD) direct expenditures abroad for major equipment, as compiled from DOD data from BEA. The goods in (a) include: complete military aircraft, all types; engines and turbines for military aircraft; military trucks, armored vehicles, etc.; military (naval) ships and boats; tanks, artillery, missiles, guns, and ammunition; military apparel and footwear; and other military goods, equipment, and parts. Data on countries other than the United States are estimates by U.S. Government sources. Arms transfer data for the Soviet Union and other communist countries are approximations based on limited information.

It should be noted that the arms transfer estimates for the most recent year, and to a lesser extent for several preceding years, tend to be understated. This applies to both foreign and U.S. arms exports. In the former case, information on transfers, which comes from a variety of sources, is acquired and processed with a considerable time lag. In the U.S. case, commercial transfer licenses are now valid for two years rather than one, causing a delay in the reporting of deliveries made on them to statistical agencies.

Close comparisons between the estimated values shown for arms transfers and for GNP and military expenditures are not warranted. Frequently, weapons prices do not reflect true production costs. Furthermore, much of the international arms trade involves offset or barter arrangements, multiyear loans, discounted prices, third-party payments, and partial debt forgiveness. Acquisition of armaments thus may not impose burden on an economy, in the same or in other years, that is implied by the estimated equivalent U.S. dollar value of the shipment. Therefore, the value of arms imports should be compared to other categories of data with care.

Total Imports and Exports:

The values for imports and exports cover merchandise transactions. Those for non-communist countries come from *International Financial Statistics*, published by the IMF. The Communist trade figures are from the CIA *Handbook of Economic Statistics*, 1989 edition.

Footnotes. NA stands for not available. A zero (0) stands for nil or negligible. The letter e is used to indicate data that are estimated or based on partial or uncertain data.

1. Data for Germany refer to West Germany only; data for Bosnia and Hercegovina, Croatia, Macedonia, Serbia and Montenegro, and Slovenia all refer to former Yugoslavia; data for the Czech Republic and Slovakia all refer to former Czechoslovakia. Data for Russia refer to the former Soviet Union.

2. Estimated by adding arms imports to data on military expenditures, which are believed to exclude arms purchases. However, it should be noted that the value of arms deliveries in a given year (converted at current exchange rates) may differ significantly from actual expenditures on arms imports that year.

3. This ratio is calculated from the two variables as expressed in dollar terms. Since in this case the two variables are converted to, or estimated in, dollars in differing ways the ratio in dollars differs from what it would be in national currency terms.

4. This series or entry probably omits a major share of total military expenditures, probably including most arms acquisitions. These tables show estimated annual arms imports; it should be kept in mind however, that data in these tables represent the estimated value of arms de-

livered in a given year, not actual expenditures on those arms.

5. Some part of estimated total military expenditures may not be included in announced central budget expenditures. The ratio of ME to CGE therefore may be somewhat overstated.

6. To avoid the appearance of excessive accuracy, arms transfer data have been independently rounded, with greater severity for large numbers. Because of this rounding and the fact that they are obtained from different sources, world arms exports do not equal world arms imports.

7. Total imports and exports usually are as represented by individual countries and the extent to which arms transfers are included is often uncertain. Imports are reported "Cif" (including cost of shipping, insurance, and freight) and exports are reported "fob" (excluding these costs). For these reasons and because of divergent sources, world totals for imports and exports are not equal.

8. Because some countries exclude arms imports or exports from their trade statistics and their "total" imports and exports are therefore understated and because arms transfers may be estimated independently of trade data, the resulting ratios of arms to total imports or exports may be overstated and may even exceed 100 percent.

9. In order to reduce distortions in grouped data trends caused by data gaps for individual countries and years (shown as "NA"), the totals for the world, regions, and organizations include rough approximations for the gaps.

10. Includes transfers to NATO agencies as such, which are not attributable to individual recipient countries.

11. The estimated dollar values of the Soviet Union's arms exports in this and the previous two editions of this report are revised upward substantially from those in previous editions.

12. Includes some equipment purchased by the U.S. Army Corps of Engineers from indeterminable supplier countries for use in construction projects in Saudi Arabia and recorded in U.S. accounts as imports.

13. The data series for US arms imports is here substantially revised upward from previous editions.

Panel 23 - Crime

Source. International Criminal Police Organization (INTERPOL), *International Crime Statistics 1989-1990*, Lyons: INTERPOL. Used with permission.

Notes. Statistics are based on data collected by the police in ICPO-Interpol member countries and are therefore *police statistics* and not *judicial statistics*.

The form adopted by resolution No. AGN/45/RES/6 at the 1976 session of the ICPO-Interpol General Assembly was used to collect information. A copy of the form is presented at the end of the introduction of the original source.

The information given is in no way intended for use as a basis for comparisons between different countries.

These statistics cannot take account of the differences that exist between the legal definitions of punishable offenses in various countries, of the different methods of calculation, or of any changes which may have occurred in the countries concerned during the reference period. All these factors obviously have repercussions on the figures supplied.

Police statistics reflect the crimes reported to or detected by the police and therefore cover only part of the total number of offenses actually committed. Moreover, the volume of unreported crimes depends to some extent on action taken by the police, and may therefore vary from one point in time to another and from one country to another.

Consequently, the figures given in these statistics must be interpreted with caution.

Footnotes

1. *West Germany.* Juveniles are defined as those between the ages of 14 and 17 years. The percentage of offenses by persons younger than age 14 was 4.3%.

2. *Argentina.* Juveniles are defined as those between the ages of 0 and 18 years.

3. *Austria.* Juveniles are defined as those between the ages of 14 and 18 years.

4. *Bahamas.* Juveniles are defined as those between the ages of 7 and 17 years.

5. *Bahrain.* Juveniles are defined as those between the ages of 3 and 15 years.

6. *Bangladesh.* Juveniles are defines as those between the ages of 12 and 17 years.

7. *Barbados.* Juveniles are defined as those between the ages of 7 and 16 years.

8. *Belgium.* Juveniles are defined as those between the ages of 0 and 18 years.

9. *Botswana.* Juveniles are defined as those between the ages of 8 and 18 years.

10. *Brunei.* Juveniles are defined as those between the ages of 7 and 18 years.

11. *Burundi.* Juveniles are defined as those between the ages of 1 and 17 years.

12. *Canada.* Juveniles are defined as those between the ages of 12 and 17 years.

13. *Central African Republic.* Juveniles are defined as those between the ages of 0 and 17 years.

14. *Chile.* Juveniles are defined as those between the ages of 0 and 17 years.

15. *China.* Juveniles are defined as those between the ages of 14 and 25 years.

16. *Cyprus.* Juveniles are defined as those between the ages of 7 and 15 years.

17. *Congo.* Juveniles are defined as those between the ages of 0 and 17 years.

18. *South Korea.* Juveniles are defined as those between the ages of 14 and 19 years.

19. *Djibouti.* Juveniles are defined as those between the ages of 12 and 18 years.

20. *Egypt.* Juveniles are defined as those between the ages of 0 and 18 years.

21. *Spain.* Juveniles are defined as those between the ages of 0 and 16 years.

22. *United States.* Juveniles are defined as those between the ages of 0 and 18 years.

23. *Finland.* Juveniles are defined as those between the ages of 0 and 20 years.

24. *Fiji.* Juveniles are defined as those between the ages of 0 and 17 years.

25. *France.* Juveniles are defined as those between the ages of 13 and 18 years.

26. *Gabon.* Juveniles are defined as those between the ages of 16 and 18 years.

27. *Greece.* Juveniles are defined as those between the ages of 7 and 17 years.

28. *Guinea.* Juveniles are defined as those between the ages of 1 and 15 years.

29. *Hong Kong.* Juveniles are defined as those between the ages of 7 and 15 years.

30. *Israel.* Juveniles are defined as those between the ages of 12 and 18 years.

31. *Italy.* Juveniles are defined as those between the ages of 14 and 18 years.

32. *Japan*. Juveniles are defined as those between the ages of 14 and 19 years.

33. *Kenya*. Juveniles are defined as those between the ages of 13 and 17 years.

34. *Libya*. Juveniles are defined as those between the ages of 14 and 18 years.

35. *Luxembourg*. Juveniles are defined as those between the ages of 0 and 18 years.

36. *Malawi*. Juveniles are defined as those between the ages of 7 and 17 years.

37. *Maldives*. Juveniles are defined as those between the ages of 0 and 16 years.

38. *Mauritius*. Juveniles are defined as those between the ages of 0 and 17 years.

39. *Monaco*. Juveniles are defined as those between the ages of 0 and 18 years.

40. *Nepal*. Juveniles are defined as those between the ages of 8 and 15 years.

41. *Papua New Guinea*. Juveniles are defined as those between the ages of 8 and 16 years.

42. *Netherlands*. Juveniles are defined as those between the ages of 12 and 17 years.

43. *Peru*. Juveniles are defined as those between the ages of 1 and 17 years.

44. *Qatar*. Juveniles are defined as those between the ages of 7 and 18 years.

45. *Romania*. Juveniles are defined as those between the ages of 0 and 18 years.

46. *Rwanda*. Juveniles are defined as those between the ages of 0 and 14 years.

47. *St. Vincent and the Grenadines*. Juveniles are defined as those between the ages of 8 and 15 years.

48. *Senegal*. Juveniles are defined as those between the ages of 0 and 18 years.

49. *Singapore*. Juveniles are defined as those between the ages of 7 and 15 years.

50. *Sri Lanka*. Juveniles are defined as those between the ages of 10 and 16 years.

51. *Sweden*. Juveniles are defined as those between the ages of 15 and 17 years.

52. *Switzerland*. Juveniles are defined as those between the ages of 0 and 20 years.

53. *Swaziland*. Juveniles are defined as those between the ages of 10 and 18 years.

54. *Syria*. Juveniles are defined as those between the ages of 7 and 18 years.

55. *Czech Republic and Slovakia*. Juveniles are defined as those between the ages of 15 and 18 years. Data refer to former Czechoslovakia.

56. *Trinidad and Tobago*. Juveniles refer to those between the ages of 14 and 16 years.

57. *Russia*. Juveniles refer to those between the ages of 0 and 18 years. Data refer to the former Union of Soviet Socialist Republics.

Panel 24 - Human Rights

Source. U.S. Department of State, *Country Reports on Human Rights Practices for 1992*, Report to the Joint Committee on Foreign Affairs, Washington, DC: USGPO, February 1993.

Notes. The following nations are non-ILO members (ILO stands for International Labor Organization): Albania, Brunei, Gambia, Kazakhstan, Liechtenstein, Maldives, Marshall Islands, Micronesia, Monaco, North Korea, Oman, South Africa, St. Kitts and Nevis, St. Vincent and Grenadines, Taiwan, Tajikistan, Turkmenistan, Vanuatu, Vietnam, and Western Samoa.

Human rights conventions, shown in tables as initialisms, are as follows:

ACHR   American Convention on Human Rights

AFL   Convention Concerning the Abolition of Forced Labor

APROBC   Convention Concerning the Application of the Principles of the Right to Organize and Bargain Collectively

ASST   Supplementary Convention on the Abolition of Slavery, the Slave Trade, and Institutions and Practices Similar to Slavery

CPR   International Covenant on Civil and Political Rights

EAFDAW   Convention on the Elimination of All Forms of Discrimination Against Women

EAFRD   International Convention on the Elimination of All Forms of Racial Discrimination

ESCR   Internation Covenant on Economic, Social, and Cultural Rights

FAPRO   Convention Concerning Freedom of Association and Protection of the Right to Organize

FL	Convention Concerning Forced Labor
MAAE	Convention Concerning the Minimum Age for Admission to Employment
PCPTW	Geneva Convention Relative to the Protection of Civilian Persons in Time of War
PHRFF	European Convention for the Protection of Human Rights and Fundamental Freedoms
PPCG	Convention on the Prevention and Punishment of the Crime of Genocide
PRW	Convention on the Political Rights of Women
PVIAC	Protocol Additional to the Geneva Conventions and Relating to the Protection of Victims of International Armed Conflicts
PVNAC	Protocol Additional to the Geneva Conventions and Relating to the Protection of Victims of Non-International Armed Conflicts
RC	Convention on the Rights of the Child
SR	Protocol Relating to the Status of Refugees
SSTS	Convention to Suppress Slavery Trade and Slavery
STPEP	Convention for the Suppression of the Traffic in Persons and of the Exploitation and Prostitution of Others
TCIDTP	Convention Against Torture and Other Cruel, Inhuman or Degrading Treatment or Punishment
TPW	Geneva Convention Relative to the Protection of Prisoners of War

Footnotes

1. Based on general declaration concerning treaty obligations prior to independence.

2. Party to 1926 convention only.

Panel 25 - Total Labor Force

Source. *The World Factbook 1993*, Central Intelligence Agency (CIA), Washington, DC: CIA, 1994.

Notes. These data show the number of persons in the labor force. The date for which these data are shown can be found in Panel 26: *Labor Force by Occupation* (below).

Panel 26 - Labor Force by Occupation

Source. *The World Factbook 1993*, Central Intelligence Agency (CIA), Washington, DC: CIA, 1994.

Notes. These data show percentage of the labor force engaged in each industry.

Panel 27 - Unemployment Rate

Source. *The World Factbook 1993*, Central Intelligence Agency (CIA), Washington, DC: CIA, 1994.

Notes. These data show the rate of unemployment in percent for the most recent available year (shown in parentheses).

Panel 28 - Energy Production

Sources

U.S. Department of Energy, Energy Information Administration, Office of Energy Markets and End Use, *International Energy Annual 1991*, Washington, DC: DOE, December 1992.

When information was unavailable from the DOE, data summarizing energy resources were obtained from:

The World Factbook 1993, Central Intelligence Agency (CIA), Washington, DC: CIA, 1994.

The following countries have CIA data:

Afghanistan	Antigua and Barbuda	Armenia
Azerbaijan	The Bahamas	Bangladesh
Barbados	Belarus	Belize
Benin	Bhutan	Bolivia
Botswana	Burkina	Burma
Burundi	Cambodia	Cameroon
Cape Verde	Central African Republic	Chad
Comoros	Congo	Costa Rica
Cote d'Ivoire	Croatia	Cyprus
Djibouti	Dominica	Dominican Republic
El Salvador	Equatorial Guinea	Estonia
Ethiopia	Fiji	The Gambia
Georgia	Ghana	Grenada
Guatemala	Guinea	Guinea-Bissau
Guyana	Haiti	Honduras
Hong Kong	Iceland	Israel
Jamaica	Jordan	Kazakhstan
Korea, North	Kyrgyzstan	Laos
Latvia	Lebanon	Lesotho
Liberia	Liechtenstein	Lithuania
Macedonia	Madagascar	Malawi
Maldives	Mali	Malta
Marshall Islands	Mauritania	Mauritius
Micronesia, Federated States of	Moldova	Monaco
Mongolia	Mozambique	Namibia
Nepal	Nicaragua	Niger
Papua New Guinea	Paraguay	Rwanda
Saint Kitts and Nevis	Saint Lucia	Saint Vincent and the Grenadines
San Marino	Sao Tome and Principe	Senegal
Seychelles	Sierra Leone	Singapore
Slovenia	Solomon Islands	Somalia
Sri Lanka	Sudan	Suriname
Swaziland	Tajikistan	Tanzania
Togo	Turkmenistan	Uganda
Ukraine	Uruguay	Uzbekistan
Vanuatu	Vietnam	Western Samoa
Yemen	Zaire	Zambia
Zimbabwe		

Notes for DOE Data. These data were extracted from the Executive Summary section of the *International Energy Annual 1991*. Data in these tables have been derived from published sources and U.S. embassy personnel in foreign posts. When official national statistical reports were not available, typical sources used by the Energy Information Administration are: the United Nations, the World Bank, the International Energy Agency, industry reports, academic studies, trade publications, and other sources.

Data for primary energy production worldwide are as follows:

Crude oil—World production of crude oil rose from 53.5 million barrels per day in 1982 to 60.2 million barrels per day in 1991. Total energy produced worldwide in 1991 was 128.79 quadrillion Btu.

Natural gas plant liquids—World production of natural gas plant liquids rose from 3.6 million barrels per day in 1982 to 5.0 million barrels per day in 1991. Total energy produced worldwide in 1991 was 7.40 quadrillion Btu.

Dry natural gas—World production of dry natural gas rose from 54.1 trillion cubic feet in 1982 to 75.1 trillion cubic feet in 1991. Total energy produced worldwide in 1991 was 74.51 quadrillion Btu.

Coal—World production of coal rose from 4.3 billion short tons in 1982 to 5.1 billion short tons in 1991. Total energy produced worldwide in 1991 was 92.11 quadrillion Btu.

Hydroelectric power—World generation of hydroelectric power rose from 1.8 trillion kilowatthours in 1982 to 2.1 trillion kilowatthours in 1991. Total energy produced worldwide in 1991 was 22.29 quadrillion Btu.

Nuclear electric power—World generation of nuclear power rose from 0.9 trillion kilowatthours in 1982 to 2.0 trillion kilowatthours in 1991. Total energy produced worldwide in 1991 was 21.23 quadrillion Btu.

Footnotes. Btu stands for British thermal units. (s) Denotes less than 5 trillion Btu.

a.   Data refer to former Yugoslavia.

b.   Data refer to former Czechoslovakia.

1.   Crude oil includes condensate.

2.   Coal includes anthracite, subanthracite, subbituminous, bituminous, lignite, and brown coal.

3.   Generation data consist of both utility and nonutility sources. Net generation excludes energy consumed by the generating unit.

4.   Includes petroleum processed from Athabasca Tar Sands.

Panel 29 - Energy Consumption

Sources

U.S. Department of Energy, Energy Information Administration, Office of Energy Markets and End Use, *International Energy Annual 1991*, Washington, DC: DOE, December 1992.

When information was unavailable from the DOE, data summarizing energy resources were obtained from:

The World Factbook 1993, Central Intelligence Agency (CIA), Washington, DC: CIA, 1994.

For a list of the countries for which CIA data were used, please see Panel 28 above.

Notes for DOE Data. These data were extracted from the Executive Summary section of the *International Energy Annual 1991*. Data in these tables have been derived from published sources and U.S. embassy personnel in foreign posts. When official national statistical reports were not available, typical sources used by the Energy Information Administration are: the United Nations, the World Bank, the International Energy Agency, industry reports, academic studies, trade publications, and other sources.

Data for primary energy consumption worldwide are as follows:

Petroleum—World consumption of petroleum was 136.17 quadrillion Btu in 1991.

Dry natural gas—World consumption of dry natural gas was 75.30 qaudrillion Btu in 1991.

Coal—World consumption of coal was 91.85 quadrillion Btu in 1991.

Hydroelectric power—World net consumption of hydroelectric power was 22.29 quadrillion Btu in 1991.

Nuclear electric power—World net nuclear power consumption was 21.23 quadrillion Btu in 1991.

Footnotes. (s) Denotes less than 5 trillion Btu.

a.   Data refer to former Yugoslavia.

b.   Data refer to former Czechoslovakia.

1.   Crude oil includes condensate.

2.   Coal includes anthracite, subanthracite, subbituminous, bituminous, lignite, and brown coal.

3.   Generation data consist of both utility and nonutility sources. Net generation excludes energy consumed by the generating unit.

4.   Includes petroleum processed from Athabasca Tar Sands.

Panel 30 - Telecommunications

Source. *The World Factbook 1993*, Central Intelligence Agency (CIA), Washington, DC: CIA, 1994.

Notes. Frequently used abbreviations in this panel are:

ARABSAT   Arab Satellite Communications Organization.

EUTELSAT European Telecommunications Satellite Organization.

INMARSAT International Maritime Satellite Organization.

INTELSAT Internationl Telecommunications Satellite Organization.

MARECS   Maritime Communications Sattelite.

MARISAT   Maritime Satellite Organization.

SEACOM   Southeast Asia Communications.

Panel 31 - Transportation

Source. *The World Factbook 1993*, Central Intelligence Agency (CIA), Washington, DC: CIA, 1994.

Notes. GRT stands for gross register tons. DWT stands for deadweight tons. Km stands for kilometers.

Panel 32 - Top Agricultural Products

Sources

U.S. Department of Agriculture, Foreign Agriculture Service, *Foreign Agriculture 1992*, Washington, DC: USDA, December 1992.

When data were not available from the USDA, data were used from:

The World Factbook 1993, Central Intelligence Agency (CIA), Washington, DC: CIA, 1994.

These data are in text fromat. The following countries have CIA data:

Afghanistan	Angola	Antigua and Barbuda
The Bahamas	Bahrain	Barbados
Belgium	Belize	Benin
Bhutan	Bosnia and Herzegovina	Botswana
Brunei	Burkina	Burundi
Cambodia	Cameroon	Cape Verde
Central African Republic	Chad	Comoros
Congo	Croatia	Cuba
Cyprus	Djibouti	Dominica
Equatorial Guinea	Estonia	Ethiopia
Fiji	Gabon	The Gambia
Georgia	Grenada	Guinea
Guinea-Bissau	Guyana	Haiti
Hong Kong	Iceland	Iran
Iraq	Jordan	Korea, North
Kuwait	Laos	Latvia
Lebanon	Lesotho	Liberia
Libya	Liechtenstein	Lithuania
Luxembourg	Macedonia	Madagascar
Malawi	Maldives	Mali
Malta	Marshall Islands	Mauritania
Mauritius	Micronesia, Federated States of	Monaco
Mongolia	Mozambique	Namibia
Nepal	Niger	Oman
Papua New Guinea	Qatar	Rwanda
Saint Kitts and Nevis	Saint Lucia	Saint Vincent and the Grenadines
San Marino	Sao Tome and Principe	Serbia and Montenegro
Seychelles	Sierra Leone	Slovakia
Slovenia	Solomon Islands	Somalia
Sudan	Suriname	Swaziland
Tanzania	Togo	Trinidad and Tobago
Uganda	United States	Vanuatu
Vietnam	Western Samoa	Yemen
Zaire	Zambia	

Notes. *Foreign Agriculture 1992* was produced by the Foreign Agriculture Service's (FAS) Information Division. Country profiles were prepared by FAS agricultural counselors, attaches, and specialists serving at embassies and trade offices around the world. The publication also drew on data and expertise provided by the FAS headquarters staff in Washington, DC, the Economic Research Service, and the World Agricultural Outlook Board of the U.S. Department of Agriculture, as well as the World Bank.

Footnotes

1.   July-June crop year.

2.   Denotes year harvested.

3.   Preliminary data.

4.   Estimate.

5.   Data for marketing years 1989/90 and 1990/91; Oct.-Sept. for coffee and soybeans; and Sept.-Aug. for sugar.

6.   Crop years are Jan.-Dec. for bananas and beef; Oct.-Sept. for coffee and sugar; July-June for rice; and Aug.-July for corn.

7.   Years are Jan.-Dec. for all commodities except cotton, coffee, and cocoa, which are Oct.-Sept.

8.   Data refer to former Yugoslavia.

9.   Crop years vary by commodity.

10. Production years are July-June for corn, rice, and sorghum; Oct.- Sept. for coffee; Aug.-July for cotton; Nov.-Oct. for sugar: and Jan.-Dec. for tobacco.

11. Crop production data for 1991 are based on planting intentions.

12. Marketing years vary by commodity. July-June for barely, peaches and nectarines, sugar beets, tomatoes and wheat; Aug.-July for corn and cotton; and Sept.-Aug. for oranges. Production period for alfalfa was not listed in the original source.

13. Production years are: July-June for corn, rubber, and sesame seed; Oct.-Sept. for coffee; Nov.-Oct. for sugar; and Dec.-Nov. for wheat.

14. Production years are Jan.-Dec. for bananas and palm oil; Oct.-Sept. for coffee; July-June for corn; Jan.-Dec. for palm oil; Jan.-Dec. for sawnwood; Sept.-Aug. for sugarcane.

15. Crop years are Oct.-Sept.

16. Includes oranges, lemons, and tangerines.

17. Crop years for coffee and citrus are June-July.

18. Crop years are Oct.-Sept. for barley, mikan oranges, raw sugar, and soybeans; Nov.-Oct. for rice; and July-June for wheat.

19. Crop year is Oct.-Sept. for coffee; July-June for corn, rice, wheat, and pyrethrum.

20. Calendar years 1990 and 1991.

21. Calendar year, except Oct.-Sept. for cocoa, palm oil, and palm kernel oil.

22. Peninsular Malaysia only.

23. Crop years are Oct.-Sept. for coffee; Aug.-July for rice, sorghum, and corn; and Jan.-Dec. for bananas.

24. Unofficial estimates.

25. Commercial production.

26. Marketing years are Aug.-July for cotton; Oct.-Sept. for rice, corn, sugarcane and cottonseed; May-April for wheat.

27. Preliminary estimate.

28. Calendar years except July-June for corn and rice; Sept.-Aug. for sugar; and Oct.-Sept. for copra.

29. Estimates except wheat.

30. Refined sugar.

31. Crop years are July-June for corn, Sept.-Aug. for soybeans, and Dec.-Nov. for sugar.

32. Harvested in first quarter 1992.

Panel 33 - Top Mining Products

Sources

The World Factbook 1993, Central Intelligence Agency (CIA), Washington, DC: CIA, 1994.

U.S. Department of Interior, Bureau of Mines, *Minerals Yearbook, Volume III: Mineral Industries Africa*, Washington, DC: USGPO, 1991.

U.S. Department of Interior, Bureau of Mines, *Minerals Yearbook, Volume III: Mineral Industries Asia and the Pacific*, Washington, DC: USGPO, 1991.

U.S. Department of Interior, Bureau of Mines, *Minerals Yearbook, Volume III: Mineral Industries of Latin America and Canada*, Washington, DC: USGPO, 1991.

U.S. Department of Interior, Bureau of Mines, *Minerals Yearbook, Volume III: Mineral Industries of Europe and Central Asia*, Washington, DC: USGPO, 1991.

U.S. Department of Interior, Bureau of Mines, *Minerals Yearbook, Volume III: Mineral Industries of the Middle East*, Washington, DC: USGPO, 1991.

When data were not available from the Bureau of Mines, the following source was used:

The World Factbook 1993, Central Intelligence Agency (CIA), Washington, DC: CIA, 1994.

These data are in text format. The following countries have CIA data:

Antigua and Barbuda	Armenia	Azerbaijan
Bangladesh	Belarus	Burma
Cape Verde	Chad	Comoros
Cote d'Ivoire	Djibouti	Equatorial Guinea
Estonia	The Gambia	Georgia
Grenada	Guinea-Bissau	Hong Kong
Kazakhstan	Kyrgyzstan	Laos
Latvia	Lesotho	Liechtenstein
Lithuania	Maldives	Marshall Islands
Micronesia, Federated States of	Moldova	Monaco
Oman	Saint Kitts and Nevis	Saint Lucia
Saint Vincent and the Grenadines	San Marino	Sao Tome and Principe
Seychelles	Sierra Leone	Singapore
Solomon Islands	South Africa	Tajikistan
Turkmenistan	Ukraine	United Arab Emirates
United States	Uzbekistan	Vanuatu
Western Samoa		

Notes. Tabular data were extracted from Bureau of Mines publications cited above. Mineral commodities were chosen by the editors on the basis of the apparent importance of the minerals in each country, based on volume or value.

Footnotes

1. Reported figure.

2. Data available through August 26, 1992.

3. Data available through December 22, 1992.

4. Data available through June 15, 1992.

5. Data available through November 20, 1992.

6. India's marketable production is 10% to 20% of mine production.

7. Includes a small amount of cobalt that is not recovered separately.

8. Includes Au content of cobalt ore and output by government-controlled foreign contractors' operations. Gold output by operators of so-called people's mines and

illegal small-scale mines is not available but may be as much as 18 tons per year.

9. Includes production from Malaya, Sabah, and Sarawak.

10. Figure includes 38,692 kg of metallic silver.

11. Officially reported figures for 1991 were 33,584 kg for major mines and 55,525 kg for independent miners (Garimpeiros).

12. Direct sales and beneficiated.

13. Includes synthetic crude (from oil shale and/or tar sands).

14. Blister copper from domestic ores plus recoverable Cu content of exportable matte and concentrates.

15. Refined nickel from domestic ores plus recoverable Ni content of exported matte.

16. Figure represents gross weight of usable iron ore as mine shipments.

17. Blocks and slabs production of marble, travertine, and granite, in thousand tons, were as follows: 4,468; 463; and 9 respectively. Figures for crushed production were 498; 56; and 2 respectively.

18. Estimated for cement manufacturing only; additional quantities are undoubtedly produced but information did not give output or was inadequate to reliably estimate output.

19. Beneficiated product, estimated as 30% P2O5.

20. Revised to zero.

21. Includes Saudi Arabian one-half share of production in the Kuwait-Saudi Arabia Divided Zone.

22. Turkish pumice production is officially reported in cubic meters and has a density reported to range from 0.5 to 1.0 ton per cubic meter. This value has been converted using 1 cubic meter = 0.75 ton.

23. Annuario Estadistico de Cuba provides figures of nickel-cobalt content of granular and powder oxide, oxide sinter, and sulfur production. Using an average cobalt content in these products of 0.9% in total granular and powder oxide, 1.1% in total oxide sinter, and 4.5% in total sulfide, the cobalt content of reported Ni-Co production was determined to be 1.16% of granular and powder oxide, 1.21% of oxide sinter, and 7.56% of sulfide. The remainder of reported figures would represent the nickel content.

24. Rock salt only.

25. Includes lease condensate.

26. Quantity of produced stones in 1991 was 140,300.

27. Sb content of ores for export plus Sb content of antimonial and impure bars plus refined materials.

28. Gross weight of white and black (impure) arsenic trioxide.

29. Refined metal plus Bi content of impure smelter products.

30. Estimate based on reported formal and legal artisanal production and estimated smuggled artisanal output.

31. Some additional gold production, mostly in the northwest, was illegally exported and not officially recorded; but information is inadequate for reliably estimating such output. This unrecorded production was reported to be substantial in 1990 but the 1991 production reported is considered realistic.

32. An unquantifiable amount of gold was recovered from placer deposits in Darin Province during the period 1987-89.

33. Output based entirely on imported clinker.

34. Does not include artisanal production smuggled out of the country.

35. Output reported by Empresa Minera del Centro del Pen S.A.

36. Reported as volume or pieces; conversions to metric tons are estimated.

37. Includes, in thousand 42-gallon barrels, associated natural gas lease condensate—14,600 and natural gasoline—not available.

38. Includes cement produced from imported clinker.

39. Gold production figures likely do not include production smuggled out of the country, for which there are no reliable data.

40. Production includes the Akwatia Mine (145,887 carats), PMMC (Precious Minerals Marketing Corp.) purchases of artisanal production (541,849 carats), and estimates of smuggled production.

41. Does not include estimate of smuggled production.

42. Figure estimated based on 3% moisture content for metal-grade bauxite.

43. Figure reported by Bureau de Strategie et de Marketing Minier of Guinea.

44. Figure includes undocumented artisanal production. Audifere de Guinea (AuG) is the only reporting gold mining company, reporting 1,453 kilograms for 1991.

45. Data are estimates of artisanal production, likely smuggled out of Liberia, but which is comparable to that hitherto reported to the government.

46. Data reported for the Kalama and Syama Mines.

47. From imported clinker.

48. Estimated 22% Ta plus 30% Cb.

49. Reported gross weight output estimated to contain 92% Au.

50. Reported gross weight output estimated to contain 70% Sn.

51. Reported gross weight output estimated to contain 54% W (68% WO3).

52. Estimated for cement manufacture only.

53. Diamond figures are estimated to represent 70% gem-quality or semigem-quality and 30% industrial-quality stones.

54. Exports.

55. In 1984, production of domestic clinker ended. Since that time, all cement has been produced from imported clinker.

56. Estimated for cement and lime manufacture only.

Panel 34 - Tourism

Source. World Tourism Organization, *Compendium of Tourism Statistics 1987-1991 (13th ed.)*, Madrid: World Tourism Organization, 1993. Used with permission.

Notes. Tourism arrival data refer to the number of arrivals of visitors and not to the number of persons. The same person who makes several trips to a given country during a given period will be counted each time as a new arrival.

For statistical purposes, the term *international visitor* describes "any person who travels to a country other than that in which (s)he has his/her usual residence but outside his/her usual environment for a period nor exceeding twelve months and whose main purpose of visit is other than the exercise of an activity renumerated from within the country visited."

International visitors include: a) *Tourists* (overnight visitors): "a visitor who stays at least one night in a collective or private accomodation in the country visited." b) *Same-day visitors*: "a visitor who does not spend the night in a collective or private accomodation in the country visited." This includes:

1) *Cruise passengers* who arrive in a country on a cruise ship and return to the ship each night to sleep on board even thought the ship remains in port for several days. Also included in this group are, by extension, owners or passengers of yachts and passengers on a group tour accomodated in a train.

2) *Crew members* who do not spend the night in the country of destination; this group also includes crews of warships on a courtesy visit to a port in the country of destination and who spend the night on board ship and not at the destination.

Unless otherwise stated, figures for *visitors* correspond to the aggregation of the figures for *tourists* and *same-day visitors*. In principle, data for *cruise passengers* are included in the *same-day visitors* category but in these tables, data are provided separately.

International tourism receipts are defined as "expenditure of international inbound visitors including their payments to national carriers for international transport. They should also include any other prepayments made for goods/services received in the destination country. They should in practice also include receipts from same-day visitors, except in cases when these are so important as to justify a separate classification. It is also recommended that, for the sake of consistency with the Balance of Payments recommendations of the International Monetary Fund, international fare receipts be classified separately."

International turism expenditure is defined as "expenditure of outbound visitors in other countries including their payments to foreign carriers for international transport. They should in practice also include expenditure of residents travelling abroad as same-day visitors, except in cases when these are so important as to justify a separate classification. It is also recommended that, for the sake of consistencey with the Balance of Payments recommendations of the International Monetary Fund, international expenditure be classified separately."

International fare receipts are defined as "any payment made to carriers registered in the compiling country of sums owed by non-resident visitors, whether or not travelling to that country." This category corresponds to "Other transportation, passenger services, credits" in the standard reporting form of the International Monetary Fund.

International fare expenditure is defined as "any payment to carriers registered abroad by any person resident in the compiling country." This category corresponds to "Other transportation, passenger services, debits" in the standard reporting form of the International Monetary Fund.

Footnotes

1. Includes nationals of the country residing abroad.

2. Air and sea arrivals, excluding nationals of the country residing abroad.

3. Includes cruise ships, windjammer cruises and yacht arrivals.

4. Excludes nationals of the country residing abroad.

5.   Includes the transport of merchandise.

6.   International tourist arrivals at all accomodation establishments.

7.   Includes international fare receipts.

8.   Includes international fare expenditure.

9.   Estimated.

10. International tourist arrivals at hotels and similar establishments.

11. Excludes returning residents.

12. Data are based on a survey conducted by EMBRATUR.

13. Excludes crew spending and international fares.

14. Includes ethnic Chinese arriving from Hong Kong, Macau, and Taiwan.

15. Excludes ethnic Chinese arriving from Hong Kong, Macau, and Taiwan.

16. Air arrivals.

17. International tourist arrivals at hotels and similar establishments in Brazzavile, Pointe, Noire, Loubomo, Owando, and Sibiti.

18. Includes cruise passengers.

19. Operational rates of exchange for United Nations programs; arrivals from market economy countries.

20. Includes cruise passengers and transit (1987=59,394; 1988=49,428; 1989=65,754; 1990=30,642; 1991=12,088); Includes same-day visitors (1987=117,550; 1988=126,204; 1989=73,068; 1990=38,691; 1991=33,978).

21. Data refer to the former Federal Republic of Czechoslovakia as it existed prior to 1 January 1993.

22. Estimated international tourist arrivals at frontier.

23. Arrivals by air only, including nationals of the country residing abroad.

24. All arrivals by sea.

25. Arrivals in hotels. Data refer to Dubai only.

26. Arrivals to Addis Ababa, Asmara, and Assab airports.

27. Change of series.

28. Data relate to the territory of the Federal Republic of Germany prior to October 3, 1990.

29. Excludes visitors from the former German Democratic Republic.

30. Data based on surveys.

31. Registered by the Central Bank.

32. Cruise passengers (incl. in international visitor arrivals).

33. Excludes servicemen, air crew members and transit passengers.

34. Departures.

35. Includes same-day visitors revenues.

36. Includes 1-day visitors (1987=22,003; 1988=22,003) & those staying more than one year (1987=4,390; 1988=7,092).

37. Excludes shore same-day visitors.

38. Includes pilgrims.

39. Includes approximately 800,000 refugees into Jordan, mainly from Asian countries due to the Gulf crisis.

40. International tourist arrivals at hotels and similar establishments, inlcuding international and domestic tourists.

41. Arrivals in hotels, inns, guest houses, and camping sites.

42. Also includes youth hostels, tourist private accomodation, and others.

43. Receipts from hotels, restaurants, and camping sites.

44. Foreign tourist departures, including Singapore residents crossing the frontier.

45. Includes visitors of the border zone with the U.S. whose duration of stay does not exceed 72 hours.

46. Includes receipts from frontier visitors.

47. Includes expenditures of residents and foreign tourists visiting the U.S. for more or less than 24 hours.

48. Includes arrivals from India.

49. International tourist arrivals at registered hotels and similar establishments.

50. Hotel sales.

51. Tocúmen International Airport, Paso Canoa frontier, the ports of Cristóbal and Balboa and IPAT statistics.

53. Arrivals by air and land, excluding nationals of the country residing abroad and crew members.

54. Visitor receipt figure.

55. Includes arrivals from abroad in the insular possesions of Madeira and the Azores.

56. Arrivals in hotels.

57.   Air and sea arrivals.

58.   Yacht and cruise ship arrivals.

59.   By air and sea, excluding same-day visitors.

60.   Arrivals at Arnos Vale airport.

61.   Excludes Italian visitors (1987=2,327,569; 1988=2,413,399; 1989=2,408,184; 1990=2,330,290).

62.   With visit visa (1987=412,000; 1988=1,566,000; 1990=1,310,000; 1991=1,921,000); and Omra visa [special visit to the two Holy Mosques at any time during the year] (1987=1,788,000; 1988=524,000; 1990=674,000; 1991=370,000).

63.   Of which pilgrims.

64.   Excludes arrivals of Malaysian citizens by land.

65.   Transit passengers.

66.   New series.

67.   Arrivals in hotels, rest camps and cottage accomodation, caravan parks, and camping sites.

68.   Revised figures.

69.   Includes persons in transit.

70.   Sea arrivals (except one land border from 1989).

71.   Departures, excluding Turkey.

72.   Includes Mexicans staying one or more nights in the U.S.

73.   Data refer to the former Union of Soviet Socialist Republics.

74.   Data refer to former Yugoslavia.

75.   Arrivals by air and road.

76.   Excludes in-transit passengers.

Panel 35 - Manufacturing

Sources

United Nations Industrial Development Organization, *Industry and Development Global Report* (various years), Vienna: UNIDO.

When data were not available from UNIDO, the following source was used to present an Industrial Summary:

The World Factbook 1993, Central Intelligence Agency (CIA), Washington, DC: CIA, 1994.

The following countries have CIA data:

Angola	Antigua and Barbuda	Armenia
Azerbaijan	Belarus	Burma
Cambodia	Comoros	Dominica

Estonia	Georgia	Grenada
Kazakhstan	Kyrgyzstan	Latvia
Lebanon	Liechtenstein	Lithuania
Maldives	Marshall Islands	Micronesia, Federated States of
Moldova	Monaco	Saint Kitts and Nevis
Saint Lucia	Saint Vincent and the Grenadines	San Marino
Solomon Islands	Tajikistan	Turkmenistan
Ukraine	Uzbekistan	Western Samoa

Notes. *Gross Domestic Product (GDP)*. Sum of all goods and services produced by people and capital located in the country. *Real GDP* measures economic activity in constant prices, that is after adjustments for inflation.

Value-added in Manufacturing. The difference between the value of shipments and the costs of purchased raw materials and components.

Notes for CIA Data. GNP/GDP dollar estimates for the OECD countries, the former Soviet republics, and East European countries are derived from *purchasing power parity (PPP)* calculations rather than from conversions at official currency exchange rates. The PPP method normally involves the use of international dollar price weights, which are applied to the quantities of goods and services produced in a given economy. In addition to the lack of reliable data from the majority of countries, there is a major difficulty in specifying, identifying, and allowing for the quality of goods and services. The division of a PPP GNP/GDP estimate in dollars by the corresponding estimate in the local currency gives the *PPP conversion rate*. One thousand dollars will buy the same basket of goods in the U.S. as one thousand dollars—converted to the local currency at the PPP conversion rate—will buy in the other country. GNP/GDP estimates for the less-developed countries, however, are based on the conversion of GNP/GDP estimates in local currencies to dollars at the official currency exchange rates. Because currency exchange rates depend on a variety of international and domestic financial forces that often have little relation to domestic output, use of these rates is less satisfactory for calculating GNP/GDP than the PPP method. Furthermore, exchange rates may suddenly go up or down by 10% or more because of market forces or official decree, while real output remains unchanged. One additional caution: the proportion of, say, defense expenditures as a percent of GNP/GDP in local currency accounts may differ substantially from the porportion when GNP/GDP accounts are expressed in PPP terms. An example might be when an observer estimates the dollar level of Russian or Japanese military expenditures. Similar problems exist when components are expressed in dollars under currency exchange rate procedures.

Panel 36 - Economic Indicators

Sources

National Trade Data Bank: The Export Connection [CD-ROM]. Prepared by the Economic and Statistics Administration, U.S. Department of Commerce. Washington, DC: Dept. of Commerce, December 1993.

When data were not available from DOC, appropriate fields were extracted from the following source:

The World Factbook 1993, Central Intelligence Agency (CIA), Washington, DC: CIA, 1994.

The following countries have CIA data:

Afghanistan	Albania	Antigua and Barbuda
Armenia	Azerbaijan	Belarus
Belize	Benin	Bhutan
Botswana	Brunei	Burkina
Burma	Burundi	Cambodia
Cameroon	Cape Verde	Central African Republic
Chad	Comoros	Congo
Cote d'Ivoire	Cuba	Cyprus
Djibouti	Dominica	Equatorial Guinea
Ethiopia	Fiji	The Gambia
Georgia	Grenada	Guinea
Guinea-Bissau	Guyana	Iceland
Iraq	Kazakhstan	Korea, North
Kyrgyzstan	Laos	Latvia
Lebanon	Lesotho	Liberia
Libya	Liechtenstein	Lithuania
Luxembourg	Madagascar	Malawi
Maldives	Mali	Malta
Marshall Islands	Mauritania	Mauritius
Micronesia, Federated States of	Moldova	Monaco
Mongolia	Mozambique	Namibia
Nepal	Niger	Papua New Guinea
Qatar	Rwanda	Saint Kitts and Nevis
Saint Lucia	Saint Vincent and the Grenadines	San Marino
Sao Tome and Principe	Senegal	Seychelles
Sierra Leone	Solomon Islands	Somalia
Sri Lanka	Sudan	Suriname
Swaziland	Tajikistan	Tanzania
Togo	Turkmenistan	Uganda
Ukraine	United States	Uzbekistan
Vanuatu	Vietnam	Western Samoa
Yemen	Zambia	Zimbabwe

National product. The total output of good and services in a given country. National product normally corresponds to Gross National Product (GNP), defined below.

Gross Domestic Product (GDP). The sum of all goods and services produced by people and capital within the political boundaries of a country. *Real GDP* measures economic activity in constant prices, that is after adjustments for inflation.

Gross national product (GNP). The sum of all goods and services produced by a country's nationals and their capital regardless of location.

Commercial interest rate. The interest rate at which loans are extended to the commercial sector.

Savings rate. The percentage of all income that is saved by consumers, government, and commercial sectors.

Investment rate. The percentage of all income that is invested by consumers, government, and commercial sectors.

Inflation rate. An increase in prices unrelated to value.

Consumer Price Index (CPI). The price of a basket of goods and services, usually based on surveys of household or family expenditures.

Wholesale Price Index (WPI). The price of goods at the bulk distribution level.

External debt. The amount of debt owed to foreign entities by the given country.

Footnotes

e.   Estimated.

p.   Projected or provisional.

r.   Revised.

1.   Angola's fiscal year is January 1 - December 31.

2.   Nominal monthly rate in percent at year end.

3.   US$ billions; including interest arrears.

4.   Sources: Australian Bureau of Statistics, Embassy Estimates.

5.   1984/85 prices, seasonally adjusted.

6.   Source: Bangladesh Export Promotion Bureau.

7.   The Bangladesh fiscal year is July 1 - June 30.

8.   All figures end of year unless otherwise noted.

9.   Estimated data (Central Bank of Bolivia and UDAPE) and/or targets set by the GOB and the IMF.

10.   External Sector/Foreign Debt of the Central Bank of Bolivia.

11.   As of September 30, 1991.

12.   Data refer to the former republic of Yugoslavia.

13.   Gross Social Product (GSP) is the total output of goods, plus services regarded as productive. This approximates, but is 10-15 percent less than, GNP.

14.   Forecast by the IPEA/Ministry of Economy.

15.   As of September 1991.

16.   Cost of money for working capital for 30 days average for October 1991.

17.   As of March 1991.

18.   Sources: U.S. Embassy Sofia; Government of Bulgaria; international financial institutions.

19.   Home mortgage rate for 1989/1990 was 2.0 percent; base lending rate from 2/1/91 to 11/7/91 moved in a range of 45 to 54 percent.

20.   SAAR: Seasonally Adjusted Annual Rate.

21.   Embassy projection.

22.   Actual for the year.

23.   National accounts available only in 1977 Chilean pesos. Translation of these figures into 1977 dollars distorts the structure of GDP (Exchange rate: USD 1 = Ch pesos 26.54).

24. As of July 1991.

25. RMB = renminbi. Source for GNP data for 1989 and 1990 is the IMF's International Financial Statistics. GNP per capita and real per capita GNP growth are calculated using this IMF data. Real GNP growth is based on PRC government statistics. Figures for 1991 are Embassy projections.

26. M1 estimates based on data released by the Central Bank.

27. Estimates of Gross National Savings as a percent of GNP and Gross Domestic Investment as a percent of GNP for 1989 and 1990 are as estimated by the IMF in January 1990. Actual savings in 1990 were much higher. Figures for 1991 are Embassy estimates.

28. Figures are for December over December inflation. They differ from official Chinese statistics which show average annual inflation of 17.8 percent for 1989.

29. Source: Central Bank, National Planning Department, National Department of Statistics.

30. Estimate based on October 1991 data.

31. All figures used are from Czechoslovak government sources or from the IMF.

32. January-July.

33. January-April.

34. September.

35. Until end-1990 M1. As of 1991 the M1 money supply figure is no longer available. The figure shown for 1991, which comes closest to the M1 definition, is made up of the nonbank sector holdings of coins and notes plus actual demand deposits in Danish banks (by contrast the M1 definition includes as demand deposits, deposits available upon one month's or less notice).

36. Savings as a Percent of Personal Disposable (after tax) Income.

37. End of Period.

38. Sucres converted at the average intervention rate for the year. Because of real appreciation of the sucre against the dollar in 1990 and 1991, dollar figures overstate growth of GDP and GDP/capita.

39. Sources: Annual Report of the Central Bank of the Arab Republic of Egypt, Central Agency for Mobilization and Statistics of the Arab Republic of Egypt, and IMF International Financial Statistics.

40. Except where otherwise noted, data years are based on Egyptian fiscal year which runs from July 1 to June 30

(i.e., 1991 = Egypt's fiscal year July 1, 1990 to June 30, 1991).

41. Includes banking system obligations.

42. M1 has been recalculated by the Bank of Finland to include currency in circulation, Finnmark check and postal checking account deposits, transactions account deposits as well as foreign exchange account deposits held by the public.

43. 3-month Helibor rate. Helibor (Helsinki interbank offered rate) is Finland's commercial banking reference rate.

44. 1991 figures are government forecasts unless otherwise specified.

45. Outstanding amount as of August 1991.

46. Second quarter 1991 figure.

47. France does not report comparable statistics.

48. Those figures reported for 1991 are necessarily estimates, drawn in most cases from Central Bank/IMF data. "N/A" indicates that no official estimate is available.

49. The Government of Gabon stopped publishing price indexes in June 1989. Estimates compiled by private sector consultants indicate rates of approximately 10 percent in 1989 and 1990. The Embassy does not believe that conditions in 1991 have been significantly different.

50. Data refer to West Germany.

51. Since July 1990 inclusion of eastern Germany.

52. As of August.

53. Dec. avg., credits over DM 1 billion and under DM 5 billion.

54. As of September.

55. January - September.

56. As of June.

57. Sources: Bank of Greece, National Statistical Service of Greece, Ministry of National Economy, and Embassy estimates.

58. Interest rate on long term loans to industry.

59. Sources: IMF, World Bank, USAID, U.S. Department of State or U.S. Department of Commerce.

60. 1991 figures are projected prior to the Sept. 30, 1991 coup.

61. Sources: Central Bank of Honduras (CBH), Ministry of Public Finance, United Nations Economic Commission for Latin America (CEPAL), U.S. Agency for International

Development (AID), and International Monetary Fund (IMF).

62. Consulate revised projection.

63. Prime lending rate.

64. Savings deposit rate.

65. 3-month time deposit rate.

66. Oct 1984-Sept 1985; CPI covers urban households with monthly expenditure of HKD 2,000-6,499 (approximately 50 percent of households).

67. Sources: Central Statistical Organization, Reserve Bank of India, U.S. Department of Commerce, World Bank.

68. Annual growth rate, except for 1991 which is first half of 1991 over first half of 1990.

69. Interbank funds rates; 1991 rate is January to June average.

70. End of period. For 1989, FY 1977/78 equals 100; starting 1990, FY 1988/89 equals 100; due to change in basis of calculating CPI, comparable data for earlier years are not available. Fiscal year is from April 1 to March 31.

71. First nine months of 1991.

72. Maximum.

73. Medium- and long-term debt.

74. Sources: Central Bank Bulletin; Central Statistics Office (CSO); Economic & Social Research Institute; IMF statistics.

75. December figure.

76. Base: mid-November 1982 as 100.

77. Mid-November figure.

78. Base: year 1985 as 100.

79. July figure.

80. Current GDP divided by average permanent population.

81. New Israeli Shekels (NIS).

82. Annual real short-term credit to the public, including overdraft facilities and exchange-rate indexed credits.

83. Net liabilities of government sector, nonfinancial private sector, and banking system.

84. 8/28/91.

85. 1991 data are estimates by Italian Government or U.S. Embassy except where data are followed by a month thereby indicating actual data through that period.

86. September.

87. This figure does not include foreign purchases of treasury securities issued domestically.

88. July.

89. August.

90. Jan-Sep seasonally adjusted annual rate (S.A.A.R.).

91. End of September non-seasonally-adjusted (N.S.A.).

92. Percent of household income.

93. Domestic fixed capital formation and inventory/GNP.

94. Jan-Nov average, non-seasonally-adjusted (N.S.A.)

95. Sources: Unless otherwise stated in the following paragraphs, all figures were obtained from the latest Central Bank of Jordan's (CBJ) Monthly Statistical Bulletin, Volume No. 27, Issue No. 8, August 1991.

96. 1990 figures are preliminary. 1991 figures are based on projections made in December 1991.

97. International Monetary Fund Report, December 1991.

98. Additional source: for 1991, Minister of Finance's budget speech to Parliament, December 1990.

99. Preliminary estimates obtained from Kenya's Ministry of Planning and Central Bank of Kenya.

100. Consumer Price Index (base — 1982).

101. Sources: Embassy estimates; Bank of Korea; Economic Planning Board; Ministry of Finance.

102. Economic Planning Board estimate.

103. July

104. December/December.

105. Sources: Central Bureau of Statistics (CBS), Netherlands Central Bank(NB), Central Planning Bureau (CPB).

106. Fiscal years ending March 31.

107. Sources: International Monetary Fund (IMF) and World Bank unless otherwise indicated. All indicators are for calendar years unless otherwise indicated. Figures in this chart differ from those used in the 1991 Trade Act Report, as they are from different sources. The numbers used here are considered to be more reliable. Any inconsistencies among the 1991 indicators are due to the fact that data were drawn from several different sources.

108. Source: World Bank. The IMF estimate of GDP is significantly lower, while the Nicaraguan Central Bank estimate is much higher.

109. The IMF projects -0.4 growth.

110. Investment rates appear unrealistically high. This may result in part form state-owned companies claiming that loans obtained from the state banks were used for investments, when they were likely used for current operating expenses.

111. Reflects average inflation. Figures included in the 1990 report were year-end.

112. Sources: Central Bank of Nigeria, IMF, IBRD, U.S. Department of Commerce, and U.S. Embassy estimates.

113. 1-Month NIBOR.

114. National Saving/National Disposable Income.

115. Gross Fixed Investment/GDP.

116. End-Year Foreign Assets Minus Foreign Liabilities.

117. Figure is M1 at the end of August 1991, the last period available (M1 at end August 1990 was 339.7).

118. Estimate by the Oman Chamber of Commerce and Industry (OCCI). OCCI estimates that 40 percent of the labor is Omani and the rest expatriate.

119. Figure is for CPI at end August, last figure available.

120. Pakistan's Fiscal Year (FY) is July 1 - June 30

121. Average annual interest rate on commercial bank loans to private sector borrowers.

122. For 1991, data based on assumption that Panama clears arrears with international financial institutions by year-end.

123. Source: Central Reserve Bank, National Institute of Statistics, Ministry of Labor and Embassy Estimates.

124. Excludes interest due on arrears.

125. Sources: National Economic and Development Authority, Central Bank of the Philippines, Department of Finance, U.S. Survey of Current Business.

126. Bank lending rates on secured loans; weighted average for all maturities.

127. Actual weighted average rate for January - September.

128. Money market rates; weighted average for all types of instruments.

129. Actual average for January - August.

130. End-October 1991.

131. The savings rate is calculated using the percentage of GDP held in personal time deposits. This does not include personal hard currency deposits in Polish banks,

which totalled $4.6 billion at the end of 1989, $6.3 billion at the end of 1990, and $5.6 billion at the end of October 1991.

132. These price indices measure the rise in average price levels recorded in a given year compared to average levels during the preceding year. Poland does not officially report price growth on a January 1 - December 31 basis.

133. Sources: National Institute of Statistics, Bank of Portugal, Government of Portugal, OECD, Portuguese Association of Banks and estimates by the Embassy.

134. 1991 figures are U.S. Embassy estimates.

135. Source: U.S. Embassy estimate or official U.S. Government source.

136. Source: Government of Romania.

137. Source: International financial institutions.

138. The ruble estimate for GNP was converted to 1982 geometric-mean (GM) U.S. dollars by multiplying the ruble value of estimated GNP by the geometric mean of two dollar-ruble-ratios—one weighted with U.S. price weights and the other with Soviet weights. The U.S. GNP deflator was then applied convert 1982 GM dollars to 1989 dollars.

139. Based estimates in 1982 rubles at factor cost.

140. U.S. estimates.

141. Sources: Official Saudi data, international financial statistics, IMF, U.S. Embassy estimates, Foreign Commercial Service estimates.

142. Statistics depending on population data are often unreliable; official black population and unemployment rates are likely understated. While the Central Statistical Services no longer attempts to quantify black unemployment, most economists believe the rate is in excess of 40 percent. Unemployment among other racial groups is lower.

143. Sources: Economic Research Institute, Central Bank, and Statistics Sweden.

144. Year end and 08/31/91. Includes treasury discount notes held by public plus accrued monies in deductible national savings scheme. Central Bank does not compile M1.

145. Industrial bonds, 30-month adjusted rates, percent. Annual averages and average for first 8 months of 1991.

146. Ratio of personal saving to disposable personal income.

147. Ratio of gross investment to GDP.

148. Change between annual CPI averages.

149. Product prices for total industry excluding shipbuilding.

150. Sources: *Die Volkswirtschaft*, Swiss National Bank Bulletin, Swiss Foreign Trade Statistics, Foreign Population Statistics.

151. Through end of September.

152. September rate as measured over the same month of the previous year.

153. Sources: All statistics except those for the free market exchange rate, aid, debt, annual debt service, gold and foreign exchange reserves are taken from the official Syrian statistical abstract or from official foreign trade statistics. The balance are Embassy estimates based on a variety of sources, none of which can be regarded as totally reliable. Discrepancies result in part from the use of different exchange rates in different periods.

154. Does not include debt under bilateral clearing arrangement with USSR nor non-civilian debt.

155. Estimated by the Directorate General of Budget, Accounting and Statistics.

156. Assume November interest rate to prevail to 1991 year-end.

157. Data as of October 29, 1991.

158. Data through September 1991.

159. End of year for 1989 and 1990, as of September 27 for 1991.

160. In 1991 the State Institute of Statistics changed the weightings and methods used to calculate national income. Previous years' statistics in this report have been restated to reflect the change. The income figures published here cannot be compared with previous years' reports.

161. Average deposit and lending rates for Oct. 7, 1991.

162. For 1989 and 1990, percent change during year; for 1991, percent change between September 1990 and September 1991.

163. As of March 31, 1991. Includes short-term debt.

164. 1991 figures are all estimates based on available monthly data in October 1990.

165. GDP at factor cost.

166. Figures are actual, average annual interest rates, not changes in them.

167. Embassy estimates.

168. Central Bank of Uruguay.

169. U.S. Embassy computation based on Central Bank and Office of Statistics and Census data.

170. U.S. Embassy computation based on Central Bank data.

171. Office of Statistics and Census.

172. Embassy forecast as of 11/12/91.

173. Factor cost, 1986/87 prices.

Panel 37 - Balance of Payments

Source. United Nations Conference on Trade and Development (UNCTAD), *Handbook of International Trade and Development Statistics*, Geneva: UNCTAD, 1993.

Notes. *Balance of payments* is the account of all international financial flows for a given country.

Balance of trade is the difference in value between merchandise imports and merchandise exports. If the value of imports is greater than the value of exports, the result is a *trade deficit*. If the value of exports is greater, the result is a *trade surplus*.

Debits are financial outflows, such as payments made to another country for goods and services, insurance and freight, interest payments. Another component of the debit total is *direct investment income*—this is considered debit because the foreign country is the investor and claims this income as a credit.

Credits are financial inflows, such as payments received from another country for goods and services or insurance and freight, interest payments, and tourism revenues.

In any relationship between two countries, if one country posts a debit the other country must post the same amount as a credit.

Private transfers include migrants' transfers, workers' remittances, gifts, dowries, inheritances, prize monies from non-governmental lotteries, etc. *Government transfers* include transactions between official sectors of two economies, such as grants, debt cancellations, and reparations; and transactions between official sectors and private non-residents, such as scholarships, licensing fees, and government-sponsored lottery tickets.

Long-term capital includes direct investment, loans extended abroad, and purchases of foreign securities.

Short-term capital include assets that are easily reversed, such as bank deposits and treasury bonds.

Errors and omissions. These figures indicate that—because of errors, omissions, and inconsistencies in reported figures—the sum of credits does not equal the sum of debits, as should be the case.

Notes. (-) Stands for nil. f.o.b. stands for free on board—this means that the value of goods does not include insurance and freight charges. c.i.f. stands for cost, insurance, and freight—this means that insurance and freight charges are included in the value of goods.

Panel 38 - Exchange Rates

Source. *The World Factbook 1993*, Central Intelligence Agency (CIA), Washington, DC: CIA, 1994.

Notes. The *exchange rate* is the value of a nation's monetary unit at a given date or over a period of time, as expressed in units of local currencey per U.S. dollar and as determined by international market forces or official decree.

Panel 39 - Top Import Origins

Source. *The World Factbook 1993*, Central Intelligence Agency (CIA), Washington, DC: CIA, 1994.

Notes.　The following abbreviations are used in this panel:

ASEAN　　Association of Southeast Asian Nations.

EC　　European Community.

ECOWAS　Economic Community of West African States.

EFTA　　European Free Trade Association.

CARICOM　Caribbean Community.

CEMA　　Council for Economic Mutual Assistance.

CIS　　Community of Independent States.

FRG　　Federal Republic of Germany (former West Germany).

GDR　　German Democratic Republic (former East Germany).

NZ　　New Zealand.

OECS　　Organization of Eastern Caribbean States.

PDR　　People's Democratic Republic of Yemen (South Yemen).

UAE　　United Arab Emirates.

UK　　United Kingdom.

US　　United States.

USSR　　former Union of Soviet Socialist Republics (Soviet Union).

YAR　　Yemen Arab Republic (North Yemen).

Panel 40 - Top Export Destinations

Source. *The World Factbook 1993*, Central Intelligence Agency (CIA), Washington, DC: CIA, 1994.

Notes. For abbreviations used in this panel, see the listing for Panel 39, immediately above.

Panel 41 - Foreign Aid

Source. *The World Factbook 1993*, Central Intelligence Agency (CIA), Washington, DC: CIA, 1994.

Notes. *ODA* stands for official development assistance. ODA refers to government grants that are administered with a main objective of the promotion of economic development and welfare of less developed countries (LDCs).

OOF stands for other official flows. These transactionsinclude official export credits (e.g. from Export-Import Bank), official equity and portfolio investment, and deby reorganization by the official sector that does not meet concessional terms.

Additional abbreviations used inclode *CEMA:* Council for Economic Mutual Assistance; *EC:* European Community; *IMF:* International Monetary Fund; *OPEC:* Organization of Petroleum Exporting Countires, and *US:* United States.

Aid has been committed when agreements are initialed by the parties involved, which constitutes a formal declaration of intent.

Panel 42 - Import and Export Commodities

Source. *The World Factbook 1993*, Central Intelligence Agency (CIA), Washington, DC: CIA, 1994.

Notes. Years for imports of commodities shown can be found in the legend of Panel 39: *Top Import Origins*.

Years for exports of commodities shown can be found in the legend of Panel 40: *Top Export Origins*.

KEYWORD INDEX

The Keyword Index provides access, by page number, to every country in *Statistical Abstract of the World*. Country names are capitalized. Subject references are also provided followed by (1) a listing of countries and page numbers separated by dashes or (2) by page numbers only. Countries are arranged alphabetically within each block. The phrase *pol.* follows index terms that are political parties or entities.

Keyword Index

Human Rights Protection Party (HRPP) *pol.*
Western Samoa 1009

Hungarian Christian Democratic Movement/Coexistence *pol.*
Slovakia 830

Hungarian Socialist Party (MSP) *pol.* Hungary 407

HUNGARY 405

Hydroelectric power

ICELAND 411

Ilmenite 41, 578, 865

Immunization *See* Children immunized

Imports

Independence Party *pol.* Iceland 413

Independent Democratic Union (UDI) *pol.* Chile 190

Independent Islamic *pol.* Jordan 473

Independent Moralization Front *pol.* Peru 725

Independent Revolutionary Party *pol.* Dominican Republic 271

Independent Smallholders *pol.* Hungary 407

Keyword Index

Keyword Index

Population - *continued*

Belize 88 – Benin 93 – Bhutan 98 – Bolivia 103 – Bosnia and Herzegovina 109 – Botswana 115 – Brazil 121 – Brunei 127 – Bulgaria 132 – Burkina 138 – Burma 144 – Burundi 149 – Cambodia 155 – Cameroon 160 – Canada 166 – Cape Verde 172 – Central African Republic 177 – Chad 183 – Chile 188 – China 194 – Colombia 200 – Comoros 206 – Congo 211 – Costa Rica 217 – Cote d'Ivoire 223 – Croatia 229 – Cuba 235 – Cyprus 241 – Czech Republic 247 – Denmark 253 – Djibouti 259 – Dominica 264 – Dominican Republic 269 – Ecuador 275 – Egypt 281 – El Salvador 287 – Equatorial Guinea 293 – Estonia 298 – Ethiopia 303 – Fiji 309 – Finland 315 – France 321 – Gabon 327 – Gambia, The 333 – Georgia 339 – Germany 344 – Ghana 350 – Greece 356 – Grenada 362 – Guatemala 367 – Guinea 373 – Guinea-Bissau 378 – Guyana 383 – Haiti 388 – Honduras 393 – Hong Kong 399 – Hungary 405 – Iceland 411 – India 417 – Indonesia 423 – Iran 429 – Iraq 435 – Ireland 441 – Israel 447 – Italy 453 – Jamaica 459 – Japan 465 – Jordan 471 – Kazakhstan 477 – Kenya 482 – Korea, North 488 – Korea, South 493 – Kuwait 499 – Kyrgyzstan 505 – Laos 510 – Latvia 515 – Lebanon 520 – Lesotho 525 – Liberia 530 – Libya 535 – Liechtenstein 541 – Lithuania 546 – Luxembourg 551 – Macedonia 557 – Madagascar 563 – Malawi 569 – Malaysia 575 – Maldives 581 – Mali 586 – Malta 591 – Marshall Islands 597 – Mauritania 602 – Mauritius 607 – Mexico 613 – Micronesia, Federated States of 619 – Moldova 624 – Monaco 629 – Mongolia 634 – Morocco 639 – Mozambique 645 – Namibia 650 – Nepal 655 – Netherlands 660 – New Zealand 666 – Nicaragua 672 – Niger 678 – Nigeria 683 – Norway 689 – Oman 695 – Pakistan 700 – Panama 706 – Papua New Guinea 712 – Paraguay 717 – Peru 723 – Philippines 729 – Poland 735 – Portugal 741 – Qatar 747 – Romania 752 – Russia 758 – Rwanda 764 – Saint Kitts and Nevis 769 – Saint Lucia 774 – Saint Vincent and the Grenadines 779 – San Marino 784 – Sao Tome and Principe 789 – Saudi Arabia 794 – Senegal 800 – Serbia and Montenegro 806 – Seychelles 812 – Sierra Leone 817 – Singapore 822 – Slovakia 828 – Slovenia 834 – Solomon Islands 840 – Somalia 845 – South Africa 850 – Spain 856 – Sri Lanka 862 – Sudan 868 – Suriname 873 – Swaziland 878 – Sweden 883 – Switzerland 889 – Syria 895 – Taiwan 901 – Tajikistan 907 – Tanzania 912 – Thailand 918 – Togo 924 – Trinidad and Tobago 930 – Tunisia 936 – Turkey 942 – Turkmenistan 948 – Uganda 953 – Ukraine 958 – United Arab Emirates 963 – United Kingdom 968 – United States 974 – Uruguay 980 – Uzbekistan 986 – Vanuatu 991 – Venezuela 996 – Vietnam 1002 – Western Samoa 1007 – Yemen 1012 – Zaire 1017 – Zambia 1023 – Zimbabwe 1029

PORTUGAL 741

Portuguese Socialist Party (PS) *pol.* Portugal 743

Potash 169, 347, 450, 456, 474, 761

Potassium nitrate 191

Potatoes 8, 13, 35, 36, 40, 52, 53, 67, 68, 79, 80, 105, 106, 134, 190, 202, 249, 255, 277, 283, 317, 346, 407, 443, 449, 450, 479, 480, 507, 508, 626, 627, 662, 691, 725, 737, 743, 754, 760, 852, 885, 891, 909, 910, 944, 960, 961, 970, 988, 989

Pozzolana 163

PPPP *pol.* Poland 737

Presbyterians *See* Religion

Progress *pol.* Norway 691

Progress Party *pol.* Denmark 255

Progressive Citizens' Party (FBP) *pol.* Liechtenstein 543

Progressive Conservative Party *pol.* Canada 168

Progressive Democrats *pol.* Ireland 443

Progressive Liberal Party (PLP) *pol.* Bahamas, The 57

Progressive National (RDNP) *pol.* Haiti 390

Progressive Party *pol.* Iceland 413

Propane 63

Protestantism *See* Religion

PSD/AP-5 *pol.* Guatemala 369

PSD/UNSP *pol.* Benin 95

Pulses 146, 147, 323, 419

Pumice 112, 141, 232, 267, 306, 414, 560, 809, 837, 945

PVP/PPC *pol.* Costa Rica 219

Pyrite 859

Pyrite and pyrrhotite 744

Pyrophyllite 468, 496

QATAR 747

Quartz 432, 566, 566, 865, 983

Quicklime 85, 135, 502

Radical Civic Union (UCR) *pol.* Argentina 29

Radical Liberal Party *pol.* Denmark 255

Radishes 495

Railroads *See* Transportation

Rally for Democracy & Labor (RDT) *pol.* Mali 588

Rally for Democracy (RDP) *pol.* Mali 588

Rally for the Republic (RPR) *pol.* France 323

R&D *See* Research and Development

Rapeseed 46, 317, 323, 737, 885, 970

Rare earths 197, 578

Refah Party (RP) *pol.* Turkey 944

Refinery products 74, 703

Religions

–Adventist 389

–Alawite 521, 896

–Anglican 23, 39, 56, 72, 89, 167, 363, 442, 667, 713, 770, 775, 780, 813, 841, 931, 969, 992

–Animist 145, 212, 304, 511

–Apostolic, 242

–Armenian Orthodox 34, 51, 340

–Assembly of God 620

–Atheism 248, 406, 823, 829

–Baha'i 430, 620

–Baptist 56, 145, 265, 389, 625, 667, 841

–Buddhism 66, 128, 145, 418, 424, 466, 489, 494, 511, 576, 635, 656, 730, 823, 863, 919, 1003

–Bulgarian Orthodox 133

–Calvinist 406

–Catholic (see also Roman Catholic) 23, 104, 110, 230, 236, 294, 363, 368, 394, 516, 521, 547, 558, 620, 718, 724, 770, 780, 785, 790, 959, 969, 992, 1003, 1008

–Chondogyo 489

–Christian 12, 66, 94, 116, 128, 145, 150, 161, 184, 212, 224, 260, 294, 310, 334, 351, 374, 379, 384, 400, 418, 430, 436, 448, 466, 472, 489, 494, 500, 521, 526, 531, 564, 576, 587, 598, 608, 620, 640, 646, 651, 679, 684, 701, 818, 823, 851, 863, 869, 879, 896, 913, 919, 925, 937, 943, 964, 1008, 1013, 1024, 1030

–Church of Christ 992

–Church of God 56

–Confucianism 489, 494, 576, 823

–Congregational 1008

–Coptic Christian 282

–Druze 448, 521, 896

–Eastern Orthodox 558, 625, 736, 949, 987

–Ethiopian Orthodox 304

–Evangelical Alliance 713

–Evangelical Lutheran 254, 316, 412, 690, 884

–Evangelical Methodist 104

–Evangelical Protestant 790

–Georgian Orthodox 340

–Greek Orthodox 7, 242, 316, 357

–Hindu 66, 99, 310, 384, 418, 424, 500, 576, 608, 656, 696, 701, 823, 851, 863, 874, 919, 931, 969, 1013, 1024